AF598479

Molecular Basis of Thrombosis and Hemostasis

Molecular Basis of Thrombosis and Hemostasis

edited by

Katherine A. High
Children's Hospital of Philadelphia
and University of Pennsylvania Medical Center
Philadelphia, Pennsylvania

Harold R. Roberts
University of North Carolina at Chapel Hill School of Medicine
Chapel Hill, North Carolina

Marcel Dekker, Inc. **New York • Basel • Hong Kong**

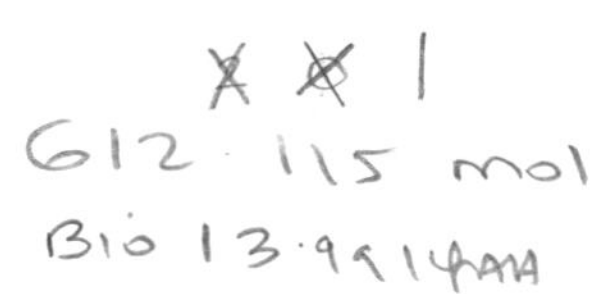

Library of Congress Cataloging-in-Publication Data

Molecular basis of thrombosis and hemostasis / edited by Katherine A. High, Harold R. Roberts.
p. cm.
Includes bibliographical references and index.
ISBN 0-8247-9501-6
1. Blood—Coagulation. 2. Blood coagulation factors. I. High, Katherine A. II. Roberts, H. R. (Harold Ross)
QP93.5.M65 1995
612.1'15—dc20 94-43114
CIP

The publisher offers discounts on this book when ordered in bulk quantities. For more information, write to Special Sales/Professional Marketing at the address below.

This book is printed on acid-free paper.

MARCEL DEKKER, INC.
270 Madison Avenue, New York, New York 10016

Current printing (last digit):
10 9 8 7 6 5 4 3 2 1

Printed in the United States of America

Preface

Students of the biochemical and cellular mechanisms responsible for normal blood coagulation and of the derangements that produce coagulopathies have benefited greatly from the introduction of recombinant DNA techniques into biomedical research. Blood clotting and its disorders are intensely molecular processes, highly susceptible to dissection by molecular strategies. The key events are mediated by soluble procoagulant and anticoagulant protein molecules that interact with one another, and with molecules on the surfaces of cells and supporting tissues to promote or inhibit hierarchical cascades of proteolytic and/or binding reactions that ultimately result in the formation of fibrin plugs, platelet aggregates, and proliferating scar tissue involved in the closure of wounds. Pathological bleeding or thrombosis results when the delicate balance and regulation of these molecular components is disrupted. Some of these pathological conditions, such as the hemophilias, have been known for many years to result from defects in individual clotting factors; others have defied understanding at the basic cellular or molecular level until recently (e. g., certain hypercoagulable states).

A major impediment that precluded precise characterization of the various coagulation pathways and their disorders was, until recently, the enormous difficulty associated with the precise purification of the responsible proteins. For the most part, these are present in minute quantities and are highly labile, making isolation a formidable challenge. The introduction of straightforward and universally applicable methods for gene cloning has revolutionized this field. Most of the known coagulation-related proteins have now been characterized at the level of their encoding genes. Isolation of the genes has permitted deduction of the amino acid sequences of these proteins (from DNA sequences), characterization of mutations responsible for inherited abnormalities, production of large quantities of protein for functional analysis, establishment of structure–function relationships by site-directed mutagenesis, and exploratory study of the biological effects of perturbing production of these proteins, using transgenic mouse and gene knockout models. Pharmacological amounts of some proteins can now be produced for use as replacements in hemophilia or intervention in acquired coagulopathies. The field stands on the threshold of gene therapy for the correction or modification of coagulation disorders.

In the first chapter of this volume we review fundamental aspects of nucleic acid chemistry, gene organization, regulation of gene expression, and basic experimental techniques. The second chapter provides a short introductory summary of the major procoagulant and anticoagulant pathways as they are currently understood. Other chapters deal with the molecular genetics of each of the clotting factors, including fibrinogen, prothrombin, tissue factor, and factors V, VII, VIII, IX, X, XI, XII, XIII, and von Willebrand factor. The molecular basis of control of the procoagulant pathways by anticoagulants is also covered in detail, including chapters on the tissue factor pathway inhibitor, protein C, protein S, antithrombin III, and the protein C inhibitor. In addition, the molecular genetics of plasminogen and tissue plasminogen activator is described. The molecular basis of control mechanisms of fibrinolysis is also discussed in the chapters on plasminogen activator inhibitors and α_2-plasmin inhibitor. The concluding chapters describe the genetics of platelet membrane glycoprotein Ib (GPIb), as well as GPIIb/IIIa, other platelet integrins, and the platelet thrombin receptor. The molecular genetics of thrombospondin and the recently described thrombopoietin round out the book and reflect the encyclopedic coverage of the molecular biology of proteins involved in blood coagulation and fibrinolysis. The authors have for the most part maintained a sharp focus on the biochemistry, molecular biology, and physiological roles of the proteins being considered. Clinical deficiency states are considered but are given a secondary emphasis, since clinical material is well treated in many other hematology and coagulation texts. This volume is intended as an authoritative and up-to-date reference about the individual proteins.

It is hoped that this volume will serve as a useful resource for several types of readers. For molecular biologists, pathologists, and other workers knowledgeable about blood coagulation and its disorders, the book will serve as an important reference work. The organization of the volume lends itself to rapid identification of a cohesive and readable summary of the structure and properties of any given coagulation protein and its encoding gene. Clinicians will find this volume useful for identifying individual mutations producing hemophilia and relating those mutations to known functional defects in the proteins. Hematologists, cardiologists, and fellows in training for these specialties and experts in the coagulation field may benefit from a careful perusal of this volume and thorough reading of chapters particularly relevant to their interest. No volume of this type can be completely up-to-date factually, nonetheless, this book is conceptually current. A thorough reading of a few chapters will allow one to develop a working knowledge of the state of the art of the coagulation proteins of interest.

As this book goes to press, new findings are emerging. The mechanism of action of γ-glutamylcarboxylase is still being worked out. Resistance to activated protein C has been traced to a specific mutation in Factor V but clinical aspects of this newly recognized defect require further assessment. The fields of coagulation and fibrinolysis continue to change rapidly and it is likely that progress in the next ten years will be even more rapid than in the past decade.

KATHERINE A. HIGH
HAROLD R. ROBERTS

Contents

Contributors

Nobuo Aoki First Department of Medicine, Tokyo Medical and Dental University, Tokyo, Japan

Joel S. Bennett, M.D. Professor, Division of Hematology-Oncology, Department of Medicine, University of Pennsylvania School of Medicine, Philadelphia, Pennsylvania

Morris A. Blajchman, M.D., F.R.C.P.(C) Professor, Departments of Pathology and Medicine, McMaster University, Hamilton, Ontario, Canada

Francis J. Castellino, Ph.D. Kleiderer-Pezold Chair Professor, Department of Chemistry and Biochemistry, and Dean, College of Science, University of Notre Dame, Notre Dame, Indiana

Dominic W. Chung, Ph.D. Research Professor, Department of Biochemistry, University of Washington, Seattle, Washington

Frank C. Church, Ph.D. Associate Professor, Departments of Pathology and Medicine, Center for Thrombosis and Hemostasis, University of North Carolina at Chapel Hill, Chapel Hill, North Carolina

Shaun R. Coughlin, M.D., Ph.D. Associate Professor, Department of Medicine, Cardiovascular Research Institute, University of California at San Francisco, San Francisco, California

Sandra J. Friezner Degen, Ph.D. Associate Professor, Division of Basic Sciences, Department of Pediatrics, Children's Hospital Research Foundation, and University of Cincinnati, Cincinnati, Ohio

William A. Dittman, M.D. Assistant Professor of Medicine and Cell Biology, Duke University Medical Center, and Durham Veterans Affairs Medical Center, Durham, North Carolina

Kazuo Fujikawa, Ph.D. Research Professor, Department of Biochemistry, University of Washington, Seattle, Washington

David Ginsburg, M.D. Professor and Chief, Division of Molecular Medicine and Genetics, Departments of Internal Medicine and Human Genetics, University of Michigan, Ann Arbor, Michigan

Charles S. Greenberg, M.D. Associate Professor, Department of Medicine, Duke University Medical Center, Durham, North Carolina

Ulla Hedner, M.D. Vice President, Biopharmaceuticals Division, Research, Novo Nordisk A/S, Gentofte, Denmark

Katherine A. High, M.D. Director, Hematology Laboratories, Children's Hospital of Philadelphia, and Associate Professor of Pediatrics and Laboratory Medicine and Member of the Institute for Human Gene Therapy, University of Pennsylvania Medical Center, Philadelphia, Pennsylvania

William H. Kane, M.D., Ph.D. Assistant Professor, Division of Hematology-Oncology, Department of Medicine, Duke University Medical Center, Durham, North Carolina

Kenneth Kaushansky, M.D. Associate Professor of Medicine, Division of Hematology, University of Washington, Seattle, Washington

Frank G. Keller, M.D. Assistant Professor, Division of Pediatric Hematology-Oncology, Department of Pediatrics, West Virginia University Children's Hospital, Morgantown, West Virginia

Tetsuhito Kojima, M.D. Assistant Professor of Medicine, First Department of Internal Medicine, Nagoya University School of Medicine, Nagoya, Japan

Thung-Shenq Lai, Ph.D. Research Associate, Department of Medicine, Duke University Medical Center, Durham, North Carolina

Peter J. Larson, M.D. Instructor, Department of Pediatrics, University of Pennsylvania Medical Center and Assistant Director, Blood Bank, Department of Clinical Laboratories, Children's Hospital of Philadelphia, Philadelphia, Pennsylvania

Jack Lawler, Ph.D. Associate Professor, Department of Pathology, Brigham and Women's Hospital and Harvard University Medical School, Boston, Massachusetts

Daniel A. Lawrence, Ph.D. Research Investigator, Department of Internal Medicine, University of Michigan, Ann Arbor, Michigan

Si Lok, Ph.D. Senior Scientist, Cytokine Research Center, ZymoGenetics, Inc., Seattle, Washington

Susan T. Lord, Ph.D. Associate Professor, Department of Pathology, University of North Carolina at Chapel Hill, Chapel Hill, North Carolina

Edwin L. Madison, Ph.D. Assistant Member, Department of Vascular Biology, The Scripps Research Institute, La Jolla, California

Daniel P. Morris Department of Biology, University of North Carolina at Chapel Hill, Chapel Hill, North Carolina

James H. Morrissey, Ph.D. Associate Member, Cardiovascular Biology Research Program, Oklahoma Medical Research Foundation, Oklahoma City, Oklahoma

Stephen C. Nelson, M.D. Fellow in Hematology-Oncology, Department of Pediatrics, Duke University Medical Center, Durham, North Carolina

Thomas L. Ortel, M.D., Ph.D. Assistant Professor, Division of Hematology-Oncology, Department of Medicine, Duke University Medical Center, Durham, North Carolina

Lars C. Petersen, D.Sc. Biopharmaceuticals Division, Research, Novo Nordisk A/S, Gentofte, Denmark

Jeanne E. Phillips, Ph.D.* University of North Carolina at Chapel Hill, Chapel Hill, North Carolina

Salvatore V. Pizzo, M.D., Ph.D. Professor and Chairman, Department of Pathology, Duke University Medical Center, Durham, North Carolina

Mortimer Poncz, M.D. Professor, Department of Pediatrics, Children's Hospital of Philadelphia, Philadelphia, Pennsylvania

Harold R. Roberts, M.D. Chief, Division of Hematology, and Sarah Graham Kenan Professor of Medicine and Pathology, University of North Carolina at Chapel Hill School of Medicine, Chapel Hill, North Carolina

Gerald J. Roth, M.D. Professor, Department of Medicine, Seattle Veterans Hospital, and University of Washington, Seattle, Washington

Zaverio M. Ruggeri, M.D. Member, Department of Molecular and Experimental Medicine, The Scripps Research Institute, La Jolla, California

Hidehiko Saito, M.D. Chairman and Professor of Medicine, First Department of Internal Medicine, Nagoya University School of Medicine, Nagoya, Japan

Samuel A. Santoro, M.D., Ph.D. Professor, Departments of Pathology and Medicine, Washington University School of Medicine, St. Louis, Missouri

William P. Sheffield, Ph.D. Assistant Professor, Department of Pathology, McMaster University, Hamilton, Ontario, Canada

Rebecca A. Shirk, Ph.D.† University of North Carolina at Chapel Hill, Chapel Hill, North Carolina

* Present affiliation: Emory University School of Medicine, Atlanta, Georgia.
† Present affiliation: Bowman Gray School of Medicine, Winston-Salem, North Carolina.

M. Sharon Stack, Ph.D. Assistant Professor, Department of Obstetrics and Gynecology, Northwestern University Medical School, Chicago, Illinois

Darrel W. Stafford, M.D. Professor, Department of Biology, University of North Carolina at Chapel Hill, Chapel Hill, North Carolina

Koji Suzuki, Ph.D. Professor, Department of Molecular Pathobiology, Mie University School of Medicine, Mie, Japan

Aldo Hugo Tabares, M.D. Coagulation Fellow, Hospital Laboratories, University of North Carolina at Chapel Hill School of Medicine, Chapel Hill, North Carolina

E. G. D. Tuddenham, M.D., F.R.C.P., F.R.C. Path. Professor of Haemostasis, Haematology and MRC Clinical Sciences Centre, Royal Postgraduate Medical School, Hammersmith Hospital, London, England

Jerry Ware, Ph.D. Assistant Member, Department of Molecular and Experimental Medicine, Scripps Research Institute, La Jolla, California

Herbert H. Watzke, M.D. Associate Professor, First Department of Medicine, University of Vienna, Vienna, Austria

Peter Wildgoose, Ph.D. Physician and Senior Scientist, Selectide Corporation, Tucson, Arizona

David J. Wright, Ph.D. Department of Biology, University of North Carolina at Chapel Hill, Chapel Hill, North Carolina

Tze-Chein Wun, Ph.D. Fellow, Department of Protein Biochemistry, Searle/Monsanto Company, St. Louis, Missouri

Ye I. Wu, M.D. Department of Pathology, McMaster University, Hamilton, Ontario, Canada

Molecular Basis of Thrombosis and Hemostasis

edited by

Katherine A. High
Children's Hospital of Philadelphia
and University of Pennsylvania Medical Center
Philadelphia, Pennsylvania

Harold R. Roberts
University of North Carolina at Chapel Hill School of Medicine
Chapel Hill, North Carolina

Marcel Dekker, Inc. New York • Basel • Hong Kong

1

Molecular Genetics: An Overview

Peter J. Larson and Katherine A. High
University of Pennsylvania Medical Center, and Children's Hospital of Philadelphia, Philadelphia, Pennsylvania

I. INTRODUCTION

This volume summarizes recent advances in the field of hemostasis and thrombosis, with an emphasis on findings that have emerged through the application of molecular biology to coagulation. Molecular biology, through a synthesis of experimental techniques derived from genetics, microbiology, cell biology, and biochemistry, has permitted the detailed characterization of proteins involved in coagulation. The unequivocal prediction of amino acid sequences from cDNA data, the synthesis in mammalian expression systems of large quantities of proteins that were difficult to purify from natural sources, and the ability to probe protein-protein interactions through the synthesis of mutant molecules are all examples of the application of molecular genetic techniques that have advanced our understanding of coagulation. In a sense, advances in the past 15 years have focused on the proteins themselves; that is, molecular genetic techniques have been employed to gain a better understanding of the structure and function of proteins involved in hemostasis and thrombosis. The molecular mechanisms that regulate the expression of these proteins have been less well explored. It is now well recognized, for example, that endothelial cells express a number of proteins that have both procoagulant and anticoagulant effects. How expression of these proteins is regulated and how gene expression may be modulated by cellular environmental factors, such as shear stress or inflammation, are poorly understood and are areas of active current investigation. It is likely that these issues related to gene regulation will be increasingly well defined by molecular genetic techniques over the next decade.

By way of introduction to the terminology and applications of molecular biology in coagulation research, this chapter provides a review of some of the basic concepts of molecular genetics. An additional section describes some of the most frequently used techniques of molecular biology, including their applications and limitations. This chapter is not intended to be encyclopedic in scope, nor is it a manual of laboratory methods, but rather it is a reference point for those with a less comprehensive knowledge base in molecular biology.

II. BIOCHEMICAL BASIS OF INHERITANCE

A. Nucleic Acid Structure

A *gene* is defined as the functional unit of inheritance, and it is now known that each gene is a nucleic acid sequence carrying the information necessary for the formation of a specific RNA or polypeptide. Early work by Mendel predicted that a factor, responsible for heredity, was passed unchanged from parent to progeny. It has been known since the 1860s that chromosomes, composed of deoxyribonucleic acid (DNA) and protein, were responsible for heredity, and Avery, McLeod, and McCarty in 1944 showed that the critical factor dictating inheritance was DNA, not protein, as was previously hypothesized (1). Using x-ray crystallographic data, Watson and Crick proposed in 1953 the double-helical structure of deoxyribonucleic acid (2). The necessary genetic information for the expression and propagation of a particular organism is contained in this biomolecule.

1. The DNA Molecule

Macromolecular DNA is an unbranched polymer consisting of purine (adenine and guanine) or pyrimidine (cytosine and thymine) bases bound to deoxyribose (Fig. 1). These bases are joined together by phosphate linkages between the 5′ carbon of one deoxyribose and the 3′ carbon of the adjacent deoxyribose. This 5′ → 3′ linkage defines a direction in the linear sequence of bases in nucleic acids. Thus a DNA sequence has directionality, and the coding information contained is critically dependent on it (5′-AGGCAT-3′ is a different molecule from 3′-AGGCAT-5′). The DNA sequence complementary to the 5′ → 3′ strand is coupled to the original strand by hydrogen bonding between a purine and a pyrimidine (base pairing), such that cytosine normally can bond only to its complementary base, guanine, and adenine to thymine (Fig. 1). RNA (ribonucleic acid) is a similar polymer in which the backbone is composed of ribose rather than deoxyribose. In addition, another pyrimidine (uracil) is substituted for thymine in RNA.

The transmission of genetic information (*inheritance*) is accomplished by the semiconservative replication of the double-stranded DNA molecule. The double helix is separated, and replication occurs by the assembly of complementary bases on templates formed by each of the separated strands (Fig. 2). This polymerization is carried out by a DNA-dependent DNA polymerase that adds bases to the elongating complementary strand in a 5′ → 3′ direction. An enzyme capable of ligating DNA molecules is also required to join discontinuous stretches of replicated DNA.

The *expression* of genetic information is accomplished by the conversion of DNA sequence information to an intermediary molecule composed of ribonucleic acid. This RNA subsequently directs the formation of polypeptides in a process called translation. The unidirectional transfer of information from DNA sequence to polypeptide via a messenger RNA template is the central dogma of molecular genetics (Fig. 3).

2. Organization of Cellular DNA

In eukaryotic cells, DNA is complexed with protein and in this form is referred to as chromatin. Euchromatin is a less densely organized DNA-protein complex; heterochromatin is DNA condensed in a form comparable to chromosomes visible at mitosis. Histones are a group of basic DNA binding proteins present in the nuclei of eukaryotes. The remarkable sequence homology between histones from widely different species implies a critical function for these proteins; they are the most prominent of the DNA binding proteins. DNA and histones are organized into nucleosomes compromised of approximately 200 bp of DNA, 146 bp of which are wrapped around a histone core of eight protein subunits (Fig. 4). Nucleosomes are further organized into a compact helical structure (solenoid) consisting of six nucleosomes per turn. The solenoid may be further condensed into large supercoiled loops. It is likely that the solenoid structure of chromatin must loosen for transcription to occur, and this is probably accomplished by modifications to the histones, such as acetylation. Scaffolding proteins are complexed with histone-associated DNA and are responsible for packing the condensed DNA into its metaphase chromosomal length. The major scaffolding protein is topoisomerase II, which also functions enzymatically to cleave and reseal double-stranded DNA. Topoisomerase II thus fulfills a critical role in DNA replication and can relieve torsional stress and tangles in the DNA molecule. Other major DNA binding proteins exist that have yet to be fully characterized. Minor DNA binding proteins, present in small amounts in the cell, function to regulate transcription and are discussed later in more detail.

B. Expression of Genetic Information Encoded by Nucleotide Sequence

1. The Genetic Code

Three-nucleotide units (codons) are the means by which nucleic acid sequence information is translated into the amino acid sequence of polypeptides. There are 64 possible trinucleotide combinations of the four bases forming RNA or DNA. Each three-base combination of nucleotides either codes for a specific amino acid or signals for the termination of translation (stop or termination codon; Table 1). Approximately 20 amino acids are utilized in most organisms in translation; therefore, some amino acids are encoded by more than one codon. Each codon, however, is specific for only one amino acid or the stop signal. Thus the genetic code is degenerate but not ambiguous.

2. Transcription

Transcription is the process by which the DNA-dependent RNA polymerase copies regions of macromolecular DNA, forming an RNA template. RNAs destined to encode proteins are known as messenger RNAs (mRNAs). This messenger RNA (mRNA) is then transported to the cytoplasm, where it directs protein synthesis. Transcription, like DNA replication, occurs in a $5' \rightarrow 3'$ direction. The DNA strand with the sequence identity to an mRNA is called the coding strand (or sense strand); the strand complementary to it and, thus, the template from which a particular mRNA is formed, is the noncoding or antisense strand (Fig. 3). In viruses and prokaryotes, there is an exact equivalence between the coding sequence and the amino acid sequence in the translated protein; however, in eukaryotes this equivalence exists for mature mRNA but usually not for the DNA sequences encoding the protein. DNA sequences that encode portions of the mature protein are known as exons; these exons are separated by intervening sequences (introns) that are not translated (Fig. 5). In eukaryotic organisms, transcription occurs in the nucleus, where the RNA polymerase produces an RNA copy that is an exact copy of the genomic DNA (with the exception that the base uracil is substituted for thymine).

3. mRNA Processing

Eukaryotic pre-mRNAs undergo a number of modifications, collectively referred to as processing, which result in generation of a mature mRNA (Fig. 5). The intronic se-

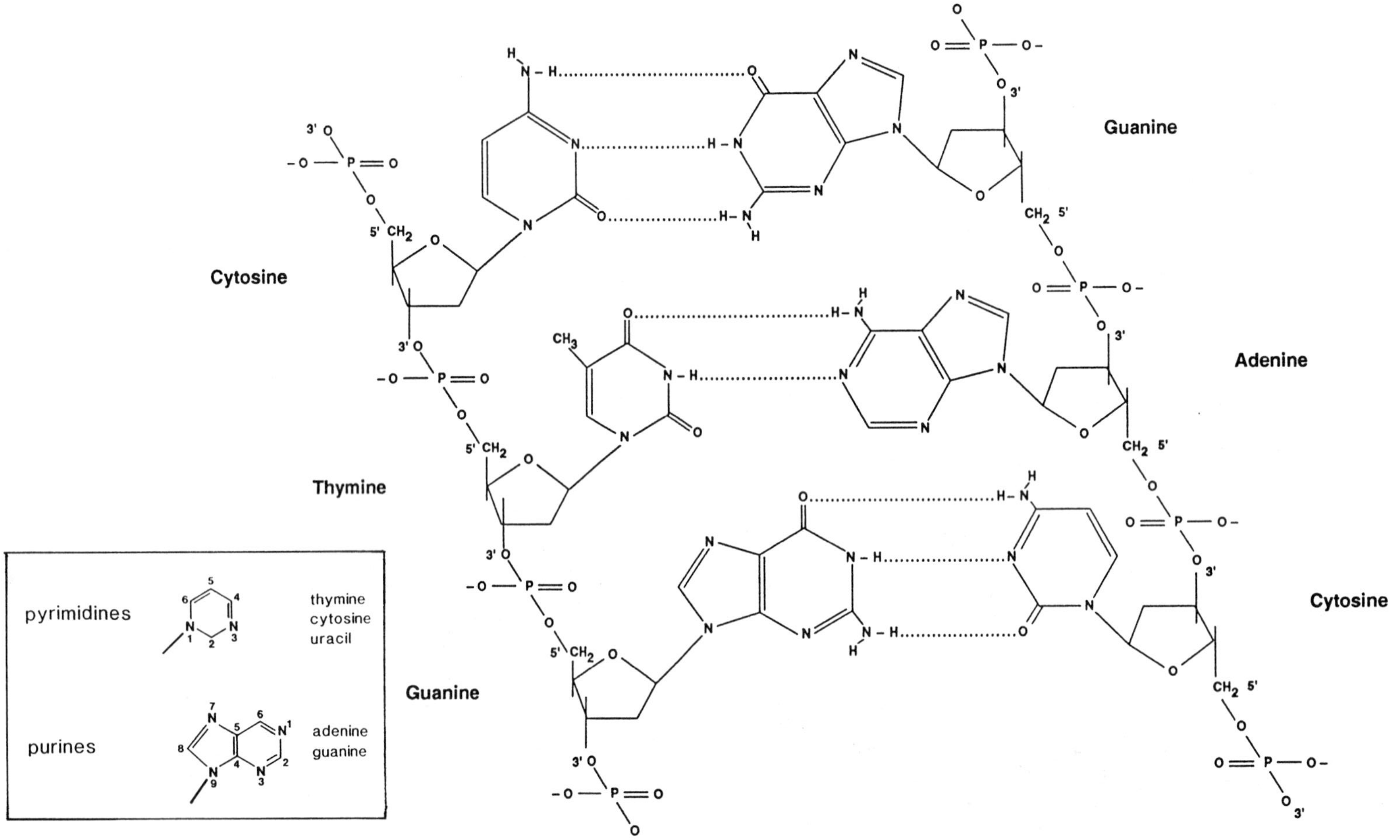

Guanine

Cytosine

Adenine

Thymine

Cytosine

Guanine

pyrimidines	thymine cytosine uracil
purines	adenine guanine

Figure 1

quences, which are not present in the mRNA, must be removed in a process called RNA splicing, producing an mRNA consisting of only the exons joined together. Splicing results in the joining of exons in the order in which they lie on the gene; thus, following processing, the linear relationship that exists in prokaryotes between the gene and protein (colinearity) also exists for eukaryotes. The intronic sequence, beginning with a GU

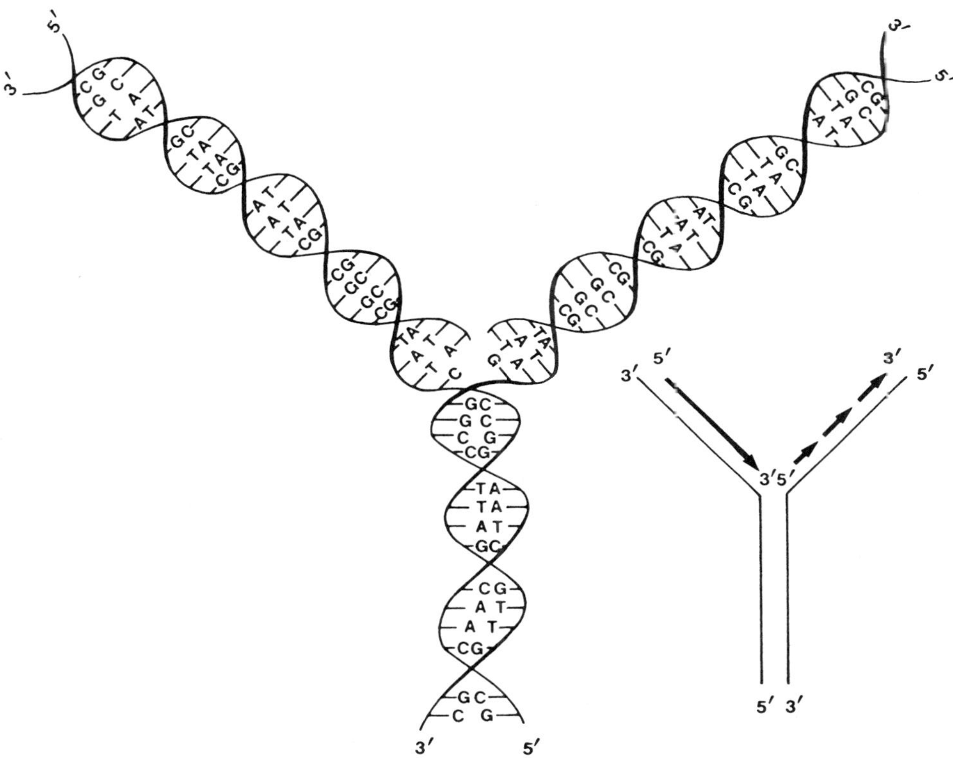

Figure 2 DNA replication occurs by the separation of the double-stranded parental molecule followed by polymerization of a new strand through complementary base pairing. The catalysis of this polymerization by DNA polymerase occurs in a 5′ to 3′ direction; shorter segments formed by DNA polymerase on the 3′ → 5′ parental strand must be joined together by a DNA ligase.

Figure 1 DNA structure. DNA is composed of a polymer of deoxyribose monomers linked by phosphate residues between the 5′ carbon of one sugar and the 3′ residue of the adjacent sugar. A purine or pyrimidine base is linked to deoxyribose at carbon 1′, and the order of these bases along the polymer conveys the genetic information contained in nucleic acids. A second polymer, oriented in the opposite direction with respect to its phosphate linkages, is coupled to the first by hydrogen bonding between complementary bases on each strand. Complementary base pairing occurs between adenine and thymine side groups and between guanine and cytosine side groups. Inset: purines and pyrimidines are the nitrogenous bases in nucleic acids. Pyrimidines are linked to deoxyribose (or ribose in RNA) at position 1; purines are linked to the sugar group at position 9. By convention, the positions in the bases are referred to by Arabic numerals and those in the sugar residues by a numeral followed by a prime.

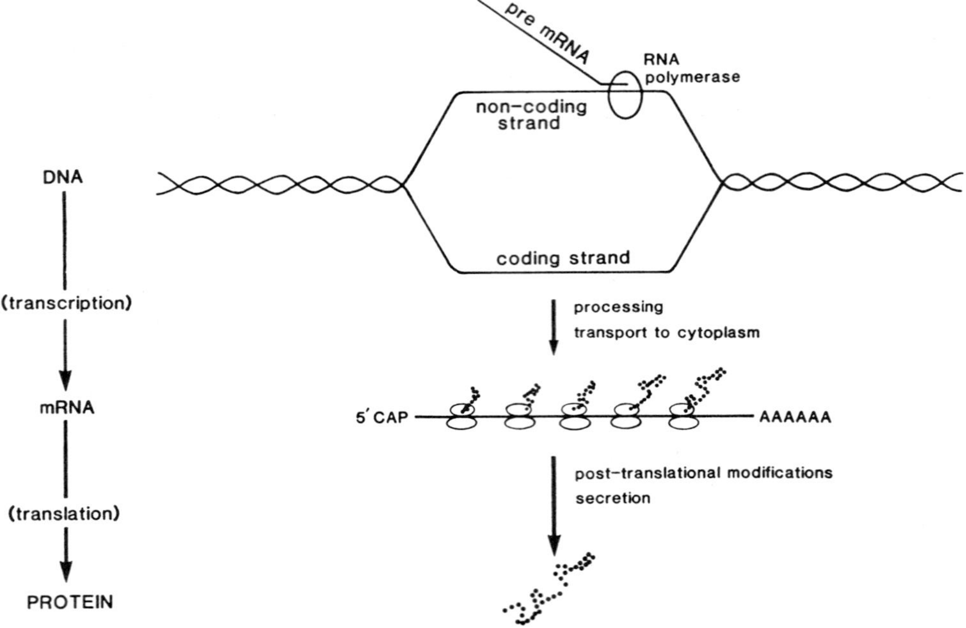

Figure 3 Expression of DNA sequence information. DNA sequence information is expressed through the production of polypeptides or RNAs. RNA polymerase produces an RNA copy of the DNA sequence in the nucleus of eukaryotic cells (transcription). Nucleic acid sequence information contained in messenger RNA is converted to the amino acid sequence of protein in a cytoplasmic process called translation. The unidirectional flow of genetic information from DNA to a messenger molecule (RNA) to protein is the central dogma of molecular genetics.

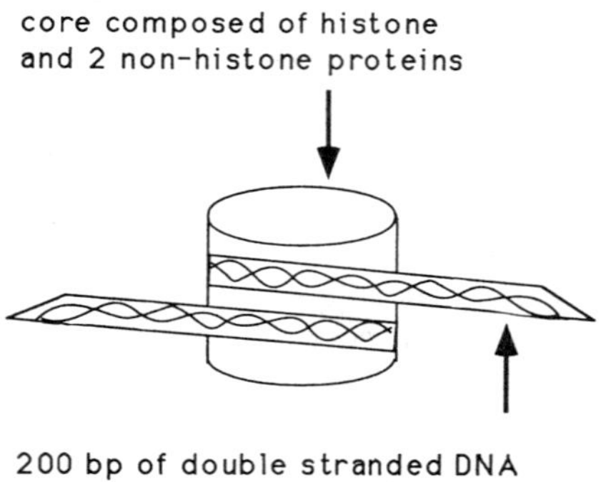

Figure 4 Nucleosome organization. Chromosomal DNA is organized into 200 bp units, 146 bp of which are wound around a protein core. Nucleosomes are further organized into solenoids, with six nucleosomes to a turn (see text).

(splice donor) at the 5′ end and ending with an AG (splice acceptor) at the 3′ end, may interrupt the coding sequence between codons (type 0 introns), between the first and second bases of a codon (type I introns), or between the second and third bases of a codon (type II introns). Other splice sites that are not used (cryptic splice sites) may be present within a transcript. The mechanisms for selection of the correct splice site during processing remains unclear. Mutations at the splice site preclude normal splicing and may unmask cryptic splice sites that result in an mRNA of a different length. This often results in functional changes in the encoded protein. Some genes may undergo alternative splicing. For example, the gene for human factor VII contains two copies of exon 1, one of which codes for an additional 22 amino acids in the pre-proleader sequence; separate species of mRNA (one that contains the additional exon) have been isolated (3). Splicing usually occurs as an intramolecular reaction, and thus mutations occurring in one allele cannot ordinarily be compensated for by a normal sequence on the other allele.

Other modifications of mRNA include "capping," in which the 5′-terminal phosphate of the maturing mRNA is coupled to the 5′ carbon of a 7-methylguanylate. This 5′ cap is required for recognition of the message by the assembling ribosome (the cap binding protein). The cap may also prevent mRNA degradation by RNAses until translation has

Table 1 The Genetic Code: mRNA Codons for Amino Acids

Alanine	Arginine	Asparagine	Aspartic acid	Cysteine
5′—GCU—3′	CGU	AAU	GAU	UGU
GCC	CGC	AAG	GAC	UGC
GCA	CGA			
GCG	AGA			
	AGG			
Glutamic acid	**Glutamine**	**Glycine**	**Histidine**	**Isoleucine**
GAA	CAA	GGU	CAU	AUU
GAG	CAG	GGC	CAC	AUC
		GGA		AUA
		GGG		
Leucine	**Lysine**	**Methionine**	**Phenylalanine**	**Proline**
UUA	AAA	AUG	UUU	CCU
UUG	AAG		UUC	CCC
CUU				CCA
CUC				CCG
CUA				
CUG				
Serine	**Threonine**	**Tryptophan**	**Tyrosine**	**Valine**
UCU	ACU	UGG	UAU	GUU
UCC	ACC		UAC	GUC
UCA	ACA			GUA
UCG	ACG			GUG
AGU				
AGC				
		Chain termination (stop) codons		
		UAA		
		UAG		
		UGA		

occurred. Multiple riboadenosines are added to the 3′ end of the message, forming a region known as the poly(A) tail; this poly(A) tail is removed before the message is degraded. These features, the cap and the poly(A) tail, identify an RNA molecule as a message or transcript. In eukaryotes, mRNAs are transported through pores in the nucleus to the cytoplasm. Other transcribed RNAs, known as ribosomal RNAs (rRNAs), become associated with cytoplasmic proteins, forming functional units known as ribosomes. Ribosomes are the cellular machinery by which mRNA is read and translated into a polypeptide. A third type of RNA, transfer RNAs (tRNAs), are RNA-amino acid complexes that are substrates for the ribosomal-mRNA complex (discussed later).

4. Translation

Translation is the process by which a polypeptide is produced from an mRNA template. During translation, a two-component ribosome assembles on the 5′ end of a mature mRNA through the binding of rRNA sequences contained within the smaller ribosomal subunit to mRNA sequences (Fig. 6). A third species of RNA involved in translation (tRNA) consists of a single aminoacyl group coupled to an RNA molecule. The amino acid bound to a particular tRNA is determined by a specific trinucleotide region in the tRNA. The ribosome-mRNA complex sequentially exposes a three-base codon on the mRNA. A complementary three-base region (the anticodon) on a tRNA binds to the codon in an energy-requiring step, bringing the first amino acid of the peptide into the ribosomal complex. In another energy-requiring step, a second tRNA bearing an amino

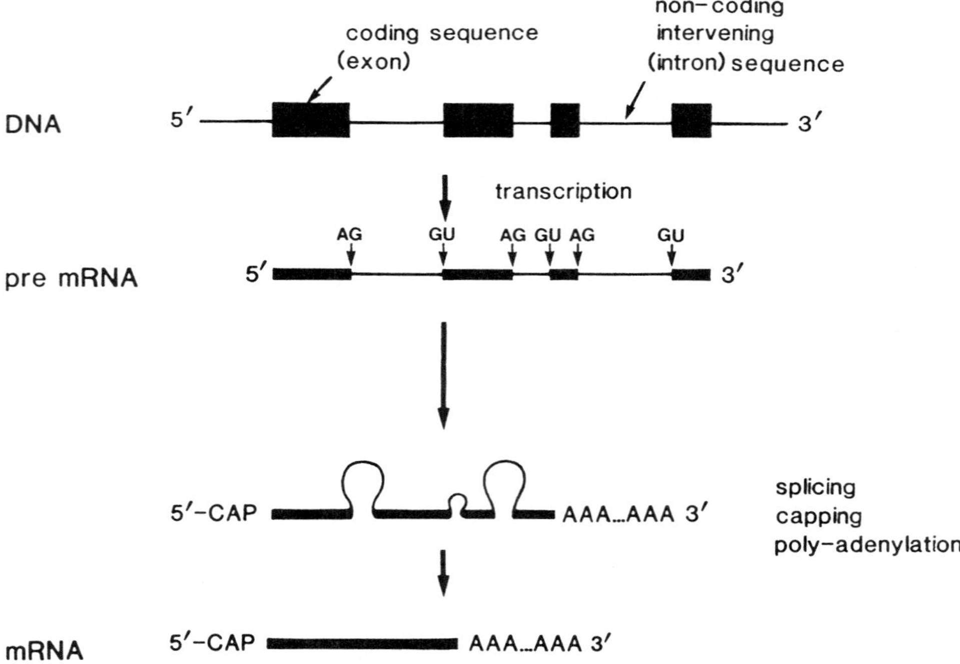

Figure 5 Transcription and mRNA processing. RNA polymerase produces a premessenger RNA from a DNA template. Intronic sequences are removed in a process called splicing (see text): a 7-methylguanylate residue is added to the 5′ end, and multiple riboadenosines are added to the 3′ end of the transcript to form a processed messenger RNA.

acid binds to the next codon exposed in the ribosomal-mRNA complex. The ribosome catalyzes the formation of a peptide bond between the two amino acids; this bond formation frees the first amino acid from its tRNA (Fig. 6). The nascent polypeptide is now tethered to the incoming tRNA at its carboxyl-terminal amino acid. In a third energy-requiring reaction, the ribosome translocates to the next codon and releases the spent tRNA from the ribosomal complex. The process is repeated until a termination codon is reached.

C. Mutation

Changes in DNA sequence can occur as a result of normal cellular processes or interactions with environmental mutagens, such as ionizing radiation and chemical exposure. These changes (mutations) include insertions and duplications, deletions, inversions, translocations, and point mutations. Mutations can occur in the coding or noncoding region of a gene and, depending on their location, can manifest their effect at the level of transcription or translation. A change in DNA sequence is passed on to future generations of dividing cells if the change does not result in cell death.

The substitution of a single nucleotide (point mutation) in the coding region of DNA may result in a codon that specifies the same amino acid; this is referred to as a silent mutation. The new codon may result in the replacement of a new amino acid in the

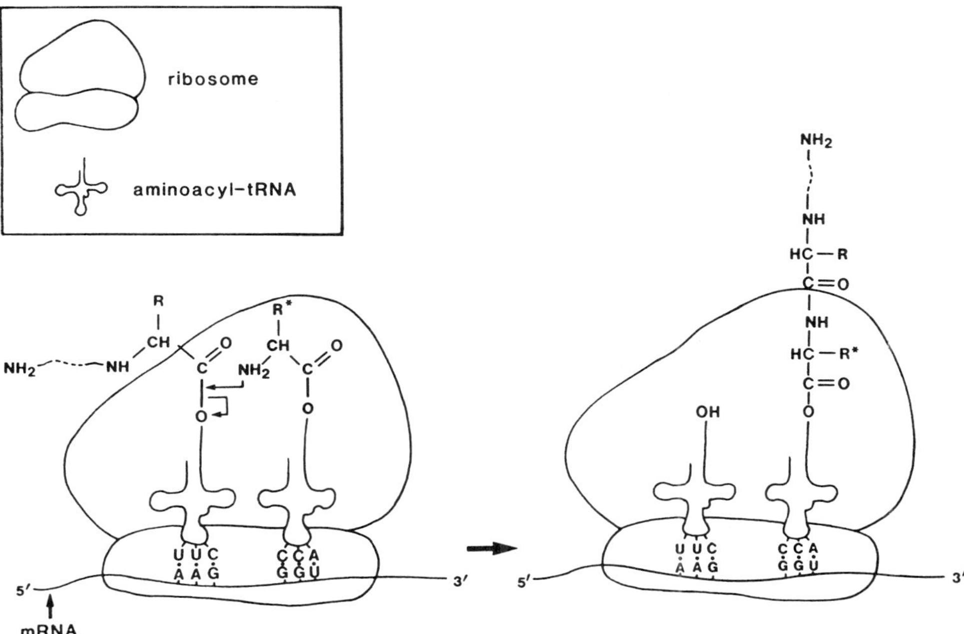

Figure 6 Translation. A two-subunit ribosome composed of protein and RNA assembles at the 5′ end of cytoplasmic mRNA. Trinucleotide codons are sequentially exposed, transfer RNAs associate with the complex by way of specific RNA sequences complementary to the mRNA codon, and the ribosomal complex catalyzes the formation of peptide bonds between the amino acid groups of the tRNAs.

subsequent translated protein (missense mutation) or may produce an early stop codon (nonsense mutation). Point mutations result when a DNA polymerase misincorporates a nucleotide (i.e., accepts a noncomplementary base) during replication or repair or when modification of any of the bases occurs as a result of exposure to environmental compounds. The error rate of *Escherichia coli* DNA polymerase is 1 wrong base in 10,000; however, cellular mechanisms have evolved that can correct these errors.

A common type of base modification involves the methylation of cytosine residues. Spontaneous deamination of these methylated cytosine residues (which occurs with a low but detectable frequency) results in a change from a methylcytosine to a thymidine; subsequent daughter strands then maintain the C → T change (Fig. 7). When the change occurs in the coding strand, it causes a C → T transition in the coding strand. If the deamination of methylcytosine occurs in the noncoding strand, a G → A transition occurs in the coding strand (Fig. 7). Of methylated cytosine residues in vertebrate genomes, 90% occur within CG dinucleotides. Thus, CG dinucleotides are "hot spots" for mutation because of the high frequency of methylated cytosines in these dinucleotides and the propensity of these methylated cytosine residues to undergo deamination.

Insertion and duplication or deletion mutations occur when fragments of DNA are either added or lost. These fragments may be very large (kb) or small (one base). Small deletions and insertions can be the result of enzymatic errors that occur during DNA polymerization or repair. Larger deletions and insertions are the result of the unequal exchange of material between or within chromosomes. If insertions or deletions occur within the DNA coding region for a protein, new amino acids are added or deleted during translation. If the number of bases added or deleted is not divisible by 3, a change in the ribosomal reading frame (or frameshift) occurs, resulting in a protein with an amino acid sequence markedly divergent from the wild-type sequence following the insertion or deletion. Frameshift mutations often result in early termination codons.

III. MAPPING THE GENOME

A. Linkage Analysis

Genetic recombination is a normal process that occurs during meiosis when homologous chromosomes become closely associated and an exchange of genetic material occurs. The frequency of this crossover reaction is a basis for determining the *linkage* of one gene to another on chromosomal DNA. The distance separating one gene or locus from another is proportional to the frequency of recombination, because longer distances result in more frequent recombination events. Shorter distances, on the other hand, result in segregation of the two loci together. Detectable phenotypic markers can sometimes be determined to segregate with a disease of interest. For example, a phenotypic marker, such as a particular red blood cell antigen, may segregate with a disease phenotype; in this case the gene encoding the marker (red cell antigen) is linked to the disease gene, and the two genes lie on the same chromosome. Linkage in human pedigrees is difficult to determine using phenotypic markers, however, because of the small number of progeny per mating and the low number of generations available for analysis.

B. Restriction Fragment Length Polymorphisms

Given the problems with phenotypic markers, a more direct approach is needed to define a detailed map of the human genome. A widely used method to generate genotypic

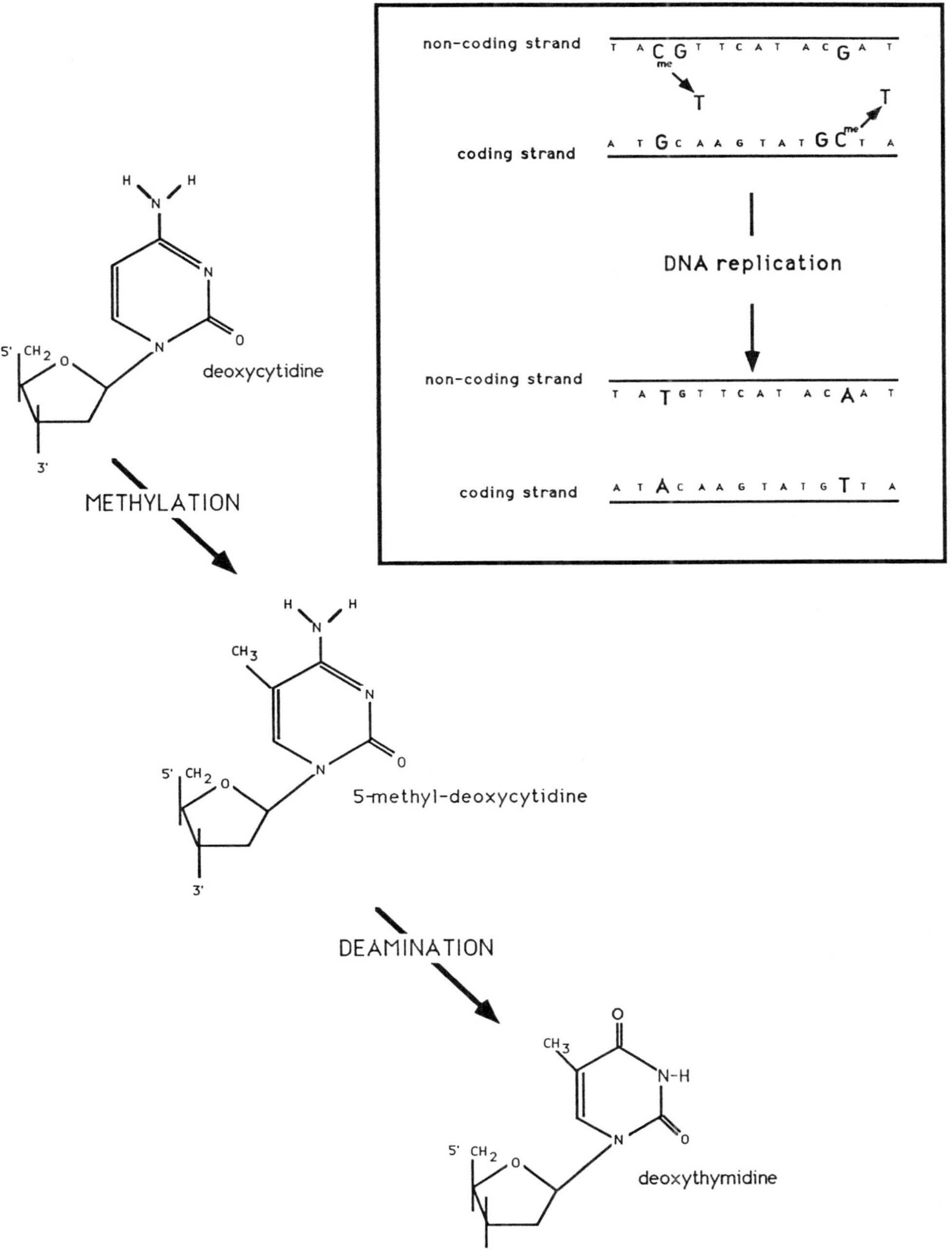

Figure 7 Deamination of methylated cytosines is responsible for some point mutations. Inset: when deamination of methylcytosine occurs in one strand, it results in a C → T transition in that strand and a G → A transition in the opposite strand.

markers is based on the variability of DNA sequences (polymorphisms) and their detection by restriction enzymes. A *polymorphism* is the occurrence of two or more genetically determined genotypes that coexist within a population at frequencies that cannot be attributed to recurrent mutation. Polymorphisms occur at a rate of 1 in 1000 base pairs and can be located in the coding region of a gene or (more often) outside the coding region. These polymorphic sequences can often be detected by digestion of DNA with certain bacterial enzymes (restriction endonucleases). Restriction endonucleases recognize and digest DNA at specific sequences, reducing large DNA molecules into a collection of smaller fragments. Restriction analysis involves determination of the lengths of these fragments generated by restriction endonuclease digestion; this is easily accomplished by electrophoretic separation. Digestions of genomic DNA from different individuals may result in differences in fragment lengths caused by the presence of polymorphic sequencesx that create or destroy endonuclease recognition sequences. Polymorphic sequences that can be detected by restriction analysis are referred to as restriction fragment length polymorphisms (RFLPs). A particular RFLP that segregates with a gene of interest can then be used as a marker for the gene (linked) by virtue of its proximity to the target gene, such that the marker and the disease gene segregate together (linkage). A more sensitive genetic map can be generated by these genotypic (RFLP) markers.

C. Variable Number of Tandem Repeats (VNTRs)

Human genomic DNA contains regions consisting of a variable number of short repeated sequences; these repeats are highly polymorphic, in that a sequence can be repeated anywhere from 2 to greater than 20 times. The function of these repeat elements remains unknown. A restriction enzyme that cuts outside the repeat element results in a fragment whose size is related to the number of repeats. These fragments can then be detected by hybridization to a probe specific for the repeated element (see Sec. VI. H). Because the number of repeats may vary widely between homologous chromosomes, these VNTRs have proven extremely useful as markers. In contrast to RFLPs, in which only the presence or absence of enzymatic cleavage can be detected (i.e., there are only two possible alleles), VNTRs present the possibility for multiple alleles that vary by the number of concatenated repeat elements. The ability to detect multiple alleles at a given locus affords a greater discrimination of genetic differences between individuals. Many of the repeat elements contain sequences that are similar, allowing the use of a single probe to detect several VNTRs. The detection of several VNTRs with one or more probes results in a "DNA fingerprint" capable of distinguishing each human. VNTRs can be used as markers for the mapping of a gene of interest and the detection of pathological mutations.

D. Assigning a Gene to a Specific Chromosome

If a gene of interest can be shown by linkage analysis to be associated with a previously mapped gene, it can be assigned to the same chromosome on which the identified gene lies. Cloned DNA can be labeled (radioactive or fluorescent) and hybridized to a metaphase chromosomal preparation. A single chromosomal site can be determined by scoring multiple metaphase preparations for probe location, thus controlling for nonspecific hybridization. Another method used to map genes to a specific chromosome with DNA probes makes use of human-rodent fusion cells (hybridomas). Formation of these hybridomas results in cell lines that contain a variable number of human chromosomes; these cell lines are then characterized for human chromosomal content by karyotype and expression markers. DNA is prepared from these hybridoma cell lines, and the presence of the gene of interest is determined by the hybridization of hybridoma DNA to a probe specific for the gene of interest (see Sec VI. G). Chromosome location is then determined by the pattern of reactivity with the panel of characterized hybridomas.

E. The Human Genome Project

In 1990, the U.S. Human Genome Project was officially funded by the National Institutes of Health and the Department of Energy to develop genetic and physical maps of the human genome, with a goal of a completely sequenced human genome by the year 2005. The goal of phase 1 of the project is to develop a map of useful genetic markers at approximately 1 centimorgan (cM) intervals (a centimorgan is defined as the nucleotide distance that results in a recombination frequency of 1% and is equal to approximately 10^6 bases in the human genome). This map will be useful in the identification of genetic loci in kindreds with heritable disorders. A genetic map, placing genes in a correct linear arrangement on chromosomes, should be available by 1994 at a resolution of 1–2 cM.

The second phase involves placing markers (phenotypic and genetic) on chromosomal maps. One type of physical marker is comprised of restriction enzyme sites that are unusual in the human genome (positioned approximately 10^6 bp apart). Sequence tagged sites (STS) constitute another set of useful genome markers. STSs are unique sequences on a particular chromosome that are identified by oligonucleotide primed DNA amplification using the polymerase chain reaction (PCR; see Sec. VI. F). This map will provide actual physical distances between the markers.

To clone a specific gene that has been localized to a region defined by this physical map, a series of overlapping segments of cloned DNA must be developed that encompass the region between the physical markers. These clones must then be ordered based on similarities in their regions of overlap. This can be accomplished by DNA fingerprinting using, for example, restriction digests or inter-Alu PCR. The development of new cloning vectors, such as yeast artificial chromosomes (Sec. VI. E. 4), capable of carrying inserts of up to 10^6 bases, has made this goal imminently achievable.

Finally, determination of the nucleotide sequence of these overlapping clones will provide a basis for searching for specific genes. For example, if a disease locus can be mapped to a region contained in a specific chromosomal clone and the sequence of that clone is available, candidate genes can be identified by analyzing the sequence for regions corresponding to potential coding regions. The identification of coding regions relies on the observation that exonic sequences of expressed genes lack stop codons and have a propensity to contain certain codons (codon usage bias). Once coding sequence is determined, protein sequence can then be deduced. Knowing the primary amino acid sequence for a protein encoded by the gene of interest will provide insight into the gene's functional relationship to the disease state. Advances in computer analysis of nucleotide sequence and improvements in nucleotide sequencing technology are necessary to allow the completion of this phase by 2005.

IV. ORGANIZATION OF THE GENE (ANATOMY OF THE GENE)

A gene consists of DNA sequences that are necessary to produce a single peptide or RNA and includes sequences that control production of the macromolecule as well as those that encode the macromolecular product itself. This section focuses on genes that encode protein.

A. Coding Region

The coding region of a gene consists of that part of the nucleotide sequence that corresponds to the amino acid sequence of the encoded protein (in eukaryotes, the exons). It is apparent that the structure of proteins has developed by concatenation and reorganization of the exons encoding structural domains. For example, in the vitamin K-dependent coagulation serine proteases, similarities exist in exon organization, suggesting that these genes arose by divergence from a primordial digestive serine protease through exon shuffling and exon duplication (Fig. 8). The functional differences among these proteins are in part a consequence of these changes in gene organization. The vitamin K-dependent serine proteases are enzymes that carry out limited proteolysis on specific substrate proteins. They require specific intermolecular interactions with cofactors. Divergence in exon sequences has arisen through point mutations and most likely accounts for the different cofactor and substrate specificities of these proteins. Phylogenetically, there is evidence that larger ancestral exons became divided into a set of smaller exons through the insertion of intronic sequences, eventually leading to a uniformity in exon length.

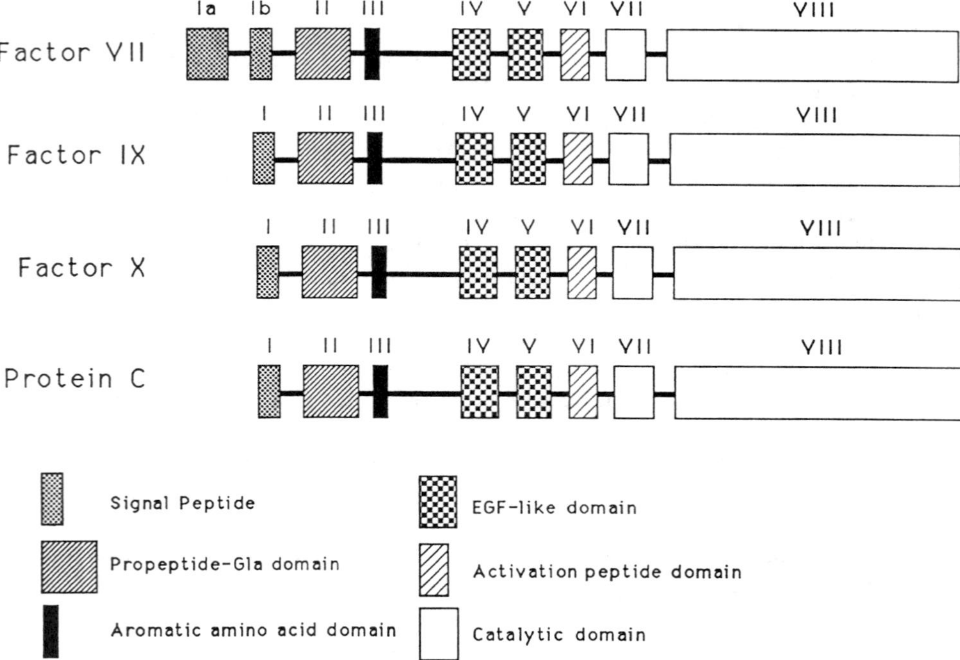

Figure 8 Gene organization of the vitamin K-dependent clotting factors. The genes encoding some of the vitamin K-dependent coagulation proteins have strikingly similar organizations. The exons correspond to the functional domains of the proteins. (Adapted from Furie and Furie. Cell 1988; 53:505–518.)

B. The 5′ Flanking Region

The sequence upstream (or 5′) from the coding region of a gene contains DNA sequences that control the proper initiation and regulation of transcription. For groups of proteins that are transcribed in a concerted fashion (e.g., genes encoding globin chains), there are regions responsible for the activation of a series of genes known as locus-activating or locus control regions. These may lie many kilobases away from the start site of transcription. Intervening regions may contain one or several DNA sequences that bind proteins and are capable of directing the initiation of RNA synthesis at a particular site and in a particular orientation. These sequences, taken together, are known as promoter elements, which in concert with their binding proteins stabilize the binding of the DNA-dependent RNA polymerase. The proteins that bind at these sites are known as transcription factors, or *trans-acting* elements, in contrast to the DNA sequences themselves, which lie in proximity to the start site of transcription and are known as *cis-acting* elements. Other sequences, usually upstream from the promoter region, bind proteins that interact with transcription factors in the vicinity of the promoter and enhance initiation. These sequences, called enhancers, are position and orientation (with respect to the promoter) independent. Enhancer elements are not confined to the 5′ flanking region; they have been found within introns and/or the 3′-flanking region.

C. The 3′ Flanking Region

This region in mRNA may contain sequences that affect mRNA stability, because of either secondary structure or the binding of elements important for delaying or enhancing mRNA degradation. Some RNA polymerase enhancer elements have been described that are present in the 3′ flanking region of genomic DNA sequence.

V. REGULATION OF EXPRESSION: TRANSCRIPTION AND TRANSLATION

Each cell in an organism has the same genetic (DNA) content. What differentiates one cell from another phenotypically is the genes that are expressed. Regulatory sequences in the DNA, combined with specific nuclear proteins that bind to these sequences and activate transcription, are the agents that control the execution of the cell-specific program. These elements together bring about both tissue-specific and temporal changes in expression. Such elements can also bring about changes in expression in response to environmental stimuli. For example, in the promoter region of the platelet-derived growth factor β chain, investigators recently identified a shear-stress response element (SSRE) (4).

A. Cis-Acting Factors

1. Promoters.

Promoters are regions of DNA involved in the binding of RNA polymerase and, therefore, are operationally defined as lying in proximity to the start site of transcription. In eukaryotic cells, the polymerase does not initiate transcription alone and requires at least six other trans-acting (accessory DNA binding) factors: TFIID, TFIIA, TFIIB, TFIIE, TFIIF, and TFIIH (for review, see Ref. 5). Although there is no conservation of sequence at the start site of transcription, there is a tendency for the first base of mRNA to be an adenine flanked on either side by pyrimidines. In comparing the promoters of mammalian genes, the only homologous sequences (in proximity to the start site) tend to be short stretches of DNA. Many promoters contain an 8 bp module called the TATA box. The TATA box almost always contains only A-T base pairs, lies approximately 25 bp upstream from the start site, and is homologous to bacterial promoter elements. The CAAT box is another common module present approximately 80 bp upstream of the start site and named for its consensus sequence. Mutagenesis studies suggest that, of all promoter elements, the CAAT box has the greatest effect on transcription initiation. The GC box contains the sequence GGGCGG, occurs in either orientation, and is often present in multiple copies. The GC and CAAT boxes are believed to bind proteins that stimulate the formation or activity of the basal transcription complex of RNA polymerase and its cofactors. The TATA box, through its binding of a component of TFIID called the TATA box binding protein (TBP), may determine the start site of transcription. Promoters that lack a TATA box often have multiple start sites of transcription.

2. Enhancers and Silencers

Enhancers are DNA sequences capable of stimulating promoter function. These cis-acting elements resemble promoter elements in that they are composed of protein binding modules, and some modules found in promoters are also found in enhancers—indeed the distinction between enhancer elements and promoter elements has become less clear as mechanisms of transcription initiation are better understood. Enhancers can act at great distances from the promoter and can lie in either orientation and on either side of the gene. They may function by binding proteins that then interact with transcription factors

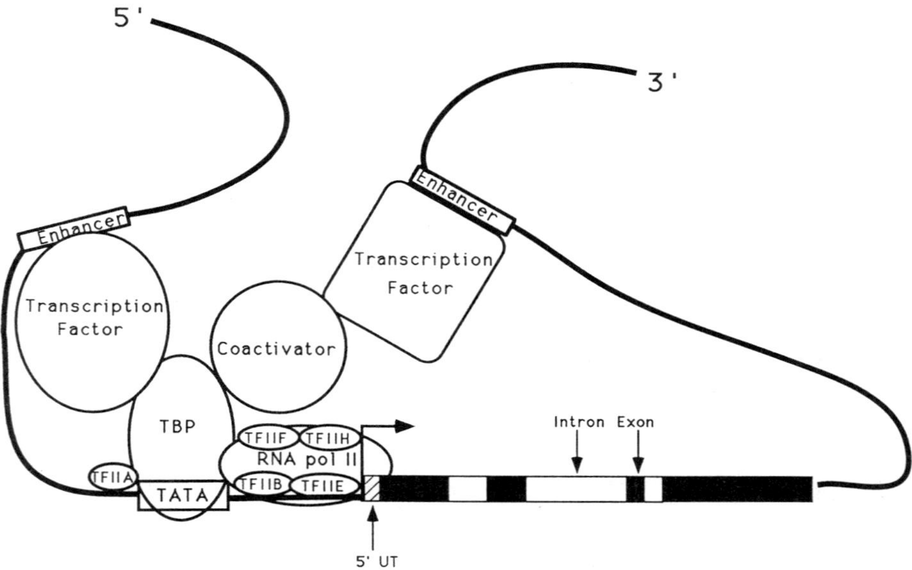

Figure 9 Transcription factors may interact with the basal transcription complex. The basal transcription complex, consisting of RNA polymerase II and its cofactors (TFIIA, TFIIB, TFIIE, TFIIF, TFIIH, and TBP), binds to DNA sequences in the 5′ region flanking the gene. Other transcription factors bind DNA at more distal regions, either 3′ or 5′ to the basal transcription complex. These transcription factors may interact directly or through intermediary proteins (coactivators).

bound to the promoter and effectively "loop out" intervening DNA (Fig. 9). Enhancers may be targets for tissue-specific or temporal regulatory elements. A novel class of tissue-specific cis-acting sequences, identified recently in the promoter region of the platelet-derived growth factor B gene, is the shear-stress–related element (4). This 12 bp sequence is required for the upregulation of transcription that occurs when responsive cells (endothelial cells) are subjected to laminar shear stress. The cognate binding protein is not yet identified. Other genes expressed in endothelial cells, including tissue plasminogen activator, intercellular adhesion molecule 1, and transforming growth factor, also contain this 12 bp element. Silencers are DNA sequences that, through their binding proteins, inhibit transcription, usually by interfering with the interaction of the basal transcription complex with promoter elements or proteins that activate the complex.

B. Transcription Factors

Proteins that bind DNA and comprise the transcription apparatus can be divided into several categories: the RNA polymerase; proteins that bind RNA polymerase when it forms the basal transcription complex but are not associated with free polymerase; and proteins that bind specific DNA sequences in the target promoter. This last class of proteins, with specificity for defined DNA sequences, confers on a cell type the property of temporal or tissue-specific transcription. A transcription factor may be tissue specific because (1) it is expressed only in certain cells; (2) its activity may be controlled by a cell-specific modification, such as phosphorylation; (3) it may be activated or imported from the cytoplasm to the nucleus by binding a tissue-specific ligand (e.g., steroid receptors); or (4) it may be available only in cells that lack inhibitory binding proteins (factors that sequester the transcription factor in the cytoplasm). These specific transcription factors are the best characterized. The number of well-characterized transcription factors has increased rapidly in the past few years; a database of vertebrate transcription factors was recently published (6).

The binding of transcription factors to DNA may either promote or block transcription. Transcription factors may act to stabilize chromatin in a more loosened configuration necessary for access of the basal transcription complex to its critical binding sequences or may stabilize the binding of the initiation complex itself. The complexity of the interaction of these transcription factors with the transcription initiation complex (RNA polymerase and its cofactors) is poorly understood and is an area of ongoing investigation. Indeed, a new class of transcription factors (coactivators) was recently identified that mediate the interaction of two DNA binding proteins (Fig. 9).

Characterization and comparison of these transcription factors have led to the identification of several common motifs in their DNA binding and protein binding (or activation) domains (Fig. 10). The zinc finger motif is comprised of a Zn^{2+} ion bound to two cysteines and two histidine residues in a characteristic 23 amino acid loop. These finger regions are usually present in tandem repeats and are a feature of the ubiquitous SP1 factor, which binds to the GC box. A modification of the zinc finger in which the Zn^{2+} binding site consists of four cysteines is present in the DNA binding domains of the glucocorticoid and estrogen receptors and the hepatic transcription factor HNF-4, which binds to the promoters of factors VII, IX, and X. Steroid receptors are zinc finger transcription factors that interact with DNA only after binding a particular steroid. Glucocorticoid, thyroid hormone, and retinoic acid receptors are examples of this type of transcription factor. A second class of DNA binding domain, known as the *homeodomain*, is a highly conserved region present in a variety of transcription factors, including DNA binding factors involved in the developmental regulation of *Drosophila*. This motif consists of three α-helical regions separated by short β turns. These proteins bind to DNA as dimers in which one α-helical region of each dimer occupies the major groove in the DNA helix. Examples of this helix-turn-helix motif in mammalian cells are the Oct general and lymphoid-specific transcription factors, the Pit1/GHF-1 pituitary transcription factors, the HOX11 oncogene, and the hepatic transcription factor HNF-1. The basic helix-loop-helix factors have an α helix followed by a loop of several helix-breaking amino acids, followed by a second α helix motif. Adjacent to the helix-loop-helix structure is a stretch of basic amino acids that bind DNA. The helical regions are crucial for the protein dimerization that occurs with this type of transcription factor. A similar motif is present in the leucine zipper, in which a leucine-rich amino acid sequence is juxtaposed to a basic stretch of amino acids capable of binding DNA. The leucine-rich regions of two proteins dimerize at these hydrophobic stretches. This motif is present in the CAAT enhancer binding protein (C/EBP) isolated from rat liver and the c-myc oncogene. The formation of heterodimers at these leucine zipper domains is known to occur; their biological significance with respect to the regulation of transcription has yet to be defined

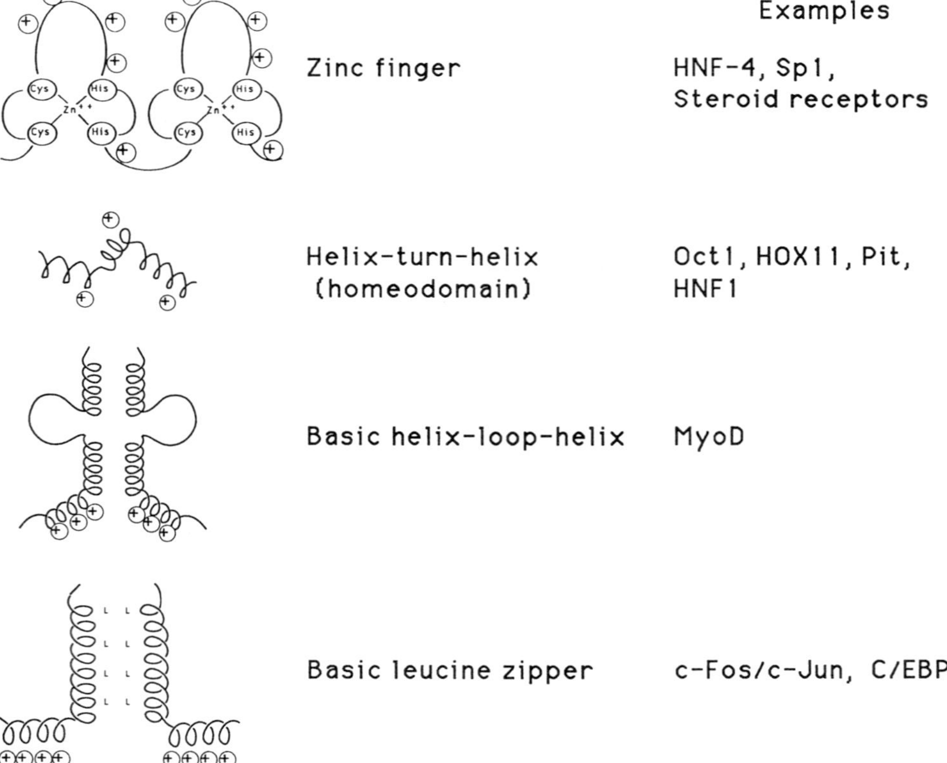

Figure 10 Classification of transcription factors may be made based on their DNA binding motifs. The plus indicates positively charged (basic) protein domains that interact with the DNA molecule.

completely. Other cloned transcription factors have been described that show no sequence homology to these DNA binding motifs. Activation domains have been characterized that have regions rich in a particular amino acid, such as the acidic residues, aspartic acid and glutamic acid; proline; glutamine, and cysteine. These regions are thought to be involved in protein-protein interactions with the basal transcription apparatus.

C. Posttranscriptional Regulation of Gene Expression

Gene expression is also affected by the rate of mRNA degradation. Some transcripts are very stable—for example, the half-life of the β-globin message is greater than 20 h; in contrast, the transcript for tissue factor has a much shorter half-life (approximately 1 h). Messenger RNAs contain a flanking untranslated (UT) sequence at the 5′ and often 3′ ends. The 5′ UT sequence contains binding sites for the small ribosomal subunits, which are important in the initiation of translation. The 3′ UT region of an mRNA may contain sequences that bind proteins responsible for stabilizing the message as it is processed in the nucleus, transported to the cytoplasm, or translated. Messenger RNAs that undergo rapid turnover have been shown to lack poly(A) tails or to share a common sequence motif that includes a high content of A and U in the 3′ UT regions. Specific sequences present in the open reading frame, as well as the 3′-terminal portion of the mRNA transcript, have been reported to destabilize the molecule. For example, the short half-life of the tissue factor transcript has been attributed to specific sequences present in the 3′ UT region (7). Rapid mRNA turnover is blocked by inhibition of translation suggesting that the activity of degrading RNAses is linked to translation. Deadenylation of the 3′ tail is an important signal for message degradation, and this deadenylation is enhanced by factors that bind to sequences in the message during translation. Some mRNAs may be stored in association with ribonuclear proteins; however, most mRNAs are processed in the nucleus, exported to the cytoplasm, and immediately undergo translation and degradation.

D. Translation: Targeting to Cellular Compartments

Translation is initiated in the cytosol of eukaryotic cells by the association of ribosomal subunits at the 5′ end of a processed mRNA. Translation begins at an AUG sequence encoding a methionine residue at the amino terminus of the nascent protein. An optimal sequence of ACCAUGG surrounding the initiation AUG has been identified for the initiation of translation by eukaryotic ribosomes (Kozak M. Point mutations define a sequence flanking the AUG initiator codon that modulates translation by eukaryotic ribosomes. Cell 1986; 44:283–292). For secreted proteins, such as the soluble factors of coagulation and fibrinolysis, the initial portion (amino terminus) of the nascent polypeptide (signal sequence) is bound to a ribonucleoprotein, SRP (signal recognition particle). Based on in vitro evidence, translation is believed to halt following binding to the SRP, until the particle binds its receptor on the cytosolic face of the endoplasmic reticulum (ER) membrane and effects ribosomal attachment to the membrane (Fig. 11). Translation then resumes, and the signal sequence is cleaved on or near the lumen of the ER. Such proteins as tissue factor, platelet glycoproteins, and endothelial cell receptors, which are destined to become integral membrane components, are first targeted to the ER by a signal sequence. Subsequent translation results in the elaboration of one or more stretches of hydrophobic amino acids called anchor sequences, which halt further translocation through the membrane. For proteins with the carboxyl terminus exposed on the outer face of the membrane, a combination signal-anchor sequence causes the protein to reverse its orientation within the membrane during translation. Vesicles composed of ER membrane and integrated proteins are further processed in the Golgi apparatus and then fuse with the plasma membrane.

E. Posttranslational Modifications of Protein

Many proteins undergo modifications following translation, which afford them unique structural and functional attributes. This is particularly true of many of the proteins involved in coagulation. In addition to the signal sequence present in a secreted protein, some proteins contain a propeptide that may be involved in directing proteins to and through the cellular compartments where modifications occur, in maintaining proper protein conformation for transport through cellular compartments, and in assuring recognition by enzymes, such as γ-carboxylase, which effect posttranslational modifications. A processing peptidase removes the amino-terminal propeptide before secretion. Virtually all proteins that pass through the secretory apparatus of the cell are glycosylated (N-linked glycosylation occurs on the $-NH_2$ group of asparagine; O-linked glycosylation occurs on the $-OH$ group of serine, threonine, or hydroxylysine). Other modifications that occur in the endo-

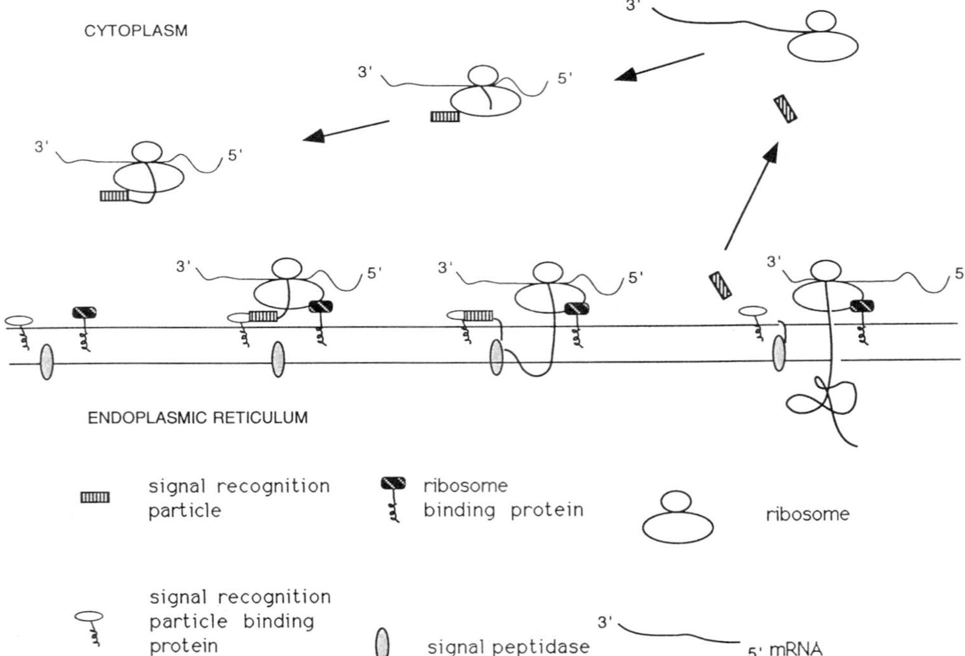

Figure 11 The signal sequence directs targeting and translocation into the ER of secreted and integral membrane proteins. Translation of proteins destined for secretion begins in the cytoplasm but is probably halted when the amino-terminal signal sequence binds to the signal recognition particle (SRP). The SRP binds a receptor on the cytosolic face of the ER; translation resumes following translocation of the nascent polypeptide through the ER membrane; the signal sequence is cleaved by signal peptidase, and the SRP is released from its receptor.

plasmic reticulum or Golgi apparatus include β-hydroxylation of specific aspartate and asparagine residues, sulfation of tyrosine residues (factor VIII), γ-carboxylation of glutamic acid residues in the amino-terminal portion of vitamin K-dependent coagulant and anticoagulant proteins, multimerization of von Willebrand factor, processing of single-chain proteins into linked multiple-chain species, and association with binding or chaperone proteins necessary for further processing, secretion, or degradation (e.g., BiP and factor VIII).

VI. TECHNIQUES OF MOLECULAR BIOLOGY

Within a cell, a gene is present as an exceedingly small quantity of DNA; it is not feasible to isolate genes directly from genomic DNA, by traditional biochemical methods, in the quantities necessary for detailed characterization. The techniques of molecular biology and recombinant DNA technology provide a means to overcome this limitation utilizing enzyme chemistry and microbiology in conjunction with biochemical procedures. These techniques have allowed the isolation and manipulation of DNA sequences and include the splicing of an isolated DNA fragment into another DNA molecule (with properties that may be exploited in the laboratory), thus forming a hybrid, or *recombinant* DNA molecule. *Cloning* is the generation of large amounts of a species of DNA from a single copy. The use of recombinant technology to isolate and characterize individual genes has led to a better understanding of the processes controlling protein synthesis, cell growth, and differentiation. In this section we describe some of the basic experimental approaches that have allowed such characterization.

A. Isolation of DNA and RNA

Nucleic acids can be isolated from cellular material, such as bacteria, eukaryotic cells grown in tissue culture, and tissues harvested from organisms; even cells without nuclei (such as platelets and red blood cells) contain RNA. To isolate nucleic acids from cells, a gentle lysis procedure using detergents, freezing, or hypertonic solutions is first employed to break down the cytoplasmic and nuclear membranes. Protein is removed by utilizing a nonspecific proteinase and/or organic extraction. Protocols for the preparation of DNA often include treatment with an RNA degrading enzyme (RNAse); likewise, RNA may be further purified with a DNA degrading enzyme (DNAse). Glassware and solutions used in the isolation of RNA must be pretreated with an RNAse inhibitor, such as diethyl pyrocarbonate, because RNA degrading enzymes are ubiquitous and very stable and the contamination of an RNA preparation by these enzymes can result in poor yields. Purified nucleic acid can be concentrated by alcohol precipitation, drying, or removal of water by treatment with unsaturated butanol. Alcohol precipitation is most commonly employed because salts contaminating the nucleic acid are also removed. Typical yields from mammalian cells are 10^{-5} μg RNA/cell and 10^{-6} μg DNA/cell. Fractionation of nucleic acids can be accomplished by chromatographic techniques. However, gel electrophoresis, which results in better resolution, is more widely used.

B. Restriction Enzymes

Restriction endonucleases probably evolved to protect bacteria from contamination by infective foreign DNA (viruses or phages). These enzymes recognize sequences, usually not present in host DNA, and cleave the foreign DNA near or within these recognition sequences (9). Useful enzymes can be produced, in quantities large enough for use in the laboratory, from bacterial culture or from recombinant expression systems. By convention, they are named for the bacterial species from which they are derived (e.g., SmaI is derived from *Serratia marcescens*). The cleavage sites and recognition sequences have been characterized for over 600 unique enzymes. Restriction enzymes from different sources that recognize the same sequence are termed *isoschizomers.* Restriction enzymes have been used singly or in combination to reproducibly reduce DNA, derived from a particular genome, into a collection of smaller fragments of variable size depending on the distribution of the recognition site in the starting material. A map of small regions of DNA (up to 20 kb) can be deduced from the sizes of these fragments. These smaller fragments are also more easily manipulated in the laboratory. Some enzymes cut DNA at the same base in the sense and antisense strand, yielding a blunt-ended fragment. Others cut at different bases in the two strands, leaving an asymmetrical terminus, with either a 5′ or a 3′ overhang (Fig. 12). These termini are referred to as cohesive or "sticky" ends because they can anneal specifically with termini possessing the comple-

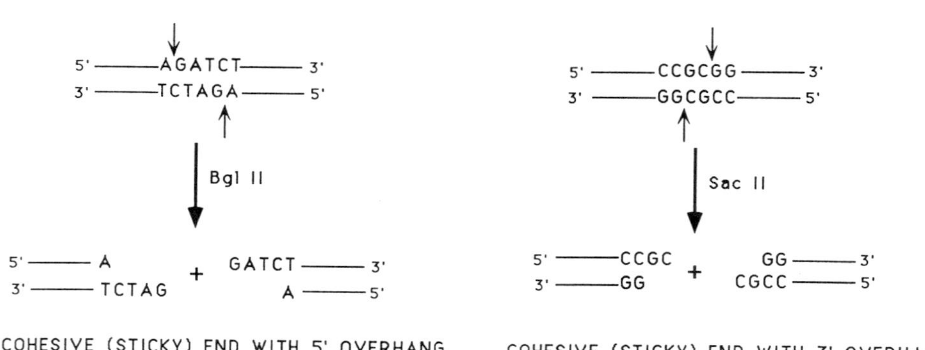

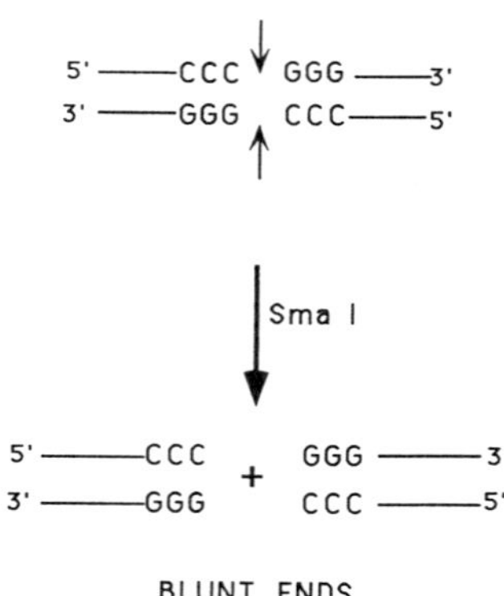

Figure 12 Restriction endonucleases. Restriction enzymes recognize specific double-stranded DNA sequences and cleave the DNA duplex at specific bases. For example, BglII digestion results in a 5′ overhanging cohesive end, SacII digestion results in a 3′ overhanging cohesive end, and SmaI digestion results in blunt ends.

mentary sequence. Using the same restriction enzyme to digest DNA from different sources allows one to recombine fragments to form a hybrid DNA molecule.

C. Other Enzymes

A variety of enzymes have been exploited in the laboratory for the manipulation of nucleic acids. DNA polymerases synthesize DNA from double-stranded or primed single-stranded templates. Exonucleases degrade DNA in a controlled fashion from either the 3′ or 5′ ends and can be used to remove overhanging termini remaining after restriction digest, to produce single-stranded DNA for sequencing, and to generate unidirectional deletions. Ligases join double-stranded annealed or blunt-ended DNA segments, allowing the generation of recombinant molecules. RNA polymerases synthesize RNA molecules from a DNA template. Kinases phosphorylate 5′-terminal hydroxyl ends of DNA and are useful for radiolabeling DNA and in preparing synthetic DNA fragments for ligation, a reaction that requires the presence of a 5′ phosphate. Phosphatases remove 5′-terminal phosphate groups and are useful for preparing DNA or RNA for radiolabeling with kinases and in preventing self-ligation of fragments of DNA during the production of recombinant molecules. RNAses degrade free RNA or the unannealed portion of an RNA molecule hybridized to DNA. Other enzymes are available for special applications, including the retrovirally derived RNA-dependent DNA polymerase, or reverse transcriptase. This enzyme produces a DNA copy (termed cDNA) complementary to single-stranded messenger RNA and has been extensively employed in the production of the cDNAs of many genes from cellularly derived mRNA.

D. Synthesis of Oligonucleotides

Advances in nucleic acid chemistry have allowed the production of small specific nucleotide sequences by employing phosphoramidite reagents to couple nucleotides sequentially onto an elongating nucleic acid polymer. Automated DNA synthesizers that can rapidly produce these chemically generated oligonucleotides are commercially available. Oligonucleotides can be labeled radioactively for use as probes in hybrization assays or used as primers for such reactions as sequencing and the polymerase chain reaction which utilize DNA polymerases.

E. Vectors

Much of the seminal work in molecular biology involved *E. coli* and the bacteriophages and plasmids that infect them. Foreign sequences of DNA (with certain size limitations) can be recombined with phage or plasmids (from which nonessential sequences have been deleted) without altering their normal life cycles. These plasmid and phage "vectors" then become hybrid molecules of DNA or chimeras that carry the foreign DNA. Vectors can be produced in large copy number by introducing them into *E. coli* and selecting and culturing bacteria that harbor the vector. Selection is made relatively simple by introducing, into the vector DNA, genes that confer antibiotic resistance on the bacteria. By exploiting differences in size, secretion, packaging, and linearity, these vectors may be effectively separated from host bacterial DNA. Thus a clone that contains the foreign DNA can be generated using these vectors. This clone can then be used to characterize the foreign sequence of DNA. The most common vectors used in the laboratory include plasmids conferring antibiotic resistance, derivatives of bacteriophage λ, and derivatives of the bacteriophage M13.

1. Plasmids

Plasmids are circular extrachromosomal DNA inclusions (episomes) in bacteria that are replicated independently of bacterial chromosomal DNA. In nature, plasmids serve the function of conferring antibiotic, heavy metal, or phage resistance or code for the expression of restriction endonucleases, synthesis of rare amino acids, or the catabolism of complicated organic molecules. Many useful recombinant plasmid vectors have been derived from these naturally occurring plasmids. Recombinant plasmids contain the following elements:

1. Replicator: DNA sequence containing the site at which replication begins (origin or ori) and sequence encoding specific plasmid RNA and proteins necessary for replication
2. Selectable marker: a gene expressed in a dominant fashion, usually encoding resistance to an antibiotic, by which bacteria containing the plasmid may be identified
3. Cloning site: a restriction endonuclease site at which a foreign DNA molecule can be inserted.

Plasmids may be present in host cells in low or high (with respect to bacterial chromosomal DNA) copy number. Low copy number plasmids usually replicate in a *stringent* fashion: they are dependent on unstable bacterial replication proteins, and their replication is synchronized with bacterial chromosomal replication. *Relaxed* plasmids are usually present in high copy number and are not dependent upon bacterial chromosomal replication. Relaxed plasmid replication can be greatly amplified by inhibiting bacterial protein synthesis with the antibiotic chloramphenicol. Most plasmids utilized in the laboratory, such as the pUC series, are derived from the naturally occurring pMB1 or ColE1 plasmids by engineering a defect in the normal negative feedback regulation of plasmid replication, thus allowing higher copy number.

2. Phage

Phages are viruses that infect bacteria, such as *E. coli*. Filamentous phage (such as M13) are composed of single-stranded circular DNA in a tubular protein coat, the size of which is determined by the length in kb of the phage DNA. Filamentous phage can only infect male *E. coli*, which possess an F pilus—a minor phage coat protein attaches to the pilus, which retracts, bringing phage DNA in contact with the bacterial cell surface. After entering the bacteria, a complementary strand is synthesized, yielding the double-stranded replicative form (RF) of the phage. RF phage functions as the transcription template for phage-encoded proteins and as the template for single-stranded phage replication. An advantage in the use of filamentous phage as a vector is the apparent lack of size limit for packaging; thus, longer inserts are packaged as longer particles. Foreign inserts up to 45 kb have been cloned using the filamentous phage M13; however, longer inserts tend to be unstable and may undergo spontaneous deletion or rearrangement. Also, filamentous phage can be recovered as double- or single-stranded molecules. Because this type of phage does not result in bacterial lysis, at steady state, a single bacteria may contain 20–40 double-stranded copies of the phage and export 100–200 particles per hour. Lambda (λ) is a linear DNA phage that has been extensively characterized. It can cause either lytic infection (whereby phage infection results in the replication of phage DNA and production of phage proteins that are assembled into infectious particles and released by bacterial host lysis) or lysogenic infection (whereby phage DNA is integrated into the bacterial host chromosome). The λ phage particles can only package a DNA molecule that is between 38 and 53 kb. The λ-derived vectors (such as λgt10 and λgt11) were designed to accept small inserts (0–4.8 kb); λEMBL3 and λ2001 accept larger inserts (10.4–20 kb). An advantage of the λgt11 vector is that it contains a cloning site within its expressed lacZ gene. Foreign DNA inserted into this cloning site is elaborated by the bacterial host as a fusion protein with part of the lacZ bacterial protein. Antibodies against the cloned protein can then be used to screen for bacterial colonies that contain the insert.

3. Cosmids

Cosmids are plasmids with features of λ phage that allow linearization and packaging into phage particles. Cosmid-containing particles can then be used to infect bacteria; after infection the particles are circulated and replicate as plasmids. Cosmids and λ phage have been used to accept the large inserts (45 kb) generated by the restriction endonuclease digestion of genomic DNA.

4. Yeast Artificial Chromosomes (YACs)

Yeast artificial chromosomes are vectors derived from cis elements important for chromosomal stability in the yeast *Saccharomyces cerevisiae.* They were developed to accommodate large inserts (200–1000 kb). The need for such vectors has become apparent with the identification of large genes, such as factor VIII (180 kb) and dystrophin (1800 kb), which cannot be cloned by λ or cosmid vectors. In addition, YAC libraries are being used to generate physical maps of chromosomes and provide the substrate for sequencing mapped regions of large genomes (see Sec. III. E). The YAC vector consists of two arm regions flanking the cloning site and contains sequences that function as telomeres (chromosomal ends). One of the arms contains a sequence that functions as a centromere and an origin of replication or autonomously replicating sequence. Each arm also contains a selectable marker. Yeasts that have taken up and stably maintain the artificial chromosome are identified on selective agar. The cloning site disrupts a suppressor gene, resulting in the formation of red (rather than white) colonies and making identification of transformants containing an insert apparent.

5. Libraries

A mixture of vector particles that contain a full complement of foreign DNA fragments from a particular source is called a library. Genomic libraries are generated by digesting genomic DNA with a restriction endonuclease. A λ phage or cosmid vector is digested with the same or a compatible enzyme. Vector and genomic fragments are mixed and ligated, yielding a collection of vector particles; the foreign inserts contained in this collection are derived from the genome of the source organism; theoretically the library should contain the entire genome of the organism. Useful cDNA libraries can be generated from mRNA molecules isolated from a cell line or tissue of interest. First-strand cDNA is produced from isolated mRNA using reverse transcriptase. The mRNA is then digested with RNAseH and second-strand DNA is synthesized with DNA polymerase. Oligonucleotide linkers containing appropriate restriction endonuclease sites are ligated to the double-stranded cDNAs; the cDNAs flanked by linkers are then digested with the appropriate restriction enzyme and inserted into an appropriate vector. Clones can be derived from these vector libraries by screening infected bacteria for the insert of interest. Screening can be accomplished by antiserum or antibody as described for λgt11 or by hybridization of a labeled DNA probe complementary to the sequence of interest.

F. Polymerase Chain Reaction

The polymerase chain reaction (PCR) is an in vitro technique for amplifying DNA sequences of interest without cloning into a vector or culturing host bacteria. The widespread introduction of PCR in the late 1980s made possible the rapid production of small linear DNA clones in amounts useful for characterization by sequence analysis or restriction enzyme digestion, recombination, and in vitro mutagenesis. This technique has become indispensable in the molecular biology laboratory. PCR involves the sequential denaturation of template double-stranded DNA, annealing of specific oligonucleotide primers to the template, and extension from the primers by a DNA polymerase to produce double-stranded replicas of the starting material. The vast majority of these amplified fragments are of a size determined by the linear separation of the primer annealing positions on the template DNA (Fig. 13). Repeated cycles of this reaction yield exponentially increasing amounts of the "target" sequence when adequate oligonucleotide primers and nucleotide substrate are included in the reaction. The discovery of a thermostable DNA polymerase from the bacteria *Thermophilus aquaticus* (Taq polymerase) has allowed the automation of PCR because, with Taq polymerase, enzyme need not be added after each high-temperature denaturation step. Taq polymerase incorporates 1–2 incorrect nucleotides for every 10,000 per cycle, and Taq-based PCR is extremely sensitive to contamination by undesired DNA. Newer thermostable polymerases (Vent polymerase) have been isolated that have a higher fidelity (sixfold lower error rate) compared with Taq.

G. Nucleotide Sequencing

The determination of the nucleotide sequence of the cDNAs and genes for many proteins active in hemostasis and thrombosis has made possible the indisputable determination of their primary amino acid sequence and the detection of mutations. The most common technique of nucleotide sequencing currently in use involves the annealing of a priming oligonucleotide upstream (5′) of the region to be sequenced. A DNA polymerase is utilized to incorporate a detectable label (e.g., radiolabeled or fluorescence-labeled nucleotides). A series of fragments with incorporated label is then generated by carrying out the polymerization reaction in the presence, separately, of small amounts of modified nucleotides (dideoxy-G, dideoxy-A, dideoxy-T, and dideoxy-C), which terminate polymerization at each of the specific residues (Fig. 14). Fragments that vary by one nucleotide in length can be separated by denaturing polyacrylamide gel electrophoresis. Each visualized fragment contains a single species of DNA, the length of which is defined by the distance from the primer at which the modified nucleotide was incorporated. Only one fragment of each size is produced from a single template. All fragments that end in the modified base (present in the termination reaction) are visualized in the lane in which that termination reaction was loaded. The concurrent separation of each of the four termination reactions generates the "sequence ladder" (Fig. 14).

H. Molecular Hybridization

A variety of hybridization techniques have been developed, the full description of which is available elsewhere. In general, hybridization relies on the specific binding of com-

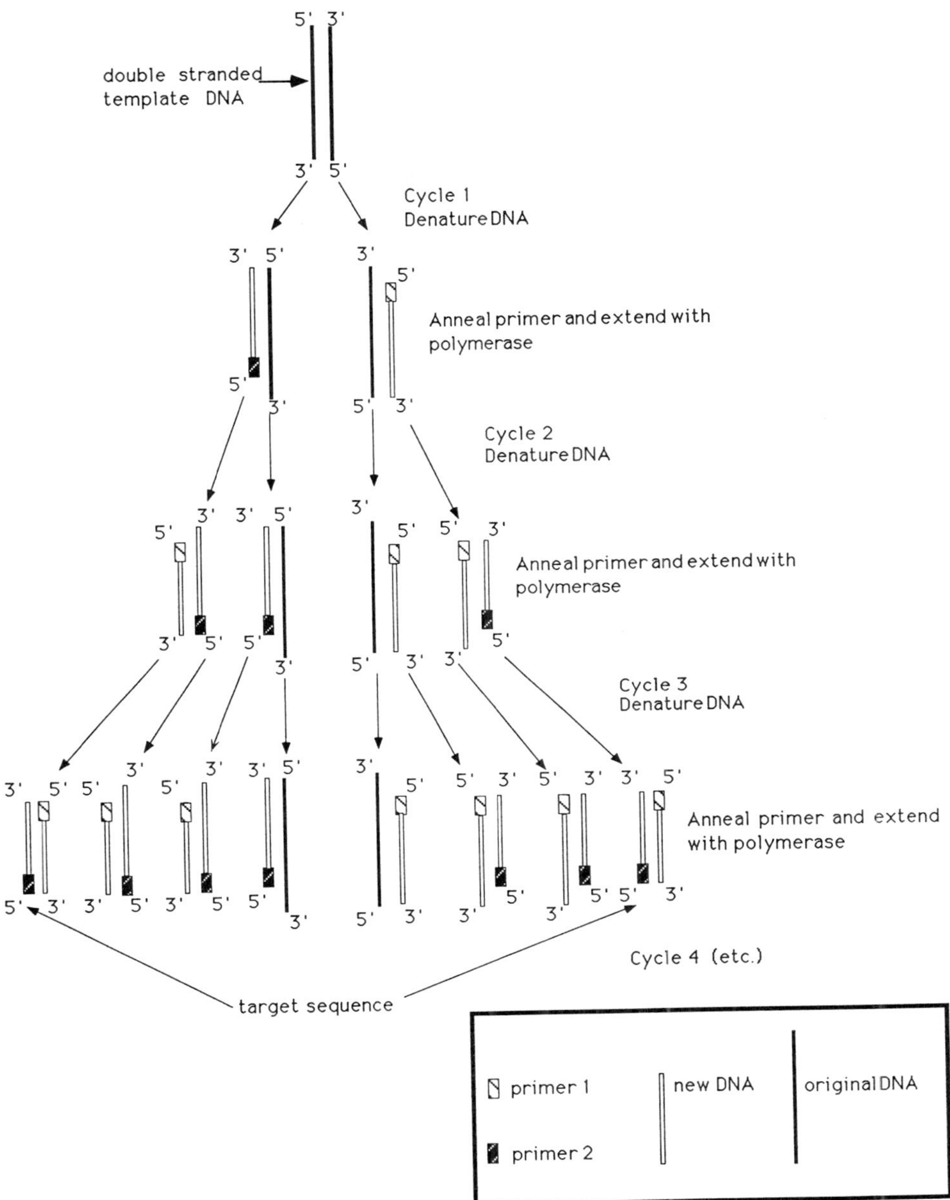

Figure 13 The polymerase chain reaction. PCR employs specific DNA oligomers to prime sequential rounds of enzymatic DNA polymerization from a specific template. The distance between the annealing positions of the oligomers on the template defines the length of the target sequence.

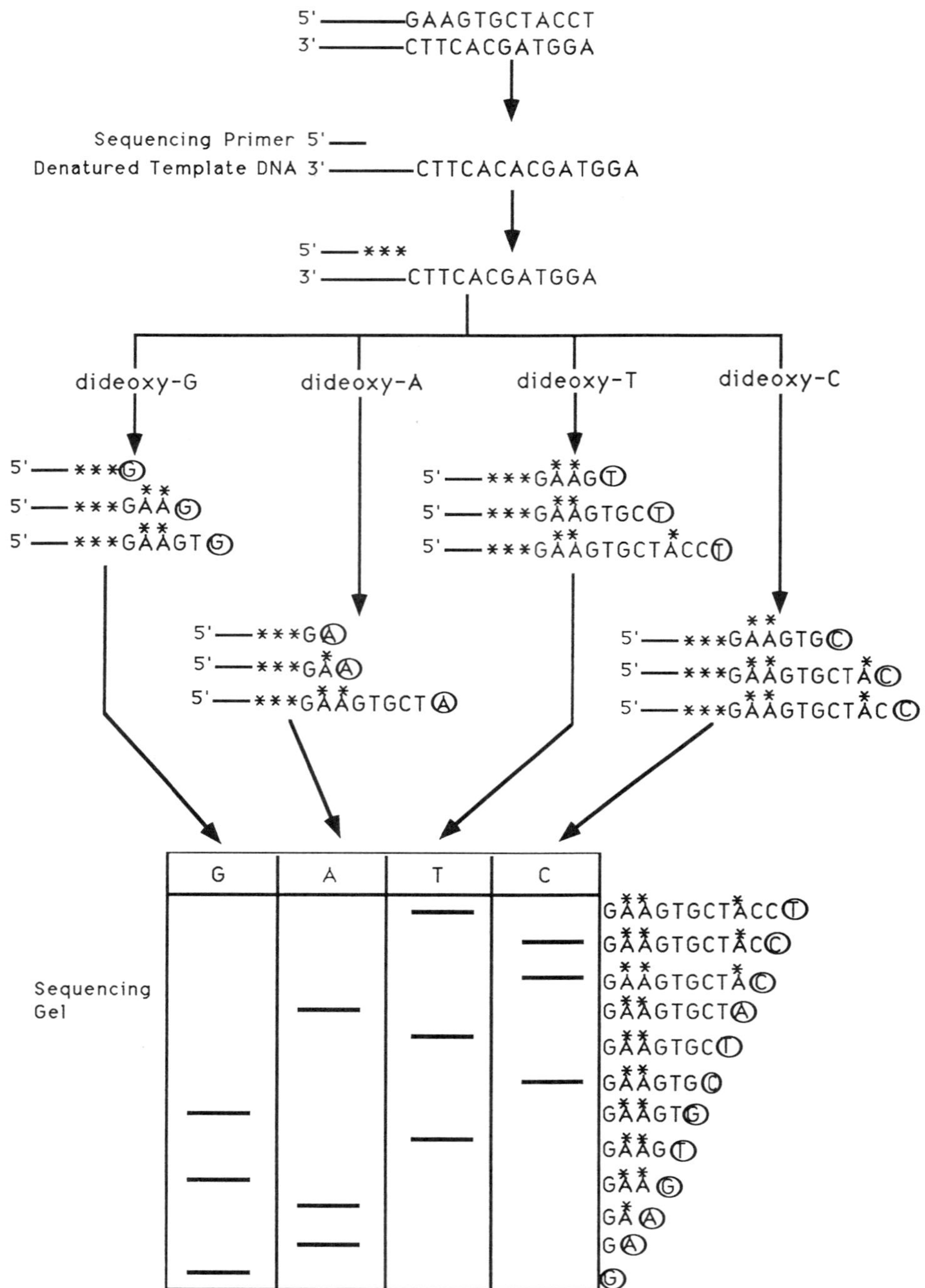

Figure 14 The dideoxy termination method of nucleotide sequencing. A single-stranded DNA oligomer (primer) is annealed to the template; enzymatic polymerization of a short region incorporates labeled nucleotide (denoted by *); polymerization is then continued in the presence of one of four dideoxy nucleotides (ddGTP, ddCTP, ddATP, or ddTTP) that terminate polymerization when incorporated into the polymer (denoted by circled residues). Because the 5′ end of the polymer is determined by the primer, DNA fragments of differing lengths (which correspond to the position at which the terminator was incorporated) are generated with each of the dideoxy terminators. These fragments can be separated electrophoretically on a denaturing polyacrylamide gel, yielding a ''ladder'' of bands that represent the nucleotide sequence of the template DNA (see text).

plementary strands of DNA or RNA. In the classic Southern blot, double-stranded DNA is prepared by restriction endonuclease digestion, separated by electrophoresis, denatured, and transferred to a DNA binding membrane (Fig. 15). The membrane or blot (typically nitrocellulose or nylon) is then incubated with a radiolabeled DNA probe of specific sequence. After allowing sufficient time for hybridization, the membrane is washed and binding of the radiolabeled probe is detected by autoradiography. A variation of this technique detects an RNA-DNA hybrid and is referred to as a northern blot. Probes may

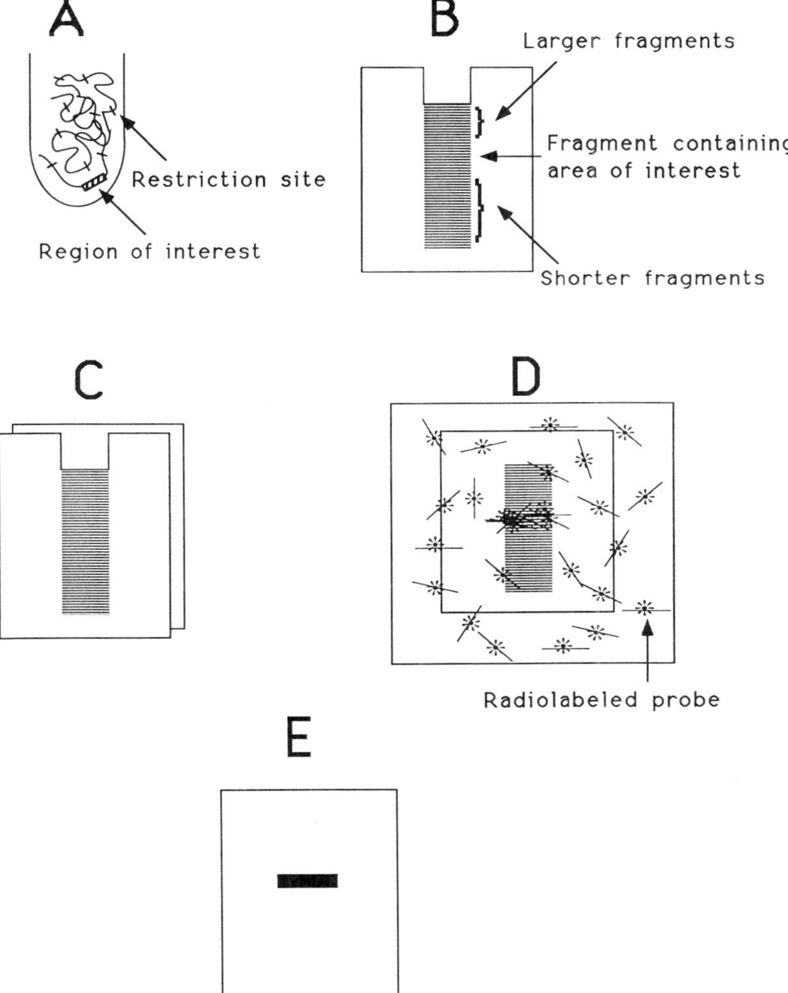

Figure 15 Molecular hybridization. (A) A nucleic acid species is generated (in this example of a Southern blot, double-stranded DNA is first digested with a restriction endonuclease). (B) The nucleic acid is separated electrophoretically by size. (C) The separated fragments are denatured and transferred from a gel matrix to a DNA binding membrane. (D) The membrane is then incubated with a labeled DNA probe complementary to the sequence of interest. (E) After washing to remove unbound probe, the band of interest is visualized by autoradiography.

be generated from cloned DNA or synthesized as oligonucleotides. Hybridization techniques have been used to detect RFLPs and specific alleles in digested genomic DNA. A clone of interest may be detected in a library by infecting a lawn of susceptible bacteria, then producing a plate replica containing phage plaque DNA on a nitrocellulose membrane. The DNA on the membrane is denatured and screened by hybridization with a labeled probe. Gene identification for proteins with a known complete or partial amino acid sequence can be accomplished by hybridization techniques. An oligonucleotide probe is synthesized based on peptide sequence from regions that are rich in amino acids with nondegenerate codons. This probe can then be used to map the gene (e.g., by in situ hybridization with chromosomes or by restriction analysis and linkage determination) or used to clone the gene by screening a library that contains genomic DNA. The probe can be used to screen a cDNA library. The unambiguous amino acid sequence of the protein can then be determined by nucleotide sequencing of a cDNA clone that hybridizes with the probe.

I. Heterologous Expression Systems

Heterologous gene expression—expression of foreign genes in cell culture systems—has allowed verification of the protein sequence of a cloned gene or cDNA, the preparation of proteins (or portions of proteins) to generate antibodies, and the production of proteins (which exist in trace amounts in homologous systems) in quantities sufficient to perform analyses of structure and function. Many of the proteins involved in coagulation are present in trace amounts in plasma, such that the analysis of structure-function relationships was extremely difficult before the advent of recombinant expression systems. Using the techniques of site-directed mutagenesis, recombinant proteins with specific changes from wild type can be expressed. The study of these engineered mutants provides a powerful means of deducing the functional importance of protein domains and specific amino acids.

Heterologous gene expression relies on the use of an expression vector that contains the sequence of interest under the control of a promoter that is responsive to the transcriptional environment of the expressing cell type. Transfection or transduction is the technique by which the heterologous expression vector is introduced into the expressing cell. A variety of cellular systems have been used to express recombinant material, including bacterial, fungal, and mammalian cells.

Bacterial expression systems, such as *E. coli*, are useful for the production of smaller mammalian polypeptides that do not undergo extensive posttranslational modifications. Recombinant protein may be expressed directly or fused to a protein that has been determined to be reliably translated in the bacterial system. The *E. coli* system has the limitation that mammalian posttranslational modifications, such as correct glycosylation, β-hydroxylation, and γ-carboxylation, are not carried out. In addition, protein folding and disulfide bond formation may not occur properly in these systems.

Mammalian expression systems overcome many of these disadvantages; however, they are more expensive to maintain. For certain applications, such as the simple detection of an expressed protein, the tracking of a protein's secretion or degradation, or functional studies that require small amounts of recombinant protein to perform, a transient transfection system often proves adequate. In transient expression, the transfected DNA is not stably integrated into the host cell genome, and protein expression falls over a period of several days to several weeks. Although expression of protein depends on the vector and cell line used in the system, it is difficult to scale up transient transfection systems for the production of large amounts of protein (>1 mg). Stably transfected cell lines can be produced by the use of a selectable marker, which allows the removal of untransfected cells, and selection of transfected cells, some of which stably integrate the foreign DNA into the host genome. Stable transfectants are expensive to produce but are essential for the production of large amounts of recombinant protein; these systems may produce up to 10 $\mu g/10^6$ cells per 24 h and may be maintained indefinitely.

VII. CONCLUSION

Molecular genetic techniques have contributed substantially to our current understanding of hemostasis and thrombosis. The cloning and sequencing of the genes encoding proteins involved in coagulation has allowed the unambiguous determination of their amino acid structures. Elucidation of the mechanisms governing the regulation of expression of these genes has led to a better understanding of the pathophysiology of certain diseases (e.g., hemophilia B Leyden; see Chap. 10). The characterization of specific genetic defects associated with disease has provided a wealth of information on protein structural anomalies and their functional consequences. Recombinant technology, employing the expression of proteins with specific mutations, has the potential to enhance the initial observations made with naturally occurring variants. In addition, the expression of biologically active recombinant protein has ushered in a new era of treatment based on non–plasma-derived proteins for individuals with clotting factor deficiencies. Finally, molecular genetic techniques have been critical to the development of gene therapy approaches for the in vivo amelioration or cure of heritable diseases of thrombosis and hemostasis.

REFERENCES

1. Avery OT, Macleod CM, McCarty M. Studies on the chemical nature of the substance inducing transformation of pneumococcal types. J Exp Med 1944; 79:137–158.
2. Watson JD, Crick FHC. A structure for deoxyribose nucleic acid. Nature 1953; 171:737–738.
3. O'Hara PJ, Grant FJ, Haldeman BA, et al. Nucleotide sequence of the gene coding for human factor VII, a vitamin K-dependent protein participating in blood coagulation. Proc Natl Acad Sci USA 1987; 84:5158–5162.
4. Resnick N, Collins T, Atkinson W, Bonthron DT, Dewey CF Jr, Gimbrone MA Jr. Platelet-derived growth factor β chain promoter contains a cis-acting fluid shear-stress–responsive element. Proc Natl Acad Sci USA 1993; 90:4591–4595.
5. Gill G, Tijian R. Eukaryotic coactivators associated with the TATA box binding protein. Curr Opin Genet Dev 1992; 2:236–242.
6. Faisst S, Meyer S. Compilation of vertebrate-encoded transcription factors. Nucleic Acids Res 1992; 20:3–26.
7. Ahern SM, Miyata T, Sadler JE. Regulation of human tissue factor expression by mRNA turnover. J Biol Chem 1993; 268:2154–2159.
8. Kozak M. Point mutations define a sequence flanking the AUG initiator codon that modulates translation by eukaryotic ribosomes. Cell 1986; 44:283–292.
9. High KA, Ware JL, Stafford DW. Isolation of genomic DNA, restriction endonuclease mapping and Southern gene blotting. In: Benz EJ, ed. Methods in hematology: molecular genetics. New York: Churchill Livingstone, 1989.

SUGGESTED READING

Biochemical Basis of Inheritance

Avery OT, Macleod CM, McCarty M. Studies on the chemical nature of the substance inducing transformation of pneumococcal types. J Exp Med 1944; 79:137–158.

Barker D, Schafer M, White R. Restriction sites containing CpG show a higher frequency of polymorphism in human DNA. Cell 1984; 36:131–138.

Cooper DN, Youssoufian H. The CpG dinucleotide and human genetic disease. Hum Genet 1988; 78:151–155.

Felsenfeld G. Chromatin as an essential part of the transcriptional mechanism. Nature 1992; 355: 219–224.

Koeberl DD, Bottema CDK, Buerstedde JM, Sommer SS. Functionally important regions of the factor IX gene have a low rate of polymorphism and a high rate of mutation in the dinucleotide CpG. Am J Hum Genet 1989; 45:448–457.

O'Hara PJ, Grant FJ, Haldeman BA, et al. Nucleotide sequence of the gene coding for human factor VII, a vitamin K-dependent protein participating in blood coagulation. Proc Natl Acad Sci USA 1987; 84:5158–5162.

Shapiro MB, Senapathy P. RNA splice junctions of different classes of eukaryotes: sequence statistics and functional implications in gene expression. Nucleic Acids Res 1987; 15:7155–7174.

Sharp PA. Speculations on RNA splicing. Cell 1981; 23:643–646.

Watson JD, Crick FHC. A structure for deoxyribose nucleic acid. Nature 1953; 171:737–738.

Mapping the Genome

Collins F, Galas D. A new five-year plan for the U.S. human genome project. Science 1993; 262: 43–46.

McKusick VA. A 40-year perspective on the evolution of a medical speciality from a basic science. JAMA 1993; 270(19):P2351–2356.

Nakamura Y, Leppert M, O'Connell P, et al. Variable number of tandem repeat (VNTR) markers for human gene mapping. Science 1987; 235:1616–1622.

Olson MV. The human genome project. Proc Natl Acad Sci USA 1993; 90:4338–4344.

Shows TB, Brown JA, Haley LL, et al. Assignment of the β-glucuronidase structural gene to the pter→q22 region of chromosome 7 in man. Cell Genet 1978; 21:99–104.

Shows TB, Sakaguchi AY, Naylor SL. Mapping the human genome, cloned genes, DNA polymorphisms, and inherited disease. In: Harris H, Hirschham K, eds. Advances in human genetics New York: Plenum Press, 1982:12:341–452, Chapter 5.

Organization of the Gene

Neurath H. Evolution of proteolytic enzymes. Science 1984; 224:350–357.

Rogers J. Exon shuffling and intron insertion in serine protease genes. Nature 1985; 315:458–459.

Regulation of Expression: Transcription and Translation

Ahern SM, Miyata T, Sadler JE. Regulation of human tissue factor expression by mRNA turnover. J Biol Chem 1993; 268:2154–2159.

Dynan WS. Modularity in promoters and enhancers. Cell 1989; 58:1–4.

Faisst S, Meyer S. Compilation of vertebrate-encoded transcription factors. Nucleic Acids Res. 1992; 20:3–26.

Gill G, Tjian R. Eurkaryotic coactivators associated with the TATA box binding protein. Curr Opin Gene Dev 1992; 2:236–242.

Kozak M. Point mutations define a sequence flanking the AUG initiator codon that modulates translation by eukaryotic ribosomes. Cell 1986; 44:283–292.

Landschulz WH, Johnson PF, McKnight SL. The DNA binding domain of the rat liver nuclear protein C/EBP is bipartite. Science 1989; 243:1681–1688.
Levine M, Manley JL. Transcriptional repression of eukaryotic promoters. Cell 1989; 59:405–408.
Lewin B. Commitment and activation at Pol II promoters: a tail of protein-protein interactions. Cell 1990; 61:1161–1164.
Maniatis T, Goodbourn S, Fischer JA. Regulation of inducible and tissue-specific gene expression. Science 1987; 236:1237–1244.
Mendel DB, Crabtree GR. HNF-1, a member of a novel class of dimerizing homeodomain proteins. J Biol Chem 1991; 266(2):677–680.
Mitchell PJ, Tjian R. Transcriptional regulation in mammalian cells by sequence-specific DNA binding proteins. Science 1989; 245:371–378.
Ptashne M, Gann AAF. Activators and targets. Nature 1990; 346:329–331.
Resnick N, Collins T, Atkinson W, Bonthron DT, Dewey CF Jr, Gimbrone MA Jr. Platelet-derived growth factor B chain promoter contains a cis-acting fluid sheer-stress-responsive element. Proc Natl Acad Sci USA 1993; 90:4591–4595.
Sachs AB. Messenger RNA degradation in eukaryotes. Cell 1993; 74:413–421.
Sladek FM, Darnell JE. Mechanisms of liver-specific gene expression. Curr Opin Genet Dev 1992; 2:256–259.
Sladek FM, Zhong W, Lai E, Darnell JE Jr. Liver-enriched transcription factor HNF-4 is a novel member of the steroid hormone receptor superfamily. Genes Dev 1990; 4:2353–2365.

Techniques of Molecular Biology

Ausubel FM, Brent R, Kingston RE, et al. Current protocols in molecular biology. Brooklyn NY: Greene Publishing, 1993.
Erhich HA, ed. PCR technology: principles and applications for DNA amplification. New York: Stockton Press, 1989.
Goeddel DV. Systems for heterologous gene expression. Methods Enzymol 1990; 185:3–7.
High KA, Ware JL, Stafford DW. Isolation of genomic DNA, restriction endonuclease mapping and Southern gene blotting. In: Benz EJ, ed. Methods in hematology: molecular genetics. New York: Churchill Livingstone, 1989.
Sambrook J, Fritsch EF, Maniatis T. Molecular cloning: a laboratory manual, 2nd ed. Cold Spring Harbor, NY: Cold Spring Harbor Laboratory Press, 1989.

General

Darnell J, Lodish H, Baltimore D. Molecular cell biology, 2nd ed. New York: Scientific American Books, 1990.
Gelehrter TD, Collins FS. Principles of medical genetics. Baltimore: Williams & Wilkins: 1990.
Lewin B. Genes V. New York: Oxford University Press; 1994.

2

Overview of the Coagulation Reactions

Harold R. Roberts and Aldo Hugo Tabares
University of North Carolina at Chapel Hill School of Medicine, Chapel Hill, North Carolina

I. INTRODUCTION

This concise review is intended only as an introduction for readers who may not be familiar with the basic clotting reactions. The brief description of the factors involved in hemostasis is augmented by a detailed description of these factors in subsequent chapters of this volume.

Blood coagulation is an important component of the host defense mechanisms. It involves the sequential activation of multiple zymogens by a process of limited proteolysis ultimately leading to thrombin generation and conversion of soluble fibrinogen to an impermeable fibrin clot. Blood clotting reactions also involve cell–cell interactions, including platelet–platelet and platelet–endothelial cell interactions, as well as interactions with other formed elements in the vessel wall. All of these components are necessary to effect hemostasis (1). The major factors involved in the blood-clotting fibrinolytic, and control reactions are listed in Table 1.

Over the past decades, dramatic advances in research have increased our knowledge of the molecular biology and the biochemistry of blood coagulation. As a result, our understanding of the process has been greatly modified. One advance was the cascade or waterfall hypothesis of blood coagulation put forward in the 1960s by Davie and Ratnoff (2) and Macfarlane (3). A modified form of this hypothesis is shown in Fig. 1. This view of coagulation proposed that fibrin formation resulted from a series of stepwise reactions involving clotting factors circulating in the blood in a precursor or zymogen form. After limited proteolysis, zymogens were converted to active two-chain serine proteinases. Early modifications of the hypothesis were necessary when it was found that some of the clotting factors acted as cofactors rather than enzymes. Thus, factor V and factor VIII, when activated by thrombin, acted as cofactors for the enzymes factors Xa and IXa, respectively, so that in the presence of a phospholipid surface and calcium ions, the rate of conversion of the substrate to active product was increased by several orders of magnitude (4,5), as shown in Table 2.

Table 1 Molecular Properties of Blood Coagulant Proteins

	Site of synthesis	Molecular weight	Protein characteristic	Plasma level	Chromosomal location	Gene organization
Fibrinogen (factor I)	Hepatocyte	330,000	Dimeric, with each monomer having three subchains	200–400 mg/dl	4	Separate genes for α, β and γ chains
Prothrombin (factor II)	Hepatocyte	72,000	Vitamin K dependent	100 μg/ml	11	14 Exons
Tissue factor (factor III)	Many cell types	37,000	Transmembrane protein	—	1	6 Exons
Factor V	Hepatocyte, megakaryocyte	330,000	Cofactor to factor Xa	10 μg/ml	1	Similar to factor VIII
Factor VII	Hepatocyte	55,000	Vitamin K dependent	0.5 μg/ml	13	9 Exons; adjacent to factor X gene
Factor VIII	Hepatocyte	330,000	Cofactor to factor IXa	0.1 μg/ml	X	26 Exons
von Willebrand factor	Endothelial cell, megakaryocyte	220,000 (multimers up to 20×10^6)	Carries factor VIII in plasma	10 μg/ml	12	52 Exons; pseudogene on chromosome 22
Factor IX	Hepatocyte	55,000	Vitamin K dependent	5 μg/ml	X	8 Exons
Factor X	Hepatocyte	55,000	Vitamin K dependent	10 μg/ml	13	8 Exons; adjacent to factor VII gene
Factor XI	Hepatocyte	160,000	Homodimeric	5 μg/ml	4	15 Exons
Factor XII	Hepatocyte	80,000	—	30 μg/ml	5	15 Exons
Factor XIII	Hepatocyte	320,000	a chains and b chains, tetrameric (a_2, b_2)	10 μg/ml	1(b) 6(a)	a and b genes on different chromosomes
Protein C	Hepatocyte	62,000	Vitamin K dependent	5 μg/ml	2	8 Exons
Protein S	Hepatocyte	80,000	Vitamin K dependent Cofactor protein C	25 μg/ml	3	15 Exons; pseudogene on chromosome 3

Thrombomodulin	Endothelial cell	60,300	Expressed on endothelial surface	—	20	No introns
Antithrombin III	Hepatocyte	60,000	Forms complex with thrombin	150 μg/ml	1	6 Exons
Tissue factor pathway inhibitor (TFPI)	Endothelial cell	33,000	Circulates in plasma; disulfide linked to apo—a, Kunitz-type inhibitor	0.01 μg/ml	2	9 Exons
Heparin cofactor II (HCII)	Hepatocyte	66,000	Forms complex with thrombin	6 μg/ml	22	5 Exons
Protein C inhibitor (PCI)	Male genital tract, liver	57,000	Trypsinlike serine protease inhibitor	5 μg/ml	14	5 Exons
Plasminogen	Hepatocyte	92,000	Single-chain glycoprotein contains 5 Kringle domains	200 μg/ml	6	19 Exons
α_2 plasmin inhibitor	Hepatocyte, endothelial cells	70,000	Serine protease inhibitor	70 μg/ml	18	10 Exons
Tissue plasminogen activator (t-PA)	Endothelial cells	68,000	Serine protease contains 2 Kringle domains	0.005 μg/ml	8	14 Exons
Single-chain urokinase plasminogen activator (scu-PA)	Kidney cells	54,000	Similar to t-PA only 1 Kringle domain	0.008 μg/ml	10	11 Exons
Plasminogen activator inhibitor 1 (PAI-1)	Endothelial cells	52,000	Serine protease inhibitor	0.2 μg/ml	7	9 Exons
Plasminogen activator inhibitor 2 (PAI-2)	Human placenta, leukocytes	47,000	Serine protease inhibitor	<0.05 μg/ml	18	8 Exons

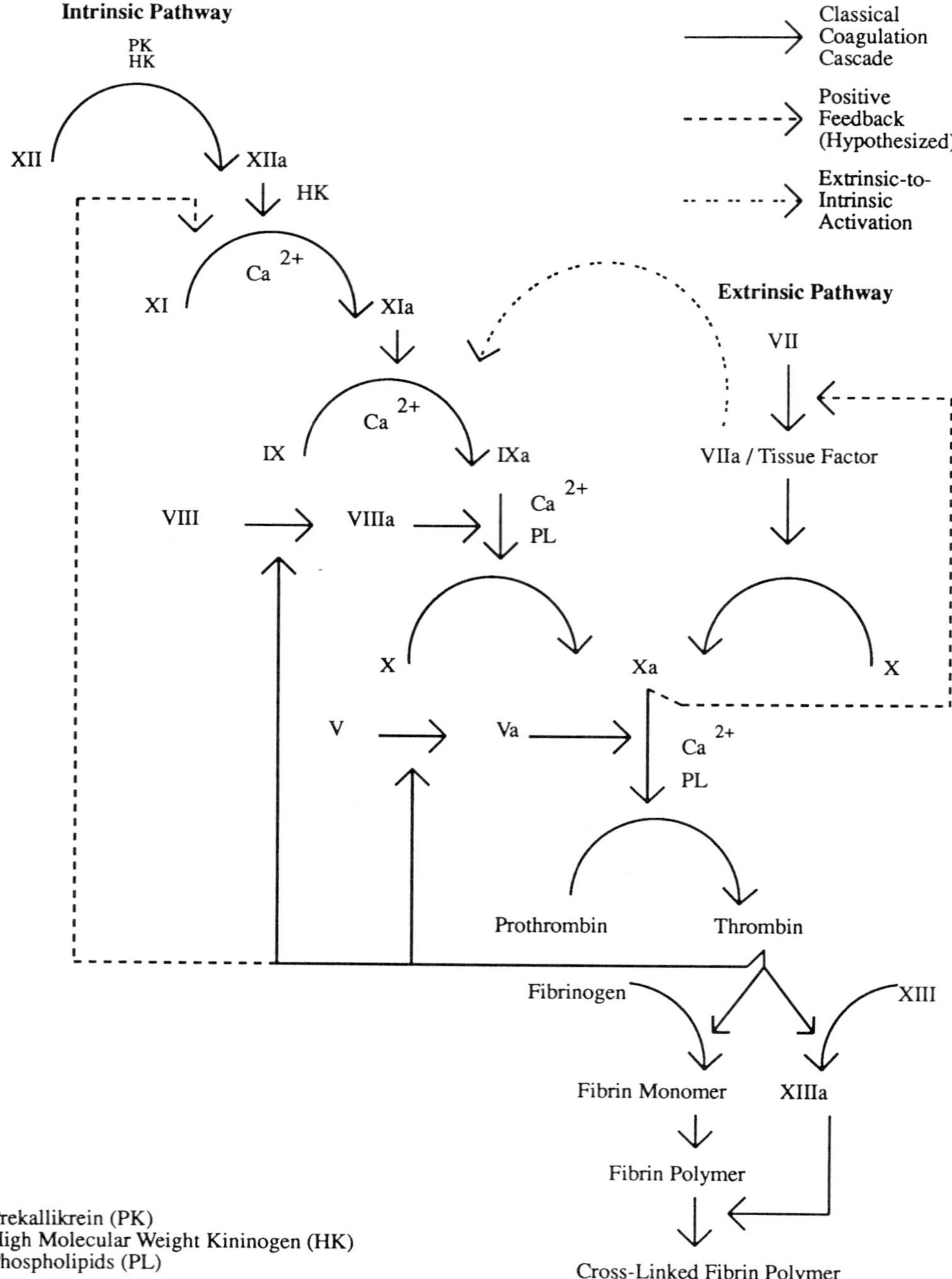

Figure 1 Waterfall hypothesis of blood coagulation. (Modified from Roberts HR, Lozier J. *Hospital Practice* 1992; 26:97; reproduced with permission of the publisher and the artist.)

Table 2 Effect of Cofactors V and VIII on the Rate of Activation of Factor X and Prothrombin

Composition of the reaction	Km (mMol/L)	VMax	Catalytic efficacy (relative)
Factor X activation			
IXa	299	0.0022	(1)
IXa, Ca^{2+}	181	0.0105	8
IXa, Ca^{2+}, PL	0.058	0.0247	58,000
IXa, Ca^{2+}, PL, VIIIa	0.063	500	1.0×10^9
Prothrombin activation			
Xa	131	0.61	(1)
Xa, Ca^{2+}	84	0.68	1.7
Xa, Ca^{2+}, PL	0.058	2.25	8300
Xa, Ca^{2+}, PL, Va	0.21	1919	2.0×10^9

The waterfall hypothesis also divided events leading to fibrin formation into extrinsic and intrinsic pathways. The extrinsic pathway consisted of factor VII and tissue factor (the latter was considered to be extrinsic to the bloodstream). Factor VII, when activated, formed a complex with tissue factor that converted factor X to activated factor X (factor Xa). The intrinsic system, consisting of factors XII, XI, IX, and X, was considered to be intrinsic to the circulating blood, leading to the formation of a complex of factor IXa with its cofactor, factor VIIIa. This complex, in the presence of phospholipid and calcium, could also convert factor X to Xa. Factor Xa, as the enzyme, formed a complex with its cofactor, factor Va, and rapidly converted prothrombin to thrombin. This reaction required a phospholipid surface and calcium. Once thrombin was formed, fibrinogen was converted to fibrin monomers that polymerized to form a fibrin clot.

Although the concept of extrinsic and intrinsic pathways of coagulation still serves as a working model for the coagulation reactions, observations by Rapaport and his group, as well as by Nemerson, Broze, and a number of others (6–9), have shown that the tissue factor–factor VIIa complex can activate factor IX in the intrinsic system. Furthermore, it was observed that patients deficient in factor XII had no bleeding tendency, although this factor presumably initiated the intrinsic pathway. On the other hand, patients severely deficient in factor VII had a severe bleeding tendency, even though the intrinsic pathway was intact. Furthermore, it was observed that thrombin could activate factors V, VIII, XI, and XIII in feedback reactions. These observations led to the concept that the extrinsic and intrinsic pathways were not independent of each other, but rather were interdependent and interrelated.

More recent observations have underscored the importance of the platelet and endothelial cell surfaces as sites of receptor-mediated clotting reactions in addition to providing the necessary phospholipids.

II. INITIATION OF BLOOD COAGULATION

A revised view of the coagulation process is shown in Fig. 2. It now appears that blood coagulation can be initiated when tissue factor, expressed after injury to the vessel wall, is exposed to factor VII inside the bloodstream (Fig. 2, panel A). Tissue factor, which

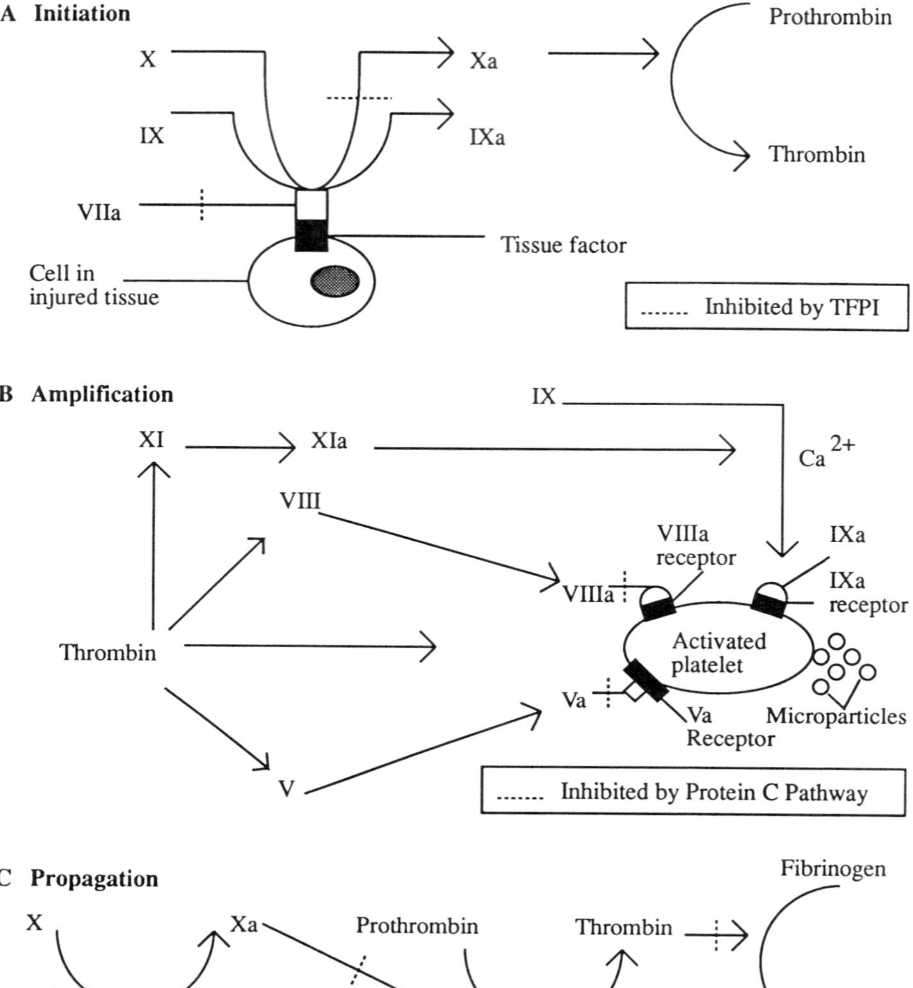

Figure 2 Schematic diagram of procoagulant reactions on cell surfaces. See text for details. TFPI = tissue factor pathway inhibitor. ATIII = antithrombin III. (Modified from Roberts HR, Lozier J. *Hospital Practice* 1992; 26:97; reproduced with permission of the publisher and the artist.)

is constitutively expressed by cells of the adventitia of blood vessels as well as in other tissues throughout the body, is not normally found either in the circulating blood or on the surface of the vascular lumen (10,11). Once expressed, tissue factor forms a complex with factor VII. Factor VII is converted to its activated form (factor VIIa) by mechanisms that are not completely understood. One prevailing hypothesis is that small amounts of factor Xa are always available to activate factor VII (12–15). Another hypothesis holds that factor VII undergoes autoactivation, that is, small amounts of factor VIIa are always available to activate the zymogen form (16,17). Nemerson also has evidence that the factor VII zymogen possesses proteolytic activity (14). Whatever the mechanism of factor VII activation, it is clear that the tissue factor–factor VIIa complex can activate both factor X and factor IX. These initial reactions, as shown in Fig. 2, panel A, take place on cell surfaces that express tissue factor. The fact that tissue factor is normally separated from factor VII by an endothelial cell barrier probably indicates that it represents a control mechanism and is required for normal hemostasis. Endothelial cells do not produce tissue factor constitutively, but only after specific stimulation by one or more substances, for example, tumor necrosis factor (TNF) or endotoxin.

III. AMPLIFICATION OF BLOOD COAGULATION

Once factors Xa and IXa are formed by the tissue factor–factor VIIa complex (TF-VIIa), subsequent reactions leading to amplification of coagulation must take place on the platelet surface, as opposed to the surface expressing tissue factor (18,19). How factors that initiate clotting on the tissue factor–bearing cells can be transported to platelets is not yet clear, but it is likely that platelets accumulate in the vicinity of TF-VIIa–bearing cells and that small amounts of thrombin leads to platelet activation. This process probably also leads to expression of specific platelet receptors that localize both the cofactors and enzymes responsible for amplification of coagulation on the platelet surface, as depicted in Fig. 2, panel B. The factor IXa formed by TF-VIIa is likely enhanced by the factor XIa pathway, which converts more factor IX to factor IXa. How factor XI is activated in vivo is not entirely clear, either. It is known that factor XIIa can activate factor XI in vitro (20–22) but, since patients with a severe factor XII deficiency exhibit no bleeding symptoms, the physiologic relevance of factor XIIa conversion of factor XI to XIa in vitro has been questioned. Recent evidence suggests that thrombin can activate factor XI (23,24) in a feedback reaction. It has been shown, however, that the thrombin activation of factor XI in plasma is not nearly as efficient as the activation of factor XI by thrombin in purified systems. Thus, the physiologic relevance of thrombin activation of factor XI has been questioned, which leaves unclear the mechanism of factor XI activation in vivo.

IV. PROPAGATION PHASE OF COAGULATION

With platelet activation, cofactors Va and VIIIa are localized and concentrated on the platelet surface, possibly by interaction with specific receptors (Fig. 2, panel C). After localization of the cofactors, factor IXa occupies a platelet binding site in the vicinity of factor VIIIa while factor Xa occupies a binding site next to factor Va. Thus the stage is set for propagation of coagulation by the *tenase complex* (IXa-VIIIa), which rapidly converts factor X to factor Xa and the formation of the *prothrombinase complex* (Va-

Xa). The prothrombinase complex then effects an explosive generation of thrombin by rapidly converting prothrombin to the enzyme thrombin (Fig. 2, panel C).

With the thrombin generation, fibrinogen is converted to the fibrin monomer, which spontaneously polymerizes to form fibrin polymers, leading to a visible fibrin clot. A visible fibrin clot that is not cross-linked is permeable to blood and before it becomes a hemostatically stable clot it must be cross-linked by the action of factor XIIIa (factor XIII zymogen is activated by thrombin). After cross-linking by factor XIIIa, the fibrin clot becomes impermeable to the blood and hemostasis is assured.

Obviously, all three phases of coagulation, that is, initiation, amplification, and propagation, must be controlled and, indeed, exquisite control mechanisms have been described for each of these steps.

V. CONTROL OF THE PROCOAGULANT PATHWAYS

The initiation phase of coagulation is controlled by tissue factor pathway inhibitor (TFPI), which is a Kunitz-type inhibitor that can block the initiation of coagulation by inhibiting the complex of tissue factor, factor VIIa, and factor Xa (25–27). The mechanism of inhibition has been intensively studied and is described elsewhere in this volume. One possible mechanism of action is depicted schematically in Fig. 3.

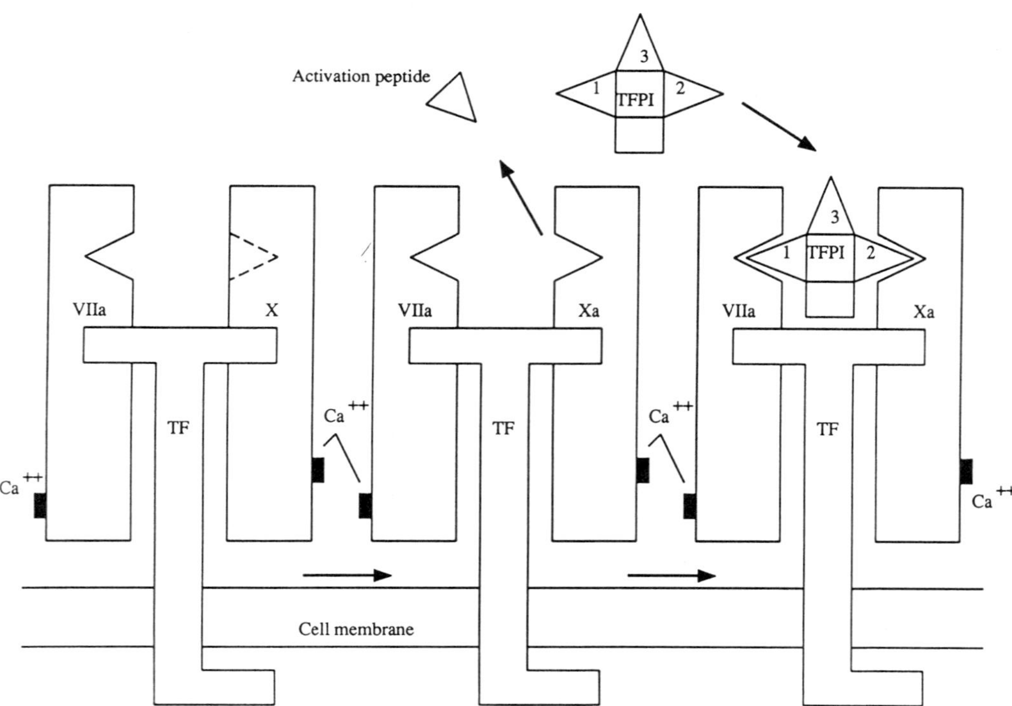

Figure 3 Schematic diagram of tissue factor pathway inhibitor (TFPI) and its interaction with the TF-VIIa-Xa complex. TFPI possesses 3 Kunitz-like domains and it acts as a specific inhibitor of a complex consisting of tissue factor (TF), factor VIIa, and factor Xa. (Modified from Broze G: *Hospital Practice* 1992; 27:71; reproduced with permission of the publisher and artist.)

The amplification phase of coagulation is controlled by the protein C pathway, shown schematically in Fig. 4. Protein C is a vitamin K–dependent zymogen, which is activated by a complex of thrombomodulin and thrombin on the endothelial cell surface. Thrombomodulin is an endothelial cell transmembrane protein that can rapidly activate protein C when complexed to thrombin. Activated protein C, in the presence of its cofactor (protein S) then inactivates factors Va and VIIIa to control the amplification phase of coagulation (28–30). Moreover, as described below, there is some evidence that activated protein C can contribute to the initiation of fibrinolysis by freeing tissue plasminogen activator from its inhibitor, plasminogen activator inhibitor-1. The importance of the protein C/protein S system is demonstrated by clinical manifestations of thrombosis when either factor is decreased in the heterozygous or homozygous state. In the heterozygous state, deep venous thrombosis is a common complaint (31–35), whereas in the homozygous state of either protein C or protein S deficiency, neonatal purpura fulminans is the invariant result and leads to early death unless adequately treated (36–38).

The thrombin formed in the propagation phase of coagulation is controlled by antithrombin III, which forms a one-to-one complex to inactivate the enzyme (39,40). All of these reactions are described in detail in other sections of this book. Deficiency of antithrombin III is also associated with a thrombotic tendency.

VI. THE FIBRINOLYTIC SYSTEM

A schematic representation of the fibrinolytic system is shown in Fig. 5. Here it can be seen that the zymogen, plasminogen, is converted to an active two-chain fibrinolytic enzyme, plasmin. Plasmin lyses fibrin clots by proteolysis of an arginyl bond, leading to the formation of fibrin fragments, as shown in Fig. 6. The final fragments resulting from plasmin dissolution of fibrin are fragments D and E, but intermediate products, fragments X and Y, are also formed (41–43).

The conversion of plasminogen to plasmin requires the presence of activators. Two physiologic activators have been identified. One is tissue plasminogen activator (t-PA) and the other is prourokinase, also known as single-chain urokinase plasminogen activator (scu-PA) scu-PA must be converted to two-chain urokinase to be active. Both activators cleave an Arg → Val bond in plasminogen, exposing the active site serine of plasmin (44–46). Urokinase activation of plasminogen is more important in cell-associated plasminogen activation, while tissue plasminogen activator is the physiologic activator within the circulation.

As in the clotting reactions, the fibrinolytic system must also be exquisitely controlled. The major control mechanism for t-PA is plasminogen activator inhibitor-1 (PAI-1). The major antiprotease that inhibits plasmin is α_2 plasmin inhibitor (also known as α_2 antiplasmin [α_2AP]), which forms a very tight bond with plasmin. Alpha-2 macroglobulin (α_2M) also inhibits plasmin, but it is not as specific for plasmin as α_2AP. The action of all the fibrinolytic components is depicted schematically in Fig. 5.

Congenital abnormalities of plasminogen or plasminogen activators lead to thromboembolic complications (47–49). Abnormalities in the inhibitors of the fibrinolytic system are accompanied by hemorrhagic symptoms (50,51).

VII. RELATIONSHIP BETWEEN COAGULATION AND FIBRINOLYSIS

Teleologically, one would expect a link between fibrin formation and fibrin dissolution, that is, procoagulant activity and fibrinolytic activity. Clearcut evidence that such a link exists in vivo has not been found. In vitro, however, it is known that factor XIIa, factor XIa, or kallikrein can convert plasminogen to plasmin under certain circumstances (52–55). Furthermore, there is in vitro evidence that kallikrein (K), formed by the action of factor XIIa on prekallikrein (PK), can convert single-chain urokinase to the two-chain species, resulting in potent activation of plasminogen to the active fibrinolytic enzyme, plasmin (44). As stated previously, it has been shown that, under certain in vitro conditions, activated protein C can contribute to the dissociation of the complex of t-PA and PAI-1 (56). The physiologic relevance of both of these putative links between coagulation and fibrinolysis has not been clearly established.

VIII. PLATELET CONTRIBUTIONS TO THE BLOOD-CLOTTING REACTIONS

It is well known that the surfaces of cells are essential for the clotting reactions to take place. Platelets have specific receptors for the clotting factors, some of which have been demonstrated, and others are suspected (57). Platelet contributions to hemostasis include the initial event of platelet adhesion to the vessel wall, followed by platelet aggregation and release. Platelet adhesion to subendothelial surfaces is mediated through platelet membrane glycoprotein Ib (GPIb) and von Willebrand factor, as shown in Fig. 7. GPIb is part of the platelet membrane glycoprotein Ib-IX complex (GPIb-IX). This complex

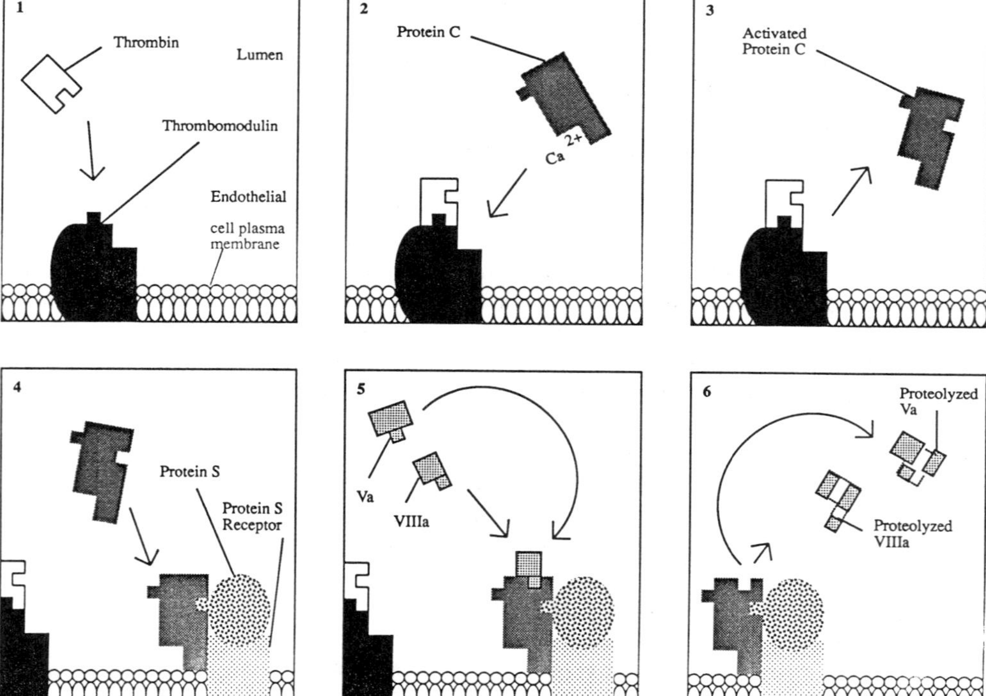

Figure 4 Protein C pathway. (Modified from Roberts HR, Lozier J: *Hospital Practice* 1992; 26: 97; and is reproduced with permission of the publisher and the artist.)

belongs to the leucine-rich glycoprotein (LRG) gene family, which is characterized by a common structural motif composed of a leucine-rich 24 amino acid sequence and is a receptor for von Willebrand factor (58,59). von Willebrand factor plays a role in the adhesion of platelets to components of the vessel wall as well as platelet–platelet interactions (60). These processes are described in detail elsewhere in this book.

After platelet adhesion, platelet activation occurs by the action of thrombin, with the result that a specific thrombin receptor is exposed on the platelet surface (61). After platelet activation by thrombin, binding sites for the tenase complex, as well as the prothrombinase complex, are formed. The platelet aggregation takes place during this period of time. Platelet aggregation is mediated largely by platelet membrane glycoproteins IIb-IIIa (GPIIb-IIIa), which are exposed only after platelet activation. The exposure of GPIIb-IIIa results in the binding of fibrinogen to this platelet integrin. Under certain circumstances GPIIb-IIIa will bind von Willebrand factor, vitronectin, or fibronectin.

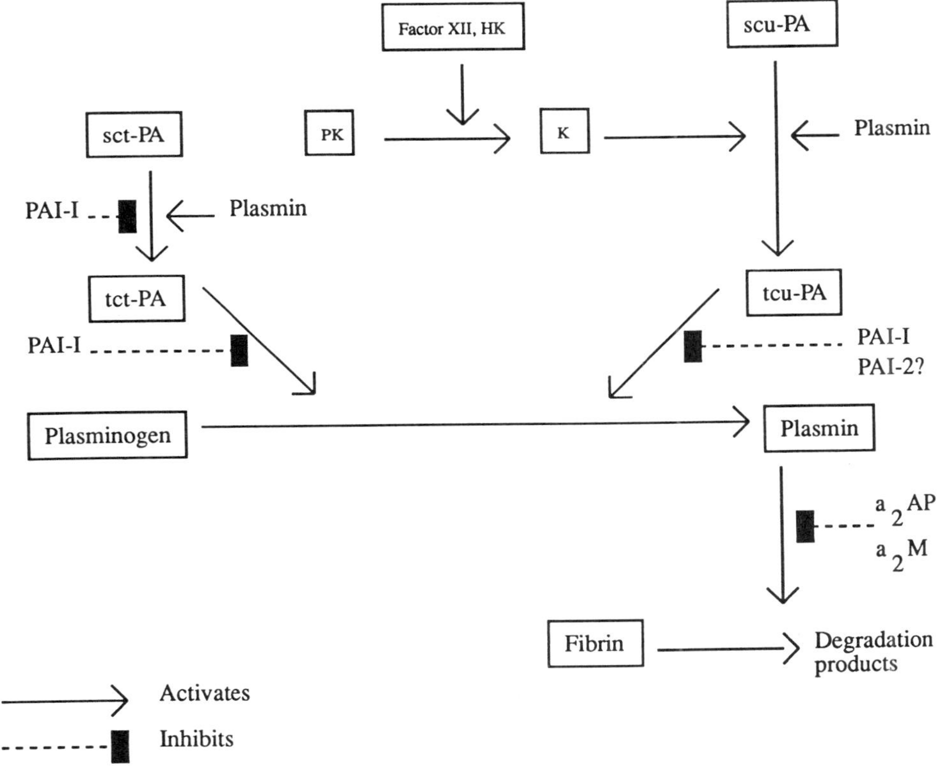

Figure 5 Fibrinolytic system. Plasminogen conversion to plasmin is mediated by tissue plasminogen activator (t-PA). t-PA exists as a single-chain t-PA (sct-PA), which is converted to two-chain t-PA (tct-PA) by plasmin. Both forms of t-PA are active and both can be inhibited by plasminogen activator inhibitor-1 (PAI-1). Single-chain urokinase plasminogen activator (scu-PA) is converted to two-chain urokinase plasminogen activator (tcu-PA) by kallikrein (K). Kallikrein is derived from prekallikrein (PK) by the action of factor XIIa in complex with high–molecular weight kininogen (HK). Urokinase is inhibited by PAI-1 and plasminogen activator inhibitor-2 (PAI-2). The major inhibitor of plasmin is α_2 plasmin inhibitor (α_2 antiplasmin [α_2AP]). Alpha-2 macroglobulin (α_2M) also inhibits plasmin, but is not as specific as α_2AP.

Fibrinogen is presumed to be most important in aggregation, probably due to its divalent structure, allowing it to form a bridge from platelet to platelet (62,63), as shown in Fig. 7.

The platelet-release reaction represents the response to appropriate agonists and is manifested by secretion of platelet constituents stored in alpha granules and dense bodies, resulting in the recruitment of additional platelets and promotion of vasoconstriction. Several platelet agonists have been well characterized and are capable of inducing platelet aggregation and release in vitro. They include thrombin, adenosine diphosphate (ADP), collagen, arachidonic acid, and epinephrine. Specific receptors exist on the platelet surface for these agonists (64), and many receptor–agonist complexes interact with coupling proteins (G proteins) whose actions require guanosine triphosphate (65). The function of G proteins is currently a matter of intense research; they appear to interact with target proteins coupled to ion-permeable channels, thus modulating calcium flux. Many reactions involved in the process of platelet activation are calcium dependent, including the phosphorylation of the light chain of myosin and the release of arachidonic acid from membrane phospholipids by the enzyme phospholipase A_2 (66,67). Arachidonic acid is converted by the enzyme cyclooxygenase to prostaglandin endoperoxides and ultimately to the potent platelet agonist, thromboxane A_2 (68). Cyclooxygenase is permanently in-

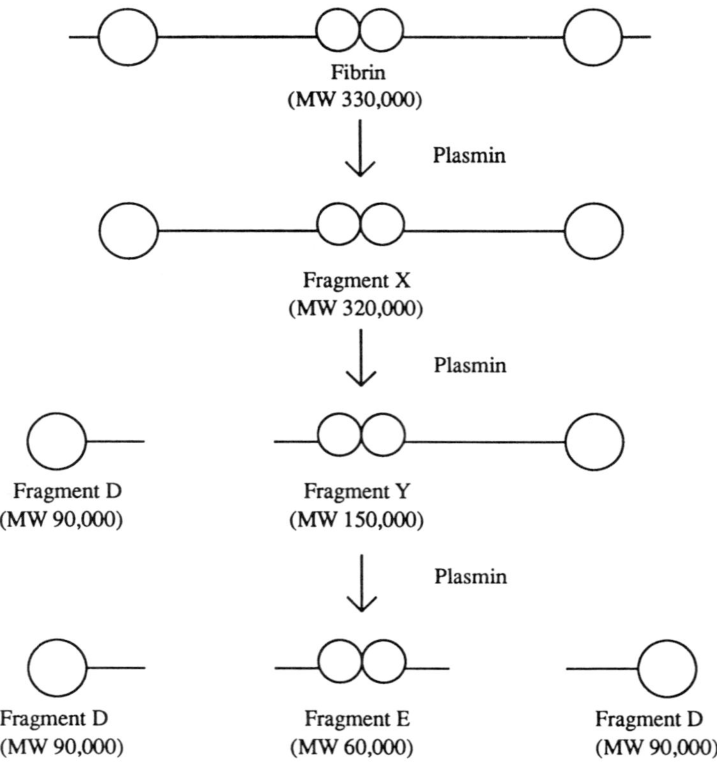

Figure 6 Schematic diagram of the plasmin digestion of fibrin. Fragment X gives rise to a molecule of fragment Y and a molecule of fragment D. Fragment Y undergoes further proteolysis to a second molecule of fragment D and a molecule of fragment E.

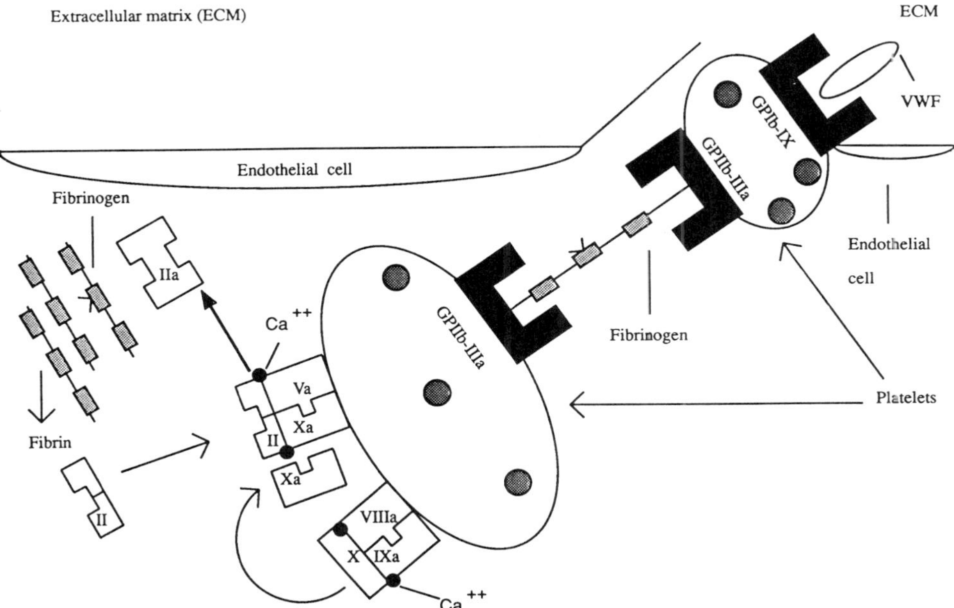

Figure 7 Schematic diagram of platelet adhesion, platelet aggregation, and assembly of tenase and prothrombinase complexes on the platelet surface. A break or injury in the vessel wall between endothelial cells exposes the extracellular matrix (ECM) of the vessel wall. Adhesion of platelets to collagen is mediated by von Willebrand factor (vWF) that couples collagen in the ECM to platelets via glycoprotein Ib. Aggregation of platelets is mediated by fibrinogen bridging between glycoproteins IIb-IIIa on the activated platelet surface. The tenase and prothrombinase complexes are shown schematically. The platelet thrombin receptor is not shown, but is described in detail elsewhere in this book.

hibited by aspirin through acetylation of a serine residue in the enzyme (69). This accounts for the major pharmacologic action of aspirin on platelets.

REFERENCES

1. Colman R, Marder V, Salzman E, Hirsh J. Overview of hemostasis. In: Colman R et al., eds. Hemostasis and thrombosis. Basic principles and clinical practice. Philadelphia: J. B. Lippincott, 1994:3–18.
2. Davie E, Ratnoff O. Waterfall sequence for intrinsic blood clotting. Science 1964; 145:1310.
3. Macfarlane RG. An enzyme cascade in the blood clotting mechanism and its function as biochemical amplifier. Nature 1964; 202:498.
4. Rosing J, Tans G, Govers-Riemslag JWP, Zwaal RFA, Hemker C. The role of phospholipids and factor Va in the prothrombinase complex. J Biol Chem 1980; 255:274.
5. van Dieijen G, Tans G, Rosing J, Hemker HC. The role of phospholipid and factor VIIIa in the activation of bovine factor X. J Biol Chem 1981; 256:3433.
6. Nemerson Y. The reaction between bovine brain tissue factor and factor VII and X. Biochemistry 1966; 5:601.

7. Osterud B, Rapaport SI. Activation of factor IX by the reaction product of tissue factor and factor VII: additional pathway for initiating blood coagulation. Proc Natl Acad Sci USA 1977; 74:5260.
8. Bach R, Nemerson Y, Konigsberg W. Purification and characterization of bovine tissue factor. J Biol Chem 1981; 256:8324.
9. Broze GJ, Jr, Leykam JE, Schwartz BD, Miletich JP. Purification of human brain tissue factor. J Biol Chem 1985; 260:10917.
10. Drake TA, Morrissey JH, Edgington TS. Selective cellular expression of tissue factor in human tissues. Implications for disorders of hemostasis and thrombosis. Am J Pathol 1989; 134:1087.
11. Fleck RA, Rao LVM, Rapaport SI. Localization of human tissue factor antigen by immunostaining with monospecific, polyclonal anti-human tissue factor antibody. Thromb Res 1990; 57:765.
12. Rao LVM, Rapaport SI, Bajaj SP. Activation of human factor VII in the initiation of tissue factor-dependent coagulation. Blood 1986; 68:685.
13. Rao LVM, Rapaport SI. Activation of factor VII bound to tissue factor: a key early step in the tissue factor pathway of blood coagulation. Proc Natl Acad Sci USA 1988; 85:6687.
14. Nemerson Y, Repke D. Tissue factor accelerates the activation of coagulation factor VII: the role of a bifunctional coagulation cofactor. Thromb Res 1985; 40:351.
15. Masys DR, Bajaj SP, Rapaport SI. Activation of human factor VII by activated factors IX and X. Blood 1982; 60:1143.
16. Pedersen AH, Lund-Hansen T, Bisgaard-Frantzen H, Olsen F, Petersen LC. Autoactivation of human recombinant factor VII. Biochemistry 1989; 28:9331.
17. Nakagaky T, Foster DC, Berkner KL, Kisiel W. Initiation of the extrinsic pathway of blood coagulation: evidence for the tissue factor dependent autoactivation of human coagulation factor VII. Biochemistry 1991; 30:10819.
18. Furie B, Furie BC. The molecular basis of blood coagulation. Cell 1988; 53:505.
19. Davie EW, Fujikawa K, Kisiel W. The coagulation cascade: initiation, maintenance, and regulation. Biochemistry 1991; 30:10363.
20. Ratnoff OD, Davie EW, Mallet DL. Studies on the action of Hageman factor: evidence that activated Hageman factor in turn activates plasma thromboplastin antecedent. J Clin Invest 1961; 40:803.
21. Griffin JH, Cochrane CG. Mechanisms for the involvement of high molecular weight kininogen in surface-dependent reactions of the Hageman factor. Proc Natl Acad Sci USA 1976; 73:2554.
22. Wiggins RC, Bouma BN, Cochrane CG, Griffin JH. Role of high-molecular-weight kininogen in surface-binding and activation of coagulation factor XI and prekallikrein. Proc Natl Acad Sci USA 1977; 74:4636.
23. Naito K, Fujikawa K. Activation of human blood coagulation XI independent of factor XII. Factor XI is activated by thrombin and factor XIa in the presence of negatively charged surfaces. J Biol Chem 1991; 266:7353.
24. Gailani D, Broze GJ, Jr. Factor XI activation in a revised model of blood coagulation. Science 1991; 253:909.
25. Rapaport SI, Rao LVM. Initiation and regulation of tissue-factor dependent blood coagulation. Arterioscler Thromb 1992; 12:111.
26. Rapaport SI. The extrinsic pathway inhibitor: a regulator of tissue-factor dependent blood coagulation. Thromb Haemost 1991; 66:6.
27. Broze GJ, Jr, Girard TJ, Novotny WF. Regulation of coagulation by a multivalent Kunitz-type inhibitor. Biochemistry 1990; 29:7539.
28. Esmon C. Regulation of natural anticoagulant pathways. Science 1987; 235:1348.
29. Broze GJ, Tollefsen DM. Regulation of blood coagulation by protease inhibitors. In: Stamatoyannopoulos G et al., eds. The molecular basis of blood diseases. Philadelphia: W. B. Saunders, 1994:629–656.

30. Esmon CT. Molecular events that control the protein C anticoagulant pathway. Thromb Haemost 1993; 70:29.
31. Griffin JH, Evatt B, Zimmerman TS, et al. Deficiency of protein C in congenital thrombotic disease. J Clin Invest 1981; 68:1370–1373.
32. Bertina RM, Broekmans AW, van der Linden IK, Mertens K. Protein C deficiency in a Dutch family with thrombotic disease. Thromb Haemost 1982; 45:237–241.
33. Bovill EG, Bauer KA, Dickerman JD, et al. The clinical spectrum of heterozygous protein C deficiency in a large New England kindred. Blood 1989; 73:712–717.
34. Comp PC, Esmon CT. Recurrent venous thromboembolism in patients with a partial deficiency of protein S. N Engl J Med 1984; 311:1525–1528.
35. Schwartz HP, Fisher M, Hopmeier P, et al. Plasma protein S deficiency in familial thrombotic disease. Blood 1984; 64:1297–1300.
36. Branson HE, Katz J, Marble R, Griffin JH. Inherited protein C deficiency and coumarin-responsive chronic relapsing purpura fulminans in a newborn infant. Lancet 1983; 2:1165–1168.
37. Mahasandana C, Suvatte V, Marlar R, et al. Neonatal purpura fulminans associated with homozygous protein S deficiency. Thromb Haemost 1989; 62:301 (abstract).
38. Marciniak E, Wilson HD, Marlar RA. Neonatal purpura fulminans: a genetic disorder related to the absence of protein C in blood. Blood 1985; 65:15–20.
39. Rosenberg RD, Damus PS. The purification and mechanism of action of human antithrombin-heparin cofactor. J Biol Chem 1973; 248:6490–6505.
40. Pratt CW, Church FC. General features of the heparin-binding serpins antithrombin, heparin cofactor II and protein C inhibitor. Blood Coag and Fibrinol 1993; 4:479–490.
41. Mihayi E, Weinberg RM, Towne DW, et al. Proteolytic fragmentation of fibrinogen. I. Comparison of the fragmentation of human and bovine fibrinogen by trypsin or plasmin. Biochemistry 1976; 15:5372.
42. Marder V, Shulman NR, Carrol WR. The importance of intermediate degradation products of fibrinogen in fibrinolytic hemorrhage. Trans Assoc Am Physicians 1967; 80:156.
43. Marder V, Shulman NR, Carrol WR. High-molecular-weight derivatives of human fibrinogen produced by plasmin. I. Physicochemical and immunological characterization. J Biol Chem 1969; 244:2111.
44. Ichinose A, Fujikawa K, Suyama I. The activation of pro-urokinase by plasma prekallikrein and its inactivation by thrombin. J Biol Chem 1986; 261:3486.
45. Ranby M, Bergsdorf N, Nilsson T. Enzymatic properties of the one- and two-chain form of tissue plasminogen activator. Thromb Res 1982; 27:175.
46. Ichinose A, Kisiel W, Fujikawa K. Proteolytic activation of tissue plasminogen activator by plasmin and tissue enzymes. FEBS (letter) 1984; 175:412.
47. Aoki NB, Moroi M, Sakata Y, et al. Abnormal plasminogen: a hereditary molecular abnormality found in a patient with recurrent thrombosis. J Clin Invest 1978; 61:1186.
48. Jorgensen M, Mortensen JZ, Madsen AG, et al. A family with reduced plasminogen activator activity in blood associated with recurrent venous thrombosis. Scand J Haematol 1982; 29:217.
49. Stead NW, Bauer KA, Kinney TR, et al. Venous thrombosis in a family with defective release of vascular plasminogen activator and elevated plasma factor VIII/von Willebrand's factor. Am J Med 1983; 74:33.
50. Schleef RR, Higgins DL, Pillemer E, Levitt LJ. Bleeding diathesis due to decreased functional activity of type I plasminogen activator inhibitor. J Clin Invest 1989; 83:1747.
51. Fay WP, Shapiro AD, Shih JL, Schleef RR, Ginsburg D. Complete deficiency of plasminogen-activator inhibitor type 1 due to a frame-shift mutation. N Engl J Med 1992; 327:1729.
52. McDonagh J, Ferguson JH. Studies on the participation of Hageman factor in fibrinolysis. Thromb Haemost 1970; 24:1.
53. Iatridis PG, Ferguson JH. Active Hageman factor. A plasma lysokinase of the human fibrinolytic system. J Clin Invest 1962; 41:1277.

54. Mandle RJ, Jr, Kaplan AP. Hageman factor dependent fibrinolysis. Generation of fibrinolytic activity by the interaction of human activated factor XI and plasminogen. Blood 1979; 54: 850.
55. Hauret J, Nicoloso G, Schleuning W-D, et al. Plasminogen activators in dextran sulfate-activated euglobulin fractions: a molecular analysis of factor XII- and prekallikrein-dependent fibrinolysis. Blood 1989; 73:994.
56. Sakata Y, Curriden S, Lawrence D, Griffin JH, et al. Activated protein C stimulates the fibrinolytic activity of cultured endothelial cells and decreases antiactivator activity. Proc Natl Acad Sci USA 1985; 82:1121.
57. Nesheim ME, Furmaniak-Kazmierczak E, Henin C, Cote G. On the existence of platelet receptors for factor Va and factor VIIIa. Thromb Haemost 1993; 70:80.
58. Takahashi N, Takahashi Y, Putman FW. Periodicity of leucine and tandem repetition of a 24 amino acid segment in the primary structure of leucine rich alpha 2 glycoprotein of human serum. Proc Natl Acad Sci USA 1985; 82:1906.
59. Weiss JH, Turitto VT, Baumgartner HR. Effect of shear rate on platelet interaction with subendothelium in citrated and native blood. I. Shear rate-dependent decrease of adhesion in von Willebrand's disease and the Bernard-Soulier syndrome. J Lab Clin Med 1978; 92:750.
60. Meyer D, Girma J-P. von Willebrand factor: structure and function. Thromb Haemost 1993; 70:99.
61. Coughlin SR. Thrombin receptor structure and function. Thromb Haemost 1993; 70:184.
62. Phillips DR, Charo IF, Parise LV, Fitzgerald LA. The platelet membrane glycoprotein IIb-IIIa complex. Blood 1988; 71:831.
63. Kieffer N, Phillips DR. Platelet membrane glycoproteins: functions in cellular interactions. Ann Rev Cell Biol 1990; 6:329.
64. Colman RW. Platelet receptors. Hematol-Oncol Clin North Am 1990; 4:27.
65. Brass LF, Hoxie JA, Manning DR. Signaling through G proteins and G protein-coupled receptors during platelet activation. Thromb Haemost 1993; 70:217.
66. Adelstein RS, Conti MA. Phosphorylation of platelet myosin increases actin-activated myosin ATPase activity. Nature 1975; 256:597.
67. Pickett WC, Jesse RL, Cohen P. Initiation of phospholipase A_2 activity in human platelets by calcium ionophore A 23187. Biochim Biophys Acta 1977; 486:209.
68. Marcus AJ. Multicellular eicosanoid and other metabolic interactions of platelets and other cells. In: Colman R et al., eds. Hemostasis and thrombosis: basic principles and clinical practice. Philadelphia: J. B. Lippincott, 1994:590–602.
69. Al-Mondhiry H, Marcus AJ, Spaet TH. On the mechanism of platelet function inhibition by acetylsalicylic acid. Proc Soc Exp Biol Med 1970; 133:632.

3

Fibrinogen

Susan T. Lord
University of North Carolina at Chapel Hill, Chapel Hill, North Carolina

I. INTRODUCTION AND BACKGROUND

Fibrinogen is a soluble plasma glycoprotein that participates in the cellular phase of coagulation by mediating platelet aggregation and in the fluid phase by conversion to insoluble fibrin fibers. It circulates at 2–4 mg/ml and is readily purified from plasma. A general review of fibrinogen and its roles in thrombosis and hemostasis can be found in Reference 1. The primary structure of fibrinogen was first determined by amino acid sequence analysis and subsequently confirmed by the cDNA sequence. The protein is a dimer of three polypeptide chains, Aα, Bβ, and γ, that are linked by 29 disulfide bridges.

High-resolution structural data are not available, but analysis by electron microscopy and calorimetry has established a model for fibrinogen (Fig. 1) that consists of five globular domains; a single, central domain is separated from pairs of peripheral domains by rod-like pieces (2–6). The central domain contains the N-terminal residues of all six chains, includes 11 of the disulfide bridges, and constitutes about 10% of the molecule. This domain can be isolated as a plasmin fragment called the E domain. The peripheral domains consist primarily of the C-terminal residues of the Bβ and γ chains, each chain forming an independent globular domain. These domains share remarkable sequence homology and are unlike the C-terminal residues of the Aα chain. They can be isolated as a plasmin fragment called the D domain.

Multiple biochemical studies on purified fibrinogen and fragments of fibrinogen have provided a basic understanding of the domains critical to fibrinogen functions. Experiments defining the role of fibrinogen in platelet aggregation have been reviewed (7–9). Platelet aggregation is readily supported by γ chains isolated from purified fibrinogen and, to a lesser extent, by isolated Aα chains (10). Although current results are not completely consistent, it appears that the C-terminal dodecapeptide of γ chain and two Arg-Gly-Asp (RGD) sequences in the Aα chain are critical for aggregation (11–15).

In the fluid phase of coagulation, fibrinogen is converted to fibrin by thrombin-catalyzed cleavage of four peptide bonds, releasing two fibrinopeptides A (FpA, Aα

1–16), two fibrinopeptides B (FpB, Bβ 1–14), and fibrin monomers. Fibrin monomers spontaneously polymerize into fibers that are subsequently cross-linked by the transglutaminase factor XIIIa. Active thrombin binds to fibrin and thereby is localized to the site of clot formation. Fibrin fibers are the substrate for plasmin, which cleaves at specific sites during fibrinolysis, releasing fibrin degradation products. Fibrin potentiates plasmin activation: the conversion of the inactive precursor plasminogen to active plasmin by tissue plasminogen activator is enhanced about 1000-fold by the presence of fibrin. Experiments defining the formation and lysis of fibrin fibers have been reviewed (1,16).

This chapter focuses on fibrinogen studies enabled by cloning and sequencing the human fibrinogen genes and cDNAs. The first section covers the organization of the fibrinogen genes and the constituents that regulate gene expression. The second section includes aspects that relate fibrinogen structure to its biological functions. Both sections consider genetic defects associated with disease. The correlation of variant proteins with disease symptoms is emphasized because these results provide insight into the in vivo roles of fibrinogen.

II. GENE STRUCTURE AND FUNCTION

A. Gene Organization

The three fibrinogen polypeptides are encoded by three single-copy, linked genes located on less than 50 kb of human chromosome 4q23–q32 (17–19). All three genes have been completely sequenced (20). They are ordered γ, Aα, and Bβ, with the Aα and γ genes transcribed in one direction and the Bβ gene transcribed in the opposite direction (19–21). The Aα gene is 5.4 kb long with 5 exons, Bβ is 8.2 kb with 8 exons, and γ is 8.4 kb with 10 exons. A variant form of the γ chain, called γ′, γA, or γ-Leu^{427}, (22), arises from alternative polyadenylation within intron 9 such that 20 amino acids encoded by the 5′ end of intron 9 are appended to the amino acids encoded by exon 9 (20,23–25). The ratio of γ to γ-Leu^{427} chains in plasma fibrinogen is typically about 9:1. Because the complete gene sequences are known, it is practical to use genomic DNA to determine the genetic mutation and thereby the amino acid abnormality in variant fibrinogens.

Analysis of the fibrinogen genes has supported the proposal, originally suggested by Doolittle on the basis of the protein sequence (26), that the three chains arose by suc-

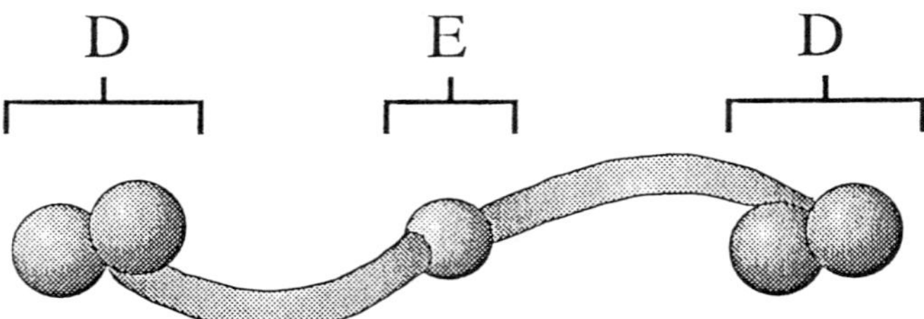

Figure 1 A current model for the fibrinogen molecule. The central sphere, or E domain, contains the N-terminal segments from all six polypeptide chains, two copies each of the Aα, Bβ, and γ chains. The two peripheral pairs of spheres, or D domains, contain the C-terminal residues of the Bβ and γ chains, with each chain represented by a separate sphere. (Adapted from Binnie CG, Lord ST. Blood 1993; 81:3186–3192.)

cessive duplication of a common ancestral gene (20,26). The closely linked arrangement of the genes is characteristic of gene duplication. In addition, the three genes share two intron junctions. A type 0 splice junction interrupts the coding region of each gene between the codons three and four residues before the first Cys of the N-terminal disulfide bonds that link all three chains, and a type I splice junction interrupts the coding sequence 62 residues downstream in all three cases (20). Although these conserved features demonstrate the evolutionary relation of the three genes, it is noteworthy that other introns within the three genes are not conserved (20,27). Analysis of fibrinogen sequences in other species has demonstrated the presence of two unexpected Aα-related sequences. Within the chicken Aα-chain gene is an open reading frame that resembles the C-terminal domains of Bβ and γ rather than the C-terminal domain of Aα (28). In lamprey, a second Aα cDNA was found that also has a C terminus more like Bβ and γ C-terminal domains than the typical Aα C-terminal domain (29). These new data support the hypothesis than an ancestral gene gave rise to all three chains but also demonstrate that the subsequent evolution of the three genes remains obscure.

B. Gene Expression

1. Regulation of Transcription

Fibrinogen is synthesized in hepatocytes and secreted into the plasma. Together the three fibrinogen messages constitute about 3% of total liver mRNA, and during the acute phase of inflammation the message levels increase approximately threefold (30). The enhanced expression in vivo can also be induced by administration of interleukin-6 (Il-6), demonstrating that this cytokine has a role in modulating fibrinogen expression (31). Extensive study of tissue specificity and acute-phase regulation of fibrinogen mRNA synthesis has produced unexpected findings. For example, the tissue-specific expression of fibrinogen γ mRNA is different from that of Aα and Bβ mRNAs. Transcription from the γ-chain promoter is regulated by ubiquitous factors, such as Sp1, a CAAT binding factor (32); transcription from the Aα and Bβ promoters requires a tissue-specific factor, hepatocyte nuclear factor 1 (HNF-1) (33,34). Hepatocyte-specific expression has been associated with specific upstream sequences (34–36). Recent results demonstrate that the Bβ-chain promoter also contains an HNF-1-dependent Il-6-responsive element (37). This Il-6 stimulation appears to be linked to the cis element motif CTGGGAA, which was previously identified in all three fibrinogen chain genes 120–150 bp upstream of the point of message initiation (38).

The differences in these promoters are reflected in the tissue distribution of the fibrinogen mRNAs. The levels of all three fibrinogen mRNAs are substantially greater in liver than in other tissues. Messanger RNAs for Aα and Bβ have also been found in kidney but not in lung tissue (39). This is consistent with the role of HNF-1 in Aα and Bβ expression, for HNF-1 is found in liver and kidney, but not in lung (39). In contrast, fibrinogen γ-chain transcripts were found in lung, but not in kidney (39–41). Haidaris and Courtney also found γ-chain transcripts in brain and marrow (40,41). Thus, γ-chain transcripts are more ubiquitous than Aα and Bβ transcripts. The significance of the expression of Aα and Bβ mRNAs without γ, or γ mRNA alone, remains unclear.

Because the α granules of platelets contain fibrinogen, it was assumed that fibrinogen was synthesized in megakaryocytes. Fibrinogen transcripts have been isolated from rat marrow enriched for megakaryocytes (42), and γ-chain transcripts were found in promegakaryocytes by in situ hybridization (43). However, other experiments demonstrated that α-granule fibrinogen is not synthesized in megakaryoctyes but results from endocytosis of plasma fibrinogen (44–47). Although the data appear contradictory, a reasonable postulate is that γ-chain transcripts are synthesized in some marrow cells that may become platelets, and the major source of α-granule fibrinogen is endocytosis.

2. *Overexpression Associated with Disease*

Abnormally high levels of plasma fibrinogen correlate strongly with increased risk of cardiovascular disease (48–52). A relation between increased fibrinogen concentrations and altered fibrin gel characteristics has been demonstrated (53), indicating that increased fibrinogen per se may predispose to cardiovascular diseases. Studies to determine whether plasma fibrinogen levels correlate with specific polymorphisms within the fibrinogen genes show varied results. An association between fibrinogen levels and a restriction enzyme (BclI) polymorphism in the Bβ gene was demonstrated by Humphries et al. (54). In this study of 91 individuals, those homozygous for one allele had a mean fibrinogen of 2.74 mg/ml, but those homozygous for a second allele had a mean concentration of 3.69 mg/ml; heterozygous individuals had 2.98 mg/ml. In similar studies, Berg and Keirulf were not able to confirm the BclI association (55). They postulate that because the number of individuals who were homozygous for the second BclI allele was small (54), the association with high plasma fibrinogen was a chance event. In a study of 292 healthy men, a HaeIII polymorphism in the Bβ gene was also associated with between-individual differences in fibrinogen levels (56). Connor et al., in a study of four polymorphic loci in 247 individuals, found no significant association of genetic variation and plasma fibrinogen levels (57). In another study of fibrinogen genotypes, Fowkes et al. found that polymorphisms within the three genes were not significantly related to fibrinogen concentrations (58). They did, however, find an association between the Bβ genotypes and an increased risk of atherosclerotic disease. They speculated that this association could be caused by a structurally variant fibrinogen associated with the polymorphism or similar linkage disequilibrium with a neighboring gene. The situation appears to be that a relation between fibrinogen genotype and fibrinogen concentration requires further analysis of larger and different populations.

3. *Underexpression Associated with Disease*

Hypofibrinogenemia is associated with both acquired defects, such as severe liver disease (59), and with inherited defects (60). Patients with inherited disease often have a hemorrhagic diathesis related to trauma or surgery but rarely bleed spontaneously. Inherited disease is rare, with approximately 150 patients reported. In some cases fibrinogen has been isolated from patient plasma and found to be abnormal, but the frequency of this association is not known. Individuals with low plasma fibrinogen are likely to be asymptomatic unless concentrations are below 0.6 mg/ml (60).

Congenital hypofibrinogenemia has been found as an obstetric complication (61). Successful gestation occurred in two cases when patients with low fibrinogen and recurrent spontaneous abortion were treated by fibrinogen infusion (61,62). Further examination of one patient and her family demonstrated that the low fibrinogen levels were associated with a unique karyotype, 46, XX t(7;12)(p15.2–3;q24.31) (62). Because Il-6 is known to influence fibrinogen synthesis (31), it is noteworthy that the chromosomal location of the Il-6 gene is 7pl5 (63).

Inherited hypofibrinogenemia has also been seen in association with defective secretion of fibrinogen from the liver (64). In these patients, analysis of liver biopsies with fibrinogen-specific antibodies demonstrated the excessive accumulation of droplets of fibrinogen in the hepatocellular cytoplasm. Electron micrographs indicated that the protein had accumulated in the rough endoplasmic reticulum. Because analysis of mammalian cell systems expressing only one or two fibrinogen chains demonstrated intracellular accumulation of fibrinogen chains (65–67), the accumulation suggests that only one or two fibrinogen chains are synthesized in these patients. The familial nature of this disease was demonstrated by analysis of fibrinogen levels within the family; the patient was 0.4 mg/ml, her father was 0.6 mg/ml, and one daughter was 0.3 mg/ml; her mother and two grandmothers had normal levels (3–5 mg/ml).

III. PROTEIN STRUCTURE AND FUNCTION

A. Evolutionary Considerations

The genes for the three fibrinogen polypeptides are thought to have arisen by duplication of a common ancestral gene (20,26,27), although the putative gene has not been identified. All vertebrate fibrinogens are dimers of three polypeptides, and the Aα, Bβ, and γ chains are clearly related to the homologous human polypeptides (68). In all species, soluble fibrinogen is cleaved by thrombin to remove short amino-terminal peptides from the Aα and Bβ chains, giving rise to fibrin monomers that spontaneously polymerize, and the polymerized fibrin is then covalently cross-linked by a transglutaminase (68).

A comparison of the structures of the lamprey and mammalian proteins permits a correlation between the conserved structures and the common functions (68). For example, bovine thrombin cleaves FpB, but not FpA, from lamprey fibrinogen even though the scissile bond in both peptides is Arg-Gly (69). Several residues within human FpA—Phe^{8}, Leu^{9}, and Val^{14}—are known to bind to bovine or human thrombin, and they are absent in lamprey FpA. Thus the resistance of lamprey fibrinogen to cleavage by bovine thrombin correlates with the known structural requirements for mammalian thrombins.

This precise difference between lamprey and human fibrinogens reinforces our detailed image of mammalian FpA-thrombin interactions and encourages utilization of such structural comparisons as a means to define regions important for function. From an overall comparison of the lamprey and human sequences, two distinctive regions are striking. The Aα-chain C terminus is striking for the cross-species variability, and the C-terminal domain of both the Bβ and γ chains is striking for the cross-species identity. Aα-chain residues 264–391 in human fibrinogen contain a 13 amino acid repeat present in 10 copies (70). The Aα chain of lamprey fibrinogen contains an 18 amino acid repeat present in more than 20 copies between residues 367 and 821 (71). The difference in polypeptide chain size, 66,124 for the human and 96,722 for the lamprey, is accounted for by the difference in these two repeat units. The two repeats are composed of similar residues, yet the segments share no sequential elements. The variation in primary structure implies that this region of the Aα chain exists as a domain of significant flexibility, consistent with the name ''free-swimming appendages'' originally coined by Doolittle (72). It remains to be determined whether this large difference in structure is associated with functional differences between lamprey and human fibrinogens.

The remarkable similarity of the Bβ and γ C-terminal domains of all fibrinogens prompted Doolittle to propose a model for the secondary structure and folding of these domains based mainly on analysis of the sequence conservation (73). When comparing the human γ chain from residue 152 to the C terminus with the lamprey γ chain and the human Bβ chain from residue 210 to the C terminus with the lamprey β chain, Doolittle found that the lamprey and human sequences are approximately 63% identical for both the Bβ and γ chains. The human γ-chain domain is 42.4% identical to the human β chain, and the lamprey γ chain is 42.5% identical to the lamprey β chain. From his analysis Doolittle proposed a model in which the C-terminal domain of each chain exists as two subdomains. The N-terminal subdomain is more compact and contains two sheets of three β strands. Some β structure and a C-terminal helix are proposed for the second domain. It will be of interest to see whether an experimentally based structure will resemble this novel high-resolution model.

Because it has long been known that the fibrinogen D domain, which includes the C-terminal domains of both Bβ and γ chains, inhibits fibrin polymerization (72), it is reasonable to conclude that the conserved structures are critical for polymerization. The D domain is presumed to participate in polymerization by binding to the E domain sites, which are exposed by removal of the fibrinopeptides. Because the first three residues of the fibrin α chain (Gly-Pro-Arg) and four of the first five residues of the fibrin β chain are also conserved between human and lamprey, a critical role for both the E domain and the D domain in fibrin polymerization is implicated. The importance of these domains is confirmed when variant proteins with substitutions in these domains are examined; they all demonstrate impaired polymerization (see later). All the residue substitutions that have been identified in variant human fibrinogens are at positions conserved between humans and lampreys, except Asn^{308}, which is changed to Ile in fibrinogen Baltimore III (74) and to Lys in Kyoto I (75). Furthermore, recent cross-linking experiments with Gly-Pro-Arg–photoactivatable probes indicate that γ residues 337–379 comprise the primary D domain polymerization site (76). The predominant labeling was at Tyr^{363} (77), a residue conserved between lamprey and human. It is surprising that there was no indication of labeling in the highly homologous Bβ-chain domain. Participation of the Bβ domain in polymerization is indicated by a Bβ-chain variant, $Ala^{335} \rightarrow$ Thr, that shows impaired polymerization (78). This substitution leads to a novel carbohydrate addition, however, and the presence of this carbohydrate moiety may cause large conformational changes resulting indirectly in impaired polymerization.

B. Analyses of Variant Fibrinogens

Inherited variant fibrinogens are rare. A significant fraction of these variants are found in clinically asymptomatic individuals and have been discovered through routine coagulation assays before surgery. In vitro studies of these abnormal proteins have been useful in associating fibrinogen structure with biochemical properties. Analyses of variant fibrinogens that are associated with clinical symptoms are potentially more informative, because they permit association of structure with in vivo biologic roles and in vitro biochemical properties. Variant fibrinogens are associated with bleeding, with thrombosis, and with both bleeding and thrombosis. The variant fibrinogens discussed here are those with known abnormalities. They are divided into four groups: (1) asymptomatic cases, (2) cases associated with bleeding, (3) cases associated with thrombosis, and (4) cases associated with both bleeding and thrombosis.

Of the 30 known defects listed in Table 1, 13 are asymptomatic, 6 are associated with bleeding, 5 with thrombosis, and 6 with both bleeding and thrombosis. To conclude that a variant fibrinogen is the pathogenic agent there must be a direct correlation between the abnormality and the clinical phenotype. Hemostasis is a multifactorial event, and a specific abnormal fibrinogen may be manifest because the variant fibrinogen activity complicates another clinical abnormality. One anticipates, for example, that in individuals predisposed to atherosclerosis plaque formation is potentiated by a variant fibrinogen with slightly impaired fibrinolysis, leading to thrombotic events that are then associated with the variant fibrinogen. In individuals who lack this predisposition, the variant fibrinogen is probably clinically silent. Although one can conclude that impaired fibrinolysis is detrimental to health, one would be incorrect to conclude that this variant fibrinogen is the etiological agent in the formation of atherosclerotic plaque. A causative role of the variant fibrinogen is therefore questionable with the four variants that are sometimes symptomatic. Similarly, cases in which a single individual serves to imply that a variant fibrinogen is the etiological agent for hemostatic disease must be assessed with

Table 1 Variant Fibrinogens

Structural defect		Number of cases	Genetic change	Clinical manifestation[a]	Allelic nature
Aα	$Asp^{7} \rightarrow Asn$	1	GAC → AAC[b]	A	Heterozygous
	$Gly^{12} \rightarrow Val$	1	GGA → GTA[b]	A	Heterozygous
	$Arg^{16} \rightarrow Cys$	14	CGT → TGT	A/B/T	Both[c]
	$Arg^{16} \rightarrow His$	27	CGT → CAT	A/B	Both[c]
	$Pro^{18} \rightarrow Leu$	1	CCA → CTA[b]	B	Heterozygous
	$Arg^{19} \rightarrow Asn$	1	AGG → AAT/C[b]	B	Unknown
	$Arg^{19} \rightarrow Gly$	2	AGG → GGG[b]	B/T	Both[c]
	$Arg^{19} \rightarrow Ser$	1	AGG → AGT/C[b]	B	Homozygous
	$Arg^{141} \rightarrow Ser$	1	AGA → AGT/C[b]	B	Homozygous
	$Ser^{434} \rightarrow Asn$	1	AGC → AAC[b]	A	Heterozygous
	$Lys^{461} \rightarrow stop$	1	AAA → TAA	B/T	Homozygous
	$Arg^{554} \rightarrow Cys$	1	CGT → TGT	T	Heterozygous
Bβ	$Arg^{14} \rightarrow Cys$	3	CGT → TGT	A/B/T	Heterozygous
	$Gly^{15} \rightarrow Cys$	1	GGT → TGT[b]	A	Heterozygous
	$Arg^{44} \rightarrow Cys$	1	CGT → TGT	T	Heterozygous
	$Ala^{68} \rightarrow Thr$	1	GCT → ACT	T	Homozygous
	Δ^{9-72}	1	Deletes exon 2[b]	T	Heterozygous
	$Ala^{335} \rightarrow Thr$	1	GCC → ACC	NA	Heterozygous
γ	$Arg^{275} \rightarrow Cys$	5	CGC → TGC	A	Heterozygous
	$Arg^{275} \rightarrow His$	5	CGC → CAC[b]	A/B/T	Heterozygous
	$Gly^{292} \rightarrow Val$	1	GGC → GTC	B/T	Heterozygous
	$Asn^{308} \rightarrow Ile$	1	AAT → ATT[b]	A	Heterozygous
	$Asn^{308} \rightarrow Lys$	1	AAT → AAG	A	Heterozygous
	$Met^{310} \rightarrow Thr$	1	ATG → ACG	B	Heterozygous
	$\Delta^{319-320}$	1	Deletes AAT GAT	T	Heterozygous
	$Gln^{329} \rightarrow Arg$	1	CAG → CGG[b]	A	Heterozygous
	$Asp^{330} \rightarrow Tyr$	1	GAT → TAT	A	Heterozygous
	$Asp^{330} \rightarrow Val$	1	GAT → GTT[b]	A	Heterozygous
	350/insert 15 aa/351[d]	1	Intron bp 6588 A → G	A	Heterozygous
	$Arg^{375} \rightarrow Gly$	1	CGG → GGG	A	Heterozygous

[a]A asymptomatic; B, bleeding; T, thrombosis; NA, not available.
[b]The genetic change is surmised from the amino acid change.
[c]Some cases are homozygous and some are heterozygous.
[d]aa, amino acids.

caution. Currently, an association between a variant fibrinogen found in more than one individual and specific clinical symptoms has been demonstrated in only a limited number of cases: Detroit I (Aα Arg^{19} → Ser) (79), Munich I (Aα Arg^{19} → Asn) (80), Mannheim I (Aα Arg^{19} → Gly) (81), New York I (Bβ Δ^{9-72} (82), Naples (Bβ Ala^{68} → Thr) (83), and Dusart (Arg^{554} → Cys) (84). An index of variant fibrinogen has been published (80).

1. Asymptomatic Cases

A total of 13 variants (17 individual cases) have been identified by routine coagulation assays and are asymptomatic. In 15 cases these individuals are known to be heterozygous, supporting the hypothesis that clinical manifestations of dysfibrinogenemia are usually recessive. In vitro analysis of these 13 variant proteins has provided important functional information, much of which is consistent with biochemical analysis and some of which has provided leads for further analysis.

Two variants within FpA, Asp^{7} → Asn (85,86) and Gly^{12} → Val (87), were associated with delayed thrombin release of FpA. Biochemical and conformational analysis of peptides analogous to normal FpA and these two variant FpAs confirmed that Val at position 12 impairs thrombin binding to this region but did not confirm that Asn at position 7 is detrimental (88,89). Similarly, analysis of engineered mutants expressed in a hybrid construct containing Aα 1–50 demonstrated that the substitution of Val at position 12 reduced the rate of FpA release by 100-fold, but a substitution of Gly at position 7 had no influence (90). The biochemical data, interpreted as suggesting that the assignment of Asp^{7} → Asn is the defect in fibrinogen Lille, may need to be reconsidered because this assignment was published only as an abstract and actual experimental data are not available (86). Together these experiments indicate that position Aα 7 is not critical, but Aα 12 is critical, for effective thrombin-catalyzed release of FpA. Furthermore, they indicate that effective release of FpA in vivo from approximately half the Aα chains is sufficient to support normal hemostasis.

A third asymptomatic Aα variant is Caracas II, a heterozygous Ser^{434} → Asn substitution that is accompanied by N-linked glycosylation (91). Analysis of purified patient fibrinogen demonstrated abnormal aggregation profiles, but normal factor XIIIa-catalyzed cross-linking and normal fibrin enhancement of tissue plasminogen activator activity. The structure of the C-terminal region of Aα chain is poorly defined, and thus the availability of residue 434 for glycosylation is not unexpected. Recent results indicate that the C-terminal region of the Aα chain participates in fibrin polymerization (92), and the data on Caracas II support a proposal that this region is important for lateral aggregation of protofibrils. Again, the asymptomatic nature of this variant indicates that the altered fibrin formed with this heterozygous molecule is a trivial defect.

Fibrinogen Ise, Bβ Gly^{15} → Cys, was also found in an asymptomatic individual (93). The Ise proband, his two sisters, and his daughter were all found to have hypofibrinogenemia when assessed by thrombin time, and all are asymptomatic. Thrombin catalyzed the release of only 50% of FpB from purified fibrinogen Ise under normal conditions, but complete release was seen after long incubation with excessive amounts of thrombin. Because the thrombin time was prolonged, the data from fibrinogen Ise support the hypothesis that the release of FpB and exposure of the β-chain amino terminus influence fibrin formation. The defective FpB release is clinically silent, however. This is anticipated from the biochemical finding that removal of only FpA by the enzyme Reptilase is sufficient to form a clot that is essentially the same as that formed when both FpA and FpB are removed by thrombin (94). Another Bβ substitution, Ala^{335} → Thr, is probably also asymptomatic, because no clinical symptoms have been reported for this patient (78). There are few biochemical data relevant to that region of the Bβ chain, so it is not unexpected that variants in this region are asymptomatic.

Eight heterozygous substitutions within the γ chain are associated with asymptomatic individuals, and all show impaired monomer polymerization. Together these variants clearly support the conclusion that this C-terminal region of the γ chain, residues γ 275–375, is important for fibrin polymerization but that such impaired polymerization is not clinically manifest. In the case Osaka II, $Arg^{275} \rightarrow$ Cys, the fate of a novel Cys was examined and found to be disulfide linked to a free cysteine molecule, an addition that probably perturbs the fibrinogen structure minimally (95). Because the C-terminal domain of the γ chain contains a high-affinity calcium binding site (96,97) and calcium binding influences fibrin polymerization (98), it is pertinent to note that the variant $Asn^{308} \rightarrow$ Lys (Kyoto I) (75) binds calcium normally, but the $Arg^{375} \rightarrow$ Gly (Osaka V) (99) substitution does not. For this latter case, defective fibrin polymerization can be reversed to normal by the addition of 5 mM calcium, demonstrating that calcium binding residues may be separated from residues important for polymerization. Another of these asymptomatic variants (Paris I) is a 15 amino acid insertion (100). In addition to defective polymerization, this variant also has defective factor XIIIa cross-linking of fibrin γ chains (101) and defective fibrinogen-mediated platelet aggregation (102). Other data demonstrate that the C-terminal 14 residues, γ 398–411, are important for these two functions (11,103), so it was unexpected that the abnormality in this variant is 50 residues removed from the C terminus. These data imply that the novel insertion alters the C-terminal conformation of the variant chain such that it cannot participate in cross-link formation or platelet aggregation (100).

2. *Cases Associated with Bleeding*

Six variants have been associated with a clinically manifest bleeding tendency, and one of these, $Arg^{16} \rightarrow$ His, is only sometimes symptomatic. Both homozygous and heterozygous individuals are found in this group.

In two cases, Asahi (104) and Lima (105), the clinical manifestations are associated only with the proband either because family studies were not reported or because no other affected family members were identified. Thus, it is possible to hypothesize an association between the abnormalities and the clinical symptoms, but further studies are needed to regard these as hereditary and to support the association with clinical manifestation. Fibrinogen Asahi is a substitution of γ Met^{310} with Thr, a change in amino acid structure with consequent glycosylation at γ Asn^{308}. This heterozygous variant was associated with markedly impaired fibrin monomer polymerization. When normal D fragments of fibrinogen Asahi were separated from the variant D fragments, the former prolonged the clotting time of normal fibrinogen in a dose-dependent manner, but the latter did not. Factor XIIIa-catalyzed γ chain cross-linking was also impaired in this variant. The association of this heterozygous defect with posttraumatic bleeding demonstrates that dysfibrinogenemias may be clinically expressed dominants. The homozygous defect in fibrinogen Lima, Aα $Arg^{141} \rightarrow$ Ser, is also associated with novel glycosylation, N-linked to Asn^{139}. This 10-year-old patient had transient hematuria but no other bleeding symptoms. Her parents, who are first cousins, are presumably heterozygous for the defect. There was no family history of bleeding or thrombosis. Thus, the association between this defect and bleeding anomalies is minimal and clinically recessive. It should be noted that novel addition of carbohydrate is not invariably associated with clinical manifestation; for example, fibrinogen Caracas II, Aα Ser^{434} → Asn, has a unique N-glycosylation in an asymptomatic girl and her father (91).

One group of variants is associated with a bleeding diathesis in multiple individuals. These are substitutions at Aα Arg^{19} → Asn, Gly, or Ser (78,81,106–108). For all three substitutions, the probands presented with bleeding symptoms and multiple affected family members displayed mild bleeding symptoms. Although these patients indicate a direct association between mild bleeding and substitutions at this residue, another case in which Aα Arg^{19} is substituted with Gly, Aarhus (107), the proband suffered from transient ischemic attacks after surgery, not bleeding symptoms. Ischemic difficulties following surgery are not uncommon, so the symptoms may not reflect the fibrinogen abnormality. In addition, the Aarhus patient is homozygous for this defect, which may also influence the clinical manifestation.

Biochemical analysis of these Arg^{19} variants demonstrated impaired fibrin polymerization associated with impaired fibrinopeptide release. Analysis of fibrinogens Detroit and Mannheim I indicated that this region of the Aα chain constitutes a critical domain in polymer formation (81,108). This conclusion is supported by the observation that a synthetic peptide that includes the first three residues of the α chain (Gly-Pro-Arg-Pro) inhibits the polymerization of fibrin monomers, but a peptide based on fibrinogen Detroit, Gly-Pro-Ser-Pro, does not (109). The association of this domain with impaired polymerization was also demonstrated by an Aα Pro^{18} → Leu variant, Kyoto II (110). In this instance the clinical manifestation was genital bleeding 2 weeks after delivery, consistent with the manifestations of the heterozygous Arg^{19} substitutions. In this case, however, an affected relative, the mother of the proband, was asymptomatic.

Two other variants, Aα Arg^{16} → His and Aα Arg^{16} → Cys, have been associated with both asymptomatic and bleeding conditions. Of the 27 cases of Aα Arg^{16} → His, 8 examples were identified as bleeders and the remainder as asymptomatic (80). Unexpectedly, it was found that thrombin can cleave the His^{16}-Gly^{17} bond (111), although at a greatly reduced rate (112). This defective cleavage, of course, impairs fibrin formation, which is expected to cause bleeding. Thus, these cases may be classified as a genetic defect with variable expression; that is, the abnormal fibrinogen is the causative agent, but manifestation of symptoms is diverse. This conclusion is probably appropriate for the cases of Aα Arg^{16} → Cys, of whom 3 (Homburg III, Ledyard I, and Metz I) of 14 are bleeders and 10 are asymptomatic (80). The remaining case (Hershey II) (113), which displays symptoms of thrombosis, is complicated because the proband suffers from diabetes, a disease known to be associated with thrombotic symptoms, and was over 60 years old at the time of her first thrombotic episode.

3. *Cases Associated with Thrombosis*

Five variants are associated with thrombotic disease, one of which is manifest only in homozygous individuals and is therefore a recessive clinical trait. For three of these variants there is a correlation of the protein abnormality and the clinical symptoms in multiple individuals.

Fibrinogen Nijmegen is a substitution, Bβ Arg^{44} → Cys, that is associated with disulfide-linked albumin (114). Studies demonstrated a decreased binding of tissue plasminogen activator and impaired tissue plasminogen activator-mediated plasminogen activation (115), characteristics expected to cause thrombosis. If it can be confirmed that this variant fibrinogen is the etiological agent for thrombosis, then this abnormality will also represent a dominant disease. Fibrinogen Vlissingen is a heterozygous deletion of γ

Asn^{319} and Asp^{320}, and this defect is associated with thrombosis (116). Analysis of purified fibrinogen demonstrated normal release of fibrinopeptides but impaired fibrin polymerization. Furthermore, this fibrinogen bound only 2 mol calcium per mol fibrinogen, rather than the 3 mol/mol found in normal fibrinogen. These results demonstrate that Asn^{319} and Asp^{320} are a crucial part of the high-affinity calcium binding site located in the D domain and that these residues are also critical to the activity of the C-terminal polymerization domains. Again, if the variant fibrinogen is responsible for the thrombotic disease, then this is a clinically dominant defect.

Fibrinogen New York I, in which Bβ residues 9–72 are deleted, was identified in a patient with thrombosis (82,117). Analysis of 13 family members demonstrated that two brothers have similar impaired coagulation assays, one of these with symptoms of thrombosis (82). A third brother, whose coagulation status was unknown, died at age 57 with suspected pulmonary embolism following cholecystectomy. The patient's parents have no history to suggest abnormal hemostasis. Further biochemical studies on the patient and one brother demonstrated that the abnormal fibrinogen represented 50% of the total fibrinogen is the patient but only 35–40% in her brother with recent symptoms. Although this large alteration in protein structure is anticipated to lead to clinically abnormal hemostasis, absence of symptoms in some affected family members indicates that expression of thrombosis is variable.

Two cases of thrombosis appear to be completely correlated with a variant fibrinogen, fibrinogen Naples (83,118) and fibrinogen Dusart (84,119,120). Naples is a homozygous substitution of Bβ $Ala^{68} \rightarrow$ Thr. This substitution is associated with delayed thrombin-catalyzed fibrinopeptide release and markedly impaired thrombin binding to fibrin. Fibrin prepared from heterozygous family members bound approximately 50% as much thrombin as fibrin from normal individuals. The association with thrombophilia is complete in that all homozygous family members, three siblings, have clinically manifest thrombotic disease, and the heterozygous family members are asymptomatic. Even this case must be interpreted with some caution, because the parents are first cousins, so that the siblings are homozygous for other genes as well. However, no data were obtained during clinical examination of these patients to indicate any other abnormal gene product. Again, this is a recessive disease associated with a variant fibrinogen.

Fibrinogen Dusart is a heterozygous substitution of Aα $Arg^{554} \rightarrow$ Cys, with disulfide-linked albumin (84). Studies with isolated fibrinogen demonstrated abnormal fibrin polymerization with altered lateral fibril association and fibers that are thinner than those formed from normal fibrinogen. Ultrastructural analysis of fibrinogen fragments prepared by limited plasmin digestion demonstrated that polymers formed from normal fibrinogen and fibrinogen Dusart were essentially equivalent for fragments lacking the C-terminal Aα regions of fibrinogen (120). Thus, the defective polymerization of fibrinogen Dusart is associated only with the abnormal Aα C-terminal domain. In addition to abnormal polymerization, functional studies showed reduced plasminogen binding and impaired plasminogen activation by tissue plasminogen activator (119). These results further demonstrate that the C-terminal region of the Aα chain is critical to normal fibrin fiber formation and, furthermore, that normal fiber formation is critical to normal fibrinolysis. The proband had a history of recurrent venous thrombosis and died from pulmonary embolism (119). His father and his two brothers also died from thrombotic disease at ages 52, 20, and 30, respectively; the father and one brother were known to have an abnormal fibrinogen. His two sons are affected but at ages 18 and 15 did not have thrombotic disease. Because the patient is heterozygous, the clinical manifestation of disease associated with this variant fibrinogen is dominant.

4. Cases Associated with Both Bleeding and Thrombosis

Six variants have been associated with both bleeding and thrombosis, and three of these also occur in asymptomatic individuals. Two variants have been associated with both bleeding and thrombosis in the same individual, indicating that a single defect may impair both fibrin formation and fibrin lysis and thus be associated with both bleeding and thrombosis, although the clinical history of one case suggests that the variant fibrinogen alone did not bring about the hemostatic disease.

The three variants that are sometimes symptomatic are Aα Arg^{16} → Cys, Bβ Arg^{14} → Cys, and γ Arg^{275} → His. As noted, in 1 of 14 identified cases (Hershey II) (113), Aα Arg^{16} → Cys is associated with thrombosis. This patient is an elderly diabetic, so that it is reasonable to conclude that this substitution is associated with only asymptomatic or bleeding cases. The heterogeneity of clinical manifestations with the Bβ Arg^{14} → Cys substitution may be associated with the fate of the odd Cys. In fibrinogen Seattle I (121), two variant chains associate in a homodimer to form a novel disulfide bond that links the two variant Bβ chains. Biochemical analysis of this variant demonstrated that only half the FpB was released, and polymerization was impaired, with the thrombin time increased about threefold. This patient was asymptomatic. In contrast, the patient Christchurch II (122) had recurrent epistaxis. Biochemical data demonstrated abnormal release of FpB and a twofold increase in the thrombin time. The status of the novel Cys was not reported. Finally, the patient IJmuiden I (114) suffered from thrombophilia. FpB release from this variant was also half that of normal, and the thrombin time was increased about 1.5-fold. In this case, the Bβ chain was shown to be disulfide linked to albumin, which, as described before, is a characteristic of several variant fibrinogens associated with thrombosis. Based on these three cases, one may speculate that the fate of a novel Cys residue in terms of its pairing partner markedly influences the clinical manifestation of these variants.

In contrast, there is no known structural difference among the five cases of the variant γ Arg^{275} → His, yet two cases, Essen I (123) and Perugia I (123), are asymptomatic. One, Saga I (124), is associated with hematuria, another, Bergamo II (123), with deep vein thrombosis and pulmonary embolism, and the fifth, Haifa I, with peripheral arterial thrombosis (125,126). It is relevant to note that in all five cases in whom this Arg is replaced by Cys, rather than His, the patients are asymptomatic. Thus, it appears that the clinical manifestations of the three symptomatic cases arise from unrelated or compounding causes. Biochemical characterization of these 10 (Cys plus His) Arg^{275} substitutions demonstrated impaired polymerization consistent with the proposal that the C-terminal domain of the γ chain contains a critical polymerization site. In perhaps the most carefully studied case, Osaka II, Arg^{275} → Cys (95), Terukina et al. isolated the defective C-terminal D domain from the heterozygous fibrinogen and demonstrated that in contrast to the normal D fragment, the defective fragment did not inhibit normal fibrin monomer polymerization. Thus, the impaired polymerization associated with these heterozygous variant proteins probably reflects the presence of an altered polymerization site in the variant D domain.

As mentioned earlier, there are two cases in which one individual manifests symptoms of both bleeding and thrombosis. One is fibrinogen Marburg, which is a homozygous case in which Aα Lys^{461} is substituted by a stop codon (127). This results in a deletion of 150 amino acids, one of which is Cys, and thereby an unpaired Cys is found in the defective chain. Koopman et al. showed that albumin is disulfide linked to this variant fibrinogen. The deletion of 150 amino acids and addition of albumin is an extensive change in fibrinogen structure, and therefore it is reasonable that this variant is associated with both bleeding and thrombosis in a homozygous case. These symptoms are not found in heterozygous family members. Immunological analysis of plasma from the heterozygous individuals demonstrated that less than 15% of plasma fibrinogen in these individuals was abnormal. This case serves as a single example of an apparently recessive trait with both thrombotic and bleeding symptoms.

The second case that manifests both thrombosis and bleeding is the heterozygous defect Baltimore I, γ $Gly^{292} \rightarrow Val$ (128). In this case it appears that the symptoms of the proband arise from causes other than the defective fibrinogen. The patient was a 29-year-old woman with a history of thrombophlebitis with pulmonary emboli and a mild hemorrhagic disorder (129). However, in a 25 year follow-up she has shown no hemostatic symptoms. In addition, she has eight asymptomatic offspring who express the variant fibrinogen (128).

5. Conclusions

In reviewing these individual cases, two things are apparent. First, most cases associated with bleeding are variants within the N terminus of the Aα chain. These alterations impair the first step of the conversion of fibrinogen to fibrin or the first step in fibrin polymer formation. It is reasonable that these activities would impair thrombus formation, leading to bleeding symptoms. Caution should be observed, however, because the N-terminal regions of proteins are more easily studied, so variants in other regions may not yet have been found and associated with bleeding. It appears that variants in the N terminus of Bβ and γ are *not* associated with bleeding, however. Second, many cases associated with thrombosis are variants with disulfide-linked albumin. This generalization also requires caution because the only cases examined to date for linked albumin have been thrombotic cases. It seems appropriate to screen routinely all variant fibrinogens for linked albumin.

It is clear that biochemical analysis of these 30 cases with known structural alterations has been informative in associating fibrinogen structure with functions assayed in vitro. In some cases the results have substantiated prior conclusions based on biochemical analysis of normal fibrinogen; for example, the deletion of γ Asn^{319} and Asp^{320} (116) confirmed that these residues are critical to calcium binding (130). In other cases the biochemical analysis of abnormal fibrinogens has provided the impetus for additional experiments, for example the preparation of α chain-like peptides Gly-Pro-Arg-Pro and Gly-Pro-Ser-Pro (109), based on fibrinogen Detroit (79,108), to determine whether these residues could inhibit fibrin polymerization as would be predicted if they formed one of a pair of binding domains essential for polymerization. Furthermore, it is clear that analysis of examples in which the variant protein is found in more than one individual can provide novel information on the in vivo functions of fibrinogen. For instance, we learned from analysis of the fibrinogen Naples family members that thrombin binding to fibrin is apparently critical in preventing life-threatening thrombosis (83,118). It is relevant that only homozygous individuals presented with clinical symptoms. Moreover, these three individuals suffered from spontaneous thrombosis in the abdominal aorta and the internal carotid artery and venous thrombosis after abdominal surgery, not from other thrombotic disease, such as disseminated intravascular coagulation. This suggests that defective thrombin binding was not pathogenic in smaller vessels, perhaps reflecting the increased levels of thrombomodulin as a result of the relatively increased surface area of endothelial cells in the smaller vessels. These results also suggest that the levels of circulating antithrombin III were not sufficient to prevent thrombosis in this abnormal situation. Clearly, further studies of variant fibrinogens associated with clinical manifestations would be enhanced if multiple affected individuals were identified, a goal that can be most easily met by family studies. Finally, several variant fibrinogens have been identified, but the structural abnormalities remain unknown.

C. Analysis of Recombinant Fibrinogens

The first expression of recombinant fibrinogen was in *Escherichia coli* (131). The three individual fibrinogen chains and domains from these chains have been expressed in separate clones (131–134), but functional, assembled fibrinogen has not been synthesized in a bacterial system. Analysis of these expressed domains and engineered variant domains have added to our understanding of fibrinogen residues critical for thrombin binding by analysis of hybrid proteins containing Aα 1–50 (90) and, for platelet aggregation, by analysis of γ-chain aggregates (135).

High-resolution structural studies of thrombin-FpA complexes demonstrate that specific residues within Aα 7–16 interact directly with thrombin; in particular, residues Phe^{8}, Leu^{9}, and Val^{15} are bound in a single hydrophobic pocket in thrombin (136–138). The juxtaposition of these three residues depends on the flexibility of the peptide chain linking them, which includes Gly at positions 12, 13, and 14. Single substitutions of Gly with Val led to a 100-fold reduced rate of thrombin cleavage for Val^{12} or Val^{13} but no change in the rate of cleavage for Val^{14}. These data support the importance of this compact structure for efficient cleavage of the Aα Arg^{16}-Gly^{17} peptide bond. In contrast, substitution of Phe^{8} with Tyr completely prevents cleavage of this bond, changing this Aα segment from a thrombin substrate to a thrombin inhibitor. It is known that Tyr can bind in the substrate hydrophobic pocket because structural analysis of thrombin complexed with the leach-derived inhibitor, hirudin, showed that Tyr^{3} of hirudin binds in this pocket in a position similar to, but with an orientation different from, Phe^{8} of FpA (139,140). If the substitution of Tyr for Phe^{8} in the Aα hybird protein alters the orientation of the bound Aα domain, then the scissile bond may not be accessible to the thrombin catalytic domain, such that the variant is not cleaved. This line of reasoning explains the observation that the Tyr^{8} variant binds to thrombin with an affinity similar to that of the normal hybird protein, but the variant is not cleaved.

Analysis of recombinant normal and variant γ chains synthesized as aggregates in *E. coli* has refined our understanding of which γ-chain residues are critical to platelet aggregation (135). In this study novel proteins were engineered to ask which γ-chain C-terminal residues are important for aggregation. Plasma fibrinogen contains about 10% of a variant chain, γ-Leu^{427}, that arises by alternative mRNA processing (24,25); γ-Leu^{427} differs from γ in that the last four amino acids of γ are replaced by a different 20 amino acid segment (23). Recombinant γ chains supported platelet aggregation to the same extent as plasma fibrinogen, but platelet aggregation with γ-Leu^{427} chains was markedly reduced (135). These results confirmed previously reported data that examined γ and γ-Leu^{427} chains isolated from plasma fibrinogen (14). Markedly reduced aggregation was also found with engineered γ chains that added γ-Leu^{427} residues 412–427 onto the C terminus of γ. Finally, truncated γ chains that were deleted for the last four residues were unable to support platelet aggregation (135). These results demonstrate that the four C-terminal residues of the γ chain, Ala-Gly-Asp-Val, are required and must be C terminal to support platelet aggregation.

Because plasma fibrinogen contains only 10% γ-Leu^{427}, it has not been possible to isolate fibrinogen molecules with two γ-Leu^{427} chains. However, a preparation of fibrinogen that is about 50% γ and 50% γ-Leu^{427}, presumably composed of heterodimers, has been prepared. Whether this preparation supports platelet aggregation is controversial (13,14). The complicated analysis of such heterodimers has been circumvented by synthesis of intact recombinant fibrinogen with only γ-Leu^{427} chains (15).

The synthesis of functional recombinant fibrinogen was first described by Farrell et al. in 1989 (141) and the details published in 1991 (142). Briefly, the three fibrinogen cDNAs were cloned into two plasmids (one containing a stable selection marker) that were cotransfected into baby hamster kidney (BHK) cells. Cells grown in selective media were screened for fibrinogen secretion and the highest producing clones isolated for further study. Initial analysis of homodimeric γ-chain fibrinogen and homodimeric γ-Leu^{427} chain fibrinogen demonstrated that sulfated tyrosine residues are found in γ-Leu^{427} chains. The presence of sulfated tyrosine in fibrinogen was described in 1963 (143) and, based on β-chain mobility on sodium dodecyl sulfate–polyacrylamide gel electrophoresis (SDS-PAGE), assigned to the β chain (144). Analysis of the recombinant protein indicates that fibrinogen is sulfated on the γ-Leu^{427} chain rather than the β chain.

Further analysis of these recombinant fibrinogens demonstrated that homodimeric γ-Leu^{427} fibrinogen is markedly defective in platelet aggregation, such that the extent of aggregation after a 5 minute incubation is about 14% of that seen with homodimeric γ-chain fibrinogen or plasma fibrinogen (15). The limited support of platelet aggregation by γ-Leu^{427} fibrinogen is similar to that seen with isolated γ-Leu^{427} chains synthesized in *E. coli* (135). Using the BHK system, Farrell et al. also altered the two RGD segments found in the Aα chain (15). Because RGD-containing peptides inhibit fibrinogen-mediated platelet aggregation, it has been hypothesized that either or both of these two RGD sequences are critical to platelet aggregation (145–147). However, changing either RGD to RGE in the recombinant protein did not alter fibrinogen-mediated platelet aggregation (15). This finding contrasts with that from von Willebrand factor, in which an RGD to RGE substitution abolished vWF binding to GPIIb/IIIa, the platelet receptor that binds both vWF and fibrinogen (148). These results indicate that fibrinogen binding to GPIIb/IIIa is mediated by the C terminus of γ chains, an interaction that is diminished for the variant γ-Leu^{427} chains, and is not mediated by the Aα RGD domains.

Recombinant human fibrinogen has also been expressed in COS-1 cells and CHO cells (65–67). The successful expression of apparently normal fibrinogen with equivalent amounts of Aα, Bβ, and γ chains, and the expected mobility on SDS-PAGE suggests that the information necessary for correct assembly and secretion of fibrinogen is inherent within the chains themselves, at least when they are expressed in mammalian cells with functioning secretory systems. Moreover, expression of one chain or pairs of chains in COS-1 and CHO cells has provided information on the chain requirements for assembly and secretion. The results are not completely consistent in that some laboratories (66,67) found that secretion of isolated chains or pairs of chains is sometimes possible, but Roy et al. (65) found that only fully assembled fibrinogen is secreted into the media.

Chain assembly has also been explored using the BHK expression system, and a novel model for chain assembly was recently proposed by Huang et al. (149). Using site-directed mutagenesis to alter the Cys residues that occur in the N terminus of all three chains, Huang et al. examined the formation of six-chain molecules. Their results appeared to be inconsistent with the current assignment of three disulfide bonds linking the fibrinogen half-molecules. They propose a new fibrinogen model in which the half-molecules are held together by five disulfide bonds. Specifically, the disulfide pair between Aα Cys^{28} and Bβ Cys^{65} is proposed to link chains from different half-molecules rather than within the same half-molecule.

Binnie et al. used directed mutagenesis of fibrinogen to examine the pattern of thrombin-catalyzed cleavage of fibrinopeptides (150). Under normal conditions, FpA is released before FpB, a result consistent with either FpA being a superior substrate or FpB being less accessible than FpA. Binnie et al. synthesized a molecule that had essentially four FpAs, two as usual on the α chains and two attached to the β chains, substituting for FpB. Analysis of this altered fibrinogen indicated that FpA is a better substrate than FpB and that FpA cleavage from the β chain is delayed relative to FpA cleavage from the α chain.

Taken together, these three studies (15,149,150) of recombinant normal and altered fibrinogens demonstrate the extensive potential for this new approach in further defining the association of fibrinogen structure with its multiple functions.

IV. FUTURE DIRECTIONS

This chapter has attempted to summarize the impact of molecular genetics on the biochemistry and biology of fibrinogen. The new data have answered many questions, such as the mechanism that regulates the level of plasma fibrinogen in the acute-phase response, but they have also introduced several new puzzles. For example, is there a role for γ chain synthesized in tissues that do not make the Aα and Bβ chains? New genetic information and techniques have enhanced the efficiency of analysis of fibrinogen variants, analyses that will continue to provide important clues for understanding of fibrinogen functions in vivo. I anticipate the production of transgenic animals whose characterization will provide further critical data about fibrinogen biology. Successful expression of recombinant fibrinogens has enabled the production of designed variants. These, in turn, will permit analysis of such complex issues as the mechanism of intracellular chain assembly. Furthermore, recombinant fibrinogen synthesis, which bypasses virus contamination, may stimulate utilization of fibrin glue in surgical procedures (151). Finally, synthesis of fibrinogen fragments or of homogeneous fibrinogen, which lacks the γ-Leu^{427} variant and the α chain heterogeneity found in plasma, may permit the long-awaited studies required to develop a high-resolution model for fibrinogen.

REFERENCES

1. Doolittle RF. Fibrinogen and fibrin. In: Bloom AL, Thomas DP, eds. Haemostasis and thrombosis, 2nd ed. Edinburgh: Churchill Livingstone, 1987:192–215.
2. Williams RC. Morphology of fibrinogen monomers and of fibrin protofibrils. Ann NY Acad Sci 1983; 408:180–193.
3. Weisel JW, Stauffacher CV, Bullitt E, Cohen C. A model for fibrinogen: domains and sequence. Science 1985; 230:1388–1391.
4. Rao SP, Poojary MD, Elliott Jr BW, Melanson LA, Oriel B, Cohen C. Fibrinogen structure in projection at 18Å resolution. Electron density by co-ordinated cryo-electron microscopy and X-ray crystallography. J Mol Biol 1991; 222:89–98.

5. Medved LV, Litvinovich SV, Privalov PL. Domain organization of the terminal parts in the fibrinogen molecule. FEBS Lett 1986; 202:298–302.
6. Medved LV. Relationship between exons and domains in the fibrinogen molecule. Blood Coag Fibrinol 1990; 1:439–442.
7. Peerschke EIB. The platelet fibrinogen receptor. Semin Hematol 1985; 22:241–259.
8. Marguerie GA, Ginsberg MH, Plow EF. Fibrinogen and platelet function. Adv Exp Med Biol 1985; 192:41–54.
9. Phillips DR, Charo IF, Scarborough RM. GPIIb-IIIa: the responsive integrin. Cell 1991; 65: 359–362.
10. Hawiger J, Timmons S, Kloczewiak M, Strong DD, Doolittle RF. γ And α chains of human fibrinogen possess sites reactive with human platelet receptors. Proc Natl Acad Sci USA 1982; 79:2068–2071.
11. Kloczewiak M, Timmons S, Bednarek MA, Sakon M, Hawiger J. Platelet receptor recognition domain on the γ chain of human fibrinogen and its synthetic peptide analogues. Biochemistry 1989; 28:2915–2919.
12. Hawiger J, Kloczewiak M, Bednarek MA, Timmons S. Platelet receptor recognition domains on the α chain of human fibrinogen: structure-function analysis. Biochemistry 1989; 28: 2909–2914.
13. Peerschke EIB, Francis CW, Marder VJ. Fibrinogen binding to human blood platelets: effect of γ chain carboxyterminal structure and length. Blood 1986; 67:385–390.
14. Amrani DL, Newman PJ, Meh D, Mosesson MW. The role of fibrinogen Aα chains in ADP-induced platelet aggregation in the presence of fibrinogen molecules containing γ′ chains. Blood 1988; 72:919–924.
15. Farrell DH, Thiagarajan P, Chung DW, Davie EW. Role of fibrinogen α and γ chain sites in platelet aggregation. Proc Natl Acad Sci USA 1992; 89:10729–10732.
16. Mosesson MW. The roles of fibrinogen and fibrin in hemostasis and thrombosis. Semin Hematol 1992; 29:177–188.
17. Olaisen B, Teisberg P, Gedde-Dahl T Jr. Fibrinogen γ-chain locus is on chromosome 4 in man. Hum Genet 1982; 62:24–26.
18. Humphries SE, Imam AMA, Robbins TP, et al. The identification of a DNA polymorphism of the α fibrinogen gene, and the regional assignment of the human fibrinogen gene to 4q26-qter. Hum Genet 1984; 68:148–153.
19. Kant JA, Fornace AJ, Saxe D, Simon MI, McBride OW, Crabtree GR. Organization and evolution of the human fibrinogen locus on chromosome four. Proc Natl Acad Sci USA, 1985; 82:2344–2348.
20. Chung DW, Harris JE, Davie EW. Nucleotide sequences of the three genes coding for human fibrinogen. In: Liu CY, Chien S, eds. Fibrinogen, thrombosis, coagulation, and fibrinolysis. New York: Plenum Press, 1990:39–48.
21. Huber P, Dalmon J, Courtois G, Laurent M, Assouline Z, Marguerie G. Characterization of the 5′-flanking region for the human fibrinogen β gene. Nucleic Acids Res 1987; 15:1615–1625.
22. Francis CW, Mosessen MW. Terminology for fibrinogen γ-chains differing in carboxyl terminal amino acid sequence. Thromb Haemost 1989; 62:813–814.
23. Wolfenstein-Todel C, Mosesson MW. Carboxy-terminal amino acid sequence of a human fibrinogen γ chain variant (γ′). Biochemistry 1981; 20:6146–6149.
24. Chung DW, Davie EW. γ and γ′ chains of human fibrinogen are produced by alternative mRNA processing. Biochemistry 1984; 23:4232–4236.
25. Fornace AJ Jr, Cummings DE, Comeau CM, Kant JA, Crabtree GR. Structure of the human γ-fibrinogen gene. Alternate mRNA splicing near the 3′ end of the gene produces γA and γB forms of γ-fibrinogen. J Biol Chem 1984; 259:12826–12830.
26. Doolittle RF. The evolution of vertebrate fibrinogen. Fed Proc 1976; 35:2145–2149.

27. Crabtree GR, Comeau CM, Fowlkes DM, Fornance AJ Jr, Malley JD, Kant JA. Evolution and structure of the fibrinogen genes: random insertion of introns or selective loss? J Mol Biol 1985; 185:1–19.
28. Weissbach L, Grieninger G. Bipartite mRNA for chicken α-fibrinogen potentially encodes an amino acid sequence homologous to β- and γ-fibrinogens. Proc Natl Acad Sci USA 1990; 87:5198–5202.
29. Pan Y, Doolittle RF. cDNA sequence of a second fibrinogen α chain in lamprey: an archetypal version alignable with full-length β and γ chains. Proc Natl Acad Sci USA 1992; 89: 2066–2070.
30. Crabtree GR, Kant JA. Coordinate accumulation of the mRNAs for the α, β, and γ chains of rat fibrinogen following defibrination. J Biol Chem 1982; 257:7277–7279.
31. Geiger T, Andus T, Klapproth J, Hirano T, Kishimoto T, Heinrich PC. Induction of rat acute-phase proteins by interleukin 6 in vivo. Eur J Immunol 1988; 18:717–721.
32. Morgan JG, Courtois G, Fourel G, et al. Sp1, a CAAT-binding factor, and the adenovirus major late promoter transcription factor interact with functional regions of the γ-fibrinogen promoter. Mol Cell Biol 1988; 8:2628–2637.
33. Mendel DB, Crabtree GR. Minireview: HNF-1, a member of a novel class of dimerizing homeodomain proteins. J Biol Chem 1991; 266:677–680.
34. Courtois G, Morgan JG, Campbell LA, Fourel G, Crabtree GR. Interaction of a liver-specific nuclear factor with the fibrinogen and α_1 antitrypsin promoters. Science 1987; 238:688–692.
35. Huber P, Laurent M, Dalmon J. Human β-fibrinogen gene expression: upstream sequences involved in its tissue specific expression and its dexamethasone and interleukin 6 stimulation. J Biol Chem 1990; 265:5696–5701.
36. Kugler W, Wagner U, Ryffel GU. Tissue-specificity of liver gene expression: a common liver-specific promoter element. Nucleic Acids Res 1988; 16:3165–3174.
37. Dalmon J, Laurent M, Courtois G. The human β fibrinogen promoter contains a HAF-1 dependent IL-6 responsive element. Mol Cell Biol 1993; 13:1183–1193.
38. Fowlkes DM, Mullis NT, Comeau CM, Crabtree GR. Potential basis for regulation of the coordinately expressed fibrinogen genes: homology in the 5′ flanking regions. Proc Natl Acad Sci USA 1984; 81:2313–2316.
39. Baumhueter S, Mendel DB, Conley PB, et al. HNF-1 shares three sequence motifs with the POU domain proteins and is identical to LF-B1 and APF. Genes Dev 1990; 4:372–379.
40. Haidaris PJ, Courtney MA. Molecular biology and regulation of the fibrinogen gene: tissue-specific and ubiquitous expression of fibrinogen γ-chain mRNA. Blood Coag Fibrinol 1990; 1:433–437.
41. Haidaris PJ, Courtney MA. Liver-specific RNA processing of the ubiquitously transcribed rat fibrinogen γ-chain gene. Blood 1992; 79:1218–1224.
42. Uzan G, Courtois G, Stanckovic Z, Crabtree GR, Marguerie G. Expression of the fibrinogen genes in rat megakaryocytes. Biochem Biophys Res Commun 1986; 140:543–549.
43. Courtney MA, Stoler MH, Marder VJ, Haidaris PJ. Developmental expression of mRNAs encoding platelet proteins in rat megakaryocytes. Blood 1991; 77:560–568.
44. Handagama PJ, Shuman MA, Bainton DF. Incorporation of intravenously injected albumin, immunoglobulin G, and fibrinogen in guinea pig megakaryocyte granules. J Clin Invest 1989; 84:73–82.
45. Harrison P, Wilbourn B, Debili N, et al. Uptake of plasma fibrinogen into the alpha granules of human megakaryocytes and platelets. J Clin Invest 1989; 84:1320–1324.
46. Handagama P, Rappolee DA, Werb Z, Levin J, Bainton DF. Platelet α-granule fibrinogen, albumin, and immunoglobulin G are not synthesized by rat and mouse megakaryocytes. J Clin Invest 1990; 86:1364–1368.
47. Louache F, Debili N, Cramer E, Breton-Gorius J, Vainchenker W. Fibrinogen is not synthesized by human megakaryocytes. Blood 1991; 77:311–316.

48. Wilhelmsen L, Svardsudd K, Korsan-Bengtsson K, Larsson B, Welin L, Tibblin G. Fibrinogen as a risk factor for stroke and myocardial infarction. N Engl J Med 1984; 311:501–505.
49. Stone MC, Thorpe JM. Plasma fibrinogen—a major coronary risk factor. J R Coll Gen Pract 1985; 35:565–570.
50. Meade TW, Mellows S, Brozovic M, et al. Haemostatic function and ischaemic heart disease: principal results of the Northwick Park heart study. Lancet 1986; 2:533–537.
51. Kannel WB, Wolf PA, Castelli WP, D'Agostino RB. Fibrinogen and risk of cardiovascular disease. The Framingham Study. JAMA 1987; 258:1183–1186.
52. Handa K, Kone S, Saku K, et al. Plasma fibrinogen levels as an independent indicator of severity of coronary atherosclerosis. Atherosclerosis 1989; 77:209–213.
53. Fatah K, Hamsten A, Blomback B, Blomback M. Fibrin gel network characteristics and coronary heart disease: relations to plasma fibrinogen concentration, acute phase protein, serum lipoproteins and coronary atherosclerosis. Thromb Haemost 1992; 68:130–135.
54. Humphries SE, Cook M, Dubowitz M, Stirling Y, Meade TW. Role of genetic variation at the fibrinogen locus in determination of plasma fibrinogen concentrations. Lancet 1987; 1: 1452–1455.
55. Berg K, Kierulf P. DNA polymorphisms at fibrinogen loci and plasma fibrinogen concentration. Clin Genet 1989; 36:229–235.
56. Thomas AE, Green FR, Kelleher CH, et al. Variation in the promoter region of the β fibrinogen gene is associated with plasma fibrinogen levels in smokers and non-smokers. Thromb Haemost 1991; 65:487–490.
57. Connor JM, Fowkes FGR, Wood J, Smith FB, Donnan PT, Lowe GDO. Genetic variation at fibrinogen loci and plasma fibrinogen levels. J Med Genet 1992; 29:480–482.
58. Fowkes FGR, Connor JM, Smith FB, Wood J, Donnan PT, Lowe GDO. Fibrinogen genotype and risk of peripheral atherosclerosis. Lancet 1992; 339:693–696.
59. Tytgat GN, Collen D, Verstraete M. Metabolism of fibrinogen in cirrhosis of the liver. J. Clin Invest 1971; 50:1690–1701.
60. Galanakis DK. Fibrinogen anomalies and disease: a clinical update. Hematol/Oncol Clin North Am 1992; 6:1171–1187.
61. Goodwin TM. Congenital hypofibrinogenemia in pregnancy. Obstet Gynecol Surv 1989; 44: 157–161.
62. Kitchens CS, Cruz AC, Kant JA. A unique 7p/12q chromosomal abnormality associated with recurrent abortion and hypofibrinogenemia. Blood 1987; 70:921–925.
63. Ferguson-Smith AC, Chen YF, Newman MS, May LT, Sehgal PB, Ruddle FH. Regional localization of the interferon-beta 2/B-cell stimulatory factor 2/hepatocyte stimulating factor gene to human chromosome 7p15-p21. Genomics 1988; 203–208.
64. Pfeifer U, Ormanns W, Klinge O. Hepatocellular fibrinogen storage in familial hypofibrinogenemia. Virchows Arch [B] 1981; 36:247–255.
65. Roy SN, Procyk R, Kudryk BJ, Redman CM. Assembly and secretion of recombinant human fibrinogen. J Biol Chem 1991; 266:4758–4763.
66. Hartwig R, Danishefsky KJ. Studies on the assembly and secretion of fibrinogen. J Biol Chem 1991; 266:6578–6585.
67. Binnie CG, Hettasch JM, Strickland E, Lord ST. Characterization of purified recombinant fibrinogen: partial phosphorylation of fibrinopeptide A. Biochemistry 1993; 32:107–113.
68. Doolittle RF. The structure and evolution of vertebrate fibrinogen: a comparison of the lamprey and mammalian proteins. In: Liu CY, Chien S, eds. Fibrinogen, thrombosis, coagulation, and fibrinolysis. New York: Plenum Press, 1990:25–37.
69. Doolittle RF: Difference in the clotting of lamprey fibrinogen by lamprey and bovine thrombins. Biochem J 1965; 94:735–741.
70. Doolittle RF, Watt KWK, Cottrell BA, Strong DD, Riley M. The amino acid sequence of the α-chain of human fibrinogen. Nature 1979; 280:464–468.

71. Wang YZ, Patterson J, Gray JE, et al. Complete sequence of the lamprey fibrinogen α chain. Biochemistry 1989; 28:9801–9806.
72. Doolittle RF. Structural aspects of the fibrinogen-fibrin conversion. Adv Protein Chem 1973; 27:1–109.
73. Doolittle RF. A detailed consideration of a principal domain of vertebrate fibrinogen and its relatives. Protein Sci 1992; 1:1563–1577.
74. Bantia S, Bell WR, Dang CV. Polymerization defect of fibrinogen Baltimore III due to a γAsn^{308} → Ile mutation. Blood 1990; 75:1659–1663.
75. Yoshida N, Terukina S, Okuma M, Moroi M, Aoki N, Matsuda M. Characterization of an apparently lower molecular weight γ-chain variant in fibrinogen Kyoto I: the replacement of γ-asparagine 308 by lysine which causes accelerated cleavage of fragment D_1 by plasmin and the generation of a new plasmin cleavage site. J Biol Chem 1988; 263:13848–13856.
76. Shimizu A, Nagel GM, Doolittle RF. Photoaffinity labeling of the primary fibrin polymerization site: isolation and characterization of a labeled cyanogen bromide fragment corresponding to γ-chain residues 337–379. Proc Natl Acad Sci USA 1992; 898:2888–2892.
77. Yamazumi K, Doolittle RF. Photoaffinity labeling of the primary fibrin polymerization site: localization of the label to γ-chain Tyr-363. Proc Natl Acad Sci USA 1992; 89:2892–2896.
78. Kaudewitz H, Henschen A, Soria J, Soria C. Fibrinogen Pontoise—a genetically abnormal fibrinogen with defective fibrin polymerization but normal fibrinopeptide release. In: Lane DA, Henschen A, Jasani MK, eds. Fibrinogen, fibrin formation and fibrinolysis. Berlin: Walter de Gruyter, 1986:91–96.
79. Blomback M, Blomback B, Mammen EF, Prasad AS. Fibrinogen Detroit—a molecular defect in the N-terminal disulphide knot of human fibrinogen? Nature 1966; 218:134–137.
80. Ebert RF. Index of variant human fibrinogens. Boca Raton, FL: CRC Press, 1991.
81. Dempfle C-EH, Henschen A. Fibrinogen Mannheim I—identification of an Aα19 Arg → Gly substitution in dysfibrinogenaemia associated with bleeding tendency. In: Matsuda M, Iwanaga S, Takada A, Henschen A, eds. Fibrinogen 4, current basic and clinical aspects. Amsterdam: Elsevier Science Publishers, 1990:159–166.
82. Liu CY, Koehn JA, Morgan FJ. Characterization of fibrinogen New York 1: a dysfunctional fibrinogen with a deletion of Bβ(9–72) corresponding exactly to exon 2 of the gene. J Biol Chem 1985; 260:4390–4396.
83. Koopman J, Haverkate F, Lord ST, Grimbergen J, Mannucci PM. Molecular basis of fibrinogen Naples associated with defective thrombin binding and thrombophilia: homozygous substitution of Bβ 68 Ala → Thr. J Clin Invest 1992; 90:238–244.
84. Koopman J, Haverkate F, Grimbergen J, et al. The molecular basis for fibrinogen Dusart (Aα 554 Arg → Cys) and its association with abnormal fibrin polymerization and thrombophilia. J Clin Invest 1993; 91:1637–1643.
85. Denninger MH, Finlayson JS, Reamer LA, Parquet-Gernez A, Goudemand M, Menache D. Congenital dysfibrinogenemia: fibrinogen Lille. Thromb Res 1978; 13:453–466. Thromb Res 1979; 15:291–293 (errata).
86. Morris S, Denninger MH, Finlayson JS, Menache D. Fibrinogen Lille: Aα7 Asp → Asn. Thrombo Haemostas 1981; 46:104.
87. Kehl M, Lottspeich F, Henschen A. Genetically abnormal fibrinogens releasing abnormal fibrinopeptides as characterized by high-performance liquid chromatography. In: Haverkate F, Henschen A, Nieuwenhuizen W, Straub PW, eds. Fibrinogen—structure, functional aspects, metabolism. Berlin: Walter de Gruyter, 1983:125–144.
88. Ni F, Konishi Y, Bullock LD, Rivetna MN, Scheraga HA. High-resolution NMR studies of fibrinogen-like peptides in solution: structural basis for the bleeding disorder caused by a single mutation of Gly(12) to Val(12) in the Aα chain of human fibrinogen Rouen. Biochemistry 1989; 28:3106–3119.
89. Zheng Z, Ashton RW, Ni F, Scheraga HA. Thrombin hydrolysis of an N-terminal peptide from fibrinogen Lille: kinetic and NMR studies. Biochemistry 1992; 31:4426–4431.

90. Lord ST, Byrd PA, Hede KL, Wei C, Colby TJ. Analysis of fibrinogen Aα-fusion proteins: mutants which inhibit thrombin equivalently are not equally good substrates. J Biol Chem 1990; 265:838–843.
91. Maekawa H, Yamazumii K, Muramatsu S-I, et al. An Aα Ser-434 to N-glycosylated Asn substitution in a dysfibrinogen, fibrinogen Caracas II, characterized by impaired fibrin gel formation. J Biol Chem 1991; 266:11575–11581.
92. Hasegawa N, Sasaki S. Location of the binding site ''b'' for lateral polymerization of fibrin. Thromb Res 1990; 57:183–195.
93. Yoshida N, Wada H, Morita K, et al. A new congenital abnormal fibrinogen Ise characterized by the replacement of Bβ glycine-15 by cysteine. Blood 1991; 77:1958–1963.
94. Hantgan R, Fowler W, Erickson H, Hermans J. Fibrin assembly: a comparison of electron microscopic and light scattering results. Thromb Haemost 1980; 44:119–124.
95. Terukina S, Matsuda M, Hirata H, et al. Substitution of γArg-275 by Cys in an abnormal fibrinogen, ''fibrinogen Osaka II'': evidence for a unique solitary cystine structure at the mutation site. J Biol Chem 1988; 27:13579–13587.
96. Haverkate F, Timan G. Protective effect of calcium in the plasmin degradation of fibrinogen and fibrin fragments D. Thromb Res 1977; 10:803–812.
97. Nieuwenhuizen W, Ruijven-Vermeer IAM, Nooijen WJ, Vermond A, Haverkate F. Recalculation of calcium-binding properties of human and rat fibrin(ogen) and their degradation products. Thromb Res 1981; 22:653–657.
98. Ratnoff OD, Potts AM. The accelerating effect of calcium and other cations on the conversion of fibrinogen to fibrin. J Clin Invest 1954; 33:206–210.
99. Yoshida N, Hirata H, Morigami Y, et al. Characterization of an abnormal fibrinogen Osaka V with the replacement of γ-arginine 375 by glycine: the lack of high affinity calcium binding to D-domains and the lack of protective effect of calcium on fibrinolysis. J Biol Chem 1992; 267:2753–2759.
100. Rosenberg JB, Newman PJ, Mosesson MW, Guillin M-C, Amrani DL. Paris I dysfibrinogenemia: a point mutation in intron 8 results in insertion of a 15 amino acid sequence in the fibrinogen γ chain. Thromb Haemost 1993; 69:217–220.
101. Mosesson MW, Amrani DL, Menache D. Studies on the structural abnormality of fibrinogen Paris I. J Clin Invest 1976; 57:782–790.
102. Denninger M-H, Jandrot-Perrus M, Elion J, et al. ADP-induced platelet aggregation depends on the conformation or availability of the terminal gamma chain sequence of fibrinogen. Study of the reactivity of fibrinogen Paris I. Blood 1987; 70:558–563.
103. Chen R, Doolittle RF. γ-γ Cross-linking sites in human and bovine fibrin. Biochemistry 1971; 10:4486–4494.
104. Yamazumi K, Shimura K, Terukina S, Takahashi N, Matsuda M. A γ methionine-310 to threonine substitution and consequent N-glycosylation at γ asparagine-308 identified in a congenital dysfibrinogenemia associated with posttraumatic bleeding, fibrinogen Asahi. J Clin Invest 1989; 83:1590–1597.
105. Maekawa H, Yamazumii K, Muramatsu S-I, et al. Fibrinogen Lima: a homozygous dysfibrinogen with an Aα-arginine-141 to serine substitution associated with extra N-glycosylation at Aα-asparagine-139: impaired fibrin gel formation but normal fibrin-facilitated plasminogen activation catalyzed by tissue-type plasminogen activator. J Clin Invest 1992; 90: 67–76.
106. Henschen A, Southan C, Kehl M, Lottspeich F. The structural error and its relation to the malfunction in some abnormal fibrinogens. Thromb Haemost 1981; 46:181.
107. Hessel B, Stenbjerg S, Dyr J, Kudryk B, Therkildsen L, Blomback B. Fibrinogen Aarhus—a new case of dysfibrinogenemia. Thromb Res 1986; 42:21–37.
108. Kudryk B, Blomback B, Blomback M. Fibrinogen Detroit—an abnormal fibrinogen with non-functional NH_2-terminal polymerization domain. Thromb Res 1976; 9:25–36.

109. Laudano AP, Doolittle RF. Studies on synthetic peptides that bind to fibrinogen and prevent fibrin polymerization. Structural requirements, number of binding sites, and species differences. Biochemistry 1980; 19:1013–1019.
110. Yoshida N, Okuma M, Hirata H, Matsuda M, Yamazumi K, Asakura S. Fibrinogen Kyoto II, a new congenitally abnormal molecule, characterized by the replacement of Aα proline-18 by leucine. Blood 1991; 78:149–153.
111. Higgins DL, Shafer JA. Fibrinogen Petoskey, a dysfibrinogenemia characterized by replacement of Arg-Aα16 by a histidyl residue: evidence for thrombin-catalyzed hydrolysis at a histidyl residue. J Biol Chem 1981; 256:12013–12017.
112. Southan C, Lane DA, Bode W, Henschen A. Thrombin-induced fibrinopeptide release from a fibrinogen variant (fibrinogen Sydney I) with an Aα Arg-16 → His substitution. J Biochem 1985; 147:593–600.
113. Galanakis DK, Henschen A, Schubach W, Lord S, Al-Mondhiry H. Two new 16Arg → Cys dysfibrinogens, Hershey II and Leogan: determination of abnormal structure and heterozygosity by amplification of genomic DNA using the polymerase chain reaction and by amino acid sequence analyses. In: Matsuda M, Iwanaga S, Takada A, Henschen A, eds. Fibrinogen 4. Current basic and clinical aspects. Amsterdam: Elsevier Science Publishers, 1990:173–178.
114. Koopman J, Haverkate F, Grimbergen J, et al. Abnormal fibrinogens IJmuiden (Bβ Arg14 → Cys) and Nijmegen (Bβ Arg44 → Cys) form disulfide-linked fibrinogen-albumin complexes. Proc Natl Acad Sci USA 1992; 89:3478–3482.
115. Engesser L, Koopman J, Munk G, et al. Fibrinogen Nijmegen: congenital dysfibrinogenemia associated with impaired t-PA mediated plasminogen activation and decreased binding of t-PA. Thomb Haemost 1988; 60:113–120.
116. Koopman J, Haverkate F, Briet E, Lord ST. A congenitally abnormal fibrinogen (Vlissingen) with a six-base deletion in the γ-chain gene, causing defective calcium binding and impaired fibrin polymerization. J Biol Chem 1991; 265:13456–13461.
117. Al-Mondhiry HAB, Bilezikian SB, Nossel HL. Fibrinogen ''New York''—an abnormal fibrinogen associated with thromboembolism: functional evaluation. Blood 1975; 46:607–612.
118. Minno GD, Martinez J, Cirillo F, et al. A role for platelets and thrombin in the juvenile stroke of two siblings with defective thrombin-adsorbing capacity of fibrin(ogen). Arterioscl Thromb 1991; 11:785–796.
119. Soria J, Soria C, Caen JP. A new type of congenital dysfibrinogenaemia with defective fibrin lysis—Dusard syndrome: possible relation to thrombosis. Br J Haematol 1983; 53:575–586.
120. Siebenlist KR, Mosesson MW, DiOrio JP, Soria J, Soria C, Caen JP. The polymerization of fibrinogen Dusart (Aα554 Arg → Cys) after removal of carboxy terminal regions of the Aα chains. Blood Coag Fibrinol 1993; 4:61–65.
121. Pirkle H, Kaudewitz H, Henschen A, Theodor I, Simmons G. Substitution of Bβ-14 arginine by cyst(e)ine in fibrinogen Seattle I. In: Lowe GDO, Douglas JT, Forbes CD, Henschen A, eds. Fibrinogen 2. Biochemistry, physiology and clinical relevance. Amsterdam: Elsevier Science Publishers, 1987:49–52.
122. Kaudewitz H, Henschen A, Soria C, Soria J, Bertrand O, Heaton D. The molecular defect of the genetically abnormal fibrinogen Christchurch II. In: Muller-Berghaus G, Scheefers-Borchen V, Selmayr E, Henschen A, eds. Fibrinogen and its derivatives. Amsterdam: Elsevier Science Publishers, 1986:31–36.
123. Reber P, Furlan M, Henschen A, et al. Three abnormal fibrinogen variants with the same amino acid substitution (γ275 Arg → His): fibrinogens Bergamo II, Essen and Perugia. Thromb Haemost 1986; 56:401–406.
124. Yamazumi K, Terukina S, Onohara S, Matsuda M. Normal plasmic cleavage of the γ-chain variant of ''fibrinogen Saga'' with an Arg-275 to His substitution. Thromb Haemost 1988; 60:476–480.

125. Brook JG, Tabori S, Tatarsky I, Hasmonai M, Schramek A. Fibrinogen "Haifa": a new fibrinogen variant. A case report. Haemostasis 1983; 13:277–281.
126. Siebenlist KR, Mosesson MW, DiOrio JP, Tavori S, Tatarsky I, Rimon A. The polymerization of fibrin prepared from fibrinogen Haifa (γ275Arg → His). Thromb Haemost 1989; 62:875–879.
127. Koopman J, Haverkate F, Grimbergen J, Egbring R, Lord ST. Fibrinogen Marburg: a homozygous case of dysfibrinogenemia, lacking amino acids Aα461–610 (Lys 461 AAA → Stop TAA). Blood 1992; 80:1972–1979.
128. Bantia S, Mane SM, Bell WR, Dang CV. Fibrinogen Baltimore I: polymerization defect associated with a γ292Gly → Val (GGG → GTC) mutation. Blood 1990; 76:2279–2283.
129. Beck EA, Charache P, Jackson DP. A new inherited coagulation disorder caused by an abnormal fibrinogen ("fibrinogen Baltimore"). Nature 1965; 5006:143–145.
130. Dang CV, Ebert RF, Bell WR. Localization of a fibrinogen calcium binding site between γ-subunit positions 311 and 336 by terbium fluorescence. J Biol Chem 1985; 260:9713–9719.
131. Lord ST. Expression of a cloned human fibrinogen cDNA in *E. coli*: synthesis of an Aα polypeptide. DNA 1985; 4:33–38.
132. Bolyard MG, Lord ST, High level expression of functional fibrinogen γ chain in *E. coli*. Gene 1988; 66:183–192.
133. Lord ST, Fowlkes DM, Expression of a fibrinogen fusion peptide in *E. coli*: a model thrombin substrate for structure/function analysis. Blood 1989; 73:166–171.
134. Bolyard MG, Lord ST. Expression in *Escherichia coli* of the human fibrinogen Bβ chain and its cleavage by thrombin. Blood 1989; 73:1202–1206.
135. Hettasch JM, Bolyard MG, Lord ST. The residues AGDV of recombinant γ chains of human fibrinogen must be carboxy-terminal to support human platelet aggregation. Thromb Haemost 1992; 68:701–706.
136. Ni F, Meinwald YC, Vasquez M, Scheraga HA. High-resolution NMR studies of fibrinogen-like peptides in solution: structure of a thrombin-bound peptide corresponding to residues 7–16 of the Aα chain of human fibrinogen. Biochemistry 1989; 28:3094–3105.
137. Stubbs MT, Oschkinat H, Mayr I, et al. The interaction of thrombin with fibrinogen. Eur J Biochem 1992; 206:187–195.
138. Martin PD, Robertson W, Turk D, Huber R, Bode W, Edwards BFP. The structure of residues 7–16 of the Aα-chain of human fibrinogen bound to bovine thrombin at 2.3-Å resolution. J Biol Chem 1992; 267:7911–7920.
139. Rydel RJ, Ravichandran KG, Tulinsky A, et al. The structure of a complex of recombinant hirudin and human α-thrombin. Science 1990; 249:277–280.
140. Grutter MG, Priestle JP, Rahuel J, et al. Crystal structure of the thrombin-hirudin complex: a novel mode of serine protease inhibition. EMBO J 1990; 9:2361–2365.
141. Farrell DH, Mulvihill ER, Chung DW, Davie EW. Expression of functional human fibrinogen from cDNA clones in baby hampster kidney cells. Blood 1989; 74:55a.
142. Farrell DH, Mulvihill ER, Huang S, Chung DW, Davie EW. Recombinant human fibrinogen and sulfation of the γ′ chain. Biochemistry 1991; 30:9414–9420.
143. Jevons FR. Tyrosine O-sulphate in fibrinogen and fibrin. Biochem J 1963; 89:621–624.
144. Liu M-C, Yu S, Sy J, Redman CM, Lipmann F. Tyrosine sulfation of proteins from the human hepatoma cell line HepG2. Proc Natl Acad Sci USA 1985; 82:7160–7164.
145. Plow EF, Pierschbacher MD, Ruoslahti E, Marguerie GA, Ginsberg MH. The effect of Arg-Gly-Asp-containing peptides on fibrinogen and von Willebrand factor binding to platelets. Proc Natl Acad Sci USA 1985; 82:8057–8061.
146. Gartner TK, Bennett JS. The tetrapeptide analogue of the cell attachment site of fibronectin inhibits platelet aggregation and fibrinogen binding to activated platelets. J Biol Chem 1985; 260:11891–11894.

147. Haverstick DM, Cowan JF, Yamada KM, Santoro SA. Inhibition of platelet adhesion to fibronectin, fibrinogen, and von Willebrand factor substrates by a synthetic tetrapeptide derived form the cell-binding domain of fibronectin. Blood 1985; 66:946–952.
148. Beacham DA, Wise RJ, Turci SM, Handin RI. Selective inactivation of the Arg-Gly-Asp-Ser (RGDS) binding site in von Willebrand factor by site-directed mutagenesis. J Biol Chem 1992; 267:3409–3415.
149. Huang S, Mulvihill ER, Farrell DH, Chung DW, Davie EW. Assembly of human fibrinogen: potential intermediates and pathways. Blood 1992; 80:263a.
150. Binnie CG, Lord ST, Strickland E, Hettasch JM. Investigation of thrombin-catalyzed fibrinopeptide cleavage from fibrinogen using an engineered recombinant fibrinogen. Blood 1992; 80:263a.
151. Brennan M. Fibrin glue. Blood Rev 1991; 5:240–244.

4

Prothrombin

Sandra J. Friezner Degen
Children's Hospital Research Foundation,
and University of Cincinnati, Cincinnati, Ohio

I. INTRODUCTION

Prothrombin and its enzymatically active form, thrombin, play roles in several biological processes. The role of thrombin as both a procoagulant and an anticoagulant in the process of blood coagulation has been well studied. The coagulation of blood is a complex series of reactions ultimately resulting in the formation of a fibrin clot (1). During this process zymogens are converted to active serine proteases by limited proteolysis. In the final activation step of this pathway, prothrombin is activated to the active protease, thrombin, by factor Xa in the presence of factor Va, calcium, and a phospholipid surface. Factor Xa cleaves two peptide bonds to form thrombin. Thrombin is responsible for proteolytically removing fibrinopeptides A and B from fibrinogen, which results in formation of the fibrin clot. Thrombin is also involved in the activation of the coagulation cofactors, factors V and VIII, and in the cross-linking of fibrin to form the insoluble fibrin clot by activation of factor XIII (1). The anticoagulant properties of thrombin include its role in the activation of protein C in the presence of thrombomodulin on the surface of vascular endothelial cells. Activated protein C proteolytically degrades factors Va and VIIIa and thereby inhibits blood coagulation (2). Thrombomodulin acts as a regulatory protein that alters the specificity of thrombin from its preferred substrate fibrinogen in the coagulation process to protein C in the inhibition of the clotting reaction.

Besides the central role of thrombin in the coagulation process, it is also involved in the stimulation of platelet aggregation by activation of the thrombin receptor, causes mitogenesis of fibroblasts, is chemotactic for macrophages, and is involved in the regulation of proliferation of endothelial cells and possibly other cell types (3–6). A receptor specific for thrombin has been characterized and found to be present on platelets and vascular endothelial cells (7).

Prothrombin is synthesized in the liver as a single-chain glycoprotein of 72,000 molecular weight (8). Vitamin K is required for the accurate biosynthesis of prothrombin, which includes the posttranslational modification of specific glutamic acid residues to γ-carboxyglutamic acid (Gla). Vitamin K is a cofactor for a microsomal carboxylase that is responsible for these modifications in prothrombin and in several other coagulation proteins, as well as several proteins found in bone (9,10). The other vitamin K-dependent proteins involved in blood coagulation are factors IX, X, and VII, protein C, and protein S. Gla residues in prothrombin enable it to bind to the cell surface at the site of injury.

The complete amino acid sequence of both human and bovine prothrombin has been determined (9,11–14). The form of prothrombin found circulating in blood contains 579 amino acids and approximately 8% carbohydrate (Fig. 1). Ten Gla residues are present at the amino-terminal end of the protein (9,15,16). The three apparent carbohydrate at-

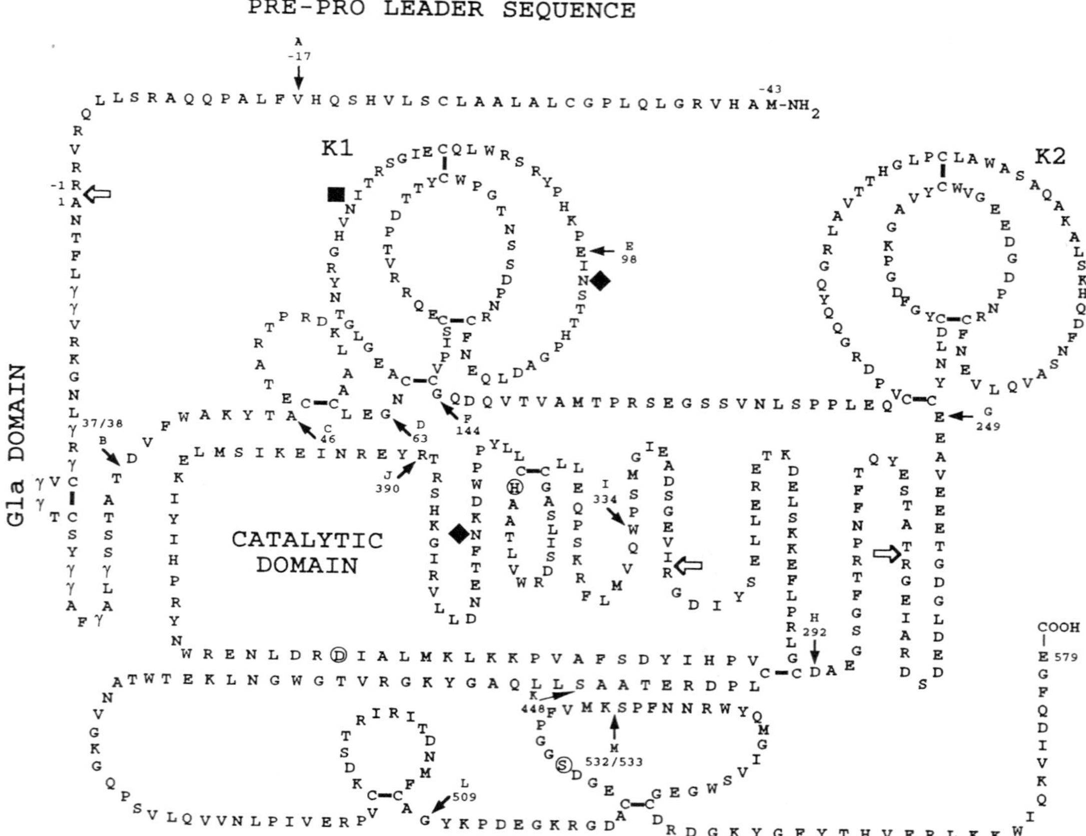

Figure 1 The amino acid sequence of human prepro-prothrombin and placement of the intervening sequences in the gene with respect to the structure of the protein. Domains are indicated. The processing protease and factor Xa cleavage sites are indicated by arrows. Thrombin is composed of two chains: a 49 amino acid light chain (A chain) disulfide linked to a 259 amino acid catalytic domain or heavy chain (B chain). After activation thrombin undergoes self-autolysis, removing 13 amino acids from the amino terminus of the light chain. The amino acids in the active site of the catalytic domain are circled (H^{363}, D^{419}, and S^{525}). The three carbohydrate attachment sites are indicated by solid diamonds. The positions of the 13 intervening sequences are indicated by arrows labeled A–M, with the corresponding residue numbered.

tachment sites are at asparagines at positions 78, 100, and 373 (Fig. 1). Prothrombin consists of several structural and functional domains, which include a Gla domain, two kringle structures, and a serine protease domain. The amino-terminal 40 amino acids contain the 10 Gla residues and are homologous to Gla-containing regions in the other vitamin K-dependent proteins. Following the Gla region are two triple disulfide-bonded regions of internal homology of approximately 80 amino acids. These structures were first identified in prothrombin and have been called kringles (9). At present, eight kringle-containing proteins have been identified that include coagulation factors (factor XII and prothrombin), fibrinolytic proteins (tissue plasminogen activator, urokinase-type plasminogen activator, and plasminogen), apolipoprotein(a), hepatocyte growth factor, and hepatocyte growth factor-like protein (17–19). Kringle structures appear to function autonomously (20,21) and fold independently (22). Kringles are over 50% identical to each other. The function of only a few of the kringle structures in these proteins has been elucidated (21). For example, the second kringle in prothrombin may bind to its cofactor, factor Va (23). The primary function of kringles appears to be in interacting with substrates, cofactors, or receptors (24).

At the carboxyl-terminal end of prothrombin is the catalytic domain, which is homologous to trypsin and other serine proteases. Factor Xa proteolytically cleaves two peptide bonds in prothrombin (Fig. 1) to form the active enzyme thrombin. After activation, thrombin is released from the Gla- and kringle-containing regions, which remain attached to the cell surface. Thrombin is composed of two polypeptide chains held together by a disulfide bond. The active site of thrombin is composed of the catalytic triad of His^{363}, Asp^{419}, and Ser^{525} (Fig. 1). The substrate specificity of thrombin is limited compared with that of trypsin, although the peptide bond following basic amino acids is cleaved by both thrombin and trypsin. This limited specificity is believed to be a result of interactions at secondary binding sites distant from the active site.

Besides its catalytic properties, thrombin binds such macromolecules as fibrinogen, thrombomodulin, and hirudin (a potent inhibitor specific for thrombin from the saliva of the leech) (25) at or near a site that has been termed the anion binding exosite (residues 382–393; Fig. 1) (26). The anion binding exosite is a positively charged surface region located some distance from the active site. Residues 367–381 have been found to be required for the chemotactic and mitogenic properties of human prothrombin for monocytes and macrophages (Fig. 1) (5).

The three-dimensional structure of the complex of α-thrombin and D-Phe-Pro-Arg chloromethyl ketone has been solved at 1.92 Å resolution (27). Thrombin was found to be structurally similar to trypsin with the addition of two extended insertions on the surface, which included residues Tyr^{367} to Trp^{370} and an enlarged loop around Trp^{468} (Fig. 1). Both these loops appear to be responsible for the strict substrate specificity of thrombin. Lys^{385} to Glu^{396} represents an arginine-rich surface loop that may be part of the anion binding exosite region. The crystallographic structure of a complex of human α-thrombin with hirudin has been solved at 2.3 and 2.95 Å resolution (28,29). Of the 65 residues of hirudin 27 are in close proximity to residues in thrombin (less than 4.0 Å apart), which probably accounts for the high affinity and specificity of thrombin toward hirudin. The carboxyl-terminal portion of hirudin interacts with the anion binding exosite.

The x-ray structure of bovine prothrombin fragment 1 (the amino-terminal 155 residues, which includes the Gla and first kringle domains; Fig. 1) has been determined at 2.25 Å resolution in the absence of calcium ions (30) and 2.2 Å resolution in the presence of calcium ions (31). In the absence of calcium, the amino-terminal 35 amino acids, which includes all 10 Gla residues, has a disordered structure. In the presence of calcium the tertiary structure of this region is well defined and binds seven calcium ions. The interaction of this region with calcium is apparently a new and different type of calcium ion-protein interaction that has not previously been observed. The structure of the kringle domain is approximately the same in the presence and absence of calcium. The 18 residues conserved in all kringle domains were found to be responsible for the tertiary structure of the kringle, half of these forming an inner core surrounding the inner loop disulfide groups and the remainder in association with the many turns of the main chain.

II. STRUCTURE OF THE cDNA FOR PROTHROMBIN

The first complementary DNA (cDNA) isolated for prothrombin was a partial cDNA coding for bovine prothrombin (32). In 1983, the first partial cDNA clones coding for human prothrombin were isolated and characterized (33). Full-length cDNAs have now been isolated by several laboratories (34–36). The full-length cDNA clone is about 2000 bp in length and has an open reading frame of 1866 bp followed by a stop codon and a 97 bp 3′ noncoding region (Fig. 2). The traditional polyadenylation signal of 5′-AATAAA-3′ starts 22 bp upstream of the poly(A) sequence. The open reading frame codes for 622 amino acids, which includes a 43 amino acid leader sequence (residues −43 to −1 in Figs. 1 and 2), followed by 579 amino acids coding for the mature protein.

Upon translation of the cDNA it was apparent that there were an additional 43 amino acids preceding the mature protein sequence (residues −43 to −1; Fig. 2). Proteins that are secreted from the cell traditionally have signal sequences at the amino-terminal end of the synthesized protein that are required for the secretory process and are removed before secretion. The signal peptide in prothrombin is probably 25 amino acids in length, followed by an 18 amino acid propeptide. It was previously shown that mature uncarboxylated proteins do not serve as substrates for vitamin K-dependent carboxylation (37) but that a slightly larger precursor could be carboxylated in vitro (38). It also has been observed that the intracellular forms of prothrombin contained a basic peptide of approximately 1500 daltons that was not present in the plasma form of the protein (39). The size of this peptide is consistent with the size predicted for the propeptide. The cloning of cDNAs coding for other vitamin K-dependent proteins also indicated the presence of pre-pro leader sequences. Although the structure of the signal sequences for various vitamin K-dependent proteins vary and show no sequence homology, the propeptides are homologous. Residues at −17, −16, −15, −12, −10, −7, −6, −4, −2, and −1 are well conserved among all vitamin K-dependent proteins in blood coagulation and in bone Gla protein. Site-directed mutagenesis of prothrombin was performed to study the residues required for γ-carboxylation in the propeptide. Mutation of residues at −18, −17, −15, or at −10 resulted in partial inhibition of carboxylation in human prothrombin (Fig. 2) (40). In contrast, mutation of residue −14 or residue −8 had no effect on carboxylation. In addition, natural human factor IX mutants, as well as site-directed mutants of factor IX and protein C, have shown specific amino acids in the pro region that are required for the complete γ-carboxylation of these proteins (41–43). Residues −18, −17, −16, −15, and −10 of the propeptide constitute part of the carboxylation signal in the vitamin K-dependent proteins (44).

Full-length cDNAs have also been isolated for mouse, rat, and bovine prothrombin (36,45,46), and a partial cDNA has been isolated for rabbit prothrombin (47). A comparison of the translated amino acid sequences from these cDNAs with the human amino acid sequence is shown in Figure 3. The length of the mature proteins is similar, with the differences primarily caused by deletions in the region between the second kringle domain and the factor Xa activation site (residues 252–276; Fig. 3) and the region between the two kringles (residues 145–172; Fig. 3). These two regions appear to be able to accommodate insertions or deletions and probably serve as spacers between domains. There is an additional insertion at residue 3 in bovine, mouse, and rat prothrombin.

Human and bovine prothrombin have 10 Gla residues in the Gla region of the protein at residues 7, 8, 15, 17, 20, 21, 26, 27, 30, and 33 in Fig. 3. Glutamic acids are coded for at each of these sites in the cDNAs for mouse and rat prothrombin and therefore are likely to be carboxylated. No additional glutamic acid residues are present in this region of the mouse and rat protein.

The cDNA sequence of the B chain of thrombin has been determined for nine vertebrate species, including rat, mouse, and rabbit (36,46,48). The amino acid sequence of thrombin has been highly conserved during evolution, as can be seen by comparison of the amino acid sequence of 11 vertebrate species (starting at residue 349 in Fig. 3). Mouse and rat are the most conserved (96.5%), and hagfish and newt are the least conserved (62.6%). The size of the B chain of thrombin is relatively invariant between species. Sequences surrounding the three active site amino acids (His^{368}, Asp^{424}, and Ser^{530} in Fig. 3; residues 363, 419, and 525 in Figs. 1 and 2) are conserved in all species of thrombin characterized to date. This conservation is also present for all cysteines and tryptophans present in the B chain (48).

Comparison of amino acid sequences of the same protein isolated from different species can provide insight into amino acid residues important for structure and/or functional properties of the protein. The anion binding exosite that apparently is involved in binding thrombomodulin, hirudin, and fibrinogen spans residues 387–398 (Fig. 3; residues 382–393 in Figs. 1 and 2) (26). This region is identical in human, bovine, rat, and mouse prothrombin. In other species in which this sequence has been determined there are several differences, primarily at residues 392 and 394–396 (Fig. 3). The region identified as involved in the chemotactic and mitogenic properties of prothrombin for macrophages and monocytes are residues 372–386 (Fig. 3; residues 367–381 in Figs. 1 and 2) (5). Human, rat, and mouse prothrombin have identical sequences in this region, and the bovine sequence is 87% identical. All other sequences are 73–87% identical. Although these functions of thrombin have not been characterized for mouse and rat prothrombin, based on their high degree of identity with the human and bovine proteins it can be inferred that the mouse and rat proteins also have these biological functions. This may also be the case with the other characterized species of thrombin, but because the sequences are not conserved to the same extent these inferences cannot be made.

III. PROTHROMBIN MUTANTS

A. Congenital Abnormalities

Congenital abnormalities associated with dysfunctional prothrombins are rare. Several variants have been characterized to date, and all have been found to be caused by the substitution of a single amino acid in regions important for the structure and/or function of the protein (Table 1). Prothrombin Quick was isolated from a patient with less than 2% of normal prothrombin activity. Dysfunctional thrombin Quick I and II were isolated after activation of the abnormal prothrombin. Thrombin Quick I had a decreased ability

```
                                                                                                                                                                      CCCAGGAGCTGACACACT  18
-43         -40                                     -30                                     -20                                     -10
Met Ala His Val Arg Gly Leu Gln Leu Pro Gly Cys Leu Ala Leu Ala Ala Leu Cys Ser Leu Val His Ser Gln His Val Phe Leu Ala Pro Gln Gln Ala Arg Ser Leu Leu Gln Arg
ATG GCG CAC GTC CGA GGC TTG CAG CTG CCT GGC TGC CTG GCC CTG GCT GCC CTG TGT AGC CTT GTG CAC AGC CAG CAT GTG TTC CTG GCT CCT CAG CAA GCA CGG TCG CTG CTC CAG CGG  138

        -1  +1                                  10                                      20                                      30
Val Arg Arg Ala Asn Thr Phe Leu Glu Glu Val Arg Lys Gly Asn Leu Glu Arg Glu Cys Val Glu Glu Thr Cys Ser Tyr Glu Glu Ala Phe Glu Ala Leu Glu Ser Ser Thr Ala Thr
GTC CGG CGA GCC AAC ACC TTC TTG GAG GAG GTG CGC AAG GGC AAC CTA GAG CGA GAG TGC GTG GAG GAG ACG TGC AGC TAC GAG GAG GCC TTC GAG GCT CTG GAG TCC TCC ACG GCT ACG  258

        40                                      50                                      60                                      70
Asp Val Phe Trp Ala Lys Tyr Thr Ala Cys Glu Thr Ala Arg Thr Pro Arg Asp Lys Leu Ala Ala Cys Leu Glu Gly Asn Cys Ala Glu Gly Leu Gly Thr Asn Tyr Arg Gly His Val
GAT GTG TTC TGG GCC AAG TAC ACA GCT TGT GAG ACA GCG AGG ACG CCT CGA GAT AAG CTT GCT GCA TGT CTG GAA GGT AAC TGT GCT GAG GGT CTG GGT ACG AAC TAC CGA GGG CAT GTG  378
                                                                                         ◆              └K1
◆       80                                      90                                      100                                     110
Asn Ile Thr Arg Ser Gly Ile Glu Cys Gln Leu Trp Arg Ser Arg Tyr Pro His Lys Pro Glu Ile Asn Ser Thr Thr His Pro Gly Ala Asp Leu Gln Glu Asn Phe Cys Arg Asn Pro
AAC ATC ACC CGG TCA GGC ATT GAG TGC CAG CTA TGG AGG AGT CGC TAC CCA CAT AAG CCT GAA ATC AAC TCC ACT ACC CAT CCT GGG GCC GAC CTA CAG GAG AAT TTC TGC CGC AAC CCC  498

        120                                     130                                     140                                     150
Asp Ser Ser Asn Thr Gly Pro Trp Cys Tyr Thr Thr Asp Pro Thr Val Arg Arg Gln Glu Cys Ser Ile Pro Val Cys Gly Gln Asp Gln Val Thr Val Ala Met Thr Pro Arg Ser Glu
GAC AGC AGC AAC ACG GGA CCC TGG TGC TAC ACT ACA GAC CCC ACC GTG AGG AGG CAG GAA TGC AGC ATC CCT GTC TGT GGC CAG GAT CAA GTC ACT GTA GCG ATG ACT CCA CGC TCC GAA  618
                                                                                                       K1┘
        160                                     170                                     180                                     190
Gly Ser Ser Val Asn Leu Ser Pro Pro Leu Glu Gln Cys Val Pro Asp Arg Gly Gln Gln Tyr Gln Gly Arg Leu Ala Val Thr Thr His Gly Leu Pro Cys Leu Ala Trp Ala Ser Ala
GGC TCC AGT GTG AAT CTG TCA CCT CCA TTG GAG CAG TGT GTC CCT GAT CGG GGG CAG CAG TAC CAG GGG CGC CTG GCG GTG ACC ACA CAT GGG CTC CCC TGC CTG GCC TGG GCC AGC GCA  738
                                               └K2
        200                                     210                                     220                                     230
Gln Ala Lys Ala Leu Ser Lys His Gln Asp Phe Asn Ser Ala Val Gln Leu Val Glu Asn Phe Cys Arg Asn Pro Asp Gly Asp Glu Glu Gly Val Trp Cys Tyr Val Ala Gly Lys Pro
CAG GCC AAG GCC CTG AGC AAG CAC CAG GAC TTC AAC TCA GCT GTG CAG CTG GTG GAG AAC TTC TGC CGC AAC CCA GAC GGG GAT GAG GAG GGC GTG TGG TGC TAT GTG GCC GGG AAG CCT  858

        240                                     250                                     260                                     270
Gly Asp Phe Gly Tyr Cys Asp Leu Asn Tyr Cys Glu Glu Ala Val Glu Glu Glu Thr Gly Asp Gly Leu Asp Glu Asp Ser Asp Arg Ala Ile Glu Gly Arg▼Thr Ala Thr Ser Glu Tyr
GGC GAC TTT GGG TAC TGC GAC CTC AAC TAT TGT GAG GAG GCC GTG GAG GAG GAG ACA GGA GAT GGG CTG GAT GAG GAC TCA GAC AGG GCC ATC GAA GGG CGT ACC GCC ACA AGT GAG TAC  978
                                         K2┘
        280                                     290                                     300                                     310
Gln Thr Phe Phe Asn Pro Arg Thr Phe Gly Ser Gly Glu Ala Asp Cys Gly Leu Arg Pro Leu Phe Glu Lys Lys Ser Leu Glu Asp Lys Thr Glu Arg Glu Leu Leu Glu Ser Tyr Ile
CAG ACT TTC TTC AAT CCG AGG ACC TTT GGC TCG GGA GAG GCA GAC TGT GGG CTG CGA CCT CTG TTC GAG AAG AAG TCG CTG GAG GAC AAA ACC GAA AGA GAG CTC CTG GAA TCC TAC ATC  1098

        320                                     330                                     340                                     350
Asp Gly Arg▼Ile Val Glu Gly Ser Asp Ala Glu Ile Gly Met Ser Pro Trp Gln Val Met Leu Phe Arg Lys Ser Pro Gln Glu Leu Leu Cys Gly Ala Ser Leu Ile Ser Asp Arg Trp
GAC GGG CGC ATT GTG GAG GGC TCG GAT GCA GAG ATC GGC ATG TCA CCT TGG CAG GTG ATG CTT TTC CGG AAG AGT CCC CAG GAG CTG CTG TGT GGG GCC AGC CTC ATC AGT GAC CGC TGG  1218

        360                                     370         ◆                           380                                     390
Val Leu Thr Ala Ala [His] Cys Leu Leu Tyr Pro Pro Trp Asp Lys Asn Phe Thr Glu Asn Asp Leu Leu Val Arg Ile Gly Lys His Ser Arg Thr Arg Tyr Glu Arg Asn Ile Glu Lys
GTC CTC ACC GCC GCC [CAC] TGC CTC CTG TAC CCG CCC TGG GAC AAG AAC TTC ACC GAG AAT GAC CTT CTG GTG CGC ATT GGC AAG CAC TCC CGC ACC AGG TAC GAG CGA AAC ATT GAA AAG  1338

        400                                     410                                     420                                     430
Ile Ser Met Leu Glu Lys Ile Tyr Ile His Pro Arg Tyr Asn Trp Arg Glu Asn Leu Asp Arg [Asp] Ile Ala Leu Met Lys Leu Lys Lys Pro Val Ala Phe Ser Asp Tyr Ile His Pro
ATA TCC ATG TTG GAA AAG ATC TAC ATC CAC CCC AGG TAC AAC TGG CGG GAG AAC CTG GAC CGG [GAC] ATT GCC CTG ATG AAG CTG AAG AAG CCT GTT GCC TTC AGT GAC TAC ATT CAC CCT  1458

        440                                     450                                     460                                     470
Val Cys Leu Pro Asp Arg Glu Thr Ala Ala Ser Leu Leu Gln Ala Gly Tyr Lys Gly Arg Val Thr Gly Trp Gly Asn Leu Lys Glu Thr Trp Thr Ala Asn Val Gly Lys Gly Gln Pro
GTG TGT CTG CCC GAC AGG GAG ACG GCA GCC AGC TTG CTC CAG GCT GGA TAC AAG GGG CGG GTG ACA GGC TGG GGC AAC CTG AAG GAG ACG TGG ACA GCC AAC GTT GGT AAG GGG CAG CCC  1578

        480                                     490                                     500                                     510
Ser Val Leu Gln Val Val Asn Leu Pro Ile Val Glu Arg Pro Val Cys Lys Asp Ser Thr Arg Ile Arg Ile Thr Asp Asn Met Phe Cys Ala Gly Tyr Lys Pro Asp Glu Gly Lys Arg
AGT GTC CTG CAG GTG GTG AAC CTG CCC ATT GTG GAG CGG CCG GTC TGC AAG GAC TCC ACC CGG ATC CGC ATC ACT GAC AAC ATG TTC TGT GCT GGT TAC AAG CCT GAT GAA GGG AAA CGA  1698

        520                                     530                                     540                                     550
Gly Asp Ala Cys Glu Gly Asp [Ser] Gly Gly Pro Phe Val Met Lys Ser Pro Phe Asn Asn Arg Trp Tyr Gln Met Gly Ile Val Ser Trp Gly Glu Gly Cys Asp Arg Asp Gly Lys Tyr
GGG GAT GCC TGT GAA GGT GAC [AGT] GGG GGA CCC TTT GTC ATG AAG AGC CCC TTT AAC AAC CGC TGG TAT CAA ATG GGC ATC GTC TCA TGG GGT GAA GGC TGT GAC CGG GAT GGG AAA TAT  1818

        560                                     570
Gly Phe Tyr Thr His Val Phe Arg Leu Lys Lys Trp Ile Gln Lys Val Ile Asp Gln Phe Gly Glu ***
GGC TTC TAC ACA CAT GTG TTC CGC CTG AAG AAG TGG ATA CAG AAG GTC ATT GAT CAG TTT GGA GAG TAG GGGGCCACTCATATTCTGGGCTCCTGGAACCAATCCCGTCAAAGAATTATTTTTGTGTTTCTAAAAC  1954

TATGGTTCCCAATAAAAGTGACTCTCAGCGAAAAAAAAAAAAAAAAAAAAAAAAAAA                                                                                          2011
```

to release fibrinopeptide A from fibrinogen and decreased activity in stimulating platelet aggregation and release of prostacyclin from endothelial cells. Amino acid sequence analysis of this variant indicated that Arg^{382} had been substituted with a Cys (Figs. 1 and 2; residue 387 in Fig. 3) (49). Thrombin Quick II is characterized by a Gly to Val substitution at position 558 and is associated with a disruption in the Arg/Lys binding pocket involved in substrate recognition (Figs. 1 and 2; residue 563 in Fig. 3) (50). The patient with prothrombin Tokushima was heterozygous for both dysprothrombinemia and hypoprothrombinemia. Amino acid sequence analysis of isolated peptides indicated that Arg^{418} was replaced by Trp (Figs. 1 and 2; residue 423 in Fig. 3) (51). This amino acid substitution appears to result in a reduced interaction of prothrombin with various substrates, including fibrinogen and platelet receptors, which accounts for the recurrent bleeding observed in the patient. Prothrombins Barcelona and Madrid were isolated from patients with normal prothrombin antigen levels but with reduced prothrombin coagulant activity because of the inability of activation by factor Xa. Substitution of Cys for Arg at residue 271 was found to be the defect in both prothrombin mutants (Figs. 1 and 2; residue 276 in Fig. 3) (52,53). Activation of prothrombin to thrombin by factor Xa occurs by proteolytic cleavage following residues 271 and 320 (Figs. 1 and 2). Substitution of a Cys for an Arg at one of these sites is consistent with the observation that factor Xa is unable to activate this dysfunctional prothrombin. Characterization of prothrombin Salakta indicated that the defect is caused by an abnormality near or at the primary substrate binding site, and not in the active site of the molecule. Substitution of Glu^{466} by Ala was determined to be the defect by amino acid sequence analysis (Figs. 1 and 2; residue 471 in Fig. 3) (54). This substitution occurs in one of the surface loops of the protein that has been implicated as responsible for the strict substrate specificity of α-thrombin (27). A patient with prothrombin Himi was found to heterozygous for two dysfunctional prothrombin molecules that are both defective in the thrombin portion of the molecule. Sequence analysis of the genomic DNA from this patient indicated that the mutations were caused by a substitution of Thr for Met^{337} and a His for Arg^{388} (Figs. 1 and 2; residues 342 and 393 in Fig. 3) (55). Arg^{388} appears to be part of the anion binding exosite. Fibrinogen, thrombomodulin, and hirudin all bind competitively to this region. The function of Met^{337} is not clear, and therefore the basis for the prothrombin Himi I defect is not understood.

It is expected that the amino acid substitutions responsible for these bleeding disorders are conserved in prothrombin from different species because of their obvious functional importance. All the substitutions that result in the dysfunctional prothrombins discussed here occur at amino acid residues that are absolutely conserved in all species of prothrombin characterized to date (Figs. 1 and 2).

The amino acid defect(s) in several other abnormal prothrombins have not yet been characterized. Like prothrombins Barcelona and Madrid, prothrombin Cardeza (56) and prothrombin Clamart (57) cannot be activated to thrombin by factor Xa. Prothrombin Metz (58) and prothrombin Molis (59) are defective in their thrombin domains.

Figure 2 The nucleotide and translated amino acid sequence of the cDNA coding for human prothrombin. The translated amino acid sequence is shown above the nucleotide sequence. Amino acids are numbered every 10 residues. Numbers in the right margin correspond to the last nucleotide of each line. Arrows indicate factor Xa cleavage sites. The kringle domains are marked. Amino acids in boxes correspond to the active site residues in the catalytic domain. The stop codon is indicated by three asterisks.

```
       -43         -40                                     -30                                     -20                                     -10
Human  Met Ala His Val Arg Gly Leu Gln Leu Pro Gly Cys Leu Ala Leu Ala Ala Leu Cys Ser Leu Val His Ser Gln His Val Phe Leu Ala Pro Gln Gln Ala Arg Ser Leu Leu Gln Arg
Bovine --- --- Arg --- --- --- Pro Arg --- --- --- --- --- --- --- --- --- --- Phe --- --- --- --- --- --- --- --- --- --- --- His --- --- --- Ser --- --- --- --- ---
Rat    --- Leu --- --- --- --- --- Gly --- --- --- --- --- --- --- --- --- --- Ala --- --- --- --- --- --- --- --- --- --- --- --- --- --- --- Leu --- --- --- --- ---
Mouse  --- Ser --- --- --- --- --- Gly --- --- --- --- --- --- --- --- --- --- Val --- --- --- --- --- --- --- --- --- --- --- --- --- --- --- Leu --- --- --- --- ---

               -1  +1                                  10                                      20                                      30
Human  Val Arg Arg Ala Asn xxx Thr Phe Leu Glu Glu Val Arg Lys Gly Asn Leu Glu Arg Glu Cys Val Glu Glu Thr Cys Ser Tyr Glu Glu Ala Phe Glu Ala Leu Glu Ser Ser Thr Ala
Bovine Ala --- --- --- --- Lys Gly --- --- --- --- --- --- --- --- --- --- --- --- --- --- Leu --- --- Pro --- --- Arg --- --- --- --- --- --- --- --- --- Leu Ser ---
Rat    --- --- --- --- --- Ser Gly --- --- --- --- Leu --- --- --- --- --- --- --- --- --- --- --- --- Gln --- --- --- --- --- --- --- --- --- --- --- --- Pro Gln Asp
Mouse  --- --- --- --- --- Ser Gly --- --- --- --- Leu --- --- --- --- --- --- --- --- --- --- --- --- Gln --- --- --- --- --- --- --- --- --- --- --- --- Pro Gln Asp

               40                                      50                                      60                                      70
Human  Thr Asp Val Phe Trp Ala Lys Tyr Thr Ala Cys Glu Thr Ala Arg Thr Pro Arg Asp Lys Leu Ala Ala Cys Leu Glu Gly Asn Cys Ala Glu Gly Leu Gly Thr Asn Tyr Arg Gly His
Bovine --- --- Ala --- --- --- --- --- --- --- --- --- Ser --- --- Asn --- --- Glu --- --- Asn Glu --- --- --- --- --- --- --- --- --- Val --- Met --- --- --- --- Asn
Rat    --- --- --- --- --- --- --- --- --- Val --- Asp Ser Val --- Lys --- --- Glu Thr Phe Met Asp --- --- --- --- Arg --- --- Met Asp --- --- Leu --- --- His --- Asn
Mouse  --- --- --- --- --- --- --- --- --- Val --- Asp Ser Val --- Lys --- --- Glu Thr Phe Met Asp --- --- --- --- Arg --- --- Met Asp --- --- Val --- --- Leu --- Thr
                                                                                                                   K1

               80                                      90                                      100                                     110
Human  Val Asn Ile Thr Arg Ser Gly Ile Glu Cys Gln Leu Trp Arg Ser Arg Tyr Pro His Lys Pro Glu Ile Asn Ser Thr Thr His Pro Gly Ala Asp Leu Gln Glu Asn Phe Cys Arg Asn
Bovine --- Ser Val --- --- --- --- --- --- --- --- --- --- --- --- --- --- --- --- --- --- --- --- --- --- --- --- --- --- --- --- --- --- Arg --- --- --- --- --- ---
Rat    --- Ser Val --- His Thr --- --- --- --- --- --- --- --- --- --- --- --- --- Arg --- Asp --- --- --- --- --- --- --- --- --- --- --- Lys --- --- --- --- --- ---
Mouse  --- --- Val --- His Thr --- --- --- Gln --- --- --- --- --- --- --- --- --- --- --- --- --- --- --- --- --- --- --- --- --- --- --- Lys --- --- --- --- --- ---

               120                                     130                                     140                                     150
Human  Pro Asp Ser Ser Thr Thr Gly Pro Trp Cys Tyr Thr Thr Asp Pro Thr Val Arg Arg Gln Glu Cys Ser Ile Pro Val Cys Gly Gln xxx Asp Gln Val Thr Val Ala xxx Met Thr Pro
Bovine --- --- Gly --- Ile --- --- --- --- --- --- --- --- Ser --- --- Leu --- --- Glu --- --- --- Val --- --- --- --- --- xxx --- Arg --- --- --- Glu xxx Val Ile ---
Rat    --- --- --- --- --- Ser --- --- --- --- --- --- --- --- --- --- --- --- --- Glu --- --- --- --- --- --- --- --- --- Glu Gly Arg Thr --- --- Lys xxx --- --- ---
Mouse  --- --- --- --- --- --- --- --- --- --- --- --- --- --- --- --- --- --- --- Glu --- --- --- Val --- --- --- --- --- xxx Glu Gly Arg --- Thr Val Val --- --- ---
Rabbit                                         --- --- --- --- --- Ser --- --- --- Glu Glu --- --- Val --- Asp --- Ser --- xxx xxx Glu --- --- Leu Glu xxx Leu Arg ---
                                                                                                           K1

               160                                     170                                     180                                     190
Human  Arg Ser Glu Gly Ser Ser Val Asn Leu Ser Pro Pro Leu Glu Gln Cys Val Pro Asp Arg Gly Gln Gln Tyr Gln Gly Arg Leu Ala Val Thr Thr His Gly Leu Pro Cys Leu Ala Trp
Bovine --- --- Gly --- --- Thr Thr Ser Gln --- --- Leu --- --- Thr --- --- --- --- --- --- Arg Glu --- Arg --- --- --- --- --- --- --- --- --- Ser Arg --- --- --- ---
Rat    --- --- Arg --- --- Lys Glu --- --- --- --- --- --- Gly Glu --- Leu Leu Glu --- --- Arg Leu --- --- --- Asn --- --- --- --- --- Leu --- Ser --- --- --- --- ---
Mouse  --- --- Gly --- --- Lys Asp --- --- --- --- --- --- Gly --- --- Leu Thr Glu --- --- Arg Leu --- --- --- Asn --- --- --- --- --- Leu --- Ser --- --- --- Pro ---
Rabbit --- ?   Gly His --- Gly Ile Thr Gln Pro --- --- --- ?   Ala --- --- Ser Arg --- --- --- Asp --- --- ?   ?   ?   ?   ?   ?   ?   ?   ?   ?   ?   ?   ?   ?   ?
                                                                   K2

               200                                     210                                     220                                     230
Human  Ala Ser Ala Gln Ala Lys Ala Leu Ser Lys His Gln Asp Phe Asn Ser Ala Val Gln Leu Val Glu Asn Phe Cys Arg Asn Pro Asp Gly Asp Glu Glu Gly Val Trp Cys Tyr Val Ala
Bovine Ser --- Glu --- --- --- --- --- --- --- Asp --- --- --- --- Pro --- --- Pro --- Ala --- --- --- --- --- --- --- --- --- --- --- --- --- Ala --- --- --- --- ---
Rat    Asp --- Leu Pro Thr --- Thr --- --- --- Tyr --- Asn --- Asp Pro Glu --- Lys --- --- Gln --- --- --- --- --- --- --- Arg --- --- --- --- Ala --- --- Phe --- ---
Mouse  Asn --- Leu Pro --- --- Thr --- --- --- Tyr --- --- --- Asp Pro Glu --- Lys --- --- --- --- --- --- --- --- --- --- Trp --- --- --- --- Ala --- --- --- --- ---
Rabbit ?   ?   ?   ?   ?   ?   ?   ?   ?   ?   ?   ?   ?   ?   ?   ?   ?   ?   ?   ?   ?   ?   ?   ?   ?   ?   ?   ?   ?   ?   ?   ?   ?   ?   ?   ?   ?   ?   ?   ?

               240                                     250                                     260                                     270
Human  Gly Lys Pro Gly Asp Phe Gly Tyr Cys Asp Leu Asn Tyr Cys Glu Glu Ala Val Glu Glu Glu Thr Gly Asp Gly Leu Asp Glu Asp Ser Asp Arg xxx xxx Ala Ile Glu Gly Arg Thr
Bovine Asp Gln --- --- --- --- Glu --- --- --- --- --- --- --- --- --- Pro --- Asp Gly Asp Leu --- --- Arg --- Gly --- --- Pro --- Pro Asp Ala --- --- --- --- --- ---
Rat    Gln Gln --- --- xxx --- Glu --- --- Ser --- --- --- --- Asp --- --- --- Gly --- --- Asn His xxx xxx xxx xxx xxx --- Gly --- Glu xxx xxx Ser --- Ala --- --- ---
Mouse  --- Gln --- --- --- --- Glu --- --- Asn --- --- --- --- --- --- --- --- Gly --- --- Asn Tyr xxx xxx xxx xxx xxx --- Val --- Glu xxx xxx Ser --- Ala --- --- ---
Rabbit ?   ?   ?   ?   ?   ?   ?   ?   ?   ?   ?   ?   ?   ?   ?   ?   ?   ?   ?   ?   ?   ?   ?   ?   ?   ?   ?   ?   ?   ?   ?   ?   ?   ?   ?   ?   ?   ?   ?   ---
                                                 K2
```

```
                    280                                     290                                     300                                     310
Human       Ala Thr Ser Glu Tyr Gln Thr Phe Phe Asn Pro Arg Thr Phe Gly Ser Gly Glu Ala Asp Cys Gly Leu Arg Pro Leu Phe Glu Lys Lys Ser Leu Glu Asp Lys Thr Glu Arg Glu Leu
Bovine      Ser Glu Asp His Phe --- Pro --- --- --- Glu Lys --- --- --- Ala --- --- --- --- --- --- --- --- --- --- --- --- --- --- Gln Val Gln --- Gln --- --- Lys --- ---
Rat         Thr Asp Ala --- Phe His --- --- --- Asp Glu --- --- --- --- Leu --- --- --- --- --- --- --- --- --- --- --- --- --- --- --- --- Thr --- --- --- --- Lys --- ---
Mouse       Thr Asp Ala --- Phe His --- --- --- --- Glu Lys --- --- --- Leu --- --- --- --- --- --- --- --- --- --- --- --- --- --- --- --- Lys --- Thr --- --- Lys --- ---
Rabbit      Thr Glu Gln --- Phe --- --- --- --- --- Gln Gln  ?  --- --- Thr --- --- --- --- --- --- --- --- --- --- --- --- ---  ?   ?   ?   ?   ?   ?   ?   ?   ?   ?   ?   ?

                    320                                     330                                     340                                     350
Human       Leu Glu Ser Tyr Ile Asp Gly Arg Ile Val Glu Gly Ser Asp Ala Glu Ile Gly Met Ser Pro Trp Gln Val Met Leu Phe Arg Lys Ser Pro Gln Glu Leu Leu Cys Gly Ala Ser Leu
Bovine      Phe --- --- --- --- Glu --- --- --- --- --- --- Gln --- --- --- Val --- Leu --- --- --- --- --- --- --- --- --- --- --- --- --- --- --- --- --- --- --- --- ---
Rat         --- Asp --- --- --- --- --- --- --- --- --- --- Trp --- --- --- Lys --- Ile Ala --- --- --- --- --- --- --- --- --- --- --- --- --- --- --- --- --- --- --- ---
Mouse       --- Asp --- --- --- --- --- --- --- --- --- --- Trp --- --- --- Lys --- Ile Ala --- --- --- --- --- --- --- --- --- --- --- --- --- --- --- --- --- --- --- ---
Rabbit       ?   ?   ?   ?   ?   ?   ?   ?   ?   ?   ?   ?   ?   ?   ?   ?   ?   ?   ?   ?   ?   ?   ?   ?   ?   ?   ?   ?   ?   ?   ?  --- --- --- --- --- --- --- --- ---
Chicken                                                                                                                                 --- --- --- --- --- --- --- --- ---
Gecko                                                                                                                                   --- Asp --- --- --- --- --- --- ---
Newt                                                                                                                                    --- --- --- Ile --- --- --- --- Ile
Rainbow Tr                                                                                                                              --- --- --- --- --- --- --- --- ---
Sturgeon                                                                                                                                --- --- --- --- --- --- --- --- ---
Hagfish                                                                                                                                 --- --- Gly Met --- --- --- --- ---

                    360                                     370                                     380                                     390
Human       Ile Ser Asp Arg Trp Val Leu Thr Ala Ala His Cys Leu Leu Tyr Pro Pro Trp Asp Lys Asn Phe Thr Glu Asn Asp Leu Leu Val Arg Ile Gly Lys His Ser Arg Thr Arg Tyr Glu
Bovine      --- --- --- --- --- --- --- --- --- --- --- --- --- --- --- --- --- --- --- --- --- --- --- Val Asp --- --- --- --- --- --- --- --- --- --- --- --- --- --- ---
Rat         --- --- --- --- --- --- --- --- --- --- --- --- Ile --- --- --- --- --- --- --- --- --- --- --- --- --- --- --- --- --- --- --- --- --- --- --- --- --- --- ---
Mouse       --- --- --- --- --- --- --- --- --- --- --- --- Ile --- --- --- --- --- --- --- --- --- --- --- --- --- --- --- --- --- --- --- --- --- --- --- --- --- --- ---
Rabbit      --- --- --- --- --- --- --- --- --- --- --- --- --- --- --- --- --- --- --- --- --- --- --- Val --- --- Ile --- --- --- --- --- --- Tyr Ala --- Ser --- --- ---
Chicken     --- --- Asn Ser --- Ile --- --- --- --- --- --- --- --- --- --- --- --- --- --- --- Leu --- Thr --- --- Ile --- --- --- Met --- Leu --- Phe --- Ala Lys --- ---
Gecko       --- --- --- --- --- --- --- --- --- --- --- --- Ile Phe --- --- --- --- --- --- --- --- --- Ala Asp --- --- Val --- --- --- --- --- --- Asn --- Arg Ile His ---
Newt        --- --- --- --- --- --- --- --- --- --- --- --- Ile Phe --- --- --- --- --- --- --- Tyr --- Thr Glu --- Ile --- --- --- --- --- --- --- Tyr --- --- Lys --- ---
Rainbow Tr  --- --- --- Glu --- Ile --- --- --- --- --- --- Ile --- --- --- --- --- Asn --- --- --- --- Ile --- --- Ile --- --- --- Leu --- --- --- Asn --- Ala Lys Phe ---
Sturgeon    --- --- --- Gln --- Ile --- --- --- --- --- --- Ile --- --- --- --- --- Asn --- --- --- --- Ala --- --- Ile --- --- --- Val --- --- --- Tyr --- Ala Lys Phe ---
Hagfish     --- --- --- Lys --- --- --- --- --- --- --- --- Ile --- --- --- --- --- Gly --- --- --- Ser His --- --- --- Val --- --- Val --- --- --- Phe --- Ala Ala His ---

                    400                                     410                                     420                                     430
Human       Arg Asn Ile Glu Lys Ile Ser Met Leu Glu Lys Ile Tyr Ile His Pro Arg Tyr Asn Trp Arg Glu Asn Leu Asp Arg Asp Ile Ala Leu Met Lys Leu Lys Lys Pro Val Ala Phe Ser
Bovine      --- Lys Val --- --- --- --- --- --- Asp --- --- --- --- --- --- --- --- --- --- Lys --- --- --- --- --- --- --- --- --- Leu --- --- --- Arg --- Ile Glu Leu ---
Rat         --- --- Val --- --- --- --- --- --- --- --- --- --- --- --- --- --- --- --- --- --- --- --- --- --- --- --- --- --- --- Leu --- --- --- --- --- --- Pro --- ---
Mouse       --- --- Val --- --- --- --- --- --- --- --- --- --- Val --- --- --- --- --- --- --- --- --- --- --- --- --- --- --- --- Leu --- --- --- --- --- --- Pro --- ---
Rabbit      --- --- Met --- --- --- --- Thr --- --- --- --- Ile --- --- --- Gly --- --- --- --- --- --- --- --- --- --- --- --- --- --- --- --- --- --- --- --- --- --- ---
Chicken     --- --- Lys --- --- --- Val Leu --- Asp --- Val Ile --- --- --- Lys --- --- --- Lys --- --- Met --- --- --- --- --- --- Leu His --- --- Arg --- --- Ile --- ---
Gecko       Lys Thr Arg --- --- --- Ala Leu --- Asp --- --- Ile --- --- --- Lys --- --- --- Lys --- --- --- --- --- --- --- --- --- Leu Arg --- Arg --- --- --- Pro --- ---
Newt        --- Gln Gln --- --- --- Arg --- --- --- Arg --- Ile --- --- --- Lys --- --- --- --- --- --- --- --- --- --- --- --- --- Ile Gln --- --- Arg --- Ile Gly --- Thr
Rainbow Tr  Lys Gly Thr --- --- --- Val Ala Ile Asp Glu --- Ile Val --- --- Lys --- --- --- Lys --- --- --- Asn --- --- --- --- --- Leu His Met Arg Arg --- Ile Thr --- Thr
Sturgeon    Lys Gln Thr --- --- --- Val Ala --- Asp Glu --- Ile Leu --- --- Lys --- --- --- Lys --- --- --- --- --- --- --- --- --- Leu His --- Arg --- --- Leu Thr --- Thr
Hagfish     Lys --- Gln --- Gln --- Ala Ala Ile Lys --- --- Ile Leu --- --- --- --- Asp --- Lys --- --- --- Asn --- --- --- --- --- Ile Leu --- --- Arg --- --- His --- Thr
```

Figure 3 Comparison of the amino acid sequences of 11 species of vertebrate prothrombin. The amino acid sequence of human prothrombin is shown on the top line (33). The amino acid sequence is numbered with respect to the human sequence and includes the deletions present in the human protein compared with the other species. Inclusion of these deletions in the numbering of the amino acid sequence results in a different number system than in Figures 1 and 2. Identity with the human sequence is indicated by dashed lines for the other sequences, and differences are indicated (36,45–48). Deletions are indicated by xxx, and insertions are represented by xxx in the other sequences. Arrows indicate the two factor Xa cleavage sites. The active site amino acids are His^{368}, Asp^{424}, and Ser^{530}. The two kringle domains are indicated.

```
                    440                                     450                                     460                                     470
Human       Asp Tyr Ile His Pro Val Cys Leu Pro Asp Arg Glu Thr Ala Ala Ser Leu Leu Gln Ala Gly Tyr Lys Gly Arg Val Thr Gly Trp Gly Asn Leu Lys Glu Thr Trp Thr Ala Asn Val
Bovine      --- --- --- --- --- --- --- --- --- --- Lys Gln --- --- --- Lys --- --- His --- --- Phe --- --- --- --- --- --- --- --- --- Arg Arg --- --- --- --- Thr Ser ---
Rat         --- --- --- --- --- --- --- --- --- --- Lys Gln --- Val Thr --- --- --- --- --- --- --- --- --- --- --- --- --- --- --- --- --- Arg --- --- --- --- Thr --- Ile
Mouse       --- --- --- --- --- --- --- --- --- --- Lys Gln --- Val Thr --- --- --- Arg --- --- --- --- --- --- --- --- --- --- --- --- --- Arg --- --- --- --- Thr --- Ile
Rabbit      --- --- --- --- --- --- --- --- --- --- Lys Gln Ile Val Thr --- --- --- --- --- --- His --- --- --- --- --- --- --- --- --- --- --- --- Met --- --- Val --- Met
Chicken     --- --- --- --- --- --- --- --- --- Thr Lys --- Leu Val Gln Arg --- Met Leu --- --- Phe --- --- --- --- --- --- --- --- --- --- --- --- --- --- Ala Thr Thr xxx
Gecko       --- --- --- Gln --- --- --- --- --- Thr Lys --- --- Val Gln --- --- --- Leu Thr --- --- --- --- --- --- --- --- --- --- --- --- Phe --- --- --- Gly Ser Ser xxx
Newt        Asn --- --- --- --- --- --- --- --- Thr Lys --- Ile Val Gln Thr --- Met Leu Asn Arg His --- --- --- --- Ser --- --- --- --- --- His --- --- --- --- Ser Gly xxx
Rainbow Tr  --- Glu --- --- --- --- --- --- --- Thr Lys Gln Val --- Lys Thr --- Met Phe --- --- --- --- --- --- --- --- --- --- --- --- --- Tyr --- --- --- Ser Ser Ser xxx
Sturgeon    Glu Asn --- Val --- Ile --- --- --- Thr Lys Lys Val --- Lys Thr --- Met Phe --- --- Val Gln --- --- --- --- --- --- --- --- --- Tyr --- --- --- --- Ser Ser xxx
Hagfish     Lys --- Val Ala --- --- --- --- --- Glu Ser Ala Val --- Arg Lys --- Met Arg --- --- --- --- --- --- --- --- --- --- --- --- --- Gln --- Met --- Ser Leu Ser xxx

                    480                                     490                                     500                                     510
Human       Gly Lys Gly Gln Pro Ser Val Leu Gln Val Val Asn Leu Pro Ile Val Glu Arg Pro Val Cys Lys Asp Ser Thr Arg Ile Arg Ile Thr Asp Asn Met Phe Cys Ala Gly Tyr Lys Pro
Bovine      Ala Glu Val --- --- --- --- --- --- --- --- --- --- --- Leu --- --- --- --- --- --- --- Ala --- --- --- --- --- --- --- --- --- --- --- --- --- --- --- --- ---
Rat         Asn Glu Ile --- --- --- --- --- --- --- --- --- --- --- --- --- --- --- --- --- --- --- Ala --- --- --- --- --- --- --- --- --- --- --- --- --- Phe --- Val
Mouse       Asn Glu Ile --- --- --- --- --- --- --- --- --- --- --- --- --- --- --- --- --- --- --- Ala --- --- --- --- --- --- --- --- --- --- --- --- --- Phe --- Val
Rabbit      Asn Glu Val --- --- --- --- --- --- Met --- --- --- --- Leu --- --- --- --- Ile --- --- Ala --- --- Gly --- --- Val --- --- --- --- --- --- --- --- --- --- ---
Chicken     Pro Glu Asn Leu --- Thr --- --- --- Gln Leu --- --- --- --- --- Asp Gln Asn Thr --- --- Ala --- --- --- Val Lys Val --- --- --- --- --- --- --- --- --- Ser ---
Gecko       Thr Pro Ala Leu --- Thr Tyr --- --- Leu --- --- --- --- --- --- Asp --- Asp Thr --- --- Ala --- --- Lys --- Lys --- --- --- --- --- --- --- --- --- --- Ser ---
Newt        --- Gln Ala Leu --- Gln --- --- --- Gln --- --- --- --- --- --- Asp Gln Glu Thr --- --- Ala --- --- Lys --- Lys Val --- Ser --- --- --- --- --- --- --- --- ---
Rainbow Tr  Pro --- Ser Leu --- Thr --- --- --- Gln Ile His --- --- --- --- --- Gln Asp Ile --- Arg --- --- --- Ser --- --- --- --- --- --- --- --- --- --- --- Phe --- ---
Sturgeon    Pro Gln Ser Leu --- Gln --- --- --- Gln Ile His --- --- --- --- Gln Gln Glu Thr --- Arg --- --- --- Lys --- --- Val --- --- --- --- --- --- --- --- Phe Ser ---
Hagfish     Ser --- Val His --- Arg --- --- --- Leu Ile --- --- --- --- --- Asp Thr Arg Thr --- His --- --- --- Thr --- Lys --- --- Arg --- --- --- --- --- --- --- Ser ---

                    520                                     530                                     540                                     550
Human       Asp Glu Gly Lys Arg Gly Asp Ala Cys Glu Gly Asp Ser Gly Gly Pro Phe Val Met Lys Ser Pro Phe Asn Asn Arg Trp Tyr Gln Met Gly Ile Val Ser Trp Gly Glu Gly Cys Asp
Bovine      Gly --- --- --- --- --- --- --- --- --- --- --- --- --- --- --- --- --- --- --- --- --- Tyr --- --- --- --- --- --- --- --- --- --- --- --- --- --- --- --- ---
Rat         Asn Asp Thr --- --- --- --- --- --- --- --- --- --- --- --- --- --- --- --- --- --- --- Tyr --- His --- --- --- --- --- --- --- --- --- --- --- --- --- --- ---
Mouse       Asn Asp Thr --- --- --- --- --- --- --- --- --- --- --- --- --- --- --- --- --- --- --- --- --- --- --- --- --- --- --- --- --- --- --- --- --- --- --- --- ---
Rabbit      Glu --- --- --- --- --- --- --- --- --- --- --- --- --- --- --- --- --- --- --- Asn --- Tyr --- --- --- --- --- --- --- --- --- --- --- --- --- --- --- --- ---
Chicken     Glu Asp Ser --- --- --- --- --- --- --- --- --- --- --- --- --- --- --- --- --- Asn --- Asp Asp --- --- --- --- --- --- --- Val --- --- --- --- --- --- --- ---
Gecko       Glu Asp Ser --- --- --- --- --- --- --- --- --- --- --- --- --- --- --- --- --- Asn --- Gln Asp --- --- --- --- --- --- --- Val --- --- --- --- --- --- --- ---
Newt        --- --- Pro Asn --- --- --- --- --- --- --- --- --- --- --- --- --- --- --- --- --- --- Asp Asp --- --- --- --- --- --- --- Val --- --- --- --- --- --- --- ---
Rainbow Tr  Glu --- Gln --- Thr --- --- --- --- --- --- --- --- --- --- --- --- --- --- --- --- --- Asp Asp --- --- --- --- --- --- --- Ile --- --- --- --- --- --- --- ---
Sturgeon    Glu Asp Ser Ile Ser --- Ser --- --- --- --- --- --- --- --- --- --- --- --- --- Asn --- Glu Asp Asp --- --- --- --- --- --- Ile --- --- --- --- --- --- --- ---
Hagfish     Glu Asp Met --- --- --- --- --- --- --- --- --- --- --- --- --- --- --- --- --- Asn --- Glu Gln --- --- --- --- --- --- --- Val --- --- --- --- --- --- --- ---

                    560                                     570                                     580
Human       Arg Asp Gly Lys Tyr Gly Phe Tyr Thr His Val Phe Arg Leu Lys Lys Trp Ile Gln Lys Val Ile Asp Gln Phe Gly Glu
Bovine      --- --- --- --- --- --- --- --- --- --- --- --- --- --- --- --- --- --- --- --- --- --- --- Arg Leu --- Ser
Rat         --- Asn --- --- --- --- --- --- --- --- --- --- --- --- --- Arg --- Met --- --- --- --- --- --- His Arg xxx
Mouse       --- Lys --- --- --- --- --- --- --- --- --- --- --- --- --- Arg --- --- --- --- --- --- --- --- --- --- xxx
Rabbit      --- --- --- --- --- --- --- --- --- --- --- --- --- --- --- --- --- --- Arg --- Met Val --- Arg --- --- xxx
Chicken     --- --- --- --- --- --- --- --- --- --- --- --- --- --- --- --- --- Met Arg --- Thr --- Glu Lys Gln --- xxx
Gecko       --- --- --- --- --- --- --- --- --- --- --- --- --- --- --- --- --- Leu Lys --- Thr Val Glu Lys His --- Asn
Newt        --- --- --- --- --- --- --- --- --- --- Leu His --- Met Arg Gln --- Met Met --- Ile --- Glu Lys Cys --- Ser
Rainbow Tr  --- --- --- --- --- --- --- --- --- --- Leu --- --- Met Arg Arg --- Met Lys --- --- --- --- Lys Thr --- Gly Asp Asp Asp Asp
Sturgeon    --- Ser --- --- --- --- --- --- --- --- Leu --- --- Met Arg --- --- Met Leu --- Thr --- Val Asp Thr Glu xxx
Hagfish     Lys --- --- --- --- --- --- --- --- --- Leu --- --- Met Leu Arg --- Leu Lys --- Ile Val Asn Arg Glu --- Ala Arg
```

Figure 3 Continued

Table 1 Naturally Occurring Dysfunctional Prothrombins

Variant	Substitution amino acid (position)[a]	DNA	Impaired function	Reference
Madrid	Arg to Cys (271)	CGT to TGT[b]	Factor Xa activation	53
Barcelona	Arg to Cys (271)	CGT to TGT[b]	Factor Xa activation	52
Himi I	Met to Thr (337)	ATG to ACG	?	55
Quick I	Arg to Cys (382)	CGC to TGC[b]	Fibrinogen binding, platelet aggregation	49
Himi II	Arg to His (388)	CGC to CAC	Fibrinogen binding	55
Tokushima	Arg to Trp (418)	CGG to TGG[b]	Fibrinogen binding, platelet aggregation	51
Salakta	Glu to Ala (466)	GAG to GCG[b]	Substrate binding site	54
Quick II	Gly to Val (558)	GGC to GTC[b]	Arg/Lys binding pocket	50
Clamart			Factor Xa activation	57
Cardeza			Factor Xa activation	56
Metz			Thrombin domain	58
Molis			Thrombin domain	59

[a]Amino acid residues are numbered as shown in Figures 1 and 2.
[b]These base substitutions have not been experimentally determined in genomic DNA from these patients but are inferred based on the sequence of the cDNA in Figure 2.

The genetic defect responsible for hypoprothrombinemia in the compound heterozygote for prothrombin Tokushima (described earlier) and hypoprothrombinemia is caused by the insertion of a T at position 4177 in exon 6 of the human prothrombin gene (60,61). Insertion of an additional base at this position causes a frameshift mutation with an in-frame stop codon at codon 173 in exon 7. Exon 7 encodes all of the second kringle domain, and therefore it is likely that this frameshift mutation causes a null prothrombin phenotype.

Preliminary results from another patient with hypoprothrombinemia indicate two mutations that may be responsible for the low circulating levels of prothrombin antigen and thrombin activity seen in the patient (62,63). A G was found to be deleted at position 7247 or 7248 in exon 8 and a T substituted for a C at position 8759 in exon 10 of the human prothrombin gene (61). The deletion is in the region immediately following the exon coding for the second kringle domain and causes a frameshift that leads to premature translation termination. The mutation at position 8759 results in the substitution of Trp at Arg^{340} (Fig. 2). Whether these two mutations are present on different alleles was not determined.

B. Site-Directed Mutants

Prothrombin has been expressed by several laboratories in several eukaryotic cell lines (35,64,65). Expressed recombinant prothrombin exhibits characteristics of the native protein, including complete γ-carboxylation. Expression of prothrombin has enabled the functions of this protein to be studied by site-directed mutagenesis analysis. Wu and coworkers (66) created several mutants to study the thrombomodulin-binding and fibrinogen clotting activities of thrombin because fibrinogen is normally the preferred substrate of thrombin, but in the presence of thrombomodulin protein C is preferentially activated. Three mutants (K372E, R388E, and R390E) replaced basic amino acids with glutamic acid in the putative substrate binding groove of the anion binding exosite (Figs. 1 and 2). These mutants established that Lys^{372} is required for fibrinogen clotting activity and platelet activation and is not required for activation of protein C; Arg^{388} was required for all three activities. Arg^{390} appears to be required only for the activation of protein C. These results indicate that the fibrinogen and thrombomodulin binding sites are not identical but overlapping.

The thrombin mutant E345K in which Lys was substituted for Glu^{345}, was made to test the requirement of this residue in determining substrate specificity (Fig. 2) (67). Results with this mutant indicate that this residue is indeed a determinant of the strict specificity of thrombin and suggest that thrombomodulin functions in part by changing this specificity from the preferred substrate of fibrinogen to protein C. The mutant E522Q, in which Glu^{522} is replaced by Gln, was also synthesized to test another aspect of the substrate specificity of thrombin (Fig. 2) (68). In trypsin and most other trypsin-like enzymes with broad specificity, the residue at this position is a Gln, but in thrombin this residue is a Glu. The activity of E522Q toward protein C was increased significantly compared with wild-type thrombin, but release of the fibrinopeptides from fibrinogen was approximately the same for the mutant and wild-type thrombin. These results indicate the critical role of Glu^{522} in restricting the specificity of thrombin and again suggest that thrombomodulin functions partly by altering the interaction of thrombin toward protein C near this residue.

Residues 466, 467, and 468 have been deleted from thrombin to form the mutant des-ETW to study the interaction of thrombin with specific serine protease inhibitors (Fig. 2) (69). These three residues are present in one of two insertion loops present in thrombin that are not present in trypsin as determined by x-ray diffraction studies and that potentially limit the access of proteins to the active site (27). Studies with this mutant suggest that these three residues either directly or indirectly prevent the interaction of thrombin with two Kunitz inhibitors specific for trypsin and are not essential for the thrombomodulin-mediated enhancement of protein C activation.

The active site Ser (residue 525 in Fig. 2) has been replaced by an Ala to produce an enzymatically inactive form of thrombin (S525A) (7,66,70). The mutant displayed no activity toward fibrinogen or synthetic chromogenic substrates. In addition, this form of thrombin could bind to the thrombin receptor but not activate it.

IV. THE GENE CODING FOR PROTHROMBIN

A. Chromosomal Localization

The chromosomal location has been determined for both the human and mouse prothrombin genes. The human gene was localized to chromosome 11 near the centromere (71). The mouse gene was localized to chromosome 2, 1.8 ± 1.3 map units proximal to the catalase locus (36). This region of mouse chromosome 2 shares homology with the proximal end of human chromosome 11p, which may indicate that the human gene is located on the human chromosome 11p side of the centromere rather than 11q.

B. Organization of the Human Prothrombin Gene

The gene coding for human prothrombin has been completely sequenced, including 5′ and 3′ flanking regions (61,72). A total of 26,929 bp continuous sequence has been determined, including 6544 bp sequence upstream from the site of initiation of transcription, 20,241 bp from the transcription start site to the polyadenylation site, and 145 bp 3′ flanking sequence. The gene is composed of 14 exons separated by 13 intervening sequences (Fig. 4). The exons range in size from 25 to 315 bp in length. Intervening sequences range from 84 to 9447 bp in length. The gene is composed of approximately 90% intervening sequence.

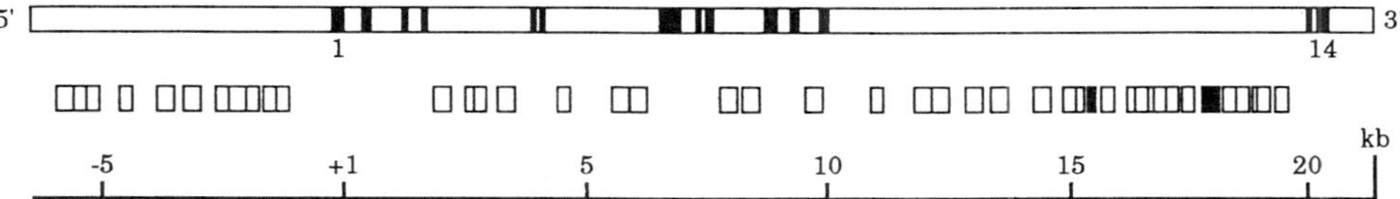

Figure 4 The human prothrombin gene. Blackened parts of the bar represent the 14 exons (exons 1 and 14 are labeled) (61). Areas between exons represent intervening sequences. The region upstream of the first exon represents the 5′ flanking sequence that has been characterized by DNA sequence analysis (61,72). The orientation of transcription is indicated by the 5′ and 3′. Boxes below the bar represent repetitive sequences; open boxes represent Alu-repetitive sequences, and blackened boxes represent Kpn-repetitive sequences. The scale in kb is shown at the bottom. The site of initiation of transcription is indicated by +1; sequence upstream is numbered in a negative fashion, with −1 the nucleotide immediately 5′ to the initiation site.

The site of initiation of transcription for the human prothrombin gene has been determined (72,73). Heterogeneous start sites are present from 3 to 38 bp upstream from the initiator methionine (Fig. 5). The site that was found to agree best with the consensus sequence for transcription start sites was 31 bp upstream (Fig. 5) (72).

The genes coding for serine proteases are the result of duplication and divergence from a common ancestral gene (74). Further diversification has been proposed to occur by recombination of exons from different ancestral genes and may include exon shuffling (75–77). It is in this way that proteins with common domains exist even when other regions of the protein are different. The organization of the prothrombin gene, as well as other coagulation and fibrinolytic proteins, is consistent with these proposals.

The placement of intervening sequences with respect to the coding sequence of human prothrombin is shown in Figure 1 (61). Intervening sequences A, B, and C are located in almost the same position in the amino acid sequence of prothrombin as the intervening sequences in the genes coding for other vitamin K-dependent proteins (78). The Gla region is the most conserved region when the vitamin K-dependent proteins in blood coagulation are compared, and the placement of intervening sequences in this region of these genes indicates that the conservation extends to the level of gene organization. The size and sequence of these three intervening sequences in the genes for all vitamin K-dependent proteins differ and indicate that there is no evolutionary constraint on these regions of the gene and therefore that they probably play no regulatory role.

The placement of the intervening sequences with respect to the two kringle domains is shown in Figure 1. The first kringle is encoded by two exons as a result of the presence of intervening sequence E within the DNA coding for this domain, and the second kringle is encoded by one exon. The lack of an internal intervening sequence in the DNA coding for the second kringle is unique among the genes coding for other kringle-containing proteins. All other kringles are encoded by two exons. The placement of the internal intervening sequence within the first kringle is also unique in that no other gene structure for a kringle has a similar placement. The evolution of kringle structures can be inferred from the placement of the internal intervening sequence. Kringles with identically placed internal intervening sequences appear to have duplicated after the insertion event. For example, tissue plasminogen activator, urokinase, and factor XII have kringles that are more similar to each other than to all other kringle domains based on amino acid sequence (77). The internal intervening sequence separating these kringles into two exons is identically placed in the genes coding for these proteins. From these data and the homology of their DNA and amino acid sequences, it is generally accepted that the kringles in tissue plasminogen activator, urokinase-type plasminogen activator, and factor XII are more closely related to each other than to the kringles in prothrombin (77).

The splice junctions of the intervening sequence within the first kringle of prothrombin is of the type II class, and the intervening sequences flanking both kringles are type I (61). Type I intervening sequences interrupt codons between the first and second base, type II sites are between the second and third base of the codon, and type 0 splice junctions are between codons (79). This arrangement of type I and II intervening sequences has been found for all other genes coding for kringle structures. It has been noted that intervening sequences flanking kringle domains are always type I (80). Type I intervening sequences appear to be dominant at domain junctions and are consistent with the importance of exon shuffling in the evolution of genes coding for multidomain-containing proteins because the reading frame remains intact. The consistent use of any

```
TGCGGGAGTC CTGGTCCCCA GTTTTTCTGC CAACACTCCC TGTTCCCACA CATGACCTGG TCCAGACCCC AAACAGCCAG GCCCAAAGGA CAGGTGAGGC   -945

GAGGCGAGAA CTTGTGCCTC CCCGTGTTCC TGCTCTTTGT CCCTCTGTCC TACTTAGACT AATATTTGCC TTGGGTACTG CAAACAGGAA ATGGGGGAGG   -845

GACAGGAGTA GGGCGGAGGG TAGGGTAGGA CCAGAAGCCT CTCTAGGCCT GCCATGGGGC AGGCAGCCAG GGAGAAGGAG GGCCCCTCAG TGGAGACCCA   -745

GGGATTTCAG TAGCCCCTGT TCCGGGACAG GCGCAGGTCC TGGGAGGTGA CAGAAGATAG ACTAAAGGCC CAAGAGTCCC TGGACCTGAC TCCTCCCAGC   -645
                             ||| ||     ||| ||||| |||   |||||||| || | |||||    |||  ||  || || ||| |||||| ||
                             CAG GCCGGCCTCC TGGGACCTGA ACGAAGATAG ACCAGAGGCC TGGGAG GCC AGGGCCCGAC TCCTCCTGGC

AGCTGCCACA CACAAACACA CCTCCAGGCA CCCTGGACAG GAAGGAGGAG AAATGGGCCC CTCCTCCAGT GGCTGAGAAG CTGGGGCAAA TGTTGGCTGT   -545
| | || ||| |||||||||  |||||  ||  ||| || |||   || || | |  |||| || ||||
AACCGCTACA CACAAACACG CCTCCCAGCT CCCAGG CAG    GGCGGGG ACGTGGGACC CTCC

TCCTATCCCT GGTGCATCCC ATGGCGAGGG GCAACTTCCA TCAGGCCACA CCTTTTATCT TTGTCTCTAT TTTTGATATC TGTGTATTAT GATTATACAA   -445

ACCCCCACAT TGGCCTATAT GTGCAGATCT GATTAAGAAC TTACGATATT CCATGGACAT TCCATTCCTA ATCTCCTTTA GTCCTCACAA CAAAGTATTA   -345

TTCCCATTGT ATAGATGAGG AAACTGAGGC ACACAGAGAT GACAAGCAAC CACCGCTATA TGTTAGGATT CGAAGGAGCT CCAGGAAAGT CTCATAGCCC   -245

CACTGGCCAG AATGGGCTAA ATCTCAGAGG GGGAGGGTGG GAGATGGGGG TGACAGTGAC CTTTTTTGTG ACTCCTCCTA GACCATCCAT CCCTGCTCCC   -145

AGGAGGACCT GTCCTCCCAG ATGGTGGAGA TGGACAGGAG GACTATCTAC CCACCCGTCC CCACGGCCCT GACCCTCTGA CCTCACCCTC TCCGCTGATT    -45
                                       || | | || || |  ||||  |||| | | |||||| ||||| |||| ||||  | ||  | |||||||
                                       GGTG GGCTCTCCAT CCACGTGTCC CTATGGCCCT GACCCGCTGA CCTCCGCTTC CCGGCTGATT

                                             *                            MetAl aHisValArg GlyLeuGlnL
TCTTCATGTT AGTTCAACAT TAACCCAGAG GGGTCAGGAC AGACAATTCC TCAGTGACCC AGGAGCTGAC ACACTATGGC GCACGTCCGA GGCTTGCAGC     57
|||||| |||  ||||||||| |||||| | | |||||||      ||| |  ||  ||| |  ||  ||||| || |||| ||||| || ||||||| |||  || ||
TCTTCACGTT GGTTCAACAT TAACCCGGTG GGGTCAG      GACCAGCCC GCAGAGTGCC  GGAGCGGAT ACACCATGGC GCGCGTCCGA GGCCCGCGGC

euProGlyCy sLeuAlaLeu AlaAlaLeuC ysSerLeuVa lHisSerGln HisV
TGCCTGGCTG CCTGGCCCTG GCTGCCCTGT GTAGCCTTGT GCACAGCCAG CATGGTAAGG GAGTGCTT                                      124
|||||||||| |||||||||| ||||||||||    ||||| || |||||||||| ||||
TGCCTGGCTG CCTGGCCCTG GCTGCCCTGT TCAGCCTCGT GCACAGCCAG CATG
```

Figure 5 DNA sequence of the 5′ flanking sequence and first exon of the human prothrombin gene. The DNA sequence of 1044 bp of 5′ flanking sequence, the first exon, and 14 bp of the 5′ end of the first intervening sequence are shown (61,72). The site of initiation of transcription is indicated by an asterisk. The translated amino acid sequence of the first exon is shown above the nucleotide sequence. Numbers in the right margin indicate the final nucleotide on that line. The sequence most similar to a TATA sequence is underlined and is present immediately upstream of the site of initiation of transcription. The putative HNF-1 binding site is in bold type and underlined. Two inverted repeat sequences flanking the HNF-1 site are underlined. The portions of the 5′ flanking region that are similar to the bovine prothrombin gene are indicated by comparison of the two sequences (87). Vertical lines between these two sequences indicate identity. Deletions are represented by spaces; insertions are not shown but occur following nucleotides -702, -697, -655, and -622.

one type of junction is favorable for exon shuffling, although type I intervening sequences appear most frequently in the genes that have been studied.

The placement of intervening sequences in the human prothrombin gene with respect to the catalytic domain is shown in Figure 1. Each of the amino acids that constitute the active site (His^{363}, Asp^{419}, and Ser^{525} is present in individual exons). This has also been found for several other genes coding for serine proteases. The genes coding for the other vitamin K-dependent proteins in blood coagulation have only the active site His in an individual exon, whereas the remainder of the catalytic domain is encoded by one exon.

C. Repetitive Sequences

An unusual feature of the human prothrombin gene and its 5′ flanking region is the presence of a high concentration of repetitive sequences (Fig. 4) (61,72). Repetitive sequences of the Alu type constitute 41% of the sequence of the gene as well as sequence upstream of the gene. Alu-repetitive DNA has been estimated to have over 500,000 copies throughout the genome, which represents approximately 5% of the total mass of human DNA (81). The majority of Alu repeats are flanked by short directly repeated sequences. Several Alu repeats have been found to be transcribed by RNA polymerase III, but as yet their function, if any, is not known. There are 41 copies of Alu sequences upstream or in the intervening sequences of the human prothrombin gene (Fig. 4). The majority of these repeats are flanked by direct repeats 5–17 bp in length. In many cases, the repeats occur in clusters. For example, intervening sequence L contains 20 Alu repeats, and 5 of these occur in head-to-tail orientation with no additional DNA between them. These 5 tandem repeats are flanked by 6 bp direct repeats.

The high content of Alu sequences within this gene is unique among genes characterized to date. None of the genes coding for other coagulation factors have a high concentration of repetitive sequences. For the majority of genes that have been characterized, the DNA sequence of entire intervening sequences are not usually determined. Therefore, it is not possible to determine whether the prothrombin gene is unique in its high content of repetitive sequences. There is no obvious reason for the high concentration of these sequences, except possibly that the gene is located near the centromere (71).

Several other types of repetitive sequences are present in the human prothrombin gene (61,78). Two copies of partial KpnI repetitive sequences are present in intervening sequence L (Fig. 4). Several copies of middle repetitive sequences (MERs) have been found in the 5′ flanking region of the gene (82). In addition, another sequence in the 5′ flanking region (nucleotides −4965 and −4898) (72) is homologous to two regions in the human adenosine deaminase gene but not to any other sequences in the database (78).

D. Polymorphisms

When the sequence of the exons in the gene were compared with the human prothrombin cDNA sequences (33–35,61), several differences were found that could be the result of polymorphisms. These are noted in Table 2. Five of these result in silent substitutions and therefore have no effect on the amino acid sequence of the protein (at residues −42, 13, 31, 274, and 389). Differences at three positions result in amino acid substitutions at residues −41 and −40 in the signal sequence and at residue 121 in the first kringle domain.

Polymorphic sites have been identified in the human prothrombin gene by analysis for restriction fragment length polymorphisms. PstI detects a polymorphism that results

Amino acid		cDNA		Gene		
Residue	Position[a]	Nucleotide	bp[b]	Nucleotide	bp[c]	References
Ala	−42	C	24	G	6	61
		G				
His	−41	A	26	A	8	61
Arg	−41	G				
Ile	−40	A	28	G	10	61
Val	−40	G				
Leu	13	G	186	A	554	61
		A				
Leu	31[d]	G	240	G	608	63
				A		
Asn	121	A	509	C	4200	61
Thr	121					
Thr	122	C	512	C	4203	85, 86
Met	122			T		
Thr	274	A	969	C	7322	33
Thr	389	C	1314	A	8909	33
(IVS E)				T	4048	86
				C		
(IVS E)				T	4096	86
				C		
(IVS E)				T	4097	86
				C		
(IVS E)				C	4125	86
				G		
(IVS F)				G	4272	86
				A		
(IVS G)				C	7177	60
				T		

[a]Numbering in Figures 1 and 2.
[b]Numbering as in Figure 2.
[c]Reference 61.
[d]The authors claim that the substitution is at nucleotide 609 in the gene in the codon for Leu[61]. For this substitution to be silent, the substitution should be at position 608 in the codon for Leu[31].

in the insertion or deletion of a 0.8 kb region, yielding fragments of 9.2 and 10 kb (83). The frequency of this polymorphism in 18 unrelated European whites was 0.72 for the 9.2 kb allele and 0.28 for the 10 kb allele. The polymorphism is between exons 4 and 5 between the TaqI site at position 1575 and the EcoRI site at 3017 in the gene (61,83). This same polymorphism was also reported by McAlpine and coworkers (84) with a frequency of 0.70 for the larger allele and 0.30 for the smaller in 20 unrelated whites. Other restriction enzymes have been used to detect this polymorphism in the same individuals (HindIII, SstI, TaqI, and BamHI); EcoRI and PvuII do not (84). NcoI detects a polymorphism in exon 6, where a T has been substituted for a C at position 4203 in the gene (Table 2) (85). This polymorphism results in substitution of Met for Thr at residue 122. In 20 unrelated Japanese, the frequency is 0.425 for the originally reported allele and 0.575 for the new allele (T substitution). The region surrounding exon 6 (including intervening sequences E and F) has been found to be highly polymorphic, with six substitutions in a 418 bp region (Table 2) (86). Only one of the polymorphisms is in an exon (exon 6) and is the same as mentioned earlier. Several of these polymorphisms, as well as an additional site at position 7177 in intervening sequence G, were also identified using polymerase chain reaction–single-strand conformation polymorphism analysis (60).

E. Comparison with the Bovine Prothrombin Gene

The gene coding for bovine prothrombin has been partially characterized by DNA sequence analysis (87). The gene is approximately 15.4 kb in length; 5.1 kb of the gene has been sequenced. The organization of the bovine prothrombin gene is identical to that of the human gene, with 14 exons separated by 13 intervening sequences (Fig. 4). All intervening sequences are in identical positions with regard to the coding sequence, as in the human gene. The approximately 5 kb difference in the size of the human and bovine genes can be accounted for by the high concentration of repetitive sequences in the human gene, although the repetitive DNA content of the bovine gene has not been determined. Bovine Alu sequences do not exist. The site of initiation of transcription in the bovine gene was found to be 24 bp upstream of the initiator methionine (87), which is similar to the placement of this site in the human gene at 31 bp (Fig. 5) (72). All the exons and seven of the intervening sequences in the bovine prothrombin gene have been completely sequenced and therefore can be compared with those in human gene. Exons are 65–96% identical, and the intervening sequences that can be compared are 52–74% identical (78).

V. TISSUE-SPECIFIC EXPRESSION OF PROTHROMBIN

The vast majority of prothrombin is synthesized in the liver (8,88). The expression of prothrombin mRNA in the rat has been determined by northern analysis of RNA isolated from various tissues obtained during development and during and after pregnancy (88). Prothrombin was found to be expressed primarily in the liver and to a smaller extent in diaphragm, stomach, intestine, kidney, spleen, and adrenal tissues during some stages of development. No expression was observed by this analysis in brain, heart, aorta, or lung at any point during development. During pregnancy, prothrombin was primarily expressed in the liver but was also found at some time points in the diaphragm, stomach, uterus, and placenta. No expression was observed as determined by this analysis in the brain, heart, lung, spleen, small or large intestine, kidney, adrenal, ovary, or urinary bladder at any time point. For all nonliver tissues analyzed during development and during and after pregnancy, the level of expression of prothrombin mRNA was less than 5% of that expressed in the liver. Northern analysis also indicated that the level of prothrombin expression in the liver increased severalfold late in gestation and reached maximal levels by 13 days after birth (88). The levels of prothrombin mRNA in the liver increased slightly during pregnancy, but immediately after delivery, the levels decreased to approximately 50% of preparturition levels.

The polymerase chain reaction and *in situ* hybridization analysis have been used to demonstrate the presence of prothrombin transcripts in the olfactory bulb, the cortex, the cerebellum, and other regions of the rat and human nervous system, as well as in neural cell lines (89). These techniques are usually more sensitive than northern analysis, which probably explains why prothrombin mRNA was not detected previously in the brain (88).

VI. REGULATORY SEQUENCES

Tissue- or cell-specific expression of genes is often regulated at the level of transcription. Specific nuclear proteins (trans-acting factors) interact with sequences in or around the gene (cis-acting sequences) to regulate its expression in either a positive or a negative manner. For example, liver-specific expression of a gene could result from the presence of positive regulatory factors (enhancers) in the liver that are not present in other tissues or the presence of negative regulatory factors (repressors or extinguishers) in other tissues that are not present in the liver. Our present state of knowledge indicates that specific proteins bind to several short DNA sequences scattered throughout the 5′ flanking region of liver-specific genes. Each of these interactions contribute to the overall transcriptional activity of the gene by interacting indirectly with the promoter of the gene, where RNA polymerase II binds and initiates transcription of that gene. These sequences or enhancers function regardless of orientation or distance from the promoter.

Cis-acting regulatory sequences have been identified and characterized for several liver-specific genes that appear to bind many different trans-acting factors (90,91). Several nuclear proteins that are involved in the liver-specific regulation of transcription of specific genes have been isolated and characterized. To date, four gene families that appear to play a major role in liver specificity have been identified. These include the C/EBP, HNF-1, HNF-3A, and HNF-4 family of proteins (92).

A. Promoter Region

Specific DNA sequences present in eukaryotic genes are required for the accurate and efficient transcription of the gene by RNA polymerase II. These promoter regions usually include 5′-TATA-3′ and 5′-CCAAT-3′ sequences located 30 and 80 bp upstream of the transcription start site, respectively. No sequences similar to these promoter consensus sequences are present upstream of the site of initiation of transcription in the human prothrombin gene (Fig. 5) (61,72). The closest sequence to a TATA box, in the appropriate position, is the sequence 5′-CATTAA-3′ between nucleotides −27 and −22. There is no obvious 5′-CCAAT-3′ sequence upstream of this site. Lack of these sequences may contribute to the heterogeneity observed in the transcription initiation site in the human prothrombin gene (discussed earlier) (72,73). Two sequences important for the promoter activity of the human factor X gene (5′-CCAAT-3′ at −120 and 5′-ACTTTG-3′ at −56) are not present in this region of the human prothrombin gene (93).

B. Identification of Liver-Specific Sequences

The 5′ flanking region of the human prothrombin gene has been examined for the presence of cis-acting sequences that determine the liver-specific expression of this gene (72,73,94). Liver-specific sequences were identified by use of hybrid constructs containing portions of the 5′ flanking region of the human prothrombin gene attached to either the chloramphenicol acetyltransferase gene or the gene for human growth hormone transiently transfected into liver and nonliver cell lines (73,94). Sequences between −919 and −849 were found to be required for liver-specific expression and, when deleted, no expression in liver and nonliver cells was observed (Fig. 5) (94).

A region between nucleotides −887 and −875 in the human prothrombin gene is highly homologous to the consensus sequence for the binding site of HNF-1 (Fig. 5) (73,94). Of 10 bases, 8 are identical. HNF-1 is a DNA binding protein that participates in the liver-specific expression of a number of genes (95).

Gel mobility shift experiments using nuclear extracts from HepG2 cells and DNA coding for nucleotides −919 to −790 in the human prothrombin gene indicate that several proteins or protein complexes bind to this region (94). Formation of these specific complexes was inhibited when an excess of an oligonucleotide identical in sequence to the HNF-1 sequence in the gene coding for the β chain of human fibrinogen was included, implying that HNF-1 protein binds to this DNA sequence in the human prothrombin gene in HepG2 cells. Linker scanning mutagenesis indicated that this region was essential for enhancer activity (73). Although these results suggest that there is a functional HNF-1 binding site in the human prothrombin gene, these experiments do not definitively demonstrate that HNF-1 binding is necessary for tissue-specific expression of prothrombin.

The sequence 5′-CCTCCC-3′ flanking the putative HNF-1 binding site as an inverted repeat was found to bind protein as determined by deoxyribonuclease I footprint analysis and to have enhancer activity (nucleotides −845 to −850 in Fig. 5) (73). Mutagenesis of the 3′ repeat eliminated the enhancer activity of this region. This sequence was also found to be present in the 5′ flanking regions of several other liver-specific genes, which suggests a potential functional role for this sequence (73). This motif appears to be significant only in the presence of other cis-acting elements because mutagenesis of the 5′ repeat (Fig. 5) had no effect on the enhancer activity. In addition, this sequence is present several other times in the 5′ flanking region of the human prothrombin gene and when deleted had no effect on liver-specific expression.

Results from deoxyribonuclease I footprint analysis and gel mobility shift assays of the region from −919 to −790 in the human prothrombin gene suggest that multiple proteins bind to this region (73,94). A non–tissue-specific enhancer appears also to be present in this region because its presence upstream of a heterologous promoter directs transcription in HeLa cells regardless of orientation (94).

C. Other Potential Regulatory Sequences

The nucleotide sequence of 583 bp of the 5′ flanking sequence has been determined upstream of the site of initiation of transcription for the bovine prothrombin gene (87). When this sequence was compared with the 6544 bp of 5′ flanking sequence of the

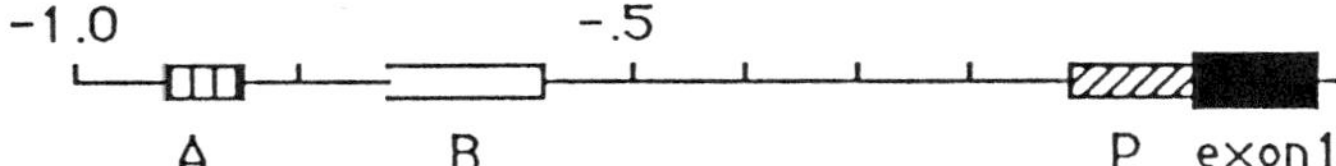

Figure 6 The potential regulatory elements in the 5′ flanking region of the human prothrombin gene. Exon 1 is indicated by the blackened box, with the promoter sequence (P) immediately upstream represented by a horizontally striped box. The putative HNF-1 binding site, as well as a non-tissue-specific enhancer, is present in region A, represented by a vertically striped box. The region conserved between the human and bovine genes (not including the promoter region) is indicated by the white box (region B). The 5′ limit of this conservation has not been determined and is indicated by the open end of the box. The scale is indicated in kb.

human prothrombin gene, two regions of identity were apparent (Fig. 5) (72). Regions from −717 to −581 and −108 to −1 in the human gene are 69 and 79% identical to nucleotides −580 to −443 and −110 to −1 in the bovine gene, respectively. The high degree of identity of both these regions between the two genes indicates a possible role in the regulation of prothrombin. The region immediately upstream of the site of initiation of transcription contains the putative promoter region of both genes. The actual length of the putative upstream regulatory region (Fig. 5) is not known because sequence has not been determined upstream of this region in the bovine gene (87). Deletion of this region has no effect on the liver-specific expression of human prothrombin as tested in HepG2 cells (94). A regulatory function for this region remains to be identified.

D. Summary of Regulatory Elements

A diagram of the putative regulatory elements present in the 5′ flanking region of the human prothrombin gene in HepG2 cells is summarized in Fig. 6 (94). Two or more enhancer sequences are present in region A, an HNF-1 binding site that is liver specific and a second sequence with non–tissue-specific enhancer activity. The enhancers in region A and the prothrombin promoter (region P in Fig. 6) are the required cis-acting elements for tissue-specific expression of this gene in HepG2 cells (73,94). Region B is the highly conserved region in the 5′ flanking regions of both the human and bovine prothrombin genes. This region does not participate in the liver-specific expression of the human prothrombin gene in HepG2 cells but may have other as yet unidentified regulatory functions. The 5′ end of this element has not been delineated. Further studies on region A and the promoter are required to define the specific sequences within these regions that direct liver-specific expression. Identification of these sequences and the proteins that bind to them will facilitate evaluation of the regulatory mechanisms of the other vitamin K-dependent coagulation proteins. The lack of obvious homology between sequences required for the liver-specific expression of several genes and the gene for human prothrombin suggests that several mechanisms must exist for liver-specific expression. Additional trans-acting factors may be required for expression of this gene.

REFERENCES

1. Davie EW, Fujikawa K, Kisiel W. The coagulation cascade: initiation, maintenance, and regulation. Biochemistry 1991; 30:10363–10370.
2. Esmon CT. Protein C: biochemistry, physiology, and clinical implications. Blood 1983; 62: 1155–1158.

3. Davey MG, Luscher EF. Actions of thrombin and other coagulant and proteolytic enzymes on blood platelets. Nature 1967; 216:857–858.
4. Chen LB, Buchanan JM. Mitogenic activity of blood components. 1. Thrombin and prothrombin. Proc Natl Acad Sci USA 1975; 72:131–135.
5. Bar-Shavit R, Kahn AJ, Mann KG, Wilner GD. Identification of a thrombin sequence with growth factor activity on macrophages. Proc Natl Acad Sci USA 1986; 83:976–980.
6. Cunningham DD, Farrell DH. Thrombin interactions with cultured fibroblasts: relationship to mitogenic simulation. Ann NY Acad Sci 1986; 485:240–248.
7. Vu T-KH, Hung DT, Wheaton VI, Coughlin SR. Molecular cloning of a functional thrombin receptor reveals a novel proteolytic mechanism of receptor activation. Cell 1991; 64:1057–1068.
8. Barnhart MI. Cellular site for prothrombin synthesis. Am J Physiol 1960; 199:360–366.
9. Magnusson S, Petersen TE, Sottrup-Jensen L, Claeys H. Complete primary structure of prothrombin: isolation, structure and reactivity of ten carboxylated glutamic acid residues and regulation of prothrombin activation by thrombin. In: Reich E, Rifkin DB, Shaw E, eds. Proteases and biological control. Cold Spring Harbor, NY: Cold Spring Harbor Laboratories, 1975:123–149.
10. Suttie JW. Vitamin K-dependent carboxylase. Annu Rev Biochem 1985; 54:459–477.
11. Butkowski RJ, Elion J, Downing MR, Mann KG. Primary structure of human prethrombin 2 and α-thrombin. J Biol Chem 1977; 252:4942–4957.
12. Thompson AR, Enfield DL, Ericsson LH, Legaz ME, Fenton JW II. Human thrombin: partial primary structure. Arch Biochem Biophys 1977; 178:356–367.
13. Walz DA, Hewett-Emmett D, Seegers WH. Amino acid sequence of human prothrombin fragments 1 and 2. Proc Natl Acad Sci USA 1977; 74:1969–1972.
14. Seegers WH. Prothrombin yesterday and today. Prog Chem Fibrinolysis Thromb 1979; 4: 241–254.
15. Stenflo J, Fernlund P, Egan W, Roepstorff P. Vitamin K-dependent modifications of glutamic acid residues in prothrombin. Proc Natl Acad Sci USA 1974; 71:2730–2733.
16. Nelsestuen GL, Zytkovicz TH, Howard JB. A mode of action of vitamin K: identification of γ-carboxyglutamic acid as a component of prothrombin. J Biol Chem 1974; 249:6347–6350.
17. McLean JW. Tomlinson JE, Kuang W-J, et al. cDNA sequence of human apolipoprotein(a) is homologous to plasminogen. Nature 1987; 330:132–137.
18. Nakamura T, Nishizawa T, Hagiya M, et al. Molecular cloning and expression of human hepatocyte growth factor. Nature 1989; 342:440–443.
19. Han S, Stuart LA, Degen SJF. Characterization of the DNF15S2 locus on human chromosome 3: identification of a gene coding for four kringle domains with homology to hepatocyte growth factor. Biochemistry 1991; 30:9768–9780.
20. Trexler M, Patthy L. Folding autonomy of the kringle 4 fragment of human plasminogen. Proc Natl Acad Sci USA 1983; 80:2457–2461.
21. Van Zonneveld A-J, Veerman H, Pannekoek H. Autonomous functions of structural domains on human tissue-type plasminogen activator. Proc Natl Acad Sci USA 1986; 83:4670–4674.
22. Tulinsky A, Park CH, Skrzypczak-Jankun E. Structure of prothrombin fragment 1 refined at 2.8 Å resolution. J Mol Biol 1988; 202:885–901.
23. Esmon CT, Jackson CM. The conversion of prothrombin to thrombin. IV. The function of the fragment 2 region during activation in the presence of factor V. J Biol Chem 1974; 249: 7791–7797.
24. Patthy L, Trexler M, Vali Z, Banyai L, Varadi A. Kringles: modules specialized for protein binding. Homology of the gelatin-binding region of fibronectin with the kringle structures of proteases. FEBS Lett 1984; 171:131–136.
25. Markwardt F. Hirudin as an inhibitor of thrombin. Methods Enzymol 1970; 19:924–932.
26. Noe G, Hofsteenge J, Rovelli G, Stone SR. The use of sequence-specific antibodies to identify a secondary binding site in thrombin. J Biol Chem 1988; 263:11729–11735.

27. Bode W, Mayr I, Baumann UL, Huber R, Stone SR, Hofsteenge J. The refined 1.9 Å crystal structure of human α-thrombin: interaction with D-Phe-Pro-Arg chloromethylketone and significance of the Tyr-Pro-Pro-Trp insertion segment. EMBO J 1989; 8:3467–3475.
28. Rydel TJ, Ravichandran KG, Tulinsky A, et al. The structure of a complex of recombinant hirudin and human α-thrombin. Science 1990; 249:277–280.
29. Grutter MG, Priestle JP, Rahuel J, et al. Crystal structure of the thrombin-hirudin complex: a novel mode of serine protease inhibition. EMBO J 1990; 9:2361–2365.
30. Seshadri TP, Tulinsky A, Skrzypczak-Jankun E, Park CH. Structure of bovine prothrombin fragment 1 refined at 2.25 Å resolution. J Mol Biol 1991; 220:481–494.
31. Soriano-Garcia M, Padmanabhan K, de Vos AM, Tulinsky A. The Ca^{2+} ion and membrane binding structure of the Gla domain of Ca-prothrombin fragment. Biochemistry 1992; 31: 2554–2566.
32. MacGillivray RTA, Degen SJF, Chandra T, Woo SLC, Davie EW. Cloning and analysis of a cDNA coding for bovine prothrombin. Proc Natl Acad Sci USA 1980; 77:5153–5157.
33. Degen SJF, MacGillivray RTA, Davie EW. Characterization of the complementary deoxyribonucleic acid and gene coding for human prothrombin. Biochemistry 1983; 22:2087–2097.
34. MacGillivray RTA, Irwin DM, Guinto ER, Stone JC. Recombinant genetic approaches to functional mapping of thrombin. Ann NY Acad Sci 1986; 458:73–79.
35. Jorgensen MJ, Cantor AB, Furie BC, Furie B. Expression of completely γ-carboxylated recombinant human prothrombin. J Biol Chem 1987; 262:6729–6734.
36. Degen SJF, Schaefer LA, Jamison CS, et al. Characterization of the cDNA coding for mouse prothrombin and localization of the gene on mouse chromosome 2. DNA Cell Biol 1990; 9: 487–498.
37. Soute BAM, Vermeer C, DeMetz M, Hemker HC, Lijnen HR. In vitro prothrombin synthesis from a purified precursor protein. Biochim Biophys Acta 1981; 67:101–107.
38. Esmon CT, Grant GA, Suttie JW. Purification of an apparent rat liver prothrombin precursor: characterization and comparison to normal rat prothrombin. Biochemistry 1975; 14:1595–1600.
39. Swanson JC, Suttie JW. Prothrombin biosynthesis: characterization of processing events in rat liver microsomes. Biochemistry 1985; 24:3890–3897.
40. Huber P, Schmitz T, Griffin J, et al. Identification of amino acids in the γ-carboxylation recognition site on the propeptide of prothrombin. J Biol Chem 1990; 265:12467–12473.
41. Duiguid DL, Rabier MJ, Furie BC, Liebman HA, Furie B. Molecular basis of hemophilia B: a defective enzyme due to an unprocessed propeptide is caused by a point mutation in the factor IX precursor. Proc Natl Acad Sci USA 1986; 83:5803–5807.
42. Jorgensen MJ, Cantor AB, Furie BC, Brown CL, Shoemaker CB, Furie B. Recognition site directing vitamin K-dependent γ-carboxylation resides on the propeptide of factor IX. Cell 1987; 48:185–191.
43. Foster DC, Rudinshi MS, Schach BG, et al. Propeptide of human protein C is necessary for γ-carboxylation. Biochemistry 1987; 26:7003–7011.
44. Furie B, Furie BC. Molecular basis of vitamin K-dependent γ-carboxylation. Blood 1990; 75:1753–1762.
45. MacGillivray RTA, Davie EW. Characterization of bovine prothrombin mRNA and its translation product. Biochemistry 1984; 23:1626–1634.
46. Dihanich M, Monard D. cDNA sequence of rat prothrombin. Nucleic Acids Res 1990; 18: 4251.
47. Karpatkin M, Tang Z, Meer J, Blei F, Samuels HH. Cloning and partial sequence of a cDNA for rabbit prothrombin. Thromb Res 1991; 62:757–763.
48. Banfield DK, MacGillivray RTA. Partial characterization of vertebrate prothrombin cDNAs: amplification and sequence analysis of the B chain of thrombin from nine different species. Proc Natl Acad Sci USA 1992; 89:2779–2783.

49. Henriksen RA, Mann KG. Identification of the primary structural defect in the dysthrombin thrombin Quick I: substitution of cysteine for arginine-382. Biochemistry 1988; 27:9160–9165.
50. Henriksen RA, Mann KG. Substitution of Val for Gly-558 in the congenital dysthrombin, thrombin Quick II, alters primary substrate specificity. Biochemistry 1989; 28:2078–2082.
51. Miyata T, Morita T, Inomoto T, Kawauchi S, Shirakami A, Iwanaga S. Prothrombin Tokushima, a replacement of arginine-418 by tryptophan that impairs the fibrinogen clotting activity of derived thrombin Tokushima. Biochemistry 1987; 26:1117–1122.
52. Rabiet M-J, Furie BC, Furie B. Molecular defect of prothrombin Barcelona. J Biol Chem 1986; 261:15045–15048.
53. Diuguid D, Rabiet MJ, Furie BC, Furie B. Molecular defects of factor IX Chicago-2 (Arg 145 → His) and prothrombin Madrid (Arg 271 → Cys): arginine mutations that preclude zymogen activation. Blood 1989; 74:193–200.
54. Miyata T, Aruga R, Umeyama H, Bezeaud A, Guillin M-C, Iwanaga S. Prothrombin Salakta: substitution of glutamic acid-466 by alanine reduces the fibrinogen clotting activity and the esterase activity. Biochemistry 1992; 31:7457–7462.
55. Morishita E, Saito M, Kumabashiri I, Asakura H, Matsuda T, Yamaguchi K. Prothrombin Himi: a compound heterozygote for two dysfunctional prothrombin molecules (Met-337 to Thr and Arg-388 to His). Blood 1992; 80:2275–2280.
56. Shapiro SS, Martinez J, Holburn RR. Congenital dysprothrombinemia—an inherited structural disorder of human prothrombin. J Clin Invest 1969; 48:2251–2259.
57. Huisse MG, Dreyfus M, Guillin MC. Prothrombin Clamart: prothrombin variant with defective Arg-320-Ile cleavage by factor Xa. Thromb Res 1986; 44:11.
58. Rabiet MJ, Jandrot-Perriss M, Boissel JP, Elion J, Josso F. Thrombin Metz: characterization of the dysfunctional thrombin derived from a variant of human prothrombin. Blood 1984; 63:927–934.
59. Rabiet MJ. Prothrombin Molis: a mutant prothrombin characterized by a defect in the thrombin domain. Thromb Haemost 1985; 54:46.
60. Iwahana H, Yoshimoto K, Shigekiyo T, Shirakami A, Saito S, Itakura M. Molecular and genetic analysis of a compound heterozygote for dysprothrombinemia of prothrombin Tokushima and hypoprothrombinemia. Am J Hum Genet 1992; 51:1386–1395.
61. Degen SJF, Davie EW. Nucleotide sequence of the gene for human prothrombin. Biochemistry 1987; 26:6165–6177.
62. Gill FM, Shapiro SS, Schwartz E. Severe congenital hypoprothrombinemia. J Pediatr 1978; 93:264–266.
63. Tamary H, Surrey S, Augustine JG, Schwartz E, Rappaport EF. Molecular characterization of hypoprothrombinemia. Blood 1991; 78:65a.
64. Russo G, Mertens K, De Magistris L, Lange H, Cortese R, Pietropaolo C. Biologically active recombinant prothrombin and antithrombin III expressed in a human hepatoma/vaccinia virus system. Biotechnol Appl Biochem 1991; 14:222–233.
65. Falkner FG, Turecek PL, MacGillivray RTA, et al. High level expression of active human prothrombin in an vaccinia virus expression system. Thromb Haemost 1992; 68:119–124.
66. Wu Q, Sheehan JP, Tsiang M, Lentz SR, Birktoft JJ, Sadler JE. Single amino acid substitutions dissociate fibrinogen-clotting and thrombomodulin-binding activities of human thrombin. Proc Natl Acad Sci USA 1991; 88:6775–6779.
67. Le Bonniec BF, MacGillivray RTA, Esmon CT. Thrombin Glu-39 restricts the P′3 specificity to nonacidic residues. J Biol Chem 1991; 266:13796–13803.
68. Le Bonniec BF, Esmon CT. Glu-192 to Gln substitution in thrombin mimics the catalytic switch induced by thrombomodulin. Proc Natl Acad Sci USA 1991; 88:7371–7375.
69. Le Bonniec BF, Guinto ER, Esmon CT. Interaction of thrombin des-ETW with antithrombin III, the Kunitz inhibitors, thrombomodulin and protein C. J Biol Chem 1992; 267:19341–19348.
70. Pei G, Baker K, Emfinger SM, Fowlkes DM, Lentz BR. Expression, isolation, and characterization of an active site (serine 528 to alanine) mutant of recombinant bovine prothrombin. J Biol Chem 1991; 266:9598–9604.

71. Royle NJ, Irwin DM, Koschinsky ML, MacGillivray RTA, Hamerton JL. Human genes encoding prothrombin and ceruloplasmin map to 11p11–q12 and 3q21–24, respectively. Som Cell Mol Genet 1987; 13:285–292.
72. Bancroft J, Schaefer LA, Degen SJF. Characterization of the Alu-rich 5′-flanking region of the human prothrombin-encoding gene: identification of a positive cis-acting element that regulates liver-specific expression. Gene 1990; 95:253–260.
73. Chow BK-C, Ting V, Tufaro F, MacGillivray RTA. Characterization of a novel liver-specific enhancer in the human prothrombin gene. J Biol Chem 1991; 266:18927–18933.
74. Neurath H. Evolution of proteolytic enzymes. Science 1984; 224:350–357.
75. Gilbert W. Genes-in-pieces revisited. Science 1985; 228:823–824.
76. Rogers J. Exon shuffling and intron insertion in serine protease genes (news). Nature 1985; 315:458–459.
77. Patthy L. Evolution of the proteases of blood coagulation and fibrinolysis by assembly from modules. Cell 1985; 41:657–663.
78. Degen SJF. The prothrombin gene and its liver-specific expression. Semin Thromb Hemost 1992; 18:230–242.
79. Sharp PA. Speculations on RNA processing. Cell 1981; 23:643–646.
80. Patthy L. Intron-dependent evolution: preferred types of exons and introns. FEBS Lett 1987; 214:1–7.
81. Schmid CW, Jelinek WR. The Alu family of dispersed repetitive sequences. Science 1982; 216:1065–1070.
82. Jurka J. Novel families of interspersed repetitive elements from the human genome. Nucleic Acids Res 1990; 18:137–141.
83. De Vetten M, Ploos van Amstel HK, Reitsma PH. RFLP for the human prothrombin (F2) gene. Nucleic Acids Res 1990; 18:5917.
84. McAlpine PJ, Dickson M, Guy C, Wiens A, Irwin DM, MacGillivray RTA. Polymorphism detected by multiple RENS in the human coagulation factor II (F2) gene. Nucleic Acids Res 1991; 19:193.
85. Iwahana H, Yoshimoto K, Itakura M. Nco I RFLP in the human prothrombin (F2) gene. Nucleic Acids Res 1991; 19:4309.
86. Iwahana H, Yoshimoto K, Itakura M. Highly polymorphic region of the human prothrombin (F2) gene. Hum Genet 1992; 89:123–124.
87. Irwin DM, Robertson KA, MacGillivray RTA. Structure and evolution of the bovine prothrombin gene. J Mol Biol 1988; 200:31–45.
88. Jamison CS, Degen SJF. Prenatal and postnatal expression of mRNA coding for rat prothrombin. Biochim Biophys Acta 1991; 1088:208–216.
89. Dihanich M, Kaser M, Reinhard E, Cunningham D, Monard D. Prothrombin mRNA is expressed by cells of the nervous system. Neuron 1991; 6:575–581.
90. Maniatis T, Goodbourn S, Fischer JA. Regulation of inducible and tissue-specific gene expression. Science 1987; 236:1237–1245.
91. Mitchell PJ, Tjian R. Transcriptional regulation in mammalian cells by sequence-specific DNA binding proteins. Science 1989; 245:371–378.
92. Di Simone V, Cortese R. Transcription factors and liver-specific genes. Biochim Biophys Acta 1992; 1132:119–126.
93. Huang M-N, Hung H-L, Stanfield-Oakley SA, High KA. Characterization of the human blood coagulation factor X promoter. J Biol Chem 1992; 267:15440–15446.
94. Bancroft JD, McDowell SA, Degen SJF. The human prothrombin gene: transcriptional regulation in HepG2 cells. Biochemistry 1992; 31:12469–12476.
95. Courtois G, Baumhueter S, Crabtree GR. Purified hepatocyte nuclear factor 1 interacts with a family of hepatocyte-specific promoters. Proc Natl Acad Sci USA 1988; 85:7937–7994.

5

Tissue Factor

James H. Morrissey
Oklahoma Medical Research Foundation, Oklahoma City, Oklahoma

I. OVERVIEW: TISSUE FACTOR AS THE TRIGGER OF BLOOD CLOTTING

In normal hemostasis, the primary triggering agent of the blood clotting cascade is a cell surface protein known as tissue factor (TF) or, sometimes, tissue thromboplastin [for an excellent review of the history and general biology of TF, see Bach (1)]. TF is a relatively small, integral membrane protein distributed in vivo in a highly cell type-specific manner. Under normal circumstances, TF cannot be detected in any cell in direct contact with plasma. Instead, TF is expressed on a variety of cell types located outside the vasculature (2–4). On those cells expressing TF, the protein is localized to the cell surface, where it serves as the high-affinity receptor for the plasma protein, factor VII (5–10).

Factor VII is the zymogen (inert precursor) form of a serine protease that is structurally related to trypsin and chymotrypsin (11). This is the predominant form in plasma; the activated form, factor VIIa, is generated during clotting. Factor VIIa is derived from factor VII by cleavage of a specific arginine-isoleucine bond (arginine 152 and isoleucine 153). As depicted in Figure 1, when zymogen factor VII is bound to TF, its conversion to factor VIIa is accelerated (12,13). In addition, TF is also the essential protein cofactor for the enzymatic activity of factor VIIa [reviewed by Nemerson (14)]. The stimulatory effect of TF on factor VIIa is reversible and is thought to result from a conformational change in factor VIIa that enhances its enzymatic activity. Thus, the complex of factor VIIa and TF (TF-VIIa) can be thought of as a two-subunit enzyme, in which factor VIIa is the catalytic subunit and TF is the positively acting regulatory subunit (Fig. 2). When this two-subunit protease is assembled on the cell surface, it constitutes the first enzyme in the clotting cascade.

The substrates for the TF-VIIa enzyme complex are two other serine protease zymogens of the plasma, factors IX and X (Figs. 1 and 2). Like factor VII, factors IX and X are activated to IXa and Xa by limited proteolysis. This arrangement (catalytic subunit

and substrates circulating in the plasma-regulatory subunit embedded in the surface of cells located outside the vasculature) ensures that the clotting cascade is triggered at sites of vasculature injury that permit plasma to contact extravascular cells (some of which express TF). This can be thought of as a kind of sensing system for notifying the blood that it is no longer circulating within a blood vessel. Because TF is an integral membrane protein, the triggering complex (TF-VIIa) remains bound to the cells expressing TF. This

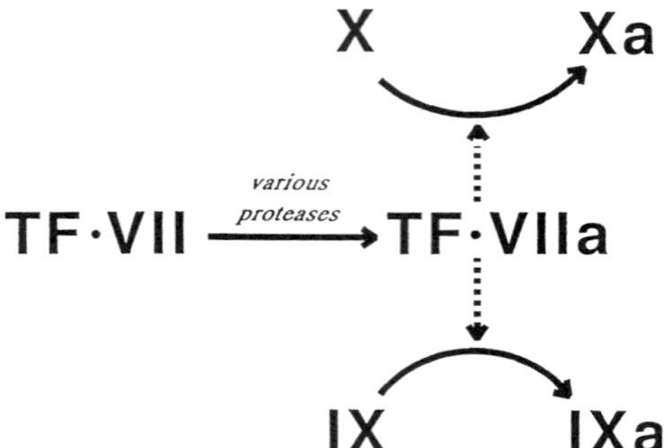

Figure 1 Initiation of the clotting cascade by TF. TF promotes the conversion of factor VII to VIIa and also acts as the essential protein cofactor for the ability of factor VIIa to activate enzymatically factors IX and X.

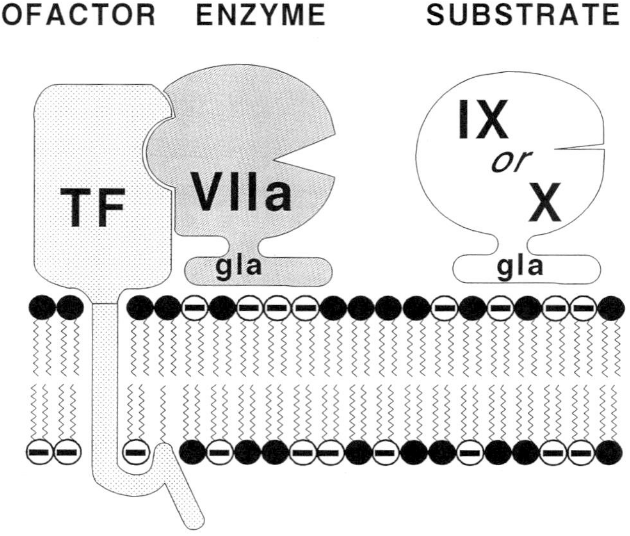

Figure 2 The complex of TF and factor VIIa constitutes a two-subunit, cell surface enzyme. TF, an integral membrane protein, functions as the regulatory subunit; factor VIIa, a serine protease, functions as the catalytic subunit. The substrates, factors IX and X, are the zymogen forms of two serine proteases of the clotting cascade. As depicted, these substrates can associate with negatively charged phospholipid molecules, such as phosphatidylserine, in a calcium-dependent manner via the Gla domains.

arrangement may be important in localizing the formation of clots to wound sites, to prevent their spreading beyond regions of local trauma.

It has long been known that blood clots spontaneously when placed in containers made of such materials as glass. Because this clotting tendency appeared to be an intrinsic property of blood, the enzymes involved in this phenomenon were termed the intrinsic pathway of blood clotting. Conversely, the TF pathway of blood clotting was called the extrinsic pathway because it required mixing an extrinsic activator (i.e., TF) with the blood. The two pathways were originally thought to converge at the level of activation of factor X. However, two factors (VIII and IX) originally assigned to the intrinsic pathway were subsequently found to be common to both pathways (15). Because some individuals who lack critical early enzymes in the intrinsic pathway of blood clotting do not exhibit bleeding disorders, the intrinsic pathway is generally not considered important for normal hemostasis. By the same token, because individuals with severe factor VII deficiency exhibit bleeding problems, TF is generally considered the primary physiological trigger of the clotting cascade in vitro. This notion has been confirmed by the fact that anti-TF monoclonal antibodies are able to block, essentially completely, the clot-promoting ability of most cells, cell lysates, and tissue extracts (16,17). In addition to its role in hemostasis, TF is thought to trigger the clotting cascade in a number of life-threatening thrombotic disorders [reviewed by Edwards and Rickles (18)].

II. TISSUE FACTOR PROTEIN

A. Domain Organization and Posttranslational Modifications

The amino acid sequence of TF has been determined from cDNA cloning and partial amino acid sequencing (19–22). Compared with other protein cofactors of the clotting cascade (23), TF is remarkably small, the mature protein consisting of a mixture of forms containing either 261 or 263 amino acids (Fig. 3) (19–21). The protein has three easily recognizable domains: an amino-terminal extracellular domain consists of 217 or 219 amino acids, a single transmembrane domain of 23 amino acids, and a small cytoplasmic tail of 21 amino acids (Fig. 3).

There are several known posttranslational modifications of TF. The precursor of TF contains a 32 amino acid signal peptide at its amino terminus, which is almost certainly involved in the initial targeting of the nascent protein to the endoplasmic reticulum. Removal of the signal peptide appears to occur at two alternative sites located two amino acids apart, resulting in microheterogeneity of the amino terminus (19–21). There are three carbohydrate chains attached to the extracellular domain at asparagines 11, 124, and 137 (24). The four cysteine residues in the extracellular domain are involved in two relatively small disulfide-bonded loops (Fig. 3) (24,25). The remaining cysteine in the cytoplasmic domain does not exist as a free sulfhydryl, but instead is thioester linked to a fatty acyl moiety (palmitate or stearate) (25). A heterodimeric form of TF identified in early reports (19) was later shown to result from artifactual mixed disulfide formation between the cytoplasmic cysteine of TF and hemoglobin during purification (26,27). Finally, the cytoplasmic domain of TF can be phosphorylated on one or more serine residues (28). Although phosphorylation is stimulated by activators of protein kinase C, it is unlikely that protein kinase C phosphorylates TF directly. Instead, Zioncheck et al. (28) proposed proline-directed protein kinases as likely candidates for phosphorylating

TF. The stoichiometry and functional significance of TF phosphorylation have not been examined in detail.

B. Factor VII/VIIa Binding

Factor VIIa binds to TF in a manner that is saturable, reversible, and calcium dependent. The affinity of this binding interaction was determined by quantifying the association of radiolabeled factor VIIa with purified TF incorporated into phospholipid vesicles (10), TF expressed on the surface of cells (5,7–9,29–32), or TF immobilized on solid supports (33,34). Reported K_d values vary from 82 pM to 21 nM. This rather broad range of reported affinities of factor VIIa for TF may reflect methodological differences among the different studies, as well as influences of the local environment of TF on its affinity for factor VIIa. When binding of radiolabeled factor VIIa to living cells is compared with the generation of cell-associated factor VIIa enzymatic activity, there are substantial differences in binding kinetics (31). These discrepancies have not been fully explained but suggest greater complexity of the system than can be explained by binding of factor VIIa to a homogeneous class of receptors (31).

Recently, equilibrium binding of factor VIIa to TF in solution was quantified using the TF-dependent increase in factor VIIa amidolytic activity (35,36). Compared with

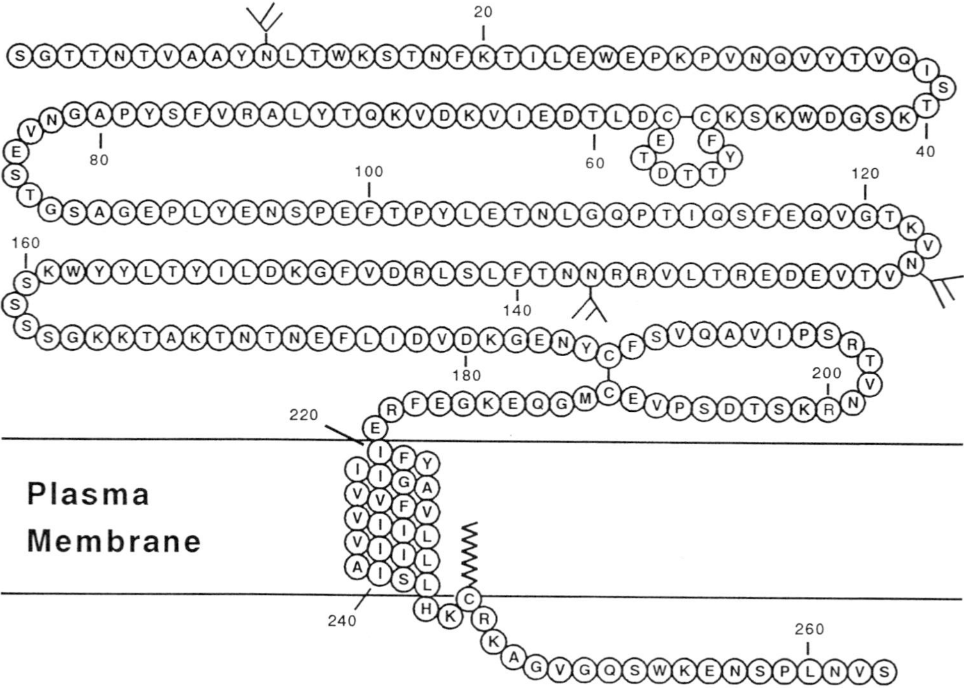

Figure 3 The primary sequence and disulfide bond assignment of TF. Amino acids are abbreviated using the standard one-letter code and numbered according to Morrissey et al. (18) Because of microheterogeneity in the amino terminus, approximately half the TF molecules lack serine 1 and glycine 2 (18–20). The branched structures attached to asparagine residues represent N-linked carbohydrate, and the jagged line attached to cysteine 245 represents a fatty acyl moiety.

measuring the binding of radiolabeled factor VIIa, this approach avoids the need to modify factor VIIa chemically, immobilize TF, or wash out unbound ligand. This approach has yielded K_d values of 180–259 pM (35,36). A caveat is that, under some circumstances, the kinetically derived K_d may differ somewhat from the true K_d (37). Pressure-induced changes in the anisotropy of a fluorescent probe covalently attached to the active site of factor VIIa have also been used to study the interaction between factor VIIa and TF (38). This approach yielded a derived K_d value for the factor VIIa-TF interaction of 7.3 pM. A limitation in interpretation is that the derived K_d requires the assumption that all of the change in anisotropy is caused by dissociation of factor VIIa from TF, not from pressure-induced local changes in dye mobility or other pressure effects, such as on the protein-phospholipid interaction. Nevertheless, these approaches yield broad agreement that the binding of factor VII to TF is a high-affinity interaction ($K_d < 1$ nM).

TF binds zymogen factor VII with an affinity about 5- to 10-fold weaker than that for factor VIIa (10,39). The plasma concentration of factor VII is approximately 10 nM (40), which is much higher than the K_d reported by most studies. Thus, when TF is exposed to plasma, most of the TF molecules are saturated with bound factor VII or VIIa.

It is thought that the calcium dependence of the interaction of factor VIIa with TF is caused by Ca^{2+} binding sites in factor VIIa rather than in TF (41). This view is supported by the lack of observable Ca^{2+} effects on nuclear magnetic resonance signals of TF (34). Recent studies using a soluble mutant form of TF (sTF; see Sec. IV.B) have indicated that, although Ca^{2+} enhances the affinity of binding of factor VIIa to sTF, it is not strictly required (Neuenschwander and Morrissey, submitted for publication). Thus, when calcium was exhaustively removed and a chelating agent was added, binding of factor VIIa to sTF was still observed. However, the K_d was substantially higher than in the presence of Ca^{2+} (3.2 μM versus 1.7 nM).

C. Factor VIIa Cofactor Function

Although factor VIIa has detectable enzymatic activity in the absence of TF (42,43), binding to TF increases all its known activities. Thus, TF dramatically enhances the ability of factor VIIa to activate proteolytically factors IX and X (42,43) and also enhances the amidolytic activity of factor VIIa (i.e., the ability to cleave small, tripeptidylamide substrates) (35,36,44–47). When activation of factor X by factor VIIa is examined, adding TF in phospholipid vesicles causes a large increase in k_{cat} and a large decrease in K_m (42,43). However, much of the decrease in K_m is probably attributable to the phospholipid surface, because negatively charged phospholipid vesicles alone cause a similar decrease in the K_m for factor X utilization by factor VIIa (42,43). A plausible explanation for this effect is that the local concentration of factor X on the vesicle surface is much higher than in solution, resulting in a pronounced decrease in the *apparent* K_m [reviewed by Mann et al. (23)].

When factor VIIa amidolytic activity is measured, TF causes a relatively large increase in k_{cat} but only a small change in K_m (35). This ability of TF to increase the amidolytic activity of factor VIIa is unusual compared with other protein cofactors in the clotting cascade, indicating that the binding event changes the catalytic center of factor VIIa. Consistent with this notion, TF enhances the reactivity of factor VIIa with a tripeptidylchloromethyl ketone active site inhibitor (36). TF also increases the susceptibility of factor VIIa to inactivation by two plasma inhibitors: tissue factor pathway inhibitor (48), and antithrombin III (49). The latter reaction is greatly stimulated by heparin.

D. Conversion of Factor VII to Factor VIIa

An essential function of TF is to bind zymogen factor VII and enhance its rate of conversion to factor VIIa (12,13). A number of proteases, including factors Xa, IXa, XIIa, and thrombin, can activate factor VII (12,13,50–53). In particular, the rate of activation of factor VII by factor Xa has been shown to be enhanced by TF (12,13,39). A plausible explanation for these results is that factor VII becomes a better substrate for activation when it is bound to TF, possibly because of binding-induced conformational changes in factor VII. This results in the preferential activation of TF-bound factor VII, rather than free factor VII, during clotting (12,13).

The proteases just mentioned, which can activate factor VII, are members of the blood-clotting cascade and are produced in the activated forms during clotting. However, these activated forms have very short plasma half-lives because of the presence of plasma protease inhibitors. Free factor VIIa, on the other hand, has no effective plasma inhibitor (54) and has a relatively long half-life (2 h) when given intravenously in humans and animals (55). For these reasons, Miller et al. (56) and others (57) hypothesized that traces of factor VIIa are present in the plasma at all times, to serve a priming role in triggering the clotting cascade. When bound to TF, low levels of factor VIIa therefore generate small amounts of factors IXa and Xa, which could feed back to activate the remainder of TF-bound factor VII molecules.

Interestingly, factor VIIa itself can activate factor VII in the presence of TF (58,59) or certain artificial surfaces (60), a process termed factor VII autoactivation (although, strictly speaking, it is factor VIIa, not factor VII, that is the activator). In this reaction, it has been hypothesized that both the enzyme (factor VIIa) and the substrate (factor VII) are bound to TF and that the reaction requires lateral diffusion of TF-VIIa and TF-VII complexes in the plane of the membrane (39). This enzyme reaction is unusual because its rate is independent of the solution concentration of any of the reactants but instead obeys strictly two-dimensional enzyme kinetics (Neuenschwander, Fiore, and Morrissey, submitted for publication). Factor VII autoactivation (via preexisting factor VIIa in the plasma) may be responsible for the very first stages of the activation of TF-bound factor VII. Therefore, this is another way in which trace levels of factor VIIa could serve a priming role in initiating the clotting cascade. Indeed, all individuals tested have been found to have low but readily measurable levels of factor VIIa in the plasma (61).

III. TISSUE FACTOR GENE ORGANIZATION AND REGULATION

A. Regulation of Expression

TF is expressed throughout the body in a highly cell type specific manner. Immunohistochemical and in situ hybridization approaches have indicated that, under normal circumstances, cells in direct contact with the plasma are devoid of detectable TF (2–4). This includes vascular endothelial cells as well as all circulating blood cells and platelets. Similarly, TF is undetectable in the vascular smooth muscle cells of most vessels. TF is prominent, however, in the adventitial cells and pericytes surrounding most blood vessels larger than capillaries. TF is also present throughout the integument, being prominently expressed in differentiating epidermal keratinocytes but nearly undetectable in the dermis (including resting dermal fibroblasts). TF is present in the epithelial cells of the gut and respiratory mucosae, as well as in cells in most organ capsules. This distribution of TF has been proposed to constitute a ''hemostatic envelope'' surrounding blood vessels, organs, and even the entire body (2). TF is also particularly abundant at sites of increased risk of bleeding or where bleeding would be particularly dangerous, such as renal glomeruli, cardiac muscle, and the brain.

Under pathological conditions, two cell types in contact with the plasma (monocytes and endothelial cells) can be induced to express TF. These cells express TF in response to a variety of inflammatory mediators, including bacterial lipopolysaccharide (LPS), tumor necrosis factor α, interleukin-1, certain viral infections, and other agents (reviewed in Refs. 18, 62, and 63). Induced TF expression on monocytes, and possibly also on endothelial cells, is thought to underlie the life-threatening disseminated intravascular coagulation that accompanies such diseases as septic shock. TF is also expressed within atherosclerotic plaques: chiefly by macrophages (foam cells), monocytes adjacent to cholesterol clefts, mesenchymal-appearing cells within the intima, and on what are probably shed vesicles or fragments of dead cells within the necrotic core (4). Expression of TF within atherosclerotic plaques is predicted to trigger thrombotic episodes (including myocardial infarction and stroke) following plaque fissure or rupture (4).

TF expression in monocytes and endothelial cells requires de novo synthesis of TF protein and is regulated by a combination of transcriptional control and mRNA turnover (64–68). The half-life of TF mRNA is relatively short (0.75–1.5 h) because of an AU-rich sequence in the 3′ noncoding region (68).

Expression of TF is highly regulated in a variety of other cell types. TF is inducible by serum or growth factors in fibroblasts and COS cells (69–71). Because TF mRNA induction is a prompt response to serum or cytokine treatment and does not require protein synthesis, TF has been categorized as an ''immediate-early'' inducible gene in fibroblasts; this has led some to speculate on possible additional roles for TF, such as in growth control (69,70). The pattern of expression of TF by epidermal keratinocytes in vivo indicates that its expression is highly regulated during the program of terminal differentiation (2,3), but this has not been studied in vitro.

B. Gene Organization

The human TF gene, designated *F3* because TF is sometimes called coagulation factor III, has been localized to segment 1p21–22 on the short arm of chromosome 1 (72). The entire DNA sequence of the TF gene has been determined and comprises six exons and five introns, spanning approximately 13 kb (73). The first exon encodes the leader peptide; the second through fifth exons encode the extracellular domain. The sixth exon comprises the transmembrane and cytoplasmic domains and the relatively long 3′ noncoding sequence of the mRNA. This 3′ noncoding sequence includes an *Alu* family repeat (21,22). In many cells, such as fibroblasts, a single mRNA species of 2.3 kb is observed (19). In other tissues and cell types, such as monocytes and endothelial cells, the predominant TF mRNA size is 2.3 kb, but substantial amounts of larger TF transcripts (chiefly around 3.1 kb) are often observed (21,22,64). These larger transcripts are a result of the incomplete removal of the first intron and are not thought to encode an active protein (74).

The 5′ end of the TF gene is embedded in a CpG island (73), a region of DNA sequence approximately 1 kb long that is extremely high in G + C content but devoid of DNA methylation (75). CpG islands are associated with many housekeeping genes and some cell type-specific genes (as with TF); however, their functional significance is not known. Unlike many genes with CpG islands at the 5′ end, the TF gene has a unique start site for transcription located 26 bp downstream from a TATA sequence (73). A number of potential transcription factor binding sites are located within the 5′ flanking sequence of the TF gene (Fig. 4).

Deletional mapping studies have defined the limits of the promoter for inducible expression in COS cells and THP-1 monocytic leukemia cells (66,71). In COS cells, a sequence extending from −383 to 121 bp (relative to the start site of transcription) contains full promoter activity (71). This sequence is located within the CpG island. A sequence capable of directing serum-responsive transcription was localized to −111 to 14, which, however, does not bear obvious sequence similarity to the serum response element in the c-*fos* gene (71) (Fig. 4). Studies of LPS inducibility of TF promoter constructs transfected into THP-1 cells indicated the presence of an LPS response element located between −172 and −227 (66) (Fig. 4). This region contains both AP-1 sites and an NFκ-B-like site, which are required for LPS responsiveness (66).

IV. INSIGHTS INTO TISSUE FACTOR FUNCTION FROM MUTANTS AND MUTAGENESIS

A. Polymorphisms and Interspecies Sequence Conservation

Unlike the situation for many of the other clotting factors, no individual with a naturally occurring deficiency in TF has been reported. This is probably because it is difficult to measure TF levels in a patient, because the protein is normally absent from the blood. TF is expressed in varying levels in a variety of tissues of the body, and this variability

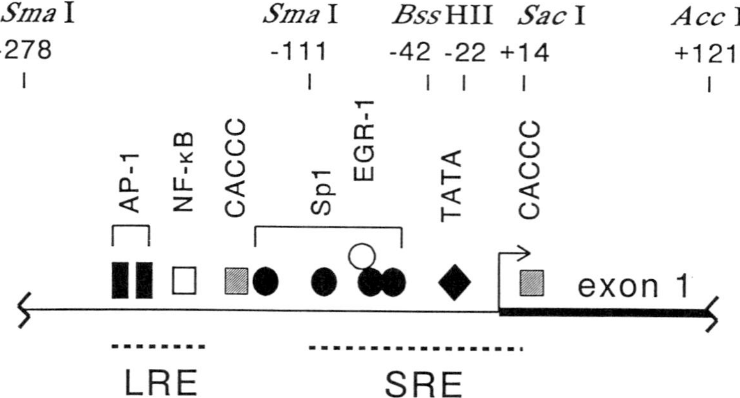

Figure 4 The promoter region of the TF gene. Nucleotide numbering is relative to the start site of transcription (72), which is indicated by the bent arrow. The positions of putative transcription factor binding sites and the regions conferring serum responsiveness in COS cells (SRE) or LPS inducibility (LRE) in THP-1 cells are from Mackman et al. (65,71).

makes it difficult to quantitate TF reliably in biopsy samples, which presumably would be required to establish the presence of a deficiency state. Given that TF is likely to be essential for normal hemostasis, it is anticipated that a total deficiency in TF would be lethal in utero. However, less severe genetic deficiencies in TF may underlie some cases of otherwise unexplained bleeding diatheses. Such patients might be expected to exhibit a phenotype similar to that of factor VII deficiency.

Another possible class of naturally occurring mutations in the TF gene are the promoter mutants or other regulatory mutants that result in overexpression of TF. If such mutations exist, they are expected to confer a tendency toward thrombosis. For the same reasons regarding deficiencies in TF, it would be technically difficult to screen such patients for expression of TF activity or antigen levels. One readily accessible cell type that can be made to express TF in vitro is the circulating monocyte, although variability in the content and reactivity of monocytes in the blood may make quantitation difficult.

A polymorphism in TF amino acid sequence was reported during the initial cloning of the TF cDNA sequence, consisting of a conservative replacement of alanine for valine in the transmembrane domain of TF (21). This substitution is likely to be silent with regard to TF function, and indeed the individual from whom the cDNA sequence was derived had no bleeding disorder. Restriction fragment length polymorphisms have been reported for the TF gene, and some of these have been mapped at the nucleotide level (73,76).

The primary sequence of the TF protein has been deduced via cDNA cloning of human, bovine, rabbit, and mouse TF (19–22,69,70,77,78). (Unless otherwise stated, the studies described in this chapter refer to human tissue factor.) For the mature protein, sequence identity between human and animal TF sequences ranges from 57% for the human-murine comparison to 72% for the human-bovine and human-rabbit comparisons. The degree of sequence identity parallels functional similarity, in that bovine and rabbit TF can promote the clotting of human plasma, but mouse TF has very little procoagulant activity when tested with human plasma (79). The basis for the species specificity in TF clotting activity is thought to be species specificity in the interaction of TF with factor VII, not the recognition of factor X as a substrate (14).

When human TF is compared with rabbit, mouse, or bovine TF, most of the sequence identity is found in the extracellular domain, including conservation of the four cysteine residues involved in disulfide bond formation (Fig. 5). When the human TF amino acid sequence was first deduced, it was noted that three of the five tryptophan residues were contained with a tripeptide sequence tryptophan-lysine-serine (WKS) (19). However, none of these WKS sequences is conserved among all four species (although all four TF amino acid sequences contain at least one WKS motif somewhere in the extracellular domain). The functional significance, if any, of the WKS motif is unclear. All TF amino acid sequences contain multiple glycosylation signals in the extracellular domain, but the only such signal conserved among all four species is at asparagine 11.

The transmembrane domains of TF from the four species, like the leader peptide and the cytoplasmic domain, show little sequence identity (Fig. 5). However, three basic residues in the cytoplasmic domain located just C terminal to the transmembrane domain (the presumptive stop-transfer sequence) are absolutely conserved. There is also conservation of the cytoplasmic cysteine residue that, in humans, was shown to be thioester linked to a fatty acid. However, this cysteine occurs two residues amino terminal in the murine sequence compared with the position in the other three species. Another conserved amino acid in the cytoplasmic domain is the proline at position 259 (human TF numbering system). In all four species, the amino acid immediately N terminal to this proline is either serine or threonine, which is consistent with the idea that a proline-directed protein kinase is responsible for phosphorylation of the cytoplasmic domain of TF (28).

Although initial database searches revealed no obvious homology between TF sequence and any previously identified protein (19–22), subsequent analysis revealed a likely distant relationship with a cytokine receptor family that includes interferon receptors (80). The amino acid identity between TF and the other members of this family is of the order of 20%. This relationship has formed part of the basis of attempts to predict secondary structures of TF (81), the three-dimensional structure of which has yet to be determined experimentally.

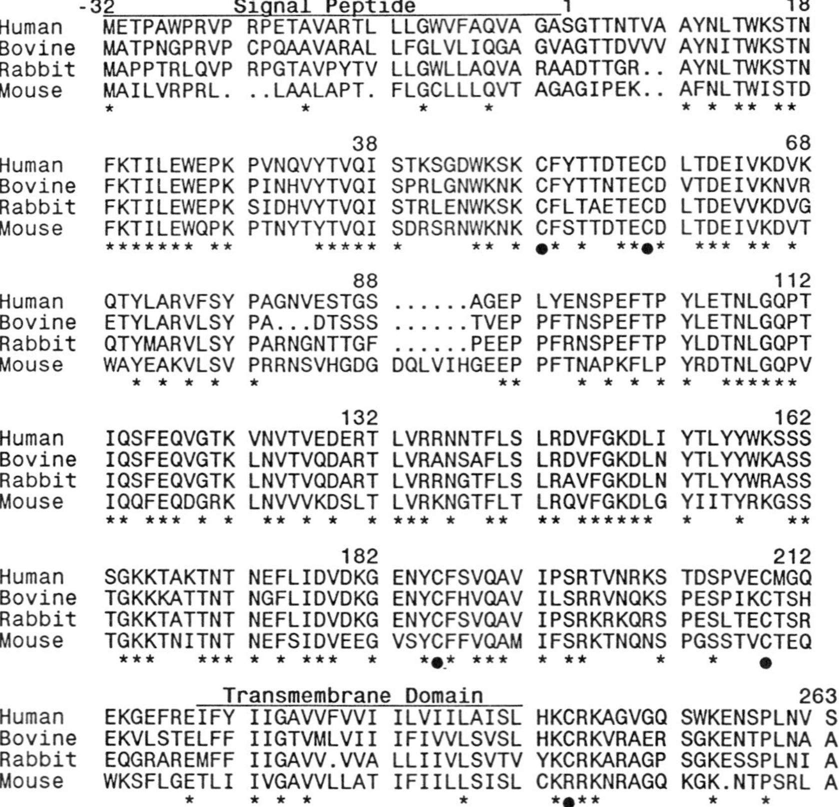

Figure 5 Alignment of human, bovine, rabbit, and mouse TF amino acid sequences. Published TF amino acid sequences (18–21,68,69,75,76) were aligned using a multiple-sequence alignment program (Pileup; Genetics Computer Group, Version 7) and numbered according to the human TF sequence (18). Negative values refer to the presumptive signal peptide, and the mature N terminus of the longer form of TF is designated 1. The signal peptide and transmembrane domains are indicated; the extracellular domain extends from positions 1 through 219, and the cytoplasmic domain extends from positions 243 through 263. Asterisks mark positions of amino acid identity between all four sequences, except cysteine residues, which are indicated by filled circles.

B. Selective Functional Deficiency of the Isolated Extracellular Domain of Tissue Factor

The isolated extracellular domain of TF, or soluble TF, consists of amino acids 1–219 and has been produced by recombinant means via expression in mammalian cells (45,82), yeast (83), and bacteria (38,84). Unlike wild-type TF, sTF is highly soluble in the absence of detergents (38) and appears to have no detectable tendency to associate with phospholipids (45). sTF retains high-affinity binding to factor VIIa, with a K_d determined from solution-phase binding of 0.6–2.0 nM for the bacterially derived protein (36,38).

The k_{cat} of the sTF-VIIa complex toward factor X has been reported to be far lower than that of TF-VIIa, suggesting that sTF-VIIa is intrinsically a much less active enzyme than TF-VIIa (45). We recently found, however, that when high concentrations of phospholipids are employed—to obtain nearly quantitative binding of sTF-VIIa to the vesicles—sTF-VIIa exhibits a k_{cat} very similar to that of TF-VIIa (Fiore, Neuenschwander, and Morrissey, submitted for publication). High phospholipid concentrations are required because factor VIIa (and hence sTF-VIIa) has a weak affinity for phospholipid surfaces (K_d 15 μM) (85). These high phospholipid concentrations are not necessary for wild-type TF, because it is anchored in the membrane. According to this view, sTF is a useful model for the protein-protein interaction component of the TF-VIIa-phospholipid complex and causes similar changes in factor VIIa upon binding.

An early report indicated that sTF is essentially devoid of activity in clotting assays (82). This discrepancy with studies of factor X activation using purified proteins was explained by the later finding that sTF fails to support factor VII activation under conditions in which TF is very active (39). Because the bulk of factor VII in plasma is the inert, zymogen form, that sTF does not support conversion to factor VIIa means that it clots plasma based solely on preexisting factor VIIa (39,61). This led to the development of a sensitive new clotting assay for measuring factor VIIa without interference from factor VII (61). Normal individuals were found to have levels of factor VIIa in their plasma ranging from 0.5 to 8.4 ng/ml, with a mean of 3.6 ng/ml (61), or approximately 0.5% of the total plasma factor VII. Epidemiological studies have implicated elevated plasma factor VII coagulant activity as a risk predictor for ischemic heart disease (86), but plasma factor VIIa could not previously be measured directly, as a result of interference by factor VII. Therefore, it has been controversial whether these results implicate higher levels of total factor VII or specifically of factor VIIa as a risk factor (87). The new assay can now be used to resolve this question.

The basis for the apparent selective loss by sTF of the ability to promote factor VII activation has been investigated. Initial studies demonstrated that both factor VII and VIIa bind to sTF, indicating the selective deficiency of sTF is *not* caused by selective binding to factor VIIa versus VII (39). We recently found that sTF promotes factor VII autoactivation provided conditions permit nearly quantitative binding of sTF-VII and sTF-VIIa to a suitable surface or template, such as polylysine or high concentrations of phospholipid vesicles (Fiore, Neuenschwander, and Morrissey, submitted for publication). Under such conditions, rates of factor VII autoactivation promoted by sTF are similar to those promoted by wild-type TF. The reason sTF fails to promote factor VII autoactivation during normal clotting assays, while promoting factor X activation, is proposed to be a combination of the relatively low affinity of factor VIIa for phospholipid surfaces, along with basic differences in the kinetics of factor X activation versus factor VII autoactivation. Specifically, factor X activation is first order in factor VIIa concentration, and factor VII autoactivation is second order in factor VII(a) concentration, the latter tending greatly to amplify the effects of low-affinity binding to the phospholipid surface.

C. Other Site-Directed Mutants

Structural features other than the transmembrane domain of TF have been modified and/or deleted to investigate their role in the function of this molecule. TF is inactivated by reducing agents, indicating a requirement for the integrity of at least one of the two disulfide bonds in the extracellular domain. Accordingly, site-directed mutagenesis showed that the more amino-terminal cysteine pair is dispensable for function, and mutation of the more carboxyl-terminal cysteine pair to serines disrupts TF function (88). However, the conversion of cysteine 186 to serine fortuitously generates a new signal for N-linked carbohydrate addition, and an ectopic carbohydrate chain could alter function in a manner unrelated to the disrupted disulfide bond. Indeed, general derangements in glycosylation were observed, but the report of this mutant neither pointed out the existence of the ectopic glycosylation signal nor reported whether the site contained carbohydrate (88). With this caveat in mind, the most likely explanation for these results is that the carboxyl-terminal disulfide-bonded loop of TF is essential for function but the amino-terminal loop is not.

Based on the observation that chemical reagents that react with lysine inactivate TF, Roy et al. (89) identified two adjacent reactive lysine residues in the TF extracellular domain (lysine 165 and lysine 166). Mutating either of these lysines to alanine resulted in decreased TF activity; mutating both lysines resulted in loss of function (89). Because mutating either lysine singly resulted in mutants with increased dependence on phosphatidyl-serine for factor X activation, these authors postulated that lysines 165 and 166 of TF may interact with negatively charged phospholipid head groups. Ruf et al. (90) expressed the same double mutant and reported the intriguing finding that the mutant retained normal cofactor function for factor VIIa amidolytic activity but had an increased K_m for factor X activation. Another study of alanine substitution mutants in the region of amino acids 157–167 of TF (which includes lysines 165 and 166) reported similar differential effects on the activation of factor X versus small peptide chromogenic substrates (91). Ruf et al. (91) proposed that this region of TF represents a factor X binding site that is disrupted when lysine 165 or 166 residues are mutated. Alternatively, the conformation of factor VIIa may be different when bound to mutant versus wild-type TF, such that its ability to activate macromolecular substrates is reduced. Loss of function mutants are inherently difficult to interpret because mutating a given amino acid may alter the folding of portions of the protein distal to the site of mutation. Demonstrating that the mutant protein retains binding to an antibody that does not bind the denatured protein (92) indicates only that the epitope for that specific antibody is intact, not that the overall protein structure is the same as wild type. It should be noted that all these studies have employed either whole cells expressing mutant TF or crude cell lysates containing the mutant TF molecules (88–92). Biophysical or biochemical studies employing purified TF with mutations in this region have yet to be reported.

The cytoplasmic domain has been deleted without any obvious effect on coagulant function (82). Thus, the cytoplasmic domain of TF does not play an essential role in factor VIIa binding or cofactor function, although it is possible that it has some other cellular role.

The transmembrane domain of TF has been replaced by the transmembrane domain of human immunodeficiency virus gp41 protein (93) and also with a phosphatidylinositol membrane anchor (82); both mutants retained coagulant activity. These results indicate that although membrane anchoring of TF is essential for full function, the nature of the membrane anchor is not important. The propensity of TF to form dimers and higher multimers within phospholipid membranes has been shown, using chimeric molecules, to be a property of the transmembrane domain (82). However, oligomerization is dispensable for TF coagulant activity, because mutants that have lost the ability to oligomerize are functional in clotting (82).

V. SUMMARY AND FUTURE PROSPECTS

Although spontaneously occurring mutations in TF have not been found to date, site-directed mutagenesis of TF has begun to provide insights into the structure-function relationship of this molecule. The isolated extracellular domain of TF (sTF) retains high affinity binding to factor VIIa and also cofactor function. Much of the differences in enzymatic activities of sTF-VIIa versus wild-type TF-VIIa can be attributed to the enzyme kinetic consequences of weak association of sTF-VIIa with phospholipid surfaces compared with membrane anchoring of TF-VIIa. These differences affect the conversion of factor VII to VIIa to a far greater degree than the activation of factor X. This unique property of sTF has permitted the development of a highly sensitive and specific means of quantifying trace levels of factor VIIa in plasma, which may have clinical significance.

Site-directed mutagenesis of other parts of TF have shown that a sequence in the transmembrane domain is essential for dimerization of TF, although this property is dispensable for coagulant activity. Also, the thioester-linked fatty acid and the entire cytoplasmic domain of TF is dispensable for function. Finally, a region in the extracellular domain of TF is dispensable for factor VIIa binding but is required for efficient activation of macromolecular substrates by the TF-VIIa complex.

Hereditary deficiencies in TF have not been reported, probably because of the relative difficulty of quantifying TF levels in patients compared with quantifying levels of the soluble plasma proteins. Although a total lack of TF is probably incompatible with life, less severe deficiencies in TF may be present in individuals with otherwise inexplicable bleeding tendencies. Furthermore, it is conceivable that individuals prone to overexpression of TF exhibit thrombotic tendencies. Screening for genetic alterations in TF by activity or antigen assays would be difficult, because TF is normally restricted to the tissues (and even within tissues it is expressed to varying degrees on different cell types). The coding sequence of the TF gene is not large, so it is possible that screening for mutations in the TF gene could be accomplished using approaches based on the polymerase chain reaction.

REFERENCES

1. Bach RR. Initiation of coagulation by tissue factor. CRC Crit Rev Biochem 1988; 23:339–368.
2. Drake TA, Morrissey JH, Edgington TS. Selective cellular expression of tissue factor in human tissues: implications for disorders of hemostasis and thrombosis. Am J Pathol 1989; 134:1087–1097.

3. Fleck RA, Rao LVM, Rapaport SI, Varki N. Localization of human tissue factor antigen by immunostaining with monospecific, polyclonal anti-human tissue factor antibody. Thromb Res 1990; 59:421–437.
4. Wilcox JN, Smith KM, Schwartz SM, Gordon D. Localization of tissue factor in the normal vessel wall and in the atherosclerotic plaque. Proc Natl Acad Sci USA 1989; 86:2839–2843.
5. Drake TA, Ruf W, Morrissey JH, Edgington TS. Functional tissue factor is entirely cell surface expressed on lipopolysaccharide-stimulated human blood monocytes and a constitutively tissue factor-producing neoplastic cell line. J Cell Biol 1989; 109:389–395.
6. Bach R, Rifkin DB. Expression of tissue factor procoagulant activity: regulation by cytosolic calcium. Proc Natl Acad Sci USA 1990; 87:6995–6999.
7. Rodgers GM, Broze GJ Jr, Shuman MA. The number of receptors for factor VII correlates with the ability of cultured cells to initiate coagulation. Blood 1984; 63:434–438.
8. Fair DS, MacDonald MJ. Cooperative interaction between factor VII and cell surface-expressed tissue factor. J Biol Chem 1987; 262:11692–11698.
9. Ploplis VA, Edgington TS, Fair DS. Initiation of the extrinsic pathway of coagulation. J Biol Chem 1987; 262:9503–9508.
10. Bach R, Gentry R, Nemerson Y. Factor VII binding to tissue factor in reconstituted phospholipid vesicles: induction of cooperatively by phosphatidylserine. Biochemistry 1986; 25: 4007–4020.
11. Hagen FS, Gray CL, O'Hara P, et al. Characterization of a cDNA coding for human factor VII. Proc Natl Acad Sci USA 1986; 83:2412–2416.
12. Nemerson Y, Repke D. Tissue factor accelerates the activation of coagulation factor VII: the role of a bifunctional coagulation cofactor. Thromb Res 1985; 40:351–358.
13. Rao LVM, Rapaport SI. Activation of factor VII bound to tissue factor: a key early step in the tissue factor pathway of blood coagulation. Proc Natl Acad Sci USA 1988; 85:6687–6691.
14. Nemerson Y. Tissue factor and hemostasis. Blood 1988; 71:1–8.
15. Østerud B, Rapaport SI. Activation of factor IX by the reaction product of tissue factor and factor VII: additional pathway for initiating blood coagulation. Proc Natl Acad Sci USA 1977; 74:5260–5264.
16. Carson SD, Ross SE, Bach R, Guha A. An inhibitory monoclonal antibody against human tissue factor. Blood 1987; 70:490–493.
17. Morrissey JH, Fair DS, Edgington TS. Monoclonal antibody analysis of purified and cell-associated tissue factor. Thromb Res 1988; 52:247–261.
18. Edwards RL, Rickles FR. Macrophage procoagulants. Prog Hemost Thromb 1984; 183–209.
19. Morrissey JH, Fakhrai H, Edgington TS. Molecular cloning of the cDNA for tissue factor, the cellular receptor for the initiation of the coagulation protease cascade. Cell 1987; 50:129–135.
20. Spicer EK, Horton R, Bloem L, et al. Isolation of cDNA clones coding for human tissue factor: primary structure of the protein and cDNA. Proc Natl Acad Sci USA 1987; 84:5148–5152.
21. Scarpati EM, Wen D, Broze G Jr, et al. Human tissue factor: cDNA sequence and chromosome localization of the gene. Biochemistry 1987; 26:5234–5238.
22. Fisher KL, Gorman CM, Vehar GA, O'Brien DP, Lawn RM. Cloning and expression of human tissue factor cDNA. Thromb Res 1987; 48:89–99.
23. Mann KG, Jenny RJ, Krishnaswamy S. Cofactor proteins in the assembly and expression of blood clotting enzyme complexes. Annu Rev Biochem 1988; 57:915–956.
24. Paborsky LR, Harris RJ. Post-translational modifications of recombinant human tissue factor. Thromb Res 1990; 60:367–376.
25. Bach R, Konigsberg WH, Nemerson Y. Human tissue factor contains thioester-linked palmitate and stearate on the cytoplasmic half-cystine. Biochemistry 1988; 27:4227–4231.

26. Morrissey JH, Revak D, Tejada P, Fair DS, Edgington TS. Resolution of monomeric and heterodimeric forms of tissue factor, the high-affinity cellular receptor for factor VII. Thromb Res 1988; 50:481–493.
27. Carson SD, Ross SE, Gramzinski RA. Protein co-isolated with human tissue factor impairs recovery of activity. Blood 1988; 71:520–523.
28. Zioncheck TF, Roy S, Vehar GA. The cytoplasmic domain of tissue factor is phosphorylated by a protein kinase C-dependent mechanism. J Biol Chem 1992; 267:3561–3564.
29. Broze GJ Jr. Binding of human factor VII and VIIa to monocytes. J Clin Invest 1982; 70: 526–535.
30. Sakai T, Lund-Hansen T, Paborsky L, Pedersen AH, Kisiel W. Binding of human factors VII and VIIa to a human bladder carcinoma cell line (J82): implications for the initiation of the extrinsic pathway of blood coagulation. J Biol Chem 1989; 264:9980–9988.
31. Le DT, Rapaport SI, Rao LVM. Relations between factor VIIa binding and expression of factor VIIa/tissue factor catalytic activity on cell surfaces. J Biol Chem 1992; 267:15447–15454.
32. Rao LVM, Tait JF, Hoang AD. Binding of annexin V to a human ovarian carcinoma cell line (OC-2008): contrasting effects on cell surface factor VIIa/tissue factor activity and prothrombinase activity. Thromb Res 1992; 67:517–531.
33. Toomey JR, Smith KJ, Stafford DW. Localization of the human tissue factor recognition determinant of human factor VIIa. J Biol Chem 1991; 266:19198–19202.
34. Ruf W, Kalnik MW, Lund-Hansen T, Edgington TS. Characterization of factor VII association with tissue factor in solution: high and low affinity calcium binding sites in factor VII contribute to functionally distinct interactions. J Biol Chem 1991; 266:15719–15725.
35. Krishaswamy S. The interaction of human factor VIIa with tissue factor. J Biol Chem 1992; 267:23696–23706.
36. Fiore MM, Neuenschwander PF, Morrissey JH. An unusual antibody that blocks tissue factor/ factor VIIa function by inhibiting cleavage only of macromolecular substrates. Blood 1992; 80:3127–3134.
37. Jesty J. The determination of enzyme-cofactor dissociation constants by kinetic methods: a correction. Thromb Res 1988; 50:745–746.
38. Waxman E, Ross JBA, Laue TM, et al. Tissue factor and its extracellular soluble domain: the relationship between intermolecular association with factor VIIa and enzymatic activity of the complex. Biochemistry 1992; 31:3998–4003.
39. Neuenschwander PF, Morrissey JH. Deletion of the membrane anchoring region of tissue factor abolishes autoactivation of factor VII but not cofactor function: analysis of a mutant with a selective deficiency in activity. J Biol Chem 1992; 267:14477–14482.
40. Fair DS. Quantitation of factor VII in the plasma of normal and warfarin-treated individuals by radioimmunoassay. Blood 1983; 62:784–791.
41. Wildgoose P, Foster D, Schiodt J, Wiberg FC, Birktoft JJ, Petersen LC. Identification of a calcium binding site in the protease domain of human blood coagulation factor VII: evidence for its role in factor VII-tissue factor interaction. Biochemistry 1993; 32:114–119.
42. Bom VJJ, Bertina RM. The contributions of Ca^{2+}, phospholipids and tissue-factor apoprotein to the activation of human blood-coagulation factor X by activated factor VII. Biochem J 1990; 265:327–336.
43. Komiyama Y, Pedersen AH, Kisiel W. Proteolytic activation of human factors IX and X by recombinant human factor VIIa: effects of calcium, phospholipids, and tissue factor. Biochemistry 1990; 29:9418–9425.
44. Pedersen AH, Nordfang O, Norris F, et al. Recombinant human extrinsic pathway inhibitor: production, isolation, and characterization of its inhibitory activity on tissue factor-initiated coagulation reactions. J Biol Chem 1990; 265:16786-16793.
45. Ruf W, Rehemtulla A, Morrissey JH, Edgington TS. Phospholipid-independent and -dependent interactions required for tissue factor receptor and cofactor function. J Biol Chem 1991; 266:2158–2166 (correction in J Biol Chem 1991; 266:16256).

46. Pedersen AH, Lund-Hansen T, Komiyama Y, Petersen LC, Oestergard PB, Kisiel W. Inhibition of recombinant human blood coagulation factor VIIa amidolytic and proteolytic activity by zinc ions. Thromb Haemost 1991; 65:528–534.
47. Lawson JH, Butenas S, Mann KG. The evaluation of complex-dependent alterations in human factor VIIa. J Biol Chem 1992; 267:4834–4843.
48. Broze GJ Jr, Girard TJ, Novotny WF. The lipoprotein-associated coagulation inhibitor: why do hemophiliacs bleed? Prog Hemost Thromb 1991; 10:243–268.
49. Lawson JH, Butenas S, Ribarik N, Mann KG. Complex-dependent inhibition of factor VIIa by antithrombin III and heparin. J Biol Chem 1993; 268:767–770.
50. Radcliffe R, Nemerson Y. Activation and control of factor VII by activated factor X and thrombin: isolation and characterization of a single chain form of factor VII. J Biol Chem 1975; 250:388–395.
51. Kisiel W, Fujikawa K, Davie EW. Activation of bovine factor VII (proconvertin) by factor XII_a (activated hageman factor). Biochemistry 1977; 16:4189–4194.
52. Seligsohn U, Østerud B, Brown SF, Griffin JH, Rapaport SI. Activation of human factor VII in plasma and in purified systems: roles of activated factor IX, kallikrein and activated factor XII. J Clin Invest 1979; 64:1056–1065.
53. Masys DR, Bajaj SP, Rapaport SI. Activation of human factor VII by activated factors IX and X. Blood 1982; 60:1143–1150.
54. Kondo S, Kisiel W. Regulation of factor VIIa in plasma: evidence that antithrombin III is the sole plasma protease inhibitor of human factor VIIa. Thromb Res 1987; 46:325.
55. Seligsohn U, Kasper CK, Østerud B, Rapaport SI. Activated factor VII: presence in factor IX concentrate and persistence in the circulation after infusion. Blood 1978; 53:828.
56. Miller BC, Hultin MB, Jesty J. Altered factor VII activity in hemophilia. Blood 1985; 65: 845–849.
57. Rapaport SI. Regulation of the tissue factor pathway. Ann NY Acad Sci 1991; 614:51–62.
58. Nakagaki T, Foster DC, Berkner KL, Kisiel W. Initiation of the extrinsic pathway of blood coagulation: evidence for the tissue factor dependent autoactivation of human coagulation factor VII. Biochemistry 1991; 30:10819–10824.
59. Yamamoto M, Nakagaki T, Kisiel W. Tissue factor-dependent autoactivation of human blood coagulation factor VII. J Biol Chem 1992; 267:19089–19094.
60. Pedersen AH, Lund-Hansen T, Bisgaard-Frantzen H, Olsen F, Petersen LC. Autoactivation of human recombinant coagulation factor VII. Biochemistry 1989; 28:9331–9336.
61. Morrissey JH, Macik BG, Neuenschwander PF, Comp PC. Quantitation of activated factor VII levels in plasma using a tissue factor mutant selectively deficient in promoting factor VII activation. Blood 1993; 81:734–744.
62. Ryan J, Geczy C. Coagulation and the expression of cell-mediated immunity. Immunol Cell Biol 1987; 65:127–139.
63. Morrissey JH, Gregory SA, Mackman N, Edgington TS. Tissue factor regulation and gene organization. In: MacLean N, ed. Oxford surveys on eukaryotic genes, Vol. 6. Oxford University Press, Oxford 1989:67–83.
64. Gregory SA, Morrissey JH, Edgington TS. Regulation of tissue factor gene expression in the monocyte procoagulant response to endotoxin. Mol Cell Biol 1989; 9:2752–2755.
65. Brand K, Fowler BJ, Edgington TS, Mackman N. Tissue factor mRNA in THP-1 monocytic cells is regulated at both transcriptional and posttranscriptional levels in response to lipopolysaccharide. Mol Cell Biol 1991; 11:4732–4738.
66. Mackman N, Brand K, Edgington TS. Lipopolysaccharide-mediated transcriptional activation of the human tissue factor gene in THP-1 monocytic cells requires both activator protein 1 and nuclear factor κB binding sites. J Exp Med 1991; 174:1517–1526.
67. Crossman DC, Carr DP, Tuddenham EGD, Pearson JD, McVey JH. The regulation of tissue factor mRNA in human endothelial cells in response to endotoxin or phorbol ester. J Biol Chem 1990; 265:9782–9787.

68. Ahern SM, Miyata T, Sadler JE. Regulation of human tissue factor expression by mRNA turnover. J Biol Chem 1993; 268:2154–2159.
69. Hartzell S, Ryder K, Lanahan A, Lau LF, Nathans D. A growth factor-responsive gene of murine BALB/c 3T3 cells encodes a protein homologous to human tissue factor. Mol Cell Biol 1989; 9:2567–2573.
70. Ranganathan G, Blatti SP, Subramaniam M, Fass DN, Maihle NJ, Getz MJ. Cloning of murine tissue factor and regulation of gene expression by transforming growth factor type β_1. J Biol Chem 1991; 266:496–501.
71. Mackman N, Fowler BJ, Edgington TS, Morrissey JH. Functional analysis of the human tissue factor promoter and induction by serum. Proc Natl Acad Sci USA 1990; 87:2254–2258.
72. Kao F-T, Hartz J, Horton R, Nemerson Y, Carson SD. Regional assignment of human tissue factor gene (F3) to chromosome 1p21-p22. Somat Cell Mol Genet 1988; 14:407–410.
73. Mackman N, Norrissey JH, Fowler B, Edgington TS. Complete sequence of the human tissue factor gene, a highly regulated cellular receptor that initiates the coagulation protease cascade. Biochemistry 1989; 28:1755–1762.
74. Van der Logt CPE, Reitsma PH, Bertina RM. Alternative splicing is responsible for the presence of two tissue factor mRNA species in LPS stimulated human monocytes. Thromb Haemost 1992; 67:272–276.
75. Bird AP. CpG-rich islands and the function of DNA methylation. Nature 1986; 321:209–213.
76. Scarpati EM, Sadler JE, O'Connell P, et al. Identification and mapping of RFLPs for human tissue factor (HTF) to chromosome 1p. Nucleic Acids Res 1987; 15:9098.
77. Takayenoki Y, Muta T, Miyata T, Iwanaga S. cDNA and amino acid sequences of bovine tissue factor. Biochem Biophys Res Commun 1991; 181:1145–1150.
78. Andrews BS, Rehemtulla A, Fowler BJ, Edgington TS, Mackman N. Conservation of tissue factor primary sequence among three mammalian species. Gene 1991; 98:265–269.
79. Kadish JL, Wenc KM, Dvorak HF. Tissue factor activity of normal and neoplastic cells: quantitation and species specificity. J Natl Cancer Inst 1983; 70:551–557.
80. Bazan JF. Structural design and molecular evolution of a cytokine receptor superfamily. Proc Natl Acad Sci USA 1990; 87:6934–6938.
81. Ruf W, Edgington TS. Two sites in the tissue factor extracellular domain mediate the recognition of the ligang factor VIIa. Proc Natl Acad Sci USA 1991; 88:8430–8434.
82. Paborsky LR, Caras IW, Fisher KL, Gorman CM. Lipid association, but not the transmembrane domain, is required for tissue factor activity: substitution of the transmembrane domain with a phosphatidylinositol anchor. J Biol Chem 1991; 266:21911–21916.
83. Shigematsu Y, Miyata T, Higashi S, Miki T, Sadler JE, Iwanaga S. Expression of human soluble tissue factor in yeast and enzymatic properties of its complex with factor VIIa. J Biol Chem 1992; 267:21329–21337.
84. Rezaie AR, Fiore MM, Neuenschwander PF, Esmon CT, Morrissey JH. Expression and purification of a soluble tissue factor fusion protein with an epitope for an unusual calcium-dependent antibody. Prot Express Purif 1992; 3:453–460.
85. Nelsestuen GL, Kisiel W, DiScipio RG. Interaction of vitamin K dependent proteins with membranes. Biochemistry 1978; 17:2134–2138.
86. Hultin MB. Fibrinogen and factor VII as risk factors in vascular disease. Prog Haemost Thromb 1991; 10:215–241.
87. Mann KG. Factor VII assays, plasma triglyceride levels, and cardiovascular disease risk. Arteriosclerosis 1989; 9:783–784.
88. Rehemtulla A. Ruf W, Edgington TS. The integrity of the cysteine 186-cysteine 209 bond of the second disulfide loop of tissue factor is required for binding of factor VII. J Biol Chem 1991; 266:10294–10299.
89. Roy S, Hass PE, Bourell JH, Henzel WJ, Vehar GA. Lysine residues 165 and 166 are essential for the cofactor function of tissue factor. J Biol Chem 1991; 266:22063–22066.

90. Ruf W, Miles DJ, Rehemtulla A, Edgington TS. Cofactor residues lysine 165 and 166 are critical for protein substrate recognition by the tissue factor-factor VIIa protease complex. J Biol Chem 1992; 267:6375–6381.
91. Ruf W, Miles DJ, Rehemtulla A, Edgington TS. Tissue factor residues 157–167 are required for efficient proteolytic activation of factor X and factor VII. J Biol Chem 1992; 267:22206–22210.
92. Rehemtulla A, Ruf W, Miles DJ, Edgington TS. The third Trp-Lys-Ser (WKS) tripeptide motif in tissue factor is associated with a function site. Biochem J 1992; 282:737–740.
93. Roy S, Paborsky LR, Vehar GA. Self-association of tissue factor as revealed by chemical cross-linking. J Biol Chem 1991; 266:4665–4668.

6

Factor V

Thomas L. Ortel and William H. Kane
Duke University Medical Center, Durham, North Carolina

Frank G. Keller
West Virginia University Children's Hospital, Morgantown, West Virginia

I. INTRODUCTION

Factor V was discovered 50 years ago by Owren (1), who identified a labile factor in normal human plasma that could correct the thrombin generation defect in a patient with a bleeding disorder. Subsequent work by many investigators established that activated factor V was a nonenzymatic cofactor that was essential for formation of the prothrombinase complex, which consists of factor Xa, factor Va, calcium, and a phospholipid or cellular surface (2). Characterization of factor V was hampered for many years because the circulating form of factor V has little or no procoagulant activity and the protein is exquisitely sensitive to proteolytic activation and inactivation. Seminal work by Nesheim, Mann, and coworkers led to the isolation and characterization of single-chain factor V from bovine plasma (3). The structure of the activated cofactor was first elucidated by Esmon (4), who demonstrated that thrombin-activated bovine factor Va was a heterodimer composed of a heavy chain and a light chain derived from a single-chain precursor by limited proteolysis. Most studies on factor V have been carried out with the bovine protein, because the factor V present in bovine plasma is considerably more stable than the human protein and is present at a 5- to 10-fold higher concentration in plasma (5–7). The availability of large quantities of bovine factor V allowed detailed study of the kinetics of the prothrombinase complex and determination of the association constants for the binding of factor V to each of the other components of the complex (8). A further advance in our understanding of this protein was achieved with the cloning of the cDNAs encoding human factor V and determination of the primary sequence of the molecule (9–11). These data demonstrated that factor V was a mosaic protein, composed of several domains, or ''modules,'' and that it was closely related to coagulation factor VIII, the defective protein in patients with hemophilia A (12,13).

Regulation of the prothrombinase complex is relevant to human disease processes because the generation of thrombin plays an important role in both hemostasis and thrombosis (14,15). Additionally, the role of factor V as a nonenzymatic cofactor in the prothrombinase complex has become a general paradigm for the role of factor VIII and other nonenzymatic cofactors in blood coagulation (8,16). This chapter focuses on recent developments in the molecular biology of coagulation factor V, including the cloning of the human factor V gene and the bovine factor V cDNA and the identification of additional members of this gene family. Additionally, we review recent data concerning structure-function relationships in this protein, including those determined by site-directed mutagenesis of recombinant factor V.

II. GENOMIC ORGANIZATION AND cDNA SEQUENCE

A. Organization of the Factor V Gene

The gene for human factor V has been localized to human chromosome 1q21–25 (17). Recent gene mapping studies have located the gene within a 300 kilobase (kb) region that also includes the genes for the selectin family of leukocyte adhesion molecules (18). Our group recently isolated and characterized genomic clones encoding human factor V from lung fibroblast and monocyte λ phage genomic libraries (19). The human factor V gene spans >80 kb of DNA and consists of 25 exons (Fig. 1). The exons range in size from 72 to 286 base pairs (bp), with the exception of exon 13, which spans 2820 bp. The 24 introns in the factor V gene range in size from 0.4 to ≥11 kb. The genomic clones include ~5.4 kb DNA 5′ to exon 1 and ~7 kb 3′ to exon 25. Cross-hybridization experiments indicate that a portion of intron 2 is not contained in our genomic clones and that the size of this intron is >10 kb (19). At the present time, there is no evidence for alternative splicing of factor V transcripts, and the mechanisms responsible for the control of factor V gene transcription and translation remain unknown.

B. Factor V cDNA and the Predicted Amino Acid Sequence

Human factor V cDNAs have been isolated from adult liver, fetal liver, and HepG2 cDNA libraries (9–11). The cDNA for human factor V is approximately 7 kb in size. The sequence characterized in our laboratory includes 76 nucleotides of 5′-untranslated sequence, 84 nucleotides that code for a leader peptide of 28 amino acids, 6588 nucleotides that code for a mature protein of 2196 amino acids, a stop codon, and 155 nucleotides of 3′ noncoding sequence (9,10). Analysis of the predicted amino acid sequence revealed that the protein contains several types of internal repeats organized with the following domain structure: A1-A2-B-A3-C1-C2. The A-type domains contain ~350 amino acids each, the B domain contains ~836 amino acids, and the C-type domains contain ~150 amino acids each. Recently, hepatic cDNA clones encoding bovine factor V were characterized by Guinto et al. (20). The predicted amino acid sequence for bovine factor V is 85% identical to the human sequence except in the B domain, where there is only 59% amino acid identity (see later discussion).

Human factor V contains 19 cysteine residues that occur in similar locations in the bovine protein, except a single free cysteine in the B domain of human factor V that is missing from the bovine protein (Fig. 1). The disulfide bonding of the cysteines in the light chain of bovine factor Va was recently determined by Xue et al. (21). Based on the conservation of the human and bovine sequences, these data indicate that the light chain of human factor V contains three disulfide-linked cysteines, corresponding to Cys^{1697} and

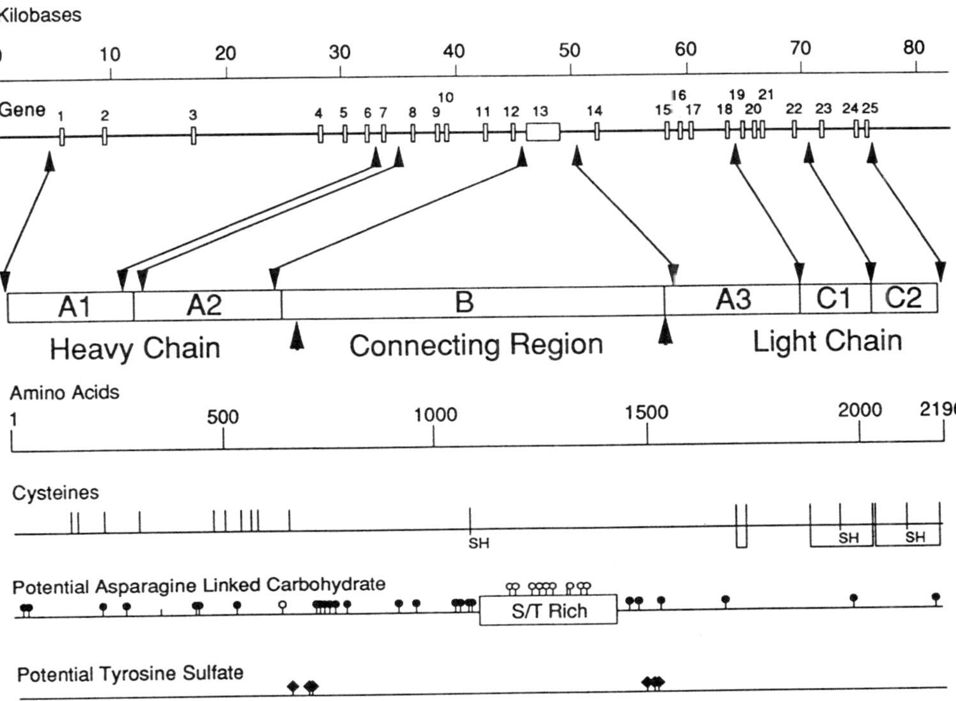

Figure 1 Features of the factor V gene and predicted amino acid sequence. The top line shows the scale for factor V genomic sequences in kilobases. The second line shows the intron-exon organization of the factor V gene, with the exons represented by the open boxes. The number of each exon is indicated, and the intron sequences are represented by the thin lines. The domain structure encoded by the factor V cDNA is shown by the boxes labeled A1, A2, B, A3, C1, and C2. The correlation between the exon-intron structure of the factor V gene and the protein domains are shown by the lines with arrows at each end. Domain A1 is encoded by exons 1–7; the A1-A2 junction is encoded by exon 7; domain A2 is encoded by exons 7–13; the A2-A3 junction, including the B domain, is encoded by exon 13; the A3 domain is encoded by exons 13–18; the C1 domain is encoded by exons 19–22; and the C2 domain is encoded by exons 23–25. The heavy chain of factor Va contains domains A1 and A2, and the light chain of factor Va contains domains A3, C1, and C2. The connecting region includes the B domain. The thrombin cleavage sites following Arg_{709} and Arg_{1545}, which release the heavy-chain and light-chain, fragments, respectively, are indicated by the arrowheads. The scale in amino acids is shown in the fourth line. The fifth line shows the positions of the cysteine residues in the predicted amino acid sequence for human factor V, indicated by the vertical lines. The predicted sulfhydryl groups in human factor V are indicated by −SH, and the disulfide bonds are indicated by the connecting horizontal lines. The sixth line indicates the presence of potential asparagine-linked carbohydrate. Strong acceptor sites are indicated by the solid circles and vertical lines; weak acceptor sites are indicated by the open circles and vertical lines. Acceptor sites that are known not to be utilized are indicated by the two vertical lines. A region within the B domain spanning >200 amino acids that contains >30% serine and threonine is indicated by the box labeled S/T Rich. The seventh line shows the location of potential tyrosine sulfation sites, which are indicated by the solid diamonds and vertical lines.

Cys_{1723}, Cys_{1879}, and Cys_{2033}, and Cys_{2038} and Cys_{2193}. The two remaining cysteines, Cys_{1960} and Cys^{2113}, exist as free sulfhydryls. Interestingly, only one of the two free sulfhydryl groups in the bovine factor V light chain is reactive with dithiobis(nitrobenzoic acid) (22).

Factor V is a glycoprotein (6,23) that contains both N-linked and O-linked carbohydrate moieties. The primary sequence of human factor V encodes 26 strong (N-*X*-S/T) and 11 weak (N-*X*-S/T-P) potential N-linked oligosaccharide attachment sites (Fig. 1) (24). Many, but not all, of these sites are clustered in the B domain. The use of individual asparagine acceptor sites has not been determined except for amino acid residues 354 and 1310, which are not glycosylated. Of the weak acceptor sites located in the B domain, 9 are not conserved in bovine factor V. The B domain of human factor V also contains O-linked carbohydrate, as judged by binding to the lectin jacalin (25). Although it is not possible to predict the exact location of O-linked glycosylation sites based on primary sequence data (26), the B domain contains a region of approximately 200 amino acids that is >30% serine and threonine, which appears to be a good candidate for O-linked glycosylation.

Factor V has also been shown to contain tyrosine sulfate (25). Consensus tyrosine sulfation sites are clustered near thrombin cleavage sites that release the heavy and light chains during activation of factor V (Fig. 1). Tyrosine sulfate has been implicated in the interaction of thrombin with a number of proteins, including the thrombin inhibitor hirudin (27). The exact locations of tyrosine sulfate in human factor V and the role of these residues, if any, in regulating activity of the molecule requires further characterization (28). Last, Kalafatis et al. (29) recently demonstrated that bovine factor V can be phosphorylated by a casein kinase II-like kinase expressed in activated platelets. This modification accelerated the inactivation of the molecule by activated protein C. Their data localized the phosphorylation site to Ser_{690} in the bovine molecule. Human factor V contains 34 potential serine phosphorylation sites, but the site corresponding to Ser_{690} is not conserved (30) and it is not known whether the human protein is phosphorylated.

C. Homologies with Other Proteins

The primary sequence of human factor V is 40% identical to that of human coagulation factor VIII, except in the B domain, where there is no similarity (12,13). The exon-intron organization of the genes encoding these coagulation factors are almost identical (19,31). The genes for factors V and VIII are composed of 25 and 26 exons, respectively. Exon 5 of factor V corresponds to exons 5 and 6 of factor VIII, accounting for the single extra exon present in factor VIII. This can be explained by either the gain or the loss of an intron after the divergence of these two genes. The gene for factor V spans ≥80 kb, whereas the gene for factor VIII is considerably larger at ~180 kb. This difference is largely because six of the introns in the factor V gene are much smaller than the corresponding introns in the factor VIII gene. Inspection of the aligned amino acid sequences and the corresponding genomic nucleotide sequences for factor V and factor VIII reveals that the exon-intron boundaries occur at precisely the same location in the aligned sequence in 21 of 24 cases. Three exon-intron boundaries that are not precisely conserved, as judged by the protein alignment, code for regions of the proteins where there is significantly weaker amino acid homology. This suggests that the discrepancy may be caused by either inaccuracies in the protein alignment or the phenomenon of intron drift (32). Last, although the mRNA of factor VIII is considerably larger than the message for factor V, this is primarily a result of the 1.8 kb of 3′-untranslated sequence in factor VIII. This striking similarity in genomic organization and cDNA sequences confirms that the genes encoding these two proteins have evolved through gene duplication.

Most of the proteins involved in blood coagulation appear to be mosaic proteins, composed of individual domains or "modules" that are shared with a variety of other proteins (33). This is also the case for factor V and factor VIII, which contain three A-type domains, approximately 350 amino acids in length, which are ~30% identical to each other and to the triplicated A-type domains in the plasma protein ceruloplasmin (Fig. 2) (34). Ceruloplasmin is a blue, multicopper oxidase that binds five to six copper atoms per molecule of protein in spectroscopically defined sites that differ in their co-ordination environments (35). Mann et al. (36) demonstrated that bovine factor V contains one copper ion, although the functional importance of this finding is unknown. The proposed copper binding ligands present in ceruloplasmin are not conserved in factor V, however, and thus the location of the copper binding site is not known (9–11).

The tandem C-type domains in factor V and factor VIII are ~150 amino acids in length and are homologous to duplicated domains present in human (BA-46) and murine (MFG-E8) milk fat globule proteins (37,38) and in a putative neuronal cell adhesion molecule (A5 antigen) from *Xenopus laevis* (Fig. 2) (39). The sequences of these C-type domains are ~30–50% identical, the cysteines at the beginning and end of each domain being conserved in all cases (Fig. 3). With the exception of factor V and factor VIII, genomic clones encoding these proteins have not yet been reported. In both these cases, the exon-intron boundaries in the genomic sequences correspond exactly to the boundaries of the C-type domains in the proteins. The murine protein MFG-E8 consists of a single epidermal growth factor-like domain followed by a proline-rich spacer region and two tandem C-type domains (38). Partial cDNAs for the human protein BA46 also dem-

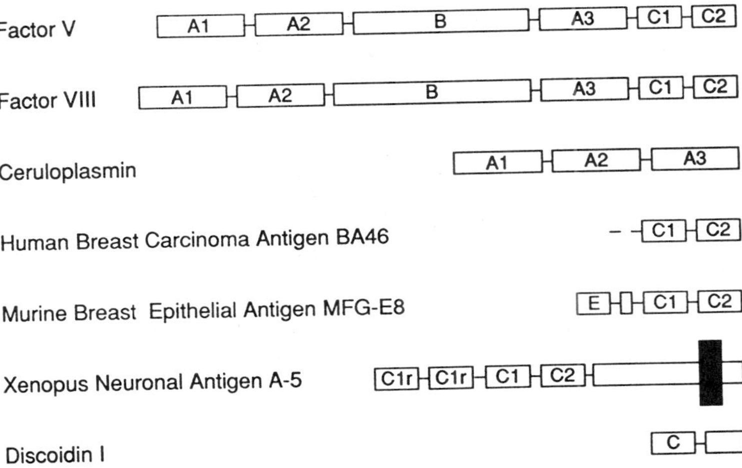

Figure 2 Protein modules in factor V and related proteins. Structural domains in factor V and homologous proteins are indicated by the labeled boxes (see text). The complete sequence for BA46 has not yet been determined. The epidermal growth factor-like domain in MFG-E8 is labeled E. In Reference 39, the domains homologous to complement regulatory proteins (C1r) were labeled a1 and a2 and the C1 and C2 domains were labeled b1 and b2. Discoidin I contains a single C-type domain.

onstrate the presence of tandem C-type domains (37). Northern blotting experiments indicate that the transcripts for MFG-E8 and BA46 are both ~2.1 kb; however, the sequence for the amino terminus of BA46 has not yet been reported. The similarity between domain C2 in human factor V and BA-46 is particularly striking because 44% of the amino acid residues are identical. As a point of comparison, the C2 domains of BA-46 and MFG-E8, both milk fat globule proteins, are only 65% identical. Little is known about the physiological function of the milk fat globule proteins, although recent evidence suggests that the murine protein MFG-E8 binds specifically to phosphatidylserine and may be involved in the process of membrane secretion during the biogenesis of milk fat globules (40). The human protein BA-46 is also isolated from milk fat globules and has been shown to circulate in the plasma of patients with metastatic breast cancer (37). The putative neuronal adhesion molecule, the A5 antigen, also contains domains that are 30–35% identical to the duplicated C-type domains present in factor V. In addition, this protein also contains tandem domains homologous to complement regulatory proteins and is unique among proteins containing C-type domains in that it appears to be anchored to cell membranes via a transmembrane domain. The A5 antigen is selectively expressed in pre- and postsynaptic sites of the visual nervous system and also possibly in the general somatosensory nervous system (39). In addition to the proteins just described, discoidin I, a protein that mediates cell adhesion and migration in the single-cell organism *Dic-*

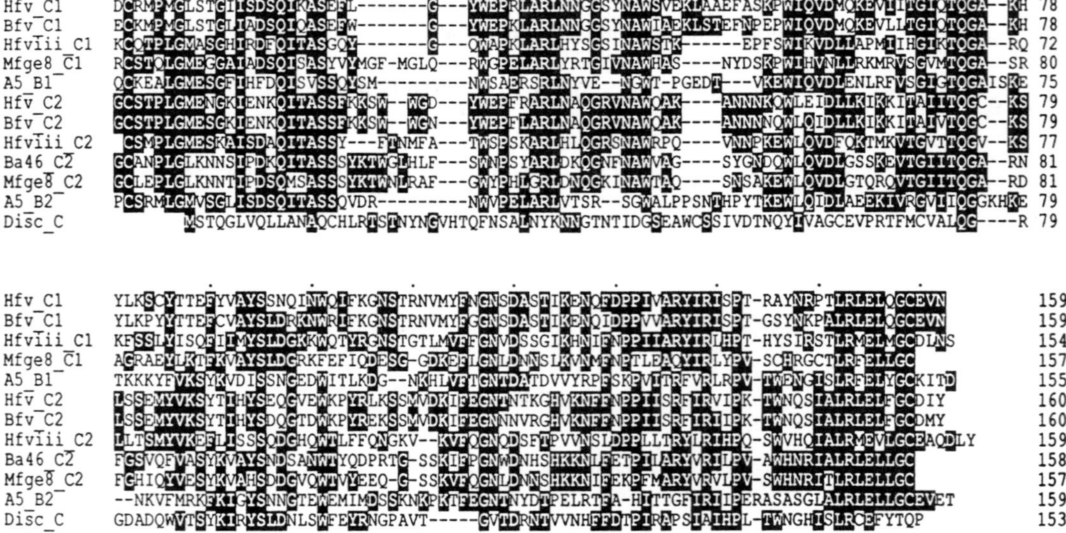

Figure 3 Amino acid sequences of C-type domains. Amino acid sequences were aligned and displayed using the programs Pileup and Psmalign (160). Gaps in the aligned sequences are shown by the dashes. Amino acids agreeing with the consensus are highlighted in black. The amino acid sequences shown here are Hfv_C1, human factor V residues 1878–2036; Bfv_C1, bovine factor V residues 1860–2018; Hfviii_C1, human factor VIII residues 2020–2173; Mfge8_C1, MFG-E8 residues 125–281; A5_B1, *Xenopus* A5 antigen residues 274–428; Hfv_C2, human factor V residues 2037–2196; Bfv_C2, bovine factor V residues 2019–2179; Hfviii_C2, human factor VIII residues 2174–2332; Mfge8_C2, MFG-E8 residues 285–441; Ba46_C2, BA-46 antigen residues 54–217; A5_B2, *Xenopus* A5 antigen, residues 430–588; and discoidin I, residues 1–153.

tyostelium discoideum, possesses a single C-type domain that is 15–20% identical to the other C-type domains (41). This similarity is most apparent in the carboxyl-terminal portion of the C domains (Fig. 3). Discoidin I is a galactose binding lectin that binds to specific cell surface receptors (42) and has also been reported to bind to and agglutinate negatively charged phospholipid vesicles (43). Although much remains to be learned about the structure and function of this family of proteins, one general observation is that they are all involved in interactions with cellular and phospholipid surfaces.

D. Evolution of the Connecting Region (B Domain)

Although the B domain does not appear to be necessary for procoagulant activity (see later), it possesses several interesting structural features. The B domain of human factor V contains two tandem repeats of a 17 amino acid sequence and 31 tandem repeats of a nine amino acid sequence (Fig. 4) (9–11). The consensus sequences for these two types of repeats are SQDTGSPSXMRPWEDXP and [T,N,P]LSPDLSQT. Comparison of the sequences of the human and bovine factor V B domains reveals that there is only 59% sequence identity, and that three gaps are present in the aligned sequences (Fig. 4) (20). Each of the three gaps in the aligned sequences corresponds to regions containing a different number of tandem repeats. First, in the bovine protein, the sequences corresponding to the 29th and a portion of the 30th nine amino acid repeat are absent. Furthermore, the consensus sequence for the bovine tandem nine amino acid repeats, [A,S]LSPD[G,S]Q[T,E], differs significantly from the human consensus sequence. Second, bovine factor V contains only a single copy of the 17 amino acid tandem repeat present in human factor V (Fig. 4). Last, bovine factor V also contains a 14 amino acid tandem repeat with a consensus sequence of XDPTHSTTAPSNRS, which is present as

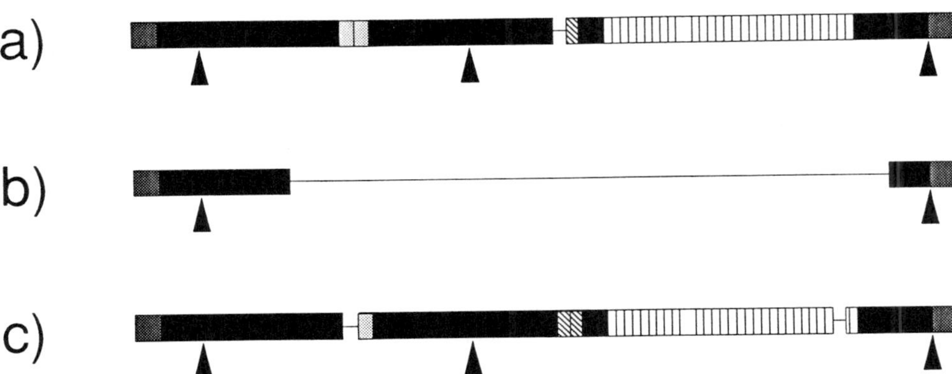

Figure 4 Structural features present in the B domains of human and bovine factor V. (a) The region encoded by exons 13 of the human factor V gene. (b) The corresponding region of recombinant human factor V des $B^{811-1491}$. (c) The corresponding region of bovine factor V. Sequences homologous to ceruloplasmin are indicated by the darkly shaded boxes. The B domains are indicated by the thick black bars. Deleted amino acid sequences are represented by the thin black lines. The 17 amino acid sequences that are duplicated in human factor V are indicated by the lightly shaded boxes. The 14 amino acid sequences that are duplicated in bovine factor V are indicated by the hatched boxes. The nine amino acid sequences that are repeated 31 times in human factor V and 29 times in bovine factor V are indicated by the white boxes.

a single copy in the human molecule (Fig. 4). These data indicate that the duplication of tandem repeats within the connecting region of factor V is a relatively recent event that has continued since the divergence of the bovine and human species. The continued evolution of the nine amino acid tandem repeats suggests there is some selective pressure to maintain these structures in factor V.

The entire B domain of human factor V is encoded within a single exon, exon 13, which also encodes for the carboxyl end of the A2 domain and the amino end of the A3 domain (19). Similarly, the entire B domain of human factor VIII is also encoded in a single large exon (31). These exons are much larger than the reported average size of exons in vertebrate genes of 133 bp (44), exon 13 of factor V spanning 2820 bp and exon 14 of factor VIII spanning 3106 bp. Recently, however, large exons ranging from 1.5 to 8.7 kb in length were described in a number of genes and gene families, including apolipoprotein B (45), the large subunit of RNA polymerase II (46), nuclear hormone receptor genes (47), collagen gene (48), and mucin genes (49,50). As with the exons encoding the A2-A3 domain boundaries, the exons encoding the A1-A2 domain boundaries of factor V and factor VIII also differ in their central portions. Specifically, exon 8 of the factor VIII gene contains a 33 amino acid insert that includes 15 acid residues, which is not present in exon 7 of the factor V gene (19,31). Because activation of factor VIII requires a proteolytic cleavage between the A1 and A2 domains at Arg_{372}-Ser_{373}, whereas activation of factor V does not, this insertion may be important in the regulation of the procoagulant activity of factor VIII.

Because of the large size of the exons encoding the B domains in factor V and factor VIII, we speculate that these exons arose through the insertion of a processed gene derived from reverse transcription of an mRNA. It is possible that the exons encoding the B domains of both factor V and factor VIII are derived from a single primordial cofactor gene. In this case, the lack of sequence similarity between the two cofactors would be a result of divergent evolution. Analysis of the amino acid sequences for factor V and factor VIII from different species suggests that there has been relatively little selective pressure to conserve amino acid sequences in the B domains. Thus the predicted amino acid sequence for the B domain of bovine factor V is only 59% identical to the corresponding sequence of the human protein, whereas there is 85% amino acid sequence identity in the heavy-chain and light-chain regions (20). Similar divergence has been observed between human and porcine factor VIII sequences. Toole et al. (51) reported ~50% amino acid identity in the B domain region, whereas there was 80–85% amino acid identity in the heavy-chain and light-chain regions. The alignment of the porcine and human B domains also included 10 deletions spanning over 200 amino acids in the porcine B domain. In contrast to the factor V B domains, neither the porcine nor the human factor VIII sequences contained tandem repeat structures. Characterization of related genes in primitive species may further clarify the evolution of this family of genes.

III. PROTEIN STRUCTURE, FUNCTION, AND INTERMOLECULAR INTERACTIONS

A. Physical Properties

Factor V circulates in the plasma as a single-chain glycoprotein with an approximate molecular weight of 330,330 kDa (3,5–7). Ultracentrifugation studies of bovine factor

V revealed a high frictional ratio (f/f_{min} = 2.01) and a sedimentation coefficient of 9.19*S*, suggesting that the molecule is highly asymmetrical and rod-like, with an axial ratio as high as 25:1 (3). In contrast, thrombin-activated factor Va, which lacks the large B domain (see later), has a much lower frictional ratio (f/f_{min} = 1.39) (52). Furthermore, Laue et al. (52) found that the frictional coefficient ratio of bovine factor Va was comparable to the ratio for the isolated heavy chain or light chain. These studies suggested that the B domain was responsible for the molecular asymmetry of factor V, and Laue et al. (52) speculated that the B domain extended like a long tail from the more globular heavy and light chains.

To explain the unusual hydrodynamic properties of factor V, several groups have investigated the ultrastructural features of the protein using electron microscopy. These studies have differed significantly in technique and interpretation. The most recent studies by Mosesson and coworkers (53), using scanning transmission electron microscopy and mass analysis, suggest that factor V consists of a globular core 10–12 nm in diameter accounting for ~70% of the mass of the molecule. The remainder of the mass of the molecule extends significantly from the core structure and, in some images, can be seen as an indistinct satellite structure connected to the globular core by a thin stalk up to 35 nm in length. Mass analysis of these images yielded a mass of ~330 kD, which is in good agreement with values obtained by ultracentrifugation and sodium dodecyl sulfate–polyacrylamide gel electrophoresis (SDS-PAGE). Images of thrombin-activated factor Va revealed a globular structure of 8–12 nm in diameter without any peripheral stalk or satellite structures. Mass analysis of these images yielded values of ~180 kD, which is consistent with estimates derived from ultracentrifugation data. Fowler et al. (54) reached similar conclusions based on an analysis of rotary-shadowed images of human factor V. These data also demonstrated core globular domains 10–12 nm in diameter. In approximately 25% of the molecules rod-like tails up to 50 nm in length were observed. In agreement with Mosesson et al. (53), the rod-like structures were not present in thrombin-activated factor Va. Furthermore, images of the connecting region fragment released following thrombin activation demonstrated rod-like structures up to 34 nm in length. These studies are consistent with a model in which the heavy chain and light chain of factor V form a globular core connected to an extended tail comprising portions of the B domain. Thrombin cleavage of factor V removes the tail structure from the globular domain, resulting in activation of the molecule and the marked change in hydrodynamic properties. This ultrastructural model differs substantially from earlier models. The larger and more complex images reported in earlier studies appear in retrospect to represent aggregates of two or more factor V molecules (55–57).

B. Activation of Factor V

As a single-chain protein, factor V expresses very little, if any, procoagulant activity (58). Procoagulant activity is expressed following limited proteolysis at discrete sites in the molecule. Thrombin activates human factor V by cleavage at three peptide bonds: Arg_{709}-Ser_{710}, Arg_{1018}-Thr_{1019}, and Arg_{1545}-Ser_{1546} (Fig. 5) (11). Thrombin activation of bovine factor V occurs by proteolytic cleavage at the corresponding arginine residues in the bovine protein (20). The activation intermediates observed during the activation of bovine factor V are different from those seen during the activation of human factor V, suggesting that the relative rates of cleavage differ between the two species (6,59,60). Thrombin-activated factor V is a calcium-dependent heterodimer consisting of a heavy chain (molecular weight ≃ 110 kD) and a light chain (molecular weight ≃ 73 kD) (60,61). The heavy chain consists of the first two A-type domains, spanning amino acids 1–709, and the light chain consists of the third A-type domain and the two smaller C-type domains, spanning amino acids 1546–2196 (9–11). The large, heavily glycosylated B domain, spanning amino acids 710–1545, is not necessary for procoagulant activity and is released during activation by thrombin (60,61).

Thrombin-activated factor Va is a calcium-dependent heterodimer: incubation of the activated cofactor with ethylenediaminetetraacetic acid (EDTA) or other chelating agents results in the loss of procoagulant activity caused by dissociation of the heavy chain and the light chain (4,60). Neither the isolated heavy chain nor the light chain exhibits significant procoagulant activity (4,60). Mixing the two chains in the presence of various divalent cations, however, results in the formation of an active cofactor. In addition to

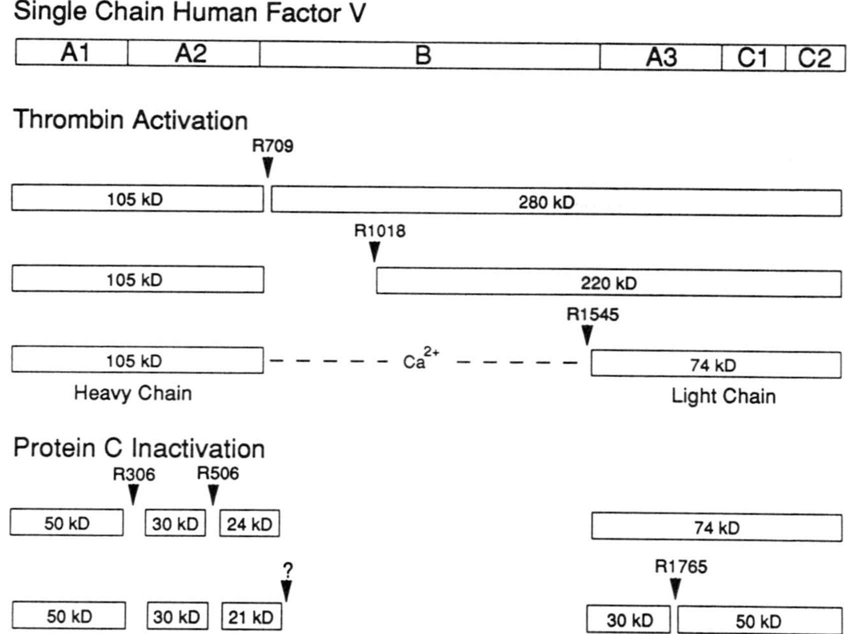

Figure 5 Proteolytic activation and inactivation of human factor V. The domains present in single-chain human factor V are indicated by the labeled boxes. The locations of thrombin cleavage sites following arginine 709, 1018, 1545 are indicated by the arrowheads. The apparent molecular mass of the activation intermediates and final activation products are indicated. The order of bond cleavage is consistent with the activation intermediates observed but does not exclude other pathways. The heavy chain and light chain of thrombin-activated factor Va form a calcium-dependent heterodimer. The activation fragments released from the connecting region are not shown. Activated protein C cleaves both the heavy chain and light chain of human factor V; however, the locations of the cleavage sites have not been identified directly. Approximate locations for the protein C cleavage sites in human factor V are indicated by the arrowheads. The two protein C cleavage sites identified correspond to known cleavage sites in bovine factor V. Cleavage at sites generating the 50, 30, and 24 kD fragments occurs more rapidly than at the remaining sites.

Ca^{2+} ions, Mn^{2+}, Co^{2+}, and Cd^{2+} ions are capable of restoring bovine factor Va activity, whereas Mg^{2+} is relatively ineffective (4,60,62). The regeneration of procoagulant activity is dependent on reassociation of the heavy chain and the light chain with a stoichiometry of 1:1 (4,60,63). This interaction involves hydrophobic interactions between the two chains and results in the formation of a tight binding site for a single calcium ion with an apparent K_d of 50 μM (52,64).

Besides thrombin, the best characterized activator of factor V is a 27 kD protein isolated from the venom of the pit viper *Vipera russelli* (RVV-V). RVV-V is capable of fully activating human factor V by a single proteolytic cleavage at Arg_{1545}-Ser_{1546}. This results in a heavy chain of ~230 kD (A1A2B) and a light chain of ~73 kD (A3C1C2) (6,9,60). Thrombin cleavage of RVV-V-activated factor Va generates a heavy chain of ~110 kD, with release of the B domain fragment(s), but no further increase in procoagulant activity. These results indicate that (1) a single proteolytic cleavage at Arg_{1545}-Ser_{1546} is *sufficient* for complete activation of factor V, and (2) release of the B domain from the heavy chain is not necessary for expression of procoagulant activity. Consequently, the functional significance of the thrombin-mediated cleavages at Arg_{709}-Ser_{710} and Arg_{1018}-Thr_{1019} require further study.

Because factor V has essentially no intrinsic procoagulant activity and because thrombin is the product of the prothrombinase complex, other blood clotting proteins have been investigated for their ability to activate factor V. Factor Xa has been reported to activate factor V in a reaction that is dependent on anionic phospholipids. Monkovic and Tracy reported that human factor Xa can activate human factor V by cleaving at Arg_{1018} and at or near Arg_{709} (65). Based on densitometric analysis of SDS–polyacrylamide gels, they concluded that generation of procoagulant activity correlated with cleavage at Arg_{1018}. These results are different from those previously obtained in the bovine system (66). In the bovine system, thrombin-catalyzed activation of factor V was 100-fold faster than factor Xa on a molar basis, whereas in the human system this difference was only 5-fold. Second, the bonds cleaved in the bovine system, corresponding to Arg_{348} and Arg_{1765}, are not the same as those cleaved in the human system. Finally, reconstitution experiments in the bovine system using fragments obtained by cleavage at the site corresponding to Arg_{1018} result in formation of a complex with minimal procoagulant activity (66). The significance of these apparent substantial differences between the human and bovine system remains to be determined. In experiments studying the activation of factor V in human plasma, inhibition of thrombin by hirudin completely blocks activation of factor V (67,68). These results, however, cannot rule out the possibility that activation of a very small number of factor V molecules by factor Xa contributes to the initial formation of the prothrombinase complex (65,66).

Several other enzymes have also been shown to activate factor V. In contrast to the activators discussed earlier, these enzymes can only partially activate factor V. The proteolytic cleavage patterns are complex, and the locations of individual cleavage sites have not been defined. In most cases, it is not clear whether partial activation reflects partial cleavage, generation of a molecule with a reduced specific activity, or the concomitant activation and inactivation of the molecule. Osterud et al. (69) reported that platelet lysates contained an activated form of factor V as well as a factor V activator. This observation was significant because up to 20% of the factor V present in blood is localized in platelet α granules and is released upon platelet activation (70). Platelet factor V appears to be heterogeneous because of proteolysis within the B domain (71). Platelet lysates contain protease(s) that cleave factor V within the B domain, which can result in partial activation of the molecule (72). Work by Colman et al (73) suggests that platelet calpain is an enzyme capable of partially activating factor V and that the proteolytic cleavage products are distinct from those obtained with thrombin. Furthermore, data reported by Monkovic and Tracy (74) suggested that the factor V released from *activated* platelets is partially activated and that it is more sensitive to activation by factor Xa than is plasma factor V. These data are in contrast to earlier reports that suggested that the factor V released from activated platelets was largely in the procofactor form (75–78). Last, plasmin has been reported to activate human factor V, but this activation is incomplete compared with thrombin activation of factor V (maximal activity $\simeq$ 15–25% of thrombin-activated factor Va) and is followed by a subsequent decline in procoagulant activity associated with further proteolysis of the molecule (79).

That factor V has minimal or no procoagulant activity and thrombin activation of factor V results in the removal of the B domain raise questions about the role of the B domain. To explore the role of the B domain further, we used the cDNA of human factor V to construct and express a recombinant factor V deletion mutant that lacks most of the connecting region (factor V des B; see Fig. 4) (80). This mutant lacks amino acids Leu_{811} through Gly_{1491}, encompassing the internal repeats as well as many of the potential glycosylation sites. Deletion of these amino acids resulted in a two- to fivefold increase in the level of expression in COS cells compared with wild-type factor V. The mutant is expressed as a single-chain protein with an apparent molecular weight of ~230 kD. Proteolytic processing of the mutant by thrombin and RVV-V proceeds in a manner similar to that by wild-type factor V. Thrombin cleavage results in the formation of the predicted heavy-chain (110 kD) and light-chain (73 kD) fragments. In contrast, RVV-V appears to cleave the mutant at a single location, resulting in the formation of a slightly larger heavy chain of ~160 kD and a light chain of ~73 kD. In contrast to wild-type factor V, however, this single-chain deletion mutant appeared to possess significant constitutive procoagulant activity. In a chromogenic assay using purified prothrombin and factor Xa, recombinant factor V possessed ~3.8 $\pm$ 1.3% of the procoagulant activity of the fully activated cofactor, whereas factor V des B expressed 38 $\pm$ 7% of the activity of the fully activated protein. Analysis by SDS-PAGE indicated that the mutant was not more sensitive to proteolysis by thrombin or factor Xa. Activation of the mutant with thrombin or RVV-V resulted in a further two- to fourfold increase in procoagulant activity. These results differed from those obtained by Pittman et al. (81), who constructed a B domain deletion mutant that lacked the entire B domain, attaching the carboxyl end of the heavy chain directly to the amino end of the light chain. This deletion mutant did not express constitutive procoagulant activity and was only slowly activated by thrombin (81).

To confirm that this deletion mutant expresses constitutive procoagulant activity, we used site-directed mutagenesis to change the arginine residues at positions 709 and 1545 (numbering from wild-type factor V) to isoleucines (82). These amino acid substitutions would be predicted to render these sites resistant to proteolysis by thrombin. As expected, the mutant factor V_{II} des B, which contains isoleucines at both sites, is expressed as a single-chain molecule and is resistant to proteolysis by thrombin or factor Xa. Nevertheless, this mutant expresses constitutive procoagulant activity with a relative specific activity comparable to that expressed by factor V des B (82). These results confirm that this deletion of amino acids 811–1492 in the B domain results in a single-chain protein that expresses constitutive procoagulant activity and suggest that one of the roles of the large B domain is to prevent the expression of constitutive procoagulant activity by the circulating procofactor. These results also confirm that the B domain is not necessary for the proper folding and interaction of the heavy- and light-chain domains.

C. Inactivation of Factor Va

The proteolytic inactivation of factor Va by activated protein C is a second mechanism involved in regulating the activity of the prothrombinase complex. Protein C is a vitamin K-dependent zymogen that is activated by thrombin in the presence of the endothelial cell cofactor thrombomodulin (83). Proteolysis of factor Va by activated protein C requires the activated form of the cofactor and results in cleavage of both the heavy and light chains (Fig. 5) (84,85). Three of the four protein C cleavage sites in bovine factor Arg_{306}-Asn_{307} have been identified by protein sequencing (84), and these most likely correspond to Arg_{506}-Gly_{507} in the heavy chain and Arg_{1765}-Leu_{1766} in the light chain of the human protein. In the presence of bovine-activated protein C, cleavage of the bovine factor Va heavy chain results in fragments of 70 and 24 kD and cleavage of the bovine factor Va light chain results in fragments of 30 and 48 kD (84). An additional cleavage at an as yet unidentified position in the carboxyl terminus of the 24 kD fragment results in formation of a 20 kD fragment during prolonged incubations. Based on reconstitution studies utilizing native and protein C-modified heavy and light chains, cleavage of the heavy chain by protein C results in inactivation of the molecule, whereas cleavage of the light chain has no effect on activity (85). Suzuki et al. (86) characterized the inactivation of human factor Va by activated protein C, demonstrating that proteolysis of the heavy chain resulted in fragments with molecular weights of 50 and 30 kD and a doublet of 24 and 21 kD, whereas proteolysis of the light chain produced a 50 kD doublet and a 30 kD fragment (Fig. 5). Cleavage of the heavy chain occurred much more rapidly than that of the light chain and was most closely associated with the loss of procoagulant activity.

Inactivation of factor Va by activated protein C is accelerated in the presence of acidic phospholipids. Kalafatis et al. (86a) have recently demonstrated that cleavage at Arg_{306} is dependent on the presence of anionic phospholipids in bovine factor V, and that this cleavage is required for complete inactivation of the cofactor. In both the human and bovine systems, formation of the prothrombinase complex by addition of factor Xa protects factor Va from inactivation by activated protein C (87,88). The ability of factor Xa to protect human factor Va from inactivation by activated protein C is overcome by the addition of the cofactor protein S (88). Protein S has also been reported to enhance the binding of bovine activated protein C to phospholipid vesicles (89), but this effect is minimal for the human protein (88,90). To characterize the interaction between activated protein C and bovine factor V and factor Va in the presence of phospholipid vesicles, Krishnaswamy et al. (91) used active site-blocked activated protein C. They found that factor V had essentially no effect on the binding of activated protein C to phospholipid vesicles but that factor Va decreased the dissociation constant from 7.3×10^{-8} to 7×10^{-9} M. Furthermore, they found that the interaction between factor Va and activated protein C was mediated by the light chain of factor Va. Subsequently, Walker et al. (92), using recombinant fragments of the factor VIII light chain expressed in *Escherichia coli*, demonstrated that two fragments that overlapped between amino acids 1974 and 2052 inhibited the anticoagulant activity of activated protein C. The binding site was localized further using synthetic peptides that spanned amino acid residues 1865–1874 of the human factor Va light chain. This synthetic peptide inhibited the anticoagulant activity of activated protein C and quenched the fluorescence of a dansyl-Glu-Gly-Arg active site-modified activated protein C. Thus, a portion of the binding site for activated protein C to the light chain of factor Va has been localized to the carboxyl end of the A3 domain.

D. Formation of the Prothrombinase Complex

The procoagulant activity of factor Va is manifested by its ability to markedly enhance the activation of prothrombin to thrombin by the serine protease factor Xa in the presence of calcium ions and a procoagulant phospholipid surface (93). Factor V contributes to the amplification of prothrombin activation (1) by stabilizing the enzyme (factor Xa)-cofactor (factor Va)-substrate (prothrombin) complex, and (2) by altering the kinetic mechanism of prothrombin activation. *In vitro* studies revealed that prothrombin activation by the fully formed prothrombinase complex proceeds exclusively through the intermediate meizothrombin (94–96). In contrast, solution-phase factor Xa slowly converts prothrombin to thrombin through the reaction intermediates fragment 1.2 and prethrombin 2 (97). The effects of factor Va on the kinetics of the prothrombinase complex *in vitro* are confirmed *in vivo* by patients with congenital deficiency of factor V, who have a significant bleeding diathesis (see later).

1. *Interactions with Factor Xa*

Factor Va interacts with factor Xa in the presence of calcium ions on a phospholipid membrane surface to accelerate the conversion of prothrombin to thrombin. This interaction requires activation of the cofactor, because factor V does not bind to factor Xa (60). In the absence of a phospholipid membrane, factor Va and factor Xa form a complex, but with a dissociation constant that is three orders of magnitude greater than the value observed in the presence of acidic phospholipids (98). When bound to a membrane surface, factor Va increases the affinity of factor Xa to the membrane (22,99). Likewise, factor Va forms at least part of the receptor for factor Xa on an activated platelet surface (100). On the membrane surface, the two proteins interact in a 1:1 stoichiometry that occurs independently of the substrate, prothrombin (97). The interaction between factor Va and factor Xa involves epitopes on both the light chain and the heavy chain of the activated cofactor (101,102) and directly results in an alteration in the height of the factor Xa active site above the membrane (103). Recent studies suggest that the amino-terminal regions of the A1 and the A3 domains of factor Va interact with factor Xa (104). Last, the ability of factor Va to bind to factor Xa is lost when factor Va is proteolytically cleaved by activated protein C (105).

2. *Interactions with Phospholipid and Cell Membrane Surfaces*

Formation of the prothrombinase complex requires the presence of a membrane surface to which the individual components bind and subsequently assemble into a functional complex. Activated platelets, platelet microparticles, and damaged vascular cells most likely provide this surface *in vivo*. The interaction of factor V with cellular surfaces appears to be mediated, at least in part, by binding to acid phospholipids (e.g., phosphatidylserine). These phospholipids are normally sequestered in the inner leaflet of the cellular membrane, facing the cytoplasm, but become exposed following platelet activation and microparticle formation (106). Other peripheral blood cells, such as neutrophils, lymphocytes, and monocytes, as well as endothelial cells, have also been shown to support assembly of the prothrombinase complex (107). Because of the difficulties in studying the binding of these proteins to living cells, however, most studies have been performed using phospholipid vesicles as a more defined model system. Using a variety of analytical techniques, several common observations have been made by a number of investigators concerning the interaction of factor V and factor Va with phospholipid vesicles. First, binding to phospholipid membranes does not require cofactor activation (108–110). Second, binding of factor V and factor Va to a phospholipid membrane surface requires the presence of an anionic phospholipid; no binding occurs to vesicles containing only the neutral phospholipid phosphatidylcholine (108,111). Third, binding of factor Va to a phospholipid membrane (as well as to a platelet membrane surface) is mediated by the light chain of the cofactor (109,111,112). Last, the light-chain–mediated binding of factor Va to a membrane surface is not dependent on calcium ions (22,109,111,112). Treatment of membrane-bound factor Va with a calcium chelator, such as EDTA, results in dissociation of the heavy chain from the light chain, but the light chain remains bound to the membrane surface.

Although all studies agree that binding to acid phospholipids is mediated through the light chain of factor Va, there are considerable differences in the reported physical characteristics of this interaction. First, the reported dissociation constants range from 10^{-7} to 10^{-11} M, most likely reflecting differences in the techniques employed (22,108–110,113,114). Second, the relative contributions of hydrophobic and ionic interactions to this binding is unclear. Two independent studies using lipophilic reagents have shown that a small portion of the light chain of factor Va is exposed to the lipid bilayer core, suggesting that hydrophobic interactions are involved (115,116). Other studies, however, have shown that ionic strength has a significant effect on the dissociation rate of the factor V-phospholipid vesicle interaction, indicating an electrostatic interaction between the protein and phospholipids (110,111,114). Increasing calcium ion concentration has been reported to result in decreased binding to phospholipid vesicles (111), although this has not been observed by others (22). Lysine modification by citraconic anhydride has been shown to prevent factor V-membrane binding, whereas tryptophan modification by 2-hydroxy-5-nitrobenzyl bromide does not (110). Last, the observation that factor V does not compete with factor VIII for binding to the platelet surface suggests that interactions with cellular surfaces may involve binding to phospholipids as well as to other, as yet unidentified, platelet membrane components (117).

To understand better the molecular mechanisms involved in phospholipid and cell membrane surface binding, we (118–120) and others (22,121,122) have begun to define discrete regions in the light chain of factor Va that mediate binding to these surfaces. Krishnaswamy and Mann (22) have identified and characterized a proteolytic fragment containing the first ~220 amino acids of the A3 domain of the light chain of bovine factor Va, which is capable of binding to phosphatidylserine-containing phospholipid vesicles. This region corresponds to amino acids 1546–1765 in human factor V (Fig. 5). Fluorescence polarization curves demonstrated that the amino-terminal fragment of the bovine factor Va light chain could displace labeled factor Va from phosphatidylserine-containing vesicles. Furthermore, the dissociation constant for this fragment was ~1.55 ± 0.55 nM, which was similar to the constants obtained for factor Va (~2.51 ± 0.44 nM) and the isolated light chain (~1.93 ± 0.49 nM) (22). The interaction of the carboxyl-terminal fragment of the bovine factor Va light chain, which contains the two tandem C-type domains, could not be studied because this fragment was insoluble in aqueous solution (22). Kalafatis et al. (121) subsequently demonstrated that several regions were protected from proteolysis by trypsin, chymotrypsin, and elastase when bound to phosphatidylserine-containing vesicles. One of these peptides, corresponding to residues 1683–1765 in human factor V, was isolated, sequenced, and shown to bind to phosphatidylserine-containing vesicles in a qualitative binding assay. More recently, the same group has used 1-azidopyrene, a fluorescent lipophilic cross-linking reagent, to identify regions of the bovine factor Va light chain that interact with the lipid bilayer. When the labeled light chain was subjected to proteolysis with activated protein C, both the 30 kD and the ~50 kD fragments were modified by the cross-linking reagent (Fig. 5). These data suggested that there is a second region of the factor Va light chain, corresponding to residues 1766–2196 in the human protein, that interacts with phospholipid (122).

To investigate the role of individual domains in the binding of factor V to acid phospholipids, we used the cDNA encoding human factor V to construct a series of deletion mutants that lack domain-sized fragments of the light chain (118). These mutants lack (1) the second C-type domain (factor V des C2); (2) both C-type domains (factor V des C1C2); (3) the entire light chain (factor V des LC); (4) the third A-type domain (factor V des A3); or (5) the third A-type domain and the first C-type domain (factor V des A3C1; see Fig. 2). The light-chain deletion mutants were expressed in COS cells as single-chain proteins and were processed normally by thrombin or RVV-V. These mutants did not express significant procoagulant activity. Using an immobilized phosphatidylserine enzyme-linked immunosorbent assay that was specific for light-chain–mediated binding to acid phospholipids, we found that those light-chain deletion mutants lacking the second C-type domain were not able to bind immobilized phosphatidylserine. In contrast, the presence of the C2 domain alone (factor V des A3C1) partially restored phosphatidylserine-specific binding. This binding was inhibited by soluble phosphatidylserine but not by phosphatidylcholine, indicating that the binding was specific for phosphatidylserine. Furthermore, binding was inhibited by increasing calcium ion concentration, suggesting that the C2 domain-mediated binding to phosphatidylserine was electrostatic in nature. In contrast, the mutant lacking only the third A-type domain (factor V des A3) did not bind to immobilized phosphatidylserine. It is unclear whether this reflects the presence of the C1 domain, which does not appear to bind to acid phospholipids, or a protein-folding abnormality introduced with this specific deletion.

The identification of a phosphatidylserine-specific binding site in the second C-type domain of factor V parallels earlier studies localizing a phosphatidylserine binding site to the homologous C-type domain of factor VIII (123,124). Factor VIII inhibitors are immunoglobulins that may arise as alloantibodies in patients with hemophilia A who receive therapeutic factor VIII or as spontaneous autoantibodies in nonhemophilic patients. Many of these inhibitors have been epitode mapped, and most have been found to bind to the A2 domain of the heavy chain or the C2 domain of the light chain of factor VIII (123). Furthermore, those inhibitors that bind to the C2 domain of factor VIII were shown to interfere with binding to phosphatidylserine, and preincubation of the activated cofactor with procoagulant phospholipids protected it from neutralization by inhibitors directed against the C2 domain (124). We therefore investigated a factor V inhibitor that was previously reported to bind to the light chain of factor V and to block factor Va-mediated binding of factor Xa to the platelet surface (125). This antibody rapidly neutralized the procoagulant activity of factor Va. Using the recombinant factor V light-chain deletion mutants, we found that this factor V inhibitor bound to the C2 domain of factor V (119). We further localized the epitope recognized by this inhibitor by using a series of recombinant factor V light-chain chimeras that contain exon-sized segments of the factor VIII C2 domain substituted for the corresponding segments of factor V (120). Because the factor V inhibitor does not bind to the C2 domain of factor

VIII, this localized the epitope to the amino-terminal region of the C2 domain of factor V (120). This antibody, as well as papain-generated F(ab) fragments, blocked binding of factor V to immobilized phosphatidylserine. Furthermore, preincubation of the activated cofactor with procoagulant phospholipids protected it from neutralization by the inhibitor (119). These results confirmed the importance of the C2 domain in the expression of procoagulant activity, as well as in mediating phosphatidylserine specific binding. In conclusion, the interaction of factor Va with phospholipid is complex with regions within domains A3 and C2 contributing to the protein phospholipid binding site. Further study is required for a more complete understanding of the interaction of factor V with artificial and natural membranes.

3. *Interaction with Prothrombin*

Factor Va has been shown to facilitate the binding of prothrombin to a phospholipid membrane (126), thereby promoting the assembly of the enzyme (factor Xa)-substrate (prothrombin) complex. The interaction between bovine factor Va and prothrombin is mediated through the heavy chain of the cofactor, and this interaction is independent of calcium ions (127). In contrast, prothrombin does not bind to factor V, indicating that the cofactor must be activated before it can bind prothrombin (105). Sedimentation equilibrium analysis revealed that the factor Va heavy chain and prothrombin form a 1:1 complex with a dissociation constant of 10 μM (127). Similar results were obtained if prethrombin 1 was substituted for prothrombin, indicating that this interaction is not mediated through the fragment 1 portion of prothrombin. Furthermore, bovine factor Va heavy chain forms a similar 1:1 complex with either bovine or human prethrombin 1, differing by only a slightly weaker dissociation constant (27 μM) for the bovine-human interaction (127). Last, proteolysis of the factor Va heavy chain by activated protein C results in the loss of its ability to bind prothrombin (105).

E. Cellular Synthesis of Factor V

Factor V circulates in plasma at a concentration of ~7 μg/mL (30 nM). In addition to circulating in the plasma compartment, approximately 20% of the factor V in blood is contained in platelet α granules (70). The storage of factor V within platelets appears to be important, because when platelets aggregate to form the primary hemostatic plug they also provide both a surface for assembly of the prothrombinase complex and high local concentrations of released platelet factor V. The main site of synthesis of plasma factor V appears to be the hepatocyte. This is consistent with the finding of factor V mRNA in liver (11) and HepG2 cells (10) and with the finding that plasma factor V levels are decreased in advanced liver disease. Megakaryocytes have been shown to contain factor V mRNA, which suggests that the factor V present in α granules is synthesized by megakaryocytes rather than taken up from the plasma compartment (128,129). The synthesis of factor V by several other cell types, including bovine aortic endothelial cells (130), aortic vascular smooth muscle cells (131), monocytes (132), and macrophages (133), has also been reported. The physiological significance of synthesis at these sites requires further investigation because the amounts of factor V released into the culture media by these cells are relatively minor compared with the concentration of factor V circulating in plasma (70).

IV. DEFICIENCY STATES

A. Parahemophilia

Congenital factor V deficiency, also referred to as parahemophilia, is a rare disorder that has an estimated incidence of 1 in 10^6. Approximately 150 cases have been reported in the literature (134–137). Parahemophilia is characterized by low levels of functional factor V activity in plasma and platelets and is inherited in an autosomal recessive manner. Clinical manifestations of this bleeding diathesis include ecchymoses, epistaxis, oral bleeding, and menorrhagia. In contrast to hemophilia A, hemarthroses are uncommon and are often traumatically induced. Excessive bleeding occurs following surgery, trauma, and dental extractions. Interestingly, there have been several reports of patients with this disorder who have had thromboembolic events, including deep and superficial venous thromboses, pulmonary embolism, stroke, and myocardial infarction (138–141). Most patients with congenital factor V deficiency have low antigenic levels of factor V, suggesting a quantitative defect. A minority of patients exhibit a discrepancy between their functional and antigenic levels, however, suggesting a circulating dysfunctional factor V molecule. Chiu et al. (142) found such a discrepancy in 4 of 14 patients studied, and Tracy and Mann (137) found it in 2 of 21 patients. The molecular basis for factor V deficiency has not been defined. In general, the functional plasma factor V levels have not correlated well with the hemorrhagic tendency (135), and several authors have suggested that the severity of bleeding symptoms may be more closely correlated with platelet rather than plasma factor V levels (100,143). The description of two members of a family with a significant bleeding diathesis associated with a dysfunctional factor V and markedly diminished platelet factor V activity (factor V Quebec) lends support to the concept that platelet factor V content is critically important in normal hemostasis (143).

B. Combined Factor V and Factor VIII Deficiency

Combined factor V and factor VIII deficiency is a fascinating disorder that occurs with too high a frequency to be attributable to the chance association of hemophilia A and homozygous factor V deficiency. The expected frequency of concurrent factor V and factor VIII deficiency has been calculated to be 6 per 100 billion, but more than 30 families with this combined disorder have been reported in the literature in the last 30 years (144). There seems to be a particularly high incidence of this disorder in non-Ashkenazi Jews, in whom the frequency of combined factor V and factor VIII deficiency is estimated to be 1 in 10^5 (145). The inheritance pattern appears to be autosomal recessive, with females affected as frequently as males and familial consanguinity appearing commonly in reported cases (144–147). Clinical manifestations are similar to those of parahemophilia (144,145). The molecular basis for combined factor V and factor VIII deficiency is not known. In most cases, antigenic determination of factor V and factor VIII levels correspond to the low functional levels seen in this disorder (137,146–150). In addition, Hultin and Eyster reported normal activation of both factors V and VIII by thrombin in a patient with the combined deficiency (146), suggesting a quantitative rather than qualitative deficiency of these factors. Marlar and Griffin presented data to suggest a defect in protein C inhibitor leading to excessive inactivation of factor V and factor VIII (151). However, this mechanism was largely discounted by subsequent reports (148,150,152,153). Garcia and colleagues reported a rise in factor VIII activity in response to DDAVP (desmopressin) in two patients with combined factor V and factor

VIII deficiency, with a rate of decline of VIII activity that did not differ significantly from that in mild hemophiliacs similarly treated (154). This suggests that the combined deficiency is not caused by excessive clearance of the factors from the plasma. It is possible that a defect in a posttranslational modification common to these structurally similar proteins is responsible for the combined deficiency. Further characterization of posttranslational modifications shared by these proteins is required to verify this hypothesis.

C. Factor V Inhibitors

Factor V inhibitors are pathological immunoglobulins that bind to the factor V molecule and result in its functional neutralization (155). These inhibitors are relatively uncommon and have been most frequently observed in the postoperative setting or in association with antibiotic therapy, especially aminoglycoside antibiotics (155). Factor V inhibitors tend to be in low titer and frequently resolve spontaneously with no specific treatment (155). Hemorrhagic manifestations are variable and range from essentially no bleeding to severe hemorrhage and death (155,156). Although most factor V inhibitors develop as autoantibodies in individuals with no prior coagulopathy, there have been several recent reports of factor V inhibitors developing following exposure to topical bovine thrombin (157,158). Topical thrombin, used extensively as a local hemostatic agent, has been shown to contain bovine factor V in addition to bovine thrombin (157). The bovine proteins presumably serve as the initial immunogen, and the resultant antifactor V antibodies have been shown to cross-react with the human molecule. As with the spontaneously acquired inhibitors, these patients exhibit a variable initial presentation and clinical course. Therapy for factor V inhibitors is largely empirical and is determined by the clinical manifestations of the individual patient. Platelet transfusions may be useful for the treatment of hemorrhagic manifestations (159), because they provide a source of factor V that may be ''sequestered'' from the circulating antibody, but platelets are not always effective (157). In contrast to factor VIII inhibitors, which have been extensively characterized and studied, only three factor V inhibitors have been characterized in detail (102,119,157). Two inhibitors have been mapped to the light chain of factor Va, including the inhibitor we have mapped to the C2 domain (119) and a second inhibitor that developed following exposure to topical bovine thrombin (157). The third inhibitor has been shown to bind the heavy chain of factor Va (102). This antibody has also been shown to compete with factor Xa for binding to factor Va, suggesting the presence of a factor Xa binding site in the heavy chain of factor Va (102).

V. FACTOR V DEFECTS AND THROMBOPHILIA

Defects in factor V that interfere with natural anticoagulant pathways have recently been demonstrated to be important in the pathogenesis of thrombosis [161,162]. Dahlbäck et al. [163] developed an assay for protein C resistance based on the prolongation of the clotting time caused by the addition of activated protein C to the activated partial thromboplastin time assay. When this assay was used to evaluate 104 consecutive Swedish patients with venous thrombosis the prevalence of protein C resistance was found to be 33% [164]. Subsequent investigations have estimated the prevalence of protein C resistance in patients with venous thrombosis to be between 21 and 52% [165,166]. The incidence of protein C resistance to various normal populations and in subcategories of thrombotic disease remains to be established. Dahlbäck et al. [161] subsequently demonstrated that resistance to activated protein C was corrected by the addition of factor V. At least two kinds of defects have been identified. First, Bertina et al. [162] demonstrated that the phenotype of protein C resistance was associated with heterozygosity or homozygosity for a single point mutation in exon 10 of the factor V gene [19]. This mutation substitutes a glutamine residue for arginine-506 at the protein C cleavage (factor V R506Q or factor V Leiden). This mutation has been found in ~80-100% of patients with protein C resistance and appears to be tenfold more common than all other known genetic risk factors for thrombosis [162]. The mechanism whereby the factor V R506Q mutation results in protein C resistance remains to be confirmed. Second, Dahlbäck et al. [163] originally postulated that protein C resistance was due to a new cofactor which was later identified to be factor V. Factor V serves as a cofactor for the inactivation of factor Va [161] and factor VIIIa [167] by activated protein C in vitro. In addition Zoller and Dahlbäck [168] have identified several patients with protein C resistance that do not express the factor V R506Q mutation. The molecular defects in these patients and the importance of factor V as a cofactor for protein C in vivo remain to be determined.

VI. CONCLUSIONS

The use of molecular approaches to complement classic protein chemistry has helped to increase our understanding of the structure and function of factor V and its role in the prothrombinase complex. Despite the progress outlined in this review, much remains to be learned about this large, complicated protein. The molecular mechanisms that convert the inactive precursor to an active cofactor receptor and the molecular interactions involved in the binding of factor V to phospholipid, factor Xa, and prothrombin remain to be defined in detail. The use of site-directed mutagenesis in conjunction with x-ray crystallographic studies may help resolve these questions. Further studies are also needed to characterize the molecular interactions involved in the assembly of the prothrombinase complex on cellular surfaces, because most data have been obtained using artificial phospholipid membranes. Finally, the identification of molecular defects in factor V deficiency and combined factor V-factor VIII deficiency may provide additional insights into the structure and function of this protein *in vivo*. Given the pivotal role of factor V in the prothrombinase complex, this knowledge may have important new diagnostic, as well as therapeutic, applications.

ACKNOWLEDGMENTS

This work was supported by Grant HL43106 from the National Institutes of Health (Kane), Searle Scholar Award 89-G-130 (Kane), and Clinician Scientist Award 91-419 from the American Heart Association and Genentech (Ortel). Dr. Kane is an Established Investigator of the American Heart Association (91-181).

REFERENCES

1. Owren PA. The coagulation of blood: investigations on a new clotting factor. Acta Med Scand (Suppl) 1947; 194:1–327.
2. Colman RW. Factor V. Prog Hemost Thromb 1976; 3:109–143.

3. Nesheim ME, Myrmel KH, Hibbard L, Mann KG. Isolation and characterization of single chain bovine factor V. J Biol Chem 1979; 254:508–517.
4. Esmon CT. The subunit structure of thrombin-activated factor V. Isolation of activated factor V, separation of subunits, and reconstitution of biological activity. J Biol Chem 1979; 254: 964–973.
5. Dahlback B. Human coagulation factor V purification and thrombin-catalyzed activation. J Clin Invest 1980; 66:583–591.
6. Kane WH, Majerus PW. Purification and characterization of human coagulation factor V. J Biol Chem 1981; 256:1002–1007.
7. Katzmann JA, Nesheim ME, Hibbard LS, Mann KG. Isolation of functional human coagulation factor V by using a hybridoma antibody. Proc Natl Acad Sci USA 1981; 78:162–166.
8. Jenny RJ, Mann KG. Factor V: a prototype pro-cofactor for vitamin K-dependent enzyme complexes in blood clotting. Clin Haematol 1989; 2:919–944.
9. Kane WH, Davie Ew. Cloning of a cDNA coding for human factor V, a blood coagulation factor homologous to factor VIII and ceruloplasmin. Proc Natl Acad Sci USA 1986; 83: 6800–6804.
10. Kane WH, Ichinose A, Hagen FS, Davie EW. Cloning of cDNAs coding for the heavy chain region and connecting region of human factor V, a blood coagulation factor with four types of internal repeats. Biochemistry 1987; 26:6508–6514.
11. Jenny RJ, Pittman DD, Toole JJ, et al. Complete cDNA and derived amino acid sequence of human factor V. Proc. Natl Acad Sci USA 1987; 84:4846–4850.
12. Toole JJ, Knopf JL, Wozney JM, et al. Molecular cloning of a cDNA encoding human antihaemophilic factor. Nature 1984; 312:342–347.
13. Vehar GA, Keyt B, Eaton DL, et al. Structure of human factor VIII. Nature 1984; 312:337–342.
14. Davie EW, Fujikawa K, Kisiel W. The coagulation cascade: initiation, maintenance, and regulation. Biochemistry 1991; 30:10363–10370.
15. Bauer KA, Rosenberg RD. The pathophysiology of the prethrombotic state in humans: insights gained from studies using markers of hemostatic system activation. Blood 1987; 70:343–350.
16. Kane WH, Davie EW. Blood coagulation factors V and VIII: structural and functional similarities and their relationship to hemorrhagic and thrombotic disorders. Blood 1988; 71:539–555.
17. Wang H, Riddell DC, Guinto ER, MacGillivray RTA. Localization of the gene encoding human factor V to chromsome 1q21-25. Genomics 1988; 2:234.
18. Watson ML, Kingsmore SF, Johnston GI, et al. Genomic organization of the selectin family of leukocyte adhesion molecules on human and mouse chromosome 1. J Exp Med 1990; 172:263–271.
19. Cripe LD, Moore KD, Kane WH. Structure of the gene for human coagulation factor V. Biochemistry 1992; 31:3777–3785.
20. Guinto ER, Esmon CT, Mann KG, MacGillivray RTA. The complete cDNA sequence of bovine coagulation factor V. J Biol Chem 1992; 267:2971–2978.
21. Xue J, Kalafatis M, Mann KG. Determination of the disulfide bridges in factor Va light chain. Biochemistry 1993; 32:5917–5923.
22. Krishnaswamy S, Mann KG. The binding of factor Va to phospholipid vesicles. J Biol Chem 1988; 263:5714–5723.
23. Bruin T, Sturk A, ten Cate JW Cath M. The function of the human factor V carbohydrate moiety in blood coagulation. Eur J Biochem 1987; 170:305–310.
24. Gavel Y, von Heijne G. Sequence differences between glycosylated and non-glycosylated Asn-X-Ser/Thr acceptor sites: implications for protein engineering. Protein Eng. 1990; 3: 433–442.
25. Hortin GL. Sulfation of tyrosine residues in coagulation factor V. Blood 1990; 76:946–952.

26. Wilson IB, Gavel Y, von Heijne G. Amino acid distributions around O-linked glycosylation sites. Biochem J 1991; 275:529–534.
27. Braun PJ, Dennis S, Hofsteenge J, Stone SR. Use of site-directed mutagenesis to investigate the basis for specificity of hirudin. Biochemistry 1988; 27:6517–6522.
28. Pittman DD, Tomkinson K, Kaufman RJ. Post-translational sulfation of factor V is required for efficient cleavage and activation mediated by thrombin. (abstract). Blood 1991; 78:63a.
29. Kalafatis M, Rand MD, Jenny RJ, Ehrlich YH, Mann KG. Phosphorylation of factor Va and factor VIIIa by activated platelets. Blood 1992; 81:704–719.
30. Pinna LA. Casein kinase 2: an "eminence grise" in cellular regulation? Biochim Biophys Acta 1990; 1054:267–284.
31. Gitschier J, Wood WI, Goralka TM, et al. Characterization of the human factor VIII gene. Nature 1984; 312:326–330.
32. Craik CS, Rutter WJ, Fletterick R. Splice junctions: association with variation in protein structure. Science 1983; 220:1125–1129.
33. Davie EW, Ichinose A, Leytus SP. Structural features of the proteins participating in blood coagulation and fibrinolysis. Cold Spring Harb Symp Quant Biol 1986; 51:509–514.
34. Ortel TL, Takahashi N, Putnam FW. Structural model of human ceruloplasmin based on internal triplication, hydrophilic/hydrophobic character, and secondary structure of domains. Proc Natl Acad Sci USA 1984; 81:4761–4765.
35. Ryden L. Cooper proteins and copper enzymes, 3rd ed. Boca Raton FL: CRC Press, 1984: 37n100.
36. Mann KG, Lawler CM, Vehar GA, Church WR. Coagulation factor V contains copper ion. J Biol Chem 1984; 259:12949–12951.
37. Larocca D, Peterson JA, Urrea R, Kuniyoshi J, Bistrain AM, Ceriani RL. A M_r 46,000 human milk fat globule protein that is highly expressed in human breast tumors contains factor VIII-like domains. Cancer Res 1991; 51:4994–4998.
38. Stubbs JD, Lekutis C, Singer KL, et al. cDNA cloning of a mouse mammary epithelial cell surface protein reveals the existence of epidermal growth factor-like domains linked to factor VIII-like sequences. Proc Natl Acad Sci USA 1990; 87:8417–8421.
39. Takagi S, Hirata T, Agata K, Mochii M, Eguchi G, Fujisawa H. The A5 antigen, a candidate for the neuronal recognition molecule, has homologies to complement proteins and coagulation factors. Neuron 1991; 7:295–307.
40. Buse P, Peck E, Stubbs JD, Gendler S, Parry G. MFG-E8: an apical phospholipid binding protein of mammary epithelia is associated with membrane secretion. (abstract). J Cell Biol 1991; 115:196a.
41. Poole S, Firtel RA, Lamar E, Towekamp W. Sequence and expression of the discoidin I gene family in *Dictyostelium discoideum*. J Mol Biol 1981; 153:273–289.
42. Gabius HJ, Springer WR, Barondes SH. Receptor for the cell binding site of discoidin I. Cell 1985; 42:449–456.
43. Bartles JR, Galvin NJ, Frazier WA. Discoidin I-membrane interactions II. Discoidin I binds to and agglutinates negatively charged phospholipid vesicles. Biochim Biophys Acta 1982; 687: 129–136.
44. Smith MW. Structure of vertebrate genes: a statistical analysis implicating selection. J Mol Evol 1988; 27:45–55.
45. Ludwig EW, Blackhart BD, Pierotti VI, et al. DNA sequence of the human apolipoprotein B gene. DNA 1987; 6:363–372.
46. Ahearn JM, Barolomei MS, West ML, Cisek LJ, Corden JL. Cloning and sequence analysis of the mouse genomic locus encoding the largest subunit of RNA polymerase II. J Biol Chem 1987; 262:10695–10705.
47. Faber PW, Kuiper GGJM, van Rooij HCJ, van der Korput JAGM, Brinkman AO, Trapman J. The N-terminal domain of the human androgen receptor is encoded by one large exon. Mol Cell Endocrinol 1989; 61:257–262.

48. Muragaki Y, Jacenko O, Apte S, Mattei M, Ninomiya Y, Olsen BR. The α_2(VIII) collagen gene, a novel member of the short chain collagen family located on the human chromosome 1. J Biol Chem 1991; 266:7721–7727.
49. Ligtenberg MJL, Vos HL, Gennissen AMC, Hilkens J. Episialin, a carcinoma-associated mucin, is generated by a polymorphic gene encoding splice variants with alternative amino termini. J Biol Chem 1990; 265:5573–5578.
50. Toribara NW, Gum JR, Culhane PJ, et al. MUC-2 human small intestinal mucin gene structure: repeated arrays and polymorphism. J Clin Invest 1991; 88:1005–1013.
51. Toole JJ, Pittman DD, Orr EC, Murtha P, Wasley LC, Kaufman RJ. A large region (approximately equal to 95 kDa) of human factor VIII is dispensable for in vitro procoagulant activity. Proc Natl Acad Sci USA 1986; 83:5939–5942.
52. Laue TM, Johnson AE, Esmon CT, Yphantis DA. Structure of bovine blood coagulation factor Va. Determination of the subunit associations, molecular weights, and asymmetries by analytical ultracentrifugation. Biochemistry 1984; 23:1339–1348.
53. Mosesson MW, Church WR, DiOrio JP, et al. Structural model of factors V and Va based on scanning transmission electron microscope images and mass analysis. J Biol Chem 1990; 265:8863–8868.
54. Fowler WE, Fay PJ, Arvan DS, Marder VJ. Electron microscopy of human factor V and factor VIII: correlation of morphology with domain structure and localization of factor V activation fragments. Proc Natl Acad Sci USA 1990; 87:7648–7652.
55. Mosesson MW, Nesheim ME, DiOrio J, Hainfeld JF, Wall JS, Mann KG. Studies on the structure of bovine factor V by scanning transmission electron microscopy. Blood 1985; 65: 1158–1162.
56. Dahlback B. Bovine coagulation factor V visualized with electron microscopy. Ultrastructure of the isolated activated forms and of the activation fragments. J Biol Chem 1986; 261: 9495–9501.
57. Dahlback B. Ultrastructure of human coagulation factor V. J Biol Chem 1985; 260:1347–1349.
58. Nesheim ME, Taswell JB, Mann KG. The contribution of bovine factor V and factor Va to the activity of prothrombinase. J Biol Chem 1979; 254:10952–10962.
59. Nesheim ME, Mann KG. Thrombin-catalyzed activation of single chain bovine factor V. J Biol Chem 1979; 254:1326–1334.
60. Suzuki K, Dahlback B, Stenflo J. Thrombin-catalyzed activation of human coagulation factor V. J Biol Chem 1982; 257:6556–6564.
61. Kane WH, Majerus PW. The interaction of human coagulation factor Va with platelets. J Biol Chem 1982; 257:3963–3969.
62. Laue TM, Lu R, Krieg UC, Esmon CT, Johnson AE. Ca^{2+}-dependent structural changes in bovine blood coagulation factor Va and its subunits. Biochemistry 1989; 28:4762–4771.
63. Krishnaswamy S, Russell GD, Mann KG. The reassociation of factor Va from its isolated subunits. J Biol Chem 1989; 264:3160–3168.
64. Guinto ER, Esmon CT. Formation of a calcium-binding site on bovine activated factor V following recombination of the isolated subunits. J Biol Chem 1982; 257:10038–10043.
65. Monkovic DD, Tracy PB. Activation of human factor V by factor Xa and thrombin. Biochemistry 1990; 29:1118–1128.
66. Foster WB, Nesheim ME, Mann KG. The factor Xa-catalyzed activation of factor V. J Biol Chem 1983; 258:13970–13977.
67. Pieters J, Lindhout T, Hemker HC. In situ-generated thrombin is the only enzyme that effectively activates factor VIII and factor V in thromboplastin-activated plasma. Blood 1989; 74:1021–1024.
68. Yang X-J, Blajchman MA, Craven S, Smith LM, Anvari N, Ofosu FA. Activation of factor V during intrinsic and extrinsic coagulation. Inhibition by heparin, hirudin and D-Phe-Pro-Arg-CH_2Cl. Biochem J 1990; 272:399–406.

69. Osterud B, Rapaport SI, Lavine KK. Factor V activity of platelets: evidence for an activated factor V molecule and for a platelet activator. Blood 1977; 49:819–834.
70. Tracy PB, Eide LL, Bowie EJ, Mann KG. Radioimmunoassay of factor V in human plasma and platelets. Blood 1982; 60:59–63.
71. Viskup RW, Tracy PB, Mann KG. The isolation of human platelet factor V [published erratum appears in Blood July 1987; 70(1):339]. Blood 1987; 69:1188–1195.
72. Kane WH, Mruk JS, Majerus PW. Activation of coagulation factor V by a platelet protease. J Clin Invest 1982; 70:1092–1100.
73. Bradford HN, Annamalai A, Doshi K, Colman RW. Factor V is activated and cleaved by platelet calpain: comparison with thrombin proteolysis. Blood 1988; 71:388–394.
74. Monkovic DD, Tracy PB. Functional characterization of human platelet-released factor V and its activation by factor Xa and thrombin. J Biol Chem 1990; 265:17132–17140.
75. Chesney CM, Pifer D, Colman RW. Subcellular localization and secretion of factor V from human platelets. Proc Natl Acad Sci USA 1981; 78:5180–5184.
76. Vicic WJ, Lages B, Weiss HJ. Release of human platelet factor V activity is induced by both collagen and ADP and is inhibited by aspirin. Blood 1980; 56:448–455.
77. Kane WH, Lindhout MJ, Jackson CM, Majerus PW. Factor Va-dependent binding of factor Xa to human platelets. J Biol Chem 1980; 255:1170–1174.
78. Baruch D, Hemker HC, Lindhout T. Kinetics of thrombin-induced release and activation of platelet factor V. Eur J Biochem 1986; 154:213–218.
79. Lee CD, Mann KG. Activation/inactivation of human factor V by plasmin. Blood 1989; 73: 185–190.
80. Kane WH, Devore-Carter D, Ortel TL. Expression and characterization of recombinant human factor V and a mutant lacking a major portion of the connecting region. Biochemistry 1990; 29:6762–6768.
81. Pittman DD, Thompkinson K, Marquette K, Jenny RJ, Mann KB, Kaufman RJ. Expression of recombinant factor V and B-domain mutants in mammalian cells. (abstract). Blood (Suppl) 1990; 76:433a.
82. Ortel TL, Quinn-Allen MA, Keller FG, Kane WH. Expression of a single-chain coagulation factor V deletion mutant that possesses constituative procoagulant activity. (abstract). Circulation 1992; 86:I–686.
83. Esmon CT. The protein C anticoagulant pathway. Arterioscler Thromb 1992; 12:135–145.
84. Odegaard B, Mann K. Proteolysis of factor Va by factor Xa and activated protein C. J Biol Chem 1987; 262:11233–11238.
85. Van de Waart P, Bruls H, Hemker HC, Lindhout T. Functional properties of factor Va subunits after proteolytic alterations by activated protein C. Biochim Biophys Acta 1984; 799:38–44.
86. Suzuki K, Stenflo J, Dahlback B, Teodorsson B. Inactivation of human coagulation factor V by activated protein C. J Biol Chem 1983; 258:1914–1920.
86a. Kalafatis M, Mann KG. Role of the membrane in the inactivation of factor Va by activated protein C. J Biol Chem 1993; 268:27246–27257.
87. Nesheim ME, Canfield WM, Kisiel W, Mann KG. Studies of the capacity of factor Xa to protect factor Va from inactivation by activated protein C. J Biol Chem 1982; 257:1443–1447.
88. Solymoss S, Tucker MM, Tracy PB. Kinetics of inactivation of membrane-bound factor Va by activated protein C. Protein S modulates factor Xa protection. J Biol Chem 1988; 263: 14884–14890.
89. Walker FJ. Regulation of activated protein C by protein S. The role of phospholipid in factor Va inactivation. J Biol Chem 1981; 256:11128–11131.
90. Bakker HM, Tans G, Janssen-Claessen T, et al. The effect of phospholipids, calcium ions and protein S on rate constants of human factor Va inactivation by activated human protein C. Eur J Biochem 1992; 208:171–178.

91. Krishnaswamy S, Williams EB, Mann KG. The binding of activated protein C to factors V and Va [published erratum appears in J Biol Chem Feb 5, 1987; 262(4):1926]. J Biol Chem 1986; 261:9684–9693.
92. Walker FJ, Scandella D, Fay PJ. Identification of the binding site for activated protein C on the light chain of factors V and VIII. J Biol Chem 1990; 265:1484–1489.
93. Mann KG, Jenny RJ, Krishnaswany S. Cofactor proteins in the assembly and expression of blood clotting enzyme complexes. Annu Rev Biochem 1988; 57:915–956.
94. Krishnaswamy S, Mann KG, Nesheim ME. The prothrombinase-catalyzed activation of prothrombin proceeds through the intermediate meizothrombin in an ordered, sequential reaction. J Biol Chem 1986; 261:8977–8984.
95. Rosing J, Zwaal RFA, Tans G. Formation of meizothrombin as intermediate in factor Xa-catalyzed prothrombin activation. J Biol Chem 1986; 261:4224–4228.
96. Tans G, Janssen Claessen T, Hemker HC, Zwaal RF, Rosing J. Meizothrombin formation during factor Xa-catalyzed prothrombin activation. Formation in a purified system and in plasma. J Biol Chem 1991; 266:21864–21873.
97. Krishnaswamy S, Church WR, Nesheim ME, Mann KG. Activation of human prothrombin by human prothrombinase. Influence of factor Va on the reaction mechanism. J Biol Chem 1987; 262:3291–3299.
98. Pryzdial ELG, Mann KG. The association of coagulation factor Xa and factor Va. J Biol Chem 1991; 266:8969–8977.
99. Krishnaswamy S, Jones KC, Mann KG. Prothrombinase complex assembly. Kinetic mechanism of enzyme assembly on phospholipid vesicles. J Biol Chem 1988; 263:3823–3834.
100. Miletich JP, Majerus DW, Majerus PW. Patients with congenital factor V deficiency have decreased factor Xa binding sites on their platelets. J Clin Invest 1978; 62:824–831.
101. Tucker MM, Foster WB, Katzmann JA, Mann KG. A monoclonal antibody which inhibits the factor Va: factor Xa interaction. J Biol Chem 1983; 258:1210–1214.
102. Annamalai AE, Rao AK, Chiu HC, et al. Epitope mapping of functional domains of human factor Va with human and murine monoclonal antibodies. Evidence for the interaction of heavy chain with factor Xa and calcium. Blood 1987; 70:139–146.
103. Husten EJ, Esmon CT, Johnson AE. The active site of blood coagulation factor Xa. Its distance from the phospholipid surface and its conformational sensitivity to components of the prothrombinase complex. J Biol Chem 1987; 262:12953–12961.
104. Kalafatis M, Xue JC, Lawler CM, Mann KG. Factor Xa-factor Va interactions involve the NH_2-termini of the A1 and A3 domains of factor Va Biochemistry 1994; 33:6538–6545.
105. Guinto ER, Esmon CT. Loss of prothrombin and of factor Xa-factor Va interactions upon inactivation of factor Va by activated protein C. J Biol Chem 1984; 259:13986–13992.
106. Zwaal RFA, Comfurius P, Bevers EM. Platelet procoagulant activity and microvesicle formation. Its putative role in hemostasis and thrombosis. Biochim Biophys Acta Mol Basis Dis 1992; 1180:1–8.
107. Tracy PB, Eide LL, Mann KG. Human prothrombinase complex assembly and function on isolated peripheral blood cell populations. J Biol Chem 1985; 260:2119–2124.
108. Bloom JW, Nesheim ME, Mann KG. Phospholipid-binding properties of bovine factor V and factor Va. Biochemistry 1979; 18:4419–4425.
109. Higgins DL, Mann KG. The interaction of bovine factor V and factor V-derived peptides with phospholipid vesicles. J Biol Chem 1983; 258:6503–6508.
110. Pusey ML, Nelsestuen GL. Membrane binding properties of blood coagulation factor V and derived peptides. Biochemistry 1984; 23:6202–6210.
111. Van de Waart P, Bruls H, Hemker HC, Lindhout T. Interaction of bovine blood clotting factor Va and its subunits with phospholipid vesicles. Biochemistry 1983; 22:2427–2432.
112. Tracy PB, Mann KG. Prothrombinase complex assembly on the platelet surface is mediated through the 74,000-dalton component of factor Va. Proc Natl Acad Sci USA 1983; 80:2380–2384.

113. Lampe PD, Pusey ML, Wei GJ, Nelsestuen GL. Electron microscopy and hydrodynamic properties of blood clotting factor V and activation fragments of factor V with phospholipid vesicles. J Biol Chem 1984; 259:9959–9964.
114. Pusey ML, Mayer LD, Wei GJ, Bloomfield VA, Nelsestuen GL. Kinetic and hydrodynamic analysis of blood clotting factor V-membrane binding. Biochemistry 1982; 21:5262–5269.
115. Lecompte MF, Krishnaswamy S, Mann KG, Nesheim ME, Gitler C. Membrane penetration of bovine factor V and Va detected by labeling with 5-iodonaphthalene-1-azide. J Biol Chem 1987; 262:1935–1937.
116. Krieg UC, Isaacs BS, Yemul SS, Esmon CT, Bayley H, Johnson AE. Interaction of blood coagulation factor Va with phospholipid vesicles examined by using lipophilic photoreagents. Biochemistry 1987; 26:103–109.
117. Nesheim ME, Pittman DD, Wang JH, Slonosky D, Giles AR, Kaufman RJ. The binding of ^{35}S-labeled recombinant factor VIII to activated and unactivated human platelets. J Biol Chem 1988; 263:16467–16470.
118. Ortel TL, Devore-Carter D, Quinn-Allen MA, Kane WH. Deletion analysis of recombinant human factor V. Evidence for a phosphatidylserine binding site in the second C-type domain. J Biol Chem 1992; 267:4189–4198.
119. Ortel TL, Quinn-Allen MA, Charles LA, Devore-Carter D, Kane WH. Characterization of an acquired inhibitor to coagulation factor V. Antibody binding to the second C-type domain of factor V inhibits the binding of factor V to phosphatidylserine and neutralizes procoagulant activity. J Clin Invest 1992; 90:2340–2347.
120. Ortel TL, Quinn-Allen MA, Keller FG, Peterson JA, Larocca LD, Kane WH. Localization of functionally important epitopes within the second C-type domain of coagulation factor V using recombinant chimeras. J Biol Chem 1994; 269:15898–15905.
121. Kalafatis M, Jenny RJ, Mann KG. Identification and characterization of a phospholipid-binding site of bovine factor Va. J Biol Chem 1990; 265:21580–21589.
122. Kalafatis M, Mann KG. Factor Va membrane interaction is mediated by 2 regions located on the light chain of the cofactor. Biochemistry 1994; 33:486–493.
123. Scandella D, Mahoney SD, Mattingly M, Roeder D, Timmons L, Fulcher CA. Epitope mapping of human factor VIII inhibitor antibodies by deletion analysis of factor VIII fragments expressed in *Escherichia coli*. Proc Natl Acad Sci USA 1988; 85:6152–6156.
124. Arai M, Scandella D, Hoyer LW. Molecular basis of factor VIII inhibition by human antibodies. Antibodies that bind to the factor VIII light chain prevent the interaction of factor VIII with phospholipid. J Clin Invest 1989; 83:1978–1984.
125. Miletich JP, Jackson CM, Majerus PW. Properties of the factor Xa binding site on human platelets. J Biol Chem 1978; 253:6908–6916.
126. Van de Waart P, Hemker HC, Lindhout T. Interaction of prothrombin with factor Va-phospholipid complexes. Biochemistry 1984; 23:2838–2842.
127. Luckow EA, Lyons DA, Ridgeway TM, Esmon CT, Laue TM. Interaction of clotting factor V heavy chain with prothrombin and prethrombin 1 and role of activated protein C in regulating this interaction: analysis by analytical ultracentrifugation. Biochemistry 1989; 28: 2348–2354.
128. Gewirtz AM, Shen YM. Effect of phorbol myristate acetate on c-myc, b-actin, and FV gene expression in morphologically recognizable human megakarocytes: a kinetic analysis employing in situ hybridization. Exp Hematol 1990; 18:945–952.
129. Gewirtz AM, Shapiro C, Shen YM, Boyd R, Colman RW. Cellular and molecular regulation of factor V expression in human megakaryocytes. J Cell Physiol 1992; 153:277–287.
130. Cerveny TJ, Fass DN, Mann KG. Synthesis of coagulation factor V by cultured aortic endothelium. Blood 1984; 63:1467–1474.
131. Rodgers GM. Vascular smooth muscle cells synthesize, secrete and express coagulation factor V. Biochim Biophys Acta 1988; 968:17–23.

132. Altieri DC, Edgington TS. Sequential receptor cascade for coagulation proteins on monocytes. Constitutive biosynthesis and functional prothrombinase activity of a membrane form of factor V/Va. J Biol Chem 1989; 264:2969–2972.
133. Rothberger H, McGee MP. Generation of coagulation factor V activity by cultured rabbit alveolar macrophages. J Exp Med 1984; 160:1880–1890.
134. Girolami A, De Marco L, Dal Bo Zanon R, Patrassi G, Cappellato MG. Rarer quantitative and qualitative abnormalities of coagulation. Clin Haematol 1985; 14:385–411.
135. Seeler RA. Parahemophilia: factor V deficiency. Med Clin North Am 1972; 56:119–125.
136. Melliger MJ, Duckert F. Major surgery in a subject with factor V deficiency: cholecystectomy in a parahaemophilic woman and a review of the literature. Thromb Diath Haemorrh 1971; 25:438–446.
137. Tracy PB, Mann KG. Abnormal formation of the prothrombinase complex: factor V deficiency and related disorders. Hum Pathol 1987; 18:162–169.
138. Miller SP. Coagulation dynamics in factor V deficiency: a family study, with a note on the occurrence of thrombophlebitis. Thromb Diath Haemorrh 1965; 13:500–515.
139. Reich NE, Hoffman GC, deWolfe VG, Van Ordstrand HS. Recurrent thrombophlebitis and pulmonary emboli in congenital factor 5 deficiency. Chest 1976; 69:113–114.
140. Manotti C, Quintavalla R, Pini M, Jeran M, Paolicelli M, Dettori AG. Thromboembolic manifestations and congenital factor V deficiency: a family study. Haemostasis 1989; 19: 331–334.
141. Petiot P, Croisile B, Confavreux C, et al. Thalamic stroke and congenital factor V deficiency. Stroke 1991; 22:1606.
142. Chiu HC, Whitaker E, Colman RW. Heterogeneity of human factor V deficiency. Evidence for the existence of antigen-positive variants. J Clin Invest 1983; 72:493–503.
143. Tracy PB, Giles AR, Mann KG, Eide LL. Hoogendoorn H, Rivard GE. Factor V (Quebec): a bleeding diathesis associated with a qualitative platelet factor V deficiency. J Clin Invest 1984; 74:1221–1228.
144. Soff GA, Levin J. Familial multiple coagulation factor deficiencies. I. Review of the literature: differentiation of single hereditary disorders associated with multiple factor deficiencies from coincidental concurrence of single factor deficiency states. Semin Thromb Hemost 1989; 7:112–148.
145. Seligsohn U, Zivelin A, Zwant E. Combined factor V and factor VIII deficiency among non-Ashkenazi Jews. N Engl J Med 1982; 307:1191–1195.
146. Hultin MB, Eyster ME. Combined factor V-VIII deficiency: a case report with studies of factor V and VIII activation by thrombin. Blood 1981; 58:983–985.
147. Brown JM, Selik NR, Voelpel MJ, Mammen EF. Combined factor V/VIII deficiency: a case report including levels of factor V and factor VIII coagulant and antigen as well as protein C inhibitor. Am J Hematol 1985; 20:401–407.
148. Suzuki K, Nishioka J, Hashimoto S, Kamiya T, Saito H. Normal titer of functional and immunoreactive protein-C inhibitor in plasma of patients with congenital combined deficiency of factor V and factor VIII. Blood 1983; 62:1266–1270.
149. Seligsohn U, Zivelin A, Zwang E. Decreased factor VIII clotting antigen levels in the combined factor V and VIII deficiency. Thromb Res 1983; 33:95–98.
150. Bern MM, Suzuki K, Mann K, et al. Response of protein C and protein C inhibitor to warfarin therapy in patient with combined deficiency of factors V and VIII. Thromb Res 1984; 36:485–495.
151. Marlar RA, Griffin JH. Deficiency of protein C inhibitor in combined factor V/VIII deficiency disease. J Clin Invest 1980; 66:1186–1189.
152. Canfield WM, Kisiel W. Evidence of normal functional levels of activated protein C inhibitor in combined factor V/VIII deficiency disease. J Clin Invest 1982; 70:1260–1272.
153. Gardiner JE, Griffin JH. Studies on human protein C inhibitor in normal and factor V/VIII deficient plasmas. Thromb Res 1984; 36:197–203.

154. Garcia VV, Silva IA, Borrasca AL. Response of factor VIII/von Willebrand factor to intranasal DDAVP in healthy subjects and mild haemophiliacs (with observations in patients with combined deficiency of factors V and VIII). Thromb Haemost 1982; 48:91–93.
155. Nesheim ME, Nichols WL, Cole TL, et al. Isolation and study of an acquired inhibitor of human coagulation factor V. J Clin Invest 1986; 77:405–415.
156. Coots MC, Muhleman AF, Glueck HI. Hemorrhagic death associated with a high titer factor V inhibitor. Am J Hematol 1978; 4:193–206.
157. Zehnder JL, Leung LLK. Development of antibodies to thrombin and factor V with recurrent bleeding in a patient exposed to topical bovine thrombin. Blood 1990; 76:2011–2016.
158. Rapaport SI, Zivelin A, Minow RA, Hunter CS, Donnelly K. Clinical significance of antibodies to bovine and human thrombin and factor V after surgical use of bovine thrombin. Am J Clin Pathol 1992; 97:84–91.
159. Chediak J, Ashenhurst JB, Garlick I, Desser RK. Successful management of bleeding in a patient with factor V inhibitor by platelet transfusions. Blood 1980; 56:835–841.
160. Genetics Computer Group. Program manual for the GCG package, Version 7, April 1991, 7th ed. Madison, WI: Genetics Computer Group, 1991.
161. Dahlbäck B, Hildebrand B. Inherited resistance to activated protein C is corrected by anticoagulant cofactor activity found to be a property of factor V. Proc Natl Acad Sci USA 1994; 91:1396–1400.
162. Bertina RM, Koeleman BPC, Koster T, et al. Mutation in blood coagulation factor V associated with resistance to activated protein C. Nature 1994; 369:64–67.
163. Dahlbäck B, Carlsson M, Svensson PJ. Familial thrombophilia due to a previously unrecognized mechanism characterized by poor anticoagulant response to activated protein C: Prediction of a cofactor to activated protein C. Proc Natl Acad Sci USA 1993; 90:1004–1008.
164. Svensson PJ, Dahlbäck B. Resistance to activated protein C as a basis for venous thrombosis. N Engl J Med 1994; 330:517–522.
165. Koster T, Rosendaal FR, De Ronde H, Briët E, Vandenbroucke JP, Bertina RM. Venous thrombosis due to poor anticoagulant response to activated protein C: Leiden Thrombophilia Study. Lancet 1993; 342:1503–1506.
166. Griffin JH, Evatt B, Wideman C, Fernández JA. Anticoagulant protein C pathway defective in majority of thrombophilic patients. Blood 1993; 82:1989–1993.
167. Shen L, Dahlbäck B. Factor V and protein S as synergistic cofactors to activated protein C in degradation of factor VIIIa. J Biol Chem 1994; 269:18735–18738.
168. Zöller B, Dahlbäck B. Linkage between inherited resistance to activated protein C and factor V gene mutation in venous thrombosis. Lancet 1994; 343:1536–1538.

7

Factor VII

Lars C. Petersen and Ulla Hedner
Novo Nordisk A/S, Gentofte, Denmark

Peter Wildgoose
Selectide Corporation, Tucson, Arizona

I. INTRODUCTION

Recent research has provided increasing evidence of the importance of factor VIIa in initiation of blood coagulation via the so-called extrinsic coagulation pathway (see Refs. 1–4 for reviews). Under normal conditions factor VII circulates in the blood, primarily in its inactive zymogen form, unable to initiate the formation of a hemostatic plug. Upon vascular injury, however, cell surface tissue factor (TF) is exposed and readily forms a one-to-one stoichiometric complex with factor VII. The formation of a TF complex has two functions. It augments the activation of factor VII to VIIa, as well as enhancing the proteolytic activity of factor VIIa toward its substrates. Once complexed to TF, the zymogen factor VII is rapidly converted to factor VIIa by cleavage of a single internal peptide bond located at Arg^{152}-Ile^{153}. A number of serine proteases, including factor Xa, factor IXa, factor XIIa, thrombin, and factor VIIa, are capable of activating factor VII under in vitro conditions. Precisely which of these proteases is responsible for the *in vivo* activation of factor VII remains unknown. Once formed, however, the factor VIIa-TF complex rapidly initiates blood coagulation by proteolytically activating its substrates factors IX and X, which eventually leads to thrombin formation and a fibrin clot.

Factor VII is unique among the vitamin K-dependent coagulation proteases in that it possesses a half-life of 2–3 h regardless of whether it is present as the zymogen or in an activated state (5). This is in sharp contrast to other vitamin K-dependent proteases, such as factors IX and X, which in their activated forms are cleared from the circulation within a few minutes. Unlike these other coagulation proteases, there are also no direct inhibitors of factor VIIa in plasma. The only physiologically significant factor VIIa inhibitor is tissue factor pathway inhibitor (TFPI), and it exerts its inhibitory properties only in the presence of factor Xa and then only when the factor VIIa molecule is bound to TF.

Increasing evidence has accumulated that blood coagulation is a permanently ongoing process (6,7) and that the state of activation in an undisturbed blood vessel, that is, the basal level, results mainly from the activity of factor VIIa-TF (8,9). Although important for the maintenance of basal coagulation activation, the TF-dependent pathway is not sufficient for normal hemostasis when it comes to more serious bleeding episodes. The clinical symptoms of hemophilia A and B patients demonstrate the mandatory role of factors VIII and IX when more excessive activation of the coagulation system is required. This is in accordance with a model in which the intrinsic pathway serves as a powerful feedback loop (2,4). It is possible to surpass this enhancement mechanism, however, by reinforcement of the extrinsic pathway with high levels of factor VIIa. Thus the use of recombinant factor VIIa has proven to be very successful in preventing abnormal bleeding in hemophilia patients with acquired inhibitors (10). This places factor VII (FVII) in a central position with respect to its role in initiation and regulation of the coagulation process. A detailed knowledge of FVII structure and function is thus pertinent to a thorough understanding of the hemostatic process as it exists in the normal and diseased states.

Because human factor VII is present in plasma in only trace concentrations (11), it was not until the cloning and production of recombinant factor VIIa in the latter half of the 1980s that it became possible to investigate thoroughly the function of this molecule. The structural and functional properties that make factor VII uniquely suited as a key enzyme in the regulation of the hemostatic process have since been the subject of much research. In many ways factor VII is unique compared with other coagulation factors. These unique properties are likely clues to an understanding of the regulatory mechanisms in which factor VII is involved. However, it is clear that much has yet to be resolved before a final and thorough description of the overall regulation can be given.

II. PROTEIN STRUCTURE

A. Factor VII Primary Structure

Because factor VII is present in human plasma in only trace concentrations (400 ng/ml), initial attempts to characterize the primary structure of factor VII by protein chemical techniques resulted in only a partial amino acid sequence (12). It was not until cloning of several factor VII cDNAs that the complete primary amino acid sequence of factor VII was determined (13). In addition to providing the sequence for the mature protein, this also provided information about its precursor structure during biosynthesis. The factor VII molecule, like the other vitamin K-dependent proteases, is initially synthesized as a precursor with a 38 amino acid pre-pro leader segment immediately preceding the mature amino terminus of the protein structure that circulates in plasma. The first 20 amino acids have a structure resembling a classic signal peptide, which directs the nascent polypeptide to associate with the rough endoplasmic reticulum (ER) during translation and subsequent translocation of the protein into the lumen of the ER and into the cellular secretory pathway. This first section is consequently referred to as the ''prepeptide'' or ''signal'' peptide. The latter 18 amino acids of the leader sequence possess a strong similarity to the corresponding structure found in the other vitamin K-dependent coagulation factors and has since been shown to be a recognition signal for an endoplasmic carboxylase enzyme that posttranslationally modifies specific glutamic acid residues present in the N terminus of the protein (14). This latter section has been called the propeptide. The remainder of the cDNA encodes a mature protein for factor VII of 406 amino acids with a predicted molecular weight of 45,500. The sequence shown in Figure 1 predicts two sites for potential N-linked carbohydrate addition, which together bring the predicted molecular weight of 50,000 into close agreement with that of the plasma protein.

The primary sequence of factor VII possesses a great deal of similarity to other vitamin K-dependent coagulation proteases and provides important clues concerning its structure and function. As shown in Figure 1, the protein sequence can be viewed as follows. The first 152 amino acids comprise a "light" (smaller) chain ($M_w \simeq 20$ kD), which is connected by a single disulfide bond to the "heavy" chain ($M_w \simeq 30$ kD), comprised of the last 254 amino acids following cleavage after Arg^{152}. The light chain contains the Gla (γ-carboxyglutamic acid) domain consisting of approximately the first 40 amino acids and probably constitutes the phospholipid binding domain of the protein. The Gla domain is followed by two repeating copies of a 36 amino acid domain with considerable sequence similarity to epidermal growth factor (EGF). These two EGF-like domains contain cysteine patterns identical to those of EGF and therefore probably fold into structures resembling the growth factor.

Directly following the FVII EGF domains is a so-called connecting region that contains the activation cleavage site (Arg^{152}-Ile^{153}) at which the two chains are eventually cleaved from each other. The final 254 amino acids of the protein comprise a domain called the catalytic or serine proteinase domain because of its striking sequence homology

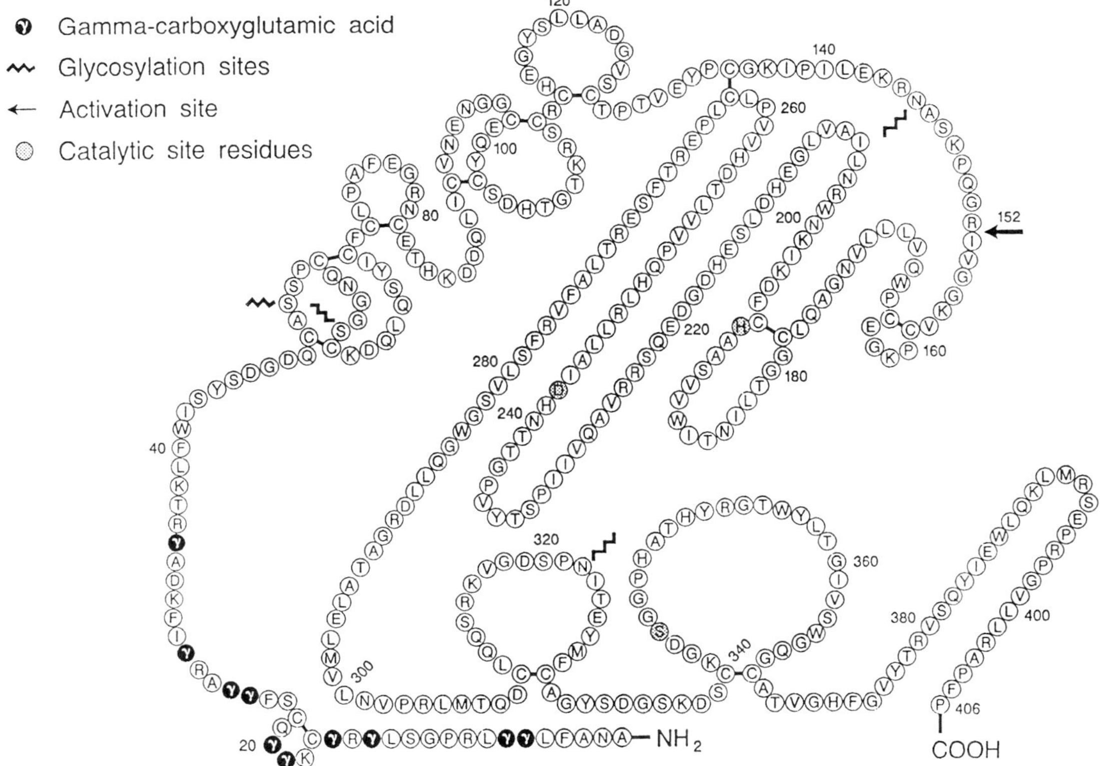

Figure 1 Human coagulation factor VII.

with the trypsin family of serine proteases. This domain contains the catalytic triad (consisting of Ser^{344}, Asp^{242}, and His^{193}) with which factor VIIa cleaves its protein substrates, factors IX and X.

B. Posttranslational Modifications

Like other coagulation proteases, factor VII undergoes several posttranslational modifications, including glycosylation as well as γ-carboxylation, of several N-terminal glutamic acid residues. This latter modification involves the action of a carboxylase enzyme, which in the presence of reduced vitamin K fixes carbon dioxide to form a new carboxyl group on the glutamic acid while converting vitamin K to vitamin K epoxide (14). This posttranslational processing of Gla residues is limited to 10 sites, all located within the first 38 residues of the factor VII light chain (generally referred to as the Gla domain). Full carboxylation of these residues is essential for calcium binding and expression of factor VIIa coagulant activity. Based on studies with other vitamin K-dependent proteases, such as factor IX and factor X, γ-carboxylation of the Gla domain is essential for calcium binding, which effects a conformational change that is crucial for interaction of the protease with the phospholipid membrane.

Factor VII also undergoes posttranslational N-glycosylation of two Asn-*X*-Thr/Ser sites located at positions 145–147 and 322–324 (15). Both sites are normally fully glycosylated, and the structures are composed of *N*-acetylglucosamine, mannose, galactose, fucose, and sialic acid. Factor VII also contains several novel O-linked glycosylations located within the EGF domains (16). These consist of approximately equal amounts of glucose, glucose-xylose, or glucose $(xylose)_2$ O-glycosidically linked to Ser^{52}, and in addition to these O-linked glycan structures factor VII also contains a single fucose covalently linked to Ser^{60} (16). Similar glycans have also been found attached to conserved serine residues in factor IX (17) and human and bovine protein Z, as well as bovine factor VII (18). Precisely what function these novel carbohydrate moieties play is unclear. It is unlikely, however, that they contribute to the biological function of the protein in light of recent findings that factor VII mutated at either position 52 and/or 60 (Ser to Ala) possesses essentially normal coagulant activity (16). Based on sequence comparison of the EGF homology regions that contain β-hydroxylated Asp or Asn, a consensus sequence that seems to be required for this modification has been identified. Although factor VII contains this consensus sequence in EGF domain 1, the Asp^{63} of human factor VII has been shown not to be β-hydroxylated (15).

III. FACTOR VII GENE STRUCTURE

The factor VII cDNA has been used as a molecular probe to isolate the human structural gene for factor VII from human genomic DNA (19). By analysis of the isolated gene with restriction enzymes and DNA sequencing, the organization of the gene has been determined. The human factor VII gene is located on chromosome 13, where factor X also resides (20). The factor VII gene spans a distance of approximately 13 kb. The gene is comprised of eight exons interrupted by seven introns and is organized in a manner highly similar to the gene organization of the other vitamin K coagulation factors, such as factor X (21), factor IX (22), and protein C (23) (Table 1).

In each of these cases, the introns occur in precisely identical positions in the gene, relative to the corresponding sections of the protein encoded by the exons, suggesting

Table 1 Comparison of the Positions, Phases, and Lengths of the Introns of Human Factor VII, Factor IX, Factor X, and Protein C[a]

	Position								Phase								Size							
Intron	A1	A	B	C	D	E	F	G	A1	A	B	C	D	E	F	G	A1	A	B	C	D	E	F	G
Factor VII	−40	−17	37/38	46	84	131	167/168	209	I	I	0	I	I	I	0	I	1067	2574	1919	68	1908	971	595	816
Factor IX		−17	38/39	47	85	128	195/196	234		I	0	I	I	I	0	I		6206	188	3689	7163	2365	9473	668
Factor X		−17	37/38	46	84	128	209/210	249		I	0	I	I	I	0	I		?	7400	850	1800	2900	3400	1700
Protein C		−19	37/38	46	92	137	184/185	224		I	0	I	I	I	0	I		1263	1462	92	102	2668	873	1129

[a]The positions are give in amino acid residue number for each protein. Phase I = the intron following the first nucleotide of a codon; phase II = the intron following the second nucleotide of a codon; phase 0 = the intron between codons. The length of the introns is given in nucleotide pairs.

Source: Adapted from O'Hara et al.

that all these enzymes evolved by gene duplication and divergent evolution from a common ancestral gene whose structural organization has been preserved. Although the precise position of introns have been strongly conserved in this family, their internal intervening sequences and their sizes have diverged markedly, to the extent that there are no recognizable similarities in intronic sequences.

Interestingly, the genes for this family of enzymes appear to be organized around functional units of the protein molecule. Regions of the protein that can be thought of as individual functional domains also correspond to sections of the amino acid sequence encoded in their entirety on individual exons in the gene structure. For example, the carboxylase recognition signal and the Gla domain reside together on an exon. Each of the EGF domains is precisely encoded by an exon, as is a very short eight amino acid ''hydrophobic stack'' located between the Gla and EGF domains. Such apparent correlation between seemingly functional domains and exons has been widely observed in the genome of mammals and has led to the speculation that one evolutionary mechanism for the design of new protein or enzyme functions may occur through the phenomenon of ''exon shuffling,'' whereby intact functional units of different proteins can be spliced together by genetic means to create new enzymes with novel properties. The general gene organization for this family of coagulation factors seems to support such an interpretation.

IV. FACTOR VII INTERACTIONS

A. Factor VII-TF Binding

As mentioned previously, factor VIIa possesses little coagulant activity in itself, and it is only upon binding to its cell surface cofactor, tissue factor, that factor VIIa possesses significant proteolytic activity to activate its substrates, factors IX and X. TF is an integral cell membrane glycoprotein that is irreversibly associated with the lipid bilayer (24). This is in sharp contrast to other coagulation cofactors, such as factors Va and VIIIa, which circulate in plasma and form reversible associations with cell surface phospholipids. TF also differs from other coagulation cofactors in that it is bifunctional in nature, accelerating the conversion of factor VII to VIIa as well as enhancing the proteolytic activity of factor VIIa toward its substrates (25). Although factor VIIa possesses full activity toward simple peptide substrates solely in the presence of TF apoprotein and calcium (26), this is not the case for physiological substrates, such as factor IX or factor X. These latter reactions also require negatively charged phospholipids for full expression of factor VIIa proteolytic activity (27). This has led to the hypothesis that anatomical localization of TF is not the only factor regulating *in vivo* factor VIIa proteolytic activity. In this regard, it has been shown that the procoagulant activity of a TF-producing cell line can be modulated by calcium-mediated changes in the asymmetrical distribution of negatively charged phospholipids, such as phosphatidylserine (28). Because phospholipid is not directly required for the formation of a factor VIIa-TF complex, the functional increase in factor VIIa activity probably reflects increased substrate binding to the phospholipid-factor VIIa-TF complex.

Numerous studies have been conducted to assess which structural regions of the factor VIIa molecule are involved in the formation of a factor VIIa-TF complex. It has long been known that γ-carboxylation of the factor VII Gla domain is crucial for manifestation of full biological activity, but it was not until more recently that the Gla domain was directly indicated in TF binding. Experiments conducted with Gla-domainless factor VIIa indicate that proteolytic removal of the first 38 residues results in a factor VIIa molecule incapable of binding to cell surface-expressed TF (29). Although these experiments clearly indicate the involvement of the FVII Gla domain, it is not clear whether this is a result of direct protein-protein interactions or a more indirect effect caused by allosteric interactions mediated by the binding of calcium ions to the factor VII Gla domain. Competition binding experiments indicate that the purified FVII Gla domain (residues 1–38) is incapable of inhibiting the specific binding of factor VIIa to cell surface TF (30), indicating that the factor VII Gla domain does not, in itself, contain the structural elements necessary for the formation of a stable binary factor VIIa-TF complex. Other studies using synthetic peptides (resides 195–206) to factor VII have implicated regions in the protease domain as important in TF binding (31). An interaction between this region and the factor VII Gla domain is suggested by the observation that antibodies raised against residues 195–206 are calcium dependent and bind tighter to full-length factor VIIa than to Gladomainless factor VIIa (31). This suggests that the factor VII Gla domain is necessary in inducing a calcium-dependent conformational change in the factor VIIa molecule, which in turn expresses one or more neoepitopes directly involved in TF binding. Other calcium binding sites, such as that recently identified in the factor VII protease domain (residues 210–220), also appear to contribute to factor VIIa-TF binding, presumably by a similar mechanism (32,33). Binding of TF to structures outside the Gla region is also indicated by studies that show that the amidolytic activity of Gla-domainless factor VIIa is stimulated by TF and calcium ions (34). It should be pointed out, however, that the affinity of TF for truncated factor VIIa, as determined by this method, is at least an order of magnitude lower than that of the full-length protein.

The EGF domains of factor VIIa have also been implicated as possible TF binding sites. It has been demonstrated that a monoclonal antibody that specifically reacts with the first EGF domain (residues 51–88) directly inhibits the binding of factor VIIa to TF (35). Others have shown that factor VII/factor IX chimeras, consisting of various portions of the factor VII light chain connected to the factor IX heavy chain, bind to TF apoprotein with an affinity essentially equal to that of wild-type recombinant factor VIIa (36). This has been used as evidence to support the notion that the high-affinity TF binding site is located in the EGF domains. A limitation in the interpretation of these latter studies is that they were conducted with factor VII variants that were undercarboxylated, making it impossible to separate the individual contributions of the factor VII EGF and Gla domains. More recent studies attempted to address this problem by studying the TF binding properties of factor VII fragments obtained by tryptic digestion of fully carboxylated wild-type VIIa (37). Fragments corresponding to residues 1–109 were purified and found to inhibit the binding of factor VIIa to cell surface TF. Interestingly, cleavage of the factor VII Gla domain by treatment with cathepsin G resulted in complete loss of inhibitory activity, suggesting that the calcium ion-dependent conformation of the Gla domain provides some degree of structural integrity necessary for the interaction of the EGF domain with TF.

Taken collectively, these studies indicate that the physical interaction of factor VIIa with TF involves multiple regions of the factor VIIa molecule, including the N-terminal Gla domain and the EGF domains, as well as residues located in the factor VII protease domain.

B. Calcium Ion Binding

Like all reactions involving vitamin K-dependent coagulation factors, the initiation of coagulation via factor VIIa-TF is strictly dependent on the presence of calcium ions. Although the binding of calcium ions to factor VII generally resembles that of the other vitamin K-dependent coagulation factors, it is important to stress that several marked differences exist. It is interesting to note, for instance, that the amidolytic activity of factor VIIa toward peptidyl substrates is strongly enhanced by calcium ions in both the presence and absence of TF.

Detailed calcium ion binding studies have been conducted on a number of vitamin K-dependent coagulation factors including, factor IX (38), factor X (39,40), and prothrombin (41). Existing evidence suggests that the Gla domain of all these vitamin K-dependent coagulation factors contains several low-affinity (>1 mM) calcium binding sites that support the binding of the vitamin K-dependent proteins to acid phospholipid vesicles. X-ray crystallographic studies conducted using bovine prothrombin fragment I indicate that calcium binding to the Gla region induces the folding of a relatively disordered protein structure into an ordered one (42). Although similar x-ray crystallographic studies have not been conducted with factor VIIa, Ca^{2+} ion-induced conformational change similar to that observed in prothrombin fragment I is indicated by a number of observations. The binding of several antibodies directed toward specific epitopes of the factor VII Gla domain is Ca^{2+} ion dependent (43), as is the cathepsin G-mediated proteolytic cleavage of the Gla region at Tyr^{44}-Ser^{45} (44). Direct analysis by equilibrium dialysis of the calcium binding properties of the isolated factor VII Gla domain indicates that it can bind five calcium ions with a K_d of 1.5 mM per site (45). Binding studies on the isolated Gla domains of factor IX and factor X have shown a reduced number of binding sites, although the affinity of calcium for these sites is similar to what is observed with factor VII.

From the striking homology among the Gla domains of the vitamin K-dependent coagulation proteases, one might assume that the affinity of the calcium-Gla domain complexes for phospholipids would also be very much the same. This is not the case, however. In the presence of Ca^{2+} ions, factor IX and in particular factor X bind much more strongly to phospholipids than factor VII (46,47). The reported values (47) suggest that the affinity of factor VII for phosphatidyl choline/phosphatidyl serine structures (K_d = 17 μM) is several orders of magnitude weaker than that of factor X (K_d = 0.25 μM). Thus direct binding of factor VIIa to phospholipids via Gla-calcium complexes is unlikely to play a role in the in vivo adherence of factor VIIa to cell surfaces.

Like other vitamin K-dependent proteases, factor VII also contains several Gla-independent calcium binding sites. Equilibrium dialysis and Ca^{2+} ion electrode measurements of calcium binding to Gla-domainless factor VIIa (residues 39–406) indicate that there are two calcium binding sites outside the Gla region (45,48). The K_d of these sites differ somewhat with respect to the method used, equilibrium dialysis estimating K_d of 0.2 and 10 mM (45) and the calcium-specific electrode measuring an affinity of 0.2 mM for both sites (48). By analogy to factor IX and factor X, one of these Ca^{2+} binding sites is probably located in the first EGF-like domain, with Asp^{46}, Asp^{48}, and Asp^{63} as putative ligands. The second Gla-independent calcium binding site appears to be located in the protease domain of factor VII. Evidence for the involvement of this region in calcium binding is provided by the observation that a synthetic peptide spanning FVII residues 206–242 binds calcium with a K_d of 0.2 mM (48). Homology considerations suggest that the calcium binding ligands in factor VIIa are probably provided by the carboxyl oxygens of Glu^{210} and Glu^{220} and the main-chain carbonyl oxygens of Asp^{212} and Glu^{215} (49). Direct support for this hypothesis was obtained using site-specific mutants of factor VII in which Glu^{220} was changed to either a lysine (E22OK FVII) or alanine (E220A) (32). Consistent with the existence of a calcium binding site involving Glu^{220}, mutagenesis of this residue resulted in a prominent decrease in the affinity of the factor VII molecule for calcium ions, as evidenced using the luminescent lanthanide series element terbium as a probe. Under normal conditions, terbium binding is independent of the factor VII Gla domain and can be directly inhibited by calcium (50). Characterization of the functional properties of these mutants indicates that calcium binding to this region is crucial for the normal interaction of factor VIIa with its cofactor, TF (32). Involvement of this calcium binding loop in enzyme-cofactor interactions appears to be a common theme, with the analogous regions of factor IX (51) and factor X (52) implicated in factor VIII and factor V binding, respectively.

Equilibrium binding studies conducted using full-length factor VII indicate that the zymogen form bonds a total of 10 calcium ions, the first 3 calcium ions binding in a cooperative manner (45). Similar results have been obtained with prothrombin fragment 1 (53), and it has been argued that this results from a direct interaction between the Gla domain and the EGF domains. Interestingly, activation of factor VII to VIIa results in loss of cooperativity and loss of a calcium binding site, implying that the binding of calcium to FVII also involves interactions between the factor VII heavy and light chains (45).

V. ZYMOGEN ACTIVATION REACTIONS

A. Activation of Factor VII

For many years zymogen factor VII was considered to possess inherent enzymatic activity, mainly because of its ability to incorporate diisopropyl fluorophosphate (isofluorophate) (54). It was thus postulated that initiation of the extrinsic coagulation pathway involved the formation of a zymogen factor VII-TF complex, which then initiated coagulation by directly activating factor X. The factor Xa, in turn, reciprocally activated the TF-bound factor VII, resulting in amplification of factor VII proteolytic activity (55). This intrinsic activity hypothesis was subsequently contradicted by experiments showing that a factor IX variant incapable of feedback-activating factor VII is not activated by a complex of zymogen factor VII-TF (56). In addition, there is direct evidence that peptide chloromethyl ketones specific for the FVII active site react exclusively with the activated form of factor VII (57). These observations are further supported by experiments showing that an activation cleavage site mutant of factor VII (R152E FVII), designed to remain in the one-chain zymogen form, possesses no intrinsic proteolytic activity toward either factor X or factor IX (58). Taken together, these findings suggest that factor VII is a true zymogen and absolutely requires activation for expression of significant biological activity.

In vitro activation studies reveal that factor VII is most rapidly activated by factor Xa (59–61), although several other proteases, including factor IXa (62), factor XIIa (63,64), and thrombin (65), are also capable of activating factor VII. Like all vitamin K-dependent zymogen activation reactions, this also requires the presence of calcium ions and phospholipid. The activation of factor VII differs from the activation of other vitamin K-dependent proteases, however, in that it is greatly enhanced by binding to its cofactor,

TF. Formation of a zymogen factor VII-TF complex renders the scissile bond at Arg^{152}-Ile^{153} extremely susceptible to cleavage by trace amounts of factor Xa or IXa. More recent studies indicate that the factor VII-TF complex is also susceptible to autoactivation by factor VIIa (66,67) and that this reaction is actually favored in the presence of plasma concentrations of antithrombin III.

Precisely which of these proteases is responsible for the activation of factor VII under in vivo conditions is less clear. Recent studies have attempted to mimic what happens in vivo by assessing the ability of normal pooled plasma and various congenitally factor-deficient plasmas to support the activation of an inactive factor VII mutant (S344A factor VII) on the surface of a TF-expressing cell line (J82 bladder carcinoma cells) (68). It was found that incubation of the S344A FVII-TF complex with a 10-fold dilution of normal pooled plasma in the presence of hirudin resulted in a time-dependent conversion of S344A FVII to its VIIa counterpart. Substitution of the normal plasma with plasma congenitally deficient in factor IX, factor VIII, factor XI, factor V, or prothrombin resulted in qualitatively similar activation rates. Incubation with factor X-deficient plasma also resulted in activation of S344A FVII, but at a reduced rate of that observed with normal pooled plasma. In sharp contrast, incubation of the [^{125}I]S344A FVII with factor VII-deficient plasma produced no detectable activation. From these studies the authors suggest that the autoactivation of factor VII by VIIa plays a meaningful role in the initiation of extrinsic blood coagulation.

The question about in vivo activation of FVII was also addressed by studying the degree of factor VII activation under basal (i.e., in the absence of thrombosis or provocative stimuli) conditions. Using the recently developed factor VIIa-specific coagulation assay (69,70), it was possible to measure and compare directly the circulating factor VIIa levels in normal plasma as well as plasma from severe factor VIII- or factor IX-deficient patients (71). The data from this study indicate that normal individuals possess an average functional factor VIIa level corresponding to 1% of their total FVII antigen (assuming an average factor VII/VIIa antigen level of 400 ng/ml). Factor VIII deficiency has a small but statistically significant effect on circulating factor VIIa levels, patients on the average possessing a functional factor VIIa level corresponding to 60% of that observed in normal individuals. Factor IX deficiency, on the other hand, has a far greater effect, factor IX-deficient patients possessing a basal factor VIIa level corresponding to <10% of that observed in normal individuals. When taken in conjunction with previous observations that factor VIII- and factor IX-deficient individuals possess essentially normal levels of factor Xa (9), these data support the notion that factor IXa is responsible for factor VII activation under in vivo conditions. It is important to stress, however, that this factor IX-dependent activation may be important only under nonstimulatory and nonthrombotic conditions, and it is quite possible that other activation mechanisms such as autoactivation and/or factor Xa-mediated activation, take over immediately following vascular injury.

B. Activation of Factors IX and X

As already discussed, it is not until formation of a FVIIa-TF complex that the factor VIIa molecule possesses sufficient proteolytic activity to activate its substrates, factor IX or factor X. Because factor VIIa-TF can directly activate factor X and the function of factor IX is also to activate factor X, there was early skepticism about the physiological importance of the factor VIIa-TF–dependent mechanism of factor IX activation. Precisely which of these proteases is the preferred substrate has been a matter of considerable debate.

Detailed kinetic parameters for the activation of factor IX and X by factor VIIa-TF have been reported by a number of groups, utilizing a variety of TF sources. With TF-producing cell lines as a source of TF, factor VIIa activates factors IX and X with a K_m corresponding to 200 and 300 nM, respectively (72–74). Significantly lower K_m values (20–250 nM) are obtained when the intact TF-expressing cells are substituted with purified reconstituted TF apoprotein or lysed and sonicated cell extracts. As expected, the reported k_{cat} values also vary widely with the source of TF used. Reported k_{cat} values vary between 0.25 and 2.0 s^{-1} for factor IX activation, with values between 0.5 and 3.0 s^{-1} reported for the activation of factor X by FVIIa-TF. In practice, a number of theoretical and practical complications can account for the observed variability in kinetic parameters just mentioned. Probably the most serious complication has to do with the effector function of TF. That TF is an essential activator of factor VIIa implies that the apparent K_m is not a single constant but rather a function of the relative concentrations of TF and factor VIIa (75). Other complications that add to the complexity of this system arise as a consequence of substrate (factor IX or X) binding to the effector (FVII-TF). The interaction of factors IX and X with the factor VIIa-TF complex depends among other things on the nature of the phospholipids surrounding the TF apoprotein. Thus direct extrapolation of kinetic data, obtained under in vitro conditions, to the physiological situation must be undertaken with extreme caution.

Of course some attempts have also been made to assess the *in vivo* contribution of the factor VIIa-TF complex to factor IX and X activation under basal conditions. An indirect assessment of plasma factor IXa levels, by measurement of factor IX activation peptide (8), reveals that plasma from factor XI-deficient patients has essentially the same basal levels of factor IX activation peptide as that from normal individuals. In contrast, factor VII-deficient patients exhibit markedly diminished levels of factor IX activation peptide, suggesting that under basal conditions factor IXa levels are entirely dependent on the action of factor VIIa-TF. Similar experiments have been conducted to assess the relative contribution of factor VIIa to factor X activation (9). Lower than normal levels of factor X activation peptide levels were observed in hereditary factor VII-deficient patients, whereas patients deficient in either factor VIII or factor IX has plasma factor X activation peptide levels similar to those of normal subjects. Collectively these clinical observations indicate that the activation and maintenance of basal levels of factor IXa and Xa are mainly dependent on the action of factor VIIa-TF. Precisely what accounts for the trace amounts of factor VIIa, however, remains uncertain.

VI. FACTOR VII DEFICIENCY AND VARIANTS

Hereditary factor VII deficiency is a rare autosomal recessive disorder with an estimated prevalence of 1 in 500,000. The clinical picture of factor VII deficiency is variable, approximately 16% of homozygotes reported to experience episodes of cerebral hemorrhage (76). Interestingly, there have also been scattered reports of thromboembolism associated with factor VII deficiency (77–79). Precisely how these latter cases occur, in light of our current knowledge of factor VIIa-TF function, is unclear.

Characterization of naturally occurring mutants has contributed significantly to the structure-function relationships of coagulation proteases, such as factor IX and factor X. To date, however, only relatively few factor VII mutants have been characterized at the molecular level. The most commonly encountered and best studied factor VII variant is an Arg^{304} to Gln variant that was first described in a homozygous individual (80). Although this mutation is not associated with any clinical bleeding tendency, a significantly reduced factor VII coagulant activity is observed when measured in the presence of human thromboplastin. Interestingly, no factor VII coagulant activity is detected when the coagulation assays are performed using rabbit brain thromboplastin. Ligand blot analysis of the patient material suggests that the reduced coagulant activity is caused by a decreased affinity for TF apoprotein. A Japanese variant in which Arg^{304} was mutated to tryptophan has also been reported (81) and appears to behave in a similar manner. Interestingly, an Arg^{304}-Gln mutation (FVII Richmond) has also been reported in a patient suffering from pulmonary embolism and deep vein thrombosis (82). Computer modeling of the serine protease domain suggests that Arg^{304} is involved in stabilization of the core structure of this domain and that abolition of the positive charge might disrupt the conformation of the protein and thus indirectly affect its interactions with substrates and cofactors. Because heterozygous forms of this mutation (83) have been reported in conjunction with von Willebrand disease and factor IX deficiency, it is possible that the thrombotic complications associated with factor VII Richmond actually stem from some other, as yet unidentified, complication.

An equally complicated picture emerges from characterization of the variant known as factor VII Charlotte. This variant is associated with two missense mutations in the factor VII gene (R79Q and R152Q). A preliminary report stated that the patient was homozygous for the Arg^{79} mutation and heterozygous for the Arg^{152} mutation (84), but subsequent work demonstrated clearly that the patient was homozygous for both mutations (85). It seems likely that the patient's severe phenotype (<1% factor VII activity) is caused by the mutation at Arg^{152}, which prevents activation of factor VII. The role of the R79Q mutation is unclear; one study with a recombinant R79Q molecule reported that the molecule was fully active (86), but other groups reported reduced activity with a recombinant R79Q molecule (Ref. 86 and John McVey, personal communication). Recently, a patient was described whose only defect is homozygosity for R79Q; this patient has a factor VII activity level of 20% (87).

The only other naturally occurring factor VII mutation to be characterized at a molecular level is a missense mutation affecting cysteine 310 (Cys^{310} to Phe) (88). This particular residue is conserved among all serine proteases, and a mutation at this position abolishes disulfide bridging to Cys^{329} and results in a molecule with 4% coagulant activity. Precisely how this mutation affects the functional properties of the molecule is not known, although a substitution at this position could be expected to destabilize the tertiary structure and induce erroneous folding.

VII. FACTOR VII AND CARDIOVASCULAR DISEASE

The importance of the thrombotic component of coronary heart disease is increasingly evident, a number of studies reporting elevated levels of FVII coagulant activity (FVIIc) in the plasma of patients at risk for cardiovascular disease (89–92). The prospectively designed Northwick Park heart study concluded that fibrinogen and factor VIIc levels are at least as powerful as total cholesterol in predicting risk for ischemic heart disease (93). Several studies have also reported a positive correlation between plasma triglyceride levels and factor VIIc. The antigenic levels of factor VII (FVII Ag) have also been implicated as a risk factor, one study reporting an increase in factor VII Ag levels in patients with established ischemic heart disease as well as their young first-degree relatives (92). The elevation in FVII Ag levels also correlated, in both groups, with increased plasma levels of cholesterol and triglycerides. Surprisingly, however, this study found no statistically significant differences in the factor VIIc levels. One can thus speculate that the FVII Ag elevation is a consequence of the lipoprotein abnormality reflected by total triglyceride levels in patients at risk for cardiovascular disease. However, two significantly correlated risk factors may still be independent risk factors. It should also be emphasized that an association between high plasma levels of FVIII and coronary heart disease clearly emerged in the epidemiological study by Rosendaal (94).

VIII. FACTOR VIIa IN THE TREATMENT OF HEMOPHILIA

As discussed earlier, a number of cleavage peptides are generated as by-products to the activation of coagulation, and measurement of basal levels of these peptides provides evidence that coagulation is a permanently ongoing process. Activated proteases, in contrast, are usually too short-lived as free entities to be measured in the plasma because of their interactions with inhibitors, such as antithrombin III. Factor VIIa is an exception to this rule, however, and measurements of the factor VIIa steady-state level in plasma is an additional monitor of (basal) coagulation activity. Severe hemophilia A and B are associated with lower than normal factor VIIa plasma levels (10), whereas, interestingly enough, these patients have essentially normal levels of prothrombin and factor X activation peptides (6,7). This suggests that the factor VIIa plasma level is a more sensitive monitor of coagulation abnormality than activation peptide levels and also that hemophilia A and B result in an impaired capacity of activating factor VIIa.

Adequate treatment of patients deficient in FVIII (hemophilia A) or FIX (hemophilia B) is usually considered the administration of the coagulation protein that is lacking. However, in a substantial fraction of hemophilia patients complicated by the development of antibodies against FVIII/FIX, such therapy is met with limited success. Much effort has thus been focused on the search for so-called FVIII bypassing agents for use in these patients. A FVIII bypassing agent should be capable of initiating hemostasis independently of the presence of FVIII/FIX. Most commonly used are so-called prothrombin complex concentrates containing variable amounts of the vitamin K-dependent coagulation factors (FIX/FIXa, FX/FXa, FVII/FVIIa, prothrombin, protein C, and protein S). However such concentrates are limited in their effectiveness, and a general supplement of activated coagulation factors is associated with thrombotic side effects (95). Because factor VIIa is proteolytically active only when complexed with TF, the administration of factor VIIa alone should minimize the risk of inducing a generalized systemic activation of the coagulation system.

An initial attempt to test this hypothesis was made in a few hemophilia A patients with antibodies who were successfully treated with factor VIIa derived from human plasma (96). Later this concept was further substantiated when recombinant factor VIIa was shown to shorten the cuticle bleeding time in hemophilia A as well as hemophilia B dogs (97). With the availability of human recombinant factor VIIa, it was then also possible to apply this treatment to 83 patients with a variety of serious bleeding complications (98). Most patients suffered from hemophilia A with inhibitors. A smaller group suffered from hemophilia B with inhibitors, and a few patients with factor VII deficiency, von Willebrand disease, or coagulopathy as a result of liver impairment have also been treated. To date, more than 500 bleeding episodes, from a variety of sources, have been treated. In approximately 90% of these instances the clinical response was judged effective (98), indicating that supplementary factor VIIa is capable of temporarily restoring the coagulation process in a variety of hemophilia patients. A plasma level of FVIIa approximately 5–7 u/ml (five times higher than the normal endogenous FVII antigen level) seems adequate to sustain hemostasis, although a higher level (20–30 u/ml; 8–12 μg/ml) seems to be needed to initiate hemostasis in moderate to severe bleeding episodes (98). In terms of factor VIIa levels, however, this means that more that a hundredfold increase relative to the plasma level of FVIIa in healthy individuals is needed for an adequate potentiation of coagulation in hemophilics. The reason for the requirement of such an excess is not clear; however, several explanations might be offered.

The inhibitory action of TFPI in vivo has not been fully elucidated. In the absence of an efficient factor Xa generation via factor IXa/VIIIa activity, it is possible that high concentrations of factor VIIa are needed to overcome the blocking of TF caused by the binding of TFPI, factor Xa, and factor VIIa to this receptor. Another possibility exists that under these conditions factor Xa is generated by a phospholipid-dependent but TF-independent pathway and that in this case the slow catalytic rate might be compensated for by the high concentration of factor VIIa. Third, a plausible explanation for this phenomenon could involve feedback inhibition of the TF pathway. It is an old and well-established but hereto unexplained in vitro observation (unpublished) that the progress curve for the factor VIIa-dependent activation of factors IX and X is characterized by some sort of product inhibition that is a serious constraint to full activation of factors IX and X. Proteolytic cleavage of factor VIIa or TF could perhaps explain this behavior. Finally, it is possible that the displacement of inactive factor VII-TF complexes by factor VIIa can account for the large excess of recombinant factor VIIa needed to restore coagulation in hemophilia A and B patients. In the absence of an efficient activator of TF-bound factor VII, this replacement may represent an alternative although sluggish route to the generation of the TF-factor VIIa complex.

REFERENCES

1. Rapaport SI, Rao LVM. Initiation and regulation of tissue factor-dependent coagulation. Arteriosclerosis Thromb 1992; 12:111–121.
2. Davie EW, Fujikawa K, Kisiel W. The coagulation cascade: initiation, maintenance, and regulation. Biochemistry 1991; 30:10363–10370.
3. Østerud B. Factor VII and hemostasis. Blood Coag Fib 1990; 1:175–181.
4. Broze GJ. Tissue factor pathway inhibitor and the revised hypothesis of blood coagulation. Trends Cardiovasc Med 1992; 2:72–77.
5. Seligsohn U, Kasper CK, Østerud B, Rapaport SI. Activated factor VII: presence in factor IX concentrates and persistence in the circulation after infusion. Blood 1978; 58:828–837.
6. Bauer KA, Kass BL, ten Cate H, Bednarek MA, Hawiger JJ, Rosenberg RD. Detection of factor X activation in humans. Blood 1989; 82:2523–2527.
7. Teitel JM, Bauer KA, Lau HK, Rosenberg RD. Studies of prothrombin activation pathways utilizing radioimmunoassay for the F2/F1+2 fragment and the thrombin:antithrombin complex. Blood 1982; 59:1086–1097.
8. Bauer KA, Kass BL, ten Cate H, Hawiger JJ, Rosenberg RD. Factor IX is activated in vivo by the tissue factor mechanism. Blood 1990; 76:731–736.
9. Bauer KA, Mannucci PM, Gringeri A, et al. Factor IXa-factor VIIIa-cell surface complex does not contribute to the basal activation of the coagulation mechanism in vivo. Blood 1992; 79:2039–2047.

10. Hedner U. Factor VIIa in the treatment of hemophilia. Blood Coag Fib 1990; 1:307–317.
11. Fair DS. Quantitation of factor VII levels in the plasma of normal and warfarin treated individuals by radioimmunoassay. Blood 1983; 62:784–791.
12. Kisiel W, McMullen BA. Isolation and characterization of human factor VIIa. Thromb Res 1981; 22:375–380.
13. Hagen FS, Gray CL, O'Hara P, et al. Characterization of a cDNA coding for human factor VII. Proc Natl Acad Sci USA 1986; 83:2412–2416.
14. Suttie JW. Vitamin K-dependent carboxylase. Annu Rev Biochem 1985; 54:459–477.
15. Thim L, Bjoern S, Christensen M, et al. Amino acid sequence and posttranslational modifications of human factor VIIa from plasma and transfected baby hamster kidney cells. Biochemistry 1988; 27:7785–7793.
16. Bjoern S, Foster DC, Thim L, et al. Human plasma and recombinant factor VI: characterization of O-glycosylations at serine 52 and 60 and effects of site directed mutagenesis of serine 52 to alanine. J Biol Chem 1991; 266:11051–11057.
17. Nishimura H, Kawabata S, Kisiel W, et al. Human factor IX has a tetra saccharide O-glycosidically linked to serine 61 through the fucose residue. J Biol Chem 1989; 264:20320–20325.
18. Hase S, Kawabata SI, Nishimura H, et al. A new trisaccharide sugar chain linked to a serine residue in bovine blood coagulation factors VII and IX. J Biochem 1988; 104:867–886.
19. O'Hara PJ, Grant FJ, Haldeman BA, et al. Nucleotide sequence of the gene coding for human factor VII, a vitamin K dependent protein participating in blood coagulation. Proc Natl Acad Sci USA 1987; 84:5158–5162.
20. Cox DR, Gedde-Dahl T. Human gene mapping 8: report of the committee on the genetic constitution of chromosomes 13, 14, 15 and 16 Cytogenet. Cell Genet 1985; 40:206.
21. Leytus SP, Foster DC, Kurachi K, Davie EW. Gene for human factor X, a blood coagulation factor whose gene organization is essentially identical to that of factor IX and protein C. Biochemistry 1986; 25:5098–5102.
22. Yoshitake S, Schach BG, Foster DC, Davie EW, Kurachi K. Nucleotide sequence of the gene for human factor IX (antihemophilic factor B). Biochemistry 1985; 24:3736–3750.
23. Foster DC, Yoshitake S, Davie EW. The nucleotide sequence of the gene for human protein C. Proc Natl Acad Sci USA 1985; 82:4673–4677.
24. Nemerson Y. Tissue factor and hemostasis. Blood 1988; 71:1–7.
25. Nemerson Y, Repke D. Tissue factor accelerates the activation of coagulation factor VII: the role of a bifunctional coagulation cofactor. Thromb Res 1985; 40:351–358.
26. Ruf W, Rehemtulla A, Edgington TS. Phospholipid-independent and -dependent interactions required for tissue factor receptor and cofactor functions. J Biol Chem 1991; 266:2158–2166.
27. Nemerson Y. The phospholipid requirement of tissue factor in blood coagulation. J Clin Invest 1968; 47:72–80.
28. Bach R, Rifkin DB. Expression of tissue factor procoagulant activity: regulation by cytosolic calcium. Proc Natl Acad Sci USA 1990; 87:6995–6999.
29. Sakai T, Lund-Hansen T, Thim L, Kisiel W. The γ-carboxyglutamic acid domain of human factor VIIa is essential for its interaction with cell surface tissue factor. J Biol Chem 1990; 265:1890–1894.
30. Wildgoose P, Jørgensen T, Komiyama Y, Nakagaki T, Pedersen A, Kisiel W. The role of phospholipids and the factor VII Gla-domain in the interaction of factor VII with tissue factor. Thromb Haemost 1992; 67:679–685.
31. Wildgoose P, Kazim AI, Kisiel W. The importance of residues 195–206 of human blood clotting factor VII in the interaction of factor VII with tissue factor. Proc Natl Acad Sci USA 1990; 87:7290–7294.
32. Wildgoose P, Foster D, Schiødt J, Wiberg FC, Petersen LC. Identification of a calcium binding site in the protease domain of blood coagulation factor VII: evidence for its implication in factor VII:tissue factor interaction. Biochemistry 1993; 32:114–119.

33. Kuman A, Blumenthan DK, Fair DS. Identification of region of factor VII mediating the assembly and function of the extrinsic pathway activation complex (abstract). Blood 1989; 74:352.
34. Ruf W, Kalnik MW, Lund-Hansen T, Edgington TS. Characterization of factor VII association with tissue factor in solution. High and low affinity calcium binding sites in factor VII contribute to functionally distinct interactions. J Biol Chem 1991; 266:15719–15725.
35. Clarke BJ, Ofosu FA, Sridhara S, Bona RD, Rickles FR, Blajchman MA. The first epidermal growth factor domain of human blood coagulation factor VII is essential for binding with tissue factor. FEBS Lett 1992; 298:206–210.
36. Toomey JR, Smith KJ, Stafford DW. Localization of the human tissue factor recognition determinant of human factor VIIa. J Biol Chem 1991; 266:19198–19202.
37. Kazama Y, Pastuszyn A, Wildgoose P, Hamamoto T, Kisiel W. Isolation and characterization of proteolytic fragments of human factor VIIa that inhibit the tissue factor-enhanced amidolytic activity of factor VIIa. J Biol Chem. In Press.
38. Bajaj SP. Cooperative Ca^{2+} binding to human factor IX: effects of Ca^{2+} on the kinetic parameters of the activation of factor IX by factor Xa. J Biol Chem 1982; 257:4127–4132.
39. Sugo T, Björk I, Holmgren A, Stenflo J. Calcium binding properties of bovine factor X lacking the γ-carboxyglutamic acid region. J Biol Chem 1984; 259:5705–5710.
40. Monroe DM, Deerfield DW, Olson DL, et al. Calcium ion binding to human and bovine factor X. Blood Coag Fib 1990; 1:633–640.
41. Pollock JS, Shepard AJ, Weber DJ, et al. Phospholipid binding properties of bovine prothrombin peptide residues 1–45. J Biol Chem 1988; 252:840–850.
42. Soriano-Garcia M, Park CH, Tulinsky A, Ravichandran KG, Skrzypczak-Jankun E. The Ca^{2+} ion and membrane binding structure of the Gla domain of Ca-prothrombin fragment 1. Biochemistry 1992; 31:2554–2566.
43. Higashi S, Kawabata S, Nishimura H, et al. Monoclonal antibody (VII-M31) to bovine factor VII: a specific epitope in the γ-carboxyglutamic acid domain. J Biochem 1990; 108:654–662.
44. Nicolaisen EM, Petersen LC, Thim L, Jacobsen JK, Christensen M, Hedner U. Generation of Gla-domainless cathepsin G-mediated cleavage. FEBS Lett 1992; 306:157–160.
45. Monroe DM, Huh NW, Wildgoose P, Jørgensen T, Roberts HR, Pedersen LG. Equilibrium dialysis analysis of calcium ion binding to factor VII/factor VIIa (abstract). Circulation 1992; 86:2710.
46. Bloom JW. The interactions of factors X and IX with phospholipid. Thromb Res 1989; 54: 261–268.
47. Nelsestuen GL, Kisiel W, Di Scipio RG. Interaction of vitamin K dependent proteins with membranes. Biochemistry 1978; 17:2134–2138.
48. Bajaj SP, Sabharwal AK, Wildgoose P, Petersen LC, Gorka JG, Birktoft JJ. Ca^{2+}-binding site in the protease domain of human factor VIIa (abstract). Circulation 1992; 86:2702.
49. Bajaj SP, Sabharwal AK, Gorka J, Birktoft JJ. Antibody probed conformational transitions in the protease domain of human factor IX upon calcium binding and zymogen activation: putative high affinity Ca^{2+}-binding site in the protease domain. Proc Natl Acad Sci USA 1992; 89:152–156.
50. Schiødt J, Harrit N, Christensen U, Petersen LC. Two different Ca^{2+} ion binding sites in factor VIIa and in des(1–38) factor VIIa. FEBS Lett 1992; 306:265–268.
51. Bajaj SP, Rapaport SI, Maki SL. A monoclonal antibody to factor IX that inhibits the factor VIII:Ca^{2+} potentiation of factor X activation. J Biol Chem 1985; 260:11574–11580.
52. Chattopadhyay A, Fair DS. Molecular recognition in the activation of human blood coagulation factor X. J Biol Chem 1989; 264:11035–11043.
53. Deerfield DW, Olson DL, Berkoqitz P, Koehler KA, Pedersen LG, Hiskey RD. Relative affinity of Ca(II) and Mg(II) ions for human and bovine prothrombin and fragment 1. Biochem Biophys Res Commun 1987; 144:520–527.

54. Radcliffe R, Nemerson Y. Mechanism of activation of bovine factor VII. Products of cleavage by factor Xa. J Biol Chem 1976; 251:4797–4802.
55. Zur M, Radcliffe R, Oberdick J, Nemerson Y. The dual role of factor VII in blood coagulation. Initiation and inhibition of a proteolytic system by a zymogen. J Biol Chem 1982; 257:5623–5631.
56. Rao LVM, Rapaport SI, Bajaj SP. Activation of human factor VII in the initiation of tissue factor-dependent coagulation. Blood 1986; 68:685–691.
57. Williams EB, Krishnaswamy S, Mann KG. Zymogen/enzyme discrimination using peptide chloromethyl ketones. J Biol Chem 1989; 264:7536–7545.
58. Wildgoose P, Berkner K, Kisiel W. Synthesis, purification and characterization of an Arg^{152}-Glu^{152} site directed mutant of recombinant human blood clotting factor VII. Biochemistry 1990; 29:3413–3420.
59. Radcliffe R, Nemerson Y. Mechanism of activation of bovine factor VII. Products of cleavage by factor Xa. J Biol Chem 1976; 251:4797–4802.
60. Broze GJ, Majerus PW. Purification and properties of human coagulation factor VII. J Biol Chem 1980; 255:1242–1247.
61. Wildgoose P, Kisiel W. Activation of human factor VII by factors IXa and Xa on human bladder carcinoma cells. Blood 1989; 73:1888–1895.
62. Masys DR, Bajaj SP, Rapaport SI. Activation of human factor VII by activated factors IX and X. Blood 1982; 60:1143–1150.
63. Kisiel W, Fujikawa K, Davie EW. Activation of bovine factor VII (proconvertin) by factor XIIa (activated hageman factor). Biochemistry 1977; 16:4189–4194.
64. Broze GJ, Majerus PW. Human factor VII. Methods Enzymol 1981; 80:228–237.
65. Broze GJ, Majerus PW. Purification and properties of human coagulation factor VII. J Biol Chem 1981; 256:8324–8331.
66. Nakagaki T, Foster DC, Berkner KL, Kisiel W. Initiation of the extrinsic pathway of blood coagulation: evidence for the tissue factor dependent autoactivation of human coagulation factor VII. Biochemistry 1991; 30:10819–10824.
67. Pedersen AH, Lund-Hansen T, Bisgaard-Frantzen H, Olsen F, Petersen LC. Autoactivation of human recombinant coagulation factor VII. Biochemistry 1989; 28:9331–9336.
68. Yamamoto M, Nakagaki T, Kisiel W. Tissue-factor dependent autoactivation of human blood coagulation factor VII. J Biol Chem 1992; 267:19089–19094.
69. Morrissey JH, Macik BG. Novel clotting assay that factor VIIa in plasma using a tissue factor mutant that does not support activation of factor VII (abstract). Atherosclerosis Thromb 1991; 11:1544a.
70. Macik BG, Morrissey JH. Determination of activated factor VII (FVIIa) levels in plasma using a clotting assay specific for FVIIa (abstract). Blood 1991; 78:61a.
71. Wildgoose P, Nemerson Y, Lyng-Hansen L, Nielsen FE, Glazer S, Hedner U. Measurement of basal levels of factor VIIa in hemophilia A & B patients. Blood 1992; 80:25–28.
72. Komiyama Y, Pedersen A, Kisiel W. Proteolytic activation of human factors IX and X by recombinant human factor VIIa: effects of calcium phospholipids, and tissue factor. Biochemistry 1990; 29:9418–9425.
73. Rao LVM, Robinson T, Hoang AD. Factor VIIa/tissue factor-catalyzed activation of factors IX and X on a cell surface and in a suspension: A kinetic study. Thromb Haemost 1992; 67: 654–659.
74. Bom VJJ, Bertina RM. The contributions of calcium, phospholipids, and tissue factor apoprotein to the activation of human blood coagulation factor X by activated factor VII. Biochem J 1990; 265:327–336.
75. Nemerson Y, Gentry R. An ordered addition, essential activation model of the tissue factor pathway of coagulation: evidence for a conformational cage. Biochemistry 1986; 25:4020–4033.

76. Hedner U, Davie EW. Introduction to hemostasis and the vitamin K-dependent coagulation factors. In: Scriver CR, Beaudet AL, Sly WS, Valle D, eds. The Metabolic Basis of Inherited Disease, 6th ed. New York: McGraw-Hill, 1989:2107–2127.
77. Godal HC, Madsen K, Nissen-Meyer R. Thrombo-embolism in patients with total proconvertin (factor VII) deficiency, a report on two cases. Acta Med Scand 1962; 171:325–327.
78. Hall CA, Rapaport SI, Ames SB, DeGroot JA. A clinical and family study of hereditary proconvertin (factor VII) deficiency. Am J Med 1964; 37:172–181.
79. Gershwin ME, Gude JK. Deep venous thrombosis and pulmonary embolism in congenital factor VII deficiency. N Engl J Med 1973; 288:141–142.
80. O'Brien DP, Gale KM, Anderson JS, et al. Purification and characterization of factor VII 304-Gln: a variant molecule with reduced activity isolated from a clinically unaffected male. Blood 1991; 78:132–140.
81. Matsushita T, Emi N, Takamatsu J, Saito H. Identification and in vitro expression analysis of a single amino acid substitution in blood clotting factor VII_{Nagoya} (Arg^{304}-Trp) (abstract). Blood 1991; 78:181a.
82. Sabharwal AK, Kuppuswamy MN, Foster DC, et al. Factor VII deficiency ($FVII_{Richmond}$, R304Q mutant) associated with thrombosis (abstract). Circulation 1992; 86:I-679.
83. Berube C, Ofosu FA, Kelton JG, Blajchman MA. A novel congenital hemostatic defect: combined factor VII and factor XI deficiency. Blood Coag Fib 1992; 3:357–360.
84. Chaing SH, High KA. Severe FVII deficiency associated with two missense mutations in the factor VII gene (abstract). Thromb Haemost 1991; 65:2048.
85. Chaing S, Clarke B, Sridhara S, et al. Severe factor VII deficiency due to mutations abolishing the cleavage site for activation and altering binding to tissue factor. Blood, in press, 1994.
86. Kazama Y, Foster DC, Kisiel W. Evidence that an $Arg^{79} \rightarrow$ Gln substitution in human factor VII is not associated with a reduction in coagulant activity. Blood Coag Fib 1992; 3:697–702.
87. Takamiya O, Kemball-Cook G, Martin DMA, et al. Detection of missense mutations by single-strand conformational polymorphisms (SSCP) analysis in five dysfunctional variants of coagulation factor VII. Hum Mol Genet 1993; 2:1355–1359.
88. Marchetti G, Patracchini P, Gemmati D, et al. Detection of two missense mutations and characterization of a repeat polymorphism in the factor VII gene. Hum Genet 1992; 89:497–502.
89. Meade TW, North WRS, Chakraarti R, Stirling Y, Haines AT, Thompson SG. Haemostatic function and cardiovascular death: early results of a prospective study. Lancet 1980; 1:1050–1014.
90. Dalaker K, Hjermann I, Prydz H. A novel form of factor VII in plasma from men at risk for cardiovascular disease. Br J Haematol 1985; 61:315–322.
91. Balleisen L, Bailey J, Epping PH, Schulte H, van de Loo J. Epidemiological study on factor VII, factor VIII, and fibrinogen in an industrial population. I. Baseline data on the relation to age, gender, body weight, smoking alcohol, pill-using, and menopause. Thromb Haemost 1985; 54:475–479.
92. Hoffman CJ, Miller RH, Lawson WE, Hultin MB. Elevation of factor VII activity and mass in young adults at risk of ischemic heart disease. J Am Coll Cardiol 1989; 14:941–946.
93. Meade TW, Brozovic M, Chakrabarti RR, et al. Haemostatic function and ischaemic heart disease: principal results of the Northwick Park Heart study. Lancet 1986; 2:533–537.
94. Rosendaal FR. Factor VIII and coronary heart disease. Eur J Epidemiol 1992; 8:71–75.
95. Lusher LM, Shapiro SS, Plascak JE. Efficacy of prothrombin complex concentrates in hemophiliacs with antibodies to factor VIII: a multicenter therapeutic trial. N Engl J Med 1980; 303:421–425.
96. Hedner U, Kisiel W. Use of human factor VIIa in the treatment of two hemophilia A patients with high titer inhibitors. J Clin Invest 1983; 71:1836–1841.

97. Brinkhous KM, Hedner U, Garris JB, Diness V, Read MS. Effect of recombinant factor VIIa on the hemostatic defect in dogs with hemophilia A, hemophilia B, and von Willebrand disease. Proc Natl Acad Sci USA 1989; 86:1382–1386.
98. Hedner U, Glazer S, Falch J. Recombinant activated factor VII (rFVIIa) in the treatment of bleeding episodes on patients with congenital and acquired bleeding disorders. Transfusion Med Rev, in press.

8

Factor VIII

E. G. D. Tuddenham
Royal Postgraduate Medical School,
Hammersmith Hospital, London, England

I. STRUCTURE AND FUNCTION OF FACTOR VIII

Factor VIII (FVIII) is a glycoprotein cofactor that accelerates the activation of FX by FIXa in the central reaction of the blood coagulation cascade. The essential role of FVIII in blood coagulation is evidenced by the severe bleeding diathesis associated with its deficiency in humans and other mammals: hemophilia A. FVIII is one of the largest, least stable coagulation factors, with a complex polypeptide composition. The plasma concentration of FVIII has been estimated as between 100 and 200 ng/ml (1). FVIII circulates in plasma in a noncovalent complex with von Willebrand factor (vWF), a multimeric glycoprotein with a plasma concentration of approximately 10 μg/ml (2). These characteristics combined to delay the purification and characterization of this molecule. Following the isolation of homogeneous preparations of FVIII and the cloning and determination of the nucleic acid sequence of FVIII cDNAs in 1984, there has been dramatic progress in the study of this clinically important coagulation factor. The molecular biochemistry of FVIII has been studied by many groups using the isolated plasma-derived and recombinant wild-type proteins, recombinant molecules altered by oligonucleotide-directed mutagenesis, and naturally occurring variant molecules isolated from hemophilia A plasmas. In addition, monoclonal antibodies to defined epitopes and synthetic peptides representing linear FVIII amino acid sequences have been used to probe the relationship between the structure of FVIII and its interaction with several other coagulation factors. The results of these studies and their contribution to our understanding of FVIII structure and function are summarized here.

A. Function of FVIII

FVIII is the protein procofactor for the serine protease FIXa. Optimum rates of activation of factor X by FIXa require the presence of activated factor VIII, calcium ions, and phospholipid, a multicomponent activator known as the tenase complex. FIXa is capable of hydrolyzing its substrate FX in the absence of calcium ions, phospholipid, and proteolytically activated FVIII (FVIIIa), but the reaction rates are extremely slow. A number of groups have carried out kinetic studies of the activation of FX in the presence of FIXa and the other components of the tenase complex. In a system using bovine components, van Dieijen et al. showed that the addition of calcium alone had little effect on the velocity of the reaction. (V_{max}) or the affinity of the enzyme for the substrate (K_m) (3). The addition of phospholipid, however, decreases the apparent K_m for the reaction by 3000-fold. This effect is probably caused by increasing the concentration of the enzyme, FIXa, and the substrate, FX, on the phospholipid surface. The addition of FVIIIa to the reaction mixture dramatically enhances the rate of the reaction by increasing the V_{max} 200,000-fold. Thus, phospholipid acts to increase the local concentration of the enzyme and the substrate, whereas FVIIIa binds to the enzyme and the substrate increasing the forward rate constant. Kinetic studies using human proteins suggest that the addition of FVIIIa to FIXa, FX, phospholipid, and calcium alters the reaction rate by modestly decreasing K_m and greatly increasing V_{max} (4,5). One possible explanation for this effect is that FVIIIa binds to FIXa and induces a conformational change in the enzyme, increasing its activity toward the scissile bond in FX. Direct evidence for such an allosteric mechanism recently came from studies using factor IXa fluorescently labeled in the active site. Factor VIIIa induced a conformational change in the active site of factor IXa without altering its distance from a phospholipid surface (6,7). Hence the effect of factor VIIIa on factor IXa resembles that of factor Va on factor Xa (8). There is other evidence that FVIII is capable of directly interacting with both FIXa (9) and FX (10). The requirement for the preactivation of FVIII before participation in tenase is absolute. FVIII is activated through selective proteolysis by thrombin or FXa, and as discussed later, variant molecules with cleavage site mutations are nonfunctional. When unactivated procofactor VIII is used in kinetic analyses, there is a considerable lag phase before maximal rates of FXa generation. This reflects the time taken for backactivation of FVIII by FXa in the reaction mixture. The initial generation of FXa in vivo probably occurs as a FVIIIa-independent process whereby tissue factor and FVIIa generate FXa directly before the pathway is damped by the tissue factor pathway inhibitor TFPI (11). Trace amounts of thrombin and FXa are capable of proteolytically activating FVIII, and so the model in which the TF/FVIIa pathway generates sufficient FXa and thrombin from subsequent full activation of the FVIII is plausible. In this model of the initiation of the coagulation cascade, FV, FVIII, and TF are the key regulatory proteins.

B. Purification and Primary Structure

Biochemical characterization of FVIII was not possible until preparations of the molecule were isolated from plasma virtually free of vWF. Homogeneous FVIII was first isolated from bovine plasma (10) and subsequently porcine (12) and human plasma (1,13). All these preparations resolved as multiple noncovalently linked polypeptides on sodium dodecyl sulfate–acrylamide gel electrophoresis (SDS-PAGE) as the result of extensive in vivo processing. The largest molecular mass species resolved at 330 kD (13), with principal polypeptide bands of 220, 116, and 90 kD and a doublet at 80 kD. Two-dimensional tryptic mapping of the 90–210 kD pool of peptides isolated by electrophoresis indicated that they were derived from the same or similar polypeptides (14). The 80 kD doublet gave different tryptic maps, and it was concluded that it was derived from a different part of the FVIII precursor molecule. Monoclonal antibody mapping confirmed that the 90–210 kD species contained several epitopes in common, whereas the anti-80 kD polypeptide monoclonal antibodies were unique to the polypeptide (13). Pools of the purified FVIII polypeptides were subjected to amino acid sequence analysis following tryptic digestion and separation on reversed-phase high-performance liquid chromatography (HPLC) by Vehar et al. (14). The sequence AWAYFSDVDLEK was obtained from digestion of the 80 kD polypeptide. This was used to prepare synthetic DNA probes that identified FVIII genomic (15) and cDNA (16) clones generated from a four X chromosome library and a lymphoblast cell line. From this the complete amino acid sequence of FVIII was deduced. About one-third of the cDNA sequence was matched to the protein-derived sequence in these studies. Toole and colleagues (17) used a sequence from porcine FVIII (12) to isolate porcine FVIII genomic clones and thence human cDNA clones, the sequence of which was nearly identical to that reported by Wood et al. (16).

The 2351 amino acid sequence of FVIII deduced from the nucleotide sequence of these clones is shown in Figure 1. The first 19 amino acids comprise a core of 10 hydrophobic residues flanked by two charged residues, a structure that conforms to that observed for the leader sequences found in most secreted proteins (18). The mature protein after cleavage of the leader contains 2332 amino acids with a mass calculated as 264,763 D. Posttranslational covalent attachment of sugar moieties to nascent polypeptide chains can occur on asparagine residues in the consensus sequence Asn-*X*-Ser or Thr (19). This sequence motif appears 25 times in the FVIII molecule, indicating that it is potentially rich in N-linked glycosylation sites. At least some of these are known to be occupied (20). The addition of carbohydrate accounts for the largest form found in plasma, the single-chain 330 kD species. All the tryptic peptide sequences determined from the 90–210 kD polypeptides were found to be derived from the N terminus of FVIII. Those sequences obtained from the 80 kD fragment were derived from the C terminus. Because the 90–210 kD species all exhibited the same N terminus, it became clear that these polypeptides were derived from C terminal processing of the 210 kD form. The protease that effects this proteolysis has not been identified. Tyrosine sulfation is a posttranslational modification associated with acid sequences in many secreted proteins (21), although the functional significance of this has not for the most part been established. Tyrosines at positions 346, 395, 407, 718, 719, and 722 on the FVIII heavy chain and residues 1664 and 1680 on the light chain have been proposed as possible sites for sulfation based on consensus sequences. Site-directed mutagenesis and expression of recombinant FVIII labeled with $^{35}SO_4$ demonstrated that the residues at 1664 and 1680 are indeed targets for sulfation (22), the latter being essential for the interaction of FVIII with vWF (see later).

C. Structural Homologies

Computer-aided analysis of the FVIII sequence revealed two types of internal homology (Fig. 2) (14). The first, the A domains, consist of a triplicated segment of about 330 amino acid residues at positions 1–329, 380–711, and 1690–2019 of the mature polypeptide. The A domains have approximately 30% amino acid homology. The second internal homology is an unrelated duplication of about 150 amino acid residues, called C domains, found at the C terminus of the molecule. Homology between the two C domains is approximately 40%. The first two A domains are separated from the third by a unique segment of 983 residues containing 19 of the 25 potential N-linked glycosylation sites, referred to as the B domain. Between the first two A domains and immediately N

```
FACTOR VIII                                                    -19       -10
                                                               MQIELSTCFFLCLLRFCFS

                   1        10        20        30        40        50        60        70        80        90
FACTOR VIII        AIRRYYLGAVELSWDYM--QSDLGELPVDARFPPRVPKSFPFNTSVVYKKTLFVEFTDHLFNIAKPRPPWMGLLGPTIQAEVYDTVVITLKNMASHPVSL
CERULOPLASMIN   1  KEKHYYIGIIETTWDYASDHGEKKLISVDTEHSNIYLQNGPDRIGRLYKKALYLQYTDETFRTTIEKPVWLGFLGPIIKAETGDKVYVHLKNLASRPYTF
CONSENSUS          ----YY-G--E--WDY-----------VD-----------P------YKK-L----TD--F------P-W-G-LGP-I-AE--D-V---LKN-AS-P---

                   100       110       120       130       140       150       160       170       180       190
FACTOR VIII        HAVGVSYWKASEGAEYDDQTSQREKEDDKVFPGGSHTYVWQVLKENGPMASDPLCLTYSYLSHVDLVKDLNSGLIGALLVCREGSLAKEKTQTLHKFILL
CERULOPLASMIN 101  HSHGITYYKEHEGAIYPDNTTDFQRADDKVYPGEQYTYMLLATEEQSPGEGDGNCVTRIYHSHIDAPKDIASGLIGPLIICKKDSLDKEKEKHIDREFVV
CONSENSUS          H--G--Y-K--EGA-Y-D-T------DDKV-PG---TY------E--P---D--C-T--Y-SH-D--KD--SGLIG-L--C---SL-KEK---------

                   200       210             220       230       240       250       260       270       280       290
FACTOR VIII        -FAVFDEGKSWHSETKNSL------MQDRDAASARAWPKMHTVNGYVNRSLPGLIGCHRKSVYWHVIGMGTTPEVHSIFLEGHTFLVRNHRQASLEISPI
CERULOPLASMIN 201  MFSVVDENFSWYLEDNIKTYCSEPEKVDKDNEDFQESNRMYSVNGYTFGSLPGLSMCAEDRVKWYLFGMGNEVDVHAAFFHGQALTNKNYRIDTINLFPA
CONSENSUS          -F-V-DE--SW--E------------D-D---------M--VNGY---SLPGL--C----V-W---GMG----VH--F--G------N-R-------P-

                             300       310       320       330       340       350       360       370       380       390
FACTOR VIII        TFLTAQTLLMDLGQFLLFCHISSHQHDGMEAYVKVDSCPEEPQLRMKNNEEAEDYDDDLTDSEMDVVRFDDDNSPSFIQIRSVAKKHPKTWVHYIAAEEE
CERULOPLASMIN 301  TLFDAYMVAQNPGEWMLSCQNLNHLKAGLQAFFQVQECNKSSSKD-------------------------------------------NIRGKHVRHYYIAAEEI
CONSENSUS          T---A-------G---L-C----H---G--A---V--C----------------------------------------------------YIAAEE-

                             400       410               420       430       440          450       460       470       480
FACTOR VIII        DWDYAPLVLAPDDRSYKSQYLN-------NGPQRIGRKYKKVRFMAYTDETFKTREAIQHE---SGILGPLLYGEVGDTLLIIFKNQASRPYNIYPHGIT
CERULOPLASMIN 363  IWNYAPSGIDIFTKENLTAPGSDSAVFFEQGTTRIGGSYKKLVYREYTDASFTNRKERGPEEEHLGILGPVIWAEVGDTIRVTFHNKGAYPLSIEPIGVR
CONSENSUS          -W-YAP-----------------------G--RIG--YKK-----YTD--F--R-----E----GILGP----EVGDT----F-N----P--I-P-G--

                                490       500       510       520       530       540       550       560       570
FACTOR VIII        DVRP----LYSR--RLPKGVKHLKDFPILPGEIFKYKWTVTVEDGPTKSDPRCLTRYYSSFVNMERDLASGLIGPLLICYKESVDQRGNQIMSDKRNVIL
CERULOPLASMIN 463  FNKNNEGTYYSPNYNPQSRSVPPSASHVAPTETFTYEWTVPKEVGPTNADPVCLAKMYYSAVDPTKDIFTGLIGPMKICKKGSLHANGRQKDVDKE-FYL
CONSENSUS          ---------YS-----------------P-E-F-Y-WTV--E-GPT--DP-CL---Y-S-V----D---GLIGP--IC-K-S----G-Q---DK----L

                      580       590       600       610       620       630       640       650       660       670
FACTOR VIII        F-SVFDENRSWYLTENIQRFLPNPAGVQLEDPEFQASNIMHSINGYVFD-SLQLSVCLHEVAYWYILSIGAQTDFLSVFFSGYTFKHKMVYEDTLTLFPF
CERULOPLASMIN 562  FPTVFDENESLLLEDNIRMFTTAPDQVDKEDEDFQESNKMHSMNGFMYGNQPGLTMCKGDSVVWYLFSAGNEADVHGIYFSGNTYLWRGERRDTANLFPQ
CONSENSUS          F--VFDEN-S--L--NI--F---P--V--ED--FQ-SN-MHS-NG--------L--C------WY--S-G---D-----FSG-T---------DT--LFP-

                        680       690       700       710       720       730       740       750       760       770
FACTOR VIII        SGETVFMSMENPGLWILGCHNSDFRNRGMTALLKVSSCDKNTGDYYEDSYEDISAYLLSKNNAIEPRSFSQNSRHPSTRQKQFNATTIPENDIEKTDPWF
CERULOPLASMIN 662  ISLTLHMWPDTEGTFNVECLTTDHYTGGMKQKYTVNQCRRQSEDST (707)
CONSENSUS          ---T--M-----G-----C---D----GM-----V--C-----D--

                        780       790       800       810       820       830       840       850       860       870
FACTOR VIII        AHRTPMPKIQNVSSSDLLMLLRQSPTPHGLSLSDLQEAKYETFSDDPSPGAIDSNNSLSEMTHFRPQLHHSGDMVFTPESGLQLRLNEKLGTTAATELKK

                        880       890       900       910       920       930       940       950       960       970
FACTOR VIII        LDFKVSSTSNNLISTIPSDNLAAGTDNTSSLGPPSMPVHYDSQLDTTLFGKKSSPLTESGGPLSLSEENNDSKLLESGLMNSQESSWGKNVSSTESGRLF

                        980       990       1000      1010      1020      1030      1040      1050      1060      1070
FACTOR VIII        KGKRAHGPALLTKDNALFKVSISLLKTNKTSNNSATNRKTHIDGPSLLIENSPSVWQNILESDTEFKKVTPLIHDRMLMDKNATALRLNHMSNKTTSSKN

                        1080      1090      1100      1110      1120      1130      1140      1150      1160      1170
FACTOR VIII        MEMVQQKKEGPIPPDAQNPDMSFFKMLFLPESARWIQRTHGKNSLNSGQGPSPKQLVSLGPEKSVEGQNFLSEKNKVVVGKGEFTKDVGLKEMVFPSSRN

                        1180      1190      1200      1210      1220      1230      1240      1250      1260      1270
FACTOR VIII        LFLTNLDNLHENNTHNQEKKIQEEIEKKETLIQENVVLPQIHTVTGTKNFMKNLFLLSTRQNVEGSYDGAYAPVLQDFRSLNDSTNRTKKHTAHFSKKGE

                        1280      1290      1300      1310      1320      1330      1340      1350      1360      1370
FACTOR VIII        EENLEGLGNQTKQIVEKYACTTRISPNTSQQNFVTQRSKRALKQFRLPLEETELEKRIIVDDTSTQWSKNMKHLTPSTLTQIDYNEKEKGAITQSPLSDC

                        1380      1390      1400      1410      1420      1430      1440      1450      1460      1470
FACTOR VIII        LTRSHSIPQANRSPLPIAKVSSFPSIRPIYLTRVLFQDNSSHLPAASYRKKDSGVQESSHFLQGAKKNNLSLAILTLEMTGDQREVGSLGTSATNSVTYK

                        1480      1490      1500      1510      1520      1530      1540      1550      1560      1570
FACTOR VIII        KVENTVLPKPDLPKTSGKVELLPKVHIYQKDLFPTETSNGSPGHLDLVEGSLLQGTEGAIKWNEANRPGKVPFLRVATESSAKTPSKLLDPLAWDNHYGT

                        1580      1590      1600      1610      1620      1630      1640      1650      1660      1670
FACTOR VIII        QIPKEEWKSQEKSPEKTAFKKKDTILSLNACESNHAIAAINEGQNKPEIEVTWAKQGRTERLCSQNPPVLKRHQREITRTTLQSDQEEIDYDDTISVEMK

                        1680      1690      1700      1710          1720                1730      1740      1750      1760
FACTOR VIII        KEDFDIYDEDENQSPRSFQKKTRHYFIAAVERLWDYGMSSSP----HVLRNRAQSGSVPQ---------FKKVVFQEFTDGSFTQPLYRGELNEHLGLLG
CERULOPLASMIN 708                   FYLGERTYYIAAVEVEWDYSPQREWEKELHHLQEQNVSNAFLDKGEFYIGSKYKKVVYRQYTDSTFRVPVERKAEEEHLGILG
CONSENSUS                          F----R-Y-IAAVE--WDY----------H-L-----S---------------KKVV----TD--F--P--R----EHLG-LG

                             1770      1780      1790      1800      1810      1820      1830      1840      1850      1860
FACTOR VIII        PYIRAEVEDNIMVTFRNQASRPYSFYSSLISYEEDQRQGAEPRKNFVKPNETKTYFWKVQHHMAPTKDEFDCKAWAYFSDVDLEKDVHSGLIGPLLVCHT
CERULOPLASMIN 791  PQLHADVGDKVKIIFKNMATRPYSIHAHGVQTESSTVTPTLPGETL-------TYVWKIPERSGAGTEDSACIPWAYYSTVDQVKDLYSGLIGPLIVCRR
CONSENSUS          P---A-V-D-----F-N-A-RPYS--------E--------P-----------TY-WK--------------C--WAY-S-VD--KD--SGLIGPL-VC--

                             1870      1880      1890      1900      1910      1920      1930      1940      1950      1960
FACTOR VIII        NTLNPAHGRQVTVQEFALFFTIFDETKSWYFTENMERNCRAPCNIQMEDPTFKENYRFHAINGYIMDTLPGLVMAQDQRIRWYLLSMGSNENIHSIHFSG
CERULOPLASMIN 884  PYLKVFNPRRKL--EFALLFLVFDENESWYLDDNIKTYSDHPEKVNKDDEEFIESNKMHAINGRMFGNLQGLTMHVGDEVNWYLMGMGNEIDLHTVHFHG
CONSENSUS          --L-----R-----EFAL-F--FDE--SWY---N-------P------D--F-E----HAING-----L-GL-M-------WYL--MG-----H--HF-G

                             1970      1980      1990      2000      2010      2020      2030      2040      2050      2060
FACTOR VIII        HVFTVRKKEEYKMALYNLYPGVFETVEMLPSKAGIWRVECLIGEHLHAGMSTLFLVYSNKCQTPLGMASGHIRDFQITASGQYGQWAPKLARLHYSGSIN
CERULOPLASMIN 982  HSFQYKHRGVYSSDVFDIFPGTYQTLEMFPRTPGIWLLHCHVTDHIHAGMETTYTVLQNEDTKSG
CONSENSUS          H-F-------Y--------PG---T-EM-P---GIW---C----H-HAGM-T---V--N------

                             2070      2080      2090      2100      2110      2120      2130      2140      2150      2160
FACTOR VIII        AWSTKEPFSWIKVDLLAPMIIHGIKTQGARQKFSSLYISQFIIMYSLDGKKWQTYRGNSTGTLMVFFGNVDSSGIKHNIFNPPIIARYIRLHPTHYSIRS

                             2170      2180      2190      2200      2210      2220      2230      2240      2250      2260
FACTOR VIII        TLRMELMGCDLNSCSMPLGMESKAISDAQITASSYFTNMFATWSPSKARLHLQGRSNAWRPQVNNPKEWLQVDFQKTMKVTGVTTQGVKSLLTSMYVKEF

                             2270      2280      2290      2300      2310      2320      2330
FACTOR VIII        LISSSQDGHQWTLFFQNGKVKVFQGNQDSFTPVVNSLDPPLLTRYLRIHPQSWVHQIALRMEVLGCEAQDLY
```

terminal to the third A domain are two short acid segments with a high proportion of Asp and Glu residues. Thus, the domain structure of FVIII can be represented as A1-a1-A2-B-a2-A3-C1-C2 (Fig. 2). The A domains of FVIII show striking homology with the A domains of a copper binding plasma protein, ceruloplasmin (23). Both these proteins have a triplicate A domain structure with a pairwise homology of about 30%. This family relationship is extended to FV, the cofactor for FXa in the prothrombinase complex of blood coagulation (24). The C domains of FVIII share 20% homology with the discoidin lectins from *Dictyostelium* (25). They are also homologous to the C domains of a milk fat globule membrane protein (26,27). The B domain of FVIII shows little homology with other proteins in available databases, but FV has a similar highly glycosylated region separating the A1-A2 domains from the A3-C1-C2 carboxyl-terminal polypeptide (28). The complex pattern of FVIII polypeptides resolved on SDS-PAGE is the result of in vivo processing of the B domain of the molecule. Product-precursor relationships were demonstrated by western blotting with monoclonal antibodies (13) and sequence analysis of bands eluted from gels (14). The band at 210 kD represents the A1-A2-B polypeptide, with removal of the A3-C1-C2 C-terminal region of the molecule following proteolytic processing at position 1648–1649 by an unknown protease. This cleavage occurs extensively in vivo because purification of FVIII in the presence of a cocktail of protease inhibitors does not result in the isolation of a predominantly single-chain preparation (13). Polypeptides less than 210 kD but greater than 90 kD represent the N-terminal A1-A2 domains, with varying extensions of the B domain generated by further proteolysis at several sites. The 90 kD fragment is the A1-A2 domain polypeptide generated by cleavage of FVIII at position 740. The 80 kD band represents the C-terminal A3-C1-C2 domains. Each molecule of FVIII therefore comprises a variable-length heavy-chain A1-A2-B noncovalently linked to the light-chain A3-C1-C2 polypeptide.

D. Interaction with Divalent Cations

It has been known for many years that strong chelating agents, such as EDTA, rapidly, and apparently irreversibly, inactivate FVIII (29). This phenomenon was further investigated by Fass et al., who showed that when porcine FVIII was immobilized on light-chain–specific monoclonal antibody columns the heterogeneous heavy chains could be selectively eluted with EDTA (12). Andersson et al., Eaton et al., and Nordfang and Ezban performed similar experiments with human FVIII (30–32). Recombinant FVIIIa that has been inactivated by chelation can also be reconstituted by recombining the heavy and light chains following dialysis into cation-containing buffers (31). In those reconstitution experiments, the most effective cation was Mg^{2+}, followed by Ca^{2+} and Co^{2+}. The nature and site of the cation interaction with FVIII is unknown but could involve either calcium-induced conformational change favoring the binding of the light and heavy chains or direct metal ion bridging of the subunits. Based on homology with plastocyanin, the residues involved in type 1 copper binding in ceruloplasmin comprise two histidines, a cysteine, and a methionine residue near the carboxyl terminus of each A domain. These

Figure 1 Amino acid sequence of human FVIII and its homology with human ceruloplasmin. Amino acid sequence of human FVIII predicted from cDNA clone; ceruloplasmin sequence determined from protein. The consensus line shows residues that are identical in the two proteins. The numbering above the line is that of FVIII. The single-letter notation is used. [Reprinted from Vehar et al. (1984) with permission.]

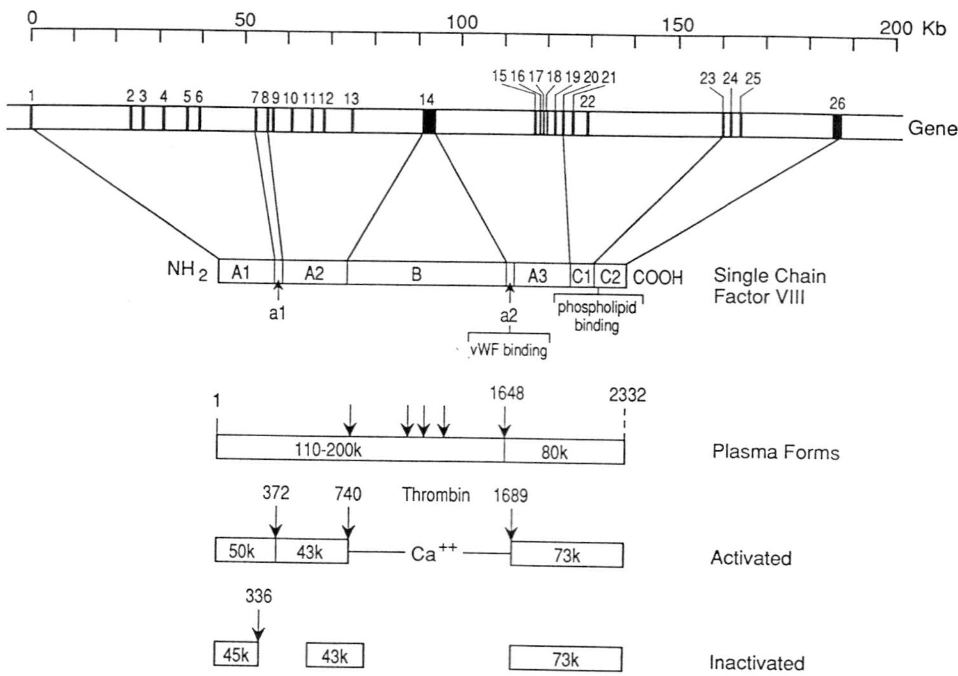

Figure 2 Factor VIII gene and protein. Line 1: Scale for gene (kilobase). Line 2: Map of gene, shaded bars exons numbered 1–26, 5′ to 3′. Line 3: Protein domains based on homology analysis. A1–3, ceruloplasmin homology; C1, 2, discoidin/milk fat globule binding protein homology; a1, 2, acid domains. Line 4: Partially processed forms found in plasma proteolyzed as indicated by arrows. Line 5: Activated factor VIII. Line 6: Inactivated factor VIII. Spontaneous dissociation of 43,000 A2 domain also occurs without additional cleavage at Arg^{336}.

residues are conserved in the first and third A domains of FVIII, but it is not known whether the molecule binds copper. One may speculate that the A1-A2 domains of FVIII are noncovalently bridged to the A3-C1-C2 complex via specific metal ion binding sites in the A1 and A3 domains.

E. Interaction of FVIII with Von Willebrand Factor

FVIII and vWF circulate in plasma in a noncovalent complex that can be dissociated with reducing agents (10) or by high ionic strength buffers (20,33,34). vWF is a multimeric glycoprotein with a role in platelet adhesion in primary hemostasis (see Chap. 8). vWF is also required to stabilize FVIII in plasma, as evidenced by the parallel decline in FVIII levels in patients suffering from severe homozygous von Willebrand disease (vWD): these patients are clinically indistinguishable from severe hemophilia A patients. The $t_{1/2}$ of purified FVIII relatively free of vWF administered to a patient with severe vWD was 2.5 h, compared with 10–12 h observed in hemophilia A patients, who have endogenous vWF capable of binding to, and stabilizing, infused purified FVIII (35). vWF therefore acts as a carrier protein, protecting FVIII from proteolysis in vivo. It appears that each vWF protomer has a single FVIII binding site (36). FVIII binds to vWF through the N-terminal region of the 80 kD light chain. Foster et al. mapped the epitope of a monoclonal antibody to FVIII that inhibited the binding of the cofactor to vWF (37). Residues 1670–1684 of FVIII bound to the monoclonal antibody, a segment that incorporates the highly acid region at the N terminus of the A3-C1-C2 polypeptide, the second acid segment. Leyte et al. produced a series of site-directed mutants of recombinant FVIII with deletions in the B domain and the second acid segment, which confirmed the results of the monoclonal antibody epitope mapping by indicating a role for residues within the sequence 1669–1689 in the formation of a vWF binding site (38). Site-directed mutants of FVIII, with conservative substitution of tyrosine residues at positions 1664 and 1680 to phenylalanine, bound poorly to vWF (22). These two tyrosine residues, which lie in the second acid segment, are sulfated. The recombinant mutant FVIII des (741–1668) produced by Leyte et al. lacked the tyrosine at position 1664 but bound vWF efficiently, and so the tyrosine at position 1680 must be directly involved in the vWF interaction (38). This was further confirmed by the lack of vWF binding by the recombinant FVIII mutant Tyr^{1680} to Phe. A naturally occurring dysfunctional FVIII molecule with replacement of tyrosine 1680 by phenylalanine produces mild hemophilia A, presumably reflecting the lack of binding to and stabilization by vWF (39). Thus the charged second acid segment, specifically the highly charged tyrosine sulfate at position 1680, forms the vWF binding site on FVIII. Fay and Smudzin investigated the physical topography of the FVIII/vWF complex using fluorescence quenching (40). Isolated FVIII subunits labeled with the fluorescence donor were used to label fluorescence acceptor vWF SPIII homodimers. Modified FVIII light chain showed an interfluorophore distance of 28 Å. Heavy chains did not interact with vWF.

F. Platelet and Phospholipid Interactions

The tenase complex comprises FVIIIa, FIXa, FX, and phospholipid. All three proteins bind to phospholipid, provided in vivo by activated platelets. FVIII binds to phospholipid without requiring calcium ions (41). Fractionated polyclonal antibodies that bind to the FVIII light-chain A3-C1-C2 polypeptide inhibit phospholipid binding (42). Arai et al. found that several alloantibodies directed against the FVIII light chain prevented the association of the cofactor and phospholipid (PL) (43). It is now clear from a number of lines of evidence that some of the FVIII sequences responsible for the FVIII-PL interaction are contained within the C domains. The C domains of FVIII show a pairwise homology of 40% and are in turn about 20% homologous to the first 150 amino acids of the slime mold lectin discoidin I (25). This tetrameric galactose binding lectin binds to phospholipid through the C domain sequence, as do the milk fat globule membrane proteins that have C domain homologies (26,27). Further evidence for the involvement of the FVIII C domains in PL binding came from peptide mimetic studies. Synthetic peptides representing FVIII sequences 2302–2332 inhibited the binding of FVIII to PL by 90% (44). Phospholipid vesicles containing 20% phosphotidylserine bind FVIII with an apparent K_d of 4 nM (45).

G. Activation and Inactivation of FVIII

A number of blood coagulation serine proteases modulate the activity of FVIII by specific limited proteolysis. The interactions between FVIII and thrombin, FXa, FIXa, and activated protein C (APC) have all been studied in detail. The activation of FVIII procoagulant activity was first reported by Rapaport et al. in 1963 (46). Switzer and McKee showed that the activation of FVIII by solid-phase thrombin was followed by a decay in activity on removal of the enzyme from the reaction mixture (47). Vehar and Davie used highly purified bovine FVIII to demonstrate a 30-fold activation of the cofactor in the presence of a 260:1 weight ratio of FVIII to thrombin (10). Stable activated FVIIIa was formed at pH 6.6, but the activity of the cofactor decayed rapidly in the presence of 0.2 M EDTA or at pH 7.0. This activation was associated with proteolysis of a triplet of polypeptides to three smaller bands that resolved at 73, 55, and 38 kD. Hoyer and Trabold also showed that the activation of purified human FVIII was associated with a reduction in molecular weight as determined by gel filtration studies (48). These experiments were repeated with purified human FVIII when sufficient material was available for reaction time course analysis (13,49). Product precursor relationships of thrombin-activated FVIII polypeptides were established by a combination of western blotting using monoclonal antibodies (13) and amino acid sequencing (14). Vehar et al. and Eaton et al. identified the specific cleavages in the FVIII molecule mediated by thrombin (14,50). Thrombin-activated FVIII was resolved by SDS-PAGE and the individual bands electroeluted and subjected to N-terminal amino acid sequencing. Thrombin first cleaves FVIII on the carboxyl side of Arg^{740} to remove the B domain from the procofactor and simultaneously cleaves adjacent to residue Arg^{1689} to remove the second acid segment from the amino terminus of the 80 kD chain, generating the 70 kD A3-C1-C2 polypeptide. This cleavage removes the putative vWF binding site, which includes the sulfated tyrosine at position 1680, thus effecting the release of the activated cofactor from the carrier protein. A cleavage at position 372 cleaves the 90 kD chain into the 50 kD A1 domain and the 40 kD A2 fragment. The first acid segment remains part of the A1 domain, accounting for the difference in the molecular weights of the 50 and 40 kD species. Little difference could be seen between the pattern of polypeptides resolved at the time of maximal procoagulant activity and that seen in the inactive product at the end of the time course.

The A1 domain cleavage of FVIII at position 336 is mediated by activated protein C. APC cleaves the procofactor 90–210 kD series of polypeptides to generate a 45 kD A1 domain fragment and a series of degradation products derived from the A2 domain (10,50–52). The 80 kD light chain of FVIII is not proteolyzed by APC. The position 336 cleavage is associated with rapid loss of FVIII procoagulant activity, suggesting that it is the cleavage responsible for the loss of activity seen when FXa activates FVIIIa. Interestingly, the activation and inactivation cleavages flank the acid region between positions 336 and 372 in FVIII. Neither ceruloplasmin nor FV, two homologous proteins with triplicated A domain structures, have acid insertions between the first two A domains. Taken together, these data strongly suggest an important role for this small acid segment in the function of FVIII. The thrombin-activated form of FVIIIa is also susceptible to inactivation by APC (53), but FVIII in complex with vWF is not (54,55). Fay et al. identified a second APC cleavage site at Arg^{562} bisecting the A2 domain (55). This site is preferentially cleaved in factor VIIIa compared with factor VIII.

Recently it was shown that incubation of physiologically relevant concentrations of highly purified factor IXa with factor VIII resulted in the inactivation of FVIII procoagulant activity in a dose-dependent fashion (9). This correlated with cleavage of the FVIII heavy chain at position 336 and the light chain at position 1719–1720. The heavy-chain cleavage was identical to that mediated by APC, and the light-chain cleavage was similar to that mediated by FXa at position 1721–1722. No cleavage was detected at position 372 or 1689, the activation cleavages mediated by thrombin and FXa (Fig. 2). FIXa was also capable of inactivating thrombin-activated FVIIIa following proteolysis of the 50 kD A1 polypeptide, and no protection of FVIII by FIXa was detected in the presence of APC, in contrast to the stabilizing effect of FIXa on thrombin-activated porcine FVIIIa (56). Purified FVIII was protected from FIXa inactivation by vWF in a manner qualitatively similar to the vWF protection of FVIII in the presence of APC (54,55,57). The potential negative feedback by FIXa-mediated inactivation of FVIIIa may explain why most protein C-deficient individuals do not suffer from a thrombotic tendency (58). In whole plasma activated by TF, only thrombin activates FV and FVIII (59). Because trace amounts of thrombin are continuously generated in vivo (60), it is likely that thrombin is the sole physiological activator of FVIII, whereas both APC and FIXa inactivate the cofactor in vivo.

By studying both the effects of site-directed mutagenesis and the properties of certain naturally occurring dysfunctional FVIII molecules, critical activation and inactivation cleavages have been determined. Transiently expressed mutants with radical substitutions at positions Arg^{740} and Arg^{1648} were resistant to cleavage at these sites but were still activated following thrombin cleavage at positions 372 and 1689 (61). The removal of the B domain from the A1-A2 polypeptide is therefore not a prerequisite for FVIII activation. Mutation of Arg^{372} or Arg^{1689} to isoleucine resulted in the expression of recombinant FVIII that was not activated by thrombin. Naturally occurring dysfunctional FVIII molecules isolated from the plasma of hemophiliacs have also been studied. A mutation in a mildly affected hemophiliac's FVIII gene led to the expression of a molecule with an Arg to Cys substitution at position 372 (62). This molecule was only threefold activated by thrombin following cleavage at positions 740 and 1689 but not 372 (Fig. 2). A patient with an Arg to His substitution had a similar phenotype (63). Bihoreau et al. have presented evidence for two-stage activation of FVIII, the first stage resulting from cleavage at positions 1689 and 740 to generate an A1-A2/A3-C1-C2 complex resulting in a fivefold activation of FVIII (64). In the second stage, full activation is concomitant with the cleavage at position 372. This is in keeping with the finding that the variant FVIII Arg^{372} to Cys molecule was partially activatable, therefore causing only a mild bleeding phenotype. Another patient was shown to have a circulating variant molecule with substitution of Arg^{1689} by Cys (65). This molecule was completely inactive, even though it was cleaved efficiently at positions 372 and 740. Gel permeation chromatography studies showed that this abnormal molecule was not released from vWF following incubation with thrombin, confirming that at least one of the functions of this cleavage is to effect the release of the cofactor from the carrier molecule. Interestingly, other cases with the same mutation sometimes have residual factor VIII activity (66). Activation of porcine FVIII in the absence of vWF by a snake venom protease that cleaved the (homologous) 372 but not the 1689 site suggested the function of the latter cleavage is simply to release FVIII from vWF (67). In contrast, the variant molecule Arg^{1689} to Cys was not activated by thrombin following purification from vWF (65), and vWF-free recombinant FVIII Arg^{1689} to Ile was also inactive (61). It is possible that the lack of activity in these molecules is the result of nonconservative amino acid substitutions at this site, sterically inhibiting the activated form of the cofactor.

H. Structure of Activated FVIII

Several studies have shown that activation of FVIII by thrombin is followed by a first-order decay with no apparent alteration in the polypeptides resolved by SDS-PAGE (13,49,56). In contrast to these reports, Vehar and Davie and Eaton et al. were able to demonstrate the presence of a stable activated form of bovine and human FVIIIa following thrombin activation (10,50). This led to controversy about the nature of the activated form of FVIII. The maximally activated form of the cofactor and the inactive species resolved as the characteristic polypeptides at 70, 50, and 40 kD on SDS-PAGE. It was not clear from these studies which polypeptide or combination of polypeptides represented the active moiety. When porcine FVIIIa was subjected to cation-exchange HPLC by Lollar and Parker, the coagulant activity was isolated as a single peak containing three polypeptides, the A1, A2, and A3-C1-C2 segments of factor VIII (68). This material remained stably activated for several weeks at pH 6.0. All three FVIII fragments were shown to be associated by analytical ultracentrifugation. Thus porcine FVIIIa was shown to be a complex of the A1, the A2, and the A3-C1-C2 domains. The stability of the active form of porcine FVIII was found to be markedly pH and concentration dependent (69), being stable at pH 6.0 but irreversibly inactivated by elevation of the pH to 8.0, which caused FVIIIa to elute from cation-exchange columns as a complex of the A1 and A3-C1-C2 fragments, with loss of the A2 domain by adsorption to the column. The stability of the active species was also markedly improved by increasing the concentration of FVIII in reaction mixtures. These data indicated that the inactivation of FVIIIa was associated with dissociation of the A2 domain from the active moiety and suggested a mechanism for the pH dependence of FVIIIa stability, a phenomenon first described by Vehar and Davie (10). Because the loss of activity and dissociation of FVIIIa occur over a narrow pH band, it seems that deprotonation leads to disruption of the quartenary structure of the complex. This hypothesis predicts that the activation of human FVIII at slightly acid pH would result in the formation of stable activated FVIIIa without the need for chromatography. Fay et al. subjected human FVIIIa to cation-exchange chromatography at pH 6.0 and isolated two pools of protein, a minimally active A1, A3-C1-C2 complex and an earlier eluting A2 polypeptide (70). The dissociation was reversible: A2 polypeptide added back to the A1, A3-C1-C2 pool effected reconstitution of the activated cofactor. The association of the A2 subunit appeared to be primarily electrostatic in that even low concentrations of NaCl prevented the reassociation, in keeping with the rapid inactivation of the activated cofactor at alkaline pH. The A2 domain appeared to bind to the A1 domain but not the A3-C1-C2 polypeptide, and this association was independent of the presence of divalent cations. Recently, a recombinant FVIII mutant was expressed in COS-1 cells that lacks the A2 domain. This molecule was nonfunctional but was cleaved by thrombin to generate an A1–A3-C1-C2 complex (71). When this molecule was coexpressed with an expression vector encoding the A2 domain alone, the two molecules assembled to generate a fully functional FVIII protein. These results confirmed those of Lollar and Parker and indicate that activated FVIIIa is a heterotrimeric A1/A2/A3-C1-C2 molecule with a highly pH-dependent quarternary structure (69).

The role of the B domain of FVIII is unknown. B-domainless mutants have been expressed that have similar properties to the wild type in vitro (72). This variant, FVIII des (797–1562), retained only 142 of the 909 amino acid residues of the B domain and was predominantly secreted as a two-chain molecule with a heavy chain of 115 kD and a light chain of 80 kD. These polypeptides comprised the A1-A2 domains (with a small extension into the B domain) and the A3-C1-C2 domains. Thus, intracellular processing to cleave the single-chain B-domainless molecule occurred in the same way as cleavage of the wild type. Some single-chain material was detected, however. The B-domainless FVIII still bound to vWF immobilized on a solid support, indicating that no essential vWF binding site is incorporated in the 797–1562 sequence. The molecule was also cleaved and activated by thrombin to generate the characteristic heterotrimer. The specific activity of the mutant was comparable to the wild-type protein, further showing that the in vitro activity of FVIII is not dependent on the presence of the B domain. A second B-domainless mutant was expressed that lacked 880 amino acids of the B domain and was fully functional in assays in vitro (73). Both these molecules have now been tested in hemophilic dogs and appear to have half-life and phamacokinetic properties identical to those of the wild-type recombinant molecule. The function of the B domain therefore remains unclear, but as discussed later, it may play a role in the control of FVIII secretion from the cell.

II. IN VIVO SYNTHESIS OF FACTOR VIII

Immunological methods were used to locate the presence of factor VIII antigen to lung, liver, placenta, endothelial cells, and lymph nodes (74–77). However, such immunological studies are not able to distinguish between synthesis and storage. The cloning and sequencing of the factor VIII gene permitted the use of gene or oligonucleotide hybridization probes in mRNA localization studies. Factor VIII mRNA was thus detected in both liver and AL-7 (T cell hybridoma) cells and also in spleen and placenta (15–17). The use of oligonucleotides in the latter study potentiated the cloning of factor VIII cDNA from a human liver cDNA library. Subsequent studies found small amounts of factor VIII mRNA in lymph nodes, pancreas, kidney, a liver-derived cell line, placenta, heart, and muscle, in addition to those tissues already mentioned (78,79). Human tissues that do not appear to express factor VIII mRNA as assessed by northern blotting include peripheral blood lymphocytes, thymus, histiocytic lymphoma (U937) cells, fetal lung, bone marrow, brain, and cultured umbilical vein endothelial cells (15,17,78,79). Despite that the hepatocyte appears to be the major site of synthesis of factor VIII, factor VIII mRNA was not detectable in the human hepatocyte cell line HepG2 (15,78,79). To date, no example of a naturally occurring human cell line has been reported that synthesizes factor VIII. However, Hellman et al. detected secretion of factor VIII from short-term cultures of rat liver sinusoidal endothelial cells but not parenchymal cells (80). Thus the liver cell types responsible for factor VIII synthesis in the rat appears to differ from that in the human.

The level of factor VIII mRNA in human liver is reported to be some 20–40 times lower than that of factor IX mRNA (17). A crude estimate of relative mRNA abundance can be made from the observed representation of factor VIII clones in a cDNA library. Toole et al. screened 10^6 recombinants and isolated a total of 55 factor VIII cDNA clones (17). This proportion (5.5×10^{-5}) is in good agreement with the estimate of 1/100,000 total mRNA molecules estimated by a dot-blotting hybridization assay (78). Hassan et al. showed that the level of factor VIII mRNA increases in the human fetus from 30 to 50% of the adult level between the age of 5 and 10 weeks (81).

Very few data are available on the initiation of transcription of the factor VIII gene. A TATA box (GATAAA) is present 5′ to the presumed mRNA start site, but no CCAAT box is apparent. The RNAase protection technique was used to determine the mRNA start site (15). Radiolabeled fragments from the 5′ end of the gene were found to be protected from RNAase digestion by hybridization with poly(A)$^+$ mRNA from AL-7 and human liver cells. It was thus inferred that transcription usually initiates −170 bp (and less frequently −172 bp) upstream of the initiation codon. An additional ATG codon is found at −131, but it cannot serve as a start codon because of the presence of numerous intervening stop codons. At the 3′ end of the gene, a polyadenylation signal (AATAAA) occurs 19 bp upstream of the polyadenylation site, followed by four other such sequences in the next 400 bp.

III. THE FACTOR VIII GENE AND cDNA

The human factor VIII gene was first cloned and mapped in 1984 (15,17). Both groups isolated genomic clones from a library made from cell lines containing four X chromosomes. Over 200 kb of DNA sequence encompassing the factor VIII gene was isolated; this distance represents some 0.1% of the length of the human X chromosome.

Toole et al. were able to localize the site of factor VIII synthesis to the liver by oligonucleotide hybridization (17). Subsequent screening of a liver cDNA library with oligonucleotides yielded 10 clones of a total of 2 million, and a full-length cDNA was pieced together as a series of overlapping clones. Similarly, the Genentech group, using factor VIII genomic clones as probes, screened nearly 80 different human tissues and cell lines to select one, T cell hybridoma cell line AL-7, as a potential source of factor VIII mRNA for cDNA cloning (16). The 9010 bp cDNA includes 150 bp 5′-untranslated sequence, 57 bp encoding a 19 amino acid signal peptide, 6996 bp protein-coding sequence, a stop codon, and an unusually long 1806 bp 3′ noncoding region.

A restriction map of the 210 kb region around the factor VIII gene has been derived using a total of 10 restriction enzymes, later augmented by the addition of restriction sites for the rare cutting enzymes SstII, NarI, SfiI, and NruI (15,82). This permitted the placement of the 26 exons within the gene (Fig. 2). Exon size ranges from 69 bp (exon 5) to an exceptionally long 3106 bp (exon 14). However, only 5% of the gene region is coding sequence; the rest, 177 kb, comprises the introns, which vary in size from 207 bp to 32.4 kb. All splice-donor acceptor sites conform to the GT/AG rule (83).

A third full-length factor VIII cDNA from human kidney has been isolated by a group at Chiron Corporation (84). Comparison of the various cloned factor VIII sequences revealed polymorphic variation within the coding region. The cDNA and genomic sequences isolated by Gitschier et al. and Wood et al. showed two differences [G/C in codon 1241 resulting in either glutamic acid (GAG) or aspartic acid (GAC); G/A at nucleotide 8728 in the 3′-untranslated region] (15,16). Comparison of the sequences reported by Truett et al. (84) and Toole et al. (17) also showed a difference in codon 1241, as well as a difference at codon 56, where either GAC (aspartic acid) or GTC (valine) may occur. The relative paucity of sequence polymorphisms within the factor VIII gene-coding sequence is consistent with the low polymorphism frequency associated with X-linked genes and may reflect strict selective constraints acting on these loci (85).

The prospect of a gene within a gene was first raised by the discovery of a CpG island within intron 22 of the factor VIII gene (86). A 1.8 kb mRNA homologous to this CpG island has now been detected in human liver and HeLa and U937 cells (87). This gene (F8A), transcribed in the opposite direction to that of factor VIII, is one of three homologs, all present on the X chromosome and all transcribed. These genes are intronless and are thought to fulfill a ''housekeeping'' role. Confirmation that the intron 22 gene is active came from the study of hemophilia A patients with deletions spanning intron 22 in whom a 60% decrease in mRNA levels was found in lymphoblastoid cell lines. Another mRNA transcript (F8B) originating near the 3′ end of F8A in intron 22 has been identified in lymphoid cells (88). This transcript appears to code for eight amino acids upstream from exon 23 and then uses the rest of the factor VIII coding sequence. Its function, if any, is unknown.

The factor VIII gene has been localized to Xq28 both by somatic cell hybrid analysis and by in situ hybridization (89,90).

A. Mutations at the Factor VIII Locus

The molecular genetic analysis of hemophilia A has yielded a large number of different lesions of the factor VIII gene. The most recent compilation of mutations is found in a database (66). Tables 1 through 5 summarize these mutations, updated to 1993.

Over 80 different point mutations have been found in the factor VIII gene by a combination of Southern blotting, oligonucleotide discriminant hybridization, denaturing gradient gel electrophoresis (DGGE), single-strand conformation polymorphism analysis, chemical cleavage, and DNA sequencing. In these cases, the patients are thought to be unrelated, although it is not always possible to distinguish formally between the alternative possibilities of recurrent mutation and identity by descent (see later).

Of the different point mutations now characterized, 27 result in severe hemophilia, 29 result in moderately severe hemophilia, and 18 result in mild hemophilia. All nonsense mutations resulted in severe hemophilia, and indeed the majority of known point mutations causing severe hemophilia A are of this type. All examples of point mutations causing moderate and mild hemophilia are missense types.

About 5% of patients with hemophilia A have an excess of factor VIII antigen over functional activity caused by the presence of a factor VIII molecule with reduced specific activity. These cases are termed cross-reacting material positive (CRM^+). They have often been specifically selected for study in the hope of gaining information on the functionally critical regions of the factor VIII protein. Unfortunately, information on CRM status is available for only a minority of the cases with identified mutations reported so far. By measuring FVIII Ag as well as FVIIIc, both the specific activity and the protein level can be determined, and those mutations affecting protein stability can be distinguished from those with a critical effect on protein function. For example, mutations producing a CRM^+ phenotype have been identified in the two arginine residues (372 and 1689) adjacent to the scissile bonds cleaved upon activation (91). The defect in function of these molecules has been clearly demonstrated to be associated with resistance to cleavage by thrombin of the heavy and light chains, respectively. This result was anticipated by site-directed in vitro mutagenesis, but other CRM^+ variants identify functionally important regions of factor VIII not previously defined in this way (61). Thus Arg^{2209} appears critical for function because its substitution by Gln produces a CRM^+ phenotype [FVIIIc = 7%; FVIII Ag (antigen) = 130%].

An interesting example of a mutation almost certainly affecting the stability of the protein in the circulation is that converting Tyr^{1680} to Phe. This residue is normally sulfated and is critical for the interaction of factor VIII with von Willebrand factor, the loss of which interaction leads to rapid clearance from the circulation (38). The lower activity (FVIIIc = 10%) than antigen (FVIII Ag = 20%) in the case reported suggests an additional effect on function.

Two examples of mutations creating new N-glycosylation sites have been reported. M1772T and I566T create carbohydrate attachment sites at N1770 and N564, respectively (Table 1). Aly and colleagues proved that these sites are occupied and that removal of carbohydrate from the variant protein by *N*-glycanase digestion restores function (92). Such a mechanism of disease has only previously been noted for antithrombin Rouen III

Table 1 Single-Base Substitutions Found in the Coding Region of the Factor VIII Gene of Patients with Hemophilia A

Exon	Codon[a]	Nucleotide change	Codon change	Number of unrelated cases	FVIIIc (U/dl)	FVIII Ag (U/dl)	Species conservation[b]	Presence of inhibitors	Comments
1	−5	CGA → TGA	Arg → term	1	<1	<1	—	No	Signal peptide
1	11	GAA → GTA	Glu → Val	1	?	?	M	No	
3	73	GGT → GTT	Gly → Val	1	?	?	M	No	
3	85	GTC → GAC	Val → Asp	1	?	?	M	No	
3	89	AAG → ACG	Lys → Thr	1	?	?	M	?	
3	91	ATG → GTG	Met → Val	1	?	?	M	?	
4	145	GGT → GTT	Gly → Val	1	?	?	M	No	
4	162	GTG → ATG	Val → Met	2	8/5	?	M	No	
4	166	AAA → ACA	Lys → Thr	1	19	?	M	?	
4	170	TCA → TTA	Ser → Leu	1	3.5	8.7	M	No	
5	205	G/gt → T/gt	Gly → Trp	1	3.2	?	M	?	−1 IVS5 donor splice site
7	255	TGG → TGA	Trp → term	1	?	?	M	?	
7	266	GTG → GGG	Val → Gly	1	?	?	CS	?	
7	272	GAA → GGA	Glu → Gly	1	2	3.5	M	No	
7	280	AAC → ATC	Asn → Ile	1	4	?	M	No	
7	280	CGC → CAC	Arg → Cys	1	<1	?	M	No	
7	282	CGC → CAC	Arg → His	3	<1	18	M	?	
7	289	TCG → TTG	Ser → Leu	1	37	106	M	?	
7	293	TTC → TCC	Phe → Ser	1	?	?	M	?	
7	295	ACT → GCT	Thr → Ala	1	14–16	?	M	?	
8	326	GTA → CTA	Val → Leu	2	?	?	M	No	
8	329	TGT → CGT	Cys → Arg	1	?	?	M	No	
8	336	CGA → TGA	Arg → term	7	0	<1	—	No	Activated protein C cleavage site
8	372	CGC → TGC	Arg → Cys	3	3	70–80	M	No	Thrombin activation site
8	372	CGC → CAC	Arg → His	2	5	325/57	M	No	Thrombin activation site
9	412	TTG → TTT	Leu → Phe	1	10.5	?	M	?	

9	425	AAA → AGA	Lys → Arg	1	<1	5	M	?	
3	427	CGA→ TGA	Arg → trm	1	<1	<1	M	No	
9	431	TAC → AAC	Tyr → Asn	1	4	?	M	No	
10	473	TAT → CAT	Tyr → His	1	?	?	M	?	
10	473	TAT → TGT	Tyr → Cys	1	2.7/3.5	?	M	?	
10	479	GGA → AGA	Gly → Arg	1	2	?	M	No	
11	504	CTG → CTT	Leu → Leu	4	?	?	—	?	Potential new acceptor site
11	527	CGG → TGG	Arg → Trp	4	17	245	M	?	
11	531	CGC → TGC	Arg → Cys	4	6.7/4.2	?	M	?	
11	531	CGC → GGC	Arg → Gly	1	9.2	?	M	?	
11	535	AGT → GGT	Ser → Gly	3	?	?	M	?	
11	542	GAT → GGT	Asp → Gly	1	<1	5	M	No	
11	557	GAA → TAA	Glu → term	1	?	?	M	No	
11	558	TCT → TTT	Ser → Phe	1	21	175	M		
11	565	CAG/gt → AAG/gt	Gln → Lys	1	?	?	M	?	−3 IVS11 donor splice site
12	566	ATA → ACA	Ile → Thr	1	<1	154	CS	?	New N-glycosylation site N564
12	577	TCT → CCT	Ser → Pro	1	?	?	M	?	
12	583	CGA → TGA	Arg → term	1	<1	<1	—	No	
12	584	AGC → ATC	Ser → Ile	1	?	?	M	?	Loss N-glycosylation site N582
12	593	CGC → TGC	Arg → Cys	8	10	?	M	No	
12	612	AAC → AGC	Asn → Ser	1	?	?	M	?	
13	634	GTG → GCG	Val → Ala	1	5	138	M	No	
13	634	GTG → ATG	Val → Met	1	<1	175	M	No	
13	644	GCA → GTA	Ala → Val	1	14	25	M	?	
13	652	TTC[c]	Phe^{652}/3[c]	1	1.4	12	M	No	
14	698	CGG → TGG	Arg → Trp	1	?	?	M	No	
14	704	[C]GCC → ACC	Ala → Thr	2	?	?	M	?	
14	795	CGA → TGA	Arg → term	1	<1	<1	—	No	
14	1038	GAG → AAG	Glu → Lys	1	2.4	10–20	M	?	
14	1441	AAT → AAA	Asn → Lys	1	6	7	MP	No	
14	1680	TAT → TTT	Tyr → Phe	2	10	20	MP	No	Tyrosine sulfation/vWF interaction
14	1680	TAT → TGT	Tyr → Cys	1	?	?	MP	No	

Table 1 Continued

Exon	Codon[a]	Nucleotide change	Codon change	Number of unrelated cases	FVIIIc (U/dl)	FVIII Ag (U/dl)	Species conservation[b]	Presence of inhibitors	Comments
14	1686	CAG → TAG	Gln → term	1	<1	?	MP	?	
14	1689	CGC → TGC	Arg → Cys	8	0–5	87–220	MP	No	Thrombin activation site
14	1689	CGC → CAC	Arg → His	1	7	165	MP	No	Thrombin activation site
14	1696	CGA → TGA	Arg → term	2	<1	?	MP	Yes 2	
14	1709	TAT → TGT	Tyr → Cys	1	18	?	MP	No	
15	1772	ATG → ACG	Met → Thr	1	<1	72	MP	?	New N-glycosylation site N1770
16	1781	CGT → CAT	Arg → His	1	?	?	M	?	
16	1784	TCC → TAC	Ser → Tyr	1	?	?	M	?	
16	1789	CTT → TTT	Leu → Phe	2	?	?	M		
16	1825	CCC → TCC	Pro → Ser	1	15	?	M	?	
16	1826	ACT → CCT	Thr → Pro	1	?	?	M	?	
16	1843	CTG/gt → CTA/gt	Leu → Leu	1	9–18	?	M	?	−1 IVS16 donor splice site
17	1848	CAC → CGC	His → Arg	1	1–5	?	M	?	
17	1874	CAG-TAG	Gln → term	1	<1	?	M	Yes	
18	1922	AAT → GAT	Asn → Asp	2	<1	?	M	Yes	
18	1922	AAT → AGT	Asn → Ser	2	?	?	M	?	
18	1941	CGA → TGA	Arg → term	8	<1	?	M	No 2 Yes 5 ? 1	
18	1941	CGA → CTA	Arg → Leu	1	7	?	M	No	
18	1941	CGA → CAA	Arg → Gln	1	2	?	M	No	
18	1966	CGA → TGA	Arg → term	1	<1	?	M	Yes	
18	1987	GAA → TAA	Glu-term	1	<1	?	M	No	
19	1997	CGG → TGG	Arg → Trp	2	3.4/2.6	?	M	?	
21	2046	TGG-CGG	Trp → Arg	1	?	?	M	No	

22	2101	TTT → TTG	Phe → Leu	1	11	?	M	?
22	2105	TAT → TGT	Tyr → Cys	1	14	?	M	Yes
22	2116	CGA → TGA	Arg → term	6	<1	?	—	No
22	2116	CGA → CCA	Arg → Pro	1	<1	?	—	?
22	2119	TCC → TAC	Ser → Tyr	1	5–8	?	M	?
23	2147	CGA → TGA	Arg → term	7	<1	<0.1	M	No 2
								Yes 3
								? 1
23	2150	CGT → CAT	Arg → His	8	5–7	?	M	No 3
								? 5
23	2159	CGC → TGC	Arg → Cys	4	7.4–12	2	M	No 4
23	2163	CGC → CAC	Arg → His	2	?	?	M	?
23	2166	GTT → GCT	Leu → Ser	1	<1	?	M	No
24	2205	CTC[c]	Pro[c]	2	?	?	M	?
24	2209	CGA → CTA	Arg → Leu	1	3	2.5	M	No
24	2209	CGA → TGA	Arg → term	10	0	?	M	Yes 3
								No 4
								?3
24	2209	CGA → CAA	Arg → Gln	12	<1/2.5	4	M	No 11
								Yes 1
25	2229	TGG → TGT	Trp → Cys	2	3	?	M	Yes 2
26	2300	CCG → CTG	Pro → Leu	1	7.5	?	M	?
26	2304	CGC → TGC	Arg → Cys	1	?	?	M	?
26	2307	CGA → TGA	Arg → term	6	0	?	M	No 6
26	2307	CGA → CTA	Arg → Leu	3	2	4	M	No
26	2307	CGA → CAA	Arg → Gln	2	10	6	M	No

[a]Codons numbered after scheme of Vehar et al. (1984): starts at mature N terminus, and 19 signal peptide residues numbered negatively.

[b]M, conserved in mouse (111); P, conserved in pig (73); CS, conservative substitution; —, not conserved.

[c]Deleted.

Source: Compiled from references cited in Tuddenham et al. (66) and from McGinness et al. (112), Naylor et al. (94), Pieneman et al. (113), Economou et al. (114), Nafa et al. (115), and Diamond et al. (116).

Table 2 Mutations Definitely or Putatively Affecting mRNA Processing in the Factor VIII Gene

Location[a]	Nucleotide[b] change	Codon change	Number of unrelated cases	FVIIIc (U/dl)	FVIII Ag (U/dl)	Presence of inhibitors	Clinical status	Comment
Intron 2300 bp 3′ to exon 2	cga → tga	—	1	<1	?	No	Severe	Possibly a neutral change
Intron 5.SA	ag/ → gg/	—	1	<1	?	No	Severe	
Intron 6 SA	ag/ → ac/	—	1	?	?	?	Severe	
Exon 11								
CA	TTCCAATTCTGCCAG/ →							
MA	TTCCAATTCTTCCAG/	L504L[c]	1	?	?	No	Mild	Possibly a rare variant
Intron 12 SD	/gtgagt → /gtgaat	—	1	?	?	?	Mild	
Exon 16 intron 16 SD -1	CTG/gt → CTA/gt	L1843L[c]	1	9–18	?		Moderate	
Intron 22	Inversion involving intra- and extragenic copies of F8A[d]	—	10	<1	<1	No 8 Yes 2	Severe	Accounts for up to half of severe cases
Intron 25 ~1.9 kb 5′ to exon 26	caa → cga	—	1	<1	?	No	Severe	Possibly a neutral change

[a]SA, splice acceptor; SD, splice donor; CA, cryptic acceptor; MA, mutant acceptor.
[b]Intron sequences, lowercase; exon sequences, uppercase.
[c]Change is neutral at protein level but could interfere with splicing.
[d]Dr. J. Gitschier, personal communication.

Table 3 Insertions in the Factor VIII Gene of Patients with Hemophilia A

Exon	Nature of insertion	Severity	Inhibitors
11	1 bp (G at codon 513)	Severe	?
14	3.8 kb LINE element	Severe	?
14	2.1 kb LINE element	Severe	Yes
14	1 bp (TCA → TCAA at codon 1395)	Severe	?
14	1 bp (A in stretch of 8 A residues at codon 1439)	Severe	?
14	1 bp (A in a stretch of 6 A residues at codons 961–963)	Severe	No
17	1 bp (A in stretch of 4 A residues in codon 1887)	Severe	?

and fibrinogen Asaki (92). Confirmation of the mutation FVIII I566T has been obtained in the author's laboratory (unpublished observations) in an unrelated case.

The few other missense mutations for which information is available produce a CRM reduced or CRM$^-$ phenotype with low amounts of residual activity and protein. This is most probably a result of effects on stability caused by interference with proper folding, often because the substituted residues are in the hydrophobic core of a domain.

Higuchi et al. highlighted an intriguing phenomenon through their attempt to screen the factor VIII genes of hemophilia A patients for mutations using DGGE (39,93). They detected the causative lesions in virtually all patients with moderately severe or mild hemophilia, but they were successful in only 16 of 30 (53%) patients with severe hemophilia. This discrepancy argues against a detection artifact associated with the screening method and instead points to either a mutation cluster in a region of the gene that was incompletely screened (introns, 3′-untranslated region, or promoter) or mutations in another gene on the X chromosome, possibly within the factor VIII gene itself or very closely linked to it.

The explanation for this has now come to light and involves intron 22. Naylor et al. found by analyzing FVIII mRNA in lymphocytes that about half the severely affected

Table 4 Small Deletions in the Factor VIII Gene of Patients with Hemophilia A

Location codons	Size (bp; nucleotides deleted)	Presence of inhibitors	Clinical status	Comments
104–111	23	?	Severe	Includes IVS3 donor splice site
340–341	4 (AATG)	No	Severe	Frameshift in exon 8
341	2 (GA)	?	Severe	Frameshift in exon 8
652–653	3 (TTC)	—	Moderate	Deletes Phe^{652}
1212	1 (C)	No	Severe	Frameshift in exon 14
1422	4 (AAGA)	No	Severe	Frameshift in exon 14
1439	1 (A)	?	Severe	A8 → A7, frameshift in exon 14
1535–1536	2 (GA)	?	Severe	Frameshift in exon 14
2136	2 (AA)	?	Severe	A4 → A2, frameshift in exon 23
2204–2205	3 (CTC)	?	Moderate	Deletes Pro^{2205}

Table 5 Large Deletions in the Factor VIII Gene of Patients with Hemophilia A

Exon(s) deleted	Size of deletion (kb)	FVIIIc (U/dl)	FVIII Ag (U/dl)	Severity	Inhibitors
1–26	>210	?	?	Severe	No
1–26	>210	?	?	Severe	Yes
1–6	>55	<1%	?	Severe	Yes
1–5	>35	<1%	?	Severe	No
1	>2	<1%	<1%	Severe	No
1	>1	<1%	?	Severe	No
1	?	?	?	Severe	?
Intron 1[a]	7	?	?	Severe	No
2–3	9–12	<1%	<1%	Severe	No
3–13	60	<1%	<1%	Severe	Yes
3–5	11.7	<1%	?	Severe	No
3	1.7–2.0	<1%	?	Severe	No
4–25	133–145	<1%	<1%	Severe	No
5–13	57	<1%	?	Severe	Yes
5, 6	?	?	?	Severe	?
5, 6	2.5–10	?	?	Severe	Yes
5, 6	?	<1%	?	Severe	No
5	2	?	?	Severe	No
5 or 6	2	?	?	Severe	?
6	?	<1%	?	Severe	No
6	10	?	?	Severe	No
6	7	<1%	<1%	Severe	No
6	3–6	?	?	Severe	No
7–22	110	?	?	Severe	Yes
7–14	40–56	<1%	<1%	Severe	No
7–9	15–20	<1%	?	Severe	Yes
11–22	60	?	?	Severe	Yes
14–22	>36	?	?	Severe	Yes
14–22	?	?	?	Severe	Yes
14–21	50	<1%	?	Severe	Yes
14	12–16	<1%	?	Severe	Yes
14	6	<1%	<0.1%	Severe	Yes
14	2.3–3.0	<1%	?	Severe	No
14	2.5	<1%	<1%	Severe	No
14	2.5	?	?	Severe	?
15–21	?	?	?	Severe	?
15–18	13	?	?	Severe	Yes
15	?	?	?	Severe	?
17–19	?	?	?	Severe	No
18, 19[b]?	?	?	?	Severe	?
19	1.9	<1%	?	Severe	No
19–21	4.7	<1%	?	Severe	Yes
22	5.5	2–5%	?	Moderate	No
Intron 22[a]	?	?	?	Severe	No
Intron 22[a]	?	?	?	Moderate	No
23–26	?	?	?	Severe	Yes
23–26	?	?	?	Severe	Yes
23–25	>16	<1%	<1%	Severe	No

Table 5 Continued

Exon(s) deleted	Size of deletion (kb)	FVIIIc (U/dl)	FVIII Ag (U/dl)	Severity	Inhibitors
23–25	39	<1%	?	Severe	Yes
23–24	?	?	?	Moderate	No
24–25	>3.4	<1%	<1%	Severe	No
26	22	<1%	?	Severe	No
26	>18	<1%	?	Severe	No
26	14	<1%	<1%	Severe	No
26	8.7	?	?	Severe	?
26	>2	?	?	Severe	?
26	>2	?	?	Severe	No
26	>2	?	?	Severe	?
26	?	?	?	Severe	?
26	?	?	?	Severe	No
26	?	?	?	Severe	No
26	?	?	?	Severe	No

[a]Not proven to be cause of disease phenotype, although segregates with disease allele.
[b]Precise extent unknown but includes at least region indicated.

patients had clearly identifiable lesions affecting the coding region and the remainder failed to splice exon 22 to exon 23 correctly (94). Therefore, about half of severe hemophilia A is caused by lesions within intron 22. The mechanism appears to be a complete inversion disrupting the gene, involving crossover between the intragenic and extragenic copies of F8A (see Fig. 3).

Surprisingly, the clinical phenotype exhibited by hemophilia A patients with the same point mutation is not always the same. Pattinson et al. noted that one patient with a C → T mutation at codon 1689 was severely affected but two others were only moderately affected (91). Several other examples of this phenomenon are known (66). The reasons for these differences are unknown. The complete coding sequence of two discordant pairs were recently screened in the authors' laboratory, and no differences other than the reported mutations (R1689C and R2209/Q) were found. Possible explanations include the effect of other genes acting in trans or some unidentified rearrangement, such as that preventing splicing across intron 22. Whatever the explanation, only missense mutations are involved, nonsense mutations always resulting in severe hemophilia.

Of known point mutations in the factor VIII gene, 38% are C → T or G → A transitions in a CpG nucleotide. This doublet is already known to be hypermutable as a consequence of methylation-mediated deamination of 5-methylcytosine; indeed, some 33% of all point mutations causing human genetic disease are CG → TG or CG → CA transitions consistent with this postulate (95). The high proportion of CpG mutations is at least in part caused by the deliberate screening of these sites, however. This notwithstanding, recurrent mutation at CpG dinucleotides appears to have occurred in at least 16 sites [codons 282, 336, 372(2), 527, 1689, 1941, 2116, 2147, 2150, 2159, 2209(2), 2228, and 2307(2)]. This assertion is made on the basis of either restriction fragment length polymorphism (RFLP) haplotyping data or extreme geographical separation. Multiple mutations have been reported at four further CpG sites at codons 531, 593, 704, and 1997. Identity by descent may be suspected, however, in the case of the Arg^{593} →

Cys mutation in (apparently unrelated) patients JH82, 83, 84, 110, and 113, all of whom originate from the Nashville area of Tennessee and all of whom possess an identical RFLP haplotype. For the other CpG mutations at codons 531, 704, and 1997, it is at present difficult to distinguish recurrent mutation from identity by descent.

There are 70 CpG dinucleotides within the 9 kb coding region of the factor VIII gene, representing the potential for 140 different base pair changes (7 of these CpG dinucleotides occur within TaqI restriction sites). Youssoufian et al. estimated that the relative mutation rate of CpG dinucleotides is 10–20 times greater than the average mutation rate at other sites in the factor VIII gene (96). Because these authors also estimated that there may be as many as 2500 different point mutations (of a total of perhaps 3400 possible lesions) causing hemophilia A, it may be that up to 50% of all point mutations occur through methylation-mediated deamination at CpG sites. This is a maximum estimate: it is likely that not all CpG mutations are deleterious. The estimate, however, is not dissimilar to the observed figure of 35% for all point mutations causing human genetic disease that are located in CpG dinucleotides (95).

For multiple mutations at non-CpG sites, identity by descent may often be more plausible. This is certainly the case for the $Tyr^{1680} \rightarrow$ Phe and $Asn^{1922} \rightarrow$ Asp substitutions, causing mild and moderate hemophilia, respectively, even although the patients are not known to be related. On the other hand, recurrent mutation at non-CpG sites apparently occurs: the $Arg^{2307} \rightarrow$ Leu substitution causing severe hemophilia A has now been reported in American, German, and Japanese patients. These lesions are extremely unlikely to be identical by descent.

A number of point mutations that putatively interfere with mRNA splicing have been detected in the factor VIII genes of hemophiliacs. Because the factor VIII gene is tran-

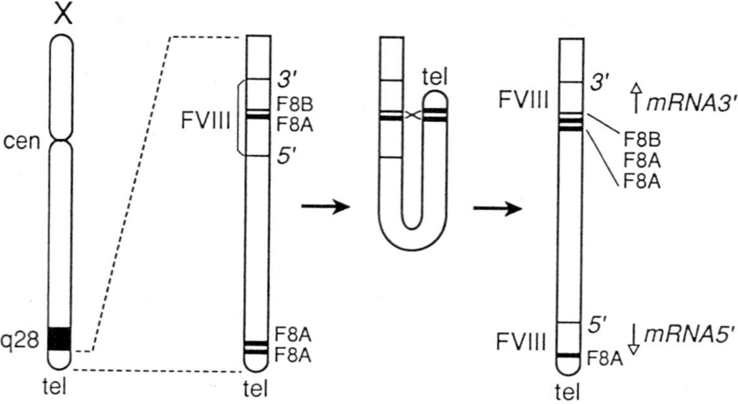

Figure 3 The inversion mechanism causing hemophilia in up to half of severely affected cases. The existence of at least two copies of the F8A gene near the telomere of the X chromosome enables a crossover to occur with the copy of the F8A gene within intron 22. As a consequence, the factor VIII gene is separated into two halves distanced from each other by several megabases. Messenger RNA corresponding to the first 22 exons can be detected, spliced to novel sequences derived from telomeric genes. Messenger corresponding to the F8B sequences and exons 23–26 is also detectable. No protein is detectable in the plasma of patients affected by this prevalent inversion, however. (From Lakich D, Kazazian HH, Antonarakis SE, Gitschier J. Inversions disrupting the factor VIII gene as a common cause of severe haemophilia A. Nature Genet 1993; 5:236–241.)

scribed in fairly inaccessible tissues, formal confirmation of a splicing defect is difficult. However, "ectopic" transcript analysis has been successfully employed to analyze aberrant splicing in lymphocyte factor VIII mRNA (97,98). The putative splicing defects reported to date may be divided into three categories: (1) Mutations of the invariant GT and AG dinucleotides at the donor and acceptor splice junctions, respectively; (2) mutations within the extended consensus sequences of the donor and acceptor splice junctions; and (3) mutations that create a novel donor or acceptor splice site. In the first category, the invariant AG dinucleotide in the IVS5 and IVS6 acceptor sites were mutated to GG and AC, respectively; both resulted in a severe phenotype. In the second category, four mutations in donor splice site consensus sequences have been reported. In two cases, the altered nucleotide is at the last base (−1) of an exon (Gly^{205} and Leu^{1943}) and in a third case is located at position −3 (Gln^{565}). These lesions result in either a mild or a moderate phenotype. Only in the Gly^{205} mutation is the encoded amino acid changed; it is not yet possible to dissect the relative contribution of amino acid substitution and altered mRNA splicing to the clinical phenotype in this patient.

Only two potential examples of the activation of cryptic splice sites have been reported, one within intron 4 (donor site) and the other at codon 504 (no amino acid substitution) in exon 11 (acceptor site). Both are reportedly associated with mild hemophilia, suggesting that use of the novel splice site is <95%.

To date, over 60 different deletions or partial deletions of the factor VIII gene have been reported. They vary in size from as little as 1 bp to over 200 kb (99–101). All but 3 are associated with the severe phenotype. That gross deletions of a portion of the factor VIII gene are associated with the severe phenotype is of course not very surprising. One of the exceptions involves exon 22 and result in moderately severe hemophilia (102,103). An in-frame deletion of exon 22 would remove 52 amino acids from the protein and presumably results in a protein with a shortened C1 domain. Similarly the exon 23-24 deletion removes 98 amino acids in frame.

Of a total of 1386 hemophilia A patients now examined, 34 exhibit a gross deletion of the factor VIII gene, a frequency of 2.5% (104). Known deletions within the factor VIII gene are fairly heterogeneous in terms of both their size and their position, proving no evidence for a deletion "hot spot." Deletion of the factor VIII gene, however, appears to be associated with a fivefold higher risk of developing inhibitors compared with other severe hemophiliacs without gene deletions (104).

Four examples of the insertion of a single base (three of them A residues) have also been noted. The introduction of an A into an existing string of A residues is consistent with slipped mispairing at the replication fork.

A new type of mutational inactivation has been found in the factor VIII gene causing severe hemophilia A, the insertion of a highly repetitive LINE-1 element into exon 14 of the gene (105). In a screen of 240 unrelated hemophiliac DNA samples, two (0.8%) such insertion events were found. Both occurred in exon 14, the largest exon in the factor VIII gene, and both probably occurred de novo.

B. Prospects for Gene Therapy

Hopes for gene therapy in hemophilia A were recently raised with the successful expression of factor VIII in human and mouse fibroblasts after retrovirus-mediated gene transfer (106,107).

It is axiomatic that such therapy would only be applied to somatic cells, isolated from a suitable tissue of the intended recipient and reintroduced after modification. The risks inherent in germ line modification are all too obvious. Before somatic gene therapy can be contemplated in humans, it should be extensively tested in animal models. An approach to this has begun with the introduction of retrovirus-modified factor VIII-expressing human skin fibroblasts into immune-deficient mice (108). The results of this study were mixed in that the cells persisted and could be shown on recovery to continue secreting factor VIII in vivo, but no human factor VIII was detectable in recipient mouse plasma. Several explanations are available, but a likely one is that it is because the human factor VIII has a very short half-life (~60 minutes) when injected into the mouse with or without von Willebrand factor. Clearly, a great deal more work needs to be done on optimizing expression vectors, the choice of cell for modification and site of reintroduction, and the long-term study of correction in appropriate animal models before clinical trials can be considered. Nevertheless a start has been made, and many laboratories are working toward the goal of gene therapy for various genetic disorders, including hemophilia (109,110).

REFERENCES

1. Fulcher C, Zimmerman T. Characterization of the human FVIII procoagulant protein with a heterologous precipitating antibody. Proc Natl Acad Sci USA 1982; 79:1648.
2. Legaz M, Schmer G, Counts R, Davie E. Isolation and characterization of human factor VIII (anti-hemophilic factor). J Biol Chem 1973; 248:3946–3955.
3. Van Dieijen G, Tans G, Rosing J, Hemker HC. The role of phospholipid and factor VIIIa in the activation of bovine factor X. J Biol Chem 1981; 256:3433–3441.
4. Griffith M, Reisner H, Lundblad R, Roberts R. Measurement of human FIXa activity in an isolated factor X activation system. Thromb Res 1982; 27:289–301.
5. Hultin M. Role of human factor VIII in factor X activation. J Clin Invest 1981; 69:950–958.
6. Duffy EJ, Parker ET, Mutucumarana VP, Johnson AE, Lollar P. Binding of factor VIIIa and factor VIII to factor IXa on phospholipid vesicles. J Biol Chem 1992; 267:17006–17011.
7. Mutucumarana VP, Duffy EJ, Lollar P, Johnson AE. The active site of factor IXa is located far above the membrane surface and its conformation is altered upon association with factor VIIIa. A fluorescence study. J Biol Chem 1992; 267:17012–17021.
8. Husten EJ, Esmon CT, Johnson AE. The active site of blood coagulation factor Xa. Its distance from the phospholipid surface and its conformational sensitivity to components of the prothrombin are complex. J Biol Chem 1987; 262:12953–12961.
9. O'Brien DP, Johnson D, Byfield P, Tuddenham EGD. Inactivation of factor VIII by factor IXa. Biochemistry 1992; 31:2805–2812.
10. Vehar G, Davie E. Preparation and properties of bovine FVIII (anti-hemophilic factor). Biochemistry 1980; 19:401–410.
11. Rapaport S. The extrinsic pathway inhibitor: a regulator of tissue factor-dependent blood coagulation. Thromb Haemost 1991; 6:6–15.
12. Fass DN, Knutson GJ, Katzmann JA. Monoclonal antibodies to porcine factor VIII coagulant and their use in the isolation of active coagulant protein. Blood 1992; 59:594–600.
13. Rotblat F, O'Brien DP, O'Brien FJ, Goodall A, Tuddenham EGD. Purification of factor VIII:C and its characterization by western blotting using monoclonal antibodies. Biochemistry 1985; 24:4294–4300.
14. Vehar G, Keyt B, Eaton D, et al. Structure of human factor VIII. Nature 1984; 313:337–342.

15. Gitschier J, Wood WI, Goralka TM, et al. Characterization of the human factor VIII gene. Nature 1984; 312:326–330.
16. Wood WI, Capon DJ, Simonsen CC, et al. Expression of active human factor VIII from recombinant DNA clones. Nature 1984; 312:330–337.
17. Toole JJ, Knopf JL, Wozney JM, et al. Molecular cloning of a cDNA encoding human antihaemophilic factor. Nature 1984; 312:342–347.
18. Perlman D, Halvorson H. A putative signal peptidase recognition site and sequence in eukaryotic and prokaryotic signal peptides. J Mol Biol 1983; 167:394.
19. Kornfeld R, Kornfeld S. Assembly of aparagine-linked oligosaccharides. Annu Rev Biochem 1985; 54:631.
20. Tuddenham E, Trabold N, Collins J, Hoyer L. The properties of FVIII coagulant activity prepared by immunoadsorbent chromatography. J Lab Clin Med 1979; 93:40.
21. Huttner W, Baeuerle P. Protein sulfation on tyrosine. In: Satiri B, ed. Modern Cell Biology. New York: Alan R. Liss, 1988; 97–140.
22. Pittman D, Walsley L, Murray B, Wang J, Kaufman R. Analysis of structural requirements for FVIII function using site-directed mutagenesis. Thromb Haemost 1987; 58:804a.
23. Ortel T, Takahashi N, Putnam F. Structural model of human ceruloplasmin based on internal triplication, hydrophobic/hydrophillic character and secondary structure of domains. Proc Natl Acad of Sci USA 1984; 81:4761–4768.
24. Kane W, Davie E. Blood coagulation factors V and VIII: structural and functional similarities and their relationships to hemorrhagic and thrombotic disorders. Blood 1988; 71:539–555.
25. Poole S, Firtel R, Lamar E, Rowekamp W. Sequence and expression of the discoidin I gene family in dictostelium discoideum. J Mol Biol 1981; 153:273.
26. Stubbs JD, Lekutis C, Singer KL, et al. cDNA cloning of a mouse mammary epithelial cell surface protein reveals the existence of epidermal growth factor-like domains linked to factor VIII-like sequences. Proc Natl Acad Sci USA 1990; 87:8417–8421.
27. Larocca D, Peterson J, Urrea R, Kuniyoshi J, Bistrain A, Ceriani L. A 46,000 human milk fat globule protein that is highly expressed in human breast tumours contains FVIII-like domains. Cancer Res 1991; 51:4994–4998.
28. Jenny R, Pittman D, Toole J, et al. Complete cDNA and derived amino acid sequence of human FV. Proc Natl Acad Sci USA 1987; 84:4846.
29. Michaelson ME, Forsman N, Oswaldsson UM. Human factor VIII: a calcium linked protein complex. Blood 1983; 62:1006–1015.
30. Andersson L-O, Forsman N, Huang K, et al. Isolation and characterization of human factor VIII: molecular forms in commercial factor VIII concentrate, cryoprecipitate and plasma. Proc Natl Acad Sci USA 1986; 83:2979–2983.
31. Eaton DL, Hass PE, Riddle L, et al. Characterization of recombinant human factor VIII. J Biol Chem 1987; 262:3285–3290.
32. Nordfang O, Ezban M. Generation of active coagulation factor VIII from isolated subunits. J Biol Chem 1987; 263:1115–1118.
33. Rick ME, Hoyer LW. Immunologic studies of antihemophilic factor (AHF, factor VIII). V. Immunologic properties of AHF subunits produced by salt dissociation. Blood 1973; 42: 737–747.
34. Hoyer LW. The factor VIII complex: structure and function. Blood 1981; 58:1–13.
35. Tuddenham EGD, Lane RS, Rotblat F, et al. Response to infusions of polyelectrolyte fractionated human factor VIII concentrate in human haemophilia A and von Willebrand's disease. Br J Haematol 1982; 52:259–267.
36. Lollar P, Parker CG. Stoichiometry of the porcine factor VIII-von Willebrand factor association. J Biol Chem 1984; 262:17572–17576.
37. Foster PA, Fulcher CA, Houghten RA, Zimmerman TS. An immunogenic region within residues Val^{1670}-Glu^{1684} of the factor VIII light chain induces antibodies which inhibit binding of factor VIII to von Willebrand factor. J Biol Chem 1988; 263:5230–5234.

38. Leyte A, van Schijndel HB, Niehrs C, et al. Sulfation of Tyr^{1680} of human blood coagulation factor VIII is essential for the interaction of factor VIII with von Willebrand factor. J Biol Chem 1991; 266:740–746.
39. Higuchi M, Antonarakis SE, Kasch L, et al. Towards complete characterization of mild-to-moderate hemophilia A: detection of the molecular defect in 25 of 29 patients by denaturing gradient gel electrophoresis. Proc Natl Acad Sci USA 1991; 88:8307–8311.
40. Fay PJ, Smudzin TM. Topography of the human factor VIII-von Willebrand factor complex. J Biol Chem 1990; 265:6197–6202.
41. Hemker HC, Kahn MJP, Devilee PP. The adsorption of coagulation factors onto phospholipids. Its role in the reaction mechanisms of blood coagulation. Thromb Diath Haemorrh 1970; 24:214–223.
42. Kemball-Cook G, Edwards SJ, Sewerin K, Andersson LO, Barrowcliffe TW. Factor VIII procoagulant protein interacts with phospholipid vesicles via its 80 kDa light chain. Thromb Haemost 1988; 60:442–446.
43. Arai M, Scandella D, Hoyer LW. Molecular basis of factor VIII inhibition by human antibodies. Antibodies that bind the factor VIII light chain prevent the interaction of factor VIII with phospholipid. J Clin Invest 1989; 83:1978–1984.
44. Foster PA, Fulcher CA, Houghton RA, Zimmerman TS. Synthetic factor VIII peptides with amino sequences contained within the C2 domain of factor VIII inhibit factor VIII binding to phosphatidylserine. Blood 1990; 75:1999–2004.
45. Gilbert GE, Sims PJ, Wiedmer T, Furie B, Furie BC, Shattil SJ. Platelet-derived microparticles express high affinity receptors for factor VIII. J Biol Chem 1991; 266:17261–17268.
46. Rapaport SI, Schiffman S, Patch MJ, Ames SB. The importance of activation of anti-haemophilic globulin and proaccelerin by traces of thrombin in the generation of intrinsic prothrombinase activity. Blood 1963; 21:221–236.
47. Switzer ME, McKee PA. Reactions of thrombin with human factor VIII/von Willebrand factor protein. J Biol Chem 1980; 255;10606–10611.
48. Hoyer LW, Trabold NC. The effect of thrombin on human factor VIII. J Lab Clin Med 1981; 97:50–64.
49. Fulcher C, Roberts J, Zimmerman T. Thrombin proteolysis of purified FVIII procoagulant protein: correlation of activation with generation of specific polypeptides. Blood 1983; 60: 807.
50. Eaton D, Rodriguez H, Vehar G. Proteolytic processing of human FVIII. Correlation of specific cleavages by thrombin FXa and activated ptoein C with activation and inactivation of factor VIII coagulant activity. Biochemistry 1986; 25:505–512.
51. Fulcher C, Gardiner J, Griffin J, et al. Proteolytic inactivation of human FVIII procoagulant protein by activated protein C and its analogy with FV. Blood 1984; 63:486.
52. Walker F, Chavin S, Fay P. Inactivation of FVIII by activated protein C and protein S. Arch Biochem Biophys 1987; 252:322–328.
53. Eaton D, Vehar G. FVIII structure and proteolytic processing. In: Progress in Hemostasis and Thrombosis. BS Coller (Ed.). Grune and Stratton Inc., Orlando. Vol 8. 1986 pp 47–70.
54. Koedam J, Meijers J, Sixma J, Bouma B. Inactivation of human factor VIII by activated protein C: cofactor activity of protein S and protective effect of von Willebrand factor. J Clin Invest 1988; 82:1236–1243.
55. Fay P, Coumans J, Walker F. Von Willebrand factor mediates protection of FVIII from activated protein C-catalysed inactivation. J Biol Chem 1991; 266:2172–2177.
56. Lollar P, Knutson G, Fass D. Stabilisation of thrombin activated porcine factor VIII:C by factor IXa and phospholipid. Blood 1984; 63:1303–1308.
57. Rick ME, Esmon N, Krizek D. Factor IXa and von Willebrand factor modify the inactivation of factor VIII by activated protein C. J Lab Clin Med 1990; 115:415–21.
58. Miletich J, Sherman L, Broze G. Absence of thrombosis in subjects with heterozygous protein C deficiency. N Engl J Med 1987; 317:991–996.

59. Pieters J, Lindhout T, Hemker C. In situ-generated thrombin is the only enzyme that effectively activates factor VIII and factor V in thromboplastin-activated plasma. Blood 1989; 74:1021–1024.
60. Bauer KA, Rosenberg RD. The pathophysiology of the prethrombotic state in humans: insights gained from studies using markers of hemostatic system activators. Blood 1987; 70:343–350.
61. Pittmann DD, Kaufman RJ. Proteolytic requirements for thrombin activation of anti-hemophilic factor (factor VIII). Proc Natl Acad Sci USA 1988; 85:2429–2433.
62. O'Brien DP, Pattinson JK, Tuddenham EGD. Purification and characterization of factor VIII 372-Cys: a hypofunctional cofactor from a patient with moderately severe haemophilia A. Blood 1990; 75:1664–1672.
63. Arai M, Inaba H, Higuchi M, et al. Direct characterization of factor VIII in plasma: detection of a mutation altering a thrombin cleavage site (arginine 372 histidine). Proc Natl Acad Sci USA 1989; 86:4277–4281.
64. Bihoreau N, Sanger A, Yon JM, de Pol HV. Isolation and characterization of different activated forms of factor VIII, the human antihemophilic A factor. Eur J Biochem 1989; 185:111–118.
65. O'Brien DP, Tuddenham EGD. Purification and characterization of factor VIII 1689 Cys: a non-functional cofactor occurring in a patient with severe haemophilia A. Blood 1989; 73: 2117–2122.
66. Tuddenham EGD, Schwaab R, Seehafer J, et al. Haemophilia A: database of nucleotide substitutions, deletions, insertions and re-arrangements of the factor VIII gene. Second edition Nucleic Acids Res 1994; In press.
67. Hill-Eubanks DC, Parker CG, Lollar P. Differential proteolytic activation of factor VIII-von Willebrand factor complex by thrombin. Proc Natl Acad Sci USA 1989; 86:6508–6512.
68. Lollar P, Parker C. Subunit structure of thrombin activated porcine factor VIII. Biochemistry 1987; 28:666–674.
69. Lollar P, Parker C. pH-dependent denaturation of thrombin-activated porcine factor VIII. J Biol Chem 1990; 265:1688–1692.
70. Fay PJ, Haidari PJ, Smudzin TM. Human factor VIIIa subunit structure. Reconstitution of factor VIIIa from the isolated A1/A3-C1-C2 dimer and A2 subunit. J Biol Chem 1991; 266: 8957–8962.
71. Pittman D, Millenson M, Marquette K, Bauer K, Kaufman R. A2 domain of human recombinant-derived FVIII is required for procoagulant activity but not for thrombin cleavage. Blood 1992; 79:389–397.
72. Eaton DL, Wood W, Eaton D, et al. Construction and characterization of an active FVIII variant lacking the central one third of the molecule. Biochemistry 1986; 25:8343–8347.
73. Toole J, Pittman D, Orr E, Murtha P, Walsey L, Kaufman R. A large region (=95kDa) of human factor VIII is dispensable for in vitro procoagulant activity. Proc Natl Acad Sci USA 1986; 83:5939–5942.
74. Exner T, Richard KA, Kronenberg H. Measurement of factor VIIIC Ag by immunoradiometric assay in human tissue extracts. Thromb Res 1983; 32:427.
75. Stel HV, van der Kwast TH, Veerman ECI. Detection of factor VIII/coagulant antigen in human liver tissue. Nature 1983; 303:530–532.
76. Stel HV, Veerman ECI, Jeijer CJLM. Immunohistological localization of factor VIII in placental endothelial cells. Br J Haematol 1986; 63:565–569.
77. Zelechowska MG, van Mourik JA, Brodniewicz-Proba T. Ultrastructural localization of factor VIII procoagulant antigen in human liver hepatocytes. Nature 1985; 317:729–730.
78. Rall LB, Bell GI, Caput D, Truett MA, et al. Factor VIII:C synthesis in the kidney. Lancet 1985; 1:44.
79. Wion KL, Kelly D, Summerfield JA, Tuddenham EGD, Lawn RM. Distribution of factor VIII mRNA and antigen in human liver and other tissues. Nature 1985; 317:726–729

80. Hellman L, Smedsrod B, Sandberg H, Pettersson U. Secretion of coagulant factor VIII activity and antigen by in vitro cultivated rat liver sinusoidal endothelial cells. Br J Haematol 1989; 73:348–355.
81. Hassan HJ, Leonardi A, Chelucci C, et al. Blood coagulation factors in human embryonic-fetal development: preferential expression of the FVII/tissue factor pathway. Blood 1990; 76:1158–1164.
82. Cutting GR, Antonarakis SE, Youssoufian H, Kazazian HH. Accuracy and limitations of pulsed field gel electrophoresis in sizing partial deletions of the factor VIII gene. Mol Biol Med 1988; 5:173–184.
83. Breathnach R, Chambon P. Organization and expression of eukaryotic split genes coding for proteins. Annu Rev Biochem 1981; 50:349–383.
84. Truett MA, Blacher R, Burke RL, et al. Characterization of the polypeptide composition of human factor VIII:C and the nucleotide sequence and expression of the human kidney cDNA. DNA 1985; 4:333–349.
85. Cooper DN, Smith BA, Cooke HJ, Niemann S, Schmidtke J. An estimate of unique DNA sequence heterozygosity in the human genome. Hum Genet 1985; 69:201–205.
86. Levinson B, Lakich D, Silvera P, Kenwrick S, Gischier J. A gene contained within a factor VIII intron is identified by a CpG island. Proc Am Soc Clin Genet New Orleans 1988; A192.
87. Levinson B, Kenwrick S, Lakich D, Hammonds G, Gitschier J. A transcribed gene in an intron of the human factor VIII gene. Genomics 1990; 7:1–11.
88. Levinson B, Kenwrick S, Gamel P, Fisher K, Gitschier J. Evidence for a third transcript from the human factor VIII gene. Genomics 1992; 14:585–589.
89. Purrello M, Alhadeff B, Esposito D, et al. The human genes for hemophilia A and hemophilia B flank the X chromosome fragile site at Xq27.3. EMBO J 1985; 4:725–729.
90. Tantravahi U, Murty VVVS, Jhanwar SC, et al. Physical mapping of the factor VIII gene proximal to two polymorphic DNA probes in human chromosome band Xq28: implications for factor VIII gene segregation analysis. Cytogenet Cell Genet 1986; 42:75–79.
91. Pattinson JK, Millar DS, Grundy CB, et al. The molecular genetic analysis of haemophilia A; a directed-search strategy for the detection of point mutations in the human factor VIII gene. Blood 1990; 76:2242–2248.
92. Aly AM, Higuchi M, Kasper CK, Kazazian HH, Antonarakis SE, Hoyer LW. Haemophilia A due to mutations that create new N-glycosylation sites. Proc Natl Acad Sci USA 1992; 89:4933–4937.
93. Higuchi M, Kazazian HH, Kasch L, et al. Molecular characterization of severe haemophilia A suggests that about half the mutations are not within the coding regions and splice junctions of the factor VIII gene. Proc Natl Acad Sci USA 1991; 88:7405–7409.
94. Naylor JA, Green PM, Rizza CR, Giannelli F. Analysis of factor VIII mRNA reveals defects in everyone of 28 haemophilia A patients. Hum Mol Genet 1993; 2:11–17.
95. Cooper DN, Youssoufian H. The CpG dinucleotide and human genetic disease. Hum Genet 1988; 78:151–155.
96. Youssoufian H, Antonarakis SE, Bell W, Griffin AM, Kazazian HH. Nonsense and missense mutations in hemophilia A: estimate of the relative mutation rate at CpG dinucleotides. Am J Hum Genet 1988; 42:718–725.
97. Berg L-P, Wieland K, Millar DS, et al. Detection of a novel point mutation causing haemophilia A by PCR/direct sequencing of ectopically transcribed factor VIII mRNA. Hum Genet 1990; 85:655–658.
98. Naylor JA, Green PM, Montandon AJ, Rizza CR, Giannelli F. Detection of three novel mutations in two haemophilia A patients by rapid screening of whole essential region of factor VIII gene. Lancet 1991; 337:635–639.
99. Higuchi M, Wong C, Kochhan L, et al. Characterization of mutations in the factor VIII gene by direct sequencing of amplified genomic DNA. Genomics 1990; 6:65–71.

100. Casarino L, Pecorara M, Mori PG, et al. Molecular basis for hemophilia A in Italians. Ric Clin Lab 1986; 16:227.
101. Schwartz C, Fitch N, Phelan MC, Richer CL, Stevenson R. Two sisters with a distal deletion at the Xq26/Xq27 interface: DNA studies indicate that the gene locus for factor IX is present. Hum Genet 1987; 76:54–57.
102. Youssoufian H, Antonarakis SE, Aronis S, Tsiftis G, Phillips DG, Kazazian HH. Characterization of five partial deletions of the factor VIII gene. Proc Natl Acad Sci USA 1987; 84:3772–3776.
103. Wehnert M, Herrmann FH, Wulff K. Partial deletions of factor VIII gene as molecular diagnostic markers in haemophilia A. Dis Markers 1989; 7:113–117.
104. Millar DS, Steinbrecher RA, Wieland K, et al. The molecular genetic analysis of haemophilia A; characterization of six partial deletions in the factor VIII gene. Hum Genet 1990; 86:219–227.
105. Kazazian HH, Wong C, Youssoufian H, Scott AF, Phillips DG, Antonarakis SE. Haemophilia A resulting from de novo insertion of L1 sequences represents a novel mechanism for mutation in man. Nature 1988; 332:164–166.
106. Hoeben RC, van der Jagt RCM, Schoute F, et al. Expression of functional factor VIII in primary human skin fibroblasts after retrovirus-mediated gene transfer. J Biol Chem 1990; 265:7318–7323.
107. Israel DI, Kaufman RJ. Retroviral-mediated transfer and amplification of a functional human factor VIII gene. Blood 1990; 75:1074–1080.
108. Hoeben RC, Miachielson AAJ, Vander Jagt RCM, Van Ormandt H, van der Erb AJ. Inactivation of the Moloney murine leukemia virus long terminal repeat in fibroblasts in vivo is associated with methylation and dependent on its chromosomal position. J Vivol 1991; 65:904–912.
109. Thompson AR. Status of gene transfer for hemophilia A and B. Thromb Haemost 1991; 66: 119–122.
110. Friedmann T. Progress toward human gene therapy. Science 1989; 224:1275–1281.
111. Elder B, Lakich D, Gitschier J. Sequence of the murine factor VIII cDNA. Genomics 1993; 16:374–379.
112. McGinnis MJ, Kazazian HH, Jr, Hoyer LW, Bi L, Inaba H, Antonarakis SE. Spectrum of mutations in CRM-positive and CRM-reduced hemophilia A. Genomics 1993; 15:392–398.
113. Pieneman WC, Reitsma PH, Briët E. Double strand conformation polymorphism (DSCP) detects two point mutations at codon 280 (AAC → ATC) and at codon 431 (TAC → AAC) of the blood coagulation factor VIII gene. Thromb Haemost 1993; 69:473–475.
114. Economou EP, Kakazian HH, Jr, Antonarakis SE. Detection of mutations in factor VIII gene using single-stranded conformational polymorphism (SSCP). Genomics 1992; 13:909–911.
115. Nafa K, Baudis M, Deburgrave N, et al. A novel mutation (Arg → Leu in exon 18) in factor VIII gene responsible for moderate hemophilia A. Hum Mutat 1992; 1:77–78.
116. Diamond C, Kogan S, Levinson B, Gitschier J. Amino acid substitutions in conserved domains of factor VIII and related proteins: study of patients with mild and moderately severe hemophilia A. Hum Mutat 1992; 1:248–257.

9

The Molecular Bases of Von Willebrand Disease

Jerry Ware and Zaverio M. Ruggeri
The Scripps Research Institute, La Jolla, California

I. INTRODUCTION

Although genetic diseases are one tragic reminder of the fragile nature of life, congenitally inherited bleeding disorders have historically provided some important and definitive "experiments of nature" documenting the physiological importance of a molecule. For many of the disorders, the symptoms and clinical recognition of the disease occurred long before an unraveling of the biochemical problem. The description of hemophilia can be traced to biblical times, yet it was only during the latter half of the twentieth century that the biochemical basis of hemophilia was defined. For von Willebrand disease, it was nearly half a century after the description of the disease by Erik von Willebrand (circa 1926) that a direct molecular link was established between the glycoprotein, now referred to as von Willebrand factor, and the disease (1). Nevertheless, studies during the past decade have led to remarkable progress in defining the molecular anatomy of von Willebrand factor and its role in hemostasis and thrombosis. Throughout, the characterization of variant molecules resulting in von Willebrand disease has provided some of the key insights into our current understanding of this molecule.

There are two essential roles for von Willebrand factor in hemostasis. The first is its ability to support platelet adhesion and platelet activation to a damaged vessel wall. The second role is to serve as the carrier protein for procoagulant factor VIII in circulating blood, in which both factor VIII and von Willebrand factor circulate as a noncovalently linked glycoprotein complex. With regard to its role in platelet attachment, von Willebrand factor has functional domains interacting with the platelet receptor complexes, glycoprotein Ib-IX and glycoprotein IIb-IIIa (integrin $\alpha_{IIB}\beta_3$) (2). The von Willebrand factor-platelet interaction is most significant under flow conditions generating high shear stress, such as occurs in the arterial circulation and stenosed arteries (3–5). Current models suggest that the two platelet binding sites are integrated into a temporal sequence of events initiated by the interaction of subendothelian-bound von Willebrand factor and the platelet glycoprotein Ib component of the Ib-IX receptor complex (6–9). This binding event establishes platelet contact with the damaged surface and initiates signals associated with platelet activation (10–12). Subsequently, von Willebrand factor interacts with the glycoprotein IIb-IIIa receptor to establish irreversible adhesion and support platelet aggregation at the site of injury (3–5).

The close association between von Willebrand factor and factor VIII resulted for a number of years in considerable confusion in the terminology and molecular identities of the two molecules (13). In the mid-1980s, the genes for von Willebrand factor and factor VIII were isolated by molecular cloning techniques and it was clearly established that each protein is synthesized as a distinct gene product (14–19). In fact, it is now well established that mutations within the former give rise to von Willebrand disease and the latter to hemophilia A. However, as discussed later, some variants of von Willebrand factor may be impaired in their ability to bind and stabilize factor VIII. In this case, the lack of factor VIII clinically disguises itself as a form of hemophilia even though the defect is within the von Willebrand factor gene (20).

In this review we examine the structural organization of the von Willebrand factor gene and protein in an attempt to appreciate how this knowledge has been used to identify the genetic bases of the clinically heterogeneous disorder, von Willebrand disease. In particular, we discuss those mutations resulting in qualitative defects that have been informative for understanding the role of von Willebrand factor in hemostasis. Given the immense structural complexity of von Willebrand factor and its gene, many of the recently utilized strategies for identifying mutations have specifically targeted restricted regions of the gene. In general, this strategy has relied upon previous studies of the normal von Willebrand factor molecule to localize a functional domain and then determine whether mutations altering the same function could be found within the same domain. Because this strategy characterizes only a limited portion of the von Willebrand factor gene, conclusive proof depends upon the generation of recombinant molecules and functional studies to validate the consequences of the mutation.

II. VON WILLEBRAND FACTOR STRUCTURE-FUNCTION RELATIONSHIPS

A. Multimeric Von Willebrand Factor

The protein referred to as von Willebrand factor is actually a series of heterogeneous oligomers ranging in molecular mass from 500 to 10,000 kD (Fig. 1) (21,22). These oligomers, more commonly referred to as multimers, are assembled from varying numbers of a 275 kD von Willebrand factor subunit (23). The multimeric process is dependent upon a large propeptide, also referred to as von Willebrand antigen II, that is cleaved from multimeric von Willebrand factor before secretion (24–26). The biological importance of multimer formation is best exemplified by the fact that the highest molecular weight multimeric species have the greatest hemostatic efficacy (27–29). Indeed, the largest von Willebrand factor multimers are stored within specific organelles of endothelial cells and platelets and are only transiently observed in the circulation in response to tissue damage and hemostasis.

An in vivo role for circulating von Willebrand antigen II after cleavage from mature von Willebrand factor has not been established. In experimental assays, the molecule can inhibit collagen-induced platelet aggregation; thus, it may participate in regulating the size of thrombi at sites of vascular injury and balance the prothrombotic action of von Willebrand factor. It is worth noting that molecular variants of the von Willebrand factor propeptide resulting in aberrant functional properties have not been found.

It has been proposed that regulating the size of von Willebrand factor multimers may be an important in vivo mechanism for regulating the hemostatic response. Protein studies have demonstrated that the multimer may be altered by a proteolytic cleavage event occurring between Tyr^{842} and Met^{843} of the constituent subunit (30,31). When a subunit is cleaved at this position, the multimer is separated into two smaller species as a result

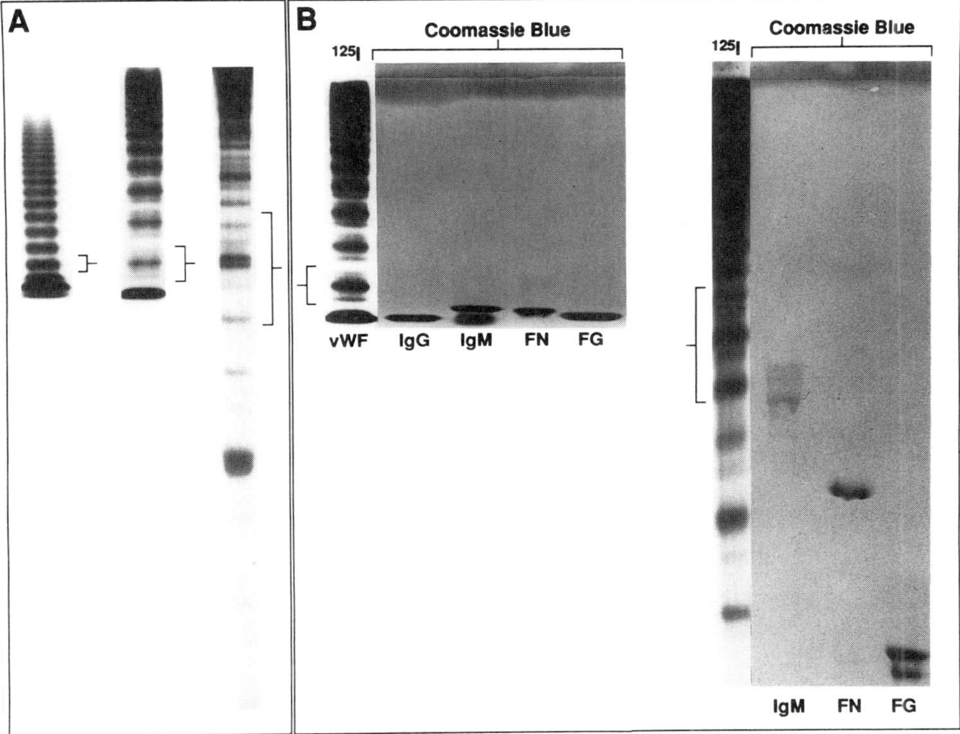

Figure 1 Von Willebrand factor multimers of plasma can be separated by electrophoresis through agarose gels in the presence of sodium dodecyl sulfate. Samples are applied at the top of the gel (cathode) and migrate toward the anode at the bottom. The von Willebrand factor species are visualized by reacting with ^{125}I-labeled anti–von Willebrand factor antibodies. (*A*) By varying the agarose concentration, the complete range of multimers can be observed. The lowest agarose concentration (left lane) separates the largest species into discernible bands, whereas the highest agarose concentration (right lane) separates the smallest species into individual bands. Brackets enclose the same molecular weight species using three different agarose concentrations. (*B*) Purified IgG (166 kD), IgM (900 kD, slowest migrating band), fibronectin (450 kD), and fibrinogen (340 kD) were electrophoresed in the same agarose gel for comparison with plasma von Willebrand factor multimers. The purified proteins were visualized by staining with Coumassie blue, and the von Willebrand factor was visualized by autoradiography as just described. The gel on the left in B is the same agarose concentration as the middle gel of A, and the gel on the right in B is the same agarose concentration as the gel on the right in A. (Reprinted with permission from Ref. 36.)

of the absence of intra- and intermolecular disulfide linkages between residues on the amino-terminal side of the cleavage and residues on the carboxyl-terminal side of the cleavage. It has been speculated that this proteolytic event may be a normal process regulating the size of circulating von Willebrand factor to reduce, in effect, the adhesive potential of von Willebrand factor by removing the highest molecular weight species (32). Even though this cleavage event is a normal process, variant molecules resulting in the subtype of von Willebrand disease, designated type IIA, have a heightened susceptibility to this proteolytic event (30,33,34). As discussed later, the localization of this cleavage site led to a targeted approach to identify mutations responsible for the type IIA phenotype (35).

B. Interactions of Von Willebrand Factor with Platelets

The von Willebrand factor subunit is composed of multiple structural and functional domains (Fig. 2). Protein studies have identified regions and isolated fragments conferring binding specificity for components of the subendothelial matrix, the platelet glycoprotein Ib-IX and IIb-IIIa receptors and factor VIII (36). However, identification of residues or domains responsible for the adhesive properties of von Willebrand factor still yields an incomplete understanding of this molecule. Von Willebrand factor circulates as a soluble component of blood with no measurable affinity for platelets, yet in the presence of vessel injury, it supports platelet attachment at the site of injury. Understanding the physiological events that transform von Willebrand factor from a noncompetent ligand to a competent ligand is paramount for understanding the role of this molecule in hemostasis. As discussed later, mutations resulting in the subtype of von Willebrand disease, designated type IIB, demonstrate that conformational changes within the von Willebrand factor subunit can generate a ligand that spontaneously binds to the platelet surface. Thus, as illustrated by molecular analysis of variant molecules, conformation changes within von Willebrand factor may be an important physiological modulator of the ligand-platelet receptor interaction.

The platelet glycoprotein Ib-IX receptor complex interacts with von Willebrand factor through a domain originally isolated as a tryptic fragment composed of Val^{449}-Lys^{728} (Fig. 2) (37,38). The original determination of the von Willebrand factor primary amino acid sequence revealed that this region, the A1 domain, is the first motif of a contiguous triplicate A repeat with the mature subunit (39). The A, or sometimes referred to as I, domains have been described in other adhesive proteins, including some integrins (Mac-1, LFA-1, p150,95, and VLA-2), components of the extracellular matrix (cartilage matrix protein and collagen VI), and components of the complement system (C2 and factor B) (40). Although sequence similarities exist among the different A domains, only the von Willebrand factor A1 domain interacts with platelet glycoprotein Ib-IX. Thus, each A domain has probably evolved to have distinct and specific binding properties through precise amino acid substitutions. Recent experiments using recombinant analogs of the domain have demonstrated the importance of residues within a loop generated by an intrachain disulfide bond between Cys^{509} and Cys^{695} (41,42). In vitro assays evaluating the structural requirements for binding to platelet glycoprotein Ib-IX have demonstrated that the formation of the intrachain disulfide bond between Cys^{509} and Cys^{695} can regulate von Willebrand factor affinity for glycoprotein Ib-IX (41–43). The residues within the von Willebrand factor A1 domain that are in direct contact with the platelet glycoprotein IB-IX receptor remain to be defined.

Once von Willebrand factor and glycoprotein Ib-IX establish platelet adhesion, platelets activate, spread, and become irreversibly bound to a thrombogenic surface. These latter processes are dependent upon von Willebrand factor binding with the platelet receptor, glycoprotein IIb-IIIa (5,44). The Arg-Gly-Asp sequence of von Willebrand factor, residues 1744–1746, is a signature recognition sequence for a variety of adhesive ligands that bind glycoprotein IIb-IIIa, including fibrinogen, fibronectin, and vitronectin (45). Although Arg-Gly-Asp is a crucial sequence within each ligand, additional sequences are important for the binding specificity of each ligand (46,47). To date, no mutations resulting in von Willebrand disease have been localized to the glycoprotein IIb-IIIa binding site within von Willebrand factor. In contrast, a number of mutations resulting in von Willebrand disease disrupt von Willebrand factor binding to platelet GP Ib-IX. However, it should not be concluded that von Willebrand factor binding to platelet glycoprotein

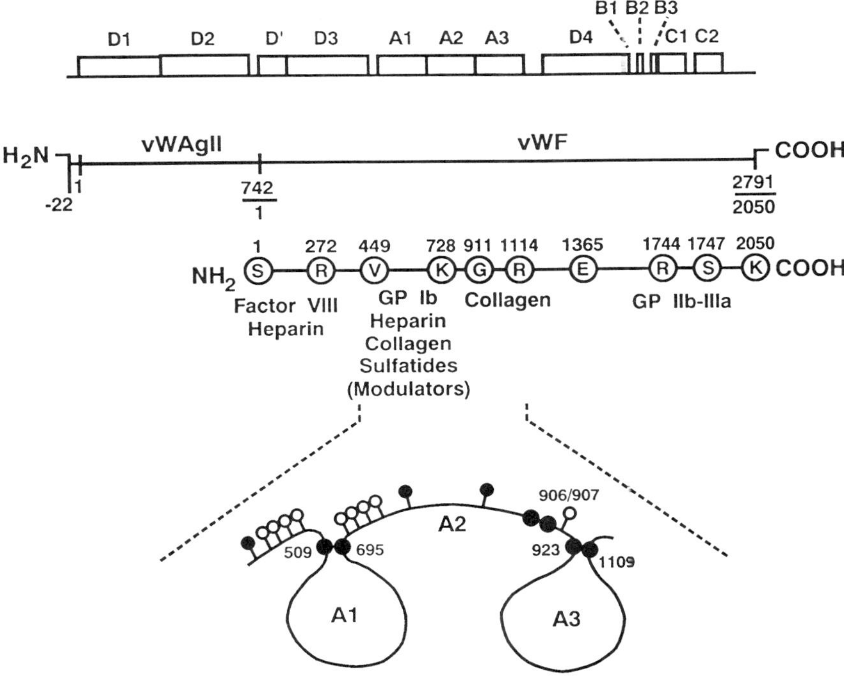

Figure 2 Von Willebrand factor is synthesized as a primary translation product of 2813 amino acids (pre-pro–von Willebrand factor). (*Top*) Pre-pro–von Willebrand factor consists of a 22-residue signal peptide (SP), a 741-residue propeptide composed of D1 and D2 domains, and a 2050 residue mature or constituent subunit. The primary amino acid sequence revealed that von Willebrand factor is composed of a series of repeated domains, A → D (39). (*Middle*) Protein analysis of von Willebrand factor has identified a number of functional fragments or domains within the mature subunit. Below each domain is the molecule(s) that binds within the indicated region. (*Bottom*) The triplicate A domains contain 6 cysteine residues (black dots) within the mature subunit that generate two large intramolecular disulfide loops (Cys^{509}-Cys^{695} and Cys^{923}-Cys^{1109}) comprising the A1 and A3 domains and a small disulfide pairing between Cys^{906} and Cys^{907}. Positions of *O*-linked carbohydrate and *N*-linked carbohydrate side chains are illustrated with open circles and filled circles, respectively, at their corresponding positions. (Reprinted with permission from Ref. 32).

Ib-IX is more important than binding to glycoprotein IIb-IIIa. The most common clinical assay for characterizing von Willebrand factor-platelet interaction is a ristocetin-mediated binding event that requires only the glycoprotein Ib-IX binding site on the platelet surface. This purely experimental assay uses the antibiotic ristocetin, a modulator of soluble von Willebrand factor binding to platelets (48,49). Modulator-independent assays that require binding sites for both platelet receptors, such as shear-dependent platelet aggregation, are starting to be used as an alternative to the modulator-dependent assay (50–52). Widespread use of such assays may identify variant molecules with an aberrant binding site for glycoprotein IIb-IIIa. Thus, the absence of described mutations within the glycoprotein IIb-IIIa binding site of von Willebrand factor may reflect only an inability to identify experimentally such variants.

C. Interactions of Von Willebrand Factor with Factor VIII and the Subendothelium

The von Willebrand factor domain that stabilizes factor VIII is within the amino terminus of the mature subunit (Fig. 2). This conclusion was originally based on the observation that a tryptic fragment comprised of residues 1–272 of von Willebrand factor could inhibit the interaction of intact multimeric von Willebrand factor and purified factor VIII (53). Supporting evidence was obtained with inhibitory monoclonal antibodies that, when bound to the amino-terminal region of von Willebrand factor, blocked the interaction of von Willebrand factor with factor VIII (53,54). That the binding site for factor VIII is exclusively within the amino terminus of the constituent subunit has not been determined. It is possible that a complete factor VIII binding site may require residues outside the von Willebrand factor domain spanning residues 1–272, but as we discuss later, the importance of the amino terminus was recently corroborated with amino-terminal mutations resulting in molecules unable to interact with factor VIII.

Because von Willebrand factor serves as a bridging molecule between platelets and the subendothelium, it must simultaneously bind both (8). Indeed, functional binding sites within von Willebrand factor for components of the subendothelial matrix have been described. Von Willebrand factor interacts with heparin, supporting a concept that it interacts with heparin-like glycosaminoglycans in the extracellular matrix (55,56). Collagens of the subendothelium, such as types I, III, and VI, bind von Willebrand factor, providing another mechanism for interaction with the subendothelium (57–59). Although a number of possible candidate molecules of the subendothelium have been described, the most physiologically relevant has not been determined with certainty. To date, no mutations resulting in von Willebrand disease have been attributed to loss of these functional binding sites. However, the observation that von Willebrand factor can interact with a number of different molecules of the subendothelial matrix and the possibility that the binding sites for each of these components are nonoverlapping may explain why mutations disrupting only one binding site do not result in impaired hemostasis.

III. VON WILLEBRAND FACTOR GENE ORGANIZATION AND STRUCTURE

The expression of the von Willebrand factor gene is restricted to endothelial cells and megakaryocytes. The first cDNA clones were originally prepared from endothelial cell mRNA, an abundant source of von Willebrand factor mRNA (14–17), and revealed the sequence of an 8500 base mature transcript (60,61). Platelets retain residual amounts of von Willebrand factor mRNA that serve as a ready source of material for amplification via the polymerase chain reaction and characterization of a mutant gene (35). To date, no differences have been described between the von Willebrand factor sequence of endothelial cell- and platelet-derived mRNA. The von Willebrand factor protein is encoded by a single copy gene (14), so it is assumed that differences, if any, between endothelial cell- and platelet-synthesized von Willebrand factor are the result of posttranslational differences, such as glycosylation, between the different cell types. Indeed, the molecule is subjected to unique processing events within platelets, as evidenced by its strict localization to platelet α granules, where it is only released by platelet activation. In contrast, von Willebrand factor synthesized by endothelial cells may be constitutively secreted after synthesis or stored within the unique organelles of endothelial cells, termed Weibel-Palade bodies.

After characterization of cDNA clones, the von Willebrand factor gene was isolated and characterized as a composite array of molecular clones spanning 180 kb (62). The nucleotide sequence for 34 kb of the gene has been determined. This analysis defined the boundaries for each of the 52 exons of the gene (62). Studies that localized the von Willebrand factor gene to the tip of the short arm of chromosome 12 also identified a highly homologous sequence on chromosome 22 (63,64). This homologous sequence is a nonprocessed von Willebrand factor pseudogene representing a duplication of exons 23–34 of the functional von Willebrand factor gene (64). The pseudogene has presumably arisen through a translocation event between chromosomes 12 and 22. The von Willebrand factor gene and pseudogene are 97% identical at the nucleotide level. This high degree of homology warrants a careful interpretation of the potential defects resulting in von Willebrand disease. However, the use of gene- or pseudogene-specific oligonucleotide primers in the polymerase chain reaction can be used to amplify selectively fragments of the von Willebrand factor gene (64).

Polymorphisms within the von Willebrand gene have been particularly informative for genetic studies of von Willebrand disease. The large size of the von Willebrand factor gene accounts, at least in part, for the description of 33 polymorphisms throughout the gene (65). Of the described polymorphisms, approximately half are present in the noncoding sequence and half are present in the coding sequence. Within the coding sequence, 8 polymorphisms result in amino acid substitutions or are the cause of naturally occurring amino acid dimorphisms within what is considered a normal von Willebrand factor protein. By definition, it is assumed that amino acid polymorphisms do not affect protein function; however, it is possible in the case of von Willebrand factor that such polymorphisms result in subtle, yet important, heterogeneity in the protein. Although unproven, such heterogeneity could explain some of the diversity observed in the hemostatic response. In combination with single amino acid substitutions representing authentic von Willebrand factor mutations, these polymorphisms may also contribute to the phenotype of a variant von Willebrand factor molecule.

One of the most powerful markers within the von Willebrand factor gene is a highly polymorphic region within intron 40 (62). Termed the variable number tandem repeat (VNTR) sequence, this sequence can be quite informative for pedigree analysis designed to follow the inheritance of a mutant allele (66). The polymorphic nature of the VNTR region is dependent upon varying numbers, as many as eight, of a repetitive element approximately 600 bp in length and composed primarily of the tetranucleotide repeat 5′-

TCTA-3′. In a recent family studies report, analysis of the VNTR sequence was informative for the inheritance of 98% of all von Willebrand factor alleles (67).

IV. GENETIC BASES OF VON WILLEBRAND DISEASE

A. Heterogeneity and Classification of Von Willebrand Disease

Under the nomenclature established in 1993 by the Scientific and Standardization Committee of the International Society on Thrombosis and Haemostasis, the absent or dysfunctional properties of von Willebrand factor form the basis for the primary division of von Willebrand disease into one of three types designated by Arabic numbers. This classification replaces an older scheme in which the multimeric structure of von Willebrand factor formed the basis for the primary division into one of three types designated by Roman numerals (22). Under the new system, quantitative defects are divided into dominantly transmitted partial deficiencies (type 1) and recessively transmitted severe deficiencies (type 3). Qualitative defects (type 2) are divided into three distinct categories: (1) those with decreased glycoprotein Ib-IX binding, type 2A; (2) those with increased affinity for platelet glycoprotein Ib-IX, type 2B; and (3) those with decreased affinity for factor VIII, type 2N.

During the last few years there have been an increasing number of publications defining the genetic lesions resulting in von Willebrand disease, and a recent review summarized this emerging database of information and the formation of a consortium to ensure the database remains updated (68). As expected, mutations resulting in type 3 von Willebrand disease have strictly been those genetic lesions that are expected to decrease protein production significantly, such as large gene deletions, nonsense mutations, defective mRNA processing, and mRNA instability (68). To date, no mutations have been identified to explain the "classic" form of type 1 von Willebrand disease, although in one case it has been linked to the von Willebrand factor locus (69). A number of genetic analyses have described the qualitative defects resulting in type 2 von Willebrand disease, and these studies have been particularly informative for understanding structure-function relationships within von Willebrand factor.

B. Qualitative Defects of Type 2 Von Willebrand Disease

Type 2 von Willebrand disease was originally recognized as containing two major subtypes, designated type IIA and type IIB (or type 2A and 2B under the newer classification scheme), both inherited in an autosomally dominant manner and lacking the highest von Willebrand factor multimers but differing in their ability to support ristocetin-mediated platelet aggregation (70). Umder the older classification scheme additional subtypes were reported, but many were single case reports (22). Experimentally, type 2A von Willebrand disease is characterized by decreased ristocetin cofactor activity, whereas type 2B is characterized by a variant von Willebrand factor that supports platelet aggregation at absent or subnormal levels of ristocetin (22). It must be emphasized that the ability of von Willebrand factor to support ristocetin-mediated platelet aggregation is an experimental assay, and no in vivo equivalent for ristocetin has been described, even though it is the most widely used test of von Willebrand factor competence. The mechanism by which ristocetin supports a von Willebrand factor-platelet glycoprotein Ib-IX interaction suggests that ristocetin dimerizes with itself and interacts with both von Willebrand factor and platelet glycoprotein Ib-IX to support platelet agglutination (71). It has been proposed that during thrombogenesis the physiological equivalent of ristocetin may be specific functional conformations acquired by von Willebrand factor or the platelet glycoprotein Ib-IX receptor. Such conformational changes may reflect intrinsic changes within von Willebrand factor resulting from binding to subendothelial matrix components, unidentified agonists, or rheological conditions affecting platelet glycoprotein Ib-IX and/or von Willebrand factor (32,72).

As discussed, protein studies on the molecules resulting in type 2A von Willebrand disease identified a cleavage site with heightened susceptibility to proteolysis within the variant von Willebrand factor subunit (30,33). The localization of this proteolytic cleavage site to Tyr^{842}-Met^{843} resulted in a hypothesis that mutations resulting in type 2A von Willebrand disease are near the cleavage (31,34,35). This hypothesis was proven to be correct with the identification of single amino acid substitutions resulting in type 2A von Willebrand disease within a region spanning residues 742–875 of the constituent subunit (Fig. 3) (35,68).

The recombinant expression of type 2A von Willebrand factor molecules further demonstrated that within this subtype, an additional classification can be made. Studies have shown that some of the type 2A missense mutations result in abnormal intracellular transport (73). Such an observation suggests that another distinct pathogenetic mechanism for the type 2A phenotype is a quantitative decrease in von Willebrand factor secretion; this factor alone may result in the selective absence of the larger multimers. Of course, an abnormal intracellular processing mechanism does not preclude increased proteolysis,

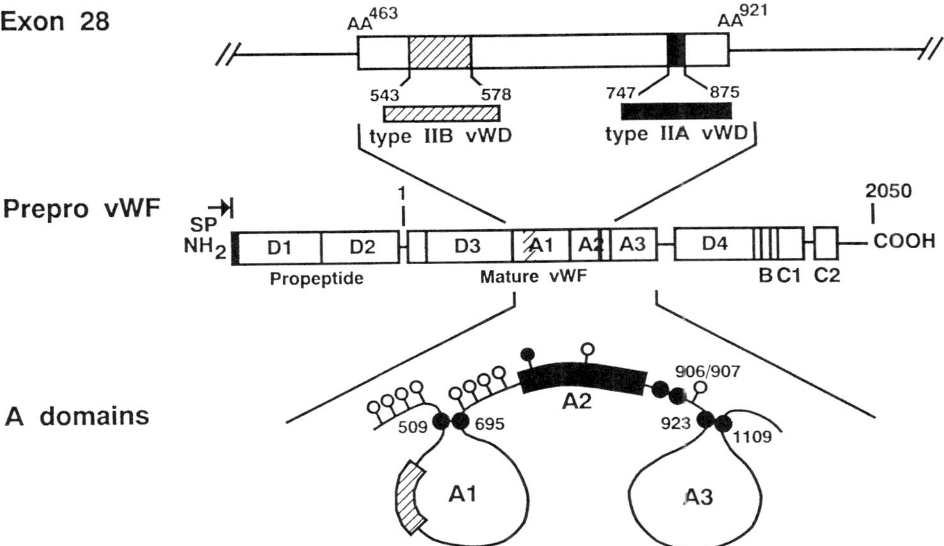

Figure 3 The majority of mutations resulting in type 2A and type 2B von Willebrand disease are clustered within two regions of exon 28 (68). This exon codes for the A1 and S2 domains of von Willebrand factor. The corresponding position of each mutation within pre-pro–von Willebrand factor (middle) and the disulfide loops of the triplicated A domains (lower) are shown.

as described earlier, but could be a separate mechanism for the selective loss of the largest von Willebrand factor multimers.

Insights into the regulation of von Willebrand factor affinity for the platelet glycoprotein Ib-IX complex have been provided by the elucidation of the molecular defects resulting in type 2B von Willebrand disease. The soluble form of the von Willebrand factor molecule resulting in the type 2B phenotype has an increased affinity for platelets and, unlike normal von Willebrand factor, can bind to platelet glycoprotein Ib-IX in the absence of a modulating substance (74). Further evidence that the circulating type 2B von Willebrand factor molecule directly interacts with platelets is the observation that patients may exhibit a transient or persistent thrombocytopenia (75). Presumably, the absence of the largest von Willebrand factor multimers is a consequence of the variant molecules' spontaneous interaction with platelets, which may, in fact, explain the thrombocytopenia as a consequence of intravascular platelet clumping mediated by the aberrant von Willebrand factor.

Studies focusing on the coding sequence for the glycoprotein Ib-IX binding domain within a type 2B von Willebrand factor molecule identified a $Trp^{550} \rightarrow Cys$ codon substitution within the A1 disulfide loop (76). By expressing the isolated domain in prokaryotic and eukaryotic cells, it became apparent that the native conformation of normal von Willebrand factor prevents binding to platelet glycoprotein Ib-IX. This conclusion was based on the observation that recombinant molecules expressed in eukaryotic cells, such as Chinese hamster ovary cells, acquired a disulfide bond-dependent conformation that is analogous to plasma von Willebrand factor and, like plasma von Willebrand factor, required ristocetin to bind platelet glycoprotein Ib-IX (41,77). When the same recombinant domain was expressed containing a $Trp^{550} \rightarrow Cys$ amino acid change, the mutant domain spontaneously interacted with platelet glycoprotein Ib-IX, the characteristic feature of the type 2B phenotype (76). This experiment verified the functional consequence of a single amino acid change within the domain responsible for binding to platelet glycoprotein Ib-IX.

An important aspect of the consequences of a type 2B mutation is how the amino acid change results in an increased affinity for platelet glycoprotein Ib-IX. One explanation is that the substituted amino acid is within the contact site for the platelet receptor and simply generates a contact site with increased affinity. Alternatively, the amino acid change may generate a conformation that exposes a previously hidden glycoprotein Ib-IX contact site, suggesting that the mutation is within a region important for the modulation of binding. To determine experimentally which explanation applies to the $Trp^{550} \rightarrow Cys$ mutation, recombinant fragments were generated in bacteria and purified after reduction and alkylation of the cysteine residues (76). These fragments lack a disulfide bond-dependent conformation and are able to bind platelet glycoprotein Ib-IX in the absence of any exogenous modulator, analogous to the reduced and alkylated tryptic fragment purified from normal plasma-derived von Willebrand factor (37). Comparative assays demonstrated that the $Trp^{550} \rightarrow Cys$ substitution had no functional consequence on affinity in the reduced and alkylated bacterial analogue (76). Thus, assuming that the contact site(s) for platelet glycoprotein Ib-IX are the same for molecules with or without disulfide bond-dependent conformation, the increase in affinity resulting in the type 2B phenotype is most likely not related to changes within the contact site(s) for platelet glycoprotein Ib-IX. Only in the context of a von Willebrand factor molecule with a disulfide bond-dependent conformation does the $Trp^{550} \rightarrow Cys$ substitution increase the affinity for platelet glycoprotein Ib. This demonstrates that subtle alterations in the conformational state of this domain are paramount in regulating the glycoprotein Ib binding site of von Willebrand factor. A number of additional amino acid changes have been associated with the variant molecules exhibiting the type 2B phenotype. These amino acids are clustered within an isolated region of the A1 domain, further suggesting that this region represents a modulating element for the interaction of von Willebrand factor with platelet glycoprotein Ib-IX (Fig. 3) (68).

Although a type 2B designation is restricted to individuals with increased affinity for glycoprotein Ib, it is commonly characterized by a lack of the highest molecular weight multimers. However, the use of multimeric structure as the sole criteria for classification can be misleading, as evidenced by one recently studied variant. This molecule has functional attributes for type 2B classification yet has a full range of plasma von Willebrand factor multimers, which is more typical of type 1 von Willebrand disease. Originally, two different laboratories reported such a variant. In one case, the full complement of von Willebrand factor multimers was used as the primary basis for the classification under the older scheme, and the variant was appropriately designated type I New York (78). In the other case, the observation that the variant molecule could aggregate platelets at low doses of ristocetin led to the designation type II Malmö (79). More recently, genetic studies of four different families, including both of the original propositi, identified a mutation resulting in a $Pro^{503} \rightarrow$ Leu amino acid substitution (80). Recombinant expression of the isolated glycoprotein Ib binding domain of von Willebrand factor provided direct evidence that a $Pro^{503} \rightarrow$ Leu mutation is responsible for enhanced platelet agglutination at subnormal ristocetin concentrations. As noted earlier, it has been suggested that the lack of high-molecular-weight multimers associated with type 2B von Willebrand disease is a result of the spontaneous binding between platelets and von Willebrand factor and the removal of the complex from the circulation. If this hypothesis is correct, then it can be further hypothesized that such variants as type I New York or type II Malmö represent aberrant von Willebrand factor molecules that have an intermediate affinity for platelet glycoprotein Ib. Accordingly, the reduced in vivo affinity is not sufficient to support in vivo platelet clumping yet can be experimentally distinguished from normal von Willebrand factor in a ristocetin-mediated assay. In support of this conclusion is that the bleeding symptoms from individuals with the type I New York or type II Malmö phenotype are relatively mild (79,80).

Another interesting observation made during the genetic characterization of type 2 Malmö was the occurrence of a second mutation within codon 500 that did not result in a change in the amino acid sequence. The intriguing aspect of this change at codon 500 is that the mutations within codons 500 and 503 are also present in the von Willebrand factor pseudogene sequence. This observation raises two possibilities. First, the mutations within codons 500 and 503 may represent corresponding ‘‘hot spots’’ for mutations. Supporting this possibility is that both nucleotide substitutions occur at CpG dinucleotides that have an increased potential for mutation (81). However, arguing against the hot spot explanation is that the silent mutation within codon 500 was not identified in a panel of more than 100 normal alleles and, as such, is a rare base change. Additionally, both mutations were present in two unrelated patients with the same phenotype (80). A second mechanism whereby both base changes could simultaneously occur is through a nonreciprocal recombination event, known as gene conversion. In this process, the normal von Willebrand factor gene located on chromosome 12 would have paired with a short homologous segment within the von Willebrand factor pseudogene during meiosis. Correction of the mismatched base pairs by excision and repair (mismatch repair) would have resulted in the transfer of pseudogene nucleotide sequence to the functional von Willebrand factor gene sequence. Gene conversion has been documented in yeast and other lower organisms (82) but may also be an unappreciated mechanism in some human diseases. Such a mechanism has been implicated as the cause of human steroid 21-hydroxylase deficiency (83,84), but definitive proof that such a mechanism can occur in higher organisms is difficult to obtain.

Another unique form of von Willebrand disease has provided some definitive evidence that the amino terminus of the von Willebrand factor mature subunit contains the binding site for the coagulation cofactor, factor VIII (85). This form of von Willebrand disease, termed type Normandy, after the French province from which the propositus originates, was originally identified as an autosomally inherited form of hemophilia A. The pathogenesis of type Normandy (termed type 2N under the more recent classification scheme) is a variant von Willebrand factor molecule with decreased affinity for factor VIII. Thus, this variant emphasizes the in vivo stabilizing role of von Willebrand factor for factor VIII. As described earlier, the region of von Willebrand factor that binds factor VIII was proposed to be within residues 1–272 of the mature subunit (53). Thus, to identify the molecular basis of the type Normandy variant, a logical place to begin genetic characterizations was within the coding sequence for the amino terminus of von Willebrand

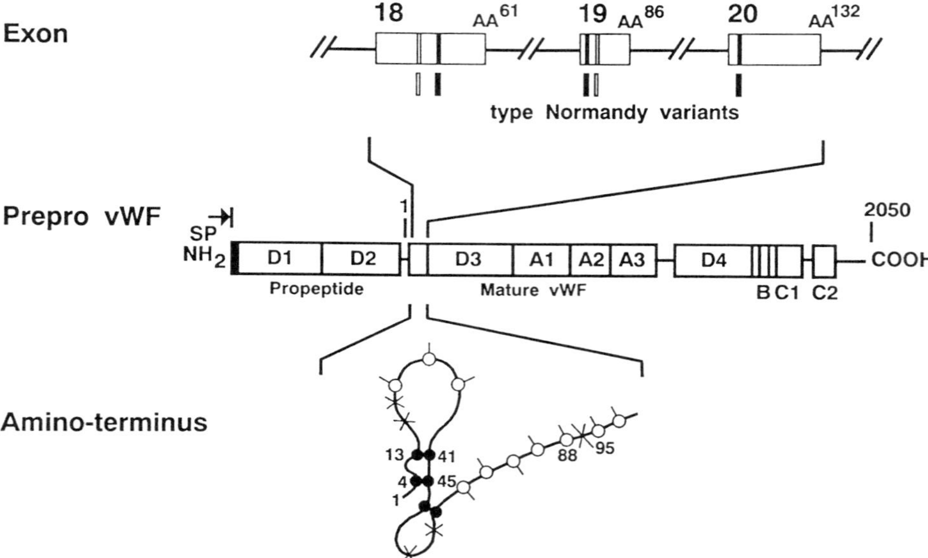

Figure 4 The mutations resulting in the type Normandy (type 2N) variant of von Willebrand factor are within the amino terminus of von Willebrand factor (68). (*Top*) Exons 18, 19, and 20 code for residues 1–132 of the von Willebrand factor mature subunit (62). The position of each mutation within three unrelated type Normandy variants is shown below the corresponding position in each exon (closed box). Another type Normandy variant contains two mutations at positions 19 and 54 (open box). (*Middle*) The amino terminus of the mature subunit is composed of a truncated D domain, D′, composed of residues 1–103 (39). This domain contains 16 cysteine residues. The intramolecular linkages between six of these residues have been defined (closed circles). The intramolecular linkages for the remaining cysteine residues have not been defined (open circles) (90). The corresponding positions of the type Normandy mutations are indicated by X.

factor. Studies to date have identified single amino acid substitutions responsible for the type Normandy phenotype at residues 28, 53, and 91 and a double mutant with changes at residues 19 and 54 of the mature subunit (Fig. 4) (68). Clearly, this is another example in which the protein characterizations provided a strong hypothesis for the localization of a binding site and the genetic studies validated the hypothesis.

V. CONCLUDING REMARKS

The heterogeneity of von Willebrand disease is one confirming example of the multiple functions of this protein in hemostasis. Unlike genetic defects in which loss of function or activity simply inhibits or blocks a biochemical pathway, von Willebrand factor activity is highly regulated, so that it plays its essential role in hemostasis only when necessary. However, it should be remembered that in pathological situations, the same activity may participate in processes leading to arterial occlusion. A greater understanding of the events surrounding the thrombogenesis recently attracted considerable interest in the design of molecules that inhibit von Willebrand factor binding to platelets, as potential antithrombotic agents (86). The evaluation of this concept has become a topic of active research using structural analogs of von Willebrand factor, such as recombinantly derived fragments of von Willebrand factor (42,87–89). Thus, understanding the mechanisms and structural bases of von Willebrand factor function may result in new and effective approaches for antithrombotic therapy. Insights provided by the variant molecules of von Willebrand disease have been a major contributing factor in the development of these therapeutic strategies.

REFERENCES

1. Zimmerman TS, Ratnoff OD, Powell AE. Immunologic differentiation of classic hemophilia (factor VIII deficiency) and von Willebrand's disease. With observations on combined deficiencies of antihemophilic factor and proaccelerin (factor V) and on an acquired circulating anticoagulant against antihemophilic factor. J Clin Invest 1971; 40:244–254.
2. Ruggeri ZM, De Marco L, Gatti L, Bader R, Montgomery RR. Platelets have more than one binding site for von Willebrand factor. J Clin Invest 1983; 72:1–12.
3. Peterson DM, Stathopoulos NA, Giorgio TD, Hellums JD, Moake JL. Shear-induced platelet aggregation requires von Willebrand factor and platelet membrane glycoproteins Ib and IIb-IIIa. Blood 1987; 69:625–628.
4. Weiss HJ, Hawiger J, Ruggeri ZM, Turitto VT, Thiagarajan P, Hoffmann T. Fibrinogen-independent platelet adhesion and thrombus formation on subendothelium mediated by glycoprotein IIb-IIIa complex at high shear rate. J Clin Invest 1989; 83:288–297.
5. Ikeda Y, Handa M, Kawano K, et al. The role of von Willebrand factor and fibrinogen in platelet aggregation under varying shear stress. J Clin Invest 1991; 87:1234–1240.
6. Baumgartner HR, Tschopp TB, Weiss HJ. Defective adhesion of platelets to subendothelium in von Willebrand's disease and Bernard Soulier syndrome. Thromb Haemost 1978; 39:782–783.
7. Stel HV, Sakariassen KS, de Groot PG, von Mourik JA, Sixma JJ. Von Willebrand factor in the vessel wall mediates platelet adherence. Blood 1985; 65:85–90.
8. Turitto VT, Weiss HJ, Zimmerman TS, Sussman II. Factor VIII/von Willebrand factor in subendothelium mediates platelet adhesion. Blood 1985; 65:823–831.
9. Weiss HJ, Turitto VT, Baumgartner HR. Platelet adhesion and thrombus formation on subendothelium in platelets deficient in glycoproteins IIb-IIIa, Ib, and storage granules. Blood 1986; 67:322–330.

10. Kroll MH, Harris TS, Moake JL, Handin RI, Schafer AI. von Willebrand factor binding to platelet GPIb initiates signals for platelet activation. J Clin Invest 1991; 88:1568–1573.
11. Savage B, Shattil SJ, Ruggeri ZM. Modulation of platelet function through adhesion receptors: a dual role for glycoprotein IIb-IIIa (integrin $\alpha_{IIB}\beta_3$) mediated by fibrinogen and glycoprotein Ib-von Willebrand factor. J Biol Chem 1992; 267:11300–11306.
12. Chow TW, Hellums JD, Moake JL, Kroll MH. Shear stress-induced von Willebrand factor binding to platelet glycoprotein Ib initiates calcium influx associated with aggregation. Blood 1992; 80:113–120.
13. Marder VJ, Mannucci PM, Firkin BG, Hoyer LW, Meyer D. Standard nomenclature for factor VIII and von Willebrand factor: a recommendation by the International Committee on Thrombosis and Haemostasis. Thromb Haemost 1985; 54:871–872.
14. Ginsburg D, Handin RI, Bonthron DT, et al. Human von Willebrand factor (vWF): isolation of complementary DNA (cDNA) clones and chromosomal localization. Science 1985; 228: 1401–1406.
15. Lynch DC, Zimmerman TS, Collins CJ, et al. Molecular cloning of cDNA for human von Willebrand factor: authentication by a new method. Cell 1985; 41:49–56.
16. Sadler JE, Shelton-Inloes BB, Sorace JM, Harlan JM, Titani K, Davie EW. Cloning and characterization of two cDNAs coding for human von Willebrand factor. Proc Natl Acad Sci USA 1985; 82:6394–6398.
17. Verweij CL, de Vries CJM, Distel B, et al. Construction of cDNA coding for human von Willebrand factor using antibody probes for colony-screening and mapping of the chromosomal gene. Nucleic Acids Res 1985; 13:4699–4717.
18. Toole JJ, Knopf JL, Wozney JM, et al. Molecular cloning of a cDNA encoding human antihaemophilic factor. Nature 1984; 312:342–347.
19. Gitschier J, Wood WI, Goralka TM, et al. Characterization of the human factor VIII gene. Nature 1984; 312:326–330.
20. Tuley EA, Gaucher C, Jorieux S, Worrall NK, Sadler JE, Mazurier C. Expression of von Willebrand factor ''Normandy'': an autosomal mutation that mimics hemophilia A. Proc Natl Acad Sci USA 1991; 88:6377–6381.
21. Ruggeri ZM, Zimmerman TS. The complex multimeric composition of factor VIII/von Willebrand factor. Blood 1981; 57:1140–1143.
22. Ruggeri ZM, Zimmerman TS. Von Willebrand factor and von Willebrand disease. Blood 1987; 70:895–904.
23. Titani K, Kumar S, Takio K, et al. Amino acid sequence of human von Willebrand factor. Biochemistry 1986; 25:3171–3184.
24. Fay PJ, Kawai Y, Wagner DD, et al. Propolypeptide of von Willebrand factor circulates in blood and is identical to von Willebrand antigen II. Science 1986; 232:995–998.
25. Wagner DD, Fay PJ, Sporn LA, Sinha S, Lawrence SO, Marder VJ. Divergent fates of von Willebrand factor and its propolypeptide (von Willebrand antigen II) after secretion from endothelial cells. Proc Natl Acad Sci USA 1987; 84:1955–1959.
26. Wise RJ, Pittman DD, Handin RI, Kaufman RJ, Orkin SH. The propeptide of von Willebrand factor independently mediates the assembly of von Willebrand multimers. Cell 1988; 52:229–236.
27. Sixma JJ, Sakariassen KS, Beeser-Visser NH, Ottenhof-Rovers M, Bolhuis PA. Adhesion of platelets to human artery subendothelium: effect of factor VIII-von Willebrand factor of various multimeric composition. Blood 1984; 63:128–139.
28. Sporn LA, Marder VJ, Wagner DD. Inducible secretion of large, biologically potent von Willebrand factor multimers. Cell 1986; 46:185–190.
29. Sporn LA, Marder VJ, Wagner DD. Von Willebrand factor released from Weibel-Palade bodies binds more avidly to extracellular matrix than that secreted constitutively. Blood 1987; 69:1531–1534.

30. Zimmerman TS, Dent JA, Ruggeri ZM, Nannini LH. Subunit composition of plasma von Willebrand factor. Cleavage is present in normal individuals, increased in IIA and IIB von Willebrand disease, but minimal in variants with aberrant structure of individual oligomers (types IIC, IID and IIE). J Clin Invest 1986; 77:947–951.
31. Dent JA, Berkowitz SD, Ware J, Kasper CK, Ruggeri ZM. Identification of a cleavage site directing the immunochemical detection of molecular abnormalities in type IIA von Willebrand factor. Proc Natl Acad Sci USA 1990; 87:6306–6310.
32. Ruggeri ZM, Ware J. The structure and function of von Willebrand factor. Thromb Haemost 1992; 67:594–599.
33. Berkowitz SD, Dent J, Roberts J, et al. Epitope mapping of the von Willebrand factor subunit distinguishes fragments present in normal and type IIA von Willebrand disease from those generated by plasmin. J Clin Invest 1987; 79:524–531.
34. Dent JA, Galbusera M, Ruggeri ZM. Heterogeneity of plasma von Willebrand factor multimers resulting from proteolysis of the constituent subunit. J Clin Invest 1991; 88:774–782.
35. Ginsburg D, Konkle BA, Gill JC, et al. Molecular basis of human von Willebrand disease: analysis of platelet von Willebrand factor mRNA. Proc Natl Acad Sci USA 1989; 86:3723–3727.
36. Ruggeri ZM, Ware J. Von Willebrand factor. FASEB J 1993; 7:308–316.
37. Fujimura Y, Titani K, Holland LZ, et al. Von Willebrand factor. A reduced and alkylated 52/48 kDa fragment beginning at amino acid residue 449 contains the domain interacting with platelet glycoprotein Ib. J Biol Chem 1986; 261:381–385.
38. Mohri H, Fujimura Y, Shima M, et al. Structure of the von Willebrand factor domain interacting with glycoprotein Ib. J Biol Chem 1988; 263:17901–17904.
39. Shelton-Inloes BB, Titani K, Sadler JE. cDNA sequences for human von Willebrand factor reveal five types of repeated domains and five possible protein sequence polymorphisms. Biochemistry 1986; 25:3164–3171.
40. Colombatti A, Bonaldo P. The superfamily of proteins with von Willebrand factor type A-like domains: one theme common to components of extracelluar matrix, hemostasis, cellular adhesion, and defense mechanisms. Blood 1991; 77:2305–2315.
41. Azuma H, Hayashi T, Dent JA, Ruggeri ZM, Ware J. Disulfide bond requirements for assembly of the platelet glycoprotein Ib binding domain of von Willebrand factor. J Biol Chem 1993; 268:2821–2827.
42. Sugimoto M, Dent J, McClintock RA, Ware J, Ruggeri ZM. Analysis of structure-function relationships in the GP Ib-binding domain of von Willebrand factor by expression of deletion mutants. J Biol Chem 1993; In Press:
43. Berndt MC, Ward CM, Booth WJ, Castaldi PA, Mazurov AV, Andrews RK. Identification of aspartic acid 514 through glutamic acid 542 as a glycoprotein Ib-IX complex receptor recognition sequence in von Willebrand factor. Mechanisms of modulation of von Willebrand factor by ristocetin and botrocetin. Biochemistry 1992; 31:11144–11151.
44. Sakariassen KS, Nievelstein PF, Coller BS, Sixma JJ. The role of platelet membrane glycoproteins Ib and IIb-IIIa in platelet adherence to human artery subendothelium. Br J Haematol 1986; 63:681–691.
45. Ruoslahti E, Pierschbacher MD. New perspectives in cell adhesion: RGD and integrins. Science 1987; 238:491–497.
46. Berliner S, Niiya K, Roberts JR, Houghten RA, Ruggeri ZM. Generation and characterization of peptide-specific antibodies that inhibit von Willebrand factor binding to GP IIb-IIIa without interacting with other adhesive molecules: selectivity is conferred by Pro^{1743} and other amino acid residues adjacent to the sequence Arg^{1744}-Gly^{1745}-Asp^{1746}. J Biol Chem 1988; 263:7500–7505.
47. Cheresh DA, Berliner AS, Vicente V, Ruggeri ZM. Recognition of distinct adhesive sites on fibrinogen by related integrins on platelets and endothelial cells. Cell 1989; 58:945–953.

48. Howard MA, Firkin BG. Ristocetin—a new tool in the investigation of platelet aggregation. Thromb Haemost 1971; 26:362–369.
49. Weiss HJ, Rogers J, Brand H. Defective ristocetin-induced platelet aggregation in von Willebrand's disease and its correction by factor VIII. J Clin Invest 1973; 52:2697–2707.
50. Moake JL, Turner NA, Stathopoulos NA, Nolasco LH, Hellums JD. Involvement of large plasma von Willebrand factor (vWF) multimers and unusually large vWF forms derived from endothelial cells in shear stress-induced platelet aggregation. J Clin Invest 1986; 78:1456–1461.
51. O'Brien JR, Salmon GP. Shear stress activation of platelet glycoprotein IIb/IIIa plus von Willebrand factor causes aggregation:filter blockage and the long bleeding time in von Willebrand's disease. Blood 1987; 70:1354–1361.
52. Fukuyama M, Sakai K, Itagaki I, et al. Continuous measurement of shear-induced platelet aggregation. Thromb Res 1989; 54:253–260.
53. Foster PA, Fulcher CA, Marti T, Titani K, Zimmerman TS. A major factor VIII binding domain resides within the amino-terminal 272 amino acid residues of von Willebrand factor. J Biol Chem 1987; 262:8443–8446.
54. Bahou WF, Ginsburg D, Sikkink R, Litwiller R, Fass DN. A monoclonal antibody to von Willebrand factor (vWF) inhibits factor VIII binding. Localization of its antigenic determinant to a nonadecapeptide at the amino terminus of the mature vWF polypeptide. J Clin Invest 1991; 84:56–61.
55. Fujimura Y, Titani K, Holland LZ, et al. A heparin-binding domain of human von Willebrand factor. Characterization and localization to a tryptic fragment extending from amino acid residue Val-449 to Lys-728. J Biol Chem 1987; 262:1734–1739.
56. Mohri H, Yoshioka A, Zimmerman TS, Ruggeri ZM. Isolation of the von Willebrand factor domain interacting with platelet glycoprotein Ib, heparin, and collagen, and characterization of its three distinct functional sites. J Biol Chem 1989; 264:17361–17367.
57. Roth GJ, Titani K, Hoyer LW, Hickey MJ. Localization of binding sites within human von Willebrand factor for monomeric type III collagen. Biochemistry 1986; 25:8357–8361.
58. Pareti FI, Niiya K, McPherson JM, Ruggeri ZM. Isolation and characterization of two domains of human von Willebrand factor that interact with fibrillar collagen types I and III. J Biol Chem 1987; 262:13835–13841.
59. Rand JH, Patel ND, Schwartz E, Zhou S, Potter BJ. 150-kD von Willebrand factor binding protein extracted from human vascular subendothelium is type VI collagen. J Clin Invest 1991; 88:253–259.
60. Bonthron D, Orr EC, Mitsock LM, Ginsburg D, Handin RI, Orkin SH. Nucleotide sequence of pre-pro-von Willebrand factor cDNA. Nucleic Acids Res 1986; 14:7125–7127.
61. Collins CJ, Underdahl JP, Levene RB, et al. Molecular cloning of the human gene for von Willebrand factor and identification of the transcription initiation site. Proc Natl Acad Sci USA 1987; 84:4393–4397.
62. Mancuso DJ, Tuley EA, Westfield LA, et al. Structure of the gene for human von Willebrand factor. J Biol Chem 1989; 264:19514–19527.
63. Shelton-Inloes BB, Chehab FF, Mannucci PM, Federici AB, Sadler JE. Gene deletions correlate with the development of alloantibodies in von Willebrand disease. J Clin Invest 1987; 79:1459–1465.
64. Mancuso DJ, Tuley EA, Westfield LA, et al. Human von Willebrand factor gene and pseudogene: structural analysis and differentiation by polymerase chain reaction. Biochemistry 1991; 30:253–269.
65. Sadler JE, Ginsburg D. A database of polymorphisms in the von Willebrand factor gene and pseudogene. Thromb Haemost 1993; 69:185–191.
66. Peake IR, Bowen D, Bignell P, et al. Family studies and prenatal diagnosis in severe von Willebrand disease by polymerase chain reaction amplification of a variable number tandem repeat region of the von Willebrand factor gene. Blood 1990; 76:555–561.

67. Mercier B, Gaucher C, Mazurier C. Characterisation of 98 alleles in 105 unrelated individuals in the F8VWF gene. Nucleic Acids Res 1993; 19:4800–4800.
68. Ginsburg D, Sadler JE. von Willebrand disease: a database of point mutations, insertions and deletions. Thromb Haemost 1993; 69:177–184.
69. Randi AM, Sacchi E, Castaman GC, Rodeghiero F, Mannucci PM. The genetic defect of type I von Willebrand disease ''Vicenza'' is linked to the von Willebrand factor gene. Thromb Haemost 1993; 69:173–176.
70. Ruggeri ZM, Zimmerman TS. Variant von Willebrand's disease: characterization of two subtypes by analysis of multimeric composition of factor VIII/von Willebrand factor in plasma and platelets. J Clin Invest 1980; 65:1318–1325.
71. Scott JP, Montgomery RR, Retzinger GS. Dimeric ristocetin flocculates proteins, binds to platelets, and mediates von Willebrand factor-dependent agglutination of platelets. J Biol Chem 1991; 266:8149–8155.
72. Roth GJ. Developing relationships: arterial platelet adhesion, glycoprotein Ib, and leucine-rich glycoproteins. Blood 1991; 77:5–19.
73. Lyons SE, Bruck ME, Bowie EJW, Ginsburg D. Impaired intracellular transport produced by a subset of type IIA von Willebrand disease mutations. J Biol Chem 1993; 267:4424–4430.
74. De Marco L, Mazzucato M, De Roia D, et al. Distinct abnormalities in the interaction of purified IIA and IIB von Willebrand factor with the two platelet binding sites GP Ib-IX and GP IIb-IIIa. J Clin Invest 1990; 86:785–792.
75. Holmberg L, Nilsson IM, Borge L, Gunnarsson M, Sjorin E. Platelet aggregation induced by 1-desamino-8-D-arginine vasopressin (DDAVP) in type IIB von Willebrand's disease. N Engl J Med 1983; 309:816–821.
76. Ware J, Dent JA, Azuma H, et al. Identification of a point mutation in type IIB von Willebrand disease illustrating the regulation of von Willebrand factor affinity for the platelet GP Ib-IX receptor. Proc Natl Acad Sci USA 1991; 88:2946–2950.
77. Azuma H, Dent JA, Sugimoto M, Ruggeri ZM, Ware J. Independent assembly and secretion of a dimeric adhesive domain of von Willebrand factor containing the glycoprotein Ib-binding site. J Biol Chem 1991; 266:12342–12347.
78. Weiss HJ, Sussman II. A new von Willebrand variant (type I, New York): increased ristocetin-induced platelet aggregation and plasma von Willebrand factor containing the full range of multimers. Blood 1986; 68:149–156.
79. Holmberg L, Berntorp E, Donner M, Nilsson IM. Von Willebrand's disease characterized by increased ristocetin sensitivity and the presence of all von Willebrand factor multimers in plasma. Blood 1986; 68:668–672.
80. Holmberg L, Dent JA, Schneppenheim R, Budde U, Ware J, Ruggeri ZM. Von Willebrand factor mutation enhancing interaction with platelets in patients with normal multimeric structure. J Clin Invest 1993; 91:2169–2177.
81. Barker D, Schafer M, White R. Restriction sites containing CpG show a higher frequency of polymorphism in human DNA. Cell 1984; 36:131–138.
82. Petes TD, Hill CW. Recombination between repeated genes in microorganisms. Annu Rev Genet 1988; 22:147–168.
83. Higashi Y, Tanae A, Inoue H, Fujii-Kuriyama Y. Evidence for frequent gene conversion in the steroid 21-hydroxylase P-450(C21) gene: implications for steroid 21-hydroxylase deficiency. Am J Hum Genet 1988; 42:17–25.
84. Urabe K, Kimura A, Harada F, Iwanaga T, Sasazuki T. Gene conversion in steroid 21-hydroxylase genes. Am J Hum Genet 1990; 46:1178–1186.
85. Mazurier C, Dieval J, Jorieux S, Delobel J, Goudemand M. A new von Willebrand factor (vWF) defect in a patient with factor VIII (FVIII) deficiency but with normal levels and multimeric patterns of both plasma and platelet vWF. Characterization of abnormal vWF/FVIII interaction. Blood 1993; 75:20–26.

86. Ruggeri ZR. Von Willebrand factor as a target for antithrombotic intervention. Circulation 1992; 86:III-26–III-29.
87. Sugimoto M, Ricca G, Hrinda ME, et al. Functional modulation of the isolated glycoprotein Ib-binding domain of von Willebrand factor expressed in *Escherichia coli*. Biochemistry 1991; 30:5202–5209.
88. Prior C, Chu V, Holt J, et al. Production and functional characterization of a recombinant fragment of von Willebrand factor (vWF): an antagonist to platelet receptor GP Ib. Biotechnology 1992; 10:66–73.
89. Gralnick HR, Williams S, McKeown L, et al. A monomeric von Willebrand factor fragment, Leu-504—Ser-728, inhibits von Willebrand factor interaction with glycoprotein Ib-IX. Proc Natl Acad Sci USA 1992; 89:7880–7884.
90. Marti T, Roesselet S, Titani K, Walsh KA. Identification of disulfide-bridged substructures within human von Willebrand factor. Biochemistry 1987; 26:8099–8109.

10

Factor IX

Katherine A. High
Children's Hospital of Philadelphia, and University of Pennsylvania Medical Center, Philadelphia, Pennsylvania

Harold R. Roberts
University of North Carolina at Chapel Hill School of Medicine, Chapel Hill, North Carolina

I. MOLECULAR BIOLOGY OF FACTOR IX

Factor IX is a vitamin K-dependent serine protease that plays a critical role in the middle phase of blood coagulation. Its existence was first recognized in 1952, when Aggeler and colleagues (1) and Biggs and coworkers (2) showed that hemophilia B was distinct from hemophilia A and was caused by a deficiency of a clotting factor different from factor VIII. In the last 40 years, work by many investigators has generated detailed knowledge of the structure and function of factor IX. Nearly 400 mutations in the factor IX gene have been reported in the literature (3); all of these, as well as a number of carefully studied recombinant mutants, have contributed to our working knowledge of factor IX. Although x-ray crystallographic structure is not available for factor IX, the structure of the gene and the protein, and its role in coagulation, are well understood based on the data discussed in this review.

II. ROLE OF FACTOR IX IN COAGULATION

Like the other enzymes in the cascade, factor IX participates in two physiologically important reactions: cleavage of specific peptide bonds to generate the activated form (factor IXa) and cleavage of a specific substrate, factor X, by the activated species factor IXa (FIXa). In the first reaction, either FVIIa-tissue factor-Ca^{2+} or FXIa-Ca^{2+} can cleave FIX to generate FIXa. In the second, FIXa in the presence of the cofactors VIIIa, Ca^{2+}, and phospholipid cleaves factor X to generate factor Xa. The first of these two reactions, the conversion of FIX to IXa, is understood in more detail than the second. Much current research focuses on determining sites of interaction of IX with its cofactors, its physiological activators, and its substrate, factor X.

III. ACTIVATION OF FACTOR IX

Activation involves the cleavage of two bonds, Arg^{145}-Ala^{146} and Arg^{180}-Val^{181}, resulting in the release of an activation peptide (molecular weight, MW, 10,000 daltons) and the generation of activated FIX, a molecule of MW 45,000, consisting of two chains (designated light and heavy) connected by a disulfide bond (Fig. 1). The Arg^{145}-Ala^{146} bond is the first to be cleaved. The resulting intermediate, designated FIXα, which has the activation peptide still covalently linked to the heavy chain, has no clotting activity. Cleavage of the Arg^{180}-Val^{181} bond generates the active form, factor IXaβ, with the activation peptide no longer covalently linked to the molecule. Russell's viper venom activates factor IX by cleaving the Arg^{180}-Val^{181} bond; this generates FIXaα, with the activation peptide attached to the light chain. FIXaα has about 20% of the clotting activity of the mature enzyme FIXa. The bonds cleaved are the same whether FIX is activated by FXIa-Ca^{2+} or FVIIa-TF-Ca^{2+}.

IV. ACTIVATION OF FACTOR X BY FACTOR IXa

Activated factor IXa specifically cleaves the Arg^{52}-Ile^{53} bond within the heavy chain of factor X to generate activated Xa. The same bond is cleaved if X is activated by two other activators, factor VIIa-tissue factor and Russell's viper venom. The activation of X by IXa requires FVIIIa, Ca^{2+}, and a phospholipid surface. Calcium ions interact with the γ-carboxyglutamic acid (Gla) domain of IX (and IXa) to induce a necessary conformational change, and a suitable phospholipid surface enhances assembly of the X activating complex. These two cofactors, Ca^{2+} and phospholipid, lower the K_m of factor IXa for its substrate factor X by over 5000-fold. Factor VIIIa has little effect on the K_m but increases the V_{max} by 200,000-fold compared with IXa alone (4). These findings are summarized in Table 1.

V. STRUCTURE OF THE FACTOR IX PROTEIN

Human factor IX is present in plasma as a 415 amino acid glycoprotein (Fig. 2). The protein has a MW of 55,000 daltons, of which about 20% is carbohydrate. Two potential

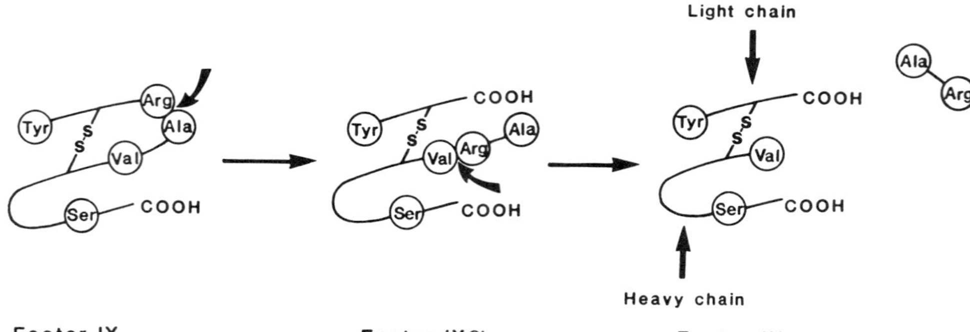

Figure 1 *Activation of factor IX.* FXIa, or VIIa-tissue factor, cleaves the Arg^{145}-Ala^{146} bond to generate FIXα, with the activation peptide still linked to the heavy chain. The Arg^{180}-Val^{181} bond is then cleaved to generate the active form, FIXαβ, consisting of a heavy chain (containing the active site) and a light chain linked by a disulfide bond.

N-linked glycosylation sites exist at residues 157 and 167 in the activation peptide, and an oligosaccharide residue is attached at Ser^{53} in the light chain (5). Carbohydrate is also found in the heavy chain, but the sites of attachment are not known.

The mature protein consists of a γ-carboxyglutamic acid region extending from residues 1 to 46. This domain contains 12 γ-carboxyglutamic acid residues, 6 of which are paired. A major function of the Gla domain is related to calcium-dependent lipid binding; the precise number of calcium binding sites in the Gla domain has not been accurately determined (6). In addition to being necessary for calcium-dependent phospholipid binding, the Gla domain is also essential for the binding of factor IX to endothelial cells. Stafford and his colleagues identified residues 3–11 as necessary for binding to endothelial cells (7). In contrast, these same residues are *not* thought to be necessary for the binding of factor IX to activated platelets.

Factor IX also contains two epidermal growth factor-like domains, EGF-1 and EGF-2 (reviewed in Ref. 8). EGF-1 extends from residues 47 through 84. The function of this domain is not entirely clear, but it contains at least one high-affinity calcium binding site that is thought to be required for conformational changes necessary for interactions with factor VIII and factor X. The three-dimensional structure of the first EGF-like domain has been determined using two-dimensional nuclear magnetic resonance (9). EGF-2 extends from residue 85 through 124. Again, the precise function of this domain is not known, although it has been postulated that it is necessary for FIXa incorporation into the tenase complex. The activation peptide, extending from residues 145 to 180, although cleaved from the zymogen during activation, appears to remain noncovalently linked to the molecule. However, no specific function is known for the activation peptide, although it contains most of the carbohydrate moieties.

The catalytic domain, extending from residues 181 to 415, contains the typical active site triad of a serine protease, namely H^{221}, D^{269}, and S^{365}. After cleavage of the activation peptide, the amino-terminal V^{181} of the heavy chain forms an ion pair with D^{364} to position the active site serine 365 for interaction with substrate FX (10).

VI. BIOSYNTHESIS OF FIX

The mature FIX protein consists of 415 amino acids, but the polypeptide is initially synthesized as a longer precursor containing a signal sequence and a propeptide. The precise length of the precursor is unknown, owing to uncertainty regarding the start site of translation. Between the start site of transcription and the amino terminus of the mature

Table 1 Activation of Bovine Factor X by Bovine Factor IXa[a]

Composition of reaction	K_m	V_{max}	Catalytic efficiency
IXa	299	0.0022	1
IXa, Ca^{2+}	181	0.0105	8
IXa, Ca^{2+}, PL	0.058	0.0247	5.8×10^4
IXa, Ca^{2+}, PL, VIIIa	0.063	500	1×10^9

[a]As indicated, reactions contained Ca^{2+}, 10 mM; phospholipid (PL), 10 μM; and factor VIIIa, 11 U/ml. Relative catalytic efficiency is V_{max}/K_m.
Source: Adapted from van Dieijen et al. (4).

protein, there are three potential methionine codons clustered at −46, −41, and −39 relative to the amino terminus of the mature protein. It is possible that all three are used as start sites of translation, but it is more likely that there is a single start site. Of these three potential methionines, only the one at −39 is conserved in macaque, dog, rat, and mouse (11), as well as human, suggesting that this third potential methionine is the start site of translation. In addition, the residue at −39 has the best match to the Kozak consensus sequence (six of nine nucleotide match; four of nine and five of nine for the −46 and −41 sites, respectively) (12). Finally, the −39 site defines a signal sequence

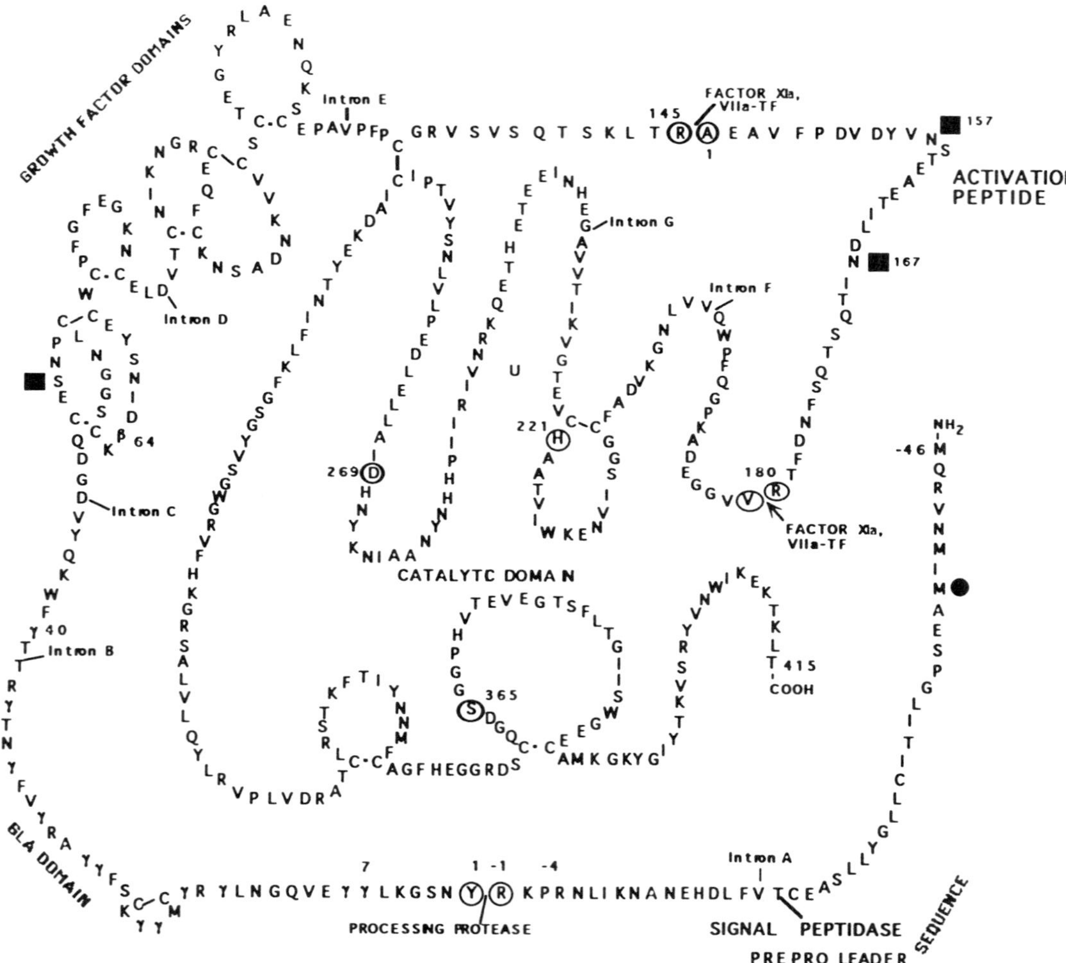

Figure 2 *Amino acid sequence of human factor IX.* The mature protein contains 415 amino acids. The domains of the protein are labeled. A solid circle denotes the probable start site of translation within the pre-pro leader sequence. Cleavage sites for signal peptidase and the processing protease are indicated. Boxes denote glycosylation sites. Positions of introns are indicated. Cleavage sites for activation by XIa or VIIa-TF are indicated. The residues of the active site triad (H^{221}, D^{269}, and S^{365}) are circled and numbered. (Adapted from Yoshitake et al. Biochemistry 1985; 24:3736.)

similar in length to that of factors VII and X, whereas the −46 start site would define a considerably longer signal sequence. Taken together, these facts suggest that the −39 residue is the start site of translation.

As is the case for other secreted proteins, the signal sequence of factor IX serves to direct the nascent polypeptide chain to the endoplasmic reticulum (and thence to the Golgi), where a number of critical posttranslational modifications, including glycosylation, carboxylation of glutamic acid residues, and β-hydroxylation, take place. Following translocation into the lumen of the endoplasmic reticulum (ER), the signal sequence is removed by signal peptidase, an enzyme in the membrane of the ER. The cleavage site for signal peptidase in factor IX is thought to be at Cys^{-19}-Thr^{-18} based on analysis of two naturally occurring variants, factor IX Oxford 3 (13) and factor IX Cambridge (14). In both these cases, mutations at basic residues near the start site of the mature protein (Arg → Gln at −4 and Arg → Ser at −1) prevent the proteolytic removal of the propeptide by the processing protease; the protein thus circulates with the propeptide attached. N-terminal sequence analysis of these variant proteins reveals the initial sequence of TVFLDH, confirming that the signal peptidase cleavage site occurs at Cys^{-19}-Thr^{-18}.

The function of the propeptide is to serve as a recognition signal for the γ-glutamylcarboxylase. The carboxylase is a membrane-bound microsomal enzyme that catalyzes the posttranslational conversion of the 12 amino-terminal glutamyl residues in the factor IX precusor to γ-carboxyglutamyl (Gla) residues in the mature protein. Uncarboxylated mature vitamin K-dependent plasma proteins (i.e., lacking the propeptide) cannot serve as substrates for carboxylation (15); the requirement for the propeptide in carboxylation is also evident from the observation that a synthetic peptide composed of the propeptide and part of the Gla domain of factor IX was used as an affinity ligand in the purification of the carboxylase (16).

The propeptides themselves are highly conserved among the vitamin K-dependent clotting factors. Using site-directed mutagenesis, several groups have defined residues within the propeptide that are absolutely required for carboxylation (17–20), including those at positions −18, −17, −16, −15, and −10. Substitutions at these residues prevent carboxylation but have no effect on cleavage of the signal sequence, removal of the propeptide, or β-hydroxylation. The recent isolation of the cDNA encoding carboxylase (21) will allow a more complete characterization of its mechanism of action (see Chap. 15).

Following carboxylation of the NH_2-terminal glutamic acid residues and before secretion of the mature protein into the circulation, the propeptide is removed. The identity of the protease that cleaves the propeptide is uncertain. Two candidate enzymes were recently isolated. PACE (paired basic amino acid cleaving enzyme), a 90 kD subtilisin-like serine protease found in the Golgi compartment, processes profactor IX correctly in an in vitro expression system (22). Cotransfection of a PACE expression vector with a profactor IX expression vector resulted in secretion of fully processed factor IX, and specific inhibition of PACE by the α_1-antitrypsin Pittsburgh mutant inhibited factor IX processing. The second candidate enzyme was reported by Kawabata and Davie (23), who isolated a 67 kD protein from rabbit liver that cleaves synthetic peptides mimicking the processing sites of the vitamin K-dependent proteins. Bristol et al. recently defined residues within the propeptide of factor IX required for proper processing of the molecule (24).

In addition to the γ-carboxyglutamic acid residues, factor IX contains an additional post-translationally modified amino acid, β-hydroxyaspartic acid, occurring at residue 64 in the mature protein. β-Hydroxy Asp was initially described in protein C (25) and subsequently, among the vitamin K-dependent clotting factors, in factors IX and X and protein S (26). This modified amino acid has since been found in other proteins containing EGF-like modules, including the complement proteins C1r and C1s, the transforming growth factor β-1 binding protein, the low-density lipoprotein receptor, and thrombomodulin. The content of β-hydroxy Asp in factor IX averages only 0.26 mol/mol protein. Factor IX purified from recombinant expression systems also contains this modification (18). The function of β-hydroxy Asp is unclear. The enzyme that catalyzes this posttranslational modification, aspartyl β-hydroxylase, has been partially purified from bovine liver (27). A putative recognition sequence for the enzyme has been described by Stenflo et al. (28).

VII. FACTOR IX cDNA

The site of factor IX synthesis is the hepatocyte. Plasma levels are approximately 4 μg/ml. The volume of distribution of infused factor IX is three times the plasma volume, and the half-life of factor IX is roughly 24 h (29).

A factor IX cDNA was first isolated in 1982 by Kurachi and Davie (30). A human liver cDNA library was screened with baboon liver cDNA that had been prepared from liver mRNA enriched for factor IX sequences and with synthetic oligonucleotides derived from a short stretch of known amino acid sequence. Factor IX cDNAs were subsequently isolated by other groups as well (31,32). All the reported human factor IX cDNA sequences contain a short (approximately 30 base pair, bp) 5′-untranslated region, followed by an open reading frame of 1383 bp and a very long (approximately 1400 bp) 3′-untranslated sequence. The open reading frame encodes a pre-pro leader sequence of 39–46 amino acids (depending upon which of three methionines is the start site of translation), followed by a mature protein of 415 amino acids. The coding region is followed by a double-stop translation terminator (UAAUGA) and then a long 3′-untranslated region. This long 3′ untranslated is also found in mouse (33) and canine (34) factor IX; its function is unknown. Yoshitake et al. pointed out several regions of potential secondary structure (35), but similar thermodynamically stable regions are not conserved in the canine cDNA (34). It is possible that these sequences contribute to mRNA stability, but definitive experiments to address this issue have not been done.

In addition to the human FIX cDNA, complete cDNA sequence is available for mouse (33), canine (34), and rabbit (36) FIX and partial sequence (primarily the heavy chain and the promoter region) is available for sheep, pig, macaque, guinea pig, and rat FIX (11,37). The amino acid sequence of the bovine protein is also known (38). Comparison of sequences among the species for which data are available is informative. The most extensive conservation of amino acid sequence occurs in the Gla domain (>90% among canine, human, bovine, and mouse) and in the carboxyl terminus of the heavy chain (>90% for the last 50 residues). Conservation is much less marked in the activation peptide and the 3′-untranslated region (39,40).

VIII. CHROMOSOMAL LOCALIZATION

Based on the pattern of inheritance for hemophilia B, the gene for factor IX was known to be located on the X chromosome. The availability of a cDNA in 1983 permitted precise regional localization of factor IX to the tip of the long arm of the X chromosome (41). The factor IX gene maps to Xq27.1; it is adjacent to the fragile X locus. The factor VIII gene is also present in this region, in which the order of loci has been determined as factor IX-fragile X site-factor VIII-Xqter (42).

IX. GENE ORGANIZATION OF FACTOR IX

The human factor IX cDNA was used to screen a genomic library, and the sequence of the complete factor IX gene was reported in 1985 (35). A total of 38,060 bp have been determined, including approximately 2.8 kb of 5′ flanking sequence, 32,742 nucleotides from the start site of transcription to the polyadenylation signal, and an additional 2.3 kb of 3′ flanking sequence. The gene is composed of eight exons and seven introns The exons range in size from 25 to 1935 nucleotides and the introns from 188 to 9473 bases.

The start site of transcription has been identified by several groups. Anson et al., using S1 nuclease mapping, have identified a start site 29 bp upstream of the first AUG (32). Reijnen et al., using primer extension and S1 nuclease mapping on human liver RNA, identified transcription start sites at −29, −26, and +1 relative to the first AUG (43). Huang et al., using human liver poly(A)$^+$ RNA as substrate for anchored polymerase chain reaction (PCR), identified a start site 26 bp upstream of the first AUG (44). These start sites are all in close agreement. Salier et al., on the other hand, using primer extension on RNA from HepG2 cells transfected with a IX promoter-bearing construct, identified a start site approximately 150 bp farther upstream (45). HepG2, a human hepatoma cell line, secretes a number of hepatic proteins, including prothrombin and factor X, but does not produce factor IX (46). Thus, some question exists about whether HepG2 cells transfected with IX promoter-bearing reporter genes initiate transcription correctly or at an aberrant site. It seems likely that the site at 26–29 bp upstream of the first AUG is the major transcription start site; the more upstream site may be a minor transcription start site.

The structure of the vitamin K-dependent clotting factors VII, IX, and X is strikingly similar at the protein level and at the level of gene organization. Each of the eight exons of the factor IX gene corresponds to a domain of the protein (Fig. 3), and a similar exon-domain correlation exists for factors VII and X. For factor IX, exon 1 encodes the signal peptide, exon 2 the propeptide and the amino-terminal glutamic acid-rich domain, the Gla domain, exon 3 an aromatic acid-rich stack, exons 4 and 5 two epidermal growth factor-like domains, exon 6 the activation peptide, and exons 7 and 8 the catalytic domain [and in factor IX, the long (1.4 kb) 3′-untranslated region].

The placement of the intervening sequences is also remarkably similar in factors VII, IX, and X, and the splice junction types are identical (see Table 2). The lengths of the introns vary considerably, and this variation in intron size accounts to a large extent for

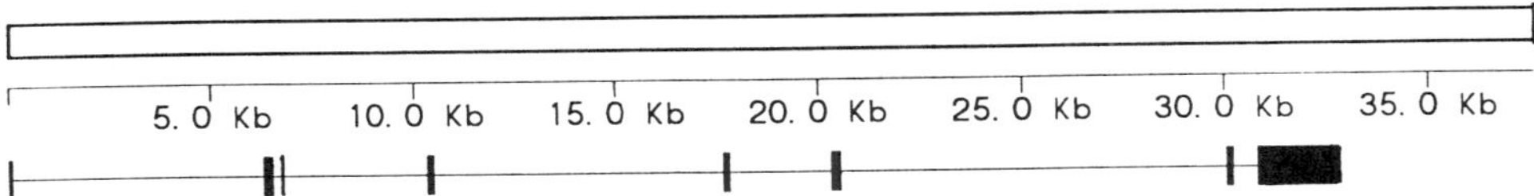

Figure 3 *Exon-intron structure of human factor IX.* Drawn to scale. The gene consists of eight exons and seven introns. Exon 1 encodes the signal sequence, exon 2 the Gla domain, exon 3 a short connecting peptide, exons 4 and 5 the EGF-like domains, exon 6 the activation peptide, and exons 7 and 8 the catalytic domain and the 3′-untranslated sequence.

the differences in the overall size of the genes (34 kb for factor IX, 20 kb for factor X, and 12.3 kb for factor VII), because the sizes of the coding regions are quite similar. The FIX gene contains a number of repetitive elements, including *Alu* repeats in intron A, intron F, and the 3′ UT and *Kpn* repeats in the 5′ flanking region and in intron D (35).

X. POLYMORPHISMS WITHIN THE HUMAN FACTOR IX GENE

Eight polymorphisms have been described within or flanking the factor IX gene. The locations and gene frequencies of these polymorphisms are described in Table 3. The gene frequencies are markedly different in whites and U.S. black populations; most of these polymorphisms are not found in the Asian population (47). Based on reported gene frequencies, >90% of white females should be heterozygous for one or more markers;

Table 2 Intron Lengths and Splice Junction Types in Human Factors VII, IX, X

Intron	IX	VII	X[a]	Splice junction type
A	6206	1067[b] 2574	Unknown	I
B	188	1919	8000	0
C	3689	68	920	I
D	7163	1908	1400	I
E	2565	971	3000	I
F	9473	595	3300	0
G	668	816	1400	I

[a]Intron lengths for factor X are approximate. The length of intron A has not been determined.
[b]The two intron lengths are for the introns following exons 1a and 1b in factor VII.

Table 3 Polymorphisms in or Near the Human Factor IX Gene

Polymorphism	Location	Gene frequency of minor allele: In white population	Gene frequency of minor allele: In U.S. black population	References
TaqI	Intron D	0.29–0.35	0.14	41, 52
XmnI	Intron C	0.29	0.12	48, 53
HinfI/DdeI	Intron A	0.24	0.36	53
Ala/Thr dimorphism in activation peptide	Amino acid 148 in mature protein	0.33	0.03–0.15	54, 59
HhaI	8 kb 3′ to exon 8	0.39		49
MspI	Intron D	0.22	0.39	55, 56
5′ BamHI	5′ Flanking sequence	0.06	0.48	57
BamHI	Intron B	0	0.13	48

the observed frequency is somewhat lower because of linkage disequilibrium between some sites (48,49). The most widely separated polymorphic sites (5′ BamHI and HhaI) do not display linkage disequilibrium with the other sites and may therefore be useful in cases in which the other sites are uninformative. For whites and U.S. blacks, antenatal diagnosis and carrier detection based on restriction fragment length polymorphism (RFLP) analysis alone should yield a diagnosis in >80% of cases (49). RFLP analysis has also proven useful for demonstrating "founder" effects (50,51), in which a common distant ancestor appears to have given rise to the same mutation in several apparently unrelated kindreds.

The Ala/Thr polymorphism within the factor IX protein, first reported by McGraw and colleagues, is of special interest (58). Unlike the other reported polymorphisms, this one occurs within the coding region of the gene, where an A or a G at the first nucleotide of codon 148 specifies either Thr or Ala. The polymorphism appears to be silent at the protein level, because there is no difference in factor IX antigen or activity levels between individuals with Thr or Ala. The minor allele (Ala) has a gene frequency of approximately 0.30 in white populations. It is much lower (0.03–0.15) in U.S. black populations and has not been detected in Asian populations analyzed (Chinese and Malay) (59). The polymorphism is readily detected using either mouse monoclonal antibodies (60) or synthetic oligonucleotides (61).

XI. REGULATION OF EXPRESSION

As for other genes expressed in a tissue-specific manner, factor IX expression results from the interplay of sequence-specific elements within the promoter (cis-acting elements) and transcription factors found in certain cell types (trans-acting factors). For factor IX, the study of gene expression has been rendered especially interesting by the existence of the hemophilia B Leyden variants. Patients with the Leyden phenotype exhibit severe factor IX deficiency during childhood (<1% normal), but following puberty, FIX levels gradually rise to near normal (30–60%) (62). Molecular characterization of the factor IX gene from patients with the Leyden variant has revealed a number of defects (see later), all clustered within a 33 bp region adjacent to the start site of transcription. The presence of a disease state associated with a promoter mutation is not unique to the hemophilias (a number of the thalassemia variants are caused by promoter mutations), but the correction of the promoter defect through the physiological changes of puberty in unique. Despite considerable study of the factor IX promoter, this tantalizing observation remains unexplained.

A. Cis-Acting Elements Required for Factor IX Expression

The promoter activity of the 5′ flanking sequence has been examined by several groups through the use of reporter gene assays in transient expression systems. The most detailed analysis is that of Salier et al. (45); similar data have been reported by Crossley and Brownlee (63). In the report of Salier et al., varying lengths of 5′ flanking DNA were linked to the chloramphenicol acetyltransferase (CAT) gene and transfected into HepG2 cells. Maximal promoter activity was obtained with a construct extending from −274 (numbering the major start site of transcription as 1) to +4. Inclusion of additional upstream sequence to −416 resulted in no change in promoter activity; addition of 5′ sequence upstream of −416 resulted in loss of promoter strength. Using a similar assay system (CAT constructs transfected into HepG2 cells), Crossley and Brownlee obtained maximal promoter activity with a −189 to 21 construct (same numbering system) but did not test longer constructs, except a single construct of −2258 to 21. This long construct had activity similar to that of the −189 construct. In the Salier et al. analysis, both these constructs had approximately 15–20% activity compared with the maximum seen with the −274 construct. The choice of constructs in the former study, then, may have skipped the fragment conferring maximal activity. The data suggest that maximal promoter activity requires the presence of the 274 bp upstream of the major transcription start site. Constructs containing as little as 80 bp upstream have little detectable promoter activity over background (see Table 4). A potential flaw in these studies is in the choice of HepG2 cells for the transient expression assays. There are no permanent cell lines that produce factor IX; HepG2 cells, however, secrete other procoagulant proteins, including factor X and prothrombin, so that it is reasonable to assume that trans-acting factors required for factor IX expression may be present in HepG2 cells. However, Friedman et al. (64) have shown that levels of the CCAAT/enhancer binding protein (C/EBP) are greatly reduced in HepG2 cells compared with normal hepatocytes (approximately 5% of hepatocyte levels by western blot), and C/EBP has been shown to bind to at least two sites in the factor IX promoter (see Fig. 4). It must be recognized, therefore, that in vitro results derived from studies with HepG2 cells (or nuclear extracts) may *not* accurately reflect the course of events in vivo.

B. Protein Binding Sites in the Factor IX Promoter

The effect of cis-acting elements on gene expression is mediated through transcription factors (TFs), proteins that bind to specific sequences and facilitate transcription (65).

Table 4 Promoter Activity in HepG2 Cells

Bases in CAT construct (kb)	Salier et al. (%)	Crossley and Brownlee (%)
−2.2	16	105
−416/4	100	Not tested
−274/4	107	Not tested
−207/4	15	Not tested
−189/21	Not tested	100
−171/4	17	Not tested
−77/21	Not tested	20
−80/4	1.6	Not tested

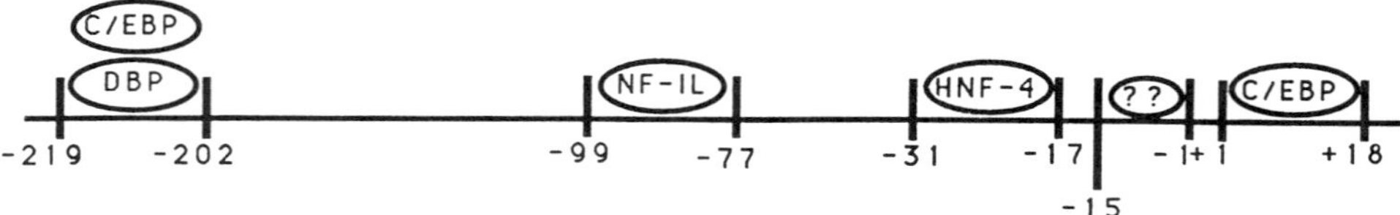

Figure 4 *Protein binding sites in factor IX promoter.* Based on DNase protection assays, the regions marked are binding sites for nuclear proteins. Tentative identification of proteins binding at some sites is as shown: C/EBP, CCAAT/enhancer binding protein (68); DBP, D-site binding protein (69); HNF-4, hepatocyte nuclear factor 4 (70); and NF-IL, nuclear factor 1-liver (71).

Using DNase protection studies, Picketts et al. identified five protein binding sites within the factor IX promoter (66): −219 → −202, −99 → −77, −31 → −17, −15 → −1, and +1 → +18 (see Fig. 4). The sites at −99 → −77 and 1 → 18 were previously identified in similar studies by Crossley and Brownlee (63). The transcription factors that bind to some of these sites have been identified through (1) recognition of the typical consensus sequence for binding; (2) DNA-protein binding studies carried out with the purified TF protein; and (3) trans-activation studies in which a reporter gene linked to the factor IX promoter shows an increase in expression following cotransfection with an expression plasmid coding for a specific TF. The cumulative results of several studies are shown in Figure 4 (63,66,67). Mutations within the binding sites for several of these liver-specific transcription factors are associated with the hemophilia B Leyden phenotype.

C. Hemophilia B Leyden Variants

As noted earlier, the Leyden phenotype is characterized by very low levels of factor IX during childhood and gradually rising levels following puberty. The Leyden phenotype was initially described in 1970 (72); a detailed longitudinal study of eight related patients was published by the same group in 1982 (62). Briet and colleagues documented that affected patients demonstrated concordance of factor IX antigen and activity levels throughout the course of the disease and that, beginning at puberty, factor IX levels rise at the rate of 4–5% per year until the age of 20 or so, when they reach a stable maximum in the 30–60% range. In subsequent work, they documented the response of a child with the Leyden variant to short-term treatment with testosterone (levels rose to 5%) and an anabolic steroid (levels rose to 10–15%) (73). Since the initial description, a number of kindreds with the Leyden phenotype have been studied. For all of these, the nucleotide defect has been determined and found to be in the promoter region of the gene. Defects at −21 (74), −20 (75,76), −6 (77,78), −5 (66), +6 (79), +8 (80), and +13 (81–83) have been reported. Note that these mutations fall into three different protein binding domains (see Fig. 4) but all display a similar phenotype.

In contrast to the other hemophilia B variants, a physiological explanation of the Leyden variants requires an explanation for two observations: the absence of factor IX expression during childhood, and the gradual rise in expression following puberty. Because all the Leyden mutations fall within protein binding sites in the promoter, the absence of expression during childhood is not surprising; however, determining the basis for the improved expression after the onset of puberty is more difficult. Experimental evidence supports several possible explanations, all of which may be correct. Brownlee and colleagues (84) have pointed out the existence of a possible androgen-responsive element (ARE) at −36 to −22 in the IX promoter and have shown that this element can bind androgen receptor in vitro. Thus, with increasing concentrations of androgen receptor in the liver following puberty, levels of factor IX are expected to rise gradually. In the absence of a promoter mutation, the ARE exerts no effect on transcription, because a binding site for the liver transcription factor HNF-4 exists at the same site. Mutations in the promoter that alter or abolish HNF-4 binding permit the modest effect of the ARE to become apparent. Picketts et al. (66) proposed another hypothesis. Based on gel mobility shift assays carried out with purified proteins, they demonstrated that the D site binding protein, which is developmentally regulated in the rat, with expression commencing at puberty, enhances the binding of C/EBP to its cognate site at −219 to −202. If the D site binding protein is similarly regulated in the human, then enhanced promoter activity from the distal C/EBP binding site may contribute to the improvement in factor IX levels that occurs postpuberty.

XII. MUTATIONS IN THE FACTOR IX GENE

An absence of functional factor IX leads to the clinical condition known as hemophilia B. Hemophilia B is indistinguishable from hemophilia A (absence of factor VIII) on clinical grounds, and indeed the two were not recognized as distinct entities until 1952, when Aggeler and colleagues (1) and Biggs and coworkers (2) demonstrated that there is a factor in the plasma of one group of patients that could correct the deficiency of the other in an in vitro clotting assay. Initial characterization of hemophilia B variants was based on factor IX antigen and activity levels; those with no detectable antigen or activity were referred to as CRM negative (cross-reacting material negative), those with normal levels of antigen but greatly reduced activity, CRM-positive, and those with similar reductions of antigen and activity, CRM reduced. The second group of patients, the CRM-positive patients, were considered of special interest because they appeared to synthesize a dysfunctional factor IX molecule. Most initial descriptions of mutations in patients with hemophilia B occurred in this group.

The distribution of disease severity differs between hemophilia A and hemophilia B. In hemophilia A, most patients (60%) have severe disease, with <1% FVIII activity. In factor IX deficiency, however, the percentage of patients with severe disease is lower. In one large U.S. study, which included only patients with severe and moderate disease (i.e., mild hemophiliacs were excluded), only 44% of patients with hemophilia B had severe disease (85). In addition, patients with hemophilia B have a much lower rate of inhibitor formation (antibodies directed against the molecule that prevent treatment with exogenous factor IX) than patients with hemophilia A (86). The reasons for this are unclear. For both hemophilia A and B, it has been suggested that inhibitors may be associated with the occurrence of large gene deletions. It is hypothesized in these cases that the therapeutically administered product is perceived as a foreign protein by the immune system and antibody formation therefore ensues. Clearly this mechanism does not explain all cases of inhibitor formation, because inhibitors occasionally arise in the setting of CRM-positive disease (87).

Determination of defects in the factor IX gene resulting in hemophilia B has generated a wealth of information about the structure and function of the molecule. The utility of the information derived from the mutations has rested in large part on the wide array of functional assays available for assessing factor IX activity. In contrast to the situation seen in a number of other genetic diseases (e.g., muscular dystrophy or cystic fibrosis), for which no straightforward functional assays exist or the effect of the mutation can only be judged in the whole organism, the functional consequences of factor IX variants, either plasma-derived or recombinant, are readily analyzed in studies in vitro. Using purified components, one can measure the rate of activation of IX to IXa, and the activity of IXa in activating X can also be measured. Thus, one can readily determine whether a particular mutation interferes with the activation of IX or with the activity of the activated species (or both).

In 1983, Noyes et al. reported the first characterization of a factor IX mutation at the molecular level, factor IX Chapel Hill (88). The mutation in this instance was determined by amino acid sequencing and showed the substitution of histidine for arginine at position 145 of the zymogen. This substitution precludes cleavage by factor XIa at the Arg^{145}-

Ala^{146} peptide bond; the activation peptide remains associated with the light chain, producing a IXaα form, resulting in a mild hemophilic phenotype.

Since the first report some 10 years ago, hundreds of mutations in the factor IX gene have been described. The initial reports based on DNA sequencing necessitated the rather laborious preparation and screening of a genomic library for each hemophilic patient studied. With the advent of PCR and rapid screening procedures based on PCR (e.g., single-strand conformation polymorphism), the pace of detection of mutations accelerated dramatically. To utilize effectively the data being generated in many laboratories throughout the world, Brownlee and colleagues organized a hemophilia B database, which collated all available information on factor IX mutations in a concise format, which is now published yearly in *Nucleic Acids Research* (3). The 1993 database contained reports of 806 mutations representing 378 unique molecular events. (The database includes point mutations, small insertions, and small deletions but does not include large gene deletions.) Mutations have been reported in every domain of the factor IX molecule and in the promoter as well (see earlier). In addition, mutations occur at donor and acceptor splice sites and within introns to create new cryptic splice sites. As one would predict, a number of the most frequently reported mutations occur at CG dinucleotides and involve a CG → TG or CA change (hot spots for mutations). Other frequently observed mutations seem to be caused by a "founder" effect; based on haplotype analysis, these mutations, described from distinct geographical locations, appear to have arisen from a single (prolific) ancestor (50,51). For a complete listing of mutations in the factor IX gene, the reader is referred to the hemophilia B database (3). The most frequently occurring mutations giving rise to hemophilia B are listed in Table 5.

The database has been of considerable utility to both scientists and clinicians. For those interested in structure-function studies involving factor IX or other vitamin K-dependent clotting factors, it serves as a rapid guide to functional studies already reported for a particular locus. For those involved in patient care, Thompson and colleagues have used the database to generate an algorithm for rapid determination of the defect in newly presenting kindreds with hemophilia B, based on the mutations that occur most commonly in CRM-positive, CRM-reduced, and CRM-negative phenotypes (89). Perhaps the strongest indication of the utility of the hemophilia B database is the recent establishment of similar databases for factor VIII and von Willebrand factor deficiencies (90,91).

One other interesting subset of hemophilia B variants merits discussion here. In 1967, Hougie and Twomey first described two brothers with an unusual phenotype (92). Patients with the usual type of hemophilia B have a prolonged activated partial thromboplastin time but a normal prothrombin time (PT); these brothers had a prolongation of the PT, but only if ox brain (as opposed to rabbit or human brain) was used as the source of tissue factor. This type of hemophilia B is referred to as hemophilia B_m. A number of kindreds with the B_m variant (after the surname of the family in the initial description) have now been described. The molecular variants giving rise to this phenotype include mutations at the carboxyl-terminal end of the activation peptide (Arg^{180}) (93) and the amino-terminal end of the catalytic domain (Val^{181}, Val^{182}) (94,95) and selected mutations in the catalytic domain that involve interactions with Val^{181}. A typical factor IX B_m patient has been described by us (93,96), and the mutant molecule has been termed factor IX Hilo. In this case, glutamine replaces the normal arginine at position 180 of the zymogen. Because this is a critical cleavage site, activation is delayed, and the activation peptide is not released but remains linked to the heavy chain, resulting in the formation of factor IXα, which has no activity. The ox brain prothrombin time of factor IX Hilo is shown

Table 5 Hemophilia B: Most Frequently Reported Mutations

Clotting (normal = 100%)	Antigen (normal = 100%)	Nucleotide position and mutation	Amino acid change	Comments	References[a]
1.8	34	6,364, C → T	−4, R → W	16 cases	111
<1	43	6,365, G → T	−4, R → L	6 cases	76
<1	57	6,365, G → A	−4, R → Q	29 cases, abnormal carboxylation, circulates with propeptide	13, 14
<1	<1	6,460, C → T	29, R → stop	16 cases including 2 inhibitors	111
22	73	6,461, G → A	29, R → Q	7 cases including 4 with double mutation	82
13	29	10,430, G → A	60, G → S	29 cases	112
10	5	17,810, A → G		6 cases; new donor splice?	113
3	41	20,413, C → T	145, R → C	16 cases; decreased XIa activation	114
6	113	20,414, G → A	145, R → H	17 cases; decreased XIa activation	88

<1	88	20,518, C → T	180, R → W	11 cases including 5 B_m	94
<1	112	20,519, G → A	180, R → Q	8 cases including 3 B_m; decreased XIa activation	93
<1	<1	30,863, C → T	248, R → stop	15 cases including 1 inhibitor	111, 113
5	5	30,864, G → A	248, R → Q	20 cases including 1 double	115, 116
<1	<1	30,875, C → T	252, R → stop	12 cases	117
10		30,992, G → A	291, A → T	6 cases	118
5	8	31,008, C → T	296, T → M	41 cases	113
2	88	31,119, G → A	333, R → Q	19 cases including 1 double	119, 120
2	2	31,127, T → C	336, C → R	4 cases	120
3.5	72	31,311, T → C	397, I → T	30 cases including 3 B_m; at least 24 have the same rare haplotype	51, 121

[a]Representative references listed. For a complete list of references for each mutation, see Reference 3.

in Table 6. As can be seen, the PT on Hilo plasma is markedly prolonged but can be normalized simply by removing the abnormal molecule by adsorption to a specific antifactor IX antibody. The reason for the prolongation of the ox brain prothrombin time is that the abnormal factor IX competes as a substrate for the activation of factor X by factor VIIa-tissue factor. The degree of prolongation of the PT is determined to some extent by the concentration of the abnormal factor IX. On the other hand, increasing concentrations of factor X shorten the ox brain PT, making it necessary to take into account concentrations of both factor IX and factor X when interpreting a prolonged ox brain prothrombin time.

XIII. HEMOPHILIA B AS A MODEL FOR GENE THERAPY

Conventional tratment for hemophilia B has rested on administration of exogenous (plasma-derived) factor concentrates during hemorrhagic episodes. An extensive discussion of therapy is beyond the scope of this review; several comprehensive reviews are available (97–99). However, it is appropriate to discuss briefly the present state of efforts aimed at establishing an experimental basis for gene therapy for hemophilia. The object of somatic cell gene therapy is to supply the organism with a normal copy of the mutant gene rather than the protein encoded by the gene. The transcriptional and translational machinery of the organism then completes synthesis of the missing or defective protein. For several reasons, the hemophilias are a likely choice for early efforts at human gene therapy. First, tissue-specific expression is not required. Work by several investigators has shown that biologically active factor IX can be synthesized in a number of different tissues, including fibroblasts (100), myoblasts (101), and endothelial cells (102). The only requirement is that the secreted protein be able to gain access to the circulation. Second, precise regulation of expression is not required. In contrast to some deficiency states (e.g., diabetes), in which the level of the replaced protein must be very precisely regulated, clotting factor replacement therapy is effective over a wide range of protein levels. Thus, levels as low as 5% of normal would result in a marked clinical improvement for patients whose endogenous levels are less than or equal to 1%, and levels as high as 200% are not associated with ill effects (the level of IXa would not rise proportionately, and the donated gene is not contaminated with other activated clotting factors). A third advantage is the existence of large-animal models for the hemophilias; canine models exist for both hemophilia A and B. For hemophilia B, the normal canine factor IX cDNA has been isolated (34) and the defect in one strain of hemophilic animals elucidated (103). The strain characterized is CRM-negative, that is, neither antigen nor activity is detectable, so that any expression of factor IX from a donated gene is readily distinguished from baseline. Finally, it should be noted that of the two most prevalent he-

Table 6 Ox Brain Prothrombin Times with Factor IX Hilo

	Ox brain PT (s)
Normal plasma	45
Patient plasma (IX Hilo)	100
Depleted patient plasma	41
Depleted patient plasma + IX Hilo	85
Depleted normal plasma + IX Hilo	91

mophilia variants (A and B), hemophilia B is more easily approached by most strategies for gene therapy because of the smaller size of the cDNA (3 kb for factor IX and 7.3 kb for factor VIII). Many virus-based vectors have limitations on the size of inserts that can be accepted; with the currently available vectors, the size of the factor IX cDNA is a distinct advantage.

A wide variety of gene delivery systems are theoretically available, ranging from liposomes to virally-based vectors to DNA-polycation conjugates (reviewed in Ref. 104). The most extensive experience by far is with retroviral vectors. A number of investigators have shown that retroviral vectors bearing a FIX cDNA can be used to transduce a variety of target cells in vitro and that the transduced cells express biologically active FIX. Target cells have included primary culture fibroblasts (100,105,106), endothelial cells (102), hepatocytes (107), and myoblasts (101,108), as well as a number of cell lines. Levels of expression in vitro have generally not exceeded 1–3 $\mu g/10^6$ cells/24 h. In vivo data are less plentiful; early experiments had been disappointing, duration of expression usually not exceeding 4 weeks. The cause of the failure of long-term expression is unknown, but one study showed that although transplanted transduced cells persisted at constant levels for >8 months, *expression* from the retroviral construct decreased to undetectable levels after 1 month. Explants cultured from the grafts also failed to express the transgene, suggesting that the donated gene sequences had been specifically inactivated (109). In one recent report, primary myoblasts obtained from 2-day-old mice were transduced with a retroviral vector containing a CMV promoter, a mouse muscle CK enhancer, and canine FIX and then injected into 6-week-old mice (108). Canine FIX was detected in mouse plasma at a level of 10 ng/ml for up to 6 months. Another recent publication (110) documents prolonged expression (≥6 months) of canine factor IX in the plasma of hemophilic dogs following in vivo transduction of hepatocytes with a retroviral vector. Again, however, levels of expression were quite low (<8 ng/ml). In addition, since retroviral vectors require a dividing target cell for integration into the host cell genome, the animals underwent partial hepatectomy before retroviral transduction. Such a requirement clearly limits the feasibility of extending this approach to humans, but similar approaches using viral vectors that do not have a requirement for dividing target cells are currently under investigation. Therapeutic efficacy will require plasma levels of factor IX closer to 500 ng/ml, but these results are encouraging in terms of prolonged expression from a viral vector and suggest that hemophilia may eventually be treatable by somatic cell gene therapy.

REFERENCES

1. Aggeler PM, White SG, Glendenning MB, Page EW, Leake TB, Bates G. Plasma thromboplastin component (PTC) deficiency: a new disease resembling hemophilia. Proc Soc Exp Biol Med 1952; 79:692–694.
2. Biggs R, Douglas AS, Macfarlane RG, et al. Christmas disease: a condition previously mistaken for hemophilia. BMJ 1952; 2:1378–1382.
3. Giannelli F, Green PM, High KA, et al. Haemophilia B: database of point mutations and short additions and deletions, fourth edition, 1993. Nucleic Acids Res 1993; 21(13):3075–3087.
4. Van Dieijen G, Tans G, Rosing J, Hemker HC. The role of phospholipid and factor VIIIa in the activation of bovine factor X. J Biol Chem 1981; 256:3433–3442.
5. Nishimura H, Kawabata S-I, Kisiel W, et al. Identification of a disaccharide (Xyl-Glc) and a trisaccharide (Xyl_2-Glc) *O*-glycosidically linked to a serine residue in the first epidermal

growth factor-like domain of human factors VII and IX and protein Z and bovine protein Z. J Biol Chem 1989; 264(34):20320–20325.
6. Bajaj SP. Co-operative Ca^{2+}-binding to human factor IX. J Biol Chem 1982; 257:4127–4132.
7. Cheung W-F, Hamaguchi N, Smith KJ, Stafford DW. The binding of human factor IX to endothelial cells is mediated by residues 3-11. J Biol Chem 1992; 267:20529–20531.
8. Stenflo J. Structure-function relationships of epidermal growth factor modules in vitamin K-dependent clotting factors. Blood 1991; 78:1637–1651.
9. Baron M, Norman DG, Harvey TS, et al. The three-dimensional structure of the first EGF-like module of human factor IX: comparison with EGF and TGF-alpha. Protein Sci 1992; 1:81–90.
10. Hamaguchi N, Roberts H, Stafford DW. Mutations in the catalytic domain of factor IX that are related to the subclass hemophilia Bm. Biochemistry 1993; 32:6324–6329.
11. Pang CP, Crossley M, Kent G, Brownlee GG. Comparative sequence analysis of mammalian factor IX promoters. Nucleic Acids Res 1990; 18(22):6731–6732.
12. Kozak M. Compilation and analysis of sequences upstream from the translational start site in eukaryotic mRNAs. Nucleic Acids Res 1984; 12(2):857–872.
13. Bentley AK, Rees DJG, Rizza C, Brownlee GG. Defective propeptide processing of blood clotting factor IX caused by mutation of arginine to glutamine at position-4. Cell 1986; 45: 343–348.
14. Diuguid DL, Rabiet MJ, Furie BC, Liebman HA, Furie B. Molecular basis of hemophilia B: a defective enzyme due to an unprocessed propeptide is caused by a point mutation in the factor IX precursor. Proc Natl Acad Sci USA 1986; 83:5803–5807.
15. Soute BAM, Vermeer C, DeMetz M, Hemker HC, Lijen HR. In vitro prothrombin synthesis from a purified precursor protein. III. Preparation of an acid-soluble substrate for vitamin K-dependent carboxylase by limited proteolysis of bovine descarboxyprothrombin. Biochim Biophys Acta 1981; 676:101–107.
16. Wu SM, Morris DP, Stafford DW. Identification and purification to near homogeneity of the vitamin K-dependent carboxylase. Proc Natl Acad Sci USA 1991; 88:2236–2240.
17. Jorgensen MJ, Cantor AB, Furie BC, Brown CL, Shoemaker CB, Furie B. Recognition site directing vitamin K-dependent γ-carboxylation residues on the propeptide of factor IX. Cell 1987; 48:185–191.
18. Rabiet M-J, Jorgensen MJ, Furie B, Furie BC. Effect of propeptide mutations on posttranslational processing of factor IX: evidence that β-hydroxylation and γ-carboxylation are independent events. J Biol Chem 1987; 262:14895–14898.
19. Huber P, Schmitz T, Griffin J, et al. Identification of amino acids in the γ-carboxylation recognition site on the propeptide of prothrombin. J Biol Chem 1990; 265(21):12467–12473.
20. Foster DC, Rudinski MS, Schach BG, et al. Propeptide of human protein C is necessary for gamma-carboxylation. Biochemistry 1987; 26:7003–7011.
21. Wu SM, Cheung WF, Frazier D, Stafford DW. Cloning and expression of the cDNA for human gamma-glutamyl carboxylase. Science 1991; 254:1634–1636.
22. Wasley LC, Rehemtulla A, Bristol JA, Kaufman RJ. PACE/furin can process the vitamin K-dependent pro-factor IX precursor within the secretory pathway. J Biol Chem 1993; 268(12):8458–8465.
23. Kawabata SI, Davie EW. A microsomal endopeptidase from liver with substrate specificity for processing proproteins such as the vitamin K-dependent proteins of plasma. J Biol Chem 1992; 267(15):10331–10336.
24. Bristol JA, Furie BC, Burie B. Propeptide processing during factor IX biosynthesis. Effect of point mutations adjacent to the propeptide cleavage site. J Biol Chem 1993; 268:7577–7584.
25. Drakenberg T, Fernlund P, Roepstorff P, Stenflo J. β-Hydroxyaspartic acid in vitamin K-dependent protein C. Proc Natl Acad Sci USA 1983; 80:1802–1806.

26. Fernlund P, Stenflo J. β-Hydroxyaspartic acid in vitamin K-dependent proteins. J Biol Chem 1983; 258(20):12509–12512.
27. Gronke RS, Welsh DJ, VanDusen WJ, et al. Partial purification and characterization of bovine liver aspartyl β-hydroxylase. J Biol Chem 1990; 265:8558–8565.
28. Stenflo J, Lundwall A, Dahlback B. β-Hydroxyasparagine in domains homologous to the epidermal growth factor precursor in vitamin K-dependent protein S. Proc Natl Acad Sci USA 84:368–372.
29. Zauber NP, Levin J. Factor IX levels in patients with hemophilia B (Christmas disease) following transfusion with concentrates of factor IX or fresh frozen plasma (FFP). Medicine (Baltimore) 1977; 56:213–224.
30. Kurachi K, Davie EW. Isolation and characterization of a cDNA coding for human factor IX. Proc Natl Acad Sci USA 1982; 79:6461–6464.
31. Jaye M, de la Salle H, Schamber F, et al. Isolation of a human anti-haemophilic factor IX cDNA clone using a unique 52-base synthetic oligonucleotide probe deduced from the amino acid sequence of bovine factor IX. Nucleic Acids Res 1983; 11(8):2325–2335.
32. Anson DS, Choo KH, Rees DJG, et al. The gene structure of human anti-haemophilic factor IX. EMBO J 1984; 3(5):1053–1060.
33. Wu SM, Stafford DW, Ware J. Deduced amino acid sequence of mouse blood-coagulation factor IX. Gene 1990; 86:275–278.
34. Evans JP, Watzke HH, Ware JL, Stafford DW, High KA. Molecular cloning of a cDNA encoding canine factor IX. Blood 1989; 74(1):207–212.
35. Yoshitake S, Schach BG, Foster DC, Davie EW, Kurachi K. Nucleotide sequence of the gene for human factor IX (antihemophilic factor B). Biochemistry 1985; 24:3736–3750.
36. Pendurthi UR, Tukey RH, Rao LV. Characterization of a rabbit factor IX cDNA. Thromb Res 1992; 65:177–186.
37. Sarkar G, Koeberl DD, Sommer SS. Direct sequencing of the activation peptide and the catalytic domain of the factor IX gene in six species. Genomics 1990; 6:133–143.
38. Katayama K, Ericsson LH, Enfield DL, et al. Comparison of amino acid sequence of bovine coagulation factor IX (Christmas factor) with that of other vitamin K-dependent plasma proteins. Proc Natl Acad Sci USA 1979; 76(10):4990–4994.
39. Frazier D, Smith KJ, Cheung WF, et al. Mapping of monoclonal antibodies to human factor IX. Blood 1989; 74:971–977.
40. McGraw RA, Frazier D, de Serres M, Reisner H, Stafford DW. Antigenic determinant in human coagulation of factor IX immunological screening and DNA sequence analysis of recombinant phage map: a monoclonal antibody to residues 111 through 132 of the zymogen. Blood 1986; 67:1344–1348.
41. Camerino G, Grzeschik KH, Jaye M, et al. Regional localization on the human X chromosome and polymorphism of the coagulation factor IX gene (hemophilia B locus). Proc Natl Acad Sci USA 1984; 81:498–502.
42. Purrello M, Alhadeff B, Esposito D, et al. The human genes for hemophilia A and hemophilia B flank the X chromosome fragile site at Xq27.3. EMBO J 1985; 4(3):725–729.
43. Reijnen M, Bertina RM, Reitsma PH. Localization of transcription initiation sites in the human coagulation factor IX gene. FEBS Lett 1990; 270(1,2):207–210.
44. Huang M-N, Hung H-L, Stanfield-Oakley SA, High KA. Characterization of the human blood coagulation factor X promoter. J Biol Chem 1992; 267(22):15440–15446.
45. Salier JP, Hirosawa S, Kurachi K. Functional characterization of the 5′-regulatory region of human factor IX gene. J Biol Chem 1990; 265(12):7062–7068.
46. Fair DS, Bahnak BR. Human hepatoma cells secrete single chain factor X, prothrombin, and antithrombin III. Blood 1984; 64:194–204.
47. Kojima T, Tanimoto M, Kamiya T, et al. Possible absence of common polymorphisms in coagulation factor IX gene in Japanese subjects. Blood 1987; 69:349–352.

48. Driscoll MC, Dispenzieri A, Tobias E, Miller CH, Aledort LM. A second *Bam*HI DNA polymorphism and haplotype association in the factor IX gene. Blood 1988; 72(1):61–65.
49. Winship PR, Rees DJG, Alkan M. Detection of polymorphisms at cytosine phosphoguanadine dinucleotides and diagnosis of haemophilia B carriers. Lancet 1989; 1:631–634.
50. Thompson AR, Bajaj SP, Chen SH, MacGillivray RTA. ''Founder'' effect in different families with haemophilia B mutation. Lancet 1990; 335:418.
51. Ketterling RP, Bottema CDK, Phillips JA III, Sommer SS. Evidence that descendants of three founders constitute about 25% of hemophilia B in the United States. Genomics 1991; 10:1093–1096.
52. Giannelli F, Choo KH, Winship PR, et al. Characterisation and use of an intragenic polymorphic marker for detection of carriers of haemophilia B (factor IX deficiency). Lancet 1984; 239–241.
53. Winship PR, Anson DS, Rizza CR, Brownlee GG. Carrier detection in haemophilia B using two further intragenic restriction fragment length polymorphisms. Nucleic Acids Res 1984; 12(23):8861–8872.
54. Graham JB, Kunkel GR, Egilmez NK, Wallmark A, Fowlkes DM, Lord ST. The varying frequencies of five DNA polymorphisms of X-linked coagulant factor IX in eight ethnic groups. Am J Hum Genet 1991; 49:537–544.
55. Camerino G, Oberle I, Drayna D, Mandel JL. A new MspI restriction fragment length polymorphism in the hemophilia B locus. Hum Genet 1985; 71:79–81.
56. Freedenberg DL, Chen SH, Kurachi K, Scott CR. MspI polymorphic site within the factor IX gene. Hum Genet 1987; 76:262–264.
57. Hay CW, Robertson KA, Yong SL, Thompson AR, Growe GH, MacGillivray RTA. Use of a *Bam*HI polymorphism in the factor IX gene for the determination of hemophilia B carrier status. Blood 1986; 67(5):1508–1511.
58. McGraw RA, Davis LM, Noyes CM, et al. Evidence for a prevalent dimorphism in the activation peptide of human coagulation factor IX. Proc Natl Acad Sci USA 1985; 82:2847–2851.
59. Wallmark A, Kunkel G, Mouhli H, et al. Population genetics of the Malmö polymorphism of coagulation factor IX. Hum Hered 1991; 41:391–396.
60. Wallmark A, Ljung R, Nilsson IM, et al. Polymorphism of normal factor IX detected by mouse monoclonal antibodies. Proc Natl Acad Sci USA 1985; 82:3839–3843.
61. Graham JB, Lubahn DB, Lord ST, et al. The Malmö polymorphism of coagulation factor IX, an immunologic polymorphism due to dimorphism of residue 148 that is in linkage disequilibrium with two other F.IX polymorphisms. Am J Hum Genet 1988; 42:573–580.
62. Briët E, Bertina RM, Van Tilburg NH, Veltkamp JJ. Haemophilia B Leyden: a sex-linked hereditary disorder that improves after puberty. N Engl J Med 1982; 306:788–790.
63. Crossley M, Brownlee GG. Disruption of a C/EBP binding site in the factor IX promoter is associated with haemophilia B. Nature 1990; 345:444–446.
64. Friedman AD, Landschulz WH, McKnight SL. CCAAT/enhancer binding protein activates the promoter of the serum albumin gene in cultured hepatoma cells. Genes Dev 1989; 3: 1314–1322.
65. Johnson PF, McKnight SL. Eukaryotic transcriptional regulatory proteins. Annu Rev Biochem 1989; 58:799–839.
66. Picketts DJ, Lillicrap DP, Mueller CR. Synergy between transcription factors DBP and C/EBP compensates for a haemophilia B Leyden factor IX mutation. Nature Genet 1993; 3:175–179.
67. Reijnen MJ, Sladek FM, Bertina RM, Reitsma PH. Disruption of a binding site for hepatocyte nuclear factor 4 results in hemophilia B Leyden. Proc Natl Acad Sci USA 1992; 89:6300–6303.
68. Landschulz WH, Johnson PF, McKnight SL. Homologous recognition of a promoter domain common to the MSV LTR and the HSV *tk* gene. Cell 1986; 44:565–576.

69. Mueller CR, Maire P, Schibler U. DBP, a liver-enriched transcriptional activator, is expressed late in ontogeny and its tissue specificity is determined post translationally. Cell 1990; 61:279–291.
70. Sladek FM, Zhong W, Lai E, Darnell JE Jr. Liver-enriched transcription factor HNF-4 is a novel member of the steroid hormone receptor superfamily. Genes Dev 1990; 4:2353–2365.
71. Paonessa G, Gounari F, Frank R, Cortese R. Purification of a NF1-like DNA-binding protein from rat liver and cloning of the corresponding cDNA. EMBO J 1988; 7:3115–3123.
72. Veltkamp JJ, Meilof J, Remmelts HG, van der Vlerk D, Loeliger EA. Another genetic variant of haemophilia B: haemophilia B Leyden. Scand J Haematol 1970; 7:82–90.
73. Briët E, Wijnands MC, Veltkamp JJ. The prophylactic treatment of haemophilia B Leyden with anabolic steroids. Ann Intern Med 1986; 103:225–226.
74. Reijnen MJ, Peerlinck K, Maasdam D, Bertina RM, Reitsma PH. Hemophilia B Leyden: substitution of thymine for guanine at position −21 results in a disruption of a hepatocyte nuclear factor 4 binding site in the factor IX promoter. Blood 1993; 82(1):151–158.
75. Reitsma PH, Bertina RM, Ploos van Amstel JK, Riemens A, Briet E. The putative factor IX gene promoter in hemophilia B Leyden. Blood 1988; 72:1074–1076.
76. Ghanem N, Costes B, Martin J, et al. Eur J Hum Genet 1993; in press.
77. Crossley M, Winship PR, Austen DEG, Rizza CR, Brownlee GG. A less severe form of haemophilia B Leyden. Nucleic Acids Res 1990; 18:4633.
78. Hirosawa S, Fahner JB, Salier JP, Wu CT, Lovrien EW, Kurachi K. Structural and functional basis of the developmental regulation of human coagulation factor IX gene: factor IX Leyden. Proc Natl Acad Sci USA 1990; 87:4421–4425.
79. Freedenberg DL, Black B. Altered developmental control of the factor IX gene: a new T to A mutation at position +6 of the F-IX gene resulting in hemophilia B Leyden (abstract). Thromb Haemost 1991; 65:964.
80. Royle G, Van De Water NS, Berry E, Ockelford PA, Browett PJ. Haemophilia B Leyden arising de novo by point mutation in the putative factor IX promoter region. Br J Haematol 1991; 77:191–194.
81. Reitsma PH, Mandalaki T, Kasper CK, Bertina RM, Briët E. Two novel point mutations correlate with an altered developmental expression of blood coagulation factor IX (hemophilia B Leyden phenotype). Blood 1989; 73:743–746.
82. Koeberl DD, Bottema CDK, Buerstedde JM, Sommer SS. Functionally important regions of the factor IX gene have a low rate of polymorphism and a high rate of mutation in the dinucleotide CpG. Am J Hum Genet 1989; 45:448–457.
83. Crossley PM, Winship PR, Black A, Rizza C, Brownlee GG. Unusual case of haemophilia B. Lancet 1989; 1:960.
84. Crossley M, Ludwig M, Stowell KM, De Vos P, Olek K, Brownlee GG. Recovery from hemophilia B Leyden: an androgen-responsive element in the factor IX promoter. Science 1992; 257:377–379.
85. Department of Health and Human Services. National Heart and Lung Institute Blood Resource Studies. Vol. 3. Pilot study hemophilia treatment in the U.S. U.S. Government Printing Office, Washington, D.C., 1972.
86. Biggs R. Jaundice and antibodies directed against factors VIII and IX in patients treated for haemophilia or Christmas disease in the United Kingdom. Br J Haematol 1974; 26:313–329.
87. Ludwig M, Schwaab R, Olek K, Brackmann HH, Egli H. Haemophilia B^{+} with inhibitor. Thromb Haemost 1988; 59:340.
88. Noyes CM, Griffith MJ, Roberts HR, Lundblad RL. Identification of the molecular defect in factor IX Chapel Hill. Substitution of a histidine for arginine at position 145. Proc Natl Acad Sci USA 1983; 80:4200–4202.
89. Thompson AR, Chen S-H. Characterization of factor IX defects in hemophilia B. Methods Enzymol 1993; 222:143–169.

90. Tuddenham EGD, Cooper DN, Gitschier J, et al. Haemophilia A: database of nucleotide substitutions, deletions, insertions and rearrangements of the factor VIII gene. Nucleic Acids Res 1991; 19:4821–4833.
91. Sadler JE, Ginsberg D. A database of polymorphisms in the von Willebrand factor gene and pseudogene. Thromb Haemost 1993; 69:185–191.
92. Hougie C, Twomey JJ. Haemophilia B_m: a new type of factor-IX deficiency. Lancet 1967; 698–700.
93. Huang M-N, Kasper CK, Roberts HR, Stafford DW, High KA. Molecular defect in factor IX_{HILO}, a hemophilia B_m variant: Arg → Gln at the carboxyterminal cleavage site of the activation peptide. Blood 1989; 73:718–721.
94. Bertina RM, van der Linden IK, Mannucci PM, et al. Mutations in hemophilia B_m occur at the Arg^{180}-Val^{181} activation site or in the catalytic domain of factor IX. J Biol Chem 1990; 265:10876–10883.
95. Sakai T, Yoshioka A, Yamamoto K, et al. Blood clotting factor IX Kashihara: amino acid substitution of $Valine^{-182}$ by phenylalanine. J Biochem 1989; 105:756–759.
96. Monroe DM, McCord DM, Huang M-N, et al. Functional consequences of an arginine 180 to glutamine mutation in factor IX Hilo. Blood 1989; 73:1540–1544.
97. Bardin JM, Sultan Y. Factor IX concentrate versus prothrombin complex concentrate for the treatment of hemophilia B during surgery. Transfusion 1990; 30:441–443.
98. Kim HC, McMillan CW, White GC, Bergman GE, Saidi P. Purified factor IX using monoclonal immunoaffinity technique: clinical trials in hemophilia B and comparison with prothrombin complex concentrates. Blood 1992; 79:568–572.
99. Mannucci PM. Modern treatment of hemophilia: from the shadows towards the light. Thromb Haemost 1993; 70(1):17–23.
100. Palmer TD, Thompson AR, Miller AD. Production of human factor IX in animals by genetically modified skin fibroblasts: potential therapy for hemophilia B. Blood 1989; 73(2): 438–445.
101. Yao SN, Kurachi K. Expression of human factor IX in mice after injection of genetically modified myoblasts. Proc Natl Acad Sci USA 1992; 89:3357–3361.
102. Yao S-N, Wilson JM, Nabel EG, Kurachi S, Hachiya H, Kurachi K. Expression of human factor IX in rat capillary endothelial cells: toward somatic gene therapy for hemophilia B. Proc Natl Acad Sci USA 1991; 88:8101–8105.
103. Evans JP, Brinkhous KM, Brayer GD, Reisner HW, High KA. Canine hemophilia B resulting from a point mutation with unusual consequences. Proc Natl Acad Sci USA 1989; 86: 10095–10099.
104. Mulligan RC. The basic science of gene therapy. Science 1993; 260:926–932.
105. Lozier JN, Palmer TD, Thompson AR, Miller AD, Brinkhous KM, High KA. Production of recombinant canine factor IX by hollow fiber tissue culture. Thromb Haemost 1991; 65: 1158. (Presented to the XIIIth Congress of the International Society on Thrombosis and Haemostasis, Amsterdam, The Netherlands, June 30 to July 6, 1991.)
106. Axelrod JH, Read MS, Brinkhous KM, Verma IM. Phenotypic correction of factor IX deficiency in skin fibroblasts of hemophilic dogs. Proc Natl Acad Sci USA 1990; 87:5173–5177.
107. Armentano D, Thompson AR, Darlington G, Woo SLC. Expression of human factor IX in rabbit hepatocytes by retrovirus-mediated gene transfer: potential for gene therapy of hemophilia B. Proc Natl Acad Sci USA 1990; 87:6141–6145.
108. Dai Y, Roman M, Naviaux RK, Verma IM. Gene therapy via primary myoblasts: long-term expression of factor IX protein following transplantation in vivo. Proc Natl Acad Sci USA 1992; 89:10892–10895.
109. Palmer TD, Rosman GJ, Osborne WRA, Miller AD. Genetically modified skin fibroblasts persist long after transplantation but gradually inactivate introduced genes. Proc Natl Acad Sci USA 1991; 88:1330–1334.

110. Kay MA, Rothenberg S, Landen CN, et al. In vivo gene therapy of hemophilia B: sustained partial correction in factor IX-deficient dogs. Science 1993; 262:117–119.
111. Green PM, Bentley DR, Mibashan RS, Nilsson IM, Giannelli F. Molecular pathology of haemophilia B. EMBO J 1989; 8:1067–1072.
112. Denton PH, Fowlkes DM, Lord ST, Reisner HM. Hemophilia B Durham: a mutation in the first EGF-like domain of factor IX that is characterized by polymerase chain reaction. Blood 1988; 72:1407–1411.
113. Koeberl DD, Bottema CDK, Ketterling RP, Bridge PJ, Lillicrap DP, Sommer SS. Mutations causing hemophilia B: direct estimate of the underlying rates of spontaneous germ-line transitions, transversions, and deletions in a human gene. Am J Hum Genet 1990; 47:202–217.
114. Winship PR, Dragon AC. Identification of haemophilia B patients with mutations in the two calcium binding domains of factor IX: importance of a β-OH Asp 64 → Asn change. Br J Haematol 1991; 77:102–109.
115. Chen SH, Thompson AR, Zhang M, Scott CR. Three point mutations in the factor IX genes of five hemophilia B patients. J Clin Invest 1989; 84:113–118.
116. Ludwig M, Sabharwal AK, Brackmann HH, et al. Hemophilia B caused by five different nondeletion mutations in the protease domain of factor IX. Blood 1992; 79:1225–1232.
117. Chen SH, Scott CR, Lovrein EW, Kurachi K. Factor $IX_{Portland}$: a nonsense mutation (CGA to TGA) resulting in hemophilia B. Am J Hum Genet 1989; 44:567–569.
118. Montandon AJ, Green PM, Giannelli F, Bently DR. Direct detection of point mutations by mismatch analysis: application to haemophilia B. Nucleic Acids Res 1989; 17:3347–3358.
119. Tsang TC, Bentley DR, Mibashan RS, Giannelli F. A factor IX mutation, verified by direct genomic sequencing, causes haemophilia B by a novel mechanism. EMBO J 1988; 7:3009–3015.
120. Chen SH, Zhang M, Lovrein EW, Scott CR, Thompson AR. CG dinucleotide transitions in the factor IX gene account for about half of the point mutations in hemophilia B patients: a Seattle series. Hum Genet 1991; 87:177–182.
121. Geddes VA, LeBonniec BF, Louie GV, Brayer GD, Thompson AR, MacGillivray RTA. A moderate form of hemophilia B is caused by a novel mutation in the protease domain of factor $IX_{Vancouver}$. J Biol Chem 1989; 264:4689–4697.

11

Factor X

Herbert H. Watzke
University of Vienna, Vienna, Austria

Katherine A. High
Children's Hospital of Philadelphia, and University of Pennsylvania Medical Center, Philadelphia, Pennsylvania

I. INTRODUCTION

Factor X (FX) is a vitamin K-dependent plasma protein that plays a central role in the process of blood coagulation (1). The presence of FX in plasma was first described in 1957 by Graham et al. (2). The presence of a hitherto unknown coagulation factor was suspected in two patients with a bleeding disorder. The unknown protein was termed factor X. The deficiency state was named, after the two initial patients, Mr. Stuart and Mrs. Prower, "Stuart/Prower deficiency" (3).

FX plays a pivotal role in blood coagulation: it is the main target for the serine proteases factor VIIa and factor IXa. Once activated by these enzymes, factor Xa is the main enzyme in the process of the conversion of prothrombin into thrombin (1).

II. THE GENE CODING FOR FX

A. Chromosomal Location

The human gene for FX is located on chromosome 13q34-qter (4–6). It spans approximately 25 kb genomic DNA: it is in close proximity to the gene encoding coagulation factor VII. Patients with a deletion of chromosome 13 have been found to be deficient in FX and FVII (4,7–10).

B. Structure

The FX gene (Fig. 1) is composed of eight exons separated by seven intervening sequences (11,12). The exons range in size from 24 bp (exon III) to 602 bp (exon VIII). The sequence of the introns is known only in part. Their sizes range from approximately 960 bp (intron C) to approximately 8100 bp (intron B) as determined by restriction enzyme analysis (13). The nucleotide sequence of the introns has been only partially determined. The nucleotides at all the intron/exon splice junction sites have been determined. Their sequence follows the CT/AG rule and includes splice junction types I and 0 (11). The 3′-untranslated region of the gene is remarkably short and comprises only 10 base pairs. The 5′ flanking region of the FX gene has been completely sequenced, including the complete intergenic region between the genes for FX and for FVII (14). The introns in the FX gene fall between various functional domains. Their location in the coding sequence is highly conserved compared with other vitamin K-dependent proteins (15). It has been proposed that the genes coding for multidomain proteins, such as the serine proteases, evolved from a common ancestral gene (16). Diversification occurred through various events, including gene duplication, carryover of genetic material, and random transposition. Because the serine proteases not only share significant structural homology but also show extensive sequence similarity, they are believed to constitute a distinct family of serine proteases. The first intron in the FX gene (intron A) occurs at a location analogous to the first introns found in the genes for FVII, FIX, and protein C (11). This first intron in the FX gene is located between the codons for Ser^{-18} and Leu^{-17} (see Fig. 2). It separates two functional domains: the N-terminal signal peptide encoded by exon I and the propeptide γ-carboxyglutamic (Gla) domain, encoded by exon II. The Gla domain is the region in the FX gene with the highest degree of homology to the other vitamin K-dependent coagulation factors. The second intron (intron B) separates the exons coding for the propeptide Gla domain from a short exon (exon III) coding for a stack of hydrophobic amino acids (aa). Intron C is located immediately upstream of exon IV, which encodes the first of two modules that have a high degree of sequence homology and structural similarity to the epidermal growth factor (EGF) (17). The exon coding for the second EGF domain is separated from that encoding the first EGF domain by intron D. Intron E is placed between the second EGF domain and the part of the gene that encodes the COOH terminus of the light chain, the FX activation peptide, and the N terminus of the heavy chain. The remainder of the sequence coding for the heavy chain (exon VII and exon VIII) is separated by intron G. Although the position of the introns is highly conserved among the vitamin K-dependent coagulation factors, the sizes of the introns vary and the sequence lacks any significant degree of homology.

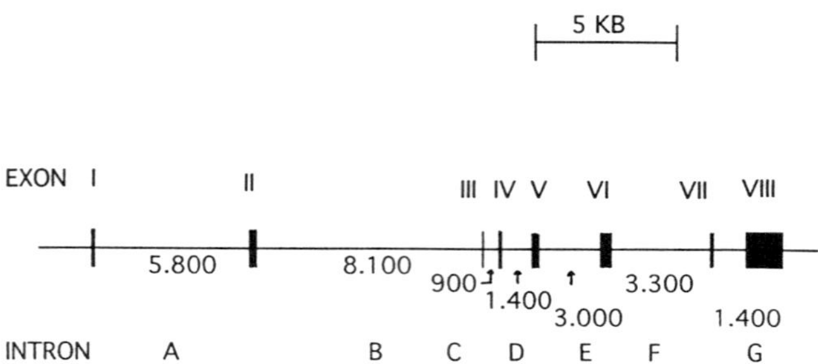

Figure 1 The gene for human factor X. The exons are named I–VIII and shown as solid boxes. Their size is 70 bp for exon I, 161 bp for exon II, 25 bp for exon III, 114 bp for exon IV, 132 bp for exon V, 245 bp for exon VI, 118 bp for exon VII, and 612 bp for exon VIII. Areas between exons represent intervening sequences and are labeled A–G. The size of the introns are estimated from restriction enzyme maps (13). The scale in kb is shown at the top.

Five Alu repetitive sequences have been identified in the gene for FX. Three of these have been found in intron B, one in intron D, and one in intron G (11).

C. Polymorphisms

A number of polymorphic sites have been identified in the FX gene and can be used for diagnostic purposes: a TaqI polymorphism identifies four invariant bands (2.5, 2.6, 1.35, and 0.42 kb) in addition to a simple two-allele polymorphism with a band at either 1.62 or 1.25 kb (18,19). The reported allele frequency in whites is 0.77 for the 1.62 kb allele and 0.23 for the 1.25 kb allele (19). This polymorphism is also detected by probing with a cDNA fragment of nt 347–1130 (18), indicating its position in the 5′ intragenic region. Other polymorphic sites have been identified in digests with EcoRI (18,20), PstI (18,20), HindIII, and PvuII (18). EcoRI identifies a simple two-allele polymorphism with fragments of 7.5 or 7.2 kb. Digestion with PstI shows three constant bands (4.4, 2.9, and 0.85 kb) and an additional band of 2.5 with a frequency of 0.10. The EcoRI polymorphism and a PstI polymorphism show co-dominant segregation and are in linkage disequilibrium. A HindIII polymorphism creates a 7.5 kb fragment with a frequency of 0.05. PvuII identifies two constant fragments (5.5 and 0.8 kb) in addition to four variant fragments (2.7, 3.3, 3.1, and 2.6 kb) with frequencies of 0.73, 0.08, 0.12, and 0.06, respectively. A combination of EcoRI, PstI, PvuII, and TaqI has been reported to be informative in the diagnosis of a family with a hereditary FX deficiency (18). An intragenic NlaIV polymorphism has been found in the coding sequence of FX (exon VII) (21). The variant is found at residue 817 of the cDNA (numbering according to Ref. 11), where either a C or a T may be present. The polymorphism is silent at the amino acid level, because both triplets (ACT and ACC) encode threonine. The frequency is 0.76 for the ACT sequence and 0.24 for the ACC sequence. Another sequence polymorphism has been observed in the coding sequence of exon V (22,23). The Glu residue at position 102 (GAG) has been found to be substituted by a Lys (AAG) residue in two independent kindreds. Both families have an additional mutation in their coding sequence that is responsible for the observed FX deficiency. However, the sequence variation at Glu^{102} has not been reported in normal control subjects. Further studies in normal individuals are thus necessary before this sequence variability can be classified as a true polymorphism.

D. Regulation and Expression of Factor X

The 5′ flanking region of human factor X has been sequenced and its promoter activity characterized (14,24). There are 2.823 bp between the polyadenylation site of the factor VII gene and the first AUG of factor X. The factor X promoter lacks a typical TAT box but contains a CCAAT sequence at −120 to −116 (where 1 is the first nucleotide of the first AUG). Based on S1 nuclease analysis and anchored polymerase chain reaction, factor X appears to have multiple start sites of transcription (24). Five start sites are clustered within a 21 bp region ranging from −13 to −33, and two other minor start sites of transcription, at −57 and −130, are also detected. The finding of multiple start sites of transcription is consistent with the results reported from other TATA-less promoters.

Using growth hormone as the reporter gene, Huang et al. analyzed the promoter activity of the factor X 5′ flanking region in HepG2 cells (a human hepatoma cell line) and HeLa cells. In this system, maximal promoter activity was achieved with the fragment containing 209 bp upstream of the first AUG; addition of the next 179 bp flanking

ATGGGGCGCCCACTGCACCTCGTCCTGCTCAGTGCCTCCCTGGCTGGCCTCCTGCTGCTCGGGGAAAGTCTGTTCATCCG 80
-40 Met Gly Arg Pro Leu His Leu Val Leu Leu Ser Ala Ser Leu Ala Gly Leu Leu Leu Leu Gly Glu Ser Leu Phe Ile Arg -14
-18 -17
signal peptidase

CAGGGAGCAGGCCAACAACATCCTGGCGAGGGTCACGAGGGCCAATTCCTTTCTTGAAGAGATGAAGAAAGGACACCTCG 160
Arg Glu Gln Ala Asn Asn Ile Leu Ala Arg Val Thr Arg Ala Asn Ser Phe Leu Glu Glu Met Lys Lys Gly His Leu 13
-1 +1 γ γ
processing protease

AAAGAGAGTGCATGGAAGAGACCTGCTCATACGAAGAGGCCCGCGAGGTCTTTGAGGACAGCGACAAGACGAATGAATTC 240
Glu Arg Glu Cys Met Glu Glu Thr Cys Ser Tyr Glu Glu Ala Arg Glu Val Phe Glu Asp Ser Asp Lys Thr Asn Glu Phe 40
γ γ γ γ γ γ γ γ γ

TGGAATAAATACAAAGATGGCGACCAGTGTGAGACCAGTCCTTGCCAGAACCAGGGCAAATGTAAAGACGGCCTCGGGGA 320
Trp Asn Lys Tyr Lys Asp Gly Asp Gln Cys Glu Thr Ser Pro Cys Gln Asn Gln Gly Lys Cys Lys Asp Gly Leu Gly Glu 67
β

ATACACCTGCACCTGTTTAGAAGGATTCGAAGGCAAAAACTGTGAATTATTCACACGGAAGCTCTGCAGCCTGGACAACG 400
Tyr Thr Cys Thr Cys Leu Glu Gly Phe Glu Gly Lys Asn Cys Glu Leu Phe Thr Arg Lys Leu Cys Ser Leu Asp Asn 93

GGGACTGTGACCAGTTCTGCCACGAGGAACAGAACTCTGTGGTGTGCTCCTGCGCCCGCGGGTACACCCTGGCTGACAAC 480
Gly Asp Cys Asp Gln Phe Cys His Glu Glu Gln Asn Ser Val Val Cys Ser Cys Ala Arg Gly Tyr Thr Leu Ala Asp Asn 120

GGCAAGGCCTGCATTCCCACAGGGCCCTACCCCTGTGGGAAACAGACCCTGGAACGCAGGAAGAGGTCAGTGGCCCAGGC 560
Gly Lys Ala Cys Ile Pro Thr Gly Pro Tyr Pro Cys Gly Lys Gln Thr Leu Glu Arg Arg Lys Arg Ser Val Ala Gln Ala 147
139 143
processing protease

CACCAGCAGCAGCGGGGAGGCCCCTGACAGCATCACATGGAAGCCATATGATGCAGCCGACCTGGACCCCACCGAGAACC 640
Thr Ser Ser Ser Gly Glu Ala Pro Asp Ser Ile Thr Trp Lys Pro Tyr Asp Ala Ala Asp Leu Asp Pro Thr Glu Asn 173
CCTTCGACCTGCTTGACTTCAACCAGACGCAGCCTGAGAGGGGCGACAACAACCTCACCAGGATCGTGGGAGGCCAGGAA 720

```
Pro Phe Asp Leu Leu Asp Phe Asn Gln Thr Gln Pro Glu Arg Gly Asp Asn Asn Leu Thr Arg Ile Val Gly Gly Gln Glu   200
                                                                              194 ↑ 195
                                                                              FIXa;FVIIa
TGCAAGGACGGGGAGTGTCCCTGGCAGGCCCTGCTCATCAATGAGGAAAACGAGGGTTTCTGTGGTGGAACTATTCTGAG   800
Cys Lys Asp Gly Glu Cys Pro Trp Gln Ala Leu Leu Ile Asn Glu Glu Asn Glu Gly Phe Cys Gly Gly Thr Ile Leu Ser   227
CGAGTTCTACATCCTAACGGCAGCCCACTGTCTCTACCAAGCCAAGAGATTCAAGGTGAGGGTAGGGGACCGGAACACGG   880
Glu Phe Tyr Ile Leu Thr Ala Ala (His) Cys Leu Tyr Gln Ala Lys Arg Phe Lys Val Arg Val Gly Asp Arg Asn Thr   253
AGCAGGAGGAGGGCGGTGAGGCGGTGCACGAGGTGGAGGTGGTCATCAAGCACAACCGGTTCACAAAGGAGACCTATGAC   960
Glu Gln Glu Glu Gly Gly Glu Ala Val His Glu Val Glu Val Val Ile Lys His Asn Arg Phe Thr Lys Glu Thr Tyr Asp   280
TTCGACATCGCCGTGCTCCGGCTCAAGACCCCCATCACCTTCCGCATGAACGTGGCGCCTGCCTGCCTCCCCGAGCGTGA   1040
Phe (Asp) Ile Ala Val Leu Arg Leu Lys Thr Pro Ile Thr Phe Arg Met Asn Val Ala Pro Ala Cys Leu Pro Glu Arg Asp   307
CTGGGCCGAGTCCACGCTGATGACGCAGAAGACGGGGATTGTGAGCGGCTTCGGGCGCACCCACGAGAAGGGCCGGCAGT   1120
Trp Ala Glu Ser Thr Leu Met Thr Gln Lys Thr Gly Ile Val Ser Gly Phe Gly Arg Thr His Glu Lys Gly Arg Gln   333
CCACCAGGCTCAAGATGCTGGAGGTGCCCTACGTGGACCGCAACAGCTGCAAGCTGTCCAGCAGCTTCATCATCACCCAG   1200
Ser Thr Arg Leu Lys Met Leu Glu Val Pro Tyr Val Asp Arg Asn Ser Cys Lys Leu Ser Ser Ser Phe Ile Ile Thr Gln   360
AACATGTTCTGTGCCGGCTACGACACCAAGCAGGAGGATGCCTGCCAGGGGGACAGCGGGGGCCCGCACGTCACCCGCTT   1280
Asn Met Phe Cys Ala Gly Tyr Asp Thr Lys Gln Glu Asp Ala Cys Gln Gly Asp (Ser) Gly Gly Pro His Val Thr Arg Phe   387
CAAGGACACCTACTTCGTGACAGGCATCGTCAGCTGGGGAGAGAGCTGTGCCCGTAAGGGGAAGTACGGGATCTACACCA   1360
Lys Asp Thr Tyr Phe Val Thr Gly Ile Val Ser Trp Gly Glu Ser Cys Ala Arg Lys Gly Lys Tyr Gly Ile Tyr Thr   413
AGGTCACCGCCTTCCTCAAGTGGATCGACAGGTCCATGAAAACCAGGGGCTTGCCCAAGGCCAAGAGCCATGCCCCGGAG   1440
Lys Val Thr Ala Phe Leu Lys Trp Ile Asp Arg Ser Met Lys Thr Arg Gly Leu Pro Lys Ala Lys Ser His Ala Pro Glu   440
GTCATAACGTCCTCTCCATTAAAG
Val Ile Thr Ser Ser Pro Leu Lys   448
```

Figure 2 cDNA and primary amino acid sequences for human factor X. Numbers in bold represent amino acid sequences: 1 is the start site of mature plasma FX. Arrows denotes sites of physiological cleavages. Encircled residues indicate the active site triad His^{236}, Asp^{282}, and Ser^{379}, γ = gamma-carboxyglutamic acid and β = β-hydroxyaspartic acid.

sequence (388 bp fragment) resulted in no change in promoter activity, and sequences farther upstream resulted in a gradual diminution of promoter activity (to 50% maximum). These findings were tissue specific; similar experiments carried out in HeLa cells showed maximal promoter activity with the 209 bp fragment, but rapid diminution of promoter activity with addition of further 5′ sequence, so that the 388 bp fragment showed 28% of maximum promoter strength. In addition, the maximum promoter strength achieved in HeLa cells was fivefold lower than seen with the HepG2 cells. Similar results were reported by Miao et al. (14) using CAT as the reporter gene. Using a DNase I footprint assay, Miao et al. identified an element at −64 to −42 that is protected by nuclear extracts from HepG2 cells. Huang et al. showed that even a single base mutation within a 6 bp sequence (ACTTTG) contained within this element can result in loss of protein binding function (using gel shift assays) and loss of promoter activity (in reporter gene assays). The protein that binds to this element has been tentatively identified as HNF-4 (25), based on the binding of the HNF-4 protein to this element in gel shift assays and the "super-shifting" of the complex following incubation with antiserum to HNF-4. Of considerable interest is the finding that the ACTTTG element is conserved in the promoters of factors VII, IX, and X. Mutations within the ACTTTG element in the factor IX promoter result in the hemophilia B Leyden phenotype (26).

Mutations within the CCAAT element at −120 to −116 also result in a marked loss of promoter activity and loss of binding function on gel shift assays (24). In contrast to the factor IX promoter, C/EBP (the CCAAT/enhancer binding protein) does not appear to bind to this CCAAT element in the FX promoter; the identity of the nuclear protein that binds in this region is unknown. Based on promoter activity in a reporter gene assay, Miao et al. identified two other elements, one at −215 to −149, the other at −457 to −351, as positive regulators of FX expression. The transcription factors that bind at these elements are unknown.

III. THE cDNA FOR FACTOR X

The cDNA of FX has been isolated by several laboratories (27–29). The FX cDNA has an open reading frame of 1467 nt. The stop codon TGA is followed by a remarkably short 3′-untranslated region of only 10 nucleotides. A putative polyadenylation signal ATTAAA is present within the coding sequence of exon VIII (nt 1458–1463 in the open reading frame). Elements corresponding to a second additional polyadenylation site with the consensus sequence CAYTG are located 3′ to the stop codon. The 5′-untranslated region varies in length. Multiple different start sites of transcription clustered within 33 nt of the first ATG were found when total human liver poly(A) RNA was analyzed by S1 nuclease digestion and anchored PCR (24). This finding of multiple start sites of transcription is consistent with the results from other TATA-less promoters. A major start site of transcription for FIX yields a 5′-untranslated region of about the same length. The open reading frame starts with a stretch of 23 aa, which comprise the signal peptide. In vitro expression of a truncated FX cDNA has identified the Ser^{-18}-Leu^{-17} bond as the cleavage site of the signal peptidase (30). Signal peptides are common features of secretory proteins (31). Although the actual homology among the signal peptides of different proteins is low, similar amino acids are found in certain regions of the signal peptide: the center of the peptide is hydrophobic, the N terminus has a strong net positive charge, and the carboxyl terminus carries small nonpolar residues (32). A propeptide 17 aa in length is located between the signal peptide and the N terminus of the mature protein.

Sequence homology exists to a great extent within the propeptide region of the vitamin K-dependent coagulation factors (33). The Gla domain, which comprises the N-terminal portion of the mature FX molecule, is also highly conserved among the vitamin K-dependent coagulation factors (34). A total of 11 γ-carboxylated glutamic acid residues are present in this portion of the FX molecule (35). Of these, 6 occur in "pairs" (6/7, 19/20, and 25/26) and 5 occur as single residues (14, 16, 29, 32, and 39). Posttranslationally, these residues undergo vitamin K-dependent carboxylation at the γ carbon. Further posttranslational modifications of the primary translational product include β-hydroxylation of residue 63 (35) and removal of a 3 aa peptide that, before secretion, creates the two-chain form of the mature FX molecule, consisting of a light chain and a heavy chain. This Arg-Lys-Arg tripeptide (codons 180–182) is invariably present in the coding sequence of the FX cDNA but is lacking in the mature FX molecule, as shown by amino acid sequence analysis (36).

IV. THE FX PROTEIN

A. Protein Structure

Human coagulation factor X is synthesized in the liver but has been expressed in vitro in other cell types, such as human embryonic kidney cells (37,38) and Chinese hamster ovary cells (39). The primary translational product of FX (pre-pro FX) is 488 amino acids in length as derived from the coding sequence of the FX gene (Fig. 2). Through the cleavage by the signal peptidase at residues Ser^{-18}-Leu^{-17}, the preportion of FX, which directs the protein to the ER, is cleaved away from the primary translational product (37). Cleavage of the propeptide at residues Arg^{-1}-Ala^{1} establishes the NH_2 terminus of the mature protein. FX circulates in plasma as a two-chain molecule with a molecular weight of 59,000 according to sedimentation equilibrium analysis and sodium dodecyl sulfate (SDS)–polyacrylamide gel electrophoresis (40). It consists of a light chain with a molecular weight (MW) of 17,000 and heavy chain of 40,000 (41) held together by a single disulfide bond. A single-chain form of FX has been found in human and rat hepatocyte culture systems (42). When human FX is expressed in the human embryonic cell line 293, it is inefficiently cleaved at position −1 Arg^{-1}-Ala^{+1} (43), in contrast to the in vivo situation.

The structural domains of FX are shared with many of the vitamin K-dependent coagulation factors. The heavy chain includes the catalytic domain of FX, which shows considerable homology to the pancreatic protease trypsin. The hydrolytically functional residues of FX are evident by comparison with these enzymes and include heavy-chain residues serine 379, histidine 236, and aspartic acid 282, all of which are involved in proton transfer during substrate hydrolysis (Fig. 2) (44). The NH_2-terminal isoleucine of the heavy chain (following activation) is also conserved in the pancreatic proteases and other coagulation factors. It participates in the formation of the substrate binding pocket by formation of a salt bridge to the catalytic domain, which in turn permits full hydrolytic activity (45). The FX light chain consists of four characteristic structural domains, each of which possesses distinct functional properties: the γ-carboxyglutamic acid-rich domain containing 11 γ-carboxyglutamic acid residues, two modules homologous to epidermal growth factor (EGF-1 and EGF-2), and the activation peptide, which is cleaved off during the activation of the proenzyme FX to the active protease factor Xa. The activation peptide is 52 amino acids in length. It has molecular weight of 11,000 and is heavily glycosylated (40).

B. Protein Function

1. *Membrane Binding of FX*

FX is the proenzyme of the vitamin K-dependent serine protease FXa. Activation of coagulation FX is accomplished through the incorporation of FX proenzyme into either of two different membrane-bound complexes: the factor IXa-factor VIIIa (intrinsic) complex and the factor VIIa-tissue factor (TF); extrinsic complex (46). The degree of binding of these complexes is tissue specific: unactivated platelets do not exhibit significant binding of either FX activation complex (47), but human blood monocytes show functional assembly of both the intrinsic and the extrinsic complexes (48). FX activation by the intrinsic complex occurs in the presence of activated platelets (49–51). Quiescent endothelial cells are not thought to facilitate proteolytic activity in coagulation (52,53); however, physically perturbed endothelium or endothelial cells stimulated with interleukin-1 and tumor necrosis factor (54–56) express TF and result in FXa generation. FXa has been demonstrated to bind to aortic and vena caval endothelium (57). The membrane binding of FX, as with other vitamin K-dependent factors, is mediated through its Gla domain (58). This binding is Ca^{2+} dependent and is required for full biological activity of the protein; FX, which lacks the Gla domain, is unable to bind to endothelial cells and shows reduced activity in a coagulation assay (59).

2. *Activation of FX*

Activation of FX is accomplished by the cleavage of the Arg^{194}-Ile^{195} bond, which releases the activation peptide. Hydrolysis at this bond can be achieved by the coagulation factors of the intrinsic complex and the extrinsic complex and by Russell's viper venom (60). When activation by the FVIIa-TF (the extrinsic) complex is followed on an SDS gel, disappearance of the band representing the heavy chain of FX is accompanied by the appearance of bands with a MW of 34,300, termed α-Xa, and a MW of 29,000, termed β-Xa (Fig. 3) (61). A third band with a molecular weight of 41,900, termed intermediate I, is seen only when a lower concentration of FVIIa is used for activation. It is generated by cleavage at an Arg-Gly bond near the carboxyl terminus of the heavy chain without prior cleavage at the Arg^{194}-Ile^{195} bond. This intermediate lacks coagulant activity. Generation of α-Xa occurs via the cleavage at the Arg^{194}/Ile^{195} bond and represents the activation of FX to FXa. The appearance of β-Xa is the result of autocatalytic cleavage at the Arg-Gly Bond in the heavy chain. β-Xa also has enzymatic activity. The light chain of FX remains unchanged during the activation process (60). The activation of FX by the intrinsic pathway is catalyzed by the FIXa-FVIIIa complex. Upon activation by this complex, the same bands are observed on an SDS gel as for the FVIIa/TF complex. However, the β-Xa fragment is generated to a greater extent than the other fragments as the reaction proceeds (61). A specific protease of Russell's viper venom is also a potent activator of FX (62). In the presence of phospholipid, both activation fragments, α-Xa and β-Xa, are generated. Trypsin also activates FX with formation of bands corresponding to α- and β-Xa and intermediate I. Continued incubation with trypsin leads to complete proteolytic degradation of FX (59).

C. Factor Xa

The primary function of FXa within the coagulation cascade is cleavage of two peptide bonds of prothrombin, which results in the formation of thrombin. In addition, FXa can also activate factor V, FVII, factor VIII, and FX. Under physiological conditions, FXa associates with its nonproteolytic cofactor (FVa) on a membrane surface in the presence of calcium to form the prothrombinase complex. This macromolecular complex cleaves prothrombin, generating the active enzyme, thrombin, and an activation fragment (prothrombin fragment 1.2) (63,64). In addition to its activity in coagulation, FXa is a potent mitogen of cultured smooth muscle cells (65). This mitogenic activity may contribute to intimal proliferation after vascular injury.

The important inhibitor of activated FX is antithrombin III (ATIII). FXa, in a reaction augmented 1000-fold in the presence of heparin (66), cleaves a specific Arg-Ser bond near the COOH terminus of ATIII (67). This results in an irreversible complex consisting of the two proteins in a 1:1 stoichiometry. This cleavage of ATIII is thought to result in its covalent bonding to FXa at the active site serine (68). The ATIII-FXa complex may be cleared through ATIII binding to endothelial cell-associated proteoglycan, resulting in endocytosis and intracellular degradation (69,70). ATIII accounts for approximately 50% of FXa inhibition in plasma (57,71); α_1-proteinase inhibitor and α_2-macroglobulin also irreversibly inhibit FXa, but to a lesser degree than ATIII (57,71). The tissue factor pathway inhibitor (TFPI) is a Kunitz-type protease inhibitor that binds FXa, with a 1:1 stoichiometry, near the FXa active site. Plasma levels (2 nM) are 1000-fold less than ATIII (3 μM), and the importance of TFPI in vivo as an inhibitor of FXa is unknown (72). TFPI is bound to glycosaminoglycans on the endothelial surface and thus may be localized in higher concentrations at sites of active coagulation. TFPI binds FXa reversibly, and it is speculated that TFPI-bound FXa is eventually transferred to the irreversible plasma inhibitors of FXa (ATIII, α_1-proteinase inhibitor, and α_2-macroglobulin; see Chap. 15) (72).

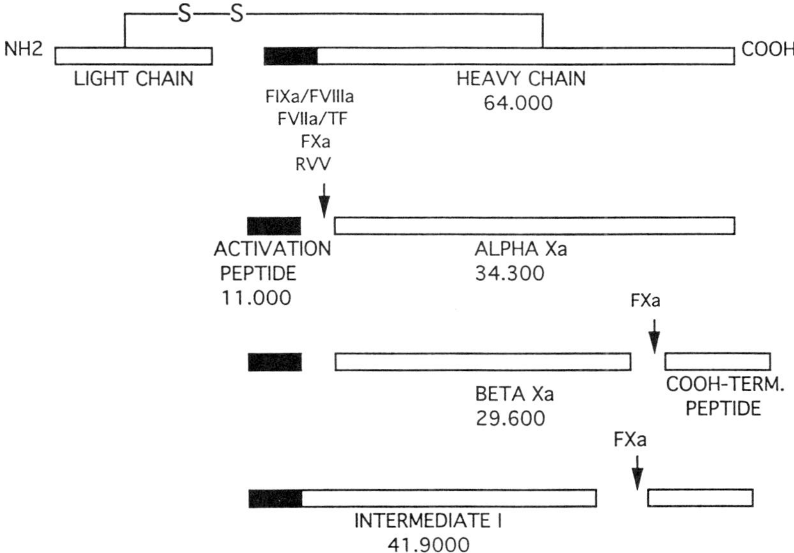

Figure 3 Activation of factor X by the FVIIa-TF complex, the FIXa-FVIIIa complex and RVV. The name of the fragments and the respective size in daltons is given below each fragment. The site of the proteolytic cleavage is shown by an arrow.

V. HEREDITARY FX DEFICIENCY

A. Genetics

Hereditary FX deficiency was first described in 1957 by Graham et al. (2). It is an autosomal recessive trait. Based on the number of phenotypically homozygous patients observed in North Carolina, Graham et al. estimated a frequency of the homozygous form of 1:1,000,000. Application of the Hardy-Weinberg equation to this observation gives an estimate of the heterozygous state of 2:1000 (73).

B. Naturally Occurring Mutants

1. *Functional Characterization*

Functional characterization of hereditary FX mutants has been carried out in several ways (74–76). First, the various defects are categorized by the amount of circulating FX in plasma (FX Ag) as cross-reacting material (CRM) positive (normal amount of FX), CRM reduced, and CRM negative. The exeptional position of FX in the coagulation cascade allows further differentiation. Physiologically, FX can be converted into its active form FXa by two enzyme complexes: the FVIIA-TF complex (the extrinsic pathway) and the FIXa-FVIIIa complex (the intrinsic pathway). Hereditary FX deficiency is characterized by a defective activation of FX in one or both pathways. In addition, a failure of activation by Russel's viper venom has been observed in some forms of hereditary FX deficiency (77). Hereditary FX deficiency can be further characterized by assessment of the FX antigen (Ag) in plasma. Thus, a rather detailed functional characterization of hereditary FX variants is easily possible, which makes the allocation of certain molecular defects to the observed functional defects particularly interesting.

2. *Molecular Characterization*

The causative mutation has been clucidated in the nine kindreds described in (Table 1). Among these are seven homozygous and one heterozygous FX deficiency, all caused by point mutations. One compound heterozygous has been reported in which one allele is affected by a point mutation and the other carries a point deletion. In another heterozygote patient, one allele carries a large deletion of exon VII and exon VIII but the defect in

Table 1 Point Mutation Causing Hereditary Factor X Deficiency

Factor X	FX-C PT	FX-C aPTT	FX-C RVV	FX Ag	Amino acid change[a]		References
Vorarlberg	7	25	15	30	$Gla^{+14} \rightarrow Lys$	Homozygous	22
Santo Domingo	1	1	1	5	$Gly^{-20} \rightarrow Arg$	Homozygous	37
Friuli	7	8	76	100	$Pro^{344} \rightarrow Ser$	Homozygous	80, 81
San Antonio	14			36	$Arg^{366} \rightarrow Cys$ del nt^{838}	Compound heterozygous	83
Malmö	35	43			$Gla^{+26} \rightarrow Asp$	Homozygous	79
Vienna	1			5	$Gly^{201} \rightarrow Glu$	Homozygous	82
Marseille	21	21	26	100	$Ser^{334} \rightarrow Pro$	Homozygous	84
Stockton	43			100	$Asp^{282} \rightarrow Asn$	Heterozygous	85
Stuart	1	1	1	1	$Val^{339} \rightarrow Met$	Homozygous	86

[a]Numbering of amino acids according to Reference 11.

the other allele has not yet been defined. Seven of the reported amino acid changes occur in the heavy chain of FX, two in the light chain, and one in the signal peptide. Six G to A transitions were found among the point mutations. Four of these occur at CG dinucleotides.

a. Point Mutations FX Vorarlberg (FXv) (22) was the first homozygous FX deficiency in which the causative mutation was identified. A G to A transition at a CpG dinucleotide in exon II of the FX gene results in the substitution of a Lys residue for the γ-carboxylated residue at position 14 in the mature protein (position 54 according to Ref. 11) Three homozygous and five heterozygous members of the kindred were detected by genetic testing. Consanguinity is suspected because the defect is present in a large kindred living in a rather remote valley in the mountain area of Austria. The (homozygous) affected family members have a marked prolongation of the prothrombin time (PT) and the (activated) partial thromboplastin time (aPTT). The index case was found to have a FX activity level of 5% using a PT-based assay. His FX Ag level in plasma is reduced to about 30% of normal. However, spontaneous bleeding was not present in either the heterozygous or the homozygous family members. The molecular defect causes an incomplete conformational change in the molecule in the presence of Ca^{2+}, as shown by incomplete intrinsic fluorescence quenching. Functional analysis of the plasma-purified defective FXv protein revealed a lower Ca^{2+} affinity compared with normal FX. It is therefore speculated, that the loss of the γ-carboxylated Gla^{+14} and its substitution by Lys interfere with the Ca^{2+} binding to the Gla domain, which impairs the proper conformational change. Under in vitro conditions only the activation of FXv by the VIIa-TF complex was severely affected; activation by the FIXa-FVIIa complex was only slightly affected, and activation by Russell's viper venom (RVV) was normal. In addition, the catalytic activity of the activated FXv was completely normal as measured by the conversion of prothrombin into thrombin and by the cleavage of a FX-specific chromogenic substrate. These data suggest that a structural change in the light chain of FXv, which is induced by the aa change from Gla^{+14} to Lys, interferes mainly with the activators in the extrinsic complex and to a much lesser extent with the FIXa-FVIIIa complex.

The corresponding Gla residue (Gla^{+15}) in FIX was changed to Asp by site-directed mutagenesis (78). In vitro expression and functional characterization of the variant FIX molecule did not reveal any apparent effects on FIX function. These data suggest that major functional differences exist between corresponding Gla residues in factors IX and X.

FX Malmö (FXM) (79) is a homozygous FX deficiency. The molecular basis of this deficiency is a G to C transversion in exon II, which leads to the substitution of Gla^{+26} (GAG) by Asp (GAC) (Glu^{66} according to Ref. 11). The functional consequences are a decreased clotting activity in the extrinsic (35%) and in the intrinsic (43%) pathway. The plasma-purified FXM is indistinguishable from normal FX when activated with FVIIa-TF or RVV. The rate of activation was only 20% of normal following activation by FIXa-FVIIIa.

FX Santo Domingo (FXSD) (37) is a homozygous FX deficiency with a severe bleeding diathesis. The proposita has a markedly prolonged PT and aPTT. Her FX activity is below 1%, and her FX antigen level is reduced to 5%. The molecular defect in FXSD is a point mutation from G to A, which leads to the substitution of Gly^{-20}(GGG) by Arg (AGG); (position 21 according to Ref. 11). The amino acid change occurs in the signal peptide region of the primary translational product of FX. Introduction of this mutation into the wild-type cDNA and in vitro expression showed normal levels of intracellular

FXSD. However, although the wild-type FX is readily secreted into the cell supernatant, no FXSD was detectable in the supernatant of the mutant construct. The amino acid substitution in FXSD is only three amino acids N terminal to the cleavage site of the signal peptidase, which posttranslationally cleaves the signal sequence and allows secretion of the pro-FX molecule. Residues at position −3 of the signal peptide are conserved as small neutral amino acids. It was therefore speculated that the introduction of the basic Arg at this position in FXSD may inhibit hydrolysis of this Ser-Leu bond by the signal peptidase, which in turn would prevent the FXSD molecule from being released from the membrane of the endoplasmic reticulum. Further studies of in vitro expression of the primary translational product of FXSD in a cell-free system (30) showed that the mutant FX protein is correctly targeted to and translocated into microsomal membranes. However, the correctly targeted and translocated FXSD cannot be cleaved by the signal peptidase, but the wild type is cleaved at the Ser^{-18}-Leu^{-17} bond. FXSD was the first example of human disease caused by a homozygous mutation in the signal peptide.

FX Friuli (FXF) (80) is a homozygous CRM^{+} FX deficiency caused by decreased clotting activity in the intrinsic and extrinsic pathways. Activation of FXF by RVV is normal. In addition, the activated FXF is severely defective in the conversion of prothrombin into thrombin and in cleavage of the chromogenic substrate S-2222. The genetic defect in FXF is a point mutation in exon VIII, which leads to a change in the amino acid sequence from Pro^{344} to Ser (81). In vitro expression of FXF revealed that the amino acid change in the catalytic domain not only causes a complete loss of catalytic activity but also impairs interaction with the coagulation factors of the extrinsic and intrinsic pathways (38).

FX Vienna (82) is a clinically severe CRM-reduced FX deficiency. The FX activity is reduced in the extrinsic and intrinsic pathways to $<1\%$. The FX Ag level is 5% of normal. The patient is homozygous for a point mutation in exon VI, resulting in a change from Gly (GGG) 201 to Glu (GAG).

FX San Antonio (FXSA) (83) is a clinically severe CRM-reduced FX deficiency. The propositus has a prolonged PT but a normal aPTT. The FX activity in the extrinsic pathway is reduced to 14%. The FX Ag is also reduced to 36%. Two mutations were detected in the FX gene of the propositus. A C to T transition occurs in exon VIII, which leads to a replacement of Arg (CGC) 366 by Cys (TGC). The base substitution creates a new restriction site for ApaLI. Using this enzyme, restriction digests of the amplified exon VIII of FXSA reveal a heterozygous state for the mutation in the propositus. A second mutation was found in exon VII. A point deletion of nt 838 of the coding sequence results in a premature stop codon. Direct sequencing of enzymatic amplified genomic DNA of exon VII demonstrated a heterozygous state for this mutation also. On the basis of the low FX Ag level of the propositus, it was concluded that the patient is a compound heterozygote.

The genetic defect of FX Marseille was published recently in a preliminary report (84). A T to C mutation in exon VIII leads to the exchange of Ser^{334} by Pro (position 374 according to Ref. 11). The homozygous patient has normal FX antigen but reduced FX activity upon activation in the extrinsic and intrinsic pathways of blood coagulation. Activation of the purified FX Marseille by RVV showed that the rate of cleavage was decreased. Once FXa was generated from FX Marseille it had normal catalytic efficiency for prothrombin. It was therefore speculated that Ser^{334} participates in the recognition of FX activators.

The active site Asp^{282} (position 322 according to Ref. 11) is replaced by Asn in a heterozygous FX deficiency, named FX Stockton (85). The affected family members have a decreased FX activity (43% of normal) but a normal FX antigen.

The genetic defect in one of the two index patients of FX deficiency, Mr. Stuart, has also been detected (86). The patient has a clinically severe CRM-FX deficiency. A G to A transversion in exon VIII results in the change in the amino acid sequence of the heavy chain from Val (GTG) 338 to Met (ATG). How this missense mutation leads to an absence of protein is not known.

b. Deletions Indirect genetic analysis was used to delineate the genetic defect in a patient with a severe CRM-reduced FX deficiency (87). Analysis of restriction fragment length polymorphisms showed that one allele of the propositus, inherited from the mother, has a deletion of exon VII and part of exon VIII. The paternal allele did not show any gross abnormalities. Because the FX Ag in the propositus plasma is <10%, it was suggested that a mutation event must also be present in the paternal allele.

C. Site-Directed Mutagenesis

Some of the naturally occurring mutants were expressed to study their functional defects, as described earlier. In addition, site-directed mutagenesis and in vitro expression were also used to study the effect of amino acid substitutions that were not previously reported to occur in hereditary FX deficiency. A catalytically inactive form of FXa was produced by the introduction of two point mutations in the part of the gene that codes for the heavy chain of FX (88). The mutations lead to the introduction of an Asn residue for the Asp^{+322} and an Ala residue for Ser^{+419}. The resulting FXa molecule is devoid of measurable proteolytic activity and completely inhibits plasma FX assembly into the prothrombinase complex.

FX variants in which a major part of the activation peptide is deleted (deletion from aa 136 to aa 182 and from aa 143 to aa 187) have completely normal activity following activation in the intrinsic or extrinsic pathway (43). The activation of these variants by RVV is defective, however. A double missense mutation within the activations peptide (Arg^{140} to Ser and Lys^{141} to Glu) prevents processing of this FX molecule to its two-chain form. The rate of activation is significantly less in the extrinsic system, the intrinsic system, and after incubation with RVV compared with normal FX.

Variants with a deletion of the COOH-terminal peptide of FX (89), generated by the introduction of a stop codon (TGA) for Gly (GGC) 430 (position 470 according to Ref. 11), have normal coagulant activity. However, this molecule is inhibited at only half the rate by antithrombin III/heparin compared with normal FX.

REFERENCES

1. Davie EW, Fujikawa K, Kurachi K, Kisiel W. The role of serine proteases in the coagulation cascade. *Adv Enzymol* 1979; 48:277.
2. Hougie C, Barrow EM, Graham JB. Stuart clotting defect. I. Segregation of an hereditary hemorrhagic state from the heterogeneous group heretofore called ''stable factor'' (SPCA, proconvertin, factor VII) deficiency. *J Clin Invest* 1957; 36:485.
3. Telfer TP, Denson KW, Wright DW. A ''new'' coagulation defect. Brit J Hematol 1956; 2: 308.

4. DeGrouchy J, Dautzemberg MD, Turleau C, Beguin S, Chavin-Colin F. Regional mapping of clotting factors VII and X to 13q34. Expression of factor VII through chromosome 8. *Hum Genet* 1984; 66:230.
5. Scambler PJ, Williamson R. The structural gene for human coagulation factor X is located on chromosome 13q34. *Cytogenet Cell Genet* 1985; 39:231.
6. Royle NJ, Fung MR, MacGillivray RTA, Hamerton JL. The gene for clotting factor X is located on chromosome 13q32qter. *Cytogenet Cell Genet* 1986; 41:185.
7. Fukushima Y, Kuroki Y, Iizuka A. Activity and antigen of coagulation factors VII and X in five patients with abnormal chromosome 13. *Jpn J Hum Genet* 1987; 32:91.
8. Pfeiffer RA, Ott R, Gilgenkrantz S, Alexander P. Deficiency of coagulation factors VII and X associated with deletion of a chromosome 13(q34). Evidence from two cases with 46 XY, t(23_Y(q11;q34). *Hum Genet* 1982; 62:358.
9. Ott R, Pfeiffer RA. Evidence that activities of coagulation factors VII and X are linked to chromosome 13(q34). *Hum Hered* 1984; 34:123.
10. Pfeiffer RA, Ott R, Taben KD. Clotting factors VII and X as useful markers of terminal deletion of chromosome 13. *Hum Genet* 1985; 69:192.
11. Leytus SP, Foster DC, Kurachi K, Davie EW. Gene for human factor X: a blood coagulation factor whose gene organization is essentially identical with that of factor IX and protein C. *Biochemstry* 1986; 25:5098.
12. Jagadeeswaran P, Reddy SV, Rao KJ, Hamsabhushanam K, Lyman G. Cloning and characterization of the 5′ end (exon 1) of the gene encoding human factor X. *Gene* 1989; 84:517.
13. Fung MR. *Molecular genetics of blood coagulation factor X*. Dissertation, University of British Columbia, Vancouver, Canada, 1988.
14. Miao CH, Leytus SP, Chung DW, Davie EW. Liver-specific expression of the gene coding for human factor X, a blood coagulation factor. *J Biol Chem* 1992; 267:7395.
15. Furie B, Furie BC. The molecular basis of blood coagulation. *Cell* 1988; 53:505.
16. Neurath H. Evolution of proteolytic enzymes. *Science* 1984; 224:350.
17. Doolittle RF, Feng DF, Johnson MS. Computer based characterization of epidermal growth factor precursors. *Nature* 1984; 307:558.
18. Hassan HJ, Guerriero R, Chelucci C, et al. Multiple polymorphic sites in factor X locus. *Blood* 1988; 71:1353.
19. Jaye R, Ricca G, Kaplan R, et al. Polymorphisms associated with the human factor X (F10) gene. *Nucleic Acids Res* 1985; 22:8268.
20. Hay CW, Robertson KA, Fung MR, MacGillivray RTA. RFLPs for PstI and EcoRI in the human blood clotting factor X gene. *Nucleic Acids Res* 1986; 14:5118.
21. Wallmark A, Rose VL, Ho C, High KA. A NlaIV polymorphism within the human factor X gene. *Nucleic Acids Res* 1993; 19:4022.
22. Watzke HH, Lechner K, Roberts HR, et al. Molecular defect (Gla^{+}14-Lys) and its functional consequences in a hereditary factor X deficiency (factor X ''Vorarlberg''). *J Biol Chem* 1990; 265:11981–11989.
23. Wallmark A. *Personal communication*, 1994.
24. Huang MN, Hung HL, Stanfield-Oakley SA, High KA. Characterization of the human coagulation factor X promoter. *J Biol Chem* 1992; 267:15440.
25. Hung HL, High KA. Hepatocyte nuclear factor 4 binds to the promoters of factors VII, IX and X. *Thromb Haemost* 1993; 69:2476.
26. Gianelli F, Green PM, High KA. Haemophilia B: database of point mutations and short deletions, second edition. *Nucleic Acids Res* 1991; 19:2193.
27. Leytus SP, Chung DW, Kisiel W, Kurachi K, Davie EW. Characterization of cDNA coding for human factor X. *Proc Natl Acad Sci* 1984; 82:3699.
28. Fung MR, Hay CW, MacGillivray RTA. Characterization of an almost full-length cDNA coding for human blood coagulation factor X. *Proc Natl Acad Sci USA* 1985; 82:3591.

29. Kaul RK, Hildebrand B, Roberts S, Jagadeeswaran P. Isolation and characterization of human blood-coagulation factor X cDNA. *Gene* 1986; 41:311.
30. Racchi M, Watzke HH, High KA, Lively MO. Human coagulation factor X deficiency caused by a mutant signal peptide that blocks cleavage by the signal peptidase but not targeting and translocation to the endoplasmic reticulum. *J Biol Chem* 1993; 268:5735.
31. Walter P, Gilmore R, Blobel G. *Protein translocation across the endoplasmic reticulum.* **Cell** *1984; 38:5.*
32. Von Heijne G. Signal sequences: the limits of variation. *J Mol Biol* 1985; 184:99.
33. Davie EW. The blood coagulation factors: their cDNAs, genes and expression. In: *Hemostasis and thrombosis. Basic principles and clinical practice.* 2nd ed. Philadelphia: JB Lippincott, 1987.
34. Jackson CM. Factor X. In: *Progress in hemostasis and thrombosis.* New York: Grune and Stratton, 1984.
35. McMullen BA, Fujikawa K, Kisiel W. Complete amino acid sequence of the light chain of human blood coagulation factor X: evidence for identification of residue 63 as β-hydroxy aspartic acid. *Biochemistry* 1983; 22:2875.
36. Titani K, Fujikawa K, Enfield D. Bovine FX1 (Stuart factor): amino acid sequence of the heavy chain. *Proc Natl Acad Sci USA* 1975; 72:3082.
37. Watzke HH, Wallmark A, Hamagichi N, Giardina P, Stafford DW, High KA. Factor X ''Santo Domingo'': evidence that the severe clinical phenotype arises from a mutation blocking secretion. *J Clin Invest* 1991; 88:1685.
38. James HL, Kumar A. Recapitulation of factor X Friuli by site directed mutagenesis of the recombinant protein. *Blood* 1992; 80(Suppl. 1):1216A.
39. Wolff DL, Sinha U, Hancock TE, et al. Design and constructs for the expression of biologically active recombinant human factors X and Xa. *J Biol Chem* 1991; 266:13726.
40. DiScipio K, Hermodson, Yates S. A comparison of human prothrombin, factor IX (Christmas factor), factor X (Stuart factor) and protein S. *Biochemistry* 1977; 16:698.
41. Miletich J, Jackson C, Majerus P. Interaction of coagulation factor Xa with human platelets. *Proc Natl Acad Sci USA* 1977; 74:4033.
42. Graves C, Munns T, Willingham A. Rat factor X is synthesized as a single chain precursor inducible by prothrombin fragments. *J Biol Chem* 1982; 257:13108.
43. Mullane MP, Eby CS, Porche-Orbet RM, Miletich JP. Deletional and processing mutants of the activation peptide of human FX. *Blood* 1992; 80(Suppl. 1):1215A.
44. Huber R, Bode W. Structural basis of the activation and action of trypsin. *Accounts Chem Res* 1978; 11:114.
45. Stroud RM, Krieger M, Koeppe RE. Structure function relationships in the serine proteases. In: Reich E, Rifkin DB, Shaw E, eds.: *Proteases and biological control, Cold Spring Harbor Conference on Cell Proliferation, Vol. 2.* Cold Spring Harbor, NY: Cold Spring Harbor Laboratories, 1975:13.
46. Davie EW, Fujikawa K, Kisiel W. The coagulation cascade: initiation, maintenance and regulation. *Biochemistry* 1991; 30:10363.
47. Miletich JP, Jackson CM, Majerus PW. Interaction of coagulation factor Xa with human platelets. *Proc Natl Acad Sci USA* 1977; 74:4033.
48. McGee MP, Li LC. Functional differences between intrinsic and extrinsic coagulation pathways. Kinetics of factor X activation on human monocytes and alveolar macrophages. *J Biol Chem* 1991; 266:8079.
49. Hultin MA. Role of human factor VIII in factor X activation. *J Clin Invest* 1982; 69:950–958.
50. Walsh PN. Different requirements for intrinsic factor-XA forming activity and platelet factor 3 activity and their relationship to platelet aggregation and secretion. *Br J Haematol* 1978; 40:311–331.

51. Rosing J, van Rijn JLML, Bevers EM, van Dieijen G, Comfurius P, Zwaal RFA. The role of activated human platelets in prothrombin and factor X activation. *Blood* 1985; 65:319–332.
52. Nemerson Y. Tissue factor and hemostasis. *Blood* 1988; 71:1–8.
53. Mann KG, Nesheim ME, Church WR, Haley P, Kroshnaswamy S. Surface-dependent reactions of the vitamin K-dependent enzyme complexes. *Blood* 1990; 76:1–16.
54. Bevilacqua MP, Pober JS, Majeal GR, Cotran RS, Gimbrone MA Jr. Interleukin 1 (IL-1) induces biosynthesis and cell surface expression of procoagulant activity in human vascular endothelial cells. *J Exp Med* 1984; 160:618–623.
55. Lindhout T, Blezer R, Schoen P, Nordfang P, Reutelingsperger C, Hemker HC. Activation of factor X and its regulation by tissue factor pathway inhibitor in small-diameter capillaries lined with human endothelial cells. *Blood* 1992; 79:2909–2916.
56. Bevilacqua MP, Pober JS, Majeau GR, Fiers W, Cotran RS, Gimbrone MA Jr. Recombinant tumor necrosis factor induces procoagulant activity in cultured human vascular endothelium: characterization and comparison with the actions of interleukin 1. *Proc Natl Acad Sci USA* 1986; 83:4533–4537.
57. Fuchs HE, Pizzo SV. Regulation of factor X_a in vitro in human and mouse plasma and in vivo in mouse. *J Clin Invest* 1983; 72:2041–2049.
58. Skogen WF, Esmon CT, Cox C. Comparison of coagulation factor Xa and des-(1–44) factor Xa in the assembly of prothrombinase. *J Biol Chem* 1984; 259:2306.
59. Morita T, Jackson CM. Preparation and properties of derivatives of bovine factor X and factor Xa from which the γ-carboxyglutamic acid containing domain has been removed. *J Biol Chem* 1986; 261:4015.
60. Jesty J, Spencer AK, Nakashima J. The activation of coagulation factor X: identity of cleavage site in the alternative activation pathways and characterization of the COOH-terminal peptide. *J Biol Chem* 1975; 250:4497.
61. Jesty J, Spencer AK, Nemerson Y. The mechanism of action of factor X: kinetic control of alternative pathways leading to the formation of activated factor X. *J Biol Chem* 1974; 249: 5614.
62. Fujikawa K, Coan HM, Legaz ME, Davie EW. The mechanism of activation of bovine factor X (Stuart factor) by intrinsic and extrinsic pathways. *Biochemistry* 1974; 13:5290.
63. Esmon CT, Owen WG, Jackson CM. The conversion of prothrombin to thrombin: II. Differentiation between thrombin- and factor Xa-catalyzed proteolyses. *J Biol Chem* 1974; 249: 606–611.
64. Mann KG. Prothrombin. *Methods Enzymol* 1976; 45:123.
65. Gasic GP, Arenas CP, Gasic TB, Gasic GJ. Coagulation factors X, Xa and protein S as potent mitogens of cultured aortic smooth muscle cells. *Proc Natl Acad Sci USA* 1992; 89:2317.
66. Jordan RE, Oosta GM, Gardner WT, Rosenberg RD. The kinetics of hemostatic enzyme-antithrombin interactions in the presence of low molecular weight heparin. *J Biol Chem* 1980; 255:10081.
67. Bjork I, Jackson CM, Jornvall H, Lavine KK, Nordling K, Salsgiver WJ. The active site of antithrombin: release of the same proteolytically cleaved form of the inhibitor from complexes with factor IXa, factor Xa, and thrombin. *J Biol Chem* 1982; 257:2406–2411.
68. Owen WG. Evidence for the formation of an ester between thrombin and heparin cofactor. *Biochim Biophys Acta* 1975; 405:380.
69. Stern D, Nawroth P, Marcum J, et al. Interaction of antithrombin III with bovine aortic segments. *J Clin Invest* 1985; 75:272–279.
70. Savion N, Farzame N. Binding uptake and degradation of antithrombin III protease complexes by cultured corneal endothelial cells. *Exp Cell Res* 1984; 153:50–60.
71. Gitel SN, Medina VM, Wessler S. Inhibition of human activated factor X by antithrombin III and α_1-proteinase inhibitor in human plasma. *J Biol Chem* 1984; 259:6890–6895.

72. Broze GJ Jr, Girard TJ, Novotny WF. Regulation of coagulation by a multivalent Kunitz-type inhibitor. *Biochemistry* 1990; 29:7539–7546.
73. Graham JB, Barrow EM, Hougie C. Stuart clotting defect. II. Genetic aspects of a ''new'' hemorrhagic state. *J Clin Invest* 1957; 36:497.
74. Fair DS, Plow EF, Edgingtown TS. Combined functional and immunochemical analysis of normal and abnormal human factor X. *J Clin Invest* 1979; 64:884.
75. Fair DS, Edgington TS. Heterogeneity of hereditary and acquired factor X deficiencies by combined immunochemical and functional analysis. *Br J Haematol* 1985; 59:235.
76. Girolami A. Tentative and update classification of factor X variants. *Acta Haematol* (Basel) 1986; 75:58.
77. Denson KWE, Lurie A, de Cataldo F, Manucci PM. The factor X defect: recognition of abnormal forms of factor X. *Br J Haematol* 1970; 18:317.
78. Liles DK, Lefkowitz J, Monroe DM, Roberts HR. Corresponding γ-carboxyglutamyl (Gla) residues of factor IX and factor X have different functional consequences. *Blood* 1992; 80(Suppl. 1):163a.
79. Wallmark A, Larson P, Ljung R, Monroe D, High KA. Molecular defect (Gla 26-Asp) and its functional consequences in a hereditary FX deficiency (factor X ''Malmo 4''). *Blood* 1991; 78(Suppl. 1):229a.
80. Girolami A, Carli A, Falomo R, DeMarco L. Factor X Friuli coagulation disorder. *Blut* 1973; 17:151.
81. James H, Girolami A, Fair DS. Molecular defect in coagulation factor X ''Friuli'' results from substitution of serine for proline at position 343. *Blood* 1991; 77:317.
82. Watzke HH, Lechner K, Larson P, High KA. Factor X ''Vienna'': molecular analysis and in vitro expression of a severe CRM-FX deficiency. *Thromb Haemost* 1993; 69:752.
83. Sakamuri VR, Zhou ZQ, Rao KJ, Scott PJ, et al. Molecular characterization of human factor X ''San Antonio.'' *Blood* 1989; 74:1486.
84. Bezeaud A, Miyata T, Zheng YZ, et al. Functional consequences of Ser^{334} to Pro mutation in a factor X variant (factor X Marseille). *Thromb Haemost* 1993; 69:617(265a).
85. Meissier TL, Wong C, Bovill EG, Long G, Church WR. Factor X ''Stockton'': a missense mutation (GAC to AAC) of the serine protease active site aspartic acid to asparagine. *Thromb Haemost* 1993; 69:751(761a).
86. Roberts HR, Lozier JN. *Other clotting factor deficiencies. In: Hematology. Basic principles and practice*, New York: Churchill Livingstone, 1991.
87. Bernardi F, Marchetti G, Ptracchini P, et al. Partial gene deletion in a family with factor X deficiency. *Blood* 1989; 73:2123.
88. Sinha U, Hancock TE, Lin PH, Hollembach S, Wolf DL. Expression, purification, and characterization of inactive human coagulation factor Xa (Asn^{322} Ala^{419}). *Prot Exp Purif* 1992; 3: 518.
89. Eby CS, Mullane MP, Porche-Sorbet RM, Miletich JP. Characterization of the structure and function of the carboxyterminal peptide of human factor X. *Blood* 1992; 80(Suppl. 1):1214a.

12

Factor XI

Kazuo Fujikawa and Dominic W. Chung
University of Washington, Seattle, Washington

I. INTRODUCTION

Factor XI was described by Rosenthal et al. in 1953 as a third type of hemophilia (distinct from hemophilia A and B): the first report described three patients in one related family who had moderate hemorrhage, mostly bleeding after tooth extraction (1). To date, several hundred cases of factor XI deficiency have been reported (2). These deficiencies are often found in certain populations, such as Ashkenazi Jews (3) and an autochthonous Swiss population (4). Marriage between close family members seems to be the reason for the high frequency of factor XI deficiency in these populations. The deficiency is inherited as an autosomal recessive trait. Patients with the homozygous defect have a low circulating factor XI activity and risk surgical bleeding, whereas heterozygotes have no significant risk of bleeding (2,3).

Decades after the first reported case of the deficiency, the protein, factor XI (also known as plasma thromboplastin antecedent), was finally purified from bovine plasma (5) and subsequently from a human source (6,7). A low content of factor XI in plasma (~5 μg/ml) and its instability during purification hampered the purification. The use of a pseudoaffinity column and addition of various protease inhibitors during purification solved most of the problems. The purification opened the door to the study of factor XI function and its interactions with other proteins at a molecular level and eventually led to the isolation of the gene. Gene analysis of affected Ashkenazi Jews was particularly interesting because of an unusually high frequency (11% or higher) in this population (3).

The liver is only the organ that is known to synthesize factor XI. At least two cases have been reported in whom the transfer of factor XI deficiency is associated with liver transplantation from deficient donors to recipients (8,9). These incidents show clearly that the liver is the sole organ of factor XI synthesis. Factor XI is synthesized as a single polypeptide chain and secreted into blood as a homomeric dimer held by a disulfide bond. The circulating homodimer is the zymogen of a serine protease. Factor XI migrates closely with IgG with an apparent molecular weight of 120,000–140,000 in nonreduced sodium dodecyl sulfate–polyacrylamide gel electrophoresis. Upon reduction, the subunits have an apparent molecular weight of 80,000, including ~5% carbohydrate. Factor XI circulates in blood as a noncovalent complex with a nonenzymatic cofactor, high M_r kininogen, (10), and participates in the early stage of the intrinsic pathway of blood coagulation (11). Factor XI is activated in plasma by contact activation; four of the contact factors, three zymogens of serine proteases (factor XII, factor XI, and plasma prekallikrein) and one non-enzymatic cofactor (high M_r kininogen), gather on a negatively charged surface, such as dextran sulfate, sulfatide, heparin, or kaolin, and the zymogens are reciprocally activated (12,13). In a purified system, factor XI is activated by activated factor XII (factor XIIa) to an active form, factor XIa, in the presence of a negatively charged surface, and high M_r kininogen stimulates the activation of factor XI by factor XIIa (6,7,12,13). However, the activation of factor XI by factor XIIa in the presence of high M_r kininogen is probably not a physiological event in normal hemostasis, because neither factor XII nor high M_r kininogen deficiencies are associated with bleeding problems. Contact activation is thought to take place in certain pathological conditions, such as sepsis, typhoid fever (14), myocardial infarction (15), and when blood comes in contact with foreign materials, as in kidney dialysis membranes or artificial heart and blood vessels.

Thrombin activation of factor XI is a possible alternative pathway of factor XI activation in normal hemostasis (16,17). Factor XI is activated by thrombin in the presence of dextran sulfate, sulfatide, or heparin. Among these negatively charged surfaces, dextran sulfate also stimulates the autoactivation of factor XI. The factor XI activation by thrombin is proposed to play a role in maintaining clot formation rather than in initiating coagulation. Regardless of how factor XIa is formed, it activates factor IX in the presence of Ca^{2+} and leads to eventual clot formation (11).

In this chapter, we review recent progress on the structure and function relationship of factor XI, the interactions of factor XI with other proteins, and the analysis of mutations in the factor XI gene. Recent review articles on factor XI and its deficiency are available (18,19).

II. PROTEIN STRUCTURE OF FACTOR XI

The primary structure of factor XI was deduced by sequencing cDNAs isolated from a human liver library. A full-length cDNA of 2097 base pairs encodes the hydrophobic signal peptide (18 amino acid residues) and the mature monomeric protein. The monomer of mature factor XI is composed of 608 amino acid residues, including 35 half-cystines, and five potential N-linked glycosylation sites (20). All half-cystine residues are involved in disulfide bonds, and the complete disulfide bond structure has been determined by sequencing isolated disulfide bond-containing peptides (Fig. 1) (21). The disulfide bond structure and the amino acid sequence of factor XI are highly homologous (58% identical) to plasma prekallikein (22), which also circulates as a noncovalent complex with high M_r kininogen. The activation site of factor XI by factor XIIa, thrombin, and factor XIa itself is identical and is located between Arg^{369} and Ile^{370}. The cleavage of this bond converts factor XI to factor XIa, which is composed of two amino-terminal heavy chains and two carboxyl-terminal light chains held together by disulfide bonds. The disulfide bond between the heavy and light chains of factor XIa can be preferentially reduced under mild conditions, and functionally active heavy and light chains can be separated.

The isolated heavy chain binds to high M_r kininogen, whereas the light chain retains full amidolytic activity but loses the procoagulant activity (23).

The heavy chain is composed of four tandem repeats, each containing 90 or 91 amino acid residues with six half-cystine residues in conserved positions. These repeats are designated "apple domains," and each domain contains three characteristic disulfide bonds formed by the six conserved half-cystines. Similarly, four apple domains are also present in the amino-terminal portion of plasma prekallikrein. In addition to the six conserved half-cystines, the first and fourth apple domains in factor XI each contains an extra half-cystine residue. The extra half-cystine residue in the first domain (Cys[11]) links with a free cysteine, and the extra half-cystine in the fourth domain (Cys[321]) forms a disulfide bond with the corresponding Cys[321] in the second subunit. The fourth apple domain is followed by a short connecting region that is composed of eight amino acid

Figure 1 Sequence and disulfide bond structure of human factor XI monomer. The four apple domains are shown as A1, A2, A3, and A4. The Asp, His, and Ser residues of the catalytic triad are circled. Asn residues marked with closed diamonds are glycosylated. The Arg-Ile bond marked with a curved arrow is the cleavage site during the activation by factor XIIa, thrombin, or factor XIa. The Cys residue marked by an asterisk forms a disulfide bond between the monomers.

residues from Cys^{362} to Arg^{369}. Cys^{362} in the connecting region links with Cys^{482} in the light chain to form an interchain disulfide bond between the heavy and light chains.

The amino acid sequence and the disulfide bond structure of the light chain (catalytic domain) are homologous to digestive enzymes, such as trypsin and chymotrypsin, as well as the catalytic domain of other coagulation or fibrinolytic enzymes. The catalytic triad is located at His^{413}, Asp^{462}, and Ser^{557}, and Asp^{551} is situated in the bottom of the binding pocket, indicating that factor XIa has a typical trypsin type of substrate specificity (20,24). The enzymatic activity of factor XIa is completely inactivated by diisopropylphosphofluoridate, an inhibitor specific for serine proteases. Antithrombin III (7) or a platelet-derived inhibitor (see later) (25) at 2 mol is required to inhibit 1 mol factor XIa completely, indicating that both the light chains of factor XIa have independent catalytic activity. Factor XIa exhibits a specific activity in converting the zymogen factor IX into its active form factor IXa. During this activation, factor XIa cleaves two internal peptide bonds between Arg^{145} and Ala^{146} and between Arg^{180} and Val^{181} in factor IX in the presence of Ca^{2+}, liberating a carbohydrate-containing peptide, called the activation peptide (Ala^{146} to Arg^{180}), from the middle of the molecule (26,27). Calcium is probably not required for the catalytic activity of factor XIa but is required for the formation of a proper conformation in factor IX susceptible to the enzyme. Factor XIa also cleaves the synthetic peptide substrates Boc-Glu(OBz)-Ala-Arg-methylcoumaryl-7-amide (28) and Glu-Pro-Arg-*p*-nitroaniline (29), which are conveniently used for kinetic analysis of factor XIa.

III. FUNCTIONS OF THE HEAVY CHAIN OF FACTOR XI

As mentioned, factor XI is complexed with high M_r kininogen. This interaction involves the heavy chain of factor XI and the carboxyl-terminal portion (light chain) of high M_r kininogen. A synthetic peptide, 31 residues in length, corresponding to residues 194–224 of the carboxyl-terminal light chain of high M_r kininogen, has full binding affinity to plasma prekallikrein (K_d = 18 nM) and retains fairly strong affinity for factor XI (K_d = 1.8 μM) (30). This binding is important for the expression of factor XI procoagulant activity when factor XI is activated by contact activation. The binding site for high M_r kininogen in factor XI was localized to the apple 1 domain with chemically produced fragments of factor XI (31,32). The structural requirement for the binding was further studied by testing the binding affinities of a number of synthetic peptides. A peptide corresponding to Phe^{56} through Ser^{86} in the apple 1 domain was found to be a core unit that blocks the binding of factor XI to high M_r kininogen. This peptide also inhibits the clotting of human plasma when clotting is initiated by kaolin. The affinity of this peptide for high M_r kininogen, however, is two to three orders of magnitude lower than that of native factor XI. As a control, peptides from corresponding regions of the other three apple domains have no effect on the binding of factor XI with high M_r kininogen. Therefore, it is proposed that the apple 1 domain in factor XI is the primary site for binding high M_r kininogen.

For the activation of factor IX, both the light chain (catalytic domain) and the heavy chain of factor XIa interact with the substrate, factor IX (33). A monoclonal antibody specific for the heavy chain of factor XI decreases the catalytic activity of factor XIa for the activation of factor IX. A synthetic peptide corresponding to a region (Asn^{145}-Ala^{176}) in the apple 2 domain competitively inhibits the activation of factor IX with a K_d of 30 nM. On the other hand, peptides from the corresponding regions of apple 1, 3, and 4 domains have much less effect on the activation. Accordingly, the binding site of factor XI with factor IX is thought to be located in the apple 2 domain (34). A proposed tertiary structure of the apple 2 domain constructed by computer modeling shows that the binding segment forms two antiparallel β structures connected by a β turn, which is often found in protein-protein binding sites (34).

A striking feature of factor XI is that, unlike other serine proteases of the coagulation cascade, it is secreted as a homodimer. Structural features important for subunit interactions have been studied in a mammalian cell culture system transfected with normal and mutant factor XI cDNAs (35). Substitution of Cys^{321} by Ser in the apple 4 domain results in a factor XI subunit that is unable to form an interchain disulfide bond. The two subunits, however, remain bound as a noncovalently linked dimer that is functionally indistinguishable from normal plasma factor XI. Therefore, it is not possible to determine whether the monomeric factor XI subunit possesses the same activity as the dimeric form. These results suggest that recognition between the two mutant subunits remains intact and the interchain disulfide bond is normally formed after dimerization that positions Cys^{321} in both subunits in the proper orientation for disulfide bond formation. To study further the contribution of each apple domain to dimerization, each of the four apple domains of factor XI is fused to tissue plasminogen activator (tPA). In these constructs, the various apple domains of factor XI replace the finger and growth factor region of native tPA. Fusion of the apple 4 domain of factor XI to tPA generates a secreted dimerized chimeric protein in which the two subunits are linked by a disulfide bond. Substitution of Cys^{321} by Ser in the apple 4 domain in this fusion protein produced a chimeric protein that lacks a covalent interchain disulfide bond. This chimeric protein, similar to the case with factor XI, is secreted as a noncovalently linked dimer. These results demonstrate that the structural information necessary for dimer formation is contained exclusively in the apple 4 domain of factor XI.

IV. STRUCTURE AND MUTATIONS OF THE FACTOR XI GENE

The human factor XI gene is 23 kb in length and consists of 15 exons (exons 1–15, introns A–N) (36). To date, the transcription initiation site has not been determined. A 5′-untranslated sequence is encoded in exon 1, and the signal peptide is encoded in exon 2. Each of the four apple domains is encoded in two exons. The conservation in the location and splice junction types is consistent with multiple exon duplications during the evolution of the heavy chain. Thus, introns B, D, F, H, and J that separate the apple domains from the signal peptide and from each other have conserved type I splice junctions. Within each apple domain, the introns C, E, G, and I occur in homologous positions of the coding sequence and have conserved type II splice junctions. The light chain is encoded in five exons with an organization similar to the genes for plasma prekallikrein, tPA, urokinase, and factor XII. In contrast, the catalytic chains of vitamin K-dependent coagulation factors are composed of three exons. This difference suggests that the genes for proteins participating in the early phase of intrinsic coagulation and fibrinolysis evolve from a common ancestor that is different from the vitamin K-dependent coagulation factors. The gene for human factor XI has been localized to chromosome 4q32–35 (37). Two polymorphic sites in the factor XI gene have been identified (38,39). An HhaI restriction fragment length polymorphic site was identified in intron E, which gives rise in whites to polymorphic fragments of 0.15 kb (A1 allele) or 0.14 kb (A2 allele) in length with frequencies of 0.4 and 0.6 (39). A dinucleotide repeat polymorphism involving variable numbers of (CA) repeats has also been observed in intron B of the factor XI gene (38). The alleles, designated D1 and D2, occur with frequencies of 0.47 and 0.53, respectively, in whites.

Using a combination of gene cloning, Southern blotting, the polymerase chain reaction, and DNA sequencing, three independent point mutations have been identified in the factor XI genes of six unrelated Ashkenazi patients with severe factor XI deficiency (40). The type I mutation occurs at the splice junction boundary of the last intron (intron N). This mutation changes the G in the obligatory 5′ splice donor dinucleotide GT to an A and either disrupts mRNA splicing or leads to premature polypeptide termination (Fig. 2). A second mutation, designated the type II mutation, was identified in exon 5 of the factor XI gene, in which the GAA codon for Glu^{117} was changed to the stop codon TAA. Premature translation termination would result in a truncated and presumably unstable factor XI molecule. A third independent missense mutation, the type III mutation, was also identified in exon 9, in which the codon TTC coding for Phe^{283} was changed to CTC coding for Leu.

Type I mutation

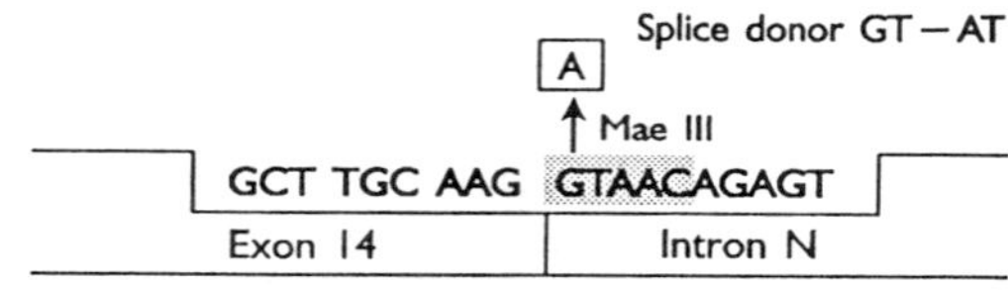

Type II mutation

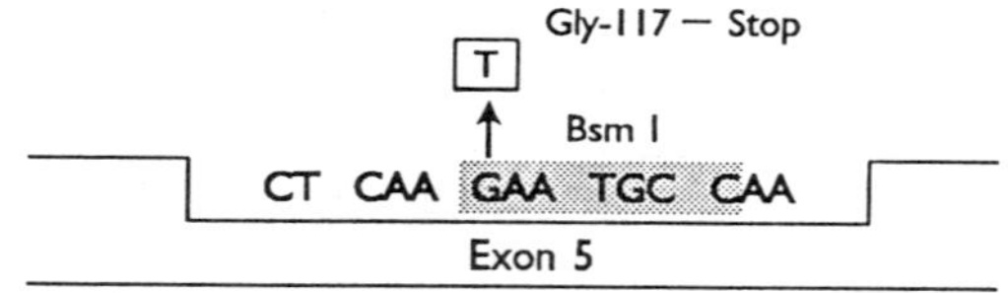

Type III mutation

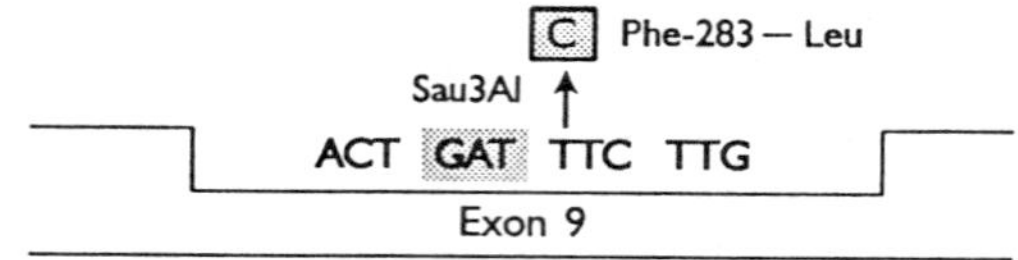

Figure 2 Mutations in the factor XI gene. Single base changes at mutation sites are shown in boxes. Shaded sequences represent restriction endonuclease recognition sequences used in the identification of each type of mutation.

To understand further the mechanism by which the type III mutation causes factor XI deficiency, this mutation was introduced into a factor XI cDNA by site-directed mutagenesis and the biosynthesis of the mutant polypeptide studied in a mammalian cell culture system (41). Cells transfected with an expression vector containing the type III mutation secreted about 8% of the level of factor XI, in contrast to cells expressing normal factor XI, a level similar to that observed in patients with homozygous type III deficiency. The secreted type III mutant factor XI is dimeric and is functionally normal, indicating that the amino acid substitution affected intracellular processing and secretion rather than function. Pulse-chase labeling experiments showed that the type III mutant protein has a prolonged intracellular transit time and is present intracellularly predominantly as a monomer; an appreciable amount of normal factor XI is usually converted to the dimeric form intracellularly. These results are consistent with the previous suggestion that the apple 4 domain is essential for dimerization and further suggest that residue Phe^{283} in apple 4 is involved in this process.

The three types of mutations in the factor XI gene occur in sequences that either abolish or create restriction endonuclease recognition sites. Accordingly, type-specific restriction digestion following the polymerase chain reaction was developed for mutant genotype identification (40,42). Thus, the abolition of a MaeIII and a BsmI site and the creation of a Sau3AI (or MboI) site were used to identify the three types of mutations. In one study of 43 unrelated Ashkenazi Jewish patients from Israel with severe factor XI deficiency, the type II and type III mutations account for 49 and 47% of the analyzed defective alleles. The proportion of the type II/II homozygotes to type II/III compound heterozygotes and to type III/III homozygotes was 10:20:10. In addition, two unrelated probands of Iraqi Jewish origin were homozygous for the type II mutation (42). This finding suggests that factor XI deficiency caused by the type II mutation is not confined to the Ashkenazi community. In a second study of 50 severe factor XI patients of Ashkenazi Jewish descent at present residing in the United Kingdom, the frequencies for the type II and III mutations were found to be 52 and 36%, respectively (43). The significant difference in the type III allele frequency may be attributed to population differences. In the second study, both the type II and type III mutations have been identified in individuals of non-Jewish origin, indicating that, although at much lower frequency, similar mutations have occurred in other ethnic groups. The high frequency of the type II and type III mutations in the Ashkenazi community is attributed to extreme genetic drifts caused by drastic changes in population size, migration, and the founder effect.

In a comparison of the circulating factor XI levels with the genotype, individuals with type II/II genotypes have 1.2 ± 0.5% factor XI levels, but type III/III homozygotes have 9.7 ± 3.8% and type II/III compound heterozygotes have an intermediate level of 3.3 ± 1.6%, respectively. These data are consistent with the interpretation that the type II mutation, which causes premature chain termination, produces practically no functional factor XI. The 1% level measured in type II/II homozygotes is at the threshold of detection of the factor XI-dependent clotting assay. The type III mutation, which results in an amino acid substitution in the factor XI molecule, produces approximately 5% of circulating factor XI; the mean activity of 10% observed in type III/III homozygotes was attributed to the presence of two copies of the type III allele, each contributing about 5% of the circulating factor XI. This interpretation is consistent with in vitro expression studies that show cells synthesizing factor XI with the type III mutation secrete about 8% of the level, in contrast to cells expressing normal factor XI (41).

Although genotype difference shows a definite correlation with circulating levels of factor XI, the correlation of bleeding tendency with genotype is not clear. A higher incidence of injury-related bleeding was observed in type II/II homozygotes (4 of 8) compared with type II/III compound heterozygotes (3 of 20) and type III/III homozygotes (1 of 10) (42). In a separate study, however, the type II/III compound heterozygotes have a greater risk of bleeding compared with either homozygous group (43). Resolution of these differences requires additional studies on larger patient populations. The possibility of other contributing or precipitating factors, including site of injury and use of aspirin, should also be examined (19).

V. INTERACTIONS OF FACTOR XI WITH CELLS

Thrombin-activated platelets bind factor XI and factor XIa with K_d = 822 pM at distinct sites in the presence of high M_r kininogen. Bound factor XIa can activate factor IX with the same efficiency as unbound factor XIa (44). Recently, quiescent human umbilical vein endothelial cells were also found to bind factor XI or factor XIa at the same site in the presence of high M_r kininogen (45). The binding is Zn^{2+} dependent, reversible, and saturable, and the K_d is 1.5 nM for factor XIa and 4.5 nM for factor XI, respectively. Bound factor XI is activated by factor XIIa, and the resulting factor XIa can activate factor IX. These results show that the activation of factor XI and subsequent activation of factor IX by the resulting factor XIa can proceed on the surface of certain cell types. The binding of factor XIa to cell surfaces does not enhance the catalytic efficiency of factor XIa, unlike the binding of factor Xa with factor Va, factor IXa with factor VIIIa, or factor VIIa with tissue factor in the presence of phospholipid vesicles and Ca^{2+}, which greatly enhances the catalytic efficiency of the proteases. However, the binding of factor XI or factor XIa to cell surfaces may serve to localize clot formation and may also prevent or protect factor XIa from inactivation by circulating plasma protease inhibitors or an inhibitor released from activated platelets.

VI. INHIBITORS OF FACTOR XIa

A number of protease inhibitors (antithrombin III, α_1-trypsin inhibitor, C_1-esterase inhibitor, α_2-macroglobulin, and others) circulate in blood at relatively high concentrations and inactivate unbound forms of activated coagulation factors to prevent disseminated clot formation. These inhibitors have a broad inhibition spectrum and inhibit factor XIa activity when factor XIa is added to plasma (46). High M_r kininogen competitively protects factor XIa from inactivation by these circulating inhibitors. α_1-Trypsin inhibitor has the highest affinity to factor XIa, followed by antithrombin III. C_1-esterase inhibitor and α_2-plasmin inhibitor have a relatively weak inhibitory activity to factor XIa. From kinetic analysis with purified proteins, inhibition by α_1-trypsin inhibitor contributes 68% of the total factor XIa inhibition in plasma, antithrombin III, 16%, and α_2-plasmin inhibitor and C_1-esterase inhibitor, 8% each. The inhibition of factor XIa by antithrombin III is enhanced by heparin, and the degree of enhancement is dependent on the number of the binding sites for antithrombin III per heparin molecule (47). Recently, activated protein C inhibitor was also found to inhibit factor XIa (48). The sensitivity of factor XIa to these inhibitors is different from that of plasma kallikrein, which is most sensitive to C_1-esterase inhibitor (49).

Stimulated platelets secrete a potent but slow-reacting inhibitor of factor XIa. This inhibitor has a molecular weight of ~112,000 and was found to be a truncated form of Alzheimer amyloid β-protein precursor. It contains a Kunitz-type structure and inhibits factor XIa and trypsin with a K_d of 450 ± 50 and 20 ± 10 pM, respectively. Interestingly, the factor XIa inhibition is enhanced severalfold by a low concentration of heparin, and its binding affinity to factor XIa is several orders of magnitude higher than the plasma inhibitors mentioned earlier. This inhibitor does not inhibit other coagulation factors, such as factor VIIa, factor VIIa/tissue factor complex, factor Xa, or thrombin. Cultured hepatoma cells (HepG2) also secrete this inhibitor. Amyloid β-protein precursor, also known as protease nexin 2 (50), is the parent molecule of amyloid β-protein that is the major constituent of cerebrovascular deposits in Alzheimer's disease and in Down syndrome. All these inhibitors, especially β-amyloid precursor protein, which has a high affinity and specificity for factor XIa, are thought to play a physiological role in maintaining hemostasis.

VII. PERSPECTIVE

Like other coagulation and fibrinolytic enzymes, the heavy chain of factor XIa is a regulatory unit for the enzyme activity. It presumably determines the substrate specificity and affects the catalytic efficiency of factor XIa by interacting with the substrate, factor IX, and a cofactor, high M_r kininogen. These specific interactions are associated with unique apple domains and are analogous to the function of growth factor domains and kringle domains in other coagulation factors. It is surprising to note, however, that although factor XI and plasma prekallikrein share extensive structural and functional similarities, the binding to high M_r kininogen was apparently mediated by different apple domain(s) in these two proteins (51). These observations suggest the possibility that in addition to the primary site of interaction, secondary or cooperative binding sites contributed by other apple domains are present. Furthermore, because factor XI is a homodimer, the presence of two possible sites of interaction with high M_r kininogen and with factor IX raises questions about the interdependence of the two sites in factor XI in these interactions. It is unclear at present whether the two sites are independent of each other, interact cooperatively, or function in a mutually exclusive manner. Resolution of these questions may depend on elucidation of the three-dimensional structure of each apple domain and how the apple domains interact with each other in the native molecule. A combination of peptide modeling and site-directed mutagenesis would be useful in these structure and function studies.

The activation mechanism of factor XI under physiological conditions is not yet clear. Although thrombin was able to activate factor XI in a purified system, this reaction requires the presence of a negatively charged surface, such as dextran sulfate, sulfatide, or heparin. This activation is greatly reduced in plasma, because circulating fibrinogen and antithrombin III compete with factor XI for the binding to thrombin. The physiological surface that supports this reaction has not been identified. Thus far, the activated platelet surface and the endothelial cell surface are the only surfaces shown to bind factor XI or factor XIa. The binding to these surfaces, however, may not be involved in normal hemostasis because this binding is dependent on high M_r kininogen and a deficiency in high M_r kininogen is not associated with bleeding problems. Moreover, unlike other surface-dependent activations in the coagulation cascade, this surface binding does not lead to an enhancement of factor IX activation. Therefore, the identification of the physiological surface and other potential activators and cofactors in the activation of factor XI remains a crucial area of future study.

REFERENCES

1. Rosenthal RL, Dreskin OH, Rosenthal N. New hemophilia-like disease caused by deficiency of a third plasma thromboplastin factor. Proc Soc Exp Biol Med 1953; 82:171–174.
2. Saito H, Ratnoff OD, Bouma BN, Seligsohn U. Failure to detect variable (CRM^+) plasma thromboplastin antecedent (PTA, factor XI) molecule in hereditary PTA deficiency: a study of 125 patients of several ethnic backgrounds. J Lab Med 1985; 106:718–722.
3. Seligsohn U. Factor XI (PTA) deficiency in Ashkenazi Jews. In: Goodman RM, Motulsky AG, eds. Genetic Diseases Among Ashkenazi Jews. New York: Raven Press, 1979:141–148.
4. Franscillon, Knecht H, Winkel L, Hauert J, Bachmann F. Hemophilia C, a deficiency of a contact phase protein which may involve in a risk of hemorrhage. Schweiz Med Wochenschr 1992; 122:351–353.
5. Koide H, Kato H, Davie EW. Isolation and characterization of bovine factor XI (plasma thromboplatin antecedent). Biochemistry 1977; 16:2279–2286.
6. Bouma BN, Griffin JH. Human coagulation factor XI: purification, properties, and mechanism of activation by activated factor XII. J Biol Chem 1977; 252:6432–6437.
7. Kurachi K, Davie EW. Activation of human factor (plasma thromboplastin antecedent) by factor XIIa (activated Hageman factor). Biochemistry 1977; 16:5831–5839.
8. Dzik WH, Arkin CF, Jenkins RL. Transfer of congenital factor XI deficiency from a donor to a recipient by liver transplantation (letter). N Engl J Med 1987; 316:1217–1218.
9. Clarkson K, Rosenfeld B, Fair J, Klein A, Bell W. Factor XI deficiency acquired by liver transplantation. Ann Intern Med 1991; 115:877–879.
10. Thompson RE, Mandel R Jr, Kaplan AP. Association of factor XI and high molecular weight kininogen in human plasma. J Clin Invest 1977; 60:1376–1380.
11. Davie EW, Fujikawa K, Kisiel W. The coagulation cascade: initiation, maintenance, and regulation. Biochemistry 1991; 30:10363–10370.
12. Colman RW. Surface-mediated defence reactions. The plasma contact activation system. J Clin Invest 1984; 73:1249–1253.
13. Griffin JH. Role of surface-dependent activation of Hageman factor (blood coagulation factor XII). Proc Natl Acad Sci USA 1978; 75:1998–2002.
14. Colman RW. Contact systems in infectious disease. Rev Infect Dis 1989; 11(Suppl. 4):S689–699.
15. Keller CC, Mitropoulos KA, Imeson J, et al. Hageman factor and risk of myocardial infarction in middle-aged men. Atherosclerosis 1992; 1:67–73.
16. Naito K, Fujikawa K. Activation of human blood coagulation factor XI independent of factor XII. Factor XI is activated by thrombin and factor XIa in the presence of negatively charged surfaces. J Biol Chem 1991; 266:7353–7358.
17. Gailani D, Broze GJ Jr. Factor XI activation in a revised model of blood coagulation. Science 1991; 253:909–912.
18. Walsh PN. Factor XI: a renaissance. Semin Hematol 1992; 29:189–201.
19. Kitchens CS. Factor XI: a review of its biochemistry and deficiency. Semin Thromb Hemost 1991; 17:55–72.
20. Fujikawa K, Chung DW, Hendrickson LE, Davie EW. Amino acid sequence of human factor XI, a blood coagulation factor with four tandem repeats that are highly homologous with plasma prekallikrein. Biochemistry 1986; 25:2417–2424.
21. McMullen BA, Fujikawa K, Davie EW. Location of the disulfide bonds in human coagulation factor XI: the presence of tandem apple domains. Biochemistry 1991; 30:2056–2060.
22. Chung DW, Fujikawa K, McMullen BA, Davie EW. Human plasma prekallikrein, a zymogen to a serine protease that contains four tandem repeats. Biochemistry 1986; 25:2410–2417.

23. Van der Graaf F, Greengard JS, Bouma BN, Kerbiriou DM, Griffin JH. Isolation and functional characterization of the active light chain of activated human blood coagulation factor XI. J Biol Chem 1983; 258:9669–9675.
24. Titani K, Fujikawa K, Enfield DL, Ericsson LH, Walsh KA, Neurath H. Bovine factor X_1 (Stuart factor): amino-acid sequence of heavy chain. Proc Natl Acad Sci USA 1975; 72: 3082–3086.
25. Raymond PS, Higuchi DA, Broze GJ Jr. Platelet coagulation factor XIa-inhibitor, a form of Alzheimer amyloid precursor protein. Science 1990; 248:1126–1128.
26. DiScipio RD, Hermodson MA, Davie EW. Activation of human factor X (Stuart factor) by a protease from Russell's viper venom. Biochemistry 1977; 16:5253–5260.
27. Kurachi K, Davie EW. Isolation and characterization of a cDNA coding for human factor IX. Proc Natl Acad Sci USA 1982; 79:6461–6464.
28. Kawabata S, Miura T, Morita T, et al. High sensitive peptide-4-methylcoumaryl-7-amide substrate for blood-clotting proteases and trypsin. Eur J Biochem 1988; 172:17–25.
29. Scott CF, Sinha D, Seaman FS, Walsh PN, Colman RW. Amidolytic assay of human factor XI in plasma: comparison with a coagulant assay and a new rapid radioimmunoassay. Blood 1984; 63:42–50.
30. Tait JF, Fujikawa K. Primary structure requirements for the binding of human high molecular weight kininogen to plasma prekallikrein and factor XI. J Biol Chem 1987; 262:11651–11656.
31. Baglia FA, Jameson BA, Walsh PN. Localization of the high molecular weight kininogen binding site in the heavy chain of human factor XI to amino acids phenylalanine 56 through serine 86. J Biol Chem 1990; 265:4149–4154.
32. Baglia FA, Jameson BA, Walsh PN. Fine mapping of the high molecular weight kininogen binding site on blood coagulation factor XI through the use of rationally designed synthetic analogs. J Biol Chem 1992; 267:4247–4252.
33. Sinha D, Seaman FS, Walsh PN. Role of calcium ions and the heavy chain of factor XIa in the activation of human coagulation factor IX. Biochemistry 1987; 26:3768–3775.
34. Baglia FA, Jameson BA, Walsh PN. Identification and chemical synthesis of a substrate-binding site for factor IX on coagulation factor XIa. J Biol Chem 1991; 266:24190–24197.
35. Meijers JC, Mulvihill ER, Davie EW, Chung DW. Apple four in human blood coagulation factor XI mediates dimer formation. Biochemistry 1992; 31:4680–4684.
36. Asakai R, Davie EW, Chung DW. Organization of the gene for human factor XI. Biochemistry 1987; 26:7221–7228.
37. Kato A, Asakai R, Davie EW, Aoki N. Factor XI gene (F11) is located on the distal end of the long arm of human chromosome 4. Cytogenet Cell Genet 1989; 52:77–78.
38. Bodish P, Warne D, Watkins C, Nyberg K, Spurr NK. Dinucleotide repeat polymorphism in the human coagulation factor XI gene, intron B (F11), detected using polymerase chain reaction. Nucleic Acids Res 1991; 19:6979.
39. Butler MG, Parsons AD. RFLP for intron E of factor XI gene. Nucleic Acids Res 1990; 18: 5327.
40. Asakai R, Chung DW, Ratnoff OD, Davie EW. Factor XI (plasma thromboplastin antecedent) deficiency in Ashkenazi Jews is a bleeding disorder that can result from three types of point mutations. Proc Natl Acad Sci USA 1989; 86:7667–7671.
41. Meijers JC, Davie EW, Chung DW. Expression of human blood coagulation factor XI: characterization of the defect in factor XI type III deficiency. Blood 1992; 79:1435–1440.
42. Asakai R, Chung DW, Davie EW, Seligsohn U. Factor XI deficiency in Ashkenazi Jews in Israel. N Engl J Med 1991; 325:153–158.
43. Hancock JF, Wieland K, Pugh RE, et al. A molecular genetic study of factor XI deficiency. Blood 1991; 77:1942–1948.
44. Sinha D, Seaman FS, Koshy A, Knight LC, Walsh PN. Blood coagulation factor XIa binds specifically to a site on activated human platelets distinct from that for factor XI. J Clin Invest 1984; 73:1550–1556.

45. Berrettini M, Raymond RS, Heeb MJ, Hopmeier P, Griffin JH. Assembly and expression of an intrinsic factor IX activator complex on the surface of cultured human endothelial cells. J Biol Chem 1992; 267:19833–19839.
46. Scott CF, Schapira M, James HL, Cohen AB, Colman RW. Inactivation of factor XIa by plasma inhibitors: predominant role of α_1-protease inhibitor and protective effect of high molecular weight kinogen. J Clin Invest 1982; 69:844–852.
47. Soons H, Tans G, Hemker HC. The heparin-catalysed inhibition of human factor XIa by antithrombin III is dependent on the heparin type. Biochem J 1988: 256:815–820.
48. Meijers JC, Kanters DH, Vlooswijk RA, van Erp HE, Hessing M, Bouma BN. Inactivation of human plasma kallikrein and factor XIa by protein C inhibitor. Biochemistry 1988; 27: 4231–4237.
49. Van der Graaf F, Koedam JA, Bouma BN. Inactivation of kallikrein in human plasma. J Clin Invest 1983; 71:149–158.
50. Van-Nostrand WE, Farrow JS, Wagner SL, et al. The predominant form of the amyloid beta-protein precursor in human brain is protease nexin 2. Proc Natl Acad Sci USA 1991; 88: 10302–10306.
51. Page JD, Colman RW. Localization of distinct functional domains on prekallikrein for interaction with both high molecular weight kininogen and activated factor XII in a 28 kDa fragment (amino acids 141–371). J Biol Chem 1991; 266:8143–8148.

13

Factor XII, Prekallikrein, and High-Molecular-Weight Kininogen

Hidehiko Saito and Tetsuhito Kojima
Nagoya University School of Medicine, Nagoya, Japan

I. INTRODUCTION

Normal blood quickly clots when it is allowed to come into contact with a variety of foreign surfaces. The initial steps in this pathway of blood coagulation are known as the contact activation reactions, because they are triggered by surface contact. The contact activation system of human plasma consists principally of the three proteins: factor XII (Hageman factor), plasma prekallikrein, and high-molecular-weight (HMW) kininogen. Like most other blood-clotting factors, each of these three factors was initially discovered as an agent functionally deficient in blood of very rare individuals with abnormal blood coagulation. They were then isolated from normal plasma, and their roles in normal blood coagulation were defined.

Factor XII, prekallikrein, and HMW kininogen are unique among blood-clotting factors in several respects. They participate not only in blood coagulation but also in kinin generation and fibrinolysis. The reactions among these proteins and surfaces do not require the presence of calcium ions. Interestingly, individuals with a severe deficiency of factor XII, prekallikrein, or HMW kininogen do not have a bleeding tendency despite markedly abnormal coagulation tests (1). They do not bleed excessively even in response to trauma or surgery. Thus, it appears that factor XII, prekallikrein, and HMW kininogen are not required for hemostasis. Paradoxically, several anecdotal case reports of thromboembolism in subjects with a hereditary deficiency of these contact factors were reported (2,3). Because contact activation stimulates fibrinolysis under certain in vitro conditions, it was speculated that thrombotic complications are caused by impaired fibrinolytic activity (4). There is no definite evidence, however, that the in vivo development of fibrinolytic activity is diminished in patients with contact factor deficiency. The true incidence of thrombosis in such individuals is difficult to assess, because most subjects are asymptomatic and are diagnosed only if routine clotting tests are performed. Therefore, the roles of the contact activation factors in the pathogenesis of thrombosis are controversial.

Recent advances in protein chemistry and recombinant DNA technology have made it possible to characterize factor XII, prekallikrein, and HMW kininogen at the molecular level (5). They have also provided a basis for identifying the structural and functional defects in hereditary deficiencies of these proteins. This chapter briefly summarizes the structure-function relationship of contact factors and the molecular genetics of their deficiencies. There are very few cases of contact factor deficiencies whose structural defects have been identified. This is not surprising, because hereditary deficiencies of the contact factor are rare. In this regard, it should be also noted that only a small percentage of patients with the most common hereditary disorder of blood coagulation, hemophilia, have been characterized at the molecular level.

II. ROLE OF FACTOR XII, PREKALLIKREIN, AND HMW KININOGEN IN COAGULATION

Table 1 shows some properties of three plasma proteins involved in contact activation reactions. Some features of the mRNA and gene encoding for these proteins are also listed. Factor XII, prekallikrein, and HMW kininogen are synthesized in the liver and secreted into the blood. Both factor XII and prekallikrein exist in the blood as an inactive precursor (zymogen) of a serine protease, whereas HMW kininogen is a nonenzymatic cofactor that circulates as a noncovalent complex with prekallikrein and factor XI (6,7). Contact activation is initiated and amplified by complex reactions that involve both protein-protein interactions and protein-surface interactions.

A large number of agents, including glass, kaolin, Celite (infusorial earth), ellagic acid, and dextran sulfate, are known to trigger the contact phase of blood coagulation in vitro. Some biological substances, such as sulfatides (cerebroside sulfates), acid phospholipid, and basement membrane components, can also support contact activation. A common property of these agents is that they are negatively charged and of a high molecular mass; certain positively charged substances, such as Polybrene (hexadimethrine bromide), lysozyme and β_2-glycoprotein I, can attach to negatively charged surfaces and inhibit contact activation. The exact nature of the clot-promoting substances is not completely understood. Contact activation was considered a result of a general surface property, highly dense negative charges, rather than to specific surface functional groups, such as a sulfate moiety (8).

Table 1 Contact Factors[a]

	Protein				mRNA		Gene		
Factor	MW (kD)	Plasma Concentration µg/ml	Plasma Concentration µM	In vivo half-life (h)	Size (kb)	No. coding aas	Size (kb)	No. exons	Chromosome localization
Factor XII	80	30	0.4	50–70	2.4	615	12	14	5q33-qter
Prekallikrein	85	50	0.6	35	2.4[b]	638	22[c]	15[c]	4q35
High-molecular-weight kininogen	120	70	0.7	150	3.5	644	27	11	3q26-qter

[a]MW, molecular weight; aas, amino acids.
[b]Expected from cDNA data.
[c]Data in rat.

Exposure of blood to negatively charged surfaces leads to rapid binding of all contact factors to surfaces. Factor XII and HMW kininogen directly bind, but prekallikrein and factor XI are bound to surfaces via HMW kininogen (9). Surface binding is assumed to serve to bring factor XII, factor XI, prekallikrein, and HMW kininogen in close proximity and to increase their chance of interacting. Binding of factor XII to surfaces also makes the molecule more susceptible to proteolytic cleavage (10). It was shown that surface-bound factor XII was 500 times more susceptible than soluble factor XII to proteolytic cleavage by kallikrein in the presence of HMW kininogen. The binding mechanism appears to be different depending on different surfaces. Factor XII binds to sulfatides by an electrostatic interaction, whereas it is bound to kaolin by a nonelectrostatic interaction (5).

Surface-bound factor XII is activated to factor XIIa by limited proteolysis. Once formed, factor XIIa converts surface-bound prekallikrein to kallikrein (Fig. 1). Kallikrein, in turn, activates surface-bound factor XII to factor XIIa. This process, called reciprocal activation, is a positive feedback loop that serves to amplify the rapid mutual activation of factor XII and prekallikrein (11). Although other mechanisms exist, such as autoactivation of factor XII and activation of factor XII by factor XIa, the kallikrein-dependent factor XII activation plays the predominant role in contact activation. This is clearly illustrated by the markedly slow activation of factor XII in prekallikrein-deficient plasma. A kinetic study demonstrated that at prekallikrein and factor XII levels equal to those in plasma the kallikrein-dependent activation is 2000-fold more rapid than autoactivation (12). Once sufficient amounts of factor XIIa are generated, factor XIIa then activates surface-bound factor XI to factor XIa, leading to the intrinsic blood coagulation pathway.

HMW kininogen functions as a cofactor in contact activation. It augments the activation of prekallikrein and factor XI with surface-bound factor XIIa by linking both

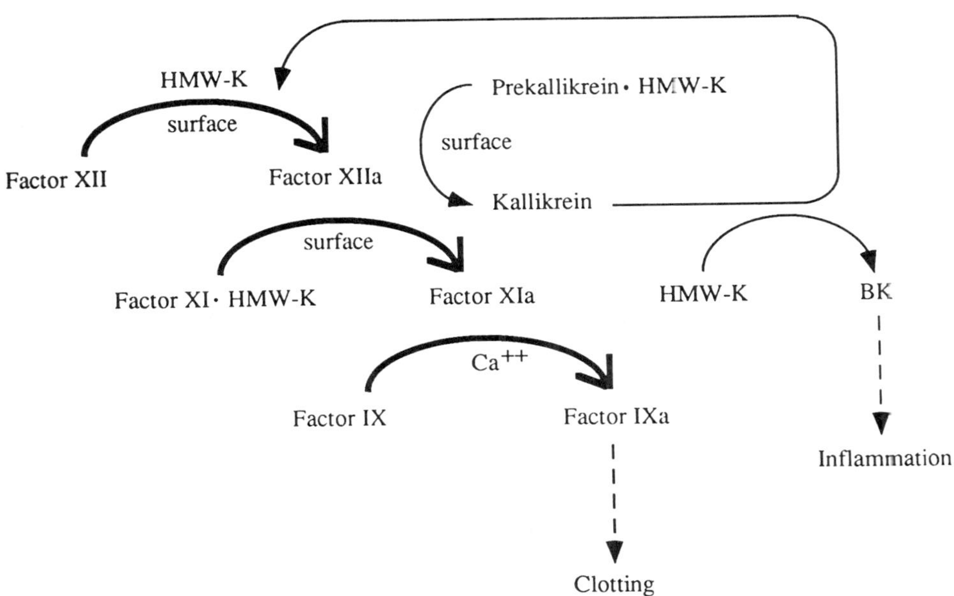

Figure 1 Contact activation reactions. HMW-K, high-molecular-weight kininogen; BK, bradykinin.

proteins to surfaces (9), and it also enhances the activation of factor XII by kallikrein. Limited proteolysis of HMW kininogen by kallikrein appears to enhance the absorption of HMW kininogen to surfaces, thereby amplifying the procoagulant activity of HMW kininogen (13). At the same time, kallikrein releases bradykinin, a potent vasodilator, from HMW kininogen. Bradykinin also has the ability to increase vascular permeability, to cause pain and to contract smooth muscle, and has been implicated as an inflammatory mediator. Thus, factor XII, prekallikrein, and HMW kininogen are shared by the intrinsic pathway of thrombin generation and kinin formation, thereby linking blood coagulation and inflammatory reactions.

The leading question in contact activation is how the first factor XII or prekallikrein is activated. The origin and nature of the initial enzymatic activity responsible for the conversion of zymogens into active serine proteases are unknown. Autoactivation of factor XII (14) or prekallikrein (15) cannot explain the origin of the first factor XIIa or kallikrein during contact activation, because autoactivation is dependent on traces of factor XIIa or kallikrein that are invariably present in zymogen preparations.

Several hypotheses have been advanced. The binding of factor XII to surfaces may induce a conformational change in the molecule that exposes hidden active sites. There is experimental evidence that conformational changes in factor XII indeed occur upon binding to ellagic acid or dextran sulfate (16,17). The results of ultraviolet difference spectroscopy and circular dichroism studies suggest that the interaction of factor XII and dextran sulfate is a biphasic process. It is initiated by a fast contraction of the molecule upon binding and is followed by a slow relaxation process (17). Factor XII and prekallikrein may have a low degree of inherent enzymatic activity that may be sufficient to initiate activation upon surface binding. Alternatively, low levels of factor XIIa or kallikrein, or both, may always circulate in the blood, and these enzymes may trigger zymogen activation (18). Furthermore, activated platelets have been shown to enhance the activation of factor XII in the presence of kallikrein and HMW kininogen (19). It is also possible that factor XII and prekallikrein are activated by cellular enzymes present in basophils, endothelial cells, and bacteria (20,21).

There are at least two mechanisms by which contact activation is controlled. First, certain plasma proteins, such as IgG, fibrinogen, β_2-glycoprotein I, and C1q, are known to compete with factor XII for surface binding and to inhibit contact activation (22,23). Second, activated contact factors (factor XIIa and kallikrein) are readily inactivated by naturally occurring plasma protease inhibitors. C1INH is a major inhibitor of both factor XIIa and kallikrein (24,25). The physiological significance of C1INH is illustrated by the fact that the activation of contact system was found in the blood of patients with hereditary angioneurotic edema, a functional deficiency of C1INH (26). During acute attacks of angioedema, in vivo cleavage of HMW kininogen was detected by a ligand blotting technique (27).

III. FACTOR XII

A. Gene Organization and Structure

The gene of factor XII, located on chromosome 5q33-qter, spans approximately 12 kb and is composed of 13 introns and 14 exons (Table 1) (28–30). Although exons 1 and 2 are separated from other exons by large introns, exons 3–14 are contained in a genomic region of only 4.2 kb with introns ranging in size 80 to 554 base pairs (Fig. 2). The coding sequence of factor XII has multiple putative domains that are homologous to putative domains found in fibronectin and tissue-type plasminogen activator. The intron-exon gene organization is similar to that of the serine protease gene family of plasminogen activators but is different from that of the clotting factor family. It was proposed that factor XII and plasminogen activators evolved from a common ancestor through exon shuffling (31). Analysis of the promoter region shows that it does not contain the typical TATA and CAAT sequences found in other genes. This is consistent with the finding that transcription of the gene is initiated at multiple sites. The coding DNA is approximately 2 kb, with codons for a polypeptide of 615 amino acids and with 150 base pairs containing a typical AATAAA polyadenylation signal for a 3′-untranslated end of mRNA. Translation begins with a methionine, 19 amino acids from the amino terminus of the mature protein, and terminates with a TGA stop codon. After translation of the mRNA, the polypeptide chain undergoes signal peptidase cleavage, removing 19 amino acids.

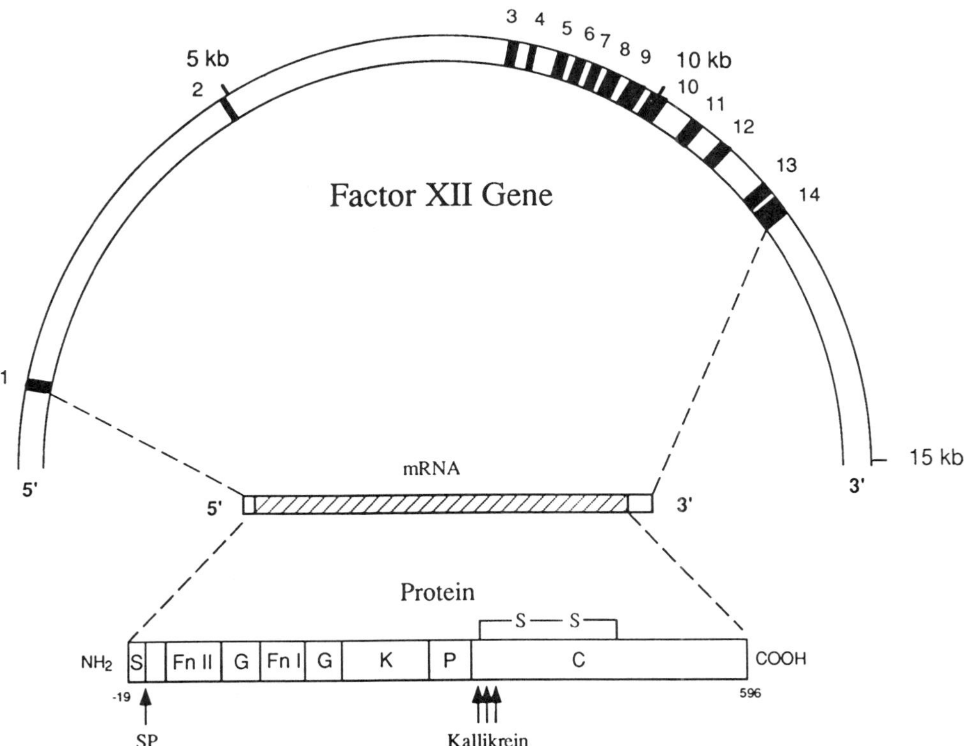

Figure 2 Factor XII gene and protein. The factor XII gene is shown in the upper portion, and the protein is shown in the lower portion, with the amino terminus at the left and the carboxyl terminus at the right. S, signal peptide; SP, signal peptidase; Fn II, fibronectin type II homology domain; G, growth factor-like domain; Fn I, fibronectin type I homology domain; K, kringle domain; P, proline-rich domain; C, catalytic domain. Arrows indicate a cleavage site by signal peptidase and three cleavage sites by kallikrein.

B. Protein Structure and Interaction

Factor XII is synthesized in the liver (32), secreted into the blood, and present in normal plasma at a level of 30 μg/ml. The primary structure of factor XII was determined by protein and cDNA sequencing (28,33,34). A mature factor XII protein (80,000 MW) consists of a single polypeptide chain of 596 amino acids containing 16.8% carbohydrate. Figure 3 illustrates the complete amino acid sequence of factor XII. The arrow indicates the Arg^{353}-Val^{354} bond that is cleaved by plasma kallikrein upon the activation of factor XII. The heavy chain, located in the amino-terminal portion of the molecule, consists of 353 amino acid residues containing one N-glycosylation site and six O-glycosylation sites and has five discrete folding domains, a fibronectin type II homology, a first growth factor-like region, a fibronectin type I homology, a second growth factor-like region, and a kringle structure. The domain organization of the heavy chain is very similar to that of tissue plasminogen activator (tPA) or urokinase (34). The exact function of the different domains is not known at present, but a putative surface binding site was localized to 28 amino acids at the amino terminus of the heavy chain (35). A possible function of the growth factor-like domains was suggested by a recent report that factor XII has the

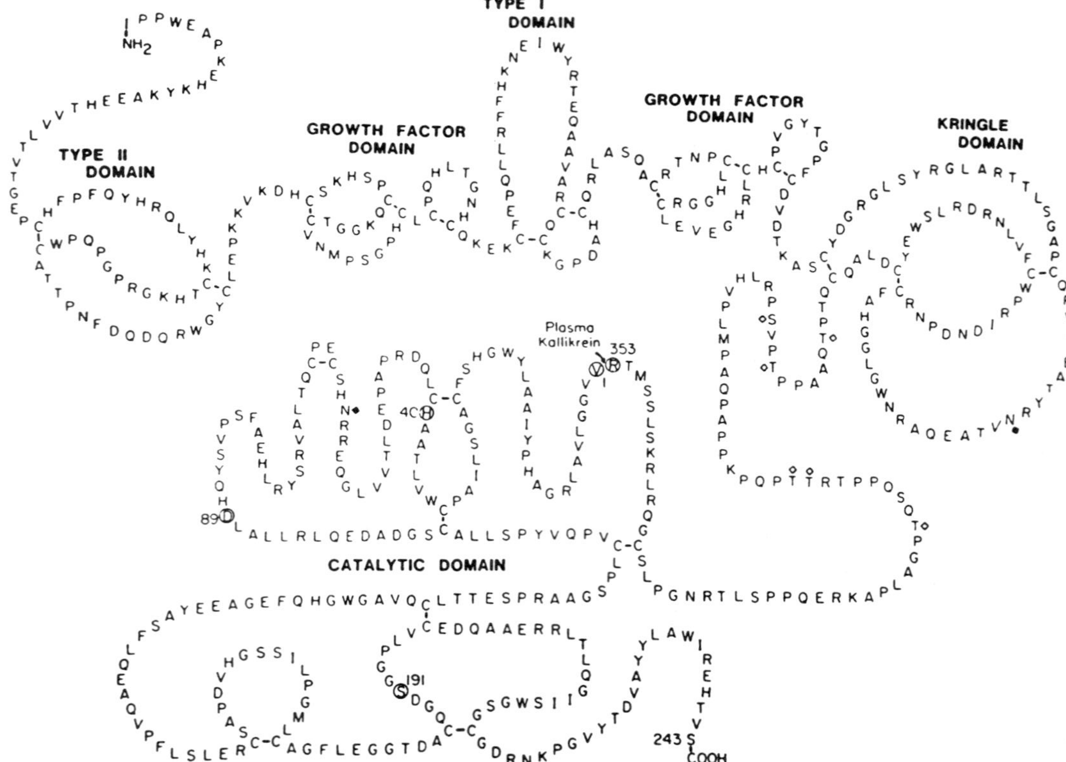

Figure 3 Complete amino acid sequence of factor XII. The curved arrow indicates the Arg-Val bond cleaved by plasma kallikrein. The catalytic triad (His^{40}, Asp^{89}, and Ser^{191}) are circled. The solid diamonds show the N-glycosylation sites, and the open diamonds represent the proposed O-glycosylation sites. (Reproduced with permission from Ref. 5.)

ability to promote proliferation of HepG2 liver cells (36). The carboxyl-terminal portion of factor XII is a light chain or a catalytic domain that expresses the catalytic activity upon activation. The light chain contains the catalytic triad of histidine 40, aspartic acid 89, and serine 191. The amino acid sequence of the light chain (243 amino acid residues) has significant homology with the corresponding regions of other serine proteases, such as trypsin, tPA, plasmin, and thrombin. The homology is the highest with plasmin (42%). Thus, the primary structure of factor XII has a higher degree of homology with fibrinolytic enzymes than with clotting enzymes.

Factor XII is activated to factor XIIa by proteolytic cleavage with kallikrein at three sites. Two forms of factor XIIa are known (37). Initial cleavage at Arg^{353}-Val^{354} results in formation of α-XIIa, a two-chain molecule composed of a heavy (52,000 MW) and a light (28,000 MW) chain linked by a disulfide bridge. α-XIIa has the ability to activate both factor XI and prekallikrein. Subsequent cleavage at Arg^{334}-Asn^{335} and at Arg^{343}-Leu^{344} converts α-XIIa to β-XIIa. β-XIIa consists of the entire light chain of α-XIIa and a small nonapeptide attached via the disulfide bridge (32). β-XIIa is a potent activator of prekallikrein but has essentially no ability to activate surface-bound factor XI, because the surface binding site in the amino-terminal region is lost.

C. Mutations in Factor XII Deficiency

Hereditary deficiency of factor XII (Hageman trait) (38) is transmitted as an autosomal recessive trait. In general, homozygous individuals have essentially undetectable factor XII activity (less than 1% of normal pooled plasma), and heterozygous subjects have 20–60% factor XII activity (1). The molecular basis of factor XII deficiency has been studied in several individuals with this disorder by DNA analysis as well as protein sequencing. Only limited data are available at present. Gross gene deletions were not detected in six Italian kindreds by Southern blot hybridization (39,40). In five families an additional *Taq*I restriction site, located in intron 2 of the factor XII gene, was found to be closely linked to the inheritance of factor XII deficiency. This *Taq*I site was not found in 120 normal control genes. It was suggested that the high frequency of the additional *Taq*I site in Italian factor XII deficiency may be explained by the founder effect or by selective advantage (40). Whether it represents a DNA alteration underlying factor XII deficiency or a very tightly linked restriction fragment length polymorphism remains to be elucidated. Thus, no definite DNA abnormality causing factor XII deficiency has yet been identified.

Protein abnormalities causing functional factor XII deficiency have been disclosed by amino acid sequencing in two cases. In the plasma of the majority of individuals with hereditary factor XII deficiency, factor XII antigen (immunoreactive factor XII) is reduced in parallel with decreased factor XII activity. Very rare subjects with this disorder, however, have been found to have nonfunctional material immunologically indistinguishable from normal factor XII (cross-reacting material-positive, CRM^+, variants) (41). Structural studies of abnormal nonfunctional factor XII protein may provide a unique opportunity to examine the structure-function relationship of this plasma protein. At least six CRM^+ cases of factor XII deficiency have been reported, as shown in Table 2: XII Washington D.C. (42,43), XII Toronto (44), XII Bern (45), XII Bari (46), XII Locarno (47), and XII Valencia (48). In most cases, only functional characterization of the abnormal protein, such as the molecular weight, the ability to bind to kaolin, limited proteolysis during contact activation, and enzymatic activities of cleaved molecules, have been achieved. Amino acid sequence analysis of XII Washington D.C. identified replacement of Cys^{571} with serine in the catalytic domain. It was proposed that this amino acid substitution destroys the formation of the disulfide linkage between Cys^{540} and Cys^{571}, giving rise to an altered conformation of the active-site serine residue or the secondary substrate binding site (43). In XII Locarno, an amino acid substitution ($Arg^{353} \rightarrow$ Pro) at the critical cleavage site by kallikrein was recently identified. Structural defects in other CRM^+ cases have not yet been determined, because sufficient amounts of abnormal protein are difficult to purify from the plasma of index subjects. DNA analysis by polymerase chain reaction (PCR) amplification and direct sequencing will be helpful in such cases.

IV. PREKALLIKREIN

A. Gene Organization and Structure

The precise gene structure of human plasma prekallikrein is not yet determined, whereas the entire rat plasma prekallikrein gene has been isolated and characterized in detail. The gene of rat plasma prekallikrein is 22 kb in length and composed of 15 exons and 14 introns, which are similar to human factor XI gene (Fig. 4) (49). The 5′ promoter region of this gene has some putative control elements that may regulate the rat prekallikrein gene expression. Exon 1 codes for the 5′ noncoding region, and exon 2 codes for the signal peptide. The following eight exons (exons 3–10) encode the four tandem repeats that are present in the heavy chain of rat plasma kallikrein. These repeats are separated by introns located at the same position. Exon 11 codes for the connecting region, and the remaining four exons (exons 12–15) encode the entire light chain (catalytic domain) of plasma kallikrein.

The gene of human plasma prekallikrein has been mapped on chromosome 4q35 by in situ hybridization using human plasma prekallikrein cDNA as a probe (49). The human factor XI gene was also localized to this region of the chromosome (50). The close linkage of the two genes on the same chromosome, the numbers and positions of introns, the similarity at the level of intron type, and the 58% identity of human factor XI and prekallikrein cDNA strongly suggest a gene duplication event from a common ancestor to both prekallikrein and factor XI (51).

Table 2 Some Properties of Abnormal Factor XII Found in Cross-Reacting Material-Positive Factor XII Deficiency[a]

Factor XII	XII (%) Activity	XII (%) Antigen	MW	Binding to surfaces	Limited proteolysis by kallikrein	Enzymatic activities	Structural defect
Washington D.C.	<1	80	Normal	Normal	Normal	−	$Cys^{571} \rightarrow$ Ser
Toronto	<1	39	Normal	Normal	ND	ND	ND
Bern	<1	11	Normal	Normal	Normal	−	ND
Bari	1.5	35–52	Normal	Normal	Delayed	+	ND
Locarno	<1	46	Normal	Normal	Delayed	−	$Arg^{353} \rightarrow$ Pro
Valencia	<1	41–70	ND	ND	ND	ND	ND

[a]ND, not determined.

Human prekallikrein cDNA, which contained a 5′ noncoding region followed by the regions coding for a signal peptide of 19 amino acids and the mature prekallikrein of 619 amino acids, was cloned from a human liver cDNA library. The 3′-untranslated region contains a TGA stop codon and a typical AATAAA polyadenylation signal (51).

B. Protein Structure and Interaction

A combination of protein sequencing and cDNA sequencing techniques unraveled the primary structure of prekallikrein (51,52). Figure 5 illustrates the complete amino acid sequence of human plasma prekallikrein. Prekallikrein (85,000 MW) consists of a single polypeptide chain of 619 amino acids and is converted to kallikrein by factor XIIa by the cleavage of a Arg^{371}-Ile^{372} bond, as indicated by an arrow. Protein sequence analysis revealed dimorphism at position 124 in the heavy chain, where Asn and Ser are found at a 7:3 ratio. The heavy chain (371 amino acid residues) is composed of four tandem repeats, called apple domains, and the light chain (248 amino acid residues) contains the catalytic domain with an active-site serine residue. Each apple domain contains 90 or 91 amino acid residues that are highly homologous to those of factor XI, and it also has characteristic disulfide bonds linking the first and sixth, second and fifth, and third and

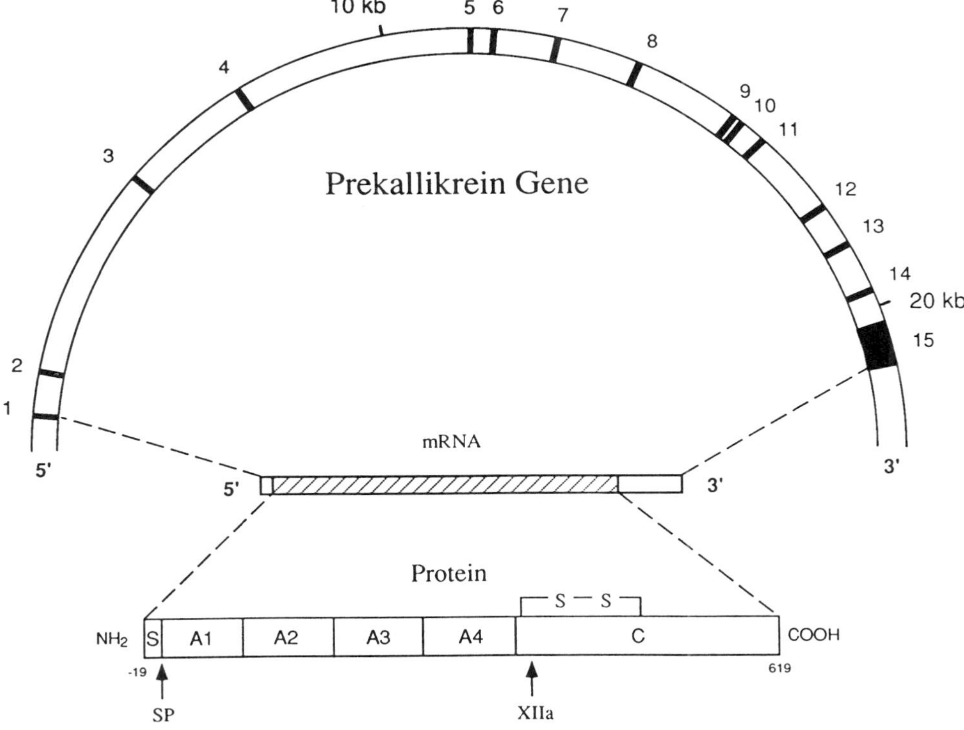

Figure 4 Prekallikrein gene and protein. The rat prekallikrein gene is illustrated in the upper portion, and the protein is shown in the lower portion. S, signal peptide; SP, signal peptidase; A1–A4, apple domains; C, catalytic domain. Arrows indicate cleavage sites by signal pepetidase and by factor XIIa.

fourth Cys residues (52). The amino acid sequence of the light chain is homologous to that of trypsin, and the catalytic triad is located at His[415], Asp[464], and Ser[559].

Prekallikrein is a zymogen of a serine protease. It circulates in blood as an equimolar noncovalent complex with HMW kininogen (6). The plasma concentration of prekallikrein is approximately 50 μg/ml. When blood comes into contact with negatively charged surfaces, the prekallikrein-HMW kininogen complex is absorbed onto the surface (9). Activation of prekallikrein to kallikrein is catalyzed by surface-bound factor XIIa. Factor XIIa cleaves a single peptide bond and changes a one-chain polypeptide to a two-chain species composed of heavy (52,000 MW) and light (33,000 MW) chains held together by a disulfide bridge. Studies using murine monoclonal antibodies against human prekallikrein localized the binding region for HMW kininogen to the COOH-terminal 231 amino acids (amino acids 141–371) of the heavy chain of prekallikrein (53). The substrate recognition site on prekallikrein for factor XIIa is also present in the separate region of the same fragment.

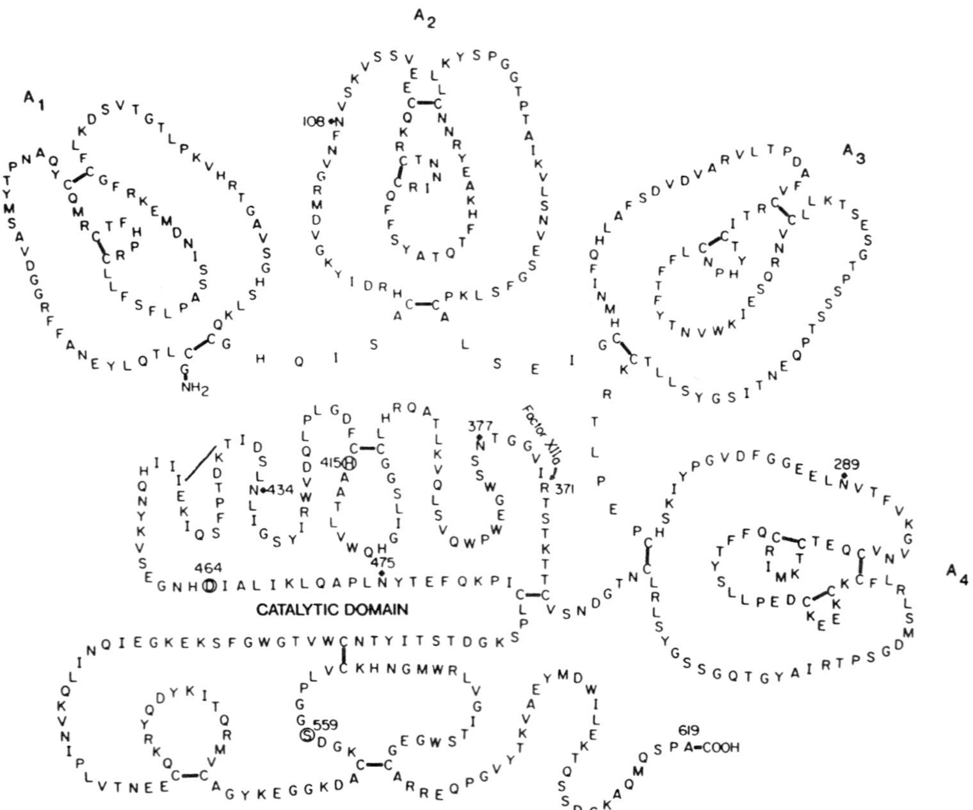

Figure 5 The complete amino acid sequence of plasma prekallikrein. The curved arrow shows the Arg-Ile bond cleaved by factor XIIa. A1, A2, A3, and A4 indicate four apple domains. The catalytic triad (His[415], Asp[464], and Ser[559]) are circled. The solid diamonds illustrate attachment sites for carbohydrate chains. (Reproduced with permission from Ref. 51.)

C. Mutations in Prekallikrein Deficiency

Hereditary prekallikrein deficiency (Fletcher trait) is a rare autosomal recessively transmitted disorder (54,55). The plasma of the homozygotes contains less than 1% prekallikrein activity; that of the heterozygotes has 20–60%. Abnormal nonfunctional prekallikrein has been found in the plasma of some patients with this disorder (56). Of 25 subjects with homozygous prekallikrein deficiency, 9 were CRM^+ variants (1). The functional characterization of abnormal prekallikrein has been reported in two cases. Prekallikrein Long Beach is similar to normal prekallikrein in molecular weight and the ability to form a complex with HMW kininogen. It is immunologically indistinguishable from the normal (57). However, the variant protein is cleaved by β-XIIa very slowly and has no enzymatic activity. Prekallikrein Zurich appears to have a functional defect similar to that of prekallikrein Long Beach (58). Further studies are needed to characterize the structural defect in these variants. No DNA alterations underlying prekallikrein deficiency have yet been reported.

V. HMW KININOGEN

A. Gene Organization and Structure

The gene of human kininogen was isolated from a human genomic library and characterized (59,60). The gene comprises approximately 27 kb and contains 11 exons divided by 10 introns. The nine 5′-terminal exons encode the 5′-untranslated region and the protein coding region for the signal peptide (18 amino acid residues) and the heavy chain, which are common for high-molecular-weight and low-molecular-weight (LMW) prekininogen mRNAs (Fig. 6). The largest exon, exon 10, consists of the common sequence for a nonapeptide of bradykinin (exon 10 bradykinin) and the adjacent 3′ unique sequence for HMW prekininogen mRNA (exon 10 HMW). Exon 11 is located following a 90-nucleotide sequence downstream of exon 10 and precisely specifies the sequence unique to LMW prekininogen mRNA. This, together with the hybridization analysis of total human cellular DNA, reveals that human HMW and LMW prekininogen mRNAs are produced from a single gene as a consequence of alternative RNA processing. The HMW prekininogen mRNA is transcribed from exons 1–10, and the LMW prekininogen mRNA is produced from exons 1–11 by splicing out the exon 10 HMW sequence and its flanking 90 nucleotide sequence. The structural analysis of the kininogen gene also demonstrates that each of nine 5′-terminal exons separately specifies the nine protein domains observed in the amino-terminal protein of the kininogens. Interestingly, these nine genetic domains are characterized by a thrice repeated pattern of the three genetic segments. Particularly, two sets of three domains, encompassing exons 3–5 and exons 6–8, are closely related to each other in nucleotide sequence homology, the sizes of paired exons, and the positions of introns. Therefore, two successive duplication mechanisms as a model for the generation of the structure of the kininogen gene were proposed (60). The kininogen gene was localized to chromosome 3q26-qter (61).

Human HMW kininogen cDNA was isolated from a human liver cDNA library and was found to be approximately 3.2 kb in length, containing a 5′ noncoding region followed by a coding region for a signal peptide of 18 amino acids and for a mature protein of 626 amino acids, a TAA stop codon, and a 1286 bp 3′-untranslated region (59).

B. Protein Structure and Interaction

HMW kininogen is present in normal plasma at a concentration of 70 μg/ml. The complete primary structure of HMW kininogen was established through cDNA sequencing and protein sequence analysis (59,62). HMW kininogen (120,000 MW) consists of a single polypeptide chain containing 626 amino acid residues. HMW kininogen contains 40% of its mass as carbohydrate, but the function of carbohydrate is not known. It is composed of a heavy-chain region, a bradykinin moiety, and a light-chain region (Fig. 6). HMW kininogen is a protein with multiple functions; it is a precursor of bradykinin, a nonenzymatic cofactor in contact activation, an inhibitor of thiol protease (63–65), and an antiadhesive protein (66). LMW kininogen is also a precursor of bradykinin and an inhibitor of thiol protease, but it has no function in blood coagulation. HMW kininogen has been divided into six domains, D1, D2, and D3 in the heavy chain, D4 as bradykinin, and D5 and D6 in the light chain (67). The light-chain region has the procoagulant activity.

A variety of methods, including mapping with monoclonal antibodies, protein fragmentation, and competitive binding using synthetic peptides, have been used to assign various functions of HMW kininogen to different domains. The amino-terminal portion

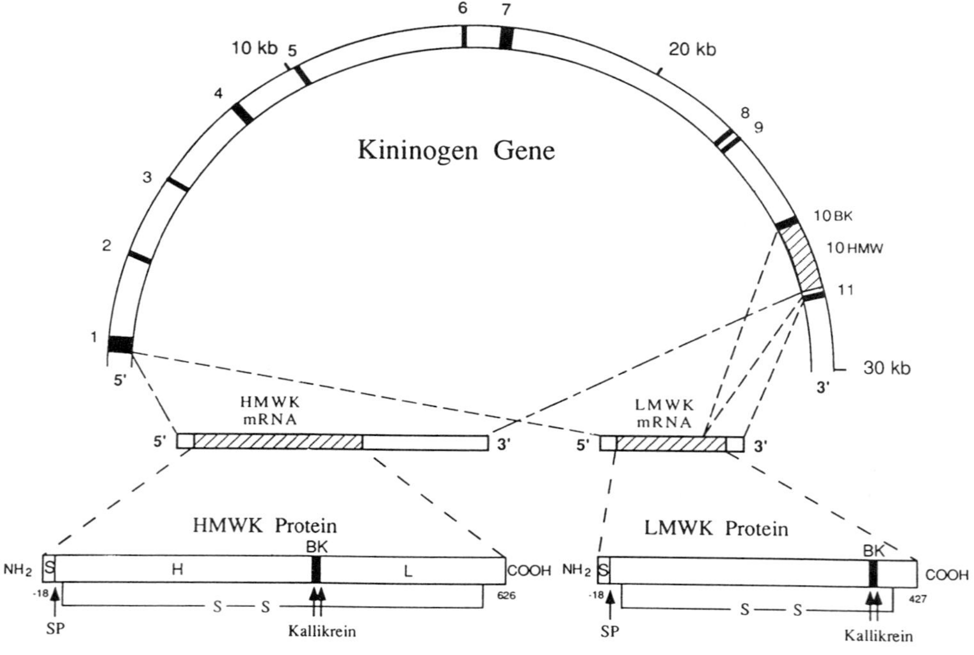

Figure 6 Kininogen gene and protein. The kininogen gene is shown in the upper portion. The lower portion demonstrates the way in which HMW prekininogen mRNA and LMW prekininogen mRNA are produced by alternative RNA splicing: 10 BK, exon 10 encoding for bradykinin; 10 HMW, exon 10 encoding for a unique sequence for HMW kininogen mRNA; S, signal peptide; SP, signal peptidase; H, heavy chain; BK, bradykinin; L, light chain. The arrows indicate a cleavage site by signal peptidase and two cleavage sites that release bradykinin from HMW kininogen and LMW kininogen.

of domain 1 appears to be a possible Ca^{2+} binding site, but the physiological role of the metal binding by kininogens is not known at present (68). Each of the reactive sites of the thiol protease inhibitor is located in domains 2 and 3 (69). As mentioned, the procoagulant activity of HMW kininogen is mostly dependent on its interaction with factor XI and prekallikrein and on its binding to foreign surfaces. Domain 5 (amino acid residues 420–502), which is rich in histidine, has been postulated to contain the anionic surface binding site (70). Recently, at least two surface binding subdomains (Lys^{420}-Asp^{474} and His^{475}-Lys^{502}) in domain 5 have been identified by site-specific mutagenesis (71). The binding sites on HMW kininogen for factor XI and prekallikrein have been localized to amino acid residues 556–613 and 565–595 in domain 6, respectively (72).

HMW kininogen is cleaved by kallikrein at Lys^{362}-Arg^{363} and Arg^{371}-Ser^{372}, liberating a nonapeptide, bradykinin. Cleaved kinin-free HMW kininogen consists of a 65,000 MW heavy chain and a 45,000 light chain linked by a disulfide bridge. Kinin-free HMW kininogen has more procoagulant activity than intact HMW kininogen (13,70). A recent report on the expression of HMW kininogen light chain in *Escherichia coli* suggests that glycosylation is not essential for the procoagulant activity (73). Kinin-free HMW kininogen has antiadhesive activity that seems to be dependent on domain 5 (66).

C. Mutations in HMW Kininogen Deficiency

Hereditary deficiency of HMW kininogen is a very rare disorder of blood coagulation that is inherited as an autosomal recessive trait (74). Two types of kininogen deficiency are known, HMW kininogen deficiency and total kininogen deficiency (75,76). HMW kininogen procoagulant activities in homozygotes and heterozygotes are less than 1 and 20–60% of normal pooled plasma, respectively. No CRM^+ variant has been described in HMW kininogen deficiency. DNA analysis has been conducted in only a few individuals with this disorder. No gross abnormalities of the kininogen gene were found by Southern blotting in four cases with total kininogen deficiency, but a partial deletion in intron 7 was detected in one subject with isolated HMW kininogen deficiency (77). The latter has not yet been characterized in detail. A recent study by PCR amplification and direct DNA sequencing identified a point mutation (C $\rightarrow$ T transition) in exon 5 in an individual with total kininogen deficiency (Williams' trait) (78). This mutation results in a CGA (Arg) to TGA (stop) change in exon 5 and is expected to prevent the synthesis of both HMW and LMW kininogens, because the stop codon occurs before the splice site in exon 10. This single base mutation is inherited in the family and is responsible for total kininogen deficiency. No mutation responsible for HMW kininogen deficiency has yet been reported.

ACKNOWLEDGMENTS

We thank Drs. Robert W. Colman and Bernhard Lämmle for providing us a chance to see their in-press and preprint papers.

REFERENCES

1. Saito H. Contact factors in health and disease. Semin Thromb Hemost 1987; 13:36–49.
2. Goodnough LT, Saito H, Ratnoff OD. Thrombosis or myocardial infarction in congenital clotting factor abnormalities and chronic thrombocytopenias: a report of 21 patients and a review of 50 previously reported cases. Medicine (Baltimore) 1983; 62:248–255.

3. Halbmayer WM, Mannhalter C, Feichtinger C, Rubi K, Fisher M. The prevalence of factor XII deficiency in 103 orally anticoagulated outpatients suffering from recurrent venous and/or arterial thromboembolism. Thromb Haemost 1992; 68:285–290.
4. Kluft C, Dooijewaard G, Emeis JL. Role of the contact system in fibrinolysis. Semin Thromb Hemost 1987; 13:50–68.
5. Fujikawa K, Saito H. Contact activation. In: Scriver CR, Beaudet AL, Sly WS, Valle D, eds. The Metabolic Basis of Inherited Disease, 6th ed. New York: McGraw-Hill, 1989:2189–2206.
6. Mandle R, Colman RW, Kaplan AP. Identification of prekallikrein and high molecular weight kininogen as a complex in human plasma. Proc Natl Acad Sci USA 1976; 73:4179–4183.
7. Thompson R, Mandle R, Kaplan AP. Association of factor XI and high molecular weight kininogen in human plasma. J Clin Invest 1977; 60:1376–1380.
8. Griep MA, Fujikawa K, Nelsestuen GL. Possible basis for the apparent surface selectivity of the contact activation of human blood coagulation factor XII. Biochemistry 1986; 25:6688–6694.
9. Wiggins RC, Bouma BN, Cochrane CG, Griffin JH. Role of high-molecular-weight kininogen in surface-binding and activation of coagulation factor XI and prekallikrein. Proc Natl Acad Sci USA 1977; 74:4636–4640.
10. Griffin JH. Role of surface in surface-dependent activation of Hageman factor (blood coagulation factor XII). Proc Natl Acad Sci USA 1978; 75:1998–2002.
11. Cochrane CG, Revak SD, Wuepper KD. Activation of Hageman factor in solid and fluid phases. A critical role of kallikrein. J Exp Med 1973; 138:1564–1583.
12. Tankersley DL, Finlayson JS. Kinetics of activation and autoactivation of human factor XII. Biochemistry 1984; 23:273–279.
13. Scott CF, Silver LD, Schapira M, Colman RW. Cleavage of human high molecular weight kininogen markedly enhances its coagulant activity. Evidence that this molecule exists as a procofactor. J Clin Invest 1984; 73:954–962.
14. Silverberg M, Dunn JT, Garen L, Kaplan AP. Autoactivation of human Hageman factor. Demonstration utilizing a synthetic substrate. J Biol Chem 1980; 255:7281–7286.
15. Tans G, Rosing J, Berrettini M, Lämmle B, Griffin JH. Autoactivation of human plasma prekallikrein. J Biol Chem 1987; 262:11308–11314.
16. McMillin CR, Saito H, Ratnoff OD, Walton AG. The secondary structure of human Hageman factor (factor XII) and its alteration by activating agents. J Clin Invest 1974; 54:1312–1322.
17. Samuel M, Pixley RA, Villanueva MA, Colman RW, Villanueva GB. Human factor XII (Hageman factor) autoactivation by dextran sulfate. Circular dichroism, fluorescence and ultraviolet difference spectroscopic studies. J Biol Chem 1992; 267:19691–19697.
18. Tans G, Rosing J. Structural and functional characterization of factor XII. Semin Thromb Hemost 1987; 13:1–14.
19. Walsh PN. Platelet-mediated trigger mechanisms in the contact phase of blood coagulation. Semin Thromb Hemost 1987; 13:86–94.
20. Wiggins RC, Loskutoff DJ, Cochrane CG, Griffin JH, Edgington TS. Activation of rabbit Hageman factor by homogenates of cultured rabbit entothelial cells. J Clin Invest 1980; 65: 197–206.
21. Molla A, Yamamoto T, Akaike T, Miyoshi S, Maeda H. Activation of Hageman factor and prekallikrein and generation of kinin by various microbial proteinases. J Biol Chem 1989; 264:10589–10594.
22. Henry ML, Everson B, Ratnoff OD. Inhibition of the activation of Hageman factor (factor XII) by β_2-glycoprotein I. J Lab Clin Med 1988; 111:519–523.
23. Rehmus EH, Greene BM, Everson BA, Ratnoff OD. Inhibition of the activation of Hageman factor (factor XII) by complement subcomponent C1q. J Clin Invest 1987; 80:516–521.
24. Pixley RA, Schapira M, Colman RW. The regulation of human factor XIIa by plasma proteinase inhibitors. J Biol Chem 1985; 260:1723–1729.

25. Van der Graaf, Koedam JA, Bouma BN. Inactivation of kallikrein in human plasma. J Clin Invest 1983; 71:149–158.
26. Schapira M, Silver LD, Scott CF, et al. Prekallikrein activation and high molecular weight kininogen consumption in hereditary angioedema. N Engl J Med 1983; 308:1050–1053.
27. Lämmle B, Zuraw BL, Heeb MJ, et al. Detection and quantitation of cleaved and uncleaved high molecular weight kininogen in plasma by ligand blotting with radiolabeled plasma prekallikrein or factor XI. Thromb Haemost 1988; 59:151–161.
28. Cool DE, Edgell C-JS, Louie GV, Zoller MJ, Brayer GD, MacGillivray RTA. Characterization of human blood coagulation factor XII cDNA. Prediction of the primary structure of factor XII and the tertiary structure of β-factor XIIa. J Biol Chem 1985; 260:13666–13676.
29. Cool DE, MacGillivray RTA. Characterization of the human blood coagulation factor XII gene. Intron/exon gene organization and analysis of the 5′-flanking region. J Biol Chem 1987; 262:13662–13673.
30. Royle NJ, Nigli M, Cool DE, MacGillivray RT, Hamerton JL. Structural gene encoding human factor XII is located at 5q33-qter. Somat Cell Mol Genet 1988; 14:217–221.
31. Patthy L. Evolutionary assembly of blood coagulation proteins. Semin Thromb Hemost 1990; 16:245–259.
32. Gordon EM, Gallagher CA, Johnson TR, Blossey BK, Ilan J. Hepatocytes express blood coagulation factor XII (Hageman factor). J Lab Clin Med 1990; 115:463–469.
33. Fujikawa K, McMullen BA. Amino acid sequence of human β-factor XIIa. J Biol Chem 1983; 258:10924–10933.
34. McMullen BA, Fujikawa K. Amino acid sequence of the heavy chain of human α-factor XIIa (activated Hageman factor). J Biol Chem 1985; 260:5328–5341.
35. Clarke BJ, Cote HCF, Cool DE, et al. Mapping of a putative surface-binding site of human coagulation factor XII. J Biol Chem 1989; 264:11497–11502.
36. Schmeidler-Sapiro KT, Ratnoff OD, Gordon EM. Mitogenic effects of coagulation factor XII and factor XIIa on HepG2 cells. Proc Natl Acad Sci USA 1991; 88:4382–4385.
37. Revak SD, Cochrane CG, Bouma BN, Griffin JH. Surface and fluid phase activities of two forms of activated Hageman factor produced during contact activation of plasma. J Exp Med 1978; 147:719–729.
38. Ratnoff OD, Colopy JE. A familial hemorrhagic trait associated with a deficiency of a clot-promoting fraction of plasma. J Clin Invest 1955; 34:602–613.
39. Bernardi F, Marchetti G, Patracchini P, et al. Factor XII gene alteration in Hageman trait detected by *Taq*I restriction enzyme. Blood 1987; 69:1421–1424.
40. Bernardi F, Marchetti G, Volinia S, et al. A frequent factor XII gene mutation in Hageman trait. Hum Genet 1988; 80:149–151.
41. Saito H, Scott JG, Movat HZ, Scialla SJ. Molecular heterogeneity of Hageman trait (factor XII deficiency). Evidence that two of 49 subjects are cross-reacting material positive (CRM^+). J Lab Clin Med 1979; 94:256–265.
42. Saito H, Scialla SJ. Isolation and properties of an abnormal Hageman factor (factor XII) molecule in a cross-reacting material-positive Hageman trait plasma. J Clin Invest 1981; 68: 1028–1035.
43. Miyata T, Kawabata S, Iwanaga S, Takahashi I, Alving B, Saito H. Coagulation factor XII (Hageman factor) Washington D.C.: Inactive factor XIIa results from Cry 571 → Ser substitution. Proc Natl Acad Sci USA 1989; 86:8319–8322.
44. Takahashi I, Saito H. A rapid purification with high recovery of factor XII (Hageman factor) on immunoaffinity column: Application to an abnormal clotting factor XII (factor XII Toronto). J Biochem 1988; 103:641–643.
45. Wuillemin WA, Huber I, Furlan M, Lämmle B. Functional characterization of an abnormal factor XII molecule (F XII Bern). Blood 1991; 78:997–1004.

46. Berrettini M, Lämmle B, Clavarella G, Clavarella N. Functional and immunological studies of abnormal factor XII in a cross reacting material positive (CRM+) factor XII deficiency. Thromb Haemost 1985; 54:120 (abstr).
47. Wuillemin WA, Furlan M, Stricker H, Lämmle B. Functional characterization of a variant factor XII (FXII Locarno) in a cross reacting material positive FXII deficient plasma. Thromb Haemost 1992; 67:219–225.
48. Aznar J, Villa P, Vayá A, Mira Y, Lorengo I, España F. Functional anomaly of factor XII. Haemostasis 1992; 22:345–347.
49. Beaubien G, Rosinski-Chupin I, Mattei MG, Mbikay M, Chretien M, Seidah NG. Gene structure and chromosomal localization of plasma kallikrein. Biochemistry 1991; 30:1628–1635.
50. Kato A, Asakai R, Davie EW, Aoki N. Factor XI gene (F11) is located on the distal end of the long arm of chromosome 4. Cytogenet Cell Genet 1989; 52:77–78.
51. Chung DW, Fujikawa K, McMullen BA, Davie EW. Human plasma prekallikrein, a zymogen to a serine protease that contains four tandem repeats. Biochemistry 1986; 25:2410–2417.
52. McMullen BA, Fujikawa K, Davie EW. Location of the disulfide bonds in human plasma prekallikrein: the presence of four novel apple domains in the amino-terminal portion of the molecule. Biochemistry 1991; 30:2050–2056.
53. Page JD, Colman RW. Localization of distinct functional domains on prekallikrein for interaction with both high molecular weight kininogen and activated factor XII in a 28-kDa fragment. J Biol Chem 1991; 266:8143–8148.
54. Hathaway WE, Belhasen LP, Hathaway HS. Evidence for a new plasma thromboplastin factor I. Case report, coagulation studies and physicochemical properties. Blood 1965; 26:521–532.
55. Wuepper KD. Prekallikrein deficiency in man. J Exp Med 1973; 138:1345–1355.
56. Saito H, Goodnough LT, Soria J, Soria C, Aznar J, España F. Heterogeneity of human prekallikrein deficiency (Fletcher trait). Evidence that five of 18 cases are positive for cross-reacting material. N Engl J Med 1981; 305:910–914.
57. Bouma BN, Kerbiriou DM, Baker J, Griffin JH. Characterization of a variant prekallikrein, prekallikrein Long Beach, from a family with mixed cross-reacting material-positive and cross-reacting material-negative prekallikrein deficiency. J Clin Invest 1986; 78:170–176.
58. Wuillemin WA, Furlan M, von Felten A, Lämmle B. Functional characterization of a variant prekallikrein (PK Zürich). Thromb Haemost 1993; 70:427–432.
59. Takagaki Y, Kitamura N, Nakanishi S. Cloning and sequence analysis of cDNAs for human high molecular weight and low molecular weight prekininogens. Primary structures of two human prekininogens. J Biol Chem 1985; 260:8601–8609.
60. Kitamura N, Kitagawa H, Fukushima D, Takagaki Y, Miyata T, Nakanishi S. Structural organization of the human kininogen gene and a model for its evolution. J Biol Chem 1985; 260:8610–8617.
61. Cheung PP, Cannizzaro LA, Colman RW. Chromosomal mapping of human kininogen gene (KNG) to 3q26 → qter. Cytogenet Cell Genet 1992; 59:24–26.
62. Kellermann J, Lottspeich F, Henschen A, Müller-Esterl W. Completion of the primary structure of human high-molecular-mass kininogen. The amino acid sequence of the entire heavy chain and evidence for its evolution by gene triplication. Eur J Biochem 1986; 154:471–478.
63. Ohkubo I, Kurachi K, Takasawa T, Shiokawa H, Sasaki M. Isolation of a human cDNA for α_2-thiol proteinase inhibitor and its identity with low molecular weight kininogen. Biochemistry 1984; 23:5691–5697.
64. Sueyoshi T, Enjyoji K, Shimada T, et al. A new function of kininogen as thiol-proteinase inhibitors: inhibition of papain and cathepsin B, H and L by bovine, rat and human kininogens. FEBS Lett 1985; 182:193–195.
65. Müller-Esterl W, Fritz H, Machleidt W, et al. Human plasma kininogens are identical with α-cysteine proteinase inhibitors. FEBS Lett 1985; 182:310–314.

66. Asakura S, Hurley RW, Skorstengaard K, Ohkubo I, Mosher DF. Inhibition of cell adhesion by high molecular weight kininogen. J Cell Biol 1992; 116:465–476.
67. Müller-Esterl W, Iwanaga S, Nakanishi S. Kininogens revisited. TIBS 1986; 11:336–339.
68. Higashiyama S, Ohkubo I, Ishiguro H, Sasaki M, Matsuda T, Nakamura R. Heavy chain of human high molecular weight and low molecular weight kininogens binds calcium ion. Biochemistry 1987; 26:7450–7458.
69. Higashiyama S, Ohkubo I, Ishiguro H, Kunimatsu M, Sawaki K, Sasaki M. Human high molecular weight kininogen as a thiol proteinase inhibitor: presence of the entire inhibition capacity in the native form of heavy chain. Biochemistry 1986; 25:1669–1675.
70. Sugo T, Ikari N, Kato H, Iwanaga S, Fujii S. Functional sites of bovine high molecular weight kininogen as a cofactor in kaolin-mediated activation of factor XII (Hageman factor). Biochemistry 1980; 19:3215–3220.
71. Kunapuli SP, Dela Cadena RA, Colman RW. Deletion mutagenesis of high molecular weight kininogen light chain. Identification of two anionic surface binding subdomains. J Biol Chem 1993; 268:2486–2492.
72. Tait JF, Fujikawa K. Primary structure requirements for the binding of human high molecular weight kininogen to plasma prekallikrein and factor XI. J Biol Chem 1987; 262:11651–11656.
73. Kunapuli SP, Dela Cadena RA, Colman RW. Bacterial expression of biologically active high molecular weight kininogen light chain. Thromb Haemost 1992; 67:428–433.
74. Saito H, Ratnoff OD, Waldmann R, Abraham JP. Fitzgerald trait. Deficiency of a hitherto unrecognized agent, Fitzgerald factor, participating in surface-mediated reactions of clotting, fibrinolysis, generation of kinins, and the property of diluted plasma enhancing vascular permeability (PF/Dil). J Clin Invest 1975; 55:1082–1089.
75. Wuepper KD, Miller DR, Lacombe MJ. Flaujeac trait: deficiency of human plasma kininogen. J Clin Invest 1975; 56:1663–1672.
76. Colman RW, Bagdasarian A, Talamo RC, et al. Williams trait: human kininogen deficiency with diminished levels of plasminogen proactivator and prekallikrein associated with abnormalities of the Hageman factor-dependent pathways. J Clin Invest 1975; 56:1650–1662.
77. Hayashi H, Ishimaru F, Fujita T, Tsurumi N, Tsuda T, Kimura I. Molecular genetic survey of five Japanese families with high-molecular-weight kininogen deficiency. Blood 1990; 75: 1296–1304.
78. Cheung PP, Kunapuli SP, Scott CF, Wachtfogel YT, Colman RW. Genetic basis of total kininogen deficiency in Williams' trait. J Biol Chem 1993; 268:23361–23365.

14

Factor XIII

Thung-Shenq Lai and Charles S. Greenberg
Duke University Medical Center, Durham, North Carolina

I. ROLE OF PLASMA FACTOR XIII IN HEMOSTASIS

A. Overview

The initial hemostatic plug does not produce a permanent barrier to blood loss and must undergo stabilization by the plasma factor XIII molecule. At the site of vascular injury, a complex set of interactions develops between thrombin, fibrin polymers, plasma factor XIII, platelets, and the adhesive glycoproteins in the vessel wall. During the assembly of fibrin polymers, thrombin cleaves plasma factor XIII to form factor XIIIa (1–3), the fibrin stabilizing factor (4). Factor XIIIa enzymatically transforms the loose fibrin network into a stable covalent collection of fibrin fibers. Factor XIIIa functions within the fibrin clot to catalyze the formation of intermolecular γ-glutamyl-ϵ-lysyl covalent bonds between fibrin molecules and between fibrin and several other plasma (1–3) and extracellular matrix proteins (5–7). The covalently modified fibrin clot is mechanically stronger, less deformable, and more resistant to lysis by plasmin than the non–cross-linked fibrin network (2).

The importance of fibrin stabilization is well documented by the serious lifelong bleeding disorder of individuals congenitally deficient in the factor XIII molecule. Congenital deficiency of plasma factor XIII was first reported in 1960 by Duckert and colleagues many years after the discovery of the in vitro fibrin stabilization process (8). Since that initial case report, over 200 cases of congenital factor XIII deficiency have been reported in the literature. A detailed historical account of the discovery of factor XIII and the biochemical basis of fibrin stabilization was published by Loewy et al. (9,10). Since Robbins first reported in 1944 a serum-derived fibrin stabilizing factor (11), the plasma factor XIII subunits were purified (10,12), the cDNA sequenced, and the gene structures established (13–16). In this chapter, we review the structure and function of plasma factor XIII and the fibrin stabilization process as currently understood through recent advances in molecular biology.

B. Plasma Factor XIII Subunits

The extracellular form of blood coagulation factor XIII, known as the plasma factor XIII molecule, is a tetramer, composed of two A chains ($M_r \simeq 75$ kD) and two B chains ($M_r \simeq 80$ kDa) that are noncovalently associated to produce the 340 kD zymogen. The A chains and B chains are distinct gene products without homology to each other. The factors regulating the synthesis, secretion, and assembly of the plasma factor XIII molecule remain very poorly understood. The plasma levels of factor XIII A chains and B chains were established by enzyme-linked immunosorbent assay determination at 0.13–0.16 and 0.26–0.28 μM respectively (17). The A chains are completely complexed with the B chains to form the plasma factor XIII tetramer (A_2B_2); half of the B chains are in a free form (17). The factor XIII A chain gene product is also found intracellularly in platelets, megakaryocytes, monocytes, and histiocytes (18–22). The intracellular form is a dimer composed of two A chains that are identical to the plasma factor XIII A chains (13–15,23). In human blood, ~50% of the total factor XIIIa activity exists inside platelets as an A chain dimer (24). The A chains are responsible for the catalytic activity (25), whereas the B chains serve as a carrier protein that stabilizes the A chains in the circulation and regulates the calcium-dependent activation of plasma factor XIII (26,27).

The assembly of the tetrameric plasma factor XIII molecule remains a mystery. Bone marrow transplantation studies demonstrate that factor XIII A chains are synthesized in the bone marrow by monocytes and megakaryocytes (21,28). The mRNA for the A chain was also detected in human monocytes and the transformed monocyte cell line U937 (29). The liver was also suggested as another site for factor XIII A chain synthesis because factor XIII A chain mRNA was detected in human liver samples (29) and the factor XIII A chain antigen was found in hepatocytes (30,31) and two hepatoma cell lines, HepG2 and PLC/PRF/5 (32). However, Grundmann et al. did not detect factor XIII A chain mRNA in human liver extracts (15). Further investigation is needed to define whether hepatocytes, *Kupffer*'s cells, or sinusoidal endothelial cells are the site(s) of A chain synthesis in the liver.

Human liver transplantation studies demonstrate that factor XIII B chains are produced by the liver (21). The studies by Nagy et al. using HepG2 cell lines also support the conclusion that hepatocytes are the major site for B chain synthesis (32). Patients with congenital factor XIII A chain deficiency usually have a reduced level of plasma factor XIII B chain (29–50%) concentrations, suggesting that the factor XIII A chain can alter the half-life or stability of the B chain. An alternative hypothesis is that the biosynthesis of the B chain is decreased in the absence of factor XIII A chain (33,34). A recent report describes a patient with a congenital B chain deficiency. Factor XIII B chains were undetectable in the patient's plasma (35). The level of factor XIII A chains in platelets of the B chain-deficient patient was normal, but the half-life of the A chain determined from the disappearance curve of infused factor XIII concentrate was reduced (35). These results support an important role for the B chain in regulating factor XIII A chain half-life in plasma (35). Because most of the A chains are apparently derived from the bone marrow and the B chains are from the liver, there must be a high degree of affinity of the A chains for the B chains in human plasma. The addition of purified A chains to purified plasma factor XIII B chains produces a tetrameric complex (2). Therefore, there is no biochemical prerequisite for these distinct subunits to be synthesized at the same site. The mechanism(s) by which the genes for these subunits are regulated and the process regulating protein secretion are also unknown.

C. Activation of Plasma Factor XIII

The activation process of the tetrameric plasma factor XIII (A2B2) and the dimeric platelet factor XIII (A2) share several similar biochemical features but are not identical. The activation of plasma factor XIII is more complex because of the presence of factor XIII B chains. Plasma factor XIII is activated by thrombin cleavage of the Arg^{37}-Gly^{38} peptide bond and the release of a 37 amino acid activation peptide (4 kD) from the N terminus of the A chains (41). A single proteolytic event is required for the dimeric platelet factor XIII zymogen to be converted to factor XIIIa (3). In the presence of calcium ions, the B chains dissociate from the tetramer and a calcium-dependent conformational change unmasks the active-site thiol group on the A chain, forming factor XIIIa (Fig. 1) (42). Factor XIIIa apparently can express only one active-site cysteine at a time, although some investigators reported that both sites can be exposed simultaneously. Kinetic studies reveal that the binding of two calcium ions is required to convert thrombin-cleaved plasma factor XIII to factor XIIIa (43). The platelet factor XIII zymogen does not require thrombin proteolysis or calcium to express the active-site cysteine (43). Apparently, the B chains prevent exposure of the active site in the plasma factor XIII molecule.

Fibrinogen plays an important role in plasma factor XIII activation (Fig. 1). Fibrinogen reduces the calcium ion concentration (1.5 mM Ca^{2+}) required for B chain dissociation (43–45). In the absence of fibrinogen, the B chains dissociate from thrombin-cleaved plasma factor XIII only at calcium ion concentrations that are 10-fold higher than exist

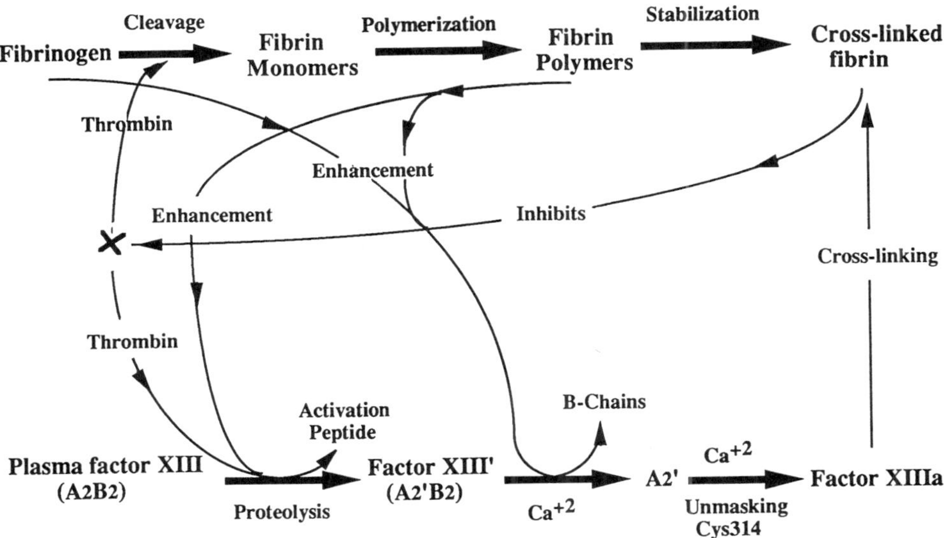

Figure 1 Regulation of fibrin cross-linking by factor XIII-fibrin(ogen) interactions. The activation of plasma factor XIII and the formation of cross-linked fibrin are processes that are very closely related to each other. Thrombin cleavage of plasma factor XIII is enhanced by fibrin polymers. Fibrinogen and fibrin promote the calcium-dependent dissociation of thrombin-cleaved factor XIII ($A_2'B_2$) to yield the active enzyme, factor XIIIa, which has the active-site Cys^{314} residue exposed for catalytic cross-linking of fibrin polymers. Once factor XIIIa cross-links fibrin, thrombin cleavage of factor XIII is inhibited (marked by X).

in plasma (43–45). In addition, thrombin cleavage of plasma factor XIII is dramatically enhanced by fibrin(ogen) (46). The polymerization of fibrin actually increases the rate of thrombin cleavage of the plasma factor XIII A chains by providing binding sites for the plasma factor XIII molecule and thrombin (Fig. 1) (44). The capacity of fibrin to promote plasma factor XIII activation ensures that factor XIIIa is formed after a critical mass of fibrin is assembled. Once fibrin polymers are cross-linked by factor XIIIa, they no longer function to accelerate thrombin cleavage of plasma factor XIII (47,48). This well-regulated series of reactions between thrombin, fibrin, and plasma factor XIII carefully limits the formation of factor XIIIa in plasma.

In the presence of high thrombin concentrations, factor XIIIa is also cleaved at the Lys^{513}-Ser^{514} position to form a Gly^{38}-Lys^{513} (51 kD) fragment and a C-terminal 24 kD peptide. The proteolysis of the A chain at the Lys^{513} site is influenced by the presence of calcium ions (49). When calcium ions are present during this secondary proteolytic event, a catalytic 51 kD fragment is produced (50). The Gly^{38}-Lys^{513} (51 kD) fragment retains factor XIIIa activity, can bind to fibrin, and behaves as a monomer in gel filtration analysis (50). These results were confirmed by thrombin activating the purified Met^{1}-Asp^{520} A chain fragment expressed in *Escherichia coli* (51). The thrombin-activated Met^{1}-Asp^{520} fragment displayed ~25% of the specific activity of full-length factor XIIIa and could also bind to fibrin (51). In preliminary studies, the Met^{1}-Asp^{520} fragment also behaved as a monomer in gel filtration analysis.

The concentration of fibrinogen in plasma is ~100-fold higher than the concentration of the plasma factor XIII molecule (3). To catalyze a substantial number of covalent bonds between polymerizing fibrin molecules and the other substrate molecules at the site of blood clotting, factor XIIIa must interact with several sites on fibrin. Specific binding sites for both factor XIII and factor XIIIa exist on fibrinogen and fibrin (3,52,53). Furthermore, the zymogen forms of platelet and plasma factor XIII do not bind to the same sites as factor XIIIa (3,52,53). The best characterized site for plasma factor XIII binding was localized to the midportion of the Aα chain of fibrinogen (residues 243–476) (54). This portion of fibrinogen promotes the calcium-dependent dissociation of the B chain from thrombin-cleaved plasma factor XIII (3). The B chains and the activation peptides from the A chains are not bound to the cross-linked fibrin matrix (3).

Factor XIIIa also catalyzes the covalent cross-linking of several other plasma proteins, including α_2-plasmin inhibitor (55–58), fibronectin (6,7), von Willebrand factor (59,60), thrombospondin (61), vitronectin (62), and lipoprotein(a) (63). The proteins that function as factor XIIIa substrates also play an important role in fibrinolysis (55–58,63), cell adhesion (64,65), and wound healing (66). These reactions are discussed in the following section.

D. Factor XIIIa Substrates Involved in Hemostasis

The fibrinogen molecule is a soluble protein composed of six polypeptide chains with a very well defined structure, which is summarized in several recent reviews (67) and in Chapter 3. Fibrinogen consists of pairs of three distinct polypeptide chains, designated Aα (66 kD), Bβ (53 kD), and γ (46 kD), with a combined M_r of 340 kD. The cleavage of the Aα chains and Bβ chains by thrombin releases fibrinopeptide A, an 18 amino acid residue, from each α chain and a 20 amino acid residue of fibrinopeptide B from each β chain, respectively. The fibrinogen molecule devoid of fibrinopeptides A and B is called a fibrin monomer and develops adhesive properties that promote fibrin polymerization

(67). The factor XIIIa cross-linking sites that yield dimeric fibrin molecules are located within the carboxyl termini of the γ chains in close proximity to the sites involved in fibrin polymerization (68). The γ dimer is formed by the cross-linking reaction between the Gln^{398} and Lys^{406} in one γ chain with Lys^{406} and Gln^{398}, respectively, of a second γ chain in an adjacent molecule. The fibrin polymerization site was located between residues 356–411 on the γ chain (68). When the tetrapeptide GPRP was used to inhibit fibrin polymerization, the γ chains lost their ability to serve as the glutamine cross-linking sites (69). This suggests that the cross-linking sites must be exposed in the proper conformation for factor XIIIa to recognize them (69). The initial rapid covalent dimerization of the γ chains from different fibrin monomers is followed by the much slower cross-linking of the α chains of fibrin to form α chain polymers (70). There are five potential cross-linking sites localized to the Aα chain at residues Gln^{328}, Gln^{366}, Lys^{508}, Lys^{556}, and Lys^{562}. However, the exact geometry and sequence of assembling the α chain cross-linked products have not been established. Recent studies demonstrate that factor XIIIa can catalyze rapid α-γ crosslinks in a γ dimer-dependent reaction (71). Fibrinogen can be cross-linked into high-molecular-weight soluble multimers by factor XIIIa, but at a much slower rate than fibrin (67). In the presence of a high fibrinogen concentration, the rate of fibrinogen cross-linking can approach rates of fibrin cross-linking (2). This suggests that the crosslink sites are exposed on fibrinogen and need to be properly aligned to facilitate the formation of the isopeptide bond between the molecules.

There is controversy regarding the orientation of the isopeptide bond in relationship to the long axis of the fibrin monomer. It was originally reported that the factor XIIIa bond formed between two fibrin molecules aligned within the long axis of adjacent fibrin molecules resulted in a *cis* pattern of cross-linking (72). However, these studies were performed using fibrinogen as the substrate, not fibrin (72). Recent studies suggest that factor XIIIa cross-links fibrin molecules that are aligned between the half-staggered overlapping fibrin monomer subunits within the fibrin protofibril, called the *trans* pattern of cross-linking (73). Additional studies are needed to establish the molecular basis of factor XIIIa catalysis of intermolecular isopeptide bonds during fibrin clot formation.

The plasma glycoprotein α_2-plasmin inhibitor (α2AP) is a 67 kD serine esterase inhibitor that is one of the major physiological inhibitors of plasmin in human plasma. The α2AP molecule is rapidly cross-linked to the Aα chains of fibrin during blood clotting (55). The α2AP cross-linking sites are localized to the N terminus of α_2-plasmin inhibitor at Gln^2 and to the carboxyl terminus of the Aα chain of fibrin(ogen) at Lys^{303} (56,57). Within the fibrin clot, ~1 in 25 fibrin molecules contains a covalent α_2-plasmin inhibitor complex capable of reacting with plasmin (58). The cross-linking of α_2-plasmin inhibitor to the Aα chains of fibrin substantially increases the resistance of fibrin clots to plasmin degradation (74) and is an important reaction for normal hemostasis (1,2). The clinical syndrome of α2AP deficiency closely resembles that of factor XIII deficiency (75). This further supports the concept that a major portion of the physiological function of factor XIIIa is dependent on α2AP cross-linking to fibrin. Platelets can also release and cross-link α2AP to fibrinogen bound to platelet aggregates. Reed et al. (76) reported that platelet factor XIIIa cross-linking of α2AP to platelet fibrinogen makes the platelet aggregate resistant to degradation by plasmin.

Fibronectin is a 400 kD glycoprotein present in plasma, on cell surfaces, and in the extracellular matrix surrounding cells that plays an important role in cell adhesion (5,6). Factor XIIIa catalyzes the cross-linking of fibronectin to the Aα chain of fibrin and to collagen types I, II, III, and V (6,7). Fibronectin is preferentially cross-linked to α chain polymers, and the α_2-plasmin inhibitor molecule is cross-linked to α monomer and α polymers at a similar rate (58,77). The cross-linking of fibronectin and α_2-plasmin inhibitor to fibrin are not competitive reactions and proceed independently (58). The cross-linking of fibronectin to fibrin occurs after α chain cross-linking, but the cross-linking of α_2-plasmin inhibitor to α chain can occur earlier than α chain cross-linking (58). The cross-linking of fibronectin to fibrin clots affects the mechanical properties of the clot and promotes cellular adherence and migration of cells into the fibrin clot (64,65). The cross-linking of fibronectin to collagen could be important for anchoring fibrin to subendothelial tissue and facilitating wound-healing processes (66). The cross-linking of fibronectin to the α chains of fibrin (58) and to collagen (78) also suggests that this reaction plays an important role in stabilizing the structures of basement membrane (18) at sites of tissue injury (67). Factor XIIIa binds to fibroblasts and catalyzes the assembly of insoluble fibronectin fibrils at the cell surface (79). Factor XIIIa bound to the fibroblast membrane is also internalized and degraded, providing a catabolic pathway for factor XIIIa in tissues (79).

Thrombospondin (TSP) is a 420–540 kD plasma glycoprotein that is secreted from the α granules of thrombin-stimulated platelets (80). During blood coagulation, it is covalently and noncovalently incorporated (61,81) into the polymerizing fibrin clot. The incorporation of TSP into a fibrin clot appears to be the result of a strong noncovalent interaction with fibrin (61,81). The cross-linking of TSP by factor XIIIa only slightly increases the amount of TSP associated with fibrin, and the physiological function of this reaction is not established (61,81).

There are a host of other plasma and platelet proteins molecule that serve as substrates for factor XIIIa, including von Willebrand factor (59,60), vinculin (82), vitronectin (62), actin (83), myosin (84), and lipoprotein (a) (63). The von Willebrand factor molecule can be cross-linked to fibrin and collagen, suggesting its function as a molecular bridge between the fibrin clot and the collagen in the subendothelium (59,60). The location of the crosslink sites in von Willebrand factor have not been identified. The extent of von Willebrand factor cross-linking to fibrin depends on the rate of fibrin formation. When fibrin is formed slowly, factor XIIIa cross-linking of von Willebrand factor to fibrin is detected (60). Vitronectin can also serve as a factor XIIIa substrate (62), although the physiological significance of this reaction remains to be established. Lipoprotein (a), a protein with striking structural homology to plasminogen, is also a factor XIIIa substrate (63). Because lipoprotein (a) can antagonize fibrinolysis, factor XIIIa cross-linking of Lp(a) could contribute to alterations in the fibrinolytic response in vascular tissues. The actin (83), myosin (84), and vinculin (82) molecules associated with the platelet cytoskeleton also serve as factor XIIIa substrates, along with platelet membrane glycoproteins IIb and IIIa (85). The cross-linking of adhesive glycoproteins to platelet membranes and cytoskeletal components could function during vascular injury to aid wound healing and to promote hemostasis. To date, molecular defects in the cross-linking sites of these proteins are unidentified. However, mutations in the cross-linking sites of these proteins could lead to an abnormal hemostatic response.

II. FACTOR XIII AND THE TRANSGLUTAMINASE GENE FAMILY

The factor XIII A chain belongs to the transglutaminase (E.C. 2.3.2.13) gene family that catalyzes the stabilization of tissues (1). The factor XIII A chain is a unique member of the transglutaminase gene family, being a plasma glycoprotein zymogen. The enzyme, factor XIIIa, originates from thrombin proteolysis of the A chains.

The transglutaminases are calcium- and thiol-dependent enzymes that are found in almost all human tissues and body fluids. These enzymes are widely distributed and highly conserved among the vertebrates and are also found in a number of invertebrates, including sea urchins, *Homarus* (36), insect (37), and *Limulus* species (38). The transglutaminases catalyze the posttranslational modification of proteins by transamidation of available glutamine residues. A modified double-displacement reaction mechanism is used to catalyze a calcium-dependent acyl transfer reaction between the γ-carboxamide group of a peptide-bound glutamine residue and the ϵ-amino group of a peptide-bound lysine or the primary amino group of a polyamine (Fig. 2). The active-site cysteine reacts with the γ-carboxamide of the glutamine, forming a γ-glutamyl thioester and releasing ammonia. The transient acyl-enzyme intermediate then reacts with any nucleophilic primary amine, yielding either an isopeptide bond or a (γ-glutamyl) polyamine bond (Fig. 2). When an amine is not available, the acyl-enzyme intermediate reacts with water to form a glutamic acid residue (39,40).

The primary amino acid sequence of the factor XIII A chain and several other transglutaminases, including tissue transglutaminase (TGc or type II) and keratinocyte transglutaminase (TGk or type I), were established by sequencing human cDNA clones (1). The erythrocyte membrane protein designated band 4.2 also belongs to the transglutaminase gene family because it shares significant amino acid sequence homology with these enzymes. However, band 4.2 does not contain an active-site cysteine and cannot function as a transglutaminase. A detailed account of the function of the other transglutaminases is reported elsewhere (1) and is beyond the scope of this review. In general, the transglutaminases are involved in specialized biological processes producing tissue stabilization through the covalent cross-linking of proteins, such as skin formation, apoptosis (programmed cell death), hair formation, and cell adhesion (1).

III. PRIMARY STRUCTURE OF THE FACTOR XIII A CHAIN

The primary structure of the human factor XIII A chain was established by a combination of cDNA and protein sequencing methods (14). The factor XIII A chain cDNA sequence was obtained by screening and sequencing cDNA clones derived from a human placental

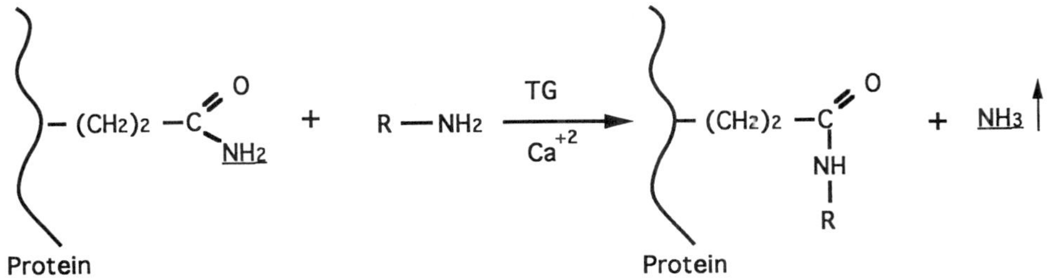

$$\text{Protein}-(CH_2)_2-C(=O)NH_2 + R-NH_2 \xrightarrow[Ca^{+2}]{TG} \text{Protein}-(CH_2)_2-C(=O)-NH-R + NH_3\uparrow$$

Figure 2 The cross-linking reaction as catalyzed by the transglutaminases. The reaction involves the exchange of primary amines (R-NH_2) for ammonia at the γ-carboxamide group of glutamine residues and results in the formation of an isopeptide bond. The details are described in the text. (From Ref. 1.)

cDNA library (15). The deduced amino acid sequence was in good agreement with the protein sequencing data from purified placental factor XIII A chain (14). Ichinose et al. also reported that the protein sequence data for 363 amino acids from the purified plasma factor XIII A chain was identical to the sequence deduced from analyzing placental cDNA (13). These data document that the A chain sequence from placental and plasma factor XIII are quite similar.

The full-length mRNA of the factor XIII A chain gene is 4 kb and contains 2196 nucleotides of protein coding sequence (13,15). The deduced amino acid sequence starting from Ser^2 contains 731 amino acid residues with a molecular weight of 83,150. There were six potential N-glycosylation sites identified; however, carbohydrate modification was not detected by protein sequencing of the A chains (14). There are nine cysteine residues present in the factor XIII A chain, and none of them form disulfide bonds (14,39,86). The mechanism of factor XIII A chain secretion into the plasma is unknown. The cDNA sequence reveals the absence of the typical N-terminal hydrophobic peptide and internal signal sequence utilized for protein secretion (13,15). Furthermore, because the mature factor XIII A chain N terminus is acetylated and the protein is not glycosylated or disulfide bonded, this suggests that the protein is predominantly a cytoplasmic protein. Immunohistochemical studies have detected factor XIII A chains within the cytoplasm of platelets (87), monocytes (88), and tissue macrophages (89).

The primary structure of the factor XIII A chain protein can be divided into several distinct functional domains defined by specific exons (Fig. 3) (90). The active site, which contains the reactive thiol group at Cys^{314}, is present in exon VII (Fig. 3) (91). The active-site sequence is conserved among the other members of the transglutaminase gene

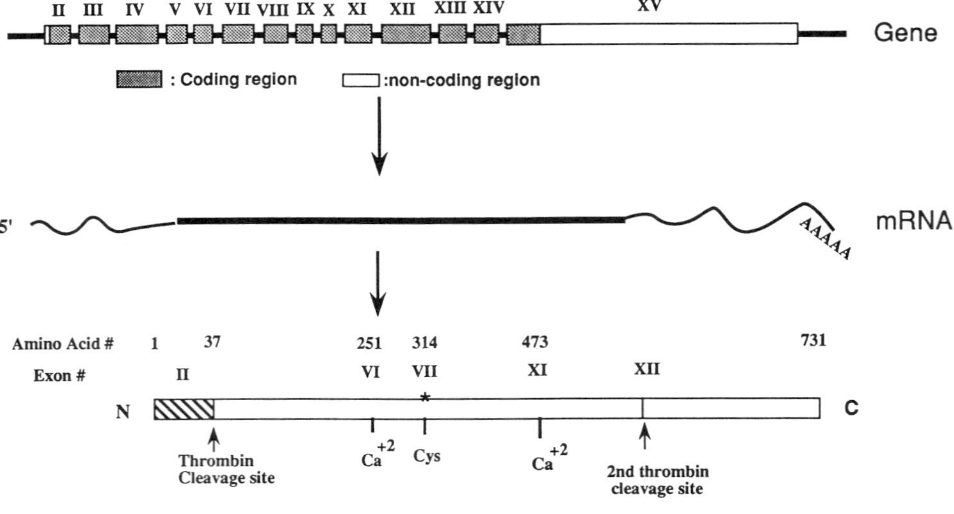

Figure 3 Gene structure, mRNA, and protein of factor XIII A chains. The exon number is shown above each exon, and its size is drawn to scale. The intron size is not drawn to scale. The various domains, activation peptide, putative calcium binding site, active site Cys^{314}, and thrombin cleavage site are indicated, and the corresponding exon and amino acid number are also shown above the factor XIII A chain diagram.

family (Fig. 4). If one assumes that these genes have similar intron-exon boundaries and align the published sequences, a high degree of identity is maintained for the human sequences, including human tissue TGc, keratinocyte TGk, and band 4.2 (Fig. 4) (92–95). In band 4.2, however, an Ala^{298} is substituted for Cys^{298} and the protein does not display transglutaminase activity. No dysfunctional factor XIII A chain molecules are reported with mutations in the active site. It is interesting to speculate that mutations in the active-site pocket could produce dysfunctional factor XIIIa and a bleeding disorder.

The fibrin binding domain covers exons II–XII and was initially localized to the Gly^{38}-Lys^{513} fragment produced by proteolysis of platelet factor XIII A chains (50). This was confirmed by using *E. coli* to express the Met^{1}-Asp^{520} sequence. The purified protein could bind and cross-link fibrin after being thrombin activated (51). Recent studies from a patient with an acquired inhibitor (antifactor XIII A chain IgG) to factor XIII A chains demonstrated that the antibody prevented the Gly^{38}-Lys^{513} fragment from binding to fibrin and inhibited fibrin cross-linking (96).

This laboratory used several synthetic peptides derived from the factor XIII A chain sequence to probe the substrate recognition sites on factor XIII. We identified two domains, Asn^{72}-Asp^{97} and Asp^{190}-Phe^{230}, derived from exons III and V (Fig. 5), respectively, that inhibited the factor XIIIa cross-linking of fibrin (97). The synthetic peptides did not block factor XIIIa binding to fibrin at concentrations that completely inhibited cross-linking, illustrating that the substrate recognition sites and the fibrin binding sites are distinct domains within the Gly^{38}-Lys^{513} fragment (97). The amino acid sequences in

A. Exon VII (Active site)

```
FXIII A 266 V N A K D D E G V L V G S W D N I Y A Y G V P P S A W T G 294
TGk     329 V N S L D D N G V L I G N W S G D Y S R G T N P S A W V G 357
TGc     227 V N C N D D Q G V L L G R W D N N Y G D G V S P M S W I G 255

FXIII A 295 S V D I L L E Y R S S E N P • V R Y G Q C W V F A G V F N T 323
TGk     358 S V E I L L S Y L R T G Y S • V P Y G Q C W V F A G V T T T 386
TGc     256 S V D I L R R W K N H G C Q R V K Y G Q C W V F A A V A C T 285
```

B. EXON XI (Calcium Binding Site)

```
FXIII A 435 V N S D L I Y I T A K K D G T H V V E N V D A T H 459
TGk     497 V N S D K V Y W Q R Q D D G S F K I V Y V E E K A 521
TGc     397 V N A D V V D W I Q Q Q D G S V H K S I H R L L V 421

FXIII A 460 I G K L I V T K Q I G G D G M M D I T D C Y K F Q E 485
TGk     522 I G T L I V T K A I S S N M R E D I T Y L Y K H P E 547
TGc     422 • G L K I S T K S V G R D E R E D I T H C Y K Y P E 446
```

Figure 4 Alignment of the amino acid sequence derived from the exon VII (active site) and XI (major calcium binding site) of factor XIII A chain gene with that of TGc and TGk. The amino acid numbers are shown at the beginning and end of each sequence. (From Refs. 16, 109, and 111.)

exons III and V of factor XIII A chain are quite conserved among other transglutaminases (Fig. 5).

A major calcium binding site flanking Gly^{473} was tentatively located by Takahashi et al. (14) by identifying partial amino acid sequence homology with the classic E-F hand structure of calmodulin (98). There are two other low-affinity sites postulated to exist in the C terminus of the Ser^{514}-Met^{730} fragment (14). Ichinose et al. also suggested that a calcium binding site was present in a negatively charged region surrounding Gly^{251} (13). However, the regions flanking the Gly^{251} and Gly^{473} do not have the typical secondary structure of a E-F hand, which usually appears as a loop flanked by two α helices (98). The absence of an E-F hand structure could explain the low-affinity constant for calcium ions (10^{-4} M) reported for the factor XIII A chains (99). The amino acid sequence surrounding Gly^{473} in exon XI is quite conserved among the various transglutaminases (Fig. 4), but the region flanking Gly^{251} is not conserved. In preliminary studies, we found that once we deleted the major calcium binding site by introducing a stop codon after Lys^{462}, we lost >99% of the transglutaminase activity. These data support a major role for the Gly^{473} site in regulating transglutaminase activity.

Several investigators reported minor differences in the cDNA nucleotide sequence for the A chain. These differences appear to represent genetic polymorphisms that exist within the normal population (13-15). DNA polymorphisms exist in both the coding and noncoding regions of the A chain gene. The DNA polymorphisms for the protein

A. Exon III.

```
FXIII A  43  E • F L N V T S V H L F K E R W D T N K V D  63
   TGk  106  E G M L V V N G V D L L S S R S D Q N R R E 127
   TGc    1  • • • A E E L V L E R C D L E L E T N G R D  19

FXIII A  64  H H T D K Y E N N K L I V R R G Q S F Y V Q  85
   TGk  128  H H T D E Y E Y D E L I V R R G Q P F H M L 149
   TGc   20  H H T A D L C R E K L V V R R G Q P F W L T  41

FXIII A  86  I D L • S R P Y D P R R D L F R V E Y V I 105
   TGk  150  L L L • S R T Y E S S • D R I T L E L L I 168
   TGc   42  L H F E G R N Y E A S V D S L T F S V V T  62
```

B. Exon V.

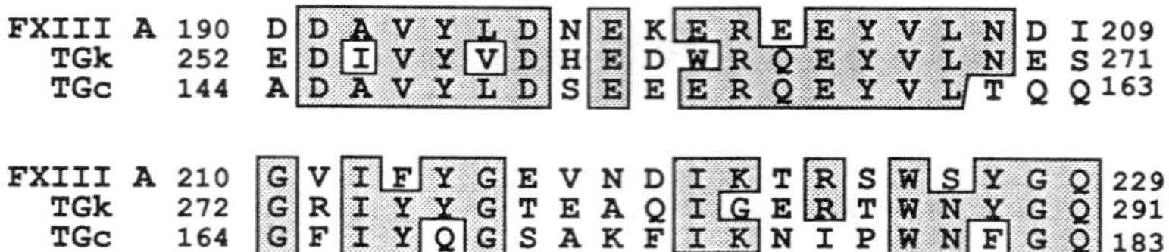

Figure 5 Amino acid sequence of the putative substrate recognition site, exons III and V, of factor XIII A chain gene is aligned with the relative amino acid sequences of TGc and TGk. The amino acid numbers are shown at the beginning and end of each sequence. (From Refs. 16, 109, and 111.)

coding region are listed in Table 1. A single nucleotide change produces the following amino acid changes: Arg^{77} to Gly^{77}, Arg^{78} to Lys^{78}, Phe^{88} to Leu^{88}, Pro^{564} to Leu^{564}, Val^{650} to Ile^{650}, and Glu^{651} to Gln^{651}. The polymorphisms at amino acid positions 77 and 78 were derived from the protein sequencing data and need to be confirmed by DNA sequencing studies (14). A polymorphism at amino acid 567 (GAA to GAG) does not change the amino acid residue; however the nucleotide change causes a loss of an EcoRI site (13,15). The EcoRI polymorphism could be a useful genetic marker in future studies. The protein charge polymorphism in the A chain allows genotype analysis by gel electrophoresis (101,102). These amino acid substitutions do not appear to produce any alteration in the protein's function. The factor XIII A chain cDNAs cloned by Ichinose et al. (Ile^{650}, Gln^{651}) and Board et al. (Val^{650}, Glu^{651}) were expressed and characterized in yeast and *E. coli* and show similar fibrin cross-linking and transglutaminase activities (51,103,104).

Electron microscopic studies revealed that isolated factor XIII A chains form dimers that are compact, slightly asymmetrical globular particles with diameters of about 6 × 9 nm (105). Sedimentation velocity experiments also suggest the dimer is a compact particle of 8.02*S* (103). Purified B chains appear as flexible rod-like molecules that are 30 nm long by ~2–3 nm in diameter (see later) (105). A model for the plasma factor XIII A2B2 tetramer is proposed to have A2 dimers forming the core and the two B chains wrapping around the outside (105).

The recent molecular cloning of several other human transglutaminases permits a direct comparison of their primary structures with the factor XIII A chain. The primary structure of the A chain shows significant amino acid sequence homology with the human tissue TGc (type II), keratinocyte TGk (type I), and erythrocyte band 4.2, especially in the regions flanking the active site and putative calcium binding site (1). The factor XIII A chain shares 45, 39, and 25–29% amino acid identity with keratinocyte transglutaminase (type I), endothelial tissue transglutaminase (type II), and erythrocyte band 4.2, respectively (93). Factor XIII A chain is more closely related to keratinocyte transglutaminase than to the other two members of the transglutaminase gene family.

Table 1 Differences in the Primary Sequence of the Factor XIII A Chains Derived from cDNA, Protein, or Gene Sequence

Amino acid no.	Kamura et al. (34)	Grundmann et al. (15)	Ichinose et al. (13,16)	Takahashi et al. (14)	Board et al. (51,104)
77	Arg (CGC)	Arg (CGC)	Arg (CGC)	Arg Gly	Arg (CGC)
78	Arg (AGA)	Arg (AGA)	Arg (AGA)	Arg Lys	Arg (AGA)
88	Phe (TTC)	Leu (CTC)	Phe (TTC)	Phe	Phe (TTC)
564	Pro (CCG)	Pro (CCG)	Leu (CTG) Pro (CCG)	Pro	Pro (CCG)
567	Glu	Glu (GAG)	Glu (GAA)	Glu	Glu (GAA)
650	Ile (ATT)	Val (GTT)	Val (GTT) Ile (ATT)	Val	Val (GTT)
651	Gln (CAG)	Gln (CAG)	Gln (CAG) Glu (GAG)	Glu	Glu (GAG)

IV. GENE STRUCTURE OF FACTOR XIII A CHAIN

The factor XIII A chain gene has 15 exons separated by 14 introns and is more than 160 kb in size (16). The size of each intron is currently unknown. The gene (F13A) was localized to chromosome 6 bands p24–p25 by *in situ* hybridization studies and shows linkage with the major histocompatibility complex (106,107). The placental and plasma factor XIII A chains are the product of the same gene, based on the concordance of the electrophoretic phenotypes of the factor XIII A chain in placental and plasma samples from 10 different subjects (106). The various exons divide the protein into distinct structural and functional domains. The A chain gene has a noncoding first exon. Exon 2 encodes the activation peptide and the first thrombin cleavage site (16). The domains proposed as calcium binding sites are localized within exons VI and XI (16). The active-site Cys^{314} region is encoded by exon VII. A second thrombin cleavage site thought to inactivate factor XIIIa is encoded by exon XII (Fig. 3) (16). Putative substrate recognition sites are coded by exons III and V (108).

The transcription initiation site for the A chain is unknown. Ichinose et al. proposed three potential transcription initiation sites, GATAA, TAGAA, and TTTAA, which were present 170, 150, and 40 bp upstream of the initiator methionine, respectively (16). Additional experiments are needed to identify which site or sites function as the transcription start site. Factors that regulate the expression of the factor XIII A chain gene are also unknown. Transfection experiments using promoter-driven expression vectors in an appropriate responsive cell are required to elucidate the nature of these regulatory elements.

Comparison of the organization and structure of the gene coding for the keratinocyte transglutaminase (TGk) and erythrocyte band 4.2 has revealed some common features with the factor XIII A chain gene (109,110). The TGk gene is similar to the factor XIII A chain gene and consists of 15 exons. Except for the first, second, and the last exons, the sizes of the remaining exons are almost identical between these two genes (Table 2) (111). Some distinct functional domains of TGk are coded by the same exons: for example, the active site is located within exon VII, and the major calcium binding site is in exon XI. The differences between these two genes are located at exons I, II, and XV. Exon II of the factor XIII A chain gene encodes 43 amino acid residues, including the activation peptide and first thrombin cleavage site, and exon II of TGk gene encodes a 106 amino acid sequence of a hydrophilic region that functions to localize TGk to the cytoplasmic side of the keratinocyte membrane. After removal of the NH_2-terminal extension of 43 and 106 amino acids, respectively, factor XIII A chain and TGk are virtually the same length as TGc (1). Thus, these enzymes could have diverged from a common transglutaminase lineage upon the addition of an NH_2-terminal extension, conferring specialized membrane cross-linking properties on the protein. Despite these similarities, the factor XIII A chain gene spans more than 140 kb, whereas the TGk gene spans only 14 kb.

The gene structure of band 4.2 differs slightly from that of factor XIII A chain and TGk. Band 4.2 gene has 13 exons and 12 introns (112). Despite that sequence identity between the corresponding exons is low, the size of the exons from exon 3 to 14 are very similar. The gene structure for the other transglutaminases remains to be determined.

V. PRIMARY STRUCTURE OF FACTOR XIII B CHAINS

The factor XIII B chain gene is a member of the complement regulatory gene family and functions to bind the A chain and stabilize its circulation in the plasma (13,26,27).

The complete amino acid sequence of the B chain of human factor XIII was obtained by a combination of cDNA cloning and partial amino acid sequence analysis (113,114). The cDNA was cloned by screening a human liver cDNA library and was found to encode a protein of 661 amino acids. Based on the amino-terminal sequence of Glu-Glu-Lys-Pro, which was determined by Takagi and Doolittle (41), the mature protein of B chain should start from Glu, which is located 20 amino acids downstream of the Met initiator. The full-length protein comprises 641 amino acid residues with a molecular weight of 73,183 (113). The addition of 8.5% carbohydrate gives a molecular weight of approximately 76,500. The first 20 amino acid residues have the consensus sequence of a signal peptide (115). This is consistent with the fact that factor XIII B chains are secreted into plasma by hepatocytes (32).

There are three potential N-glycosylation sites and 20 potential disulfide bonds that match the 16–17 disulfide bonds initially reported by Chung et al. (86). An Arg-Gly-Asp tripeptide (RGD) is located at amino acids 597–599, which is often found in proteins involved in cell adhesion (116). Analysis of the amino acid sequence of the B chains has revealed 10 short consensus repeat units (SCR, also known as GP-1, or Sushi domains). The 10 SCR units in the B chain were further classified into four subgroups based on an alignment of their amino acid sequences (117). The degree of amino acid identity within each subgroup varies between 34 and 42%; however, the identities among the 10 SCR units is much lower (117). The short consensus repeat was first described at the amino terminus of factor B, a protein involved in the alternative pathway of human complement system (118). The SCR is also a common structural feature of other proteins associated with the regulation of the complement system. This SCR family includes the six complement control proteins factor H, C4 binding protein (C4bp), CR1, CR2, decay accelerating factor (DAF), the membrane cofactor proteins (MCP), and the complement

Table 2 Comparison of Exon Size Among Genes Of Factor XIII A Chain, TGk, and Band 4.2

	Exon size (bases)		
Exon number	FXII A chain	TGk	Band 4.2 (exon no.)
1	63	89	—[a]
2	148	321	
3	189	189	186 (2)
4	252	249	234 (3)
5	119	119	119 (4)
6	108	108	105 (5)
7	175	175	178 (6)
8	139	139	139 (7)
9	104	104	104 (8)
10	89	89	
11	154	154	242 (9)
12	288	282	300 (10)
13	161	161	161 (11)
14	137	137	134 (12)
15	1688	403	—[a] (13)

[a]The exact size of the exon is unknown.

Source: From References 16, 109, 111, and 112.

factors, such as C1r, C1s, C2, factor B, C6, and C7. There are also several noncomplement proteins with sequence homology to the SCR unit, including lymph node homing receptor (interleukin-2 receptor), endothelial leukocyte adhesion molecule 1, and β_2-glycoprotein I (119). The SCR consists of 60–70 amino acids in each unit, each containing four conserved Cys residues with a characteristic disulfide bonding pattern. The first and third and the second and fourth cysteinyl residues are connected in this SCR unit. The same disulfide bonding pattern probably exist in the factor XIII B chain. The SCR units also have highly conserved amino acid residues including Pro, Trp, Tyr/Phe, and Gly (90). A major function of the regulators of complement activation proteins, factor B, C2, C1r, and C1s, is the binding to the C3b/C4b proteins. A major function of the factor XIII B chain is to bind to factor XIII A chains and regulate the function of the A chain. It is unknown which SCR in the B chain regulates A chain binding. The B chain has not been expressed, and structure-function studies of this protein are not reported.

Because the B chain shares the same SCR structural motif, one can envision that the factor XIII B chain will have secondary structure features common to the other SCR-containing proteins. To date, most of the secondary structure information for the SCR-containing proteins has been obtained from the studies of C4bp and factor H. C4bp is an abundant plasma glycoprotein with seven identical polypeptide chains. Each chain is composed of eight SCRs and a carboxyl-terminal non-SCR domain (120). Electron microscopic studies have revealed that each chain is a flexible arm 33 nm long and 3 nm in diameter (121). Based on the Fourier transform spectroscopic analysis of factor H and studies of other SCR-containing proteins, the likely secondary structure of the SCR is a globular polypeptide domain consisting mostly of an antiparallel β sheet structure that is folded into a sandwich and held in place by the disulfide linkage in the hydrophilic center (122). The structure of a protein with many SCRs is probably similar to a long string of beads, each bead representing SCR. The recent electron microscopic studies of the isolated factor XIII B chain indicate that it appears as a monomer instead of a dimer, and the micrographic images show a filamentous, flexible strand with kinks, with a contour length of 30 nm and a diameter of ~2–3 nm, very similar to the C4bp subunit structure (105).

VI. GENE STRUCTURE OF FACTOR XIII B CHAIN

The factor XIII B chain gene is composed of 12 exons separated by 11 introns and is 28 kb in size (117). The gene was located on chromosome 1 band q31–q32.1 based on *in situ* hybridization studies (123). The genes for the other SCR-containing proteins, including factor H, C4bp, MCP, CR1, CR2, and DAF, are also localized to band q32 on the long arm of chromosome 1 (124–127). The first exon and the last exon encode the leader sequence and carboxyl-terminal region of the B chain, respectively, and the remaining exons (II–XI) each encodes a short consensus repeat (117). It is evident that gene duplication has played an important part in the evolutionary development of the factor XIII B chain gene. One important feature of these 10 SCR-containing exons is that the splice junction occurs after the first nucleotide of a codon (117). This feature allows an ancestral SCR to duplicate within a gene and still encode a mature mRNA with the correct reading frame. Similarly, the genes for other SCR-containing proteins were sequenced and, with few exceptions, were found to have each SCR encoded by a single exon (128).

The regulation of factor XIII B chain expression, synthesis, and release is poorly understood. A total of 30 potential TATA box-like sequences and 1 ''CCAAT'' box (129) are located at the 5′ upstream position of the gene and may function as regulatory elements to control the expression of factor XIII B chains. Examination of the intron sequences has identified four *Alu* (130), three *Kpn*I (131), and one O (132) family of repetitive sequences. Bottenus et al. (117) suggested that exons II–V were duplicated to give exons VI–IX based on the position of the *Alu* sequences that are known to undergo homologous recombination (133).

VII. MOLECULAR GENETICS OF FACTOR XIII DEFICIENCY

Since the first description of inherited factor XIII A chain deficiency in 1960 (8), more than 200 cases of congenital factor XIII deficiency have been reported. Most of the cases reported have no detectable A chains and only ~50% B chain levels. The consequence of this deficiency is a serious bleeding diathesis that can lead to death as a result of intracranial hemorrhage (134). Some of the factor XIII-deficient patients also experience difficulty in wound healing (134). Pregnant factor XIII-deficient women experience recurrent spontaneous abortions unless factor XIII A chains are infused throughout pregnancy (135). These clinical observations suggest that factor XIII A chains have an important function in maintaining the hemostatic integrity of the placental tissue (134).

A. Factor XIII A Chain Deficiency

Congenital deficiency of the factor XIII A chain is a rare autosomal recessive disorder, with a frequency reported to be only 1 case per 2 million in the population (134). Inherited deficiency of factor XIII A chains is not associated with any particular ethnic group (134). Only two cases of factor XIII A chain deficiency are characterized at the molecular level. The first case was found to have a normal size mRNA, but subsequent DNA sequence analysis revealed a dinucleotide (AG) deletion at the 5′ end of exon III (amino acids 210–213), which resulted in frameshift mutation with a new termination codon (TAA) six bases downstream of the deletion site (34). The new truncated protein does not contain the active site Cys^{314}.

The second case studied at the molecular level also had no detectable factor XIII A chain and reduced plasma levels of factor XIII B chains (33). Subsequent DNA sequence analysis documented a point mutation at the last nucleotide of exon 14 (G to A) (33). This mutation could result in an amino acid substitution (Arg^{681} to His^{681}) or alter the pre-mRNA splicing because the mutation is located at the exon/intron junction (33). We found that the C terminus of Ser^{514}-Met^{730} fragment is not absolutely required for fibrin cross-linking activities (50); therefore, the Arg^{681} to His^{681} mutant is unlikely to cause a complete loss of enzyme activity. Additional studies are needed to establish the effect of this mutation on protein synthesis, stability, and secretion.

B. Factor XIII B Chain Deficiency

Factor XIII B chain deficiency is also a very rare autosomal recessive disorder, with only one case reported (35). The complete absence of the B chain, together with a very short half-life of factor XIII A chains, was a characteristic of this unique patient (35). Interestingly, the B chain-deficient patient had symptoms similar to those of the factor XIII A chain-deficient patients (35). DNA sequence analysis detected two defects in the factor

XIII B chain from this patient, a deletion of adenosine at the acceptor splice junction of intron A and a substitution of Cys (TGC) in exon VIII to Phe (TTC). The deletion with a loss of the AG consensus splicing sequence could result in the incorporation of the intron sequence into the mRNA and expression of an abnormal protein. The amino acid substitution (Cys to Phe) in exon VIII could cause a conformational change caused by a loss of disulfide bonding in the seventh SCR domain of the B chain (100). The B chain of factor XIII has not been expressed, and detailed biochemical studies on this mutation remain to be performed.

VIII. CONCLUDING REMARKS

The cloning, sequencing, and expression of recombinant factor XIII A chains have advanced our understanding of the catalytic subunit of the plasma factor XIII molecule. The molecular genetic defects responsible for factor XIII deficiency are just beginning to be defined. In future years, mutant factor XIII molecules with defects in thrombin activation, calcium binding, active-site catalysis, and substrate recognition should be detected. Furthermore, the B chain of factor XIII is now established to play a critical role in regulating factor XIII A chain circulatory half-life, activation by thrombin, and binding to fibrin. Additional molecular defects in the B chain that alter A chain binding, plasma half-life, or other important functions of this plasma glycoprotein remain to be discovered.

It is interesting that there are very few examples of dysfunctional factor XIII A chain molecules. This could be because of insensitive assays or failure to detect clinical disorders associated with the protein defects and/or because the molecule can tolerate deletions and mutations without causing serious bleeding. Future molecular biology studies will address many unanswered questions regarding the synthesis, secretion, assembly, and function of the plasma factor XIII molecule.

REFERENCES

1. Greenberg CS, Birckbichler PJ, Rice RH. Transglutaminases: multifunctional cross-linking enzymes that stabilize tissue. FASEB J 1991; 5:3071–3077.
2. McDonagh JA. Structure and function of factor XIII. In: Colman RW, Hirsh J, Marder VJ, Salzman EW, eds. Hemostasis and Thrombosis. Philadelphia: J.B. Lippinscott, 1987:289–300.
3. Lorand L. Activation of blood coagulation factor XIII. Ann NY Acad Sci 1986; 485:144–158.
4. De Vreker RA. Fibrin stabilizing factor (FSF). Thromb Diath Haemorrh (Suppl) 1964; 13: 411–428.
5. Hynes RO. Fibronectins. Sci Am 1986; 254(6):42–51.
6. Mosher DF. Physiology of fibronectin. Annu Rev Med 1984; 35:561–575.
7. Mosher DF, Schad PE, Vann JM. Cross-linking of collagen and fibronectin by factor XIIIa. J Biol Chem 1980; 255:1181–1188.
8. Duckert F, Jung E, Sherling DH. A undescribed congenital haemorrhagic diathesis probable due to fibrin stabilizing factor deficiency. Thromb Diath Haemorrh 1960; 5:179–186.
9. Loewy AG, Dunathan K, Gallant JA, Gardner B. Fibrinase III. Some enzymatic properties. J Biol Chem 1961; 236:2644–2647.
10. Loewy AG, Dunathan K, Kriel R, Wolfinger J Jr. Fibrinase I. Purification of substrate and enzyme. J Biol Chem 1961; 236:2625–2633.

11. Robbins KC. A study on the conversion of fibrinogen to fibrin. Am J Physiol 1944; 142: 581–588.
12. Loewy AG, Veneziale C, Forman M. Purification of the factor involved in the formation of urea-insoluble fibrin. Biochim Biophys Acta 1957; 26:670–671.
13. Ichinose A, Hendrickson LE, Fujikawa K, Davie EW. Amino acid sequence of the a subunit of human factor XIII. Biochemistry 1986; 25:6900–6906.
14. Takahashi N, Takahashi Y, Putnam FW. Primary structure of blood coagulation factor XIIIa (fibrinoligase, transglutaminases) from human placenta. Proc Natl Acad Sci USA 1986; 83: 8019–8023.
15. Grundmann U, Amann E, Zettlmeissl G, Kupper HA. Characterization of cDNA coding of human factor XIIIa. Proc Natl Acad Sci USA 1986; 83:8024–8028.
16. Ichinose A, Davie EW. Characterization of the gene for the a subunit of human factor XIII (plasma transglutaminase), a blood coagulation factor. Proc Natl Acad Sci USA 1988; 85: 5829–5833.
17. Yorifuji H, Anderson K, Lynch GW, VanDe Water L, McDonagh J. B Protein of factor XIII: differentiation between free B and complexed B. Blood 1988; 72(5):1645–1650.
18. Lorand L, Losowsky MS, Miloszewski KJM. Human factor XIII: fibrin-stabilizing factor. Prog Hemost Thromb 1980; 5:245–290.
19. Henrikson P, Becker S, Lynch G, McDonagh J. Identification of intracellular factor XIII in human monocytes and macrophages. J Clin Invest 1985; 76:528–534.
20. Adany R, Belkin A, Vasilevskaya T, Muszbek L. Identification of blood coagulation factor XIII in human peritoneal macrophages. Eur J Cell Biol 1985; 38:171–173.
21. Wolpl A, Lattke H, Board PG, et al. Coagulation factor XIII A and B subunits in bone marrow and liver transplantation. Transplantation 1987; 43:151–153.
22. Adany R, Muszbek L. Immunohistochemical detection of factor XIII subunit a in histiocytes of human uterus. Histochemistry 1989; 91:169–174.
23. Schwartz ML, Pizzo SV, Hill RL, McKee PL. Human factor XIII from plasma and platelets. J Biol Chem 1973; 248:1395–1407.
24. Lopaciuk S, Lovette KM, McDonagh J, Chuang HYK, McDonagh RP. Subcellular distribution of fibrinogen and factor XIII in human blood platelets. Thromb Res 1976; 8:453–465.
25. Chung SI, Lewis MC, Folk JE. Relationships of the catalytic properties of human plasma and platelet transglutaminases (activated blood coagulation factor XIII) to their subunit structures. J Biol Chem 1974; 249:940–950.
26. Lorand L, Gray AJ, Brown K, et al. Dissociation of subunit structure of fibrin stabilizing factor. Biochem Biophys Res Commun 1974; 56:914–922.
27. Mary A, Achyuthan KE, Greenberg CS. B-chains prevent the proteolytic inactivation of the A-chains of plasma factor XIII. Biochim Biophys Acta 1988; 966:328–335.
28. Poon MC, Russell JA. Low S, et al. Hemopoietic origin of factor XIII A subunits in platelets, monocytes, and plasma. J Clin Invest 1989; 84:787–792.
29. Weisberg LJ, Shiu DT, Conkling PR, Shuman MA. Identification of normal peripheral blood monocytes and liver as sites of synthesis of coagulation factor XIII A-chain. Blood 1987; 70:579–582.
30. Lee SY, Chung ST. Biosynthesis and degradation of plasma protransglutaminase (factor XIII). Fed Proc 1976; 35:1486.
31. Ikematsu S. An approach to the metabolism of factor XIII. Acta Haematol (Basel) 1981; 33:299–301.
32. Nagy JA, Kradin RL, McDonagh J. Biosynthesis of factor XIII A and B subunits. Adv Exp Med Biol 1988; 231:29–49.
33. Board PG, Coggan M, Miloszewskik K. Identification of a point mutation in factor XIII A subunit deficiency. Blood 1992; 80(4):937–941.
34. Kamura T, Okamura T, Murakawa M, et al. Deficiency of coagulation factor XIII A subunit caused by the dinucleotide deletion at the 5′-end of exon III. J Clin Invest 1992; 90:315–319.

35. Saito M, Asakura H, Yoshida T, et al. A familial factor XIII subunit B deficiency. Br J Haematol 1990; 74:290–294.
36. Myhrman R, Bruner-Lorand J. Lobster muscle transpeptidase. Methods Enzymol 1970; 19: 765–770.
37. Singer MA, Hortsch M, Goodman CS, Bentley D. Annulin, a protein expressed at limb segment boundaries in the grasshopper embryo, is homologous to protein cross-linking transglutaminases. Dev Biol 1992; 154:143–159.
38. Tokunaga F, Yamada M, Miyata T, et al. Limulus hemocyte transglutaminases. J Biol Chem 1993; 268(1):252–261.
39. Folk JE, Finlayson JS. The ϵ-(γ-glutamyl) lysine crosslink and the catalytic role of transglutaminases. Adv Protein Chem 1977; 31:1–133.
40. Folk JE. Transglutaminases. Annu Rev Biochem 1980; 49:517–531.
41. Takagi T, Doolittle RF. Amino acid sequence studies on factor XIII and the peptide released during its activation by thrombin. Biochemistry 1974; 13:750–756.
42. Curtis CG, Brown KL, Credo RB, et al. Calcium-dependent unmasking of active center cysteine during activation of fibrin stabilizing factor. Biochemistry 1974; 13:3774–3779.
43. Hornyak TJ, Shafer JA. Role of calcium ion in the generation of factor XIII activity. Biochemistry 1991; 30:6175–6182.
44. Greenberg CS, Achyuthan KE, Fenton JW. Factor XIIIa formation promoted by complexing of α-thrombin, fibrin and plasma XIII. Blood 1987; 69:867–871.
45. Credo RB, Curtis CG, Lorand L. Ca^{2+}-related regulatory function of fibrinogen. Proc Natl Acad Sci USA 1978; 75:4234–4237.
46. Janus TJ, Lewis SD, Lorand L, Shafer JA. Promotion of thrombin-catalyzed activation of factor XIII by fibrinogen. Biochemistry 1983; 22:6269–6272.
47. Lewis ESD, Janus TJ, Lorand L, Shafer JA. Regulation of formation of factor XIIIa by its fibrin substrate. Biochemistry 1985; 24:6772–6777.
48. Naski MC, Lorand L, Shafer JA. Characterization of the kinetic pathway for fibrin promotion of alpha-thrombin-catalyzed activation of plasma factor XIII. Biochemistry 1991; 30:934–941.
49. Mary A, Achyuthan KE, Greenberg CS. The binding of divalent metal ions to platelet factor XIII modulates its proteolysis by trypsin and thrombin. Arch Biochem Biophys 1988; 216: 112–121.
50. Greenberg CS, Enghild JJ, Mary A, Dobson JV, Achyuthan KE. Isolation of a fibrin-binding fragment from blood coagulation factor XIII capable of cross-linking fibrin(ogen). Biochem J 1988; 256:1013–1019.
51. Lai TS, Santiago MA, Achyuthan KE, Greenberg CS. Expression and characterization of recombinant human factor XIII A-chains and a fibrin binding mutant in E. coli. Blood (Suppl) 1991; 78(10):391a.
52. Greenberg CS, Dobson JV, Miraglia CC. Regulation of plasma factor XIII binding to fibrin in vitro. Blood 1985; 66:1028–1034.
53. Hornyak TJ, Shafer JA. Interaction of factor XIII with fibrin as substrate and cofactor. Biochemistry 1992; 31:423–429.
54. Credo RB, Curtis CG, Lorand L. A-chain domain of fibrinogen controls generation of fibrinoligase (coagulation factor XIIIa). Calcium ion regulatory aspects. Biochemistry 1981; 20:3770–3778.
55. Sakata Y, Aoki N. Cross-linking of α_2-plasmin inhibitor to fibrin by fibrin-stabilizing factor. J Clin Invest 1980; 65:290–297.
56. Tamaki T, Aoki N. Cross-linking of α_2-plasmin inhibitor to fibrin catalyzed by activated fibrin-stabilizing factor. J Biol Chem 1982; 257:24767.
57. Kimura S, Aoki N. Cross-linking site in fibrinogen for α_2-plasmin inhibitor. J Biol Chem 1986; 261:15591–15595.
58. Tamaki T, Aoki N. Cross-linking of α_2-plasmin inhibitor and fibronectin to fibrin by fibrin-stabilizing factor. Biochim Biophys Acta 1981; 661:280–286.

59. Bockenstedt P, McDonagh J, Handin RI. Binding and covalent cross-linking of purified von Willebrand factor to native monomeric collagen. J Clin Invest 1986; 78:551.
60. Hada M, Kaminski M, Bockenstedt P, McDonagh J. Covalent cross-linking of von Willebrand factor to fibrin. Blood 1986; 68:95–101.
61. Bale MD, Westrick LG, Mosher DF. Incorporation of thrombospondin into fibrin clots. J Biol Chem 1985; 260:7502–7508.
62. Sane DC, Moser TL, Pippen AMM, Parker CJ, Achyuthan KE, Greenberg CS. Vitronectin is a substrate for transglutaminases. Biochem Biophy Res Commun 1988; 157:115–120.
63. Borth W, Chang V, Bishop P, Harpel PC. Lipoprotein (a) is a substrate for factor IIIa and tissue transglutaminase. J Biol Chem 1991; 266:18149–18153.
64. Grinnell F, Feld M, Minter, D. Fibroblast adhesion to fibrinogen and fibrin substrata: requirement for cold-insoluble globulin (plasma fibronectin). Cell 1980; 19:517–525.
65. Kamykowski GW, Mosher DF, Lorand L, Ferry JD. Modification of shear modulus and creep compliance of fibrin clots by fibronectin. Biophys Chem 1981; 13:25–28.
66. Grinnell F. Fibronectin and wound healing. J Cell Biochem 1984; 26:107.
67. Shafer JA, Higgins DL. Human fibrinogen. CRC Crit Rev Clin Lab Sci 1988; 26:1–41.
68. Varadi A, Scheraga HA. Localization of segments essential for polymerization and for calcium binding in the γ-chain of human fibrinogen. Biochemistry 1986; 25:519–528.
69. Achyuthan KE, Dobson JV, Greenberg CS. GPRP modifies the glutamine residues in the alpha- and gamma-chains of fibrinogen: inhibition of transglutaminase cross-linking. Biochim Biophys Acta 1986; 872:261–268.
70. McKee PA, Mattock P, Hill RL. Subunit structure of human fibrinogen, soluble fibrin and crosslinked insoluble fibrin. Proc Natl Acad Sci USA 1970; 66:738–744.
71. Shainoff JR, Urbanic DA, Dibello PM. Immunoelectrophoretic characterization of the cross-linking of fibrinogen and fibrin by factor XIIIa and tissue transglutaminase. J Biol Chem 1991; 266:6429–6437.
72. Fowler WE, Erickson HP, Hantgan R, McDonagh J, Hermans J. Cross-linked fibrinogen dimers demonstrate a feature of the molecular packing in fibrin fibers. Science 1981; 211: 287–289.
73. Selmayr E, Deffner M, Bachmann L, Muller-Berghaus G. Chromatography and electron microscopy of cross-linked fibrin polymers: a new model describing the cross-linking at the DD-trans contact of the fibrin molecules. Biopolymers 1988; 27:1733–1748.
74. Tamaki T, Aoki N. Cross-linking of α_2-plasmin inhibitor to fibrin catalyzed by activated fibrin-stabilizing factor. J Biol Chem 1982; 257:14767–14772.
75. Aoki N, Saito H, Kamiya T, Koie K, Sakata Y, Kobakura M. Congenital deficiency of α_2-plasmin inhibitor associated with severe hemorrhagic tendency. J Clin Invest 1979; 63:877–884.
76. Reed GL, Matsueda GR, Haber E. Platelet factor XIII increases the fibrinolytic resistance of platelet-rich clots by accelerating the crosslinking of alpha 2-antiplasmin to fibrin. Thromb Haemost 1992; 68:315–320.
77. Okada M, Blomback B, Chang MD, Horowitz B. Fibronectin and fibrin gel structure. J Biol Chem 1985; 260:1811–1820.
78. Mosher DF, Schad PE, Kleinman HK. Cross-linking of fibronectin to collagen by blood coagulation factor XIIIa. J Clin Invest 1979; 64:781–000.
79. Barry EL, Mosher DF. Binding and degradation of blood coagulation factor XIII by cultured fibroblasts. J Biol Chem 1990; 265:9302–9307.
80. Lawler JW, Slayter HS, Coligan JE. Isolation and characterization of a high molecular weight glycoprotein from human blood platelets. J Biol Chem 1978; 253:8609–8616.
81. Lynch GW, Slayter HS, Miller BE, McDonagh J. Characterization of thrombospondin as a substrate for factor XIII transglutaminase. J Biol Chem 1987; 262:1772–1778.
82. Asijee GM, Muszbek L, Kappelmayer J, Polgar J, Horvath A. Platelet vinculin: a substrate of activated factor XIII. Biochim Biochys Acta 1988; 954:303–308.

83. Mui PTK, Ganguly P. Cross-linking of actin and fibrin by fibrin-stabilizing factor. Am J Physiol 1977; 233:H346.
84. Cohen I, Young-Bandala L, Blankenberg TA. Fibrinoligase-catalyzed cross-linking reaction of myosin from platelet and skeletal muscle. Arch Biochem Biophys 1979; 192:100.
85. Cohen I, Glaser T, Veis A, Bruner-Lorand J. Ca^{2+}-dependent cross-linking processes in human platelets. Biochim Biophys Acta 1981; 676:1137.
86. Chung SI, Lewis MS, Folk JR. Relationship of the catalytic properties of human plasma and platelet transglutaminase (activated blood coagulation factor XIII) to their subunit structure. J Biol Chem 1974; 249:940–950.
87. Sixma JJ, van den Berg A, Schiphorst M, Geuze HJ, McDonagh J. Immunocytochemical localization of albumin and factor XIII in thin cryo sections of human blood platelets. Thromb Haemost 1984; 51:388–391.
88. Muszbek L, Adany R, Szegedi G, Polgar J, Kavai M. Factor XIII of blood coagulation in human monocytes. Thromb Res 1985; 37:401–410.
89. Henriksson P, Becker S, Lynch G, McDonagh J. Identification of intracellular factor XIII in human monocytes and macrophages. J Clin Invest 1985; 76:528–534.
90. Ichinose A, Bottenus RE, Davie EW. Structure of transglutaminase. J Biol Chem 1990; 265(23):13411–13414.
91. Holbrook JJ, Cooke RD, Kingston IB. The amino acid sequence around the reactive cysteine residue in human plasma factor XIII. Biochem J 1973; 135:901–903.
92. Gentile V, Saydak M, Chiocca EA, et al. Isolation and characterization of cDNA clones to mouse macrophage and human endothelial cell tissue transglutaminases. J Biol Chem 1991; 266:478–483.
93. Phillips MA, Stewart BE, Qin Q, et al. Primary structure of keratinocyte transglutaminase. Proc Natl Acad Sci USA 1990; 97:9333–9337.
94. Sung LA, Chien S, Chang LS, et al. Molecular cloning of human protein 4.2: a major component of the erythrocyte membrane. Proc Natl Acad Sci USA 1990; 87:955–959.
95. Ikura K, Nasu C, Yokota H, Tsuchiya Y, Sasaki R, Chiba H. Amino acid sequence of guinea pig liver transglutaminase from cDNA sequence. Biochemistry 1988; 27:2898–2905.
96. Fukue H, Anderson K, McPhedran P, Clyne L, McDonagh J. A unique factor XIII inhibitor to a fibrin-binding site on factor XIII A. Blood 1992; 79(1):65–74.
97. Achyuthan KE, Lai TS, Santiago MA, Enghild JJ, Mannun YA, Greenberg CS. Isolation and characterization of synthetic and recombinant peptide inhibitors of factor XIIIa. Blood (Suppl) 1991; 78(10):184a.
98. Klee CB, Vanaman TC. Calmodulin. Adv Protein Chem 1982; 35:213–321.
99. Lewis BA, Freyssinet JM, Holbrook JJ. An equilibrium study of metal ion binding to human plasma coagulation factor XIII. Biochem J 1978; 169:397–402.
100. Hashiguchi T, Ichinose A, Saito M, Marishita E, Mutsuda T. Two genetic defects, in a patient with complete deficiency of the B subunit for coagulation factor XIII. Blood (Suppl) 1992; 80(10):307a.
101. Board PG. Genetic polymorphism of the A-subunit of human coagulation factor XIII. Am J Hum Genet 1979; 31:116–124.
102. Castle S, Board PG, Anderson RAM. Genetic heterogeneity of factor XIII deficiency: first description of unstable A-subunits. Br Haematol 1981; 48:337–342.
103. Bishop PD, Teller DC, Smith RA, Lasser GW, Gilbert T, Seale RL. Expression, purification and characterization of human factor XIII in *Saccharomyces cerevisiae.* Biochemistry 1990; 29:1861–1869.
104. Board PG, Pierce K, Coggan M. Expression of functional coagulation factor XIII in E. coli. Thromb Haemost 1990; 63:235–240.
105. Carrell NA, Erickson HP, McDonagh J. Electron microscopy and hydrodynamic properties of factor XIII subunits. J Biol Chem 1989; 264(1):551–556.

106. Board PG, Webb GC, McKee J, Ichinose A. Localization of the coagulation factor XIII A subunit gene (F13A) to chromosome bands 6p24–p25. Cytogenet Cell Genet 1988; 28:25–27.
107. Weisberg LJ, Shiu DT, Greenberg CS, Kan YW, Shuman MA. Localization of the gene for coagulation factor XIII A-chain to chromosome 6 and identification of sites of synthesis. J Clin Invest 1987; 79:649–652.
108. Achyuthan KE, Lai TS, Santiago MA, Enghild JJ, Hannun YA, Greenberg CS. Isolation and characterization of synthetic and recombinant peptide inhibitors of factor XIIIa. Blood 1991; 78(Suppl.1):184a.
109. Phillips MA, Stewart BE, Rice RH. Genomic structure of keratinocyte transglutaminase. J Biol Chem 1992; 267:2282–2286.
110. Korsgren C, Cohen CM. Organization of the gene for human erythrocyte membrane protein 4.2: structural similarities with the gene for the a subunit of factor XIII. Proc Natl Acad Sci USA 1991; 88:4840–4844.
111. Polakowska RR, Eickbush T, Falciano V, Razvi F, Goldsmith LA. Organization and evolution of the human epidermal keratinocyte transglutaminase I gene. Proc Natl Acad Sci USA 1992; 89:4476–4480.
112. Kersgren C, Cohen CM. Organization of the gene for human erythrocyte membrane protein 4.2: structural similarities with the gene for the a subunit of factor XIII. Proc Natl Acad Sci USA 1991; 88:4840–4844.
113. Ichinose A, McMullen BA, Fujikawa K, Davie EW. Amino acid sequence of the b subunit of human factor XIII, a protein composed of ten repetitive segments. Biochemistry 1986; 25:4633–4638.
114. Grundmann U, Nerlich C, Rein T, Zettlmeissl G. Complete cDNA sequence encoding the B subunit of human factor XIII. Nucleic Acids Res 1990; 18(9):2817–2818.
115. Perlman D, Halvorson HO. A putative signal peptidase recognition site and sequence in eukaryotic and prokaryotic signal peptides. J Mol Biol 1983; 167:391–409.
116. Ruoslahti E, Pierschbacher MD. Arg-Gly-Asp: a versatile cell recognition signal. Cell 1986; 44:517–518.
117. Bottenus RE, Ichinose A, Davie EW. Nucleotide sequence of the gene for the B subunit of human factor XIII. Biochemistry 1990; 29:11195–11209.
118. Morley BJ, Campbell RD. Internal homologies of the Ba fragment from human complement component factor B, a class III MHC antigen. EMBO J 1984; 3:153–157.
119. Lozier J, Takahashi N, Putnam FW. Complete amino acid sequence of human plasma beta 2-glycoprotein I. Proc Natl Acad Sci USA 1984; 81:3640–3644.
120. Chung LP, Gagnon J, Reid KB. Amino acid sequence studies of human C4b-binding protein: N-terminal sequence analysis and alignment of the fragments produced by limited proteolysis with chymotrypsin and the peptides produced by cyanogen bromide treatment. Mol Immunol 1985; 22:427–435.
121. Dahlback B, Smith CA, Muller-Eberhard HJ. Visualization of human C4b-binding protein and its complexes with vitamin K-dependent protein S and complement protein C4b. Proc Natl Acad Sci USA 1983; 80:3461–3465.
122. Perkins SJ, Haris IP, Sim RB, Chapman D. A study of the structure of human complement component factor H by Fourier transform infrared spectroscopy and secondary structure averaging methods. Biochemistry 1988; 27:4004–4012.
123. Webb GC, Coggan M, Ichinose A, Board PG. Localization of the coagulation factor XIII B subunit gene (F13B) to chromosome bands 1q31–32.1 and restriction fragment length polymorphism at the locus. Hum Genet 1989; 81:157–160.
124. Eiberg H, Nielsen LS, Mohr J. Human complement factor H (HF) linked to coagulation factor XIII B (F13B). Cytogenet Cell Genet 1987; 46:610.
125. Weis JH, Morton CC, Bruns GAP, et al. A complement receptor locus: genes encoding C3b/C4b receptor and C3d/Epstein-Barr virus receptor map to 1q32. J Immunol 1987; 138:312–315.

126. Lublin DM, Lemons RS, LeBeau MM, et al. The gene encoding decay-accelerating factor (DAF) is located in the complement-regulatory locus on the long arm of chromosome 1. J Exp Med 1987; 165:1731–1736.
127. Lublin DM, Liszewski MK, Post TW, et al. Molecular cloning and chromosomal localization of human membrane cofactor protein (MCP). Evidence for inclusion in the multigene family olf complement-regulatory proteins. J Exp Med 1988; 168:181–194.
128. Hourcade D, Holers VM, Atkinson JP. The regulation of complement activation (RCA) gene cluster. Adv Immunol 1989; 45:381–416.
129. Graves BJ, Johnson PF, Mcknight SL. Homologous recognition of a promoter domain common to the MSV LTR and the HSV tk gene. Cell 1986; 44:565–576.
130. Deininger PL, Jolly DJ, Rubin CM, Friedmann T, Schmid CW. Base sequence studies of 300 nucleotide renatured repeated human DNA clones. J Mol Biol 1981; 151:17–33.
131. Adams JW, Kaufman RE, Kretschmer PJ, Harrison M, Nienhuis AW. A family of long reiterated DNA sequences, one copy of which is next to the human beta globin gene. Nucleic Acids Res 1980; 8:6113–6128.
132. Sun L, Paulson KE, Schmid CW, Kadyk L, Leinwand L. Non-Alu family interspersed repeats in human DNA and their transcriptional activity. Nucleic Acids Res 1984; 12:2669–2690.
133. Huang LS, Ripps RE, Korman SH, Deckelbaum RJ, Breslow JL. Hypobetalipoproteinemia due to an apolipoprotein B gene exon 21 deletion derived by Alu-Alu recombination. J Biol Chem 1989; 264:11394–11400.
134. Miloszewski KJM, Losowsky MS. Fibrin stabilisation and factor XIII deficiency. In: Francis JL, ed., Ellis Horwood, Fibrinogen, Fibrin Stabilization, and Fibrinolysis. Chichester; 1988: 175–202.
135. Fisher S, Rikover M, Naor S. Factor XIII deficiency with severe hemorrhagic diathesis. Blood 1966; 28:34–39.

15

Vitamin K-Dependent γ-Glutamyl Carboxylase

David J. Wright, Daniel P. Morris, and Darrel W. Stafford
University of North Carolina at Chapel Hill, Chapel Hill, North Carolina

I. INTRODUCTION: THE ROLE OF VITAMIN K IN BLOOD COAGULATION

While studying the biological effects of dietary cholesterol, Dam (1) observed that chickens fed a lipid-depleted diet developed severe bleeding disorders that were not prevented by supplementing with pure cholesterol. He concluded that some component of the extracted lipids, which he termed vitamin K, was required for proper blood coagulation activity. A decade later, Doisy and coworkers (2) and Dam and colleagues (3) independently identified the antihemorrhagic lipid factor from green plants as 2-methyl-3-phytyl-1,4-naphthoquinone (vitamin K_1, also known as phylloquinone or phytonadione; Fig. 1A). Similar compounds with vitamin K activity, the menaquinones (vitamins K_2; Fig. 1A), were subsequently identified in fish meal (4).

Success in identifying the chemical nature of vitamin K did not lead to immediate understanding of its role in blood coagulation: 35 years passed before Stenflo and colleagues (5) and Nelsestuen and coworkers (6) discovered that the hemorrhagic state brought about by vitamin K deficiency was accompanied by synthesis of an abnormal form of the coagulation factor, prothrombin (factor II). This aberrant prothrombin lacked the unusual amino acid, γ-carboxyglutamic acid (Gla), which is found in normal prothrombin and in several other coagulation factors (7). Treatment with vitamin K resulted in recovery of normally carboxylated prothrombin and normal blood-clotting activity, verifying that the vitamin is required for this essential posttranslational modification. Shortly thereafter, Esmon et al. (8) reported the discovery of a vitamin K-dependent activity in microsomal fractions from rat liver that catalyzed the incorporation of radiolabeled bicarbonate into endogenous prothrombin precursors. The enzyme catalyzing this reaction, γ-glutamyl carboxylase, is the subject of this chapter.

Before we focus on the carboxylase, it is worthwhile to review the role of γ-carboxyglutamate in blood coagulation. The serum coagulation factors II, VII, IX, and X, and proteins C, S, and Z each contain 9–12 Gla residues within regions of about 44 amino acid residues at the amino termini of the mature zymogens (9). In the presence of Ca^{2+}, these ''Gla domains'' undergo conformational transitions that result in the ability of the zymogens to bind to negatively charged phospholipid surfaces, such as the membrane of an activated platelet (10–12). This Ca^{2+}-dependent, Gla-dependent membrane binding activity is correlated with dramatic increases in rates of activation of each of these zymogens by their activating proteases (13,14). These rate increases may result from restriction of the diffusional search process between the zymogen and protease to the two dimensions of the membrane surface. In fact, this has been the accepted explanation for many years. However, recent studies have revealed the existence of specific cell surface

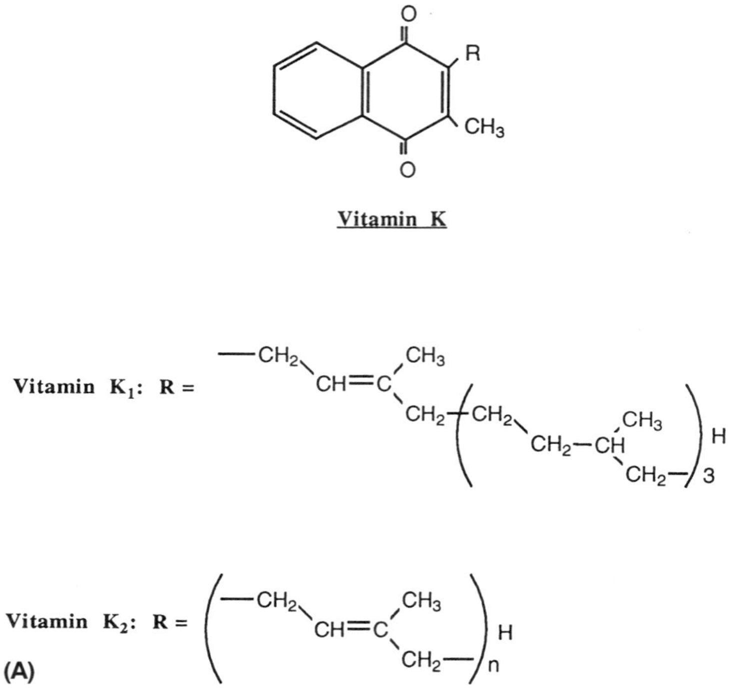

Vitamin K

Vitamin K_1: R = $-CH_2-CH=C(CH_3)-CH_2-(CH_2-CH_2-CH(CH_3)-CH_2)_3-H$

Vitamin K_2: R = $(-CH_2-CH=C(CH_3)-CH_2-)_n-H$

(A)

Figure 1 Vitamins K and the vitamin K cycle. (**A**) Two biologically active forms of vitamin K are shown. Vitamin K_1 is a 2-methyl-1,4-naphthoquinone substituted at the 3 position with a monounsaturated isoprenoid chain known as a phytyl group. This form of the vitamin is found in green plants. Vitamins K_2, or menaquinones, are of bacterial origin. These compounds differ from vitamin K_1 only in that the 3 substituent is a polyunsaturated isoprenoid chain of variable length. (**B**) The vitamin K cycle. Reduced vitamin K (vitamin KH_2) is oxidized to vitamin K epoxide by γ-glutamyl carboxylase (**1**). Vitamin K epoxide is reduced to vitamin K by a reductase activity (**2**) that is dependent in vitro on the presence of the nonphysiological reducing agent, dithiothreitol (DTT). A second DTT-dependent enzyme (**3**) reduces vitamin K to vitamin KH_2, completing the cycle. These reductase activities are the physiologically important site of action of the vitamin K antagonist anticoagulants, warfarin and dicumarol. An alternative, NAD(P)H-dependent pathway for the reduction of vitamin K to vitamin KH_2 also exists in hepatic microsomes. This latter pathway is clinically important: it allows the effects of vitamin K antagonists to be alleviated by administration of vitamin K.

receptors for several of the coagulation factors on the surfaces of platelets (15,16) and endothelial cells (17–20), two cell types of unquestionable relevance to the coagulation process. Because the precise roles of nonspecific membrane binding and specific receptor binding are yet to be determined, it is probably safest to say only that Gla residues are involved in the Ca^{2+}-dependent activation of the coagulation factors on which they are found.

We should note, in the interest of not being discriminatory, that Gla is widespread in nature and is found in many proteins having nothing whatsoever to do with blood coagulation. However, a full discussion of the occurrences and functions of Gla is beyond the scope of this chapter.

With this as introduction, we now focus on the γ-glutamyl carboxylase itself. First we describe the current state of knowledge concerning the biochemical properties of the enzyme, focusing on the bovine liver enzyme, the only form that has been satisfactorily purified. We then discuss possible chemical mechanisms for vitamin K-dependent γ-carboxylation of glutamate. There has been considerable success in identifying recognition determinants on peptide substrates, and we present a summary of this work. We then turn to the molecular biology and genetics of the enzyme, focusing on ongoing studies of the gene encoding the human carboxylase. Finally, we present a hypothetical model for the structural and functional organization of γ-glutamyl carboxylase within the endoplasmic reticulum. For discussion of issues beyond the scope of this chapter, many reviews are available (94–99).

II. PURIFICATION AND PROPERTIES OF BOVINE γ-GLUTAMYL CARBOXYLASE

There have been many attempts to purify the γ-glutamyl carboxylase polypeptide (21). Because the enzyme is an integral membrane protein (22), most efforts met with little

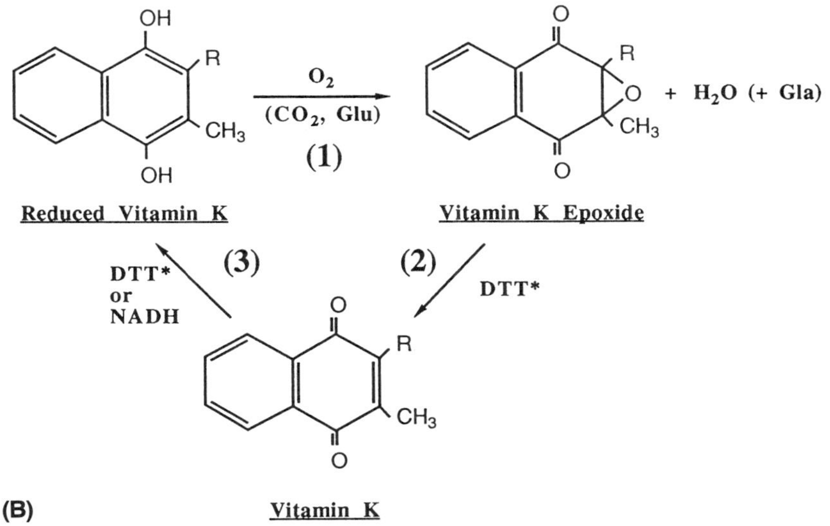

(B)

Figure 1 *Continued*

success. Several groups were able to partially purify the activity by virtue of the high affinity of the enzyme for the endogenous substrates (23) or for peptides based on the sequences of various coagulation factors (24). However, other proteins within the same lipid/detergent vesicles copurified with the carboxylase, resulting in highly impure preparations in which the carboxylase polypeptide could not be identified. A successful method for purification of γ-glutamyl carboxylase from bovine liver was recently devised by Stafford and coworkers (25). Microsomes solubilized in the presence of 1% 3-[(3-choleamidopropyl)diethylammonio]-1-propanesulfonate (CHAPS) were sonicated and loaded on an affinity matrix. This matrix was made with the propeptide-containing substrate FIXQ/S (26), derived from factor IX. The column was then extensively washed with Triton X-100 and CHAPS gradients and eluted with a factor IX propeptide/CHAPS gradient at high ionic strength. This method resulted in a 7000-fold purification of the carboxylation activity over solubilized microsomes, and sodium dodecyl sulfate–polyacrylamide gel electrophoresis analysis showed that the activity coeluted with a 94 kD polypeptide. This protein displayed both vitamin K epoxidase and γ-glutamyl carboxylase activities (27), verifying that both activities are contained within a single polypeptide chain (28).

Purified bovine γ-glutamyl carboxylase was retained on concanavalin A (29) and lentil lectin (30) adsorbents, indicating that the enzyme is a glycoprotein.

The characteristics of the purified carboxylase are for the most part identical to those of the activity in detergent-solubilized microsomes (27). The carboxylation reaction is depicted in Figure 1B as part of the vitamin K cycle. γ-Carboxyglutamate is formed in a reaction requiring the reduced form of vitamin K (hereafter referred to as vitamin KH_2) O_2, CO_2, and a glutamate-containing peptide or protein substrate. The carboxylation reaction is coupled to oxidation of vitamin KH_2 to vitamin K epoxide. Unlike the microsomal system, the purified enzyme is devoid of reductase activities (27) capable of reducing vitamin K epoxide or vitamin K to vitamin KH_2, so the vitamin K cycle (Fig. 1B) that is responsible for recycling oxidized forms of vitamin K into the active reduced form does not operate in the purified system.

Steady-state turnover numbers cannot be calculated for the activity in solubilized microsomes because the concentration of carboxylase in these preparations is unknown. Purified carboxylase with substrate concentrations yielding maximal activity incorporates one mol CO_2 into the peptide substrate FLEEL (Phe-Leu-Glu-Glu-Leu) per second per mol of enzyme (27). As shown in Table 1, the K_m values (or concentrations at half-

Table 1 Kinetic Parameters for the Bovine Carboxylase

Substrates	Solubilized[a] microsomes, K_m	Purified carboxylase[b] K_m	Purified carboxylase[b] Concentration at half-maximal activity
FLEEL	1.6–14 mM	—	1 mM
Vitamin KH_2	10-100 μM	—	30–40 μM
CO_2	0.2–2 mM	0.25–0.35 mM	—
Oxygen	40–90 μM	Not determined	—

[a]Value depends upon cosubstrate concentrations and also varies with the investigator.
[b]The substrates FLEEL and vitamin KH_2 did not display Michaelis-Menten kinetics.

maximal activity for substrates displaying non–Michaelis-Menten behavior) for the various substrates are similar in the crude and purified systems (27).

A number of compounds have been reported to stimulate carboxylase activity in microsomes (7,9), and several of these have also been tested for effects on the purified carboxylase (27). [Ammonium sulfate (0.8–1.0 M) stimulates carboxylation of FLEEL by the purified enzyme (27), while phospholipid is required for activity (27,31).] Dithiothreitol stimulates carboxylation of the FLEEL about 1.5-fold (27), but this may reflect protection of vitamin KH_2 from oxidation rather than direct effects on the reaction.

Of the many inhibitory compounds that have been reported (7,9), only the anticoagulant warfarin and the sulfhydryl-modifying reagent *N*-ethylmaleimide have been tested for effects on the purified enzyme. Both compounds irreversibly inhibit the purified carboxylase (27). It must be stressed that the carboxylase is not thought to be the physiological target of warfarin action; the vitamin K epoxide reductase is apparently inhibited by warfarin at much lower concentrations (~1 μM) than those required to achieve significant inhibition of the carboxylase (~1 mM; see Ref. 27).

III. CHEMISTRY OF γ-CARBOXYGLUTAMATE FORMATION

The chemical course of γ-carboxyglutamate formation has been studied extensively using solubilized microsomes and partially purified carboxylase preparations from bovine and rat liver. The carboxylation pathway is depicted in Figure 1 as part of the vitamin K cycle. In the presence of reduced vitamin K and the gaseous substrates O_2 and CO_2, a proton or hydrogen atom on a glutamate residue in the protein (or peptide) substrate is abstracted and replaced by CO_2, yielding γ-carboxyglutamate. In the process, vitamin KH_2 is oxidized to vitamin K epoxide and O_2 is reduced to H_2O. In microsomes, the carboxylation and epoxidation reactions are closely coupled when adequate amounts of all substrates are present (32–34). The copurification of these activities to near homogeneity (25,27) has verified that both activities reside within a single polypeptide chain. Carboxylation is accompanied by inversion of stereochemical configuration about the 4-methylene carbon of the glutamate side chain (35).

Coupling of carboxylation to epoxidation can be disrupted by selective alteration of reaction conditions. In the absence of CO_2, the epoxidation reaction is accompanied by the exchange of hydrogen between the γ-methylene groups of glutamate residues and solvent (32–34). The carboxylase can also catalyze excess vitamin K epoxide formation at low concentrations of peptide or protein substrate (36) or in the presence of cyanide (37). The latter could reflect the involvement of heme iron in the carboxylation reaction (but not in vitamin K epoxide formation). Alternatively, HCN, which appears to compete with CO_2, could deactivate the activated glutamate by donating hydrogen (38).

These observations are all consistent with formation of an activated 4-methylene carbon by abstraction of a proton or hydrogen atom, followed by attack on the activated carbon by CO_2 or, in its absence, by water. Figure 2 depicts two competing hypotheses concerning the activated glutamate intermediate and the chemical mechanism of its formation: (a) a vitamin K semiquinone radical is produced, initiating a chain of radical reactions that culminates in abstraction of a hydrogen atom from the glutamate side chain to leave a γ-methylene radical (39,40), or (b) an ionic form of oxidized vitamin K abstracts a proton to leave a glutamate carbanion (41,42).

Based on their observations that glutathione peroxidase inhibited carboxylation by rat liver microsomes and that *t*-butyl hydroperoxide displayed weak vitamin K activity, Larson and Suttie (39) proposed various mechanisms involving vitamin K hydroperoxide. It was later shown, however, that highly purified glutathione peroxidase preparations did not inhibit carboxylation (43). Reports of inhibition of carboxylation by other scavengers of active oxygen species, such as superoxide dismutase (44) and copper complexes (45), were attributed to rapid oxidation of vitamin KH_2 by these agents (46). In addition, *t*-butyl hydroperoxide-dependent fixation of CO_2 was shown to result from artifactual incorporation of CO_2 into impurities in the reaction rather than into the peptide substrate (47,48). Despite the lack of experimental evidence, involvement of a vitamin K hydroperoxide radical or anion intermediate is reasonable.

Several possible mechanisms using the vitamin K hydroperoxy radical have been described (49). A related mechanism suggested by Vermeer and colleagues (40) involves a hydroperoxy radical initiation, followed by a concerted electron rearrangement between the enzyme-bound hydroperoxide, glutamate, and CO_2.

Pulse radiolysis and laser flash photolysis techniques have been used in attempts to observe vitamin K radicals in oxygenated vitamin KH_2 solutions and in carboxylation reactions. Rao and Olson (50) found electron pulse-induced transient absorption spectra in oxygenated vitamin K solutions attributable to the vitamin K hydroperoxy radical. These results are interesting, but they do not show that this radical participates in enzymatic catalysis. Canfield and Ramelow (51) used laser flash photolysis of vitamin KH_2 to generate the radical, vitamin K semiquinone. In a crude microsomal system containing carboxylase, the lifetime of the semiquinone transient absorption spectrum increased somewhat with increasing concentrations of either microsomal protein or peptide substrate. However, it is unlikely that the carboxylase could stabilize an observable portion of the radical as the concentration of the carboxylase in their reaction is 4 to 5 orders of magnitude below the concentration of vitamin K.

Suttie and colleagues (52) attempted to demonstrate the formation of a γ-methylene radical using a 4-cyclopropane analog of the peptide substrate, Boc-EEL-OMe. Unfortunately, the enzyme had little or no affinity for the peptide probe, as judged by its failure to act as a competitive inhibitor of carboxylation of Boc-EEL-OMe.

In summary, the available evidence does not allow us to confirm or deny the involvement of free radicals in carboxylation by γ-glutamyl carboxylase.

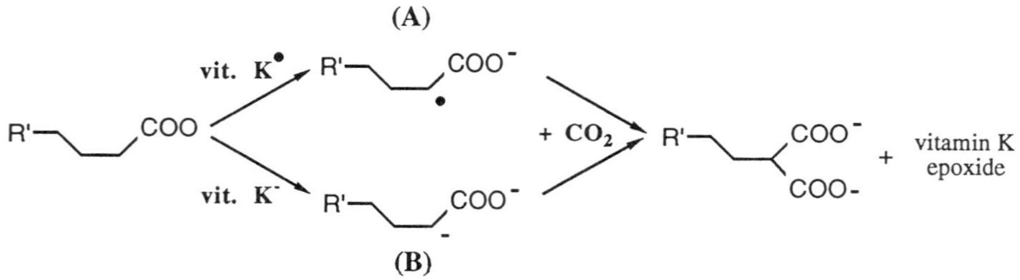

Figure 2 Alternative mechanisms for γ-glutamyl carboxylation. Two competing hypotheses concerning the pathway for formation of γ-carboxyglutamate are shown. In pathway **A**, a series of free radical reactions culminates in the production of a γ-methylene radical on the glutamic acid side chain, which then reacts with CO_2 to produce Gla. In pathway **B**, the activated glutamate contains a γ-methylene carbanion rather than a free radical, and reaction with CO_2 is a simple nucleophilic attack on CO_2 by this carbanion.

The evidence for formation of a glutamate carbanion during carboxylation is more convincing. Gaudry and coworkers (42) used peptide substrates containing 3-fluoroglutamic acid to study the identity of the activated glutamate intermediate. The carboxylase stereospecifically reacted with a 3(*R*)-fluoroglutamate-containing peptide, converting the glutamate side chain to dehydroglutamate by elimination of HF. This type of β elimination reaction is consistent with a carbanion rather than a radical intermediate. As described earlier, several groups have reported that the carboxylase catalyzes hydrogen exchange between glutamate side chains and water when CO_2 is omitted from the reaction (32–34). This was also interpreted as more consistent with the formation of a glutamate carbanion. The best evidence available at present thus supports the carbanion hypothesis, but involvement of radicals at some point in the reaction has not been ruled out. In fact, the two hypotheses are not mutually exclusive, because a radical could be converted to a carbanion by one-electron reduction.

The first mechanism suggested for the generation of the glutamate carbanion involved abstraction of the γ hydrogen by the vitamin K hydroperoxy anion (39). However, the peroxy anion, with a pK_a near 13, may not be a strong enough base to abstract the γ hydrogen of glutamate, which may have a pK_a of 26–28 if the carboxy group is ionized (41). On the other hand, it seems possible that neutralization of the charged group and polarization of the carbonyl by the enzyme may sufficiently lower the pK_a of the γ hydrogen to allow its abstraction by the peroxy anion.

Based on studies of model compounds and the apparent need for a strong base, Dowd and coworkers (41,53) proposed an interesting pathway for formation of a glutamate carbanion that features a dioxygenated form of vitamin K as an intermediate (Fig. 3). In this ‘‘basicity enhancement’’ model, vitamin KH_2 combines with O_2 to form an unstable cyclic dioxetane intermediate that decays to form a vitamin K epoxy alkoxide anion. This anion abstracts a proton from the glutamate side chain, yielding a carbanion that then reacts with CO_2. This reaction scheme predicts that one oxygen atom should be incorporated in the epoxide ring and that one oxygen atom should be incorporated in the quinone position(s). Mass spectrometric analyses of ^{18}O incorporation into vitamin K epoxide by carboxylase have been performed (54,55). One ^{18}O atom is clearly incorporated into the epoxide ring (54,55), but the results for the quinone oxygens have been inconclusive, reported levels of ^{18}O at these positions ranging from 5% (55) to as much as 20% of the total incorporated ^{18}O (53). Dowd and coworkers (53) pointed out that the low levels of ^{18}O incorporation in the quinone positions could reflect stereospecific elimination of $H_2{}^{18}O$ from the alkoxide rather than lack of alkoxide formation, and Walsh and colleagues (55) argued that the appearance of any ^{18}O at all in these positions proves that such an alkoxide species can be formed during the reaction. One argument against this model is that the dioxetane intermediate proposed by Dowd is a disfavored resonance form of the hydroperoxy anion (56); consequently, the hydroperoxy anion should predominate. Of course, enzymes specialize in stabilizing disfavored structures, so this argument may be misleading. Dowd's hypothesis and most other proposed mechanisms include a role for some form of oxygenated vitamin K in hydrogen abstraction. However, there is as yet no direct evidence that any oxygenated vitamin K species is involved in activation of the glutamate substrate.

As described in more detail later, the carboxylase contains protein sequence homology with the iron-containing proteins, protozoan cytochrome b and soybean seed lipoxygenase. If carboxylase contains a mechanistically important iron, as suggested by these homologies, many roles for iron are possible. In keeping with a common role of iron in

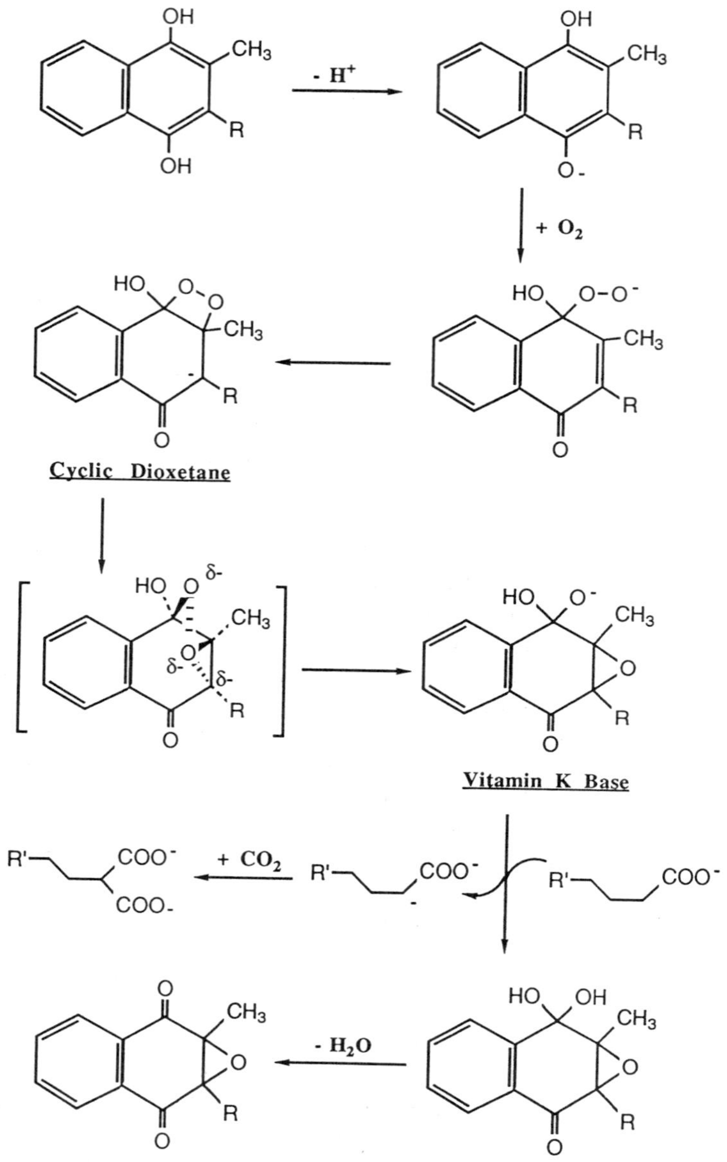

Figure 3 "Basicity enhancement" model of vitamin K action during carboxylation. The readily produced vitamin K anion reacts with molecular oxygen. The resulting peroxy anion undergoes rearrangement to produce a cyclic dioxetane intermediate which presumably decays through a high energy transition state, yielding a vitamin K (epoxy alkoxide anion) base. This base then abstracts a hydrogen from the γ carbon of glutamate, and the resulting carbanion reacts with CO_2.

biochemical reactions, the oxidation of vitamin K hydroquinone and the reduction of molecular oxygen to water could involve iron in the electron transfers required for these reactions. Iron could also chelate the glutamate carboxyl group either by salt bridge formation or by charge transfer between iron and the anionic carboxyl oxygen, resulting in a lower pK_a for the hydrogens of the γ-carbon. The sequence similarities mentioned above suggest that iron is involved in the carboxylase mechanism. However, convincing experimental evidence is not yet available.

IV. RECOGNITION AND CARBOXYLATION OF NATURAL AND SYNTHETIC SUBSTRATES

Studies of γ-glutamyl carboxylase have been greatly facilitated by the development of alternative peptide and protein substrates. Carboxylation of tightly bound endogenous precursor proteins, tremendously important in the discovery of carboxylase, has been less useful for studying the biochemical properties of the enzyme. One problem with using endogenous substrates to study carboxylase involves substrate heterogeneity. In one study, about 60–70% of the bound endogenous protein was factor X precursor (23), the other vitamin K-dependent coagulation factors presumably comprising the remaining 30–40%. Furthermore, the reaction with endogenous precursors was substrate limited and was essentially complete within a few minutes (8).

To overcome these problems, Suttie and coworkers (57) developed synthetic peptide substrates for the carboxylase. They found that the pentapeptide Phe-Leu-Glu-Glu-Val (FLEEV), based on the sequence containing the first two Gla residues of bovine prothrombin, yielded a relatively rapid rate of CO_2 incorporation, with a Michaelis constant (K_m) of about 10 mM. A similar pentapeptide, Phe-Leu-Glu-Glu-Leu, proved to be slightly superior to FLEEV. This peptide, derived from the sequence of factor VII, displayed a K_m of 2–3 mM. Apparently, only the first Glu in these pentapeptides is converted to Gla (58,59). In contrast, the corresponding Glu residues in the endogenous prothrombin precursors are both carboxylated (8). The K_m for FLEEL was also much too high to reflect the K_m for intact protein substrates, which are unlikely to be present in millimolar concentrations within the cell. Despite these drawbacks, FLEEL quickly became the standard carboxylase substrate, and much of the work done with the enzyme has been done with FLEEL.

An alternative to synthetic substrate development is to adapt naturally occurring protein substrates. In this vein, Soute and colleagues (60) reported that a small proteolytic fragment from uncarboxylated prothrombin precursor, encompassing amino acid residues 13–29 of the mature protein, was an excellent carboxylase substrate, with a K_m in the micromolar range. However, a synthetic version of this peptide was not significantly different from FLEEL in terms of kinetics of carboxylation (61), a discrepancy that has not been satisfactorily resolved. Another ''adapted'' natural substrate, thermally decarboxylated osteocalcin (also known as bone Gla protein), was also reported to have a K_m in the micromolar range (62,63). In contrast, heat-decarboxylated prothrombin and des-γ-carboxyprothrombin isolated from blood plasma were very poor substrates (60,64–66), although a slightly larger precursor form of prothrombin isolated from liver tissue was a much better substrate (67).

Clearly, something was missing from FLEEL and from most of the decarboxylated protein substrates that was required for lowering K_m to a level nearer the expected in vivo concentrations of the normal carboxylase substrates, the precursor proteins. When the genes encoding the vitamin K-dependent proteins were cloned and sequenced (68), a likely candidate for the missing carboxylase recognition site was identified (69). With a single exception, all the DNA sequences encoding vitamin K-dependent proteins also encoded amino-terminal signal sequences, directing them into the secretory pathway, and 19–25 amino acid residue propeptides between the signal sequences and the amino termini of the mature proteins. An exception to this pattern is the matrix Gla protein (MGP), a calcium binding protein found in bone and cartilage. The propeptide of MGP is actually within the sequence of the mature protein, and Glu residues on both sides of the propeptide-like sequence are carboxylated (70).

The amino acid sequences of the propeptide and Gla domains of the major vitamin K-dependent proteins are shown in Figure 4. There is a high degree of sequence similarity in the Gla domains of the blood coagulation factors. These coagulation proteins form a functional and presumably structural family distinct from bone and cartilage proteins. The propeptide sequences are more divergent, but a basic pattern of alternation of short stretches of hydrophobic and hydrophilic amino acids is maintained in the propeptides of both families of proteins.

Analyses of the effects of deletions and point mutations within the propeptide-encoding regions of the factor IX (71,72), prothrombin (73), and protein C (74) cDNAs on expression of carboxylated protein in cultured mammalian cells verified that the propeptides of these proteins are required for efficient carboxylation *in vivo.* These studies have identified several conserved amino acid residues that appear to play major roles in recognition by the carboxylase. For example, substitutions for Phe^{-16} or for Ala/Gly^{-10}

	Propeptide		Gla Domain
	-18 -10 -1		+1 +10 +20 +30 +40
h-PT	HVFLAPQQARSLLQRVRR	-	ANT-FLEEVRKGNLERECVEETCSYEEAFEALESSTATDVFW
b-PT	HVFLAHQQASSLLQRARR	-	ANKGFLEEVRKGNLERECLEEPCSREEAFEALESLSATDAFW
h-FX	SLFIRREQANNILARVTR	-	ANS-FLEEMKKGHLERECMEETCSYEEAREVFEDSDKTNEFW
b-FX	SVFLPRDQAHRVLQRARR	-	ANS-FLEEVKQGNLERECLEEACSLEEAREVFEDAEQTDEFW
h-FIX	TVFLDHEMANKILNRPKR	-	YNSGKLEEFVQGNLERECMEEKCSFEEAREVFENTEKTTEFW
b-FIX			YNSGKLEEFVRGNLERECKEEKCSFEEAREVFENTEKTTEFW
h-FVII	RVFVTQEEAHGVLHRRRR	-	AN-AFLEELRPGSLERECKEEQCSFEEAREIFKDAERTKLFW
b-FVII			AN-GFLEELLPGSLERECREELCSFEEAHEIFRNEERTRQFW
h-PC	SVFSSSERAHQVLRIRKR	-	ANS-FLEELRHSSLERECIEEICDFEEAKEIFQNVDDTLAFW
b-PC	SVFSSSQRAHQVLRIRKR	-	ANS-FLEELRPGNVERECSEEVCEFEEAREIFQNTEDTMAFW
h-PS	ANLLSKQQASQVLVRKRR	-	ANS-LLEETKQGNLERECIEELCNKEEAREVFENDPETDYFY
b-PS	ANFLSRQHASQVLIRRRR	-	ANT-LLEETKKGNLERECIEELCNKEEAREIFENNPETEYFY
b-PZ			AGSYLLEELFEGHLEKECWEEICVYEEAREVFEDDETTDEFW
h-OST	KAFVSKQEGSEVVKRPRR	-	YLYQWLGAPVPYPDPLEPRREVCELNPDCDELADHIGFQEAY
b-OST			YLDHWLGAPAPYPDPLEPKREVCELNPDCDELADHIGFQEAY
b-MGP	NPFINRRNANSFISPQQR	W	RAKAQERIRELNKPQYELNREACDDFKLCERYAMVYGYNAAY

Figure 4 Propeptides and Gla domains of the major vitamin K-dependent proteins. The amino acid sequences of the propeptides and Gla domains of human (h) and bovine (b) forms of prothrombin (PT), coagulation factors VII (FVII), IX (FIX), and X (FX), anticoagulant proteins C (PC), S (PS), and Z (PZ), and osteocalcin (OST) and matrix Gla protein (MGP) are shown in single-letter code. With the exception of the underscored residues, all glutamate residues (E) within the Gla domains of these proteins are γ-carboxylated in vivo.

caused significant decreases in production of carboxylated factor IX (71). Substitutions for several other hydrophobic amino acids in this region have somewhat less severe effects, and substitutions nearer the amino-terminal end of the mature protein have little or no effect. Substitutions for Arg^{-4} or Arg^{-1}, which are found naturally in the factor IX genes of some hemophilia B patients, may result in undercarboxylation in vivo (75,76), although this has been disputed (77). These factor IX variants are found in circulation with their propeptides still attached, so these Arg residues are presumably important for proteolytic removal of the propeptide after carboxylation. The role of these residues in carboxylase recognition is unclear, as similar substitutions in peptide substrates have no effect on carboxylation *in vitro* (26,78).

Several groups have created synthetic peptide substrates that include propeptide sequences at their amino termini. Propeptide-containing substrates based on the prothrombin (79) and factor IX (26) sequences were found to be excellent carboxylase substrates, with K_m values three to four orders of magnitude lower than that for FLEEL. Addition of an unattached synthetic propeptide to FLEEL carboxylation reactions decreases the K_m for FLEEL from six- (80) to ninefold (81) in rat and threefold (79) in bovine microsomes. Although significant, this decrease is orders of magnitude less than required to approach the K_m values of the propeptide-containing substrates. Ulrich and coworkers (79) verified the importance of covalent attachment of the propeptide by comparing carboxylation of proPT28 (residues −18 to 10 of prothrombin) with that of a two-peptide mixture consisting of a prothrombin propeptide (−18 to −1) and a prothrombin Gla domain peptide (1–10). The apparent K_m for proPT28 was in the micromolar range, but the two-peptide mixture gave a millimolar K_m for the Gla domain peptide. Turnover of the Gla domain 10 mer was faster than turnover of proPT28, suggesting that covalently attached propeptide decreases the dissociation rates of the enzyme-substrate and/or enzyme-product complexes. The major carboxylase recognition site thus appears to be the hydrophobic amino-terminal end of the propeptide.

A second potential carboxylase recognition site was identified by Price and coworkers (70), who observed that all the vitamin K-dependent proteins whose amino acid sequences were known contained the sequence Glu-*X*-*X*-*X*-Glu-*X*-Cys, where *X* is a nonconserved amino acid. As shown in Figure 4, there is an even greater degree of conservation in this region in the vitamin K-dependent coagulation factors, which all contain the sequence Glu-Cys-*X*-Glu-Glu-*X*-Cys. All the Glu residues in these sequences are normally carboxylated in the mature proteins (7,9,21). A recent study suggests that amino acid residues within this sequence play a role in determining carboxylation efficiency. Castellino and coworkers (82,83) created mutant forms of the vitamin K-dependent anticoagulant enzyme, protein C, bearing substitutions of Asp for Glu^{20} (E20D) or Glu^{21} (E21D) or of Ser for Cys^{22} (C22S). Both the E20D and C22S substitutions resulted in expression of undercarboxylated protein in human kidney 293 cells, but the E21D mutant was normally carboxylated. However, we found that substituting Ser for both Cys residues in this region of the 59-mer peptide substrate FIXQ/S (26) has no effect on the kinetics of in vitro carboxylation by the purified bovine carboxylase (Morris, Wu, and Stafford, unpublished results). Much additional work is needed to clarify the roles of specific amino acid residues within the vitamin K-dependent proteins in determining carboxylase specificity.

The natural carboxylase substrates are proteins, and all of the known Gla-containing proteins contain multiple Gla residues. Liska and Suttie (84) examined the levels of carboxylation of individual glutamate residues in the Gla domains of undercarboxylated prothrombin isolated from cattle treated with the vitamin K antagonist, dicumarol. The data appeared consistent with a preferential addition at particular positions within the Gla domain. In particular, the most amino-terminal Glu residue in the Gla domain (Glu^6) appeared to be highly carboxylated regardless of the overall extent of Gla domain carboxylation. Other preferentially carboxylated residues include Glu^7, Glu^{25}, and Glu^{26}. A similar examination of naturally occurring undercarboxylated variants of prothrombin from a human patient revealed a form that was not carboxylated at Glu^{17} but was carboxylated at Glu residues on either side (85). These results suggest that carboxylation does not occur in a linear sequence along the polypeptide chain.

V. γ-CARBOXYLATION OF GLUTAMATE AS A STEP IN PROTEIN TRAFFICKING

The vitamin K-dependent coagulation factors are synthesized primarily in the liver (86,87), posttranslationally processed through the cellular protein-sorting compartments, and secreted into circulation. Grinnell and coworkers (88) examined the secretion of protein C, a vitamin K-dependent anticoagulant, with emphasis on the timing of the γ-glutamyl carboxylation events. They found γ-carboxylation of glutamate residues in protein C to be a relatively early posttranslational event, occurring within the first 10 minutes after completion of the polypeptide chain. Core glycosylation events preceded carboxylation under normal conditions, but blocking core glycosylation with an antibiotic did not prevent carboxylation. Both γ-glutamyl carboxylation and processing of the core polysaccharide chains in the endoplasmic reticulum (ER) were required for efficient secretion of protein C.

A slow step in this pathway occurs after carboxylation, because fully carboxylated precursors accumulate in the ER lumen. This slow step may be the further processing of the core polysaccharide chains before departure from the ER. There appear to be two calcium-dependent steps in the secretory pathway for protein C. One of these calcium-dependent steps has not been identified but appears to occur early in the secretory pathway. The other step, which occurs late in the secretory pathway, is the proteolytic cleavage of an internal Lys-Arg sequence, which generates a two-chain disulfide-linked form of the zymogen. Propeptide removal probably occurs late in the pathway, although this has not been experimentally verified.

In short, with the exception of the carboxylation events themselves, which occur early in the secretory pathway and are not rate limiting for secretion, the processing and transport of protein C, and probably most of the vitamin K-dependent proteins, is very much like that of any secretory protein produced by mammalian cells.

VI. MOLECULAR GENETICS OF γ-GLUTAMYLCARBOXYLASE

We now switch from consideration of the details of carboxylation to the molecular biology and genetics of γ-glutamyl carboxylase. A cDNA encoding the entire human γ-glutamylcarboxylase was recently cloned and its DNA sequence determined (29). Preliminary data from genomic clones encoding human carboxylase are also available (Solera, Wu, Frazier, and Stafford, unpublished result). The gene is 18–20 kb and contains 15 exons. In addition, the human carboxylase gene has been mapped to the p12 region of chromosome 2 (Gray et al., personal communication). A 2.2 kb DNA fragment con-

```
5' Flanking Region of the Human Gamma-Glutamyl Carboxylase Gene

-460
CCCAGGCAACATTTCCAGGCGCGTCCTCAACTCGGCGTCACTCTG
-415
ACAAAAAGTTTGCCCTTTGCGCCCGATCCCTCACACCAGAGAGCA
-370
CCGCAGGGACGCAGGGACGGAGAGGACTAAGGGGTCGGGGAGGAC
-325
CCAAGGAAGTTGGGGAACTCAGACTCTTTTCAGTGTTTAAAAAGC
-280
AGAGGCCCCCTCGGAAGGCATAACAGGCGAGTCTAACAATGCCCC
-235
TCGAGATCAACGGCACTAAGGGGTACCTTTGCCCCGCTCCAAGGC
-190                                 HNF-4
TCCTTGTTCCGCCCCCTGCCGCCAAGCAAGGAATCGCCGCCGGTC
-145          Sp-1
GTGGGTAGTATTTACGACAGTGCTGGCGCTCTGTGCCGGCCGCAC
-100                           HRE                 AP-2
ACCTCGACGCGTCGGGAGCGGAGCCTAGGGAAGCAAATTCTCCCT
-55                                             HNF-3
GGCGGCCTCCGTTCAGACGCGGCAGCTGTGACCCACCTGCCTCCT
-10           +1                       In-frame stop
CCGCAGAGCAATGGCGGTGTCT
          M  A  V  S
```

Figure 5 DNA sequences upstream of the human γ-glutamyl carboxylase coding region. The DNA sequence of a 500 bp region immediately 5′ to the carboxylase coding region is shown. This region contains potential binding sites for two general transcription factors (Ap-2 and Sp-1), as well as sites for two liver-enriched transcription factors (HNF-3 and HNF-4) and a potential hormone response element (HRE). This region may act as a liver-specific transcriptional enhancer.

taining approximately 2 kb of the 5′ flanking sequence upstream of the human carboxylase coding region has also been isolated, and 500 bp immediately upstream of the protein start codon have been sequenced (Fig. 5). There are potential binding sites for the general transcription factors Sp-1 and Ap-2 and for the liver-enriched factors HNF-3 and HNF-4 within this region, but we have not yet shown that transcription factors bind to these sites, nor have we verified that these sequences can direct transcription.

Northern analyses of poly(A)$^+$ RNA and of total cellular RNA from human liver using the cloned human carboxylase cDNA as a probe show two major transcripts of 2.7 and 3.5 kb (Huisse et al., unpublished result). At least 2277 bp is required to encode the entire 758 amino acid carboxylase polypeptide, so the 2.7 kb transcript can contain only about 400 untranslated nucleotides. Because this transcript also contains a 3′ poly(A) tract, its 5′ end must be very near the protein initiation codon.

VII. STRUCTURAL FEATURES OF γ-GLUTAMYLCARBOXYLASE

The amino acid sequence of human γ-glutamyl carboxylase deduced from the DNA sequence of the cloned cDNA is shown in Figure 6. The enzyme is a 758 amino acid polypeptide with a predicted molecular weight of 87,542. The higher molecular weight calculated from electrophoretic mobility (94,000) presumably reflects glycosylation of the protein. Hydropathy analysis of the carboxylase amino acid sequence (29) predicts three helical transmembrane segments within the amino-terminal third of the polypeptide; the remainder of the polypeptide is relatively hydrophilic.

We have conducted amino acid identity searches using all available protein sequences in both GenBank and SwissProt databases and have found some interesting and perhaps significant amino acid sequence similarities with other proteins. As we previously reported (29), the region between residues 100–340 of the carboxylase is 16% identical with a protozoan cytochrome b. The similarity is much greater when conservative amino acid substitutions are taken into account, suggesting that this portion of the carboxylase folds into a cytochrome b-like structure. This region of the carboxylase, particularly the first putative transmembrane segment, also shares similar levels of identity with several of the mitochondrially encoded NADH-ubiquinone oxidoreductase subunits. The first transmembrane segment contains the amino acid sequence YLVYT, whose only other occurrence in the protein sequence databases is in the ubiquinone binding site of a subunit of ubiquinol-cytochrome c oxidoreductase (89). It may be that this sequence comprises part of the as yet unidentified vitamin KH_2 binding site on the carboxylase.

The central region of the carboxylase between residues 400 and 660 shares 19% amino acid identity with the c-terminal third of soybean seed lipoxygenase 1 (29). This is

```
MAVSAGSARTSPSSDKVQKDKAELISGPRQDSRIGKLLGF
EWTDLSSWRRLVTLLNRPTDPASLAVFRFLFGFLMVLDIP
QERGLSSLDRKYLDGLDVCRFPLLDALRPLPLDWMYLVYT
IMFLGALGMMLGLCYRISCVLFLLPYWYVFLLDKTSWNNH
SYLYGLLAFQLTFMDANHYWSVDGLLNAHRRNAHVPLWNY
AVLRGQIFIVYFIAGVKKLDADWVEGYSMEYLSRHWLFSP
FKLLLSEELTSLLVVHWGGLLLDLSAGFLLFFDVSRSIGL
FFVSYFHCMNSQLFSIGMFSYVMLASSPLFCSPEWPRKLV
SYCPQRLQQLLPLKAAPQPSVSCVYKRSRGKSGQKPGLRH
QLGAAFTLLYLLEQLFLPYSHFLTQGYNNWTNGLYGYSWD
MMVHSRSHQHVKITYRDGRTGELGYLNPGVFTQSRRWKDH
ADMLKQYATCLSRLLPKYNVTEPQIYFDIWVSINDRFQQR
IFDPRVDIVQAAWSPFQRTSWVQPLLMDLSPWRAKLQEIK
SSLDNHTEVVFIADFPGLHLENFVSEDLGNTSIQLLQGEV
TVELVAEQKNQTLREGEKMQLPAGEYHKVYTTSPSPSCYM
YVYVNTTELALEQDLAYLQELKEKVENGSETGPLPPELQP
LLEGEVKGGPEPTPLVQTFLRRQQRLQEIERRRNTPFHER
FFRFLLRKLYVFRRSFLMTCISLRNLILGRPSLEQLAQEV
TYANLRPFEAVGELNPSNTDSSHSNPPESNPDPVHSEFU
```

Figure 6 Amino acid sequence of human γ-glutamyl carboxylase. The amino acid sequence of the human γ-glutamyl carboxylase deduced from the sequence of a cDNA encoding the enzyme is shown in single-letter code. Putative transmembrane domains identified by hydropathy analysis are underscored.

interesting because lipoxygenases catalyze the hydroperoxidation of unsaturated fatty acids, a reaction similar to that proposed for vitamin K hydroperoxide formation (39). Lipoxygenases contain a nonheme, nonsulfur iron that is essential for activity (90). Two conserved histidine residues within a central histidine-rich region in lipoxygenase 1, as well as a third histidine in the carboxyl-terminal portion of the enzyme, have been implicated in iron binding (90,91). The portion of carboxylase that is homologous to lipoxygenase 1 does not include the histidine-rich region of the latter enzyme, but the carboxylase has a cluster of three histidine residues just before the lipoxygenase-like domain, in a similar position as the histidine-rich region in the lipoxygenase. The carboxylase has not been definitively shown to contain iron, although one group (92) has claimed that ^{59}Fe-containing heme copurifies with carboxylase activity through several chromatographic steps. Large quantities of purified carboxylase can now be obtained, so that the metal content of the enzyme can now be examined directly by chemical and spectroscopic methods.

The successful purification of carboxylase has allowed us to begin identifying functionally important sites on the enzyme. The carboxylase enzyme is cleaved by mild

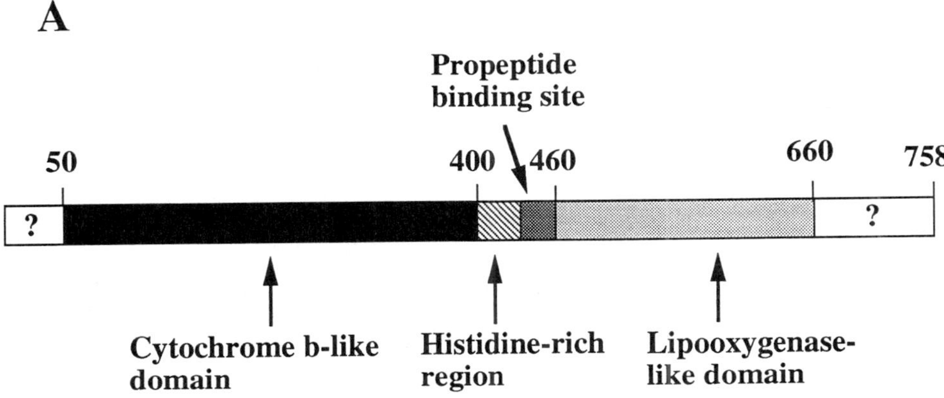

Figure 7 Functional and structural organization of γ-glutamyl carboxylase. A model for the organization of functional domains of γ-glutamyl carboxylase is depicted. (**A**) Linear arrangement of functional domains of γ-glutamyl carboxylase. A unique amino-terminal domain of about 50 amino acid residues is followed by a cytochrome b-like domain of approximately 300 amino acid residues. The cytochrome b-like domain is followed by a lipoxygenase-like domain of approximately 250 residues. A histidine-rich region and the putative propeptide binding site lie at the junction between these two domains. The enzyme ends in a unique, highly hydrophilic carboxyl-terminal domain of about 100 residues. (**B**) Organization of γ-glutamyl carboxylase within the endoplasmic reticulum. The carboxylase is oriented within the ER membrane, with the amino terminus on the cytoplasmic side and the bulk of the protein within the ER lumen. The cytochrome b domain is predicted to form three membrane-spanning helices and a luminal globular domain containing a heme iron. The lipoxygenase domain is predicted to form a luminal globular domain containing a nonheme, nonsulfur iron. The propeptide binds within the lipoxygenase domain, orienting the protein substrate within the active site, depicted here as lying between the two iron atoms. Vitamin KH_2 is depicted as interacting with the first transmembrane helix by the phytyl group and with the cytochrome b and/or lipoxygenase domain(s) by the naphthoquinone group.

proteolysis into two distinct species. Chemical cross-linking experiments (100) show that the propeptide binding site resides in the larger, C-terminal proteolytic fragment. A peptide to which the propeptide was cross-linked has been isolated, and its amino acid sequence has been determined (Wu and Stafford, manuscript in preparation). This peptide is located between the three-histidine cluster and the lipoxygenase 1-like domain of the carboxylase. Walsh and colleagues (93) reported cross-linking of an *N*-bromosuccinimide derivative of the glutamate-containing substrate FLEELY to the carboxylase, but the site on the enzyme that was cross-linked to this peptide was not identified.

VIII. STRUCTURAL AND FUNCTIONAL MODEL OF γ-GLUTAMYL CARBOXYLASE

Based on the homologies noted earlier, we have devised a tentative model for the structural and functional organization of the carboxylase (Fig. 7). The enzyme is organized into an amino-terminal cytoplasmic domain, a cytochrome b-like domain containing the membrane spanning region, and a lipoxygenase-like domain in the ER lumen. Because of its lipophilic nature, vitamin KH_2 is likely to be found largely within the membrane bilayer. However, the structural and functional similarities between the carboxylase and

B

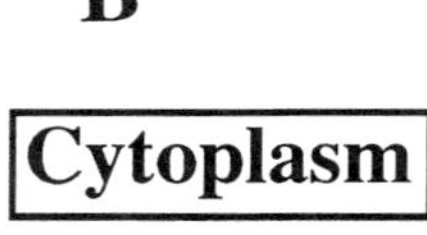

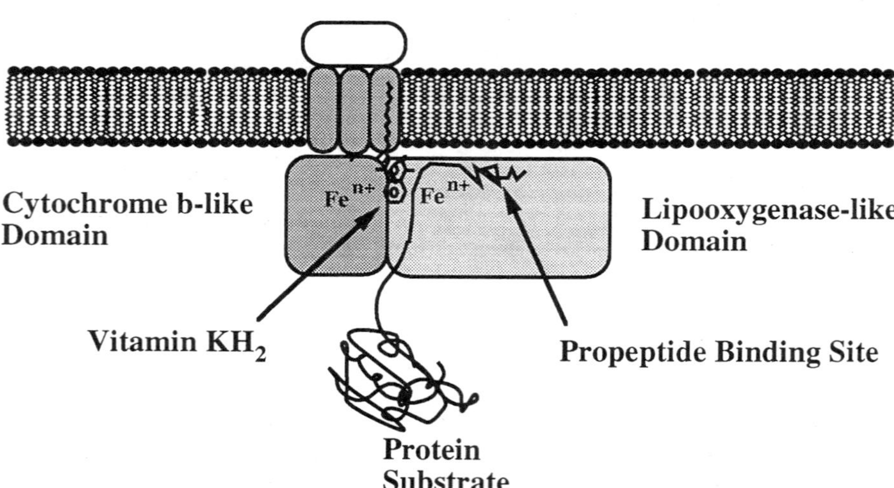

ER Lumen

Figure 7 *Continued*

soybean seed lipoxygenase 1 suggest that the catalytically relevant vitamin K binding site is in the relatively hydrophilic lipoxygenase-like domain. One possibility is that the active site of the enzyme is adjacent to the membrane and can thus easily interact with membrane-bound vitamin KH_2. The propeptide is envisioned to bind in a hydrophobic cleft, positioning the Gla domain in the active site. It is likely that binding of the propeptide anchors the substrate to the enzyme, permitting multiple carboxylation events during a single substrate binding event. We suggest that one or two iron atoms may be involved in the carboxylation/epoxidation reaction, largely on the basis of the similarities between the carboxylase and the iron-containing cytochrome b and lipoxygenase 1 proteins. Clearly, a great deal of further work is needed to verify the presence of iron and to elucidate the chemistry of the carboxylation and epoxidation reactions. The availability of large quantities of purified carboxylase places these goals within reach.

REFERENCES

1. Dam H. Cholesterinstoffwechsel in Huhnereiern und Huhnchen. Biochem Z 1929; 215:475–492.
2. Binkley SB, MacCorquodale D, Thayer SA, Doisy EA. The isolation of vitamin K_1. J Biol Chem 1939; 130:219–234.
3. Dam H, Geiger A, Glavind J, et al. Isolierung des vitamins K in hochgereinigter form. Helv Chim Acta 1939; 22:310.
4. McFarlane WD, Graham WR, Richardson F. The fat-soluble vitamin requirements of the chick. Biochem J 1931; 25:358–366.
5. Stenflo J, Fernlund P, Egan W, Roepstorff P. Vitamin K-dependent modifications of glutamic acid residues in prothrombin. Proc Natl Acad Sci USA 1974; 71:2730–2733.
6. Nelsestuen GL, Zytkovicz TH, Howard JB. The mode of action of vitamin K. J Biol Chem 1974; 249:6347–6350.
7. Suttie JW. Vitamin K-dependent carboxylase. Annu Rev Biochem 1985; 54:459–477.
8. Esmon CT, Sadowski JA, Suttie JW. A new carboxylation reaction. The vitamin K-dependent incorporation of $H^{14}CO_3$- into prothrombin. J Biol Chem 1975; 250:4744–4748.
9. Suttie JW. Vitamin K-dependent carboxylation of glutamyl residues in proteins. Biofactors 1988; 1:55–60.
10. Nelsestuen GL. Role of gamma-carboxyglutamic acid. An unusual transition required for calcium-dependent binding of prothrombin to phospholipid. J Biol Chem 1976; 251:5648–5656.
11. Prendergast FG, Mann KG. Differentiation of metal ion-induced transitions of prothrombin fragment 1. J Biol Chem 1977; 252:840–850.
12. Borowski M, Furie BC, Bauminger S, Furie B. Prothrombin requires two sequential metal-dependent conformational transitions to bind phospholipid. J Biol Chem 1986; 261:14969–14975.
13. Esmon CT, Owen WG, Jackson CM. A plausible mechanism for prothrombin activation by factor Xa, factor Va, phospholipid, and calcium ions. J Biol Chem 1974; 249:8045–8047.
14. Rosing J, Tans G, Govers-Reimslag J, et al. The role of phospholipids and factor Va in the prothrombinase complex. J Biol Chem 1980; 255:274–283.
15. Ahmad SS, Rawala-Sheikh R, Walsh PN. Comparative interactions of factor IX and factor IXa with human platelets. J Biol Chem 1989; 264:3244–3251.
16. Vu T-K, Hung DH, Wheaton VI, Coughlin SR. Molecular cloning of a functional thrombin receptor reveals a novel proteolytic mechanism of receptor activation. Cell 1991; 64:1057–1068.
17. Heimark RL, Schwartz SM. Binding of coagulation factors IX and X to the endothelial cell surface. Biochem Biophys Res Commun 1983; 111:723–731.

18. Stern DM, Drillings M, Nossel HL, Hurlet-Jensen A, LaGamma KS, Owen J. Binding of factors IX and IXa to cultured vascular endothelial cells. Proc Natl Acad Sci USA 1983; 80:4119–4123.
19. Cheung W-F, Hamaguchi N, Smith KJ, Stafford DW. The binding of human factor IX to endothelial cells is mediated by residues 3–11. J Biol Chem 1992; 267:20529–20531.
20. Toomey JR, Smith KJ, Roberts HR, Stafford DW. The endothelial cell binding determinant of human factor IX resides in the gamma-carboxyglutamic acid domain. Biochemistry 1992; 31:1806–1808.
21. Vermeer C. Gamma-carboxyglutamate-containing proteins and the vitamin K-dependent carboxylase. Biochem J 1990; 266:625–636.
22. Carlisle TL, Suttie JW. Vitamin K-dependent carboxylase: subcellular location of the carboxylase and enzymes involved in vitamin K metabolism in rat liver. Biochemistry 1980; 19:1161–1167.
23. DeMetz M, Vermeer C, Soute BAM, van Scharrenburg GJM, Slotboom AJ, Hemker HC. Partial purification of bovine liver vitamin K-dependent carboxylase by immunospecific adsorption onto antifactor X. FEBS Lett 1981; 123:215–218.
24. Hubbard BR, Ulrich MMW, Jacobs M, et al. Vitamin K-dependent carboxylase: affinity purification from bovine liver by using a synthetic propeptide containing the gamma-carboxylation recognition site. Proc Natl Acad Sci USA 1989; 86:6893–6897.
25. Wu S-M, Morris DP, Stafford DW. Identification and purification to near homogeneity of the vitamin K-dependent carboxylase. Proc Natl Acad Sci USA 1991; 88:2236–2240.
26. Wu S-M, Soute BAM, Vermeer C, Stafford DW. *In vitro* gamma-carboxylation of a 59-residue recombinant peptide including the propeptide and the gamma-carboxyglutamic acid domain of coagulation factor IX. J Biol Chem 1990; 265:13124–13129.
27. Morris DP, Soute BAM, Vermeer C, Stafford DW. Characterization of the purified vitamin K-dependent gamma-glutamyl carboxylase. J Biol Chem 1993; 268:8735–8742.
28. Wallin R, Suttie JW. Vitamin K-dependent carboxylase: evidence for cofractionation of carboxylase and epoxidase activities, and for carboxylation of high molecular weight microsomal protein. Arch Biochem Biophys 1982; 214:155–163.
29. Wu S-M, Cheung W-F, Frazier D, Stafford DW. Cloning and expression of the cDNA for human gamma-glutamyl carboxylase. Science 1991; 254:1634–1636.
30. Berkner KL, Harbeck M, Lingenfelter S, Bailey C, Sanders-Hinck CM, Suttie JW. Purification and identification of bovine liver gamma-glutamyl carboxylase. Proc Natl Acad Sci USA 1992; 89:6242–6246.
31. DeMetz M, Vermeer C, Soute BAM, Hemker HC. Identification of phospholipid as an essential part of bovine vitamin K-dependent carboxylase. J Biol Chem 1981; 256:10843–10846.
32. Anton DL, Friedman PA. Fate of the activated gamma-carbon-hydrogen bond in the uncoupled vitamin K-dependent gamma-glutamyl carboxylation reaction. J Biol Chem 1983; 258:14084–14087.
33. McTigue JJ, Suttie JW. Vitamin K-dependent carboxylase. Demonstration of a vitamin K- and O_2-dependent exchange of ^3H from ^3H_2O into glutamic acid residues. J Biol Chem 1983; 258:12129–12131.
34. Friedman PA, Shia MA, Gallop PM, Griep AE. Vitamin K-dependent gamma-carbon-hydrogen bond cleavage and nonmandatory concurrent carboxylation of peptide-bound glutamic acid residues. Proc Natl Acad Sci USA 1979; 76:3126–3129.
35. Marechal PD-L, Ducrocq C, Marquet A, Azerad R. The stereochemistry of hydrogen abstraction in vitamin K-dependent carboxylation. J Biol Chem 1984; 259:15010–15012.
36. Wood GM, Suttie JW. Vitamin K-dependent carboxylase: stoichiometry of vitamin K epoxide formation, gamma-carboxyglutamyl formation, and gamma-glutamyl ^3H cleavage. J Biol Chem 1988; 263:3234–3239.

37. DeMetz M, Soute BAM, Hemker HC, Vermeer C. The inhibition of vitamin K-dependent carboxylase by cyanide. FEBS Lett 1982; 137:253–256.
38. Dowd P, Ham S-W. Mechanism of cyanide inhibition of the blood-clotting, vitamin K-dependent carboxylase. Proc Natl Acad Sci USA 1991; 88:10583–10585.
39. Larson AE, Suttie JW. Vitamin K-dependent carboxylase: evidence for a hydroperoxide intermediate in the reaction. Proc Natl Acad Sci USA 1978; 75:5413–5416.
40. DeMetz M, Soute BAM, Hemker HC, Fokkens R, Lugtenberg J, Vermeer C. Studies on the mechanism of the vitamin K-dependent carboxylation reaction. J Biol Chem 1982; 257: 5326–5329.
41. Ham SW, Dowd P. On the mechanism of action of vitamin K. A new nonenzymic model. J Am Chem Soc 1990; 112:1660–1661.
42. Dubois J, Gaudry M, Bory S, Azerad R, Marquet A. Vitamin K-dependent carboxylation. Study of the hydrogen abstraction stereochemistry with gamma-fluoroglutamic acid containing peptides. J Biol Chem 1983; 258:7897–7899.
43. Hall AL, Turner PM, Dunkle BF, Wing DA, Olson RE. Vitamin K-dependent carboxylase for peptide-bound glutamate. In: Suttie JW, ed. Vitamin K metabolism and vitamin K-dependent proteins. New York: Academic Press, 1980:433–442.
44. Esnouf MP, Green MR, Hill HAO, Irvine G, Walter SJ. Evidence for the involvement of superoxide in vitamin K-dependent carboxylation of glutamic acid residues in prothrombin. Biochem J 1978; 174:345–348.
45. Esnouf MP, Green MR, Hill HAO, Walter SJ. The inhibition of the vitamin K-dependent carboxylation of glutamic acid residues in prothrombin by some copper complexes. FEBS Lett 1979; 107:146–150.
46. Kanabus-Kaminska JM, Girardot J-M. Inhibition of vitamin K-dependent carboxylase by metal ions and metal complexes: a reassessment. Arch Biochem Biophys 1984; 228:646–652.
47. Hall AL, Kloepper R, Zee-Cheng RK-Y, Chiu YJD, Lee FC, Olson RE. Mechanism of action of *t*-butyl hydroperoxide in the inhibition of vitamin K-dependent carboxylation. Arch Biochem Biophys 1982; 214:45–50.
48. Wallin R, Suttie JW. Vitamin K-dependent carboxylase: possible artifact of analysis due to a pyridine nucleotide-dependent carboxylation. Biochem Biophys Res Commun 1980; 94: 1374–1380.
49. Gallop PM, Friedman PA, Henson E. A radical-radical c-carboxylation reaction as a model related to the vitamin K-dependent carboxylation. In: Suttie JW, ed. Vitamin K metabolism and vitamin k-dependent proteins. New York: Academic Press, 1980:408–412.
50. Rao PS, Olson RE. Identity of peroxy radicals produced from vitamin K in oxygenated solutions as studied by pulse radiolysis technique. Biochem Biophys Res Commun 1982; 105:231–235.
51. Canfield LM, Ramelow U. Vitamin K-dependent carboxylase evidence for a semiquinone radical intermediate. Arch Biochem Biophys 1984; 230:389–399.
52. Slama JT, Satsangi RK, Simmons A, Lynch V, Bolger RE, Suttie J. An approach to trapping gamma-glutamyl radical intermediates proposed for vitamin K-dependent carboxylase: alpha/beta-methyleneglutamic acid. J Med Chem 1990; 33:824–832.
53. Dowd P, Ham SW, Geib SJ. Mechanism of action of vitamin K. J Am Chem Soc 1991; 113:7734–7743.
54. Sadowski JA, Schnoes HK, Suttie JW. Vitamin K epoxidase: properties and relationship to prothrombin synthesis. Biochemistry 1977; 16:3856–3863.
55. Kuliopulos A, Hubbard BR, Lam Z, et al. Dioxygen transfer during vitamin K-dependent carboxylase catalysis. Biochemistry 1992; 31:7722–7728.
56. Walsh C. Enzymatic reaction mechanisms. San Francisco: W. H. Freeman, 1979:513.
57. Suttie JW, Hageman JM, Lehrman SR, Rich DH. Vitamin K-dependent carboxylase: development of a peptide substrate. J Biol Chem 1976; 251:5827–5830.

58. Decottignies-LeMarechal P, Rikong-Adie H, Azerad R, Gaudry M. Vitamin K-dependent carboxylation of synthetic substrates. Nature of the products. Biochem Biophys Res Commun 1979; 90:700–707.
59. Finnan JL, Suttie JW. Carboxylation of low-molecular-weight substrates by the rat liver vitamin K-dependent carboxylase: characterization of products. In: Suttie JW, ed. Vitamin K Metabolism and Vitamin K-Dependent Proteins. Baltimore: University Park Press, 1980: 480–483.
60. Soute BAM, Vermeer C, DeMetz M, Hemker HC, Lijnen HR. *In vitro* prothrombin synthesis from a purified precursor protein. III. Preparation of an acid-soluble substrate for vitamin K-dependent carboxylase by limited proteolysis of bovine descarboxyprothrombin. Biochim Biophys Acta 1981; 676:101–107.
61. Pottorf RS, Rich DH, Engelke J, Suttie JW. Synthetic bovine prothrombin precursor 13–29 for studies of vitamin K-dependent carboxylase. J Med Chem 1987; 30:445–448.
62. Vermeer C, Soute BAM, Hendrix H, de Boer-van den Berg MAG. Decarboxylated bone Gla-protein as a substrate for hepatic vitamin K-dependent carboxylase. FEBS Lett 1984; 165:16–20.
63. Ulrich MMW, Soute BAM, van Haarlem LJM, Vermeer C. Substrate recognition by vitamin K-dependent carboxylase. Thromb Haemost 1987; 57:17–19.
64. Malhotra OP. Partially carboxylated prothrombins. III. Carboxylation studies. Biochim Biophys Acta 1983; 746:81–86.
65. Shah DV, Swanson JC, Suttie JW. Vitamin K-dependent carboxylation: effect of detergent concentrations, vitamin K status, and added protein precursors on activity. Arch Biochem Biophys 1983; 222:216–221.
66. Dubin A, Suen ET, Delaney R, Chiu A, Johnson BC. Regulation of vitamin K-dependent carboxylation. J Biol Chem 1980; 255:349–352.
67. Esmon CT, Grant GA, Suttie JW. Purification of an apparent rat liver prothrombin precursor: characterization and comparison to normal rat prothrombin. Biochemistry 1975; 14:1595–1600.
68. Davie EW. The blood coagulation factors: their cDNAs, genes, and expression. In: Colman RW, Hirsh J, Marder VJ, Salzman EW, ed. Hemostasis and Thrombosis: Basic Principles and Clinical Practice, 2nd ed. New York: J. B. Lippincott, 1987:
69. Pan LC, Price PA. The propeptide of rat bone gamma-carboxyglutamic acid protein shares homology with other vitamin K-dependent protein precursors. Proc Natl Acad Sci USA 1985; 82:6109–6113.
70. Price PA, Fraser JD, Metz-Virca G. Molecular cloning of the matrix Gla-protein: implications for substrate recognition by the vitamin K-dependent carboxylase. Proc Natl Acad Sci USA 1987; 84:8335–8339.
71. Jorgensen MJ, Cantor AB, Furie BC, Brown CL, Shoemaker CB, Furie B. Recognition site directing vitamin K-dependent gamma-carboxylation resides on the propeptide of factor IX. Cell 1987; 48:185–191.
72. Rabiet M-J, Jorgensen MJ, Furie B, Furie BC. Effect of propeptide mutations on post-translational processing of factor IX. J Biol Chem 1987; 262:14895–14898.
73. Huber P, Schmitz T, Griffin J, et al. Identification of amino acids in the gamma-carboxylation recognition site on the propeptide of prothrombin. J Biol Chem 1990; 265:12467–12473.
74. Foster DC, Rudinski MS, Schach BG, et al. Propeptide of human protein C is necessary for gamma-carboxylation. Biochemistry 1987; 26:7003–7011.
75. Ware J, Diuguid DL, Liebman HA, et al. Factor IX San Dimas: substitution of glutamine for arginine-4 in the propeptide leads to incomplete gamma-carboxylation and altered phospholipid binding properties. J Biol Chem 1989; 264:11401–11406.
76. Diuguid DL, Rabiet M-J, Furie BC, Liebman HA, Furie B. Molecular basis of hemophilia B: a defective enzyme due to an unprocessed propeptide is caused by a point mutation in the factor IX precursor. Proc Natl Acad Sci USA 1986; 83:5803–5807.

77. Sugimoto M, Miyata T, Kawabata F, Yoshioka H, Fukui H, Iwga S. Factor IX Kawachinagano: impaired function of the Gla-domain caused by attached propeptide region due to substitution of arginine by glutamate at position −4. Br J Haematol 1989; 72:216–221.
78. Hubbard BR, Jacobs M, Ulrich MMW, Walsh C, Furie B, Furie BC. Vitamin K-dependent carboxylation: *in vitro* modification of synthetic peptides containing the gamma-carboxylation recognition site. J Biol Chem 1989; 264:14145–14150.
79. Ulrich MMW, Furie B, Jacobs MR, Vermeer C, Furie BC. Vitamin K-dependent carboxylation. A synthetic peptide based upon the gamma-carboxylation recognition site sequence of the prothrombin propeptide is an active substrate for the carboxylase *in vitro* J Biol Chem 1988; 263:9697–9702.
80. Knobloch JE, Suttie JW. Vitamin K-dependent carboxylase: control of enzyme activity by the ''propeptide'' region of factor X. J Biol Chem 1987; 262:15334–15337.
81. Cheung A, Engelke JA, Sanders C, Suttie JW. Vitamin K-dependent carboxylase: influence of the ''propeptide'' region on enzyme activity. Arch Biochem Biophys 1991; 274:574–581.
82. Zhang L, Castellino FJ. Role of the hexapeptide disulfide loop present in the gamma-carboxyglutamic acid domain of human protein C in its activation properties and in the in vitro anticoagulant activity of activated protein C. Biochemistry 1991; 30:6696–6704.
83. Zhang L, Jhingan A, Castellino FJ. Role of individual gamma-carboxyglutamic acid residues of activated human protein C in defining its in vitro anticoagulant activity. Blood 1992; 80: 942–952.
84. Liska DJ, Suttie JW. Location of gamma-carboxyglutamyl residues in partially carboxylated prothrombin preparations. Biochemistry 1988; 27:8636–8641.
85. Borowski M, Furie BC, Furie B. Distribution of gamma-carboxyglutamic acid residues in partially carboxylated prothrombins. J Biol Chem 1986; 261:1624–1628.
86. Vermeer C, Hendrix H, Daemen M. Vitamin K-dependent carboxylase from non-hepatic tissues. FEBS Lett 1982; 148:317–320.
87. Roncaglioni MC, Soute BAM, van der Berg MAGdB, Vermeer C. Warfarin-induced accumulation of vitamin K-dependent proteins. Comparison between hepatic and nonhepatic tissues. Biochem Biophys Res Commun 1983; 114:991–997.
88. McClure DB, Walls JD, Grinnell BW. Post-translational processing events in the secretion pathway of human protein C, a complex vitamin K-dependent antithrombotic factor. J Biol Chem 1992; 267:19710–19717.
89. Usui S, Yu L, Yu C. Small molecular mass ubiquinone binding protein in mitochondrial ubiquinol cytochrome c reductase: isolation, ubiquinone binding domain, and immunoinhibition. Biochemistry 1990; 29:4618–4626.
90. Steczko J, Donoho GP, Clemens JC, Axelrod B. Conserved histidine residues in soybean lipooxygenase: functional consequences of their replacement. Biochemistry 1992; 31:4053–4057.
91. Steczko J, Axelrod B. Identification of the iron binding histidine in soybean lipooxygenase L-1. Biochem Biophys Res Commun 1992; 186:686–689.
92. Olson RE, Hall AL, Lee FC, Kappel WK, Meyer RG, Bettger WJ. Vitamin K-dependent carboxylase: a heme protein? In: Johnson BC, ed. Posttranslational Covalent Modification of Proteins. New York: Academic Press, 1983:295–316.
93. Kuliopulos A, Cieurzo CE, Furie B, Furie BC, Walsh CT. *N*-bromoacetyl-peptide substrate affinity labeling of vitamin K-dependent carboxylase. Biochemistry 1992; 31:9436–9444.
94. Suttie JW and Nelsestuen, GL. CRC Crit Rev Biochem 1980 8:191–223.
95. Suttie JW. Ann Rev Biochem 1985; 54:459–477.
96. Suttie JW. Biofactors 1988; 1:55–60.
97. Price PA. Ann Rev Nutr 1988; 8:565–583.
98. Vermeer C. Biochem J. 1990; 266:625–636.
99. Furie B and Furie BC. Blood 1990; 75:1753–1762.
100. Wu S-M and Stafford DW. Am Soc Biochem Mol Biol 1994; 8:A1440 (Abstract).

16

Tissue Factor Pathway Inhibitor

Tze-Chein Wun
Searle/Monsanto Company, St. Louis, Missouri

I. INTRODUCTION

The blood coagulation cascade is by convention classified into extrinsic and intrinsic pathways. The extrinsic pathway is initiated when plasma comes into contact with tissue factor (TF, previously also known as tissue thromboplastin), leading to the formation of a TF-factor VIIa complex, which activates factors IX and X into IXa and Xa. The intrinsic pathway is initiated when plasma is exposed to a negatively charged surface, which activates factor XII via a contact activation reaction involving prekallikrein and high-molecular-weight kininogen. This contact activation reaction results in the conversion of factor XI into XIa, which activates factor IX into IXa. Factor IXa forms a complex with factor VIIIa and phospholipid, which converts factor X into Xa. Factor Xa, generated via the extrinsic or intrinsic pathway, forms a prothrombinase complex with factor Va and phospholipid, which converts prothrombin into thrombin. Although contact activation reactions are required for blood to clot normally in a glass test tube, persons lacking factor XII, prekallikrein, or high-molecular-weight kininogen are without hemorrhagic symptoms. This suggests that contact activation reactions are not required for hemostasis. In contrast, patients with severe factor VII deficiency may bleed seriously. Therefore, the formation of TF-factor VIIa complex (tissue factor pathway) appears to be the primary mechanism for initiating blood coagulation during hemostasis.

Studies of the regulation of thromboplastin-induced coagulation were first reported in 1947, when Thomas (1) and Schneider (2) independently showed that the preincubation of tissue thromboplastin with serum prevented the lethal intravascular coagulation that occurred following thromboplastin infusion into animals. Later, Hjort (3) showed that an inhibitor in the serum recognized the TF-factor VIIa-Ca^{2+} complex, which he called convertin. Subsequently, several groups investigated the inhibitory moieties against factor Xa and TF-factor VIIa complex present in the plasma lipoproteins. Marciniak (4) demonstrated that the plasma low-density lipoproteins (LDL) carried an inhibitor of factor

Xa, and Barrowcliffe et al. showed that this inhibitor was present in LDL > high-density lipoprotein (HDL) >> very low density lipoprotein (VLDL) (5,6) and that its plasma concentration was increased following in vivo administration of a heparin analog (7). Carson (8) noted that plasma lipoproteins inhibited TF-factor VIIa activity and that this inhibitory activity was contained in the protein component of the lipoproteins. Morrison and Jesty (9) found that factors X and IX were only partially activated following the addition of a low concentration of TF to plasma and that this apparent inhibition of TF-factor VIIa activity required the presence of factor X or brief pretreatment of the plasma with factor Xa. Sanders et al. (10) further documented that both factor X and an inhibitor present in the plasma lipoprotein fraction were needed to produce this TF-factor VIIa inhibition. Additional reports by Hubbard and Jennings (11), Broze and Miletich (12), and Rao and Rapaport (13) demonstrated that the functional properties of the inhibitor were the same as those described for "anticonvertin" by Hjort. This inhibitor was later called extrinsic pathway inhibitor (13) or lipoprotein-associated coagulation inhibitor (12). A new name, tissue factor pathway inhibitor (TFPI), was recommended by the International Society on Thrombosis and Hemostasis in 1991.

A number of excellent reviews on tissue factor pathway inhibitor (14–17) have been published in the past few years. This chapter emphasizes more recent progress in the understanding of the structure and function of TFPI, the mechanism of inhibitory action, the gene structure and its regulation, and potential therapeutic uses of recombinant TFPI.

II. ISOLATION AND CHARACTERIZATION OF TFPI PROTEIN

TFPI was first isolated from the serum-free conditioned media of HepG2 hepatoma cells using a four-step procedure including $CdCl_2$ precipitation, diisopropylphosphoryl (DIP)-factor Xa affinity chromatography, Sephadex G-75 gel filtration, and Mono Q ion-exchange chromatography (18). It was later found that a trace quantity of active Xa remaining on DIP-Xa-agarose was responsible for binding TFPI, and a simplified three-step purification (omitting the last step) was employed using bovine Xa-agarose affinity chromatography (19). In a subsequent study, TFPI proteins were purified from the serum-free conditioned media of HepG2 hepatoma, Chang liver, and SK hepatoma cells by immunoaffinity chromatography using polyclonal antibodies raised against a synthetic peptide corresponding to amino acids 3–25 of TFPI (20). The purified TFPIs varied in molecular weight ($M_r \simeq$ 38,000 and $\simeq$ 35,000) depending on whether the media were concentrated before chromatography. It was suggested that proteolysis of the carboxyl-terminal region of the larger molecule gave rise to the lower molecular weight species, because both molecules were found to have the same amino-terminal sequences. These two species of molecules possessed the same specific activity in an end-point anti-Xa assay. Despite the combined use of immunoaffinity purification and preparative sodium dodecyl sulfate–polyacrylamide gel electrophoresis (SDS-PAGE) to isolate the high-molecular-weight species, the inhibition of tissue factor-induced coagulation of plasma by the isolated $M_r \simeq$ 38,000 molecules from these three cell lines was not equivalent in a kinetic type of assay. The isolated SK hepatoma and Chang liver TFPIs were found to possess 6- and 1.4-fold higher activity, respectively, than the isolated HepG2 TFPI on an anti-Xa unit basis. The cause for the differences was not clear, but one possible reason is the variable proteolysis of the carboxyl terminus of the molecules (see the following section), which did not allow a clear-cut separation of the full-length molecule. Whether other posttranslational modifications also contributed to the activity differences was not clear.

The majority of plasma TFPI was shown to circulate in complex with plasma lipoproteins, although the functional significance of this interaction was not understood. As determined by density ultracentrifugation (21) and size exclusion gel chromatography (22), the distribution of plasma TFPI was as follows: LDL (50–60%) > HDL (26–44%) >> VLDL (~9%), and 6% remained free. Plasma TFPI was isolated using a combination of Pheny-Sepharose CL-4B, Q-Sepharose, and factor Xa-Affi-gel chromatography. When analyzed by SDS-PAGE under nonreducing conditions, the isolated plasma TFPI consisted of multiple forms, with two prominent proteins of M_r 34,000 (uncomplexed TFPI) and 40,000 (TFPI linked by disulfide to apolipoprotein AII) and less abundant bands of higher apparent molecular weight (TFPI linked by disulfide to unidentified proteins) (15,22). Upon reduction, the uncomplexed TFPI had an apparent M_r of 36,000. All these forms of molecules were recognized by antibodies directed against the amino and carboxyl termini of TFPI and were capable of inhibiting the TF-factor VIIa complex and factor Xa. Heparin administration in vivo increased the concentration of TFPI in plasma 2- to 10-fold in the human (23,24). This heparin-releasable TFPI was purified from postheparin plasma by heparin-agarose chromatography, monoclonal antibody immunoaffinity chromatography, and size exclusion chromatography (25). On SDS-PAGE under reducing condition, the purified heparin-releasable TFPI was reported to have an apparent M_r 40,000, which was the same as HepG2 TFPI but larger than plasma TFPI (M_r 36,000). The isolated lipoprotein-associated and heparin-releasable TFPIs combined with lipoproteins in vitro but HepG2 TFPI did not. When injected into rabbits, the heparin-releasable TFPI was found to be cleared more slowly than HepG2 TFPI. The slower clearance of heparin-releasable TFPI might be a result of the attachment to lipoproteins in vivo (25). Thus, there are subtle structural and functional differences among the cell culture-, plasma- and heparin-releasable TFPIs.

Cloning of the cDNA coding for TFPI was first reported in 1988 (26), and recombinant TFPIs were subsequently expressed in several mammalian cell hosts, including C127 mouse fibroblast cells (27), baby hamster kidney (BHK) cells (28,29), human SK hepatoma cells (29), and Chinese hamster ovary (CHO) cells (29). In one study, BHK TFPI protein was purified from the serum-free conditioned medium in four steps, including heparin affinity chromatography, Mono Q anion exchange, Mono S cation exchange, and reversed-phase high-performance liquid chromatography (28), and the isolated protein was found to possess much lower anticoagulant activity than plasma TFPI (30). In another study, four recombinant TFPIs were isolated by one-step immunoaffinity chromatography on a monoclonal antibody linked to Sepharose 4B (29). Like the nonrecombinant TFPIs isolated from the HepG2 hepatoma, Chang liver, and SK hepatoma cells (20), the recombinant TFPIs expressed in SK hepatoma, C127, BHK, and CHO cells exhibited a wide variation in their anticlotting activity when measured by one-stage tissue factor-induced coagulation assays, despite that all these proteins inhibited factor Xa with 1:1 stoichiometry (29). The immunoaffinity-isolated TFPIs were shown to have the following relative activities: 28:15:2.1:1, SK-C127-BHK-CHO. By SDS-PAGE, the SK and C127 TFPIs had somewhat higher apparent molecular weight than the BHK and CHO proteins. Further separation by gradient elution on Mono S cation exchanger (29,31) or heparin-agarose (32) resolved these purified proteins into a high-activity fraction that eluted around 0.6 M NaCl [TFPI(0.6)] and a low-activity fraction around 0.3 M NaCl [TFPI(0.3)], respectively. Amino acid sequencing and western blotting probed with antibodies against the amino-terminal peptide of TFPI showed that both TFPI(0.3) and TFPI(0.6) fractions possessed intact amino termini. In contrast, the carboxyl terminus of full-length TFPI was detected in the 0.6 M fraction but not the 0.3 M fraction when western blot was probed with antibodies against the carboxyl-terminal peptide of TFPI (29,31,32). Comparison of the elution patterns of the SK, C127, and CHO TFPIs in the Mono S chromatography showed that the anticoagulant activity of the immunoaffinity-purified proteins roughly correlated with the the abundance of the TFPI(0.6) molecules (29). Peptide mapping of a V8 protease digest of BHK protein confirmed that the carboxyl-terminal peptide was present in the high-activity TFPI(0.6) fraction but was lacking in the low-activity TFPI(0.3) fraction and showed that heterogeneous degradation of the carboxyl terminus of the low-activity fraction had taken place (31). These reports taken together suggest that variations in the anticoagulant activity of isolated TFPIs are caused, at least in part, by the proteolysis of the carboxyl terminus of the TFPI molecule and that the carboxyl terminus was essential for its anticoagulant activity. The full-length C127 TFPI was shown to be more potent than the truncated molecule in the inhibition of plasma coagulation induced by tissue factor, factor Xa, or factor X coagulant protein from Russell's viper venom (32). Using purified components, it was shown that the initial inhibition of factor Xa produced by the full-length TFPI was severalfold greater than that produced by the truncated molecule and that the two forms of molecules did not show a difference in the initial rates of TF-VIIa inhibition in the presence of factor X (32). Thus, the differential effect of these forms of TFPI on tissue factor-induced coagulation in normal plasma appeared to be directly related to their ability to inhibit factor Xa. Although these results suggested that proteolysis of the carboxyl terminus of TFPI could cause a large decrease in its anticoagulant activity, alternative reasons for the variation in activity could not be excluded. Among various cell cultures, SK hepatoma cells consistently secreted mostly intact TFPI, and the molecules were largely preserved in the full-length form throughout the purification process (20,29). It had been shown that non-recombinant and recombinant SK hepatoma-derived TFPI possessed similar anticoagulant activity to that of the TFPI present in human plasma (20,29). In view of the difficulty in obtaining large quantities of plasma TFPI, recombinant SK hepatoma TFPI is potentially an attractive material for use as a reference standard in future studies.

Aside from the proteolysis already described, several posttranslational modifications appeared to occur on the TFPI molecule. TFPI produced by HepG2 cells and recombinant TFPI produced by mouse C127 fibroblasts were partially phosphorylated at serine 2 (33). Casein kinase II was likely responsible for the phosphorylation at this site because a serine is followed by a cluster of acid residues, which is characteristic of sites phosphorylated by this enzyme. Furthermore, phosphorylation of purified TFPI by bovine casein kinase II has been shown to occur in vitro. Using site-directed mutagenesis, an altered form of TFPI was produced in which the serine 2 residue was changed to alanine. This altered molecule was not phosphorylated and was shown to be as potent as the partially phosphorylated wild-type molecule in the inhibition of Xa and TF-VIIa activities in end-point assays. Thus the phosphorylation of this site is not essential for the function of TFPI. However, whether phosphorylation of serine 2 affect the kinetics of inhibition, the association with plasma lipoprotein, or the in vivo clearance is not known. Human TFPI contains N-linked oligosaccharides, but which of the three possible sites of attachment are utilized and what types of carbohydrate are present have not been reported. Sulfation of N-linked oligosaccharides occurred in TFPIs produced by human umbilical endothelial cells, kidney carcinoma cells (Caki) (15), and recombinant human kidney 293 cells, but not in those produced by HepG2 or recombinant CHO cells (34). It was determined that >70% of the oligosaccharides on recombinant TFPI expressed in 293 cells terminated with the sequence SO_4-4GalNAcβ1,4GlcNAcβ1,2Manα. Oligosaccharides terminating with this sequence were described on lutropin, thyrotropin, and proopiomelanocortin, and this sulfated structure was shown to be responsible for their rapid clearance from plasma by a receptor in hepatic reticuloendothelial cells (35). However, despite the demonstration of recognition of 293 TFPI by the hepatic reticuloendothelial cell in vitro, the clearance of this protein in vivo was not detectably accelerated. It was speculated that a serum component or the association of TFPI with lipoproteins prevented the sulfated molecule from interacting with the receptor in vivo. The effect of sulfation on the functional properties of TFPI and its association with lipoproteins remains to be clarified. A rabbit aortic endothelial cell line also synthesized a N-glycan sulfated TFPI (36), which was shown to be functionally and immunologically similar to rabbit plasma TFPI (37). The rabbit TFPI was reported to have a molecular mass of 47 kD and was susceptible to proteolysis of the carboxyl terminus, which resulted in a decrease in the molecular mass to 45 kD and a sharp reduction in its ability to inhibit TF-VIIa (38).

Human TFPI is glycosylated and contains nine pairs of cysteine disulfide linkages. Recently this protein was expressed in *Escherichia coli* as a nonglycosylated protein (unpublished results). The expressed TFPI molecules accounted for 5–10% of the total *E. coli* cell protein and formed disulfide-linked polymers in inclusion bodies. The inclusion bodies containing the TFPI protein were sulfonated, purified by anion-exchange chromatography, refolded through a disulfide exchange reaction, and further fractionated by cation-exchange chromatography to separate the active molecules from the misfolded species. The active *E. coli*-derived TFPI was shown to be twofold more active, on a molar basis, than the full-length mammalian TFPI. Thus, glycosylation was not required for the inhibitory function of TFPI, and it may impede, to some extent, the inhibitory activity of TFPI.

III. TFPI cDNA AND mRNA

cDNA for human TFPI was first cloned by screening placental and fetal liver λgt11 cDNA libraries with a polyclonal antibody (26). The longest cDNA insert of these isolates consisted of 1431 bases that included a 5′ noncoding sequence of 132 nucleotides, an open reading frame of 912 nucleotides, and a 3′ noncoding region of 387 nucleotides. Northern blot analysis indicated that there were two major species of mRNA (1.4 and 4 kb) that hybridized with TFPI cDNA. The open reading frame encoded a signal peptide of 28 residues, followed by a 32 kD protein of 276 residues. The predicted sequence of mature TFPI contained 18 cysteines and three potential N-linked glycosylation sites. The translated amino acid sequence of TFPI showed several domains, including a negatively charged amino terminus, three tandem Kunitz-type inhibitory domains, and a highly positively charged carboxyl terminus. The three Kunitz-type domains are each homologous to the sequences of bovine pancreatic trypsin inhibitor and other basic protease inhibitors. Three pairs of disulfide bonding structure in each Kunitz domain were conserved. In a subsequent study (39), a 4023 nucleotide sequence was compiled from overlapping cDNAs isolated from a human endothelial cell library, which likely spanned the full length of the 4 kb TFPI message (Fig. 1). This included 382 bases of 5′-untranslated sequence, 912 bases of coding sequence identical that of the 1.4 kb clone, and 2730 bases of 3′-untranslated sequence that included 48 bases of poly(A) tail (39).

The occurrence of five potential polyadenylation signals at the 3′ end and alternative processing appeared to account for the occurrence of 1.4 and 4 kb messages. Internal deletion of 121 bases in the 5′-untranslated region was also observed in some other clones, suggesting alternative splicing. The 3′-untranslated region of the 4 kb TFPI message contained 46 AUUU and 2 UUAUUUAU sequences, motifs thought to be involved in the selective degradation of transiently expressed lymphokine, cytokine, and protooncogene messages and in the expression of inflammatory mediators (40,41). The presence

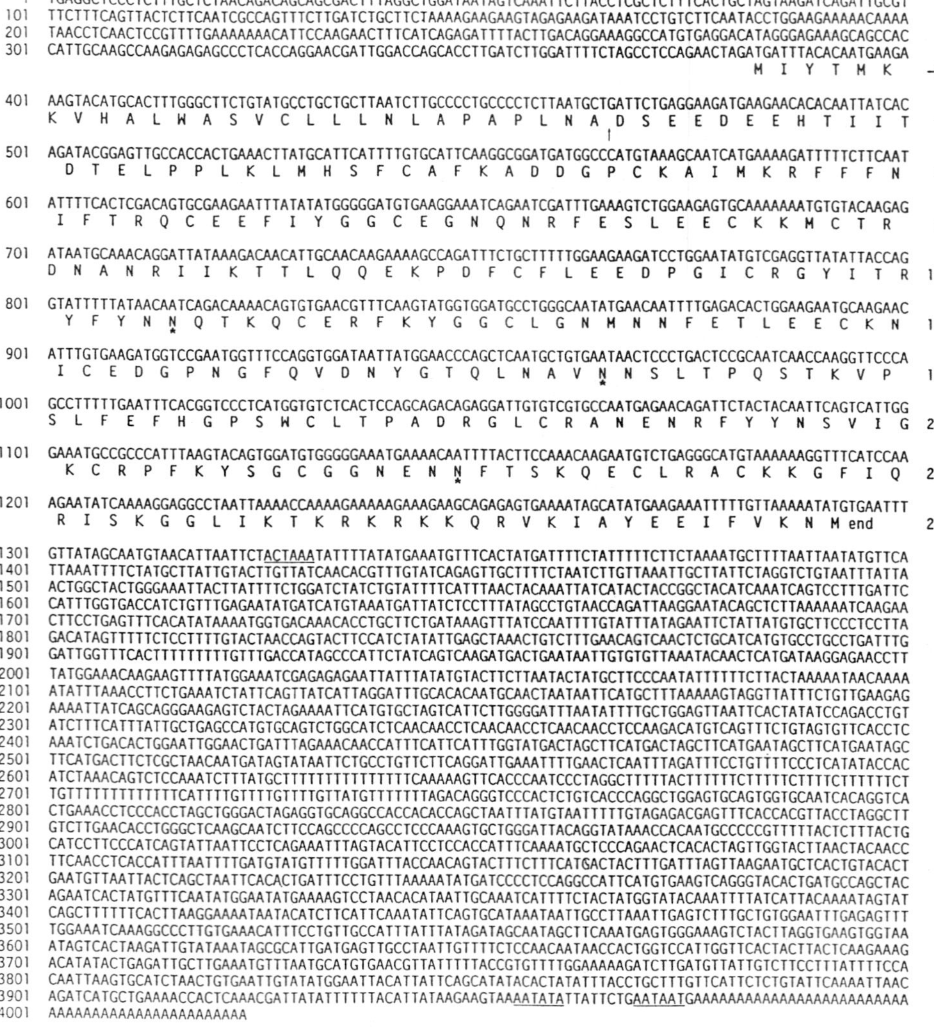

Figure 1 Nucleotide sequence and deduced amino acid sequence of human TFPI. Nucleotides are numbered on the left. Amino acids are numbered on the right. The signal peptidase cleavage site is denoted by an arrow. Potential N-linked glycosylation sites are marked by asterisks. Potential polyadenylation signals for the 1.4 and 4.0 kb messages are underscored. (Adapted from Refs. 26 and 39.)

of these motifs predicted that the 4 kb mRNA would be unstable. However, northern blots of RNA isolated from actinomycin D-treated HepG2 cells indicated that both the 1.4 and 4 kb messages are stable, greater than 80% remaining 7 h following the addition of actinomycin D (39).

cDNA sequences for rabbit (42) and rat (43) TFPIs were reported recently. Deduced amino acid sequences showed that, similar to human TFPI, rabbit and rat TFPIs both contained a cluster of negatively charged amino acid residues at the amino terminus, followed by three tandem Kunitz-type inhibitory domains and a cluster of positively charged residues near the carboxyl terminus. Three potential N-linked glycosylation sites were present. Comparison of the sequences of the three species revealed that a high degree of homology in the Kunitz-type domains (57, 86, and 69% in the Kunitz 1, 2, and 3 domains, respectively) is present, but the sequences of the polypeptide regions linking the Kunitz domains diverged considerably. Notably, rabbit TFPI contained a glutamine at the P1 residue of the active-site cleft of the third Kunitz domain instead of the arginine in the human TFPI and lysine in the rat TFPI.

IV. THE TFPI GENE

A. Gene Organization

The human TFPI gene was assigned to chromosome 2 by measuring the hybridization of a TFPI cDNA probe to DNA isolated from a panel of human-mouse somatic cell hybrids containing different human chromosomes (44). In situ hybridization to metaphase chromosome further mapped the gene to the region 2q31 → 2q32.1 (44). Genomic clones containing the human TFPI gene were isolated from genomic or chromosome 2-specific phage libraries (44,45). The TFPI gene spanned about 70 kb and consists of nine exons separated by eight introns. Exons 1 and 2 encode the 5′-untranslated portion of the TFPI messages. Alternative splicing may occur, resulting in the absence of exon 2 in some TFPI mRNAs. Exon 3 encodes the signal peptide and amino-terminal domain of mature TFPI. The three Kunitz-type inhibitor domains of TFPI are individually encoded by exons 4, 6, and 8. Exons 5 and 7 encoded the two peptide sequences interspersed between the three Kunitz-type inhibitor domains. Exon 9 encoded the carboxyl-terminal domain of TFPI and contained the entire 3′-untranslated sequence of the 1.4 and 4.0 kb TFPI messages. The eight introns each began with a GT dinucleotide and ended with an AG dinucleotide, in agreement with the consensus intron-exon splice junction sequence formulated by Shapiro and Senapathy (46). The secondary structure for TFPI and the location of introns with respect to the protein sequence are shown in Figure 2.

B. Sequence Polymorphism

A single restriction fragment length polymorphism for the TFPI gene has been reported (47). Using a cDNA fragment spanning nucleotides 458–1208 as probe, PstI digestion of genomic DNA reveals a two-allele polymorphism, with bands at either 6.4 or 6.9 kb. The allele frequencies calculated from 25 unrelated whites were 0.34 and 0.66 for 6.4 and 6.9 kb, respectively. The functional significance of this polymorphism is not known.

C. Regulatory Regions of the TFPI Gene

Primer extension and S1 nuclease protection analysis of HepG2 (44) and liver (45) mRNAs indicated that multiple transcription initiation sites for TFPI messages were present. This was thought to be related to the absence of a prototypical TATA box or CCAAT sequence in the 5′ flanking region of the human TFPI gene. Comparison of 498 bp DNA sequence upstream of the transcriptional start sites with consensus sequences for transcription factor binding sites provided matches for a GATA consensus element, two AP-1-like motifs, and two NF-1 sequences (44,45,48). The GATA motif had been shown to be essential for expression of certain genes by erythroid cells, megakaryocytes, and endothelial cells, and it was thought that the binding of GATA-2 transcription factor to the

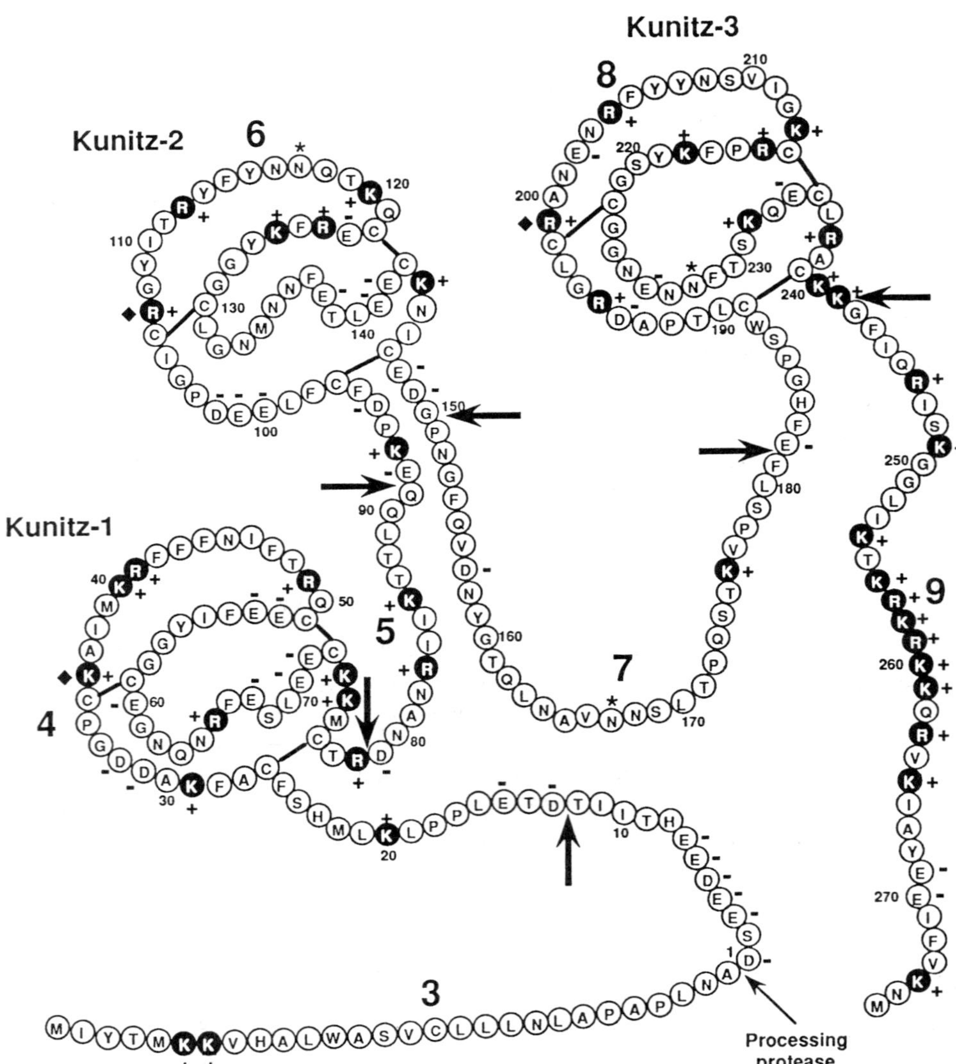

Figure 2 Secondary structure of human TFPI and location of introns. The thin arrow indicates a signal peptidase processing site. The bold arrows indicate the location of introns, and the exons are numbered. Asterisks indicate the potential N-linked glycosylation sites. Diamonds indicate the active-site P1 residues of the Kunitz domains. The charges of the amino acid side chains are shown.

GATA sequence is necessary for the constitutive expression of TFPI by the endothelium (48). Multiple copies of AP-1 binding sites in the 5′ flanking regions of other genes have been shown to act as phorbol ester-responsive elements (49), and it was thought that the two AP-1-like motifs in the 5′ flanking region of the TFPI gene are responsible for the 20-fold induction of TFPI synthesis by phorbol 12-myristate 13-acetate in U937 cells (50) and a slight increase in TFPI synthesis (~1.5-fold) in endothelial cells (48). The NF-1 motif was shown to be a transforming growth factor β (TGF-β) responsive element (51), and it was thought that the presence of two NF-1-like sequences in the 5′ flanking region of TFPI is responsible for the modest upregulation of TFPI synthesis by whole-blood serum, which contains TGF-β released from platelets (48). However, whether the GATA, AP-1-like, and NF-1 sequences are actually involved in the regulation of TFPI gene expression remains to be experimentally tested.

V. MECHANISM OF INHIBITION: STRUCTURE AND FUNCTION OF TFPI PROTEIN

Little was known about the mechanism of TFPI action until 1985–1988, when it was reported that factor X was required for a plasma inhibitor present in barium-adsorbed plasma to inhibit TF-factor VIIa activation of factor IX (10) and that the inhibitor could bind factor Xa and neutralize its enzymatic activity in a reaction independent of calcium ions (52). A two-step model (Fig. 3, left path) was proposed (52,53). In the first step, factor Xa binds to TFPI in a Ca^{2+}-independent reaction requiring the active site of factor Xa. In the second step, the TFPI-factor Xa complex binds to factor VIIa-TF by a mechanism requiring the participation of the factor Xa molecule in its calcium-dependent conformation. An alternative model (Fig. 3, right path) was also proposed that involved the binding of TFPI to a preformed Xa-VIIa-TF complex to form the quaternary complex (52). A recent study using a continuous flow reactor coated with TF-phospholipids showed that TF contained a binding site for factor Xa and that the bound factor Xa provided a TFPI binding site (54). In a similar flow system lined with endothelial cells expressing TF, it was shown that preperfusion with factor Xa-TFPI complex in the absence of factor VIIa caused much less inhibitory effect, suggesting that TFPI-mediated neutralization of TF activity required the presence of factor VIIa in addition to factor Xa (55). These data taken together indicated that the inhibition involved the formation of a quaternary Xa-TFPI-VIIa-TF complex but did not allow one to draw a conclusion on the order in which the reactants formed the inhibitory complex.

The primary structure of TFPI predicted by the sequences of cDNA clones shows that, after a signal peptide, the mature protein contains an acid amino-terminal region, followed by three tandem domains with homology to Kunitz-type protease inhibitor connected by two distinctive peptides and a highly basic carboxyl-terminal region. The secondary structure of human TFPI could be predicted based on the known Kunitz domain structure, as shown in Figure 2. The functions of the Kunitz domains were studied by mutational analysis. By site-directed mutagenesis, altered forms of TFPI were produced in which the residues at the active-site cleft of each Kunitz domain were individually changed (56). Mutants with an altered active-site residue for Kunitz 1 [TFPI(K36 → I)] and Kunitz 3 [TFPI(R199 → L)], respectively, retained the ability to inhibit Xa, but the mutant with an altered active-site residue in Kunitz 2 [TFPI(R107 → L)] did not inhibit Xa. Mutation of the active-site residue in either Kunitz 1 or 2 abrogated the ability of

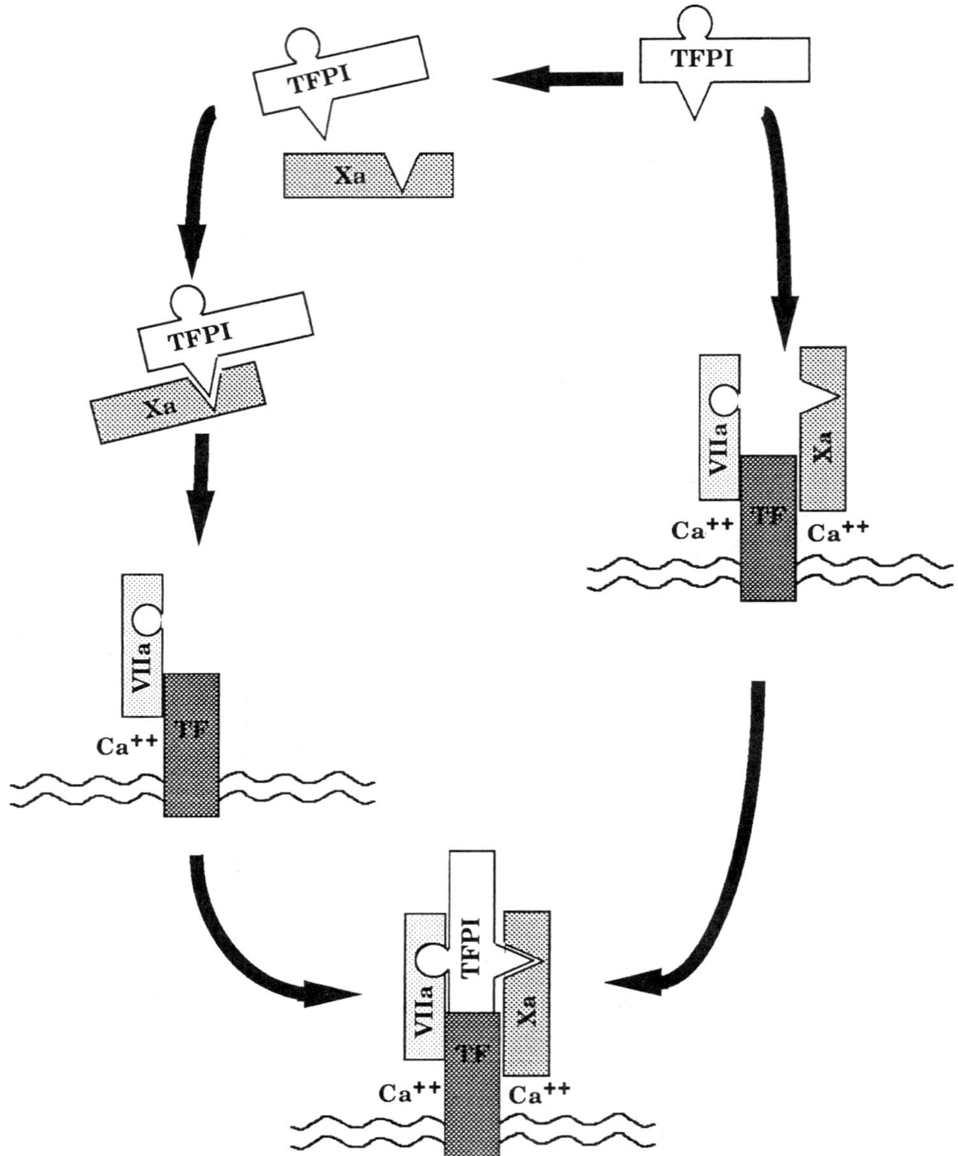

Figure 3 Formation of the putative quaternary inhibitory complex (factor Xa-TFPI-TF-factor VIIa). Left path: TFPI forms a complex with Xa followed by the binding of the complex to TF-VIIa. Right path: TFPI binds to the preformed Xa-TF-VIIa complex.

TFPI to inhibit VIIa-TF activity, whereas mutation of the Kunitz 3 domain had no effect. These results indicated that the Kunitz 2 domain is responsible for the binding to and inhibition of factor Xa and that simultaneous binding of Kunitz 1 domain to factor VIIa and Kunitz 2 domain to factor Xa within the TFPI molecule is required for the formation of the Xa-TFPI-VIIa-TF inhibitory complex. The function of the third Kunitz domain of TFPI remained unknown.

As mentioned, the carboxyl-terminal region of TFPI contains a cluster of basic amino acids, and this region is susceptible to proteolysis, resulting in truncation of the molecules. Both full-length and carboxyl-truncated TFPIs were shown to bind heparin (32). Because of the high charge density, the carboxyl-terminal region was predictably involved in the high-affinity binding to heparin, as evidenced from the higher concentration of salt required to elute the full-length molecules from heparin-agarose than the truncated species (32). The truncated and full-length TFPIs both inhibited factor Xa with 1:1 stoichiometry and inhibited factor VIIa-TF equally in preincubated end-point assays (32). However, the full-length molecule was found to be kinetically more potent than the truncated molecule in the inhibition of TF-induced coagulation of plasma (29,31,32). Using purified components, it was further shown that the initial rate of inhibition of factor Xa by the full-length TFPI was severalfold faster than that by the truncated species. Thus, the differential effect of these forms of TFPI on TF-induced plasma coagulation may be in part related to their ability to inhibit factor Xa, and the carboxyl terminus of TFPI may be involved in the acceleration of the formation of the putative Xa-TFPI-VIIa-TF complex. The mechanism by which the carboxyl terminus of TFPI enhances the inhibition of factor Xa- and TF-induced plasma coagulation remains to be defined.

Human leukocyte elastase proteolytically cleaved TFPI between threonine 87 and threonine 88 within the polypeptide that links the first and the second Kunitz domains, producing a truncated TFPI that lacks the amino-terminal peptide and the first Kunitz domain (57,58). Elastase treatment not only affected the ability of TFPI to inhibit factor VIIa-TF but also dramatically reduced its inhibition of factor Xa (57). Kinetic analysis suggested that elastase cleavage does not affect the affinity of the initial encounter between factor Xa and TFPI, but it reduced the affinity of the final factor Xa-TFPI complex, with K_i (final) values for untreated and elastase-treated TFPI of 58 pM and 4.4 nM, respectively (57). This result was opposite that produced by carboxyl-terminal truncation of the TFPI molecule, which was associated with a dramatic reduction in the affinity of the initial encounter between Xa and TFPI (32). Based on these data, it was suggested that an epitope in the amino-terminal region of TFPI or a conformation of the TFPI molecule that required the presence of this region was needed in concert with the second Kunitz domain to produce optimal inhibition of factor Xa by TFPI (57). Leukocyte cathepsin G was found to cleave TFPI into several fragments, but the reaction was slower than that by elastase (58). A weak inhibition by TFPI of both elastase and cathepsin G was also observed (58). The role of the leukocyte elastase and cathepsin G in the modulation of the coagulation associated with inflammation in vivo remained unknown.

A genetically engineered hybrid protein consisting of the light chain of factor Xa and the first Kunitz domain of TFPI was found to inhibit efficiently the activity of the factor VIIa-TF catalytic complex in the absence of factor Xa (59). This suggested that the formation of the factor Xa-TFPI complex through the binding of the Kunitz 2 domain in TFPI serves to juxtapose the factor Xa light chain and TFPI Kunitz 1 domains, thereby producing a complex with much greater affinity for VIIa/TF than TFPI alone (59).

VI. INTERACTION OF TFPI WITH SULFATED POLYSACCHARIDES

In 1980, it was found that the concentration of an inhibitor of factor Xa associated with plasma lipoproteins was increased following the *in vivo* administration of a heparin analog (7). Although the identity of this inhibitor was not known at that time, in retrospect it was likely to be TFPI. In 1988, Sandset et al. reported that intravenous injection of 7500 units heparin increases the TFPI activity about 200% over the baseline value (23). Subsequently, Novotny et al. showed that intravenous injection of 2500 units heparin into 23 patients caused increases in plasma TFPI to 150–650% of baseline levels and that as much as a 10-fold increase following the injection of 7500 units of heparin occurred in some normal individuals when measured by a competitive immunoassay (24). The source of the heparin-releasable TFPI was not known, but it was hypothesized that TFPI is released from the endothelial cell surface glycosaminoglycan (23). In the post-heparin plasma, there was a large increase (~14-fold) in TFPI with high affinity for heparin. This high-affinity fraction was eluted at 0.55–1.0 M NaCl in heparin affinity chromatography (60). The heparin-released TFPI was shown to prolong both the dilute tissue thromboplastin time and the activated partial thromboplastin time of the ex vivo plasma (30), and it contributed significantly to the anticoagulant effect of heparin ex vivo (61).

Several decades of research have led to the conclusion that the anticoagulant effect of heparin is largely a result of its catalytic action on the antithrombin III-mediated inhibition of thrombin, factor Xa, and other coagulation proteases of the intrinsic pathway, including factors XIIa, XIa, IXa, and kallikrein (62). In contrast, little was known about the effect of heparin on the components of the extrinsic pathway. Using purified components, it was first shown that heparin accelerated the inhibition of factor Xa up to 40-fold in a reaction mixture containing factor Xa and TFPI (15). In a reaction mixture containing TF, factors VIIa and Xa, TFPI, and Ca^{2+}, the inhibition of tissue factor activity was accelerated 3-fold in the presence of heparin. However, the simultaneous presence of antithrombin III with heparin was found to abrogate the TF-VIIa inhibition in the purified system, and this was possibly caused by the effective competition of heparin-antithrombin III for factor Xa, which was a required component in the formation of the TF-VIIa-Xa-TFPI inhibitory complex (63). In contrast to the results obtained from the purified system, studies using whole plasma showed that heparin enhanced the inhibition of TF-induced coagulation despite the presence of antithrombin III (64–66). To examine whether heparin directly affected the ability of TFPI to inhibit TF-induced plasma coagulation, comparative studies were made in a normal plasma and the same plasma depleted of endogenous TFPI (64). Using a certain concentration of thromboplastin, the TFPI-depleted plasma clotted within 10 minutes in the presence of pharmacologically relevant concentrations of heparin, but the normal plasma did not clot for more than 1 h in the presence of the same concentration of heparin. These results indicated that the endogenous TFPI could serve as a cofactor for heparin in the inhibition of TF-induced coagulation (64). TFPI-depleted plasma was supplemented with purified recombinant TFPI and/or heparin, and the effects on TF-induced clotting were studied. A combination of TFPI and heparin greatly enhanced anticoagulation compared with TFPI or heparin alone (64,66). Drug interaction analysis using the isobole (isoeffective curve) method (67) showed that heparin and TFPI in combination exerted an increasing synergistic anticoagulant effect with increasing concentrations of both compounds (64). Other sulfated polysaccharides were also found to enhance the TFPI-dependent inhibition of TF-induced clotting. By weight, the relative potencies are heparin > pentosan polysulfate > dermatan sulfate > dextran sulfate > heparan sulfate (64). Low-molecular-weight heparin preparations were found to be synergistic with TFPI as well (64,65).

VII. TFPI IN VIVO AND ITS LEVEL IN PLASMA

TFPI is present in vivo in at least three pools: sequestered in platelets (~8 ng/ml) (68), associated with plasma lipoproteins (~100 ng/ml) (22), and released into plasma by heparin, possibly from vascular endothelium (23,24). TFPI was shown to be synthesized by cultured tumor cells derived from liver (18,20), endothelium (69,70), and the monocyte-like cell line U937 (50). However, because poly(A)$^+$ RNA obtained from normal human liver did not contain detectable TFPI mRNA and primary hepatocyte culture from normal human liver was essentially devoid of TFPI activity and mRNA, it was proposed that endothelium is the principle site of synthesis in vivo and the source of plasma TFPI (71). Study of the clearance and the fate of recombinant TFPI after injection into rabbits was recently reported (72). When a purified recombinant TFPI labeled with ^{125}I was injected intravenously into rabbits, part of the radioactivity became unprecipitable by trichloroacetic acid after more than 10 minutes circulation in vivo, suggesting partial degradation of the molecules. When unlabeled TFPI was injected, the clearance of TFPI activity from the plasma exhibited biexponential elimination kinetics, with a rapid α-phase half-life $t_{1/2\alpha}$ of 2.3 minutes and a terminal β-phase half-life $t_{1/2\beta}$ of 79 minutes. The plasma clearance and total area under the curve were 4.2 ml/kg per minute and 129 μg/ml per minute, respectively (72). These values were 22 times smaller and 72 times larger, respectively, than those of recombinant tissue plasminogen activator (tPA) (73). This indicated that the clearance of recombinant TFPI was much slower than that of recombinant tPA, despite that the values of $t_{1/2\alpha}$ for these two proteins were similar. The slower clearance of TFPI compared with tPA may be partly related to the fact that the area under the β phase was dominant (80.6%) for TFPI (72), whereas the area under the α phase was dominant (82%) for tPA (74). The tissue distribution of intravenously administered ^{125}I-TFPI in the rabbit was studied by whole-body autoradiography (72). A high level of radioactivity was found in liver and kidney. The localization within the liver demonstrated a mottled appearance, suggesting regions of higher uptake within the liver. In the kidney, the outer cortex consistently revealed the highest activity.

Several groups have studied plasma TFPI levels using functional (69,75,76) and immunological (24) assays. The TFPI concentration in an adult control population was found to vary from 60 to 160% of the mean, with a mean value corresponding to 89 ng/ml or 2.25 nM (24). TFPI activity was found to be significantly correlated with age (77). Mean TFPI activity in adults greater than 60 years of age was slightly higher than in adults less than 60 years of age (75). Administration of 1-desamino-8-*d*-arginine vasopressin did not alter plasma TFPI levels (75). Plasma TFPI levels were not decreased in patients with warfarin therapy, primary pulmonary hypertension, deep vein thrombosis, or the lupus anticoagulant (24,69). TFPI levels were moderately elevated in normal pregnancy (24,75,78) and pregnancies complicated with hypertension (78), in acute myocardial infarction, angina pectoris, and heart failure (79), and in advanced cancer (80–82). In cancer patients, there was also an enhanced release of TFPI in response to heparin injection (82). The reason for the enhanced heparin response in cancer patients was not clear. Plasma TFPI levels fell slightly after surgical procedures (75,83–85), and the decrease was attributed mainly to hemodilution (84). Very low levels of plasma TFPI were found in patients with homozygous abetalipoproteinemia and hypobetalipoproteinemia, diseases associated with low plasma levels of apolipoprotein B-containing lipoproteins (24). Following the injection of heparin into one patient with homozygous abetalipoproteinemia, the plasma TFPI level increased to a level comparable to that in normal individuals after heparin treatment (24). Because individuals with abetalipoproteinemia have very low plasma TFPI levels and yet do not have an increased risk of thrombosis, it was suggested that plasma is not the most important reservoir of TFPI and that the heparin-releasable pool of TFPI, which is the largest pool, is of greater physiological importance (24). The plasma TFPI level was found to be correlated with total cholesterol (77). Treatment of individuals with heterozygous familial hypercholesterolemia with simvastatin (3-hydroxy-3-methylglutaryl coenzyme A reductase inhibitor) caused decreases in total cholesterol, LDL cholesterol, and apolipoprotein B, with a concomitant reduction in plasma TFPI activity (86), suggesting that plasma TFPI activity is associated with LDL. Several studies reported findings of near normal levels of plasma TFPI in hepatocellular disease (24,69,83). However, in a group of patients with decompensated chronic hepatocellular disease and acute fulminant liver failure, plasma TFPI activity was found to distribute in a bimodal fashion that varied from <20 to 194% (87). Somewhat conflicting data were reported on plasma TFPI levels in disseminated intravascular coagulation (DIC). One report showed that TFPI levels were uniformly decreased in patients with DIC as a result of sepsis and trauma (69); other investigators found that patients with DIC caused by sepsis or carcinoma had normal to elevated levels of TFPI (24,87–89). The results obtained from endotoxin-induced DIC were also inconsistent. One study (69) showed that the plasma TFPI fell, but another study (90) found no significant decrease. The physiological relevance of the differences between the plasma TFPI concentration in health and disease was not clear. The severalfold increase in plasma TFPI following heparin injection suggested that TFPI in plasma represents only a fraction of the total TFPI that can be mobilized. The amount of tissue factor present was also likely to be small, such that significant consumption of plasma TFPI might not occur even in pathological states (90). These findings raised the question of whether measurement of plasma TFPI levels truly reflects pathological changes in vivo.

Although many in vitro experiments suggested that TFPI is the sole physiologically important inhibitor of factor VIIa-TF activity, the in vivo evidence of its role in dampening TF-dependent coagulation was reported only recently. It was first shown that infusion of a concentration of TF fourfold lower than that required to induce DIC in normal rabbits induced DIC in rabbits immunodepleted of plasma TFPI (91). Further, it was found that immunodepletion of TFPI sensitized rabbits to endotoxin-induced DIC and the generalized Shwartzman reaction (92). These results supported a physiological role for TFPI as a natural anticoagulant protecting individuals from thrombotic complications of illness and minor injuries of normal living.

VIII. HEMOPHILIA AS A DEFECT OF THE TISSUE FACTOR PATHWAY: A REVISED HYPOTHESIS OF BLOOD COAGULATION

In 1964, the cascade (93) and waterfall hypotheses (94) of coagulation emphasized the amplifying effect of sequential steps of zymogen activation and separated the known coagulation factors into extrinsic and intrinsic pathways that converge at the activation of factor X with subsequent generation of thrombin. That hemorrhage occurs in patients with factor VII deficiency but not in persons lacking factor XII, prekallikrein, or high-molecular-weight kininogen suggests that the TF-VIIa, rather than the contact factors of the intrinsic pathway, is the primary initiator of blood coagulation during hemostasis. If coagulation is initiated by TF-VIIa, one must then account for the clinical manifestations of hemophilia (a deficiency of factor VIII or IX) based on the tissue factor pathway. Data have now accumulated that allow one to formulate a revised hypothesis that integrates hemophilia into a scheme in which TF is the essential initiator of coagulation.

It had long been known that the clotting times of hemophilic plasma exceeded those of normal plasma only when dilute suspensions of TF were used (95). This was interpreted later as resulting from the large molar excess of factor IX over factor VIII. Thus, at high concentrations of TF, factor IXa would be formed in excess of factor VIII in the plasma and much of the IXa would be kinetically inert. Simultaneously, TF-VIIa would rapidly convert factor X into Xa. Therefore, there would be little difference in the clotting time of hemophilic and normal plasmas when a high concentration of TF was present. Conversely, when low concentrations of TF were present, all the factor IXa generated through the action of TF-VIIa would complex with VIIIa and its contribution to factor X activation would be significant compared with the direct activation of factor X by TF-VIIa. Using dilute TF, it was shown in 1982 that the factor X activation in factor VIII- or IX-deficient plasma was only 10% of the activation rate seen for normal plasma, and reconstitution of the deficient plasmas with factor VIII or IX, respectively, restored normal factor X activation (96). These results suggested that the alternative extrinsic pathway involving TF, VII, IX, and VIII is a major physiological pathway relevant to the bleeding diathesis of hemophilia (97). In 1987, it was first postulated that TFPI-induced dampening of the direct TF-VIIa activation of factor X could account for the need of IXa-VIIIa-phospholipid activator for continuing generation of factor Xa during hemostasis and the bleeding of hemophilia (13). Using dilute TF or cell monolayers expressing TF, it was recently shown that TFPI is a more effective inhibitor of TF-VIIa-mediated activation of factor X in the absence than in the presence of factors VIII and IX (28). In a continuous flow reactor coated with TF, it was demonstrated that the direct activation of factor X by TF-VIIa could be virtually eliminated by TFPI; however, factor Xa production rates were enhanced 15- to 20-fold when factors VIII and IX were present at their approximate plasma concentrations (97). These results suggested that the tissue factor pathway, mediated through factors VIII and IX, produced significant levels of factor Xa even in the presence of TFPI, and that the bleeding of hemophilia might be attributed to the efficient inhibition of the tissue factor pathway by TFPI in the absence of factor VIII or IX. Using dilute TF, it was recently demonstrated that addition of anti-TFPI Ig shortened the coagulation time of hemophilic plasma to that of normal plasma (98,99), indicating that TFPI is, at least in part, responsible for the prolonged clotting time of hemophilic plasma.

Based on this information and the propositions provided in the previous reviews (14–17), one may revise the original cascade-waterfall hypotheses as follows: TF-VIIa is responsible for the initial burst of factor Xa generation, which provides sufficient thrombin to activate cofactors V and VIII. Once a small amount of factor Xa is generated, further direct activation of factor X is dampened by the formation of Xa-TFPI-VIIa-TF complex in a feedback fashion. However, indirect activation of factor X through the action of factors IXa and VIIIa persists, which results in ultimate hemostasis. This revised scheme differs from the original hypotheses in that (1) it integrates all the factors into a single pathway that is initiated by TF-VIIa and in which the contact factors are not required; (2) it provides alternative pathways for initial and later, progressive generation of factor Xa and thrombin, both required for ultimate hemostasis; and (3) it explains why patients with factors VIII or IX bleed but persons with deficiency in factor XII, kallikrein, or high-molecular-weight kininogen do not.

IX. POTENTIAL THERAPEUTIC USES OF RECOMBINANT TFPI

As a potent initiator of coagulation, TF is believed to have a critical function in hemostasis and thrombogenesis (100–102). Upregulation of TF expression in lipid-laden macrophages (103) and deposition of TF in the necrotic core of the atherosclerotic plaque (104) may contribute to the hyperthrombotic state of human atherosclerotic vessels. Endotoxin, cytokines, immune complexes, and activated complement factors can induce TF expression in monocytes and endothelial cells, suggesting the involvement of TF in the pathological processes of infection, cell-mediated immunity, and inflammation (105,106). As a potent inhibitor of TF function, TFPI possesses the potential to modulate these pathological processes.

Recombinant TFPI (rTFPI) was produced in large quantity to allow testing in a number of animal models of disease. Prior bolus injection or coinfusion of rTFPI was shown to prevent thromboplastin-induced fibrin deposition in the lungs and disseminated intravascular coagulation in rabbits (27). Thrombosis was induced in the femoral arteries of anesthetized dogs with the use of an anodal current to elicit extensive vascular injury and the formation of platelet-rich thrombi. The vessel was recanalized by the infusion of tPA, and the subsequent reocclusion was observed. In this model, rTFPI was found to prevent the cyclic flow variations and the reocclusion of the vessel after thrombolysis (107). These results suggested that rTFPI might be useful as a conjunctive agent to prevent reocclusion following thrombolytic therapy. Venous thrombosis was produced in the vena cava of rabbits by vascular damage and stasis. Given as an intravenous bolus 20 minutes before ligation, rTFPI was found to block formation of the thrombus completely or greatly to reduce its size in five of six animals tested (108). Using a rabbit ear artery model of crush-avulsion injury and microvascular repair, the efficacy of rTFPI as a topically applied antithrombotic agent was studied and compared to heparin (109). Traumatized arteries treated through luminal irrigation with saline (controls) achieved patency rates of 8 and 0% at 1 and 7 postoperative days, respectively. Heparin irrigation (10 units/ml) resulted in patencies of 40% at both evaluation times. In contrast, oral rTFPI at a dose of 20 μg/ml (0.2 ml; 10 minute exposure) yielded a 91% patency rate at 1 day and 73% at 7 days. Scanning electron microscopy revealed dramatically inhibited thrombogenesis upon the injured surfaces of TFPI-treated vessels. These results suggest that TFPI used as a topically applied antithrombotic agent is effective for the prevention of thrombosis in microvascular anastomosis. In a baboon *Escherichia coli* septic shock model (110), administration of rTFPI at 30 minutes after the start of a lethal *E. coli* infusion resulted in (1) permanent 7 day survivors (five of five) with significant improvement in quality of life; the mean survival time for the control (five of five) was 40 h (no survivors); and (2) significant attenuations of coagulation response and significant reductions in pathology in target organs. Delayed administration of rTFPI at 4 h after the start of bacterial infusion resulted in prolongation of survival time, with 40% survival rate (two of five) and some attenuations of the coagulopathic response.

In summary, recombinant TFPI has been shown to be highly potent in a number of preclinical animal models simulating disseminated intravascular coagulation, arterial and venous thrombosis, postsurgical thrombosis, and septic shock. The list of potential applications of TFPI is likely to expand as more tests are carried out in other disease models involving various forms of vascular injury.

X. CONCLUDING REMARKS

In the past few years, our knowledge of the role of TFPI in the regulation of the tissue factor pathway, the structure and function of the TFPI protein, and the molecular biology of the TFPI gene has increased dramatically. However, many questions remain to be answered. For example, how does TFPI associate with lipoprotein, and what is the effect of this association on TFPI function? What is the function of the third Kunitz domain of TFPI? What is the mechanism for the formation of the Xa-TFPI-VIIa-TF inhibitory complex? How is TFPI gene expression regulated? How is the TFPI level in plasma maintained? What is the relative importance of the three pools (lipoprotein associated, heparin releasable, and platelet associated) of TFPI in vivo? How is TFPI catabolized in vivo? What is the mechanism of heparin-induced TFPI release in vivo? What is the mechanism of the enhancement of TFPI activity by sulfated polysaccharides? Will suppression of TFPI activity in vivo lead to attenuation of bleeding in hemophilia? Future research efforts will provide answers to these and other questions.

ACKNOWLEDGMENTS

I am grateful to my collaborators for their contributions to our studies, especially Dr. G. J. Broze, Jr. and his colleagues at the Washington University and K. K. Kretzmer, M. O. Palmier, and J. Diaz-Collier in my group.

REFERENCES

1. Thomas L. Studies on the intravascular thromboplastin effect of tissue suspensions in mice. II. A factor in normal rabbit serum which inhibits the thromboplastin effect of the sedimentable tissue component. Bull Johns Hopkins Hosp 1947; 81:26–42.
2. Schneider CL. The active principle of placental toxin: thromboplastin; its inactivator in blood: antithromboplastin. Am J Physiol 1947; 149:123–129.
3. Hjort PF. Intermediate reactions in the coagulation of blood with tissue thromboplastin. Scand J Clin Lab Invest 1957; 9:1–182.
4. Marciniak E. Anticoagulant properties of low density lipoproteins (abstract). Fed Proc 1974; 33:217.
5. Barrowcliffe TW, Eggleton CA. Studies of anti-Xa activity in human plasma. I. Comparison of a fast-acting inhibitor with antithrombin III. Thromb Res 1981; 21:399–411.
6. Barrowcliffe TW, Eggleton CA, Stocks J. Studies of anti-Xa activity in human plasma II: the role of lipoproteins. Thromb Res 1982; 27:185–195.
7. Thomas DP, Barrowcliffe TW, Merton RE, Stocks J, Dawes J, Pepper DS. In vivo release of anti-Xa clotting activity by a heparin analogue. Thromb Res 1980; 17:831–840.
8. Carson SD. Plasma high density lipoproteins inhibit the activation of coagulation factor X by factor VIIa and tissue factor. FEBS Lett 1981; 132:37–40.
9. Morrison SA, Jesty J. Tissue factor-dependent activation of tritium-labeled factor IX and factor X in human plasma. Blood 1984; 63:1338–1347.

10. Sanders NL, Bajaj SP, Zivelin A, Rapaport SI. Inhibition of tissue factor/factor VIIa activity in plasma requires factor X and an additional plasma component. Blood 1985; 66:204–212.
11. Hubbard AR, Jennings CA. Inhibition of the tissue thromboplastin-mediated blood coagulation. Thromb Res 1986; 42:489–498.
12. Broze GJ Jr, Miletich JP. Characterization of the inhibition of tissue factor in serum. Blood 1987; 69:150–155.
13. Rao LVM, Rapaport SI. Studies of a mechanism inhibiting the initiation of extrinsic pathway of coagulation. Blood 1987; 69:645–651.
14. Rapaport SI. Inhibition of factor VIIa/tissue factor-induced blood coagulation: with particular emphasis upon a factor Xa-dependent inhibitory mechanism. Blood 1989; 73:359–365.
15. Broze GJ Jr, Girard TJ, Novotny WF. Regulation of coagulation by a multivalent Kunitz-type inhibitor. Biochemistry 1990; 29:7539–7546.
16. Rapaport SI. The extrinsic pathway inhibitor: a regulator of tissue factor-dependent blood coagulation. Thromb Haemost 1991; 66:6–15.
17. Broze GJ Jr. The role of tissue factor pathway inhibitor in a revised coagulation cascade. Semin Hematol 1992; 29:159–169.
18. Broze GJ Jr, Miletich JP. Isolation of the tissue factor inhibitor produced by HepG2 hepatoma cells. Proc Natl Acad Sci USA 1987; 84:1886–1890.
19. Broze GJ Jr, Warren LA, Girard JJ, Miletich JP. Isolation of the lipoprotein associated coagulation inhibitor produced by HepG2 (human hepatoma) cells using bovine factor Xa affinity chromatography. Thromb Res 1987; 48:253–259.
20. Wun T-C, Huang MD, Kretzmer KK, et al. Immunoaffinity purification of lipoprotein-associated coagulation inhibitors from HepG2 hepatoma, Chang liver, and SK hepatoma cells. J Biol Chem 1990; 265:16096–16101.
21. Hubbard AR, Jennings CA. Inhibition of the tissue factor-factor VII complex: involvement of factor Xa and lipoproteins. Thromb Res 1987; 46:527–537.
22. Novotny WF, Girard TJ, Miletich JP, Broze GJ Jr. Purification and characterization of the lipoprotein-associated coagulation inhibitor from human plasma. J Biol Chem 1989; 264: 18832–18837.
23. Sandset PM, Abildgaard U, Larsen ML. Heparin induces release of extrinsic coagulation pathway inhibitor (EPI). Thromb Res 1988; 50:803–813.
24. Novotny WF, Brown SG, Miletich JP, Rader DJ, Broze GJ Jr. Plasma antigen levels of the lipoprotein-associated coagulation inhibitor in patient samples. Blood 1991; 78:387–393.
25. Novotny WF, Palmier M, Wun T-C, Broze GJ Jr, Miletich JP. Purification and properties of heparin-releasable lipoprotein-associated coagulation inhibitor. Blood 1991; 78:394–400.
26. Wun T-C, Kretzmer KK, Girard TJ, Miletich JP, Broze GJ Jr. Cloning and characterization of a cDNA for the lipoprotein-associated coagulation inhibitor shows that it consists of three tandem Kunitz-type inhibitory domains. J Biol Chem 1988; 263:6001–6004.
27. Day KC, Hoffman LC, Palmier MO, et al. Recombinant lipoprotein-associated coagulation inhibitor inhibits tissue thromboplastin-induced intravascular coagulation in the rabbit. Blood 1990; 76:1538–1545.
28. Pedersen AH, Nordfang O, Norris F, et al. Recombinant human extrinsic pathway inhibitor: production, isolation, and characterization of its inhibitory activity on tissue factor-initiated coagulation reactions. J Biol Chem 1990; 265:16786–16793.
29. Wun TC, Kretzmer KK, Palmier MO, et al. Comparison of recombinant tissue factor pathway inhibitors expressed in human SK hepatoma, mouse C127, baby hamster kidney, and Chinese hamster ovary cells. Thromb Haemost 1992; 68:54–59.
30. Lindahl AK, Abildgaard U, Larsen ML, et al. Extrinsic pathway inhibitor (EPI) released to the blood by heparin is a more powerful coagulation inhibitor than is recombinant EPI. Thromb Res 1991; 62:607–614.
31. Nordfang O, Bjorn SE, Valentin S, et al. The C-terminus of tissue factor pathway inhibitor is essential to its anticoagulant activity. Biochemistry 1991; 30:10371–10376.

32. Wesselschmidt R, Likert K, Girard T, Wun T-C, Broze GJ Jr. Tissue factor pathway inhibitor: the carboxy-terminus is required for optimal inhibition of factor Xa. Blood 1992; 2004–2010.
33. Girard TJ, McCourt D, Novotny WF, MacPhail LA, Likert KM, Broze GJ Jr. Endogenous phosphorylation of the lipoprotein-associated coagulation inhibitor at serine-2. Biochem J 1990; 270:621–625.
34. Smith PL, Skelton TP, Fiete D, et al. The asparagine-linked oligosaccharides on tissue factor pathway inhibitor terminate with SO_4-4GalNAcβ1,4GlcNAcβ1,2Manα. J Biol Chem 1992; 267:19140–19146.
35. Baenziger JU, Kumar S, Brodbeck RM, Smith PL, Beranek MC. Circulatory half-life but not interaction with the lutropin/chorionic gonadotropin receptor is modulated by sulfation of bovine lutropin oligosaccharides. Proc Natl Acad Sci USA 1992; 89:334–338.
36. Colburn P, Buonassisi V. Identification of an endothelial cell product as an inhibitor of tissue factor activity. In Vitro Cell Dev Biol 1988; 24:1133–1136.
37. Warn-Cramer BJ, Maki SL, Rapaport SI. A sulfated rabbit endothelial cell glycoprotein that inhibits factor VIIa/tissue factor is functionally and immunologically identical to rabbit extrinsic pathway inhibitor (EPI). Thromb Res 1991; 61:515–527.
38. Colburn P, Crabb JW, Buonassisi V. Enhanced inhibition of tissue factor by the extended form of an endothelial cell glycoprotein (an extrinsic pathway inhibitor). J Cell Physiol 1991; 148:320–326.
39. Girard TJ, Warren LA, Novotny WF, Bejcek BE, Miletich JP, Broze GJ Jr. Identification of the 1.4 kb and 4.0 kb messages for the lipoprotein associated coagulation inhibitor and expression of the encoded protein. Thromb Res 1989; 55:37–50.
40. Shaw G, Kamen R. A conserved AU sequence from the 3′ untranslated region of GM-CSF mRNA mediates selective mRNA degradation. Cell 1986; 46:659–667.
41. Caput D, Beutler B, Hartog K, Thayer R, Brown-Shimer S, Cerami A. Identification of a common nucleotide sequence in the 3′ untranslated region of mRNA molecules specifying inflammatory mediators. Proc Natl Acad Sci USA 1986; 83:1670–1674.
42. Wesselschmidt RL, Girard TJ, Broze GJ Jr. cDNA sequence of rabbit lipoprotein-associated coagulation inhibitor. Nucleic Acids Res 1990; 18:6440.
43. Enjyoji K, Emi M, Mukai T, Kato H. cDNA cloning and expression of rat tissue factor pathway inhibitor (TFPI). J Biochem 1992; 111:681–687.
44. Girard TJ, Eddy R, Wesselschmidt RL, et al. Structure of the human lipoprotein-associated coagulation inhibitor gene—intron/exon gene organization and localization of the gene to chromosome 2. J Biol Chem 1991; 266:5036–5041.
45. Van der Logt CPE, Reitsma PH, Bertina RM. Intron-exon organization of the human gene coding for the lipoprotein-associated coagulation inhibitor: the factor Xa dependent inhibitor of the extrinsic pathway of coagulation. Biochemistry 1991; 30:1571–1577.
46. Shapiro MB, Senapathy P. RNA splice junctions of different classes of eukaryotes: sequence statistics and functional implications in gene expression. Nucleic Acids Res 1987; 15:7155–7174.
47. Van der Logt CPE, Reitsma PH, Bertina RM. A PstI RFLP of the LACI gene. Nucleic Acids Res 1990; 18:5920.
48. Ameri A, Kuppuswamy MN, Basu S, Bajaj SP. Expression of tissue factor pathway inhibitor by cultured endothelial cells in response to inflammatory mediators. Blood 1992; 79:3219–3226.
49. Lee W, Mitchell P, Tjian R. Purified transcription factor AP-1 interacts with TPA-inducible enhancer elements. Cell 1987; 49:741–752.
50. Rana SV, Reimers HJ, Pathikonda MS, Bajaj SP. Expression of tissue factor and factor VIIa/tissue factor inhibitor activity in endotoxin or phorbol ester stimulated U937 monocyte-like cells. Blood 1988; 71:259–262.

51. Rossi P, Karsenty G, Roberts AB, Roche NS, Sporn MB, de Crombrugghe B. A nuclear factor 1 binding site mediates the transcriptional activation of a type 1 collagen promotor by transforming growth factor-β. Cell 1988; 52:405–414.
52. Broze GJ Jr, Warren LA, Novotny WF, Higuchi DA, Girard TJ, Miletich JP. The lipoprotein-associated coagulation inhibitor that inhibits factor VII-tissue factor complex also inhibits Xa: insight into its possible mechanism of action. Blood 1988; 71:335–343.
53. Warn-Cramer BJ, Rao LVM, Maki SL, Rapaport SI. Modifications of extrinsic pathway inhibitor (EPI) and factor Xa that affect their ability to interact and to inhibit factor VIIa/tissue factor: evidence for a two-step model of inhibition. Thromb Haemost 1988; 60:453–456.
54. Gemmell CH, Broze GJ Jr, Turitto VT, Nemerson Y. Utilization of a continuous flow reactor to study the lipoprotein-associated coagulation inhibitor (LACI) that inhibits tissue factor. Blood 1990; 76:2266–2271.
55. Lindhout T, Blezer R, Schoen P, Nordfang O, Reutelingsperger C, Hemker HC. Activation of factor X and its regulation by tissue factor pathway inhibitor in small-diameter capillaries lined with human endothelial cells. Blood 1992; 79:2909–2916.
56. Girard TJ, Warren LA, Novotny WF, et al. Functional significance of the Kunitz-type inhibitory domains of lipoprotein-associated coagulation inhibitor. Nature 1989; 338:518–520.
57. Higuchi DA, Wun T-C, Likert KM, Broze GJ Jr. The effect of leukocyte elastase on tissue factor pathway inhibitor. Blood 1992; 79:1712–1719.
58. Petersen LC, Bjorn SE, Nordfang O. Effect of leukocyte proteinases on tissue factor pathway inhibitor. Thromb Haemost 1992; 67:537–541.
59. Girard TJ, MacPhail LA, Likert KM, Novotny WF, Miletich JP, Broze GJ Jr. Inhibition of factor VIIa/tissue factor activity by a factor X-lipoprotein-associated coagulation inhibitor hybrid protein. Science 1990; 248:1421–1424.
60. Lindahl AK, Jacobsen PB, Sandset PM, Abildgaard U. Tissue factor pathway inhibitor with high anticoagulant activity is increased in post-heparin plasma and in plasma from cancer patients. Blood Coag Fib 1991; 2:713–721.
61. Lindahl AK, Abildgaard U, Larsen ML, Aamodt L-M, Nordfang O, Beck TC. Extrinsic pathway inhibitor (EPI) and the post-heparin anticoagulant effect in tissue thromboplastin induced coagulation. Thromb Res (Suppl) 1991; 14:39–48.
62. Choay J. Structure and activity of heparin and its fragments: an overview. Semin Thromb Hemost 1989; 15:359–364.
63. Broze GJ Jr, Warren LA, Novotny WF, Higuchi DA, Girard JJ, Miletich JP. The lipoprotein-associated coagulation inhibitor that inhibits the factor VII-tissue factor complex also inhibits factor Xa: insight into its possible mechanism of action. Blood 1988; 71:335–343.
64. Wun T-C. Lipoprotein-associated coagulation inhibitor (LACI) is a cofactor for heparin: synergistic anticoagulant action between LACI and sulfated polysaccharides. Blood 1992; 79:430–438.
65. Valentin S, Ostergaard P, Kristensen H, Nordfang O. Simultaneous presence of tissue factor pathway inhibitor (TFPI) and low molecular weight heparin has a synergistic effect in different coagulation assays. Blood Coag Fib 1991; 2:629–635.
66. Valentin S, Ostergaard P, Kristensen H, Norfang O. Synergism between full length TFPI and heparin: evidence for TFPI as an important factor for the antithrombotic activity of heparin. Blood Coag Fibrinol 1992; 3:221–222.
67. Berenbaum MC. What is synergy? Pharmcol Rev 1989; 41:93–141.
68. Novotny WF, Girard TJ, Miletich JP, Broze GJ Jr. Pletelets secrete a coagulation inhibitor functionally and antigenically similar to the lipoprotein associated coagulation inhibitor. Blood 1988; 72:2020–2025.
69. Bajaj MS, Rana SV, Wysolmerski RB, Bajaj SP. Inhibitor of the factor VIIa-tissue factor complex is reduced in patients with disseminated intravascular coagulation but not in patients with severe hepatocellular disease. J Clin Invest 1987; 79:1874–1878.

70. Warn-Cramer BJ, Almus FE, Rapaport SI. Studies of the factor Xa-dependent inhibitor of factor VIIa/tissue factor (extrinsic pathway inhibitor) from cell supernates of cultured human umbilical vein endothelial cells. Thromb Haemost 1989; 61:101–105.
71. Bajaj MS, Kuppuswamy MN, Saito H, Spitzer SG, Bajaj SP. Cultured normal human hepatocytes do not synthesize lipoprotein-associated coagulation inhibitor: evidence that endothelium is the principle site of its synthesis. Proc Natl Acad Sci USA 1990; 87:3869–8873.
72. Palmier MO, Hall LJ, Reisch CM, Baldwin MK, Wilson AGE, Wun T-C. Clearance of recombinant tissue factor pathway inhibitor (TFPI) in rabbits. Thromb Haemost 1992; 68: 33–36.
73. Einarsson M, Häggroth L, Mattsson C. Elimination of native and carbohydrate-modified tissue plasminogen activator in rabbits. Thromb Haemost 1989; 62:1088–1093.
74. Hotchkiss A, Refino CJ, Leonard CK, et al. The influence of carbohydrate structure on the clearance of recombinant tissue-type plasminogen activator. Thromb Haemost 1988; 60:255–261.
75. Warr TA, Warn-Cramer B, Rao LVM, Rapaport SI. Human plasma extrinsic pathway inhibitor activity. I. standardization of assay and evaluation of physiologic variables. Blood 1989; 74:201–206.
76. Sandset PM, Abildgaard U, Pettersen M. A sensitive assay of extrinsic coagulation pathway inhibitor (EPI) in plasma and plasma fractions. Thromb Res 1987; 47:389–400.
77. Sandset PM, Larsen ML, Abildgaard U, Lindahl AK, Ødegaard OR. Chromogenic substrate assay of extrinsic pathway inhibitor (EPI): levels in the normal population and relation to cholesterol. Blood Coag Fib 1991; 2:425–433.
78. Sandset PM, Hellgren M, Uvebrandt M, Bergström H. Extrinsic coagulation pathway inhibitor and heparin cofactor II during normal and hypertensive pregnancy. Thromb Res 1989; 55:665–670.
79. Sandset PM, Sirnes PA, Abildgaard U. Factor VII and extrinsic pathway inhibitor in acute coronary disease. Br J Haematol 1989; 72:391–396.
80. Lindahl AK, Sandset PM, Abildgaard U, Andersson TR, Harbitz TB. High plasma levels of extrinsic pathway inhibitor and low levels of other coagulation inhibitors in advanced cancer. Acta Chir Scand 1989; 155:389–393.
81. Lindahl AK, Abildgaard U, Stokke G. Relase of extrinsic pathway inhibitor after heparin injection: increased response in cancer patients. Thromb Res 1990; 59:651–656.
82. Lindahl AK, Jacobsen PB, Sandset PM, Abildgaard U. Tissue factor pathway inhibitor with high anticoagulant activity is increased in post-heparin plasma and in plasma from cancer patients. Blood Coag Fib 1991; 2:713–721.
83. Abildgaard U, Sandset PM, Andersson TR, Ødegaard OR, Rosen S. The inhibitor of F VIIa in plasma measured with a sensitive chromogenic substrate assay: comparison with antithrombin, protein C and heparin cofactor II in a clinical material. Folia Haematol Leipzig 1988; 115:3,S.274–277.
84. Sandset PM, Hogevold HE, Lyberg T, Andersson TR, Abildgaard U. Extrinsic pathway inhibitor in elective surgery: a comparison with other coagulation inhibitors. Thromb Haemost 1989; 62:856–860.
85. Carson SD, Haire WD, Broze GJ Jr, Novotny WF, Pirrucello SJ, Duggan MJ. Lipoprotein associated coagulation inhibitor, factor VII, antithrombin III, and monocyte tissue factor following surgery. Thromb Haemost 1991; 66:534–539.
86. Sandset PM, Lund H, Norseth J, Abildgaard U, Ose L. Treatment with hydroxymethylglutaryl-coenzyme A reductase inhibitors in hypercholesterolemia induces changes in the components of the extrinsic coagulation system. Arteriosclerosis Thromb 1991; 11:138–145.
87. Warr TA, Rao LVM, Rapaport SI. Human plasma extrinsic pathway inhibitor activity. II. Plasma levels in disseminated intravascular coagulation and hepatocellular disease. Blood 1989; 74:994–998.

88. Sandset PM, Roise O, Aasen AO, Abildgaard U. Extrinsic pathway inhibitor in postoperative/posttraumatic septicemia: increased levels in fatal cases. Haemostasis 1989; 19:189–195.
89. Brandtzaeg P, Sandset PM, Joo GB, Øvstebo R, Abildgaard U, Kierulf P. The quantitative association of plasma endotoxin, antithrombin, protein C, extrinsic pathway inhibitor and fibrinopeptide A in systemic meningococcal disease. Thromb Res 1989; 55:459–470.
90. Warr TA, Rao LVM, Rapaport SI. Disseminated intravascular coagulation in rabbits induced by administration of endotoxin or tissue factor: effect of anti-tissue factor antibodies and measurement of plasma extrinsic pathway inhibitor activity. Blood 1990; 75:1481–1489.
91. Sandset PM, Warn-Cramer BJ, Rao LVM, Maki SL, Rapaport SI. Depletion of extrinsic pathway inhibitor (EPI) sensitizes rabbits to disseminated intravascular coagulation induced with tissue factor: evidence supporting a physiologic role for EPI as a natural anticoagulant. Proc Natl Acad Sci USA 1991; 88:708–712.
92. Sandset PM, Warn-Cramer BJ, Maki SL, Rapaport SI. Immunodepletion of extrinsic pathway inhibitor sensitizes rabbits to endotoxin-induced intravascular coagulation and the generalized Shwartzman reaction. Blood 1991; 78:1496–502.
93. MacFarlane RG. An enzyme cascade in the blood clotting mechanism, and its function as a biochemical amplifier. Nature 1964; 202:498–499.
94. Davie EW, Ratnoff OD. Waterfall sequence for intrinsic blood clotting. Science 1964; 145: 1310–1312.
95. Biggs R, MacFarlane RG. Reaction of haemophilic plasma to thromboplastin. J Clin Pathol 1951; 4:445–449.
96. Marlar RA, Kleiss AJ, Griffin JH. An alternative extrinsic pathway of human blood coagulation. Blood 1982; 60:1353–1358.
97. Repke D, Gemmell CH, Guha A, Turitto VT, Broze GJ Jr, Nemerson Y. Hemophilia as a defect of the tissue factor pathway of blood coagulation: effect of factors VIII and IX on factor X activation in a continuous-flow reactor. Proc Natl Acad Sci USA 1990; 87:7623–7627.
98. Nordfang O, Valentin S, Beck TC, Hedner U. Inhibition of extrinsic pathway inhibitor shortens the coagulation time of normal plasma and of hemophilia plasma. Thromb Haemost 1991; 66:464–467.
99. Welsch DJ, Novotny WF, Wun T-C. Effect of lipoprotein-associated coagulation inhibitor (LACI) on thromboplastin-induced coagulation of normal and hemophiliac plasmas. Thromb Res 1991; 64:213–222.
100. Nemerson Y. The tissue factor pathway of blood coagulation. Semin Haematol 1992; 29: 170–176.
101. Rapaport SI, Rao LVM. Initiation and regulation of tissue factor-dependent blood coagulation. Arteriosclerosis Thromb 1992; 12:1111–1121.
102. Drake TA, Morrissey JH, Edginton TS. Selective cellular expression of tissue factor in human tissues: implications for disorders of hemostasis and thrombosis. Am J Pathol 1989; 134:1087–1097.
103. Lesnik P, Rouis M, Skarlatos S, Kruth HS, Chapman MJ. Uptake of exogenous free cholesterol induces upregulation of tissue factor expression in human monocyte-derived macrophages. Proc Natl Acad Sci USA 1992; 89:10370–10374.
104. Wilcox JN, Smith KM, Schwartz SM, Gordon D. Localization of tissue factor in the normal vessel wall and in the atherosclerotic plaque. Proc Natl Acad Sci USA 1989; 86:2839–2843.
105. Ryan J, Geczy C. Coagulation and the expression of cell-mediated immunity. Immunol Cell Biol 1987; 65:127–135.
106. Edwards RL, Rickles FR. The role of leukocytes in the activation of blood coagulation. Semin Hematol 1992; 29:202–212.

107. Haskel EJ, Torr SR, Day KC, et al. Prevention of arterial reocclusion after thrombolysis with recombinant lipoprotein-associated coagulation inhibitor. Circulation 1991; 84:821–827.
108. Spokas EG, Wun T-C. Venous thrombosis produced in the vena cava of rabbits by vascular damage and stasis. J Pharmacol Toxicol Methods 1992; 27:225–232.
109. Khouri RK, Koudsi B, Ornberg RL, Wun T-C. Prevention of thrombosis by topical application of tissue factor pathway inhibitor in a rabbit model of vascular trauma. Ann Plast Surg 1993; 30:398–404.
110. Creasey AA, Chang ACK, Feigen L, Wun T-C, Taylor FB, Hinshaw LB. Tissue factor pathway inhibitor (TFPI) reduces mortality from *E. coli* septic shock. J Clin Invest 1993; 91:2850–2860.

17

Antithrombin: Structure and Function

William P. Sheffield, Ye I. Wu, and Morris A. Blajchman
McMaster University, Hamilton, Ontario, Canada

I. INTRODUCTION

Few processes in biology proceed unopposed. Accordingly, circulating serine protease inhibitors aid in the control of activated coagulation factors and serve to prevent inappropriate or extended clotting. The most abundant of these inhibitors is a plasma glycoprotein known as antithrombin (AT), or antithrombin III (1). Current interest in AT stems from both the clear link between inherited deficiency of this protein in patients and their increased risk of thrombosis (2,3) and from its therapeutic role as the plasma cofactor mediating the effect of heparin (1) and related compounds. A combination of molecular biological, protein chemical, and structural biological approaches over the last decade has yielded a wealth of information concerning how AT functions. In this review we examine what is known about the synthesis, action, and structure of AT, especially at the molecular level.

II. PROPERTIES OF ANTITHROMBIN

A. Complex Formation with Serine Proteases

Antithrombin is a 60 kilodalton (kD) glycoprotein found primarily in plasma at a concentration of 2–3 μM (approximately 150 μg/ml) (4). Small quantities of AT partition onto the endothelium and into subendothelial spaces (5). Three disulfide bonds link cysteine residues 8 and 128, 21 and 95, and 247 and 430 (6). The protein is capable of inhibiting the activity of many of the serine proteases of the coagulation cascade, including thrombin (1), factors IXa, Xa (7), XIa (8), and XIIa (9), kallikrein (10), and plasmin (11). When the rates of inactivation of the various proteases are taken into account, it is likely that only the reactions with thrombin and, to a lesser extent, factor Xa have physiological relevance (12,13).

Antithrombin traps thrombin, or other serine protease targets, by presenting an ideal substrate region on its surface. Attack by the protease results in cleavage of AT at its reactive center by scission of the Arg^{393}-Ser^{394} bond (termed by convention the the P1-P1′ bond). The protease-inhibitor complex fails to dissociate but instead is stabilized by formation of an acyl ester bond between the carboxyl moiety of Arg^{393} and the hydroxyl side chain of the reactive serine of the protease (14,15). Figure 1 illustrates this point, showing by sodium dodecyl sulfate–polyacrylamide electrophoresis (SDS-PAGE) the consumption of thrombin with time and the formation of a 97 kD complex between the 37 kD α-thrombin with the 60 kD AT, which is stable to boiling in SDS. Under nonreducing conditions, the 38-residue carboxyl-terminal fragment is not released but remains tethered to the rest of the cleaved AT by a disulfide bridge (not shown). In vivo, formation of the thrombin-antithrombin (TAT) complex reduces the half-life of AT by at least an order of magnitude. This rapid clearance of TAT complexes is thought to be mediated by a putative, as yet unidentified, hepatic receptor (reviewed in Ref. 16).

B. Interaction with Heparin

1. *Reaction Acceleration*

In the absence of heparin, AT inhibits thrombin in a relatively slow, progressive manner. Addition of the heterogeneous mucopolysaccharide heparin to this reaction can result in an almost instantaneous inhibition of thrombin. Under conditions selected to maximize this acceleration, a several thousandfold increase in reaction rate is observable (17). This effect explains the clinically observed anticoagulant properties of heparin. That AT contributes to the majority of the thrombin-inhibitory properties of plasma accounts for the widespread clinical use of heparin (1). The dramatic acceleration of AT action by heparin originally led researchers to believe that at least two separate coagulation inhibitors existed, one slow and one fast. A numbering scheme for antithrombin I–VI was thus devised (18). Purification and characterization of a single 60 kD polypeptide responsible for the majority of thrombin inhibition in plasma, in the presence or absence of heparin, allowed it to be called simply antithrombin, although the term ‘‘antithrombin III’’ is still used in some quarters.

2. *Physiological Significance*

It seems unlikely that AT evolved the ability to be accelerated by heparin so that clinicians could productively treat patients with this drug. What, then, is the physiological significance of this property, given that heparin is not found at plasma concentrations sufficient to affect AT function? One hypothesis is that related glycosaminoglycans of the vessel wall, such as heparan sulfate, provide the in vivo cofactor requirement (19). A number of lines of evidence support this notion. Heparan sulfate proteoglycans extracted from bovine microvascular capillary endothelial cells and from aortic endothelium have been shown to accelerate AT-thrombin reactions (19,20). Perfusion of intact rat hindlimb vasculature leads to a heparinase-sensitive acceleration of complex formation (21). Binding of rabbit AT to subendothelial basement membranes has similarly been shown to be reduced by the enzymatic removal of heparan sulfate (22). A recent electron microscopic examination of rat aortic explants extended these findings and revealed that although some AT binding takes place on the luminal surface of endothelial cells, the bulk of the binding is subendothelial (23).

3. Mechanism of Action

How does heparin, or heparan sulfate, exert its effect upon AT? Although heparin is a heterogeneous mixture of sulfated glyscosaminoglycans varying in composition and length, AT binds to a subset of heparin containing a unique pentasaccharide sequence containing four sulfate groups (24). This pentasaccharide is sufficient to accelerate substantially the inhibition of factor Xa by AT, but it has little effect on the reaction of AT with thrombin. Significant acceleration of the latter reaction appears to require at least 18 saccharide units and more than 20 such residues to achieve acceleration similar to that of unfractionated heparin (25). In a detailed comparison of the effects of the pentasaccharide and a 26 mer, it was shown that both polysaccharides bound AT with high ($K_D \simeq$ nM) affinity and induced similar fluorescence, ultraviolet, and circular dichroism changes in the inhibitor (26). Kinetic analysis, expressed as second-order rate constants, showed that the pentasaccharide enhanced AT reactions with factor Xa by 270-fold, but only by 1.7-fold with thrombin. The longer polysaccharide gave full enhancement of 580-fold for factor Xa and 4300-fold for thrombin. Pentasaccharide-related enhancement of inhibitor function was salt independent, but full enhancement by the longer polysaccharide was ionic strength dependent. The study's authors proposed that conformational

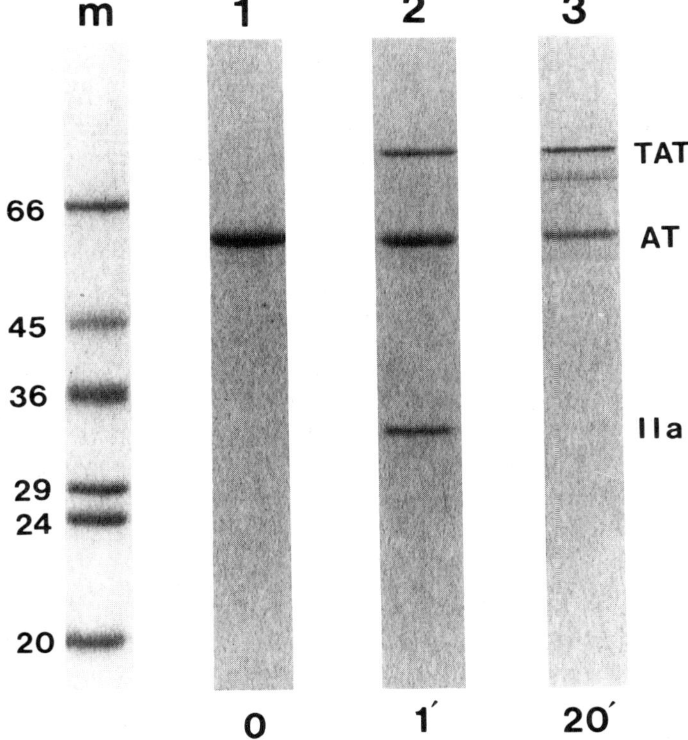

Figure 1 Complex (TAT) formation between antithrombin (AT) and thrombin (IIa) is shown on 10% SDS–polyacrylamide gels. A 2:1 molar excess of AT to IIa was reacted at 37°C and aliquots removed at 1 (lane 2) and 20 (lane 3) minutes. Lane 1 shows AT alone; molecular weight standards (m), with their weights in kD, are shown at left.

changes in AT induced by heparin are responsible primarily for the acceleration of factor Xa inhibition and bridging of AT and thrombin by the same polysaccharide chain is responsible for accelerated thrombin inhibition (26). This idea is also consistent with the finding that near maximal acceleration of AT inhibition of thrombin can be accomplished by providing a combination of pentasaccharide and high-molecular-weight, low-affinity heparin (27).

III. MOLECULAR GENETICS OF ANTITHROMBIN

A. Primary Structure and cDNA Cloning

The relative abundance of AT and the ease of its purification by immobilized heparin chromatography meant that the majority of its primary structure was solved by protein sequencing techniques before cDNA clones were isolated (28). Comparison of this information with other available protein sequences revealed unexpected homologies between α_1-antitrypsin, also known as α_1-antiproteinase (α1AP), ovalbumin, and AT (29). This family of proteins was subsequently dubbed the ''serpins'' because many members were serine protease inhibitors (30). Recent advances in our understanding of the molecular details of AT interactions with both thrombin and heparin have been contributed to by comparisons with, and modeling of, other serpins (discussed in detail later).

The isolation and sequencing of clones corresponding to human AT from liver cDNA libraries filled in gaps and corrected ambiguities in the protein sequence (31). An open reading frame of 1392 nucleotides was found, followed by 84 bases of untranslated region, and preceded by 70 untranslated bases. The deduced primary structure of the polypeptide indicated a secretory signal sequence of 32 amino acids and a mature plasma form of 432 amino acids. Four consensus sites for N-linked glycosylation were found at asparagine residues 96, 135, 155, and 192. Subsequently, it was reported that although the majority of circulating AT is glycosylated at these four positions (α-AT), approximately 10% of the circulating AT is unglycosylated at Asn^{135} (β-AT). Interestingly, β-AT binds heparin with greater avidity than the fully glycosylated α-AT (32).

The primary structure of AT from three nonhuman species has been determined. Our group isolated cDNA clones for rabbit (33) and mouse (34) AT, and Medjoub and colleagues determined the amino acid sequence of bovine AT by protein sequencing (35). When the sequences corresponding to plasma AT are compared, a remarkable degree of homology is noted, with greater than 85% conservation of residues between all nonhuman forms and the human molecule. The sequences adjacent to the reactive center of the ATs of all four species and the corresponding region of selected serpins are shown in Figure 2.

B. Gene Structure and Expression

Human cDNA probes were used to isolate the gene encoding AT, including several hundred bases of upstream nontranscribed region, as well as the margins of the introns (37,38). Protein encoding information in the human AT gene is interrupted by six introns that range in size from several hundred to several thousand base pairs; the exons range in size from 65 to 391 base pairs (37). A diagram of the AT gene and the way in which its exons encode the polypeptide is presented in Figure 3. The AT gene is located on the long arm of chromosome 1 (38). Recently, the entire sequence of the human AT gene was reported (39). The gene spans 13,480 bp from the transcriptional start site to the last

	P_{12}	P_{11}	P_{10}	P_9	P_8	P_7	P_6	P_5	P_4	P_3	P_2	P_1	Reactive Center	P_1'	P_2'	P_3'	P_4'	P_5'
	382	383	384	385	386	387	388	389	390	391	392	393	↓	394	395	396	397	398
1. Antithrombin (Human)	Ala	Ala	Ala	Ser	Thr	Ala	Val	Val	Ile	Ala	Gly	Arg		Ser	Leu	Asn	Pro	Asn
2. Antithrombin (Rabbit)	Ala	Ala	Ala	Ser	Thr	Val	Ile	Gly	Ile	Ala	Gly	Arg		Ser	Leu	Asn	Leu	Asn
3. Antithrombin (Mouse)	Ala	Ala	Ala	Ser	Thr	Ser	Val	Val	Ile	Thr	Gly	Arg		Ser	Leu	Asn	Pro	Asn
4. Antithrombin (Bovine)	Ala	Ala	Ala	Ser	Thr	Val	Ile	Ser	Ile	Ala	Gly	Arg		Ser	Leu	Asn	Ser	Asp
5. Alpha-1 Antiproteinase	Ala	Ala	Gly	Ala	Met	Phe	Leu	Glu	Ala	Ile	Pro	Met		Ser	Ile	Pro	Pro	Glu
6. Heparin Cofactor II	Ala	Thr	Thr	Val	Thr	Thr	Val	Gly	Phe	Met	Pro	Leu		Ser	Thr	Gln	Val	Arg
7. Alpha-1-Antiplasmin	Ala	Ala	Ala	Ala	Thr	Ser	Ile	Ala	Met	Ser	–	Arg		Met	Ser	Leu	Ser	Ser
8. Plasminogen Activator Inhibitor	Ala	Ser	Ser	Ser	Thr	Ala	Val	Ile	Val	Ser	Ala	Arg		Met	Ala	Pro	Glu	Glu
9. Protein C Inhibitor	Ala	Ser	Ser	Asp	Thr	Ala	Ile	Thr	Leu	Ile	Pro	Arg		Asn	Ala	Leu	Thr	Ala
10. Chymotrypsin Inhibitor	Ala	Ser	Ala	Ala	Thr	Ala	Val	Lys	Ile	Thr	Leu	Leu		Ser	Ala	Leu	Val	Glu
11. Cl-Inhibitor	Ala	Ala	Ala	Ala	Ser	Ala	Ile	Ser	Val	Ala	Arg	Thr		Leu	Leu	Val	Phe	Glu

Figure 2 The amino acid sequences near the reactive centers (P1-P1′) of antithrombin from four species, together with seven other members of the serpin family that are active serine protease inhibitors. The amino acid residue numbers are that of wild-type human antithrombin, and the sequences of the other serpins are aligned for maximum homology.

nucleotide of the poladenylation signal. Interestingly, Alu repetitive elements are present in the introns of the human AT gene at a significantly higher frequency (22%) than in the genome as a whole (5%). The significance of this observation is unclear, although in at least a few kindreds, it appears that homologous recombination between Alu repeats, specifically those in intron 4 and 5, is responsible for a gene deletion resulting in inherited AT deficiency (40).

Little is known concerning the regulation of transcription of the AT gene. Unlike the archetypal serpin α1AP, AT is not an acute-phase reactant, in that its synthesis is not upregulated during inflammatory responses (41). Like a number of other "housekeeping" or constitutively expressed genes, the 5′-non-transcribed portions of the AT gene lack recognizable TAATA or CCAAT boxes. Developmental regulation of AT expression occurs, in that AT levels in the plasma of newborns are approximately half that of adults and rise with age (42). Human AT probes have been employed on northern blots of mouse RNA to demonstrate that this increase correlates with an increase in RNA levels (43). In the same study, it was demonstrated that both rat liver and rat kidney contain mRNA species that hybridize to human AT probes. Our findings with same-species probes suggest that neither rabbit (33) nor human kidneys (Sheffield, Castillo, Blajchman, unpublished observations) contribute significant amounts of AT mRNA or protein to the total synthesized by the organism. Although treatment of individuals with such steroids as danazol (44) or cultured human hepatoma cells with TAT complexes (45) has been

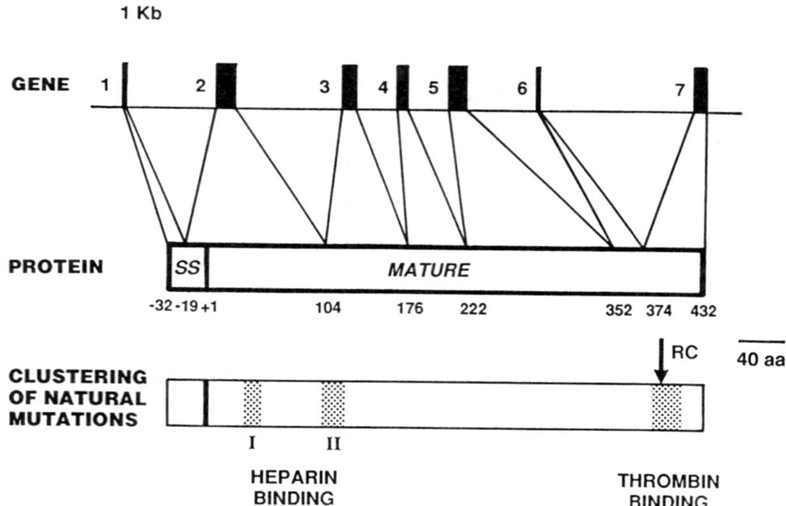

Figure 3 Gene and protein organization of AT. The spacing and relative sizes of the seven exons (numbered, at left of solid boxes) and six introns are shown. Diagonal lines show how the exons map onto the protein sequence, with the codons at intron exon boundaries numbered, below. Stippled areas on bottom panel show how the naturally occurring mutations affecting either heparin binding or thrombin binding cluster on the polypeptide chain. SS, signal sequence; mature, processed mature form of AT found in plasma (although four sites of glycosylation are not shown); RC, reactive center (Arg^{393}-Ser^{394}). The scale of the diagrams is indicated by kb, indicating 1 kilobase, for DNA, and 40 aa, indicating 40 amino acids, for the protein.

reported to increase the production of AT, it is not known whether these phenomena are mediated at the transcriptional level.

In terms of the promoter of the AT gene, no study has directly addressed the question of what constitutes the minimal upstream control region necessary for downstream transcription. A potential enhancer role for the first 400 bp of upstream region has been proposed on the basis of homology to immunoglobulin enhancers and on increased marker gene activity in transfection experiments (37). The region −89 to −68 has also been implicated in the binding of transcription factors from nuclear extracts prepared from rat liver (46).

C. Inherited Antithrombin Deficiency

A genetic deficiency of AT manifests itself as an increased risk of thrombosis (primarily venous) in affected families (2,3). The mode of transmission is autosomal dominant, with affected individuals typically having only half-normal levels of functional AT, and the deficit arising from one dysfunctional allele (47). A quantitative deficiency (type I) arises from the failure of the affected allele to produce AT protein, for instance because of frameshift (48) or termination (49) mutations or, less commonly, large deletions (50,51). Qualitative deficiencies (type II) are marked by the presence of a nonfunctional mutant AT molecule in the circulation.

The risk of a thromboembolic event in AT-deficient individuals has been estimated as approximately 1% per year of life, such that by middle age perhaps 50% of such individuals have suffered such an attack (3). We recently performed a cross-sectional study to estimate the prevalence of objectively proven thrombotic events in individuals with AT deficiency. In a large kindred with a type II deficiency having a molecular defect affecting the thrombin binding region of AT, less than 20% had evidence of having had a thrombotic event. Moreover, we found that the initial thrombotic event in those who had such an episode usually occurred in association with the presence of a predisposing factor (52). Such events are often triggered by some form of trauma or prothrombotic factor (e.g., commencement of oral contraceptive use, surgery, or childbirth).

The presence of AT activity is evidently essential for survival. Two infants, the products of a consanguineous marriage between AT-deficient parents, who lacked detectable AT in their plasma, did not survive more than 2 months (53). At the other end of the spectrum, individuals with a mutant form of AT lacking heparin cofactor activity are usually unaffected clinically and detected only in population screening programs, unless they are homozygous (54).

With the advent of recombinant DNA approaches, it has become possible to elucidate the molecular defect in kindreds with inherited AT deficiency. Although this initially involved cloning the gene from an affected individual, it is now possible to use the polymerase chain reaction to amplify any region of the gene and rapidly determine by restriction enzyme digestion, sequencing, or hybridization whether a mutant form is present. Because this topic was extensively reviewed recently (55), in this chapter we focus on selected naturally occurring mutants from which information about the interaction of AT with thrombin or heparin can be inferred. For completeness, however, Tables 1 and 2 list the AT codons involved in either type I (Table 1) or type II (Table 2) deficiencies; a thorough and exhaustive listing of all the known mutations involving AT in humans can be found in Reference 56, a database report.

Table 1 Quantitative AT Deficiency[a]

Large deletion	Frameshift	Small deletion	Termination	Unknown
Whole gene	46	49–56	129	Tyr^{166}-Cys
3′ Deletion (exon 7)	48	55	140	Ser^{349}-Pro
5′ Deletion (exon 1, 2)	81	76–77	197	Ala^{387}-Val
Internal (exon 6)	102	106–108	359	Ala^{427}-Asp
	119	120–123		
	208	311–313		
	228	427–429		
	239–240			
	244			
	244–245			
	290–291			
	295			
	308–309			
	370			
	408			
	410–411			
	412–414			
	421			
	423–424			

[a]The numbers indicate the specific codon(s) involved in each mutation.

Table 2 Qualitative AT Deficiency Associated with a Defect in Heparin Binding or Thrombin Binding or Having Multiple Effects

Heparin binding	Thrombin binding	Multiple effects
Ile^{7}-Asn	Asn^{187}-Lys, Asp	Phe^{402}-Cys, Ser, Lys
Met^{20}-Thr	Ile^{284}-Asn	Ala^{404}-Thr
Arg^{24}-Cys	Ala^{382}-Thr	Asn^{405}-Lys
Pro^{41}-Cys	Gly^{392}-Asp	Arg^{406}-Met
Arg^{47}-Cys, Ser, His	Arg^{393}-Cys, His, Pro	Pro^{407}-Thr, Leu
Cys^{99}-Phe	Ser^{394}-Leu	Pro^{429}-Leu
Arg^{129}-Gln		
Glu^{237}-Lys		

IV. EXPRESSION OF RECOMBINANT ANTITHROMBIN

A. Nonglycosylated Expression Products

The availability of AT cDNA clones has allowed the production of recombinant AT molecules from a variety of sources, including bacteria (31), yeast (57), mammalian cells (58,59), insect cells (60), and reticulocyte lysate cell-free systems (33,34,60). The characteristics of these recombinant forms of AT are discussed here and summarized in Table 3. Once the recombinant form of the protein has been produced and compared to plasma-derived AT, site-directed mutagenesis can be used to alter the AT reading frame and the new AT expressed and characterized.

Table 3 Production of Recombinant Antithrombin

Expression system	Glycosylation status	Activity	Remarks	References
E. coli	—	Not determined	—	30
Reticulocyte lysate	—	+	30% Complexes IIa; binds heparin	32, 33, 59
Yeast	+ (high-mannose type)	+	Heparin cofactor activity diminished	55
COS-1 (cultured cells)	+	+	Transient expression	56
CHO (cultured cells)	+	+	Permanent, amplified cell lines; extensive physicochemical characterization	57
Insect cells	+	+	Unexplained diminution in apparent molecule weight	58

In the initial report of the cDNA cloning of human AT, Bock et al. expressed both mature and pre-AT in *Escherichia coli* but did not report the functionality of the product (31). It is curious that bacterial expression of functional AT has not been reported. Enzymatic deglycosylation of plasma AT has been used to demonstrate that the glycosidic side chains are not required for either heparin cofactor activity or progressive thrombin inhibition (61). The inability of bacteria to glycosylate the AT product should therefore not be an issue. It is possible that the reducing intracellular environment of *E. coli* and its propensity for sequestering foreign proteins in insoluble inclusion bodies are obstacles to the production of functional AT, with essential disulfide bonding of its six cysteine residues.

Austin (60) and others from our group (33,34) have employed another system in which the recombinant AT product is not glycosylated. Positioning of the AT cDNA, altered to insert a Met codon before His[1] downstream of a bacteriophage T7 promoter, in an expression plasmid, allowed the in vitro transcription of AT mRNA. This AT mRNA was then translated in vitro in a rabbit reticulocyte lysate system supplemented with oxidized glutathione and [^{35}S]methionine. Following overnight dialysis, the product assumed a disulfide-bonded conformation, demonstrated by a mobility shift, between reduced and nonreduced samples on SDS-PAGE. In addition, cell-free translated AT was capable of forming SDS-stable covalent complexes with human α-thrombin in a time-dependent manner that is accelerated in the presence of heparin. When site-directed mutagenesis was employed to produce a cell-free recombinant AT corresponding to the Denver mutation (S394L, i.e., Ser[394] to Leu), the antithrombin activity was lost. This biological similarity between cell-free and plasma-derived forms of AT has allowed us to use the system to characterize mutants of human AT (60,62–65) and, most recently, to confirm that putative murine and rabbit AT cDNA clones encoded bona fide antithrombins (33,34). Another strength of this system is the capability of producing truncated proteins that would probably fail to be secreted from cultured cell systems. On the other hand, shortcomings of the system include likely conformational heterogeneity (only about 30% of the material can be complexed), somewhat less avid heparin binding than plasma-derived AT, and low yields (ng levels per reaction) (60).

B. Glycosylated Expression Products

1. Yeast

Although all expression systems are by definition artificial, a more ''natural'' way of producing AT is to allow it to follow the normal secretory pathway of a eukaryotic cell. Broker et al. used expression in yeast to produce recombinant human AT (57). However, this product, although demonstrating near normal progressive antithrombin activity, showed heparin cofactor activity that was reduced by an order of magnitude compared with plasma AT. The authors surmised that this impairment was likely a result of the characteristic high-mannose glycosylation pattern of yeast, as opposed to the more complex processed sugars added to mammalian ATs.

2. Cultured Mammalian Cells

Expression in cell culture systems of mammalian origin represents a further step toward the production of AT in a near native environment. However, cell lines that produce AT cannot be used if a background of nonrecombinant protein is to be avoided. Transient expression of AT in transformed African green monkey kidney cells (COS) gave yields of 50 ng/10^6 cells per 24 h of both functional normal recombinant and mutant ATs modified by mutagenesis at position 394 (P1′) (58,66). When permanently transfected Chinese hamster ovary (CHO) cell lines were established by cotransfection of AT expression plasmids with those expressing mouse dihydrofolate reductase (DHFR), 200-fold higher levels of the recombinant AT protein were obtained (58). In this latter system, amplification of AT minigenes inserted into the cell's genome was achieved by selection in increasing doses of methotrexate, a DHFR inhibitor. The high levels of protein obtained allowed a thorough physicochemical and kinetic assessment of the recombinant AT. The CHO-derived AT and plasma AT did not differ significantly when analyzed by circular dichroism, ultraviolet absorbance, fluorescence spectroscopy, heparin binding, or the kinetics of thrombin inhibition. The material contained complex sugar side chains (as does plasma AT), as judged by insensitivity to endoglycosidase F digestion. However, the sialation of the recombinant product was more heterogeneous than that of its plasma counterpart, in that tri- and tetrasialated side chains are found. An additional difference was the presence of fucose on the proximal GlcNAc residue. Despite these fine differences, the clearance behavior of the recombinant material was almost identical to that of its plasma counterpart, when injected into rabbits. Recently, this approach was employed to compare the properties of W49K mutant AT with its nonmutated recombinant counterpart (67).

3. Insect Cells

Expression of human AT in insect cells completes the list of systems employed to date for the production of recombinant AT. This approach employs a recombinant baculovirus, in which the polyhedrin gene is replaced by a cDNA encoding AT, to infect *Spodoptera frugiperda* (Sf9) insect cells. Yields of 10–35 μg/10^6 cells per 24 h were obtained, consisting of a glycosylated product that complexed thrombin in a reaction capable of being accelerated by heparin. However, the apparent molecular mass of the baculovirus-derived AT was only 50 kD, as judged by SDS-PAGE and immunoblotting. The reasons for this size difference were not determined, even though evidence was provided indicating that both the mRNA and amino-terminal sequences of the recombinant protein were intact (60).

V. MUTATIONAL ANALYSIS OF ANTITHROMBIN STRUCTURE AND FUNCTION

A. Natural Versus Artificial Mutants

Currently available molecular biological approaches allow investigators both to identify the affected amino acid in naturally occurring AT mutants and to engineer and express mutations at any point in the molecule. What do these mutations tell us about the structure and function of AT?

It is useful, at this point, to acknowledge the differences between mutations identified in human patients and those generated by directed mutagenesis. The former come to light only when phenotypic reductions in AT activity are detected. Such mutations usually give information only on what mutational changes have a negative impact clinically. In contrast, artificial mutagenesis can give additional information on what changes at a given position have minimal effect (or conceivably may be beneficial). Characterization of naturally occurring AT mutants can be complicated by the presence of equal or greater quantities of normal AT in the affected individuals' plasma. Moreover, it may not always be feasible to separate the normal and dysfunctional protein species. Expression of mutant forms in systems in which they then form the only AT-related protein produced avoids this problem, although care must be taken that the characteristics of the expression system are taken into account in interpreting the properties of a recombinantly produced mutant. The nature and effects of mutant engineered AT molecules are discussed here and summarized in Table 4.

B. Mutations Affecting Thrombin Binding

1. *Reactive Center Mutants*

Known point mutations that affect the ability of AT to form complexes with thrombin and factor Xa cluster near the reactive center scissile bond (Arg^{393}-Ser^{394}). This region, between residues 382 and 407, is presumed to be the primary site of thrombin binding. Alterations to the reactive center can abrogate or severely impair AT function, as seen in the Sheffield (R393H) (68) and Denver (S394L) (69) kindreds. At least for the P1′ residue, expression of artificial mutants in COS cells has shown that the Ser residue is not essential for complex formation (66). Glycine, alanine, or threonine substitutions yield AT molecules with near normal complexing activity, but changes to leucine, cysteine, or valine produce a poor inhibitor. Those with proline or methionine substitutions result in AT moieties with absolutely no detectable activity. Thus, both size and hydrophobicity appear to be involved in forming an optimal P1′ residue. Reproduction of the Denver mutation in a cell-free system also showed that this mutant is a poor inhibitor, in that it complexes thrombin slowly and inefficiently, and that it is a poor substrate, in that it is not cleaved efficiently by protease (61). We recently described a variant AT, from a Stockholm kindred, that involves a P2 substitution of Asp for Gly. Cell-free–derived G392D AT resembled the Denver variant: both cleavage and complex formation were greatly reduced relative to normal AT (70). These findings are consistent with an estimated 4000-fold reduction in the second-order reaction constant describing the interaction between AT Denver and thrombin, compared with normal AT (69).

Recently we have begun to characterize a family of mutant AT molecules that differ from the wild-type recombinant species only at residue 392. Utilizing cell-free expression, we demonstrated that Gly or Pro residues are required at this position for optimal inhibitor activity. Indeed, using recombinant AT generated in COS cells, it was possible to show that the G392P form was a more effective progressive inhibitor of thrombin than the wild type, as reflected in a 2.3-fold increase in its second-order rate constant (71).

An unusual combination of recombinant mutant production and chemical modification was employed by Gettins et al. to analyze the transmission of conformational change from the heparin binding site to the reactive center of AT. Mutant AT R393C was produced in baby hamster kidney (BHK) cells. Unlike the naturally occurring variant, which circulates bound to albumin via a disulfide linkage, the recombinant R393C could be purified unlinked to other proteins and then chemically modified with a nitrobenzofuran-fluorescent reporter group. In this way the binding of heparin species, which was found to be indistinguishable from that of plasma AT, could be directly demonstrated to transmit a change to the reactive center of the mutant, tagged molecule, despite the inability of this polypeptide to inhibit either factor Xa or thrombin (72).

2. *Thrombin Binding Mutations Not at the Active Center*

In contrast to mutations at the reactive center, such as AT Denver or AT Stockholm, both plasma-derived and cell-free recombinant AT Hamilton (A382T) exhibited a different phenotypic expression. AT Hamilton is cleaved efficiently by either thrombin or factor Xa but does not participate significantly in complex formation (64,73). Randomization of the 382 residue by site-directed mutagenesis revealed that neutral or hydrophobic residues gave near normal complexing reactions, but polar (Thr and Gln) or charged residues (Lys) did not (65). These observations lend strength to the idea that residue 382 lies in a region that is important in allowing movement of a portion of AT closer to the underlying β sheet following cleavage, thus facilitating complex formation (74). Attempts

Table 4 Engineered Mutants of Antithrombin[a]

Expression system	Alteration	Effects on: Cleavage	Effects on: Complexing	Effects on: Heparin binding	Reference
COS cells	S394 P/M/L/C/V	↓	↓	—	64
	S394 G/A/T	—	—	—	64
Reticulocyte lysate	S394 L	↓	↓	—	59
	G392 D	↓	↓	—	68
	A382 T/Q/K	—	↓	—	63
	A382 G/V	—	—	—	63
	C430 A/G/S	—	↓	—	61
	219–432 Truncation	—	↓	↓	61
	Rabbit 1–210 Truncation	↓	↓	—	32
	Rabbit 1–369 + human 369–432 fusion	↓	↓	—	32
	41–49 Deletion	—	—	—	79
BHK cells	W49K	—	—	—	65

[a]↑, increased activity; ↓, decreased activity; —, no change in activity.

such as these, to describe the complexing reaction at the molecular and three-dimensional level, are described later in more detail (see Sec. VI).

Mutations found at a similar distance from the reactive center as those exemplified by AT Hamilton, but in the carboxyl-terminal direction along the chain, demonstrate not only impaired serine protease inhibition but also greatly decreased antigen levels in the circulation. Whether mutant AT molecules, such as AT Utah (P407L) (37) and AT Oslo (A404T) (75), are less stable in plasma than their normal counterpart or fail to transit the secretory pathway efficiently is not known at present. This issue could be addressed readily by the expression of suitable cDNA constructs in transfected cell systems. AT Budapest (P429L), although demonstrated to be partially unfolded through its failure to form the Cys^{247}-Cys^{430} disulfide bond, and impaired in both its interaction with thrombin and heparin, is nonetheless present at normal levels in the circulation (76). Recombinant forms of AT produced in a cell-free system, in which Cys^{430} was altered to Ala, Gly, or Ser, retained 25–50% of the thrombin complexing activity of normal AT, suggesting that the P429L disrupts structural motifs within the AT molecule other than the adjacent disulfide bond alone (63).

Synthetic peptides have been used to investigate the role of the carboxyl-terminal region of AT. Peptides consisting of residues 404–412 and 420–429 have been produced and shown to inhibit the interaction of AT with both thrombin and factor Xa. The effect of the 420–429 peptide was diminished in the presence of heparin (77). Both peptides also enhanced the amidolytic activity of the two proteases on synthetic substrates. The use of a chemical modification of thrombin by a dansyl compound suggested that a direct effect on thrombin by the two peptides may be involved. The authors of this study hypothesize that the carboxyl-terminal region of the AT molecule interacts with thrombin to increase the interaction between the protease and the inhibitor.

The portion of the molecule carboxyl-terminal to the reactive center has been implicated as a determinant for TAT complexes to be recognized by their cognate hepatic receptors. Perlmutter and colleagues (78) reported that a synthetic peptide corresponding to the P1′–P15′ of α1AP bound specifically to the surface of HepG2 cells and that this binding was reduced in the presence of both TAT complexes and complexes of elastase and α1AP. The latter result is consistent with the competition that has been observed between such complexes in clearance studies performed in vivo (16). It is therefore possible that a number of serpin-enzyme complexes are removed by either similar or overlapping pathways.

3. *Deletion Mutants*

Experience with deletion mutants of AT to date has been limited to experiments in the cell-free system. Expression of a truncated form of human AT corresponding to residues 219–432 resulted in the production of a polypeptide that could be used as a substrate for cleavage by thrombin but that exhibited neither complex formation nor heparin binding (63). Conversely, expression of truncated forms of rabbit AT corresponding to residues 1–210 or 1–369 yielded proteins that retained much of the affinity of the full-length cell-free rabbit AT for heparin, but were unaffected by incubation with thrombin (33). In the latter study, it was also found that despite the high degree of identity between the rabbit and human forms of the protein, interspecies fusion proteins made up of rabbit residues from 1 to 369, fused to human residues 369–433 (rabbit AT numbering), demonstrated reduced thrombin inhibition compared with underivatized recombinant rabbit AT.

C. Mutations Affecting Heparin Binding

Most naturally occurring variants of AT with diminished heparin affinity cluster toward the amino terminus of the molecule (residues 7–47). Kindreds with amino acid substitutions P41L and R47C/H/S have been described repeatedly (reviewed in Refs. 55 and 56). Only the Budapest 3 (L99F) (79) and Geneva (R129N) (80) mutations represent reported mutations, in locations carboxyl terminal to residue 47, associated with impaired heparin binding. In the Geneva variant, it was possible to purify the product of the mutant allele by immunoaffinity chromatography of material that failed to bind to immobilized heparin. Chemical modification experiments have implicated a variety of residues within the first 150 amino acids of the AT molecule as either important for heparin binding or lying close to such critical residues. Reaction of plasma AT with dimethyl-(2-hydroxy-5-nitrobenzyl)sulfonium bromide (DMSB) results in the addition of a bulky hydroxynitrobenzyl group to the indole ring of the tryptophan residue. Although still able to complex thrombin in a progressive manner, DMSB-modified AT is no longer a heparin cofactor (81). Others, using the same chemical modification, proposed that the impaired heparin cofactor activity of the DMSB-modified AT resulted not so much from the chemical modification of Trp^{49} but from the stabilization of a particular AT conformation (82).

In an effort to clarify the contribution of the Trp^{49} residue to heparin binding, Gettins and colleagues (67) produced recombinant AT mutant W49K. This nonconservative mutant was produced in milligram quantities through the establishment of permanent baby hamster kidney cell lines resistant to both methotrexate and neomycin (G418). Purified mutant and normal recombinant AT products were then compared by proton nuclear magnetic resonance. The spectra obtained were highly similar, suggesting no change in global protein folding. The affinity for heparin pentasaccharide of the W49K mutant was somewhat reduced, but neither the heparin enhancement of thrombin nor factor Xa inhibition was affected. These results suggest that Trp^{49} is not an essential part of the heparin binding domain of AT and are consistent with the data we obtained with the cell-free system. Deletion of codons corresponding to residues 41–49 and ensuing transcription and translation from the mutant template yielded a form of AT that retained the ability both to form complexes with thrombin and to bind immobilized heparin. In addition, fusion of AT encoding fragments to the carboxyl terminus of staphlococcal protein A expressed in *E. coli* conferred heparin binding properties on the bacterial protein if residues 90–160 were appended, but not if residues 22–60 were employed (83).

Regions carboxyl terminal to residues 41–49 have also been implicated as heparin binding sites. Chemical modification studies have shown that heparin can protect Arg^{129} and Arg^{145} from reaction with (*p*-hydroxyphenyl)glyoxal and that derivatization of these amino acids correlates with severe impairment of heparin cofactor activity (84). A similar approach, using a colored sulfonic acid derivative, identified lysine residues 107, 125, and 136 to be involved in heparin binding (85). In addition, heparin cofactor activity has been reported to correlate with the intactness of the Cys^{8}-Cys^{128} bond (6). Synthetic peptides corresponding to AT residues 124–140 bind heparin more tightly than peptides made up of the same sequence, but scrambled (86). Their ability to interfere with heparin enhancement of the rate of AT inhibition of thrombin was similar, however, suggesting additional structural or conformational effects to be present within the normal heparin binding site. Moreover, a peptide corresponding to AT residues 124–145 has been used to generate an antibody that both blocked the interaction of AT with heparin and mimicked heparin by accelerating complex formation with thrombin in the absence of heparin

(87). With the exception of residue 129, the site of the mutation in the Geneva kindred, none of these positions have been investigated by mutagenesis.

VI. SERPINS AND INSIGHTS INTO ANTITHROMBIN STRUCTURE AND FUNCTION

A. Conservation of Structure and Properties

To what extent does AT resemble other members of the serpin super-family of proteins? More than 40 proteins are now known to be members of the group, including mammalian, insect, plant, and viral polypeptides (30,88). Most are approximately 400 amino acids in length. AT shares approximately 30% homology with plasma protein serpins, such as heparin cofactor II, α-$_1$-antiplasmin, plasminogen activator inhibitor (PAI), protein C inhibitor, and, arguably the most studied serpin, α1AP. Further similarities are apparent when deduced secondary structures are compared. Typical serpins exhibit 30% helical structure and 40% β-pleated structure, arranged in three sheets. One of these β sheets, the five- or six-stranded A sheet, is a characteristic feature of serpins, with the reactive site predicted to lie on a peptide loop extending from the fourth strand. The various serpins are hypothesized to have diverged from a common ancestral protein some 500 million years ago (29), with some family members developing specialized features, such as putative heparin binding domains, and others losing protease inhibitory activity (such as ovalbumin and angiotensinogen).

B. Information from X-ray Crystallography

α_1-Antitrypsin, also known as α_1-antiproteinase, to reflect its ability to inhibit both trypsin-like proteases and its natural target neutrophil elastase, was the first serpin for which crystallographic information became available (89). The form of α1AP that was successfully crystallized had been proteolytically modified at its reactive center. The cleaved loop preceding the reactive center had been completely incorporated back into the A sheet, with consequent separation of the P1′ and P1 residues to opposite poles of the molecule, a distance of some 67 Å. This dramatic "springing open" of the molecule at the reactive center implied that a major structural rearrangement had taken place, one predicted to increase the stability of the molecule greatly. This idea was confirmed experimentally in heat stability experiments (90). Exposure of either native α1AP or native AT to 60°C temperatures resulted in 50% precipitation after 2 h but practically no loss in solubility of their respective cleaved forms. Both these serpins, as well as α_1-antichymotrypsin and the C1 inhibitor, exhibit large shifts in circular dichroism (CD) spectra when cleaved, as opposed to native molecules, when tested for spectral CD changes in increasing concentrations of guanidine HCl denaturant (91).

Although ovalbumin is not a protease inhibitor, it shares 30% amino acid identity with AT and α1AP. A comparison of the crystal structures of intact (92) and cleaved (93) ovalbumin (called plakalbumin) indicates that cleavage of this serpin, at a position homologous to the active center of its inhibitory cousins, causes little structural perturbation. The pseudoreactive center loop is found on a somewhat mobile three-turn α helix, held out from the molecule by two stalks (94). It has been proposed that rigidity of the preceding hinge region in ovalbumin locks the pseudoreactive center into place, preventing the insertion into the A sheet, as seen for α1AP. Such mutations as AT Hamilton

(A383T) (64,73) and C1 inhibitor Ma (Glu to Ala at the homologous P12 residue) (94) replace one of the mobile Ala residues in this region, causing them to be unable to form complexes with their cognate proteases. This implies that the ability of the reactive center to insert fully into the A sheet, as seen in the cleaved structure, is required for inhibitor-protease complex formation. Independent evidence to support this idea is provided by a monoclonal antibody raised against TAT (95) and found to be directed against residues 382–386 (96). This agent interfered with complex formation between AT and thrombin, causing the production of cleaved AT, as was seen in the hinge mutants just described.

The circulating plasma inhibitor of tissue plasminogen activator (PAI-1) exists in plasma as an inactive latent form. It is thought to be activated through interaction with vitronectin (97). The recently reported crystal structure of latent PAI-1 (98) shows a complete insertion of the reactive center loop into the A sheet, with accompanying structural changes predicted to allow its emergence when activated.

An interesting peptide annealing technique has been used by two groups (99,100) to demonstrate the applicability of the proposals on reactive center mobility, made as the result of the evidence provided by the structures of ovalbumin and PAI-1. Incubation of AT with a peptide corresponding to P14-P2 (AT residues 380–392) conferred heat stability and spectral properties of cleaved AT to this intact but peptide-associated AT. Like cleaved AT, the peptide-complexed material displayed decreased affinity for heparin (99).

Crystals of bovine AT cleaved at the reactive center have been obtained and the structure solved using α1AP as the model (101). In terms of global conformation, it appears very similar to α1AP, particularly in terms of the separation of P1 and P1′ residues. Further refinements of our understanding of the structure of serpin are required to obtain information about the heparin binding regions, because the amino-terminal regions of AT do not align with the smaller α1AP. Moreover, the four serpins known to interact with heparin (AT, heparin cofactor II, PAI-1, and protease nexin 1) exhibit different heparin binding characteristics. Based on the deduced structure of human AT, which must be modeled on α1AP (102), it has been proposed that a zone of lysines and arginines starting with Arg^{47}, continuing through the D helix (Lys^{125}, Arg^{129}, and Arg^{132}), and extending to the neighborhood of the reactive center, is the site of heparin binding (30,103). In this hypothesis, the pentasaccharide is thought to bind to the amino-terminal regions of this zone, longer oligosaccharides binding to increasingly carboxyl-terminal portions. Crystallization of intact AT, and also of AT cocrystallized with oligosaccharides, would serve to clarify this and many other aspects of the molecular interaction of AT with heparin and other macromolecules. The currently available crystal structures of the serpins (at this writing, October 1993) are listed in Table 5.

Table 5 Available Serpin Crystal Structures

Serpin	Status	Species	Reference
α_1-Antiproteinase (α1AP)	Cleaved	Human	85
Ovalbumin	Cleaved	Chicken	89
Ovalbumin	Intact	Chicken	88
Tissue plasminogen activator inhibitor (PA1-1)	Latent	Chicken	94
Antithrombin (AT)	Cleaved	Bovine	97
α_1-Antichymotrypsin (ACT)	Cleaved	Human	104

The most recent serpin crystal structure (at this writing) to be reported is cleaved human α_1-antichymotrypsin (ACT) (104). Although α1AP and ACT are 45% homologous in terms of primary structure, their three-dimensional structures are highly similar. With a single exception, the conserved residues of the serpin family (30) are in identical positions. Importantly, strand s4a is inserted as the middle strand in the A sheet, as found for α1AP and inferred to be of general importance in serpins on the basis of the thermal stability experiments already discussed. ACT is unusual among serpins in that it possesses the capacity to bind DNA. The region thought to be involved in DNA binding is one of the major sites of local difference with the α1AP structure.

C. Serpin Mutants with Thrombin Inhibitory Activity

Both natural and artificial mutants of members of the serpin family (other than AT) have been described that have acquired the ability to inhibit thrombin. One of the first such reports was the description of the Pittsburgh (M358R) variant of α1AP (105). This mutation caused alteration of the natural reactive center of α1AP (Met-Ser) to that of AT (Arg-Ser). The mutant protein has been demonstrated to have altered its cognate protease recognition, exhibiting potent antithrombin activity but little antielastase ability. α1AP Pittsburgh association constants with thrombin approach those seen with heparin-AT. Similar results were obtained with recombinant nonglycosylated forms of this mutant protein expressed in yeast (106). The α1AP Pittsburgh mutant clearly demonstrates that P1′ alterations could significantly change the specificity of a serpin. The pathophysiological consequence of the presence of this variant in vivo is somewhat more complex: it was realized after the original description (105) that α1AP can inhibit activated protein C and that the Pittsburgh variant is particularly potent as an inhibitor of activated protein C (107). Transfection studies have indicated that the expression of the variant protein can have pleiotropic effects on protein processing (108). A second individual with α1AP Pittsburgh was recently described in whom the bleeding tendency is only mild and episodic (109), as opposed to the severe and ultimately lethal bleeding observed in the first patient reported. In the second case, protein C processing is also impaired. Thus, although the α1AP Pittsburgh variant is instructive in that it showed the importance of the reactive center for serpin specificity, its effects in vivo are quite complex and at this point incompletely understood.

The specificity of a given serpin was reported to be either redirected or enhanced in a number of other investigations. Recombinant α_1-antichymotrypsin has been produced in *E. coli* and shown to inhibit thrombin with kinetics similar to that of native AT when the P1 residue is substituted from the normal Leu to Arg (110). The ability of this engineered variant to interact with chymotrypsin is reduced 30-fold. Similarly, the conversion of recombinant heparin cofactor II from a wild-type reactive center (Leu-Ser) to that of AT (Arg-Ser) is associated with a 100-fold increase in the progressive reaction of the inhibitor with thrombin (111). However, alteration of serpin specificity by P1 mutagenesis is not always sufficient to change the target of the inhibitor, because mutant α_1-antichymotrypsin L358M does not convert it to an elastase inhibitor but displays normal specificity (112). Additionally, recombinant forms of PAI-1 expressed in *E. coli*, in which the 19 amino acids from P17 to P2′ were replaced with those of AT, maintained the ability to inhibit plasminogen activators and exhibited only minimal thrombin complexing ability (113). In addition, conversion of the active site dipeptide of α-$_2$-antiplasmin to Arg-Ser by deletion of the natural P1′ residue produced a mutant with natural specificity and no antithrombin activity (114). Finally, although the re-creation of the P10 proline mutation of AT in a recombinant α1AP molecule had the same effect, that is, conversion of the inhibitor to a substrate, re-creation of the Hamilton mutation (p12 threonine) had no effect on the ability of the recombinant α1AP to inhibit either trypsin or elastase (115).

VII. CONCLUSION

Determination of the DNA and the deduced amino acid sequence of antithrombin has spawned two very productive lines of investigation into the relationship between AT structure and function: an analysis of natural mutants and the generation and characterization of artificial ones, and informative comparisons to the crystal structures and properties of other serpins. Furthermore, the ability to generate engineered recombinant AT molecules and to perform detailed structural analyses of such proteins can be expected to bring progress to this area of knowledge. Such work could eventually lead to the development of engineered forms of antithrombin with therapeutically useful properties (e.g., increased potency, extended half-lives, or reduced susceptibility to inactivation by elastase or venoms).

Note added in proof: The crystal structure of intact, functional AT was reported as this chapter was in the final stages of editing (Carrell RW, Stein PE, Fermi G, Werdell MR. Biological implications of a 3Å structure of dimeric antithrombin. Structure 1994; 2:257–270).

REFERENCES

1. Rosenberg RD, Damus PS. The purification and mechanism of action of human antithrombin-heparin cofactor. J Biol Chem 1973; 248:6490–6505.
2. Egeberg O. Inherited antithrombin deficiency causing thrombophilia. Thromb Diath Haemorrh 1965; 13:516–530.
3. Thaler E, Lechner K. Antithrombin III deficiency and thromboembolism. Clin Haematol 1981; 10:369–390.
4. Chan V, Chan TK, Wang V, Tso SC, Todd D. The determination of antithrombin III by radioimmunoassay and its clinical application. Br J Hematol 1979; 41:563–572.
5. Carlson TH, Simon TL, Atencio AC. In vivo behaviour of human radioiodinated antithrombin III: distribution between three physiologic pools. Blood 1985; 66:13–19.
6. Sun X-J, Chang J-Y. Heparin binding domain of human antithrombin III inferred from the sequential reduction of its three disulfide linkages. J Biol Chem 1989; 264:11288–11293.
7. Kurachi K, Fujikawa K, Schmer G, Davie EW. Inhibition of bovine factor IXa and factor Xαβ by antithrombin III. Biochemistry 1976; 15:373–377.
8. Scott CF, Colman RW. Factors influencing the acceleration of human factor XIa inactivation by antithrombin III. Blood 1989; 73:1873–1879.
9. Stead N, Kaplan AP, Rosenberg RD. Inhibition of activated factor XII by antithrombin-heparin cofactor. J Biol Chem 1976; 251:6481–6488.
10. Lahiri B, Rosenberg RD, Talamo RC, Mitchell B, Bagdasarian A, Colman RW. Antithrombin III: an inhibitor of human plasma kallikrein. Fed Proc 1974; 33:642a.
11. Highsmith RF, Rosenberg RD. The inhibition of human plasmin by human antithrombin-heparin cofactor. J Biol Chem 1974; 249:4335–4338.
12. Buchanan MR, Boneu B, Ofosu F, Hirsh J. The relative importance of thrombin inhibition and factor Xa inhibition to the antithrombotic effects of heparin. Blood 1985; 65:198–201.

13. Ofosu FA, Sie P, Modi GJ, et al. The inhibition of thrombin-dependent positive-feedback reactions is critical to the expression of the anticoagulant effect of heparin. Biochem J 1987; 243:579–588.
14. Bjork I, Jackson CM, Jornvall H, Lavine KK, Nordling K, Salsgiver WJ. The active site of antithrombin. J Biol Chem 1982; 257:2406–2411.
15. Owen WG. Evidence for the formation of an ester linkage between thrombin and heparin cofactor. Biochim Biophys Acta 1975; 405:380–387.
16. Pizzo SV. Serpin receptor 1: a hepatic receptor that mediates the clearance of antithrombin III protease complexes. Am J Med 87 1989; (Suppl. 3B):10S–14S.
17. Jordan RE, Oosta GM, Gardner WT, Rosenberg RD. The kinetics of hemostatic enzyme-antithrombin interactions in the presence of low molecular weight heparin. J Biol Chem 1980; 255:10081–10090.
18. Seegers WH, Johnson JF, Fall C. An antithrombin reaction related to prothrombin activation. Am J Physiol 1954; 176:97–103.
19. Marcum JA, Rosenberg RD. Anticoagulantly active heparin-like molecules from vascular tissue. Biochemistry 1984; 23:1730–1737.
20. Marcum JA, Rosenberg RD. Heparin-like molecules with anticoagulant activity are synthesized by cultured endothelial cells. Biochem Biophys Res Commun 1985; 126:365–372.
21. Marcum JA, McKenney JB, Rosenberg RD. Acceleration of thrombin-antithrombin complex formation in rat hindquarters via heparin-like molecules from vascular tissue. J Clin Invest 1984; 74:341–350.
22. Hatton MWC, Moar SL, Richardson M. Evidence that rabbit I125-antithrombin binds to proteoheparan sulphate at the subendothelium of the rabbit aorta in vivo. Blood Vessels 1988; 67:878–886.
23. de Agostini AI, Watkins SC, Slayter HS, Youssoufian H, Rosenberg RD. Localization of anticoagulantly active heparan sulfate proteoglycans in vascular endothelium: antithrombin binding on cultured endothelial cells and perfused rat aorta. J Cell Biol 1990; 111:1293–1304.
24. Choay J, Petitou M, Lormeau JC, Sinay P, Casu B, Gatti G. Structure-activity relationship in heparin: a synthetic pentasaccharide with high affinity for antithrombin III and eliciting high anti-factor Xa activity. Biochem Biophys Res Commun 1983; 116:492–499.
25. Olson ST, Choay J. Mechanism of high molecular weight-kininogen stimulation of the heparin-accelerated antithrombin/kallikrein reaction. Thromb Haemost 1989; 62:326a.
26. Olson ST, Bjork I, Sheffer R, Craig PA, Shore JD, Choay J. Role of the antithrombin-binding pentasaccharide in heparin acceleration of antithrombin-protease reactions: resolution of the antithrombin conformational change contribution to heparin rate enhancement. J Biol Chem 1992; 267:12528–12538.
27. Evans DL, Marshall CJ, Christey PB, Carrell RW. Heparin binding site, conformational change, and activation of antithrombin. Biochemistry 1993; 31:12629–12642.
28. Petersen TE, Dudek-Wojciechowska G, Sottrup-Jensen L, Magnusson S. Primary structure of antithrombin III (heparin cofactor). Partial homology between α-1-antitrypsin and antithrombin III. In: Collen D, Wiman B, Verstraete M eds. The Physiological Inhibitors of Coagulation and Fibrinolysis. Amsterdam: Elsevier-North Holland, 1979:43–54.
29. Hunt LT, Dayhoff MO. A surprising new protein superfamily containing ovalbumin, antithrombin-III, and alpha-proteinase inhibitor. Biochem Biophys Res Commun 1982; 95:866–871.
30. Huber R, Carrell RW. Implications of the three-dimensional structure of α-1-antitrypsin for structure and function of serpins. Biochemistry 1989; 28:8951–8961.
31. Bock SC, Wion KL, Vehar GA, Lawn RM. Cloning and expression of the cDNA for human antithrombin III. Nucleic Acids Res 1982; 10:8113–8125.
32. Brennan SO, George PM, Jordan RE. Physiological variant of antithrombin III lacks carbohydrate side chain at Asn 135. FEBS Lett 1987; 219:431–436.

33. Sheffield WP, Brothers AB, Wells MJ, Hatton MWC, Clarke BJ, Blajchman MA. Molecular cloning and expression of rabbit antithrombin III. Blood 1992; 79:2330–2339.
34. Wu JK, Sheffield WP, Blajchman MA. Molecular cloning and cell-free expression of murine AT-III. Thromb Haemost 1992; 68:291–296.
35. Medjoub H, Leret M, Boulanger Y, Mamman M, Choay J, Reinbolt J. The complete amino acid sequence of bovine antithrombin (AT-III). J Protein Chem 1991; 10:205–212.
36. Bock SC, Marrinan JA, Radziejewska E. Antithrombin III Utah: proline-407 to leucine mutation in a high conserved region near the inhibitor reactive site. Biochemistry 1988; 27: 6171–6178.
37. Prochownik EV. Relationship between an enhancer element in the human antithrombin III gene and an immunoglobulin light-chain gene enhancer. Nature 1985; 316:845–848.
38. Winter JH, Bennett B, Watt JL, et al. Confirmation of linkage between antithrombin III and Duffy blood group antigen and assignment of AT3 to 1q22-25. Ann Hum Genet 1982; 26: 29–34.
39. Olds RJ, Lane DA, Chowdury V, De Stefano V, Leone G, Thein SL. Complete nucleotide sequence of the antithrombin gene: evidence for homologous recombination causing thrombophilia. Biochemistry 1993; 32:4216–4224.
40. Olds RJ, Chowdury V, Lane DA, et al. Gene rearrangement within the antithrombin locus as the basis for an inherited thrombophilic tendency. Thromb Haemost 1993; 69:576a.
41. Owens MR, Miller LL. Net biosynthesis of antithrombin-III by the isolated rat liver perfused for 12–24 hours compared with rat fibrinogen and α-2-(acute phase)-globulin: antithrombin III is not an acute phase protein. Biochim Biophys Acta 1980; 627:30–38.
42. Schmidt B, Ofosu FA, Mitchell L, Brooker LA, Andrew M. Anticoagulant effects of heparin in neonatal plasma. Pediatr Res 1989; 25:405–408.
43. D'Souza SE, Mercer JFB. Antithrombin III mRNA in adult rat liver and kidney and in rat liver during development. Biochem Biophys Res Commun 1987; 142:417–421.
44. Manoharan A, Hewitt B. Danazol therapy in familial antithrombin III deficiency. Clin Lab Haematol 1990; 12:357–362.
45. Wells MJ, Sheffield WP, Blajchman MA. Characterization of a putative hepatic receptor for thrombin-antithrombin III complexes. Blood 1992; 80(Suppl. 1):313a.
46. Ochoa A, Brunel F, Mendelzon D, Cohen GN, Zakin MM. Different liver nuclear proteins bind to similar DNA sequences in the 5′-flanking regions of three hepatic genes. Nucleic Acids Res 1989; 17:116–133.
47. Sas G, Peto I, Banhegyi D, Blasko G, Domjan G. Heterogeneity of the ''classical'' antithrombin III deficiency. Thromb Haemost 1980; 43:133–136.
48. Olds RJ, Lane DA, Finazzi G, Barbui T, Thein S-L. A frameshift mutation leading to type 1 antithrombin deficiency and thrombosis. Blood 1990; 76:2182–2186.
49. Gandrille S, Vidaud D, Emmerich J, et al. Molecular basis for hereditary antithrombin quantitative deficiencies: a stop codon in exon IIIa and a frameshift in exon VI. Br J Hematol 1991; 78:414–420.
50. Bock SC, Prochownik EV. Molecular genetic survey of sixteen kindreds with hereditary antithrombin III deficiency. Blood 1987; 70:1273–1278.
51. Fernandez-Rachubinski F, Rachubinski RA, Blajchman MA. Partial deletion of an antithrombin III allele in a kindred with a type 1 deficiency. Blood 1992; 80:1476–1485.
52. Demers C, Ginsberg JS, Hirsh J, Henderson P, Blajchman MA. Thrombosis in antithrombin-III-deficient persons: report of a large kindred and literature review. Ann Intern Med 1992; 116:754–761.
53. Hakten M, Deniz U, Ozbay G, Ulutin ON. Two cases of homozygous antithrombin III deficiency in a family with congenital deficiency of AT-III. In: Senzinger H, Vinazzer H, eds. Thrombosis and Haemorrhagic Disorders. Proceedings of the 6th International Meeting of the Danubian League Against Thrombosis and Haemorrhagic Disorders. Wurzburg, Germany: Schmitt and Meyer, 1989:177–181.

54. Boyer C, Wolf M, Vedrinni J, Mer D, Lamieu MJ. Homozygous variant of antithrombin III: AT Fountainebleau. Thromb Haemost 1986; 56:18–22.
55. Blajchman MA, Austin RC, Fernandez-Rachubinski F, Sheffield WP. Molecular basis of inherited human antithrombin deficiency. Blood 1992; 80:2159–2171.
56. Lane DA, Olds RJ, Boisclair M, et al. Antithrombin III mutation database: first update. Thromb Haemost 1993; 70:361–369.
57. Broker M, Bragg H, Karges HE. Expression of human antithrombin III in *Saccharomyces cerevisiae* and *Schizosaccharomyces pombe*. Biochim Biophys Acta 1987; 908:203–213.
58. Stephens AW, Siddiqui A, Hirs CHW. Expression of functionally active human antithrombin III. Proc Natl Acad Sci USA 1987; 84:3886–3890.
59. Zettlemeissl G, Conrad HS, Nimtz M, Karges HE. Characterization of recombinant human antithrombin III synthesized in chinese hamster ovary cells. J Biol Chem 1989; 264:21153–21159.
60. Gillespie LS, Hillesland KH, Knauer DJ. Expression of biologically active human antithrombin III by recombinant baculovirus in *Spodoptera frugiperda* cells. J Biol Chem 1991; 266:3995–4001.
61. Austin RC, Rachubinski RA, Fernandez-Rachubinksi F, Blajchman MA. Expression in a cell-free system of normal and variant forms of human antithrombin III: ability to bind heparin and react with α-thrombin. Blood 1990; 76:1521–1529.
62. Rosenfeld L, Danishefsky I. Effects of enzymatic doglycosylation on the biological activities of human thrombin and antithrombin. Arch Biochem Biophys 1984; 229:359–367.
63. Austin RC, Sheffield WP, Rachubinski RA, Blajchman MA. The NH_2-terminal domain of human antithrombin III is essential for both heparin binding and cleavage by α-thrombin of human antithrombin III. Biochem J 1992; 282:345–351.
64. Austin RC, Rachubinski RA, Ofosu FA, Blajchman MA. Antithrombin-III-Hamilton, Ala 382 to Thr: an antithrombin III variant that acts as a substrate but not an inhibitor of thrombin and factor Xa. Blood 1991; 77:2185–2189.
65. Austin RC, Rachubinski RA, Blajchman MA. Site-directed mutagenesis of alanine-382 of human antithrombin III. FEBS Lett 1991; 280:254–258.
66. Stephens AW, Siddiqui A, Hirs CHW. Site-directed mutagenesis of the reactive center (serine 394) of antithrombin III. J Biol Chem 1988; 263:15849–15852.
67. Gettins PG, Choay J, Crew BC, Zettlemeissl G. Role of tryptophan 49 in the heparin cofactor activity of human antithrombin III. J Biol Chem 1992; 267:21946–21953.
68. Lane DA, Erdjument H, Flynn A, et al. Antithrombin Sheffield: amino acid substitutions at reactive site (Arg 393 to His) causing thrombosis. Br J Hematol 1989; 71:91–96.
69. Stephens AW, Thalley BS, Hirs CHW. Antithrombin-III Denver, a reactive site variant. J Biol Chem 1987; 262:1044–1048.
70. Blajchman MA, Fernandez-Rachubinski F, Sheffield WP, Austin RC, Schulman S. Antithrombin III Stockholm: a codon 392 (Gly → Asp) mutation with normal heparin binding and impaired serine protease reactivity. Blood 1992; 79:1428–1434.
71. Sheffield WP, Blajchman MA. Site-directed mutagenesis of human antithrombin residue 392 (P2) reveals a variant with enhanced biological activity. Thromb Haemost 69:1140a.
72. Gettins PGW, Fan B, Crews BC, Turko IV, Olson ST, Streusand VJ. Transmission of conformational change from the heparin binding site to the reactive center of antithrombin. Biochemistry 1993; 32:8385–8389.
73. Devraj-Kizuk R, Chui DK, Prochownik EV, Carter CJ, Ofosu FA, Blajchman MA. Antithrombin III-Hamilton: a gene with a point mutation (guanine to adenine) in codon 382 causing impaired serine protease reactivity. Blood 1988; 72:1518–1523.
74. Bock SC. Structures and models of native serpins. Protein Eng 1990; 4:107–108.
75. Bock SC, Silbermann JA, Wikoff W, Abildgaard U, Hultin MB. Identification of a threonine for alanine substitution at residue 404 of antithrombin III Oslo suggests integrity of the 404–407 region is important for maintaining normal plasma inhibitor levels. Thromb Haemost 1989; 62:494a.

76. Olds RJ, Lane DA, Caso R, et al. Antithrombin III Budapest: a single amino acid substitution (429Pro to Leu) in a region highly conserved in the serpin family. Blood 1992; 79:1206–1212.
77. Nishioka J, Suzuki K. The role of the COOH-terminal region of antithrombin III: evidence that the COOH-terminal region of the inhibitor enhances the reactivity of thrombin and factor Xa with the inhibitor. J Biol Chem 1992; 267:22224–22229.
78. Perlmutter DH, Glover GI, Rivetna M, Schasteen CS, Fallon RJ. Identification of a serpin-enzyme complex receptor on human hepatoma cells and human monocytes. Proc Natl Acad Sci USA 1990; 87:3753–3757.
79. Olds RJ, Lane DA, Boisclair M, Sas G, Bock SC, Thein S-L. Antithrombin Budapest 3: antithrombin variant with reduced heparin affinity resulting from the substitution L99F. FEBS Lett 1992; 300:241–246.
80. Gandrille S, Aiach M, Lane DA, et al. Important role of arginine 129 in heparin-binding site of antithrombin III: identification of a novel mutation arginine 129 to glutamine. J Biol Chem 1990; 265:18997–19001.
81. Blackburn MN, Sibley CC. The heparin binding site of antithrombin III: evidence for a critical tryptophan residue. J Biol Chem 1980; 255:824–826.
82. Shah N, Scully MF, Ellis V, Kakkar VV. Influence of chemical modification of tryptophan residues on the properties of human antithrombin III. Thromb Res 1990; 57:343–352.
83. Wu YI, Sheffield WP, Austin RC, Blajchman MA. Mapping the heparin binding domain of human antithrombin III. Blood 1992; 80(Suppl. 1):312a.
84. Sun XJ, Chang JY. Evidence that arginine-129 and arginine-145 are located within the heparin binding site of human antithrombin III. Biochemistry 1990; 29:8957–8961.
85. Chang JY. Binding of heparin to human antithrombin III activates selective chemical modification at lysine 236: Lys-107, Lys-125, and Lys-136 are situated within the heparin-binding site of antithrombin III. J Biol Chem 1989; 264:3111–311.
86. Pratt CW, Whinna HC, Church FC. A comparison of three heparin-binding serine protease inhibitors. J Biol Chem 1992; 267:8795–8801.
87. Smith JW, Dey N, Knauer DJ. Heparin binding domain of antithrombin III: characterization using a synthetic peptide directed polyclonal antibody. Biochemistry 1990; 29:8950–8957.
88. Carrell RW, Evans DLI. Serpins: mobile conformations in a family of inhibitors. Curr Opin Struct Biol 1992; 2:438–446.
89. Loeberman H, Tokuoka R, Deisenhofer J, Huber R. Human α-1-proteinase inhibitor crystal structure analysis of two crystal modifications, molecular model and preliminary analysis of the implications for function. J Mol Biol 1984; 177:531–556.
90. Carrell RW, Owen MC. Plakalbumin, α-1-antitrypsin, antithrombin and the mechanism of inflammatory thrombosis. Nature 1985; 317:730–732.
91. Bruch M, Weiss V, Engel J. Plasma serine proteinase inhibitors (serpins) exhibit major conformational changes and a large increase in conformational stability upon cleavage at their reactive sites. J Biol Chem 1988; 263:16626–16630.
92. Stein PE, Leslie AGW, Finch JT, Turnell WG, McLaughlin PJ, Carrell RW. Crystal structure of ovalbumin as a model for the reactive centre of serpins. Nature 1990; 347:99–102.
93. Wright HT, Qian HX, Huber R. Crystal structure of plakalbumin, a proteolytically nicked form of ovalbumin: its relationship to the structure of the cleaved α-1-antiproteinase inhibitor. J Mol Biol 1990; 213:513–528.
94. Skriver K, Wikoff W, Patson PA, et al. Substrate properties of C1 inhibitor Ma (alanine 434 to glutamic acid): genetic and structural evidence suggesting that the P12-region contains critical determinants of serine protease inhibitor inhibitor/substrate status. J Biol Chem 1991; 266:9216–9221.
95. Asakura S, Matsuda M, Yoshida N, Terukina S, Kihara H. A monoclonal antibody that triggers deacylation of an intermediate thrombin-antithrombin III complex. J Biol Chem 1989; 264:13736–13739.

96. Asakura S, Hirata H, Okazaki H, Hashimoto-Gotoh, Matsuda M. Hydrophobic residues 382–386 of antithrombin III, Ala-Ala-Ala-Ser-Thr, serve as the epitope for an antibody which facilitates hydrolysis of the inhibitor by thrombin. J Biol Chem 265:5135–5138.
97. Keijer J, Ehrlich HJ, Linders M, Preissner KT, Pannekoek H. Vitronectin governs the interaction between plasminogen activator inhibitor 1 and tissue-type plasminogen activator. J Biol Chem 1991; 266:10700–10707.
98. Ehrlich HJ, Keijer J, Preissner KT, Gebbink RK, Pannekoek H. Functional interaction of plasminogen activator inhibitor type 1 (PAI-1) and heparin. Biochemistry 1991; 30:1021–1028.
99. Carrell RW, Evans DLI, Stein PE. Mobile reactive centre of serpins and the control of thrombosis. Nature 1991; 353:576–578.
100. Bjork I, Ylinenjarvi K, Olson ST, Bock PE. Conversion of antithrombin from an inhibitor of thrombin to a substrate with reduced heparin affinity and enhanced conformational stability by binding of a tetradecapeptide corresponding to the P1 to P14 region of the putative reactive bond loop of the inhibitor. J Biol Chem 1992; 267:1976–1982.
101. Delarue M, Samama JP, Mourey L, Moras D. Crystal structure of bovine AT-III. Acta Crystallogr B 1990; 46:550–556.
102. Engh RA, Wright HT, Huber R. Modeling the intact form of the α-1-proteinase inhibitor. Protein Eng 1990; 3:469–477.
103. Carrell RW. Antithrombin III and heparin. Br J Hematol 1992; 82:189–193.
104. Baumann U, Huber R, Bode W, Grosse D, Lesjak M. Crystal structure of α-1-antichymotrypsin at 2.7 A resolution and its comparison with other serpins. J Mol Biol 1991; 218:595–606.
105. Owen MC, Brennan SO, Lewis JH, Carrell RW. Mutation of antitrypsin to antithrombin. Alpha 1-antitrypsin Pittsburgh (358 Met leads to Arg) a fatal bleeding disorder. N Engl J Med 1983; 390:694–698.
106. George PM, Pemberton P, Bathurst IC, et al. Characterization of antithrombins produced by active site mutagenesis of human α-1-antitrypsin expressed in yeast. Blood 1989; 73:490–496.
107. Heeb MJ, Bischoff R, Courtney M, Griffin JH. Inhibition of activated protein C by recombinant α-1-antitrypsin variants with substitution of arginine or leucine for methionine 358. J Biol Chem 1990; 265:2365–2369.
108. Misumi Y, Ohkubo K, Sohda M, Takami N, Oda K, Ikehara Y. Intracellular processing of complement pro-C3 and proalbumin is inhibited by rat α-1-antiprotease variant (Met^{352} to Arg) in transfected cells. Biochem Biophys Res Commun 1990; 171:236–242.
109. Vidaud D, Emmerich J, Alhenc-Gelas M, Yvart J, Fiessinger JN, Aiach M. Met^{358} to Arg mutation of alpha-1-antitrypsin is associated with protein C deficiency in a patient with a mild bleeding tendency. J Clin Invest 1992; 89:1537–1543.
110. Rubin H, Wang Z, Nickbarg EB, et al. Cloning, expression, purification, and biological activity of recombinant native and variant human α-1-antichymotrypsins. J Biol Chem 1990; 265:1199–1207.
111. Derechin VM, Blinder MA, Tollefsen DM. Substitution of arginine for Leu^{444} in the reactive site of heparin cofactor II enhances the rate of thrombin inhibition. J Biol Chem 1990; 265:5623–5628.
112. Lawrence DA, Strandberg L, Ericson J, Ny T. Structure-function studies of the serpin plasminogen activator inhibitor type 1: analysis of chimeric strained loop mutants. J Biol Chem 1990; 265:20293–20301.
113. Holmes WE, Lijnen HR, Collen D. Characterization of recombinant human α-2-antiplasmin and of mutants obtained by site-directed mutagenesis of the reactive site. Biochemistry 1987; 26:5133–5140.
114. Hopkins PCR, Carrell RW, Stone SR. Effects of mutations in the hinge region of serpins. Biochem 1993; 32:7650–7657.

18

Heparin Cofactor II

Frank C. Church, Rebecca A. Shirk,* and Jeanne E. Phillips†
University of North Carolina at Chapel Hill, Chapel Hill, North Carolina

I. INTRODUCTION, HISTORICAL PERSPECTIVE, AND CHARACTERISTICS OF THE PROTEIN

The normal and pathological process of blood coagulation depends on the activity of proteinases (1,2). A variety of blood plasma proteinase inhibitors have been found to be regulators of hemostasis. Heparin is a highly negatively charged glycosaminoglycan that has been widely used as an anticoagulant drug (Fig. 1) (3–6). Heparin interacts with a number of plasma proteins, including two plasma serine proteinase inhibitors: antithrombin (historically known as antithrombin III) and heparin cofactor II (also called heparin cofactor A, antithrombin BM, dermatan sulfate cofactor, and human leuserpin-2; for further review see Refs. 7–9). Heparin cofactor II and antithrombin are members of the serine proteinase inhibitor (serpin) family of proteins, of which α_1-proteinase inhibitor is the prototype (10,11).

The therapeutic action of unfractionated heparin is believed to depend primarily on heparin-accelerated inhibition of coagulation proteinases by heparin binding serpins, especially antithrombin (3–6). Inhibition of α-thrombin by both heparin cofactor II and antithrombin is greatly accelerated (in some instances > 1000-fold) in the presence of heparin, but several differences are apparent. Antithrombin inhibits most of the coagulation pathway enzymes, and its activity can be accelerated by heparin, heparan sulfate, and heparan sulfate proteoglycans (Fig. 2) (8,9). In contrast, heparin cofactor II inhibits only thrombin in coagulation, and a variety of glycosaminoglycans, including heparin and dermatan sulfate, dermatan sulfate proteoglycans (decorin and biglycan), and other polyanions are capable of enhancing this inhibition (Fig. 2) (7,9). The significance of antithrombin is documented by the incidence of thrombotic disease in patients with antithrombin deficiencies or abnormal antithrombin molecules (12–14). A similar clinical

* Present affiliation: Bowman Gray School of Medicine, Winston-Salem, North Carolina.
† Present affiliation: Emory University School of Medicine, Atlanta, Georgia.

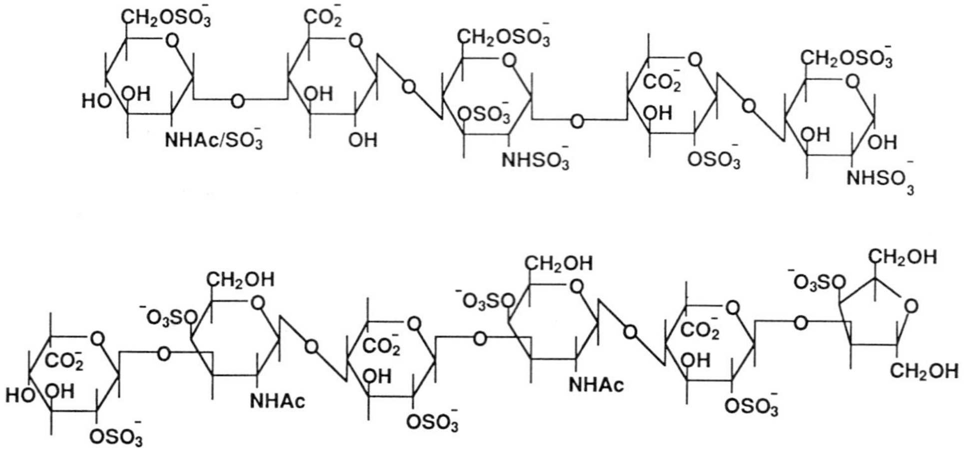

Figure 1 High-affinity oligosaccharide fragment sequence for (top) heparin, which binds antithrombin, and (bottom) dermatan sulfate, which binds heparin cofactor II. Please refer to References 3, 4, and 73 for further details concerning these glycosaminogycan fragments.

correlation is much weaker for heparin cofactor II; thus, it is difficult to assign a discrete physiological role for this serpin (15–20).

In 1974, Briginshaw and Shanberge first isolated two noninterconvertible proteins from human plasma that were identified as heparin cofactor B (which was antithrombin) and a new protein designated heparin cofactor A (21). Subsequently, in 1981–1983, Tollefsen et al. (22,23), Griffith et al. (24), and Wunderwald et al. (25) showed that

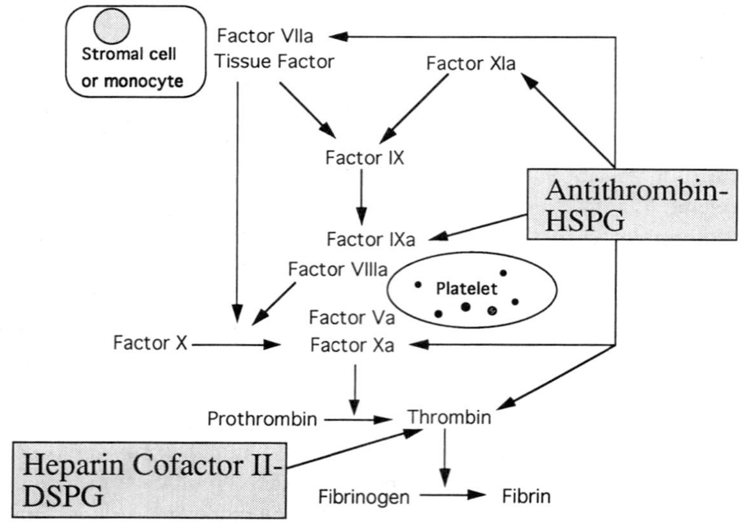

Figure 2 Regulation of blood coagulation by heparin cofactor II and antithrombin. DSPG, dermatan sulfate proteoglycan; HSPG, heparan sulfate proteoglycan. Please refer to References 1 and 2 for a detailed overview of blood coagulation.

heparin cofactor II was immunologically distinct and had unique reaction properties that distinguished it from antithrombin.

The molecular weight of plasma-derived heparin cofactor II was found to be 65,600 daltons by sedimentation equilibrium measurement (23). By sodium dodecyl sulfate–polyacrylamide gel electrophoresis, the apparent molecular weight can range from about 66,000 to 78,000, depending upon electrophoresis conditions (7,23,26–28). Heparin cofactor II is larger than its homolog antithrombin because of an amino-terminal extension (11,29,30). The mature protein has 480 amino acid residues (29,30). Two tyrosine residues in the amino-terminal acidic domain are sulfated during biosynthesis in hepatocytes (31). Heparin cofactor II contains three cysteine residues (29,30), which were shown to be accessible to sulfhydryl reagents and thus are not involved in disulfide pairing (32). Heparin cofactor II is a glycoprotein, containing at least 10% carbohydrate by weight (23,33). cDNA sequence analysis indicates there are three probable sites for N-glycosylation (29,30). The isoelectric point ranges from 4.9 to 5.3, which suggests microheterogeneity as a result of differences in carbohydrate content (23,33). An amino acid sequence alignment of heparin cofactor II and antithrombin compared with α_1-proteinase inhibitor is shown in Figure 3. There is at least 25% overall homology between heparin cofactor II compared with antithrombin and α_1-proteinase inhibitor.

II. PROTEINASE INHIBITION

Like other serpins, heparin cofactor II contains a sequence of residues termed the reactive-site loop that is accessible to proteolytic attack (Fig. 4) (10,11). A target proteinase recognizes the reactive-site bond, termed P1-P1′ using the nomenclature of Schechter and Berger (34). Leucine 444-serine 445 is the reactive-site bond in heparin cofactor II and is a potential substrate for thrombin. However, proteolysis does not proceed to completion, and the proteinase remains associated with the inhibitor in a complex that was recently shown to be a tetrahedral intermediate (using α_1-proteinase inhibitor and elastase) (35). The complex is stable to denaturants and renders the inhibitor and proteinase inactive. The reactive-site sequences of heparin cofactor II and three other serpins are shown in Figure 4.

The dissimilarity of reactive-site sequences among serpins is a well-documented example of sequence hypervariability in a protein family (10,11,36). The Leu-Ser reactive-site bond in heparin cofactor II is an unlikely substrate for thrombin, which has a preference for Arg-*X*-amino acids bonds (27,29,30). This sequence in heparin cofactor II is expected to be a suitable substrate for an enzyme, such as chymotrypsin or cathepsin G, and this has in fact been shown (37–39). Interestingly, the reactive-site sequence in α_1-antichymotrypsin is also Leu-Ser (11), yet this serpin does not exhibit any inhibitory activity for thrombin. When the reactive-site leucine of recombinant heparin cofactor II is replaced with arginine, the thrombin inhibition rate is increased about 100-fold (40). The ability of heparin cofactor II to react with proteinases of different specificity may reflect the distinct physiological role of the inhibitor.

III. GLYCOSAMINOGLYCAN BINDING SITE

The heparin binding site of heparin cofactor II was predicted and identified primarily by its homology to antithrombin (17,29,30,41–46). Heparin cofactor II contains a region around the “D helix,” lysine 165 to phenylalanine 195, that shows extensive homology to the lysine 107 to lysine 136 region of antithrombin (Figs. 5 and 6) (17,29,30,43–46). This region is populated with positively charged amino acids that can interact with various glycosaminoglycans and polyanions. Collectively, these studies implicate lysyl residues 173 and 185 and arginyl residues 184 and 193 in heparin binding, and lysyl residue 185 and arginyl residues 189, 192, 193, and possibly 184 in dermatan sulfate binding to heparin cofactor II (Fig. 6). It appears from these studies that the heparin cofactor II binding sites for heparin and dermatan sulfate are not identical but are overlapping.

IV. EFFECT OF HEPARIN AND GLYCOSAMINOGLYCAN AND POLYANION SPECIFICITY

The mechanism of heparin action appears to be different for heparin cofactor II than for antithrombin. As discussed elsewhere in this volume, antithrombin exhibits high-affinity

```
h_α1pi : MPSSVS.............................WGILLLAGLCCLVPVSLAEDPQGDAA
h_hc2  : -KH-LNalliflliitsawggskgpldqlekgg..ETAQSADPQWEQLNNKNLSM-LLP-D
h_at3  : -Y-N-IgtvtsgkrkvyllslllIigfwdcvtchgSPVDICTAKPRDI-MNPMCIYRSPEK

h_α1pi : QKTDTSHHDQDHPT..............................................
h_hc2  : FHKENTVTNDWI-Egeedddyldlekifsedddyidivdslsvsptdsdvsagnilqlfh
h_at3  : KA-EDEGSE-KI-Eatnrr.........................................

h_α1pi : ....FNKITPNLAEFAFSLYRQLAHQSN.STNIFFSPVSIATAFAMLSLGTKADTHDEIL
h_hc2  : gksrIQRLNILN-K---N---V-KD-V-tFD---IA--G-S--MG-I---L-GE--EQVH
h_at3  : ....VWELSKANSR--TTF-QH--DSK-dND---L--L--S-----TK--ACN--LQQLM

h_α1pi : EGLNFNLTEIPEAQIHEGFQELL..RTLNQPDSQLQLTTGNGLFLSEGLKLVDKFLEDVK
h_hc2  : SI-H-KDFVNASSKYEITTIHN-frKLTHRLFRRNF...-YT-RSVND-YIQKQ-PILLD
h_at3  : -VFK-DTISEKTSDQIHF-FAK-nc-LYRKANKSSK-VSA-R--GDKS-TFNETYQDISE

h_α1pi : KLYHSEAFTVNFGDTEEAKK.....QINDYVEKGTQGKIVDLVKELDRDTVFALVNYIFF
h_hc2  : FRTKVREYYFAEAQIADFSDpafisKT-NHIM-L-K-L-K-ALENI-PA-QMMIL-C-Y-
h_at3  : LV-GAKLQPLD-KENA-QSRaainkWVSNKT-GRITDV-PS..EAINEL--LV---T-Y-

h_α1pi : KGKWERPFEVKDTEEEDFHVDQVTTVKVPMMKRLGMFNIQHCKKLSSWVLLMKYLGNATA
h_hc2  : --S-VNK-P-EM-HNHN-RLNEREV---S--QTK-N-LAANDQE-DCDI-QLE-V-GISM
h_at3  : --L-KSK-SPEN-RK-L-YKADGESCSAS--YQE-K-RYRRVAEGTQVLE-PFKGDDI-M

h_α1pi : IFFLPDEGK.LQHLENELTHDIITKFLENEDRRSASLHLPKLSITGTYDLKSVLGQLGIT
h_hc2  : LIVV-HKMSgMKT--AQ--PRVVERWQKSMTN-TREVL---FKLEKN-N-VES-KLM--R
h_at3  : VLI--KPE-s-AKV-K---PEVLQEW-DELEEMMLVV-M-RFR-EDGFS--EQ-QDM-LV

h_α1pi : KVFSNGADLSGVTEEAPLK...LSKAVHKAVLTIDEKGTEAAGAMFLEAIPMSIPPE...
h_hc2  : ML-DKNGNMA-ISDQRIAIdlf....K-QGTI-VN-E--Q-TTVTTVGFM-L-TQVR...
h_at3  : DL--PEKSKLPGIVAEGRDdlyV-D-F---F-EVN-E-S---ASTAVVIAGR-LN-Nrvt

h_α1pi : VKFNKPFVFLMIEQNTKSPLFMGKVVNPTQK
h_hc2  : FTVDR--L--IY-HR-SCL----R-A--SRS
h_at3  : F-A-R--LVFIR-VPLNTII---R-A--CV-
```

Figure 3 Amino acid sequence alignment of heparin cofactor II (h_hc2) and antithrombin (h_at3) with α_1-proteinase inhibitor (h_α1pi): (dashed line) aligned identical residues; (uppercase) aligned nonidentical residues; (lowercase) unaligned residues; (dotted line) gap. Sequences were obtained from Genbank and aligned using the method of Altschul in the program EuGene (115,116). Heparin cofactor II has a 19 amino acid signal peptide, and the mature protein begins with the sequence GSKG (single-letter amino acid abbreviations); antithrombin has a 32 amino acid signal peptide, and the mature protein begins with the sequence HGSP; and α_1-proteinase inhibitor has a 24 amino acid signal peptide, and the mature protein begins EDPQ.

Reactive Site	P10					P5				P1-P1′ ↓				P5′
HC2	A	T	T	V	T	T	V	G	F	M	P	L-S	T	Q	V	R
AT3	A	A	A	S	T	A	V	V	I	A	G	R-S	L	N	P	N
PN1	A	S	A	A	T	T	A	I	L	I	A	R-S	S	P	P	W
α1AC	A	S	A	A	T	A	V	K	I	T	L	L-S	A	L	V	E

Figure 4 Reactive-site sequences of the serpins heparin cofactor II (HC2), antithrombin (AT3), α_1-antichymotrypsin (α_1AC), and protease nexin 1 (PN1). Sequences are from Reference 11. The arrow indicates the reactive-site peptide bond, termed P1-P1′, using the nomenclature of Schechter and Berger (34), that is attacked by the target proteinase.

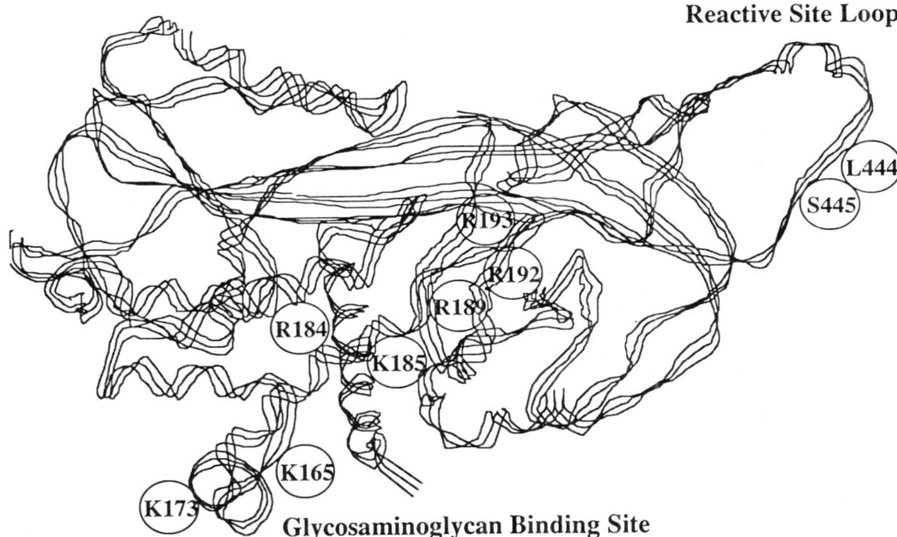

Figure 5 A molecular model of the glycosaminoglycan binding site of heparin cofactor II and its reactive-site loop. A three-dimensional model of heparin cofactor II was derived by computer-assisted mutagenesis using the α_1-proteinase inhibitor backbone determined from its crystal structure (117) and by substituting heparin cofactor II-specific residues. The first 19 residues of α_1-proteinase inhibitor lack coordinates in the crystal structure, and heparin cofactor II contains ~80 more N-terminal amino acids, such that the first 100 residues of heparin cofactor II are ignored in the simulation. It has been suggested that glutamic acid 342 (α_1-proteinase inhibitor numbering and a conserved residue in heparin cofactor II) is a critical pivot in the movement of the reactive site after cleavage (11,117), and recent studies of the structure of cleaved and intact ovalbumin agree with this hypothesis (118,119). The reactive site was reannealed by pivoting the strand of β sheet from this Glu to the P1 Leu back toward the P1′ Ser, as in ovalbumin, and the molecule was minimized using SYBYL (120). The molecule was further minimized using AMBER (Version 3.1, Singh UC, Weiner PK, Caldwell JW, Kollman PA. Department of Pharmaceutical Chemistry, University of California, San Francisco, 1988) (46,52,120). Two insertions of four and one residues, respectively, occur in HC (compared with α_1-proteinase inhibitor) at the beginning of the glycosaminoglycan binding site (120). These were made using the LOOP command in SYBYL that searches a database of molecular structures for similar regions. After insertion, the resulting molecule was minimized using AMBER (120). (This molecular model was graciously provided by Dr. Herbert C. Whinna.)

heparin binding for a pentasaccharide ($K_d \simeq 2 \times 10^{-8}$ M) (47–49). Heparin binding induces a conformational change in antithrombin that is marked by changes in intrinsic fluorescence and is accompanied by an increase in antithrombin reactivity toward factor Xa (47–50). However, the pentasaccharide alone is insufficient for accelerated inhibition of thrombin by antithrombin (51). A requirement for a longer heparin molecule, at least 16 saccharide units (52), suggests that additional heparin binding may involve subtle conformational changes or neutralization of charges near the reactive site (53). It is also possible that a larger heparin molecule, which can bind to both inhibitor and proteinase, is important for optimizing the alignment of the proteinase active site with the inhibitor reactive site as the reactants approach each other (51,52). The ability of heparin to bind both inhibitor and proteinase is the basis for the ''ternary complex'' model of heparin-enhanced proteinase inhibition (54–59). At low concentrations of heparin, the reactants are brought into closer proximity, thus increasing the reaction rate. At very high heparin concentrations, heparin sequesters the reactants and the reaction rate decreases. Thus, the heparin concentration dependence of the inhibition rate is a bimodal curve. The ternary complex model applies to heparin-enhanced antithrombin inhibition of thrombin and factors IXa and XIa, but not factor Xa.

Heparin cofactor II also appears to obey the ternary complex mechanism for heparin-accelerated thrombin inhibition (7,43,52,60). Additionally, the narrow proteinase specificity of heparin cofactor II can now be explained as a result of its unique amino terminus (29,30,44,61), which has a sequence of acidic residues including two sulfated tyrosine residues (31,62): G_{54}EEDDDYLDLEKIFSEDDDYID_{75}. The heparin cofactor II amino-terminal acid sequence is similar to the carboxyl terminus of the leech anticoagulant hirudin from residues 56 to 65 (FEEIPEEYLQ), which also contains a sulfated tyrosine residue. Hortin et al. (62) found that a synthetic peptide derived from this amino acid sequence in heparin cofactor II inhibited the fibrinogen clotting activity of thrombin. The acidic region of heparin cofactor II may interact with its own heparin binding site and then be displaced by heparin or another glycosaminoglycan or polyanion and serve as a ligand for anion binding exosite I of thrombin (44,45,61,63–65). Heparin cofactor II mutants that lack portions of the amino terminus (44,45,61) or thrombin derivatives altered in anion binding exosite I (63–70), cannot participate in this reactive-site–independent binding and exhibit virtually no glycosaminoglycan-enhanced proteinase inhibition.

Heparin cofactor II is unique among the heparin binding serpins in its ability to bind dermatan sulfate in a manner that accelerates thrombin inhibition (28,71). McGuire and Tollefsen (72) showed that dermatan sulfate from fibroblasts and smooth muscle cells accelerate heparin cofactor II activity. A dermatan sulfate hexasaccharide has been iden-

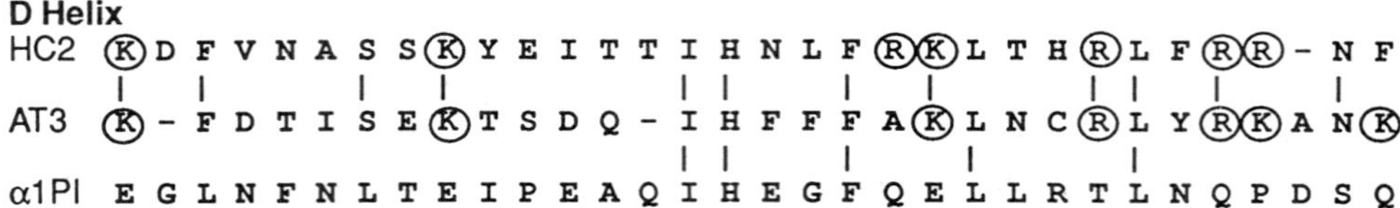

Figure 6 Putative glycosaminoglycan binding sequences in the serpin heparin cofactor II (HC2) compared with antithrombin (AT3) and α_1-proteinase inhibitor (α_1PI). The amino acid sequence contained around the D helix of heparin cofactor II 165–195 and antithrombin 107–136 is aligned with the corresponding regions in α_1-proteinase inhibitor (a nonheparin binding serpin). Basic residues in heparin cofactor II and antithrombin are circled.

tified as the high-affinity binding sequence (Fig. 1) (73). Additionally, two dermatan sulfate proteoglycans, biglycan and decorin, have the ability to bind heparin cofactor II (74). Data from numerous sources suggest that the physiological activity of heparin cofactor II is mediated through dermatan sulfate, probably in the form of a dermatan sulfate proteoglycan (7,9,28,72–77). This is further supported by evidence for dermatan sulfate proteoglycan-collagen interactions in many connective tissues (74,78–84). For instance, type V collagen is found in a wide variety of tissues, both adjacent to basement membranes and in fibrous stroma, and has been shown to be adjacent to peripheral capillaries and at the surface of blood vessels denuded of their endothelial lining (85). Local in vivo concentrations of dermatan sulfate proteoglycans may be significantly larger because of a greater ratio of extracellular matrix and cell surface to extracellular fluid (72). Therefore, following endothelial injury, plasma heparin cofactor II may be specifically ''anchored'' to dermatan sulfate chains in the extracellular matrix and regulate the multiple biological activities of thrombin. Additionally, thrombin interacts with dermatan sulfate in basement membrane derived from endothelial cells (86), which further implies a role for heparin cofactor II in regulating thrombin in the presence of dermatan sulfate proteoglycans.

The glycosaminoglycan specificity of heparin cofactor II is much broader than that of antithrombin. In fact, almost any polyanion (7), including polyphosphates (87), polysulfates (60), and polycarboxylates (88), can accelerate thrombin inhibition by heparin cofactor II. This broad specificity can be partially explained by the secondary binding mechanism of heparin cofactor II just described, in which polyanion binding to the glycosaminoglycan binding site displaces the acid amino terminus of heparin cofactor II, which then binds thrombin.

V. LEUKOCYTE CHEMOATTRACTANT ACTIVITY

Neutrophils and monocytes are important cellular components of tissue injury and play obvious roles in infection, inflammation, and wound healing (89,90). Proteolytic enzymes and oxidants are released during activation of neutrophils (89,90). A potential physiological role for heparin cofactor II, unrelated to proteinase inhibition, may depend on the generation of peptides that modulate leukocyte functions. Limited cleavage of heparin cofactor II by activated neutrophils, purified neutrophil elastase, or cathepsin G releases peptides with discrete biological activity for leukocytes (39,91–95). Proteolysis of heparin cofactor II liberates the amino-terminal acid peptide, aspartic acid 39 to isoleucine 66, with chemoattractant activity for both neutrophils and monocytes (96). Furthermore, the products of a heparin cofactor II-neutrophil elastase digest stimulate neutrophil migration in vivo in mice (96). Henry (97) demonstrated that leukocytes accumulate in thrombi in vivo. Plow (98) showed that active elastase is released from neutrophils during blood coagulation. Carp and Janoff (99) noted that there are several ways for neutrophil elastase to escape its physiological inhibitor, α_1-proteinase inhibitor. Thus, an in vivo scenario may exist in which active neutrophil elastase within a thrombus can hydrolyze heparin cofactor II further to recruit additional leukocytes to the site of injury.

VI. MOLECULAR GENETICS

A. Heparin Cofactor cDNA

A 2.1 kb cDNA clone encoding heparin cofactor II (termed human leuserpin 2) was first isolated from a human liver cDNA library by Ragg and coworkers in 1986 (29). It was shown to be 25–28% homologous with antithrombin, α_1-proteinase inhibitor, and α_1-antichymotrypsin. Inhorn and Tollefsen also reported a partial cDNA clone encoding heparin cofactor II in 1986 (100). A full-length cDNA was isolated from a human liver λgt11 cDNA library by Blinder et al. (30), consisting of 1525 bp and a 3′ noncoding region of 654 bp, and a signal peptide of 19 amino acids residues was deduced from the nucleotide sequence. Three differences were observed when the published cDNA sequences were compared (29,30,100): there is a T to C discrepancy at position 1608 in the 3′-untranslated region, a C to T at position 532 resulting in no amino acid change, and a G to A at 738 changing an arginine to a lysine. Recombinant heparin cofactor II has been expressed in *Escherichia coli* (17,40,43,46,61) and COS cells (44,45), although fully active material appears to be difficult to obtain.

B. Heparin Cofactor II Gene

The heparin cofactor II gene located on chromosome 22 was mapped to 22q11 (101). The gene includes 1749 bp 5′ flanking sequence, five exons, four introns, and 476 bp

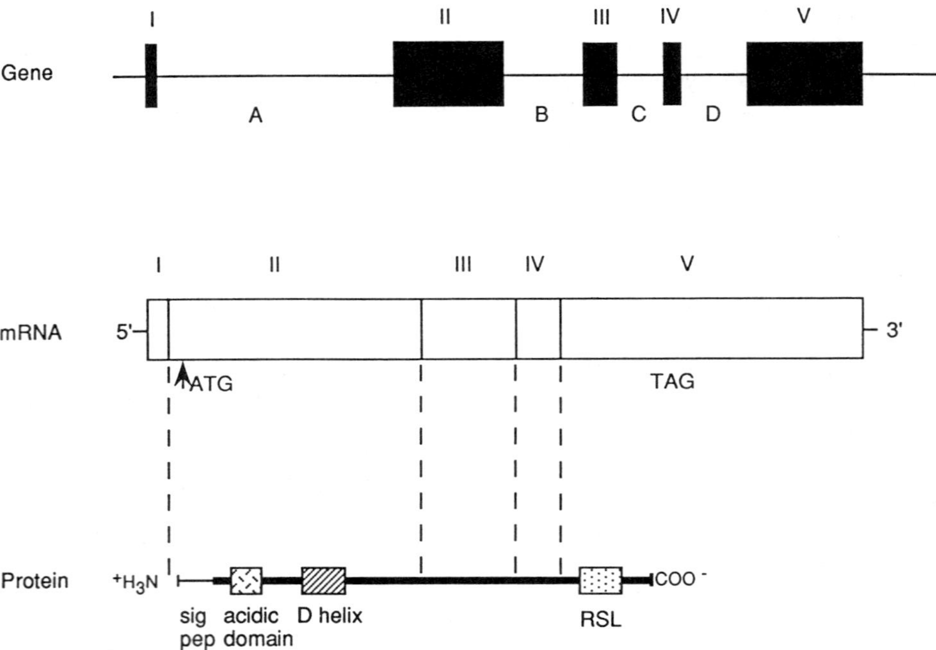

Figure 7 Heparin cofactor II gene and mRNA organization with corresponding protein domain structure. The gene for heparin cofactor II is located on chromosome 22 (30) and has been mapped to position 22q11 (101). It contains 1749 bp 5′ flanking sequence, five exons (designated I–V) and four introns (A–D), and 476 bp 3′ to the polyadenylation site (101). Exon I is composed of a 5′-untranslated sequence. Exon II contains a 16 bp untranslated region, the ATG start codon, sequences that encode a 19 amino acid signal peptide (**sig pep**), and residues 1–277 of the mature protein (which includes the acid domain and D helix). Exons III and IV encode residues 278–369 and 370–417, respectively. Exon V encodes residues 418–480 of the carboxyl terminus (which includes the reactive-site loop, **RSL**), followed by the TAG stop codon and a 3′ noncoding sequence.

3′ to the polyadenylation site (Fig. 7). Ten complete and one partial Alu repeat were also identified. Ragg and Prelbisch (102) also reported a partial gene sequence for heparin cofactor II with the same proposed exon-intron organization. These investigators indicated the positions of the introns to be highly conserved in the genes for heparin cofactor II, α_1-proteinase inhibitor, and angiotensinogen. A TATA-like sequence at −67 to −61 and an inverted CCAAT sequence at −75 to −71 were identified, although a transcription initiation site was not determined (101). Restriction fragment length polymorphisms were identified with *Bam*HI, *Hind*III and *Msp*I by Southern blot analysis (30,101,103).

C. Hereditary Heparin Cofactor II Deficiency

The normal concentration of heparin cofactor II in adult plasma is approximately 1.2 ± 0.2 μM (104). This level may decrease in patients with disseminated intravascular coagulation and liver disease (15,104). Heparin cofactor II deficiency is apparently inherited as an autosomal dominant disorder. Several people have been identified with antigen and activity levels near 50%. Heparin cofactor II deficiency has been reported to result in a thrombotic episode approximately 36% of the time (14 of 52 patients identified). Although some hereditary deficiencies or abnormalities of heparin cofactor II have been described, other studies of healthy volunteers and patients revealed that this deficiency is prevalent in both populations (15–20,105–114). Some clinical data suggest that heparin cofactor II deficiency is a high-risk factor for a thrombotic episode, but further detailed studies that eliminate the possibility of other protein abnormalities or deficiencies (such as protein C, protein S, or APC cofactor 2) are needed to support this hypothesis fully.

One of the described heparin cofactor II variants that has been characterized at the molecular level is heparin cofactor II Oslo (17,107). In this mutant, a G to A transition results in a substitution of histidine for arginine at position 189 (17). This mutation was introduced into the cDNA for heparin cofactor II and expressed in *E. coli* (17). Both recombinant wild-type heparin cofactor II and the Olso variant required similar concentrations of heparin to accelerate thrombin inhibition. In contrast, recombinant heparin cofactor II Oslo requires 100 times more dermatan sulfate than wild-type recombinant heparin cofactor II to produce equal thrombin inhibition. Characterization of heparin cofactor II Oslo was useful in identifying the function of arginine 189 in dermatan sulfate interaction.

ACKNOWLEDGMENTS

Our work described in this chapter was supported by Research Grants HL-32656 (to F. C. Church) and F32-HL-08413 (to J. E. Phillips) from the National Institutes of Health and by a Grant-in-Aid (to F. C. Church) from the American Heart Association.

REFERENCES

1. Roberts HR, Lozier JN. New perspectives on the coagulation cascade. Hosp Pract 1992; 97–111.
2. Furie B, Furie BC. Molecular and cellular biology of blood coagulation. N Engl J Med 1992; 326:800–806.
3. Casu B. Structure of heparin and heparin fragments. Ann NY Acad Sci 1989; 556:1–17.
4. Conrad HE. Structure of heparin sulfate and dermatan sulfate. Ann NY Acad Sci 1989; 556: 18–28.

5. Hemker HC, Beguin S. The mode of action of heparins in vitro and in vivo. Adv Exp Med Biol 1992; 313:221–230.
6. Samama M-M. Treatment of deep vein thrombosis (DVT) with low molecular weight heparins (LMWH). Adv Exp Med Biol 1992; 313:275–281.
7. Pratt CW, Whinna HC, Meade JM, Treanor RE, Church FC. Physicochemical aspects of heparin cofactor II. Ann NY Acad Sci 1989; 556:104–114.
8. Pratt CW, Church FC. Antithrombin: structure and function. Semir Hematol 1991; 28:3–9.
9. Pratt CW, Church FC. General features of the heparin-binding serpins antithrombin, heparin cofactor II, and protein C inhibitor. Blood Coag Fib 1993; 4:479–490.
10. Carrell RW, Travis J. α_1-Antitrypsin and the serpins: variations and countervariations. Trends Biochem Sci 1985; 10:20–24.
11. Huber R, Carrell RW. Implications of the three-dimensional structure of α_1-antitrypsin for structure and function of serpins. Biochemistry 1989; 28:8951–8966.
12. Egeberg O. Inherited antithrombin III deficiency and thromboembolism. Thromb Diath Haemorrh 1965; 13:516–530.
13. Abildgaard U. Antithrombin and related inhibitors of blood coagulation. Recent Adv Blood Coag 1981; 3:151–173.
14. Lane DA, Olds R, Thein S-L. Antithrombin and its deficiency states. Blood Coag Fib 1992; 3:315–341.
15. Tran TH, Marbet GA, Duckert F. Association of heredity heparin cofactor II deficiency with thrombosis. Lancet 1985; 2(8452):413–414.
16. Sie P, Dupouy D, Pichon J, Boneu B. Constitutional heparin co-factor II deficiency associated with recurrent thrombosis. Lancet 1985, 2(8452):414–416.
17. Blinder MA, Andersson TR, Abildgaard U, Tollefsen DM. Heparin cofactor II Oslo. Mutation of Arg-189 to His decreases the affinity for dermatan sulfate. J Biol Chem 1989; 264: 5128–5133.
18. Simioni P, Lazzaro AR, Coser E, Salmistraro G, Girolami A. Heredity heparin cofactor II deficiency and thrombosis: report of six patients belonging to two separate kindreds. Blood Coag Fib 1990; 1:351–356.
19. Andersson TR, Larsen ML, Handeland GF, Abildgaard U. Heparin cofactor II activity in plasma: application of an automated assay method to the study of a normal adult population. Scand J Haematol 1986; 36:96–102.
20. Bertina RM, van der Linden IK, Engesser L, Muller HP, Brommer EJ. Hereditary heparin cofactor II deficiency and the risk of development of thrombosis. Thromb Haemost 1987; 57:196–200.
21. Briginshaw GF, Shanberge JN. Identification of two distinct heparin cofactors in human plasma. II. Inhibition of thrombin and activated factor X. Thromb Res 1974; 4:463–477.
22. Tollefsen DM, Blank MK. Detection of a new heparin-dependent inhibitor of thrombin in human plasma. J Clin Invest 1981; 68:589–596.
23. Tollefsen DM, Majerus DW, Blank MK. Heparin cofactor II: purification and properties of a heparin-dependent inhibitor of thrombin in human plasma. J Biol Chem 1982; 257:2162–2169.
24. Griffith MJ, Carraway T, White GC, Dombrose FA. Heparin cofactor activities in a family with hereditary antithrombin III deficiency: evidence for a second heparin cofactor in human plasma. Blood 1983; 61:111–118.
25. Wunderwald P, Schrenk WJ, Port H. Antithrombin BM from human plasma: an antithrombin binding moderately to heparin. Thromb Res 1982; 25:177–191.
26. Griffith MJ, Noyes CM, Church FC. Reactive site peptide structural similarity between heparin cofactor II and antithrombin III. J Biol Chem 1985; 260:2218–2225.
27. Griffith MJ, Noyes CM, Tyndall JA, Church FC. Structural evidence for leucine at the reactive site of heparin cofactor II. Biochemistry 1985; 24:6777–6782.
28. Tollefsen DM, Pestka CA, Monafo WJ. Activation of heparin cofactor II by dermatan sulfate. J Biol Chem 1983; 258:6713–6716.

29. Ragg H. A new member of the plasma protease inhibitor gene family. Nucleic Acids Res 1986; 14:1073–1087.
30. Blinder MA, Marasa JC, Reynolds CH, Deaven LL, Tollefsen DM. Heparin cofactor II: cDNA sequence, chromosome localization, restriction fragment length polymorphism, and expression in *Escherichia coli*. Biochemistry 1988; 27:752–759.
31. Hortin G, Tollefsen DM, Strauss AW. Identification of two sites of sulfation of human heparin cofactor II. J Biol Chem 1986; 261:15827–15830.
32. Church FC, Meade JB, Pratt CW. Structure-function relationships in heparin cofactor II: spectral analysis of aromatic residues and absence of a role for sulfhydryl groups in thrombin inhibition. Arch Biochem Biophys 1987; 259:331–340.
33. Kim Y, Lee K, Linhardt RJ. Microheterogeneity of plasma glycoproteins heparin cofactor II and antithrombin III and their carbohydrate analysis. Thromb Res 1988; 51:97n104.
34. Schechter I, Berger A. On the size of the active site in proteases I. Papain. Biochem Biophys Res Commun 1967; 27:157–162.
35. Matheson NR, van Halbeek H, Travis J. Evidence for a tetrahedral intermediate complex during serpin-proteinase interactions. J Biol Chem 1991; 266:13489–13491.
36. Creighton TE, Darby NJ. Functional evolutionary divergence of proteolytic enzymes and their inhibitors. Trends Biochem Sci 1989; 14:319–324.
37. Church FC, Noyes CM, Griffith MJ. Inhibition of chymotrypsin by heparin cofactor II. Proc Natl Acad Sci USA 1985; 82:6431–6434.
38. Parker KA, Tollefsen DM. The protease specificity of heparin cofactor II. Inhibition of thrombin generated during coagulation. J Biol Chem 1985; 260:3501–3505.
39. Pratt CW, Tobin RL, Church FC. Interaction of heparin cofactor II with neutrophil elastase and cathepsin G. J Biol Chem 1990; 265:6092–6097.
40. Derechin VM, Blinder MA, Tollefsen DM. Substitution of arginine for Leu^{444} in the reactive site of heparin cofactor II enhances the rate of thrombin inhibition. J Biol Chem 1990; 265:5623–5628.
41. Church FC, Griffith MJ. Evidence for essential lysines in heparin cofactor II. Biochem Biophys Res Commun 1984; 124:745–751.
42. Church FC, Villanueva GB, Griffith MJ. Structure-function relationships in heparin cofactor II: chemical modification of arginine and tryptophan and demonstration of a two-domain structure. Arch Biochem Biophys 1986; 246:175–184.
43. Blinder MA, Tollefsen DM. Site-directed mutagenesis of arginine 103 and lysine 185 in the proposed glycosaminoglycan-binding site of heparin cofactor II. J Biol Chem 1990; 265:286–291.
44. Ragg H, Ulshofer T, Gerewitz J. On the activation of human leuserpin-2, a thrombin inhibitor, by glycosaminoglycans. J Biol Chem 1990; 265:5211–5218.
45. Ragg H, Ulshofer T, Gerewitz J. Glycosaminoglycan-mediated leuserpin-2/thrombin interaction. Structure-function relationships. J Biol Chem 1990; 265:22386–22391.
46. Whinna HC, Blinder MA, Szewczyk M, Tollefsen DM, Church FC. Role of lysine 173 in heparin binding to heparin cofactor II. J Biol Chem 1991; 266:8129–8135.
47. Nordenman B, Danielsson A, Bjork I. The binding of low-affinity and high-affinity heparin to antithrombin. Fluorescence studies. Eur J Biochem 1978; 90:1–6.
48. Olson ST, Srinivasan KR, Bjork I, Shore JD. Binding of high affinity heparin to antithrombin-III. J Biol Chem 1981; 256:11073–11079.
49. Thunberg L, Backstrom G, Lindahl U. Further characterization of the antithrombin-binding sequence in heparin. Carbohydr Res 1982; 100:393–410.
50. Chang J-Y. Binding of heparin to human antithrombin II activates selective chemical modification at lysine 236. Lys-107, Lys-125, and Lys-136 are situated within the heparin-binding site of antithrombin III. J Biol Chem 1989; 264:3111–3115.
51. Olson ST, Bjork I, Sheffer R, Craig PA, Shore JD, Choay J. Role of the antithrombin-binding pentasaccharide in heparin acceleration of antithrombin-proteinase reactions. Resolution of the antithrombin conformational change contribution to heparin rate enhancement. J Biol Chem 1992; 267:12528–12538.

52. Pratt CW, Whinna HC, Church FC. A comparison of three heparin-binding serine proteinase inhibitors. J Biol Chem 1992; 267:8795–8801.
53. Beresford CH, Owen MC. Antithrombin III. Int J Biochem 1990; 22:121–128.
54. Griffith MJ. Kinetics of the heparin-enhanced antithrombin III/thrombin reaction. J Biol Chem 1982; 257:7360–7365.
55. Griffith MJ. Heparin-catalyzed inhibitor/protease reactions: kinetic evidence for a common mechanism of action of heparin. Proc Natl Acad Sci USA 1983; 80:5460–5465.
56. Nesheim ME. A simple rate law that describes the kinetics of the heparin-catalyzed reaction between antithrombin III and thrombin. J Biol Chem 1983; 258:14708–14717.
57. Nesheim M, Blackburn MN, Lawler CM, Mann KG. Dependence of antithrombin III and thrombin binding stoichiometries and catalytic activity on the molecular weight of affinity-purified heparin. J Biol Chem 1986; 261:3214–3221.
58. Olson ST, Shore JD. Transient kinetics of heparin-catalyzed protease inactivation by antithrombin III: the reaction step limiting heparin turnover in thrombin neutralization. J Biol Chem 1986; 261:13151–13159.
59. Olson ST. Transient kinetics of heparin-catalyzed protease inactivation by antithrombin III: linkage of protease-inhibitor-heparin interactions in the reaction with thrombin. J Biol Chem 1988; 263:1698–1708.
60. Church FC, Meade JB, Treanor RE, Whinna HC. Antithrombin activity of fucoidan: the interaction of fucoidan with heparin cofactor II, antithrombin III, and thrombin. J Biol Chem 1989; 264:3618–3623.
61. Van Deerlin VMD, Tollefsen DM. The N-terminal acidic domain of heparin cofactor II mediates the inhibition of α-thrombin in the presence of glycosaminoglycans. J Biol Chem 1991; 266:20223–20231.
62. Hortin GL, Tollefsen DM, Benutto BM. Antithrombin activity of a peptide corresponding to residues 54–75 of heparin cofactor II. J Biol Chem 1989; 264:13979–13982.
63. Rogers SJ, Pratt CW, Whinna HC, Church FC. Role of thrombin exosites in inhibition by heparin cofactor II. J Biol Chem 1992; 267:3613–3617.
64. Phillips JE, Shirk RA, Whinna HC, Henriksen RA, Church FC. Inhibition of dysthrombins Quick I and II by heparin cofactor II and antithrombin. J Biol Chem 1993; 268:3321–3327.
65. Sheehan JP, Wu Q, Tollefsen DM, Sadler JE. Mutagenesis of thrombin selectively modulates inhibition by serpins heparin cofactor II and antithrombin III. Interaction with the anion-binding exosite determines heparin cofactor II specificity. J Biol Chem 1993; 268:3639–3645.
66. Bode W, Mayr I, Baumann U, Huber R, Stone SR, Hofsteenge J. The refined 1.9 ° structure of human α-thrombin: interaction with D-Phe-Pro-Arg chloromethyl ketone and significance of the Tyr-Pro-Pro-Trp insertion segment. EMBO J 1989; 8:3467–3475.
67. Rydel TJ, Ravichandran KG, Tulinsky A, et al. The structure of a complex of recombinant hirudin and human α-thrombin. Science 1990; 249:277–280.
68. Rydel TJ, Tulinsky A. Refined structure of the hirudin-thrombin complex. J Mol Biol 1991; 221:583–601.
69. Bode W, Turk D, Karshikov A. The refined 1.9-° x-ray crystal structure of D-Phe-Pro-Arg chloromethylketone-inhibited human α-thrombin: structural analysis, overall structure, electrostatic properties, detailed active-site geometry, and structure-function relationships. Protein Sci 1992; 1:426–471.
70. Whinna HC, Church FC. Interaction of thrombin with antithrombin, heparin cofactor II and protein C inhibitor. J Protein Chem 1993; 12:677–688.
71. Griffith MJ, Marbet GA. Dermatan sulfate and heparin can be fractionated by affinity for heparin cofactor II. Biochem Biophys Res Commun 1983; 112:663–670.
72. McGuire EA, Tollefsen DM. Activation of heparin cofactor II by fibroblasts and vascular smooth muscle cells. J Biol Chem 1987; 262:169–175.
73. Maimone MM, Tollefsen DM. Structure of a dermatan sulfate hexasaccharide that binds to heparin cofactor II with high affinity. J Biol Chem 1990; 265:18263–18271.

74. Whinna HC, Choi HU, Rosenberg LC, Church FC. Interaction of heparin cofactor II with biglycan and decorin. J Biol Chem 1993; 268:3920–3924.
75. Tollefsen DM, Peacock ME, Monafo WJ. Molecular size of dermatan sulfate oligosaccharides required to bind and activate heparin cofactor II. J Biol Chem 1986; 261:8854–8858.
76. Tollefsen DM, Maimone MM, McGuire EA, Peacock ME. Heparin cofactor II activation by dermatan sulfate. Ann NY Acad Sci 1989; 556:116–122.
77. Tollefsen DM. The interaction of glycosaminoglycans with heparin cofactor II: structure and activity of a high-affinity dermatan sulfate hexasaccharide. Adv Exp Med Biol 1992; 313: 167–176.
78. Scott JE. Proteoglycan-fibrillar collagen interactions. Biochem J 1988; 252:313–323.
79. Bidanset DJ, Guidry C, Rosenberg LC, Choi HU, Timpl R, Hook M. Binding of the proteoglycan decorin to collagen type VI. J Biol Chem 1992; 267:
80. Schmidt G, Hausser H, Kresse H. Interaction of the small proteoglycan decorin with fibronectin. Biochem J 1991; 280:411–414.
81. Fleischmajer R, Fisher LW, MacDonald ED, Jacobs LJ, Perlish JS, Termine JD. Decorin interacts with fibrillar collagen of embryonic and adult human skin. J Struct Biol 1991; 106: 82–90.
82. Rosenberg LC, Choi HU, Tang L-H, et al. Isolation of dermatan sulfate proteoglycans from mature bovine articular cartilages. J Biol Chem 1985; 260:6304–6313.
83. Rosenberg LC, Choi HU, Poole AR, Lewandowska K, Culp LA. Biological roles of dermatan sulphate proteoglycans. Ciba Found Symp 1986; 124:46–68.
84. Rosenberg L, Tang L, Pal S, Johnson TL, Choi HU. Proteoglycans of bovine articular cartilage. Studies of the direct interaction of link protein with hyaluronate in the absence of proteoglycan monomer. J Biol Chem 1988; 263:18071–18077.
85. Fessler JH, Fessler LI. In: Mayne R, Burgeson RE, eds. Type V Collagen. New York: Academic Press, 1987; pp. 81–103.
86. Bar-Shavit R, Eldor A, Vlodavsky I. Binding of thrombin to subendothelial extracellular matrix. J Clin Invest 1989; 84:1096–1104.
87. Church FC, Pratt CW, Treanor RE, Whinna HC. Antithrombin action of phosvitin and other phosphate-containing polyanions is mediated by heparin cofactor II. FEBS Lett 1988; 237:26–30.
88. Church FC, Treanor RE, Sherril GB, Whinna HC. Carboxylate polyanions accelerate inhibition of thrombin by heparin cofactor II. Biochem Biophys Res Commun 1987; 148:362–368.
89. Henson PM, Johnston J, R.B. Tissue injury in inflammation: oxidants, proteinases and cationic proteins. J Clin Invest 1987; 79:669–674.
90. Weiss SJ. Tissue destruction by neutrophils. N Engl J Med 1989; 320:365–376.
91. Sie P, Dupouy D, Dol F, Boneu B. Inactivation of heparin cofactor II by polymorphonuclear leukocytes. Thromb Res 1987; 47:657–664.
92. Hoffman M, Pratt CW, Brown RL, Church FC. Heparin cofactor II-proteinase reaction products exhibit neutrophil chemoattractant activity. Blood 1989; 73:1682–1685.
93. Hoffman M, Pratt CW, Corbin LW, Church FC. Characteristics of the chemotactic activity of heparin cofactor II proteolysis products. J Leuk Biol 1990; 48:156–162.
94. Hoffman M, Faulkner KA, Iannone MA, Church FC. The effects of heparin cofactor II-derived chemotaxins on neutrophil actin conformation and cyclic AMP levels. Biochim Biophys Acta 1991; 1095:78–82.
95. Corbin LW, Church FC, Hoffman M. Production of chemotactic peptides by neutrophil degradation of heparin cofactor II. Thromb Res 1990; 57:77–85.
96. Church FC, Pratt CW, Hoffman M. Leukocyte chemoattractant peptides from the serpin heparin cofactor II. J Biol Chem 1991; 266:704–709.
97. Henry RL. Leukocytes and thrombosis. Thromb Diath Haemorrh 1965; 13:35–46.

98. Plow EF. Leukocyte elastase release during blood coagulation. A potential mechanism for activation of the alternative fibrinolytic pathway. J Clin Invest 1982; 69:564–572.
99. Carp H, Janoff A. Modulation of inflammatory cell protease-tissue antiprotease interactions at sites of inflammation by leukocyte-derived oxidants. Adv Inflamm Res 1983; 5:173–201.
100. Inhorn RC, Tollefsen DM. Isolation and characterization of a partial cDNA clone for heparin cofactor II. Biochem Biophys Res Commun 1986; 137:431–436.
101. Herzog R, Lutz S, Blin N, Marasa JC, Blinder MA, Tollefsen DM. Complete nucleotide sequence of the gene for human heparin cofactor II and mapping to chromosomal band 22q11. Biochemistry 1991; 30:1350–1357.
102. Ragg H, Preibisch G. Structure and expression of the gene coding for the human serpin hLS2. J Biol Chem 1988; 263:12129–12134.
103. Turner J, Grundy CB, Kakkar VV, Cooper DN. MspI RFLP in the human heparin cofactor II (HCF2) gene. Nucleic Acids Res 1990; 18:1664.
104. Tollefsen DM, Pestka CA. Heparin cofactor II activity in patients with disseminated intravascular coagulation and hepatic failure. Blood 1985; 66:769–774.
105. Andersson TR, Berner NS, Larsen ML, Odegaard OR, Abildgaard U. Plasma heparin cofactor II, protein C and antithrombin in elective surgery. Acta Chir Scand 1987; 153:291–296.
106. Andersson TR. Consumption of heparin cofactor II (dermatan sulfate cofactor) and antithrombin during coagulation. Thromb Res 1987; 46:355–362.
107. Andersson TR, Larsen ML, Abildgaard U. Low heparin cofactor II associated with abnormal crossed immunoelectrophoresis. Thromb Res 1987; 47:243–248.
108. Andersson TR, Bell H. Plasma heparin cofactor II in alcoholic liver disease. J Hepatol 1988; 7:79–84.
109. Andersson TR, Sie P, Pelzer H, Aamodt LM, Nustad K, Abildgaard U. Elevated levels of thrombin-heparin cofactor II complex in plasma from patients with disseminated intravascular coagulation. Thromb Res 1992; 66:591–598.
110. Vinazzer H, Pangraz U. Heparin cofactor II: a simple assay method and results of its clinical application. Thromb Res 1987; 48:153–160.
111. Vinazzer H, Stocker K. Heparin cofactor II: experimental approach to a new assay and clinical results. Thromb Res 1991; 61:235–241.
112. Simioni P, Zanardi S, Prandoni P, Girolami A. Combined inherited protein S and heparin co-factor II deficiency in a patient with upper limb thrombosis: a family study. Thromb Res 1992; 67:23–30.
113. Weisdorf DJ, Edson JR. Recurrent venous thrombosis associated with inherited deficiency of heparin cofactor II. Br J Haematol 1991; 77:125–126.
114. Matsuo T, Kario K, Sakamoto S, et al. Hereditary heparin cofactor II deficiency and coronary artery disease. Thromb Res 1992; 65:495–505.
115. Needleman SB, Wunsch CD. A general method applicable to the search for similarities in the amino acid sequences of two proteins. J Mol Biol 1970; 48:443–453.
116. Altschul SF, Erickson BW. Optimal sequence alignment using affine gap costs. Bull Math Biol 1986; 48:603–616.
117. Loebermann H, Tukuok R, Deisenhofer J, Huber R. Human α_1-proteinase inhibitor crystal structure analysis of two crystal modifications, molecular model and preliminary analysis of the implications for function. J Mol Biol 1984; 107:531–556.
118. Wright HT, Qian HX, Huber R. Crystal structure of plakalbumin, a proteolytically nicked form of ovalbumin. J Mol Biol 1990; 213:513–528.
119. Stein PE, Leslie AGW, Finch JT, Turnell WG, Mclaughlin PJ, Carrell RW. Crystal structure of ovalbumin as a model for the reactive centre of serpins. Nature 1990; 347:99–102.
120. Whinna HC. Structure function studies on heparin cofactor II. Dissertation, University of North Carolina at Chapel Hill, School of Medicine, Department of Pathology, 1992.

19

Protein C

Koji Suzuki
Mie University School of Medicine, Mie, Japan

I. INTRODUCTION

Intravascularly generated thrombin activates platelets, and then negatively charged phospholipids are exposed on the cells. The phospholipids serve as the sites for binding of several vitamin K-dependent blood coagulation factors. Thrombin further proteolytically activates coagulation cofactor proteins, factor V and factor VIII, which lead to the significant acceleration of the blood coagulation cascade. Thrombin finally converts fibrinogen to fibrin, which is stabilized by factor XIIIa that is also activated by thrombin. The procoagulant actions of thrombin are, in general, limited to sites of injured tissues (1,2). On the other hand, on normal endothelium, thrombin acts as an anticoagulant by means of specific activation of the protein C pathway (Fig. 1) (3–5).

Protein C, so-called because it was isolated from chromatography fraction C during the initial purification, is a vitamin K-dependent plasma serine protease zymogen containing γ-carboxyglutamic acid (Gla) residues (6,7). It is activated by thrombin bound to thrombomodulin on endothelial cells (4,5). Activated protein C proteolyzes both factor Va and factor VIIIa on platelets and endothelial cells in the presence of two cofactors, protein S and factor V (recently identified). The activation of factors Va and VIIIa serves to regulate intravascular blood coagulation negatively. Activated protein C also shows profibrinolytic (9), antiischemic, and antiinflammatory (10) activities. Moreover, thrombomodulin blocks the procoagulant activities of thrombin, such as fibrinogen clotting and activation of factor V and platelets (4,5). Thus, thrombin is converted from a procoagulant into an anticoagulant by thrombomodulin on the vascular wall.

Activated protein C is finally neutralized by complex formation with specific plasma serine protease inhibitors, protein C inhibitor (PCI) (11), and α_1-antitrypsin (α1AT) (12). These complexes are catabolized in the reticuloendothelial system.

This review summarizes recent advances on the molecular biology of human protein C and related proteins and also covers the genetic abnormalities of protein C deficiency. Earlier reviews on the biochemistry and mechanism of action of protein C are available (3–5).

II. PROTEIN C

A. Production and Gene Structure of Protein C

Protein C is produced mainly in the liver (3) and also in the endothelial cells (13). It is secreted into plasma after undergoing several posttranslational modifications (described later).

The gene of human protein C is located on chromosome 2p13–14 (14), and the size of the gene, from the starting site of transcription to a more distant polyadenylation site, is approximately 11.2 kilobases (kb) (15,16). It consists of nine exons and eight introns (a–h). The corresponding mRNA is composed of a 75-nucleotide 5′-noncoding segment, a 1383-nucleotide region encoding a 461 amino acid precursor protein, a TAA termination codon, a 296-nucleotide 3′-noncoding segment, and a 38-nucleotide polyadenylation sequence. There are several DNA sequence polymorphisms in the protein C gene, which are potentially useful in gene tracking or haplotype analysis (described later).

Figure 2 shows the gene organization of human protein C, as well as the domain structures of the protein that correspond to the respective exons (15–17). The gene is very similar in structure to other vitamin K-dependent coagulation factors, except for the first exon corresponding to the 5′-noncoding segment (18–20), and the second exon also contains a 21-nucleotide noncoding segment. The second exon encodes a signal peptide region, the third a propeptide region and a Gla domain, the fourth a region containing hydrophobic amino acids, the fifth the first epidermal growth factor (EGF)-like domain, the sixth the second EGF-like domain, the seventh a region containing a connecting peptide between the light chain and the heavy chain and an activation peptide region, and the eighth and ninth exons a serine protease catalytic domain.

The half-life of protein C in blood is 6–8 h and approximates the half-life of factor VII (3–5 h). Other vitamin K-dependent factors have much longer half-lives: prothrombin (60–100 h); factor IX (20–24 h); and factor X (24–48 h).

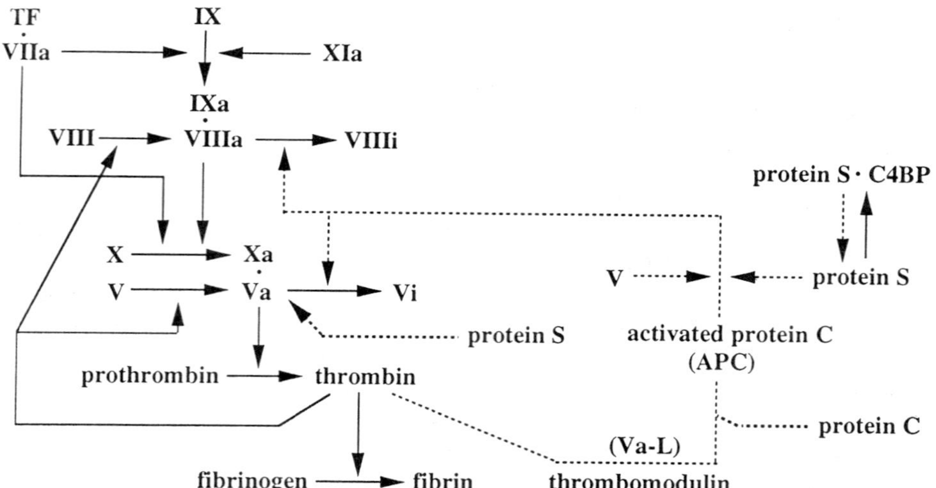

Figure 1 Anticoagulant protein C pathway. Solid lines and dotted lines indicate procoagulant actions and anticoagulant actions, respectively.

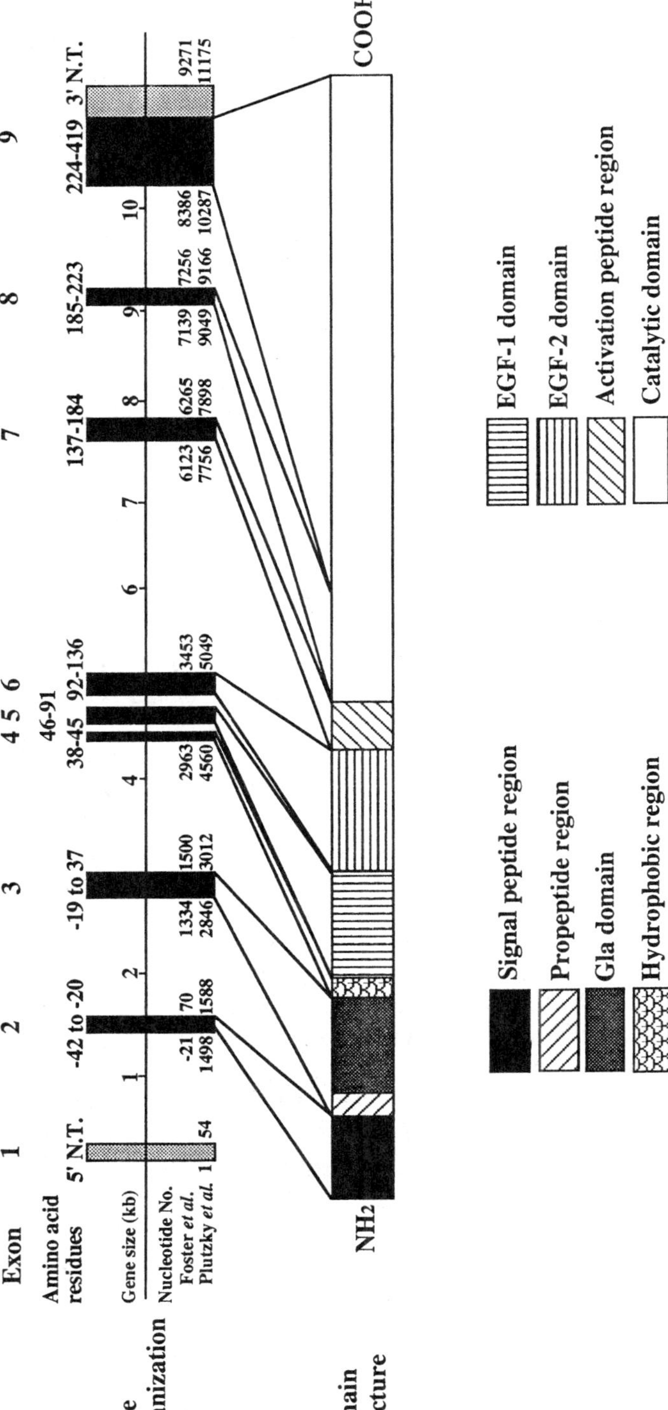

Figure 2 Gene organization and domain structures of human protein C. [Modified from Long (17).] Nucleotide number was derived from Foster et al. (15) and Plutzky et al. (16).

B. Primary Structure of Protein C

Human protein C has a molecular weight of about 62,000 daltons. The human protein C contains 9 Gla residues in the NH_2-terminal region (Gla domain) (3,7), whereas bovine protein C has 11 Gla residues (3,6). The molecule is initially synthesized in the liver as a single-chain form, but most of the protein (85–90%) is converted into a two-chain molecule consisting of a light chain and a heavy chain in the circulating plasma. Figure 3 shows the primary structure of the precursor of human protein C, deduced from its gene structure (15,16). The structure of the molecule resembles that of the precursors of the other vitamin K-dependent factors with about 40% conservation of amino acid sequence among these proteins.

The signal peptide (−42 to −20 residues) is essential to the translocation of the precursor protein from cytoplasm into endoplasmic reticulum by binding to the signal recognition particle, which targets the polypeptide in the endoplasmic reticulum. The signal peptide is then cleaved by a peptidase before transport to the endoplasmic reticulum.

The propeptide sequence (−19 to −1 residues) is typical of vitamin K-dependent proteins. The Glu residues of the precursor protein are carboxylated at the γ carbon by the vitamin K-dependent carboxyglutaminase (3,21). Recently, cDNA of human γ-carboxyglutaminase was isolated (see Chap. 14) (22). The propeptide region also plays a role in recognizing the processing protease that cleaves the propeptide from the precursor protein. The propeptidase also converts the single-chain protein C into the disulfide-linked two-chain form by cleaving the Lys-Arg dipeptide that connects the light and heavy chains (23).

The Gla domain located in the NH_2 terminus of mature protein C is a Ca^{2+}-binding region. The affinity constant K_d of Ca^{2+} ions to this domain is 5×10^{-4} M (3). In circulating plasma, about 10 Ca^{2+} ions are estimated to bind to this region. The Gla domain is believed to play a role in the binding of protein C to phospholipids on the cell membrane in the presence of Ca^{2+} ions (3). Calcium binding to the Gla domain is also thought to play a role in converting the protein into a conformationally suitable structure (24–27) for interaction with the thrombin-thrombomodulin complex. Once activated, protein C binds to negatively charged phospholipids via the Gla protein (27–29). A disulfide loop structure between Cys^{17} and Cys^{22} in the Gla domain is crucial in Ca^{2+} binding (30). Zhang et al. (31) prepared a series of modified recombinant protein C species by substituting other amino acids (mostly Asp residues) for nine Gla residues to deduce the role of each Gla and found that Gla residues at 7, 16, and 26 play an important role in expression of the Ca^{2+}-dependent anticoagulant activity. The Gla residues at 25 and 29 had a slight influence on the activity. Substitution of the Gla residues at 6, 14, and 19 had no influence on activity. Based on these findings and the relation between the Gla residues in prothrombin (whose three-dimensional structure was determined by x-ray analysis), it is thought that calcium binding to Gla residues 7, 16, 20, 26, and 29 is more important for maintenance of protein conformation than for binding to phospholipids.

The two epidermal growth factor (EGF)-like domains contain a high-affinity Ca^{2+}-binding site (K_d = 5–10 $\times 10^{-5}$ M) (3,32–34), believed to be critical to achieving a tertiary conformation conducive to activation and interaction with protein S. Recently, Rezaie et al. (35) demonstrated the existence of a high-affinity Ca^{2+}-binding site distinct from the first EGF-like domain, which was also necessary for activation of protein C by thrombomodulin. The first EGF-like domain of protein C contains a β-hydroxyaspartic acid residue (β-Hya) (3,34). Whether β-Hya plays a role in calcium binding is controversial (34,36).

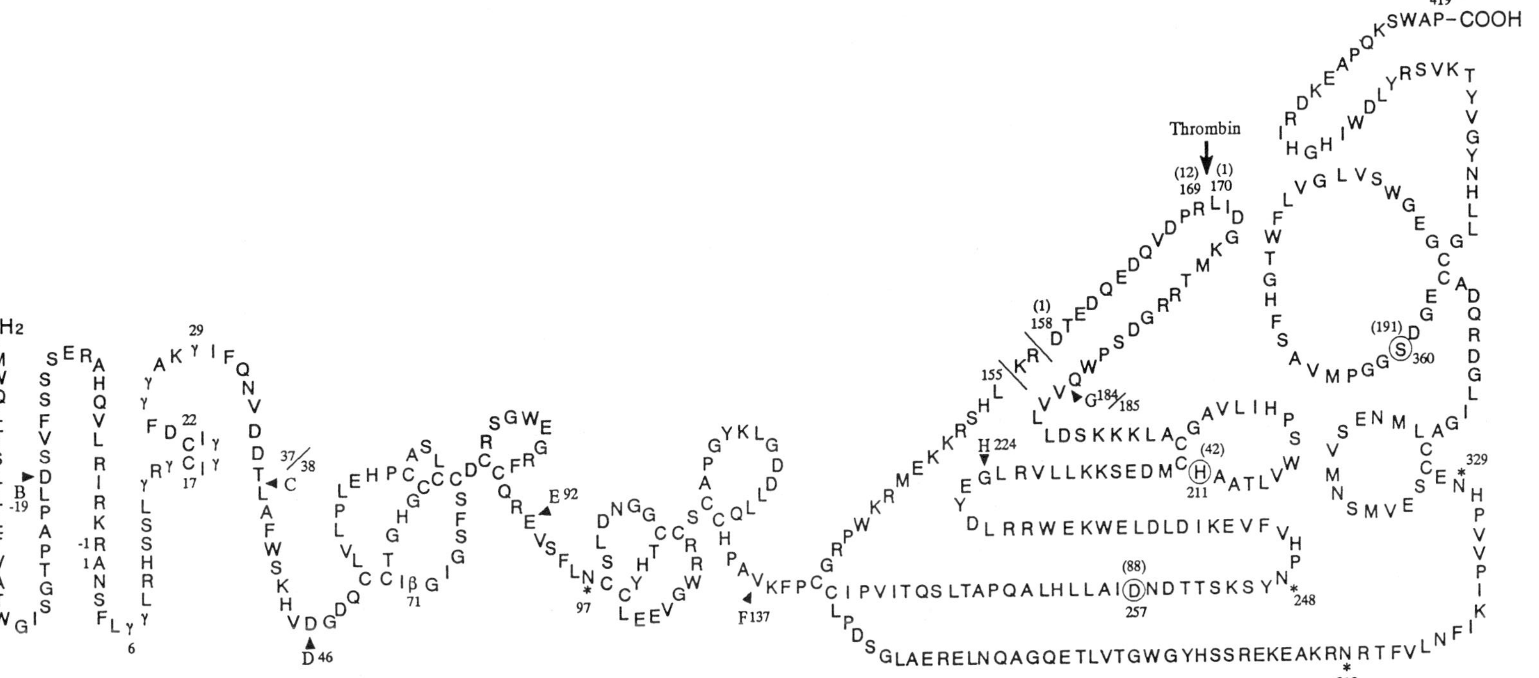

Figure 3 Primary structure of precursor of human protein C. The sequence was deduced from Foster et al. (15). The two-chain protein C lacks the K-R dipeptide (residues 156 and 157). The positions of the introns (B–H) are indicated by wedges at residues −19, 37/38, 46, 92, 137, 184/185, and 224. The active residues in the charge-relay system in the catalytic domain of activated protein C, that is, His^{42}, Asp^{88}, and Ser^{191}, are circled. *Carbohydrate binding sites; γ, γ-carboxyglutamic acid; β, β-hydroxyaspartic acid.

The activation peptide (Fig. 2) is released after cleavage of the Arg^{12}-Leu^{13} bond by thrombin-thrombomodulin. A conformational change of this region is also induced by Ca^{2+} binding (37).

The catalytic domain contains a proton relay system of the active center consisting of His^{42}, Asp^{88}, and Ser^{191} (3). In addition to its inhibition by PCI and α1AT, it is also inhibited by plasminogen activator inhibitor 1 (PAI-1) (38,39).

C. Posttranslational Processing of Protein C

The precursor of protein C undergoes posttranslational processing, such as Asn-linked glycosylation, γ-carboxylation of glutamic acid residues, β-hydroxylation of aspartic acid, and processing of several peptides before secretion from the liver cells. McClure et al. (40) determined posttranslational modifications of the precursor of recombinant human protein C using the adenovirus-transformed embryonic kidney cell line 293 and proposed the following series of events. Initial core glycosylation is necessary before γ-carboxylation of glutamic acids in the precursor protein in the early endoplasmic reticulum. The subsequent processing of the glycosyl core in the endoplasmic reticulum is rate limiting to the secretion of γ-carboxylated protein C. Subsequent events, including cleavage of the Lys-Arg dipeptide and, probably, propeptide cleavage, occur in the Golgi before secretion. Nonglycosylated recombinant human protein C is secreted inefficiently. Glycosylation and γ-carboxylation are not coupled and need not proceed sequentially. The secretion rate of recombinant human protein C is also Ca^{2+} dependent but is independent of the Ca^{2+} effect on the internal Lys-Arg dipeptide removal.

As described earlier, core glycosylation influences γ-carboxylation in the endoplasmic reticulum, and Asn-linked core glycosylation is required for secretion of protein C. Protein C has four Asn-linked glycosylation sites, Asn^{97} in the light chain and Asn^{248}, Asn^{313}, and Asn^{329} in the heavy chain. The function of these carbohydrate chains was studied using recombinant protein C variants that substituted Gln for Asn residues and thus could not be glycosylated (41). These studies showed that the Asn^{97}-linked carbohydrate chain is related to secretion of protein C from the cell and affects core glycosylation at Asn^{329}. The Asn^{248}-linked carbohydrate chain is related to cleavage of the Lys-Arg dipeptide connecting the light chain and the heavy chain. Lack of glycosylation at Asn^{313} of protein C significantly accelerates its activation by the thrombin-thrombomodulin complex. Elimination of glycosylation sites in the heavy chain of activated protein C results in a two- to three-fold increase in the anticoagulant activity. None of the carbohydrate chains relate to inactivation of activated protein C by α1AT.

Lower molecular weight β-protein C, which comprises 30% of plasma protein C, is a protein lacking the carbohydrate chain at Asn^{329} (42). A Cys^{331} residue located two residues from Asn^{329} acts for Ser or Thr of the recognition sequence of glycosylation, Asn-*X*-Ser/Thr, and functions as a hydrogen bond acceptor required for glycosylation until the disulfide loop structure is formed. O-linked fucose present in the first EGF-like domain of several other serine proteases is not contained in protein C (45–47).

D. Activation of Protein C

1. *Thrombomodulin-Independent Activation*

Protein C can be activated *in vitro* by thrombin (7,48), factor Xa (49), snake venom enzyme derived from *Agkistorodon contortrix contortrix* (50,51) or *Agkistorodon bilineatus* (52), and the factor X activator found in Russell's viper venon (RVV) (53). All the activators are serine proteases, except for the RVV activator, which is a metalloprotease requiring Ca^{2+} ions. The *A. contortrix controtrix* venom activator possesses high sequence homology to batroxobin, flavoxobin, and RVV factor V activator (54). The RVV factor X activator contains a disintegrin (platelet aggregation inhibitor)-like sequence and a Ca^{2+}-dependent lectin-like sequence in addition to the protease sequence (55). Activation of protein C by thrombin alone is decreased by Ca^{2+} ions but increased in the presence of thrombomodulin (56,57). The concentration of Ca^{2+} ions (250 μM) required to raise the activation rate to half-maximal effect in the presence of thrombomodulin is the same as that to decrease it to half-maximal effect in the absence of thrombomodulin (56). This suggests that Ca^{2+} ions induce a conformational change in protein C and that Ca^{2+}-bound protein C does not behave as a thrombin substrate in the absence of thrombomodulin. Chemical modification of thrombin has been studied to increase the substrate specificity for protein C. Treatment of thrombin with acetyl anhydride decreased the fibrinogen clotting activity of thrombin without any effect on the amidolytic activity. However, the acetylated thrombin showed conservation of the specificity for protein C activation (58). Tryptamine and its analogs, like thrombomodulin, enhance the thrombin-catalyzed activation of protein C several fold at concentrations below 10 mM (59). Factor Xa activates protein C in the presence of anionic phospholipids, and it is enhanced by sulfated polysaccharides (49,60).

Recombinant modified protein C, in which Asp^{167} is replaced with Gly or Phe, is activated by thrombin 5- to 6-fold more easily than the wild-type protein C. This suggests that Asp^{167} is related to the Ca^{2+}-dependent conformational change in protein C (61). Similarly, recombinant protein C, in which Asp^{167}, Asp^{172}, and Asn^{313} were respectively substituted by Phe, Asn, and Gln, was found to be activated by thrombin 60-fold more easily than wild type protein C. In contrast, its anticoagulant activity was increased only about 1.7-fold in comparison with that of the wild-type molecule (62).

2. *Thrombomodulin-Dependent Activation*

Protein C is physiologically activated by thrombin bound to thrombomodulin, which is a thrombin receptor on endothelial cells (4,5,56,63,66). Thrombomodulin enhances the thrombin-catalyzed activation of protein C 1000- to 2000-fold in the presence of Ca^{2+} ions.

Human thrombomodulin is a single-chain glycoprotein with a total molecular weight of 78,000, consisting of 557 amino acid residues of 55,662 daltons and carbohydrate chains of about 22,000 daltons (see Chap. 19) (67–71).

The structure-function relationship of thrombomodulin in protein C activation has been determined using proteolytic fragments and recombinant proteins of thrombomodulin (72–79). Briefly, thrombin binds to thrombomodulin with a K_d of 5 nM, primarily via the fifth EGF-like domain of thrombomodulin. The sixth EGF-like domain of thrombomodulin is thought to enhance the interaction. As a result of binding to thrombomodulin, the active center of thrombin is conformationally changed so that selective proteolysis of the activation peptide of protein C occurs (80,81). Protein C activation by thrombin shows an optimal Ca^{2+} concentration at 300–400 μM in the presence of thrombomodulin fragments that contain four to six EGF-like domains (72,79). However, the rate of activation by thrombin is unaffected by Ca^{2+} concentration when protein C lacking the Gla domain is used (72,79). These findings suggest the existence of an interaction between the Gla domain of protein C and thrombomodulin via Ca^{2+} ions. Ca^{2+}-dependent direct interactions between protein C, thrombin, and an elastase-digested fragment from thrombomodulin were determined by ultracentrifugal analysis (82). Zushi et al. (83) reported that an Asp^{349} residue in the fourth EGF-like domain of thrombomodulin plays a role in the interaction with the Gla domain of protein C (Fig. 4). Critical residues essential for protein C activation by thrombin-thrombomodulin complex in the region comprising the fourth through sixth EGF-like domains have also been studied using recombinant mutant proteins of thrombomodulin (84,85).

The molecular mechanism of the conversion of thrombin substrate specificity by thrombomodulin has also been studied. Le Bonniec and Esmon (86) prepared a recombinant mutant human thrombin in which Glu^{192} was replaced with a Gln residue located three residues from the active Ser^{195} residues and found that it activated protein C 22-fold more rapidly in the absence of thrombomodulin. In the presence of thrombomodulin, the activating activity was increased only 2-fold. Based on these observations, it was suggested that thrombomodulin, in part, alters the conformation of the Glu^{192} residue of thrombin and alleviates its interaction with the Asp^{167} at the P3 site and the Asp^{172} at P3′ site from the thrombin cleavage site of protein C. Furthermore, thrombomodulin was

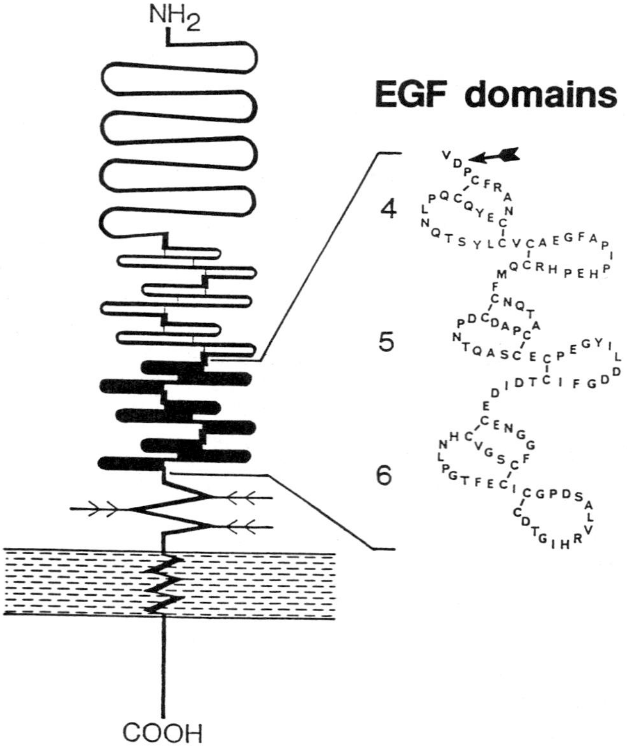

Figure 4 Thrombomodulin-dependent protein C activation (79,83). Thrombomodulin consists of an NH_2-terminal lectin-like domain, six consecutive EGF-like domains, an O-glycosylation site-rich domain, a transmembrance domain and a COOH-terminal cytoplasmic domain. Thrombin is suggested to interact with a region involving the fifth and sixth EGF-like domains of thrombomodulin. Protein C is suggested to interact with a region containing Asp^{349} residue in the fourth EGF-like domain via Ca^{2+} ions (indicated as an arrow).

suggested to alleviate the inhibitory effect of the residue Glu^{39} of thrombin, which contributes to the substrate specificity, on the proximity of the Asp^{172} at the P3′ site of protein C (87). Studies using synthetic peptides corresponding to the sequence residues located around the active center of thrombin also suggested that thrombomodulin binds to the sites near the active center of thrombin, resulting in hindrance of the thrombin interaction with fibrinogen and other procoagulants (88–90).

Besides the functions mainly dependent on the amino acid sequence, several carbohydrate-dependent functions have been reported in thrombomodulin (91). Chondroitin sulfate bound to the O-linked glycosylation site-rich domain of rabbit thrombomodulin has heparin-like activity (92). The chondroitin sulfate in recombinant human thrombomodulin enhances the binding of thrombomodulin to thrombin, resulting in the increase in protein C-activating cofactor activity (93).

E. Anticoagulant Activities of Activated Protein C

The predominant action of activated protein C is to inhibit the blood coagulation cascade and to stimulate fibrinolysis.

1. *Inhibition of Coagulation by Activated Protein C*

Activated protein C strongly inhibits blood coagulation on the phospholipid membrane of platelets and endothelial cells (3–5,94,95). The inhibition results from selective proteolysis of factor Va and factor VIIIa (3–5,96–99). Factors Va and VIIIa are inactivated 5- to 10-fold faster than nonactivated factors V and VIII (3,98,100).

Factors V and VIII share similarities in structure and function. The basic structure is composed of the following domains A1-A2-B-A3-C1-C2 (reading from the NH_2 terminus). The A domain in factors V and VIII has high homology to three domains involved in ceruloplasmin, a copper-binding protein in plasma (101).

In the process of the inactivation of factors Va and VIIIa, activated protein C first binds to the low-molecular-weight forms of factor V or VIII (3,100,102,103). Activated protein C recognizes a region located at residues Arg^{1865} to Ile^{1874} consisting of the RAGMQTPFLI sequence of factor Va and also a region located at residues His^{2009} to Val^{2018} consisting of the HAGMSTLFIV sequence of factor VIIIa (103). Ceruloplasmin contains a similar sequence HAGMETTYTV, His^{1028} to Val^{1037}. Synthetic peptides containing these sequences were found to inhibit the inactivation of factors Va and VIIIa by activated protein C, suggesting that ceruloplasmin plays a role in the regulation of activated protein C (104). Another experiment using the synthetic peptide suggested that the low-molecular-weight fragment of factor Va binds to a region containing Tyr^{390}-His^{404} residues in a protease catalytic domain of activated protein C (105). Activated protein C then cleaves in a limited fashion both the high- and low-molecular-weight fragments of factor Va and factor VIIIa. This results in the loss of interactions with factor Xa and prothrombin for factor Va and with factor IXa and factor X for factor VIIIa (2,106,107).

The site cleaved by activated protein C in human factor Va is estimated to be located at Arg^{506}-Gly^{507} bond in the A2 domain, since activated protein C cleaves bovine factor Va at the Arg^{505}-Gly^{506} bond in the A2 domain and also the Arg^{1753}-Ala^{1754} bond in the A3 domain (108). Activated protein C cleaves human factor VIIIa at the Arg^{336}-Met^{337} bond in the A1 domain and the Arg^{562}-Gly^{563} bond in the A2 domain (Fig. 5) (100).

Both factor Xa and factor X can protect factor Va from proteolytic inactivation by activated protein C (97,109). Similarly, factor IXa protects the inactivation of factor VIIIa by activated protein C (110). Jane et al. (111) suggested that the Gla domain of factor

X (Xa) plays an important role in factor Va protection, because factor X (Xa) binds to factor Va with the same affinity as activated protein C, but factor X (Xa) without the Gla domain does not protect factor Va from activated protein C. They also showed that protein S diminishes protection of factor Va by factor X (Xa). Therefore, activated protein C alone is considered insufficient for inactivation of factor Va in the prothrombinase complex, composed of factor Xa, factor Va, phospholipids, and Ca^{2+} ions, and also for inactivation of factor VIIIa in the tenase complex (7,112,113).

Factor VIII is present as a complex with the von Willebrand factor (vWF) in plasma. The levels of factor VIII activity and antigen are decreased in proportion to the reduction in vWF antigen values in patients with von Willebrand disease. This observation has now been explained by the fact that vWF protects factor VIII from proteolysis by activated protein C (114). Factor VIIIa after release from vWF is not protected.

2. *Profibrinolysis by Activated Protein C*

Activated protein C, added to a fibrin clot in vitro, shortens the clot lysis time (9,115). Infusion of activated protein C into dogs results in an increase in tissue type plasminogen activator (tPA) activity (116). This effect was also observed in cats (117) but not in squirrel monkeys (118). The difference in fibrinolytic activity in plasma for patients with congenital protein C deficiency and normal humans is slight, suggesting that there is not much physiological significance to the profibrinolytic activity of activated protein C in humans.

The effect of activated protein C on the fibrinolytic activity of cultured bovine vein endothelial cells has been examined. Activated protein C was found to form a complex with PAI-1 secreted from the endothelial cells, resulting in an increase in fibrinolytic activity (38,39). Activated protein C was also found to neutralize PAI-1 released from platelets, thus enhancing profibrinolytic activity (119,120). On the other hand, in a sep-

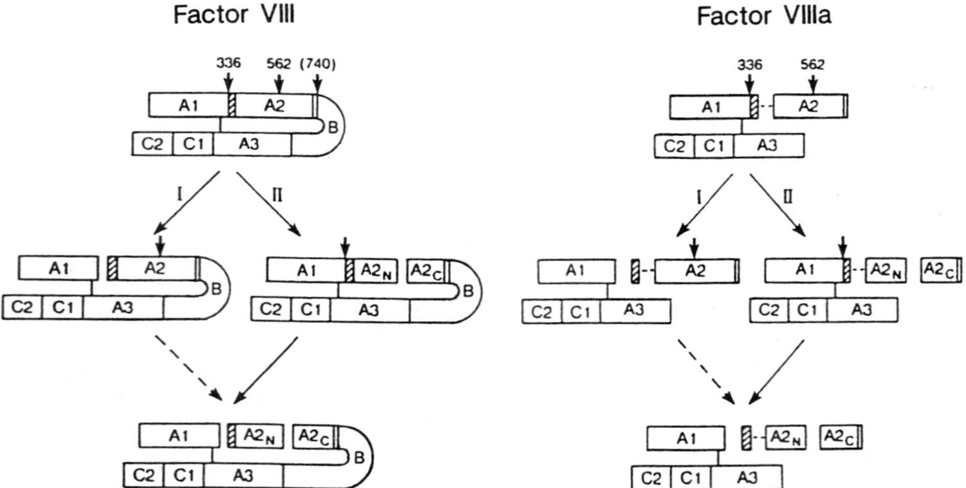

Figure 5 Proteolytic inactivation of factor VIII (left) and factor VIIIa (right) by activated protein C. This schematic model was modified from Fay et al. (100). Residues 336 and 562 and a tentative residue 740 indicate cleavage sites by activated protein C in factors VIII and VIIIa. Pathway I represents initial cleavage at residue 336 and pathway II initial cleavage at residue 562.

arate experiment, activated protein C required protein S, phospholipids, and Ca^{2+} ions for expression of profibrinolysis (121,122). Both in vitro and in vivo studies in several laboratories have revealed that thrombolysis induced by tPA is promoted in the presence of activated protein C, presumably by neutralization of PAI-1.

F. Inhibition of Activated Protein C by PCI

Activated protein C is inhibited dominantly by PCI (11,123,124) and α1AT (12,125) in plasma. The inhibition of the enzyme by PCI is heparin dependent, but that by α1AT is not. PCI has been detected not only in plasma and urine but also in seminal plasma (126), suggesting that PCI acts as a multifunctional inhibitor in body fluids. For a more extensive discussion, protein C inhibitor, see Chapter 20.

1. Mechanism of Inhibition of Activated Protein C by PCI

PCI inhibits thrombin, factor Xa, plasma and tissue kallikrein, tPA and urinary-type PA, as well as activated protein C, but not factor XIIa or plasmin (11,124,132,133). The inhibition rate of PCI for activated protein C and thrombin is significantly accelerated by the presence of a relatively higher concentration of heparin (1–5 U/ml) or negatively charged dextran sulfate (11,124). However, inhibition of plasma kallikrein by PCI is less stimulated by heparin (132,134). The site for heparin binding in PCI is presumed to be a region from Ser^{264} to Tyr^{283} located near the middle of the sequence (135). This differs from the heparin binding site in the related proteins, antithrombin III having the putative sites involving residues from Ile^{40} to Arg^{57} and Ala^{124} to Leu^{140} and heparin cofactor II having the putative site involving residues from Phe^{183} to Arg^{200}, both of which are located relatively near the NH_2-terminal side from the middle (136). Heparin does not stimulate α1AT inhibition of activated protein C. The inhibition constant K_i of PCI to activated protein C is approximately 5×10^{-8} M in the presence of heparin, but the K_i of α1AT is likely 10^{-5} M. Because the concentration of α1AT in plasma is 1000-fold higher than that of PCI (5 μg/ml), α1AT probably works physiologically. The mechanism by which activated protein C is inhibited by PCI and α1AT has been determined by experiments *in vitro* and *in vivo* using small animals and also by clinical studies (137–140).

III. CONGENITAL PROTEIN C DEFICIENCY

Congenital protein C deficiency is one of the most common hereditary causes of recurrent thromboembolic disease (142–147). The crucial role of protein C in hemostasis is demonstrated by the extensive thrombosis occurring in patients who are homozygotes or compound heterozygotes for protein C deficiency. Such patients often suffer from neonatal *purpura fulminans* and life-threatening massive venous thrombosis usually beginning at or soon after puberty (142–144). Heterozygotes, with low but detectable protein C levels, may remain free of thrombotic symptoms at least into early adulthood (146,147). Heterozygotes commonly present with deep venous thrombosis, pulmonary embolism, superficial thrombophlebitis, or mesenteric venous thrombosis. Thromboembolic episodes increase with age. Of affected patients, 78% experience their first thromboembolic event by 40 years of age. Congenital protein C deficiency is inherited as an autosomal dominant trait. The incidence of the asymptomatic heterozygous protein C deficiency is believed to be about 1 in 200–300 individuals (144), and the incidence of thrombosis to be about 1 in 15,000 individuals (142). The incidence of the homozygote is 1 in 500,000 individuals (142,143).

Of patients with protein C deficiency reported to date, mutations have been found (145): (1) promoter mutations, (2) splice site abnormalities, (3) in-frame deletions, (4) frameshift deletions, (5) in-frame insertions, (6) frameshift insertions, (7) nonsense mutations, (8) missense mutations, and (9) silent mutations. Patients with congenital protein C deficiency are phenotypically classified into two types based on concentration of plasma protein C antigen and activity as follows (142,143).

Type I protein C deficiency. Most of the patients with congenial protein C deficiency have type I deficiency. These patients have both reduced protein C antigen and activity in the plasma; thus the ratio of the activity to the antigen is nearly 1. A majority of patients are heterozygous, but a few are homozygous or compound heterozygous. The plasma concentrations of protein C antigen and activity in the heterozygous patients are approximately 50% of the normal concentration.

Type II protein C deficiency. These patients have a dysfunctional protein C molecule with markedly reduced protein C activity but a relatively normal antigen level. Thus, the ratio of the activity to the antigen is less than 1 and in the range of 0.5 in the heterozygous patients. Most patients with type II deficiency are heterozygous, rarely homozygous or compound heterozygous.

Gene analysis has been performed in both types of protein C deficiency, and nearly 150 families with abnormal genes have been reported (Table 1) (145–170). Type I deficiency has been found to result from all kinds of mutations; type II deficiencies all result from missense mutations in the protein C gene. Mutations result in amino acid substitutions throughout the molecule. The dinucleotide CpG is a ''hot spot'' for mutation thought to result from the methylation-mediated deamination of cytosine (171,172). As shown in Figure 6 (145), the distribution of CpG site mutations in the protein C gene is nonrandom: 75% (9 of 12) of single-nucleotide mutations in exon 7 and 64% (16 of 25) in exon 9, but none of the 13 point mutations in exons 5 and 6, occur in CpG dinucleotides. CpG suppression, determined by the ratio of GpC to CpG frequencies (173), is at a minimum for the region in exons 4–6 (145). The mutations at residues Arg^{15}, Gln^{132}, Arg^{157}, Arg^{169}, Arg^{178}, Arg^{229}, Arg^{230}, and Arg^{306} all occur at such hot spots as a result of $C \rightarrow T$ transitions (145).

Although most type I and type II deficiencies are caused by missense mutations, the difference between the expression modes of the mutated proteins from the liver into plasma does not always depend on the mutated position, except that proteins mutated in the Gla domain (exon 3) mostly appear to belong to type II deficiency (157,158,168–170). Sugahara et al. (164) suggested that most of the mutant protein C molecules belonging to type I deficiency undergo degradation within the liver cells instead of being secreted from the cells, and their transport is retarded in the intracellular secretory pathways.

We recently found that a male infant with neonatal purpura fulminans had a compound heterozygous deficiency derived both from the mother, with heterozygous type I deficiency, and the father, with heterozygous type II deficiency. These classifications were determined by measuring the protein C antigen and the activity levels of the family (170). From the gene analysis, it was deduced that the mother's abnormal gene had a deletion of one of four consecutive G nucleotides encoding Trp^{380} (TGG)-Gly^{381} (GGT) in exon 9, resulting in a frameshift mutation and an abnormal sequence of the 81 amino acid residues that follow Gly^{381}. This abnormality caused impaired secretion of mutant protein C from the liver cells (167a, 167b). The father's abnormality was a missense mutation

Table 1 Mutation in Protein C[a]

Pedigree and (activity %/antigen %)	Nucleotide position and mutations	Amino acid change and mutations	Comments	Reference
1. Putative promoter region				
PC-1 La Jolla IV	−1533, A → G	None		159
PC-43-010 (19c/22)	−1528, T → A	None		145
PC-1 La Jolla V	−1511, C → T	None		159
2. Exon 2 (signal peptide region)				
PC-43-001 (59c/58)	40, T → G	−29, W → G		145
PC-44-006 (38c/42)	41, G → A	−29, W → stop		145
3. Exon 3 (propeptide region and Gla domain)				
PC-1 Vermont IV	1317, C → T	None	Near 3′ end of intron b	145
PC-34-003 (57c, 50a/62)	1380–86, del 7 nt;	−3, −2, del R, K	Frameshift, stop at codon 16	158
PC-33-001 (120c, 48a/114)	1388, G → A[b]	−1, R → H	Type II (PC Malakoff)	157
PC-34-006 (45c, 4a/47)	1414, C → T[b]	9, R → C	Type II in a compound heterozygote for type I and type II (see 6246)	158
PC-1-L01 (<5/45)	1432, C → T[b]	15, R → W	Type II in a compound heterozygote (see 8790)	155
PC-81 Yonago (44c, 74a/83)	1432, C → G	15, R → G	Type II	169
PC-33-003 (60a/58)	1433, G → A[b]	15, R → Q		154
PC-1 Vermont I (115c, 42a/113)	1448, A → C	20, E → A	Type II, double mutant (see 1489)	168

Table 1 Continued

Pedigree and (activity %/antigen %)	Nucleotide position and mutations	Amino acid change and mutations	Comments	Reference
PC Mie (12c, <5a/20, 4Gla)	1465, G → A	26, E → K	Type II in a compound heterozygote for type I and type II (see 8857)	170
PC-1 Vermont I (115c, 42a/113)	1489, G → A	34, V → M	Type II, double mutant (see 1448)	168
4. Exon 5 (first EGF-like domain)				
PC-34-021 (28c/10)	3082, G → T	47, G → C		145
PC-1 La Jolla I	3139, C → A	66, H → N	Type II	159
PC-1-L03	3142, G → C	67, G → R	Compound heterozygote for type I (see 7158)	145
PC-43-005 (31c/41)	3156–70 or 3157–71, del 15 nt	72–76, del G, I, G, S, F	In-frame deletion	145
PC-31-090 (54c/50)	3169, T → C	76, F → L	Present 2 more families	153
PC-1 St. Louis II (–/42)	3172–89, del 18 nt	77–82, del S, C, D, C, R, S	In-frame delection: asymptomatic kindred	160
PC-1 St. Louis III (–/37)	3173–90, del 18 nt	78–83, del C, D, C, R, S, G	In-frame delection: asymptomatic kindred	160
5. Exon 6 (second EGF-like domain)				
PC-49-L04 (<5/<5)	3217, G → T	92, E → stop	Homozygote for type I: mutation also affects splice donor e	145
PC-31-012 (58c/58)	3222, G → A	None	In splice donor e	153
PC-31-063 (22c/37)	3222, G → T	None	In splice donor e	153
PC-31-152 (22c/35)	3222, G → C	None	In splice donor e	153
PC-31-020 (51c/50)	3359, G → A	105, C → Y	Present 1 more family	153
PC-31-046 (–/36)	3360, C → A	105, C → stop		153

PC-1-L13 (2/2)	3363/4, ins C	107, H → P	Compound heterozygote for type I (see 6246); frameshift stop at codon 119	155
PC-31-111 (–/23)	3402, C → G	119, S → R		153
PC-31-006 (42c/45)	3439, C → T	132, Q → stop	Present 11 more families	153
PC Hong Kong-1 (<1/<1)	3447–51, del 5 nt	135, 136, del P, A	Compound heterozygote for type I (see 8842), stop at codon 170	164
PC-33 Clamart (–/<1)	3451, G → C	136, A → P	Homozygote for type I	161
6. Exon 7 (COOH-terminal region of the light chain and activation peptide region)				
PC-31-160	6134, T → C	141, C → R		153
PC-34-009 (21/49, 22)	6139, ins TT	143, R → L	Frameshift, stop at codon 156	145
PC-31-033 (73c/74)	6152, C → T[b]	147, R → W	Type II, present 2 more families	145
PC-49-048 (33, 59c/50)	6153, del G	147, del R	Frameshift, stop at codon 155	162
PC-33-004 (58a/85)	6167, C → T[b]	152, R → C		154
PC-33-005 (28a/30)	6182, C → T[b]	157, R → stop	Present 3 more families	151
PC-33-L06	6216, C → T[b]	168, P → L	Homozygote for type I	147
PC-81 Tochigi (<1/19)	6218, C → T[b]	169, R → W	Type II PC London 1 (type I)	149
PC-43-011 (23/36)	6219, G → A[b]	169, R → Q	Type II	145
PC-31-056 (23c/<10)	6245, C → T[b]	178, R → W	Present 4 more families	145
PC-44-012 (30c/36)	6246, G → A[b]	178, R → Q	Present 9 more families	165
PC-31-044 (–/24)	6247, G → A	178, R → R	Cosegregates with the normal allele; 3439, C → T causes disease	153
PC-1-L08	6265, G → A	184, Q → Q	Compound heterozygote for type I (see 8524); In splice donor g	145

Table 1 Continued

Pedigree and (activity %/antigen %)	Nucleotide position and mutations	Amino acid change and mutations	Comments	Reference
PC-63-021 (<1/<1)	6265, G → C	184, Q → H	Severe homozygote for type I; in splice donor g	163
7. Exon 8 (catalytic domain)				
PC-1 St. Louis I (–/50)	6274, C → T[b]	None	Putative novel splice donor g, frameshift, stop at codon 214; asymptomatic kindred Present 1 more family	160
PC-1-L03	7158, ins A	191, K → K	Compound heterozygote for type I (see 3142); frameshift, stop at codon 207	145
PC-43-003 (35/41)	7190, C → T	202, H → Y	Double mutant (see 8751)	145
PC-43-008 (43/146)	7219, C → A	211, H → Q	Type II	145
PC-31-025 (54c/61)	7253, C → T	223, L → F	Present 2 more families	153
8. Exon 9 (catalytic domain)				
PC-34-005 (40c, 15a, 70c, 32a/75)	8400, C → T[b]	229, R → W	Type II	158
PC-33-011 (57a, 96c/90)	8401, G → A[b]	229, R → Q	Type II, known as PC Marseilles	156
PC-46-L09 (8/7)	8403, C → T[b]	230, R → C	Compound heterozygote for type I; second mutation has not been identified Present 15 more families	145
PC-43-014 (42/38)	8432, ins CTGGAC	239, ins L, D	In-frame insertion	145
PC-44-010 (6c/<10)	8455, C → T	247, P → L		165
PC-33-012 (70a, 147c/140)	8470, G → A[b]	252, S → N	Type II, known as PC Paris	156
PC-31-060 (–/35)	8481, A → G	256, N → D	Present 1 more family	145

PC-974-004 (<5/<5)	8491, C → T	259, A → V	Homozygote for type I	146
PC-33-L10 (19/9)	8514, G → A[b]	267, A → T	Homozygote for type I	147
PC-1-L08	8524, C → T[b]	270, S → L	Compound heterozygote for type I (see 6265)	145
PC-43-004 (37/38)	8551, C → T[b]	279, P → L		145
PC-43-006 (33/39)	8571, C → T[b]	286, R → C		145
PC-49 München (<1c/3)	8572, G → A[b]	286, R → H	Homozygote for type I	152
PC-31-005 (53c/55)	8589, G → A[b]	292, G → S		153
PC-1 Vermont II (20c/20)	8608, C → T[b]	298, T → M	One arm of a large kindred	145
PC-33-L11 (23/47)	8617, G → A	301, G → D	Homozygote for type I	147
PC-39-F5 (50/47)	8617, del G	301, del G	Frameshift, stop at codon 335	166
PC-31-009 (57bc/55)	8631, C → T[b]	306, R → stop	Present 5 more families	148
PC-34-002 (46sc, 46sa/53)	8678–80, del 3 nt	321–2, del I K → M	In-frame deletion	158
PC-43-002 (15/40)	8744, G → A	343, M → I	Type II	145
PC-43-003 (35/41)	8751, G → A	346, A → T	Double mutant (see 7190)	145
PC-43-009 (10/17)	8752, C → T[b]	346, A → V		145
PC-33-L12 (51/70)	8769, C → T[b]	352, R → W	Type II	145
PC-1-L01 (<5/45)	8790, G → A[b]	359, D → N	Compound heterozygote for type II (see 1432) Present 1 more family	145
PC Hong Kong-2 (<1/<1)	8842, G → A	376, G → D	Compound heterozygote for type I (see 3447)	164
PC-39-F6 (58/156)	8856, G → A	381, G → S	Type II	145

Table 1 Continued

Pedigree and (activity %/antigen %)	Nucleotide position and mutations	Amino acid change and mutations	Comments	Reference
PC Mie (38c, 43a/45)	8857, del G	381, del G	Type I; frameshift, stop at codon 462, Present 2 more families	170
PC-31 Purmerend (63c/135)	8886, G → A[b]	391, G → S	Type II	145
PC-34-ML (42c/35)	8921, G → C	402, W → C		148
PC-31-019 (53c/45)	8924, C → G	403, I → M	Present 1 more family	153

[a]Protein C activity: c, chromogenic assay; a, anticoagulant clotting assay. Nucleotide and amino acid numbering are according to Foster et al. (15).
[b]Indicates mutation of a CG to either TG or CA homology.
[c]Details in this table; see Reference 145.
Source: Modified from Table A in Reference 145.

in exon 3 resulting in a substitution of Glu^{26} by Lys. Glu^{26} is normally converted to a Gla residue necessary for Ca^{2+} binding to induce a specific conformation suitable for interaction with thrombomodulin and phospholipids.

For antenatal diagnosis and carrier detection, several techniques for gene analysis have been used. Most widely used is the restriction fragment length polymorphism (RFLP) assay for the gene of the patient and family members. Several DNA sequence polymor-

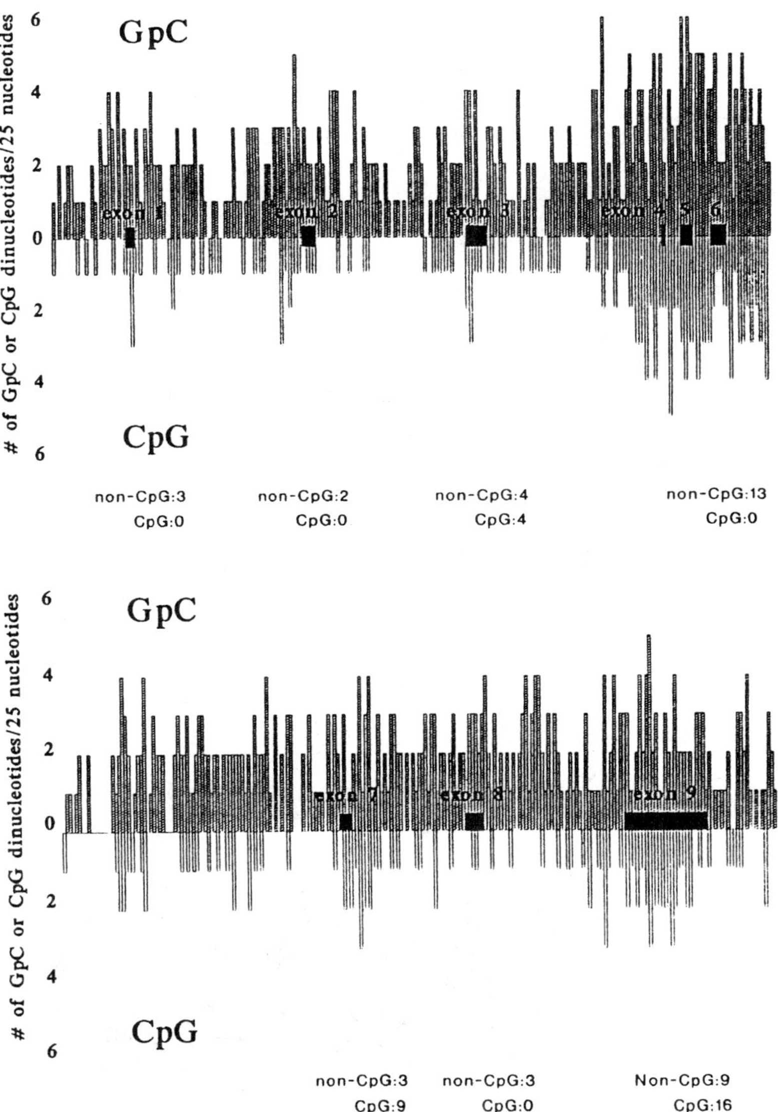

Figure 6 Histogram of the distribution of GpC and CpG dinucleotides in the protein C gene, illustrated by Reitsma et al. (145). The histogram shows the number of GpC and CpG dinucleotides in the successive stretches of 25 nucleotides. The positions and relative sizes of the exons are indicated by the black rectangles. The number of CpG and non-CpG base pair substitutions is indicated below each exon, except exon 4, 5, and 6, for which they are pooled.

Table 2 DNA Sequence Polymorphisms in the Protein C Gene

Nucleotide sequence polymorphism (position)	Amino acid residue	Frequency	Reference
T/A (−1476)	—	0.60/0.40	153
CGC/CGT (3204)	Arg^{87}	0.87/0.13	175
TCT/TCG(3342)	Ser^{99}	0.59/0.41	153
		0.82/0.18	174
AAA/AAG (6181)	Lys^{156}	0.64/0.36	175
GAT/GAC (7228)	Asp^{214}	0.67/0.33	153
		0.64/0.36	176

Source: Reproduced with permission from Reitsma et al. (145).

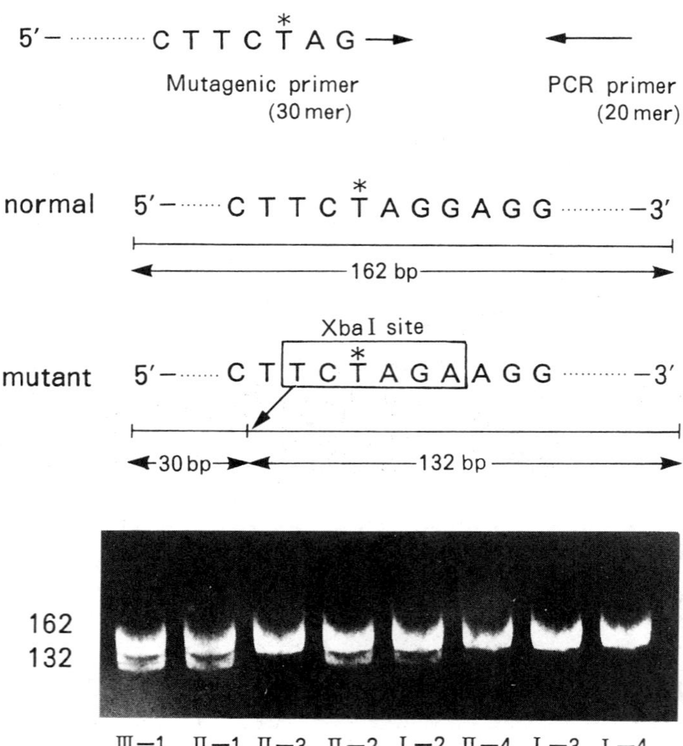

Figure 7 Detection of a point mutation in exon 3 of protein C gene (170). DNA fragments, including the mutation site in exon 3, were amplified from each family member with a mutagenic primer, in which T was introduced instead of G to create a new *Xba*I site (TCTAGA) in the mutant allele. Electrophoresis (7.5% polyacrylamide gel) of *Xba*I-digested PCR fragments yielded two fragments (132 and 30 bp) in the mutant allele, whereas the normal allele was uncut (162 bp).

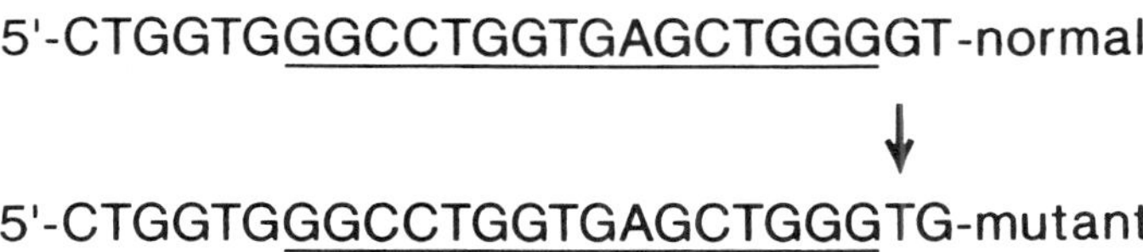

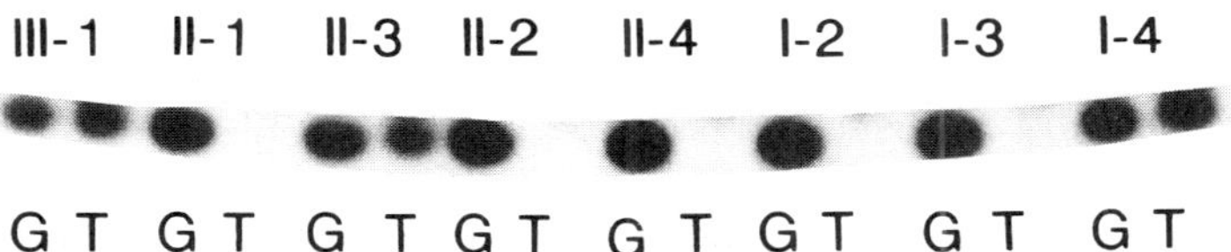

Figure 8 Single-nucleotide primer extension to detect a mutation in exon 9 of protein C gene (170). Exon 9 from family members was amplified from the genomic DNA by PCR using two primers corresponding to nucleotides 10178–10197 and 10934–10953 of the protein C gene (16). The isolated amplified fragments were used for the template and the sequences underscored for the primers. The ^{32}P-labeled single nucleotide was either G (normal) or T (mutant). When only G is incorporated into the primer, the subjects are normal at this site of mutation. When both G and T are incorporated, the subjects are heterozygous for this mutation.

phisms in the mRNA homologous region of the protein C gene have been analyzed (Table 2), none of which result in a change in the amino acid sequence (153,174–176). In addition to these, several RFLP sites within or near the protein C gene have been reported. Te Lintel Hekkert et al. (177) reported two RFLPs detected by *Msp*I and *Apa*I, respectively, at approximately 7 kb 5′ end to the first exon of the protein C gene. The frequency of these polymorphisms among the unrelated white European population was 0.31. The *Msp*I DNA polymorphisms were found to be useful for the analysis of congenital type I protein C deficiency (178). Koenhen et al. (179) reported the *Msp*I RFLP in intron 8 of the protein C gene showing a frequency of 0.01 in the healthy white population. Yamamoto et al. (174) utilized a nucleotide variation (T or G) coding for Ser^{99} (TC*T* or TC*G*) at exon 6 of the protein C gene for DNA polymorphism analysis. They created a de novo polymorphic *Xba*I site (T/CTAGA) by polymerase chain reaction (PCR) amplification using mutagenic primers and found that the *Xba*I RFLP is useful for tracing the affected protein C gene among protein C-deficient family members.

We used RFLP analysis by creation of new restriction sites using the mutagenic primers (174) and single-nucleotide primer extension analysis using PCR (180) to detect the mutations in the family members just described (170). Because the nucleotide substitution from G to A at nucleotide number 2977 identified in exon 3 (16) [corresponding to nucleotide number 1465 in exon 2 (15) of the patient gene] neither created nor destroyed any restriction sites, a nucleotide T instead of G was newly introduced at nucleotide 2974, three nucleotides upstream from the mutation site (Fig. 7). Amplification of the 162 bp fragment derived from exon 3 by PCR with a mutagenic primer followed by

*Xba*I digestion of the PCR product yielded two fragments of 132 and 30 bp in the mutant allele, whereas the wild-type product remained uncut. The patient (III-1), his father (II-1), paternal uncle (II-2), and paternal grandmother (I-2) all had this type of mutation. Thus, this mutant allele is derived from the paternal pedigree. On the other hand, for detection of a single G deletion at nucleotide 10758 in exon 9 (16) [corresponding to nucleotide 8857 in exon 8 (15) in the same family], single-nucleotide primer extension was performed. We synthesized a primer whose sequence was identical to the coding sequence of the normal gene immediately flanking the 5′ end of the mutation site and also DNA fragments containing the mutation site in exon 9 from each family member. A mixture containing the primer, DNA fragment, and either [α-^{32}P]dGTP or [α-^{32}P]dTTP was then subjected to only one cycle of PCR. The patient (III-1), his mother (II-3), and maternal grandmother (I-4) showed the incorporation of T instead of G in the mutant allele (Fig. 8), indicating that this mutant allele is derived from the maternal pedigree. These techniques are also useful for prenatal diagnosis.

Note added in proof: Recently, factor V was found to play an important role as a cofactor in the degradation of factor Va and factor VIIIa by activated protein C (APC) (181). Factor V and protein S are suggested to work in synergy as phospholipid-bound cofactors to APC (182). Congenital deficiency of the cofactor activity of factor V (APC cofactor 2) was shown first by Dahlbäck et al. (183), then by several research grounds (184–187) in a high proportion of patients with venous thromboembolism. They found patients whose plasma response to APC was defective. In these patients, that they grouped as having ‘‘APC resistance,’’ the addition of APC causes little prolongation of the activated partial thromboplastin time. It has now been shown that the APC resistance phenomenon is due to an unsusceptibility of factor V to APC, that is, the factor V molecule of the patients with APC resistance has a substitution of Arg^{506} (CGA), the cleavage site of APC, by Gln (CAA) caused by a point mutation (188–190). The relationship between this mutation of factor V and the deficiency of the APC cofactor 2 still remains unsolved.

ACKNOWLEDGMENTS

The author thanks Drs. Masaru Ido, Michiaki Ohiwa, and Tatsuya Hayashi for allowing me to include their unpublished observations and Dr. Hiroshi Takeya and Junji Nishioka and Masane Kume for preparing the manuscript.

REFERENCES

1. Davie EW, Fujikawa K, Kisiel W. The coagulation cascade: initiation, maintenance, and regulation. Biochemistry 1991; 30:10363–10370.
2. Mann KG, Nesheim ME, Church WR, Haley P, Krishnaswamy S. Surface-dependent reactions of the vitamin K-dependent enzyme complexes. Blood 1990; 76:1–16.
3. Stenflo J. The biochemistry of protein C. In: Bertina RM, ed. Protein C and Related Proteins. Longman Group UK: Churchill Livingstone, 1988:21–54.
4. Esmon CT. The roles of protein C and thrombomodulin in the regulation of blood coagulation. J Biol Chem 1989; 264:4743–4746.
5. Esmon CT. The regulation of natural anticoagulant pathways. Science 1987; 235:1348–1352.
6. Stenflo J. A new vitamin K-dependent protein. J Biol Chem 1976; 251:355–363.

7. Kisiel W. Human plasma protein C. Isolation, characterization and mechanism of activation by α-thrombin. J Clin Invest 1979; 64:761–769.
8. Dahlbäck B. Protein S and C4b-binding protein: components involved in the regulation of the protein C anticoagulant system. Thromb Haemost 1991; 66:49–61.
9. De Fouw NJ, Bertina RM, Haverkate F. Activated protein C and fibrinolysis. In: Bertina RM, ed. Protein C and Related Proteins. Longman Group UK: Churchill Livingstone, 1988: 71–90.
10. Esmon CT, Taylor FB, Snow TR. Inflammation and coagulation: linked processes potentially regulated through a common pathway mediated by protein C. Thromb Haemost 1991; 66:160–165.
11. Suzuki K, Deyashiki Y, Nishioka J, Toma K. Protein C inhibitor: structure and function. Thromb Haemost 1989; 61:337–342.
12. Heeb MJ, Griffin JH. Physiologic inhibition of human activated protein C by α_1-antitrypsin. J Biol Chem 1988; 263:11613–11636.
13. Tanabe S, Sugo T, Matsuda M. Synthesis of protein C in human umbilical vein endothelial cells. J Biochem (Tokyo) 1991; 109:924–928.
14. Patracchini P, Aiello V, Palazzi P, Calzolari E, Bernardi F. Sublocalization of the human protein C gene on chromosome 2q13–q14. Hum Genet 1989; 81:191–192.
15. Foster DC, Yoshitake S, Davie EW. The nucleotide sequence of the gene for human protein C. Proc Natl Acad Sci USA 1985; 82:4673–4677.
16. Plutzky J, Hoskins JA, Long GL, Crabtree GR. Evolution and organization of the human protein C gene. Proc Natl Acad Sci USA 1986; 83:546–550.
17. Long GL. The human protein C gene. In: Bertina RM, ed. Protein C and Related Proteins. Longman Group UK: Churchill Livingstone, 1988:117–129.
18. O'Hara PJ, Grant FJ, Haldeman BA, et al. Nucleotide sequence of the gene coding for human factor VII, a vitamin K-dependent protein participating in blood coagulation. Proc Natl Acad Sci USA 1987; 84:5158–5162.
19. Yoshitake S, Schach BG, Foster DC, Davie EW, Kurachi K. Nucleotide sequence of the gene for human factor IX (anti-hemophilic factor B). Biochemistry 1985; 24:3736–3750.
20. Leytus SP, Foster DC, Kurachi K, Davie EW. Gene for human factor X: a blood coagulation factor whose gene organization is essentially identical to that of factor IX and protein C. Biochemistry 1986; 25:5098–5102.
21. Suttie JW, Hoskins JA, Engelke J, et al. Vitamin K-dependent carboxylase—possible role of the substrate "propeptide" as an intracellular recognition site. Proc Natl Acad Sci USA 1987; 84:634–637.
22. Wu S-M, Cheung W-F, Frazier D, Stafford DW. Cloning and expression of the cDNA for human γ-glutamyl carboxylase. Science 1991; 254:1634–1636.
23. Foster DC, Sprecher CA, Holly RD, Gambee JE, Walker KM, Kumar AA. Endoproteolytic processing of the dibasic cleavage site in the human protein C precursor in transfected mammalian cells; effects of sequence alterations on efficiency of cleavage. Biochemistry 1990; 29:347–354.
24. Orthner CL, Madurawe RD, Velander WH, Drohan WN, Battey FD, Strickland DK. Conformational changes in an epitope localized to the NH_2-terminal region of protein C; evidence for interaction of protein C domains. J Biol Chem 1989; 264:18781–18788.
25. Church WR, Bhushan FH, Mann KG, Bovill EG. Discrimination of normal and abnormal prothrombin and protein C in plasma using a calcium ion-inhibited monoclonal antibody to a common epitope on several vitamin K-dependent proteins. Blood 1989; 74:2418–2425.
26. Zhang L, Castellino FJ. A γ-carboxyglutamic acid (γ) variant (γ^6D, γ^7D) of human activated protein C displays greatly reduced activity as an anticoagulant. Biochemistry 1990; 29:10828–10834.

27. Freyssinet J-M, Beretz A, Klein-Soyer C, Gauchy J, Schuhler S, Cazenave J-P. Interference of blood coagulation vitamin K-dependent proteins in the activation of human protein C—involvement of the 4-carboxyglutamic acid domain in two distinct interactions with the thrombin-thrombomodulin complex and with phospholipids. Biochem J 1988; 256:501–507.
28. Hogg PJ, Ohlin A-K, Stenflo J. Identification of structural domains in protein C involved in its interaction with thrombin-thrombomodulin on the surface of endothelial cells. J Biol Chem 1992; 267:703–706.
29. Olsen PH, Esmon NL, Esmon CT, Laue TM. Ca^{2+} dependence of the interactions between protein C, thrombin, and the elastase fragment of thrombomodulin. Analysis by ultracentrifugation. Biochemistry 1992; 31:746–754.
30. Zhang L, Castellino FJ. Role of the hexapeptide disulfide loop present in the γ-carboxyglutamic acid domain of human protein C in its activation properties and in the in vitro anticoagulant activity of activated protein C. Biochemistry 1991; 30:6696–6704.
31. Zhang L, Jhingan A, Castellino J. Role of individual γ-carboxyglutamic acid residues of activated human protein C in defining its in vitro anticoagulant activity. Blood 1992; 80: 942–952.
32. Ohlin A-K, Stenflo J. Calcium-dependent interaction between the epidermal growth factor precursor-like region of human protein C and a monoclonal antibody. J Biol Chem 1987; 262:13798–13804.
33. Ohlin A-K, Linse S, Stenflo J. Calcium binding to the epidermal growth factor homology region of bovine protein C. J Biol Chem 1988; 263:7411–7417.
34. Stenflo J. Structure-function relationships of epidermal growth factor modules in vitamin K-dependent clotting factors. Blood 1991; 78:1637–1651.
35. Razaie AR, Esmon NL, Esmon CT. The high affinity calcium-binding site involved in protein C activation is outside the first epidermal growth factor homology domain. J Biol Chem 1992; 267:11701–11704.
36. Ohlin A-L, Landes G, Bourdon P, Oppenheimer C, Wydro R, Stenflo J. β-Hydroxyaspartic acid in the first epidermal growth factor-like domain of protein C—its role in Ca^{2+} binding and biological activity. J Biol Chem 1988; 263:19240–19248.
37. Stearns DJ, Kurosawa S, Sims PJ, Esmon NL, Esmon CT. The interaction of a Ca^{2+}-dependent monoclonal antibody with the protein C activation peptide region—evidence for obligatory Ca^{2+} binding to both antigen and antibody. J Biol Chem 1988; 263:826–832.
38. Sakata Y, Curriden S, Lawrence D, Griffin JH, Loskutoff DJ. Activated protein C stimulates the fibrinolytic activity of cultured endothelial cells and decreases antiactivator activity. Proc Natl Acad Sci USA 1985; 82:1121–1125.
39. Van Hinsbergh VWM, Bertina RM, van Wijngaarden A, van Tilburg NH, Emeis JJ, Haverkate F. Activated protein C decreases plasminogen activator-inhibitor activity in endothelial cell-conditioned medium. Blood 1985; 65:444–451.
40. McClure DB, Walls JD, Grinnell BW. Post-translational processing events in the secretion pathway of human protein C, a complex vitamin K-dependent antithrombotic factor. J Biol Chem 1992; 267:19710–19717.
41. Grinnell BW, Walls JD, Gerlitz B. Glycosylation of human protein C affects its secretion, processing, functional activities, and activation by thrombin. J Biol Chem 1991; 266:9778–9785.
42. Miletich JP, Broze GJ Jr. β-Protein C is not glycosylated at asparagine 329; The rate of translation may influence the frequency of usage at asparagine-*X*-cysteine sites. J Biol Chem 1990; 265:11397–11404.
43. Bjoern S, Foster DC, Thim L, et al. Human plasma and recombinant factor VII. Characterization of O-glycosylations at serine residues 52 and 60 and effects of site-directed mutagenesis of serine 52 to alanine. J Biol Chem 1991; 266:11051–11057.

44. Nishimura H, Takao T, Hase S, Shimonishi Y, Iwanaga S. Human factor IX has a tetrasaccharide O-glycosidically linked to serine 61 through the fucose residue. J Biol Chem 1992; 267:17520–17525.
45. Harris RJ, Ling VT, Spellman MW. O-linked fucose is present in the first epidermal growth factor domain of factor XII but not protein C. J Biol Chem 1992; 267:5102–5107.
46. Kentzer EJ, Buko A, Menon G, Sarin VK. Carbohydrate composition and presence of a fucose-protein linkage in recombinant human pro-urokinase. Biochem Biophys Res Commun 1990; 171:401–406.
47. Harris RJ, Leonard CK, Guzzetta AW, Spellman MW. Tissue plasminogen activator has an O-linked fucose attached to threonine-61 in the epidermal growth factor domain. Biochemistry 1991; 30:2311–2314.
48. Kisiel W, Ericsson LH, Davie EW. Proteolytic activation of protein C from bovine plasma. Biochemistry 1976; 15:4893–4900.
49. Freyssinet J-M, Wiesel M-L, Grunebaum L, et al. Activation of human protein C by blood coagulation factor Xa in the presence of anionic phospholipids: enhancement by sulphated polysaccharides. Biochem J 1989; 261:341–348.
50. Kisiel W, Kondo S, Smith KJ, McMullen BA, Smith LF. Characterization of a protein C activator from *Agkistrodon contortrix contortrix* venom. J Biol Chem 1987; 262:12607–12613.
51. Orthner CL, Bhattacharya P, Strickland DK. Characterization of a protein C activator from the venom of *Agkistrodon contortrix contortrix*. Biochemistry 1988; 27:2558–2564.
52. Nakagaki T, Kazim AL, Kisiel W. Isolation and characterization of a protein C activator from tropical moccasin venom. Thromb Res 1990; 58:593–602.
53. Kisiel W, Canfield WM, Ericsson LH, Davie EW. Anticoagulant properties of bovine plasma protein C following activation by thrombin. Biochemistry 1977; 16:5824–5831.
54. McMullen BA, Fujikawa K, Kisiel W. Primary structure of a protein C activator from *Agkistrodon contortrix contortrix* venom. Biochemistry 1989; 28:674–679.
55. Takeya H, Nishida S, Miyata T, et al. Coagulation factor X activating enzyme from Russell's viper venom (RVV-X). A novel metalloproteinase with disintegrin (platelet aggregation inhibitor)-like and C-type lectin-like domains. J Biol Chem 1992; 267:14109–14117.
56. Esmon NL, Owen WG, Esmon CT. Isolation of a membrane-bound cofactor for thrombin-catalyzed activation of protein C. J Biol Chem 1982; 257:859–864.
57. Suzuki K, Kusumoto H, Hashimoto S. Isolation and characterization of thrombomodulin from bovine lung. Biochim Biophys Acta 1986; 882:343–352.
58. Nakagomi K, Ajisaka K, Yokota I. Effects of acetylthrombin on protein C activation and fibrinogen clotting. Thromb Res 1990; 59:713–722.
59. Musci G, Berliner KJ. Ligands which effect human protein C activation by thrombin. J Biol Chem 1987; 262:13889–13891.
60. Freyssinet J-M, Toti-Orfanoudakis F, Ravanat C, et al. The catalytic role of anionic phospholipids in the activation of protein C by factor Xa and expression of its anticoagulant function in human plasma. Blood Coag Fib 1991; 2:691–698.
61. Ehrlich HJ, Grinnell BW, Jaskunas SR, Esmon CT, Yan SB, Bang NU. Recombinant human protein C derivatives; altered response to calcium resulting in enhanced activation by thrombin. EMBO J 1990; 9:2367–2373.
62. Richardson MA, Gerlitz B, Grinnell BW. Enhancing protein C interaction with thrombin results in a clot-activated anti-coagulant. Nature 1992; 360:261–264.
63. Freyssinet F-M, Cazenave F-P. Thrombomodulin. In: Bertina RM, ed. Protein C and Related Proteins. Longman Group UK: Churchill Livingstone, 1988:91–105.
64. Esmon CT, Esmon NL, Harris KW. Complex formation between thrombin and thrombomodulin inhibits both thrombin-catalyzed fibrin formation and factor V activation. J Biol Chem 1982; 257:7944–7947.

65. Esmon NL, Carroll RC, Esmon CT. Thrombomodulin blocks the ability of thrombin to activate platelets. J Biol Chem 1983; 258:12238–12242.
66. Polgar J, Lerant I, Muszbek L, Machovich R. Thrombomodulin inhibits the activation of factor XIII by thrombin. Thromb Res 1986; 43:585–590.
67. Suzuki K, Kusumoto H, Deyashiki Y, et al. Structure and expression of human thrombomodulin, a thrombin receptor on endothelium acting as a cofactor for protein C activation. EMBO J 1987; 6:1891–1897.
68. Jackman RW, Beeler DL, Fritze L, Soff G, Rosenberg RD. Human thrombomodulin gene is intron depleted: nucleic acid sequences of the cDNA and gene predict protein structure and suggest sites of regulatory control. Proc Natl Acad Sci USA 1987; 84:6425–6429.
69. Wen D, Dittman WA, Ye RD, Deaven LL, Majerus PW, Sadler JE. Human thrombomodulin: complete cDNA sequence and chromosome localization of the gene. Biochemistry 1987; 26:4350–4357.
70. Bourin M-C, Ohlin A-K, Lane DA, Stenflo J, Lindahl U. Relationship between anticoagulant activities and polyanionic properties of rabbit thrombomodulin. J Biol Chem 1988; 263:8044–8052.
71. Bourin M-C, Lindahl U. Functional role of the polysaccharide component of rabbit thrombomodulin proteoglycan; effects on inactivation of thrombin by antithrombin, cleavage of fibrinogen by thrombin and thrombin-catalysed activation of factor V. Biochem J 1990; 270:419–425.
72. Kurosawa S, Galvin JB, Esmon NL, Esmon CT. Proteolytic formation and properties of functional domains of thrombomodulin. J Biol Chem 1987; 262:2206–2212.
73. Kurosawa S, Stearns DJ, Jackson KW, Esmon CT. A 10-kDa cyanogen bromide fragment from the epidermal growth factor homology domain of rabbit thrombomodulin contains the primary thrombin binding site. J Biol Chem 1988; 263:5993–5996.
74. Suzuki K, Hayashi T, Nishioka J, et al. A domain compound of epidermal growth factor-like structures of human thrombomodulin is essential for thrombin binding and for protein C activation. J Biol Chem 1989; 264:4872–4876.
75. Stearns DJ, Kurosawa S, Esmon CT. Microthrombomodulin—residues 310–486 from the epidermal growth factor precursor homology domain of thrombomodulin will accelerate protein C activation. J Biol Chem 1989; 264:3352–3356.
76. Ye J, Liu L-W, Esmon CT, Johnson AE. The fifth and sixth growth factor-like domains of thrombomodulin bind to the anion-binding exosite of thrombin and alter its specificity. J Biol Chem 1992; 267:11023–11028.
77. Zushi M, Gomi K, Yamamoto S, Maruyama I, Hayashi T, Suzuki K. The last three consecutive epidermal growth factor-like structures of human thrombomodulin comprise the minimum functional domain for protein C-activating cofactor activity and anticoagulant activity. J Biol Chem 1989; 264:10351–10353.
78. Tsiang M, Lentz SR, Sadler JE. Functional domains of membrane-bound human thrombomodulin: EGF-like domains four to six and the serine/threonine-rich domain are required for cofactor activity. J Biol Chem 1992; 267:6164–6170.
79. Hayashi T, Zushi M, Yamamoto S, Suzuki K. Further localization of binding sites for thrombin and protein C in human thrombomodulin. J Biol Chem 1990; 265:20156–20159.
80. Musci G, Berliner LJ, Esmon CT. Evidence for multiple conformation changes in the active center of thrombin induced by complex formation with thrombomodulin: an analysis employing nitroxide spin-labels. Biochemistry 1988; 27:769–773.
81. Ye J, Esmon NL, Esmon CT, Johnson AE. The active site of thrombin is altered upon binding to thrombomodulin: two distinct structural changes are detected by fluorescence, but only one correlates with protein C activation. J Biol Chem 1991; 266:23016–23021.
82. Olsen PH, Esmon NL, Esmon CT, Laue TM. Ca^{2+} dependence of the interactions between protein C, thrombin, and the elastase fragment of thrombomodulin. Analysis by ultracentrifugation. Biochemistry 1992; 31:746–754.

83. Zushi M, Gomi K, Honda G, Kondo S, Yamamoto S, Suzuki K. Aspartic acid 349 in the fourth epidermal growth factor-like structure of human thrombomodulin plays a role in its Ca^{2+}-mediated binding to protein C. J Biol Chem 1991; 266:19886–19889.
84. Nagashima M, Lundh E, Leonard JC, Morser J, Parkinson JF. Alanine-scanning mutagenesis of the epidermal growth factor-like domains of human thrombomodulin identifies critical residues for its cofactor activity. J Biol Chem 1993; 268:2888–2892.
85. Clarke JH, Light DR, Blasko E, et al. The short loop between epidermal growth factor-like domains 4 and 5 is critical for human thrombomodulin function. J Biol Chem 1993; 268:6309–6315.
86. Le Bonniec BF, Esmon CT. Glu-192-Gln substitution in thrombin mimics the catalytic switch induced by thrombomodulin. Proc Natl Acad Sci USA 1991; 88:7371–7375.
87. Le Bonniec BF, MacGillivray RTA, Esmon CT. Thrombin Glu-39 restricts the P′3 specificity to nonacidic residues. J Biol Chem 1991; 266:13796–13803.
88. Suzuki K, Nishioka J, Hayashi T. Localization of thrombomodulin-binding site within human thrombin. J Biol Chem 1990; 265:13263–13267.
89. Suzuki K, Nishioka J. A thrombin-based peptide corresponding to the sequence of the thrombomodulin-binding site blocks the procoagulant activities of thrombin. J Biol Chem 1991; 266:18498–18501.
90. Nishioka J, Taneda H, Suzuki K. Estimation of the possible recognition sites for thrombomodulin, procoagulant, and anticoagulant proteins around the active center of α-thrombin. J Biochem 1993; 114:148–155.
91. Preissner KT, Koyama T, Muller D, Tschopp J, Muller-Berghaus G. Domain structure of the endothelial cell receptor thrombomodulin as deduced from modulation of its anticoagulant functions; evidence for a glycosaminoglycan-dependent secondary binding site for thrombin. J Biol Chem 1990; 265:4915–4922.
92. Koyama T, Parkinson JF, Sie P, Bang NU, Muller-Berghaus G, Preissner KT. Different glycoforms of human thrombomodulin: their glycosaminoglycan-dependent modulatory effects on thrombin inactivation by heparin cofactor II and antithrombin III. Eur J Biochem 1991; 198:563–570.
93. Nawa K, Sakano K, Fujiwara H, et al. Presence and function of chondroitin-4-sulfate on recombinant human soluble thrombomodulin. Biochem Biophys Res Commun 1990; 171: 729–737.
94. Comp PC, Esmon CT. Activated protein C inhibits platelet prothrombin-converting activity. Blood 1979; 54:1272–1281.
95. Stern D, Brett J, Harris K, Nawroth P. Participation of endothelial cells in the protein C-protein S anticoagulant pathway: the synthesis and release of protein S. J Cell Biol 1986; 102:1971–1978.
96. Marlar RA, Kleiss AJ, Griffin JH. Mechanism of action of human activated protein C, a thrombin-dependent anticoagulant enzyme. Blood 1982; 59:1067–1072.
97. Suzuki K, Stenflo J, Dahlbäck B, Teodorsson B. Inactivation of human coagulation factor V by activated protein C. J Biol Chem 1983; 258:1914–1920.
98. Fulcher CA, Gardiner JE, Griffin JH, Zimmerman TS. Proteolytic inactivation of human factor VIII procoagulant protein by activated human protein C and its analogy with factor V. Blood 1984; 63:486–489.
99. Eaton D, Rodriguez H, Vehar GA. Proteolytic processing of human factor VIII. Correlation of specific cleavages by thrombin, factor Xa, and activated protein C with activation and inactivation of factor VIII coagulant activity. Biochemistry 1986; 25:505–512.
100. Fay JP, Smudzin TM, Walker FJ. Activated protein C-catalyzed inactivation of human factor VIII and factor VIIIa: identification of cleavage sites and correlation of proteolysis with cofactor activity. J Biol Chem 1991; 266:20139–145.
101. Kane WH, Davie EW. Blood coagulation factors V and VIII: structural and functional similarities and their relationship to hemorrhagic and thrombotic disorders. Blood 1988; 71:539–555.

102. Fay PJ, Walker FJ. Inactivation of human factor VIII by activated protein C—evidence that the factor VIII light chain contains the activated protein C binding site. Biochim Biophys Acta 1988; 994:142–148.
103. Walker FJ, Scandella D, Fay PJ. Identification of the binding site for activated protein C on the light chain of factors V and VIII. J Biol Chem 1990; 265:1484–1489.
104. Walker FJ, Fay PJ. Characterization of an interaction between protein C and ceruloplasmin. J Biol Chem 1990; 265:1834–1836.
105. Mesters RM, Houghten RA, Griffin JH. Identification of a sequence of human activated protein C (residues 390–404) essential for its anticoagulant activity. J Biol Chem 1991; 266:24514–24519.
106. Guinto ER, Esmon CT. Loss of prothrombin and of factor Xa-factor Va interactions upon inactivation of factor Va by activated protein C. J Biol Chem 1984; 259:13986–13992.
107. Luckow EA, Lyons DA, Ridgeway TM, Esmon CT, Laue TM. Interaction of clotting factor V heavy chain with prothrombin and prethrombin-1 and role of activated protein C in regulating this interaction: analysis by analytical ultra-centrifugation. Biochemistry 1989; 28:2348–2354.
108. Odegaard B, Mann KG. Proteolysis of factor Va by factor Xa and activated protein C. J Biol Chem 1987; 262:11233–11238.
109. Nesheim ME, Canfield WM, Kisiel W, Mann KG. Studies of the capacity of factor Xa to protect factor Va from inactivation by activated protein C. J Biol Chem 1982; 257:1443–1447.
110. Walker FJ, Chavin SI, Fay PJ. Inactivation of factor VIII by activated protein C and protein S. Arch Biochem Biophys 1987; 252:322–328.
111. Jane SM, Hau L, Salem HH. Regulation of activated protein C by factor Xa. Blood Coag Fib 1991; 2:723–729.
112. Walker FJ. Regulation of activated protein C by protein S. The role of phospholipid in factor Va inactivation. J Biol Chem 1981; 256:11128–11131.
113. Suzuki K, Nishioka J, Matsuda M, Murayama H, Hashimoto S. Protein S is essential for the activated protein C-catalyzed inactivation of platelet-associated factor Va. J Biochem 1984; 96:455–460.
114. Koedam JA, Meijers JCM, Sixma JJ, Bouma BN. Inactivation of human factor VIII by activated protein C—cofactor activity of protein S and protective effect of von Willebrand factor. J Clin Invest 1988; 82:1236–1243.
115. Taylor FB, Esmon CT, Lockhart MS. Optimum conditions for lysis of whole blood clots. In: Mann KG, Taylor FB Jr, eds. The Regulation of Coagulation. Amsterdam: Elsevier-North Holland, 1980:573–581.
116. Comp PC, Esmon CT. Generation of fibrinolytic activity by infusion of activated protein C into dogs. J Biol Chem 1981; 68:1221–1228.
117. Burdick MD, Schaub RG. Human protein C induces anticoagulation and increased fibrinolytic activity in the cat. Thromb Res 1987; 45:413–419.
118. Colucci M, Stassen JM, Collen D. Influence of protein C activation on blood coagulation and fibrinolysis in squirrel monkeys. J Clin Invest 1984; 74:201–204.
119. Taylor FB Jr, Lockhart MS. A new function for activated protein C: activated protein C prevents inhibition of plasminogen activators by released from mononuclear leukocytes-platelet suspensions stimulated by phorbol diester. Thromb Res 1985; 37:155–164.
120. De Fouw NJ, de Jong YF, Haverkate F, Bertina RM. Activated protein C increases fibrin clot lysis by neutralization of plasminogen activator inhibitor—no evidence for a cofactor role of protein S. Thromb Haemost 1988; 60:328–333.
121. D'Angelo A, Lockhart MS, D'Angelo SV, Taylor FB Jr. Protein S is a cofactor for activated protein C neutralization of an inhibitor of plasminogen activation released from platelets. Blood 1987; 69:231–237.

122. Taylor FB Jr, Hoogendoorn H, Chang ACK, et al. Anticoagulant and fibrinolytic activities are promoted, not retarded, in vivo after thrombin generation in the presence of a monoclonal antibody that inhibits activation of protein C. Blood 1992; 79:1720–1728.
123. Marlar RA, Griffin JH. Deficiency of protein C inhibitor in combined factor V/VIII deficiency disease. J Clin Invest 1980; 66:1186–1189.
124. Suzuki K. Protein C inhibitor. In: Bertina RM, ed. Protein C and Related Proteins. Longman Group UK: Churchill Livingstone, 1988:106–116.
125. Van der Meer FJM, van Tilburg NH, van Wijngaarden A, van der Linden IK, Briet E, Bertina RM. A second plasma inhibitor of activated protein C; α_1-antitrypsin. Thromb Haemost 1989; 62:756–762.
126. Laurell M, Christensson A, Abrahamsson P-A, Stenflo J, Lilja H. Protein C inhibitor in human body fluids: seminal plasma is rich in inhibitor antigen deriving from cells throughout the male reproductive system. J Clin Invest 1992; 89:1094–1101.
127. Suzuki K, Nishioka J, Hashimoto S. Protein C inhibitor. Purification from human plasma and characterization. J Biol Chem 1983; 258:163–168.
128. Suzuki K, Deyashiki Y, Nishioka J, et al. Characterization of a cDNA for human protein C inhibitor. A new member of the plasma serine protease inhibitor superfamily. J Biol Chem 1987; 262:611–616.
129. Heeb MJ, Espana F, Geiger M, Collen D, Stump DC, Griffin JH. Immunological identity of heparin-dependent plasma and urinary protein C inhibitor and plasminogen activator inhibitor-3. J Biol Chem 1987; 262:15813–15816.
130. Stief TW, Padtke K-P, Heimburger N. Inhibition of urokinase by protein C inhibitor (PCI)—evidence for identity of PCI and plasminogen activator inhibitor 3. Biol Chem Hoppe Seyler 1987; 368:1427–1433.
131. Meijers JCM, Chung DW. Organization of the gene coding for human protein C inhibitor (plasminogen activator-3): assignment of the gene to chromosome 14. J Biol Chem 1991; 266:15028–15034.
132. Meijers JCM, Kanters DHA, Vlooswijk RAA, van Erp HE, Hessing M, Bouma BN. Inactivation of human plasma kallikrein and factor XIa by protein C inhibitor. Biochemistry 1988; 27:4231–4237.
133. Ecke S, Geiger M, Resch I, et al. Inhibition of tissue kallikrein by protein C inhibitor: evidence for identity of protein C inhibitor with the kallikrein binding protein. J Biol Chem 1992; 267:7048–7052.
134. Laurell M, Stenflo J. Protein C inhibitor from human plasma: characterization of native and cleaved inhibitor and demonstration of inhibitor complexes with plasma kallikrein. Thromb Haemost 1989; 62:885–891.
135. Pratt CW, Church FC. Heparin binding to protein C inhibitor. J Biol Chem 1992; 267: 8789–8794.
136. Pratt CW, Whinna HC, Church FC. A comparison of three heparin-binding serine proteinase inhibitors. J Biol Chem 1992; 267:8795–8801.
137. Heeb MJ, Griffin JH. Physiologic inhibition of human activated protein C by α_1-antitrypsin. J Biol Chem 1988; 263:11613–11616.
138. Hoogendoorn H, Nesheim ME, Giles AR. A qualitative and quantitative analysis of the activation and inactivation of protein C in vivo in a primate model. Blood 1990; 75:2164–2171.
139. Espana F, Vicente V, Tabernero D, Scharrer I, Griffin JH. Determination of plasma protein C inhibitor and of two activated protein C-inhibitor complexes in normals and in patients with intravascular coagulation and thrombotic disease. Thromb Res 1990; 59:593–608.
140. Laurell M, Stenflo J, Carlson TH. Turnover of *I-protein C inhibitor and *I-α_1-antitrypsin and their complexes with activated protein C. Blood 1990; 76:2290–2295.
141. Suzuki K, Nishioka J, Kusumoto H, Hashimoto S. Mechanism of inhibition of activated protein C by protein C inhibitor. J Biochem 1984; 95:187–195.

142. Broekmans AW, Conard F. Hereditary protein C deficiency. In: Bertina RM, ed. Protein C and Related Proteins. Longman Group UK: Churchill Livingstone, 1988:160–181.
143. Marlar RA, Mastovich S. Hereditary protein C deficiency: a review of the genetics, clinical presentation, diagnosis and treatment. Blood Coag Fib 1990; 1:319–330.
144. Miletich J, Sherman L, Broze G Jr. Absence of thrombosis in subjects with heterozygous protein C deficiency. N Engl J Med 1987; 317:991–996.
145. Reitsma PH, Poort SR, Bernardi F, et al. Protein C deficiency: a database of mutations. Thromb Haemost 1993; 69:77–84.
146. Grundy CB, Melissari E, Lindo V, Scully MF, Kakkar VV, Cooper DN. Late onset homozygous protein C deficiency. Lancet 1991; 338:575–576.
147. Conard J, Horellou MH, Van Dreden P, et al. Homozygous protein C deficiency with late onset and recurrent coumarin-induced skin necrosis. Lancet 1992; 339:743–744.
148. Romeo G, Hassan HJ, Staempfli S, et al. Hereditary thrombophilia: identification of nonsense and missense mutation in the protein C gene. Proc Natl Acad Sci USA 1987; 84: 2829–2832.
149. Matsuda M, Sugo T, Sakata Y, et al. A thrombotic state due to an abnormal protein C. N Engl J Med 1988; 319:1265–1268.
150. Grundy C, Chitolie A, Talbot S, Bevan D, Kakkar V, Cooper DN. Protein C London 1: recurrent mutation at Arg 169 (CGG → TGG) in the protein C gene causing thrombosis. Nucleic Acids Res 1989; 17:10513.
151. Gandrille S, Vidaud M, Aiach M, et al. Quantitative hereditary protein C deficiency: identification of two novel mutations at CpG sites in exon VI of the protein C gene. Blood 1990; 76:2018.
152. Schwartz HP, Linnau Y, Schramm W, Dreyfus M, Bauer KA. Treatment of protein C deficiency with a monoclonal antibody purified protein C concentrate. Circulation 1990; 82:134a.
153. Reitsma PH, Poort SR, Allaart CF, Briet E, Bertina RM. The spectrum of genetic defects in a panel of 40 Dutch families with symptomatic protein C deficiency type I: heterogeneity and founder effects. Blood 1991; 78:890–894.
154. Gandrille S, Vidaud M, Aiach M, et al. Six previously undescribed mutations in 9 families with protein C quantitative deficiency. Thromb Haemost 1991; 65:646.
155. Reitsma PH, Poort SR, Bertina RM. Genetic abnormalities in the protein C genes of homozygous and compound heterozygotes for protein C deficiency. Thromb Haemost 1991; 65:808.
156. Gandrille S, Alhenc-Gelas M, Juhan-Vague I, Fiessinger JN, Goossens M, Aiach M. Two qualitative protein C (PC) deficiencies due to Arg 229-Gln (PC Marseille) and Ser 252-Asn (PC Paris) mutations. Thromb Haemost 1991; 65:1196.
157. Gandrille S, Alhenc-Gelas M, Goossens M, Aiach M. PC Malakoff: an abnormal protein C due to mutation at the cleavage site of the propeptide. Thromb Haemost 1991; 65: 1196.
158. Sala N, Poort SR, Bertina RM, Soria JM, Fontcuberta J, Reitsma PH. Identification of the two deletions and four point mutations in the protein C gene of 6 unrelated Spanish patients with hereditary protein C deficiency. Thromb Haemost 1991; 65:1197.
159. Tsay W, Greengard JS, Montgomery R, Griffin JH. Five previously undescribed mutations in protein C (PC) that identify elements critical for gene and protein activity. Blood 1991; 78:184a.
160. Tsuda S, Reitsma PH, Miletich J. Molecular defects causing heterozygous protein C deficiency in three asymptomatic kindreds. Thromb Haemost 1991; 65:647.
161. Dreyfus M, Magny JF, Bridey F, et al. Treatment of homozygous protein C deficiency and neonatal purpura fulminans with a purified protein C concentrate. N Engl J Med 1991; 325:1565–1568.
162. Grundy C, Plendl H, Grote W, Zoll B, Kakkar VV, Cooper DN. A single base-pair deletion in the protein C gene causing recurrent thromboembolism. Thromb Res 1991; 61:335–340.

163. Brito D, Botero L, Barcelo J, et al. Deficit total de proteina C en un recien nacido. Sangre 1991; 36(Suppl. 2):216.
164. Sugahara Y, Miura O, Yuen P, Aoki N. Protein C deficiency Hong Kong 1 and 2: hereditary protein C deficiency caused by two mutant alleles, a 5-nucleotide deletion and a missense mutation. Blood 1992; 126–133.
165. Grundy CB, Schulman S, Tengborn L, Kakkar VV, Cooper DN. Two different missense mutations at Arg_{178} of the protein C (PROC) gene causing recurrent venous thrombosis. Hum Genet 1992; 89:685–686.
166. Bernardi F, Patracchini P, Gemmati D, et al. Rapid detection of a protein C mutation present in the asymptomatic and not in the thrombosis-prone lineage. Br J Haematol 1992; 81: 277–282.
167. Yamamoto K, Matsushita T, Sugiura I, et al. Homozygous protein C deficiency: identification of a novel missense mutation that causes impaired secretion of the mutant protein C. J Lab Clin Med 1992; 119:682–689.
167a. Yamamoto K, Tanimoto M, Emi N, et al. Impaired secretion of the elongated mutant of protein C (Protein C Nagoya). Molecular and cellular basis for hereditary protein C deficiency. J Clin Invest 1992; 90:2439–2446.
168. Bovill EG, Tomczak JA, Grant B, et al. Protein $C_{Vermont}$: symptomatic type II protein C deficiency associated with two Gla domain mutations. Blood 1992; 79:1456–1465.
169. Mimuro J, Muramatsu S, Kaneko M, et al. An abnormal protein C (protein C Yonago) with an amino acid substitution of Gly for Arg-15 caused by a single base mutation of C to G in codon 57 (CGG → GGG). Int J Hematol 1993; 57:9–14.
170. Ido M, Ohiwa M, Hayashi T, et al. A compound heterozygous protein C deficiency with a single nucleotide G deletion encoding Gly-381 and an amino acid substitution of Lys for Gla-26. Thromb Haemost 1993; 70:636–641.
171. Cooper DN, Youssoufia H. The CpG dinucleotide and human genetic disease. Hum Genet 1988; 78:151–155.
172. Cooper DN, Krawczak M. The mutational spectrum of single base-pair substitutions causing human genetic disease: patterns and predictions. Hum Genet 1990; 85:55–74.
173. Bird AP. CpG-rich islands and function of DNA methylation. Nature 1986; 321:209–213.
174. Yamamoto K, Tanimoto M, Matsushita T, et al. Genotype establishments for protein C deficiency by use of a DNA polymorphism in the gene. Blood 1991; 77:2633–2636.
175. Yamamoto K, Takamatsu J, Saito H. Two novel sequence polymorphisms of the human protein C gene. Nucleic Acids Res 1991; 19:6973.
176. Gandrille S, Alach M. Polymorphism in the protein C gene detected by denaturing gradient gel electrophoresis. Nucleic Acids Res 1991; 19:6982.
177. Te Lintel Hekkert W, Bertina RM, Reitsma PH. Two RFLPS −7kb 5′ of the human protein C gene. Nucleic Acids Res 1988; 16:11849.
178. Reitsma PH, te Lintel Hekkert W, Koenhen E, et al. Application of two neutral *Msp*I DNA polymorphisms in the analysis of hereditary protein C deficiency. Thromb Haemost 1990; 64:239–244.
179. Koenhen E, Bertina RM, Reitsma PH. *Msp*I RFLP in intron 3 of the human protein C gene. Nucleic Acids Res 1989; 17:8401.
180. Kuppuswamy MN, Hoffman JW, Kasper CK, Spitzer SG, Groce SL, Bajaj SP. Single nucleotide primer extension to detect genetic disease: experimental application to hemophilia B (factor IX) and cystic fibrosis genes. Proc Natl Acad Sci USA 1991; 88:1143–1147.
181. Dahlbäck B, Hildebrand B. Inherited resistance to activated protein C is corrected by anticoagulant cofactor activity found to be a property of factor V. Proc Natl Acad Sci USA 1994; 91:1396–1400.
182. Shen L, Dahlbäck B. Factor V and protein S as synergistic cofactors to activated protein C in degradation of factor VIIIa. J Biol Chem 1994; 269:18735–18738.

183. Dahlbäck B, Carlsson M, Svensson PJ. Familial thrombophilia due to a previously unrecognized mechanism characterized by poor anticoagulant response to activated protein C. Proc Natl Acad Sci USA 1993; 90:1004–1008.
184. Griffin JH, Evatt B, Wideman C, Fernandez JA. Anticoagulant protein C pathway defective in majority of thrombophilia patients. Blood 1993; 82:1989–1993.
185. Koster T, Rosendaal FR, de Ronde H, Briët E, Vandenbroucke JP, Bertina RM. Venous thrombosis due to poor anticoagulant response to activated protein C: Leiden thrombophilia study. Lancet 1994; 342:1503–1506.
186. Halbmayer W-M, Hanshofer A, Schön R, Fisher M. The prevalence of poor anticoagulant response to activiated protein C (APC resistance) among patients suffering from stroke or venous thrombosis and among healthy subjects. Blood Coag Fibrinol 1994; 5:51–57.
187. Svensson PJ, Dahlbäck B. Resistance to activated protein C as a basis for venous thrombosis. New Engl J Med 1994; 330:517–522.
188. Bertina RM, Koeleman BPC, Koster T, Rosendaal FR, Dirven RJ, de Ronde H, van der Velden PA, Reitsma PH. Mutation in blood coagulation factor V associated with resistance to activated protein C. Nature 1994; 369:64–67.
189. Voorberg J, Roelse J, Koopman R, Büller, H, Berends F, ten Cate JW, Mertens K, van Mourik JA. Association of idiopathic venous thromboembolism with single point-mutation at Arg^{506} of factor V. Lancet 1994; 343:1535–1536.
190. Zöller B, Dahlbäck B. Linkage between inherited resistance to activated protein C and factor V gene mutation in venous thrombosis. Lancet 1994; 343:1536–1538.

20

Thrombomodulin

William A. Dittman
Duke University Medical Center, and Durham Veterans Affairs Medical Center, Durham, North Carolina

Stephen C. Nelson
Duke University Medical Center, Durham, North Carolina

I. INTRODUCTION

Maintenance of blood fluidity is a critical component of hemostatic balance. The endothelial cells lining the vasculature are crucial in this process. Under normal circumstances they represent a continuous antithrombotic surface; however, a variety of stimuli can rapidly cause the endothelium to become prothrombotic. Anticoagulant functions of endothelium include production of prostacyclin (1), secretion of plasminogen activators (2), production of cell surface heparin-like proteoglycans that can stimulate coagulation factor inhibitors (3), and the expression of thrombomodulin, an endothelial cell surface glycoprotein that enhances the activation of protein C (4,5). Upon binding of thrombin to thrombomodulin, the inherent procoagulant nature of thrombin is modified. Thrombin bound to thrombomodulin is able to activate protein C with over a 1000-fold greater efficiency but can no longer clot fibrinogen or activate platelets as efficiently as unbound thrombin.

An activity later identified as protein C was first described in thrombin-treated plasma in 1960 (6) but was not characterized or purified until much later (7–9). Protein C circulates as an inactive zymogen with a plasma concentration of 4–5 μg/ml (10–14). Activated protein C in conjunction with calcium, a phospholipid surface, and protein S is able to suppress coagulation by proteolytically inactivating factors Va and VIIIa (15–17). The essential role of the protein C pathway in controlling coagulation is demonstrated by neonatal purpura fulminans in infants with homozygous protein C deficiency (18).

Efforts to detect an efficient protein C activator were unsuccessful until Esmon and Owen identified thrombomodulin in 1981 (19). They postulated that if a protein C activator is not circulating in the bloodstream, it must be present on the vascular endothelium. Using the Langendorff rabbit heart model, perfusion of thrombin and protein C simultaneously yielded a significant increase in protein C activation compared with the effect of thrombin without endothelium present. This activity was confirmed to be on endothelial cells in culture (20) and was isolated as a detergent-solubilized protein from rabbit lungs (21). Thrombomodulin has since been isolated from several sources, including human placenta (22,23) and bovine, mouse, and rat lung (24–26).

Thrombin forms a 1:1 complex with thrombomodulin with a K_d of 0.5 nM thrombin (Fig. 1). This complex activates protein C with a K_m of 0.5 μM protein C in phospholipid vesicles or on cell surfaces. Once complexed with thrombomodulin, thrombin is no longer able to catalyze procoagulant reactions efficiently. Fibrinogen clotting (27,28) and activation of platelets (29), factor V, and factor X (30), as well as inactivation of protein S (31), by thrombin are inhibited as the result of thrombomodulin binding to thrombin.

Thrombomodulin and the protein C pathway are crucial in regulating the production and function of thrombin helping to maintain an appropriate procoagulant-anticoagulant balance. This chapter describes the structure of thrombomodulin and its gene, in which thrombomodulin is normally found, and the interaction between thrombomodulin and other important factors in thrombosis and hemostasis. Regulation of protein expression and potential clinical applications are also addressed.

II. STRUCTURE OF THROMBOMODULIN

Figure 2 shows a model of the structure of thrombomodulin as determined by cDNA cloning and protein structure analysis. Human thrombomodulin has an apparent molecular weight of 75,000 when analyzed on sodium dodecyl sulfate–polyacrylamide gel electrophoresis. The molecular weight increases to 105,000 following the reduction of disulfide bonds representing secondary structure involving cystine bridges. The secondary structure of thrombomodulin provides stability under denaturing conditions as well as extremes of pH. Human (32,33), murine (34,35), and bovine (36) thrombomodulin cDNAs have been isolated and sequenced.

The amino terminus is followed by an 18 amino acid hydrophobic leader sequence. Residues 19–179 in the mature protein exhibit homology to lectin-like regions in such

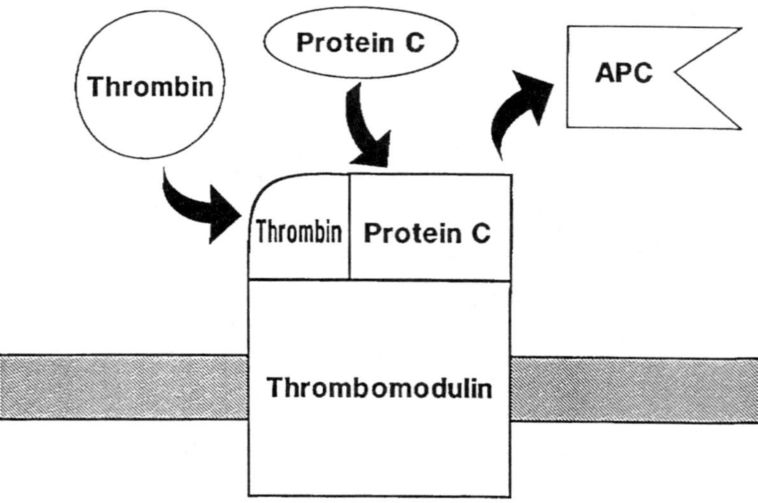

Figure 1 Model of thrombin-thrombomodulin activation of protein C. APC, activated protein C. (Courtesy of Dittman WA, Majerus PW. Blood 1990; 75:329; W. B. Saunders Company, Philadelphia.)

proteins as the asialoglycoprotein receptor (37,38). Within this lectin-like domain there is significant homology between the human and murine cDNAs; the functional significance of this conservation is unclear because thrombomodulin has no known lectin activity. Thrombomodulin contains six epidermal growth factor (EGF)-like domains comprised of residues 238–481 (39). Thrombin binding was initially localized to a peptide consisting of the fifth and sixth EGF-like regions (40), but this segment alone was not adequate for protein C activation. A peptide including the fourth, fifth, and sixth repeats exhibits both thrombin binding and protein C-activating capabilities (41). Confirmation of the protein C activation site was accomplished by studying thrombomodulin activity in cells transiently expressing truncated cDNA constructs of thrombomodulin (42,43).

Although optimal protein C-activating activity requires EGF regions 4–6 (44,45), recent studies suggest that part of the linker region between the third and fourth EGF-like domains may play a role in maximizing protein C activation (45,46). Within the region necessary for protein C activation, Asp^{349} immediately before the fourth EGF-like domain of human thrombomodulin is important in the calcium-dependent protein C activation by thrombomodulin, because mutants with Ala substituted for Asp demonstrated decreased protein C activation (47).

In addition to forming the binding sites for thrombin and protein C, the EGF-like domains may also signal posttranslational modification of thrombomodulin. Some coagulation factors with EGF-like regions show β-hydroxylation of aspartic acid and asparagine (48,49). This is also seen in the low-density lipoprotein receptor (50). β-Hydroxyl amino acids in protein C bind calcium ions (51), and these amino acids may

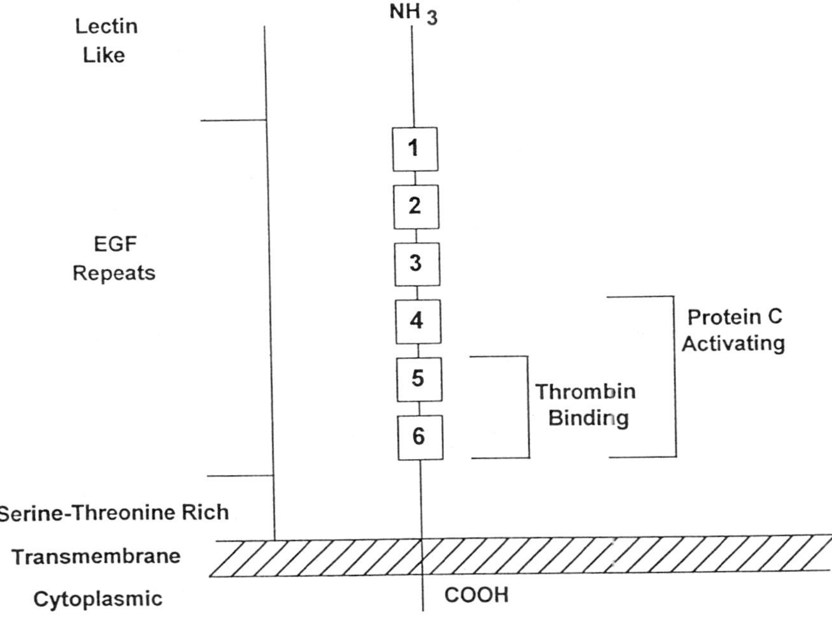

Figure 2 Structure of thrombomodulin determined by a combination of protein structure analysis and cDNA cloning. (Courtesy of Dittman WA, Majerus PW. Blood 1990; 75:329; W. B. Saunders Company, Philadelphia.)

play a similar role in thrombomodulin. The sequence signaling β-hydroxylation is conserved in thrombomodulin from all known species in EGF regions 4 and 6. Also, β-hydroxyaspartic acid has been identified in bovine thrombomodulin (50).

The serine-threonine–rich region follows the EGF domains and is the most heterogeneous region among species. This is the presumptive site of O-linked glycosylation. In some studies, heparin cofactor activity has been reported in rabbit thrombomodulin, as demonstrated by enhanced antithrombin III inactivation of thrombin in the presence of rabbit thrombomodulin (52,53). Not all studies support this finding, however (54). The role of this region in thrombin inhibition by thrombomodulin is being elucidated (55–57). Studies of recombinant human (58–62) and rabbit (63–65) thrombomodulin suggest the presence of a sulfated glycosaminoglycan moiety. Treatment of rabbit thrombomodulin with chondroitin ABC lyase decreases its ability to inhibit cleavage of fibrinogen by thrombin (63,64), decreases its effect on antithrombin III inactivation of thrombin (63–65), and decreases its inhibitory effect on factor V activation by thrombin (64). The protein C-activating ability of rabbit thrombomodulin was not affected by chondroitinase treatment (63). In contrast, surface thrombomodulin on human umbilical vein endothelial cells treated with β-D-xyloside (an inhibitor of glycosaminoglycan attachment to proteoglycan core proteins) demonstrates decreased protein C-activating ability (58). Recombinant human thrombomodulin treated with chondroitin ABC lyase lacks the ability to inhibit fibrinogen clotting efficiently by thrombin (59,62), shows decreased inhibition of thrombin by antithrombin III (59,61), and demonstrates decreased inhibition of platelet activation by thrombin (60). Protein C activation by recombinant human thrombomodulin is decreased following treatment of the thrombomodulin with chondroitinase (61). The posttranslational modification of thrombomodulin by attachment of a chondroitin sulfate glycosaminoglycan appears to be of major importance in some of the anticoagulant functions of thrombomodulin. The chondroitin sulfate glycosaminoglycan attachment site has been suggested to be Ser^{492} within the serine-threonine–rich region of human thrombomodulin (62). This residue is conserved in bovine and mouse thrombomodulin and lies within the sequence Ser-Gly-Ser^{492}-Gly-Glu-Pro, which is strongly similar to chondroitin sulfate attachment sites in other proteoglycans (32,34,36).

After the serine-threonine–rich region comes 23 hydrophobic amino acids, forming the transmembrane domain. This is the region of greatest homology between species, in which 20 amino acids are identical and the remaining 3 are conservative substitutions. In contrast, the low-density lipoprotein receptor has a heterogeneous transmembrane region among species (66), whereas, like thrombomodulin, the platelet-derived growth factor receptor exhibits strong homology between species. Platelet-derived growth factor, but not low-density lipoprotein, receptors are involved in transmembrane signaling, implying that the conserved transmembrane region of thrombomodulin may also have signaling activity (67), although no such activity has yet been identified.

Last, the short cytoplasmic tail of thrombomodulin contains several potential phosphorylation sites. Phosphorylation has been associated with increased endocytosis and degradation of thrombomodulin (34). The degree to which the cytoplasmic tail regulates the expression of thrombomodulin is not known.

III. THROMBOMODULIN GENE

A single copy of the human thrombomodulin gene is located on chromosome 20 (32). The gene has been sublocalized to the centromere of chromosome 20 at the p12 region:

20p12-cen (68). A C/T dimorphism in the thrombomodulin gene has been described that predicts an Ala^{455}-Val replacement in the sixth EGF-like domain of thrombomodulin (69). In a normal white population, the dimorphism had an allelic frequency of 82% Ala and 18% Val. This frequency was identical in a population of protein C-deficient patients and in a group of patients with thrombophilia of unknown etiology (69). Therefore, this dimorphism is neutral with respect to thrombosis.

Both the human and the mouse thrombomodulin genes are intronless (70,71). Intronless genes are uncommon and include rhodopsin, angiogenin, mitochondrial genes, and interferon-α and β, and the β-adrenergic receptors. Why some genes lack introns is unknown; however, the absence of introns, with the associated decreased mRNA processing, may allow quicker increases in protein synthesis following increased transcription.

Northern blot hybridization studies reveal a single mRNA of approximately 3700 bases, consistent with the size of cDNAs for mouse (3658 bases) and human (3693 bases) thrombomodulin. The 5′-untranslated region in humans is 150 bases; the coding region, comprised of 1725 bases, encodes 575 amino acids; and the 3′-untranslated region has 1779 bases. The 3′-untranslated region exhibits strong conservation among species, but its function is unknown. Human, murine, and bovine thrombomodulin all contain the sequence TTATTTAT in the 3′-untranslated region; this sequence has been associated with a short mRNA half-life in some RNAs, such as c-myc and interferon (72). This does not appear to be the case with thrombomodulin the mRNA half-life has been reported to be from 3 to 9 h (73,34).

IV. LOCATION OF THROMBOMODULIN

Expression of thrombomodulin both in vivo and in vitro is summarized in Table 1. Thrombomodulin was initially described as an endothelial cell surface protein, where its function as a natural anticoagulant can easily be understood. The majority of vascular thrombomodulin is found in capillaries, comprising more than 99% of the endothelial surface area. Early immunohistochemical studies of human tissues showed thrombomodulin on the endothelium of blood and lymphatic vessels. Initial studies suggested thrombomodulin was absent from the central nervous system (74,75). However, recent, more sensitive investigations suggest the presence of thrombomodulin in human brain vasculature, as well as the leptomeninges and lining of the ventricles (76,77). There appears

Table 1 Location of Thrombomodulin

In vivo	In vitro
Endothelial cells	A549 lung cancer cells
Syncytiotrophoblasts	NIH 3T3 cells
Leptomeninges	Chinese hamster ovary cells
Epidermis	MEG-01 cells
Megakaryocytes, platelets	Human smooth muscle cells
Neutrophils	Human umbilical vein endothelial cells
Monocytes, macrophages	Human saphenous vein endothelial cells
Soluble in plasma and urine	Bovine saphenous vein endothelial cells
Mesothelioma	
Vascular tumors	

to be a similar density of thrombomodulin in the cortex, cerebellum, medulla, and hippocampus, whereas there is much less thrombomodulin present in the putamen, pons, and mesencephalon (76). The significance of this regional distribution in the development of central nervous system thrombosis is unclear. There may be regional differences in distribution of thrombomodulin in the remainder of the body. Endothelial cells from different sites are known to differ in their expression of thrombomodulin (78).

The syncytiotrophoblast of the human placenta contains thrombomodulin, as does the syncytiotrophoblast of hydatidiform mole and uterine choriocarcinoma (79,80). Other neoplasms expressing thrombomodulin include mesothelioma and squamous cell carcinoma of the lung, but not adenocarcinoma (81,82). Various vascular tumors contain thrombomodulin, including angiosarcoma (80) and infantile hemangioendothelioma (83).

Normal human tissues other than the vasculature express thrombomodulin. The protein has been identified by immunohistochemistry on the squamous epithelium of the epidermis, except in the basal layer cells, mainly in the stratum malpighii (84). Its function here is not known. Thrombomodulin has also been detected in human megakaryocytes and platelets, in which approximately 60 molecules of thrombomodulin are found per platelet, compared with almost 100,000 molecules per endothelial cell (85,86). Human neutrophils synthesize thrombomodulin, but, surprisingly, neutrophil thrombomodulin lacks the ability to activate protein C (87). Thrombomodulin is expressed on human blood monocytes and human tissue macrophages (88). Thrombomodulin on monocytes suggests a role of thrombomodulin in inflammatory reactions in addition to its natural anticoagulant functions.

Smaller, soluble forms of thrombomodulin have been isolated from human urine and blood (89). There appears to be no circadian fluctuation in plasma-soluble thrombomodulin levels, although urinary excretion is highest during the day (90). Immunoassay shows that there is approximately 20 ng/ml of thrombomodulin in normal human serum. This soluble serum protein has a higher K_m for protein C activation than that of tissue thrombomodulin (91). The structure of soluble thrombomodulin is not known but is likely a cleaved form of tissue thrombomodulin with loss of a portion of the serine-threonine–rich region, the transmembrane domain, and the cytoplasmic tail. Other circulating proteins, such as macrophage colony-stimulating factor, epidermal growth factor, and transforming growth factor α, are synthesized as transmembrane glycoproteins that are proteolytically cleaved (92,93). It is conceivable that circulating plasma thrombomodulin also has some important physiological role.

Thrombomodulin has also been found in a variety of cell lines in culture, including A549 lung cancer cells (94), NIH 3T3 cells (34), Chinese hamster ovary cells, MEG-01 human megakaryoblastic cells (86), and human smooth muscle cells (95). Interestingly, thrombomodulin expression and mRNA levels may increase in human umbilical vein endothelial cells with successive cell passages (78).

V. INTERACTION OF THROMBOMODULIN WITH OTHER FACTORS

Thrombin is a disulfide-bonded two-chain protease with the active-site serine located on the larger B chain. Elucidating the sites on thrombin responsible for interaction with thrombomodulin has been an area of great interest. Studies chemically modifying the B chain of thrombin led to loss of thrombomodulin binding affinity, suggesting that the thrombomodulin binding site is on the B chain of thrombin (96,97). Indeed, a thrombomodulin binding site has been located within the sequence between Thr^{147} and Ser^{158} of the B chain by mapping of an inhibitory monoclonal antibody (98). Thrombomodulin may also alter the enzyme-substrate interaction near residue 192 on thrombin; a mutant thrombin that activates protein C in the absence of thrombomodulin can be made by a Glu^{192}-Gln substitution (99).

Thrombomodulin competes with hirudin and fibrinogen for thrombin binding at the anionic exosite (52,100–102). As expected, thrombomodulin and fibrinogen share macromolecular specificity sites on thrombin (103), as demonstrated by competition with each other but not with a smaller thrombin substrate. These binding sites may overlap but are not identical as determined by dissociation of fibrinogen cleavage and protein C activation in recombinant mutant thrombins (104). The thrombomodulin binding site on thrombin is also shared with factor V and platelets (105). These findings help to explain the inhibition by thrombomodulin of thrombin-induced procoagulant activities.

Several studies have suggested that thrombomodulin binding induces conformational changes in thrombin (106–108). After binding of thrombin to thrombomodulin, surface thrombomodulin localizes the thrombin active site on the endothelial cell surface at a height of 66 ° above the plane of the membrane (109), a distance favoring the cleavage of membrane-bound protein C.

The thrombin-thrombomodulin complex also binds to protein C. Unlike thrombin-thrombomodulin binding, protein C binding requires calcium ions. The γ-carboxyglutamic acid (Gla) domains in protein C bind Ca^{2+}; however, Gla domains are not required for protein C activation by thrombomodulin (110). Calcium binding to Gla-domainless protein C elicits a conformational change that reduces activation by thrombin but enhances activation by the thrombin-thrombomodulin complex. There is evidence for a high-affinity Ca^{2+} binding site located in the EGF homology region of protein C (111). This site appears to be the β-hydroxyaspartic acid at residue 71 in the NH_2-terminal EGF domain (112,113). Additional studies have identified a critical Ca^{2+} binding site located outside the NH_2-terminal EGF domain (114). The precise location of this high-affinity binding site within the protein C molecule remains to be discovered.

Factor V also interacts with the thrombomodulin-protein C system. Activated factor V and isolated factor Va light chain can serve as direct cofactors for protein C activation (115,116), although less efficiently than thrombomodulin (117). At low concentrations, the factor Va light chain has been shown to stimulate protein C activation by the thrombin-thrombomodulin complex on cell surfaces, with an opposite effect at higher concentrations (118). A more recent study suggests that the factor Va light chain does not compete with thrombomodulin for thrombin binding (101).

The effects of various other clotting factors on protein C activation have been investigated, with conflicting results. Factors II, VII, IX, and XI and protein S are unable to activate protein C (119,120). Whether factor X can activate protein C is less clear. Haley et al. showed that factor Xa in the presence of phospholipid vesicles and calcium ions was able to activate bovine protein C with thrombomodulin as a cofactor (121). This result has not been confirmed by other groups. In contrast, they have been unable to demonstrate thrombomodulin-dependent protein C activation by Xa (119,120,122,123). Furthermore, equilibrium dialysis studies show no binding between thrombomodulin and factor Xa (124). The zymogen factor X may actually inhibit protein C activation by the thrombin-thrombomodulin complex (120,125,126). The γ-carboxyglutamic acid domain of factor X appears to be important in this inhibition of protein C activation (125,126).

Thrombin is able to cleave single-chain urokinase-type plasminogen activator (scuPA), rendering it fibrinolytically inactive. Rabbit thrombomodulin has been shown to accelerate this reaction in vitro (127,128). Based on competition with truncated forms of scuPA, the amino-terminal region of scuPA, with its kringle and EGF-like domains, appears to be integral in the effect of thrombomodulin on thrombin-induced scuPA cleavage and inactivation (128). Thus, thrombomodulin may possess not only natural anticoagulant properties but antifibrinolytic properties as well.

VI. REGULATION OF EXPRESSION

Thrombomodulin expression is regulated by a variety of mechanisms (Table 2). As discussed earlier, thrombomodulin is limited to a small number of cell types in vivo; this tissue-specific expression is presumably regulated by the promoter in the thrombomodulin gene. Also, the presence of an intrinsic glycosaminoglycan moiety suggests that variant glycoforms of thrombomodulin may be expressed differently in a species-, organ-, or tissue-specific manner as a means to regulate thrombomodulin function (61). Changes in transcription rates with subsequent alterations in translation of thrombomodulin also regulate expression, as do variations in the amount of thrombomodulin expressed at the cell surface. Endothelial cells in culture have different properties when obtained from different areas within the vasculature or from different species (78).

Thrombomodulin expression is also regulated by internalization and degradation of the cell surface molecule, with subsequent loss of protein C-activating ability. Once bound, the thrombin-thrombomodulin complex induces internalization and transport to the lysosomes, where there is release and degradation of thrombin and return of thrombomodulin to the cell surface (94). Protein C inhibits this endocytosis, but not activated protein C, implying that internalization and degradation do not occur until protein C activation is adequate (129). Not all endothelial cells internalize thrombomodulin. Neither the endothelial cell line EA.hy 926 nor human saphenous vein endothelial cells internalize the thrombin-thrombomodulin complex in vitro (130). However, endocytosis appears to be a function in human umbilical vein endothelial cells in culture (131). An activator of protein kinase C, phorbol myristate acetate, induces internalization and degradation of thrombomodulin (34,131). Thrombomodulin is known to be phosphorylated on serine residues, and the extent to which this phosphorylation plays a role in signaling internalization and degradation is unknown.

Table 2 Regulation of Protein Expression[a]

Decreased levels	Increased levels
Endocytosis	Cycloheximide
PMA	cAMP
TNF	Interleukin-4
Interleukin-1	Pentoxifylline
Endotoxin	Retinoic acid
Hypoxia	
Homocysteine	

[a]PMA, phorbol myristate acetate; TNF, tumor necrosis factor; cAMP, cyclic adenosine monophosphate.

Cytokines, mediators of inflammatory responses, can also affect thrombomodulin expression. Tumor necrosis factor (TNF) can rapidly change the usually anticoagulant endothelium by stimulating expression of the procoagulant tissue factor while decreasing thrombomodulin activity (132–134). Earlier studies suggested that the reduction in thrombomodulin by TNF is the result of internalization and lysosomal degradation (135). Subsequent investigations have shown that TNF inhibits thrombomodulin transcription (73,136) and that degradation of thrombomodulin by TNF may be less important in regulating expression (73,137). Other factors, such as bacterial endotoxin (138–140), interleukin-1 (134,139,141), and hypoxia (142), also cause decreased thrombomodulin activity on the surface of endothelial cells, as well as decreased thrombomodulin mRNA levels.

Although inflammatory mediators appear to decrease thrombomodulin activity, a more useful change may be the stimulation of protein C activation by the thrombin-thrombomodulin complex. The protein synthesis inhibitor cycloheximide induces thrombomodulin mRNA transcription (143), with subsequent two- to fourfold increases in mRNA levels (136,144). Increased transcription in the face of protein synthesis inhibition implies the presence of a labile inhibitor of thrombomodulin mRNA transcription. Potential mediators in this setting may be members of the coagulation pathway. In mouse hemangioma cells, thrombin augments the rate of thrombomodulin transcription, levels of mRNA, and the amount of total thrombomodulin protein (144). A study in human saphenous vein endothelial cells failed to show any effect of thrombin on thrombomodulin expression (134).

Interestingly, pentoxifylline (145), by increasing intracellular cyclic adenosine monophosphate (cAMP), and interleukin-4 (139) were both able to neutralize the effect of TNF on thrombomodulin expression in human umbilical vein endothelial cells. Histamine is also able to enhance activity in endothelial cells by increasing thrombomodulin mRNA via H_1 receptors (146). The significance of these findings in the protection of the endothelium against pyrogen-induced procoagulant changes is uncertain.

Thrombomodulin is identical to the gene induced by cAMP in murine F9 cells referred to as fetomodulin (147,148). In two different human megakaryoblastic cell lines, dibutyryl-cAMP caused increases in thrombomodulin antigen, activity, and mRNA levels (149,150). Studies in human umbilical vein endothelial cells have yielded similar results (151–154). Of note, dibutyryl-cAMP was able to prevent interleukin-1- or TNF-induced suppression of thrombomodulin expression (154). Retinoic acid increases thrombomodulin expression in F9 embryonal carcinoma cells (155) and human umbilical vein endothelial cells (156). The significance of this is unknown.

The thrombomodulin gene has been identified and is now being characterized (156a). Analysis of the transcription initiation start has revealed two distinct sites 5 bases apart. Multiple potential SP1 sites have been identified by sequence homology in the 5′ flanking region of the gene. A 400-base segment from the 5′ flanking region has been more carefully evaluated by functional characterization of reporter plasmids containing the thrombomodulin promoter. In these studies, the bases from −72 to −29 appear essential for thrombomodulin transcription. Bases from −343 to −277 and −167 to −84 appear to interact at previously uncharacterized cis elements based on comparison with known transcription elements. The −167 to −84 region appears to contain an element resulting in suppressor activity; deletion of this region results in increased expression of reporter plasmids. The more 5′ region from −343 to −277 contains two distinct enhancing elements, with decreased reporter expression upon their deletion. One of these may be similar to a portion of the endothelin 1 promoter, which is important in endothelium-specific expression. Thrombomodulin transcription has been demonstrated to be inhibited in cells exposed to tumor necrosis factor. A region of the gene that appears to be required for this inhibition has been mapped to the first 51 bases 5′ to the transcription initiation site. TNF is able to inhibit the transcription of constructs containing this region but has no effect on heterologous promoters.

VII. CLINICAL OBSERVATIONS

Variation in circulating thrombomodulin levels may be of clinical significance, manifested as either thrombophilia in deficiency states or anticoagulation when present in increased amounts. Circulating thrombomodulin levels have been measured in a number of disease states (Table 3). It appears that increases in plasma thrombomodulin levels reflect endothelial damage (157–161). Elevated levels of circulating thrombomodulin have been detected in disseminated intravascular coagulation (DIC) (162–164), pulmonary embolism, adult respiratory distress syndrome, chronic renal failure, and hepatic failure (162). In one study, increasing plasma thrombomodulin found in DIC and thrombotic thrombocytopenic purpura was correlated with a worse prognosis (165). In patients with systemic lupus erythematosus, increased circulating thrombomodulin was found in those patients with active lupus nephritis (166). This is believed to be a marker of endothelial damage, perhaps secondary to circulating antiphospholipid antibodies (167). Elevations in plasma thrombomodulin have also been found in patients with chronic myelogenous leukemia. This may be caused by leukocyte elastase releasing soluble fragments of thrombomodulin into the circulation. Plasma thrombomodulin levels returned to normal following treatment with chemotherapy and the ensuing fall in white blood cell count (168). Interestingly, patients with essential hypertension and ischemic heart disease do not show elevations in circulating thrombomodulin (169). Plasma levels are higher in diabetic patients with nephropathy or retinopathy than in those without renal or retinal disease (170–172). This may serve as an early marker for endothelial damage before overt nephropathy or retinopathy in these patients.

The role of thrombomodulin in the rejection of allo- and autografts is receiving more attention (173). TNF is a mediator of renal allograft rejection through its effect of decreasing endothelial thrombomodulin expression, leading to intragraft fibrin deposition and loss of the functioning graft (174). Similar findings were noted in a rat cardiac transplant model in which increased TNF led to decreased thrombomodulin but increased tissue factor expression in the graft with resultant thrombosis (175). Thrombomodulin was reduced or absent in occluded human saphenous vein coronary artery bypass grafts

Table 3 Diseases Associated with Elevated Plasma Thrombomodulin[a]

DIC	CML
Pulmonary embolism	Diabetic nephropathy
ARDS	Diabetic retinopathy
Chronic renal failure	TTP
Hepatic failure	Lupus nephritis

[a]DIC, disseminated intravascular coagulation; ARDS, adult respiratory distress syndrome; CML, chronic myelogeneous leukemia; TTP, thrombotic thrombocytopenic purpura.

(176). A potentially confounding variable in this setting is the use of cyclosporine to prevent graft rejection. Cyclosporine is associated with vascular injury and increases in circulating thrombomodulin levels (177). Bovine endothelial cells treated with cyclosporine showed decreases in thrombomodulin surface activity with loss of protein C activation, increasing the risk of thrombosis (178). The exact role of thrombomodulin in graft rejection and cyclosporine therapy is unknown; however, the prospect of enhancing thrombomodulin expression to improve graft survival is exciting.

Certain infectious agents may have a direct effect on the tissue expression of thrombomodulin. Human umbilical vein endothelial cells infected with herpes simplex virus 1 (HSV-1), HSV-2 (179), and *Rickettsia* species (180) showed loss of thrombomodulin antigen, surface activity, and decreased mRNA levels in conjunction with elevated tissue factor expression. The significance of these findings in vitro is not clear.

Homocystinuria is a congenital metabolic disorder wherein patients are deficient in cystathionine β-synthase and have a propensity for thrombosis. The etiology of thrombosis in this disease is unclear. Protein C activation by endothelial cells is inhibited by homocysteine (181). In vitro studies using human umbilical vein endothelial cells and CV-1 cells exposed to homocysteine show depressed surface activity of thrombomodulin (182), perhaps by reducing the disulfide bond-rich EGF-like domains of the molecule (183).

Protein C activation by thrombomodulin in vitro is inhibited by plasma from some patients with the lupus anticoagulant and other antiphospholipid antibodies (184–188). Evidence for an autoantibody to thrombomodulin is lacking in these patients, implying a direct inhibitory effect on protein C activation by the lupus anticoagulant (189). A number of patients with the lupus anticoagulant demonstrate thromboses and/or recurrent fetal loss. Abnormalities in thrombomodulin function may be partly responsible.

Modulating thrombomodulin in patients may provide a means for decreasing hypercoagulability or for therapeutic anticoagulation. As noted, several agents can stimulate thrombomodulin activity in tissue culture, including cycloheximide, cAMP, and retinoic acid. Continued investigation of other pharmacological agents is needed.

It is known that infusions of activated protein C are protective against *Escherichia coli*-induced DIC (190). Similar infusions also reduce platelet and fibrin deposition in a baboon thrombosis model (191). Intravenously administered rabbit thrombomodulin circulates in a biologically active form with a half-life of approximately 200 minutes (192). Infusions of solubilized thrombomodulin into mice prevented the lethal thrombotic effects of injected thrombin (34,193,194). Recombinant human thrombomodulin infused into rats prevented both stasis-induced venous thrombosis (195) and endotoxin-induced DIC (196). Indeed, infused thrombomodulin may be a safer agent because there was less prolongation of partial thromboplastin time and bleeding time than seen with heparin (195). Also, infused thrombomodulin may be less toxic, because it only generates activated protein C in the presence of thrombin, potentially restricting its anticoagulant effect to the region of thrombosis. With further investigation, soluble forms of thrombomodulin may prove to have therapeutic benefit.

VIII. SUMMARY

Thrombomodulin is a potent activator of protein C acting by a unique mechanism whereby the procoagulant thrombin is converted to an anticoagulant. This protein may have a crucial role in the maintenance of blood fluidity. Much has been learned about thrombomodulin in the short time since its discovery. Clearly, continued study of the function of thrombomodulin on the endothelial cell, as well as its role in other recently described sites, is needed.

REFERENCES

1. Weksler BB, Ley CW, Jaffe EA. Stimulation of endothelial cell prostacyclin production by thrombin, trypsin, and the ionophore A23187. J Clin Invest 1978; 62:923.
2. Loskutoff DJ, Edgington TS. Synthesis of a fibrinolytic activator and inhibitor by endothelial cells. Proc Natl Acad Sci USA 1977; 74:3903.
3. Bauer KA, Rosenberg RD. The pathophysiology of the prethrombotic state in humans: insights gained from studies using markers of hemostatic system activation. Blood 1987; 70:343.
4. Esmon CT. The roles of protein C and thrombomodulin in the regulation of blood coagulation. J Biol Chem 1989; 264:4743.
5. Esmon CT. The regulations of natural anticoagulant pathways. Science 1987; 235:1348.
6. Mammen EF, Thomas WR, Seegers WH. Activation of purified prothrombin to autoprothrombin I or autoprothrombin II (platelet cofactor II) or autoprothrombin IIa. Thromb Diath Haemorrh 1960; 5:218.
7. Seegers WH, McCoy LE, Groben HD, Sakuragawa N, Agrawal BBL. Purification and some properties of autoprothrombin IIa: an anticoagulant perhaps related to fibrinolysis. Thromb Res 1972; 1:443.
8. Seegers WH, Novoa E, Henry RL, Hassouna HI. Relationship of ''new'' vitamin K dependent protein C and ''old'' autoprothrombin IIa. Thromb Res 1976; 8:543.
9. Stenflo J. A new vitamin K dependent protein. Purification from bovine plasma and preliminary characterization. J Biol Chem 1976; 251:355.
10. Mammen EF. Protein C. Semin Thromb Hemost 1984; 10:109.
11. Esmon CT. Protein C. In: Spaet T, ed. Progress in Hemostasis and Thrombosis, Vol. 7. New York: Grune and Stratton, 1984:25.
12. Laemmli B, Griffin JH. Formation of the fibrin clot: the balance of procoagulant and inhibitory factors. Clin Hematol 9185; 14:281.
13. Clouse LH, Comp PC. The regulation of hemostasis: the protein C system. N Engl J Med 1986; 314:1298.
14. Comp PC. Hereditary disorders predisposing to thrombosis. In: Spaet T, ed. Progress in Hemostasis and Thrombosis, Vol. 9. New York: Grune and Stratton, 1986.
15. Kisiel W, Canfield WM, Ericsson EH, Davie EW. Anticoagulant properties of bovine plasma protein C following activation of thrombin. Biochemistry 1977; 16:5824.
16. Suzuki K, Stenflo J, Dahlback B, Teodorsson B. Inactivation of human coagulation factor V by activated protein C. J Biol Chem 1983; 258:1914.
17. Fulcher CA, Gardiner JE, Griffin JH, Zimmermen TS. Proteolytic inactivation of human factor VIII procoagulant protein by activated protein C and its analogy with factor V. Blood 1984; 63:486.
18. Seligsohn Y, Berger A, Abend M, et al. Homozygous protein C deficiency manifested by massive venous thrombosis in the newborn. N Engl J Med 1984; 310:559.
19. Esmon CT, Owen WG. Identification of an endothelial cell cofactor for the thrombin-catalyzed activation of protein C. Proc Natl Acad Sci USA 1981; 78:2249.
20. Owen WG, Esmon CT. Functional properties of an endothelial cell cofactor for thrombin-catalyzed activation of protein C. J Biol Chem 1981; 256:5532.
21. Esmon NL, Owen WG, Esmon CT. Isolation of a membrane-bound cofactor for thrombin-catalyzed activation of protein C. J Biol Chem 1982; 257:859.
22. Salem HH, Maruyama I, Ishii H, Majerus PW. Isolation and characterization of thrombomodulin from human placenta. J Biol Chem 1984; 259:12246.

23. Kurosawa S, Aoki N. Preparation of thrombomodulin from human placenta. Thromb Res 1985; 37:353.
24. Jakubowski HV, Owen WG. The effect of bovine thrombomodulin on the specificity of bovine thrombin. J Biol Chem 1986; 264:11117.
25. Kumada T, Dittman WA, Majerus PW. A role for thrombomodulin in the pathogenesis of thrombin induced thromboembolism in mice. Blood 1988; 71:728.
26. Esmon CT, Esmon NL, Saufstad J, Owen WG. Activation of protein C by a complex between thrombin and an endothelial cell protein. Pathobiology of the inhibits both thrombin-catalyzed fibrin formation and factor V activation. J Biol Chem 1982; 257:7944.
27. Esmon CT, Esmon NL, Harris KW. Complex formation between thrombin and thrombomodulin inhibits both thrombin-catalyzed fibrin formation and factor V activation. J Biol Chem 1982; 257:7944.
28. Maruyama I, Salem HH, Ishii H, Majerus PW. Human thrombomodulin is not an efficient inhibitor of procoagulant activity of thrombin. J Clin Invest 1985; 75:987.
29. Esmon NL, Carroll RC, Esmon CT. Thrombomodulin blocks the ability of thrombin to activate platelets. J Biol Chem 1983; 258:12268.
30. Polgar J, Lerant I, Muszbek L, Machovich R. Thrombomodulin inhibits the activation of factor XIII by thrombin. Thromb Res 1986; 43:585.
31. Mitchell CA, Hau L, Salem HH. Control of thrombin mediated cleavage of protein S. Thromb Haemost 1986; 56:151.
32. Wen D, Dittman WA, Ye RD, Deaven LL, Majerus PW, Sadler JE. Human thrombomodulin: complete cDNA sequence and chromosome localization of the gene. Biochemistry 1987; 26:4350.
33. Suzuki K, Kusumoto H, Deyashiki Y, et al. Structure and expression of human thrombomodulin, a thrombin receptor on endothelium acting as a cofactor for protein C activation. EMBO J 1987; 6:1891.
34. Dittman WA, Kumada T, Sadler JE, Majerus PW. The structure and function of mouse thrombomodulin: phorbol myristate acetate stimulates degradation and synthesis of thrombomodulin without affecting mRNA levels in hemangioma cells. J Biol Chem 1988; 263: 15815.
35. Dittman WA, Majerus PW. Structure of a cDNA for mouse thrombomodulin and comparison of the predicted mouse and human amino acid sequences. Nucleic Acids Res 1989; 17:802.
36. Jackman RW, Beeler DL, van de Water L, Rosenberg RD. Characterization of a thrombomodulin cDNA reveals structural similarity to the low density lipoprotein receptor. Proc Natl Acad Sci USA 1986; 83:8834.
37. Petersen TE. The amino-terminal domain of thrombomodulin and pancreatic stone protein are homologous with lectins. FEBS Lett 1988; 231:51.
38. Patthy L. Detecting distant homologies of mosaic proteins: analysis of thrombomodulin, thrombospondin, complement components C9, C8 alpha, and C8 beta, vitronectin and plasma cell membrane glycoprotein PC-1. J Mol Biol 1988; 202:689.
39. Doolittle RF, Feng DF, Johnson MS. Computer-based characterization of epidermal growth factor precursor. Nature 1984; 307:558.
40. Kurosawa J, Stearns DJ, Jackson KW, Esmon CT. A 10 kDa cyanogen bromide fragment from the epidermal growth factor homology domain of rabbit thrombomodulin contains the primary thrombin binding site. J Biol Chem 1988; 263:5993.
41. Stearns DJ, Kurosawa S, Esmon CT. Microthrombomodulin. Residues 310–486 from the epidermal growth factor precursor homology domain of thrombomodulin will accelerate protein C activation. J Biol Chem 1989; 264:3352.
42. Suzuki K, Hayashi T, Nishioka J, et al. A domain composed of epidermal growth factor-like structures of human thrombomodulin is essential for thrombin binding and for protein C activation. J Biol Chem 1989; 264:4872.

43. Zushi M, Gomi K, Yamamoto S, Maruyama I, Hayashi T, Suzuki K. The last three consecutive epidermal growth factor-like structures of human thrombomodulin comprise the minimum functional domain for protein C activating cofactor activity and anticoagulant activity. J Biol Chem 1989; 264:10351.
44. Hayashi T, Zushi M, Yamamoto S, Suzuki K. Further localization of binding sites for thrombin and protein C in human thrombomodulin. J Biol Chem 1990; 265:20156–20159.
45. Tsiang M, Lentz S, Sadler JE. Functional domains of membrane bound thrombomodulin: EGF-like domains four to six and the serine/threonine rich domain are required for cofactor activity. J Biol Chem 1992; 267:6164–6170.
46. Parkinson JF, Nagashima M, Kuhn I, Jeonard J, Morser J. Structure-function studies of the epidermal growth factor domains of human thrombomodulin. Biochem Biophys Res Commun 1992; 185:567–576.
47. Zushi M, Gomi K, Honda G, et al. Aspartic acid 349 in the fourth epidermal growth factor-like structure of human thrombomodulin plays a role in its Ca^{2+}-mediated binding to protein C. J Biol Chem 1991; 266:19886–19889.
48. Fernlund P, Stenflo J. Beta-hydroxyaspartic acid in vitamin K-dependent proteins. J Biol Chem 1983; 258:12509.
49. McMullen BA, Fujikawa K, Kisiel W. The occurence of beta-hydroxyaspartic acid in the vitamin K-dependent blood coagulation zymogens. Biochem Biophys Res Commun 1983; 115:8.
50. Stenflo J, Ohlin AK, Owen WG, Schneider WJ. Beta-hydroxyaspartic acid or beta-hydroxyasparagine in bovine low density lipoprotein receptor and in bovine thrombomodulin. J Biol Chem 1988; 263:21.
51. Ohlin AK, Landes G, Bourdon P, Oppenheimer C, Wydro R, Stenflo J. Beta-hydroxyaspartic acid in the first epidermal growth factor-like domain of protein C. Its role in Ca^{2+} binding and biological activity. J Biol Chem 1988; 263:19240.
52. Hofsteenge JH, Taguchi H, Stone SR. Effect of thrombomodulin on the kinetics of the interaction of thrombin with substrates and inhibitors. Biochem J 1986; 237:243.
53. Bourin M-C, Boffa M-C, Bjork I, Lindahl U. Functional domains of rabbit thrombomodulin. Proc Natl Acad Sci USA 1986; 83:5924.
54. Esmon CT, Esmon NL, Harris KW. Complex formation between thrombin and thrombomodulin inhibits both thrombin catalyzed fibrin formation and factor V activation. J Biol Chem 1983; 258:12238.
55. Bourin M-C, Ohlin AK, Lane DA, Stenflo J, Lindahl U. Relationship between anticoagulant activities and polyanionic properties of rabbit thrombomodulin. J Biol Chem 1988; 263: 8044–8052.
56. Bourin M-C. Effect of rabbit thrombomodulin on thrombin inhibition by antithrombin in the presence of heparin. Thromb Res 1989; 54:27–39.
57. Parkinson J, Grinnell B, Moore R, Hoskins J, Vlahos C, Bang N. Stable expression of a secretable deletion mutant of recombinant human thrombomodulin in mammalian cells. J Biol Chem 1990; 265:12602–12610.
58. Parkinson J, Garcia J, Bang N. Decreased thrombin affinity of cell surface thrombomodulin following treatment of cultured endothelial cells with β-D-xyloside. Biochem Biophys Res Commun 1990; 169:177–183.
59. Nawa K, Sakano K, Fujiwara H, et al. Presence and function of chondroitin-4-sulfate on recombinant human soluble thrombomodulin. Biochem Biophys Res Commun 1990; 171: 729–737.
60. Koyama T, Parkinson J, Aoki N, Bang N, Muller-Berghaus G, Preissner K. Relationship between post-translational glycosylation and anticoagulant function of secretable recombinant mutants of human thrombomodulin. Br J Haemotol 1991; 78:515–522.
61. Koyama T, Parkinson J, Sie P, Bang N, Muller-Berghaus G, Preissner K. Different glycoforms of human thrombomodulin: their glycosaminoglycan-dependent modulatory effects

on thrombin inactivation by heparin cofactor II and antithrombin III. Eur J Biochem 1991; 198:563–570.

62. Parkinson J, Vlahos C, Yan S, Bang N. Recombinant human thrombomodulin: regulation of cofactor activity and anticoagulant function by a glycosaminoglycan side chain. Biochem J 1992; 283:151–157.
63. Preissner K, Koyama T, Muller D, Tschopp J, Muller-Berghaus G. Domain structure of the endothelial cell receptor thrombomodulin as deduced from modulation of its anticoagulant functions: evidence for a glycosaminoglucan-dependent secondary binding site for thrombin. J Biol Chem 1990; 265:4915–4922.
64. Bourin M-C, Lindahl U. Functional role of the polysaccharide component of rabbit thrombomodulin proteoglycan: effects on inactivation of thrombin by antithrombin, cleavage of fibrinogen by thrombin and thrombin-catalyzed activation of factor V. Biochem J 1990; 270:419–425.
65. Bourin M-C, Lundgren-Akerlund E, Lindahl U. Isolation and characterization of the glycosaminoglycan component of rabbit thrombomodulin proteoglycan. J Biol Chem 1990; 265:15424–15431.
66. Goldstein JL, Brown MS, Anderson TG, Russell DW, Schneider WJ. Receptor mediated endocytosis: concepts emerging from the LDL receptor system. Annu Rev Cell Biol 1985; 1:1.
67. Escobedo JA, Barr PJ, Williams LT. Role of tyrosine kinase and membrane-spanning domains in signal transduction by the platelet-derived growth factor receptor. Mol Cell Biol 1988; 13:5126.
68. Espinosa R, Sadler JE, LeBeau M. Regional localization of the human thrombomodulin gene to 20p12-cen. Genomics 1989; 5:649–650.
69. Van der Velden P, Krommenhoek-Van Es T, Allaart C, Bertina R, Reitsma P. A frequent thrombomodulin amino acid dimorphism is not associated with thrombophilia. Thromb Haemost 1991; 65:511–513.
70. Jackman RW, Beeler DL, Fritze L, Soff G, Rosenberg RD. Human thrombomodulin gene is intron depleted: nucleic acid sequences of the cDNA and gene predict protein structure and suggest sites of regulatory control. Proc Natl Acad Sci USA 1987; 84:6425.
71. Shira T, Shiojiri S, Ito H, et al. Gene structure of human thrombomodulin, a cofactor for thrombin catalyzed activation of protein C. J Biochem 1988; 103:281.
72. Shaw K, Kamen R. A conserved AU sequence from the 3′ untranslated region of GM-CSF mRNA mediates selective mRNA degradation. Cell 1986; 46:659.
73. Lentz SR, Tsiang M, Sadler JE. Regulation of thrombomodulin by tumor necrosis factor-α: comparison of transcriptional and posttranscriptional mechanisms. Blood 1991; 77:542–550.
74. Maruyama I, Bell CE, Majerus PW. Thrombomodulin found on endothelium of arteries, veins, capillaries, and lymphatics and on syncytiotrophoblast of human placenta. J Cell Biol 1985; 101:363.
75. DeBault LE, Esmon NL, Olson JR, Esmon CT. Distribution of the thrombomodulin antigen in the rabbit vasculature. Lab Invest 1986; 54:179.
76. Wong V, Hofman F, Ishii H, Fisher M. Regional distribution of thrombomodulin in human brain. Brain Res 1991; 556:1–5.
77. Boffa MC, Jackman RW, Peyri N, Boffa JF, George B. Thrombomodulin in the central nervous system. Nouv Rev Fr Hematol 1991; 33:423–429.
78. Dichek D, Quertermous T. Variability in mRNA levels in HUVECs of different lineage and time in culture. In Vitro Cell Dev Biol 1989; 25:289–292.
79. Yonezawa S, Maruyama I, Tanaka S, Nakamura T, Sato E. Immunohistochemical localization of thrombomodulin in chorionic diseases of the uterus and choriocarcinoma of the stomach. Cancer 1988; 62:569–576.
80. Maruyama I, Yonezawa S, Sato E, Igata A. Endothelium and thrombomodulin: expression and distribution of thrombomodulin on altered endothelium and neoplastic syncytiotrophoblast. Acta Haematol Jpn 1989; 51:150–155.

81. Collins C, Fink L, Hsu S, Schaefer R, Ordonez N. Thrombomodulin staining of mesothelioma cells. Hum Pathol 1992; 23:966.
82. Tamura A, Matsubara O, Hirokawa K, Aoki N. Detection of thrombomodulin in human lung cancer cells. Am J Pathol 1993; 142:79–85.
83. Yasunaga C, Sueishi K, Ohgami H, Suita S, Kawanami T. Heterogeneous expression of endothelial cell markers in infantile hemangioendothelioma: immunohistochemical study of two solitary cases and one multiple one. Am J Clin Pathol 1989; 91:673–681.
84. Yonezawa S, Maruyama I, Sakae K, Igata A, Majerus PW, Sato E. Thrombomodulin as a marker for vascular tumors: comparative study with factor VIII and *Ulex europaeus* I lectin. Am J Clin Pathol 1987; 88:405.
85. Suzuki K, Nishioka J, Hayashi T, Kosaka Y. Functionally active thrombomodulin is present in human platelets. J Biochem 1988; 104:628–632.
86. Ogura M, Ito T, Maruyama I, et al. Localization and biosynthesis of functional thrombomodulin in human megakaryocytes and a human megakaryoblastic cell line (MEG-01). Thromb Haemost 1990; 64:297–301.
87. Conway E, Nowakowski B, Steiner-Mosonyi M. Human neutrophils synthesize thrombomodulin that does not promote thrombin-dependent protein C activation. Blood 1992; 80: 1254–1263.
88. McCachren SS, Diggs J, Weinberg JB, Dittman WA. Thrombomodulin expression by human blood monocytes and by human synovial tissue lining macrophages. Blood 1991; 78: 3128–3132.
89. Ishii H, Majerus PW. Thrombomodulin is present in human plasma and urine. J Clin Invest 1985; 76:2178.
90. Ishii H, Nakano M, Tsubouchi J, et al. Establishment of enzyme immunoassay of human thrombomodulin in plasma and urine using monoclonal antibodies. Thromb Haemost 1990; 63:157–162.
91. Ishii H, Nakana M, Tsubouchi J, Kazama M, Majerus PW. Distribution of thrombomodulin in human tissues and characterization of thrombomodulin in plasma. Acta Haematol Jpn 1988; 51:1218.
92. Steele RE. Membrane-anchored growth factors work. Trends Biochem Sci 1989; 14:201.
93. Kawosaki ES, Lodner MB, Wang AM, et al. Molecular cloning of a cDNA encoding human M-CSF. Science 1985; 230:291.
94. Maruyama I, Majerus PW. The turnover of thrombin: thrombomodulin complex in cultured HUVECs and A549 lung cancer cells. Endocytosis and degradation of thrombin. J Biol Chem 1985; 260:15432.
95. Soff, G, Jackman R, Rosenberg RD. Expression of thrombomodulin by smooth muscle cells in culture: different effects of tumor necrosis factor and cyclic adenosine monophosphate on thrombomodulin expression by endothelial cells and smooth muscle cells in culture. Blood 1991; 77:515–518.
96. Thompson EA, Salem HH. Modification of human thrombin: effect on thrombomodulin binding. Thromb Haemost 1988; 59:415.
97. Hofsteenge J, Braun PJ, Stone R. Enzymatic properties of proteolytic derivatives of human alpha-thrombin. Biochemistry 1988; 27:144.
98. Suzuki K, Nishioka J, Hayashi T. Localization of thrombomodulin-binding site within human thrombin. J Biol Chem 1990; 265:13263–13267.
99. LeBonniec B, Esmon CT. Glu-192-Gln substitution in thrombin mimics the catalytic switch induced by thrombomodulin. Biochemistry 1991; 88:7371–7375.
100. Jakubowski HV, Kline MD, Owen WG. The effect of bovine thrombomodulin on the specificity of bovine thrombin. J Biol Chem 1986; 261:3876.
101. Tsiang M, Lentz SR, Dittman WA, Wen D, Scarpati EM, Sadler JE. Equilibrium binding of thrombin to recombinant human thrombomodulin: effect of hirudin, fibrinogen, factor Va, and peptide analogues. Biochemistry 1990; 29:10602–10612.

102. Rydel TJ, Ravichandran KG, Tulinsky A, et al. The structure of a complex of recombinant hirudin and human α-thrombin. Science 1990; 249:277–280.
103. Jakubowski HV, Owen WG. Macromolecular specificity determinants on thrombin for fibrinogen and thrombomodulin. J Biol Chem 1989; 264:11117–11121.
104. Wu Q, Sheehan J, Tsiang M, Lentz S, Birktoft J, Sadler J. Single amino acid substitutions dissociate fibrinogen-clotting and thrombomodulin-binding activities of human thrombin. Proc Natl Acad Sci USA 1991; 88:6775–6779.
105. Suzuki K, Nishioka J. A thrombin-based peptide corresponding to the sequence of the thrombomodulin-binding site blocks the procoagulant activities of thrombin. J Biol Chem 1991; 266:18498–18501.
106. Musci G, Berliner L, Esmon CT. Evidence for multiple conformational changes in the active center of thrombin induced by complex formation with thrombomodulin: an analysis employing nitroxide spin-labels. Biochemistry 1988; 27:769–773.
107. Ye J, Liu L, Esmon CT, Johnson AE. The fifth and sixth growth factor-like domains of thrombomodulin bind to the anion-binding exosite of thrombin and alter its specificity. J Biol Chem 1992; 267:11023–11028.
108. Ye J, Esmon N, Esmon C, Johnson A. The active site of thrombin is altered upon binding to thrombomodulin: two distinct structural changes are detected by fluorescence, but only one correlates with protein C activation. J Biol Chem 1991; 266:23016–23021.
109. Lu R, Esmon NL, Esmon CT, Johnson AE. The active site of the thrombin-thrombomodulin complex: a fluorescence energy transfer measurement of its distance above the membrane surface. J Biol Chem 1989; 264:12956–12962.
110. Esmon NL, DeBault LE, Esmon CT. Proteolytic formation and properties of carboxyglutamic acid-domainless protein C. J Biol Chem 1983; 258:5548.
111. Ohlin AK, Linse S, Stenflo J. Calcium binding to the epidermal growth factor homology region of bovine protein C. J Biol Chem 1988; 263:7411–7417.
112. Ohlin AK, Landes G, Bourdon P, Oppenheimer C, Wydro R, Stenflo J. Beta-hydroxyaspartic acid in the first epidermal growth factor-like domain of protein C: Its role in Ca^{++} binding and biological activity. J Biol Chem 1988; 263:19240.
113. Ohlin AK, Stenflo J. Calcium-dependent interaction between the EGF precursor-like region of human protein C and a monoclonal antibody. J Biol Chem 1987; 262:13798–13804.
114. Rezaie A, Esmon N, Esmon C. The high affinity calcium-binding site involved in protein C activation is outside the first epidermal growth factor homology domain. J Biol Chem 1992; 267:11701–11704.
115. Salem H, Broze GJ, Miletich JP, Majerus PW. Human coagulation factor Va is a cofactor for the activation of protein C. Proc Natl Acad Sci USA 1983; 80:1584.
116. Salem HH, Broze GJ, Miletich JP, Majerus PW. The light chain of factor Va contains the activity of factor Va that accelerates protein C activation by thrombin. J Biol Chem 1983; 258:8531.
117. Salem HH, Esmon NL, Esmon CT, Majerus PW. The effects of thrombomodulin and coagulation factor Va light chain on protein C activation in vitro. J Clin Invest 1984; 73:968.
118. Maruyama I, Salem HH, Majerus PW. Coagulation factor Va binds to human umbilical vein endothelial cells and accelerates protein C activation. J Clin Invest 1984; 74:224.
119. Thompson EA, Salem HH. Factors IXa, Xa, XIa and activated protein C do not have protein C activating ability in the presence of thrombomodulin. Thromb Haemost 1988; 59:339.
120. Freyssinet J, Beretz A, Klein-Soyer C, Gauchy J, Schuler S, Cazenave J. Interference of blood-coagulation vitamin K-dependent proteins in the activation of human protein C: involvement of the 4-carboxyglutamic acid domain in two distinct interactions with the thrombin-thrombomodulin complex and with phospholipids. Biochem J 1988; 256:501–507.
121. Haley PE, Doyle MF, Mann KG. The activation of bovine protein C by factor Xa. J Biol Chem 1989; 264:16303–16310.

122. Freyssinet J, Wiesel M, Grunebaum L, et al. Activation of human protein C by blood coagulation factor Xa in the presence of anionic phospholipids: enhancement by sulphated polysaccharides. Biochem J 1989; 261:341–348.
123. Freyssinet J, Toti-Orfanoudakis F, Ravanat C, et al. The catalytic role of anionic phospholipids in the activation of protein C by factor Xa and expression of its anticoagulant function in human plasma. Blood Coag Fib 1991; 2:691–698.
124. Wu Q, Tsiang M, Lentz S, Sadler J. Ligand specificity of human thrombomodulin: equilibrium binding of human thrombin, meizothrombin, and factor Xa to recombinant thrombomodulin. J Biol Chem 1992; 267:7083–7088.
125. Hogg P, Ohlin AK, Stenflo J. Identification of structural domains in protein C involved in its interaction with thrombin-thrombomodulin on the surface of endothelial cells. J Biol Chem 1992; 267:703–706.
126. Jane SM, Hau L, Salem HH. Regulation of activated protein C by factor Xa. Blood Coag Fib 1991; 2:723–729.
127. De Munk G, Groeneveld E, Rijken D. Acceleration of the thrombin inactivation of single-chain urokinase-type plasminogen activator (prourokinase) by thrombomodulin. J Clin Invest 1991; 88:1680–1684.
128. Molinari A, Giorgetti C, Lansen J, et al. Thrombomodulin is a cofactor for thrombin degradation of recombinant single-chain urokinase plasminogen activator ''in vitro'' and in a perfused rabbit heart model. Thromb Haemost 1992; 67:226–232.
129. Maruyama I, Majerus PW. Protein C inhibits endocytosis of thrombin: thrombomodulin complexes in A549 lung cancer cells and human umbilical vein endothelial cells. Blood 1987; 69:1481.
130. Beretz A, Freyssinet J, Gauchy J, et al. Stability of the thrombin-thrombomodulin complex on the surface of endothelial cells from human saphenous vein and from the cell line EA.hy 926. Biochem J 1989; 259:35.
131. Hirokawa K, Aoki N. Regulatory mechanisms for thrombomodulin expression in human umbilical vein endothelial cells in vitro. J Cell Physiol 1991; 147:157–165.
132. Naworth PP, Stern DM. Modulation of endothelial cell hemostatic properties by tumor necrosis factor. J Exp Med 1986; 163:740.
133. Scarpati E, Sadler JE. Regulation of endothelial cell coagulant properties: modulation of tissue factor, plasminogen activator inhibitors, and thrombomodulin by phorbol 12-myristate 13-acetate and tumor necrosis factor. J Biol Chem 1989; 264:20705–20713.
134. Archipoff G, Beretz A, Freyssinet J, Klein-Soyer C, Brisson C, Cazenave J. Heterogeneous regulation of constitutive thrombomodulin or inducible tissue-factor activities on the surface of human saphenous-vein endothelial cells in culture following stimulation by interleukin-1, tumor necrosis factor, thrombin or phorbol ester. Biochem J 1991; 273:679–684.
135. Moore KL, Esmon CT, Esmon NL. Tumor necrosis factor leads to the internalization and degradation of thrombomodulin from the surface of bovine aortic endothelial cells in culture. Blood 1989; 73:159.
136. Conway EM, Rosenberg RD. Tumor necrosis factor suppresses transcription of the thrombomodulin gene in endothelial cells. Mol Cell Biol 1988; 8:5588.
137. Yu K, Morioka H, Fritze L, Beeler D, Jackman R, Rosenberg RD. Transcriptional regulation of the thrombomodulin gene. J Biol Chem 1992; 267:23237–23247.
138. Moore KL, Andreoli SP, Esmon NL, Esmon CT, Bang NU. Endotoxin enhances tissue factor and suppresses thrombomodulin expression of human vascular endothelium in vitro. J Clin Invest 1987; 79:124.
139. Kapiotis S, Besemer J, Bevec D, et al. Interleukin-4 counteracts pyrogen-induced down-regulation of thrombomodulin in cultured human vascular endothelial cells. Blood 1991; 78:410–415.
140. He C, Kanfer A. Quantification and modulation of thrombomodulin activity in isolated rat and human glomeruli. Kidney Int 1992; 41:1170–1174.

141. Naworth PP, Handley DA, Esmon CT, Stern DM. Interleukin-1 induces endothelial cell procoagulant while suppressing cell-surface anticoagulant activity. Proc Natl Acad Sci USA 1989; 83:3460.
142. Ogawa S, Gerlach H, Esposito C, Pasagian-Macaulay A, Brett J, Stern D. Hypoxia modulates the barrier and coagulant function of cultured bovine endothelium: increased monolayer permeability and induction of procoagulant properties. J Clin Invest 1990; 85:1090–1098.
143. Ziff EB, Hermonowsky NL, Greenberg ME. Effect of protein synthesis inhibitors on growth factor activation of c-fos, c-myc and actin gene transcription. Mol Cell Biol 1986; 6:1050.
144. Dittman WA, Kumada T, Majerus PW. Transcription of thrombomodulin mRNA in mouse hemangioma cells is enhanced by cycloheximide and thrombin. Proc Natl Acad Sci USA 1989; 86:7179.
145. Ohdama S, Takano S, Ohashi K, Miyake S, Aoki N. Pentoxifylline prevents tumor necrosis factor-induced suppression of endothelial cell surface thrombomodulin. Thromb Res 1991; 62:745–755.
146. Hirokawa K, Aoki N. Up-regulation of thrombomodulin by activation of histamine H_1-receptors in human umbilical-vein endothelial cells in vitro. Biochem J 1991; 276:739–743.
147. Imada M, Imada S, Iwaski H, Kume A, Yamaguchi H, Moore EE. Fetomodulin: marker surface protein of fetal development which is modulatable by cyclic AMP. Dev Biol 1987; 122:483.
148. Imada M, Imada S, Nagumo M, Yamaguchi H, Katayanagi S. Identification of fetomodulin, a surface marker of development, as thrombomodulin by gene cloning and functional assay. J Cell Biochem 1989; 13E:200.
149. Ito T, Ogura M, Morishita Y, et al. Enhanced expression of thrombomodulin by intracellular cyclic AMP-increasing agents in two human megakaryoblastic leukemia cell lines. Thromb Res 1990; 58:615–624.
150. Ohashi K, Hirokawa K, Komatsu N, Aoki N. The biosynthesis of thrombomodulin and its enhancement by dibutyryl cAMP in a human megakaryoblastic cell line, UT-7. Int J Hematol 1991; 54:341–349.
151. Ishii H, Kizaki K, Uchiyama H, Horie S, Kazama M. Cyclic AMP increases thrombomodulin expression on membrane surface of cultured human umbilical vein endothelial cells. Thromb Res 1990; 59:841–850.
152. Kainoh M, Maruyama I, Nishio S, Nakadate T. Enhancement by beraprost sodium, a stable analogue of prostacyclin, in thrombomodulin expression on membrane surface of cultured vascular endothelial cells via increase in cyclic AMP level. Biochem Pharmacol 1991; 41: 1135–1140.
153. Hirokawa K, Aoki N. Up-regulation of thrombomodulin in human umbilical vein endothelial cells in vitro. J Biochem 1990; 108:839–845.
154. Maruyama I, Soejima Y, Osame M, et al. Increased expression of thrombomodulin on the cultured human umbilical vein endothelial cells and mouse hemangioma cells by cyclic AMP. Thromb Res 1991; 61:301–310.
155. Weiler-Guettler H, Yu K, Soff G, Gudas L, Rosenberg RD. Thrombomodulin gene regulation by cAMP and retinoic acid in F9 embryonal carcinoma cells. Proc Natl Acad Sci USA 1992; 89:2155–2159.
156. Horie S, Kizaki K, Ishii H, Kazama M. Retinoic acid stimulates expression of thrombomodulin, a cell surface anticoagulant protein, on human endothelial cells: differences between up-regulation of thrombomodulin by retinoic acid and cyclic AMP. Biochem J 1992; 281: 149–154.
156a. Yu K, Morioka H, Fritze LMS, Beeler DL, Jackman RW, Rosenberg RD. Transcriptional regulation of the thrombomodulin gene. J Biol Chem 1992; 267:23237–23247.
157. Ishii H, Uchiyama H, Kazama M. Soluble thrombomodulin antigen in conditioned medium is increased by damage of endothelial cells. Thromb Haemost 1991; 65:618–623.

158. Boffa MC, Karochkine M, Berard M. Plasma thrombomodulin as a marker of endothelial damage. Nouv Rev Fr Hematol 1991; 33:529–530.
159. Sawada K, Yamamoto H, Matsumoto K, et al. Changes in thrombomodulin level in plasma of endotoxin-infused rabbits. Thromb Res 1992; 65:199–209.
160. Uchiyama H, Hiraishi S, Ohtani H, Ishii H, Kazama M. Plasma thrombomodulin is originated by damage of endothelial cells. Thromb Haemost 1989; 62:276.
161. Giddings JC, Coles P, Williams BD. Comparison of thrombomodulin and von Willebrand factor antigen in human plasma in various diseases. Thromb Haemost 1989; 62:333.
162. Takano S, Kimura S, Ohdama S, Aoki N. Plasma thrombomodulin in health and disease. Blood 1990; 76:2024–2029.
163. Asakura H, Jokaji H, Saito M, et al. Plasma level of soluble thrombomodulin increases in cases of disseminated intravascular coagulation with organ failure. Am J Hematol 1991; 38:281–287.
164. Uchiyama H, Ohtani H, Hiraishi S, Horie S, Ishii H, Kazama M. Changes in plasma thrombomodulin antigen in rabbit developing endotoxin-induced disseminated intravascular coagulation and the effect of heparin. Thromb Res 1992; 65:593–604.
165. Wada H, Ohiwa M, Kaneko T, et al. Plasma thrombomodulin as a marker of vascular disorders in thrombotic thrombocytopenic purpura and disseminated intravascular coagulation. Am J Hematol 1992; 39:20–24.
166. Takaya M, Ichikawa Y, Kobayashi N, et al. Serum thrombomodulin and anticardiolipin antibodies in patients with systemic lupus erythematosus. Clin Exp Rheumatol 1991; 9: 495–499.
167. Karmochkine M, Boffa MC, Piette J, et al. Increase in plasma thrombomodulin in lupus erythematosus with antiphospholipid antibodies. Blood 1991; 78:837–838.
168. Morishita E, Saito M, Asakura H, et al. Increased levels of plasma thrombomodulin in chronic myelogenous leukemia. Am J Hematol 1992; 39:183–187.
169. Naruse M, Kawana M, Hifumi S, et al. Plasma immunoreactive endothelin, but not thrombomodulin, is increased in patients with essential hypertension and ischemic heart disease. J Carciovasc Pharmacol 1991; 17:S471–S474.
170. Tanaka A, Ishii H, Hiraishi S, Kazama M, Maezawa H. Increased thrombomodulin values in plasma of diabetic men with microangiopathy. Clin Chem 1991; 37:269–272.
171. Iwashima Y, Sato T, Watanabe K, et al. Elevation of plasma thrombomodulin level in diabetic patients with early diabetic nephropathy. Diabetes 1990; 39:983–988.
172. Oida K, Takai H, Maeda H, et al. Plasma thrombomodulin concentration in diabetes mellitus. Diabetes Res Clin Pract 1990; 10:193–196.
173. Labarrere C, Pitts D, Halbrook H, Faulk W. Natural anticoagulant pathways in normal and transplanted human hearts. J Heart Lung Transplant 1992; 11:342–347.
174. Tsuchida A, Salem H, Thomsom N, Hancock W. Tumor necrosis factor production during human renal allograft rejection is associated with depression of plasma protein C and free protein S levels and decreased intragraft thrombomodulin expression. J Exp Med 1992; 175:81–90.
175. Hancock W, Tanaka K, Salem H, Tilney N, Atkins R, Kupiec-Weglinski J. TNF as a mediator of cardiac transplant rejection, including effects on the intragraft protein C, protein S, thrombomodulin pathway. Transplant Proc 1991; 23:235–237.
176. Brody J, Pickering N, Fink G. Abnormalities of thrombomodulin and tissue plasminogen activator in occluded aortocoronary grafts detected by immunohistochemistry. Trans Assoc Am Physicians 1988; 101:79:87.
177. Mihatsh M, Thiel G, Basler V. Morphological patterns in cyclosporine A treated renal transplant recipients. Transplant Proc 1985; 17:101.
178. Garcia-Maldonado M, Kaufman C, Comp P. Decrease in endothelial cell-dependent protein C activation induced by thrombomodulin by treatment with cyclosporine. Transplantation 1991; 51:701–705.

179. Key N, Vercellotti G, Winkelmann J, et al. Infection of vascular endothelial cells with herpes simplex virus enhances tissue factor activity and reduces thrombomodulin expression. Proc Natl Acad Sci USA 1990; 87:7095–7099.
180. Teysseire N, Arnoux D, George F, Sampol J, Raoult D. Von Willebrand factor release and thrombomodulin and tissue factor expression in *Rickettsia conorii*-infected endothelial cells. Infect Immun 1992; 60:4388–4393.
181. Rodgers G, Conn M. Homocysteine, an atherogenic stimulus, reduced protein C activation by arterial and venous endothelial cells. Blood 1990; 75:895.
182. Lentz SR, Sadler JE. Inhibition of thrombomodulin surface expression and protein C activation by the thrombogenic agent homocysteine. J Clin Invest 1991; 88:1906–1914.
183. Hayashi T, Honda G, Suzuki K. An atherogenic stimulus homocysteine inhibits cofactor activity of thrombomodulin and enhances thrombomodulin expression in human umbilical vein endothelial cells. Blood 1992; 79:2930–2936.
184. Comp P, DeBault L, Esmon N, Esmon C. Human thrombomodulin is inhibited by IgG from two patients with nonspecific anticoagulants. Blood 1983; 62:299a.
185. Freyssinet J, Wiessel M, Gauchy J, Boneu B, Cazenave J. An IgM lupus anticoagulant that neutralizes the enhancing effect of phospholipid on purified endothelial thrombomodulin activity: a mechanism for thrombosis. Thromb Haemost 1986; 55:309.
186. Cariou R, Tobelem G, Bellucci S, et al. Effect of lupus anticoagulant on antithrombogenic properties of endothelial cells: inhibition of thrombomodulin-dependent protein C activation. Thromb Haemost 1988; 60:54.
187. Ruiz-Argulles G, Ruiz-Argulles A, Deleze M, Alarcon-Segovia D. Acquired protein C deficiency in a patient with primary antiphospholipid syndrome. Relationship to reactivity of anticardiolipin antibody with thrombomodulin. J Rheumatol 1989; 16:381.
188. Tsakiris D, Settas L, Makris P, Marbet G. Lupus anticoagulant-antiphospholipid antibodies and thrombophilia. Relation to protein C-protein S-thrombomodulin. J Rheumatol 1990; 17:785–789.
189. Gibson J, Nelson M, Brown R, Salem H, Kronenberg H. Autoantibodies to thrombomodulin: development of an enzyme immunoassay and a survey of their frequency in patients with the lupus anticoagulant. Thromb Haemost 1992; 67:507–509.
190. Taylor F, Chang A, Esmon CT, D'Angelo A, Vigano-D'Angelo S, Blick K. Protein C prevents the coagulopathic and lethal effects of *E. coli* infusion in the baboon. J Clin Invest 1987; 79:918.
191. Gruber A, Griffin J, Harker L, Hanson S. Inhibition of platelet-dependent thrombus formation by human activated protein C in primate model. Blood 1989; 73:639.
192. Ehrlich H, Esmon N, Bang N. In vivo behavior of detergent-solubilized purified rabbit thrombomodulin on intravenous injection into rabbits. J Lab Clin Med 1990; 115:182–189.
193. Kumada T, Dittman WA, Majerus PW. A role for thrombomodulin in the pathogenesis of thrombin-induced thromboembolism in mice. Blood 1987; 71:728–733.
194. Gomi K, Zushi M, Honda G, et al. Antithrombotic effect of recombinant human thrombomodulin on thrombin-induced thromboembolism in mice. Blood 1990; 75:1396–1399.
195. Solis M, Cook C, Cook J, et al. Intravenous recombinant soluble human thrombomodulin prevents venous thrombosis in a rat model. J Vasc Surg 1991; 14:599–604.
196. Maruyama I, Okadome T, Sinmyouzu K. Consumption of endothelial cell-surface thrombomodulin in LPS-induced DIC rats and its improvement by infusion of recombinant thrombomodulin. Blood 1990; 76:429a.

21

Protein C Inhibitor

Rebecca A. Shirk,* Jeanne E. Phillips,† and Frank C. Church
University of North Carolina at Chapel Hill,
Chapel Hill, North Carolina

I. INTRODUCTION AND BIOLOGICAL SIGNIFICANCE

The protein C system is an important anticoagulant regulator of blood coagulation. Protein C is a vitamin K-dependent zymogen that, when activated by thrombin bound to thrombomodulin on the endothelial cell surface, proteolytically inactivates cofactors factor V/Va and factor VIII/VIIIa (Fig. 1) (1,2). The presence in plasma of an inhibitor to activated protein C was first described by Marlar and Griffith in 1980 (3). In 1983, Suzuki et al. (4) purified protein C inhibitor from human plasma. Protein C inhibitor is thought to be the primary regulator of the protein C system, but α_1-proteinase inhibitor (historically called α_1-antitrypsin) and α_2-macroglobulin also appear to play a role when large amounts of activated protein C (APC) are generated in a pathological setting, such as disseminated intravascular coagulation (5,6). Protein C inhibitor is a better inactivator of APC than α_1-proteinase inhibitor and α_2-macroglobulin (comparing inhibition rate constants), but the latter two are present in much higher plasma concentrations than protein C inhibitor. Furthermore, primate models of protein C activation suggest that, in vivo, protein C inhibitor is the preferred APC inhibitor until its concentration becomes limiting, and then α_1-proteinase inhibitor becomes the predominant inhibitor (7–9). However, direct evidence of the role of protein C inhibitor in APC regulation is absent because patients deficient in the protein have not been described. Previous reports of a protein C inhibitor deficiency linked to congenital combined factor V/VIII deficiency (3,10) were shown to be because of the instability of the protein in stored frozen plasma (11–13).

Protein C inhibitor, also called plasminogen activator inhibitor 3 (14–16), is a member of the serine proteinase inhibitor (serpin) superfamily of proteins (4,17–19). Serpins inhibit their target proteinases by acting as pseudosubstrates and reacting with the proteinase active site to form an essentially irreversible 1:1 stoichiometric complex. Protein C

* Present affiliation: Bowman Gray School of Medicine, Winston-Salem, North Carolina.
† Present affiliation: Emory University School of Medicine, Atlanta, Georgia.

inhibitor is a glycosaminoglycan binding serpin, with structural and functional homology to antithrombin and heparin cofactor II (20–22). Its rate of proteinase inhibition is increased by heparin and, to a lesser degree, by other glycosaminoglycans and polyanions (20). In addition to APC, protein C inhibitor can also inhibit thrombin and several other procoagulant and profibrinolytic proteinases, such as factor Xa, factor XIa, kallikrein, urokinase, and tissue plasminogen activator (4,14,22–27). However, the ability to inhibit various proteinases in an in vitro environment does not establish a significant biological role in their regulation.

Protein C inhibitor is synthesized in the liver and is found in blood plasma at a concentration of 3.6–6.8 μg/ml (~90 nM) in normal individuals (24). A protein similar in antigenicity, molecular weight, and APC inhibition activity was detected in the HepG2 liver cell line (28). Protein C inhibitor is also present at much lower concentrations in urine and several other body fluids (e.g., tears, saliva, cerebral spinal fluid, and amniotic fluid) (26,27,29–32). Interestingly, protein C inhibitor is also present in synovial fluid at plasma levels and in seminal fluid at ~200 μg/ml, roughly 40 times the concentration in plasma (30,33). It may play a physiological role in some of these sites. For example, protein C inhibitor has been shown to form complexes with plasma kallikrein upon activation of the contact system (25,29), and urinary protein C inhibitor has been purified in complex with urokinase (14,16,27,32). Most recently the protein was found in seminal fluid complexed with acrosin and prostate-specific antigen, both serine proteinases. These studies further suggest that seminal protein C inhibitor is synthesized by cells in the male reproductive tract (31,33). From this broad tissue and fluid distribution, it appears that protein C inhibitor may play a broader role in proteinase regulation than formerly believed.

II. STRUCTURE AND FUNCTION OF PROTEIN C INHIBITOR

A. Physical Properties

Suzuki and coworkers used heparin-Sepharose chromatography to purify protein C inhibitor from plasma (4) and subsequently obtained a cDNA clone from a human liver

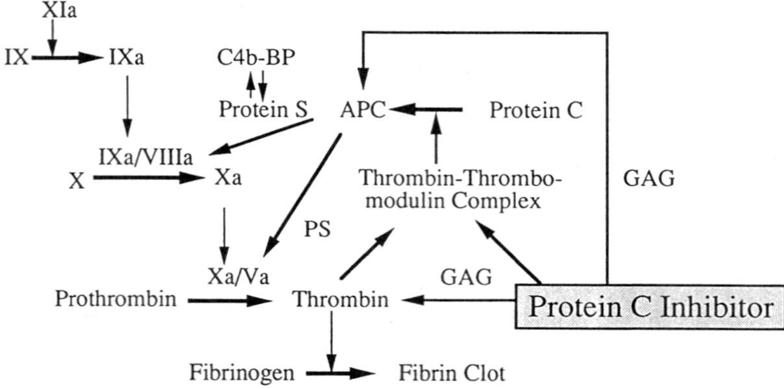

Figure 1 Activity and regulation of the protein C system. APC, activated protein C; PS, protein S; C4b-BP, C4b binding protein; GAG, glycosaminoglycan. Please refer to References 1 and 2 for a detailed overview.

cDNA library (19). Protein C inhibitor is a single-chain glycoprotein with an apparent molecular weight of 57,000 daltons (4). The mature protein contains 387 amino acids, from which a hydrophobic 19 amino acid signal peptide has been released by cleavage between a Leu-His bond. There is only one cysteine per molecule and, therefore, no intramolecular disulfide bonds. The exact carbohydrate composition and sites of attachment have not yet been characterized. However, there are five potential glycosylation sites (three Asn-*X*-Ser/Thr consensus sequences and two Thr/Ser-*X*-*X*-Pro sequences) in the amino acid sequence, and digestion with glycosidases suggests that human protein C inhibitor contains N-linked carbohydrates (Shirk and Church, unpublished observations) and that both N-linked and O-linked carbohydrates are present on the bovine protein (34). Protein C inhibitor exhibits microheterogeneity on isoelectric focusing, the multiple bands likely caused by carbohydrate composition. Three conflicting reports exist in the literature of the isoelectric point: a pI range of 7.4–8.6 for six major bands (4), a pI value of 4.5–6.0 (24), and a pI range of 6–7 for seven major bands (35). Protein C inhibitor binds glycosaminoglycans, such as heparin, and dissociates from heparin at >0.4 M NaCl (20–22).

B. Proteinase Inhibition

Serpins contain a surface-exposed sequence of residues, termed the reactive-site loop, which is accessible to proteolytic attack (Fig. 2) (17,18). A target serine proteinase recognizes the reactive-site bond, termed P1-P1′ using the nomenclature of Schechter and Berger (36), as a potential substrate. Proteolysis does not proceed to completion. Instead, the serpin and proteinase remain associated in a bimolecular complex that renders both proteins inactive. The complex remains stable in a tetrahedral intermediate (37) and is resistant to denaturants but not to strong nucleophilic agents. Within the serpin protein family, amino acid sequence variability is by far the greatest in the reactive-site loop (38), which is thought to be a primary determinant of proteinase specificity. The protein C inhibitor reactive-site peptide bond is arginine-serine and is located near the carboxyl terminus (Arg^{354}-Ser^{355}) (17–19). The P1 and P1′ residues are crucial but alone do not determine specificity (10,15,18,39). For example, protein C inhibitor can react with APC but antithrombin, which has the same Arg-Ser reactive-site sequence and inhibits a wide range of proteolytic enzymes, cannot (22). Clearly, additional molecular determinants are involved.

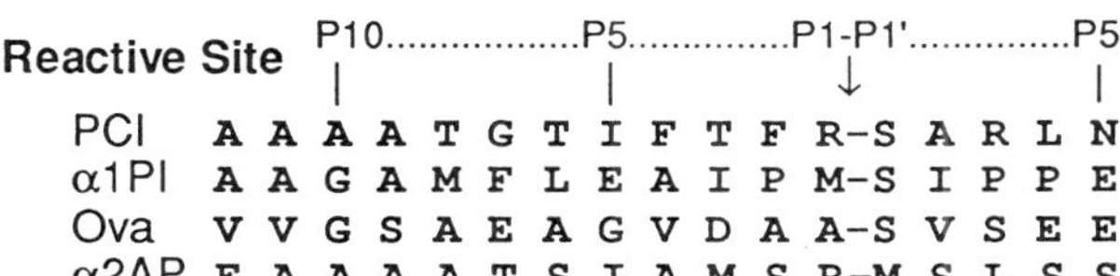

Figure 2 Reactive-site sequences of the serpins protein C inhibitor (PCI), α_1-proteinase inhibitor (α1PI), ovalbumin (Ova), and α_2-antiplasmin (α2AP). Sequences are from Reference 18. The arrow indicates the reactive-site peptide bond, termed P1-P1′ using the nomenclature of Schechter and Berger (36), that is attacked by the target proteinase. α_2-Antiplasmin and ovalbumin are two other members of the serpin superfamily. Ovalbumin does not have inhibitory activity, but the reactive-site peptide bond can be hydrolyzed by pancreatic elastase.

C. Glycosaminoglycan Interactions

The research to date suggests that the mechanism of heparin acceleration of proteinase inhibition by protein C inhibitor is comparatively simple (17–19,21,22). There is no evidence for a heparin-induced conformational change, as in antithrombin, or exposure of a secondary binding site for the proteinase, as with heparin cofactor II (21,22). Although proteinase inhibition appears to follow the ternary complex model postulated for antithrombin-heparin-thrombin (40–47), heparin-induced enhancement of the inhibition rate is modest (20) compared with that seen with antithrombin and heparin cofactor II. Interestingly, inhibition of APC is accelerated to a greater degree than inhibition of thrombin, suggesting that heparin directly alters proteinase activity or may expose ''exosites'' on the proteinase that can bind to as yet unrecognized regions of protein C inhibitor (20,21).

Like heparin cofactor II, protein C inhibitor has a broad glycosaminoglycan specificity. A variety of glycosaminoglycans and polyanions, with the exception of dermatan sulfate (when treated to remove contaminating traces of heparin and heparin sulfate), accelerate thrombin and APC inhibition by protein C inhibitor (20). Heparin fractionated by high or low affinity for antithrombin functions equally well to accelerate protein C inhibitor activity (20). Geiger et al. reported that a dermatan sulfate-containing proteoglycan (sensitive to chondroitinase ABC treatment) in an epithelial kidney cell line accelerated urokinase inhibition by protein C inhibitor (32). There are no reports of specific glycosaminoglycan sequences that bind with high affinity to protein C inhibitor. To date, the physiological proteoglycan responsible for the acceleration of the activity of protein C inhibitor in vivo has not been clearly identified.

The heparin binding site in protein C inhibitor appears to be localized not to the D helix, as in antithrombin and heparin cofactor II (18,48–51), but to the H helix, with possible contributions from the amino-terminal A^+ helix (Fig. 3) (20,21,52). Both regions have sequences of basic residues consistent with a general heparin binding motif (Fig. 4) (53). Furthermore, secondary structural predictions of the H helix region from Ser^{264} to Glu^{281} and of the A^+ helix from His^1 to Leu^{15} predict α helices. Kuhn et al. (52) predicted that heparin binds to an intensely positive electrostatic surface formed by two adjacent α helices, the H and A^+. However, there is no direct evidence for the A^+ helix being positioned near the H helix, and a synthetic peptide corresponding to the A^+ helix failed to bind heparin but an H helix peptide did bind and competed with protein C inhibitor for heparin binding (20). Studies that used a recombinant fragment of protein C inhibitor and full-length active recombinant protein C inhibitor further indicated a primary role for the H helix in heparin binding but did not rule out some contribution of the A^+ helix (54).

III. MOLECULAR GENETICS

The gene for protein C inhibitor is located on chromosome 14 in a gene cluster at 14q32.1, which also contains the genes for the serpins α_1-proteinase inhibitor, α_1-antichymotrypsin, and corticosteroid binding globulin (55). The protein C inhibitor gene, from the putative transcription initiation site to the polyadenylation signal, spans 11.5 kilobases and contains 879 bp of 3′ noncoding sequence (56). The overall gene organization resembles that of α_1-proteinase inhibitor and α_1-antichymotrypsin. Specifically, there are five exons and four introns (Fig. 5). The exon/intron boundaries follow

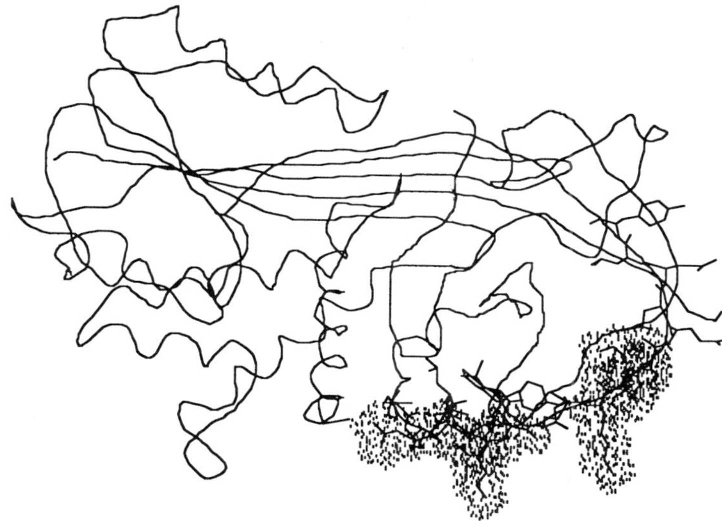

Figure 3 Molecular model of the putative glycosaminoglycan binding domain of protein C inhibitor. A three-dimensional model of protein C inhibitor was derived by computer-assisted mutagenesis using the α_1-proteinase inhibitor backbone determined from its crystal structure (18) and substituting protein C inhibitor-specific residues. The structure was energy minimized using the programs SYBYL and AMBER (21). Lysine and arginine residues of the H helix (K266, R269, K270, K273, K276, K277, and R278) are shown with their van der Waals radii highlighted. The amino-terminal A$^+$ helix is not depicted because its counterpart in α_1-proteinase inhibitor was not resolved in the crystal structure.

the "GT-AG" rule. There are no obvious TATA or CCAAT regulatory boxes near the putative transcription start site. A comparison between protein C inhibitor, α_1-proteinase inhibitor, and α_1-antichymotrypsin shows conservation of intron position for the latter three introns but a lack of homology for the first.

There are 11 nucleotide differences between the published gene sequence (56) and cDNA sequence (19) of protein C inhibitor. Only 1 of these is in the coding region and involves a G/C discrepancy at nucleotide position 1049 (cDNA numbering), resulting in

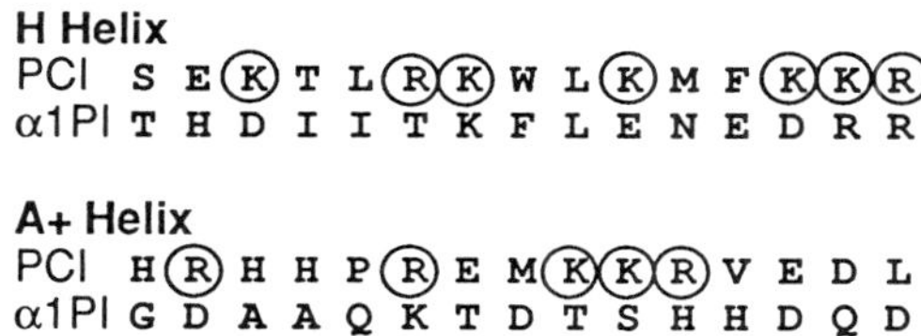

Figure 4 Putative glycosaminoglycan binding sequences in the serpin protein C inhibitor (PCI) compared with α_1-proteinase inhibitor (α1PI). The amino acid sequences of protein C inhibitor 264–278 (H helix) and 1–15 (A$^+$ helix) are aligned with the corresponding regions in α_1-proteinase inhibitor (a nonheparin binding serpin). Basic residues in protein C inhibitor are circled. Structural designations are based on either the three-dimensional structure of α_1-proteinase inhibitor deduced from x-ray crystallography (18) or from modeling studies of protein C inhibitor (52).

either a glycine or arginine at amino acid position 316. A small population study suggested that the cDNA sequence reported for that codon represents a rare polymorphism or was caused by a laboratory artifact (57). There are several other reports of genetic polymorphisms for human protein C inhibitor. Using isoelectric focusing to detect various protein isoforms, Yasuda and coworkers (35) reported an allelic polymorphism characterized by two phenotypes that represent homozygosity or heterozygosity for two autosomal codominant alleles. A Japanese population study revealed frequencies of 0.988 and 0.012 for these two alleles. Another study of cDNA clones reported a genetic polymorphism that showed ethnic frequency variability and was characterized by the less common allele possessing four nucleotide substitutions, of which two caused amino acid changes (58). There is also a report of a restriction fragment length polymorphism obtained with *Eco*RI but not with eight other restriction enzymes tested for a 3 kb *Hind*III fragment of the protein C inhibitor gene (59).

When the amino acid sequences of members of the serpin superfamily are aligned (Fig. 6), protein C inhibitor has the highest homology with α_1-proteinase inhibitor and α_1-antichymotrypsin (both 42% identity) but also shows considerable homology to the other heparin binding serpins antithrombin (28%) and heparin cofactor II (26%) (24). Little information is available about protein C inhibitor from other species. Analysis of

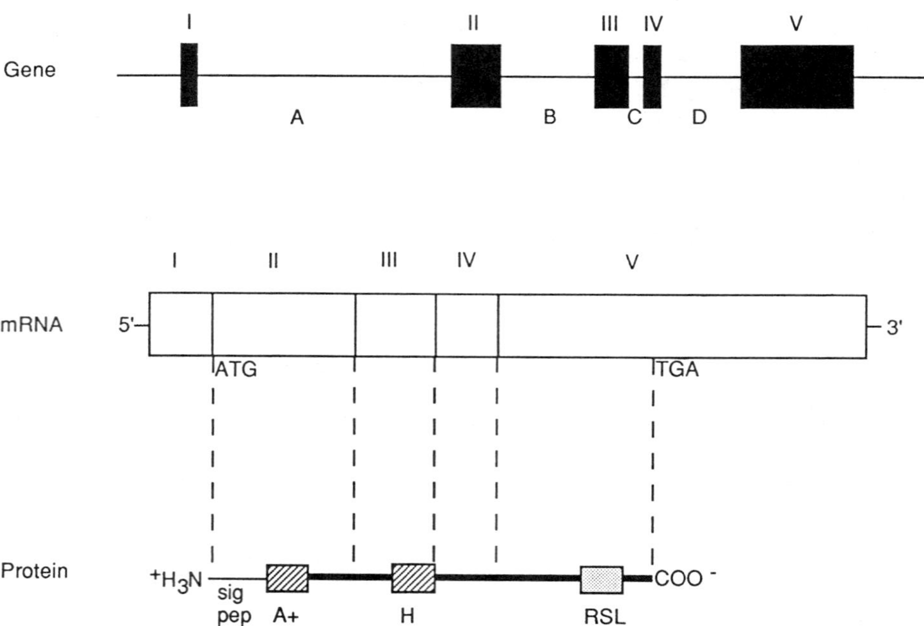

Figure 5 Protein C inhibitor gene and mRNA organization with corresponding protein domain structure (56). The gene is composed of five exons (I–V) and four introns (A–D). Exon I (98 bp) contains the putative transcription initiation site and 5′-untranslated sequences. Exon II (636 bp) contains the translation start codon and encodes the initiator methionine, the signal peptide, and the amino-terminal portion of the mature protein (residues 1–187), including the A^+ helix. Exon III (271 bp) encodes residues 188–278, which includes the H helix. Exon IV (148 bp) encodes residues 279–327. Exon V (1061 bp) encodes the carboxyl-terminal residues 328–387 (including the reactive-site loop, denoted RSL), the TGA stop codon, and 879 bp 3′-untranslated sequence.

a rhesus monkey cDNA for protein C inhibitor shows 95.7% identity with the human cDNA and predicts a protein of the same size but with 25 amino acid differences (60). These differences include 4 altered residues in the reactive-site loop (P3, P5′, P8′, and P9′) and 2 changes in the A^+ helix at positions 8 and 13. Bovine protein C inhibitor has been purified and the first 49 amino acids sequenced (34). The bovine protein lacks the first four amino-terminal residues present in human protein C inhibitor and shows very poor sequence homology for the first 15 residues, but for residues 16–49 it shows over 70% identity to the corresponding human sequence.

No abnormal human variants of protein C inhibitor have been reported. The approach of using site-directed mutagenesis to characterize protein structure-function relationships further requires a recombinant protein expression system. However, few investigators have successfully expressed recombinant protein C inhibitor. There are no reports of successful expression of the full-length molecule in a prokaryotic system. This is somewhat surprising because the protein lacks disulfide bonds and does not require carbohydrate for function (24) and because other serpins have been expressed in *E. coli.* Perhaps the unusually high susceptibility of protein C inhibitor to inactivation by limited proteolysis accounts for this phenomenon. However, recombinant protein C inhibitor has been expressed in several eukaryotic cell types, by Suzuki and coworkers in simian COS-1 cells (24), by Meijers in baby hamster kidney cells (61), and by this laboratory in insect cells infected by recombinant baculovirus (62). Molecular weight estimates for the COS-1- and baculovirus-derived proteins, based on sodium dodecyl sulfate–polyacrylamide gel electrophoresis, suggest that those recombinant proteins are not fully glycosy-

```
h_a1pi : MPSSVSWGILLLAGLCCLVPVSLAEDPQGDAAQKTDTSHHDQDHPTFNKITPNLAEFAFS
 h_pci : -.........Q-FL-L---LL-PQGASLHRHHPREMKKRVEDL-VGATVAPSSRRD-T-D

h_a1pi : LYRQLAHQSNSTNIFFSPVSIATAFAMLSLGTKADTHDEILEGLNFNLTEIPEAQIHEGF
 h_pci : ---A--SAAP-Q---------SMSL------AGSS-KMQ-----GL--QKSS-KEL-R--

h_a1pi : QELLRTLNQPDSQLQLTTGNGLFLSEGLKLVDKFLEDVKKLYHSEAFTVNFGDTEEAKKQ
 h_pci : -Q--QE----RDGF--SL--A--TDLVVD-Q-T-VSAM-T--LADT-PT--R-SAG-M--

h_a1pi : INDYVEKGTQGKIVDLVKELDRDTVFALVNYIFFKGKWERPFEVKDTEEEDFHVDQVTTV
 h_pci : -----A-Q-K------L-N--SNA-VIM-------A---TS-NH-G-Q-Q--Y-TSE-V-

h_a1pi : KVPMMKRLGMFNIQHCKKLSSWVLLMKYLGNATAIFFLPDEGKLQHLENELTHDIITKFL
 h_pci : R----S-EDQYHYLLDRN--CR-VGVP-Q-----L-I--S---M-QV--G-SEKTLR-W-

h_a1pi : ENEDRRSASLHLPKLSITGTYDLKSVLGQLGITKVFSNGADLSGVTEEAPLKLSKAVHKA
 h_pci : KMFKK-QLE-Y---F--E-S-Q-EK--PS---SN--TSH-----ISNHSNIQV-EM----

h_a1pi : VLTIDEKGTEAAGAMFLEAIPMSIPPE...VKFNKPFVFLMIEQNTKSPLFMGKVVNPTQ
 h_pci : -VEV--S--R--A-TGTIFTFR-ARLNsqrLV--R--LMFIVDN-......ILFLGKVNR

h_a1pi : K
 h_pci : P
```

Figure 6 Amino acid sequence alignment of protein C inhibitor (h_pci) with α_1-proteinase inhibitor (h_a1pi): (dashed line) aligned identical residues; (uppercase) aligned nonidentical residues; (lowercase) unaligned residues; (dotted line) gap. Sequences were obtained from Genbank and aligned using the method of Altschul in the program EuGene (63,64). The protein C inhibitor has a 19 amino acid signal peptide, and the mature protein begins with the sequence HRHH (single-letter amino acid abbreviations). The α_1-proteinase inhibitor has a 24 amino acid signal peptide, and the mature protein begins EDPQ.

lated, yet they are functionally active with specific activities the same as or higher than plasma-derived protein C inhibitor. The availability of recombinant expression systems for this serpin, in conjunction with site-directed mutagenesis techniques, will generate more knowledge of structure-function relationships in protein C inhibitor.

ACKNOWLEDGMENTS

Our work described in this chapter was supported by Research Grants HL-06530 (to F. C. Church) and F32-HL-08413 (to J. E. Phillips) from the National Institutes of Health and by a Grant-in-Aid (to F. C. Church) from the American Heart Association.

REFERENCES

1. Esmon CT. The regulation of natural anticoagulant pathways. Science 1987; 235:1348–1352.
2. Esmon CT. The protein C anticoagulant pathway. Arteriosclerosis Thromb 1992; 12:135–145.
3. Marlar RA, Griffin JH. Deficiency of protein C inhibitor in combined factor V/VIII deficiency disease. J Clin Invest 1980; 66:1186–1189.
4. Suzuki K, Nishioka J, Hashimoto S. Protein C inhibitor: purification from human plasma and characterization. J Biol Chem 1983; 258:163–168.
5. Francis RB Jr, Thomas W. Behaviour of protein C inhibitor in intravascular coagulation and liver disease. Thromb Haemost 1984; 52:71–74.
6. Marlar RA, Endres-Brooks J, Miller C. Serial studies of protein C and its plasma inhibitor in patients with disseminated intravascular coagulation. Blood 1985; 66:59–63.
7. Espana F, Gruber A, Heeb M, Hanson S, Harker L, Griffin J. In vivo and in vitro complexes of activated protein C with two inhibitors in baboons. Blood 1991; 77:1754–1760.
8. Hoogendoorn H, Nesheim ME, Giles AR. A qualitative and quantitative analysis of the activation and inactivation of protein C in vivo in a primate model. Blood 1990; 75:2164–2171.
9. Hoogendoorn H, Toh CH, Nesheim M, Giles A. Analysis of the partitioning of endogenously generated activated protein C (APC) with its inhibitors suggests the presence of heparin-like cofactor activity for protein C inhibitor (PCI) in vivo. Circulation 1992; 86:I–815 (abstract 3243).
10. Giddings JC, Sugrue A, Bloom AL. Quantitation of coagulant antigens and inhibition of activated protein C in combined factor V/VIII deficiency. Br J Haematol 1982; 52:495–502.
11. Canfield WM, Kisiel W. Evidence of normal functional levels of activated protein C inhibitor in combined factor V/VIII deficiency disease. J Clin Invest 1982; 70:1260–1272.
12. Gardiner JE, Griffin JH. Studies on human protein C inhibitor in normal and factor V/VIII deficient plasmas. Thromb Res 1984; 36:197–203.
13. Suzuki K, Nishioka J, Hashimoto S, Kamiya T, Saito H. Normal titer of functional and immunoreactive protin C inhibitor in plasma of patients with congenital combined deficiency of factor V and factor VIII. Blood 1983; 62:1266–1270.
14. Stump DC, Thienpont M, Collen D. Purification and characterization of a novel inhibitor of urokinase from human urine. Quantitation and preliminary characterization in plasma. J Biol Chem 1986; 261:12759–12766.
15. Stief TW, Radtke K-P, Heimburger N. Inhibition of urokinase by protein C inhibitor (PCI)—evidence for identity of PCI and plasminogen activator inhibitor 3. Biol Chem Hoppe Seyler 1987; 368:1427–1433.
16. Heeb MJ, Espana F, Geiger M, Collen D, Stump DC, Griffin JH. Immunological identity of heparin-dependent plasma and urinary protein C inhibitor and plasminogen activator inhibitor-3. J Biol Chem 1987; 262:15813–15816.

17. Carrell RW, Travis J. α_1-Antitrypsin and the serpins: variations and countervariations. Trends Biochem Sci 1985; 10:20–24.
18. Huber R, Carrell RW. Implictions of the three-dimensional structure of α_1-antitrypsin for structure and function of serpins. Biochemistry 1989; 28:8951–8966.
19. Suzuki K, Deyashiki Y, Nishioka J, et al. Characterization of a cDNA for human protein C inhibitor. A new member of the plasma serine protease inhibitor superfamily. J Biol Chem 1987; 262:611–616.
20. Pratt CW, Church FC. Heparin binding to protein C inhibitor. J Biol Chem 1992; 267:8789–8794.
21. Pratt CW, Whinna HC, Church FC. A comparison of three heparin-binding serine proteinase inhibitors. J Biol Chem 1992; 267:8795–8801.
22. Pratt CW, Church FC. General features of the heparin-binding serpins antithrombin, heparin cofactor II, and protein C inhibitor. Blood Coag Fib 1993; 4:479–490.
23. Suzuki K. Activated protein C inhibitor. Semin Thromb Hemost 1984; 10:154–161.
24. Suzuki K, Deyashiki Y, Nishioka J, Toma K. Protein C inhibitor: structure and function. Thromb Haemost 1989; 61:337–342.
25. Meijers JCM, Kanters DHAJ, Vlooswijk RAA, van Erp HE, Hessing M, Bouma BN. Inactivation of human plasma kallikrein and factor XIa by protein C inhibitor. Biochemistry 1988; 27:4231–4237.
26. Ecke S, Geiger M, Resch I, et al. Inhibition of tissue kallikrein by protein C inhibitor. Evidence for identity of protein C inhibitor with the kallikrein binding protein. J Biol Chem 1992; 267:7048–7052.
27. Geiger M, Heeb MJ, Binder BR, Griffin JH. Competition of activated protein C and urokinase for a heparin-dependent inhibitor. 1988; 2:2263–2267.
28. Morito F, Saito H, Suzuki K, Hashimoto S. Synthesis and secretion of protein C inhibitor by the human hepatoma-derived cell line, HepG2. Biochim Biophys Acta 1985; 844:209–215.
29. Laurell M, Stenflo J. Protein C inhibitor from human plasma: characterization of native and cleaved inhibitor and demonstration of inhibitor complexes with plasma kallikrein. Thromb Haemost 1989; 62:885–891.
30. Laurell M, Christensson A, Abrahamsson P-A, Stenflo J, Lilja H. Protein C inhibitor in human body fluids. Seminal plasma is rich in inhibitor antigen deriving from cells throughout the male reproductive system. J Clin Invest 1992; 89:1094–1101.
31. Zheng XL, Ecke S, Geiger M, Bielek E, Schleuning W-D, Binder BR. Physiological inhibition of acrosin by protein C inhibitor (PCI). Thromb Haemost 1993; 69:895(abstract 1275).
32. Geiger M, Priglinger U, Griffin JH, Binder BR. Urinary protein C inhibitor. Glycosaminoglycans synthesized by the epithelial kidney cell line TLC-598 enhance its interaction with urokinase. J Biol Chem 1991; 266:11851–11857.
33. Espana F, Gilabert J, Estelles A, Romeu A, Aznar J, Cabo A. Functionally active protein C inhibitor/plasminogen activator inhibitor-3 (PCI/PAI-3) is secreted in seminal vesicles, occurs at high concentrations in human seminal plama and complexes with prostate-specific antigen. Thromb Res 1991; 64:309–320.
34. Suzuki K, Kusumoto H, Nishioka J, Komiyama Y. Bovine plasma protein C inhibitor with structural and functional homologous properties to human plasma protein C inhibitor. J Biochem 1990; 107:381–388.
35. Yasuda T, Nadano D, Iida R, Tanaka Y, Nakanaga M, Kishi K. Discovery of a genetic polymorphism of human plasma protein C inhibitor (PCI): genetic survey utilizing isoelectric focusing followed by immunoblotting, immunological and biochemical characterization. Hum Genet 1992; 89:265–269.
36. Schechter I, Berger A. On the size of the active site in proteases I. Papain. Biochem Biophys Res Commun 1967; 27:157–162.
37. Matheson NR, van Halbeek H, Travis J. Evidence for a tetrahedral intermediate complex during serpin-proteinase interactions. J Biol Chem 1991; 266:13489–13491.

38. Creighton TE, Darby NJ. Functional evolutionary divergence of proteolytic enzymes and their inhibitors. Trends Biochem Sci 1989; 14:319–324.
39. Stein PE, Leslie AGW, Finch JT, Turnell WG, Mclaughlin PJ, Carrell RW. Crystal structure of ovalbumin as a model for the reactive centre of serpins. Nature 1990; 347:99–102.
40. Griffith MJ. The heparin-enhanced antithrombin III/thrombin reaction is saturable with respect to both thrombin and antithrombin II. J Biol Chem 1982; 257:13899–13902.
41. Griffith MJ. Heparin-catalyzed inhibitor/protease reactions: kinetic evidence for a common mechanism of action of heparin. Proc Natl Acad Sci USA 1983; 80:5460–5465.
42. Griffith MJ. Kinetics of the heparin-enhanced antithrombin III/thrombin reaction. J Biol Chem 1982; 257:7360–7365.
43. Nesheim ME. A simple rate law that describes the kinetics of the heparin-catalyzed reaction between antithrombin III and thrombin. J Biol Chem 1983; 258:14708–14717.
44. Nesheim M, Blackburn MN, Lawler CM, Mann KG. Dependence of antithrombin III and thrombin binding stoichiometries and catalytic activity on the molecular weight of affinity-purified heparin. J Biol Chem 1986; 261:3214–3221.
45. Olson ST, Shore JD. Transient kinetics of heparin-catalyzed protease inactivation by antithrombin III: the reaction step limiting heparin turnover in thrombin neutralization. J Biol Chem 1986; 261:13151–13159.
46. Olson ST. Transient kinetics of heparin-catalyzed protease inactivation by antithrombin III: linkage of protease-inhibitor-heparin interactions in the reaction with thrombin. J Biol Chem 1988; 263:1698–1708.
47. Olson ST, Bjork I, Sheffer R, Craig PA, Shore JD, Choay J. Role of the antithrombin-binding pentasaccharide in heparin acceleration of antithrombin-proteinase reactions. Resolution of the antithrombin conformational change contribution to heparin rate enhancement. J Biol Chem 1992; 267:12528–12538.
48. Borg JY, Owen MC, Soria C, Soria J, Caen J, Carrell RW. Proposed heparin binding site in antithrombin based on arginine 47. A new variant Rouen-II, 47 Arg to Ser. J Clin Invest 1988; 81:1292–1296.
49. Blinder MA, Andersson TR, Abildgaard U, Tollefsen DM. Heparin cofactor II_{Oslo}: mutation of Arg-189 to His decreases affinity to dermatan sulfate. J Biol Chem 1989; 264:5128–5133.
50. Blinder MA, Tollefsen DM. Site-directed mutagenesis of Arg-103 and Lys-185 in the proposed glycosaminoglycan-binding site of heparin cofactor II. J Biol Chem 1990; 265:286–291.
51. Whinna HC, Blinder MA, Szewczyk M, Tollefsen DM, Church FC. Role of lysine 173 in heparin binding to heparin cofactor II. J Biol Chem 1991; 266:8129–8135.
52. Kuhn LA, Griffin JH, Fisher CL, et al. Elucidating the structural chemistry of glycosaminoglycan recognition by protein C inhibitor. Proc Natl Acad Sci USA 1990; 87:8506–8510.
53. Cardin AD, Weintraub HJR. Molecular modeling of protein-glycosaminoglycan interactions. Arteriosclerosis 1989; 9:21–32.
54. Shirk RA, Elisen MGLM, Meijers JCM, Church FC. Role of the H helix in heparin binding to protein C inhibitor. J Biol Chem 1994; in press.
55. Billingsley GD, Walter MA, Hammond GL, Cox DW. Physical mapping of four serpin genes: α1-antitrypsin, α1-antichymotrypsin, corticosteroid-binding globulin, and protein C inhibitor, within a 280-kb region on chromosome 14q32.1. Am J Hum Genet 1993; 52:343–353.
56. Meijers JCM, Chung DW. Organization of the gene coding for human protein C inhibitor (plasminogen activator inhibitor-3). Assignment of the gene to chromosome 14. J Biol Chem 1991; 266:15028–15034.
57. Meijers JC, Chung DW. Evidence for a glycine residue at position 316 in human protein C inhibitor. Thromb Res 1990; 59:389–393.
58. Radtke K-P, Greengard JS, Villoutreix B, Griffin JH. Protein C inhibitor (PAI-3): an allelic polymorphism with different frequencies in different ethnic groups. Thromb Haemost 1993; 69:896(abstract 1278).

59. Ashbourne KJ, Byth BC, Meijers JCM, Cox DW. Polymorphism of the protein C inhibitor (PCI) gene on chromosome 14. Hum Mol Genet 1993; 2:92.
60. Radtke K-P, Greengard JS, Griffin JH. Protein C inhibitor in rhesus monkey plasma. Circulation 1992; 86:I–736(abstract 2930).
61. Meijers JCM. Identification of a binding site for activated protein C on protein C inhibitor. Circulation 1992; 86:I–736(abstract 2931).
62. Phillips JE, Cooper ST, Potter EE, Church FC. Mutagenesis of recombinant protein C inhibitor reactive site residues alters target proteinase specificity. J Biol Chem 1994; 269: 16696–16700.
63. Needleman SB, Wunsch CD. A general method applicable to the search for similarities in the amino acid sequences of two proteins. J Mol Biol 1970; 48:443–453.
64. Altschul SF, Erickson BW. Optimal sequence alignment using affine gap costs. Bull Math Biol 1986; 48:603–616.

22

Protein S

Koji Suzuki
Mie University School of Medicine, Mie, Japan

I. INTRODUCTION

Protein S, a protein containing γ-carboxyglutamic acid (Gla) residues, was first isolated and characterized from human plasma by DiScipio et al. in 1977 (1,2). "S" stands for Seattle, where their laboratory was situated. In 1980, Walker showed that protein S enhances the anticoagulant activity of activated protein C (APC) by acting as a cofactor (3,4). Later, Dahlbäck and Stenflo (5) found that protein S forms a reversible complex with a plasma complement component, C4b binding protein (C4BP). Successively, Comp et al. (6,7) and others (8,9) reported patients with a congenital deficiency of protein S who showed thrombotic tendencies. The physiological importance of protein S in the regulation and control of blood coagulation was thus recognized. Comp et al. (6) also found that free-form protein S, which was significantly decreased in the plasma of most patients with congenital protein S deficiency, possessed APC cofactor activity, whereas protein S complexed with C4BP did not. Dahlbäck (10) confirmed this by showing that C4BP inhibits the cofactor activity of free-form protein S. Recently, C4BP in plasma was found to form a complex with the serum amyloid P component (SAP) without any effect on protein S (11). The physiological significance of this observation has not been determined. The biochemistry of protein S and C4BP was reviewed recently by Dahlbäck (12) and Hessing (13). In this chapter, molecular biological aspects, including congenital deficiencies, of protein S and related proteins are described.

II. PROTEIN S

A. Production and Gene Structure of Protein S

Protein S is a single-chain plasma glycoprotein with an M_r of approximately 75,000, which contains 11 Gla residues in its NH_2-terminal region (2,14). It is synthesized in hepatocytes (15), vascular endothelial cells (16,17), megakaryocytes (18), Leydig cells (12), and osteoblasts (19). The plasma concentration of protein S is 20–25 μg/ml (0.26–0.33 μM).

The gene of human protein S (α gene) is located on chromosome 3p11.1–q11.2 (20). It is greater than 80 kilobases (kb) in length and contains 15 exons and 14 introns and 6 repetitive *Alu* sequences in the introns (21–23). Figure 1 shows the gene organization and the relationship between the exons and the domains that they encode. Exons I–VIII encode protein segments (residues −41 to 242) that are homologous to the Gla-containing clotting proteins and are bound by introns whose positions and types are identical with other members of this family. Exons IX–XV encode protein segments (residues 243–635) homologous to sex hormone binding globulin (SHBG) and are bound by introns of identical positions and types as in the SHBG gene. There is a pseudogene of protein S (β gene) near the α gene, which covers a size greater than 55 kb and contains regions corresponding to amino acid residues 46–635 of the mature protein. The two genes have 97% identity for coding and 3′ noncoding and 95.4% for intron regions (24).

B. Primary Structure of Protein S

Figure 2 shows the primary structure of the precursor of human protein S determined by nucleotide sequencing of the cDNA (25,26). It has high homology to that of bovine protein S, determined by a combination of amino acid and cDNA nucleotide sequencing (27). Mature protein S consists of multiple domain structures: an NH_2-terminal Gla domain, a thrombin-sensitive domain, four consecutive epidermal growth factor (EGF)-like domains, and a COOH-terminal SHBG homologous domain.

Before mature protein S is secreted into the circulation, at least four posttranslational modifications occur as studied in protein C (28). The signal peptide and propeptide are cleaved off, 11 glutamic acid residues are converted to γ-carboxyglutamyl residues, hydroxed group is added to specific aspartic acid and asparagine residues, and carbohydrate side chains are linked. Protein S has a sequence highly homologous to that of other vitamin K-dependent proteins in the propeptide region, which is required for posttranslational modifications (29).

The Gla domain of human protein S, consisting of residues 1–37, contains 11 Gla residues and is believed to participate in the interaction with the phospholipid membrane. Schwalbe et al. (30) showed that a proteolytic fragment (1–52), containing the Gla domain itself, can bind to the phospholipid membrane in the presence of a higher concentration of Ca^{2+} ions than the concentration required for intact protein S binding. Furthermore, a much higher concentration of Ca^{2+} ions dissociates the Gla-containing fragment from the membrane. These authors suggested that the Gla domain directly interacts with the phospholipid membrane via Ca^{2+} ions. The necessity of Ca^{2+} binding to the Gla domain both for binding to phospholipid vesicles and for expression of the cofactor activity was also demonstrated by Grinnell et al. (31). They found that a recombinant protein S containing 8 mol Gla residues per molecule reduced the binding ability to the vesicles and the APC cofactor activity compared with protein S containing 10 mol Gla residues per molecule. Malm et al. (32) reported the successful expression of completely γ-carboxylated and β-hydroxylated recombinant protein S having full biological activity.

The Gla domain is followed by a short connecting peptide region (residues 38–45) and a thrombin-sensitive domain (residues 46–74). In the thrombin-sensitive domain, two

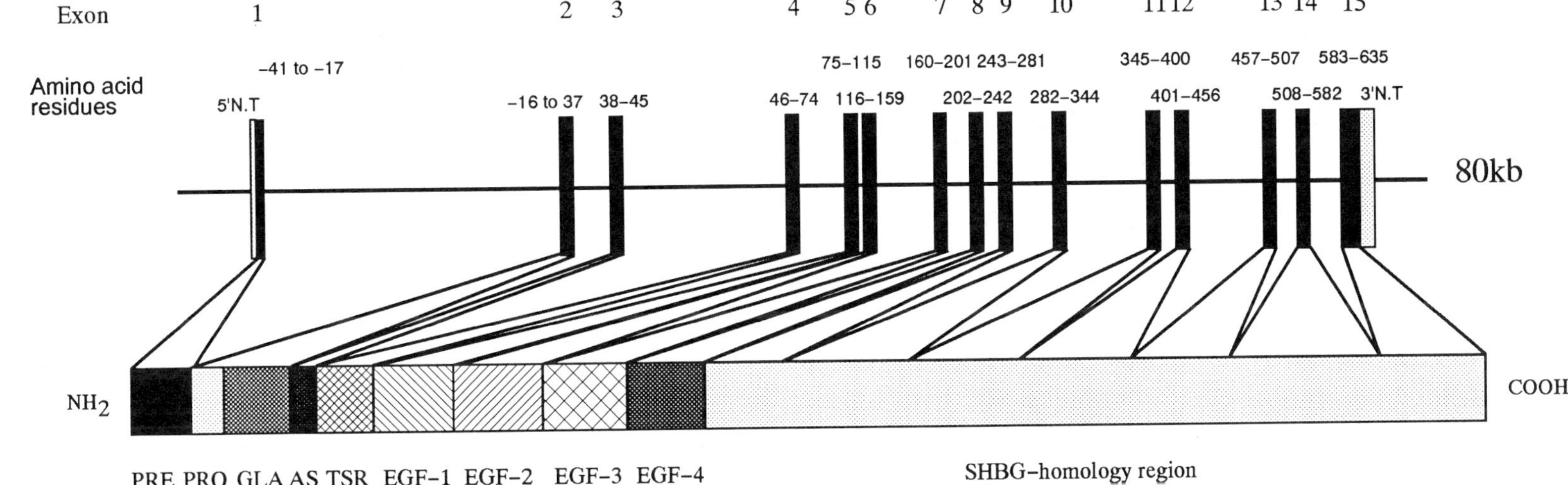

Figure 1 Gene organization of human protein S and the domain structures of protein S precursor encoded by the respective exons (21–23). 5′N.T, 5′-nontranslated region; PRE, prepeptide (signal peptide); PRO, propeptide; GLA, Gla domain; AS, aromatic stack; TSR, thrombin-sensitive region; EGF, epidermal growth factor; SHBG, sex hormone binding globulin; 3′N.T, 3′-nontranslated region.

Figure 2 Primary structure of precursor of human protein S. The sequence was deduced from Lundwall et al. (25) and Hoskins et al. (26). The positions of the introns (A–N) are indicated by wedges. *Carbohydrate binding sites; γ, γ-carboxyglutamic acid. Asp[95], Asn[136], Asn[178], and Asn[217] are β-hydroxylated.

peptide bonds Arg^{49}-Ser^{50} and Arg^{70}-Ser^{71} are easily cleaved by thrombin, particularly in the absence of Ca^{2+} ions (15,33,34). In the presence of Ca^{2+} ions, these sites are resistant to cleavage by thrombin. Thrombin-modified protein S reduced the affinity for binding to calcium ions, and as a result, the affinity for the negatively charged phospholipid membranes is also reduced (35,36). The thrombin-sensitive domain seems to be essential for Ca^{2+} binding to the Gla domain. Calcium binding, in turn, is essential for effecting a specific conformational change required for the expression of protein S cofactor activity for APC (30,36). It has been suggested that a region containing the thrombin-sensitive domain as well as the first EGF-like domain is necessary for interaction with APC (30,37,38).

The thrombin-sensitive domain is followed by four consecutive EGF-like domains (residues 75–242) in which other high-affinity Ca^{2+} binding sites are located (39). The binding constants (K_1−K_4) of Ca^{2+} ions for the first to fourth EGF-like domains are $K_1 > 1 \times 10^8$ M^{-1}, $K_2 = 1–5 \times 10^7$ M^{-1}, $K_3 = 2–6 \times 10^6$ M^{-1}, and $K_4 = 5–13 \times 10^5$ M^{-1}, respectively, in 0.15 M NaCl, pH 7.5. Each EGF-like domain contains a modified amino acid, β-hydroxyaspartic acid (Hya) or β-hydroxyasparagine (Hyn) (40). Hya is located at residue 95 in the first EGF-like domain, and Hyns are located at residues 136, 178, and 217 in the second, third, and fourth EGF-like domains of human protein S, respectively (41). Hya or Hyn has also been determined in the EGF-like domains of other plasma or cellular proteins both related and unrelated to blood coagulation, such as factor X, factor IX, protein C, protein Z, complement C1r and C1s, thrombomodulin, low-density lipoprotein receptor, uromodulin, and transforming growth factor β_1 binding protein (42). Based on the alignment of amino acid sequences of the EGF-like domains in these proteins, the following consensus sequence is suggested: *X*-*X*-Cys-*X*-Asp/Asn-*X*-*X*-*X*-*X*-Phe/Tyr-*X*-Cys-*X*-Cys-*X*-*X* (*X* is an unspecified amino acid), where Asp or Asn are hydroxylated (41). This sequence is estimated to be the recognition site for the hydroxylating enzyme(s) (41). The consensus sequence for Hyn has also been found in the EGF precursor protein and proteins encoded by genes of the Notch, Crumb, and Delta loci in *Drosophila melanogaster* and the Lin-12 locus in *Caenorhabditis elegans* and a sea urchin, although Hyn has not been actually determined in these proteins (42). The functional roles of Hya and Hyn have not been clarified. Hya is thought to relate to Ca^{2+} binding because Hya is present in the first EGF-like domains of factor IX, factor X, protein C, and protein S, whose domains all have high-affinity Ca^{2+} binding sites (42). Nelson et al. (43) reported that both Hya and Hyn in protein S are not required for anticoagulant activity or for binding to C4BP based on studies using recombinant protein S in which Hya and Hyn were replaced by other residues. Recently, the second EGF-like domain was suggested to be required for interaction with APC, factor Xa, and factor Va, but not with C4BP (our unpublished observation).

The amino acid sequence of the COOH-terminal region occupying approximately 60% of protein S is not observed in other vitamin K-dependent proteins but is homologous to the sequence of a family of androgen binding proteins, such as the SHBG and an androgen binding protein in rat testis (44,45). Androgen does not appear to bind to this region, but C4BP binds to this region with high affinity. Linear sequences, Ser^{420}-Cys^{434} (46) and Gly^{605}-Ile^{614} (47,48), in the two small loop structures near the COOH terminus of protein S were reported as candidates for binding to C4BP, based on studies using synthetic peptides corresponding to the sequences in these regions or recombinant mutant proteins. The peptide corresponding to residues Gly^{605}-Ile^{614} inhibited complex formation between protein S and C4BP and enhanced the protein S anticoagulant cofactor activity *in vivo* (49). However, Chang et al. (50) reported that replacement of residue Leu^{608}, Asp^{609}, Asp^{611}, or Glu^{612}, respectively, with Val, Asn, Asn, or Gln, in recombinant protein S did not affect the ability for binding the C4BP.

C. Function of Protein S

The functional role of protein S in the anticoagulant protein C pathway is to serve as a cofactor for APC-catalyzed inactivation of coagulation cofactors, factor Va and factor VIIIa (3,4,6–9). Although the mechanism for expression of the cofactor activity is not understood completely, it is currently accepted that, first, protein S serves as a receptor for APC on the phospholipid membrane of platelets and/or endothelial cells (4,51–54). Recently, Dahlbäck et al. (55) reported that human protein S bound specifically to platelet-derived microparticles and stimulated binding of protein C and APC, whereas protein S did not bind to the surface of activated or unactivated platelets. Second, protein S enhances the rate of APC-catalyzed inactivation of factor Va in the ''prothrombinase complex'' formed with factor Xa, factor Va, and Ca^{2+} on negatively charged phospholipids (51,52,54,56) and also enhances the rate of inactivation of factor VIIIa in the ''tenase complex'' formed with factor IXa, factor VIIIa, and Ca^{2+} on the phospholipid surface (57,58). Like factor IXa binding to factor VIIIa, factor Xa strongly interacts with factor Va bound to the phospholipid membrane with a very low dissociation constant K_d below 0.5 nM (59). By these interactions between the two proteins, factor IXa protects factor VIIIa from inactivation by APC (58,60); similarly, factor Xa protects factor Va from APC in the prothrombinase complex (61,62). Jane et al. (63) reported that factor X as well as factor Xa can protect factor Va inactivation by APC. Solymoss et al. (56) showed that protein S reduces the factor Xa protection of factor Va without any effect on the interaction between factor Xa and factor Va. Similarly, protein S reduces the protective effect of prothrombin on factor Va from APC (64). Protein S bound to APC may modulate the ability of factor Xa and prothrombin to protect factor Va and then enhances the rate of inactivation of factor Va in the prothrombinase complex by APC. Similarly, protein S may act on factor IXa and factor X to enhance the rate of inactivation of factor VIIIa in the tenase complex by APC. The cofactor activity of protein S for APC has been shown to be present in ''miniprotein S,'' which lacks the COOH-terminal SHBG-homologous domain (65), suggesting that APC interacts with a region of the amino-terminal half of protein S.

On the other hand, Heeb et al. (66) showed that protein S has an anticoagulant function independent of APC, presumably because of its competition with prothrombin for direct binding to factor Va. Hackeng et al. (67) also indicated the direct inhibition by protein S of the prothrombinase complex activity on the surface of endothelial cells and platelets in the absence of APC. They showed that both free protein S and protein S-C4BP complex interact with factor X (Xa) and inhibit the Factor Xa activity directly, but thrombin-modified protein S does not. This suggests that the site of interaction for factor Xa is located in the NH_2-terminal region of protein S. Furthermore, the binding site for factor Va is suggested to be involved in the COOH-terminal residues Arg^{621}-Ser^{635} of protein S (68).

Elastase, derived from neutrophils (69), like thrombin, cleaves protein S at the thrombin-sensitive domain and reduces its cofactor activity. These proteases may regulate the protein S cofactor activity.

The cofactor activity of protein S in vitro, observed in a system consisting of protein S, APC, factor Va, Ca^{2+} ions, and phospholipids, is very weak in comparison with that observed in the system consisting of protein S, APC, protein S-depleted plasma, Ca^{2+} ions, and phospholipids. Bakker et al. (70) demonstrated that the presence of an unknown plasma factor(s) may be required to obtain the full physiological cofactor activity of protein S. Jane et al. (71) suggested that protein S negates the inhibitor activity of an unknown factor(s) in plasma against APC. Recently, Dahlbäck et al. (72) reported a familial thrombophilia caused by a previously unrecognized mechanism characterized by poor anticoagulant response to APC and suggested the presence of a cofactor other than protein S for APC. Reports (73–75) support that the poor response to APC, called as patients with "APC resistance," is the most important congenital cause of venous thrombosis. Recently, it has been revealed that the APC resistance phenomenon is due to an unsusceptibility of factor V to APC, that is, the factor V molecule of the patients with APC resistance has a substitution of Arg^{506}, the cleavage site of APC, by Gln caused by a point mutation (133–135). On the other hand, very recently, Dahlbäck and Hildebrand found that factor V plays an important role as a cofactor of APC (APC cofactor 2) (136). Shen and Dahlbäck then demonstrated that both factor V and protein S, in a synergistic fashion, function as cofactors to APC in the degradation of factor VIIIa (137). The relationship between the mutation of factor V as the APC resistance and the deficiency of the APC cofactor 2 remains to be elucidated.

III. REGULATION OF PROTEIN S ACTIVITY BY C4BP

A. Production, Primary Structure, and Gene Organization of C4BP

C4BP is a high-molecular-weight plasma glycoprotein, synthesized in the hepatocyte, that acts as a regulator in the complement system (76). The role of C4BP in the complement system is to facilitate the dissociation of C3 convertase, a complex formed with C4b and C2a, and to serve as a cofactor in the proteolytic degradation of C4b by the serine protease factor I. C4BP is one of the acute-phase reactants. The plasma level of which is increased significantly (over four- to five-fold) during inflammatory reactions (77). It is becoming clear that C4BP may act as a regulator of protein S in the anticoagulant protein C pathway.

At least three types of C4BP are present in plasma; all are able to interact with C4b (12,13). Approximately 70% of C4BP has a high molecular weight of M_r 570,000 (C4BP-high), which has the ability to bind to protein S. The remaining 30% of C4BP has a lower molecular weight with a M_r of 500,000–530,000 (C4BP-low), in which approximately half the molecules have the ability to bind to protein S, whereas the remaining half do not (78). C4BP-high consists of seven identical α subunits with M_r 75,000 and one β subunit with M_r 45,000 (79). The C4BP-low, lacking the binding ability to protein S, is presumed to lack the β subunit from the C4BP-high, whereas the remaining half of the C4BP-low, possessing the ability to bind to protein S, is speculated to contain the β subunit, but one of the α subunits is deleted (12,13,71,80).

The primary structure of human C4BP has been elucidated based on amino acid and cDNA nucleotide sequencing (81,82). The α subunit contains 549 amino acid residues and consists of eight potential short consensus repeat sequences (SCR domains; also called Sushi domains), composed of approximately 60 residues each, and a 58 amino acid nonrepeat COOH-terminal region (81). The β subunit contains 235 amino acid residues and is comprised of three SCR domains and a 60 amino acid nonrepeat COOH-terminal region (82). Seven α subunits and a β subunit are linked to each other by disulfide bonds near the COOH-terminal end (Fig. 3) (82,83). Two cysteine residues probably involving the formation of the interchain disulfide bridges are located at the COOH-terminal nonrepeat region of both the α and β subunits (82). On the other hand, murine C4BP appears to consist of only α subunits, and these subunits are noncovalently connected together, because there are no cysteine residues at the COOH-terminal region (84).

The SCR domain has now been found in more than 30 complement and noncomplement proteins (12,76,85). Of the complement proteins, C1r, C1s, C2, factor B, C6, and C7 contain two or three SCR domains. Factor H, complement receptors 1 (CR1) and 2 (CR2), membrane cofactor protein, decay accelerating factor, and C4BP contain multiple SCR domains (76). The noncomplement proteins include the B subunit of coagulation factor XIII, β_2-glycoprotein I, haptoglobin, and cell adhesion molecules in the selectin family, ELAM-1, MEL-14, and GMP 140 (85). The function of SCR domains in these proteins has not been determined.

The gene for the α subunit of human C4BP is located on chromosome 1q32 (86) and is about 40 kb in length and composed of 12 exons (87). The second SCR domain from the NH_2 terminus of the α subunit is encoded by 2 exons, but all other SCR domains are encoded by a single exon. The gene of the β subunit is located nearby, only 4 kb apart from the α subunit gene in a head-to-tail orientation, and contains 6 exons (88). It is believed to have evolved from the α subunit gene through gene duplication. C4BP gene expression is regulated by the acute-phase mediators; tumor necrosis factor α, interleukin-6 (IL-6), and IL-1 (89). Plasma levels of C4BP are elevated during the inflammatory response, for example following infusion of the lipopolysaccharide of *Escherichia coli* (90,91).

B. Interactions of C4BP with C4b and with Protein S

Because seven α subunits are identical in structure, C4BP theoretically provides a maximum of seven binding sites for C4b. However, ultracentrifugal analysis has shown that four C4b molecules bind to a C4BP at physiological ionic strength with a dissociation constant of 8.3×10^{-8} M, and six C4b molecules bind at a lower ionic strength (92). Electron microscope studies suggest that the binding site for C4b in C4BP is located at the peripheral end of the C4BP tentacle, the NH_2-terminal region of the α subunit (83). However, studies using proteolytic fragments of C4BP indicate that the region including residues 177–322 (corresponding to the third to sixth SCR domains) is important for the cofactor activity of factor I and that residues 332–395 (corresponding to the sixth to seventh SCR domains) are required for C4b binding (93). The precise interaction mechanism between C4BP and C4b remains to be elucidated.

The interaction between C4BP and protein S is dependent on the Ca^{2+} concentration. In the presence of 5 mM Ca^{2+} ions, C4BP binds to protein S with a very low dissociation constant K_d of 5×10^{-10} M (94,95), whereas in the absence of Ca^{2+} ions the K_d increases to 10^{-7} M (96). Studies using solution-phase equilibrium showed that a K_d of 6×10^{-10} M in the presence of Ca^{2+} ions and a stoichiometry of approximately 1.7 molecules of protein S bound to each C4BP molecule and that the K_d was increased to 6×10^{-9} M in the presence of EDTA (97). Therefore, in citrated or EDTA-treated plasma, the concentration of protein S-C4BP complex appears to decrease and the free form of protein S and C4BP to increase significantly. Griffin et al. (98) determined the concentrations of free and bound forms of both proteins in plasma treated with the anticoagulant hirudin and showed that free and bound forms of protein S account for 40 and 60%, respectively, of the total protein S, whereas only 14% of the total C4BP is free and the remainder of the C4BP is in the protein S-C4BP complex. They also showed that the concentration of the protein S-C4BP complex likely corresponds to the concentration of C4BP-high containing the β subunit.

It is well-known that, in plasma of acute-phase patients, even though total C4BP concentration is increased significantly, free protein S is not elevated. The mechanism of this stable levels of free protein S was recently studied by García de Frutos et al. (138), who demonstrated that even though total C4BP level increases significantly, the level of total protein S increases only to the same extent as the level of C4BP with β subunit, whereas free protein S does not. Based on this finding, they suggested that the stable level of free protein S during acute phase is the result of different types of regulation of the C4BP α and β subunit expressions, and the free protein S is the result of the presence of a molar excess of protein S concentration over C4BP with β subunit.

The binding site for protein S in C4BP could not be localized until the β subunit was found in C4BP as an eighth tentacle, although several groups studied it by biochemical characterization (99–102) and by electron microscope analysis (83,103,104). Some experimental data suggested that the β subunit serves as a primary binding site for protein S (78–80), and recently, a recombinant β subunit of C4BP was shown to bind to protein S directly (105).

C. Regulation of Protein S Cofactor Activity by C4BP

Free protein S is thought to enhance the proteolytic degradation of factors Va and VIIIa by APC. C4BP inhibits the cofactor activity of protein S in proportion to its concentration in vitro (10,106). The mechanism whereby the protein S complexed with C4BP loses cofactor activity has not yet been elucidated. Most patients with congenital protein S

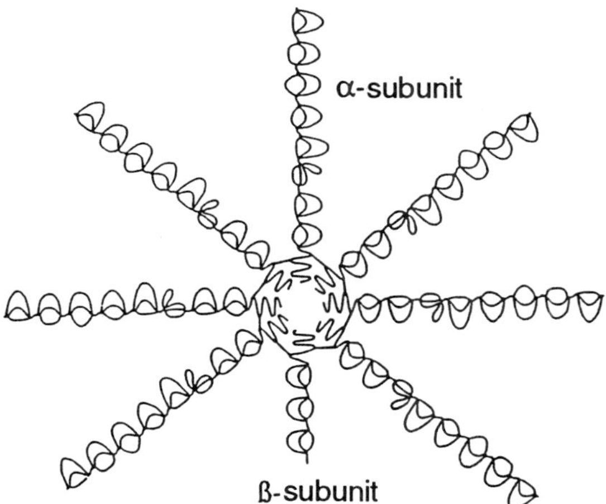

Figure 3 Structure and subunit assembly of C4BP proposed by Hillarp and Dahlbäck (79).

deficiency show a low concentration of free protein S but normal levels of the total and bound forms of the protein. We suggested that the protein S-C4BP complex competitively inhibits the interaction of APC with free protein S, based on the data that APC could bind to both the free and complexed forms of protein S equally (K_d = 10^{-10} M) and complexed protein S could competitively inhibit both the cofactor activity of free-form protein S and APC binding to free protein S (106). This idea is supported by the finding that functional protein S activity in plasma, after addition of C4BP, was lower than the residual level of free protein S antigen (107). Dahlbäck et al. (55) indicated that free protein S binds to platelet-derived microparticles and stimulates the binding of protein C and APC, whereas the protein S-C4BP complex does not bind to the microparticles or activated platelets, even though it can bind to negatively charged phospholipid vesicles. The difference between the phospholipid vesicles and microparticles in the binding of the protein S-C4BP complex may also explain the physiological significance of C4BP in the regulation of protein S cofactor activity. Dahlbäck et al. also suggested the presence of a third component that regulates the interaction between protein S and C4BP in the patient's plasma (94).

In other studies, it has been shown that the protein S-C4BP complex does not bind to factor Va, suggesting that the binding of factor Va on protein S is lost upon the interaction of C4BP with protein S (67). This C4BP binding may result in the loss of the anticoagulant activity of protein S in both the APC-dependent (10) and APC-independent (66,67) anticoagulant systems.

D. C4BP and Serum Amyloid P Component

C4BP interacts with another protein, serum amyloid P component (SAP), as well as with C4b and protein S (108). SAP is a multimeric protein consisting of five identical non-covalently bound subunits, each consisting of 204 amino acid residues and two Asn-linked carbohydrate chains with a total M_r of approximately 25,000 (109,110). The gene organization as well as the primary amino acid sequence of SAP is highly homologous to that of plasma C-reactive protein (CRP) (110,111). CRP is an acute-phase reactant in humans, but SAP is not. However, in mice, SAP, like CRP, behaves as an acute-phase reactant (112). In the 5′-noncoding region of mouse SAP and CRP genes and human CRP gene (113–116), there is a heat-shock consensus sequence, an HNF-1 recognition sequence for liver-specific expression, and TATA and CAAT box sequences.

SAP self-associates in the presence of Ca^{2+} and binds to negatively charged glycosaminoglycans, including heparin, DNA, phospholipids, and elastin fibers of extracellular tissues (109,117). SAP also forms an equimolar complex with C4BP in the presence of Ca^{2+}, and the formation of this complex appears to prevent the self-association of SAP in plasma (11). SAP-C4BP complex formation does not inhibit the formation of C4b-C4BP or protein S-C4BP complexes (11). Indeed, a part of C4BP in plasma is present as a complex composed of protein S, C4BP, SAP, and C4b, and this complex may bind to negatively charged phospholipids on the cell membrane (11,12). Although the physiological significance of this multicomponent complex has not been determined, the C4BP bound onto the cell membrane surface via protein S may stimulate or regulate the complement pathway (12). Monocytes and neutrophils migrating to the site of an inflammatory reaction release proteases that trigger both coagulation and complement cascades. At these sites, the protein S-C4BP complex, perhaps involving SAP, may control or regulate both coagulation and complement systems. Recently, García de Frutos and Dahlbäck (139) demonstrated that SAP inhibits the binding of C4BP to C4b, which regulates the stimulation by C4BP of the inhibition of factor I-mediated degradation of C4b.

IV. CONGENITAL PROTEIN S DEFICIENCY

The physiological significance of protein S in blood coagulation has been established by the identification of families with a congenital deficiency of protein S (6–9,118,119). Affected heterozygotes in these families may develop venous thromboembolic disease at a relatively young age (120). The clinical manifestations of this disorder are very similar to those in affected heterozygotes with protein C deficiency. Protein S deficiency is believed to be inherited as an autosomal dominant disorder. The prevalence of this disorder is estimated to be 1 in 20,000 in the population. Protein S deficiency accounts for about ~10% of the families with congenital thrombophilia (120).

Two distinct factors may complicate the diagnosis of protein S deficiency. First, because protein S undergoes γ-carboxylation in a vitamin K-dependent reaction, treatment with coumarin anticoagulants reduces the functional levels of the protein. Second, the concentration of the functionally active free form of protein S may be affected by the concentration of plasma C4BP, which may rise during inflammation, pregnancy, or disseminated intravascular coagulation (thus reducing levels of free protein S). On the other hand, the free protein S levels are elevated in plasma of patients with familial C4BP deficiency (121) or in the plasma of full-term newborns (122). At present, congenital protein S deficiency is tentatively classified into three types, based on the diagnostic criteria suggested by Comp et al. (123,124) and Mannucci et al. (125), which separate types based on the plasma levels of total and free protein S antigen and the cofactor activity of protein S for the APC-induced prolongation of plasma clotting (Table 1).

A. Type I Deficiency

Most patients with congenital protein S deficiency [about 94% of the patients examined by Comp (124)] are classified as type I, defined as a decreased level of free protein S antigen and a correspondingly decreased level of protein S activity. Although the free antigen and activity levels are decreased, the levels of protein S-C4BP complex are almost normal. The mutant protein S may have an increased affinity for C4BP in this type, but the detailed molecular mechanism resulting in type I deficiency has not been determined. This type of deficiency is likely a heterozygous deficiency, because while the concentra-

Table 1 Classification of Congenital Protein S Deficiency Based on Protein S Antigen and Activity

Phenotype	Antigen concentration: Total form	Antigen concentration: Free form	Antigen concentration: C4BP complex form	APC cofactor activity
Type I deficiency	Normal (or decreased)	Decreased	Normal	Decreased
Type IIa deficiency	Decreased	Decreased	Decreased	Decreased
Type IIb deficiency (type III)	Normal	Normal	Normal	Decreased

tion of protein S-C4BP complex does not decrease when there is a stable concentration of C4BP β subunit (138), a moderate decrease of total protein S concentration result in the decrease of the level of free protein S. Interestingly, the protein S complexed to C4BP has normal cofactor activity after release by treatment of the complex with detergent. This indicates that the ability to express the cofactor activity of the protein S molecule itself is unlikely to be abnormal. There may be an unknown factor(s) that regulates the interaction between protein S and C4BP, and this type of deficiency may be caused by a deficiency or overexpression of the unknown factor(s). Ploos van Amstel et al. (126) reported that 2 of the probands from 15 unrelated families with this type deficiency, having approximately 50% of total protein S antigen, showed an altered DNA fragment pattern in restriction fragment length polymorphism using protein S cDNA as a probe after genomic DNA was digested by the restriction enzyme, MspI, and that this mutation was located on the protein S β gene but not on the α gene. This suggests that the mutation in the protein S β gene is linked to type I protein S deficiency.

B. Type IIa Deficiency

A few families have been reported to belong to this type of protein S deficiency, defined as decreased levels of free and total protein S antigen and protein S activity. This type of deficiency is likely to a homozygous deficiency and basically caused by a lack of protein S in plasma, which may be caused by a significant decreased expression of the protein S gene, a large or small deletion in the gene, or abnormal secretion of protein S from the liver cells into plasma. Another possible mechanism for this type of deficiency is that the mutation results in a protein that is abnormally sensitive to degradation by plasma proteases. Ploos van Amstel et al. (127) reported a family with thrombophilia belonging to this type of deficiency caused by deletion of a large part of the middle portion of the protein S gene. Schmidel et al. (128) also reported two families with this type of deficiency who had a deletion of approximately 5.3 kb, including exon XIII, which contains all the potential Asn-linked glycosylation sites, of the protein S α gene.

C. Type IIb Deficiency

This type of deficiency may also be called type III deficiency. It is characterized by a decreased level of protein S activity in comparison with the levels of total and free protein S antigen. That is, the ratio of protein S activity per free protein S antigen in this type of deficiency is significantly lower than that in normal individuals, indicating that the protein S is probably structurally and functionally abnormal like a type II deficiency of protein C. Mannucci et al. (125) and Shigekiyo et al. (129) reported patients with dysfunctional protein S, which probably led to thrombotic disease directly. The former group showed that the distribution of protein S between the free form and the form complexed with C4BP was normal and that the level of the cofactor activity was low but the level of total and free protein S antigen were within the normal range, albeit at the low end of normal. The genes of these patients have not been studied. The patients described by the latter group had the protein S Tokushima variant, characterized by a slightly higher molecular weight than normal protein S and lacking affinity for two distinct monoclonal antibodies, one recognizing the Ca^{2+}-dependent conformation of protein S and the other recognizing the thrombin-sensitive domain of the molecule. Gene analysis revealed that the protein S Tokushima molecule had a missense mutation in the second EGF-like domain [Lys^{155} (AAG) → Glu (GAG)] (130). High-affinity Ca^{2+}-binding sites have been demonstrated in the EGF-like domains of protein S, and Ca^{2+} ions are essential for the conformational stability of these domains, which may be crucial for expressing cofactor activity, probably through interaction with APC and factor Xa (37,38,67). It has been suggested that the thrombin-sensitive domain is important for the expression of cofactor activity (35,36,38). Thus, the amino acid substitution in the second EGF-like domain may cause a conformational change in both EGF-like domains and in the thrombin-sensitive domain and thus lead to a decrease in the APC-dependent cofactor activity and also APC-independent anticoagulant activity in the patient with the protein S Tokushima variant.

Previously, Schwarz et al. (131) reported a variant protein S with a slightly decreased molecular weight, in comparison to normal protein S, in the plasma of patients with thrombotic disease. This protein S appeared to lack an Asn-linked carbohydrate site chain. However, this variant protein S showed normal cofactor activity for APC and normal C4BP binding ability, and the patients had normal molecular weight protein S in their platelets. Furthermore, Bertina et al. (132) reported a protein S polymorphism (protein S Heerlen) that was not detected by a monoclonal antibody but was detected by a polyclonal antibody. Protein S Heerlen showed a slightly lower molecular weight in comparison with normal protein S but had normal cofactor activity and C4BP binding activity. The mutation was demonstrated to be A to G change in the nucleotide sequence of exon XIII of the protein S α gene, resulting in a $Ser^{460} \rightarrow$ Pro substitution. This substitution occurred in the consensus sequence for a potential Asn-linked glycosylation site, which results in a reduction of the Asn^{458}-linked carbohydrate side chain, accounting for the reduction in the molecular weight.

Accumulation of further information on the type IIb deficiency would clarify the structure-function relationship of protein S both in vitro and in vivo in relation to the pathogenesis of thrombotic disease.

ACKNOWLEDGMENTS

The author thanks Drs. Tatsuya Hayashi and Hiroshi Takeya and also Junji Nishioka and Masane Kume for preparing the manuscript.

REFERENCES

1. DiScipio RG, Hermodson MA, Yates SG, Davie EW. A comparison of human prothrombin, factor IX (Christmas factor), factor X (Stuart factor), and protein S. Biochemistry 1977; 16: 698–706.
2. DiScipio RG, Davie EW. Characterization of protein S, a γ-carboxyglutamic acid containing protein from bovine and human plasma. Biochemistry 1979; 18:899–904.
3. Walker FJ. Regulation of activated protein C by a new protein. J Biol Chem 1980; 255: 5521–5524.
4. Walker FJ. Regulation of activated protein C by protein S: the role of phospholipid in factor Va inactivation. J Biol Chem 1981; 256:11128–11131.
5. Dahlbäck B, Stenflo J. High molecular weight complex in human plasma between vitamin K-dependent protein S and complement component C4b-binding protein. Proc Natl Acad Sci USA 1981; 78:2512–2516.
6. Comp PC, Nixon RR, Cooper MR, Esmon CT. Familial protein S deficiency is associated with recurrent thrombosis. J Clin Invest 1984; 74:2082–2088.
7. Comp PC, Esmon CT. Recurrent thromboembolism in patients with a partial deficiency of protein S. N Engl J Med 1984; 311:1525–1528.

8. Schwarz HP, Fischer M, Hopmeier P, Batard MA, Griffin JH. Plasma protein S deficiency in familial thrombotic disease. Blood 1984; 64:1297–1300.
9. Kamiya T, Sugihara T, Ogata K, et al. Inherited deficiency of protein S in a Japanese family with recurrent venous thrombosis: a study of three generations. Blood 1986; 67:406–410.
10. Dahlbäck B. Inhibition of protein C, a cofactor function of human and bovine protein S by C4b-binding protein. J Biol Chem 1986; 261:12022–12027.
11. Schwalbe RA, Dahlbäck B, Nelsestuen GL. Independent association of serum amyloid P component, protein S, and complement C4b with complement C4b-binding protein and subsequent association of the complex with membranes. J Biol Chem 1990; 265:21749–21757.
12. Dahlbäck B. Protein S and C4b-binding protein: components involved in the regulation of the protein C anticoagulant system. Thromb Haemost 1991; 66:49–61.
13. Hessing M. The interaction between complement component C4b-binding protein and the vitamin K-dependent protein S forms a link between blood coagulation and the complement system. Biochem J 1991; 277:581–592.
14. Stenflo J, Jonsson M. Protein S, a new vitamin K-dependent protein from bovine plasma. FEBS Lett 1979; 101:377–381.
15. Fair DS, Marlar RA. Biosynthesis and secretion of factor VII, protein C, protein S, and the protein inhibitor from a human hepatoma cell line. Blood 1986; 6:64–70.
16. Fair DS, Marlar RA, Levin EG. Human endothelial cells synthesize protein S. Blood 1986, 67:1168–1171.
17. Stern D, Brett J, Harris K, Nawroth P. Participation of endothelial cells in the protein C-protein S anticoagulant pathway. The synthesis and release of protein S. J Cell Biol 1986; 102:1971–1978.
18. Ogura M, Tanabe N, Nishioka J, Suzuki K, Saito H. Biosynthesis and secretion of functional protein S by a human megakaryoblastic cell line (MEG-01). Blood 1987; 70:301–306.
19. Maillard C, Berruyer M, Serre CM, Dechavannee M, Delmas PD. Protein S, a vitamin K-dependent protein, is a bone matrix component synthesized and secreted by osteoblasts. Endocrinology 1992; 130:1599–1604.
20. Watkins PC, Eddy R, Fukushima Y, et al. The gene for protein S maps near the centromere of human chromosome 3. Blood 1988; 71:238–241.
21. Schmidel DK, Tatro AV, Phelps LG, Tomczak JA, Long GL. Organization of the protein S genes. Biochemistry 1990; 29:7845–7852.
22. Ploos van Amstel HK, Reitsma PH, van der Logt PE, Bertina R. Intron-exon organization of the active human protein S gene PSα and its pseudogene PSβ: duplication and silencing during primate evolution. Biochemistry 1990; 29:7853–7861.
23. Edenbrandt C-M, Lundvall A, Wydro R, Stenflo J. Molecular analysis of the gene for vitamin K-dependent protein S and its pseudogene: cloning and partial characterization. Biochemistry 1990; 29:7861–7868.
24. Ploos van Amstel HK, Reitsma PH, Bertina RM. The human protein S locus—identification of the PSα gene as a site of liver protein S messenger RNA synthesis. Biochem Biophys Res Commun 1988; 157:1033–1038.
25. Lundwall A, Dackowski W, Cohen E, et al. Isolation and sequence of the cDNA for human protein S, a regulator of blood coagulation. Proc Natl Acad Sci USA 1986; 83:6717–6720.
26. Hoskins J, Norman DK, Beckmann RJ, Long GL. Cloning and characterization of human liver cDNA encoding a protein S precursor. Proc Natl Acad Sci USA 1987; 84:349–353.
27. Dahlbäck B, Lundwall A, Stenflo J. Primary structure of bovine vitamin K-dependent protein. Proc Natl Acad Sci USA 1986; 83:4199–4203.
28. McClure DB, Walls JD, Grinnell BW. Post-translational processing events in the secretion pathway of human protein C, a complex vitamin K-dependent antithrombotic factor. J Biol Chem 1992; 267:19710–19717.

29. Ploos van Amstel HK, van der Zanden AL, Reitsma PH, Bertina RM. Human protein S cDNA encodes Phe-16 and Tyr-222 in the consensus sequences for the post-translational processing. FEBS Lett 1987; 222:186–190.
30. Schwalbe RA, Ryan J, Stern DM, Kisiel W, Dahlbäck B, Nelsestuen GL. Protein structural requirements and properties of membrane binding by γ-carboxyglutamic acid-containing plasma proteins and peptides. J Biol Chem 1989; 264:20288–20296.
31. Grinnell BW, Walls JD, Marks C, et al. γ-Carboxylated isoforms of recombinant human protein S with different biologic properties. Blood 1990; 76:2546–2554.
32. Malm J, Cohen E, Dackowski W, Dahlbäck B, Wydro R. Expression of completely γ-carboxylated and β-hydroxylated recombinant human vitamin K-dependent protein S with full biological activity. Eur J Biochem 1990; 187:737–743.
33. Dahlbäck B. Purification of human vitamin K-dependent protein S and its limited proteolysis by thrombin. Biochem J 1983; 209:837–846.
34. Dahlbäck B, Lundwall A, Stenflo J. Localization of thrombin cleavage sites in the amino-terminal region of bovine protein S. J Biol Chem 1986; 261:5111–5115.
35. Walker FJ. Regulation of vitamin K-dependent protein S: inactivation with thrombin. J Biol Chem 1984; 259:10335–10339.
36. Suzuki K, Nishioka J, Hashimoto S. Regulation of activated protein C by thrombin-modified protein S. J Biochem 1983; 94:699–705.
37. Malm J, Persson U, Dahlbäck B. Inhibition of human vitamin K-dependent protein S-cofactor activity by a monoclonal antibody specific for Ca^{2+}-dependent epitope. Eur J Biochem 1987; 165:39–45.
38. Dahlbäck B, Hildebrand B, Malm J. Characterization of functionally important domains in vitamin K-dependent protein S using monoclonal antibodies. J Biol Chem 1990; 265:8127–8135.
39. Dahlbäck B, Hildebrand B, Linse S. Novel type of very high affinity calcium-binding sites in β-hydroxyasparagine-containing epidermal growth factor-like domains in vitamin K-dependent protein S. J Biol Chem 1990; 265:18481–18489.
40. Fernlund P, Stenflo J. β-Hydroxyaspartic acid in vitamin K-dependent proteins. J Biol Chem 1983; 258:12509–12512.
41. Stenflo J, Lundwall A, Dahlbäck B. β-Hydroxyasparagine in domains homologous to the epidermal growth factor precursor in vitamin K-dependent protein. Proc Natl Acad Sci USA 1987; 84:368–372.
42. Stenflo J. Structure-function relationships of epidermal growth factor modules in vitamin K-dependent clotting factors. Blood 1991; 78:1637–1651.
43. Nelson RM, Van Dusen WJ, Friedman PA, Long GL. β-Hydroxyaspartic acid and β-hydroxyasparagine residues in recombinant human protein S are not required for anticoagulant cofactor activity or for binding to C4b-binding protein. J Biol Chem 1991; 266:20586–20589.
44. Gershagen S, Fernlund P, Lundwall A. A cDNA coding for human sex hormone binding globulin: homology to vitamin K-dependent protein S. FEBS Lett 1987; 220:129–135.
45. Baker ME, French FS, Joseph DR. Vitamin K-dependent protein S is similar to rat androgen-binding protein. Biochem J 1987; 243:293–296.
46. Griffin J. Report in subcommittee on protein C and S. The 38th Scientific and Standardization Committee (SSC) of International Society on Thrombosis and Haemostasis (ISTH), Munich, July 7, 1992.
47. Walker FJ. Characterization of a synthetic peptide that inhibits the interaction between protein S and C4b-binding protein. J Biol Chem 1989; 264:17645–17648.
48. Nelson RM, Long GL. Binding of protein S to C4b-binding protein: mutagenesis of protein S. J Biol Chem 1992; 267:8140–8145.
49. Weinstein RE, Walker FJ. Enhancement of rabbit protein S anticoagulant cofactor activity in vivo by modulation of the protein S. J Clin Invest 1990; 86:1928–1935.

50. Chang GTG, Ploos van Amstel HK, Hessing M, Reitsma PH, Bertina RM, Bouma BN. Expression and characterization of recombinant protein S in heterologous cells: studies of the interaction of amino acid residues Leu-608 to Glu-612 with human C4b-binding protein. Thromb Haemost 1992; 67:526–532.
51. Suzuki K, Nishioka J, Matsuda J, Murayama M, Hashimoto S. Protein S is essential for the activated protein C-catalyzed inactivation of platelet-associated factor Va. J Biochem 1984; 96:455–460.
52. Harris KW, Esmon CT. Protein S is required for bovine platelets to support activated protein C binding and activity. J Biol Chem 1985; 260:2007–2010.
53. Walker FJ. Interactions of protein S with membranes. Semin Thromb Hemost 1988; 14: 216–221.
54. Stern DM, Nawroth PP, Harris K, Esmon CT. Cultured bovine aortic endothelial cells promote activated protein C-protein S-mediated inactivation of factor Va. J Biol Chem 1986; 261:713–718.
55. Dahlbäck B, Wiedmer T, Sims PJ. Binding of anticoagulant vitamin K-dependent protein S to platelet-derived microparticles. Biochemistry 1992; 31:12769–12777.
56. Solymoss S, Tucker MM, Tracy PB. Kinetics of inactivation of membrane-bound factor Va by activated protein C—protein S modulates factor Xa protection. J Biol Chem 1988; 263: 14884–14890.
57. Koedam JA, Meijers JCM, Sixma JJ, Bouma BN. Inactivation of human factor VIII by activated protein C—cofactor activity of protein S and protective effect of von Willebrand factor. J Clin Invest 1988; 82:1236–1243.
58. Walker FJ, Chavin SI, Fay PJ. Inactivation of factor VIII by activated protein C and protein S. Arch Biochem Biophys 1987; 252:322–328.
59. Mann KG, Nesheim ME, Church WR, Haley P, Krishnaswamy S. Surface-dependent reactions of the vitamin K-dependent enzyme complex. Blood 1990; 76:1–16.
60. Fulcher CA, Gardiner JH, Griffin JH, Zimmerman TS. Proteolytic inactivation of human factor VIII procoagulant protein by activated protein C and its analogy with factor V. Blood 1984; 63:486–489.
61. Nesheim ME, Canfield WM, Kisiel W, Mann KG. Studies of the capacity of factor Xa to protect factor Va from inactivation by activated protein C. J Biol Chem 1982; 257:1443–1447.
62. Luckow EA, Lyons DA, Ridgeway TM, Esmon CT, Laue TM. Interaction of clotting factor V heavy chain with prothrombin and prethrombin-1 and role of activated protein C in regulating this interaction. Biochemistry 1989; 28:2348–2354.
63. Jane SM, Hau L, Salem HH. Regulation of activated protein C by factor Xa. Blood Coag Fib 1991; 2:723–729.
64. Mitchell CA, Jane SM, Salem HH. Inhibition of the anticoagulant activity of protein S by prothrombin. J Clin Invest 1988; 82:2142–2147.
65. Chang G, Hackeng TM, Bertina RM, Bouma BN. Mini protein S, a recombinant human variant molecule that lacks the sex hormone binding globulin-like domain has cofactor activity to activated protein C and fails to bind to human C4b-binding protein (abstract). Thromb Haemost 1993; 69:789.
66. Heeb MJ, Mesters RM, Tans G, Rosing J, Griffin JH. Binding of protein S to factor Va associated with inhibition of prothrombinase that is independent of activated protein C. J Biol Chem 1993; 268:2872–2877.
67. Hackeng TM, van't Veer C, Meijers JCM, Bouma BN. Human protein S inhibits prothrombinase complex activity on endothelial cells and platelets via direct interactions with factors Va and Xa. J Biol Chem 1994; 269:21051–21058.
68. Heeb MJ, Chang G, Bouma BN, Rosing J, Tans G, Griffin JH. C-terminal residues 621–635 of protein S are essential for binding to factor Va but not factor Xa (abstract). Thromb Haemost 1993; 69:789.

69. Oates AM, Salem HH. The binding and regulation of protein S by neutrophils. Blood Coag Fib 1991; 2:601–607.
70. Bakker HM, Tans G, Janssen-Claessen T, et al. The effect of phospholipids, calcium ions and protein S on rate constants of human factor Va inactivation by activated protein C. Eur J Biochem 1992; 208:171–178.
71. Jane SM, Hau L, Salem HH. Protein S negates the activated protein C inhibitory activity of plasma. Blood Coag Fib 1992; 3:257–261.
72. Dahlbäck B, Carlsson M, Svensson PJ. Familial thrombophilia due to a previously unrecognized mechanism characterized by poor anticoagulant response to activated protein C: prediction of a cofactor to activated protein C. Proc Natl Acad Sci USA 1993; 90:1004–1008.
73. Griffin JH, Evatt B, Wideman C, Fernandez JA. Anticoagulant protein C pathway defective in majority of thrombophilic patients. Blood 1993; 82:1989–1993.
74. Koster T, Rosendaal FR, de Ronde H, Briet E, Vandenbroucke JP, Bertina RM. Venous thrombosis due to poor anticoagulant response to activated protein C: Leiden thrombophilia study. Lancet 1993; 342:1503–1506.
75. Faioni EM, Franchi F, Asti D, Sacchi E, Bernardi F, Mannucci PM. Resistance to activated protein C in nine thrombophilic families: interference in a protein S functional assay. Thromb Haemost 1993; 70:1067–1071.
76. Law SKA, Reid KBM. Activation and control of the complement. In: Male D, ed. Complement. Oxford: IRL Press, 1988:9–27.
77. Barnum SR, Dahlbäck B. C4b-binding protein, a regulatory component of the classical pathway of complement, is an acute phase protein and is elevated in systemic lupus erythematosus. Complement Inflamm 1990; 7:71–77.
78. Hillarp A, Hessing M, Dahlbäck B. Protein S binding in relation to the subunit composition of human C4b-binding protein. FEBS Lett 1989; 259:53–56.
79. Hillarp A, Dahlbäck B. Novel subunit in C4b-binding protein required for protein S binding. J Biol Chem 1988; 263:12759–12764.
80. Hessing M, Kanters D, Hackeng TM, Bouma BN. Identification of different forms of human C4b-binding protein lacking β-chain and protein S binding ability. Thromb Haemost 1990; 64:245–250.
81. Chung LP, Bentley DR, Reid KBM. Molecular cloning and characterization of the cDNA coding for C4b-binding protein: a regulatory protein of the classical pathway of the human complement system. Biochem J 1985; 230:133–141.
82. Hillarp A, Dahlbäck B. Cloning of cDNA coding for the β-chain of human complement component C4b-binding protein: sequence homology with the α chain. Proc Natl Acad Sci USA 1990; 87:1183–1187.
83. Dahlbäck B, Muller-Eberhard HJ. Ultrastructure of C4b-binding protein fragments formed by limited proteolysis using chymotrypsin. J Biol Chem 1984; 259:11631–11634.
84. Kristensen T, Ogata RT, Chung LP, Reid KBM, Tack BF. cDNA structure of murine C4b-binding protein, a regulatory component of the serum complement system. Biochemistry 1987; 26:4668–4674.
85. Baron M, Norman DG, Campbell ID. Protein modules. TIBS 1991; 16:13–17.
86. Andersson A, Dahlbäck B, Hansen C, et al. Genes for C4b-binding protein α- and β-chains (C4BPA and C4BPB) are located on chromosome 1, band 1q32, in humans and on chromosome 13 in rats. Somat Cell Mole Genet 1990; 16:493–500.
87. Aso T, Okamura S, Matsuguchi T, Sakamoto N, Sata T, Niho Y. Genomic organization of the α chain of the human C4b-binding protein gene. Biochem Biophys Res Commun 1991; 174:222–227.
88. Padro-Manuel F, Rey-Campos J, Hillarp A, Dahlback B, Rodriguez de Cordoba S. Human genes for the α and β-chains of complement C4b-binding protein are closely linked in a head-to-tail arrangement. Proc Natl Acad Sci USA 1990; 87:4529–4532.

89. Moffat GJ, Tack BF. Regulation of C4b-binding protein gene expression by the acute-phase mediators tumor necrosis factor-α, interleukin-6, and interleukin-1. Biochemistry 1992; 31: 12376–12384.
90. Taylor FB Jr, Chang A, Ferrell G, et al. C4b-binding protein exacerbates the host response to *Escherichia coli*. Blood 1991; 78:357–363.
91. Esmon CT, Taylor FB Jr, Snow TR. Inflammation and coagulation: linked processes potentially regulated through a common pathway mediated by protein C. Thromb Haemost 1991; 66:160–166.
92. Ziccardi RJ, Dahlbäck B, Muller-Eberhard HJ. Characterization of the interaction of human C4b-binding protein with physiological ligands. J Biol Chem 1984; 259:13674–13679.
93. Chung LP, Reid KBM. Structural and functional studies on C4b-binding protein, a regulatory component of the human complement system. Biosci Rep 1985; 5:855–865.
94. Dahlbäck B, Frohm B, Nelsestuen G. High affinity interaction between C4b-binding protein and vitamin K-dependent protein S in the presence of calcium: suggestion of a third component in blood regulating the interaction. J Biol Chem 1990; 265:16082–16087.
95. Schwalbe R, Dahlbäck B, Hillarp A, Nelsestuen G. Assembly of protein S and C4b-binding protein on membranes. J Biol Chem 1990; 265:16074–16081.
96. Dahlbäck B. Purification of human C4b-binding protein and formation of its complex with vitamin K-dependent protein S. Biochem J 1983; 209:847–856.
97. Nelson RM, Long GL. Solution-phase equilibrium binding interaction of human protein S with C4b-binding protein. Biochemistry 1991; 30:2384–2390.
98. Griffin JH, Gruber A, Fernandez JA. Reevaluation of total, free and bound protein S and C4b-binding protein levels in plasma anticoagulated with citrate or hirudin. Blood 1992; 79:3203–3211.
99. Nagasawa S, Unno H, Ichihara C, Koyama J, Koide T. Human C4b-binding protein, C4bp: chymotryptic cleavage and location of the 48 kDa active fragment within C4bp. FEBS Lett 1983; 164:135–138.
100. Chung LP, Gagnon J, Reid KBM. Amino acid sequence studies of human C4b-binding protein: N-terminal sequence analysis and alignment of the fragments produced by limited proteolysis with chymotrypsin and the peptides produced by cyanogen bromide treatment. Mol Immunol 1985; 22:427–435.
101. Suzuki K, Nishioka J. Binding site for vitamin K-dependent protein S on complement C4b-binding protein. J Biol Chem 1988; 263:1734–1739.
102. Hillarp A, Dahlbäck B. The protein S-binding site localized to the central core of C4b-binding protein. J Biol Chem 1987; 262:11300–11307.
103. Dahlbäck B, Smith CA, Muller-Eberhard HJ. Visualization of human C4b-binding protein and its complexes with vitamin K-dependent protein S and complement protein C4b. Proc Natl Acad Sci USA 1983; 80:3461–3465.
104. Dahlbäck B. Interaction between vitamin K-dependent protein S and the complement protein, C4-binding protein: a link between coagulation and the complement system. Semin Thromb Hemost 1984; 10:139–148.
105. Hardig Y, Rezaie A, Dahlbäck B. Protein S binds to the C4b-binding protein β-chain (abstract). Thromb Haemost 1993; 69:789.
106. Nishioka J, Suzuki K. Inhibition of cofactor activity of protein S by a complex of protein S and C4b-binding protein. J Biol Chem 1990; 265:9072–9076.
107. Espana F, Hendl S, Aznar J, Gilbert J, Estelles A. Determination of total, free and complexed protein S in plasma by ELISA, and comparison with a standard electroimmunoassay. Thromb Res 1991; 62:615–624.
108. De Beer FC, Baltz ML, Holford S, Feinstein A, Pepys MB. Fibronectin and C4-binding protein are selectively bound by aggregated amyloid P component. J Exp Med 1981; 154: 1134–1149.
109. Mantzouranis EC, Dowton SB, Whitehead AS, Edge MD, Bruns GAP, Colten HR. Human serum amyloid P component. cDNA isolation, complete sequence of pre-serum amyloid P

component, and localization of the gene to chromosome 1. J Biol Chem 1985; 260 7752–7756.

110. Skinner M, Cohen AS. Amyloid P component. Methods Enzymol 1988; 163:523–536.
111. Ohnishi S, Maeda S, Shimada K, Arao T. Isolation and characterization of the complete complementary and genomic DNA sequences of human serum amyloid P component. J Biochem 1986; 100:849–858.
112. Murakami T, Ohnishi S, Nishiguchi S, Maeda S, Araki S, Shimada K. Acute-phase response of mRNAs for serum amyloid P component, C-reactive protein and prealbumin. Biochem Biophys Res Commun 1988; 155:554–560.
113. Nishiguchi S, Maeda S, Araki S, Shimada K. Structure of the mouse serum amyloid P component gene. Biochem Biophys Res Commun 1988; 155:1366–1373.
114. Ohnishi S, Maeda S, Nishiguchi S, Arao T, Shimada K. Structure of the mouse C-reactive protein gene. Biochem Biophys Res Commun 1988; 156:814–822.
115. Lei K-J, Liu T, Zon G, Soravia E, Liu T-Y, Goldman ND. Genomic DNA sequence for human C-reactive protein. J Biol Chem 1985; 260:13377–13383.
116. Woo P, Korenberg JR, Whitehead AS. Characterization of genomic and complementary DNA sequence of human C-reactive protein, and comparison with the complementary DNA sequence of serum amyloid P component. J Biol Chem 1985; 260:13384–13388.
117. Baltz ML, De Beer FC, Feinstein A, Pepys MB. Calcium-dependent aggregation of human serum amyloid P component. Biochim Biophys Acta 1982; 701:229–236.
118. Broekmans AW, Bertina RM, Reinalda-Poot J, et al. Hereditary protein S deficiency and venous thromboembolism: a study in three Dutch families. Thromb Haemost 1985; 53:273–277.
119. Bertina RM. Hereditary protein S deficiency. Haemostasis 1985; 15:241–246.
120. Briet E, Broekmans AW, Engesser L. Hereditary protein S deficiency. In: Bertina RM, ed. Protein C and Related Proteins. Edinburgh: Churchill Livingstone, Longman Group UK, 1988:203–212.
121. Comp PC, Forristall J, West CD, Trapp RG. Free protein S levels are elevated in familial C4b-binding protein deficiency. Blood 1990; 76:2527–2529.
122. Fernandez JA, Estelles A, Gilabert J, Espana F, Aznar J. Functional and immunologic protein S in normal pregnant woman and in full-term newborns. Thromb Haemost 1989; 61:474–478.
123. Comp PC, Doray D, Patton D, Esmon CT. An abnormal plasma distribution of protein S occurs in functional protein S deficiency. Blood 1986; 67:504–508.
124. Comp PC. Laboratory evaluation of protein S status. Semin Thromb Haemost 1990; 16: 177–181.
125. Mannucci PM, Valsecchi C, Krachmalnicoff A, Faioni EM, Tripodi A. Familial dysfunction of protein S. Thromb Haemost 1989; 62:763–766.
126. Ploos van Amstel HK, Reitsma PH, Hamulyak K, de Die-Smulders CEM, Mannucci PM, Bertina RM. A mutation in the protein S pseudogene is linked to protein S deficiency in a thrombophilic family. Thromb Haemost 1989; 62:897–901.
127. Ploos van Amstel HK, Huisman MV, Reitsma PH, ten Cate JW, Bertina RM. Partial protein S gene deletion in a family with hereditary thrombophilia. Blood 1989; 73:479–483.
128. Schmidel DK, Nelson RM, Broxson EH Jr, Comp PC, Marlar RA, Long GL. A 5.3-kb deletion including exon XIII of the protein S α gene occurs in two protein S-deficient families. Blood 1991; 77:551–559.
129. Shigeliyo T, Uno Y, Tomonari A, et al. Protein S Tokushima: an abnormal protein S found in Japanese family with thrombosis. Thromb Haemost 1993; 67:244–246.
130. Hayashi T, Nishioka J, Shigekiyo T, Saito S, Suzuki K. Protein S Tokushima: abnormal molecule with a substitution of Glu for Lys-155 in the second epidermal growth factor-like domain of protein S. 1994; Blood 83:683–690.
131. Schwarz HP, Heeb MJ, Lottenberg R, Roberts H, Griffin JH. Familial protein S deficiency with a variant protein S molecule in plasma and platelets. Blood 1989; 74:213–221.

132. Bertina RM, Ploos van Amstel HK, van Wijngaarden A, et al. Heerlen polymorphism of protein S, an immunologic polymorphism due to dimorphism of residue 460. Blood 1990; 76:538–548.
133. Bertina RM, Koeleman BPC, Koster T, Rosendaal FR, Dirven RJ, de Ronde H, van der Velden PA, Reitsma PH. Mutation in blood coagulation factor V associated with resistance to activated protein C. Nature 1994; 369:64–67.
134. Voorberg J, Roelse J, Koopman R, Büller H, Berends F, ten Cate JW, Mertens K, van Mourik JA. Association of idiopathic venous thromboembolism with single point-mutation at Arg^{506} of factor V. Lancet 1994; 343:1535–1536.
135. Zöller B. Dahlbäck B. Linkage between inherited resistance to activated protein C and factor V gene mutation in venous thrombosis. Lancet 1994; 343:1536–1538.
136. Dahlbäck B, Hildebrand B. Inherited resistance to activated protein C is corrected by anti-coagulant cofactor activity found to be a property of factor V. Proc Natl Acad Sci USA 1994; 91:1396–1400.
137. Shen L, Dahlbäck B. Factor V and protein S as synergistic cofactors to activated protein C in degradation of factor VIIIa. J Biol Chem 1994; 269:18735–18738.
138. García de Frutos P, Alim RIM, Härdig Y, Zöller B, Dahlbäck B. Differential regulation of a and b chains of C4b-binding protein during acute-phase response resulting in stable plasma levels of free anticoagulant protein S. Blood 1994; 84:815–822.
139. García de Frutos P, Dahlbäck B. Interaction between serum amyloid P component and C4b-binding protein associated with inhibition of factor I-mediated C4b degradation. J Immunol 1994; 152:2430–2437.

23

Tissue-Type Plasminogen Activator

M. Sharon Stack
Northwestern University Medical School, Chicago, Illinois

Edwin L. Madison
Scripps Research Institute, La Jolla, California

Salvatore V. Pizzo
Duke University Medical Center, Durham, North Carolina

I. INTRODUCTION

A. Role of Tissue-Type Plasminogen Activator in Fibrinolysis

The fibrinolytic system is responsible for maintenance of vascular patency via a complex series of molecular interactions that result in the proteolytic degradation of fibrin. Localized thrombolysis requires the coordinated activity of zymogens, activators, effectors, and inhibitors that regulate plasmin-mediated fibrinolysis (Fig. 1). Generation of the active proteinase plasmin from the circulating zymogen plasminogen is mediated by specific proteinases, known as plasminogen activators, that initiate cleavage of the Arg^{560}-Val^{561} bond in plasminogen, resulting in the formation of the active two-chain serine proteinase plasmin (1). In addition to initiation of fibrinolysis, plasmin and plasminogen activators have been implicated in numerous physiological and pathological processes, including embryogenesis, inflammation, ovulation, neural development, tissue remodeling, and tumor cell migration (2–5).

There are two immunologically distinct physiological plasminogen activators that differ from each other with respect to molecular properties and tissue distribution. Urinary-type plasminogen activator (uPA, or urokinase) is primarily localized in fibroblast-like cells; tissue-type plasminogen activator (tPA) is found in the vascular endothelium (6). Whereas both enzymes are serine proteinases with similarly restricted substrate specificity, uPA and tPA are distinct gene products with unique amino acid sequences and resulting domain structures. Because tPA binds specifically to fibrin, induces fibrinolysis at lower concentrations than uPA, and is synthesized in the vascular endothelium, tPA is believed to be the primary physiological plasminogen activator in the fibrinolytic system (6–8).

B. Regulation of tPA Activity

1. Role of Fibrin

Because tPA catalyzes the rate-limiting step in fibrinolysis, its activity is tightly controlled in vivo. tPA is synthesized and secreted by vascular endothelial cells as a single-chain glycoprotein with a reported molecular mass of 66–72 kDa (9), which is subsequently cleaved at Arg^{275}-Ile^{276} into a two-chain disulfide-bonded form. Although single-chain tPA is catalytically active, conversion to the two-chain form results in a threefold increase in turnover rate k_{cat} against peptide substrates (10).

The activity of tPA is closely regulated by interaction with fibrin. Although tPA-catalyzed plasminogen activation follows Michaelis-Menten kinetics in both the absence and presence of fibrin, numerous studies have reported that the Michaelis constant K_m for the activation reaction is significantly decreased in the presence of fibrin (7,8,11,12). Evidence from kinetic and binding studies indicates that both tPA and plasminogen bind to the surface of the fibrin clot, forming a ternary complex that markedly enhances the efficiency of clot lysis (7,12). Estimates of the dissociation constant K_d for binding of tPA to fibrin range from 60 to 1400 nM, depending on variations in fibrin structure (13).

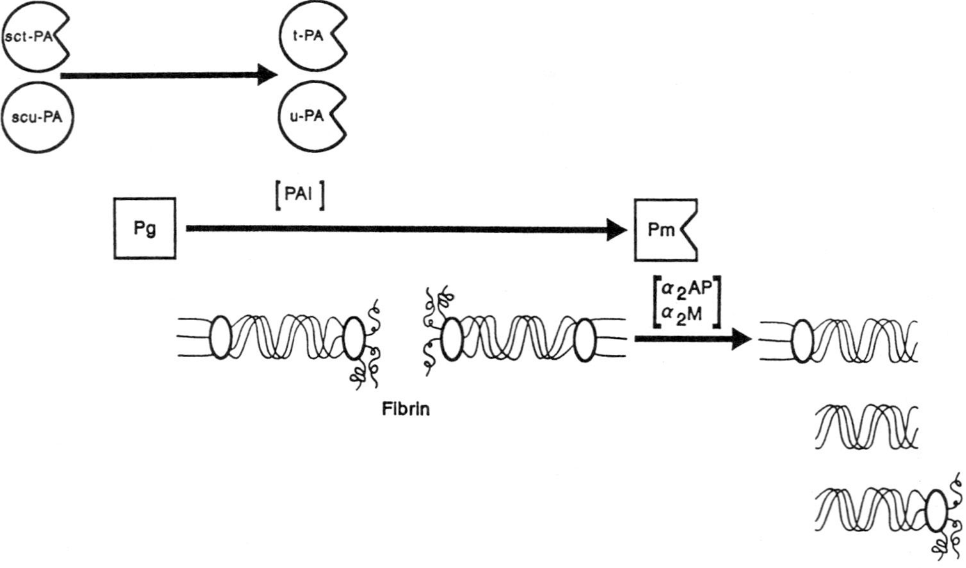

Figure 1 The fibrinolytic pathway. Brackets denote proteinase inhibitors. sct-PA, single-chain tissue-type plasminogen activator; scu-PA, single-chain urinary-type plasminogen activator; t-PA, two-chain tissue-type plasminogen activator; u-PA, two-chain urinary-type plasminogen activator; PAI, plasminogen activator inhibitor; Pg, plasminogen; Pm, plasmin; α_2AP, α_2-antiplasmin; α_2M, α_2-macroglobulin.

In addition, fibrin binding and stimulation effectively restrict plasminogen activation to sites of fibrin deposition in vivo, thereby limiting plasmin activity to the clot surface.

2. *Interaction of tPA with Other Macromolecules*

Recent studies suggested that the activity of tPA is also regulated by interaction with macromolecules other than fibrin. For example, the activation of plasminogen by tPA was shown to be promoted by heparin, which bound to both plasminogen and tPA and mediated a slight decrease in K_m for the activation reaction (14,15). In contrast, low-molecular-weight heparin does not bind tPA or plasminogen or stimulate tPA activity (16). However, it remains controversial whether heparin can compete with fibrin for the binding of tPA (14,15,17–19). Furthermore, more recent studies indicate that heparin does not stimulate tPA-catalyzed plasminogen activation in a plasma milieu (17,18). These studies suggest that the effect of heparin on tPA-mediated fibrinolysis remains to be established.

In addition to heparin, other extracellular matrix-associated macromolecules have been shown to regulate tPA activity. For example, thrombospondin and histidine-rich glycoprotein bind both tPA and plasminogen, resulting in an increase in activation rate (20). Intact extracellular matrices synthesized and secreted in vitro by endothelial cells also bind plasminogen and stimulate its activation by tPA (21). Kinetic studies using isolated extracellular matrix proteins suggest that type IV collagen is responsible for this stimulatory effect, although no direct interaction between tPA and collagen IV was detected (22). Fibronectin, a matrix-associated adhesive glycoprotein, also binds tPA, resulting in a 12-fold increase in tPA amidolytic activity, although no stimulation of tPA-catalyzed plasminogen activation was detected (22,23). However, a 55 kDa polypeptide derived from the amino terminus of fibronectin was identified that specifically binds both tPA and plasminogen (24), and a synthetic octapeptide derived from this binding region causes a 15-fold increase in the catalytic efficiency (k_{cat}/K_m) of plasminogen activation (25). Binding of tPA to the globular domain at the carboxyl terminus of the laminin A chain and stimulation of plasminogen activation by a laminin A chain-derived peptide was also reported (24,26). Together these data suggest that interaction of tPA and/or plasminogen with components of the extracellular matrix represents an additional biochemical mechanism for the regulation of tPA activity. These interactions may have significant implications in physiological and pathological processes involving plasmin-mediated proteolysis of the extracellular matrix.

3. *Regulation by Inhibition*

Specific inhibition of the catalytic activity of tPA is essential to prevent uncontrolled plasminogen activation. The most important endogenous inhibitor of tPA is endothelial cell plasminogen activator inhibitor type 1 (PAI-1) (27). PAI-1 is a 50 kDa single-chain glycoprotein that is a member of the serine proteinase inhibitor (serpin) gene family (27,28). Interaction of tPA with PAI-1 rapidly leads to the formation of a 1:1 inactive, highly stable tPA-PAI-1 complex. Because the concentration of PAI-1 is much greater than that of tPA in human plasma, circulating tPA is present predominantly in an inactive complex with PAI-1 (29).

4. *Regulation of tPA Activity by Clearance from the Circulation*

In addition to inhibitory mechanisms, negative regulation of tPA activity is also accomplished by clearance of tPA from the circulation. Clearance in humans is biphasic, phase 1 having a half-life of 5.3 minutes, followed by a second phase with a half-life of 46

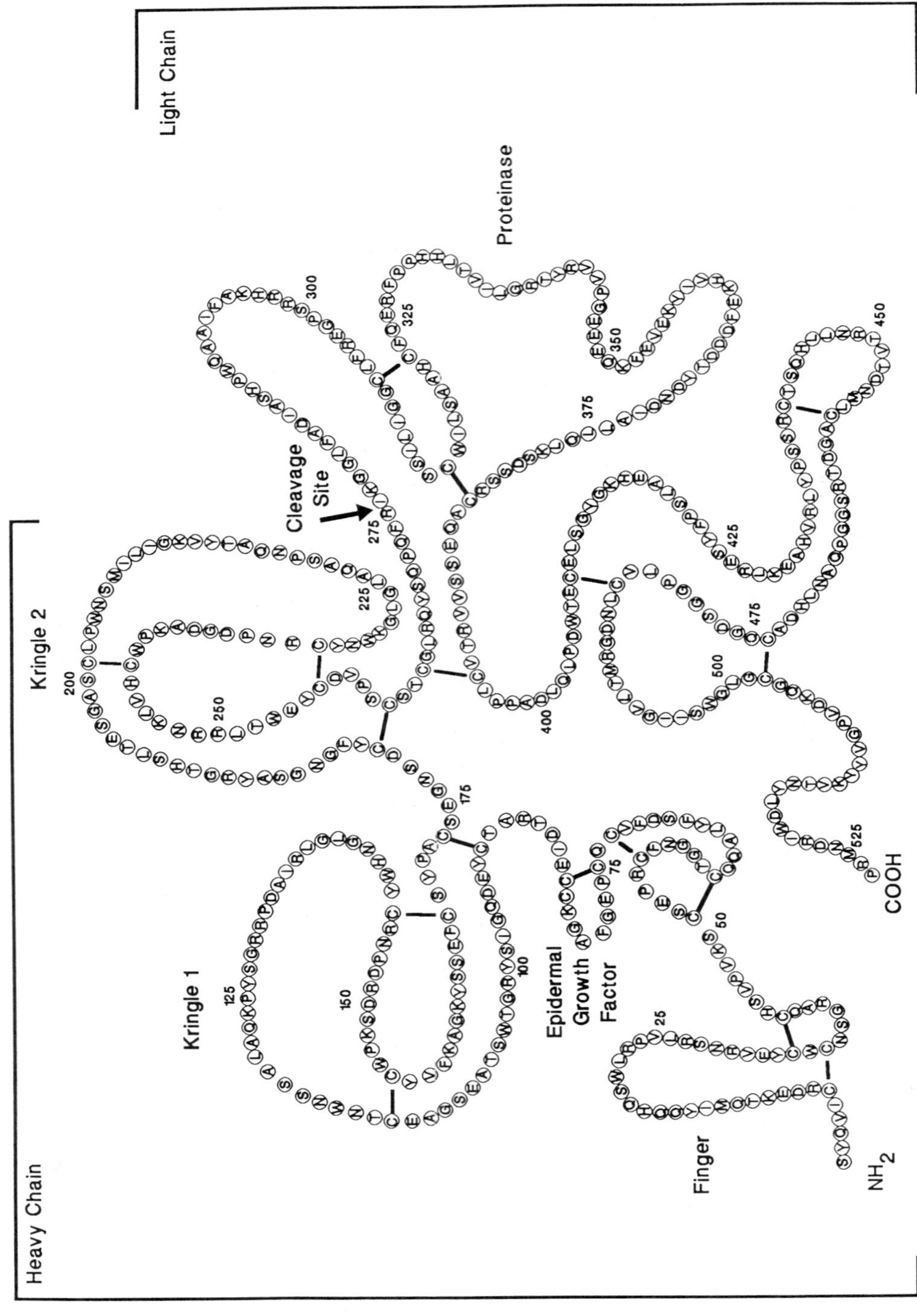
Light Chain
Heavy Chain
Proteinase
Cleavage Site
Kringle 2
Kringle 1
Epidermal Growth Factor
Finger
NH2
COOH

minutes (30). Clearance of tPA is mediated by several populations of hepatic receptors that recognize distinct portions of the tPA molecule (31). Receptors present on hepatic endothelial cells recognize the mannose-rich oligosaccharide chain present on Asn^{117} of tPA. A distinct receptor on hepatic parenchymal cells binds a determinant located in the amino terminus of the molecule (32). A third receptor recognizes only tPA-PAI-1 complexes (33–35). Together these receptors act to regulate tPA activity by limiting the half-life of tPA in circulation.

II. PROTEIN STRUCTURE OF tPA

A. Domain Structure

The precursor form of mature tPA is composed of 563 amino acids as deduced from the cDNA sequence (36,37). tPA is synthesized and secreted in vivo primarily by endothelial cells as a single-chain glycoprotein of approximately 66–70 kDa (9,38–40). Both a hydrophobic signal sequence and a hydrophilic prosequence are removed before secretion by specific endopeptidases (41–43). Secreted tPA is a 527 amino acid glycoprotein organized by 16 disulfide bonds into discrete structural domains (Fig. 2). Amino acids 4–50 are homologous to the finger domains originally described in fibronectin, which are believed to be involved in binding of fibronectin to fibrin (44). Residues 50–87 share homology with human epidermal growth factor (45). tPA also contains two triple-loop, disulfide-bonded bridge structures, known as ''kringles,'' formed by residues 88–175 and 176–263, respectively (46–48). These structural domains are followed by the catalytic or serine proteinase domain composed of residues 276–527. The serine proteinase domain of tPA contains the essential catalytic triad characteristic of serine proteinases (His^{322}, Asp^{371}, and Ser^{478}) (36). Following secretion, cleavage of single-chain tPA at the Arg^{275}-Ile^{276} bond results in the formation of the two-chain mature form of the enzyme, consisting of a 34 kDa amino-terminal heavy chain and a 31 kDa carboxyl-terminal light chain joined by a disulfide bond between Cys^{264} and Cys^{395} (49–51).

B. Glycosylation Variants

There are three potential N-glycosylation sites on tPA (Asn^{117}, Asn^{184}, and Asn^{448}), and studies using recombinant enzyme indicate that N-glycosylation appears to be cell type specific (52,53). Asn^{117} contains high-mannose oligosaccharides, whereas Asn^{184} and Asn^{448} carry diantennary, branched triantennary, and tetraantennary oligosaccharides (54). Two major tPA glycoforms are naturally occurring: type I tPA, in which Asn^{117}, Asn^{184}, and Asn^{448} are occupied, and type II tPA, in which only Asn^{117} and Asn^{448} are glycosylated (41). Analysis of cell type-specific glycosylation of the tPA types I and II produced by four different cell lines suggests that tPA is secreted by individual cell lines as a set of glycoforms, each of which contains a unique pattern of oligosaccharide side chains (52). Discernible differences among glycoforms were detected with respect to the kinetics of

Figure 2 tPA protein structure. Proposed structural domains of tPA are indicated. The cleavage site for the conversion of single-chain tPA to two-chain tPA is denoted by an arrow. Amino acid numbering is as proposed by Pennica et al. (36), with the amino-terminal serine residue designated 1. [The model is adapted from Pennica et al. (36) and Banyai et al. (44).]

fibrin-dependent activation of plasminogen, with up to five-fold differences in k_{cat} observed (53,55). Furthermore, glycosylation at Asn^{184} decreases the rate of plasmin-mediated conversion of single-chain tPA to two-chain tPA (56). These data suggest that differential glycosylation may provide an additional mechanism for regulation of tPA activity. In addition to N-linked carbohydrates, an O-linked fucose residue linked to Thr^{61} in the tPA growth factor domain has also been described (57). This novel glycosylation does not appear to be cell type specific and is of unknown functional significance.

III. CHARACTERIZATION OF tPA GENE STRUCTURE

A. Isolation of tPA cDNA

Isolation of a cDNA encoding human tPA was first reported using mRNA obtained from the human Bowes melanoma cell line (36). Agarose gel electrophoresis was used to size fractionate poly(A) mRNA, and fractions containing tPA mRNA were identified and used to prepare a cDNA library. A mixture of degenerate oligonucleotides encoding segments of tPA (determined by amino acid sequencing of the purified protein) was used to probe the library and identify cDNAs encoding tPA. Using this procedure, a 2304 bp tPA cDNA was isolated that contained a partial-length tPA cDNA insert lacking the coding sequences for the amino terminus of the protein. Therefore, a restriction fragment derived from the 5′ end of this cDNA was used to probe a human genomic library, and a genomic clone was isolated that contained a portion of the human tPA gene. A fragment from this genomic clone was used to probe a new cDNA library, and a tPA cDNA clone was isolated that overlapped the original clone and contained the amino-terminal as well as the presequence regions. From these two cDNA clones, the nucleotide sequence and corresponding amino acid sequence of tPA were determined. The 2530 bp tPA cDNA encoded a 563 amino acid polypeptide. Based on protein sequencing of purified Bowes melanoma tPA, 35 amino acids precede the serine residue designated as the amino terminus of the mature protein (49). It was suggested that this amino-terminal extension contains a 20–23 amino acid hydrophobic signal peptide followed by a hydrophilic pro sequence. Similar results were reported by other investigators (43,58,59,60).

B. Structure of the tPA Gene

The availability of tPA cDNA facilitated characterization of the tPA gene. The gene encoding human tPA has been localized to chromosome 8, bands 8.p.12 → q.11.2 (61–63). Current evidence suggests that there is only a single tPA gene in the human haploid genome (51,64,65). The structure of the tPA gene was originally elucidated from a human genomic cosmid library by Ny et al. by restriction mapping, Southern blotting, and DNA sequence analysis (51). Using tPA cDNA to probe the genomic library constructed with placental DNA, a cosmid was isolated that spanned the entire coding region of tPA mRNA as well as the 3′-untranslated region. The 5′ end of the DNA corresponded to the 5′-untranslated region of tPA mRNA 58 nucleotides downstream of the transcription initiation site. Using a similar approach, Degen and coworkers (64) sequenced 36,594 bp of the human tPA gene, including 32,720 bp corresponding to the site of transcription initiation and ending with the polyadenylation site. A further 3530 bp of 5′ and 3′ flanking sequences were also reported.

Both groups reported the presence of 13 introns dividing the gene into 14 coding regions (Fig. 3). The exons range in size from 43 to 914 bp; the size range for introns is 111–14,257 bp. Transcription is initiated at an A residue, with upstream "TATA" (−22 to −29) and "CAAT" (−112 to −116) boxes. Located within the gene and the 5′ flanking region are a KpnI repeat unit as well as 28 copies of Alu-repetitive DNA (64). A polyadenylation signal was also described (32,688–32,693).

C. Mosaic Structure of tPA

Elucidation of the tPA gene structure led to the interesting observation that the structural domains of the tPA protein correlate well with the intron/exon boundaries of the gene (51,64). These data suggest that the assembly of the tPA gene likely results from the process of exon shuffling, in which structural domains behave as independent modular units that serve as building blocks for protein assembly (66–68). Ny and coworkers (50) reported that the leader sequence, signal peptide, and pro segment are encoded by the first three exons. The fibronectin-like finger domain is encoded by exon 4, and exon 5 codes for the growth factor-like domain. The two triple-loop, disulfide-bonded kringle structures of tPA are each encoded by two exons. An identical intron-exon pattern is observed for the two kringles, suggesting a common evolutionary ancestor. The serine proteinase domain, or tPA light chain, is encoded by five exons in positions similar to those observed in the gene structure of other serine proteinases, including chymotrypsin, trypsin, and elastase (69).

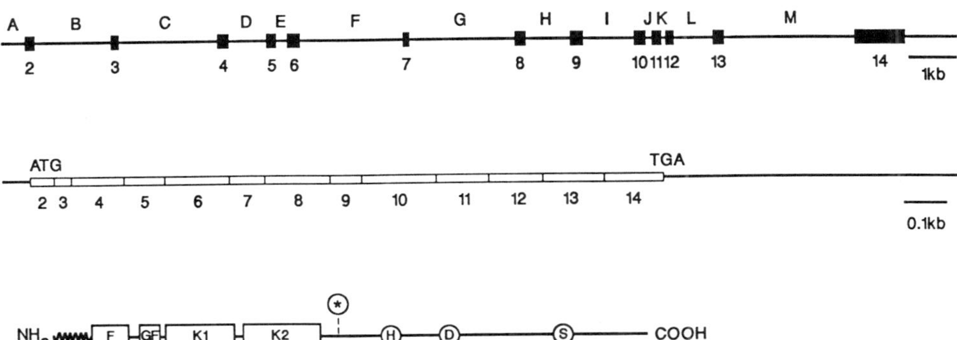

Figure 3 Organization of the human tPA gene. DNA encoding the tPA gene is shown on the first line. Introns are designated A–M; exons are numbered 1–14. The mRNA structure is depicted on the second line, with coding sequences represented by open boxes. Putative structural domains are shown on the third line, with arrows marking the location of introns. Note that the first intron is contained within the 5′ noncoding region. The asterisk indicates the cleavage site for conversion of single-chain tPA to two-chain tPA. Domain designations: F, finger; GF, epidermal growth factor; K, kringle. H, D, and S represent His^{322}, Asp^{371}, and S^{478} residues in the essential catalytic triad of the proteinase domain. [Data are adapted from Pennica et al. (36), Ny et al. (51), and Degen et al. (64).]

IV. ELUCIDATION OF STRUCTURE-FUNCTION RELATIONSHIPS USING MUTANT tPAs

A. Mutants with Altered Fibrin Binding or Catalytic Efficiency

1. *tPA Interaction with Fibrin*

Early kinetic studies using purified tPA demonstrated that the ability of tPA to activate plasminogen was dramatically enhanced in the presence of fibrin, CNBr-generated fibrinogen fragments, or polylysine (7,11,70,71). These analyses indicated that stimulation of tPA activity by fibrin results from the formation of a ternary complex involving fibrin, tPA, and plasminogen, thereby allowing efficient plasminogen activation on the clot surface. Binding of tPA to fibrin fragment FCB-2 was completely inhibited by 6-aminohexanoic acid, suggesting the involvement of a lysine binding site in the interaction (72). However, studies in which point mutations were utilized to change single amino acids in tPA demonstrated that fibrin stimulation can be dissociated from lysine binding activity, suggesting that two different sites in tPA are involved (73).

The fibrin binding activity of tPA was subsequently localized to the heavy chain of the molecule by experiments utilizing mild reduction to separate the two tPA chains (74,75). Additional experiments utilizing limited proteolysis, domain-specific monoclonal antibodies, and deletion mutants provided evidence that both the finger domain and the second kringle impart fibrin affinity to tPA (9,44,76,77).

2. *Domain Deletion Mutants*

Alignment of the gene structure of tPA with the amino acid sequence of the mature protein suggested that the position of exons or sets of exons coincide with the proposed structural domains of the protein. This led to the hypothesis that mosaic proteins, such as tPA (as well as many additional proteins involved in coagulation and fibrinolysis), are composed of functional modules resulting from exon shuffling during evolution (66). Since this initial observation, many investigators have attempted to determine whether the individual structural domains of tPA have autonomous functions in the protein.

Early experiments utilized transient expression of mutants in which entire structural domains were deleted, followed by a variety of assays to determine the effect(s) of the deletion on fibrin binding and fibrin-dependent catalytic activity. For example, van Zonneveld et al. (9) constructed mutants lacking the first kringle domain (K1), second kringle domain (K2), or the finger and epidermal growth factor domains (F/EGF). The mutants were then assayed for fibrin stimulation of tPA-catalyzed plasminogen activation. Fibrin stimulation was found to be mediated primarily by the K2 domain and, to a lesser extent, by the F/EGF region. However, Gething et al. (78) reported that mutants in which either K1 or K2 was deleted were stimulated by fibrin to the same extent as wild-type tPA, suggesting that the presence of either kringle can be both necessary and sufficient for fibrin stimulation. In similar experiments, Kalyan et al. (79) deleted the F and EGF domains of tPA and found reduced fibrin binding and stimulation, although the amidolytic activity of the mutant tPA remained comparable to that of wild-type tPA. However, mutants lacking either the F or EGF or both domains produced by Larsen and colleagues (80) were indistinguishable from wild-type tPA with respect to activity and fibrin stimulation but were found to be deficient in fibrin binding ability, suggesting that both domains are required for fibrin binding. Furthermore, additional studies by Gething et al. (78) and Larsen et al. (80) using deletion mutants lacking both the F and EGF domains showed unaltered fibrin stimulation and attributed fibrin stimulation to the kringle domain(s). Mediation of fibrin stimulation by K2 was also reported by Urano et al. (81). Although studies by Verheijen et al. (62) and Johannessen et al. (82) provided additional evidence that both the F and EGF domains are required for high-affinity fibrin binding, these conflicting results suggest that fibrin binding and fibrin stimulation of activity are mediated by multiple regions of the tPA molecule.

To define further the role of structural domains in fibrin binding and stimulation by tPA, mutants containing both deletions and duplication of structural domains were analyzed. Lijnen et al. (83) created a set of deletion mutants lacking the F, EGF, or F/EGF domains, as well as mutants constructed by replacement of K1 by an additional copy of K2. All mutants bound to fibrin with the exception of the F/EGF deletion mutant. Fibrin binding was partially restored in this mutant by the additional replacement of K1 by a second copy of K2. Results obtained from different groups suggest that both the F and K2 domains mediate fibrin binding and stimulation of tPA activity. However, conflicting data obtained using deletion mutagenesis suggest that loss of function studies should be carefully interpreted, because functional alterations may also result from conformational changes in the molecule rather than deletion of a functionally essential domain.

3. Additional Mutations Affecting Fibrin Binding and Stimulation of tPA Activity

Point mutations of individual amino acids have been used primarily to alter amino acids present at putative N-glycosylation sites. However, mutants in which proteolytic conversion of single-chain tPA to two-chain tPA is prevented have been created by mutation of Arg^{275} to Glu or Gly (84,85). The resultant single-chain mutants exhibited enhanced fibrin binding. To distinguish further amino acids within the F domain responsible for fibrin binding and stimulation, mutants containing substitution or deletion of amino acids within only the F domain were constructed (86). As was observed in some previous studies, deletion of the entire domain resulted in impaired fibrinolytic activity and fibrin binding. Furthermore, amino acids 14–323 were found to be required for full fibrinolytic activity, and amino acids 7–10 could accommodate major changes in amino acid size, charge, or deletion without loss of fibrinolytic potency.

A second alternative to domain deletion was presented by Bennett et al. (87), who used clustered charge-to-alanine scanning to create a series of clustered point mutations in which charged amino acids were replaced by Ala in groups of one to four residues. The objective of this approach was to create mutants with an altered protein surface in a restricted region of an individual domain rather than a gross alteration or deletion of the entire domain. Results from this study confirmed previous observations indicating that mutations in the F and/or EGF domain affect fibrin binding, whereas alterations in the K1, K2, or proteinase domains do not alter fibrin affinity. However, in contrast to domain deletion experiments that suggested that fibrin stimulation of tPA activity was mediated solely by nonproteinase domains, this study suggested that fibrin stimulation of activity was also affected by mutations in the proteinase domain. Mutations at charged residues throughout the proteinase domain decreased both the unstimulated and fibrin-stimulated activities of tPA. In addition, fibrin-stimulated activity greater than that of wild-type tPA was observed for mutants containing Ala substitutions at amino acids 298, 304, or 332. These results were surprising because previous studies indicated that fibrin binding and stimulation were regulated solely by residues in nonproteinase domains.

B. Mutants with Altered Zymogenicity

Unlike the majority of serine proteinase zymogens, single-chain tPA has appreciable enzymatic activity and, as such, is not a true zymogen. Both single- and two-chain tPA have similar levels of activity in the presence of fibrin and are readily inhibited by PAI. However, the two-chain form of the enzyme has greater amidolytic and plasminogen activator activity in the absence of fibrin than the single-chain form of the enzyme. Single-chain tPA is converted to the two-chain form by plasmin-catalyzed cleavage at Arg^{275}-Ile^{276}. To determine the functional consequences of Arg^{275}-Ile^{276} cleavage, site-directed mutagenesis was used to construct mutants in which Arg^{275} was converted to either Glu or Gly (81,84,85). In the absence of fibrin, the activity of these mutants was greatly reduced relative to that of wild-type tPA, whereas in the presence of fibrin full plasminogen-activating potential was attained without conversion to the two-chain enzyme.

In the proposed mechanism of activation of serine proteinase zymogens, as exemplified by chymotrypsinogen, activation occurs as a result of cleavage of an Arg^{15}-Ile^{16} bond. The new N-terminal Ile^{16} is then free to interact with an Asp^{194} residue adjacent to the active site Ser^{195}, forming a salt bridge that stabilizes the catalytic center. Cleavage of single-chain tPA at Arg^{275}-Ile^{276} is believed to follow this general mechanism. Single-chain tPA is also an active proteinase rather than an inactive zymogen, however, suggesting the existence of additional interactions stabilizing the active enzyme. It has been proposed that alternative salt bridge formation occurs between an appropriately positioned Asp^{477} and either Lys^{277} (immediately adjacent to the activation site) or Lys^{416}, thus stabilizing the active conformation of single-chain tPA (88). To test this hypothesis, site-directed mutagenesis was used to generate mutants in which Lys^{277} or Lys^{416} were replaced by neutral amino acids (89). Although the amidolytic activities of the two-chain forms were unaffected by the substitution, replacement of Lys^{416} resulted in a substantial decrease in the activity of the single-chain tPA mutant. These data suggest that Lys^{416} contributes to the stabilization of the active conformation of single-chain tPA. Additional experiments performed using charge-to-alanine scanning mutants demonstrated that mutations in the nonproteinase domains had little effect on the enzymatic activity of either the single- or two-chain form of the enzyme. However, mutations in the proteinase domain reduced the activity of single-chain variants, particularly in variants of charged residues near Arg^{275}, as well as residues 339–351 and 410–440 (87).

C. Mutants with Altered Susceptibility to PAI-1

The serine proteinase inhibitor PAI-1 rapidly inhibits both single-chain and two-chain tPA. PAI-1 binds to the active site of tPA and is cleaved at its Arg^{346}-Met^{347} bond to form a stable, inactive complex. An intact active-site serine residue in tPA is essential for stable complex formation with PAI-1, as evidenced by point mutations in which the active site Ser^{478} was changed to a Gly or Ala residue (90,91). To identify amino acids likely to be involved in interaction with PAI-1, the known structure of the complex between the serine proteinase trypsin and the serpin bovine pancreatic trypsin inhibitor was used to model the interaction between tPA and PAI-1 (92,93). Amino acids in tPA predicted to make contact with PAI-1 were then altered using site-directed mutagenesis. These studies demonstrated that mutations or deletions in amino acids 296–304 do not affect the interaction of tPA with plasminogen but confer resistance to inhibition by PAI-1.

These results were confirmed by Bennett et al. (87) using alanine scanning mutagenesis. Mutants in which alanines were substituted beginning at amino acid 298 for the sequence KHRR exhibited considerable resistance to inhibition by PAI-1.

In addition to direct interactions involving the serine proteinase domain of tPA, non-proteinase regions may also modulate the interaction of tPA with PAI-1. To determine whether exosites on the noncatalytic chain of tPA contribute to the inhibitory capacity of PAI-1, the rate of association of PAI-1 with nonproteinase domain deletion mutants of tPA was determined (94). Expression of wild-type tPA in four different cell lines that produced differentially glycosylated forms of tPA did not significantly alter the association rate with PAI-1, demonstrating that the carbohydrate moieties do not regulate the tPA-PAI-1 interaction. However, mutants in which one or both kringle domains were deleted displayed an enhanced association rate relative to wild-type tPA, suggesting that the kringles protect tPA from inactivation by PAI-1. Interestingly, whereas the mutant containing only kringle 1 and the proteinase domain (K1-P) had an enhanced association rate relative to wild-type tPA, association of the K2-P mutant was slower than that of wild type. A further decrease in association rate was obtained by substitution of a second copy of K2 for K1. Together these data suggest that the K2 domain has a role in controlling the rate of tPA association with PAI-1.

D. Summary

Structural modification of the tPA molecule has provided considerable information on the role of specific domains, groups of amino acids, or individual amino acid residues involved in tPA function (summarized in Fig. 4). Based on current data, the function of the signal sequence is to mediate entry of tPA into the cellular secretory pathway, but no known function for the pro sequence has been described. The finger domain has been implicated in fibrin binding as well as in hepatic clearance. The epidermal growth factor domain may also function in tPA clearance via interaction with hepatic parenchymal receptors. A third mechanism of tPA clearance involves recognition of the high-mannose oligosaccharide side chain on kringle 1 by receptors on hepatic endothelial cells. Kringle 2 of tPA contains a lysine binding site that mediates both fibrin binding and stimulation. The proteinase domain contains the active site and functions directly in endoproteolytic cleavage of plasminogen and PAI-1. Structure-function analysis of tPA has proven useful in understanding basic mechanisms involved in the regulation of physiological fibrinolysis.

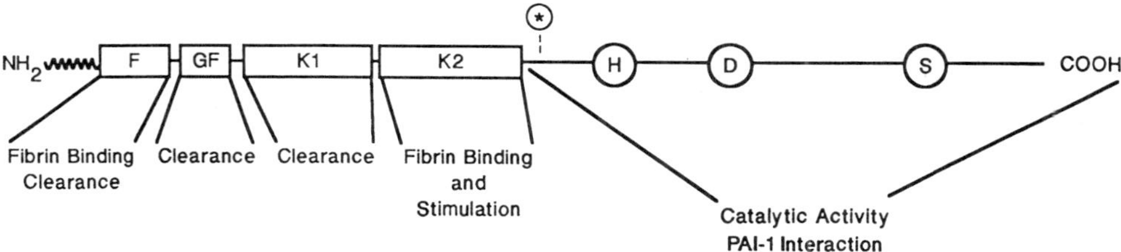

Figure 4 Proposed functions of tPA structural domains. Domains are as defined in Figure 3.

REFERENCES

1. Castellino FJ. Recent advances in the chemistry of the fibrinolytic system. Chem Rev 1981; 81:431–436.
2. Strickland S, Reich E, Sherman MI. Plasminogen activator in early embryogenesis: enzyme production by trophoblast and parietal endoderm. Cell 1976; 9:231–240.
3. Reich R, Thompson EW, Iwamoto Y, et al. Effects of inhibitors of plasminogen activator, serine proteinases and collagenase IV on the invasion of basement membranes by metastatic cells. Cancer Res 1988; 48:3307–3312.
4. Highsmith RF. Isolation and properties of a plasminogen activator derived from canine vascular tissue. J Biol Chem 1981; 256:6788–6795.
5. Ossowski L, Vassilli JD. Biological Markers of Neoplasia: Basic and Applied Aspects. Amsterdam: Elsevier, 1978.
6. Dano K, Andreasen PA, Grondahl-Hansen J, Kristensen P, Nielsen LS, Skriver L. Plasminogen activators, tissue degradation and cancer. Adv Cancer Res 1985; 44:139–266.
7. Hoylaerts M, Rijken D, Lijnen HR, Collen D. Kinetics of the activation of plasminogen by human tissue plasminogen activator. Role of fibrin. J Biol Chem 1982; 257:2912–2919.
8. Ranby M. Studies on the kinetics of plasminogen activation by tissue plasminogen activator. Biochim Biophys Acta 1982; 704:461–469.
9. Van Zonneveld A-J, Veerman H, Pannekoek H. Autonomous functions of structural domains on human tissue-type plasminogen activator. Proc Natl Acad Sci USA 1986; 83:4670–4674.
10. Boose JA, Kuismanen E, Gerard R, Sambrook J, Gething MJ. The single-chain form of tissue-type plasminogen activator has catalytic activity: studies with a mutant enzyme that lacks the cleavage site. Biochemistry 1988; 28:635–643.
11. Nieuwenhuizen W, Verheijen JH, Vermond A, Chang GTG. Identification of a site in fibrin(ogen) which is involved in the acceleration of plasminogen activation by tissue-type plasminogen activator. Biochim Biophys Acta 1983; 755:531–533.
12. Rijken DC, Groenveld E. Isolation and functional characterization of the heavy and light chains of human tissue-type plasminogen activator. J Biol Chem 1986; 261:3098–3102.
13. Fears R. Binding of plasminogen activators to fibrin: characterization and pharmacological consequences. Biochem J 1989; 261:313–324.
14. Fears R, Esmail AF, Greenwood HC. Effect of heparin on the fibrinolytic response to plasminogen activators. Semin Thromb Hemost 1991; 17:389–393.
15. Andrade-Gordon P, Strickland S. Interaction of heparin with plasminogen activators and plasminogen: effects on the activation of plasminogen. Biochemistry 1986; 25:4033–4040.
16. Andrade-Gordon P, Strickland S. Anticoagulant low molecular weight heparin does not enhance the activation of plasminogen by tissue plasminogen activator. J Biol Chem 1989; 264: 15177–15181.
17. Fry ETA, Sobel BE. Lack of interference by heparin with thrombolysis or binding of tissue-type plasminogen activator to thrombi. Blood 1988; 71:1347–1352.
18. Grailhe P, Angles-Cano E. The activation of plasminogen by tissue plasminogen activator is not enhanced by different heparin species as assessed in vitro with a solid phase fibrin method. Fibrinolysis 1991; 5:61–69.
19. Paques EP, Stohr HA, Heimburger N. Study on the mechanism of action of heparin and related substances on the fibrinolytic system: relationship between plasminogen activators and heparin. Thromb Res 1986; 42:797–807.
20. Silverstein RL, Nachman RL, Leung LLK, Harpel PC. Activation of immobilized plasminogen by tissue activator. Multimolecular complex formation. J Biol Chem 1985; 260:10346–10352.
21. Knudsen BJ, Silverstein RL, Leung LL, Harpel PC, Nachman RL. Binding of plasminogen to extracellular matrix. J Biol Chem 1986; 261:10765–10771.
22. Stack MS, Gonzalez-Gronow M, Pizzo SV. Regulation of plasminogen activation by components of the extracellular matrix. Biochemistry 1990; 29:4966–4970.

23. Salonen EM, Zitting A, Vaheri A. Laminin interacts with plasminogen and its tissue-type activator. FEBS Lett 1984; 172:29–32.
24. Moser TL, Enghild JJ, Pizzo SV, Stack MS. The extracellular matrix proteins laminin and fibronectin contain binding domains for human plasminogen and tissue plasminogen activator. J Biol Chem 1993; 268:18917–18923.
25. Stack MS, Pizzo SV. Modulation of tissue plasminogen activator-catalyzed plasminogen activation by synthetic peptides derived from the amino-terminal heparin binding domain of fibronectin. J Biol Chem 1993; 268:18924–18928.
26. Stack MS, Gray RD, Pizzo SV. Modulation of murine B16F10 melanoma plasminogen activator production by a synthetic peptide derived from the laminin A chain. Cancer Res 1993; 53:1998–2004.
27. Pannekoek H, Veerman H, Lambers H, et al. Endothelial plasminogen activator inhibitor (PAI): a new member of the serpin gene family. EMBO J 1986; 5:2539–2544.
28. Loskutoff DJ, Van Mourik JA, Erickson LA, Lawrence D. Detection of an unusually stable fibrinolytic inhibitor produced by bovine endothelial cells. Proc Natl Acad Sci USA 1983; 80:2956–2960.
29. Lucore CL, Sobel BE. Interactions of tissue-type plasminogen activator with plasma inhibitors and their pharmacologic implications. Circulation 1988; 77:660–669.
30. Garabedian HD, Gold HK, Leinbach RC, Yasuda T, Johns JA, Collen D. Dose-dependent thrombolysis, pharmacokinetics and hemostatic effects of recombinant human tissue-type plasminogen activator for coronary thrombosis. Am J Cardiol 1986; 58:673–679.
31. Korninger C, Stassen JM, Collen D. Turnover of human extrinsic (tissue-type) plasminogen activator by the human hepatoma cell line HepG2. Thromb Haemost 1981; 46:658–661.
32. Krause J, Seydel W, Heinzel G, Tanswell P. Different receptors mediate the hepatic catabolism of tissue-type plasminogen activator and urokinase. Biochem J 1990; 267:647–652.
33. Owensby DA, Sobel BE, Schwartz AL. Receptor-mediated endocytosis of tissue-type plasminogen activator by the human hepatoma cell line HepG2. J Biol Chem 1988; 263:10587–10594.
34. Morton PA, Owensby DA, Sobel BE, Schwartz AL. Catabolism of tissue-type plasminogen activator by the human hepatoma cell line HepG2. J Biol Chem 1989; 264:7228–7235.
35. Morton PA, Owensby DA, Wun TC, Billadello JJ, Schwartz AL. Identification of determinants involved in binding of tissue-type plasminogen activator-plasminogen activator inhibitor type 1 complexes to HepG2 cells. J Biol Chem 1990; 265:14093–14099.
36. Pennica D, Holmes WE, Kohr WJ, et al. Cloning and expression of human tissue-type plasminogen activator cDNA in *E. coli*. Nature 1983; 301:214–221.
37. Harris TJR, Patel T, Martson RAO, Little S, Emtage JS. Cloning of cDNA coding for human tissue-type plasminogen activator and its expression in *E. coli*. Mol Biol Med 1986; 3:279–292.
38. Allen RA, Pepper DS. Isolation and properties of human vascular plasminogen activator. Thromb Haemost 1981; 45:43–50.
39. Levin E. Latent tissue plasminogen activator produced by human endothelial cells in culture: evidence for an enzyme-inhibitor complex. Proc Natl Acad Sci USA 1983; 80:6804–6808.
40. Levin EG, Loskutoff DJ. Cultured bovine endothelial cells produce both urokinase and tissue-type plasminogen activators. J Cell Biol 1982; 94:631.
41. Pohl G, Kallstrom M, Bergsdorf N, Wallen P, Jornvall H. Tissue plasminogen activator: peptide analyses confirm an indirectly derived amino acid sequence, identify the active site serine residue, establish glycosylation sites and localize variant differences. Biochemistry 1984; 23:3701–3707.
42. Vehar GA, Spellman MW, Keyt BA, et al. Characterization studies of human tissue-type plasminogen activator produced by recombinant DNA technology. Cold Spring Harbor Symp Quant Biol 1986; LI:551–562.
43. Sambrook J, Hanahan D, Gething MJ. Expression of human tissue-type plasminogen activator from lytic viral vectors and in established cell lines. Mol Biol Med 1986; 3:459–481.

44. Banyai L, Varadi A, Patthy L. Common evolutionary origin of the fibrin-binding structures of fibronectin and tissue-type plasminogen activator. FEBS Lett 1983; 163:37–41.
45. Verde P, Stoppelli MP, Galeffi P, Nocera PD, Blasi F. Identification and primary sequence of an unspliced human urokinase poly(A)$^+$ RNA. Proc Natl Acad Sci USA 1984; 81:4727–4731.
46. Olsson G, Andersen L, Lindqvist O. A low resolution model of fragment 1 from bovine prothrombin. FEBS Lett 1982; 145:317–322.
47. Holland SK, Harlos K, Blake CCF. Deriving the generic structure of the fibronectin type II domain from the prothrombin kringle 1 crystal structure. EMBO J 1987; 6:1875–1880.
48. Park CH, Tulinsky A. Three-dimensional structure of the kringle sequence: structure of prothrombin fragment 1. Biochemistry 1986; 25:3977–3982.
49. Rijken DC, Collen D. Purification and characterization of the plasminogen activator secreted by human melanoma cells in culture. J Biol Chem 1981; 256:7035–7041.
50. Wallen P, Pohl G, Bergsdorf N, Ranby M, Ny T, Jornvall H. Purification and characterization of a melanoma cell plasminogen activator. Eur J Biochem 1983; 132:681–686.
51. Ny T, Elgh F, Lund B. The structure of the human tissue-type plasminogen activator gene: correlation of intron and exon structures to functional and structural domains. Proc Natl Acad Sci USA 1984; 81:5355–5359.
52. Parekh RB, Dwek RA, Rudd PM, et al. N-glycosylation in in vitro enzymatic activity of human recombinant tissue plasminogen activator expressed in Chinese hamster ovary cells and a murine cell line. Biochemistry 1989; 28:7670–7679.
53. Parekh RB, Dwek RA, Thomas JR, et al. Cell type specific and site specific N-glycosylation of type I and type II human tissue plasminogen activator. Biochemistry 1989; 28:7644–7662.
54. Spellman MW, Basa LJ, Leonard CK, et al. Carbohydrate structures of human tissue plasminogen activator expressed in Chinese hamster ovary cells. J Biol Chem 1989; 264:14100–14111.
55. Wittwer AJ, Howard SC, Carr LS, et al. Effects of N-glycosylation on in vitro activity of Bowes melanoma and human colon fibroblast derived tissue plasminogen activator. Biochemistry 1989; 28:7662–7669.
56. Wittwer AJ, Howard SC. Glycosylation at Asn184 inhibits the conversion of single-chain to two-chain tissue-type plasminogen activator by plasmin. Biochemistry 1990; 29:4175–4180.
57. Harris RJ, Leonard CK, Guzetta AW, Spellman MW. Tissue plasminogen activator has an O-linked fucose attached to threonine-61 in the epidermal growth factor domain. Biochemistry 1991; 30:2311–2314.
58. Edlund T, Ny T, Ranby M, et al. Isolation of cDNA sequences coding for a part of human tissue plasminogen activator. Proc Natl Acad Sci USA 1983; 80:349–352.
59. Harris TJR, Patel T, Marston RAO, et al. Cloning of cDNA coding for human tissue-type plasminogen activator and its expression in *E. coli*. Mol Biol Med 1986; 3:279–292.
60. Sambrook J, Hanahan D, Rodgers L, Gething MJ. Expression of human tissue-type plasminogen activator from lytic viral vectors and in established cell lines. Mol Biol Med 1986; 3: 459–481.
61. Rajput B, Degen SF, Reich E, et al. Chromosomal locations of human tissue plasminogen activator and urokinase genes. Science 1985; 230:672–674.
62. Verheijen JH, Visse R, Wijnen JT, Chang GTG, Kluft C, Meera Khan P. Assignment of the human tissue-type plasminogen activator gene to chromosome 8. Hum Genet 1986; 72:153–156.
63. Yang-Feng TL, Opdenakker G, Volckaert G, Francke U. Human tissue-type plasminogen activator gene located near chromosomal breakpoint in myelo-proliferative disorder. Am J Hum Genet 1986; 79–87.
64. Degen SJF, Rajput B, Reich E. The human tissue plasminogen activator gene. J Biol Chem 1986; 261:6972–6985.
65. Browne MJ, Tyrrell AWR, Chapman CG, et al. Isolation of a human tissue-type plasminogen activator genomic DNA clone and its expression in mouse L cells. Gene 1985; 33:279–284.

66. Patthy L. Evolution of the proteases of blood coagulation and fibrinolysis by assembly from modules. Cell 1985; 41:657–663.
67. Gilbert W. Why genes in pieces? Nature 1978; 271:501.
68. Blake CCF. Do genes-in-pieces imply proteins-in pieces? Nature 1978; 273:267–271.
69. Craik CS, Rutter WJ, Fletterick R. Splice junctions: association with variations in structure. Science 1983; 220:1125–1129.
70. Radcliffe R, Henize T. Stimulation of tissue plasminogen activator by denatured proteins and fibrin clots: a possible additional role for plasminogen activator. Arch Biochem Biophys 1981; 211:750–761.
71. Rijken DC, Hoylaerts M, Collen D. Fibrinolytic properties of one-chain and two-chain human extrinsic (tissue-type) plasminogen activator. J Biol Chem 1982; 257:2920–2925.
72. De Munk GAW, Caspers MPM, Chang GTG, Pouwels PH, Enger-Valk BE, Verheijen JH. Binding of tissue-type plasminogen activator to lysine, lysine analogues and fibrin fragments. Biochemistry 1989; 28:7318–7325.
73. Weening-Verhoeff EJD, Quax PHA, van Leeuwen RTJ, Rehberg EF, Marotti KR, Verheijen JH. Involvement of aspartic and glutamic residues in kringle-2 of tissue-type plasminogen activator in lysine binding, fibrin binding and stimulation of activity as revealed by chemical modification and oligonucleotide-directed mutagenesis. Protein Eng 1990; 4:191–198.
74. Rijken DC, Groenveld E. Isolation and functional characterization of the heavy and light chains of human tissue-type plasminogen activator. J Biol Chem 1986; 261:3098–3102.
75. Holvoet P, Lijnen HR, Collen D. Characterization of functional domains in human tissue-type plasminogen activator with the use of monoclonal antibodies. Eur J Biochem 1986; 158: 173–177.
76. Wojta J, Beckmann R, Turcu L, Wagner OF, van Zonnenveld A, Binder BR. Functional characterization of monoclonal antibodies directed against fibrin binding domains of tissue-type plasminogen activator. J Biol Chem 1989; 264:7957–7961.
77. De Vreis C, Veerman H, Pannekoek H. Identification of the domains of tissue-type plasminogen activator involved in the augmented binding to fibrin after limited digestion with plasmin. J Biol Chem 1989; 264:12604–12610.
78. Gething MJ, Adler B, Boose JA, et al. Variants of human tissue-type plasminogen activator that lack specific structural domains of the heavy chain. EMBO J 1988; 7:2731–2740.
79. Kalyan NK, Lee SG, Wilhelm J, et al. Structure-function analysis with tissue-type plasminogen activator. J Biol Chem 1988; 263:3971–3975.
80. Larsen GR, Henson K, Blue Y. Variants of human tissue-type plasminogen activator. Fibrin binding, fibrinolytic and fibrinogenolytic characterization of genetic variants lacking the fibronectin finger-like and/or the epidermal growth factor domains. J Biol Chem 1988; 263: 1023–1029.
81. Urano S, Metzger AR, Castellino FJ. Plasmin-mediated fibrinolysis by variant recombinant tissue plasminogen activators. Proc Natl Acad Sci USA 1989; 86:2568–2571.
82. Johannessen M, Diness V, Pingel K, et al. Fibrin affinity and clearance of t-PA deletion and substitution analogues. Thromb Haemost 1990; 63:54–59.
83. Lijnen HR, Nelles L, VanHoef B, DeCock F, Collen D. Biochemical and functional characterization of human tissue-type plasminogen activator variants obtained by deletion and/or duplication of structural/functional domains. J Biol Chem 1990; 265:5677–5683.
84. Tate KM, Higgins DL, Holmes WE, Winkler ME, Heyneker HL, Vehar GA. Functional role of proteolytic cleavage at arginine-275 of human tissue plasminogen activator as assessed by site-directed mutagenesis. Biochemistry 1897; 26:338–343.
85. Petersen LC, Johannessen M, Roster D, Kumar A, Mulvihill E. The effect of polymerized fibrin on the catalytic activities of one-chain tissue-type plasminogen activator as revealed by an analogue resistant to plasmin cleavage. Biochim Biophys Acta 1988; 952:245–254.
86. Ahern TJ, Morris GE, Barone KM, et al. Site-directed mutagenesis in human tissue-plasmiogen activator. J Biol Chem 1990; 265:5540–5545.

87. Bennett WF, Paoni NF, Keyt BA, et al. High resolution analysis of functional determinans on human tissue-type plasminogen activator. J Biol Chem 1991; 266:5195–5201.
88. Wallen P, Pohl G, Bergsdorf N, Ranby M, Ny T, Jornvall H. Purification and characterization of a melanoma cell plasminogen activator. Eur J Biochem 1983; 132:681–686.
89. Peterson LC, Boel E, Johannessen M, Roster D. Quenching of the amidolytic activity of one-chain tissue-type plasminogen activator by mutation of lysine 416. Biochemistry 1990; 29: 3451–3457.
90. Monge JC, Lucore CL, Fry ETA, Sobel BE, Billadello JJ. Characterization of interaction of active-site serine mutants of tissue-type plasminogen activator with plasminogen activator inhibitor-1. J Biol Chem 1989; 264:10922–10925.
91. Lijnen HR, VanHoef B, Collen D. On the reversible interaction of plasminogen activator inhibitor-1 with tissue-type plasminogen activator and with urokinase-type plasminogen activator. J Biol Chem 1991; 266:4041–4044.
92. Madison EL, Goldsmith EJ, Gerard RD, Gething MJH, Sambrook JF. Serpin-resistant mutants of human tissue-type plasminogen activator. Nature 1989; 338:721–724.
93. Madison EL, Goldsmith EJ, Gerard RD, Gething MJH, Sambrook JF, Basel-Duby RS. Amino acid residues that affect interaction of tissue-type plasminogen activator with plasminogen activator inhibitor-1. Proc Natl Acad Sci USA 1990; 87:3530–3533.
94. De Serrano VS, Castellino FJ. Structural determinants of the noncatalytic chain of tissue-type plasminogen activator that modulate its association rate with plasminogen activator inhibitor-1. J Biol Chem 1990; 265:10473–10478.

24

Plasminogen

Francis J. Castellino
University of Notre Dame, Notre Dame, Indiana

I. FUNCTIONS OF HUMAN PLASMINOGEN AND PLASMIN

The terminal event in activation of the human fibrinolytic system is generation of the enzyme plasmin (HPm),* a serine protease possessing a variety of functional properties, the most notable of which is clearance by proteolytic degradation of fibrin deposits. The formation of HPm is regulated by a variety of agents present in cells and in extracellular fluids in a manner that serves to localize its presence. HPm is formed upon activation of its zymogen, HPg, as a result of cleavage of a single peptide bond. This latter event is catalyzed by serine proteases with narrow specificity, termed plasminogen activators. One such activator, tPA, binds to the fibrin clot and activates the HPg that is also bound to the clot, thus directing HPm toward clot dissolution. On the clot surface, both HPm and tPA are protected from inactivation by their major inhibitors, namely, α2PI and α2M for HPm and PAI-1 for tPA (1). After degradation of the thrombus and release of HPm and tPA, these proteases are rapidly inactivated by these circulating inhibitors.

A system for cell-mediated fibrinolysis also exists and is grounded in the presence of a specific receptor (uPAR) for another plasminogen activator, uPA (2), on a variety of normal (3–5) and neoplastic (6–8) cells. Cells of these types enriched in surface-bound uPA promote HPm formation from the HPg, which is also bound to normal and carcinoma

* HPm, any molecular form of human plasmin; HPg, any molecular form of human plasminogen; Glu^1-Pg, human plasminogen containing glutamic acid 1 as the amino-terminal amino acid; Lys^{78}-Pg, human plasminogen containing lysine 78 as the amino-terminal amino acid; Glu^1-Pm, human plasmin containing glutamic acid 1 as the amino-terminal amino acid of the plasmin heavy chain; Lys^{78}-Pm, human plasmin containing lysine 78 as the amino-terminal amino acid of the plasmin heavy chain; tPA, tissue-type plasminogen activator; α2PI, α_2-antiplasmin; α2M, α_2-macroglobulin; PAI-1, plasminogen activator inhibitor 1; PAI-2, plasminogen activator inhibitor 2; uPA, urokinase-type plasminogen activator; pro-uPA, prourokinase-type plasminogen activator (used synonymously with single-chain urokinase-type plasminogen activator); tc-uPA, two-chain utokinase formed from PTO-uPA by cleavage; uPAR, utokinase receptor; SK, streptokinase; r, recombinant.

cell surfaces (9–12), and from HPg in surrounding fluids. Epicellular HPm is also resistant to inactivation by α2PI (13), and receptor-bound uPA is slightly more resistant to PAI-1 and PAI-2 than solution-phase uPA (14). These cell-associated fibrinolysis mechanisms, which can generate epicellular HPm, provide a basis for the proposed roles of HPg activators, via HPm formation, in providing a protease for normal processes involving cell migration in tissue remodeling. In this regard, HPm is believed to function in processes in which cell movement is essential, such as macrophage invasion in inflammation (15,16), mammary cell involution after lactation (17), breakdown of the follicular wall for ovulation (18), trophoblast invasion into the endometrium during embrogenesis (19–21), angiogenesis (22), and keratinocyte accumulation after wound healing (23). Further, HPm has also been strongly implicated as an important mediator in pathological processes of cell migration that are involved in tumor cell growth and invasion of surrounding tissue and, perhaps, metastases (24,25). Involvement of the HPg/uPA/HPm system in these latter processes is supported by the ability of HPm to degrade extracellular matrix proteins directly (26), such as proteoglycans (27), fibronectin (28), laminin (29,30), and type IV collagen (31), and/or to be indirectly responsible for the degradation of matrix proteins through activation of metalloprotease zymogens, such as stromolysin and procollagenase (32,33). As a result of degradation of the extracellular matrix, cell migration into surrounding areas becomes more facile.

Solution-phase activation of HPg also occurs, providing HPm that plays a role as an extracellular or pericellular protease. In this regard, solution-phase HPm may be active in such pathways as zymogen conversions in the alternate complement pathway (34) and in the contact phase of blood coagulation (35), in proinsulin to insulin conversion (36), in bradykinin generation from kininogen (37), and in proteolytic destruction of other plasma proteins (38–41). Solution-phase activation of HPg is catalyzed by soluble two-chain uPA (42) and two-chain uPA bound to uPAR (43). Additionally, tPA released from vasculature by such stimuli as venous occlusion also provides a low level of activation of plasma HPg (44). HPg in plasma is also directly activated to HPm by plasma proteases, namely, factor XIa (45), factor XIIa (46), and kallikrein (47), resulting from contact activation of the clotting cascade. However, these latter activators may play only minor roles in physiological activation of HPg, because they constitute a total of <15% of the HPg activator activity of plasma (48). Indirect activation of HPg by plasma proteases, such as HPm (49,50) and kallikrein (51), which are capable of converting pro-uPA to tc-uPA, can also occur. Finally, a factor XIIa-dependent mechanism for HPg activation exists (52), which is based upon the kallikrein-catalyzed activation of a HPg proactivator that shares some homology with uPA (53).

The physiological importance of extracellular proteolytic mechanisms involving HPm are uncertain because of the presence of relatively high levels of α2PI in plasma, which should rapidly inactivate circulating HPm. However, it is possible that some of the extracellular processes just mentioned, known to be catalyzed *in vitro* by HPm, also occur epicellularly, where HPm would be resistant to such inactivation, or pericellularly, wherein local HPm concentrations may be sufficiently high to overcome local levels of α2PI. In this manner, sufficient levels of HPm could be present to serve as a catalyst for these reactions.

Another HPg activator of importance exists, especially in situations wherein advantage is taken of the HPg/HPm system in thrombolytic therapy. This involves the bacterial protein SK. SK does not activate HPg directly but first must participate in formation of a HPg activator consisting of a stoichiometric complex of SK-HPg′ that has evolved an active site in the HPg moiety of the complex *via* a conformational rearrangement of this zymogen within the SK complex, or a stoichiometric complex of SK-HPm. Both these SK complexes serve as activators of HPg (for a review, see Ref. 54).

A summary of these HPg activation systems is present in Figure 1.

II. PRIMARY STRUCTURE OF HUMAN PLASMINOGEN

The primary structure of HPg, illustrated in Figure 2, has been determined by amino acid sequence analysis (55–59) and has also been deduced from the nucleotide sequence from the cDNA (60,61) and genomic DNA (62) that encode this protein. HPg is synthesized as an 810-residue single polypeptide chain, from which is excised a 19-residue signal peptide during secretion (61). The mature form of HPg thus contains 791 amino acid residues.

Activation of HPg to HPm occurs consequent to cleavage of the Arg^{561}-Val^{562} peptide bond (55). The HPm thus formed possesses a heavy chain of 561 amino acid residues and a light chain of 230 amino acid residues, the latter of which is homologous to serine proteases, such as trypsin and elastase. These polypeptide chains are covalently linked by two disulfide bonds, between Cys^{548} of the heavy chain and Cys^{666} of the light chain and between Cys^{558} of the heavy chain and Cys^{566} of the light chain. HPm is a serine protease with the typical catalytic triad present at His^{603}, Asp^{646}, and Ser^{741}. One consensus Asn-linked glycosylation site is present at Asp^{289}, which in human plasma HPg, contains biantennary oligosaccharide in approximately one-half of the HPg molecules (63) and is not glycosylated in the remaining HPg molecules (64), despite the integrity of the consensus sequence in the non-N-glycosylated HPg (65). HPg also contains one site containing O-linked glycan at Thr^{346} (66), which is occupied on all HPg molecules. The two N-linked glycoforms of HPg can be resolved on Sepharose-lysine affinity chromatography columns (67).

Another major cleavage site with a very significant functional consequence is that between Lys^{77} and Lys^{78}. Cleavage of this peptide bond is catalyzed by HPm (68), providing a form of HPg (Lys^{78}-Pg) that differs conformationally from native HPg (Glu^{1}-Pg) (69) and activates at a much faster rate than Glu^{1}-Pg in the absence of positive activation effectors (70,71).

The heavy chain of HPm consists of repeating homologous triple disulfide-linked peptide regions, approximately 80 amino acid residues in length, termed kringles (59). Five such repeats are present within the amino-terminal 541 residues of HPg: residues Cys^{84} to Cys^{162}, Cys^{166} to Cys^{243}, Cys^{256} to Cys^{333}, Cys^{358} to Cys^{435}, and Cys^{462} to Cys^{541}. These structures are present in several other clotting and fibrinolytic proteins, such as factor XII (72), prothrombin (73), tPA (74), and urokinase (75). Some of these kringles are responsible for interactions with regulators of these pathways. In HPg, kringles that display functional interactions with effector molecules are kringles 1, 4, and 5.

III. MECHANISMS OF ACTIVATION OF HUMAN PLASMINOGEN

Essential to the activation of HPg to HPm is the cleavage of the Arg^{561}-Val^{562} peptide bond, which alone is sufficient to convert the HPg zymogen to the HPm active enzyme

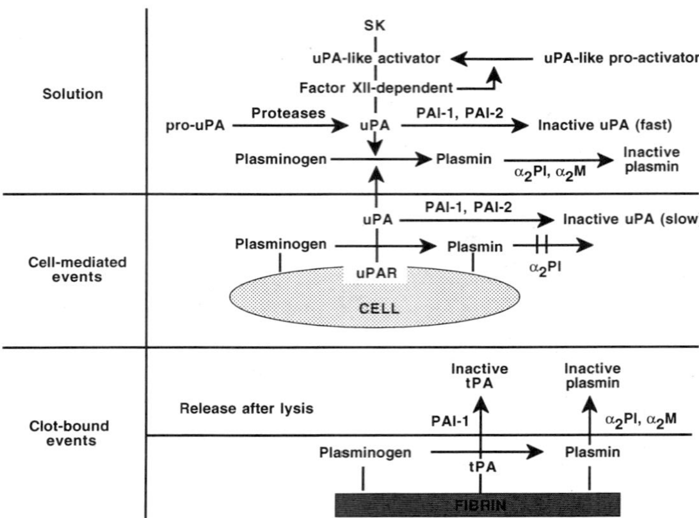

Figure 1 The interactions involved in the activation of human plasminogen (HPg). (Top) Solution-phase activation of HPg readily occurs with two-chain urokinase (uPA). uPA bound to its receptor (uPAR) can also catalyze this activation. The presence of circulating inhibitors of both uPA, that is, plasminogen activator inhibitor 1 (PAI-1) and plasminogen activator inhibitor 2 (PAI-2), and of plasmin (HPm), that is, α_2-antiplasmin (α_2PI) and α_2-macroglobulin (α_2M), probably limits the physiological effectiveness of true extracellular activation pathways but may allow effective pericellular activation of HPg to be important, wherein local concentrations of HPm may be sufficiently high to overcome local levels of inhibitors. Solution-phase activation of HPg may also occur directly from proteases, that is, kallikrein, that evolve from the contact phase of blood coagulation or indirectly from these and other proteases by way of activation of pro-uPA (single-chain urokinase), or from activation of a uPA-related proactivator. The HPm formed in solution may be responsible for various proteolytic events. Exogenously added streptokinase (SK), which is used in thrombolytic therapy, also activates HPg through an indirect mechanism by first complexing with HPg and HPm to form the actual HPg activator. In these cases, SK is added in sufficient amounts to allow levels of HPm to form that overcome the circulating concentration of α_2PI. (Middle) HPg is also activated on various cell surfaces by uPA bound to its cellular receptor (uPAR). The HPm thus formed is protected from inactivation by α_2PI and probably functions directly or indirectly in extracellular matrix degradation, perhaps locally after dissociation from the cell surface. The cell-bound uPA appears to be susceptible to inactivation by PAI-1 and PAI-2, but at a slightly slower rate than free uPA. (Bottom) Activation of HPg also occurs on the surface of the blood clot. Here, the clot-bound activator, tissue-type plasminogen activator (tPA), activates fibrin-bound HPg. The HPm formed on the clot surface is resistant to inhibition by α_2PI, as is clot-bound tPA toward PAI-1. After dissolution of the clot and release of HPm and tPA, these proteases are inhibited by their fast-acting circulating inhibitors.

(55). All HPg activators catalyze this peptide bond cleavage. However, important regulation of this activation takes place, which is at the basis of use of the fibrinolytic system for fibrinolytic and antifibrinolytic therapies.

The overall scheme for conversion of circulating Glu^1-Pg to the activation end product, Lys^{78}-Pm, is shown below (68):

PRIMARY STRUCTURE OF HUMAN PLASMINOGEN

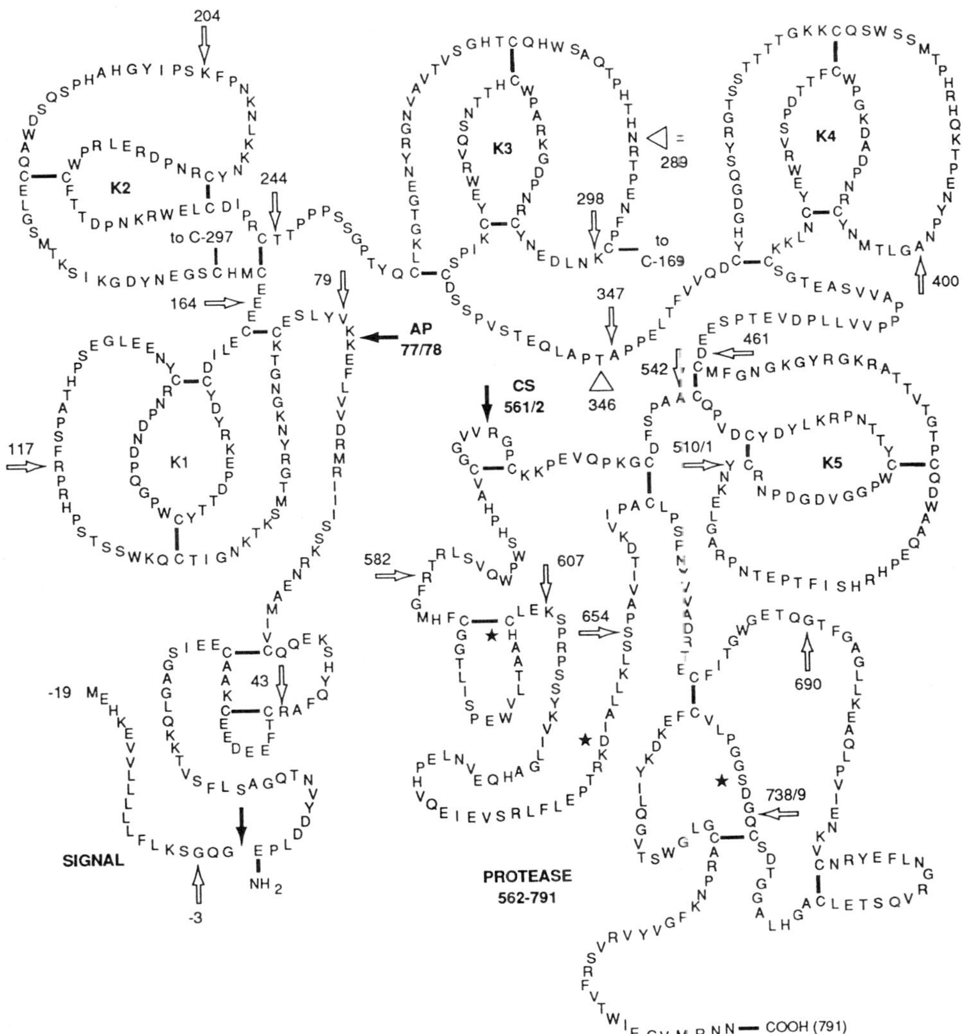

Figure 2 Primary structure of human plasminogen. The cleavage sites of the signal peptide between residues −1 and 1, that between residues 77 and 78, required for release of the activation peptide (AP) consequent to activation of Glu^{1}-plasminogen (Glu^{1}-Pg) to Lys^{78}-plasmin (Lys^{78}-Pm), and that between residues 561 and 562, needed for activation of HPg to HPm (CS), are indicated by filled arrows. Positions of introns in the gene sequence are represented by unfilled arrows. The locations of the N-linked oligosaccharide at sequence position 289 and the O-linked glycan at position 346 are also provided. *Members of the catalytic triad of plasmin consisting of His^{603}, Asp^{646}, and Ser^{741}.

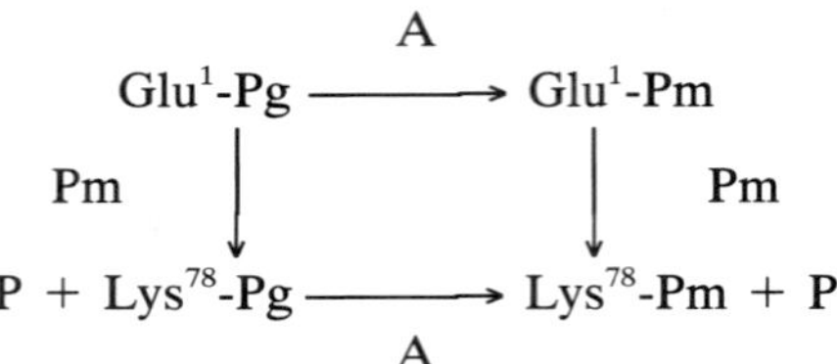

Here, circulating Glu^1-Pg is activated to Glu^1-Pm by activator (A)-catalyzed cleavage of the Arg^{561}-Val^{562} peptide bond in Glu^1-Pg. This step most likely provides the first molecules of plasmin (Pm). This reaction is slow in the presence of Cl^-, where Glu^1-Pg exists in a compact and relatively nonactivatable form (69,76–78). However, ω-amino acids, such as lysine, enhance the rate of this reaction when Cl^- is present because of the "loosening" of the Cl^-independent conformation of Glu^1-Pg (69,76,77). The initial plasmin formed catalyzes cleavage of a 77 amino acid peptide from the amino terminus of the Glu^1-Pm heavy chain, providing Lys^{78}-Pm, or from the amino terminus of Glu^1-Pg (68,79), yielding Lys^{78}-Pg. In the presence of Cl^-, the activation rate of Lys^{78}-Pg is substantially faster than that of Glu^1-Pg (70). In general, the activation rate and conformation of Lys^{78}-Pg $\pm$ Cl^- $\pm$ lysine are approximately the same as those of Glu^1-Pg in the absence of Cl^-, or Glu^1-Pg + Cl^- + lysine. Thus, Cl^- is a negative effector of Glu^1-Pg activation and lysine analogs are positive effectors of Glu^1-Pg activation in the presence of Cl^- (76–78). Fibrin is also a positive effector of HPg activation with certain activators, such as tPA. In the presence of fibrin, a large increase in the rate of tPA-catalyzed HPg activation occurs, mainly because of a decrease in the K_m of the activation reaction (80). Fibrin does not stimulate the activation rate of the uPA-catalyzed activation of HPg.

The preceding pathway occurs with all known HPg activators, except the bacterial activator SK. Because SK is not itself a protease, the question of how a protein without proteolytic activity catalyzes cleavage of the peptide bond required to activate HPg was one that commanded a great deal of attention during the past two decades. Several laboratories have contributed to the solution of this question (for a review, see Ref. 54). In the first phase of the activation SK must participate in the formation of a HPg activator *via* the series of steps outlined here:

$$\begin{array}{l} \text{SK + HPg} \leftrightarrow \text{SK-HPg} \rightarrow \text{SK-HPg}^* \rightarrow \text{SK-HPg}' \rightarrow \text{SK-HPm} \\ \hspace{22em}\uparrow \\ \hspace{21em}\text{SK + HPm} \end{array}$$

Initially, a stoichiometric complex of SK and HPg forms (SK-HPg), within which a conformational rearrangement of the HPg takes place, allowing an active site to form within the HPg moiety of the complex (SK-HPg*). This active site in HPg* is sensitive to inhibition by Cl^- and stimulation by fibrin(ogen) (81–83). With time, another complex (SK-HPg′) forms, which possesses a diminished ability to be regulated by Cl^- and fibrin(ogen), finally yielding the most stable of these complexes, SK-HPm. This latter complex also forms from SK and HPm. The SK-HPg*, SK-HPg′, and SK-HPm complexes are catalytic activators of remaining HPg, according to

$$\text{HPg} \xrightarrow{\substack{\text{SK-HPg}^* \\ \text{SH-HPg}' \\ \text{SK-HPm}}} \text{HPm}$$

with SK-HPg′ a more potent activator than SK-HPm (84). Detailed kinetic characteristics of SK-HPg* have not been evaluated. HPm alone cannot directly catalyze activation of HPg (68,79).

IV. STRUCTURE-FUNCTION RELATIONSHIPS OF HUMAN PLASMINOGEN

Both HPg and HPm interact with a variety of molecules that serve to regulate the activation of HPg and the activity of HPm. The most widely investigated of these binding interactions is that of HPg with ω-amino acids, typified by L-lysine and EACA. The strongest ω-amino acid binding site on HPg is on the kringle 1 domain of this protein (85–87), with weaker sites present on kringle 4 (85,86,88–91) and kringle 5 (89,91,92). Based on analyses by chemical modification (86,93–96), nuclear magnetic resonance spectroscopy (88,92,93,97–99), x-ray crystallography (100–104), and site-directed mutagenesis (105–110), with a variety of isolated kringle domains, a consensus regarding the kringle amino acid residues needed for direct stabilization of ω-amino acids has been developed. These essential amino acids are shown in Figure 3 with reference to HPg kringle 1.

The amino group of the ligand is stabilized by both Asp^{54} and Asp^{56} in this kringle. This is clearly inferred from the x-ray structure of the tPA kringle 2-lysine complex (103) and the HPg kringle 4-EACA complex (102). Results from chemical modification (96) and site-directed mutagenesis studies (96,110) are in complete agreement with this view. Aspartic acid residues are rigorously conserved in homologous positions in all other

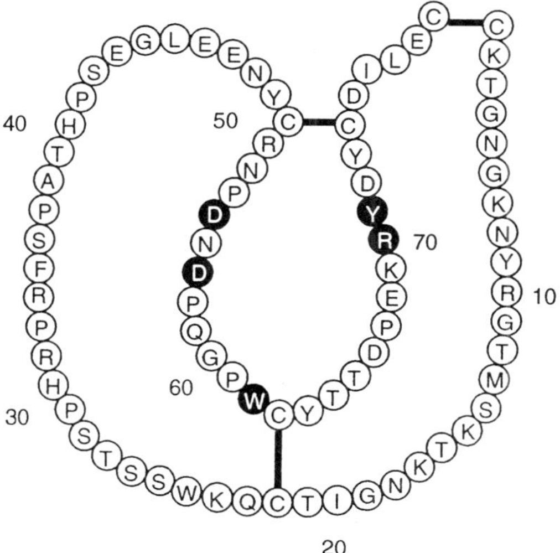

Figure 3 Primary structure of the kringle 1 domain of human plasminogen. The numbering system begins at the amino-terminal cysteine residue. The amino acid residues that have been directly implicated in ω-amino acid binding or implicated by homology considerations from direct studies with other kringle domains are indicated by black-white color reversals.

lysine binding kringles, and even simple changes of Asp to Glu at these positions in tPA kringle 2 severely disrupt ω-amino acid binding (110). The carboxylate group of the ω-amino acid ligand is stabilized by Arg^{34} and Arg^{70}. Although the x-ray structure of HPg kringle 1 clearly shows that both these Arg residues are candidates for such a ligand binding stabilization, the relative contributions of these two residues are not as clear. In this regard, chemical modification studies with HPg kringle 4 suggest the importance of Arg^{70} (93). However, mouse plasminogen kringle 1 does not contain a basic residue at a position homologous to Arg^{34} in HPg kringle 1 (111) but nonetheless binds ω-amino acids with efficacy equal to that of HPg kringle 1 (112). Further, site-directed mutagenesis studies reveal that only Lys^{33} stabilizes the binding of ω-amino acids to tPA kringle 2 (108,109), despite the presence of a residue equivalent to Arg^{70} of HPg kringle 1. Thus, it seems as though either (or both) of these residues can function for this purpose and that the particular cationic locus employed depends upon the conformation of the kringle under consideration.

Several aromatic amino acid residues are perturbed as a result of ω-amino acid binding to kringles (92,98,113–115). Shown to be of direct importance to stabilizing the binding of ω-amino acids is Trp^{74} of tPA kringle 2 (106). An aromatic residue (Tyr^{71}) is present in a homologous location in HPg kringle 1, HPg kringle 4 (Trp^{70}), and HPg kringle 5 (Tyr^{72}). This aromatic residue forms one wall of an aromatic cluster that stabilizes the hydrophobic core of the ω-amino acid ligand, the other wall being formed by Trp^{61} of HPg kringle 1 and the homologous aromatic residue found in that location in all other kringles that bind ω-amino acids (Trp^{63} in tPA kringle 2, Trp^{60} in HPg kringle 4, and Trp^{62} in HPg kringle 5) (101–104).

These kringle regions of HPg are also implicated in stabilizing the binding of HPg with macromolecules, in many cases seemingly through the ω-amino acid binding sites. The increasing ability of HPg to interact with HPm-digested fibrin, compared with virgin fibrin, most likely occurs through kringle domain interactions with the carboxyl-terminal lysine residues that are liberated consequent to HPm catalysis (116). HPg binding, mediated through the kringle domains, also occurs with other plasma proteins, such as fibrinogen (117), histidine-rich glycoprotein (118), and α_2-antiplasmin (119). This latter interaction, as well as the strong binding of HPg to fibrin fragment E (120), is also dependent upon carboxyl-terminal lysine residues and perhaps other internal lysine residues (121). In all these cases, this binding is inhibited by ω-amino acids. HPg interactions with a variety of cells have been examined, and it was found that a region of the HPg molecule comprising its first three kringle domains serves as the primary cellular binding sites (12). ω-Amino acids also displace HPg in this binding interaction.

V. STRUCTURE AND ORGANIZATION OF THE HUMAN PLASMINOGEN GENE

cDNA clones encoding HPg have been isolated and sequenced (60,61,122), and more recently the genomic DNA coding for this protein was characterized (62). The gene for HPg encompasses 52.5 kb (62) and has been mapped to chromosome 6q26–6q27 (123). Its coding sequence includes a 57 bp signal sequence and a total of 2373 nucleotides for the mature protein. The gene consists of a total of 19 exons, which range in size from 75 to 387 bp, with 18 introns, of type I, type II, and type 0 (124), interrupting the exons. All intron-exon splice junctions are in agreement with the GT-AG (125) and consensus sequence (126) rules.

The locations of the exons in the translated sequence of HPg are shown in Figure 2. The first exon (amino acids −19 to −3) comprises most of the signal sequence of the protein; exons II and III (amino acids −3 to 43 and 43 to 79, respectively) code for the amino-terminal peptide that is liberated consequent to activation of HPg to HPm. Each of the five kringles (amino acids 79–461) is encoded by two exons (exons IV–XIII). Exon XIV (amino acids 542–582) contains the Arg^{561}-Val^{562} peptide bond that is cleaved by HPg activators, as well as one of the cysteine residues (Cys^{548}) that connects the two HPm chains. Exons XV and XVI (amino acids 582–607 and 607–654) consist of the coding regions for the HPm active-site residues His^{603} and Asp^{646}, along with each of the other pair of cysteine residues (Cys^{558} and Cys^{566}) that connects these two peptide chains. A stretch of amino acids (654–690) is contained in exon XVII, within which exists the partner (Cys^{666}) of Cys^{548} that covalently stabilizes the two-chain structure of HPm. Exon XVIII encodes a sequence of amino acids (690–738) that contains a disulfide loop of unknown functional significance. However, the additional Cys residue in this exon at position 737, which pairs with Cys^{765}, is located proximal to the active center residue Ser^{741} and may be relevant to the specific functioning of the active site through its importance to the folding of this region of the molecule. Last, this active-site serine residue of HPm (Ser^{741}) is contained in exon XIX, beginning at amino acid residue 739 and terminating at an undetermined location 3′ of the coding stop sequence. These considerations, which clearly demonstrate that functional regions of HPg are contained in different exons, support the modular evolutionary theories for proteins of this type (127).

The regulatory portions of the HPg gene contained in the nucleotide sequences 5′ and 3′ of the coding sequence have been partly identified (62). Transcription regulatory nucleotide sequences 5′ of the signal sequence-initiating methionine codon include TATAA promoter elements and the forward and reverse CCAAT proximal upstream promoter element boxes (128). Nucleotide sequences (CTGGGA), found in several acute-phase reactant proteins, including fibrinogen (129), human haptoglobin (130), α_1-antitrypsin (131), and transferrin (132), are also located upstream of the HPg methionine signal-initiation codon. Regulatory ‘‘GC boxes’’ were not observed in the 5′ region of the HPg gene (62). In the 3′ noncoding region of the cDNA (61), a primary mRNA polyadenylation recognition sequence, AATAAA, is found 46 bp upstream of the poly(A) tail, and a second polyadenylation recognition unit, CTTTG (133), is positioned 13 bp downstream of this primary consensus sequence. In a cDNA isolated from a HepG2 library, which contained an additional 750 nucleotides of 3′ noncoding DNA, another AATAAA primary polyadenylation signal was found, as well as a CATTG secondary polyadenylation recognition unit 43 bp downstream of this primary recognition site (62). The YGTGTTYY consensus sequence, also needed for efficient polyadenylation of mRNA (134), is present 32 bp downstream of this latter alternative polyadenylation site in the HPg cDNA. This is close to the location of this sequence within 24–30 bp of the AATAAA sequence in a variety of mammalian gene structures (134). No such sequence is found within this distance from the first polyadenylation signal.

VI. POLYMORPHISMS OF THE PLASMINOGEN GENE IN NORMAL HUMAN POPULATIONS

A variety of electrophoretic forms of HPg exist. As shown originally with canine and rabbit plasminogens (135,136) and confirmed with HPg (137,138), at least 10 electrophoretic forms of plasminogen are resolved by isoelectric focusing methods. Each major glycoform of HPg (67) contains at least five electrophoretic isoforms with overlapping pI values (136,138). The molecular distinctions of some of these isoforms are caused by variabilities in their sialic acid contents (139), which reduce to two to three isoforms after treatment of plasminogen with neuraminidase (139,140). Since these findings were reported, it has been discovered by electrophoretic analysis that two major codominant allelic variants of HPg, that is, PLG*A (or PLG*1) and PLG*B (or PLG*2), exist (140,141), with at least 12 additional rare alleles as well as a silent allele, PLG*QD (142–147). Frequency distributions of these HPg polymorphs in a variety of human normal populations have been studied, and distinct differences have been observed (140,141,143,144,148–151). For example, the gene frequencies of PLG*A and PLG*B in Japanese were found to be 0.98 and 0.003, respectively, whereas the same alleles were present in gene frequencies of 0.69 and 0.3, respectively, in American caucasians (149). The lowest gene frequency, 0.46–0.56, of the A allele of HPg so far discovered (and thus the corresponding highest frequency of the B allele, 0.44–0.54) of HPg was found in Arab populations (Israeli Druze, Israeli Moslems, and Jordanians) (151).

The molecular bases of the polymorphisms have not been elucidated, but comparisons of the published amino acid sequence of HPg (59) with the amino acid sequence deduced from nucleic acid sequences of several cDNA (60,61) and genomic DNA (62) clones indicate some possible locations of protein polymorphisms in the normal human population. The cDNA sequences show the following deviations from the amino acid sequence: a Gln for a Glu at position 53, an extra Ile at position 85, an Asn for an Asp at 88, a Gln for a Glu at 342, and an Asp for an Asn at 453. Any of these substitutions that result in charge alterations may provide electrophoretically distinct forms of HPg. In addition to the nucleotide substitutions that result in amino acid substitutions, others have been found that yield restriction fragment length polymorphisms. For example, TGT and TGC codons for Cys^{238} have been found in different cDNA and genomic DNA clones, the former providing a *Mae*III restriction endonuclease site that does not occur when the latter codon is used (this is incorrectly labeled a *Mae*II site in Ref. 62). Similarly, in Phe^{295}, a TTC codon has been found in the cDNA and a TTT codon is used for this amino acid in the genomic DNA clone. In this latter codon, a recognition sequence for the *Xmn*I restriction endonuclease is provided. In Gly^{743}, both GGT and GGG codons have been observed in various clones. When GGT is the codon employed, an *Ava*II restriction site is present with corresponding loss of the *Hae*III restriction site that is observed when GGG is used as the codon.

VII. ABNORMAL PLASMINOGENS

Although these genetic variations in HPg are present in normal populations, other genetic alterations occur that can result in HPg deficiencies. Some of these have been associated with thrombotic conditions caused by the resulting defects in the normal balance between clot formation and clot dissolution. The first abnormal HPg (HPg-Tochigi I) described was isolated from a patient with a history of thrombotic occurrences (152). The patient possessed normal levels of HPg antigen, with 37% functional HPg, suggestive of a heterozygous condition. The inheritance followed an autosomal dominant pattern (153). Subsequent isolation and characterization of this HPg (154) demonstrated (1) that it was converted to the two-chain HPm by uPA, but the resulting HPm did not possess any proteolytic activity, nor does HPm incorporate the active-site serine inhibitor, diisopropylfluorophosphate; (2) that it interacted normally with lysine-Sepharose; and (3) that it formed a stoichiometric complex with SK, but this complex also did not incorporate diisopropyphospholfluoridate. These results suggested that HPg Tochigi I possessed an active-site defect when converted to HPm. The molecular nature of this defect was a result of replacement of Ala^{601} by a Thr residue at that location, resulting from a G to A transition in exon XV of the HPg gene (155). This alteration of Ala^{601} most likely influenced the nearby active center residue, His^{603}, in such a manner that HPm activity was lost.

The $Ala^{601} \rightarrow$ Thr mutation has also been found in individuals from unrelated families, namely, HPg Tochigi II, HPg Nagoya I_{d3} (this patient was originally called HPg Nagoya, but after family studies were conducted a revised nomenclature was adopted and the patient is a daughter of HPg Nagoya I), HPg Nagoya II, as well as HPg Kagoshima (155,156). HPg Kagoshima is a genotypic homozygote for the $Ala^{601} \rightarrow$ Thr mutation, and HPg Nagoya II is a genotypic heterozygote for this same mutation (155). Considering this same mutation, family studies have led to identification of a genotypic heterozygote daughter of HPg Kagoshima (HPg $Kagoshima_d$), as well as other genotypic heterozygote relatives of HPg Nagoya I, that is, HPg Nagoya I_{d2}, HPg Nagoya I_{d3}, HPg Nagoya I_{gs1} (a grandson of HPg Nagoya I), and HPg Nagoya I_{gs2}. Other genotypic heterozygotes for the $Ala^{601} \rightarrow$ Thr mutation were HPg Nagoya II and relatives of this patient, HPg Nagoya II_{d2} and HPg Nagoya II_{d3} (155). The DNAs have not been characterized, but from consideration of the properties of the purified HPgs, it is likely that HPg Tochigi I and HPg Tochigi II are also genotypic heterozygotes for the $Ala^{601} \rightarrow$ Thr mutation. A similar situation may exist with another HPg deficiency, HPg Jichi, a patient with a combined phenotypic homozygous protein C deficiency and heterozygous dysplasminogenemia (HPg antigen, 86%; HPg functional activity, 48–50%) (157). The HPg from this patient was activated normally with uPA to a two-chain HPm structure but possessed only 50% HPm activity, suggestive of an active-site defect in one-half of the HPm molecules and a normal HPm activity for the other half of the HPm subpopulation. This particular $Ala^{601} \rightarrow$ Thr mutation appears to be a possible genetic marker of the Japanese population because the electrophoretic pattern characteristic of the HPg Tochigi I heterozygote is present in 3.6% of a pool of Japanese subjects but is not found in a corresponding sample of American whites (149). That a finite frequency for this mutation may nonetheless exist in the white population may be evidenced by the observation of a European Jewish patient, HPg Paris I, who possessed 48% functional HPg and 100% antigenic HPg (158). Characteristics of the HPg purified from this patient suggested a similar abnormality to that of HPg Tochigi I, perhaps because of the same active-site defect.

Many, but not all, of these patients and their affected relatives had clinical histories of venous thrombosis. Interestingly, the only as yet observed genotypically homozygous patient for the $Ala^{601} \rightarrow$ Thr mutation, HPg Kagoshima, has not yet been afflicted with a thrombotic episode. This suggests that some alternative fibrin degradation mechanism is present (155).

An interesting genetic defect in HPg occurs in HPg Nagoya I (155). This patient is a phenotypic heterozygote for a HPg deficiency but yet does not contain the $Ala^{601} \rightarrow$ Thr mutation (155), as do the affected daughters of this patient. Another genetic mutation was found in the DNA of HPg of this patient. Here a G $\rightarrow$ T transition in exon X, which eliminated an *Ava*II restriction endonuclease site, led to an amino acid alteration of $Val^{355} \rightarrow$ Phe. Although 17 other nucleotide alterations occur in the genomic DNA of this patient, these were generally in introns or other noncoding regions of the molecule. The only amino acid alteration in the exons was the change $Val^{355} \rightarrow$ Phe. Thus, this change is probably responsible for the dysfunction in HPg.

Finally, a Japanese patient has been identified containing a single T to C transition in exon 14 of HPg that would translate into a $Ser^{572} \rightarrow$ Pro mutation (159). The HPg activity and antigen levels in this subject were approximately 54% (160), suggesting a heterozygous condition. The patient clinically presented with cerebral infarction. Family studies confirmed the genetic transmission of this genotype along with the thrombotic complications. Studies with normal control subjects demonstrated that this genetic was not a common polymorphism (159). The molecular basis for the phenotype observed with this mutation has not been elucidated.

A variety of other HPg dysfunctional molecules exist that have kinetic defects and charge defects. These are not discussed here, however, because the molecular bases of these dysfunctional HPgs have not been characterized. A recent detailed review of the properties of these HPgs has appeared (161).

VIII. IN VITRO MUTAGENESIS OF HPg

It is difficult to express plasminogen in mammalian cells because of the ubiquitous presence of HPg activators present therein. In such cases, the rHPg is activated to rHPm, which autodegrades. We first expressed the cDNA for HPg in a variety of recombinant baculovirus-infected lepidopteran insect cells (162–167), because cells of these types do not contain HPg activators (162). Normal functional HPg was obtained from this system, with some alterations in the Asn^{289}-linked glycan structures. Since that time, rHPg has been expressed in vaccinia virus-infected mammalian cells (166). This also represents a special case of expression because the vaccinia virus encodes and expresses a serine protease inhibitor that most likely inhibits HPg activators (167). We employed the insect cell system to express a recombinant variant HPg with a $Pro^{611} \rightarrow$ Ile mutation (168). This mutation, which is also in the vicinity of the active site His^{603}, produces a product that is activated to the two-chain form of HPm by uPA but does not possess HPm amidolytic activity. Thus, the behavior of this mutant is identical to that of the $Ala^{601} \rightarrow$ Thr HPg genotype. Thus, mutations in the vicinity of His^{603} appear to inactivate HPm, likely *via* conformational effects that result in realignment of the catalytic triad of this serine protease.

On additional HPg recombinant mutant that has been generated is an nonactivatable form of this zymogen, HPg[$Arg^{561} \rightarrow$ Ser] (84). This recombinant variant HPg has been employed to assess the HPg activation capabilities of the SK-HPg′ complex (using HPg[$Arg^{561} \rightarrow$ Ser] to generate this activator complex) *versus* the SK-HPm complex (generated with wild-type HPm). This would normally be a very difficult project to conduct because of the instability of the SK-HPg complex and its rapid conversion to the SK-HPm complex. The results of this investigation demonstrated that the SK-HPg′ complex was indeed a significantly more efficient plasminogen activator than the SK-HPm complex (84).

IX. CONCLUDING REMARKS

The last decade has been marked with great extensions of our knowledge of the structure-function relationships of components of the fibrinolytic system. It is now clear that with the methods available for the discovery of genotypic disorders in the proteins of this system, coupled with the ability to produce these mutations in vitro, our understanding of the functional requirements of these proteins will increase even more rapidly in the near future. Given the importance of this system not only in maintenance and/or generation of a fluid state of plasma, but also in other normal and disease states, such advances will be welcomed by many types of investigators in a variety of disciplines.

REFERENCES

1. Collen D. On the regulation and control of fibrinolysis. Thromb Haemost 1980; 43:77–89.
2. Vassalli J-D, Baccino D, Belin D. A cellular binding site for the M_r 55,000 form of the human plasminogen activator, urokinase. J Cell Biol 1985; 100:86–92.
3. Stoppelli MP, Corti A, Stoffientini A, Cassani G, Basi F, Assoian RK. Differentiation-enhanced binding of the amino-terminal fragment of human urokinase plasminogen activator to a specific receptor on U937 monocytes. Proc Natl Acad Sci USA 1985; 82:4939–4943.
4. Bajpai A, Baker JB. Cryptic urokinase binding sites on human foreskin fibroblasts. Biochem Biophys Res Commun 1985; 133:475–482.
5. Lazarus GS, Jensen PJ. Plasminogen activators in epithelial biology. Semin Thromb Hemost 1991; 17:210–216.
6. Boyd D, Florent G, Kim P, Brattain M. Determination of the levels of urokinase and its receptor in human colon carcinoma cell lines. Cancer Res 1988; 48:3112–3116.
7. Nielsen LS, Kellerman GM, Behrendt N, Picone R, Dano K, Blasi F. A 55,000–65,000 M_r receptor for urokinase-type plasminogen activator. Identification in human tumor cell lines and partial purification. J Biol Chem 1988; 263:2358–2363.
8. Stoppelli MP, Tacchetti C, Cubellis MV, et al. Autocrine saturation of prourokinase receptors on human A431 cells. Cell 1986; 45:675–684.
9. Plow EF, Freany F, Plescia J, Miles LA. The plasminogen system and cell surfaces: evidence for plasminogen and urokinase receptors on the same cell type. J Cell Biol 1986; 103:2411–2430.
10. Hajjar KA, Harpel PC, Jaffe EA, Nachman RL. Binding of plasminogen to cultured human endothelial cells. J Biol Chem 1986; 261:11656–11662.
11. Miles LA, Plow EF. Receptor mediated binding of the fibrinolytic components, plasminogen and urokinase to peripheral blood cells. Thromb Haemost 1987; 58:936–942.
12. Miles LA, Dahlberg CM, Plow EF. The cell binding domains of plasminogen and their function in plasma. J Biol Chem 1988; 263:11928–11934.
13. Miles LA, Plow EF. Plasminogen receptors: ubiquitous sites for cellular regulation of fibrinolysis. Fibrinolysis 1988; 2:61–71.
14. Ellis V, Wun T-C, Behrendt N, Ronne E, Dano K. Inhibition of receptor-bound urokinase by plasminogen activator inhibitors. J Biol Chem 1990; 265:9904–9908.
15. Unkeless JC, Gordon S, Reich E. Secretion of plasminogen activator by stimulated macrophages. J Exp Med 1974; 139:834–850.
16. Wohlwend A, Belin D, Vassali J-D. Plasminogen activator-specific inhibitors produced by human monocytes/macrophages. J Exp Med 1987; 165:320–339.
17. Ossowski L, Biegel D, Reich E. Mammary plasminogen activator: correlation with involution hormonal modulation and comparison between normal and neoplastic tissue. Cell 1979; 16:929–940.
18. Reich R, Miskin R, Tsafriri A. Follicular plasminogen activator: involvement in ovulation. Endocrinology 1985; 116:516–521.
19. Strickland S, Reich E, Sherman MZ. Plasminogen activator in early embryogenesis: enzyme production by trophoblast and parietal endotherm. Cell 1976; 9:231–240.
20. Sappino AP, Huarte J, Belin D, Vassalli JD. Plasminogen activators in tissue remodeling and invasion: mRNA localization in mouse ovaries and implanting embryos. J Cell Biol 1989; 109:2471–2479.

21. Feinberg RF, Kao LC, Haimowitz JE, et al. Plasminogen activator inhibitor types 1 and 2 in human trophoblasts. PAI-1 is an immunocytochemical marker for invading trophoblasts. Lab Invest 1989; 61:20–26.
22. Gross JL, Moscatelli D, Rifkin DB. Increased capillary endothelial cell protease activity in response to angiogenic stimuli in vitro. Proc Natl Acad Sci USA 1983; 80:2623–2627.
23. Morioka S, Lazarus GS, Baird JL, Jensen PJ. Migrating keratinocytes express urokinase-type plasminogen activator. J Invest Dermatol 1991; 88:418–423.
24. Ossowski L, Reich E. Antibodies to plasminogen activator inhibit human tumor metastasis. Cell 1983; 35:611–619.
25. Hearing V, Law L, Corti A, Appella E, Blasi F. Modulation of metastatic potential by surface urokinase of murine melanoma cells. Cancer Res 1988; 48:1270–1278.
26. Mullin DE, Rohrlich ST. The role of proteinases in cellular invasiveness. Biochim Biophys Acta 1983; 695:177–214.
27. Edmonds-Alt X, Quisquarter E, Vaes G. Eur J Cancer 1980; 16:1257–1261.
28. Jilek F. Cold-insoluble globulin III. Cyanogen bromide and plasminolysis fragments containing a label introduced by transamidation. Hoppe Seylers Z Physiol Chem 1977; 358: 1165–1168.
29. Schlechte W, Murano G, Boyd D. Examination of the role of the urokinase receptor in human colon cancer mediated laminin degradation. Cancer Res 1989; 49:6064–6069.
30. Schlechte W, Brattain M, Boyd D. Invasion of extracellular matrix by cultured colon cancer cells: dependence on urokinase receptor display. Cancer Commun 1990; 2:173–179.
31. Mackay AR, Corbitt RH, Hartzler JL, Thorgeirsson UP. Basement membrane type IV collagen degradation: evidence for the involvement of a proteolytic cascade independent of metalloproteinases. Cancer Res 1990; 50:5997–6001.
32. Stricklin GP, Bauer EA, Jeffrey JJ, Eisen A. Human skin collagenase: isolation of precursor and active forms from both fibroblast and organ cultures. Biochemistry 1977; 16:1607–1615.
33. He C, Wilhelm SM, Pentland AP, et al. Tissue cooperation in a proteolytic cascade activating human interstitial collagenase. Proc Natl Acad Sci USA 1989; 86:2632–2636.
34. Brade V, Nicholson A, Bitter-Suermann D, Hadding V. Formation of the C-3 cleaving properdin enzyme on zymosen. J Immunol 1974; 113:1735–1743.
35. Cochrane CG, Revak SD, Wuepper WG. Activation of Hageman factor in solid and fluid phases. J Exp Med 1974; 138:1564–1583.
36. Virgi MAG, Vassalli JD, Estensen RD, Reich E. Plasminogen activator of islets of Langerhans: modulation by glucose and correlation with insulin production. Proc Natl Acad Sci, USA 1980; 77:875–879.
37. Habal FM, Burrowes CE, Movat HZ. Generation of kinin by plasma kallikrein and plasmin and the effect of α_1-antitrypsin and antithrombin III on the kininogenases. Adv Exp Med Biol 1976; 70:23–36.
38. Mirsky IA, Perisutti G, Davis NC. The destruction of glucagon, adrenocorticotropin and somatotropin by human blood plasma. J Clin Invest 1959; 38:14–20.
39. Janeway CA, Merler E, Rosen FS, Salmon S, Crain JO. Intravenous gamma-globulin: mechanism of gamma-globulin fragments in normal and agamma-globulinemic persons. N Engl J Med 1968; 278:919–923.
40. Pizzo SV, Schwartz ML, Hill RL, McKee PA. The effect of plasmin on the subunit structure of human fibrinogen. J Biol Chem 247:636–645.
41. Omar MN, Mann KG. Inactivation of factor Va by plasmin. J Biol Chem 1987; 262:9750–9755.
42. Kluft C, Wijngaards G, Jie AFH. The factor XII-independent plasminogen proactivator system includes urokinase-related activators. Thromb Haemost 1981; 46:343–350.
43. Ellis V, Behrendt N, Dano K. Plasminogen activation by receptor-bound urokinase. A kinetic study with both cell-associated and isolated receptor. J Biol Chem 1991; 266: 12572–12578.

44. Rijken DC, Wijngaards G, Welbergen J. Relationship between tissue plasminogen activator and activators in blood and vessel wall. Thromb Res 1980; 18:815–830.
45. Mandle RJ, Kaplan AP. Generation of fibrinolytic activity by the interaction of activated factor XI and plasminogen. Blood 1979; 54:850–861.
46. Goldsmith GH, Saito H, Ratnoff OD. The activation of plasminogen by Hageman factor (factor XII) and Hageman factor fragments. J Clin Invest 1978; 62:54–60.
47. Colman RW. Activation of plasminogen by human plasma kallikrein. Biochem Biophys Res Commun 1980; 35:273–279.
48. Kluft C, Dooijewaard G, Emeis JJ. Role of the contact system in fibrinolysis. Semin Thromb Hemost 1987; 13:50–68.
49. Ellis V, Scully MF, Kakkar VV. Plasminogen activation by single-chain urokinase in functional isolation. J Biol Chem 1987; 262:14998–15003.
50. Urano T, De Serrano VS, Gaffney PJ, Castellino FJ. The activation of human [Glu1] plasminogen by human single-chain urokinase. Arch Biochem Biophys 1988; 264:222–230.
51. Ichinose A, Fujikawa K, Suyama T. The activation of pro-urokinase by plasma kallikrein and its activation by thrombin. J Biol Chem 1986; 261:3486–3489.
52. Ogston D, Ogston DM, Ratnoff OD, Forbes CD. Studies on a complex mechanism for the activation of plasminogen by kaolin and by chloroform: the participation of Hageman factor and additional cofactors. J Clin Invest 1969; 48:1786–1801.
53. Binnema DJ, Dooijewaard G, van Iersel JJL, Turion PNC, Kluft C. The contact-system dependent plasminogen activator from human plasma: identification and characterization. Thromb Haemost 1990; 64:390–397.
54. Castellino FJ. Plasminogen activators. Bioscience 1983; 33:647–650.
55. Robbins KC, Summaria L, Hsieh B, Shah RJ. The peptide chains of human plasmin. Mechanism of activation of human plasminogen to plasmin. J Biol Chem 1967; 242:2333–2342.
56. Wiman B. Primary structure of peptides released during activation of human plasminogen by urokinase. Eur J Biochem 1973; 39:1–9.
57. Wiman B, Wallen P. Amino-acid sequence of the cyanogen-bromide fragment from human plasminogen that forms the linkage between the plasmin chains. Eur J Biochem 1975; 58: 539–547.
58. Wiman B, Collen D. Purification and characterization of human antiplasmin. The fast acting plasmin inhibitor in plasma. Eur J Biochem 1977; 78:19–26.
59. Sottrup-Jensen L, Claeys H, Zajdel M, Petersen TE, Magnusson S. The primary structure of human plasminogen: isolation of two lysine-binding fragments and one "mini" plasminogen (MW, 38000) by elastase-catalyzed-specific limited proteolysis. Prog Chem Fib Thrombol 1978; 3:191–209.
60. Malinowski DP, Sadler JE, Davie EW. Characterization of a complementary DNA coding for human and bovine plasminogen. Biochemistry 1984; 23:4243–4250.
61. Forsgren M, Raden B, Israelsson M, Larsson K, Heden L-O. Molecular cloning and characterization of a full-length cDNA clone for human plasminogen. FEBS Lett 1987; 213: 254–260.
62. Petersen TE, Martzen MR, Ichinose A, Davie EW. Characterization of the gene for human plasminogen, a key proenzyme in the fibrinolytic system. J Biol Chem 1990; 265:6104–6111.
63. Hayes ML, Castellino FJ. Carbohydrate of human plasminogen variants. II. Structure of the asparagine-linked oligosaccharide unit. J Biol Chem 1979; 254:8772–8776.
64. Hayes ML, Castellino FJ. Carbohydrate of human plasminogen variants. I. Carbohydrate composition and glycopeptide isolation and characterization. J Biol Chem 1979; 254:8768–8771.
65. Powell JR, Castellino FJ. Amino acid sequence analysis of the Asn288 region of the carbohydrate variants of human plasminogen. Biochemistry 1983; 22:923–927.

66. Hayes ML, Castellino FJ. Carbohydrate of human plasminogen variants. III. Structure of the O-glycosidically-linked oligosaccharide unit. J Biol Chem 1979; 254:8777–8780.
67. Brockway WJ, Castellino FJ. Measurement of the binding of antifibrinolytic amino acids to various plasminogens. Arch Biochem Biophys 1972; 151:194–199.
68. Violand BN, Castellino FJ. Mechanism of urokinase-catalyzed activation of human plasminogen. J Biol Chem 1976; 251:3906–3912.
69. Violand BN, Sodetz JM, Castellino FJ. The effect of epsilon amino caproic acid on the gross conformation of plasminogen and plasmin. Arch Biochem Biophys 1975; 170:300–305.
70. Claeys H, Vermylen J. Physicochemical and proenzyme properties of amino-terminal glutamic acid and amino-terminal lysine human plasminogen. Biochim Biophys Acta 1974; 342:351–359.
71. Violand BN, Byrne R, Castellino FJ. The effect of α-ω-amino acids on human plasminogen structure and activation. J Biol Chem 1978; 253:5395–5401.
72. McMullen BA, Fujikawa K. Amino acid sequence of the heavy chain of human a-factor XIIa (activated Hageman factor). J Biol Chem 1985; 260:5328–5341.
73. Magnusson S, Petersen TE, Sottrup-Jensen L, Claeys H. Complete primary structure of prothrombin: isolation and reactivity of ten carboxylated glutamic residues and regulation of prothrombin activation by thrombin. In: Reich E, Rifkin DB, Shaw E, eds. Proteases and Biological Control. New York: Cold Spring Harbor Laboratories, 1975:123–149.
74. Pennica D, Holmes WE, Kohr WJ, et al. Cloning and expression of human tissue-type plasminogen activator cDNA in *E. coli*. Nature 1983; 301:214–221.
75. Steffens GJ, Gunzler WA, Otting F, Frankus E, Flohe L. The complete amino acid sequence of low molecular mass urokinase from human urine. Hoppe Seylers Z Physiol Chem 1982; 363:1043–1058.
76. Urano T, Chibber BAK, Castellino FJ. The reciprocal effects of ϵ-aminohexanoic acid and chloride ion on the activation of human [Glu^1]plasminogen by human urokinase. Proc Natl Acad Sci USA 1987; 84:4031–4034.
77. Urano T, De Serrano VS, Chibber BAK, Castellino FJ. The control of the urokinase-catalyzed activation of human glutamic acid 1-plasminogen by positive and negative effectors. J Biol Chem 1987; 262:15959–15964.
78. Urano T, de Serrano VS, Gaffney PJ, Castellino FJ. Effectors of the activation of human [Glu^1]plasminogen by human tissue plasminogen activator. Biochemistry 1988; 27:6522–6528.
79. Gonzalez-Gronow M, Violand BN, Castellino FJ. Purification and some properties of the Glu^- and Lys^- human plasmin heavy chains. J Biol Chem 1977; 252:2175–2177.
80. Hoylaerts M, Rijken DC, Lijnen HR, Collen D. Kinetics of the activation of plasminogen by human tissue plasminogen activator. Role of fibrin. J Biol Chem 1982; 257:2912–2919.
81. Chibber BAK, Morris JP, Castellino FJ. Effects of human fibrinogen and its cleavage products on activation of human plasminogen by streptokinase. Biochemistry 1985; 24: 3429–3434.
82. Chibber BAK, Castellino FJ. Regulation of the streptokinase-mediated activation of human plasminogen by fibrinogen and chloride ions. J Biol Chem 1986; 261:5289–5295.
83. Chibber BAK, Radek JT, Morris JP, Castellino FJ. Rapid formation of an anion sensitive active site in stoichiometric complexes of streptokinase and human [Glu1]plasminogen. Proc Natl Acad Sci USA 1986; 83:1237–1241.
84. Davidson DJ, Higgins DL, Castellino FJ. Plasminogen activator activities of equimolar complexes of streptokinase with variant recombinant plasminogens. Biochemistry 1990; 29:3585–3590.
85. Lerch PG, Rickli EE, Lergier W, Gillessen D. Localization of the lysine-binding regions in human plasminogen and investigations of their complex-forming properties. Eur J Biochem 1980; 107:7–13.

86. Lerch PG, Rickli EE. Studies on the chemical nature of lysine-binding sites and on their localization in human plasminogen. Biochim Biophys Acta 1980; 625:374–378.
87. Menhart N, Sehl LC, Kelley RF, Castellino FJ. Construction, expression and purification of recombinant kringle 1 of human plasminogen and analysis of its interaction with ω-amino acids. Biochemistry 1991; 30:1948–1957.
88. De Marco A, Petros AM, Laursen RA, Llinas M. Analysis of ligand-binding to the kringle 4 fragment from human plasminogen. Eur Biophys J 1987; 14:359–368.
89. Novokhatny VV, Matsuka YV, Kudinov SA. Analysis of ligand binding to kringles 4 and 5 fragments of human plasminogen. Thromb Res 1989; 53:243–252.
90. Sehl LC, Castellino FJ. Thermodynamic properties of the binding of α-, ω-amino acids to the isolated kringle 4 region of human plasminogen as determined by high sensitivity titration calorimetry. J Biol Chem 1990; 265:5482–5486.
91. Sehl LC. Thermodynamic and structural considerations for the interaction of kringle domains with ω-amino acids. Ph.D. dissertation, University of Notre Dame, 1991.
92. Thewes T, Constantine K, Byeon I-JL, Llinas M. Ligand interactions with the kringle 5 domain of plasminogen. A study by ^{1}H NMR spectroscopy. J Biol Chem 1990; 265:3906–3915.
93. Trexler M, Vali Z, Patthy L. Structure of the ω-aminocarboxylic acid-binding sites of human plasminogen. Arginine 70 and aspartic acid 56 are essential for binding of ligand by kringle 4. J Biol Chem 1982; 257:7401–7406.
94. Vali Z, Patthy L. The fibrin-binding site of human plasminogen. Arginines 32 and 34 are essential for fibrin affinity of the kringle 1 domain. J Biol Chem 1984; 259:13690–13694.
95. Trexler M, Banyai L, Patthy L, Pluck ND, Williams RJP. Chemical modification and nuclear magnetic resonance studies of human plasminogen kringle 4—assignment of tyrosine and histidine resonances to specific residues in the sequence. Eur J Biochem 1985; 152: 439–446.
96. Weening-Verhoeff EJD, Quax PHA, van Leeuwen RTJ, Rehberg EF, Mariotti KR, Verheijen JH. Involvement of aspartic and glutamic residues in kringle-2 of tissue-type plasminogen activator in lysine binding, fibrin binding and stimulation of activity as revealed by chemical modification and oligonucleotide-directed mutagenesis. Protein Eng 1990; 4:191–198.
97. Llinas M, Motta A, De Marco A, Laursen RA. Kringle 4 from human plasminogen: ^{1}H-NMR study of the interaction between ω-amino acid ligands and aromatic residues at the lysine binding site. J Biosci 1985; 8:121–129.
98. De Marco A, Motta A, Llinas M, Laursen RA. ^{1}H-NMR spectroscopic manifestations of ligand-binding to the kringle 4 domain of human plasminogen. Arch Biochem Biophys 1986; 244:727–741.
99. Petros AM, Ramesh V, Llinas M. ^{1}H NMR studies of aliphatic ligand binding to human plasminogen kringle 4. Biochemistry 1989; 28:1368–1376.
100. Mulichak AM, Park CH, Tulinsky A, Petros AM, Llinas M. Human plasminogen kringle 4. Crystallization and preliminary diffraction data of two different crystal forms. J Biol Chem 1989; 264:1922–1923.
101. Mulichak AM, Tulinsky A. Structure of the lysine-fibrin subsite of human plasminogen kringle 4. Blood Coag Fib 1990; 1:673–679.
102. Mulichak AM, Tulinsky A, Ravichandran KG. Crystal and molecular structure of human plasminogen kringle 4 refined at 1.9-Å resolution. Biochemistry 1991; 30:10576–10588.
103. De Vos AM, Ultsch MH, Kelley RF, et al. The crystal structure of the kringle-2 domain of tissue-plasminogen activator at 2.43 Å resolution. Biochemistry 1991; 31:270–279.
104. Wu T-P, Padmanabhan K, Tulinsky A, Mulichak AM. The refined structure of the ϵ-aminocaproic acid complex of human plasminogen kringle 4. Biochemistry 1991; 30: 10589–10594.
105. Kelley RF, Cleary S. Effect of residue 65 substitutions on thermal stability of tissue plasminogen activator kringle-2 domain. Biochemistry 1989; 28:4047–4054.

106. De Serrano VS, Castellino FJ. The role of tryptophan-74 of the recombinant kringle 2 domain of tissue-type plasminogen activator in its ω-amino acid binding properties. Biochemistry 1992; 31:3326–3335.
107. De Serrano VS, Menhart N, Castellino FJ. The expression, purification and characterization of the recombinant kringle 1 domain from tissue-type plasminogen activator. Arch Biochem Biophys 1992; 294:282–290.
108. De Serrano VS, Sehl LC, Castellino FJ. Direct identification of lysine-33 as the sole cationic center of the ω-amino acid binding site of the recombinant kringle 2 domain of tissue-type plasminogen activator. Arch Biochem Biophys 1992; 292:206–212.
109. De Serrano VS, Castellino FJ. The cationic locus on the recombinant kringle 2 domain of tissue-type plasminogen activator that stabilizes its interaction with ω-amino acids. Biochemistry 1992; 31:11698–11706.
110. De Serrano VS, Castellino FJ. The specific anionic residues of the recombinant kringle 2 domain of tissue-type plasminogen activator that are responsible for stabilization of its interaction with ω-amino acid ligands. Biochemistry 1993; 32:3540–3547.
111. Degen SJF, Bell SM, Schaefer LA, Elliott RW. Characterization of the cDNA coding for mouse plasminogen and localization of the gene to mouse chromosome 17. Genomics 1990; 8:49–61.
112. Menhart N, Hoover GJ, Castellino FJ. Unpublished observations, 1993.
113. De Marco A, Pluck ND, Banyai L, et al. Analysis and identification of aromatic signals in the proton magnetic resonance spectrum of the kringle 4 fragment from human plasminogen. Biochemistry 1985; 24:748–753.
114. De Marco A, Petros AM, Llinas M, Kaptein R, Boelens R. Ligand-binding effects on the kringle 4 domain from human plasminogen: a study by laser photo-CIDNP ^{1}H-NMR spectroscopy. Biochim Biophys Acta 1989; 994:121–137.
115. Petros AM, Gyenes M, Patthy L, Llinas M. Analysis of the aromatic ^{1}H NMR spectrum of chicken plasminogen kringle 4. Arch Biochem Biophys 1988; 264:192–202.
116. Bok RA, Mangel WF. Quantitative characterization of the binding of plasminogen to intact fibrin clots, lysine-Sepharose, and fibrin cleaved by plasmin. Biochemistry 1985; 24:3279–3286.
117. Wiman B, Wallen P. The specific interaction between plasminogen and fibrin. A physical role of the lysine binding site in plasminogen. Thromb Res 1977; 10:213–222.
118. Lijnen HR, Hoylaerts M, Collen D. Isolation and characterization of a human plasma protein with affinity for the lysine binding sites in plasminogen. Role in the regulation of fibrinolysis and identification as a histidine-rich glycoprotein. J Biol Chem 1980; 255: 10214–10222.
119. Moroi M, Aoki N. Isolation and characterization of alpha$_2$-plasmin inhibitor from human plasma. A novel proteinase inhibitor which inhibits activator-induced clot lysis. J Biol Chem 1976; 251:5956–5965.
120. Varadi A, Patthy L. Location of the plasminogen-binding site in human fibrin(ogen). Biochemistry 1984; 22:2440–2446.
121. Hortin GL, Trimpe BL, Fok KF. Plasmin's peptide-binding specificity: Characterization of ligand sites in α_2-antiplasmin. Thromb Res 1989; 54:621–632.
122. McLean JW, Tomlinson JE, Kuang W-J, et al. cDNA sequence of human apolipoprotein(a) is homologous to plasminogen. Nature 1987; 330:132–137.
123. Murray JC, Buetow KH, Donovan M, et al. Linkage disequilibrium of plasminogen polymorphisms and assignment of the gene to human chromosome 6q26–6q27. Am J Hum Genet 1987; 40:338–350.
124. Sharp PA. Speculations on RNA splicing. Cell 1981; 23:643–646.
125. Breathnach R, Benoist C, O'Hare K, Gannon F, Chambon P. Ovalbumin gene: evidence for a leader sequence in mRNA and DNA sequences at the exon-intron boundaries. Proc Natl Acad Sci 1978; 75:4853–4857.

126. Mount SM. A catalogue of splice junction sequences. Nucleic Acids Res 1982; 10:459–472.
127. Gilbert W. Why genes in pieces? Nature 1978; 271:501.
128. Breathnach R, Chambon P. Organization and expression of eukaryotic split genes coding for proteins. Annu Rev Biochem 1981; 50:349–383.
129. Fowlkes DM, Mullis MT, Comeau CM, Crabtree GR. Potential basis for regulation of the coordinately expressed fibrinogen genes: homology in the 5′ flanking regions. Proc Natl Acad Sci USA 1984; 81:2313–2316.
130. Maeda M. Nucleotide sequence of the haptoglobin and haptoglobin-related gene pair. J Biol Chem 1985; 260:6698–6709.
131. Long GL, Chandra T, Woo SLC, Davie EW, Kurachi K. Complete sequence of the cDNA for human alpha 1-antitrypsin and the gene for the S variant. Biochemistry 1984; 23:4828–4837.
132. Adrian GS, Korinek BW, Bowman GH, Yang F. The human transferrin gene: 5′ region contains conserved sequences which match the control elements regulated by heavy metals, glucocorticoids and acute phase reaction. Gene 1986; 49:167–175.
133. Berget SM. Are U4 small nuclear ribonucleoproteins involved in polyadenylation. Nature 1984; 309:179–182.
134. McLauchlan J, Gaffney D, Whitton JL, Clements JB. The consensus sequence YGTGTTYY located downstream from the AATAAA signal is required for efficient formation of mRNA 3′ termini. Nucleic Acids Res 1985; 13:1347–1368.
135. Heberlein PJ, Barnhart MI. Canine plasminogen: purification and demonstration of multimolecular forms. Biochim Biophys Acta 1968; 168:195–206.
136. Sodetz JM, Brockway WJ, Castellino FJ. Multiplicity of rabbit plasminogen. Physical characterization. Biochemistry 1972; 11:4451–4458.
137. Wallen P, Wiman B. Characterization of human plasminogen. II. Separation and partial characterization of different molecular forms of human plasminogen. Biochim Biophys Acta 1972; 257:122–134.
138. Summaria L, Arzadon L, Bernabe P, Robbins KC. Studies on the isolation of the multiple molecular forms of human plasminogen and plasmin by isoelectric focusing methods. J Biol Chem 1972; 247:6757–6762.
139. Siefring GE, Castellino FJ. The role of sialic acid in the distinct properties of the isozymes of rabbit plasminogen. J Biol Chem 1974; 249:7742–7746.
140. Raum D, Marcus A, Alper CA. Genetic polymorphism of human plasminogen. Am J Hum Genet 1980; 32:681–689.
141. Hobart MJ. Genetic polymorphism of human plasminogen. Ann Hum Genet 1979; 42:419–423.
142. Kuhnl P, Spielmann W. Erfahrungen mit neueren Blutgrupensystemen fur die Vaterschaftbegutachtung: FXIIA and FXIIIB, PLGN, AMY2, GAA, GDH, PGM1-Thermostabilitatsvarianten. Arztl Lab 1982; 28:367–377.
143. Nakamura S, Abe K. Genetic polymorphism of human plasminogen in the Japanese population: new plasminogen variants and relationship between plasminogen phenotypes and their biological activities. Hum Genet 1982; 60:57–59.
144. Nishigaki T, Omoto K. Genetic polymorphism of human plasminogen in Japanese: correspondence of alleles thus far reported in Japanese and difference of activity among phenotypes. Jpn J Hum Genet 1982; 27:341–348.
145. Nishimukai H, Kera Y, Sakata K, Yamasawa K. Three new variants in the plasminogen system. Hum Hered 1982; 32:130–132.
146. Bissbort S, Bender K, Mayerova A, Wienker TF, Mauff G. Genetic linkage relations of the human plasminogen gene. Hum Genet 1983; 63:126–131.
147. Skoda U, Bertrams J, Dykes D, et al. Proposal for the nomenclature of human plasminogen (Plg) polymorphism. Vox Sang 1986; 51:244–248.

148. Dykes D. Distribution of plasminogen allotypes in eight populations of the Western hemisphere. Electrophoresis 1983; 4:417–440.
149. Aoki N, Tateno K, Sakata Y. Differences of frequency distributions of plasminogen phenotypes between Japanese and American populations: new methods for the detection of plasminogen variants. Biochem Genet 1984; 22:871–881.
150. Skoda U, Klein A, Lubcke I, Mauff G, Pulverer G. Application of plasminogen polymorphism to forensic hemogenetics. Electrophoresis 1988; 9:422–426.
151. Nevo S, Cleve H, Koller A, et al. Serum protein polymorphisms in Arab Moslems and Druze of Israel—BF, F13B, AHSG, GC, PLG, PI, and TF. Hum Biol 1992; 64:587–603.
152. Aoki N, Moroi M, Sakata Y, Yoshisa N, Matsuda M. Abnormal plasminogen. A hereditary molecular abnormality found in a patient with recurrent thrombosis. J Clin Invest 1978; 61:1186–1195.
153. Sakata Y, Aoki N. Molecular abnormality of plasminogen. J Biol Chem 1980; 255:5442–5447.
154. Miyato T, Iwanaga S, Y. S, Aoki N. Plasminogen Tochigi: inactive plasmin resulting from replacement of alanine-600 by threonine at the active site. Proc Natl Acad Sci USA 1982; 79:6132–6136.
155. Ichinose A, Espling EE, Takamatsu J, et al. Two types of abnormal genes for plasminogen in families with a predisposition for thrombosis. Proc Natl Acad Sci USA 1991; 88:115–119.
156. Miyata T, Iwanaga S, Sakata Y, Aoki N, Takamatsu J, Kamiya T. Plasminogens Tochigi II and Nagoya: two additional molecular defects with Ala-600 → Thr replacement found in plasmin light chain variants. J Biochem 1984; 96:277–287.
157. Manabe S-I, Matsuda M. Homozygous protein C deficiency combined with heterozygous dysplasminogenemia found in a 21-year-old thrombophilic male. Thromb Res 1985; 39: 333–341.
158. Soria J, Soria C, Bertrand O, Dunn F, Drouet L, Caen J. Plasminogen Paris I: congenital abnormal plasminogen and its incidence in thrombosis. Thromb Res 1983; 32:229–238.
159. Azuma H, Uno Y, Shigekiyo T, Saito S. Congenital plasminogen deficiency caused by a Ser^{576} to Pro mutation. Blood 1993; 82:475–480.
160. Shigekiyo T, Uno Y, Tomonari A, et al. Type I congenital plasminogen deficiency is not a risk factor in thrombosis. Thromb Haemost 1992; 67:189.
161. Robbins KC. Dysplasminogenemias. Prog Cardiovasc Dis 1992; 34:295–308.
162. Whitefleet-Smith J, Rosen E, McLinden J, et al. Expression of human plasminogen cDNA in a baculovirus-infected insect cell system. Arch Biochem Biophys 1989; 271: 390–399.
163. Davidson DJ, Fraser MJ, Castellino FJ. Oligosaccharide processing in the expression of human plasminogen cDNA by lepidopteran insect (*Spodoptera frugiperda*) cells. Biochemistry 1990; 29:5584–5590.
164. Davidson DJ, Castellino FJ. Structures of the Asn289-linked oligosaccharides assembled on recombinant human plasminogen expressed in a *Mamestra brassicae* cell line (IZD-MBO503). Biochemistry 1991; 30:6689–6696.
165. Davidson DJ, Castellino FJ. Asparagine-linked oligosaccharide processing in lepidopteran insect cells. Temporal dependence of the nature of the oligosaccharides assembled on asparagine 289 of recombinant human plasminogen produced in baculovirus vector-infected *Spodoptera frugiperda* (IPLB-SF-21AE) cells. Biochemistry 1991; 30:6167–6174.
166. Davidson DJ, Castellino FJ. Oligosaccharide structures present on asparagine 289 of recombinant human plasminogen expressed in a Chinese hamster ovary cell line. Biochemistry 1991; 30:625–633.
167. Hink WF, Thomsen DR, Davidson DJ, Meyer AL, Castellino FJ. Expression of three recombinant proteins using baculovirus vectors in twenty-three insect cell lines. Biotechnol Prog 1991; 7:9–14.

168. Horrevoets AGJ, Fredenburgh JC, Nesheim ME, Pannekoek H. Vaccinia-virus produced mutants of human plasminogen to establish the significance of the conversion of Glu-1 plasminogen to Lys-77 plasminogen during fibrinolysis. Abstracts of the Third International Workshop on the Molecular and Cellular Biology of Plasminogen Activation. 1991:127.
169. Kotwal GJ, Blasco R, Isaacs S, Moss B. Serine protease inhibitors encoded by vaccinia virus inhibit tissue plasminogen activator activity. Abstracts of the Third International Workshop on the Molecular and Cellular Biology of Plasminogen Activation. 1991:34.
170. Mhashikar A, Viswanatha T, Chibber BAK, Castellino FJ. Breaching the conformational integrity of the catalytic triad of the serine protease plasmin: localized disruption of a side chain of His-603 strongly inhibits the amidolytic activity of human plasmin. Proc Natl Acad Sci USA 1993; 90:5374–5377.

25

Plasminogen Activator Inhibitors

Daniel A. Lawrence and David Ginsburg
University of Michigan,
Ann Arbor, Michigan

I. INTRODUCTION

A. Plasminogen Activators

Plasminogen activators (PAs) are specific serine proteases that activate the proenzyme plasminogen, by cleavage of a single Arg-Val peptide bond, to the broad-specificity enzyme plasmin (1,2). The study of plasminogen activation has a long and interesting history that was recently reviewed (3). Two plasminogen activators are found in mammals, tissue-type PA (tPA) and urokinase-type PA (uPA) (4). These enzymes are thought to influence critically many biological processes, including vascular fibrinolysis (5), ovulation (6,7), inflammation (8), tumor metastasis (9), angiogenesis (10), and tissue remodeling (Fig. 1) (4).

The regulation of PAs is a complex process that is controlled on many levels. The synthesis and release of PAs are governed by various hormones, growth factors, and cytokines (4,9). Following secretion, PA activity can be regulated both positively and negatively by a number of specific protein-protein interactions. Activity can be enhanced or concentrated by interactions with fibrin (11), the uPA receptor (12), the tPA receptor (13), or the plasminogen receptor (14). In contrast, PA activity can be downregulated by the presence of specific PA inhibitors (PAIs) (15,16) or by direct plasmin inhibition (17). In addition, PA activity is dependent on its location, or microenvironment, with significant differences observed in solution (as in circulating blood) or on a solid-phase support (as on the surface of a cell or in the extracellular matrix). The overall activity of the PA system is determined by the interactions among these various elements and the balance between the opposing activities of enzymes and inhibitors.

During the last 10 years, the recognition of PAIs as critical regulators of the PA system has gained broad acceptance. In 1968, Kawano and coworkers identified an inhibitor of

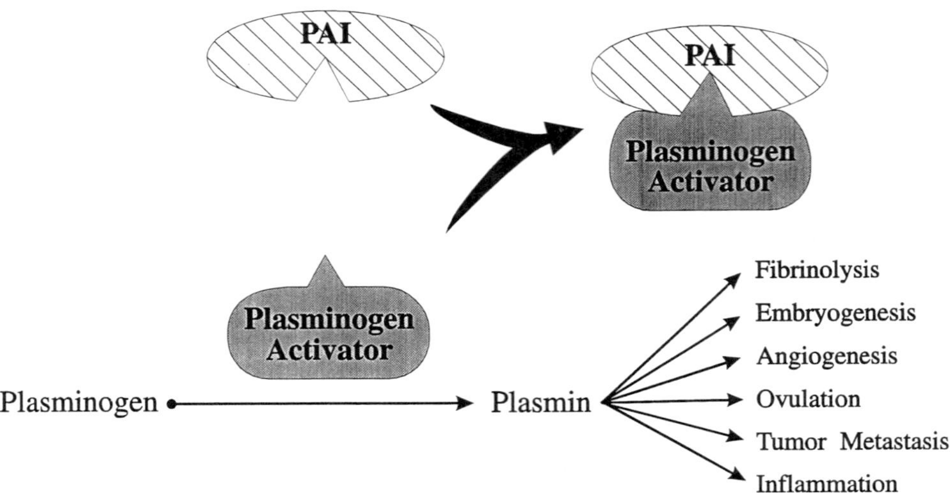

uPA in placental extracts that is generally acknowledged as the first unambiguous report of PAI activity (18). The identification of an efficient inhibitor of tPA in endothelial cells and plasma was first reported in 1983 (19–21). Four kinetically relevant PAIs are currently recognized: PAI type 1 (PAI-1), initially described as the endothelial cell PAI, PAI type 2 (PAI-2), also referred to as the placental PAI, PAI type 3 (PAI-3), also known as the activated protein C (APC) inhibitor (see Chap. 20), and protease nexin 1 (PN-1), also called glia-derived neurite-promoting factor (Table 1).

B. Serpins

All four of the major PAIs belong to the serine protease inhibitor superfamily (serpins) (22). This gene family includes many of the protease inhibitors found in blood, as well as other proteins with unrelated or unknown functions (23,24). The serpins are thought to share a common tertiary structure (25) and to have evolved from a common ancestor at least 500 million years ago (26). Proteins with recognizable sequence homology have been identified in vertebrates, plants, insects, and viruses but not, thus far, in prokaryotes (23,27–29). The evolutionary relationship among the serpins has been the subject of some debate (27,30–38). Of note, the reported serpin gene structures are very different from each other, and evolutionary relationships based on protein sequence similarity often contradict those based on gene structure. Several authors have suggested that phylogeny based on gene structure may give a better representation of relatedness than one based on protein sequence similarities (35,36).

1. Serpin Protein Structure

Current models of serpin structure are based largely on x-ray crystallographic studies of one member of the family, α 1-antitrypsin (α1AT), also called α1-protease inhibitor (reviewed in Ref. 24). The structure of a modified form of α1AT, cleaved in its reactive center, was solved by Loebermann and coworkers in 1984 (39). An interesting feature of this cleaved inhibitor is that the two amino acid residues that normally constitute the reactive center (Met-Ser bond), are found on opposite ends of the molecule, separated

Table 1 Summary of Characteristics for the Four PAIs[a]

PAI	Other Names	Sources	Second-order rate constants (M^{-1} S^{-1})	Other ligands
PAI-1	Endothelial PAI Platelet PAI Fast-acting PAI β-Migrating PAI	In vivo platelets (76) smooth muscle cells (79) LPS-activated endothelial cells (82) In vitro: many cells in culture (78)	uPA 9×10^6 (65) tctPA 2.7×10^7 (65) sctPA 4.5×10^6 (65) APC 1.1×10^4 (69) Plasmin 6.6×10^5 (66) Trypsin 7.0×10^6 (66) Thrombin 1.1×10^3–2×10^5 (70)	Vitronectin (84) Heparin (65) Fibrin (151)
PAI-2	Placental PAI	In vivo Trophoblast epithelium of placenta (178) Leukocytes (232) In Vitro Monocytes/macrophages (172) Some cell lines (171)	uPA 9×10^5 (173) tctPA 2.5×10^5 (173) sctPA 9×10^2 (173) Plasmin 1×10^2 (173)	Vitronectin (179)
PAI-3	Activated protein C inhibitor	In vivo: liver (202 In vitro: some hepatoma cell lines (202)	uPA 8×10^3–9×10^4 (205) tctPA $<10^3$ (205) APC 6.5×10^3–1.5×10^7 (207) Thrombin 4×10^4–2×10^5 (205) Factor XIa 9×10^4–7×10^5 (207) Factor Xa 2×10^4–9×10^4 (207) Kallikrein 6.5×10^4–2×10^5 (207)	Heparin (202)
PN-1	Glia-derived neurite promoting factor Glia-derived nexin	In vivo Brain (215) Olfactory system (225) In vitro: many cells in culture (218)	uPA 1.5×10^5 (214) tctPA 3.0×10^4 (214) sctPA 1.5×10^3 (214) Thrombin 8×10^5–1×10^9 (226) Plasmin 1.3×10^5 (214) Factor Xa 7.3×10^3 (214) Trypsin 4.2×10^6 (214) Mouse γ NGF 4.0×10^3 (214)	Heparin (219) Vitronectin (229)

[a] LPS, lipopolysaccharide; sctPA, single-chain tPA; tctPA, two-chain tPA; APC, activated protein C; NGF, nerve growth factor.

by almost 70 Å (Fig. 2). Loebermann and coworkers proposed that relaxation of a strained configuration takes place upon cleavage of the reactive-center peptide bond, rather than a major rearrangement of the inhibitor structure. In this model, the reactive center is part of an exposed loop, also called the strained loop (22,39). Upon cleavage this loop moves or "snaps back," becoming one of the central strands in a major β sheet structure as seen in the crystal structure. This transformation is accompanied by a large increase in thermal stability, presumably as a result of the reconstitution of the six-stranded β sheet A present in the α1AT crystal structure (41–42).

In native serpins the reactive-center loop may be analogous to the reactive-site loop of the small Kazal- and Kunitz-type inhibitors (43–46). Support for this hypothesis has come from the structure of the noninhibitory serpin, ovalbumin. The overall structure of ovalbumin is very similar to that of cleaved α1AT, but the region homologous to the reactive-center loop is exposed as an α helix above the plane of the molecule (Fig. 2) (47). In addition, several groups recently showed that synthetic peptides homologous to the reactive-center loops of α1AT and antithrombin III, when added in trans, incorporate into their respective molecules, presumably as a central strand of the major β sheet structure. This results in an increase in thermal stability of the inhibitor molecule similar to that observed following cleavage of a serpin at its reactive center. This structural change also converts the serpin from an inhibitor to a substrate for its target protease (48–50). A proposed model explaining these results suggests that active serpins have mobile reactive-center loops that can partially insert into β sheet A (49). This partial

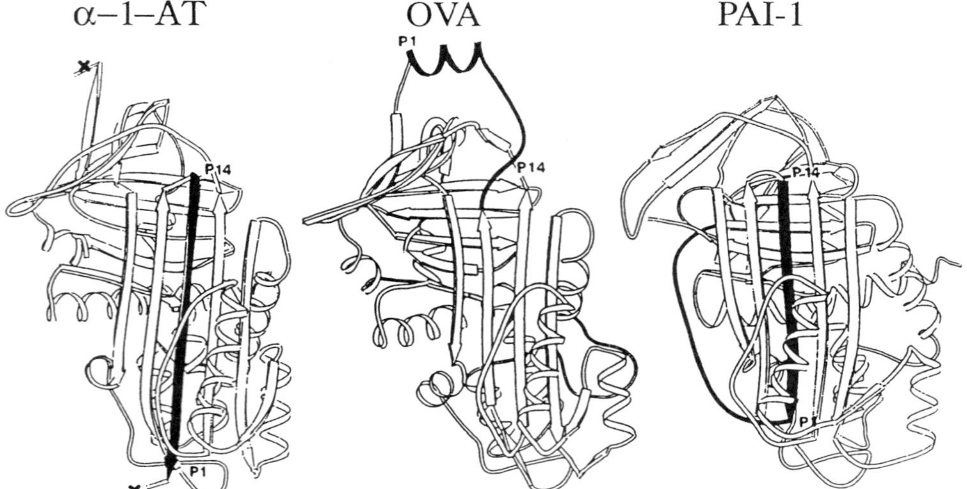

Figure 2 Backbone conformations of α_1-antitrypsin (α-1-AT), ovalbumin (OVA), and latent PAI-1. The residues corresponding to the reactive-center loop from P14 to P1 in each protein are shown as black ribbons. α1AT is shown in the relaxed conformation with the reactive-center P14–P1 residues fully inserted as the central strand in β sheet A. Ovalbumin is shown with the loop fully removed from β sheet A. The PAI-1 structure is the latent conformation, with the intact loop fully inserted and residues to the C terminus of the reactive center (up to approximately P10′) also shown as a black ribbon. The approximate position of the P1 and P1′ residues in α1AT are marked by an X. The P1 and P14 residues are labeled in each structure. (Adapted from S. R. Sprang (231), with permission.)

insertion results in the observed thermal instability of serpins but is necessary for function. Further insertion yields a latent inhibitor that is no longer reactive with the protease but has an increased thermal stability (51). Complete removal of the loop, as in ovalbumin (Fig. 2), or when a synthetic loop peptide is bound, also results in an increase in thermal stability but converts the inhibitor to a substrate.

2. *Mechanism of Inhibition*

The serpins act as so-called suicide inhibitors that react only once with their target protease to form sodium dodecyl sulfate (SDS)-stable complexes. Under certain conditions, such as hydroxylamine treatment at high pH, these complexes can dissociate to yield the free active enzyme together with a cleaved inhibitor similar to that seen in the α1AT crystal structure (52–55). The serpins interact with their target protease by providing a so-called bait amino acid residue in the reactive center. This bait residue is thought to mimic the normal substrate of the enzyme and to associate via its side-chain atoms with the specificity crevice, or S1 site, of the enzyme (24,55–58). The bait amino acid is called the P1 residue, with the amino acids toward the amino-terminal side of the scissile reactive-center bond labeled in order P1, P2, P3, and so on, and the amino acids on the carboxyl side labeled P1′, P2′, and so on (Fig. 3) (55).

The exact configuration of the complex between serpins and their target proteases is not known. However, some evidence suggests that the complex is covalently linked via an ester bond between the active-site serine residue of the protease and the new carboxyl-terminal end of the P1 residue, forming an acyl-enzyme complex (52,54,55,59). It is also possible that the complex is frozen in a transition state configuration and that the acyl-enzyme forms only when the enzyme-inhibitor complex is denatured or in some other way disturbed (55,60). Regardless of which model is correct, it is clear that the associations between the inhibitors and proteases also involve other regions, in addition to the P1 residue. This was demonstrated by the characterization of two recombinant PA mutants that have either six or seven amino acids in their catalytic domains deleted. These residues are located some distance from the active sites of the enzymes and are not involved in catalysis, because both mutants are still efficient activators of plasminogen. However, the mutant PAs were almost completely refractory to inhibition by PAI-1, suggesting that these residues distant from the active site are critical for the interaction with PAI-1 (61,62). Interestingly, despite these dramatic results, reciprocal mutations at the predicted sites of interaction in PAI-1 (P4′-P5′) had only very small effects on the rate of tPA inhibition (63).

PAI — REACTIVE CENTER

	P15	P14	P13	P12	P11	P10	P9	P8	P7	P6	P5	P4	P3	P2	P1	P1'	P2'	P3'	P4'	P5'
PAI-1	Gly	Thr	Val	Ala	Ser	Ser	Ser	Thr	Ala	Val	Ile	Val	Ser	Ala	Arg	Met	Ala	Pro	Glu	Glu
PAI-2	Gly	Thr	Glu	Ala	Ala	Ala	Gly	Thr	Gly	Gly	Val	Met	Thr	Gly	Arg	Thr	Gly	His	Gly	Gly
PAI-3	Gly	Thr	Arg	Ala	Ala	Ala	Ala	Thr	Gly	Thr	Ile	Phe	Thr	Phe	Arg	Ser	Ala	Arg	Leu	Asn
PN-1	Gly	Thr	Lys	Ala	Ser	Ala	Ala	Thr	Thr	Ala	Ile	Leu	Ile	Ala	Arg	Ser	Ser	Pro	Pro	Trp

Figure 3 Alignment of the reactive-center loop regions of the four PAIs. The amino acid sequence is given in the three-letter code, and the reactive-center scissile bond is indicated by an arrow.

II. PLASMINOGEN ACTIVATOR INHIBITOR TYPE 1

A. Background

PAI-1, formerly called the endothelial cell PAI, the fast-acting PAI, and the β-migrating PAI, is thought to be one of the principal regulators of the PA system. It is a single-chain glycoprotein with a molecular weight of 50 kDa (64) and is the most efficient inhibitor known of the single- and two-chain forms of tPA and of uPA, with second-order rate constants ranging between 0.45 and 2.7×10^7 M^{-1} s^{-1} (65). PAI-1 is also reported to inhibit plasmin and trypsin, with second-order rate constants of about 10^6 M^{-1} s^{-1} (66), as well as thrombin and activated protein C. However, these latter reactions are much less efficient, ranging from 10^3 to 10^5 M^{-1} s^{-1} (67–70). PAI-1 is present in plasma at very low concentrations, ranging from 0 to 60 ng/ml, with an average level of about 20 ng/ml (0.5 nM) (71,72) and a reported half-life of approximately 6–7 minutes (73,74). In another study comparing the clearance of two distinct forms of PAI-1 (active and latent; see later), the active form was cleared in a biphasic manner, with half-lives of 6 and 25 minutes, but latent PAI-1 was cleared with a half-life of only 1.7 minutes (75). PAI-1 is present in platelets and many other tissues and is produced by many cells in culture (76–78). In vivo the primary extravascular source of PAI-1 appears to be vascular smooth muscle cells (79). In response to endotoxemia or other pathological conditions, however, endothelial cells become a major site of PAI-1 synthesis (80–82). In plasma PAI-1 is present as a complex with vitronectin or S protein (83–85). PAI-1 is also associated with vitronectin in the extracellular matrix in culture and may be involved in maintaining the integrity of the cell substratum in vivo (86–94). The major source of plasma PAI-1 is not known but is likely to be the vascular smooth muscle cell. However, a contribution from the platelet pool cannot be excluded. PAI-1 functions efficiently in solution as well as when bound to surfaces, and it is likely that it regulates fibrinolysis in both environments.

B. The PAI-1 Gene

The gene for human PAI-1 is located on chromosome 7 bands q21.3–22 (95). It is 12.3 kb in length, composed of nine exons separated by eight introns, and is similar in structure to protease nexin 1 (30,34,36,96–98). In humans, alternative polyadenylation sites result in two different mRNA species of 2.3 and 3.1 kb, differing only in the length of their 3′-untranslated sequences (99–101). Although two mRNAs of similar sizes have been observed in other primate species (36), only a single 3.1 kb message has been detected in nonprimates (bovine, rat, and porcine) (36,102,103). Trace quantities of the smaller mRNA were noted in murine fibroblasts by one group (104,105), although we have observed exclusively the larger mRNA in murine cells (Datta, Yang, and Ginsburg unpublished data). It has been suggested that an AT-rich sequence present only in the larger message may have a regulatory function (100). Consistent with this hypothesis, insulin and insulin-like growth factor type I were recently shown to increase PAI-1 mRNA levels without affecting PAI-1 gene transcription, presumably via an increase in PAI-1 mRNA stability (106). Interestingly, although IGF-1 stabilized both messages, insulin appeared specifically to stabilize the 3.1 kb mRNA. In another study, specific induction of the 3.1 kb mRNA by cycloheximide was reported (107). Whether these effects were mediated through the AT-rich sequence was not shown.

PAI-1 is a highly regulated gene, with its synthesis and release in cell culture enhanced or depressed by such agents as endotoxin, serum, transforming growth factor β, basic fibroblast growth factor, interleukin-1, tumor necrosis factor, thrombin, dexamethasone, hydrocortisone, and phorbol esters (78,108). A more detailed discussion of PAI-1 gene expression is beyond the scope of this review; however, this subject was recently reviewed extensively (79,109,110).

C. PAI-1 Protein Structure and Function

The PAI-1 cDNA encodes a protein of 402 amino acids that includes a typical secretion signal sequence (99–101). Mature human PAI-1 isolated from cell culture is composed of two variants of 381 and 379 amino acids in approximately equal proportions. The two forms of PAI-1 likely arise from alternative cleavage of the secretion signal sequence and give rise to proteins with overlapping amino-terminal sequences of Ser-Ala-Val-His-His and Val-His-His-Pro-Pro (65,111). This latter sequence is generally referred to as mature PAI-1. PAI-1 is a glycoprotein with three potential N-linked glycosylation sites containing between 15 and 20% carbohydrate (64,99–101). In addition, mature PAI-1 contains no cysteine residues: the single cysteine present in the signal peptide is removed during membrane translocation (100,101). The lack of cysteines in PAI-1 has facilitated the efficient expression and isolation of recombinant PAI-1 from *Escherichia coli* (58,65,112–115). PAI-1 produced in *E. coli*, although nonglycosylated, is functionally very similar to native PAI-1 (65). In contrast to PAI-1 purified from mammalian cell culture (65,66,116–120), however, several groups have shown that recombinant PAI-1 can be isolated from *E. coli* in an inherently active form (see later) (58,65,113,115).

1. *Active and Latent Conformation*

PAI-1 exists in at least two conformations, an active form that is produced by cells and secreted into the culture medium and an inactive or latent form with an apparently smaller Stokes radius that accumulates in the media over time (121,122). The active form spontaneously coverts to the latent form with a half-life of about 1 h at 37°C at neutral or slightly alkaline pH (65,66,122,123). The latent form can be converted into the active form by treatment with denaturants, negatively charged phospholipids, or vitronectin, although the last reaction is very slow (118,121,124). In one report, analysis of latent PAI-1 infused into rabbits suggested that latent PAI-1 is also physiologically reactivated in vivo, although the exact mechanism remains unknown (74). This reversible interconversion between the active and latent structures, presumably on the basis of a conformational change, is a unique feature of PAI-1 that distinguishes it from other serpins. The latent form appears to be more energetically favorable, because the conversion to latent is spontaneous and the conversion from latent to active requires more extreme treatment (121,125). Recently, the three-dimensional structure of the latent form of PAI-1 was solved (Fig. 2). In this structure the entire amino-terminal side of the reactive-center loop is inserted as the central strand into β sheet A (51). This accounts for the increased stability of latent PAI-1 (1126), as well as the lack of inhibitory activity. The structure of active PAI-1 is still unknown. It has been proposed that the reactive center in active PAI-1 is exposed as a surface loop, in contrast to its position in the latent structure (see discussion of serpin structure).

2. *Oxidative Inactivation and Other Conformations*

In addition to the latent form of PAI-1, a second inactive form has also been identified that results from oxidation of one or more critical methionine residues within active

PAI-1 (127). This form differs from latent PAI-1 in that it can be partially reactivated by treatment with an enzyme that specifically reduces oxidized methionine residues. Circular dichroism analysis of oxidized PAI-1 indicates that the inactivation is correlated with a rapid conformational change to a structure that is distinct from both active and latent PAI-1 (128). Oxidative inactivation of PAI-1 may be an important mechanism for the regulation of the PA system. Oxygen radicals produced locally by neutrophils or other cells could inactivate PAI-1 and thus allow the generation of plasmin activity at sites of infection or in areas of tissue remodeling (129). Finally, a fourth conformational form of PAI-1 was recently reported that acts as a noninhibitory substrate (130,131). This conformation can be induced by the addition of SDS and can be converted back to the active or latent forms by treatment with 4 M guanidine-HCl. As a result of the unique labile structure of PAI-1, immunological methods for determining PAI-1 concentrations can vary by more than 10-fold, depending on the specific mix of conformations present and the relative specificities of the indicator antibodies. Thus, PAI-1 antigen measurements should be interpreted with caution (132).

3. *Interaction with Vitronectin*

As mentioned, PAI-1 in plasma or in the subcellular matrix is stabilized by vitronectin. Vitronectin-bound PAI-1 in solution is approximately twice as stable as unbound PAI-1 (84), and on extracellular matrix the half-life is reported to be greater than 24 h (87). Given the high concentration of vitronectin in plasma [200–400 μg/ml (133)] and the reported clearance half-life of PAI-1 in plasma of only 6–7 minutes (73–75), the conversion from active to latent PAI-1 seen in cell culture or with purified preparations may be an artifact of these in vitro systems. Consistent with this hypothesis, very little latent PAI-1 can be demonstrated in normal, fresh plasma. However, most of the PAI-1 found in platelets appears to be latent (71,72), although this point is controversial (69,134). Of note, platelets contain vitronectin (135), which could potentially function to reactivate latent platelet PAI-1 (124). However, this interaction has not yet been reported. Thus the physiological relevance of the large pool of latent PAI-1 within platelets [>90% of the total PAI-1 antigen in blood (71,72)] remains unclear. Recently, the effects of normal platelets and PAI-1-deficient platelets (see later) on clot lysis were compared in an in vitro assay. The results suggest that platelet PAI-1 may be a major factor in the resistance of platelet-rich thrombi to thrombolysis (136). Consistent with these observations, treatment of platelet-rich thrombi with anti-PAI-1 antibodies has also been shown to enhance clot lysis in vitro (137,138).

The interaction of PAI-1 with vitronectin has generated considerable debate. In one study only active PAI-1 was shown to bind vitronectin (139). In another study, however, no apparent difference in the binding was seen between active and latent PAI-1 (140). The reported dissociation constant is also controversial, one group reporting a K_d of 0.3 nM for active PAI-1 and vitronectin (141) and another reporting a major dissociation constant of 55–190 nM with a second, low-capacity but high-affinity binding site ($K_d <$ 0.1 nM) (142). Additional controversy surrounds the vitronectin binding site for PAI-1. One report, utilizing ligand blotting of vitronectin cyanogen bromide fragments, localizes the binding site to the somatomedin B domain at the N terminus of vitronectin (143). In contrast, a second study using monoclonal antibodies, localizes the PAI-1 binding site to the C terminus of vitronectin, between residues 348 and 370 (140). Some of these conflicting results may be explained by the interaction of both the active and latent forms of PAI-1 with vitronectin, but with markedly different affinities, along with differences in the relative composition of PAI-1 conformers present in alternative PAI-1 preparations. The localization of the vitronectin binding domain within PAI-1 has not been reported.

In addition to stabilizing active PAI-1, vitronectin has been shown to alter its specificity, converting it to an efficient inhibitor of thrombin (70,144). Vitronectin-bound PAI-1 has a 270-fold greater second-order rate constant toward thrombin than free PAI-1. However, this increase depends upon the source of vitronectin used. Although all forms of vitronectin appear to bind PAI-1, only vitronectin isolated under physiological conditions is able to stimulate the PAI-1 inhibition of thrombin (145). Although this enhanced rate of inhibition is similar to that of antithrombin III, in the absence of heparin, the concentration of antithrombin III in plasma is approximately 10,000 times greater than that of PAI-1. These observations suggest that PAI-1 does not contribute significantly to thrombin inhibition in plasma in vivo, although under some conditions local concentrations of PAI-1 may contribute a significant effect on thrombin. Vitronectin has also been shown to stimulate the inhibition of tPA by PAI-1, but to a much less dramatic extent (144,146). In other studies, reactive-center mutants of PAI-1 that have greatly reduced activity toward tPA were shown to have their activity partially restored in the presence of vitronectin (114).

4. *Binding to Heparin and Other Ligands*

PAI-1 also binds to heparin with high affinity, and this property has been used as a convenient method of purification (65,120). Heparin binding does not appear to affect the interaction of PAI-1 with uPA or tPA or to alter stability; however, an enhancement of the interaction with thrombin has been reported (144,147). The heparin binding domain on PAI-1 has been mapped to a region homologous to the heparin binding domain of antithrombin III, in and around α helix D (148). Critical residues appear to include lysines 65, 69, 80, and 88 arginine 76. It has also been reported that PAI-1 binds to fibrin in vitro, with a K_d of 3.8 μM and, while bound, remains able to inhibit uPA and tPA (138,149–151). PAI-1 has also been shown to interact with free arginine, resulting in increased stability (152). The significance in vivo of these interactions is unknown.

5. *Reactive-Center Loop*

The reactive-center region of PAI-1 has been the subject of extensive mutational analysis (56–59,63,70). These studies have generally demonstrated the importance of the P1 bait residue in inhibitor function, the surrounding amino acids playing a less critical role. York and coworkers and Sherman et al. performed random mutagenesis of the P3, P2, and P1 residues and the P1 and P1′ residues, respectively, and clearly demonstrated that either arginine or lysine at P1 is essential for PAI-1 to function as an effective inhibitor of uPA (57,58). These studies also demonstrated that the residues surrounding P1 can modulate PAI-1 inhibitor activity by up to two orders of magnitude and can also alter target protease specificity. P1′ was surprisingly tolerant of amino acid substitutions, with the exception of proline, which resulted in nearly total loss of function (58). This is most likely a result of a conformational change in the reactive-center scissile bond. In another study an 18 amino acid segment of PAI-1, encompassing most of the reactive-center loop, was replaced with the same region from PAI-2, antithrombin III, or a serpin consensus sequence (59). This report demonstrated that most of the requirements for PAI-1 specificity, apart from the P1 residue, lie outside the reactive-center loop sequence. All three chimeras remained very efficient inhibitors of tPA and uPA, and the antithrombin III chimera was not significantly improved as an inhibitor of thrombin. Furthermore, it was shown that the specific sequence of the reactive-center loop, the region inserted into β sheet A in the latent PAI-1 structure (Fig. 2; see preceding discussion), was not critical for the conversion between the active and latent conformations of PAI-1 because all the inhibitors had similar half-lives. This observation suggests that loop insertion is dependent more on the flexibility of β sheet A than on the specific amino acid residues in the loop. Finally, this study also demonstrated that binding to vitronectin was not affected by these substitutions in the reactive-center loop. The P4′ and P5′ residues on the C-terminal side of the reactive-center bond have also been replaced; however, these changes had only a small effect on PAI-1 activity (see preceding discussion of serpin function) (63).

D. Clinical Significance

Increased levels of PAI-1 in the circulation are associated with thrombotic disease, including myocardial infarction and deep vein thrombosis (153–159). In addition, elevated preoperative levels of PAI-1 have been correlated with postoperative deep vein thrombosis (160–162). Reduced postoperative fibrinolytic activity has also been correlated with an increase in PAI-1 activity immediately following surgery (163). This latter elevation seems to be mediated by a plasma factor that stimulates PAI-1 production and secretion from vascular endothelial cells (164). Consistent with these observations, the overproduction of PAI-1 in transgenic mice has been shown to result in venous thrombosis, primarily in the extremities (165). In contrast to these reports, a recent prospective study found no correlation between PAI-1 levels and vascular disease (166).

Three cases of either partial or complete PAI-1 deficiency have been reported in humans, and all were associated with abnormal bleeding. In one case, normal PAI-1 antigen was detected but PAI-1 activity was significantly reduced (167), whereas in another, PAI-1 antigen and activity levels in plasma were markedly reduced, with normal levels observed in platelets (168). In the most recent report, complete deficiency of platelet and plasma PAI-1 was identified in a 9-year-old Amish girl and associated with a moderate bleeding disorder. The patient was shown to be homozygous for a 2 base pair insertion at the end of exon 4 of the PAI-1 gene (169). This defect results in a frameshift, leading to a truncated PAI-1 protein and an unstable mRNA. The deficiency is inherited as an autosomal recessive disorder. Although heterozygous parents and siblings all have plasma PAI-1 activity and antigen in the normal range, they are consistently decreased compared with homozygous normal family members (Fig. 4). The lack of developmental and other abnormalities in this patient is surprising because plasminogen activation and its regulation by PAI-1 have been implicated in a large number of physiological and pathological processes (Fig. 1). The correlation of complete PAI-1 deficiency with abnormal bleeding clearly demonstrates that PAI-1 is critical for the regulation of hemostasis. Given the young age of the patient, however, an additional important in vivo role of PAI-1 in the control of ovulation or tumor metastasis cannot yet be excluded (8,170).

III. PLASMINOGEN ACTIVATOR INHIBITOR TYPE 2

A. Background

Plasminogen activator inhibitor type 2, previously called placental PAI, is a single-chain protein that exists in both extracellular glycosylated and intracellular nonglycosylated

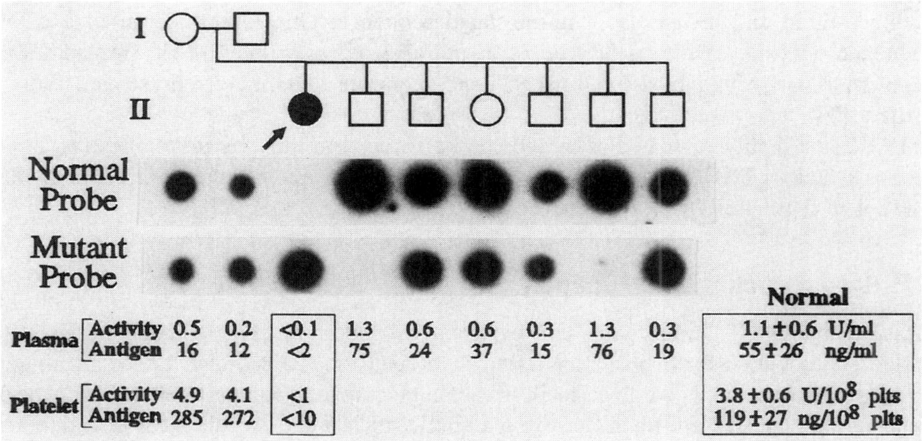

Figure 4 Pedigree of the PAI-1-deficient patient. The family pedigree is shown at the top, with the proband indicated by an arrow. Males are represented by boxes and females by circles. Analysis by allele-specific oligonucleotide hybridization of amplified DNA from exon 4 of each individual is shown beneath the pedigree. The proband is homozygous for the TA insertion mutation. The parents and four siblings are heterozygous for the mutation and normal sequence, and two siblings are homozygous normal. Plasma PAI-1 antigen and activity levels are shown at the bottom for each individual. Platelet PAI-1 levels are also indicated for the parents and proband. (Reprinted by permission of the *New England Journal of Medicine*, 1992; 327:1729–1733.)

forms (171,172). The molecular weight of glycosylated PAI-2 is approximately 60 kDa, and the nonglycosylated form is 46 kDa. PAI-2 is an efficient inhibitor of uPA and two-chain tPA, with second-order rate constants of 9×10^5 and 2×10^5 M^{-1} s^{-1}, respectively (173). This is about 20 times lower than PAI-1 toward uPA and about two orders of magnitude lower than PAI-1 toward two-chain tPA. The reaction with single-chain tPA is very slow, with a rate constant of 9×10^2 M^{-1} s^{-1}, or approximately 5000 times lower than PAI-1. The concentration of PAI-2 in plasma is very low and generally cannot be detected (171). During pregnancy, the levels of PAI-2 in plasma rise to about 100–300 ng/ml. Immediately following birth the levels decrease and are usually undetectable within 1 week (174,175). PAI-2 is produced by the placenta and by several different cell lines in vitro (18,172,173,176,177). Immunohistochemical studies have localized PAI-2 to the tophoblastic epithelium of the placenta (178). PAI-2 has been reported to be associated with vitronectin in placental extracts (179). However, in vitro binding of active PAI-2 to vitronectin could not be demonstrated (59). The physiological function of PAI-2 is not known. However, production of PAI-2 by monocyte/macrophage cells suggests a possible role in immune function (172,173).

B. The PAI-2 Gene

The gene for PAI-2, located on chromosome 18 bands q21–23, is 16.5 kb in length and consists of eight exons (33,180). The gene structure is most similar to chicken ovalbumin and gene Y and shows no obvious relationship to PAI-1. Northern blot analysis detects a single message of approximately 2 kb, consistent with the published cDNA sequence (32,181–183). The 5′-untranslated region is 74 bp in length, followed by a 1245 bp coding region and a 581 bp 3′-untranslated sequence. One report identified a cDNA sequence with an additional 581 bp of 3′-untranslated sequence (184). This additional segment is not found in the published gene sequence and likely represents a cloning artifact (33).

PAI-2 is a highly regulated gene and can be induced in vitro by phorbol esters, tumor necrosis factor (TNF), endotoxin, cholera toxin and dexamethasone (172,177,182,185–187). The regulation of PAI-2 was recently reviewed (109,188,189).

C. PAI-2 Structure and Function

PAI-2 contains 415 amino acids and exists in two alternatively processed forms, a secreted glycosylated form primarily found in pregnancy plasma, and a second, nonglycosylated species that is confined to an intracellular cytosolic compartment (172,174,183). Both forms of PAI-2 are translated from a single mRNA and are identical in amino acid sequence and protease inhibitor function (172,190). PAI-2 shows 39% sequence similarity with ovalbumin, its closest relative among the serpins (181). Like ovalbumin, but in contrast to other serpins, PAI-2 lacks a cleavable signal peptide and instead appears to utilize an internal, noncleaved secretion signal (183,190,191). This sequence is near the N terminus of PAI-2 and is similar to the ovalbumin secretion signal (192–194). However, this internal signal sequence does not function efficiently in all cell types (183,195), providing a potential additional mechanism for the regulation of PAI-2 gene expression (172). Mutations in the PAI-2 internal secretion signal sequence that increase hydrophobicity also increase the efficiency of secretion (194). By fusing the PAI-1 signal peptide sequence onto the PAI-2 N terminus, more than 90% of the PAI-2 protein can be efficiently targeted for secretion (194). Native human PAI-2 expressed in yeast is not secreted. However, fusion of the N-terminal secretion signal sequence of human serum albumin to the PAI-2 cDNA results in efficient glycosylation and secretion of PAI-2 from yeast (195). The high degree of protein sequence similarity between PAI-2 and ovalbumin, along with the similar gene structures and secretion mechanisms, suggests a close evolutionary relationship between these two proteins. The precise functions of ovalbumin and PAI-2 are unknown. A role for PAI-2 in the regulation of cellular interactions with the immune system has been proposed. Support for this hypothesis comes from a study demonstrating a protective effect of intracellular PAI-2 on cytolysis in cells treated with TNF and cycloheximide (196). Previous studies demonstrated that TNF-mediated cytolysis involves a serine protease (197) and can be blocked by exogenous protease inhibitors (198). Although PAI-2 is classified as an inhibitor of uPA, its in vivo target may actually be another protease, perhaps the protease involved in cytolysis during the TNF-mediated immune response. The reason for elevated PAI-2 levels during pregnancy is unknown but could relate to regulation of immune function, possibly the prevention of placental rejection.

D. Clinical Significance

Deficiency of PAI-2 has not yet been reported. The large increase in PAI-2 in plasma during normal pregnancy has not been specifically correlated with thrombosis. Two studies have reported decreased PAI-2 levels in plasma during severe preeclampsia (199,200) In other studies, however, no correlation between PAI-2 levels and thrombotic problems was observed (201).

IV. PLASMINOGEN ACTIVATOR INHIBITOR TYPE 3

A. Background

Plasminogen activator inhibitor type 3 (PAI-3), also called activated protein C inhibitor (see Chap. 21), is a glycoprotein with a molecular weight of approximately 57 kDa (31,202,203). PAI-3 is produced in the liver and present in plasma at a concentration of 2–5 μg/ml. PAI-3 is also found in urine and is produced by some hepatoma cell lines in culture (203,204). PAI-3 inhibits APC, uPA, thrombin, factor Xa, factor XIa, plasma kallikrein, trypsin, and chymotrypsin. Its inhibitory activity toward all these enzymes appears to be stimulated by heparin (202,205–208). The physiological function of PAI-3 is not certain, but it is likely that PAI-3 is the major inhibitor of APC in plasma. The role of PAI-3 as an inhibitor of PAs is not well defined. In urine, a significant portion of uPA is complexed with PAI-3, suggesting that PAI-3 has a physiological function as a PA inhibitor in this environment (209). The second-order rate constant for PAI-3 toward uPA is reported to be between 1×10^3 and 8×10^3 M^{-1} s^{-1} in the absence of heparin and 3×10^3 to 9×10^4 M^{-1} s^{-1} in the presence of heparin. The second-order rate constant for PAI-3 toward tPA is less than 10^3 M^{-1} s^{-1} (203,205).

B. The PAI-3 Gene

The gene for PAI-3 is located on human chromosome 14, although its precise localization has not yet been determined. It is composed of five exons and is similar in structure to α1AT and α1-antichymotrypsin. These latter genes are also located on chromosome 14, at band q31–32.3. Together these genes may represent a gene cluster that has arisen via a process of gene duplication (210). There is a single PAI-3 mRNA with a 115 bp 5′-untranslated sequence, 1218 bp coding sequence, and 839 bp 3′-untranslated region (31,210). The regulation of PAI-3 gene expression has not yet been characterized in detail.

C. PAI-3 Structure and Function

The PAI-3 cDNA encodes a 406 amino acid protein that includes a 19-residue N-terminal cleavable signal peptide (31). There are five potential N-linked glycosylation sites. The PAI-3 protein sequence shows over 40% similarity with α1AT and α_1-antichymotrypsin (204). PAI-3 binds to heparin, resulting in stimulation of inhibitory function by up to 50-fold toward all enzymes tested (APC, uPA, two-chain tPA, thrombin, factor Xa, Factor XIa, and chymotrypsin), with the exception of plasma kallikrein (207,211). The enhancing effect of heparin on PAI-3 function is similar to its effect on antithrombin III and heparin cofactor II. In contrast, heparin binding to PAI-1 shows no significant effect on inhibitory function, except with thrombin (144,147). Characterization of the heparin binding site on PAI-3, using competition with synthetic peptides, indicates that this site is distinctly different from the analogous functional domains on other serpins (208). Heparin appears to stimulate PAI-3 function via a template mechanism. In this model heparin binds to both the enzyme and the inhibitor, providing a molecular scaffold for the assembly of the protease-inhibitor complex (211).

D. Clinical Significance

The function of PAI-3 in vivo is currently unknown. Deficiency of PAI-3 was previously proposed as the molecular mechanism for combined factors V and VIII deficiency, the parallel decrease in the latter two proteins resulting from the unregulated action of APC (212). This hypothesis was later proven to be incorrect (213).

V. PROTEASE NEXIN 1

A. Background

Protease nexin 1, also called glia-derived neurite-promoting factor or glia-derived nexin, is a glycoprotein with a molecular weight of 50 kDa (214–216). PN-1 exists in two forms, α and βPN-1, that differ by only two amino acid residues centered at position 310–311 (97,217). In plasma, PN-1 levels are very low or undetectable. However, PN-1 is produced by several different cell lines in culture and is expressed in rat brain (215,218). PN-1 is a rapid inhibitor of thrombin, uPA, plasmin, trypsin, and plasma kallikrein, with second-order rate constants greater than 10^5 M^{-1} s^{-1} for each enzyme (214,219). PN-1 is a much less efficient inhibitor of both the two-chain and single-chain forms of tPA (10^3–10^4 M^{-1} s^{-1}) (214). Although PN-1 associated with the surface of platelets can inhibit thrombin, the low concentration of PN-1 in plasma suggests that it is not a functionally important inhibitor in this environment (220). Immunocytochemistry has localized PN-1 to discrete foci at the sites of neuromuscular junctions, and it has been shown to promote neurite outgrowth (221,222). This latter effect can be abolished by adding exogenous thrombin (223). PN-1 has also been found associated with the extracellular matrix of different cells in culture (224). Although these results suggest that PN-1 has an important role in regulating local extracellular proteolytic events, its exact physiological function in vivo remains unknown.

B. The PN-1 Gene

The gene for human PN-1 spans at least 36 kb, is composed of nine exons, and is similar in structure to PAI-1 (97). This observation, together with the 40% amino acid sequence similarity between these two proteins, suggests that PN-1 is the closest evolutionary relative of PAI-1 among known serpins. Northern blot analysis detects a single 2.3 kb PN-1 mRNA (97,217). Both the α and β forms of PN-1 are produced from a single gene by alternative splicing at the 3′ border of exon VI. Intron VI contains a second splice acceptor site that, when used, results in a three-base insertion at the end of exon VI, converting Arg^{310} to Thr as well as inserting a Gly codon. The resulting protein sequences are αPN-1 Ile-Thr-*Arg*-Ser-Glu and βPN-1 Ile-Thr-*Thr-Gly*-Ser-Glu (97). Regulation of the PN-1 gene has not been studied in detail. Fibroblasts and cells of neuronal origin have been shown to express PN-1, however, and induction of PN-1 mRNA has been demonstrated following injury to the sciatic nerve in the rat (97,215,225).

C. PN-1 Structure and Function

The cDNA for PN-1 encodes a protein of 397 amino acids (αPN-1) or 398 amino acids (βPN-1) (215–217). These include a 19-residue signal peptide and two potential N-linked glycosylation sites. The activity of PN-1 toward thrombin, factor Xa, and trypsin is significantly stimulated by heparin, although plasmin and uPA inhibition appear to be unaffected (224,226). Stimulation by heparin is dependent on the molecular weight of the heparin preparations, suggesting that heparin provides a template and does not alter the conformation of the PN-1 molecule (226). The heparin binding site on PN-1 is thought to be similar to those within antithrombin III and PAI-1 (148,208,227). In addition to binding heparin, PN-1 also binds to the extracellular matrix of fibroblasts in culture (219,224). Although this interaction stimulates PN-1 inhibition of thrombin, similar to the effect of heparin, a significant decrease in inhibition of uPA and plasmin is observed. These results suggest that cell surface or matrix-bound PN-1 is not a significant inhibitor of PAs (219). A second site of interaction between PN-1 and thrombin has also been proposed, involving the anion binding site on thrombin (228). Finally, it has been shown that thrombin-PN-1 complexes and, to a lesser extent, free PN-1 can bind to vitronectin (229). This interaction may be similar to the binding of antithrombin III-thrombin complexes to vitronectin and unlike the binding of free PAI-1.

D. Clinical Significance

No clinical abnormalities of PN-1 expression have yet been described. Immunostaining for PN-1 in brain specimens from Alzheimer's patients have detected high levels of PN-1 over the neuritic periphery surrounding the Alzheimer's plaque (230). Thus a role has been proposed for PN-1 in the pathogenesis of this and potentially other central nervous system disorders.

VI. FUTURE DIRECTIONS

Plasminogen activation has been postulated to play a major role in a diverse group of biological processes, and PAIs may function as central regulators in many of these systems. The recent observation of a significant hemorrhagic diathesis in a patient with complete PAI-1 deficiency demonstrates the critical role for PAI-1 in the regulation of vascular fibrinolysis in vivo. The functions of the other known PAIs in vivo are currently unknown. With the advent of new technologies for the targeted disruption of specific genes in mice, dramatic new insights will undoubtedly emerge in the coming years. Continued advances in the understanding of serpin three-dimensional structure and its relationship to function can also be anticipated. The combined advances in these two exciting areas of research should rapidly lead to new and better understandings of the importance of PAIs in biological systems.

REFERENCES

1. Wallén P. Biochemistry of plasminogen. In: Kline DL, Reddy KNN, eds. Fibrinolysis. Boca Raton, FL: CRC Press, 1980:1–25.
2. Saksela O. Plasminogen activation and regulation of pericellular proteolysis. Biochim Biophys Acta 1985; 823:35–65.
3. Astrup T. Fibrinolysis: past and present, a reflection of fifty years. Semin Thromb Hemost 1991; 17:161–174.
4. Saksela O, Rifkin DB. Cell-associated plasminogen activation: regulation and physiological functions. Annu Rev Cell Biol 1988; 4:93–126.
5. Bachmann F. Fibrinolysis. Thromb Haemost 1987; 10:227–265.
6. Hsueh AJW, Liu YX, Cajander SB, Ny T. Molecular mechanisms in the hormone regulation of plasminogen activator activity in ovarian granulosa cells and cumulus-oocyte complexes. In: Haseltine FP, First NL, eds. Meiotic Inhibition: Molecular Control of Meiosis. New York: Liss, 1988:227–258.

7. Ohlsson M, Peng XR, Liu YX, Jia XC, Hsueh AJ, Ny T. Hormone regulation of tissue-type plasminogen activator gene expression and plasminogen activator-mediated proteolysis. Semin Thromb Hemost 1991; 17:286–290.
8. Pöllänen J, Stephens RW, Vaheri A. Directed plasminogen activation at the surface of normal and malignant cells. Adv Cancer Res 1991; 57:273–328.
9. Dano K, Andreasen PA, Grondahl-Hansen J, Kristensen P, Nielsen LS, Skriver L. Plasminogen activators, tissue degradation, and cancer. Adv Cancer Res 1985; 44:139–266.
10. Moscatelli D, Rifkin DB. Membrane and matrix localization of proteinases: a common theme in tumor cell invasion and angiogenesis. Biochim Biophys Acta 1988; 948:67–85.
11. Hoylaerts M, Rijken DC, Lijnen HR, Collen D. Kinetics of the activation of plasminogen by human tissue plasminogen activator. J Biol Chem 1982; 257:2912–2919.
12. Ellis V, Dano K. Plasminogen activation by receptor-bound urokinase. Semin Thromb Hemost 1991; 17:194–200.
13. Hajjar KA, Hamel NM. Identification and characterization of human endothelial cell membrane binding sites for tissue plasminogen activator and urokinase. J Biol Chem 1990; 265:2908–2916.
14. Plow EF, Felez J, Miles LA. Cellular regulation of fibrinolysis. Thromb Haemost 1991; 66: 32–36.
15. Kruithof EKO. Plasminogen activator inhibitors—a review. Enzyme 1988; 40:113–121.
16. Hart DA, Rehemtulla A. Plasminogen activators and their inhibitors: regulators of extracellular proteolysis and cell function. Comp Biochem Physiol [B] 1988; 90B:691–708.
17. Aoki N. Hemostasis associated with abnormalities of fibrinolysis. Blood Rev 1989; 3:11–17.
18. Kawano T, Morimoto K. Uemura Y. Urokinase inhibitor in human placenta. Nature 1968; 217–253–254.
19. Loskutoff DJ, van Mourik JA, Erickson LA, Lawrence D. Detection of an unusually stable fibrinolytic inhibitor produced by bovine endothelial cells. Proc Natl Acad Sci USA 1983; 80:2956–2960.
20. Chmielewska J, Rånby M, Wiman B. Evidence for a rapid inhibitor to tissue plasminogen activator in plasma. Thromb Res 1983; 31:427–436.
21. Kruithof EKO, Ransijn A, Bachmann F. Inhibition of tissue plasminogen activator by human plasma. In: Davidson JF, Bachmann F, Bouvier CA, Kruithof EKO, eds. Progress in Fibrinolysis, Vol. VI. Edinburgh: Churchill Livingstone, 1983:365–369.
22. Carrell R, Travis J. α_1-Antitrypsin and the serpins: variation and countervariation. Trends Biochem Sci 1985; 10:20–24.
23. Boswell DR, Carrell RW. Genetic engineering and the SERPINs. Bioessays 1988; 8:83–87.
24. Huber R, Carrell RW. Implications of the three-dimensional structure of alpha 1-antitrypsin for structure and function of serpins. Biochem 1989; 28:8951–8966.
25. Doolittle RF. Angiotensinogen is related to the antitrypsin-antithrombin-ovalbumin family. Science 1983; 222:417–419.
26. Hunt LT, Dayhoff MO. A surprising new protein superfamily containing ovalbumin, antithrombin III, and alpha1-proteinase inhibitor. Biochem Biophys Res Commun 1980; 95: 864–871.
27. Kanost MR, Prasad SV, Wells MA. Primary structure of a member of the SERPIN superfamily of proteinase inhibitors from an insect, *Manduca sexta*. J Biol Chem 1989; 264:965–972.
28. Sasaki T. Patchwork-structure serpins from silkworm (*Bombyx mori*) larval hemolymph. Eur J Biochem 1991; 202:255–261.
29. Lomas DA, Evans DL, Upton C, McFadden G, Carrell RW. Inhibition of plasmin, urokinase, tissue plasminogen activator, and C_{1S} by a myxoma virus serine proteinase inhibitor. J Biol Chem 1993; 268:516–521.

30. Loskutoff DJ, Linders M, Keijer J, Veerman H, van Heerikhuizen H, Pannekoek H. Structure of the human plasminogen activator inhibitor 1 gene: nonrandom distribution of introns. Biochem 1987; 26:3763–3768.
31. Suzuki K, Deyashiki Y, Nishioka J, et al. Characterization of a cDNA for human protein C inhibitor. A new member of the plasma serine protease inhibitor superfamily. J Biol Chem 1987; 262:611–616.
32. Ye RD, Wun T-C, Sadler JE. cDNA cloning and expression in *Escherichia coli* of a plasminogen activator inhibitor from human placenta. J Biol Chem 1987; 262:3718–3725.
33. Ye RD, Ahern SM, Le Beau MM, Lebo RV, Sadler JE. Structure of the gene for human plasminogen activator inhibitor-2. J Biol Chem 1989; 264:5495–5502.
34. Strandberg L, Lawrence D, Ny T. The organization of the human plasminogen-activator-inhibitor 1 gene. Eur J Biochem 1988; 176:609–616.
35. Ragg H, Preibisch G. Structure and expression of the gene coding for the human serpin hLS2. J Biol Chem 1988; 263:12129–12134.
36. Follo M, Ginsburg D. Structure and expression of the human gene encoding plasminogen activator inhibitor, PAI-1. Gene 1989; 84:447–453.
37. Inglis JD, Hill RE. The murine Spi-2 proteinase inhibitor locus: a multigene family with a hypervariable reactive site domain. EMBO J 1991; 10:255–261.
38. Borriello F, Krauter KS. Multiple murine α_1-protease inhibitor genes show unusual evolutionary divergence. Proc Natl Acad Sci USA 1991; 88:9417–9421.
39. Loebermann H, Tokuoka R, Deisenhofer J, Huber R. Human α_1-proteinase inhibitor. Crystal structure analysis of two crystal modifications, molecular model and preliminary analysis of the implications for function. J Mol Biol 1984; 177:531–557.
40. Carrell RW, Owen MC. Plakalbumin, alpha-1-antitrypsin, antithrombin and the mechanism of inflammatory thrombosis. Nature 1985; 317:730–732.
41. Gettins P, Harten B. Properties of thrombin- and elastase-modified human antithrombin III. Biochemistry 1988; 27:3634–3639.
42. Bruch M, Weiss V, Engel J. Plasma serine proteinase inhibitors (serpins) exhibit major conformational changes and a large increase in conformational stability upon cleavage at their reactive sites. J Biol Chem 1988; 263:16626–16630.
43. Sweet RM, Wright HT, Janin J, Chothia CH, Blow DM. Crystal structure of the complex of porcine trypsin with soybean trypsin inhibitor (Kunitz) at 2.6-Å resolution. Biochem 1974; 13:4212–4228.
44. Huber R, Kukla D, Bode W, et al. Structure of the complex formed by bovine trypsin and bovine pancreatic trypsin inhibitor. II. Crystallographic refinement at 1.9 Å resolution. J Mol Biol 1974; 89:73–101.
45. Read RJ, James MNG. Introduction to the protein inhibitors: x-ray crystallography. In: Barrett AJ, Salvesen G, eds. Proteinase Inhibitors. New York: Elsevier Science Publishers, 1986:301–336.
46. Bode W, Meyer E. Jr, Powers JC. Human leukocyte and porcine pancreatic elastase: x-ray crystal structures, mechanism, substrate specificity, and mechanism-based inhibitors. Biochem 1989; 28:1951–1963.
47. Stein PE, Leslie AGW, Finch JT, Turnell WG, McLaughlin PJ, Carrell RW. Crystal structure of ovalbumin as a model for the reactive centre of serpins. Nature 1990; 347:99–102.
48. Schulze AJ, Baumann U, Knof S, Jaeger E, Huber R, Laurell C. Structural transition of α_1-antitrypsin by a peptide sequentially similar to β-strand s4A. Eur J Biochem 1990; 194:51–56.
49. Carrell RW, Evans DL, Stein PE. Mobile reactive centre of serpins and the control of thrombosis. Nature 1991; 353:576–578.
50. Björk I, Ylinenjärvi K, Olson ST, Bock PE. Conversion of antithrombin from an inhibitor of thrombin to a substrate with reduced heparin affinity and enhanced conformational sta-

bility by binding of a tetradecapeptide corresponding to the P_1 to P_{14} region of the putative reactive-bond loop of the inhibitor. J Biol Chem 1992; 267:1976–1982.
51. Mottonen J, Strand A, Symersky J, et al. Structural basis of latency in plasminogen activator inhibitor-1. Nature 1992; 355:270–273.
52. Cohen AB, Gruenke LD, Craig JC, Geczy D. Specific lysine labeling by $^{18}OH^-$ during alkaline cleavage of the α-1-antitrypsin-trypsin complex. Proc Natl Acad Sci USA 1977; 74:4311–4314.
53. Wiman B, Collen D. On the mechanism of the reaction between human alpha 2-antiplasmin and plasmin. J Biol Chem 1979; 254:9291–9297.
54. Levin EG. Latent tissue plasminogen activator produced by human endothelial cells in culture: evidence for an enzyme-inhibitor complex. Proc Natl Acad Sci USA 1983; 80:6804–6808.
55. Carrell RW, Boswell DR. Serpins: the superfamily of plasma serine proteinase inhibitors. In: Barrett AJ, Salvesen G, eds. Proteinase Inhibitors. Amsterdam: Elsevier Science Publishers (Biomedical Division), 1986:403–420.
56. Shubeita HE, Cottey TL, Franke AE, Gerard RD. Mutational and immunochemical analysis of plasminogen activator inhibitor 1. J Biol Chem 1990; 265:18379–18385.
57. York JD, Li P, Gardell SJ. Combinatorial mutagenesis of the reactive site region in plasminogen activator inhibitor I. J Biol Chem 1991; 266:8495–8500.
58. Sherman PM, Lawrence DA, Yang AY, et al. Saturation mutagenesis of the plasminogen activator inhibitor-1 reactive center. J Biol Chem 1992; 267:7588–7595.
59. Lawrence DA, Strandberg L, Ericson J, Ny T. Structure-function studies of the SERPIN plasminogen activator inhibitor type 1: analysis of chimeric strained loop mutants. J Biol Chem 1990; 265:20293–20301.
60. Matheson NR, van Halbeek H, Travis J. Evidence for a tetrahedral intermediate complex during serpin-proteinase interactions. J Biol Chem 1991; 266:13489–13491.
61. Madison EL, Goldsmith EJ, Gerard RD, Gething MH, Sambrook JF. Serpin-resistant mutants of human tissue-type plasminogen activator. Nature 1989; 339:721–724.
62. Adams DS, Griffin LA, Nachajko WR, Reddy VB, Wei C. A synthetic DNA encoding a modified human urokinase resistant to inhibition by serum plasminogen activator inhibitor. J Biol Chem 1991; 266:8476–8482.
63. Madison EL, Goldsmith EJ, Gething M-JH, Sambrook JF, Gerard RD. Restoration of serine protease-inhibitor interaction by protein engineering. J Biol Chem 1990; 265:21423–21426.
64. Van Mourik JA, Lawrence DA, Loskutoff DJ. Purification of an inhibitor plasminogen activator (antiactivator) synthesized by endothelial cells. J Biol Chem 1984; 259:14914–14921.
65. Lawrence D, Strandberg L. Grundström T, Ny T. Purification of active human plasminogen activator inhibitor 1 from *Escherichia coli*. Comparison with natural and recombinant forms purified from eucaryotic cells. Eur J Biochem 1989; 186:523–533.
66. Hekman CM, Loskutoff DJ. Bovine plasminogen activator inhibitor 1: specificity determinations and comparison of the active, latent, and guanidine-activated forms. Biochemistry 1988; 27:2911–2918.
67. Sakata Y, Curriden S, Lawrence D, Griffin JH, Loskutoff DJ. Activated protein C stimulates the fibrinolytic activity of cultured endothelial cells and decreases antiactivator activity. Proc Natl Acad Sci USA 1985; 82:1121–1125.
68. Sakata Y, Loskutoff DJ, Gladson CL, Hekman CM, Griffin JH. Mechanism of protein C-dependent clot lysis: role of plasminogen activator inhibitor. Blood 1986; 68:1218–1223.
69. Fay WP, Owen WG. Platelet plasminogen activator inhibitor: purification and characterization of interaction with plasminogen activators and activated protein C. Biochemistry 1989; 28:5773–5778.
70. Ehrlich HJ, Gebbink RK, Keijer J, Linders M, Preissner KT, Pannekoek H. Alteration of serpin specificity by a protein cofactor. J Biol Chem 1990; 265:13029–13035.

71. Declerck PJ, Verstreken M, Kruithof EKO, Juhan-Vague I, Collen D. Measurement of plasminogen activator inhibitor 1 in biologic fluids with a murine monoclonal antibody-based enzyme-linked immunoabsorbent assay. Blood 1988; 71:220–225.
72. Booth NA, Simpson AJ, Croll A, Bennett B, MacGregor IR. Plasminogen activator inhibitor (PAI-1) in plasma and platelets. Br J Haematol 1988; 70:327–333.
73. Colucci M, Páramo JA, Collen D. Generation in plasma of a fast-acting inhibitor of plasminogen activator in response to endotoxin stimulation. J Clin Invest 1985; 75:818–824.
74. Vaughan DE, Declerck PJ, Van Houtte E, De Mol M, Collen D. Studies of recombinant plasminogen activator inhibitor-1 in rabbits. Pharmacokinetics and evidence for reactivation of latent plasminogen activator inhibitor-1 in vivo. Circ Res 1990; 67:1281–1286.
75. Mayer EJ, Fujita T, Gardell SJ, Shebuski RJ, Reilly CF. The pharmacokinetics of plasminogen activator inhibitor-1 in the rabbit. Blood 1990; 76:1514–1520.
76. Erickson LA, Ginsberg MH, Loskutoff DJ. Detection and partial characterization of an inhibitor of plasminogen activator in human platelets. J Clin Invest 1984; 74:1465–1472.
77. Sawdey MS, Loskutoff DJ. Regulation of murine type 1 plasminogen activator inhibitor gene expression in vivo. Tissue specificity and induction by lipopolysaccharide, tumor necrosis factor-α, and transforming growth factor-β. J Clin Invest 1991; 88:1346–1353.
78. Krishnamurti C, Alving BM. Plasminogen activator inhibitor type 1: Biochemistry and evidence for modulation of fibrinolysis in vivo. Semin Thromb Hemost 1992; 18:67–80.
79. Loskutoff DJ. Regulation of PAI-1 gene expression. Fibrinolysis 1991; 5:197–206.
80. Pyke C, Kristensen P, Ralfkiaer E, Eriksen J, Dano K. The plasminogen activation system in human colon cancer: messenger RNA for the inhibitor PAI-1 is located in endothelial cells in the tumor stroma. Cancer Res 1991; 51:4067–4071.
81. Schneiderman J, Sawdey MS, Keeton MR, et al. Increased type 1 plasminogen activator inhibitor gene expression in atherosclerotic human arteries. Proc Natl Acad Sci USA 1992; 89:6998–7002.
82. Keeton M, Eguchi Y, Sawdey M, Ahn C, Loskutoff DJ. Cellular localization of type 1 plasminogen activator inhibitor messenger RNA and protein in murine renal tissue. Am J Pathol 1993; 142:59–70.
83. Wiman B, Chmielewska J, Ranby M. Inactivation of tissue plasminogen activator in plasma. Demonstration of a complex with a new rapid inhibitor. J Biol Chem 1984; 259:3644–3647.
84. Declerck PJ, De Mol M, Alessi MC, et al. Purification and characterization of a plasminogen activator inhibitor 1 binding protein from human plasma. J Biol Chem 1988; 263:15454–15461.
85. Wiman B, Almquist Å, Sigurdardottir O, Lindahl T. Plasminogen activator inhibitor 1 (PAI) is bound to vitronectin in plasma. FEBS Lett 1988; 242:125–128.
86. Laiho M, Saksela O, Andreasen PA, Keski-Oja J. Enhanced production and extracellular deposition of the endothelial-type plasminogen activator inhibitor in cultured human lung fibroblasts by transforming growth factor-β. J Cell Biol 1986; 103:2403–2410.
87. Mimuro J, Schleef RR, Loskutoff DJ. Extracellular matrix of cultured bovine aortic endothelial cells contains functionally active type 1 plasminogen activator inhibitor. Blood 1987; 70:721–728.
88. Knudsen BS, Harpel PC, Nachman RL. Plasminogen activator inhibitor is associated with the extracellular matrix of cultured bovine smooth muscle cells. J Clin Invest 1987; 80: 1082–1089.
89. Levin EG, Santell L. Association of a plasminogen activator inhibitor (PAI-1) with the growth substratum and membrane of human endothelial cells. J Cell Biol 1987; 105:2543–2549.
90. Pöllänen J, Saksela O, Salonen EM, et al. Distinct localizations of urokinase-type plasminogen activator and its type 1 inhibitor under cultured human fibroblasts and sarcoma cells. J Cell Biol 1987; 104:1085–1096.

91. Knudsen BS, Nachman RL. Matrix plasminogen activator inhibitor: modulation of the extracellular proteolytic environment. J Biol Chem 1988; 263:9476–9481.
92. Mimuro J, Loskutoff DJ. Binding of type 1 plasminogen activator inhibitor to the extracellular matrix of cultured bovine endothelial cells. J Biol Chem 1989; 264:5058–5063.
93. Schleef RR, Podor TJ, Dunne E, Mimuro J, Loskutoff DJ. The majority of type 1 plasminogen activator inhibitor associated with cultured human endothelial cells is located under the cells and is accessible to solution-phase tissue-type plasminogen activator. J Cell Biol 1990; 110:155–163.
94. Schleef RR, Loskutoff DJ, Podor TJ. Immunoelectron microscopic localization of type 1 plasminogen activator inhibitor on the surface of activated endothelial cells. J Cell Biol 1991; 113:1413–1423.
95. Klinger KW, Winqvist R, Riccio A, et al. Plasminogen activator inhibitor type 1 gene is located at region q21.3–q22 of chromosome 7 and genetically linked with cystic fibrosis. Proc Natl Acad Sci USA 1987; 84:8548–8552.
96. Bosma PJ, van den Berg EA, Kooistra T, Siemieniak DR, Slightom JL. Human plasminogen activator inhibitor-1 gene: promoter and structural gene nucleotide sequences. J Biol Chem 1988; 263:9129–9141.
97. McGrogan M, Kennedy J, Golini F, et al. Structure of the human protease nexin gene and expression of recombinant forms of PN-1. In: Festoff BW, ed. Serine Proteases and Their Serpin Inhibitors in the Nervous System. New York: Plenum Press, 1990: 147–161.
98. Bosma PJ, Kooistra T, Siemieniak DR, Slightom JL. Further characterization of the 5′-flanking DNA of the gene encoding human plasminogen activator inhibitor-1. Gene 1991; 100:261–266.
99. Ny T, Sawdey M, Lawrence D, Millan JL, Loskutoff DJ. Cloning and sequence of a cDNA coding for the human beta-migrating endothelial-cell-type plasminogen activator inhibitor. Proc Natl Acad Sci USA 1986; 83:6776–6780.
100. Ginsburg D, Zeheb R, Yang AY, et al. cDNA cloning of human plasminogen activator-inhibitor from endothelial cells. J Clin Invest 1986; 78:1673–1680.
101. Pannekoek H, Veerman H, Lambers H, et al. Endothelial plasminogen activator inhibitor (PAI): a new member of the serpin gene family. EMBO J 1986; 5:2539–2544.
102. Sawdey M, Ny T, Loskutoff DJ. Messenger RNA for plasminogen activator inhibitor. Thromb Res 1986; 41:151–160.
103. Zeheb R, Gelehrter TD. Cloning and sequencing of cDNA for the rat plasminogen activator inhibitor-1. Gene 1988; 73:459–468.
104. Prendergast GC, Cole MD. Posttranscriptional regulation of cellular gene expression by the c-myc oncogene. Mol Cell Biol 1989; 9:124–134.
105. Prendergast GC, Diamond LE, Dahl D, Cole MD. The c-myc-regulated gene mrl encodes plasminogen activator inhibitor 1. Mol Cell Biol 1990; 10:1265–1269.
106. Fattal PG, Schneider DJ, Sobel BE, Billadello JJ. Post-transcriptional regulation of expression of plasminogen activator inhibitor type 1 on mRNA by insulin and insulin-like growth factor 1. J Biol Chem 1992; 267:12412–12415.
107. Van den Berg EA, Sprengers ED, Jaye M, Burgess W, Maciag T, van Hinsbergh VWM. Regulation of plasminogen activator inhibition-1 mRNA in human endothelial cells. Thromb Haemost 1988; 60:63–67.
108. Loskutoff DJ, Sawdey M, Mimuro J. Type 1 plasminogen activator inhibitor. Prog Hemost Thromb 1989; 9:87–115.
109. Andreasen PA, Georg B, Lund LR, Riccio A, Stacey SN. Plasminogen activator inhibitors: hormonally regulated serpins. Mol Cell Endocrinol 1990; 68:1–19.
110. Ginsburg D. Regulation of gene expression in the fibrinolytic system. In: Haber E and Braunwald E, eds. Thrombolysis: Basic Contributions and Clinical Progress. St. Louis: Mosby-Year Book, 1991:17–26.

111. Andreasen PA, Riccio A, Welinder KG, et al. Plasminogen activator inhibitor type-1: reactive center and amino-terminal heterogeneity determined by protein and cDNA sequencing. FEBS Lett 1986; 209:213–218.
112. Franke AE, Danley DE, Kaczmarek FS, et al. Expression of human plasminogen activator inhibitor type-1 (PAI-1) in *Escherichia coli* as a soluble protein comprised of active and latent forms. Isolation and crystallization of latent PAI-1. Biochim Biophys Acta 1990; 1037: 16–23.
113. Reilly TM, Seetharam R, Duke JL, et al. Purification and characterization of recombinant plasminogen activator inhibitor-1 from *Escherichia coli*. J Biol Chem 1990; 265:9570–9574.
114. Keijer J, Ehrlich HJ, Linders M, Preissner KT, Pannekoek H. Vitronectin governs the interaction between plasminogen activator inhibitor 1 and tissue-type plasminogen activator. J Biol Chem 1991; 266:10700–10707.
115. Seetharam R, Dwivedi AM, Duke JL, et al. Purification and characterization of active and latent forms of recombinant plasminogen activator inhibitor 1 produced in *Escherichia coli*. Biochemistry 1992; 31:9877–9882.
116. Nielsen LS, Andreasen PA, Grondahl-Hansen J, Huang JY, Kristensen P, Dano K. Monoclonal antibodies to human 54,000 molecular weight plasminogen activator inhibitor from fibrosarcoma cells—inhibitor neutralization and one-step affinity purification. Thromb Haemost 1986; 55:206–212.
117. Booth NA, MacGregor IR, Hunter NR, Bennett B. Plasminogen activator inhibitor from human endothelial cells. Purification and partial characterization. Eur J Biochem 1987; 165: 595–600.
118. Lambers JW, Cammenga M, Konig BW, Mertens K, Pannekoek H, van Mourik JA. Activation of human endothelial cell-type plasminogen activator inhibitor (PAI-1) by negatively charged phospholipids. J Biol Chem 1987; 262:17492–17496.
119. Alessi MC, Declerck PJ, De Mol M, Nelles L, Collen D. Purification and characterization of natural and recombinant human plasminogen activator inhibitor-1 (PAI-1). Eur J Biochem 1988; 175:531–540.
120. Lindahl T, Wiman B. Purification of high and low molecular weight plasminogen activator inhibitor 1 from fibrosarcoma cell-line HT 1080 conditioned medium. Biochim Biophys Acta 1989; 994:253–257.
121. Hekman CM, Loskutoff DJ. Endothelial cells produce a latent inhibitor of plasminogen activators that can be activated by denaturants. J Biol Chem 1985; 260:11581–11587.
122. Levin EG, Santell L. Conversion of the active to latent plasminogen activator inhibitor from human endothelial cells. Blood 1987; 70:1090–1098.
123. Lindahl TL, Sigurdardóttir O, Wiman B. Stability of plasminogen activator inhibitor 1 (PAI-1). Thromb Haemost 1989; 62:748–751.
124. Wun T-C, Palmier MO, Siegel NR, Smith CE. Affinity purification of active plasminogen activator inhibitor-1 (PAI-1) using immobilized anhydrourokinase. J Biol Chem 1989; 264: 7862–7868.
125. Katagiri K, Okada K, Hattori H, Yano M. Bovine endothelial cell plasminogen activator inhibitor. Purification and heat activation. Eur J Biochem 1988; 176:81–87.
126. Munch M, Heegaard C, Jensen PH, Andreasen PA. Type-1 inhibitor of plasminogen activators: distinction between latent, activated and reactive centre-cleaved forms with thermal stability and monoclonal antibodies. FEBS Lett 1991; 295:102–106.
127. Lawrence DA, Loskutoff DJ. Inactivation of plasminogen activator inhibitor by oxidants. Biochemistry 1986; 25:6351–6355.
128. Strandberg L, Lawrence DA, Johansson LB, Ny T. The oxidative inactivation of plasminogen activator inhibitor type 1 results from a conformational change in the molecule and does not require the involvement of the P1′ methionine. J Biol Chem 1991; 266:13852–13858.

129. Weiss SJ, Regiani S. Neutrophils degrade subendothelial matrices in the presence of alpha-1-proteinase inhibitor. Cooperative use of lysosomal proteinases and oxygen metabolites. J Clin Invest 1984; 73:1297–1303.
130. Declerck PJ, De Mol M, Vaughn DE, Collen D. Identification of a conformationally distinct form of plasminogen activator inhibitor-1, acting as a noninhibitory substrate for tissue-type plasminogen activator. J Biol Chem 1992; 267:11693–11696.
131. Urano T, Strandberg L, Johansson LB, Ny T. A substrate-like form of plasminogen-activator-inhibitor type 1. Conversions between different forms by sodium dodecyl sulphate. Eur J Biochem 1992; 209:985–992.
132. Seiffert D, Podor TJ. Immunological detection of conformational changes of type 1 plasminogen activator inhibitor associated with activation. Fibrinolysis 1991; 5:225–231.
133. Tomasini BR, Mosher DF. Vitronectin. Prog Hemost Thromb 1991; 10:269–305.
134. Lang IM, Marsh JJ, Moser KM, Schleef RR. Presence of active and latent type I plasminogen activator inhibitor associated with porcine platelets. Blood 1992; 80:2269–2274.
135. Preissner KT, Holzhüter S, Justus C, Muller-Berghaus G. Identification and partial characterization of platelet vitronectin: evidence for complex formation with platelet-derived plasminogen activator inhibitor-1. Blood 1989; 74:1989–1996.
136. Fay WP, Shapiro AD, Ginsburg D. Role of plasminogen activator inhibitor-1 in the resistance of platelet-rich thrombi to lysis (abstract). Circulation 1992; 86(Suppl. I):I-812.
137. Levi M, Biemond BJ, van Zonneveld A-J, Wouter Ten Cate J, Pannekoek H. Inhibition of plasminogen activator inhibitor-1 activity results in promotion of endogenous thrombolysis and inhibition of thrombus extension in models of experimental thrombosis. Circulation 1992; 85:305–312.
138. Braaten JV, Handt S, Jerome WG, Kirkpatrick J, Lewis JC, Hantgan RR. Regulation of fibrinolysis by platelet-released plasminogen activator inhibitor 1: light scattering and ultrastructural examination of lysis of a model platelet-fibrin thrombus. Blood 1993; 81:1290–1299.
139. Sigurdardóttir O, Wiman B. Complex formation between plasminogen activator inhibitor 1 and vitronectin in purified systems and in plasma. Biochim Biophys Acta 1990; 1035:56–61.
140. Kost C, Stüber W, Ehrlich HJ, Pannekoek H, Preissner KT. Mapping of binding sites for heparin, plasminogen activator inhibitor-1, and plasminogen to vitronectin's heparin-binding region reveals a novel vitronectin-dependent feedback mechanism for the control of plasmin formation. J Biol Chem 1992; 267:12098–12105.
141. Seiffert D, Loskutoff DJ. Kinetic analysis of the interaction between type 1 plasminogen activator inhibitor and vitronectin and evidence that the bovine inhibitor binds to a thrombin-derived amino-terminal fragment of bovine vitronectin. Biochim Biophys Acta 1991; 1078: 23–30.
142. Salonen E-M, Vaheri A, Pollanen J, et al. Interaction of plasminogen activator inhibitor (PAI-1) with vitronectin. J Biol Chem 1989; 264:6339–6343.
143. Seiffert D, Loskutoff DJ. Evidence that type I plasminogen activator inhibitor binds to the somatomedin B domain of vitronectin. J Biol Chem 1991; 266:2824–2830.
144. Keijer J, Linders M, Wegman JJ, Ehrlich HJ, Mertens K, Pannekoek H. On the target specificity of plasminogen activator inhibitor 1: the role of heparin, vitronectin, and the reactive site. Blood 1991; 78:1254–1261.
145. Naski MC, Lawrence DA, Mosher DF, Podor TJ, Ginsburg D. Kinetics of inactivation of α-thrombin by plasminogen activator inhibitor-1: comparison of the effects of native and urea-treated forms of vitronectin. J Biol Chem 1993; 268:12367–12372.
146. Edelberg JM, Reilly CF, Pizzo SV. The inhibition of tissue type plasminogen activator by plasminogen activator inhibitor-1. J Biol Chem 1991; 266:7488–7493.
147. Ehrlich HJ, Keijer J, Preissner KT, Gebbink RK, Pannekoek H. Functional interaction of plasminogen activator inhibitor type 1 (PAI-1) and heparin. Biochemistry 1991; 30:1021–1028.

148. Ehrlich HJ, Gebbink RK, Keijer J, Pannekoek H. Elucidation of structural requirements on plasminogen activator inhibitor 1 for binding to heparin. J Biol Chem 1992; 267:11606–11611.
149. Wagner OF, de Vries C, Hohmann C, Veerman H, Pannekoek H. Interaction between plasminogen activator inhibitor type 1 (PAI-1) bound to fibrin and either tissue-type plasminogen activator (t-PA) or urokinase-type plasminogen activator (u-PA). J Clin Invest 1989; 84: 647–655.
150. Keijer J, Linders M, van Zonneveld A-J, Ehrlich HJ, de Boer J-P, Pannekoek H. The interaction of plasminogen activator inhibitor 1 with plasminogen activators (tissue-type and urokinase-type) and fibrin: localization of interaction sites and physiologic relevance. Blood 1991; 78:401–409.
151. Reilly CF, Hutzelmann JE. Plasminogen activator inhibitor-1 binds to fibrin and inhibits tissue-type plasminogen activator-mediated fibrin dissolution. J Biol Chem 1992; 267: 17128–17135.
152. Keijer J, Linders M, Ehrlich H, Gebbink RK, Pannekoek H. Stabilisation of plasminogen activator inhibitor type 1 (PAI-1) activity by arginine: possible implications for the interaction of PAI-1 with vitronectin. Fibrinolysis 1990; 4:153–159.
153. Juhan-Vague I, Moerman B, De Cock F, Aillaud MF, Collen D. Plasma levels of a specific inhibitor of tissue-type plasminogen activator (and urokinase) in normal and pathological conditions. Thromb Res 1984; 33:523–530.
154. Hamsten A, Wiman B, de Faire U, Blombäck M. Increased plasma levels of a rapid inhibitor of tissue plasminogen activator in young survivors of myocardial infarction. N Engl J Med 1985; 313:1557–1563.
155. Wiman B, Ljungberg B, Chmielewska J, Urden G, Blombäck M, Johnsson H. The role of the fibrinolytic system in deep vein thrombosis. J Lab Clin Med 1985; 105:265–270.
156. Paramo JA, Colucci M, Collen D, van de Werf F. Plasminogen activator inhibitor in the blood of patients with coronary artery disease. BMJ 1985; 291:573–574.
157. Nilsson IM, Ljungner H, Tengborn L. Two different mechanisms in patients with venous thrombosis and defective fibrinolysis low concentration of plasminogen activator or increased concentration of plasminogen activator inhibitor. BMJ 1985; 290:1453–1456.
158. Aznar J, Estelles A, Tormo G, et al. Plasminogen activator inhibitor activity and other fibrinolytic variables in patients with coronary artery disease. Br Heart J 1988; 59:535–541.
159. Angles-Cano E, Gris JC, Loyau S, Schved JF. Familial association of high levels of histidine-rich glycoprotein and plasminogen activator inhibitor-1 with venous thromboembolism. J Lab Clin Med 1993; 121:646–653.
160. Páramo JA, Alfaro MJ, Rocha E. Postoperative changes in the plasmatic levels of tissue-type plasminogen activator and its fast-acting inhibitor—relationship to deep vein thrombosis and influence of prophylaxis. Thromb Haemost 1985; 54:713–716.
161. Lijnen HR, Collen D. Congenital and acquired deficiencies of components of the fibrinolytic system and their relation to bleeding or thrombosis. Fibrinolysis 1989; 3:67–77.
162. Francis RBJ. Clinical disorders of fibrinolysis: a critical review. Blut 1989; 59:1–14.
163. Kluft C, Verheijen JH, Jie AF, et al. The postoperative fibrinolytic shutdown: a rapidly reverting acute phase pattern for the fast-acting inhibitor of tissue-type plasminogen activator after trauma. Scand J Clin Lab Invest 1985; 45:605–610.
164. Kassis J, Hirsh J, Podor TJ. Evidence that postoperative fibrinolytic shutdown is mediated by plasma factors that stimulate endothelial cell type I plasminogen activator biosynthesis. Blood 1992; 80:1758–1764.
165. Erickson LA, Fici GJ, Lund JE, Boyle TP, Polites HG, Marotti KR. Development of venous occlusions in mice transgenic for the plasminogen activator inhibitor-1 gene. Nature 1990; 346:74–76.
166. Ridker PM,Vaughan DE, Stampfer MJ, et al. Baseline fibrinolytic state and the risk of future venous thrombosis: a prospective study of endogenous tissue-type plasminogen activator and plasminogen activator inhibitor. Circulation 1992; 85:1822–1827.

167. Schleef RR, Higgins DL, Pillemer E, Levitt LJ. Bleeding diathesis due to decreased functional activity of type 1 plasminogen activator inhibitor. J Clin Invest 1989; 83:1747–1752.
168. Diéval J, Nguyen G, Gross S, Delobel J, Kruithof EKO. A lifelong bleeding disorder associated with a deficiency of plasminogen activator inhibitor type I. Blood 1991; 77:528–532.
169. Fay WP, Shapiro AD, Shih JL, Schleef RR, Ginsburg D. Complete deficiency of plasminogen-activator inhibitor type 1 due to a frame-shift mutation. N Engl J Med 1992; 327:1729–1733.
170. Liu Y-X, Peng X-R, Ny T. Tissue-specific and time-coordinated hormone regulation of plasminogen-activator-inhibitor type 1 and tissue-type plasminogen activator in the rat ovary during gonadotropin-induced ovulation. Eur J Biochem 1991; 195:549–555.
171. Åstedt B, Lecander I, Ny T. The placental type plasminogen activator inhibitor, PAI 2. Fibrinolysis 1987; 1:203–208.
172. Wohlwend A, Belin D, Vassalli J-D. Plasminogen activator-specific inhibitors produced by human monocytes/macrophages. J Exp Med 1987; 165:320–339.
173. Kruithof EKO, Vassalli J-D, Schleuning W-D, Mattaliano RJ, Bachmann F. Purification and characterization of a plasminogen activator inhibitor from the histiocytic lymphoma cell line U-937. J Biol Chem 1986; 261:11207–11213.
174. Lecander I, Astedt B. Isolation of a new specific plasminogen activator inhibitor from pregnancy plasma. Br J Haematol 1986; 62:221–228.
175. Kruithof EKO, Tran-Thang C, Gudinchet A, et al. Fibrinolysis in pregnancy: a study of plasminogen activator inhibitors. Blood 1987; 69:460–466.
176. Holmberg L, Lecander I, Persson B. The inhibitor from placenta specifically binds urokinase and inhibits plasminogen activator released from ovarian carcinoma in tissue culture. Biochim Biophys Acta 1978; 544:128–137.
177. Medcalf RL, Kruithof EKO, Schleuning W-D. Plasminogen activator inhibitor 1 and 2 are tumor necrosis factor/cachectin-responsive genes. J Exp Med 1988; 168:751–759.
178. Åstedt B, Hagerstrand I, Lecander I. Cellular localisation in placenta of placental type plasminogen activator inhibitor. Thromb Haemost 1986; 56:63–65.
179. Radtke KP, Wenz KH, Heimburger N. Isolation of plasminogen activator inhibitor-2 (PAI-2) from human placenta. Evidence for vitronectin/PAI-2 complexes in human placenta extract. Biol Chem Hoppe Seyler 1990; 371:1119–1127.
180. Kruithof EK, Cousin E. Plasminogen activator inhibitor 2. Isolation and characterization of the promoter region of the gene. Biochem Biophys Res Commun 1988; 156:383–388.
181. Webb AC, Collins KL, Snyder SE, et al. Human monocyte Arg-serpin cDNA: sequence, chromosomal assignment, and homology to plasminogen activator-inhibitor. J Exp Med 1987; 166:77–94.
182. Schleuning W-D, Medcalf RL, Hession C, Rothenbuhler R, Shaw A, Kruithof EKO. Plasminogen activator inhibitor 2: regulation of gene transcription during phorbol ester-mediated differentiation of U-937 human histiocytic lymphoma cells. Mol Cell Biol 1987; 7:4564–4567.
183. Ny T, Hansson L, Lawrence D, Leonardsson G, Åstedt B. Plasminogen activator inhibitor type 2 cDNA transfected into Chinese hamster ovary cells is stably expressed but not secreted. Fibrinolysis 1989; 3:1989–1196.
184. Antalis TM, Clark MA, Barnes T, et al. Cloning and expression of a cDNA coding for a human monocyte-derived plasminogen activator inhibitor. Proc Natl Acad Sci USA 1988; 85:985–989.
185. Wohlwend A, Belin D, Vassalli JD. Plasminogen activator-specific inhibitors in mouse macrophages: in vivo and in vitro modulation of their synthesis and secretion. J Immunol 1987; 139:1278–1284.
186. Medcalf RL, Van den Berg E, Schleuning W-D. Glucocorticoid-modulated gene expression of tissue- and urinary-type plasminogen activator and plasminogen activator inhibitor 1 and 2. J Cell Biol 1988; 106:971–978.

187. Schwartz BS, Monroe MC, Levin EG. Increased release of plasminogen activator inhibitor type 2 accompanies the human mononuclear cell tissue factor response to lipopolysaccharide. Blood 1988; 71:734–741.
188. Grant PJ, Medcalf RL. Hormonal regulation of haemostasis and the molecular biology of the fibrinolytic system. Clin Sci 1990; 78:3–11.
189. Schleuning W-D, Medcalf RL. Signal transduction chains involved in the control of the fibrinolytic enzyme cascade. In: Festoff BW, ed. Serine Proteases and Their Serpin Inhibitors in the Nervous System. New York: Plenum Press, 1990:127–135.
190. Belin D, Wohlwend A, Schleuning W-D, Kruithof EKO, Vassalli J-D. Facultative polypeptide translocation allows a single mRNA to encode the secreted and cytosolic forms of plasminogen activators inhibitor 2. EMBO J 1989; 8:3287–3294.
191. Ye RD, Wun TC, Sadler JE. Mammalian protein secretion without signal peptide removal. Biosynthesis of plasminogen activator inhibitor-2 in U-937 cells. J Biol Chem 1988; 263: 4869–4875.
192. Palmiter RD, Gagnon J, Walsh KA. Ovalbumin: a secreted protein without a transient hydrophobic leader sequence. Proc Natl Acad Sci USA 1978; 75:94–98.
193. Meek RL, Walsh KA, Palmiter RD. The signal sequence of ovalbumin is located near the NH_2 terminus. J Biol Chem 1982; 257:12245–12251.
194. von Heijne G, Liljestrom P, Mikus P, Andersson H, Ny T. The efficiency of the uncleaved secretion signal in the plasminogen activator inhibitor type 2 protein can be enhanced by point mutations that increase its hydrophobicity. J Biol Chem 1991; 266:15240–15243.
195. Steven J, Cottingham IR, Berry SJ, et al. Purification and characterisation of plasminogen activator inhibitor 2 produced in *Saccharomyces cerevisiae*. Eur J Biochem 1991; 196:431–438.
196. Kumar S, Baglioni C. Protection from tumor necrosis factor-mediated cytolysis by overexpression of plasminogen activator inhibitor type-2. J Biol Chem 1991; 266:20960–20964.
197. Suffys P, Beyaert R, Van Roy F, Fiers W. Involvement of a serine protease in tumournecrosis-factor-mediated cytotoxicity. Eur J Biochem 1988; 178:257–265.
198. Ruggiero V, Johnson SE, Baglioni C. Protection from tumor necrosis factor cytotoxicity by protease inhibitors. Cell Immunol 1987; 107:317–325.
199. De Boer K, Lecander I, ten Cate JW, Borm JJ, Treffers PE. Placental-type plasminogen activator inhibitor in preeclampsia. Am J Obstet Gynecol 1988; 158:518–522.
200. Estellés A, Gilabert J, Aznar J, Loskutoff DJ, Schleef RR. Changes in the plasma levels of type 1 and type 2 plasminogen activator inhibitors in normal pregnancy and in patients with severe preeclampsia. Blood 1989; 74:1332–1338.
201. Kruithof EKO, Gudinchet A, Bachmann F. Plasminogen activator inhibitor 1 and plasminogen activator inhibitor 2 in various disease states. Thromb Haemost 1988; 59:7–12.
202. Suzuki K. Protein C inhibitor: structure and complex formation with activated protein C. In: Takada A, Samama MM, Collen D, eds. Protease Inhibitors. Amsterdam: Elsevier Science Publishers, 1990:91–104.
203. Geiger M. Protein C inhibitor/plasminogen activator inhibitor 3. Fibrinolysis 1988; 2:183–188.
204. Suzuki K, Deyashiki Y, Nishioka J, Toma K. Protein C inhibitor: structure and function. Thromb Haemost 1989; 61:337–342.
205. Stump DC, Thienpont M, Collen D. Purification and characterization of a novel inhibitor of urokinase from human urine. Quantitation and preliminary characterization in plasma. J Biol Chem 1986; 261:12759–12766.
206. Heeb MJ, Espana F, Geiger M, Collen D, Stump DC, Griffin JH. Immunological identity of heparin-dependent plasma and urinary protein C inhibitor and plasminogen activator inhibitor-3. J Biol Chem 1987; 262:15813–15816.
207. Heeb MJ, España F, Geiger M, Griffin JH. Regulation of the protein C pathway by plasma protease inhibitors. In: Takada A, Samama MM, Collen D, eds. Protease Inhibitors. Amsterdam: Elsevier Science Publishers, 1990:105–116.

208. Pratt CW, Whinna HC, Church FC. A comparison of three heparin-binding serine proteinase inhibitors. J Biol Chem 1992; 267:8795–8801.
209. Stump DC, Thienpont M, Collen D. Urokinase-related proteins in human urine. Isolation and characterization of single-chain urokinase (pro-urokinase) and urokinase-inhibitor complex. J Biol Chem 1986; 261:1267–1273.
210. Meijers JC, Chung DW. Organization of the gene coding for human protein C inhibitor (plasminogen activator inhibitor-3). Assignment of the gene to chromosome 14. J Biol Chem 1991; 266:15028–15034.
211. Pratt CW, Church FC. Heparin binding to protein C inhibitor. J Biol Chem 1992; 267:8789–8794.
212. Marlar RA, Griffin JH. Deficiency of protein C inhibitor in combined factor V/VIII deficiency disease. J Clin Invest 1980; 66:1186–1189.
213. Gardiner JE, Griffin JH. Studies on human protein C inhibitor in normal and factor V/VIII deficient plasmas. Thromb Res 1984; 36:197–203.
214. Scott RW, Bergman BL, Bajpai A, et al. Protease nexin. Properties and a modified purification procedure. J Biol Chem 1985; 260:7029–7034.
215. Gloor S, Odink K, Guenther J, Nick H, Monard D. A glia-derived neurite promoting factor with protease inhibitory activity belongs to the protease nexins. Cell 1986; 47:687–693.
216. Sommer J, Gloor SM, Rovelli GF, et al. cDNA sequence coding for a rat glia-derived nexin and its homology to members of the serpin superfamily. Biochem 1987; 26:6407–6410.
217. McGrogan M, Kennedy J, Ping Li M, et al. Molecular cloning and expression of two forms of human protease nexin I. Biotechnology 1988; 6:172–177.
218. Hiramoto SA, Cunningham DD. Effects of fibroblasts and endothelial cells on inactivation of target proteases by protease nexin-1, heparin cofactor II, and C1-inhibitor. J Cell Biochem 1988; 36:199–207.
219. Cunningham DD, Wagner SL, Farrell DH. Regulation of protease nexin-1 activity by heparin and heparan sulfate. In: Lane DA, Björk I, and Lindahl U, eds. Heparin and Related Polysaccharides. New York: Plenum Press, 1992:297–306.
220. Gronke RS, Bergman BL, Baker JB. Thrombin interaction with platelets. Influence of a platelet protease nexin. J Biol Chem 1987; 262:3030–3036.
221. Monard D. Cell-derived proteases and protease inhibitors as regulators of neurite outgrowth. Trends Neurosci 1988; 11:541–544.
222. Festoff BW, Rao JS, Hantai D. Plasminogen activators and inhibitors in the neuromuscular system: III. The serpin protease nexin I is synthesized by muscle and localized at neuromuscular synapses. J Cell Physiol 1991; 147:76–86.
223. Cunningham DD, Gurwitz D. Proteolytic regulation of neurite outgrowth from neuroblastoma cells by thrombin and protease nexin-1. J Cell Biochem 1989; 39:55–64.
224. Wagner SL, Lau AL, Cunningham DD. Binding of protease nexin-1 to the fibroblast surface alters its target proteinase specificity. J Biol Chem 1989; 264:611–615.
225. Monard D, Reinhard E, Meier R, et al. Steps in establishing a biological relevance for glia-derived nexin. In: Festoff BW, ed. Serine Proteases and Their Serpin Inhibitors in the Nervous System. New York: Plenum Press, 1990:275–281.
226. Evans DL, McGrogan M, Scott RW, Carrell RW. Protease specificity and heparin binding and activation of recombinant protease nexin I. J Biol Chem 1991; 266:22307–22312.
227. Evans DLL, Christey PB, Carrell RW. The heparin binding site and activation of protease nexin I. In: Festoff BW, ed. Serine Proteases and Their Serpin Inhibitors in the Nervous System. New York: Plenum Press, 1990:69–78.
228. Chang AC, Detwiler TC. The reaction of thrombin with platelet-derived nexin requires a secondary recognition site in addition to the catalytic site. Biochem Biophys Res Commun 1991; 177:1198–1204.

229. Rovelli G, Stone SR, Preissner KT, Monard, D. Specific interaction of vitronectin with the cell-secreted protease inhibitor glia-derived nexin and its thrombin complex. Eur J Biochem 1990; 192:797–803.
230. Rosenblatt DE, Geula C, Mesulam M-M. Protease nexin I immunostaining in Alzheimer's disease. In: Festoff BW, ed. Serine Proteases and Their Inhibitors in the Nervous System. New York: Plenum Press, 1990:329–336.
231. Sprang SR. The latent tendencies of PAI-1. Trends Biochem Sci 1992; 17:49–50.
232. Kopitar M, Rozman B, Babnik J, Turk V, Mullins DE, Wun T-C. Human leucocyte urokinase inhibitor-purification, characterization and comparative studies against different plasminogen activators. Thromb Haemost 1985; 54:750–755.

26

α_2-Plasmin Inhibitor

Nobuo Aoki
Tokyo Medical and Dental University, Tokyo, Japan

I. INTRODUCTION

Two major enzymes are involved in the fibrinolytic system. One is the plasminogen activator and the other, plasmin. Tissue plasminogen activator in the blood is mainly synthesized in the vascular endothelial cells and released into the circulation. Plasmin is formed by a limited proteolysis of plasminogen by plasminogen activator. The plasmin, once formed, degrades fibrin. Physiologically, this proteolytic process can hardly proceed in a fluid phase, such as circulating blood, but proceeds efficiently on the surface of solid-phase fibrin. When fibrin is formed, plasminogen activator and plasminogen are bound to it. Fibrin-bound plasminogen activator efficiently activates fibrin-bound plasminogen. Fibrin-associated plasminogen activation is considered the physiological mechanism of fibrinolysis and is checked or retarded at two different levels by two protease inhibitors present in plasma, plasminogen activator inhibitor (PAI) and α_2-plasmin inhibitor (α_2PI). The former, PAI, as well as plasminogen activator, may be synthesized in the vascular endothelial cells and released into the circulation. The latter, α_2PI, is synthesized in the liver (1,2). Congenital deficiency of α_2PI results in a lifelong severe hemorrhagic tendency caused by premature degradation of hemostatic plugs by the physiologically occurring fibrinolytic process (3–6).

II. PROTEIN STRUCTURE

α_2-Plasmin inhibitor, also called α_2-antiplasmin, is a plasma glycoprotein belonging to the α_2-globulin fraction. Its concentration in human plasma is estimated as 6.9 ± 0.6 mg per 100 ml (7), which is calculated to be ~1.2 μM on the assumption of molecular weight 58 kD. The amino acid sequence of α_2PI was deduced from the cDNA (Fig. 1) (8–10).

```
  M A L L W G L L V L S W S C L Q G P C   -21
S V F S P V S A M E P L G R Q L T S G P    -1
N Q E Q V S P L T L L K L G N Q E P G G    20
Q T A L K S P P G V C S R D P T P E Q T    40
H R L A R A M M A F T A D L F S L V A Q    60
T S T C P N L I L S P L S V A L A L S H    80
L A L G A Q N H T L Q R L Q Q V L H A G   100
S G P C L P H L L S R L C Q D L G P G A   120
F R L A A R M Y L Q K G F P I K E D F L   140
E Q S E Q L F G A K P V S L T G K Q E D   160
D L A N I N Q W V K E A T E G K I Q E F   180
L S G L P E D T V L L L L N A I H F Q G   200
F W R N K F D P S L T Q R D S F H L D E   220
Q F T V P V E M M Q A R T Y P L R W F L   240
L E Q P E I Q V A H F P F K N N M S F V   260
V L V P T H F E W N V S Q V L A N L S W   280
D T L H P P L V W E R P T K V R L P K L   300
Y L K H Q M D L V A T L S Q L G L Q E L   320
F Q A P D L R G I S E Q S L V V S G V Q   340
H Q S T L E L S E V G V E A A A A T S I   360
A M S R M S L S S F S V N R P F L F F I   380
F E D T T G L P L F V G S V R N P N P S   400
A P R E L K E Q Q D S P G N K D F L Q S   420
L K G F P R G D K L F G P D L K* L V P P  440
M E E D Y P Q F G S P K*
```

Figure 1 Amino acid sequence of α_2PI deduced from the cDNA. Amino acid residues are numbered on the right, and the residues upstream of the amino terminus of the mature plasma α_2PI are identified by a negative number. The reactive-site peptide bond is indicated by an arrow. The cross-linking site residue is indicated by a short, thick bar. The carboxyl-terminal sequence of 50 amino acids extends beyond the carboxyl-terminal ends of the other serpin family members when the sequences are aligned, and the extended sequence is underscored. Lysine residues involved in the binding to plasmin(ogen) are indicated by asterisks. Potentially glycosylated asparagine residues are indicated by closed triangles. Cysteine residues are circled. (From the data in Refs. 8–10.)

α_2PI is produced in the liver in a precursor form (pre-pro-α_2) with a prepeptide (signal peptide) of 27 amino acids and a propeptide of 12 amino acids (11). The mature plasma protein is composed of 452 amino acids (9,10). Pro-α_2PI, which has 12 more amino-terminal residues (propeptide) than the mature α_2PI, is present in plasma as 30–40% of the total plasma α_2PI (12). The molecular weight of mature α_2PI, deduced from the cDNA sequence (9,10) and the carbohydrate content (14%) (13), is ~58 kD, whereas the molecular weight estimated by sodium dodecyl sulfate (SDS)-gel electrophoresis is 67 kD (13). The cause of this discrepancy is not known. When homologous amino acid sequences of α_2PI and the other members of serine protease inhibitor family (antithrombin III, α_1-antitrypsin, plasminogen activator inhibitor 1, and others) are aligned, α_2PI extends 50–52 amino acids beyond the carboxyl-terminal ends of the other members of the family (8). This extra 50–52 carboxyl-terminal amino acid sequence is therefore specific to α_2PI and contains the plasminogen binding site (8,14,15). A degraded form of α_2PI, lacking the carboxyl-terminal peptide containing the plasminogen binding site, is present in normal plasma as approximately 30% of the total α_2PI (16).

III. FUNCTIONS

α_2PI is a serine protease inhibitor (serpin) that is able to inhibit several different "serine" proteases, including plasmin, trypsin, chymotrypsin, and proteases participating in the blood coagulation and kinin forming systems (17). Because of its high affinity for plasmin, however, the major target of α_2PI is plasmin. α_2PI is the primary inhibitor of plasmin-mediated fibrinolysis (18) because of its characteristic three functions: inhibition of plasmin(ogen) binding to fibrin, inhibition of plasmin proteolytic activity, and covalent binding to fibrin. The α_2PI molecule has the sites responsible for each of these three functions: plasmin(ogen) binding site, reactive site, and cross-linking site (Fig. 1). α_2PI has a strong affinity for plasmin(ogen) and noncovalently binds to the sites in plasmin(ogen) molecule, called lysine binding sites (LBS), located in the kringle structures (19). LBS are the sites where fibrin is also noncovalently bound. Hence, α_2PI competitively inhibits the binding of plasminogen to fibrin (20,21). The naturally occurring fibrinolytic process is caused by fibrin-associated plasminogen activation, and it depends on the amount of plasminogen and plasminogen activator bound to fibrin (22). Therefore, inhibition of plasminogen binding to fibrin by α_2PI results in retardation of the initiation of the fibrinolytic process (22). The plasmin(ogen) binding site in α_2PI is located within 26 amino acid residues of the carboxyl-terminal end (8,14). Two lysine residues, Lys^{436} and Lys^{452}, located at the carboxyl-terminal end, are most likely involved in the binding (15).

The plasmin(ogen) binding site also plays an important role in the inhibition of plasmin. The inhibition reaction seems to proceed in two steps. As the first step, α_2PI rapidly forms a reversible complex with plasmin through noncovalent binding between LBS in plasmin and the plasmin(ogen) binding site in α_2PI (19). This step of the reaction can be competitively inhibited by plasminogen fragments containing LBS or by ϵ-amino caproic acid, which binds to LBS (19). A partially degraded form of α_2PI (65 kD), which lacks the plasmin(ogen) binding site, reacts less readily with plasmin than native α_2PI (67 kD) (23), indicating a significant contribution of the plasmin(ogen) binding site reaction to the efficient inhibition of plasmin. In the second step, a covalent bond is formed between the active-site serine of plasmin and the reactive site of α_2PI (Arg^{364}), resulting in a loss of plasmin proteolytic activity (10).

Another important function of α_2PI is its cross-linking to fibrin. When blood clots, part of the α_2PI present in plasma is rapidly cross-linked to the fibrin α chain by activated coagulation factor XIII (24). While the clot retraction is progressively taking place, the fibrin-bound α_2PI becomes occluded in the clot and contributes to the resistance of the clot against fibrinolysis (25). Fibrin-fibrin cross-linking is only of minor importance in endowing the clot with resistance to fibrinolysis (26). The cross-linking site in α_2PI is located at the second glutamine residue from the amino terminus (27). The glutamine residue of α_2PI is cross-linked to Lys^{303} of the Aα chain of fibrin(ogen) (28). Among these three functional sites in α_2PI, the cross-linking site and plasmin(ogen) binding site are specific to α_2PI. No other serine protease inhibitor has a factor XIII-catalyzed cross-linking activity under physiological conditions.

IV. GENE

Studies from our laboratories (8) and those of others (9,10) have led to the isolation of the cDNA coding for human α_2PI. The cDNA was subsequently used for the determination of the chromosomal location and the gene organization.

A. Chromosomal Location of the Gene

Southern blot analysis of human × mouse somatic cell hybrids and fluorescence in situ hybridization studies using the cDNA fragments indicated that the α_2PI gene, whose genetic symbol is PLI, is located at chromosome 17p13 (29).

B. Gene Organisation

Overlapping genomic clones were isolated by using the cDNA as a probe from a λ phage library (30). The α_2PI gene contains 10 exons and 9 introns distributed over ~16 kb DNA (Fig. 2). The number of introns is the highest among those of the serine protease inhibitor gene family ever reported. A schematic diagram of the gene, the mRNA, and the protein

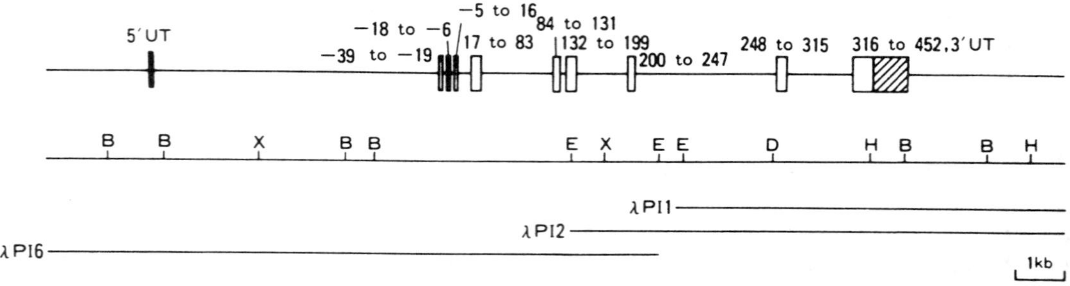

Figure 2 Organization of the human α_2PI inhibitor gene. The first line shows the positions of exons as rectangles, and the numbers above the line indicate the amino acids at which intron-exon junctions occur. Untranslated regions (UT) are shown as hatched areas. A small 5′-untranslated region exists in the second exon. The second line indicates the positions of restriction endonuclease recognition sites. Straight lines at bottom indicate the region of the three phage clones. B, BamHI; D, DraI; E, EcoRI; H, HindIII; X, XbaI. (From Ref. 30.)

product are depicted in Figure 3. The leader sequence, consisting of 39 amino acids, is encoded by a part of exon II, exon III, and a part of exon IV. The leader sequence consists of the prepeptide (signal peptide) of 27 amino acids and the propeptide of 12 amino acids (11). α_2PI is the only pre-pro type of processing protein so far reported among the serine protease inhibitor family members. The NH_2-terminal region, which contains the cross-linking site, is encoded by a part of exon IV. The COOH-terminal region, which contains the reactive site and plasmin(ogen) binding site, is encoded by exon X. A TATA box sequence, multiple GC box sequences, and a CCAAT box-like sequence are found in the 5′ flanking region. Particularly interesting is the presence of the 16 bp sequence that is 88% similar to the 17 bp sequence in the hepatitis B virus enhancer element (31). This element displays tissue-specific activity and shows high homology with sequences in the promoter region of several liver-specific genes that encode the following proteins: α-fetoprotein, α_1-antitrypsin, and albumin.

C. Restriction Fragment Length Polymorphism

A restriction fragment length polymorphism (RFLP) was found in the α_2PI gene (32). This RFLP can be attributed to the presence of two alleles, A and B. The minor allele B is a result of a deletion of about 720 bp in intron 8 of the gene (Fig. 4). Sequence analysis of the deletion junction in allele B and the corresponding regions of allele A demonstrated the presence of oppositely oriented Alu sequences at the 5′ and 3′ deletion boundaries (Fig. 5). The data suggest that the deletion was caused by intrastrand recombination between Alu sequences (Fig. 6). The two alleles, A and B, are distributed with frequencies of 73.5 and 26.5%, respectively, in 66 unrelated white individuals or with frequencies of 51.0 and 49.0%, respectively, in 50 unrelated Japanese individuals (32).

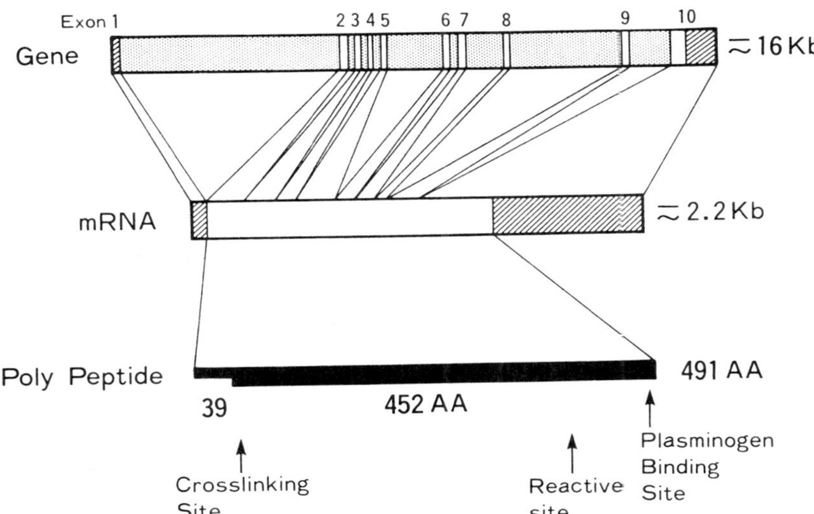

Figure 3 The gene, the mRNA, and the precursor protein of α_2PI. Dotted areas, introns; hatched areas, non-coding regions; horizontal thick bar, leader sequence and mature protein of α_2PI. Kb, kilobases; AA, amino acids.

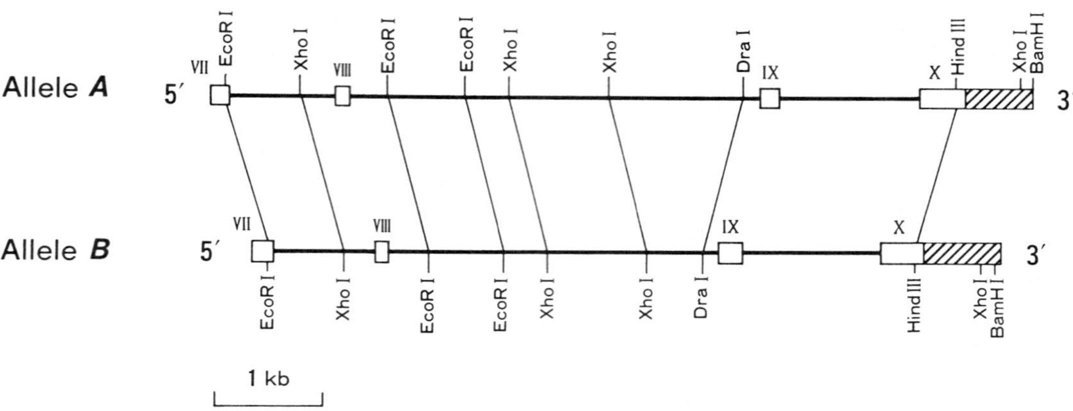

Figure 4 Comparison of restriction maps of the 3′ terminal halves of the human α_2PI gene alleles A and B. Coding regions of the exons are expressed by the open boxes and 3′-nontranslated regions by the hatched area. The exon numbers are indicated by Roman numerals. (From Ref. 32.)

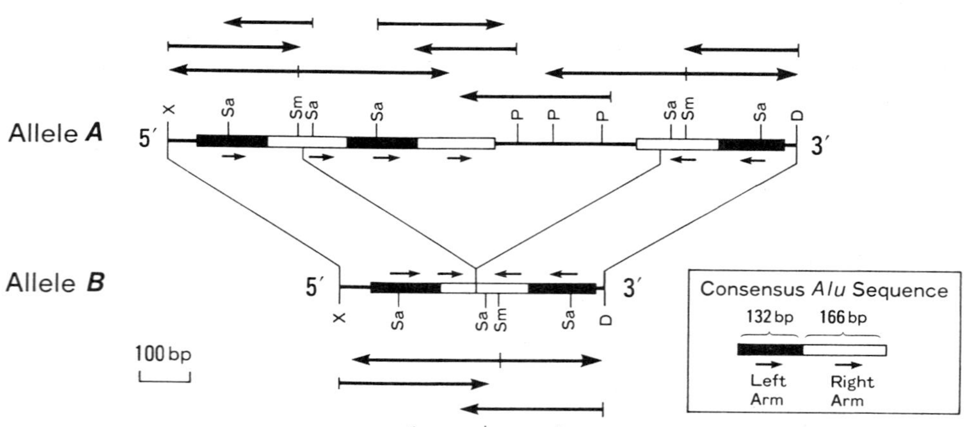

Figure 5 Strategy for sequencing the deletion boundaries in allele A and the deletion joint in allele B and orientation of the relevant Alu-repetitive sequences. The upper and lower parts of the figure show the relevant restriction fragments in intron 8 of alleles A and B derived from λαPN4 and λPI2, respectively. The direction and extent of the sequence established in a given experiment are indicated by the long arrows above or below the relevant DNA sequences. The long arrow not coinciding with a restriction endonuclease site indicates the sequence obtained by unidirectional digestion by exonuclease III. The boxed area shows an Alu sequence with left and right tandem repeats (open and closed, respectively), and short arrows indicate their orientation. The structure of a consensus Alu repeat is shown in the box. Segments of allele A and B that correspond to each other are connected by diagonal lines. D, DraI; P, PvuII; Sa, Sau3AI; Sm, SmaI; X, XhoI. (From Ref. 32.)

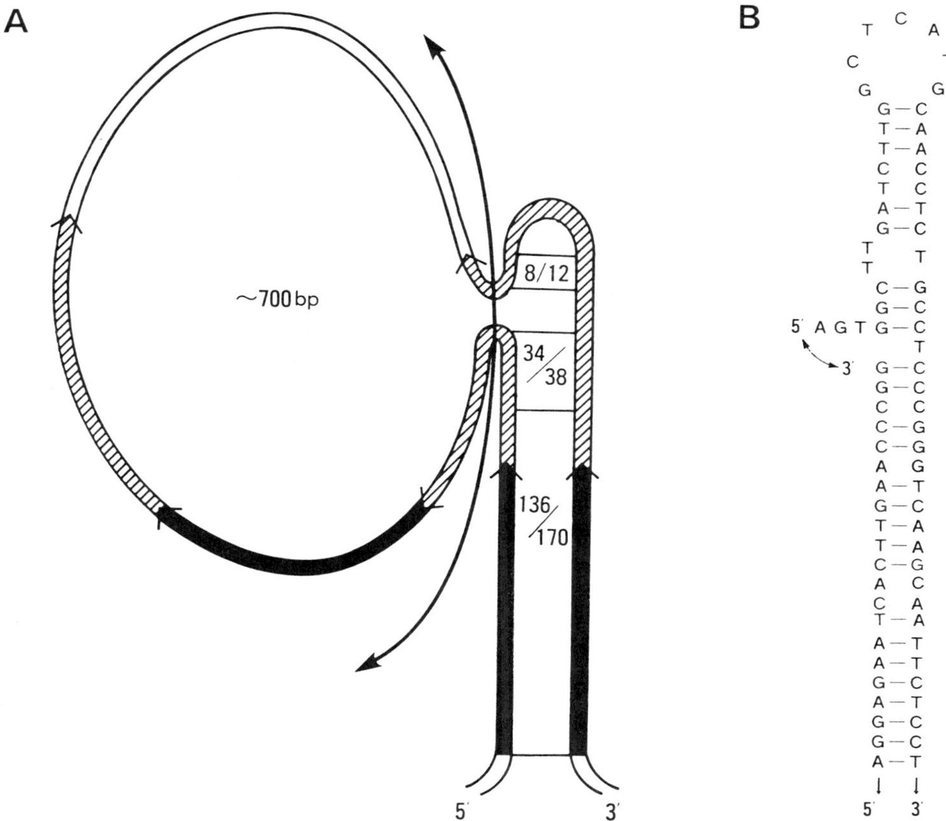

Figure 6 Potential mechanism for the deletion generating the RFLP in the α_2PI gene. (A) A potential double stem-loop structure of the XhoI/DraI region in intron 8 of allele A is shown, with the left and the right arms of the Alu sequences as closed and striped arrows. The lengths of the DNA sequences are not drawn precisely to scale. The positions of inverted repeats are shown by boxed areas, and the fractions inside the boxes denote the number of paired nucleotides divided by the number of paired plus unpaired nucleotides. The size of the deleted segment is indicated in the center of the loop, and the position of the deletion is denoted by a double arrow. (B) Potential homologous base pairings near the probable deletion points. (From Ref. 32.)

V. MUTATIONS

A. Mutation Leading to the Dysfunction

One pedigree with dysfunctional α_2PI has been reported (33). The dysfunctional α_2PI was found in two siblings living in Enschede, the Netherlands. These individuals have 3% of normal functional activity and 100% of normal antigen levels in their blood plasma (33). Using the cDNA as a probe, Holmes et al. (34) cloned a portion of the α_2PI gene from these individuals and identified a GCG in-frame insertion. This mutation results in the addition of an alanine to the consecutive alanine residues (residues 354–357), which are located only seven positions on the NH_2-terminal side of the P1 residue (Arg^{364}) in the reactive site. This alanine insertion probably results in some structural perturbation that abolishes the plasmin inhibitory activity of α_2PI.

B. Mutations Leading to the Deficiency in Plasma

More than 10 pedigrees with a congenital deficiency of α_2PI have been reported (6). Among these were two Japanese families, both of which were found to have alterations in the nucleotide sequence. One family, α_2PI Nara, has a single-nucleotide insertion in exon X (35), and the other, α_2PI Okinawa, has a trinucleotide deletion in exon VII (36).

In α_2PI Nara, a single-nucleotide insertion in a region coding for the carboxyl-terminal portion causes a frameshift mutation and results in the production of a mutant protein with alteration and elongation of the carboxyl-terminal part of the molecule (Fig. 7) (35). Synthetic oligonucleotide probes confirmed this mutation in all the affected family members. The mature normal protein consists of 452 amino acids, whereas the mature mutant protein has an extra 166 amino acids. The elongation of amino acid sequence was con-

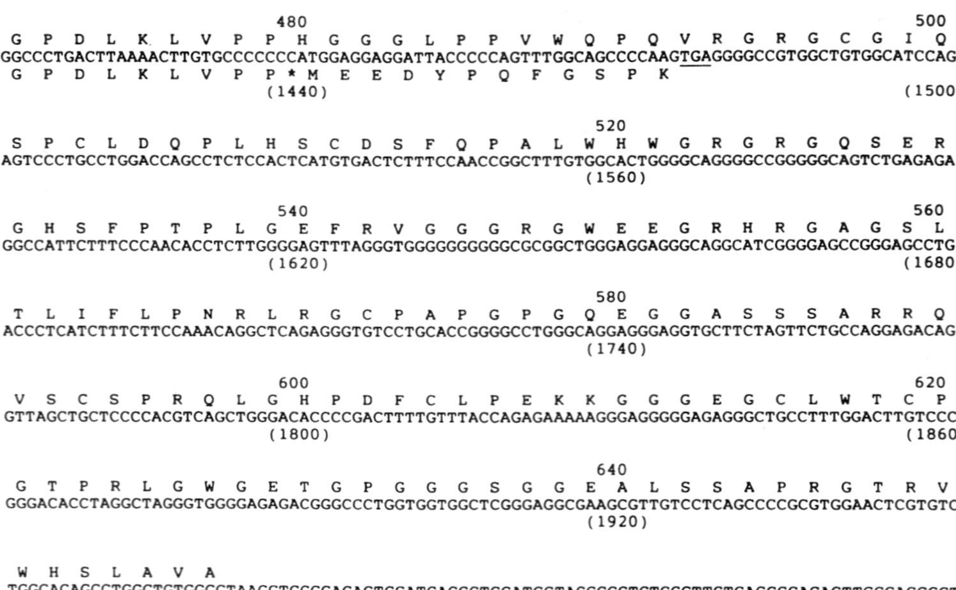

Figure 7 Effect of the single-nucleotide insertion on the deduced amino acid sequence of the carboxyl-terminal portion of the α_2PI. A portion of the nucleotide sequence of the mutated α_2PI gene is shown with nucleotides numbered in parentheses. The numbering of the nucleotides was started at the initiation codon, and only the bases in the exons have been numbered. The deduced amino acid sequence of the mutated gene (α_2PI Nara) is displayed above the nucleotide sequence with amino acid numbers, which also started with the initiator methionine. The amino acid sequence of the normal α_2PI is shown beneath the nucleotide sequence. The termination codons are underscored. As a result of the cytidine nucleotide insertion indicated by an asterisk, a frameshift occurs that leads to replacement of the carboxyl-terminal 12 amino acid residues of the normal α_2PI with 178 amino acid residues entirely unrelated to the normal amino acid sequence. (From Ref. 35.)

firmed by an analysis of the α_2PI expressed by the COS-7 cells transfected with the expression vector for the mutant α_2PI (35,37).

In α_2PI Okinawa, a trinucleotide deletion in exon VII gives rise to the deletion of Glu^{137} as identified by nucleotide sequence analysis of the mutant gene cloned from an individual homozygous for the deficiency (36). Using the DNA samples amplified with the polymerase chain reaction, hybridization analysis by oligonucleotide probes confirmed the presence of this mutation in all affected family members.

VI. MOLECULAR BASIS FOR THE DEFICIENCY

To determine the molecular basis of α_2PI Nara and Okinawa, we carried out experiments in which the mutated genes were transfected into cultured cells and expressed (36,37).

The expression vector for α_2PI, pSV2PI, was constructed by inserting a cDNA fragment coding for the carboxyl-terminal half of mature α_2PI and a fragment of the gene that codes for the leader (signal) peptide and the amino-terminal half of mature α_2PI into the pSV2 vector, which contains the replication origin, the enhancer, and the early promotor of SV40 (Fig. 8) (35). The expression vectors for α_2PI Nara and α_2PI Okinawa, pSV2PN and pSV2PO, respectively, were constructed by substituting a cloned fragment containing each mutation for the corresponding normal fragment in pSV2PI (35,36).

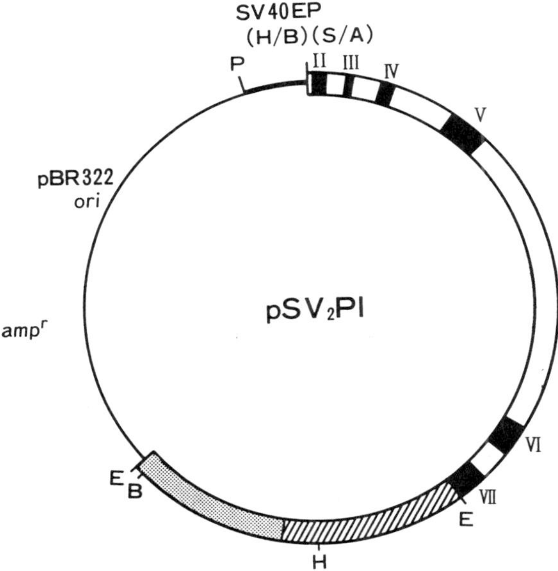

Figure 8 Structure of the α_2PI expression vector pSV2PI. The exons (solid black regions) in the genomic segment code the NH_2-terminal half of α_2PI, and the cDNA covers the rest of the coding sequence (hatched segment), stop signal, 3′ noncoding region (dotted segment), and polyadenylation signal. The exon numbers are indicated by the Roman numerals. The open segments represent the introns of the α_2PI gene. The thin and thick lines reveal parts of the pSV2 vector derived from pBR322 and the SV40 DNA, respectively. (H/B) and (S/A) indicate fusions of the blunt-ended restriction sites shown in parentheses. pBR322 ori, the origin of replication of pBR322; amp, ampicillin resistance gene; SV40EP, SV40 early promotor regions; A, AluI; B, BamHI; E, EcoRI; H, HindIII; P, PvuII; S, SmaI. (From Ref. 35.)

Each of these vectors was transfected into monkey COS-7 cells for transient expression analysis (35–37). RNA blot analysis of the cytoplasmic RNA of the transfected COS-7 cells did not reveal any significant difference between the transcript levels of a normal α_2PI expression vector, pSV2PI, and mutant expression vectors, pSV2PN and pSV2PO. Recombinant α_2PIs in the cell extracts or in the culture media of the transfected COS-7 cells were assayed by enzyme-linked immunosorbent assay (ELISA). The results indicated that ~99% of the normal recombinant α_2PI that was synthesized in the COS-7 cells was secreted into the culture medium 72 h after the transfection, whereas the levels of the mutant α_2PIs were only 5–18% of the normal protein secreted (Fig. 9) (35,36). Despite minute secretion of the mutant α_2PIs, their cellular contents did not differ much from those of the normal α_2PIs in the cells (Fig. 9). These results suggest that the secretions of these mutant proteins were retarded and the mutants retained in the cells were rapidly degraded.

To confirm that the secretory process of the mutants were impaired in COS-7 cells, the recombinant α_2PIs were immunoprecipitated from the cell extracts or the media of the transfected COS-7 cells that had been pulse labeled with [^{35}S] methionine for 15 minutes and then chased for varying periods up to 8 h with unlabeled methionine (36,37). To examine the state of the asparagine-linked oligosaccharides that undergo processing along the intracellular transport pathway, part of the immunoprecipitated α_2PI was treated with endo-β-*N*-acetylglucosaminidase (Endo H) or neuraminidase before being analyzed by SDS-polyacrylamide gel electrophoresis and fluorography. As shown in Figure 10, immediately after labeling, the cells transfected with pSV2PI contained a 64 kD form of

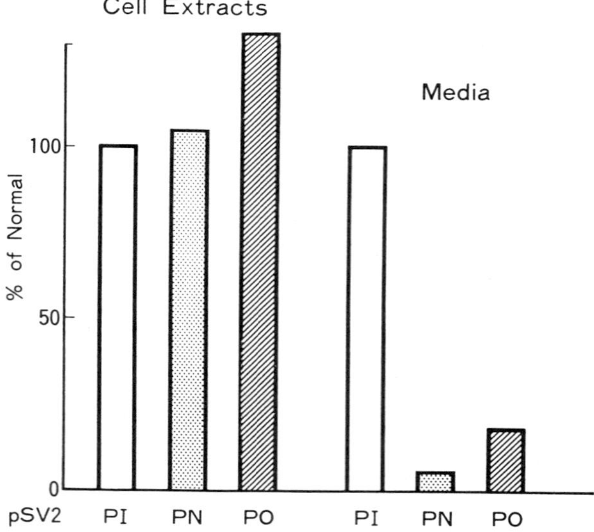

Figure 9 Expression and secretion of mutated recombinant α_2PI from COS cells. Culture media and cells were harvested 72 h after transfection with α_2PI expression plasmids. α_2PIs in the cell extracts and in the culture media were measured by ELISA, and the amounts of mutated α_2PIs in the cell extracts and in the media were expressed as a percentage of those of normal α_2PI. Approximately 99% of the total α_2PI was present in the media and only 1% in the cells when normal α_2PI was expressed. PI, normal α_2PI; PN, α_2PI Nara; PO, α_2PI Okinawa.

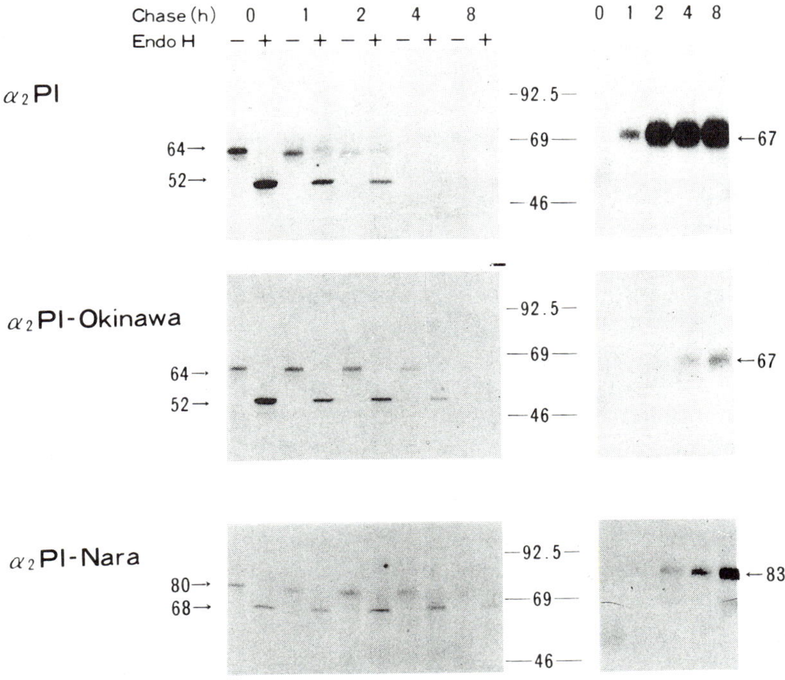

Figure 10 Pulse-chase analysis of α_2PI in transfected COS-7 cells. The expression plasmid for the normal α_2PI (pSV2PI), for α_2PI Okinawa (pSV2PO), or for α_2PI Nara (pSV2PN) was transfected into COS-7 cells for transient expression. At 48 h after transfection, the cells were incubated for 60 minutes in methionine-free medium and then pulse-labeled with [^{35}S]methionine for 15 minutes, followed by chasing with an excess of unlabeled methionine for the indicated period. Cells and media were harvested at appropriate intervals, and α_2PI was immunoprecipitated from each sample with goat antihuman α_2PI immunoglobulin. Immunoprecipitates derived from cell lysates were incubated for 16 h at 37°C in 10 μl of 10 mM Tris (pH 6.8) and 0.1% SDS in the absence or presence of 1 milliunit Endo H, as indicated. Subsequently, they were subjected to SDS-polyacrylamide gel electrophoresis (10% gels) and fluorography. The size standards used were ^{14}C-methylated myosin (200 kD), phosphorylase-b (92.5 kD), bovine serum albumin (69 kD), and ovalbumin (46 kD). (From Refs. 36 and 37.)

α_2PI. This species should represent the precursor form of α_2PI, because it is susceptible to digestion with Endo H; this enzyme cleaves the immature high-mannose form of asparagine-linked oligosaccharides but not the mature complex form (38). During the 4 h chase period, essentially all of this 64 kD form was secreted into the media after being converted to a 67 kD mature form. This higher molecular mass species is also slightly visible as the Endo H-resistant form at 1 or 2 h after labeling in the cell extracts.

The recombinant α_2PI Okinawa appeared in the cells as a 64 kD species immediately after labeling. This species was indistinguishable in size from the precursor form of the normal α_2PI and was susceptible to digestion with Endo H. Compared with normal α_2PI, however, the recombinant α_2PI Okinawa remained within the cells for prolonged periods as an Endo H-susceptible form, which is a form characteristic of processing intermediates in the endoplasmic reticulum (ER) or the Golgi apparatus (38). Moreover, α_2PI Okinawa was observed in the medium at a much lower level even 8 h after labeling, when it became almost undetectable within the cells.

The recombinant α_2PI Nara appeared in the cells as an 80 kD species immediately after labeling. This species should represent the precursor form, because it is sensitive to digestion by Endo H. The difference in size between the precursor forms of the normal α_2PI and α_2PI Nara is in agreement with the elongation of the amino acid sequence predicted by the frameshift mutation observed in the sequence analysis of the gene. Compared with the normal α_2PI, α_2PI Nara remained within the cells as an Endo H-susceptible form for a prolonged period; the mutant protein was still detected in the cells 4–8 h after labeling, when the normal protein was not detected. Moreover, only a small portion of the mutant molecule synthesized in the cells had been secreted into the medium even when the intracellular precursor form became almost undetectable.

To rule out the possibility that the small amounts of recombinant mutant α_2PIs that appeared late in the medium leaked out of the COS-7 cells injured by the transfection procedures and the subsequent transient expression of the introduced plasmid, the α_2PIs immunoprecipitated from the medium were examined for their susceptibilities to Endo H or neuraminidase. Endo H affected neither the normal nor the mutant α_2PIs in the medium. In contrast, neuraminidase decreased the apparent molecular mass of both the normal and the mutant α_2PIs in the medium, but it had no effect on either precursor form within the cells. The results indicate that the recombinant α_2PIs immunoprecipitated from the medium were mature and possessed sialylated oligosaccharides; hence, they must have been secreted from the cells after being processed along the intracellular secretory pathway.

These results suggest that most of the mutant molecules are retained within the ER and undergo degradation instead of being secreted into the medium, because Endo H removes N-linked oligosaccharides from polypeptides during its transit through the ER but has no effect on the processed oligosaccharides after transfer of the proteins to the medial stacks of the Golgi compartment (38). However, the COS-7 cells used in these experiments are fibroblast-like cells that originated from monkey kidney tissue, and the secretory process may be quite different from that of liver cells. Therefore, we examined the secretory process of α_2PI Nara in HepG2, a hepatoma cell line synthesizing α_2PI, by taking advantage of the rare opportunity that the mutant molecule can be distinguished from the native molecule by the difference in their sizes (37). For this purpose, the expression plasmid for α_2PI Nara was transfected into HepG2. The results of the pulse-chase labeling experiments using one of the HepG2 clones stably expressing α_2PI Nara were in agreement with those obtained with COS-7 cells (37); therefore, the impaired intracellular transport of the mutant molecules within liver cells should be the primary cause of the α_2PI deficiencies.

VII. INTRACELLULAR TRANSPORT-DEFICIENT MUTANTS

As discussed, molecular biological studies of two families with hereditary deficiency of α_2PI revealed that the deficiencies were caused by an impairment of intracellular transport and subsequent secretion of the mutant α_2PIs. Impaired intracellular transport, which was originally reported for the null type of α_1-antitrypsin deficiency (39), was also suggested as the cause of the absence of protein C in plasma in two families with hereditary protein C deficiency (40,41). Hence, impaired intracellular transport may be one of the prevalent causes of the deficiencies of factors involved in coagulation and fibrinolysis (42).

Although the biochemical mechanisms for the impaired transport of these proteins have not been precisely defined, it has been hypothesized that a newly synthesized secretory or membrane protein would not be transported from the ER to the Golgi apparatus unless it folds into a native or near native conformation (43,44). α_2PI Nara and protein C Nagoya (41) are substantially larger than normal; protein C Hong Kong 1 (40) is truncated to a half of normal. It is conceivable that these mutants have severely distorted folding and impaired intracellular transport. In contrast, α_2PI Okinawa and protein C Hong Kong 2 (40) have only a single amino acid deletion or replacement, respectively, and are hardly expected to have grossly aberrant folding. In α_2PI, deletion of any single amino acid in a certain limited region of the molecule would result in impaired intracellular transport and secretion (45). How these amino acids play a role in the intracellular transport of the protein is an intriguing question, and further studies are needed to answer the question. Some of the various mutants identified in hereditary deficiencies of hemostatic and fibrinolytic factors may prove to be useful for these studies and for further efforts directed toward better understanding of the biochemical basis of the intracellular transport mechanisms and its impairment.

REFERENCES

1. Aoki N, Yamanaka T. The α_2-plasmin inhibitor levels in liver diseases. Clin Chim Acta 1978; 84:99–105.
2. Saito H, Goodnough LT, Knowles BB, Aden DP. Synthesis and secretion of α_2-plasmin inhibitor by established human liver cell lines. Proc Natl Acad Sci USA 1982; 79:5684–5687.
3. Aoki N, Saito H, Kamiya T, Koie K, Sakata Y, Kobakura M. Congenital deficiency of α_2-plasmin inhibitor associated with severe hemorrhagic tendency. J Clin Invest 1979; 63:877–884.
4. Aoki N. Genetic abnormalities of the fibrinolytic system. Semin Thromb Hemostas 1984; 10: 42–50.
5. Saito H. α_2-Plasmin inhibitor and its deficiency states. J Lab Clin Med 1988; 112:671–678.
6. Aoki N. Hemostasis associated with abnormalities of fibrinolysis. Blood Rev 1989; 3:11–17.
7. Sakata Y, Aoki N. Cross-linking of α_2-plasmin inhibitor to fibrin by fibrin-stabilizing factor. J Clin Invest 1980; 65:290–297.
8. Sumi Y, Nakamura N, Aoki N, Sakai M, Muramatsu M. Structure of the carboxylterminal half of human α_2-plasmin inhibitor deduced from that of cDNA. J Biochem 1986; 100:1339–1402.
9. Tone M, Kikuno R, Kume-Iwaki A, Hashimoto-Gotoh T. Structure of human α_2-plasmin inhibitor deduced from the cDNA sequence. J Biochem 1987; 102:1033-1041.
10. Holmes WE, Nelles L, Lijnen HR, Collen D. Primary structure of human α_2-antiplasmin, a serine protease inhibitor (serpin). J Biol Chem 1987; 262:1659–1664.
11. Sumi Y, Ichikawa Y, Nakamura Y, Miura O, Aoki N. Expression and characterization of pro α_2-plasmin inhibitor. J Biochem 1989; 106:703–707.
12. Bangert K, Johnsen AH, Christensen U, Thorsen S. Different NH_2-terminal forms of α_2-plasmin inhibitor in human plasma. Biochem J 1993; 291:623–625.
13. Moroi M, Aoki N. Isolation and characterization of α_2-plasmin inhibitor from human plasma. A novel proteinase inhibitor which inhibits activator-induced clot lysis. J Biol Chem 1976; 251:5956–5965.
14. Sasaki T, Morita T, Iwanaga S. Identification of the plasminogen-binding site of human α_2-plasmin inhibitor. J Biochem 1986; 99:1699–1705.

15. Sasaki T, Sugiyama N, Iwamoto M, Isoda S. Studies on α_2-plasmin inhibitor fragment T-11. Chem Pharm Bull 1987; 35:2810–2818.
16. Kluft C, Los N. Demonstration of two forms of α_2-antiplasmin in plasma by modified crossed immunoelectrophoresis. Thromb Res 1981; 21:65–71.
17. Moroi M, Aoki N. Inhibition of proteases in coagulation, kinin-forming and complement systems by α_2-plasmin inhibitor. J Biochem 1977; 82:969–972.
18. Aoki N, Moroi M, Matsuda M, Tachiya K. The behavior of α_2-plasmin inhibitor in fibrinolytic states. J Clin Invest 1977; 60:361–369.
19. Wiman B, Lijnen HR, Collen D. On the specific interaction between the lysine-binding sites in plasmin and complementary sites in α_2-antiplasmin and in fibrinogen. Biochim Biophys Acta 1979; 579:142–154.
20. Moroi M, Aoki N. Inhibition of plasminogen binding to fibrin by α_2-plasmin inhibitor. Thromb Res 1977; 10:851–856.
21. Ichinose A, Mimuro J, Koide T, Aoki N. Histidine-rich glycoprotein and α_2-plasmin inhibitor in inhibition of plasminogen binding to fibrin. Thromb Res 1984; 33:401–407.
22. Aoki N, Sakata Y, Ichinose A. Fibrin associated plasminogen activation in α_2-plasmin inhibitor deficiency. Blood 1983; 62:1118–1122.
23. Clemmensen I, Thorsen S, Müllertz S, Peterson LC. Properties of three different molecular forms of the α_2-plasmin inhibitor. Eur J Biochem 1981; 120:105–112.
24. Sakata Y, Aoki N. Cross-linking of α_2-plasmin inhibitor to fibrin by fibrin-stabilizing factor. J Clin Invest 1980; 65:290–297.
25. Sakata Y, Aoki N. Significance of cross-linking of α_2-plasmin inhibitor to fibrin in inhibition of fibrinolysis and in hemostasis. J Clin Invest 1982; 69:536–542.
26. Jansen JWC, Haverkate F, Koopman J, Nieuwenhuis HK, Kluft C, Boschman TAC. Influence of factor XIIIa activity on human whole blood clot lysis in vitro. Thromb Haemost 1987; 57:171–175.
27. Tamaki T, Aoki N. Cross-linking of α_2-plasmin inhibitor to fibrin catalyzed by activated fibrin-stabilizing factor. J Biol Chem 1982; 257:14767–14772.
28. Kimura S, Aoki N. Cross-linking site in fibrinogen for α_2-plasmin inhibitor. J Biol Chem 1986; 261:15591–15595.
29. Kato A, Hirosawa S, Toyota S, et al. Localization of the human α_2-plasmin inhibitor gene (PLI) to 17p13. Cytogenet Cell Genet 1993; 62:190–191.
30. Hirosawa S, Nakamura Y, Miura O, Sumi Y, Aoki N. Organization of the human α_2-plasmin inhibitor gene. Proc Natl Acad Sci USA 1988; 85:6836–6840.
31. Shaul Y, Ben-Levy R. Multiple nuclear proteins in liver cells are bound to hepatitis B virus enhancer element and its upstream sequences. EMBO J 1987; 6:1913–1920.
32. Miura O, Sugahara Y, Nakamura Y, Hirosawa S, Aoki N. Restriction fragment length polymorphism caused by a deletion involving Alu sequences within the human α_2-plasmin inhibitor gene. Biochemistry 1989; 28:4934–4938.
33. Kluft C, Nieuwenhuis HK, Rijken DC, et al. α_2-Antiplasmin Enschede: dysfunctional α_2-antiplasmin molecule associated with an autosomal recessive hemorrhagic disorder. J Clin Invest 1987; 80:1391–1400.
34. Holmes WE, Lijnen HR, Nelles L, et al. α_2-Antiplasmin Enschede: alanine insertion and abolition of plasmin inhibitory activity. Science 1987; 238:209–211.
35. Miura O, Hirosawa S, Kato A, Aoki N. Molecular basis for congenital deficiency of α_2-plasmin inhibitor. J Clin Invest 1989; 83:1598–1604.
36. Miura O, Sugahara Y, Aoki N. Hereditary α_2-plasmin inhibitor deficiency caused by a transport-deficient mutation (α_2PI-Okinawa). J Biol Chem 1989; 264:18213–18219.
37. Miura O, Aoki N. Impaired secretion of mutant α_2-plasmin inhibitor (α_2PI-Nara) from COS-7 and HepG2 cells: molecular and cellular basis for hereditary deficiency of α_2-plasmin inhibitor. Blood 1990; 75:1092–1096.

38. Kornfeld R, Kornfeld S. Assembly of asparagine-linked oligosaccharides. Annu Rev Biochem 1985; 54:631–664.
39. Brantly M, Nukiwa T, Crystal RG. Molecular basis of α_1-antitrypsin deficiency. Am J Med 1988; 84:13–31.
40. Sugahara Y, Miura O, Yuen P, Aoki N. Protein C deficiency Hong Kong 1 and 2: hereditary protein C deficiency caused by two mutant alleles, a 5-nucleotide deletion and a missence mutation. Blood 1992; 80:126–133.
41. Yamamoto K, Tanimoto M, Emi N, Matsushita T, Takamatsu J, Saito H. Impaired secretion of the elongated mutant of protein C (protein C-Nagoya). J Clin Invest 1992; 90:2439–2446.
42. Miura O, Sugahara Y, Aoki N. Intracellular transport-deficient mutants causing hereditary deficiencies of factors involved in coagulation and fibrinolysis. Thromb Haemost 1993; 69: 296–297.
43. Lodish HF. Transport of secretory and membrane glycoprotein from the rough endoplasmic reticulum to the Golgi. A rate-limiting step in protein maturation and secretion. J Biol Chem 1988; 263:2107–2110.
44. Pfeffer SR, Rothman JE. Biosynthetic protein transport and sorting by the endoplasmic reticulum and Golgi. Annu Rev Biochem 1987; 56:829–852.
45. Toyota S, Hirosawa S, Aoki N. Secretion of α2-plasmin inhibitor is impaired by amino acid deletion in a small region of the molecule. J Biochem 1994; 115:293–297.

27

Human Platelet Glycoprotein Ib and the Ib-V-IX System: The Receptor for Platelet Adhesion in Arteries

Gerald J. Roth
Seattle Veterans Hospital, and University of Washington, Seattle, Washington

I. STRUCTURAL AND FUNCTIONAL ASPECTS OF GLYCOPROTEIN Ib AND THE Ib-V-IX SYSTEM

Glycoprotein (GP) Ib is a major element on the surface of human platelets (1), serving as the receptor for von Willebrand factor (vWF) and mediating platelet adhesion in the arterial circulation (2,3). GPIb is linked within a series of three separate glycoproteins that are grouped together and referred to as the Ib-V-IX system (4,5). The system consists of four distinct polypeptides: GPIbα, M_r 143,000; GPIbβ, M_r 22,000; GPV, M_r 83,000; and GPIX, M_r 20,000 (1,4,6). The four proteins share several features: physical associations, a structural motif (leucine-rich glycoprotein segments, LRGs), and a common congenital deficiency state (Bernard-Soulier syndrome, BSS) (7–10). The GPIbα chain contains a discrete binding site for vWF and appears to be the major contributor to the adhesive interaction between receptor and ligand (11). All of the proteins within the Ib-V-IX system are interrelated, however, and all appear to contribute to both the surface expression and the function of the receptor (12). A sketch of the system is given in Figure 1, showing the transmembrane nature of the four individual glycoproteins, along with their single or tandem LRG repeats and oligosaccharide chains. The critical vWF binding site is noted, lying in the intervening region ("hinge") between the LRG and O-carbohydrate (O-CHO) domains of the GPIbα chain.

A. Historical Background

Two French physicians, Bernard and Soulier, reported an unusual bleeding disorder in 1948 involving a boy with a congenital platelet disorder marked by low platelet counts, large circulating platelets, and a severe (later fatal) bleeding diathesis (10). Over the next three decades, much of the progress in understanding GPIb came from observations of the platelets of Bernard-Soulier syndrome patients. These platelets were observed to lack surface sialic acid, to adhere poorly to vascular surfaces in flow systems, and to exhibit

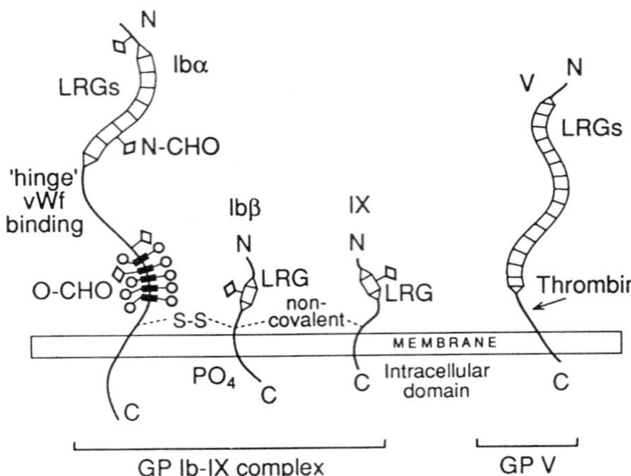

Figure 1 The Ib-V-IX system. The four transmembrane polypeptide chains, GPIbα, M_r 143,000, GPIbβ M_r 22,000, GPV M_r 83,300, and GPIX, M_r 20,000, are shown with large extracellular NH_2-terminal (N) and small intracellular (C) regions. Leucine-rich glycoprotein (LRG) segments are indicated by single or tandem boxes along with flanking regions (triangles). The tandem O-linked carbohydrate (O-CHO) repeats (solid rectangles attached to open circles) in Ibα are adjacent to a vWF binding region (hinge). Sites of thrombin cleavage (arrow), noncovalent association (dashed lines), and N-linked glycosylation (diamonds) are noted. (Reprinted with permission from Ref. 45.)

subnormal or absent agglutination with ristocetin (13,14). The biochemical basis of the syndrome was discovered in 1975, when BSS platelets were shown to lack surface glycoprotein I (later amended to be a specific deficiency of GPIb), the sialated receptor responsible for platelet adhesion and ristocetin reactivity (13–16).

B. Structural and Functional Features of GPIb

Categorization of platelet surface glycoproteins began with a rough division, on the basis of size, into three main groups, I, II, and III, which were later subdivided into individual proteins (Ia, b, and c) through the application of improved electrophoretic techniques (1). Glycoprotein Ib was identified as a heterodimeric molecule with a larger α chain, M_r 143,000, linked through disulfide bond(s) to a smaller β chain, M_r 22,000 (1). The protein is cleaved by calpain to release its large (M_r 120,000) extracellular portion, a hydrophilic carbohydrate-rich (50% by weight) peptide termed glycocalicin (17). Subsequent work showed a ''tight'' noncovalent association between GPIb and a second smaller glycoprotein, GPIX (M_r 20,000), forming the Ib-IX complex (4,7). The last member of the system, GPV, is a midsized surface glycoprotein (M_r 83,000) that is cleaved by thrombin and is ''loosely'' associated with the GPIb-IX (6,8). A normal platelet possesses about 25,000 copies of the Ibα-Ibβ-IX complex with 1:1:1 stoichiometry, and approximately one-half of the Ib-IX complexes carry an additional GPV molecule (8,18). The Ib-IX complex is an elongated rod-like structure, 60 nm in length, with an extracellular region extending approximately 50 nm from the nodular COOH terminus that includes the transmembrane and intracellular portions of the three chains (19).

With regard to function, the studies of Weiss, Nurden, Hutton, and their collaborators identified GPIb as a critically important receptor for vWF that mediated platelet adhesion in the arterial circulation (2,3,14,15). Within the receptor, the actual binding site for vWF lies in the central portion of GPIbα in close proximity to an anionic site that binds thrombin (11,20). The Ib-V-IX system also contributes to the normal shape of the platelet, perhaps through its interaction with actin binding protein and the peripheral membrane skeleton (21,22). In addition, the Ib-V-IX system provides surface antigen that can interact with antiplatelet antibodies (23).

C. Physiology of Arterial Platelet Adhesion

A variety of studies have contributed to our current knowledge of the GPIb-vWF interaction. The two critical features of the interaction are its shear dependence and activation independence, referring to fact that GPIb and vWF react with each other only under specific conditions of blood flow (shear dependence) and only on the surface of unactivated, resting platelets (activation independence).

1. Shear Dependence and Activation Independence

Under static conditions, in the absence of blood flow, platelet surface GPIb does not interact with plasma vWF, and no affinity is exhibited between the two molecules. One can induce a specific interaction with exogenous, nonphysiological agents, such as ristocetin and botrocetin (14). However, these agents have no in vivo correlate because the effects of ristocetin bear no clear relationship to the forces of flow or vascular damage that are present under living conditions and that lead to a specific interaction of the receptor-ligand pair (24). The activation independence aspect of GPIb-vWF relates to the fact that the surface receptor disappears upon platelet activation and becomes internalized within the intracellular compartment of the platelet, making the receptor unavailable for surface adhesion reactions (25).

2. Implications of Shear Dependence: Localization and Reversibility

The shear dependence of the Ib-vWF interaction implies that the two molecules will develop an affinity for one another primarily at interfaces between flowing blood and stationary vascular surfaces, where shear forces are maximal. As a result, the Ib-vWF interaction is "localized" to a specific point within the normal vasculature, and the interaction is also "reversible" because it arises only at this interface and is otherwise absent (5). Careful biophysical work has measured the shear forces generated in the arterial circulation at blood/blood vessel interfaces, and the forces are expressed as shear rates of 600–3000 s^{-1} (24). These shear rates are sufficient to "activate" the GPIb-vWF interaction from its usual "off" mode to its highly specific "on" mode (2,3,5,24).

3. Potential Mechanisms Responsible for the Off-On Quality of Ib-vWF

The molecular mechanisms that underpin the off-on aspect of Ib-vWF are unknown and present a major challenge for researchers in this field (5). Either GPIb or vWF alone, or both molecules together, appears to undergo a conformational change(s) in the presence of shear forces (Fig. 2). Some workers in this field argue that shear forces are not the primary or major influence affecting Ib-vWF and that the binding of vWF to vascular subendothelium (regardless of flow or shear) is the main stimulant for switching Ib-vWF from off to on (26). The validity of such a formulation depends on demonstrating that vWF, when bound to subendothelial elements, such as collagen, develops a specific af-

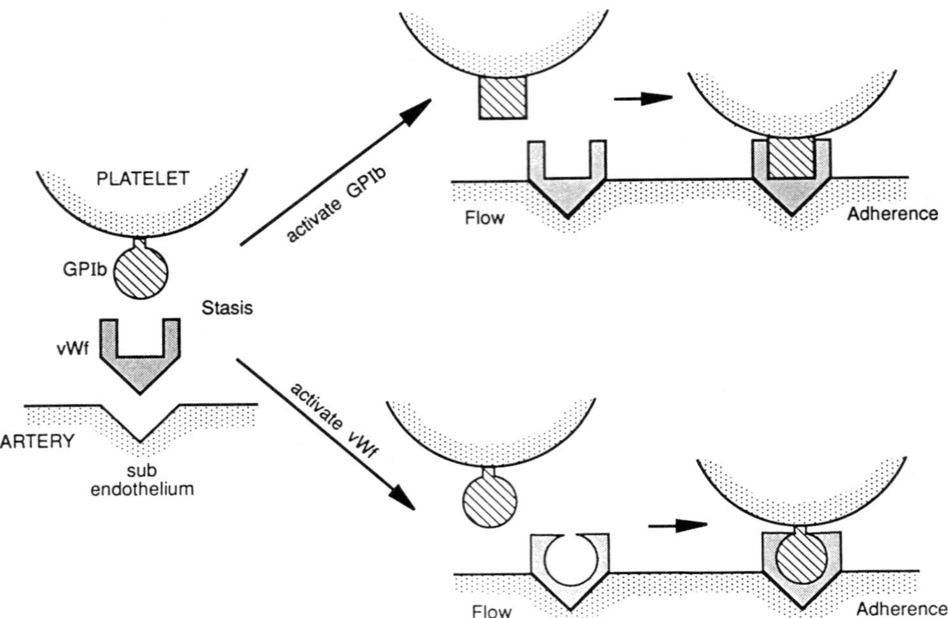

Figure 2 Potential interactions between GPIb and vWF. Under low flow (venous) conditions, platelet GPIb and plasma vWF do not interact (stasis, left). In the presence of rapidly flowing, arterial blood (flow, right), a conformational change occurs in either the GPIb receptor or the vWF ligand (or both) that ''activates'' the molecule(s), leading to a specific receptor-ligand binding event and subsequent platelet adherence to the vascular subendothelium.

finity for GPIb in the absence of flow. Such data are not readily available or reproducible, and the bulk of data point to flow or shear as the essential stimulant to Ib- and vWF-dependent platelet adhesion (5,24).

4. GPIb in Relation to Other Adhesion Receptors

Several other receptors on the surface of human platelets, such as GPIa-IIa, Ic-IIa, IIb-IIIa, and IV, contribute to the adhesion of platelets to vascular surfaces (27–30). However, these additional receptors do not appear to function under conditions of rapid blood flow in the presence of high shear forces. The important contribution of Ib-vWF to normal hemostasis is demonstrated by the fact that individuals who lack these molecules (BSS-Ib and von Willebrand disease, vWD vWF) suffer significant and even fatal bleeding (31). The other receptors appear to work under ''venous'' conditions of reduced blood flow, or if they function in the arterial circulation, they appear to act after adhesion mediated initially by Ib-vWF.

II. ANALYSIS OF THE Ib-V-IX SYSTEM THROUGH cDNA CLONING

A. Primary Structures Predicted from cDNA Sequences

Current understanding of the Ib-V-IX system includes knowledge of gene, transcript, and protein sequences (32–36). Future insight into molecular aspects of the Ib receptor will depend on sequence information. To obtain the materials needed for detailed study of gene expression and receptor function, a series of investigations was launched to clone the cDNAs encoding each of the polypeptides of the Ib-V-IX system (32–36). The work began with successive screenings of a λgt 11 cDNA expression library with polyclonal antibody probes directed against the individual chains. The nucleotide sequences of the cloned cDNA inserts were verified by comparing predicted amino acid sequences with those determined chemically by Edman degradation of the purified glycoproteins from human platelets (37–39). Because platelets lack a nucleus and are incapable of gene transcription, platelet RNA was not utilized for construction of the cDNA library (40). Instead, transcripts for the proteins of interest were obtained from human erythroleukemia (HEL) cells that had been induced to a megakaryocytic phenotype by phorbol myristate acetate (32–34). In the cDNAs encoding GPIbα, Ibβ, and IX, "full-length" cDNA sequences were obtained that provide both the primary structures of the individual polypeptides and useful probes for additional studies (Fig. 3). Several common features are present in the primary structures of the three glycoproteins. First, all three are transmembrane proteins with larger extracellular and smaller intracellular regions (Fig. 4). Second, all three possess one or more leucine-rich glycoprotein segments (Fig. 5) in a flank-LRG center-flank array, and third, all are synthesized with signal peptides of approximately 20 amino acids that precede the polypeptide chain of the mature glycoprotein (9,32–36).

1. *GPIbα cDNA*

The described 2.4 kb cDNA for GPIbα encodes a mature protein of 610 amino acids, with 485 extracellular, 96 intracellular, and 29 transmembrane residues (32). The extracellular portion has a complex domain structure consisting of an LRG region (seven tandem 24 amino acid repeats), an O-linked carbohydrate region (five tandem "mucin-like" repeats, O-CHO), and an intervening hinge binding region (residues 220–310) lying between the two. The hinge domain includes a discrete binding site for von Willebrand factor (ristocetin-induced binding) and an anionic site for thrombin that resembles similar sequences found in the thrombin receptor and in leech hirudin (11,20). An intracellular sequence in GPIbα binds to actin binding protein and may participate in transmembrane signaling through the platelet plasma membrane following interaction of the Ib receptor with vWF (41).

2. *GPIbβ cDNA*

A 0.97 kb cDNA for GPIbβ codes for a mature protein of 181 residues: 122 extracellular, 34 intracellular, and 25 transmembrane (33). The few features of Ibβ include a single extracellular LRG segment and an intracellular phosphorylation site for cAMP-dependent protein kinase(s) (42). Cysteine(s) located near the transmembrane domain are likely to be involved in the disulfide link(s) between Ibβ and Ibα.

3. *GPIX cDNA*

The 0.84 kb GPIX cDNA encodes a 160 amino acid polypeptide (134 extracellular, 6 intracellular, and 20 transmembrane residues) with a single LRG segment and an unpaired, possibly palmitoylated cysteine in the transmembrane domain that appears to contribute to membrane binding of the Ib-IX complex (34,43).

4. *GPV*

The cDNA sequence of GPV has been published recently, and considerable GPV peptide sequence has been reported (36,39). GPV is known to be a thrombin substrate and to contain multiple LRG repeats (6,39).

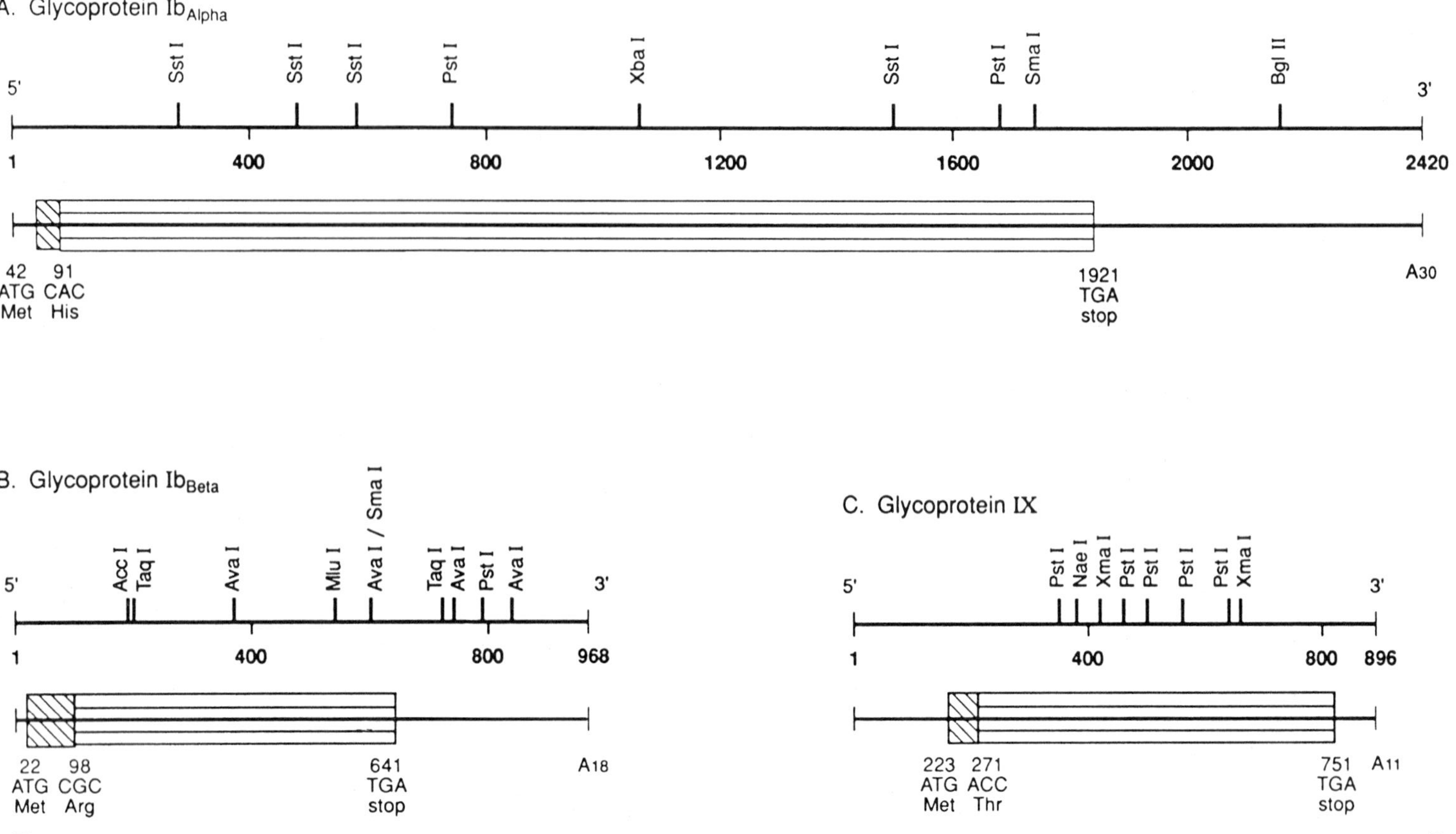

Figure 3 Diagrams and restriction maps of the GPIbα/β,IX cDNAs (32–34). (A) The 2420-nucleotide GPIbα sequence encompasses an open reading frame (626 amino acids—16 signal and 610 mature—ATG start/TGA stop codons, and CAC-HIS NH_2-terminal codon) and ends in a poly(A) tail. (B) The 968-nucleotide GPIbβ sequence encompasses an open reading frame (206 amino acids—25 signal and 181 mature—ATG start/TGA stop codons, and CGC-Arg NH_2-terminal codon) and ends in a poly(A) tail. (C) The 896-nucleotide GPIX sequence encompasses an open reading frame (176 amino acids—16 signal and 160 mature—ATG start/TGA stop codons, and ACC-Thr NH_2-terminal codon) and ends in a poly(A) tail.

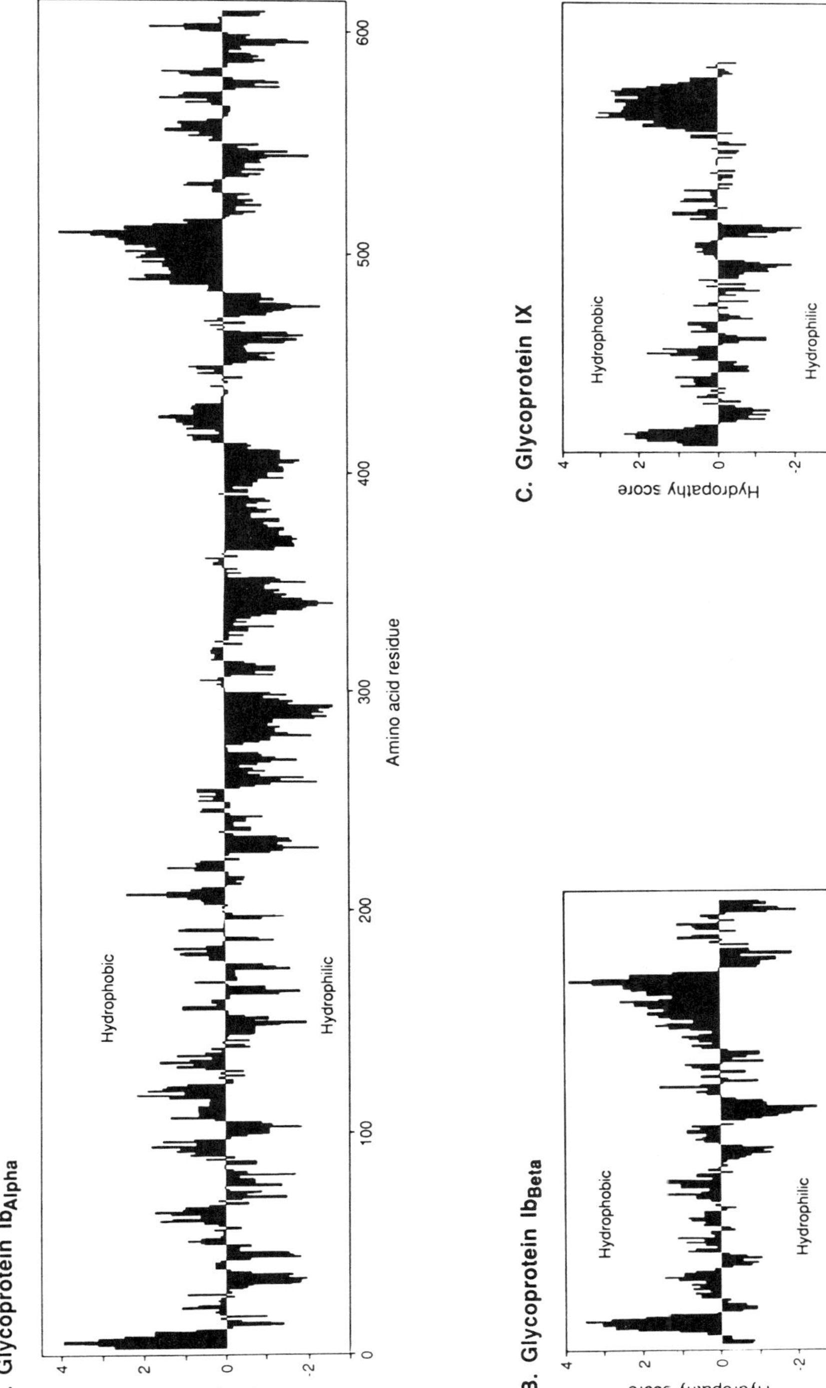

Figure 4 Hydropathy plots of the predicted primary sequences of GPIbα/β,IX indicate both signal and transmembrane domains (32–34). (A) GPIbα; (B) GPIbβ; and (C) GPIX.

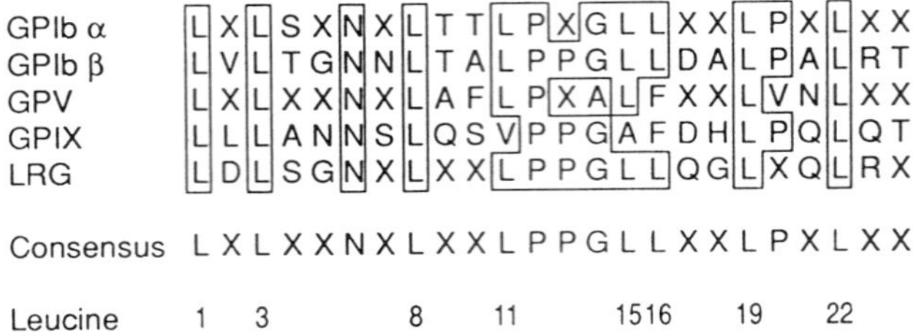

Figure 5 Leucine-rich glycoprotein (LRG) sequences in the Ib-V-IX system. The single or consensus 24 amino acid LRG sequences of GPIbα (consensus of seven tandem segments), GPIbβ (single), GPV (consensus of six segments from amino acid sequence data), GPIX (single), and LRG (consensus of 10 tandem repeats in the "proband" LRG protein from human serum reported in 1985) are shown with conserved residues boxed. An overall consensus sequence is given at the bottom, along with the position of conserved leucines. Amino acids are noted in single-letter code, and nonconserved residues are denoted by X. (Reprinted with permission from Ref. 45.)

B. Leucine-Rich Glycoprotein and Mucin Repeats

The first LRG protein was reported in 1985 and termed leucine-rich α_2-glycoprotein of human serum (9). This observation initiated a series of additional reports on a new and rapidly growing family of proteins, the LRG family, whose members are found in virtually every species (bacteria, yeast, fruit flies, cows, and humans) and in a variety of organs and tissues (retina, pancreas, nerves, embryo, and platelets) (5,34). The only common thread among the different members of the family appears to be their role in protein-protein interactions, serving a variety of functions, such as adhesion, orientation, position, and enzyme inhibition. The exact function of the LRG structure remains unknown, but the LRG sequence bears some similarity to that of "leucine zipper" proteins, such as c-jun and c-fos, that influence gene transcription (44). As shown in Figure 5, the LRG sequence includes conserved leucines at seven-residue intervals: positions 1, 8, 15, and 22 in a single repeat. This spacing and the fact that leucines are conserved, as contrasted with other hydrophobic residues (isoleucine and valine), implies that the leucine zipper mechanism (leucines along one ridge of an α helix intercalated with a complementary set of leucines on a second α helix) are involved in the function of the LRG structure (5,44,45).

A second system of tandem repeats is found within the GPIbα chain to the NH_2-terminal side of the transmembrane domain. These repeats are termed mucin-like because they are found in a number of glycoproteins and are associated with O-linked (serine/threonine) carbohydrate (32,46). The core sequence consists of a 13 amino acid repeat that is replicated in different Ibα alleles to produce a Ib size polymorphism, as recently reported (46). The function of the O-CHO domain may be to extend the vWF binding site out from the platelet membrane, where it can function in adhesion events (19,46). GPIbα contains approximately 30 hexasaccharide chains that account for 40–50% of the weight of the Ibα chain (47).

An additional polymorphism has been described in the Ibα chain, a methionine/threonine dimorphism that accounts for the HPA-2 platelet antigen system (Sib/Ko). In one kindred, an HPA-2 antigen was linked with an atypical form of Bernard-Soulier syndrome (23,48).

III. THE OFF-ON NATURE OF THE GPIb-vWF INTERACTION

A. Gain of Function Mutations

Direct experimentation on the mechanism of the in vivo Ib-vWF interaction is complicated by the fact that shear forces are not easily incorporated into studies at a molecular level. In such an instance, one can turn to congenital disorders (experiments of nature) that affect a reaction of interest; in this case, GPIb-vWF. Two types of relevant "gain of function" mutations have been described, one affecting the platelet receptor, termed pseudo-von Willebrand disease (pvWD), and the other affecting the plasma ligand, termed type IIB von Willebrand disease (49,50). In both cases, an essential quality of the receptor-ligand pair; namely, its off-on nature, has been altered and the system is switched to a permanently on mode, a gain of function condition. By understanding the molecular defect in the mutant molecule(s) (Fig. 6), one may gain insight into the change induced by shear forces because the mutation and shear exert similar effects, namely activation of the receptor-ligand interaction.

The phenotype seen in the two shear-independent mutations (pvWD and IIB-vWD) is essentially the same. Because of the on mode of either the receptor or ligand, plasma vWF is adsorbed excessively to the platelet surface, independently of shear forces, leading to premature destruction of the protein-laden platelets and to a fall in the level of plasma vWF. A bleeding disorder arises because of the drop in platelet levels (thrombocytopenia) and the low levels of plasma vWF (a vWD type of defect) (49,50).

1. Pseudo-von Willebrand Disease

First described by Weiss and coworkers in 1982, pseudo (platelet-type) von Willebrand disease is the phenotypic expression of an abnormal receptor with an altered binding site

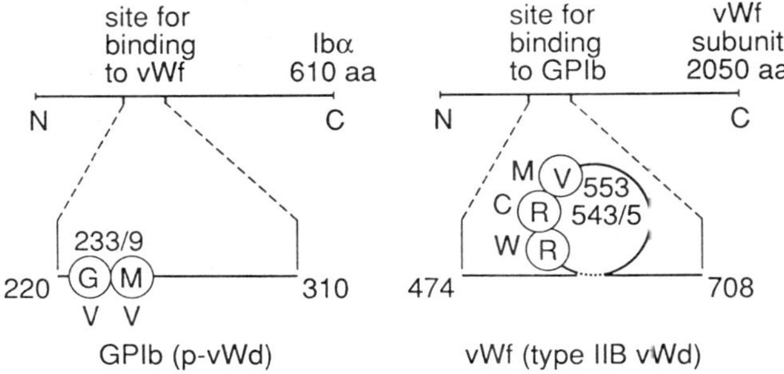

Figure 6 Point mutations in "shear-independent" mutations. Two reported mutations in pseudo-von Willebrand disease (pvWd) kindreds are shown: a valine for glycine substitution in codon 233 and a valine for methionine substitution in codon 239. Both lie within the portion of GPIbα that binds to vWF (220–310) (51–53). Three reported mutations in type IIB von Willebrand disease (vWd) kindreds are shown: a tryptophan for arginine 543, a cysteine for arginine 545, and a methionine for valine 553. All lie within the portion 474–708 of the vWF subunit that binds to platelet GPIbα. (From Refs. 54 and 55.)

for vWF. Molecular genetic defects have been defined in three different kindreds, one with a valine for glycine substitution in codon 233 and two with valine for methionine substitutions in codon 239 (51–53). Apparently, subtle alterations in primary structure can dramatically change the function of the receptor. The mutations lie in the portion of the receptor molecule that governs binding to vWF. The described mutations will now be studied with site-directed mutation and expression work.

2. *Type IIB von Willebrand disease*

The first kindred with type IIB von Willebrand disease was reported in 1980, marked by hyperfunction of the vWF ligand, which is capable of binding to platelet GPIb under static conditions in the absence of ristocetin. A series of different point mutations in different kindreds have now been reported, along with several site-directed mutation and expression studies. Similar to the findings in pvWD, the IIB-vWD mutations lie in the region of the affected molecule that governs binding to other member of the receptor-ligand pair. The mutations are clustered near codons 543–553, indicating that a subtle alteration in primary structure can have a dramatic effect on ligand function (54,55). Shear-independent (p vWD and IIB vWD) mutations are shown in Figure 6.

3. *A Unifying Hypothesis for Shear-Dependent Receptor Function*

In the absence of direct data, a hypothesis for shear-dependent receptor function has been developed that includes a functional contribution by the conserved LRG structures (5,45). In this model, shear forces physically alter the conformation of the LRG region of Ibα, and the on mode conformation is maintained through a regulated protein-protein interaction involving LRG segments. In turn, the vWF binding site in platelet GPIbα is exposed, and the receptor switches to an on mode with a highly specific affinity for the plasma and subendothelial vWF ligand. The fate of this hypothesis will depend on data from expression studies. The hypothesis itself is outlined in Figure 2 (top).

B. Expression of Ib-V-IX cDNA

The GPIb receptor requires multiple cDNAs for its expression on the surface of a transfected cell. A sharp distinction is drawn between *surface* versus *soluble* expressed proteins. Initial work with a single GPIbα cDNA reported the expression of a truncated soluble fragment, residues 1–302, from transfected CHO or COS cells (56,57). Additional studies have been reported using this system, including site-specific deletion work on the vWF binding characteristics of the expressed Ibα fragment (56). To study surface Ibα receptor function, however, one must transfect at least two cDNAs, Ibα and Ibβ, and greater extents of surface expression require the GPIX cDNA (12). The role of the GPV cDNA is currently unknown, but this cDNA may make an additional contribution to surface expression of a functional Ibα receptor that is capable of binding to vWF. Lopez and his coworkers have made a systematic study of Ibα surface expression in three transfected cells and have shown the requirement for multiple cDNAs (Ibα/β and IX) in mouse L cells (Ref. 12 and Fig. 7). These demanding studies hold promise for unraveling the complexities of the Ib receptor. For example, expression of functional receptor in adequate amounts can lead to work on the function of the LRG segments and to studies that incorporate shear forces into the function of the expressed receptor.

A single study has been reported on the individual functions of LRG segments, utilizing site-specific deletion of LRG repeats in the pancreatic RNase inhibitor (58). Indi-

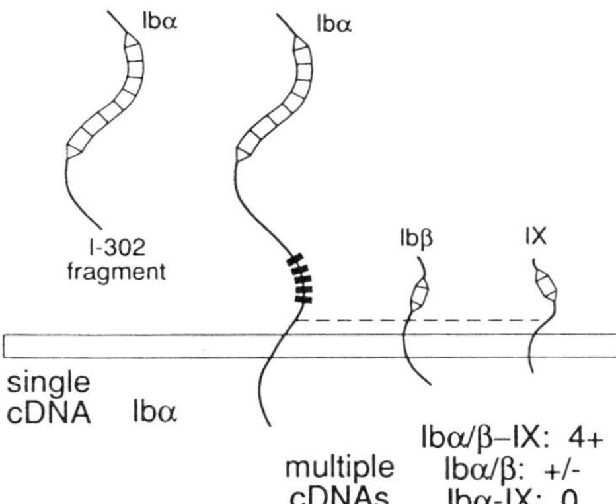

Figure 7 Current information derived from transfection studies of the GPIbα, Ibβ, and IX c-DNAs (12,56,57). Transfection of the GPIbα cDNA alone (left, single cDNA) results in the secretion of a soluble, truncated amino-terminal fragment, amino acids 1–302. No surface receptor is produced from the single Ibα transcript. Transfection of multiple cDNAs gives variable surface expression of the Ib receptor (right, multiple cDNAs). If all three cDNAs are utilized, one observes consistent expression of GPIbα receptor activity by ristocetin-induced vWF binding (Ibα/β-IX: 4+). With only two cDNAs, either minimal (Ibα/β: +/−) or absent (Ibα-IX: 0) surface expression is observed, depending on the choice of the transfected cDNAs.

vidual LRG segments were shown to subserve either inhibitory or binding functions (58). The Ib-V-IX system awaits similar study with the availability of the multiple cDNAs.

IV. THE GENES OF THE Ib-V-IX SYSTEM

A. Genomic Structures and Chromosomal Localizations

Two genes, those encoding GPIbα and GPIX, have been characterized in detail with transcriptional start sites and intron/exon boundaries (59–61). The GPIbα gene is located on chromosome 17 (60); that for GPIX is found on chromosome 3 (35). Clearly, the genes for the Ib-V-IX system are dispersed on different chromosomes and do not form a gene cluster. The genes for GPIbβ and GPV have not been cloned or localized to date.

The major feature of the Ibα and IX genes is their small size and related "intron-depleted" structure, similar to that of other described LRG genes, such as that of the oligodendrocyte-myelin glycoprotein (62). As shown in Figure 8, the Ibα and IX genes share exon sequences to the 5' side of their ATG start codons, and both contain short introns (233 or 108 bp) in this region. In view of the close relationship between the Ibα and IX genes, one can expect that the Ibβ and V genes are also related in structure to those already described.

The Ib-V-IX genes are all expressed in megakaryocytes, but their overall tissue distribution is still unclear. GPIX appears to be restricted primarily to marrow megakaryocytes, but GPIbα is expressed in endothelium exposed to interferon-γ or tumor necrosis factor α and GPV antigens are found in serum (6,36,63,64). Recent work on the 5' promoter region of the GPIX gene reveals two conserved consensus regulatory sequences in the reported megakaryocyte genes (GPIIb, PF4, BTG, and GPIX) (59). The two sites, GATA and *ets*,

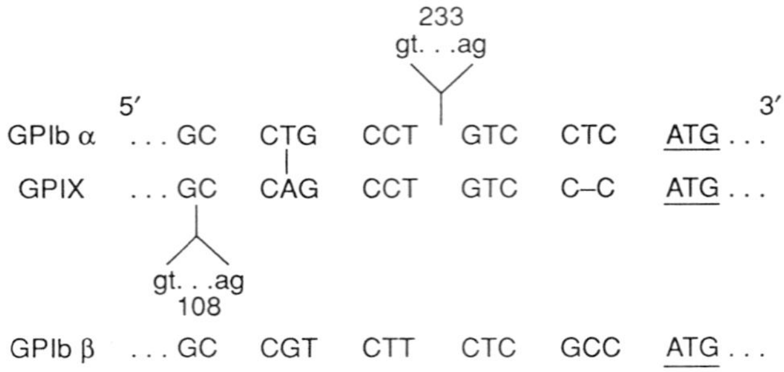

233
gt. . .ag

	5′					3′
GPIb α	. . . GC	CTG	CCT	GTC	CTC	ATG . . .
GPIX	. . . GC	CAG	CCT	GTC	C–C	ATG . . .

gt. . .ag
108

GPIb β	. . . GC	CGT	CTT	CTC	GCC	ATG . . .

Figure 8 Analysis of exon and intron structures near the *ATG* start codons of the Ibα/β,IX genes (59). The GPIbα and GPIX genes share exon sequences similar to the 5′ side of their *ATG* start codons, with 12 of 14 identical bases shown. One gap (−) is inserted, and one mismatch is present (T/A). Each gene has an intron inserted in this region: a 233 bp intron 6 bases 5′ to ATG and a 108 bp intron 12 bases 5′ to ATG in the Ibα and IX genes, respectively. The GPIbβ gene does not share the same structure. (Reprinted with permission from Ref. 59.)

are located within 90 bases of the transcriptional start sites and are juxtaposed, suggesting that they function together in regulating the expression of megakaryocyte/platelet genes (59). The relative positions of the consensus sequences are shown in Figure 9. The transcriptional start site for the GPIX gene differs between cell lines (HEL, Dami, and K562) and megakaryocytes, suggesting that gene expression differs significantly (start sites, consensus regulatory sequences, and transacting factors) between authentic megakaryocytes and cell lines with megakaryocytic features (59). Differences in reported promoter sequences in the IIb gene may reflect this difference in megakaryocyte gene regulation in various cells and cell lines (65–67). For example, in MEG-01 cells, the *ets* site in the GPIIb gene was found to influence expression but the same site does not appear to contribute to the gene's expression in HEL cells (66,67).

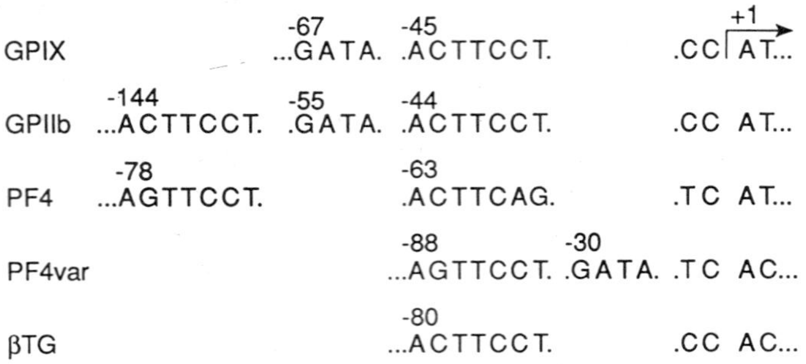

					+1
GPIX		-67 ...GATA.	-45 .ACTTCCT.		.CC AT...
GPIIb	-144 ...ACTTCCT.	-55 .GATA.	-44 .ACTTCCT.		.CC AT...
PF4	-78 ...AGTTCCT.		-63 .ACTTCAG.		.TC AT...
PF4var			-88 ...AGTTCCT.	-30 .GATA.	.TC AC...
βTG			-80 ...ACTTCCT.		.CC AC...

Figure 9 Potential consensus regulatory sequences in described megakaryocyte/platelet genes (59). Two potential regulatory sequences are present in human megakaryocyte genes, a GATA and an ACTTCCT (*ets*) sequence. The 5′ promoter regions of the four genes, GPIX, GPIIb, platelet factor 4 (PF4 and variant), and β-thromboglobulin (βTG), are noted by superscripts for the first base of each sequence to the 5′ side of the transcriptional start site, indicated as +1. Three genes contain both sequences, and two genes contain two copies of the *ets* sequence. Similar sequences are present in the related rodent genes for GPIIb and PF4. (Reprinted with permission from Ref. 59.)

The developmental regulation of the Ib-V-IX genes is not well described at this point, but the genes appear to be expressed either concomitantly or slightly later than those encoding platelet GPIIb-IIIa (68). Considerable additional work is required to characterize fully the structure and function of the Ib-V-IX genes.

B. Bernard-Soulier Syndrome

Bernard-Soulier syndrome can be roughly divided into two broad categories, typical and atypical. ''Typical'' BSS is marked by virtual absence of GPIb on the surface of the affected platelets and a marked increase in platelet size; atypical or variant BSS platelets are generally similar in size to normal platelets and possess easily detectable surface receptor and minor degrees of functional impairment (31). To date, molecular genetic defects have been reported in patients with atypical or variant BSS (48,69,70), but no typical cases have been reported. The available information on the BSS variants is summarized in Figure 10. All the defects are located in the GPIbα gene, and all three are point mutations, two of which result in amino acid substitutions, the third in a stop codon. All three described individuals are heterozygotes with one normal Ibα allele that appears to contribute to relatively normal GPIb function and preserved platelet structure.

Because expression work suggests that all four genes are required for normal surface receptor production, the search for the molecular defects in typical BSS with more severe functional and structural defects will entail extensive study of as many as four individual genes. These individuals may prove to be homozygotes for a gene or mixed heterozygotes

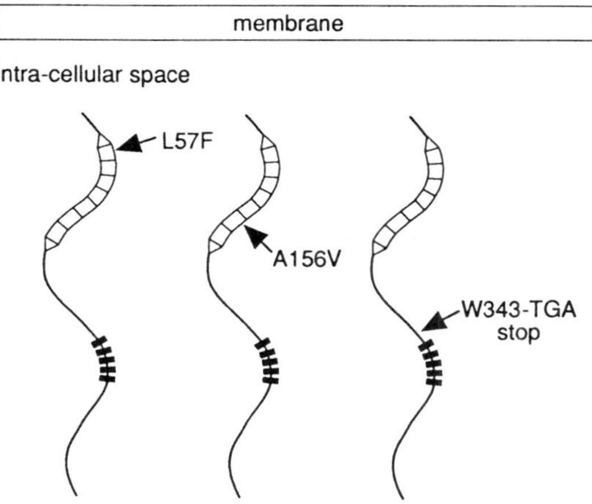

Figure 10 Described mutations associated with variant forms of Bernard-Soulier syndrome (48,69,70). Three point mutations have been described in the GPIbα genes of patients with mild, variant forms of Bernard-Soulier syndrome. One is found within the first LRG segment, a phenylalanine for leucine substitution in codon 57. A second is present in the sixth LRG, replacing valine for alanine in codon 156, associated with the Bolzano variant of BSS. The third is a stop codon mutation at codon 343 that results in a truncated Ibα chain. All are heterozygous conditions with one apparently normal allele and one mutant allele.

in whom two different genetic defects affect both alleles of a single gene. In all likelihood, BSS will prove to be a diverse disorder at the molecular genetic level, with abnormalities in any one of the four genes that result in a variety of phenotypes.

V. CURRENT QUESTIONS CONCERNING THE Ib-V-IX SYSTEM

Considerable unfinished business remains with regard to the molecular biology of the Ib-V-IX system, ranging from structural studies of genes to functional studies of the receptor interaction with vWF. With the availability of all four cDNAs, expression work can incorporate these cDNAs into work on receptor function. The critical questions to be answered by cDNA expression are the molecular basis of the off-on nature of the receptor-ligand interaction and the functional contribution of the LRG segments. The available cDNAs also provide essential probes for cloning and characterizing the Ibβ and V genes. With a complete set of gene structures, work can proceed to understand the promoter or enhancer elements and transacting factors that govern the tissue and developmental expression of the system. With regard to congenital disorders, confirmation of the functional consequences of the described pvWD mutations and work with BSs mutations will provide useful insights into the system.

In summary, the Ib-V-IX system produces the multi-component receptor for vWF that mediates platelet adhesion in the arterial circulation. The system is an essential link between extra- and intracellular events, providing for extracellular adhesion under high flow conditions and, in turn, activating the platelet and forming a nidus for further hemostatic and/or thrombotic events.

REFERENCES

1. Phillips DR, Agin PP. Platelet plasma membrane glycoproteins. Evidence for the presence of nonequivalent disulfide bonds using nonreduced-reduced two-dimensional electrophoresis. J Biol Chem 1977; 252:2121–2126.
2. Weiss HJ, Tschopp TB, Baumgartner HR, Sussman II, Johnson MM, Egan JJ. Decreased adhesion of giant (Bernard-Soulier) platelets to subendothelium. Am J Med 1974; 57:920–925.
3. Tschopp TB, Weiss HJ, Baumgartner HR. Decreased adhesion of platelets to subendothelium in von Willebrand's disease. J Lab Clin Med 1974; 83:296–300.
4. Clemetson KJ, McGregor JL, Dechavanne M, Luscher EF. Characterization of the platelet membrane glycoprotein abnormalities in Bernard-Soulier syndrome and comparison with normal by surface-labeling techniques and high-resolution two-dimensional electrophoresis. J Clin Invest 1982; 70:304–311.
5. Roth GJ. Developing relationships: arterial platelet adhesion, glycoprotein Ib and leucine-rich glycoproteins. Blood 1991; 77:5–19.
6. Berndt MC, Phillips DR. Purification and preliminary physiochemical characterization of human platelet membrane glycoprotein V. J Biol Chem 1981; 256:59–65.
7. Coller BS, Peerschke EI, Scudder IE, Sullivan CA. Studies with a murine monoclonal antibody that abolished the ristocetin-induced binding of von Willebrand factor to platelets: additional evidence in support of GP Ib as a platelet receptor for von Willebrand factor. Blood 1983; 61:99–110.
8. Modderman PW, Admiraal LG, Sonnenberg A, von dem Borne AEGK. Glycoproteins V and Ib-IX form a noncovalent complex in the platelet membrane. J Biol Chem 1992; 267:364–369.

9. Takahashi N, Takahashi Y, Putnam FW. Periodicity of leucine and tandem repetition of a 24-amino acid segment in the primary structure of leucine-rich alpha$_2$-glycoprotein of human serum. Proc Natl Acad Sci USA 1985; 82:1906–1910.
10. Bernard J, Soulier J-P. Sur une nouvelle variete de dystrophie thrombocytaire-hemorragipare congenitale. Semin Hop Paris 1948; 24:3217–3223.
11. Vicente V, Houghten RA, Ruggeri ZM. Identification of a site in the alpha chain of platelet glycoprotein Ib that participates in von Willebrand factor binding. J Biol Chem 1991; 265: 274–280.
12. Lopez JA, Leung B, Reynolds CC, Li CQ, Fox JEB. Efficient plasma membrane expression of a functional platelet glycoprotein Ib-IX complex requires the presence of its three subunits. J Biol Chem 1992; 267:12851–12859.
13. Grottum KA, Solum NO. Congenital thrombocytopenia with giant platelets: a defect in the platelet membrane. Br J Haematol 1969; 16:277–290.
14. Howard MA, Hutton RA, Hardisty RM. Hereditary giant platelet syndrome: a disorder of a new aspect of platelet function. Br M J 1973; 2:586–588.
15. Nurden AT, Caen JP. Specific roles for platelet surface glycoproteins in platelet function. Nature 1975; 255:720–722.
16. Nurden AT, Dupuis D, Kunicki TJ, Caen JP. Analysis of the glycoprotein and protein composition of Bernard-Soulier platelets by single and two-dimensional sodium dodecyl sulfate-polyacrylamide gel electrophoresis. J Clin Invest 1981; 67:1431–1440.
17. Okumura T, Lombart C, Jamieson GA. Platelet glycocalicin II: purification and characterization. J Biol Chem 1976; 251:5950–5955.
18. Berndt MC, Gregory C, Kabral A, Zola H, Fournier D, Castaldi PA. Purification and preliminary characterization of the GP Ib complex on the human platelet membrane. Eur J Biochem 1985; 151:637–649.
19. Fox JEB, Aggerbeck LP, Berndt MC. Structure of the glycoprotein Ib-IX complex from platelet membranes. J Biol Chem 1988; 263:4882–4890.
20. Katagiri Y, Hayashi Y, Yamamoto K, Tanoue K, Kosaki G, Yamazaki K. Localization of von Willebrand and thrombin-interactive domains on human platelet glycoprotein Ib. Thromb Haemost 1990; 63:122–128.
21. White JG, Burris SM, Hasegawa D, Johnson MM. Micropipette aspiration of human blood platelets. A defect in Bernard-Soulier syndrome. Blood 1985; 63:1249–1252.
22. Andrews RK. Fox JEB. Identification of a region in the cytoplasmic domain of the platelet membrane glycoprotein Ib-IX complex that binds to purified actin-binding protein. J Biol Chem 1992; 267:18605–18611.
23. Kuijpers RWAM, Faber NM, Cuypers HTM, Ouwehand WH, von dem Borne AEGK. NH_2-terminal globular domain of human platelet glycoprotein Ibα has a methionine^{145}/threonine^{145} amino acid polymorphism, which is associated with the HPA-2 (Ko) alloantigens. J Clin Invest 1992; 89:381–384.
24. Turitto VT, Baumgartner HR. Platelet-surface interactions. In: Colman RW, Hirsh J, Marder VJ, Salzman EW, eds. Hemostasis and Thrombosis, 2nd ed. Philadelphia: Lippincott, 1987:555–571.
25. Hourdille P, Heilmann E, Combrie R, Winckler J, Clemetson KJ, Nurden AT. Thrombin induces a rapid redistribution of glycoprotein Ib-IX complexes within the membrane systems of activated human platelets. Blood 1990; 76:1503–1513.
26. De Groot PG, Ottenhof-Povers M, van Mourik JA, Sixma JJ. Evidence that the primary binding site of von Willebrand factor that mediates platelet adhesion on subendothelium is not collagen. J Clin Invest 1988; 82:65–73.
27. Parmentier S, Catimel B, McGregor L, Leung LLK, McGregor JL. Role of glycoprotein IIa (β_1 subunit of very late activation antigens) in platelet functions. Blood 1991; 78:2021–2026.
28. Piotrowicz RS, Orchekowski RP, Nugent DJ, Yamada KY, Kunicki TJ. Glycoprotein Ic-IIa functions as an activation-independent fibronectin receptor on human platelets. J Cell Biol 1988; 106:1359–1367.

29. Kieffer N, Phillips DR. Platelet membrane glycoproteins: functions in cellular interactions. Annu Rev Cell Biol 1990; 6:329–357.
30. Tandon NN, Kralisz U, Jamieson GA. Identification of GP IV (CD 36) as a primary receptor for platelet-collagen adhesion. J Biol Chem 1989; 264:7576–7582.
31. George JN, Nurden AT. Inherited disorders of the platelet membrane. In: Colman RW, Hirsh J, Marder VJ, Salzman EW, eds. Hemostasis and Thrombosis, 2nd ed. Philadelphia: Lippincott, 1987:726–740.
32. Lopez JA, Chung DW, Fujikawa K, Hagen FS, Papayannopoulou T, Roth GJ. Cloning of the α chain of human platelet glycoprotein Ib: a transmembrane protein with homology to leucine-rich α_2-glycoprotein. Proc Natl Acad Sci USA 1987; 84:5615–5619.
33. Lopez JA, Chung DW, Fujikawa K, Hagen FS, Davie EW, Roth GJ. The α and β chains of human platelet glycoprotein Ib are both transmembrane proteins containing a leucine-rich amino acid sequence. Proc Natl Acad Sci USA 1988; 85:2135–2139.
34. Hickey MJ, Williams SA, Roth GJ. Human platelet glycoprotein IX: an adhesive prototype of leucine-rich glycoproteins with flank-center-flank structures. Proc Natl Acad Sci USA 1989; 86:6773–6777.
35. Hickey MJ, Deaven LL, Roth GJ. Human platelet glycoprotein IX: characterization of ''full-length'' cDNA and localization to chromosome 3. FEBS Lett 1990; 274:189–192.
36. Hickey MJ, Hagen FS, Yagi M, Roth GJ. Human platelet glycoprotein V: Characterization of the polypeptide and the related Ib-V-IX receptor system of adhesive, leucine-rich glycoproteins. Proc Natl Acad Sci USA 1993; 90:8327–8331.
37. Canfield VA, Ozols J, Nugent DJ, Roth GJ. Isolation and purification of the alpha and beta chains of human platelet glycoprotein Ib. Biochem Biophys Res Commun 1987; 147:526–534.
38. Roth GJ, Ozols J, Nugent DJ, Williams SA. Isolation and characterization of human platelet glycoprotein IX. Biochem Biophys Res Commun 1988; 157:931–993.
39. Roth GJ, Church TA, McMullen BA, Williams SA. Human platelet glycoprotein V: a leucine-rich glycoprotein related to adhesion. Biochem Biophys Res Commun 1990; 170:153–161.
40. Roth GJ, Hickey MJ, Chung DW, Hickstein DD. Circulating human blood platelets retain appreciable amounts of poly $(A)^+$ RNA. Biochem Biophys Res Commun 1989; 166:705–710.
41. Andrews RK, Fox JEB. Identification of a region in the cytoplasmic domain of the platelet membrane glycoprotein Ib-IX complex that binds to purified actin-binding protein. 1992; 267:18605–18611.
42. Wardell MR, Reynolds CC, Berndt MC, Wallace RW, Fox JEB. Platelet glycoprotein Ibβ is phosphorylated on serine 166 by cyclic AMP-dependent protein kinase. J Biol Chem 1989; 264:15656–15661.
43. Muszbek L, Laposata M. Glycoprotein Ib and glycoprotein IX in human platelets are acylated with palmitic acid through thioester linkages. J Biol Chem 1989; 264:9716–9719.
44. Landschulz WH, Johnson PF, McKnight SL. The leucine-zipper: a hypothetical structure common to a new class of DNA binding proteins. Science 1988; 240:1759–1764.
45. Roth GJ. Platelets and blood vessels: the adhesion event. Immunol Today 1992; 13:100–105.
46. Lopez JA, Ludwig EH, McCarthy BJ. Polymorphism of human glycoprotein Ibα results from a variable number of tandem repeats of a 13-amino acid sequence in the mucin-like macroglycopeptide region. J Biol Chem 1992; 267:10055–10061.
47. Tsuji T, Tsunehisa S, Watanabe Y, Yamamoto K, Tohyama H, Osawa T. The carbohydrate moiety of human platelet glycocalicin. J Biol Chem 1983; 258:6335–6339.
48. Ware J, Russell S, Murata M, Mazzucato M, De Marco L, Ruggeri ZM. $Ala^{156} \rightarrow$ Val substitution in platelet glycoprotein IBα impairs von Willebrand factor binding and is the molecular basis of Bernard-Soulier syndrome type Bolzano. (abstract). Blood 1991; 78(Suppl. 1):278a.

49. Weiss HJ, Meyer D, Rabinowitz R, et al. Psuedo-von Willebrand's disease. An intrinsic platelet defect with aggregation by unmodified human factor VIII/von Willebrand factor and enhanced absorption of its high-molecular weight multimers. N Engl J Med 1982; 306:326–333.
50. Ruggeri ZM, Pareti FI, Mannucci PM, Ciavavella N, Zimmerman TS. Heightened interaction between platelets and factor VIII/von Willebrand factor in a new subtype of von Willebrand's disease. N Engl J Med 1980; 302:1047–1051.
51. Miller JL, Cunningham D, Lyle VA, Finch CN. Mutation in the gene encoding the α chain of platelet glycoprotein Ib in platelet-type von Willebrand disease. Proc Natl Acad Sci USA 1991; 88:4761–4765.
52. Russell SD, Roth GJ. Pseudo-von Willebrand disease: a mutation in the platelet glycoprotein Ibα gene associated with a hyperactive surface receptor. Blood 1993; 81:(in press).
53. Takahashi H, Murata M, Furukawa T, Handa M, Watanabe K, Ikeda Y. A point mutation in the gene encoding glycoprotein Ibα in a Japanese family with platelet-type von Willebrand disease. (abstract). Blood 1992; 80(Suppl. 1):130a.
54. Randi AM, Rabinowitz I, Mancuso DJ, Mannucci PM, Sadler JE. Molecular basis of von Willebrand disease type IIB. J Clin Invest 1991; 87:1220–1226.
55. Cooney KA, Nichols WC, Bruck ME, et al. The molecular defect in type IIB von Willebrand disease. J Clin Invest 1991; 87:1227–1233.
56. Murata M, Ware J, Ruggeri ZM. Site-directed mutagenesis of a soluble recombinant fragment of platelet glycoprotein Ibα demonstrating negatively charged residues involved in von Willebrand factor binding. J Biol Chem 1991; 266:15474–15480.
57. Cruz MA, Petersen E, Turci SM, Handin RI. Functional analysis of a recombinant glycoprotein Ibα polypeptide which inhibits von Willebrand factor binding to the platelet glycoprotein Ib-IX complex and to collagen. J Biol Chem 1992; 267:1303–1309.
58. Lee FS, Vallee BL. Modular mutagenesis of human placental ribonuclease inhibitor, a protein with leucine-rich repeats. Proc Natl Acad Sci USA 1990; 87:1879–1883.
59. Hickey MJ, Roth GJ. Characterization of the gene encoding human platelet glycoprotein IX. J Biol Chem 1993; 268:3438–3443.
60. Wenger RH, Wicki AN, Kieffer N, Adolph S, Hameister H, Clemetson KJ. The 5′ flanking region and chromosomal localization of the gene encoding human platelet membrane glycoprotein Ibα. Gene 1989; 85:517–527.
61. Rajagopalan V, Konkle BA. Regulation of the GPIbα gene—evaluation of its transcription start site and promoter function. (abstract). Blood 1992; 80(Suppl. 1): 264a.
62. Mikol DD, Alexakos MJ, Bayley CA, Lemons RS, LeBeau MM, Stefansson K. Structure and chromosomal localization of the gene for the oligodendrocyte-myelin glycoprotein. J Cell Biol 1990; 111:2673–2679.
63. Konkle BA, Shapiro SS, Asch AS, Nachman RL. Cytokine-enhanced expression of glycoprotein Ibα in human endothelium. J Biol Chem 1990; 265:19833–19838.
64. Favaloro EJ, Maraitis N, Bradstock K, Koutts J. Co-expression of haemopoietic antigens on vascular endothelial cells: a detailed phenotypic analysis. Br J H Haematol 1990; 74:385–394.
65. Martin DIK, Zon LI, Mutter G, Orkin SH. Expression of an erythroid transcription factor in megakaryocytic and mast cell lineages. Nature 1989; 344:444–447.
66. Romeo PH, Prandini M-H, Joulin V, et al. Megakaryocytic and erythrocytic lineages share specific transcription factors. Nature 1989; 344:447–449.
67. Uzan G, Prenant M, Prandini M-H, Martin F, Marguirie G. Tissue-specific expression of the platelet GP IIb gene. J Biol Chem 1991; 266:8932–8939.
68. Breton-Gorius J, Vainchenker W. Expression of platelet proteins during the in vitro and in vivo differentiation of megakaryocytes and morphological aspects of their maturation. Semin Hematol 1986; 23:43–67.

69. Ware J, Russell SR, Vicente V, et al. Nonsense mutation in the glycoprotein Ibα coding sequence associated with Bernard-Soulier syndrome. Proc Natl Acad Sci USA 1990; 87: 2026–2030.

70. Miller JL, Lyle VA, Cunningham D. Mutation of leucine-57 to phenylalanine in a platelet glycoprotein Ibα leucine tandem repeat occurring in patients with an autosomal dominant variant of Bernard-Soulier disease. Blood 1992; 79:439–446.

28

Platelet Glycoproteins IIb and IIIa

Joel S. Bennett
University of Pennsylvania School of Medicine, Philadelphia, Pennsylvania

Mortimer Poncz
Children's Hospital of Philadelphia, Philadelphia, Pennsylvania

1. INTRODUCTION

Primary hemostasis requires platelet aggregation, a process in which blood platelets circulating in the vicinity of a damaged blood vessel adhere to one another to form a plug that occludes the wound (1). Because platelet aggregation in the absence of trauma can produce obstructive thrombi, the process is tightly controlled. To aggregate, platelets must be activated by specific agonists, such as thrombin or ADP, that are generated at the site of the wound (2). Platelet activation is followed by the binding of soluble fibrinogen and/or von Willebrand factor (vWF) to receptors on the platelet surface (3–5). The receptor-bound macromolecules then cross-link adjacent activated platelets into aggregates to form the occlusive platelet plug. The receptor for fibrinogen and vWF on activated platelets is the glycoprotein IIb-IIIa heterodimer (GPIIb-IIIa) (6,7). Normal platelet aggregation requires the presence of a sufficient number of functional GPIIb-IIIa heterodimers on the platelet surface, and quantitative and qualitative abnormalities in GPIIb-IIIa result in the hemorrhagic disorder Glanzmann thrombasthenia (8).

GPIIb-IIIa, the most abundant protein on the platelet surface, is a calcium-dependent heterodimer composed of the integral membrane proteins glycoprotein IIb and glycoprotein IIIa (9). There are approximately 50,000 heterodimers on the surface of unactivated platelets (10), and additional heterodimers, present in the membrane of the platelet α granules, can be translocated to the surface by platelet activation (11,12). GPIIb-IIIa resides on the platelet in either of two conformational states. On circulating platelets, GPIIb-IIIa is in a conformation that is incapable of interacting with soluble fibrinogen or vWF (2,13). On activated platelets, however, GPIIb-IIIa is in a conformation in which a binding site for soluble proteins is exposed. Under physiological circumstances, the conversion of GPIIb-IIIa from its resting to its activated conformation involves the transmission of signals from receptors for platelet agonists to the cytoplasmic extensions of GPIIb-IIIa. The identity of these signals and the nature of the changes in GPIIb-IIIa responsible for its activated conformation have yet to be determined. Pursuit of these fundamental questions has resulted in an extensive body of information regarding GPIIb-IIIa structure, biosynthesis, and molecular pathology, which are the subjects of this review.

II. GLYCOPROTEIN IIb AND GLYCOPROTEIN IIIa STRUCTURE

The GPIIb-IIIa heterodimer can be isolated from platelets or platelet membranes by detergent extraction and purified by gel filtration chromatography or affinity chromatography on monoclonal antibody columns. When purified by gel filtration chromatography in the presence of calcium and the detergent Triton X-100, GPIIb-IIIa has a Stokes radius of 71 Å, a sedimentation coefficient of 8.6*S*, and a calculated molecular weight of 265,000 (14). When examined by electron microscopy in the presence of the detergent octyl glucoside and after rotary shadowing with platinum and tungsten, purified GPIIb-IIIa appears as a globular head with two tails extending from one side (Fig. 1) (15,16). The average dimensions of the globular head are 12 by 8 nm, and each tail has a contour length of 18 and a diameter of ~2 nm (15). Many of the tails also terminate in a small globular region ~4 nm in diameter. Following removal of the octyl glycoside by dialysis, GPIIb-IIIa forms rosettes, with the ends of the tails interacting and pointing inward and the globular head at the outer perimeter of the rosette. These images suggest that the ends of the tails are predominantly hydrophobic and the globular head is hydrophilic, a conclusion confirmed by the morphology of GPIIb-IIIa incorporated into phospholipid vesicles. Under these conditions, the globular head extends ~20 nm from the vesicle surface and the tips of the tails are inserted in the vesicle (17). Electron microscopy of purified GPIIb-IIIa bound to well-characterized GPIIb-IIIa monoclonal antibodies has also been used to resolve the topography of the heterodimer (15). A monoclonal antibody that recognizes an epitope located near the carboxyl terminus of GPIIb heavy chain bound to GPIIb-IIIa near the tip of one of its tails, and a monoclonal antibody that reacts with a 66,000 molecular weight chymotrypsin-derived fragment of GPIIIa, composed predom-

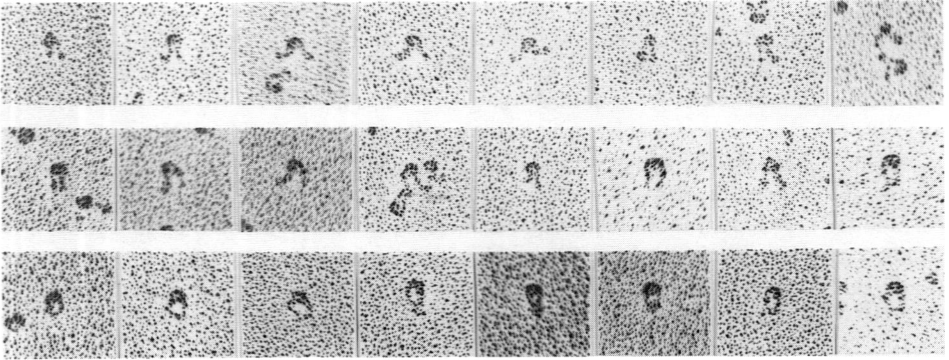

Figure 1 Electron micrographs of GPIIb-IIIa in the presence of the detergent octyl glucoside. A gallery of electron microscope images of individual GPIIb/IIIa heterodimers rotary shadowed with platinum and tungsten. Each heterodimer has a globular head and two long tails extending from one side. (Adapted from Ref. 15, with permission.)

inantly of the carboxyl-terminal portion of the molecule, also bound to one of the tails. Thus, each tail of GPIIb-IIIa contains the carboxyl terminus of one of its subunits. On the other hand, a monoclonal antibody that binds only to the heterodimer and inhibits platelet aggregation and fibrinogen binding to activated platelets bound to the globular head of the GPIIb-IIIa complex on the side opposite the tails. Thus, this region of the heterodimer likely contains its ligand binding domains. The GPIIb-IIIa heterodimer dissociates into GPIIb and GPIIIa monomers in the presence of calcium chelators. Each monomer appears as comma-shaped structure consisting of a small globular head and a single tail, with a total length of 24–27 nm (15). This appearance is consistent with the high degree of asymmetry of the monomers as inferred from frictional ratios calculated from gel filtration experiments (14).

The studies just described provide a representation of the overall shape of the GPIIb-IIIa, but further detail of GPIIb-IIIa structure has been furnished from analyses of the primary structures of GPIIb and GPIIIa as deduced from the nucleotide sequences of cloned cDNAs.

A. Glycoprotein IIb

Glycoprotein IIb has a molecular weight of 136,000, as determined by sodium dodecyl sulfate (SDS)-polyacrylamide gel electrophoresis, and is composed of disulfide-linked heavy (GPIIbα) and light (GPIIbβ) chains with molecular weights of 125,000 and 23,000, respectively (14). The complete amino acid sequence of GPIIb has been deduced from the sequence of cDNA obtained from a human erythroleukemia (HEL) cDNA library (3) and has been confirmed by sequence analysis of the human GPIIb gene (18). The features of the GPIIb sequence are shown in Figure 2A. The reading frame of the GPIIb gene encodes a 1039 amino acid GPIIb precursor, pro-GPIIb, that contains a 31 residue signal peptide and amino acids for both the GPIIb heavy and light chains. Following removal of the signal peptide in the endoplasmic reticulum and pairing with GPIIIa (see later), pro-GPIIb is transported to the Golgi complex, where it undergoes endoproteolytic cleavage into heavy and light chains. Although amino acid sequencing of purified platelet GPIIbβ suggested that endoproteolysis occurred after arginine 871, in vitro studies using recombinant GPIIb mutants demonstrated that cleavage occurred after the dibasic sequence Arg^{858}-Arg^{859} and that the endoprotease responsible for GPIIb cleavage recognizes the sequence Arg-*X*-Arg-Arg (19). The latter corresponds to the sequence recognized by enzymes of the furin family of endoproteases (20), and it is likely that a member of this family, present in the Golgi complex of megakaryocytes, performs the GPIIb cleavage. GPIIb is anchored in membranes by a stretch of 26 amino acids located near the carboxyl terminus of GPIIβ. The transmembrane domain is followed by a 20 amino acid cytoplasmic tail. GPIIb contains five potential sites for Asp-linked glycosylation, four in the heavy chain and one in the light chain (3). Each of these sites likely contains carbohydrate (Ref. 21 and Bennett and Poncz, unpublished observations). There are 18 cysteine residues in GPIIb, 15 in GPIIbα, and 3 in GPIIbβ (3). Each Cys forms a disulfide bond with its nearest neighbor, except for the most carboxyl terminal Cys in the heavy chain and the most amino terminal Cys in the light chain, which form the interchain disulfide bond (21). The GPIIb heavy chain also contains a polymorphism, isoleucine 843 → Ser, that is associated with the Bak (Lek) alloantigens involved in neonatal alloimmune thrombocytenia and posttransfusion purpura (22,23).

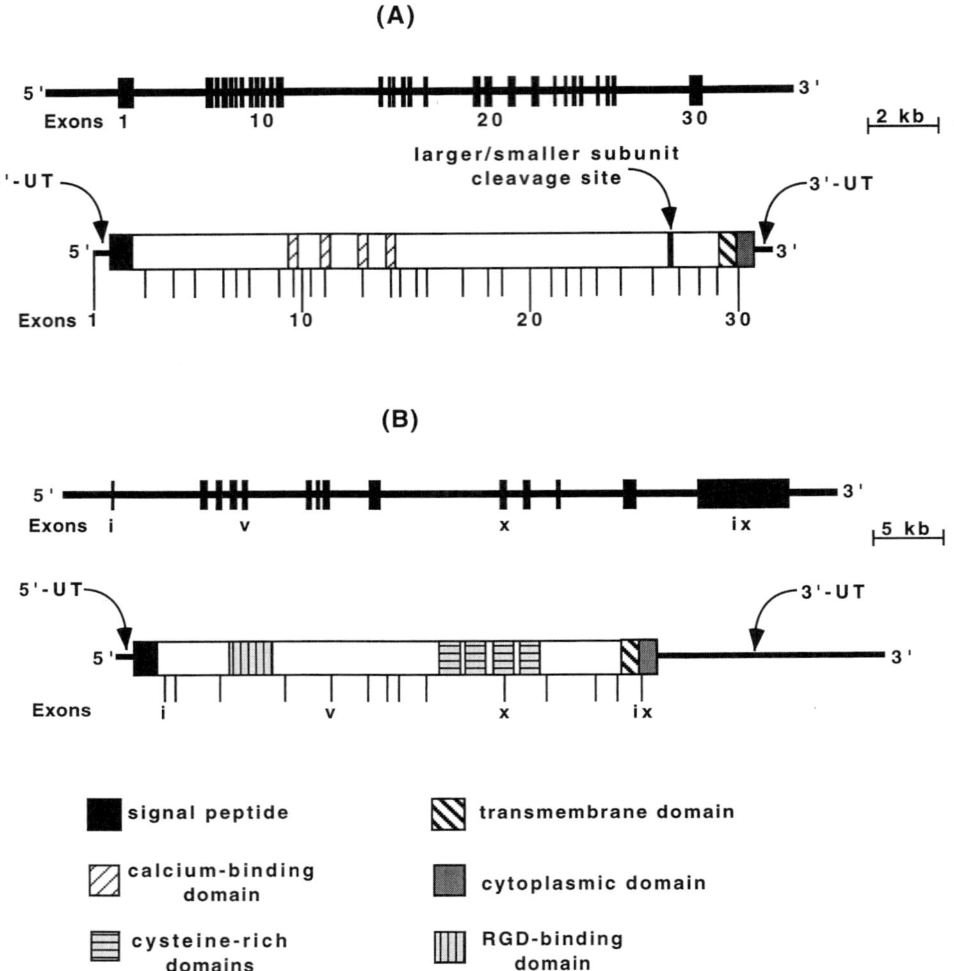

Figure 2 Comparison of the organization of the GPIIb and GPIIIa genes with the structure of the mature GPIIb and GPIIIa mRNAs. The organization of the GPIIb gene and the structure of its mRNA are shown in A, and the organization of the GPIIIa gene and its mRNA structure are shown in B. Each upper diagram depicts the exon distribution along the gene, with the exons denoted by solid black boxes. A size marker for each gene in kb is shown below and to the right. Each lower diagram depicts the structure of the mature mRNA of the corresponding gene. The reading frame of the mRNA is indicated as a box, and the vertical lines below define the contribution of the gene's exons to the reading frame. Superimposed on the reading frame is the position of the known functional domains of the protein. Of note, there is little, if any, correspondence between the exon structure of the gene and the location of functional domains. UT, untranslated.

The most interesting feature of the GPIIb sequence is the presence of four stretches of 12 amino acids each in the GPIIb heavy chain (residues 274–285, 328–339, 395–406, and 457–468) having the consensus sequence aspartic acid-*X*-Asp-*X*-Asp-Gly-*X*-*X*-Asp-*X*-*X*-valine and homology with the calcium binding regions of troponin C and calmodulin (3). Secondary structure analysis of these stretches suggests they are located at either the boundary of an α helix and a turn or within a region of turns and that they are available to coordinate with ionic calcium (3,9). Moreover, this portion of the GPIIb heavy chain has been expressed in bacteria and found to contain four binding sites that can be occupied by calcium (24). Two classes of calcium binding sites were identified, two higher affinity sites with a K_d for calcium of 30 μM and two lower affinity sites with a K_d for calcium of 120 μM. The role of the sites, if any, in GPIIb-IIIa structure and function remains to be determined. The GPIIb-IIIa heterodimer is calcium dependent, and the portion of the GPIIb heavy chain containing the calcium binding domains is clearly involved in heterodimer formation (25). Thus, it is possible that one or all of the calcium binding domains may be directly involved in heterodimer assembly. In support of this possibility, it has been reported that isolated GPIIb-IIIa heterodimers dissociate by 50% at a calcium concentration of 60 μM (26). However, other experiments indicate that heterodimer stability requires a calcium concentration of only 0.4 μM (27) and that recombinant heterodimers assemble despite the deletion of each of the calcium binding domains (Bennett and Poncz, unpublished observations). A region of GPIIb extending from residues 294 to 314, proximal to the second calcium binding domain, can be chemically cross-linked to a peptide corresponding to the 16 amino-terminal amino acids of the fibrinogen γ chain (28). Because this peptide inhibits fibrinogen binding to activated platelets (29) and because fibrinogen binding to GPIIb-IIIa requires the presence of divalent cations (2,30), it has been postulated that the second calcium binding domain is involved in or may modulate ligand binding to the heterodimer (31).

B. Glycoprotein IIIa

Glycoprotein IIIa is a cysteine-rich single-chain protein with an apparent molecular weight of 90,000 on unreduced SDS gels that increases to 110,000 following disulfide bond reduction (14). The amino acid sequence of GPIIIa has been deduced from the nucleotide sequences of cDNAs isolated from both endothelial cell and HEL cell cDNA libraries (32–34). The open reading frame of the cDNAs encodes a protein of 762 amino acids whose features are illustrated in Figure 2B. GPIIIa contains a 26 amino acid amino-terminal signal peptide, a 29 amino acid transmembrane domain located near its carboxyl terminus, and 41 amino acid cytoplasmic domain. There are seven asparagine residues in the proper context (Asn-*X*-serine/threonine) to undergo glycosylation, six of which are present in the extracellular domain of the protein. Although it has not yet been determined whether each of the six sites contains carbohydrate, glycosylation of each would account for the carbohydrate content of the protein (35). Interestingly, the Asn-linked carbohydrate appended to mature GPIIIa remains sensitive to the enzyme endo-β-*N*-acetylglucosaminidase H (Endo H), despite the fact that GPIIIa traverses the Golgi complex, where sensitivity to Endo H is lost (36). There is no obvious explanation for this observation, but it could be related to the complex tertiary structure of the protein. GPIIIa is highly polymorphic and is associated with at least six alloantigens responsible for neonatal alloimmune thrombocytopenia and posttransfusion purpura. The PIA^1/PIA^2 alloantigens, most frequently responsible for these disorders in white populations, are caused by a leucine 33 → proline polymorphism (23,37); the Pen[a]/Pen[b] alloantigens, most frequently responsible in Asian populations, are caused by a glycine 143 → alanine polymorphism (38). Several private GPIIIa polymorphisms have also been described: Mo is caused by a proline 407 → alanine substitution (39), but the polymorphisms responsible for the Va[a] (40), Tu[a] (41), and Sr[a] (42) alloantigens have not been determined. The location of the alloantigens associated with GPIIIa are of particular interest because they likely specify amino acids exposed on the surface of the protein.

GPIIIa contains 56 Cys residues and 28 disulfide bonds (32,33,43). The distribution of the disulfide bonds in GPIIIa is reflected in the complex folding of the protein and is its most distinctive feature (43). The distribution defines three regions in the extracellular portion of the molecule: a Cys-rich, protease-resistant amino terminus (residues 1–62), a protease-sensitive central region (residues 101–422), and a disulfide-rich, protease-resistant core (residues 423–622) (43). The latter contains four cysteine-rich tandem repeats of approximately 40 amino acids each, with 7 (first repeat) or 8 (three remaining repeats) cysteines in similar positions (32,33). Moreover, there appears to be a long-range disulfide bond between cysteines at positions 5 and 435 that results in a protease-sensitive loop in the molecule (43–45). Proteolysis of this loop by chymotrypsin accounts for the 66,000 molecular weight fragment of GPIIIa that remains after platelets are exposed to this enzyme (45). It is noteworthy that the 66,000 molecular weight fragment does not interact with fibrinogen, suggesting that at least a portion of the ligand binding domain of GPIIIa is located in the chymotrypsin-sensitive loop (45).

C. Assembly and Intracellular Transport of the GPIIb-IIIa Heterodimer

GPIIb and GPIIIa are the products of separate genes (46,47). Consequently, the assembly of GPIIb-IIIa heterodimers is a posttranslational process. Studies of endogenous GPIIb-IIIa expression in HEL cells (48) and cultured human megakaryocytes (49) and of recombinant GPIIb-IIIa expression in a several different in vitro systems (36,50,51) indicate that heterodimer assembly occurs in the calcium-rich environment of the endoplasmic reticulum. Furthermore, pulse-chase analyses using these expression systems reveal a lag of 2–4 h (depending on the study) before maximal heterodimer assembly is detected (36,48,49), suggesting that this period of time is required for GPIIb and GPIIIa monomers to attain competence to assemble into heterodimers. During this period, GPIIb and GPIIIa undergo folding, disulfide bond formation, and Asn-linked glycosylation. Which of these steps is rate limiting for heterodimer assembly is not known. Studies of the fate of recombinant GPIIb and GPIIIa monomers truncated proximal to their transmembrane domains reveal that each associates with the intracellular chaperone BiP (52) BiP is 78,000 molecular weight protein that resides in the endoplasmic reticulum (ER) and transiently binds to nascent and unassembled proteins to assist in their folding (53). BiP is not associated with the intracellular heterodimers formed by the truncated proteins, however, suggesting that either heterodimer formation occludes the BiP binding site or the chaperone function of BiP is no longer necessary once heterodimer formation occurs. Following heterodimer assembly, GPIIb-IIIa is transported from the ER to the Golgi complex. In the absence of heterodimer formation, however, GPIIb and GPIIIa are retained in the ER and eventually degraded (36). The signal responsible for retention of the monomers, but transport of the heterodimers, is not known. Recombinant GPIIIa lacking a transmembrane domain is secreted from COS cells (52), indicating that a signal for GPIIIa retention is present in its transmembrane domain. In contrast, recombinant truncated GPIIb is retained by the cells and is secreted only in the presence of truncated GPIIIa (52), indicating that additional signals for GPIIb retention are present in the extracellular domain of the protein. Nevertheless, heterodimer assembly alone is not sufficient to guarantee the egress of GPIIb-IIIa from the ER. Mutations that distort the conformation of the heterodimer, such as alternative splicing of GPIIb exon 28 (36) or a $Gly^{273} \rightarrow$ Asp mutation in the extracellular domain of GPIIb (54), also result in ER retention.

In the Golgi complex, the carbohydrate appended to GPIIb is modified by enzymes resident in the various Golgi compartments, from high-mannose to complex sugars (55), and, in the process, loses its sensitivity to cleavage by the enzyme Endo H (56). As mentioned, however, the carbohydrate appended to GPIIIa retains its Endo H sensitivity despite exposure to the Golgi enzymes. As discussed, GPIIb also undergoes cleavage in the Golgi complex into heavy and light chains. It is noteworthy that GPIIb cleavage is not to required for GPIIb-IIIa expression on the cell surface, nor is cleavage required for the interaction of GPIIb-III with fibrinogen (19). The intracellular processing and transport of GPIIb-IIIa are summarized in Figure 3.

D. Integrins

Heterodimers similar to GPIIb-IIIa are present on the surface of cells of mammalian, avian, amphibian, and invertebrate origin (57). Like GPIIb-IIIa, these heterodimers mediate cell-cell and cell-matrix interactions and are involved in diverse processes, such as tissue migration during embryogenesis, cell adhesion, and lymphocyte helper and killer cell functions. Comparison of the cDNA sequences of these complexes with those of GPIIb and GPIIIa indicates that GPIIb-IIIa is a member of family of adhesion receptors named the integrins (32,58). Integrins are α/β heterodimers with α subunits similar to GPIIb and β subunits similar to GPIIIa. At present, 14 distinct integrin α subunits and 8 distinct β subunits are recognized, plus several alternatively spliced α and β subunit variants (59). Because integrin β subunits associate with a restricted number of α subunits, integrins can be classified into several subfamilies (57). The largest subfamily contains the β_1 subunit and is also termed the VLA (very late antigen) subfamily because it was originally described as antigens appearing on T lymphocytes several weeks after stimulation in vitro (60). β_1 Integrins mediate the interaction of cells with the extracellular matrix by acting as receptors for fibronectin, laminin, and collagen (59). There are six members of the β_1 subfamily, three of which are found on platelets. These include $\alpha_2\beta_1$ (VLA-2 or GPIa-IIa), $\alpha_5\beta_1$ (VLA-5 or GPIc-IIa), and $\alpha_6\beta_1$ (VLA-6), receptors for collagen, fibronectin, and laminin, respectively (61). A second subfamily, the β_2 integrins, is found on leukocytes and consists of the LFA-1 antigen ($\alpha_L\beta_2$ or CDIIa-CD18) present on all leukocytes, but predominantly on lymphocytes; the Mac-1 antigen ($\alpha_M\beta_2$ or CD11b-CD18), found predominantly on phagocytes and large granular lymphocytes; and p150,95 ($\alpha_X\beta_2$ or CD11c-CD18), expressed by these cells plus activated lymphocytes, such as cytolytic T cells and hairy leukemia cells (62). In contrast to GPIIb, the α subunits of these integrins are single-chain proteins and contain a 187 amino acid segment (I domain) that is homologous to the A domain of vWF (63,64). Like GPIIb-IIIa, however, the activity of the β_2 integrins can be upregulated by intracellular signals that interact with their cytoplasmic domains (65). β_2 Integrins mediate leukocyte functions that involve cell adhesion by interacting with specific ligands, such as ICAM-1 (LFA-1) (66)

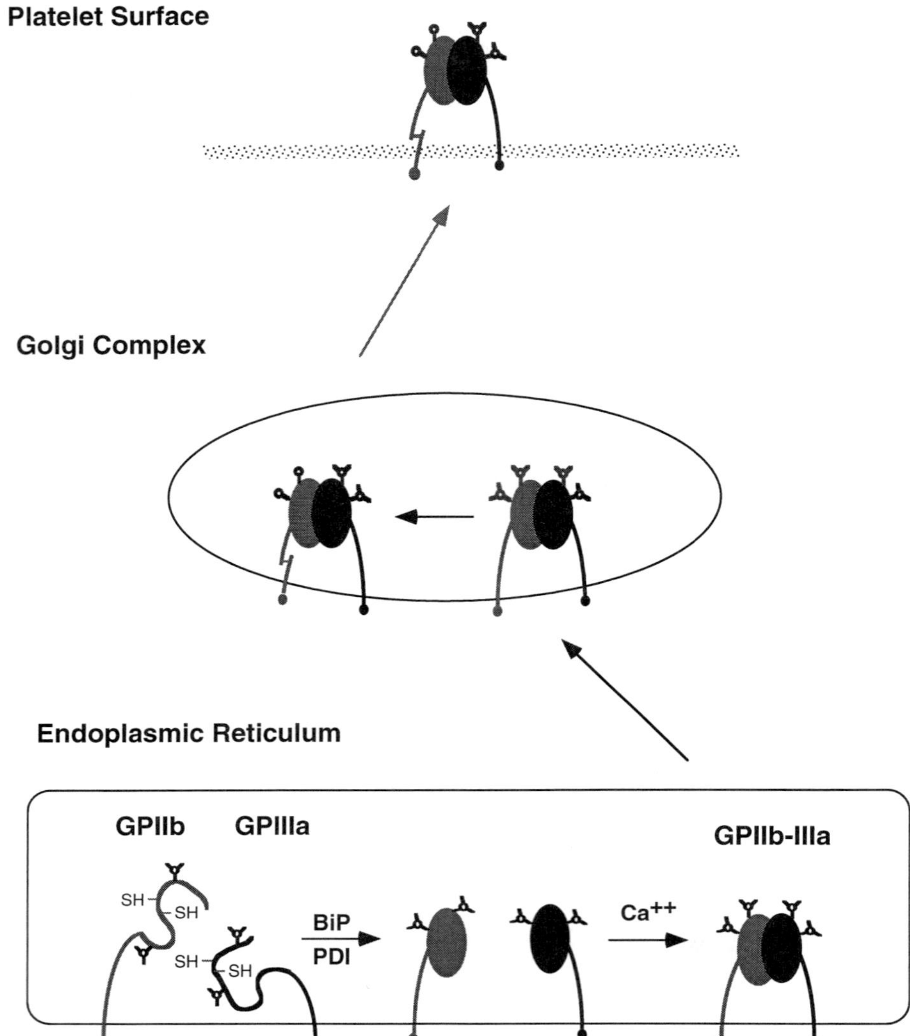

Figure 3 Assembly and intracellular transport of the GPIIb-IIIa heterodimer. Nascent GPIIb and GPIIIa polypeptides undergo cotranslational Asn-linked glycosylation, folding aided by the chaperone BiP (53), and disulfide bond formation catalyzed by the enzyme protein disulfide isomerase (PDI) (166) in the endoplasmic reticulum. Following maturation of GPIIb and GPIIIa monomers, calcium-dependent GPIIb-IIIa heterodimers assemble and are transported from the endoplasmic reticulum to the Golgi complex. In the Golgi, GPIIb undergoes cleavage into heavy and light chains and its carbohydrate is processed to Endo H-resistant forms. The processed heterodimers then appear on the platelet surface.

and iC3b (MAC-1) (67). Congenital deficiency of these protein complexes results in a disorder, leukocyte adhesion deficiency, manifested by persistent leukocytosis and recurrent, severe, and occasionally fatal bacterial and fungal infections (65). A third subfamily, the β_3 integrins, consists of GPIIb-IIIa ($\alpha_{IIb}\beta_3$) and a receptor for vitronectin $\alpha_V\beta_3$. Although $\alpha_V\beta_3$ is found on a variety of cells, including endothelial cells, GPIIb-IIIa is normally only present on cells of the megakaryocyte lineage (33). Moreover, although GPIIb-IIIa function requires cell activation, $\alpha_V\beta_3$ is constitutively active (68,69).

There is significant homology among the various integrin α and the various integrin β subunits, suggesting that each type of subunit diverged from its own common ancestral protein. For example, alignment of the amino acid sequence of GPIIb with those of α_5 and α_V reveals an overall sequence identity of ~40% (58). The three sequences are also ~25% identical to those of α_M and α_X (63), with regions of greatest identity surrounding the calcium binding and transmembrane domains. Comparison of the various β subunits also reveals significant homology with 40–48% sequence identity. A high degree of similarity is noted in their transmembrane and cytoplasmic domains, suggesting that evolutionary pressure was exerted to conserve these sequences. Moreover, the position of the cysteine residues in each of the sequences has been conserved, suggesting that the tertiary structures of the proteins are similar (33).

III. ORGANIZATION OF THE GENES FOR GPIIb AND GPIIIa

A. Location of the GPIIb and GPIIIa Genes

Because α and β integrin subunits are not evolutionarily related, it might be expected that the GPIIb and IIIa genes would not be closely linked. In fact, the genes for other integrin α and β subunits appear to be randomly distributed in the genome (70). However, somatic cell hybrid analysis of various cell lines, including lines containing various truncated versions of human chromosome 17, and in situ hybridization of human chromosomes has demonstrated that genes for both GPIIb and GPIIIa genes are localized to the long arm of chromosome 17 band 21 $\rightarrow$ 23 (34,70,71). Furthermore, pulse-field gel electrophoresis suggests that the two genes are located to within 260 kb of each other (72). The implication of the physical linkage of the GPIIb and IIIa genes, if any, is not clear, but it is possible that their coordinated expression during megakaryocytopoiesis is related to their physical proximity.

B. Physical Organization of the GPIIb and GPIIIa Genes

The organization of the entire human GPIIb gene and the genomic region encoding the reading frame for GPIIIa have been published (18,73).

The gene for GPIIb gene spans ~18 kilobase pairs (kb) and is divided into 30 exons ranging in size from 46 base pairs (bp) to 220 bp (Fig. 2A). The gene is located in a region that is rich in Alu repeats and contains at least seven complete and three incomplete repeats. Of interest, exons 5 and 8 are unusual in that they have a GC instead of a GT splice donor sequence. Whether the presence of multiple Alu repeats or the unusual splice donor sequences affect GPIIb biology, such as the frequency of mutations leading to Glanzmann thrombasthenia, requires an analysis of a greater number of thrombasthenic patients than have been studied to date (see later). Nevertheless, one patient with thrombasthenia has a GPIIb deletion (74) in which an Alu repeat was involved in the recombinant event.

Although the organization of exons in a gene often reflects the distribution of functional domains of the protein it encodes (75), this does not appear to be true for the GPIIb gene. A possible exception involves the GPIIb calcium binding domains (18). The boundaries of the exons encoding these domains and the intervening intron are located near the center of the each domain, a situation that is also seen in the gene for calmodulin.

Localization of the transcriptional start site for GPIIb by S_1-nuclease and primer extension analyses defined a 5′-untranslated region of 32 bp (18,76). Surprisingly, the GPIIb gene lacks TATA or CAAT boxes in its immediate 5′ flanking region, much like known housekeeping genes (77). Of note, the gene for the α chain of the β_2 integrin Mac-1 ($\alpha_L\beta_2$ or CDIIb-CD18) also lacks a TATA box (78). On the other hand, the sequence around the transcriptional start site of GPIIb conforms to the known cap site consensus sequence (79) and to a known weak promoter sequence (80).

The gene for GPIIIa spans over 40 kb and contains 14 exons (Fig. 2B) (73,81). The organization of the gene published to date begins with the sequence encoding the reading frame for the mature protein; exon(s) encoding the 5′-untranslated region and GPIIIa signal peptide have yet to be identified. A relationship between the organization of the GPIIIa exons and known functional domains in the protein appears to be partially preserved. Thus, the transmembrane domain and the cytoplasmic domain and 3′-untranslated regions are encoded by single exons. As the organization of more integrin α and β genes are described and as the structural and functional relationships of the proteins are defined in greater detail, insight into the evolution of the exon organization of the GPIIb and IIIa genes may be possible.

C. Regulation of the Expression of the GPIIb Gene

Expression of GPIIb-IIIa is normally restricted to platelets and their precursors, megakaryocytes, although it was recently reported that GPIIb-IIIa is also present in B16a murine melanoma cells (82). The restricted expression of GPIIb-IIIa is a result of the limited expression of GPIIb: GPIIIa is expressed by many tissues, including megakaryocytes, endothelial cells (32), fibroblasts (83), and macrophages (84). Consequently, understanding the basis for the restriction in GPIIb expression may provide insight into the mechanism for the megakaryocyte-specific expression of a number of platelet-specific proteins.

Studies on the regulated expression of genes during megakaryocytopoiesis have focused on the 5′ flanking regions of the GPIIb and platelet factor 4 (PF4) genes. Like GPIIb, PF4 expression is normally restricted to developing megakaryocytes (85), although expression of the PF4 gene appears to occur at a later stage of megakaryocyte development (86).

Megakaryocytes contain two related trans-acting nuclear factors that recognize the consensus sequence (T/A)GATA(G/A), GATA-1, and GATA-2 (87). These single-zinc finger proteins were originally discovered in erythroid cells and were thought to be erythroid specific. Recent studies have shown, however, that these factors are also expressed in megakaryocytes and mast cells (88,89), and GATA-2 is expressed in such tissues as developing brain and endothelium (87,90). The observation that GATA-1 and GATA-2 are expressed in developing erythrocytes, megakaryocytes, and mast cells suggests that these factors are active in the hematopoietic precursor cell CFU-EMMeg that gives rise to all three cell lines. It also suggests that although these transcription factors may be important for protein expression in all three cell lines, other factors must contribute to the unique protein complement of each. For example, GPIIb expression is normally megakaryocyte specific, but expression of β-globin is limited to erythroid cells. On the other hand, recognition of common transcription factors may explain why megakaryocyte-like cell lines often also have erythroid and/or mast cell-like properties. Whether GATA-1 is critical for megakaryocyte differentiation, as mice with GATA-1 gene knockouts have shown that it is important in erythroid differentiation, remains to be determined (91).

Mobility shift and DNase 1 footprinting experiments have defined potential regulatory regions in the 5′ flanking region of the human GPIIb gene (92). These regions, termed domains A–E, are shown in Figure 4. The experiments indicate that domains A (−45 to −63 bp upstream from the transcriptional start site), D (−456 to −475 bp), and E (−502 to −524 bp) are megakaryocyte specific. Furthermore, mutation or deletion of either domain D or E decreased expression to ~30% (93). The region at −406 to −598 appears to have enhancer-like qualities in that its presence results in a fivefold increase in expression independent of its orientation. Interestingly, domain D contains an inverted GATA consensus sequence, and a point mutation that changes the sequence from GATA to CATA results in a decrease in expression to 30%.

Studies focused on the 75 bp immediately upstream of the transcriptional start site have defined the minimal promoter for the GPIIb gene (94). They revealed two regulatory regions, one involving domain A and another centered at −40 bp and consisting of the sequence 5′-ACTTCCTG-3′, matching the consensus Ets binding sequence (94,95). Coexpression of a minimal promoter construct and Ets-1 (but not Ets-2) in HeLa cells resulted in efficient activation of a reporter gene, suggesting that Ets-1 is important in GPIIb expression. Moreover, northern blot analysis of different hematopoietic cell lines suggested that Ets-1 expression is restricted to megakaryocyte-like cell lines.

Comparison of the 5′ flanking region of the GPIIb gene with that of the gene for rat PF4 has also been used to determine the genetic elements required for megakaryocyte-specific gene expression (Fig. 4). Transgenic mice containing 1.1 kb of the 5′ flanking region of the rat PF4 gene placed upstream of the β-galactosidase reporter gene demonstrated that this region of the PF4 gene was sufficient to regulate tissue-specific expression (96). The only other tissue expressing β-galactosidase activity was the adrenal medulla, but at a level 2% of that seen in megakaryocytes. Transfection studies using constructs containing the flanking region defined a promoter region in the 97 bp immediately upstream of the transcriptional start site (97). This region contains a GATA consensus sequence at −30 bp. However, mutation of the GATA to AATA or TATA did not affect the expression of the constructs in megakaryocytes as much as it increased expression of the gene in megakaryocyte-depleted cells and in fibroblasts. Further, deletional constructs defined three promoter regions, P1, P2, and P3, in the −444 to −112 bp region. Deletions of the P3 region decreased megakaryocyte-specific expression by ~50% but markedly increased expression in nonmegakaryocytic cells. In addition, a stretch of 26 thymidine residues in the 5′ flanking region of the PF4 gene appears to be a negative regulator of PF4 expression. Thus, these studies suggest a complicated picture in which the GATA element plays a role as a tissue-specific promoter; other elements are either positive or negative promoters of gene expression in either megakaryocytes or heterologous tissues.

Although both the human GPIIb and the rat PF4 genes contain biologically relevant GATA consensus sequences in the 5′ flanking regions and the rat PF4 gene also contains an Ets consensus sequence, the general organization of the two genes is different. In the GPIIb gene, the biologically significant GATA sequences are ~54 and 460 bp upstream

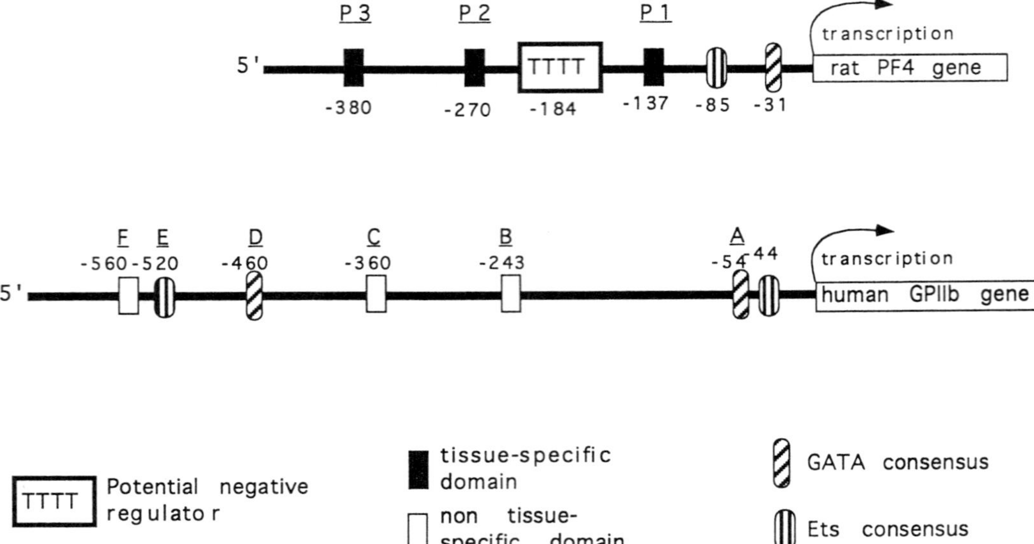

Figure 4 Comparison of the organization of regulatory elements in the 5′ flanking regions of the human GPIIb genes and the rat PF4 gene.

of the transcriptional start site; in the rat PF4 gene, the GATA sequence is only 30 bp upstream. Moreover, there is no obvious homology between the P1, P2, and P3 promoters in the PF4 gene and similar regions of the GPIIb gene. Nevertheless, the general organization and utilization of a common tissue-specific gene regulatory region, such as the muscle-specific promoter Myo D1 (98), can vary among different genes expressed in the same tissue, and the differences in the GPIIb and PF4 regulatory regions could be a reflection of the temporal difference in GPIIb and PF4 expression during megakaryocytopoiesis.

IV. GLYCOPROTEIN IIb-IIIa FUNCTION

A. Interaction of GPIIb-IIIa with Immobilized Fibrinogen

The major function of GPIIb-IIIa is to serve as a receptor for soluble fibrinogen and vWF on activated platelets (2,99). However, GPIIb-IIIa on unactivated platelets can interact with immobilized fibrinogen and to a variable extent with immobilized vWF and fibronectin (100,101), resulting in platelet spreading on the protein-coated surface. Furthermore, this interaction leads to GPIIb-IIIa and protein kinase C activation (100), and its absence may account for the decrease in platelet spreading on deendothelialized blood vessels observed in Glanzmann thrombasthenia (102). The basis for the ability of unactivated GPIIb-IIIa to interact with immobilized fibrinogen and its inability to bind soluble fibrinogen is not known. Because the binding site for soluble fibrinogen in GPIIb-IIIa requires a conformational change in the heterodimer for its exposure, it is likely that conformational changes in fibrinogen that accompany immobilization are responsible.

B. Signal Transduction in Platelets

Binding of fibrinogen, vWF, fibronectin, or vitronectin to GPIIb-IIIa requires a conformational change in the heterodimer that is induced by the intracellular signals generated during platelet activation (103). These signals interact with either or both of cytoplasmic tails of heterodimer (104) and are transmitted over the length of the molecule to induce the required change in the conformation of its globular head (105). The specific signals producing this change have not been identified, but signal transduction in platelets is an area of intensive study, and it is likely that their identity will be forthcoming shortly. Currently, it is known that platelet stimulation results in the activation of the phospholipases C and A2 and the serine/threonine kinases protein kinase C, calcium-calmodulin–dependent protein kinase, and myosin light-chain kinase (106). Moreover, platelets contain high levels of the protein tyrosine kinase $pp60^{c-src}$ (107), the src-related kinases Fyn, Lyn, Hck, and Yes (108), and the protein kinases $p72^{syk}$ and CPTK71 (109). Phospholipase C activation in platelets is a G protein-mediated process (106) and generates diacylglycerol and inositol 1,4,5-triphosphate (IP3) from phosphotidylinositol 4,5-bisphosphate (110). Diacylglycerol is a protein kinase C activator, and IP3 is responsible for the rise in intracellular calcium that accompanies platelet activation induced by most platelet agonists (111). Platelet activation is associated with the phosphorylation of a number of intracellular proteins on serine/threonine and tyrosine residues (112–114). Whether any of these phosphorylated proteins are directly in involved in GPIIb-IIIa activation remains to be determined. It has been observed that threonine residues in the cytoplasmic domain of GPIIIa are phosphorylated after platelet stimulation by thrombin, phorbol ester, or the thromboxane analog U46619 (115). However, the extent of phosphorylation is limited to 2–4%, making it unlikely that the phosphorylation is of physiological significance (116). Nevertheless, protein kinase C and tyrosine kinase antagonists inhibit GPIIb-IIIa function in platelets, clearly implicating the activity of these kinases in GPIIb-IIIa activation (117,118).

Occupation of GPIIb-IIIa by ligands also appears to generate signals in platelets. Fibrinogen binding to GPIIb-IIIa on thrombin- or collagen-stimulated platelets results in the phosphorylation of a number of platelet proteins, including the protein tyrosine kinase $pp125^{FAK}$ (119). In addition, fibrinogen binding to GPIIb-IIIa activated by monoclonal antibodies in the absence of platelet activation results in the phosphorylation of a subset of these proteins, but not $pp125^{FAK}$ (120). These phosphorylation reactions are inhibited by preincubating the platelets with cytochalasin D, suggesting that they are sensitive to the state of actin polymerization in the platelet. Studies of fibroblast adhesion mediated by adhesion receptors related to GPIIb-IIIa (see later) indicate that these receptors cluster in regions termed focal contacts and, in doing so, organize the cytoskeleton of the cell (121–123). Focal contacts contain a number of phosphorylated proteins, including $pp125^{FAK}$ (121,124,125). One consequence of fibrinogen binding to GPIIb-IIIa is the association of the cytoplasmic domains of the heterodimer with submembrane actin filaments (126,127), perhaps via intermediary cytoskeletal proteins (128,129). Thus, the tyrosine phosphorylation observed following GPIIb-IIIa occupancy may be a result of the interaction of GPIIb-IIIa with the platelet cytoskeleton in a process analogous to that seen in adherent fibroblasts.

C. Interaction of GPIIb-IIIa with Soluble Fibrinogen

The regions of soluble fibrinogen that interact with GPIIb-IIIa following platelet activation have been identified, at least to a first approximation, by studies of the effect of fibrinogen-derived peptides on GPIIb-IIIa function. Peptides corresponding to the carboxyl-terminal 10–15 amino acids of the fibrinogen γ chain or to the tetrapeptide sequences Arg-Gly-Asp-Ser (RGDS) and Arg-Gly-Asp-Phe (RDGF) found at positions 572–575 and 95–98 of the fibrinogen α chain, respectively, inhibit platelet aggregation and the interaction of GPIIb-IIIa with fibrinogen, vWF, fibronectin, and vitronectin (29,130,131). The ability of the RGD-containing peptides to inhibit vWF, fibronectin, and vitronectin binding to GPIIb-IIIa is understandable because each of these proteins contains an RGD sequence (132). However, the inhibitory effect of the γ-chain peptide on the binding of these proteins remains problematical because the γ-chain sequence is unique to fibrinogen (133). Nevertheless, these studies imply that fibrinogen can bind to GPIIb-IIIa either via the carboxyl terminus of its γ chain or two sites in its α chain. On the other hand, electron microscope images of soluble fibrinogen bound to individual GPIIb-IIIa heterodimers indicate that the globular head of GPIIb-IIIa binds to the distal ends of the fibrinogen molecule (15). Because this is the location of carboxyl terminus of the γ chain, these images suggest that the carboxyl terminus of the γ chain is the predominant region in fibrinogen that interacts with GPIIb-IIIa. This conclusion is also consistent with the ability of fibrinogen mutants lacking either RGD sequence to support platelet aggregation; fibrinogen containing a mutation at the carboxyl terminus of its γ chain is unable to do so (134).

The inhibitory effect of RGD and γ-chain peptides is mutually exclusive, suggesting that either the peptides bind to GPIIb-IIIa at the same or overlapping sites or that binding of one peptide induces a change in the heterodimer that precludes binding of the other (135). This issue has been addressed by cross-linking studies that suggest that RGD-containing peptides bind to GPIIIa between residues 109 and 171 (136) and γ-chain peptides bind to GPIIb between residues 294 and 314 (28). Thus, the binding sites for RGD and γ-chain peptides on GPIIb-IIIa appear to be spatially distinct. Additional observations that peptides corresponding to GPIIb residues 296–306 (31) and residues 309–312 (137) inhibit platelet aggregation and bind to fibrinogen strengthen the possibility that fibrinogen actually binds to this region of GPIIb. Similar experiments have also identified a region of GPIIIa encompassing residues 211–222 that may represent a second ligand binding domain in this protein (138). Taken together, these studies suggest that fibrinogen binds to a surface on GPIIb-IIIa that is composed of segments of both GPIIb and GPIIIa. Whether these segments are actually in proximity in the folded configuration of the heterodimer and the nature of the allosteric change that exposes the surface to soluble proteins remain to be determined. The speculation that peptide binding to GPIIb-IIIa induces a conformational change in the heterodimer has also been tested and has been confirmed. Incubation of GPIIb-IIIa with peptides containing RGD or RGD-like sequences converts the heterodimer to an activated conformation (139), permits the proteolytic generation of active GPIIb-IIIa fragments (140,141), and exposes neoepitopes (142,143). Whether these effects of ligand on GPIIb-IIIa conformation also account for the mutually exclusive nature of peptide binding has not been determined.

D. GPIIb-IIIa as a Calcium Channel

The initial suggestion that GPIIb-IIIa is involved in the transport of ionic calcium was derived from the observation of a 50% reduction in steady-state calcium exchange in the platelets of patients with Glanzmann thrombasthenia (144). A similar reduction could be induced in normal platelets by dissociating GPIIb-IIIa with EDTA.

Subsequently, it was found that incorporation of GPIIb-IIIa into liposomes resulted in significant facilitation of calcium movement across the lipid bilayer (145,146). Moreover, it was observed that monoclonal antibodies against GPIIb-IIIa inhibit agonist-stimulated calcium influx into platelets, although thrombin-stimulated influx was inhibited only at thrombin concentrations < 0.1 unit/ml (147). Despite these observations, the role played by GPIIb-IIIa in calcium transport remains unclear. In the latter study, no impairment of calcium influx was seen when thrombasthenic platelets were stimulated by ADP and thrombin. Moreover, it was noted that dissociation of GPIIb-IIIa had no effect on thrombin-stimulated calcium influx into HEL cells (148). Finally, the finding that voltage-independent calcium currents observed when membranes from thrombin-stimulated platelets are incorporated into planar lipid bilayers diminish, but do not cease, after dissociation of GPIIb-IIIa with EGTA or when thrombasthenic membranes are used, suggests that GPIIb-IIIa has an indirect role in calcium channel activation (149).

V. CONSEQUENCES OF NATURALLY OCCURRING GPIIb AND GPIIIa MUTATIONS

Mutations resulting in deficiency or dysfunction of GPIIb-IIIa are responsible for Glanzmann thrombasthenia, a rare autosomal recessive hemorrhagic disorder manifested by a prolonged bleeding time, a normal platelet count, and absent macroscopic platelet aggregation (8). Because of their GPIIb-IIIa abnormalities, platelets from patients with thrombasthenia are unable to bind fibrinogen (2) and consequently are unable to aggregate. Because GPIIb-IIIa also appears to be involved in the packaging of fibrinogen in platelet α granules (150,151), α granule fibrinogen is either decreased or absent in thrombasthenic platelets (152). Finally, GPIIb-IIIa is thought to transmit the force of cytoskeletal contraction to adherent fibrin strands, and platelet-mediated clot retraction has been noted to be absent or diminished in the presence of thrombasthenic platelets (153,154).

In the absence of genetic information, thrombasthenia was classified into types on the basis of the amount of GPIIb-IIIa present per platelet and the presence or absence of α granule fibrinogen and clot retraction (152,155). In type I thrombasthenia, platelets contain 1–5% of the normal amount of GPIIb-IIIa, contain little or no α granule fibrinogen, and usually fail to retract clots. In type II thrombasthenia, platelets contain 10–20% of the normal amount of GPIIb-IIIa, α granule fibrinogen is detectable, and clot retraction occurs but is diminished. In a third type of thrombasthenia termed variant thrombasthenia, 50–100% of the normal amount of GPIIb-IIIa is present and the GPIIb-IIIa abnormality is qualitative rather than quantitative.

Because there are no consistent phenotypic differences between the various types of thrombasthenia (152), the descriptive classification of the disorder is sufficient for clinical purposes. Nevertheless, cloning of the GPIIb and GPIIIa genes has made it possible to determine the genetic basis for thrombasthenia in an increasing number of cases. The information derived from these studies has proven useful in further understanding normal GPIIb-IIIa biosynthesis and function.

Seven different GPIIb and seven different GPIIIa mutations have been reported. Of the GPIIb mutations, five produce abnormal transcription or translation of the GPIIb gene. One of these mutations consists of a deletion of ~4.5 kb of the GPIIb gene containing exons 2-9 and results in the transcription of abnormal GPIIb mRNA that terminates GPIIb translation shortly after translation of the signal peptide (156). Two other mutations result from the mutation (157) or deletion (158) of a splice acceptor site leading to aberrant mRNA splicing; the remaining two mutations result from the introduction of a premature termination codon into the GPIIb reading frame (156,157). Two point mutations in GPIIb (G^{818} → A and G^{1347} → A) are of particular interest (54,159). Each results in the substitution of charged for an uncharged amino acid (Gly^{273} → Asp and Gly^{418} → Asp) adjacent to a calcium binding domain in GPIIb. Although the mutations do not prevent the assembly of GPIIb-IIIa heterodimers, they disrupt the epitopes for heterodimer-specific monoclonal antibodies and prevent the transport of the heterodimers from the ER to the Golgi. Thus, these mutations emphasize the importance of correct quaternary conformation for the normal intracellular trafficking of GPIIb-IIIa. Of the seven reported GPIIIa mutations, three produce alterations in the transcription or translation of the GPIIIa gene. One of these mutations is the result of a complex rearrangement of the GPIIIa gene (160), a second results form an 11-base deletion within exon XII of the GPIIIa gene, producing a frameshift and a premature termination codon (158), and a third is the result of the deletion of exon I of the GPIIIa gene by abnormal splicing and the subsequent introduction of a premature termination codon (161). The four remaining GPIIIa mutations consist of single amino acid substitutions that produce variant thrombasthenia. Thus, despite the inability of platelets expressing these GPIIIa mutants to aggregate, the platelets contain nearly normal amounts of GPIIb-IIIa on their surface. In one mutant, a Ser^{752} → Pro substitution in the cytoplasmic domain of GPIIIa prevents activation of the heterodimer by cellular agonists (162). The three other point mutations are similar in that they involve the extracellular domain of GPIIIa and result from the replacement of a charged amino acid by an uncharged residue (Asp^{119} → Tyr and Arg^{214} → Trp/Gln) (163–165). Interestingly, each of these mutations is located in a region of GPIIIa (residues 109–171 and 211–222) that has been implicated in ligand binding (136,138). Why such changes would disrupt ligand binding remains to be determined.

REFERENCES

1. Sixma JJ, Wester J. The hemostatic plug. Semin Hematol 1977; 14:265–299.
2. Bennett JS, Vilaire G. Exposure of platelet fibrinogen receptors by ADP and epinephrine. J Clin Invest 1979; 64:1393–1401.
3. Poncz M, Eisman R, Heidenreich R, et al. Structure of the platelet membrane glycoprotein IIb. J Biol Chem 1987; 262:8476–8482.
4. Schullek J, Jordan J, Montgomery RR. Interaction of von Willebrand factor with human platelets in the plasma milieu. J Clin Invest 1984; 73:421–428.
5. Marguerie GA, Plow EF, Edgington TS. Human platelets possess an inducible and saturable receptor specific for fibrinogen. J Biol Chem 1979; 254:5357–5363.
6. Bennett JS, Vilaire G, Cines DB. Identification of the fibrinogen receptor on human platelets by photoaffinity labeling. J Biol Chem 1982; 257:8049–8054.
7. Ruggeri ZM, Bader R, DeMarco L. Glanzmann thrombasthenia; deficient binding of von Willebrand factor to thrombin-stimulated platelets. Proc Natl Acad Sci USA 1982; 79:6038–6041.
8. George JN, Nurden AT, Phillips DR. Molecular defects in interactions of platelets with the vessel wall. N Engl J Med 1984; 311:1084–1098.
9. Phillips DR, Charo IF, Parise LV, Fitzgerald LA. The platelet membrane glycoprotein IIb-IIIa complex. Blood 1988; 71:831–843.
10. Bennett JS, Hoxie JA, Leitman F, Vilaire G, Cines DB. Inhibition of fibrinogen binding to stimulated human platelets by a monoclonal antibody. Proc Natl Acad Sci USA 1983; 80: 2417–2421.

11. Niiya K, Hodson E, Bader R, et al. Increased surface expression of the membrane glycoprotein IIb/IIIa complex induced by platelet activation. Relationship to the binding of fibrinogen and platelet aggregation. Blood 1987; 70:475–483.
12. Woods VL Jr, Wolff LE, Keller DM. Resting platelets contain a substantial centrally located pool of glycoprotein IIb-IIIa complex which may be accessible to some but not other extracellular protein. J Biol Chem 1986; 261:15242–15251.
13. Fujimoto T, Ohara S, Hawiger J. Thrombin-induced exposure and prostacyclin inhibition of the receptor for factor VIII/von Willebrand factor on human platelets. J Clin Invest 1982; 69:1212–1222.
14. Jennings LK, Phillips DR. Purification of glycoproteins IIb and III from human platelet plasma membranes and characterization of a calcium-dependent glycoprotein IIb-III complex. J Biol Chem 1982; 257:10458–10466.
15. Weisel JW, Nagaswami C, Vilaire G, Bennett JS. Examination of the platelet membrane glycoprotein IIb/IIIa complex and its interaction with fibrinogen and other ligands by electron microscopy. J Biol Chem 1992; 267:16637–16643.
16. Carrell NA, Fitzgerald LA, Steiner B, Erickson HP, Phillips DR. Structure of human platelet membrane glycoproteins IIb and IIIa as determined by electron microscopy. J Biol Chem 1985; 260:1743–1749.
17. Parise LV, Phillips DR. Reconstitution of the purified platelet fibrinogen receptor. J Biol Chem 1985; 260:10698–10707.
18. Heidenreich R, Eisman R. Surrey S, et al. Organization of the gene for platelet glycoprotein IIb. Biochemistry 1990; 29:1232–1244.
19. Kolodziej MA, Vilaire G, Gonder D, Poncz M, Bennett JS. Study of the endoproteolytic cleavage of platelet glycoprotein IIb using oligonucleotide-mediated mutagenesis. J Biol Chem 1991; 266:23499–23504.
20. Steiner DF, Smeekens SP, Ohagai S, Chan SJ. The new enzymology of precursor processing endoproteases. J Biol Chem 1992; 267:23435–23438.
21. Calvette JJ, Henschen A, Gonzalez-Rodriguez J. Complete localization of the intrachain disulphide bonds and the N-glycosylation points in the α-subunit of human platelet glycoprotein IIb. Biochem J 1989; 261:561–568.
22. Lyman S, Aster RH, Visentin GP, Newman PJ. Polymorphism of human platelet membrane glycoprotein IIb associated with the Bak^a/Bak^b alloantigen system. Blood 1990; 75:2343–2348.
23. Goldberger A, Kolodziej M, Poncz M, Bennett JS, Newman PJ. Effect of single amino acid substitutions on the formation of the PI^A and Bak alloantigenic epitopes. Blood 1991; 78: 681–687.
24. Guilino D, Boudignon C, Zhang L, Concord E, Rabiet M-J, Marguerie G. Ca^{2+}-binding properties of the platelet glycoprotein IIb ligand-interacting domain. J Biol Chem 1992; 267:1001–1007.
25. Lam SC-T. Isolation and characterization of a chymotryptic fragment of platelet glycoprotein IIb-IIIa retaining Arg-Gly-Asp binding activity. J Biol Chem 1992; 267:5649–5655.
26. Steiner B, Parise LV, Leung B, Phillips DR. Ca^{2+}-dependent structural transitions of the platelet glycoprotein IIb-IIIa complex. J Biol Chem 1991; 266:14986–14991.
27. Brass LF, Shattil SJ, Kunicki TJ, Bennett JS. Effect of calcium on the stability of the platelet membrane glycoprotein IIb-IIIa complex. J Biol Chem 1985; 260:7875–7881.
28. D'Souza SE, Ginsberg MH, Burke TA, Plow EF. The ligand binding site of the platelet integrin receptor GPIIb-IIIa is proximal to the second calcium binding domain of its α subunit. J Biol Chem 1990; 265:3440–3446.
29. Kloczewiak M, Timmons S, Lukas TJ, Hawiger J. Platelet receptor recognition site on human fibrinogen. Synthesis and structure-function relationships of peptides corresponding to the carboxy-terminal segment of the γ chain. Biochemistry 1984; 23:1767–1774.
30. Marguerie GA, Plow EF. Interaction of fibrinogen with its platelet receptor: kinetics and effect of pH and temperature. Biochemistry 1981; 20:1074–1080.

31. D'Souza SE, Ginsberg MH, Matsueda GR, Plow EF. A discrete sequence in a platelet integrin is involved in ligand recognition. Nature 1991; 350:66–68.
32. Fitzgerald LA, Steiner B, Rall SC Jr, Lo S-S, Phillips DR. Protein sequence of endothelial glycoprotein IIIa derived from a cDNA clone. J Biol Chem 1987; 262:3936–3939.
33. Zimrin AB, Eisman R, Vilaire G, Schwartz E, Bennett JS, Poncz M. Structure of platelet glycoprotein IIIa. J Clin Invest 1988; 81:1470–1475.
34. Rosa J-P, Bray PF, Gayet O, et al. Cloning of glycoprotein IIIa cDNA from human erythroleukemia cells and localization of the gene to chromosome 17. Blood 1988; 72:593–600.
35. Tsuji T, Osawa T. Structures of the carbohydrate chains of membrane glycoproteins IIb and IIIa of human platelets. J Biochem 1986; 100:1387–1398.
36. Kolodziej MA, Vilaire G, Rifat S, Poncz M, Bennett JS. Effect of deletion of glycoprotein IIb exon 28 on the expression of the platelet glycoprotein IIb/IIIa complex. Blood 1991; 78:2344–2353.
37. Newman PJ, Derbes RS, Aster RH. The human platelet alloantigens, Pl^1 and Pl^2, are associated with a $leucine^{33}$/$proline^{33}$ amino acid polymorphism in membrane glycoprotein IIIa, and are distinguishable by DNA typing. J Clin Invest 1989; 83:1778–1781.
38. Wang R, Furihata K, McFarland JG, Freidman K, Aster RH, Newman PJ. An amino acid polymorphism within the RGD binding domain of platelet membrane glycoprotein IIIa is responsible for the formation of the Pen^a/Pen^b alloantigen system. J Clin Invest 1992; 90: 2038–2043.
39. Kuijpers AWAM, Simsek S, Faber NM, Goldschmeding R, van Wermerkerken RKV, von dem Borne AEGK. Single point mutation in human glycoprotein IIIa is associated with a new platelet-specific alloantigen (Mo) involved in neonatal alloimmune thrombocytopenia. Blood 1993; 81:70–76.
40. Kekomaki R, Raivio R, Kero P. A new low-frequency platelet alloantigen, Va^a, on glycoprotein IIb/IIIa associated with neonatal alloimmune thrombocytopenia. Trans Med 1992; 2:27–33.
41. Kekomaki R, Jouhikainen T, Ollikainen J, Westman P, Laes M. A new platelet alloantigen, Tu^a, on glycoprotein IIIa associated with neonatal alloimmune thrombocytopenia in two families. Br J Haematol 1993; 83:306–310.
42. Kroll H, Kiefel V, Santoso S, Mueller-Eckhardt C. Sr^a, a private platelet antigen on glycoprotein IIIa associated with neonatal alloimmune thrombocytopenia. Blood 1990; 76:2296–2302.
43. Calvette JJ, Henschen A, Gonzalez-Rodriguez J. Assignment of disulphide bonds in human platelet GPIIIa. Biochem J 1991; 274:63–71.
44. Beer J, Coller BS. Evidence that platelet glycoprotein IIIa has a large disulfide-bonded loop that is susceptible to proteolytic cleavage. J Biol Chem 1989; 264:17564–17573.
45. Niewiarowski S, Norton KJ, Eckardt A, Lukasiewicz H, Holt JC, Kornecki E. Structural and functional characterization of major platelet membrane components derived by limited proteolysis of glycoprotein IIIa. Biochim Biophys Acta 1989; 983:91–99.
46. Silver SM, McDonough MM, Vilaire G, Bennett JS. The in vitro synthesis of polypeptides for the platelet membrane glycoproteins IIb and IIIa. Blood 1987; 69:1031–1037.
47. Bray PF, Rosa J-P, Lingappa VR, Kan YW, McEver RP, Shuman MA. Biogenesis of the platelet receptor for fibrinogen: evidence for separate precursors for glycoproteins IIb and IIIa. Proc Natl Acad Sci USA 1986; 83:1480–1488.
48. Rosa J-P, McEver RP. Processing and assembly of the integrin, glycoprotein IIb-IIIa, in HEL cells. J Biol Chem 1989; 264:12596–12603.
49. Duperray A, Berthier R, Chagnon E, et al. Biosynthesis and processing of platelet GPIIb-IIIa in human megakaryocytes. J Cell Biol 1987; 104:1665–1673.
50. O'Toole TE, Loftus JC, Plow EF, Glass AA, Harper JR, Ginsberg MH. Efficient surface expression of platelet GPIIb-IIIa requires both subunits. Blood 1989; 74:14–18.

51. Bodary SC, Napier MA, McLean JW. Expression of recombinant platelet glycoprotein IIbIIIa results in a functional fibrinogen-binding complex. J Biol Chem 1989; 264:18859–18862.
52. Bennett JS, Kolodziej MA, Vilaire G, Poncz M. Determinants of the intracellular fate of truncated forms of the platelet glycoproteins IIb and IIIa. J Biol Chem 1993; 268:3580–3585.
53. Gething M-J, Sambrook J. Protein folding in the cell. Nature 1992; 355:33–45.
54. Poncz M, Salahandrin R, Coller BS, et al. Glanzmann thrombasthenia secondary to a GLY273-ASP mutation adjacent to the first calcium-binding domain of platelet glycoprotein IIb. J Clin Invest 1993; 93:172–179.
55. Kornfeld R, Kornfeld S. Assembly of asparagine-linked oligosaccharides. Annu Rev Biochem 1985; 54:631–634.
56. Hibbs ML, Wardlaw AJ, Stacker SA, et al. Transfection of cells from patients with leukocyte adhesion deficiency with an integrin β subunit (CD18) restores lymphocyte function-associated antigen-1 expression and function. J Clin Invest 1990; 85:674–681.
57. Hynes RO. Integrins: a family of cell surface receptors. Cell 1987; 48:549–554.
58. Fitzgerald LA, Poncz M, Steiner B, Rall SC Jr, Bennett JS, Phillips DR. Comparison of cDNA-derived protein sequences of the human fibronectin and vitronectin receptor α-subunits and platelet glycoprotein IIb. Biochemistry 1987; 26:8158–8165.
59. Hynes RO. Integrins: versatility modulation, and signaling in cell adhesion. Cell 1991; 69: 11–25.
60. Takada Y, Strominger JL, Hemler ME. The very late antigen family of heterodimers is part of a superfamily of molecules involved in adhesion and embryogenesis. Proc Natl Acad Sci USA 1987; 84:3239–3243.
61. Hemler ME, Crouse C, Takada Y, Sonnenberg A. Multiple very late antigen (VLA) heterodimers on platelets. J Biol Chem 1988; 263:7660–7665.
62. Anderson DC, Springer TA. Leukocyte adhesion deficiency: an inherited defect in the Mac-1, LFA-1, and p150,95 glycoproteins. Annu Rev Med 1987; 38:175–194.
63. Larson RS, Corbi AL, Berman L, Springer T. Primary structure of the leukocyte function-associated molecule-1 α subunit: an integrin with an embedded domain defining a protein superfamily. J Cell Biol 1989; 108:703–712.
64. Colombatti A, Bonaldo P. The superfamily of proteins with von Willebrand factor type A-like domains: one theme common to components of extracellular matrix, hemostasis, and cellular adhesion, and defense mechanisms. Blood 1991; 77:2305–2315.
65. Hibbs ML, Xu H, Stacker SA, Springer TA. Regulation of adhesion to ICAM-1 by the cytoplasmic domain of LFA-1 integrin β subunit. Science 1991; 251:1611–1613.
66. Staunton DE, Marlin SD, Stratowa C, Dustin ML, Springer TA. Primary structure of ICAM-1 demonstrates interaction between members of the immunoglobulin and integrin supergene families. Cell 1988; 52:925–933.
67. Arnout MA. Structure and function of the leukocyte adhesion molecules CD11/CD18. Blood 1990; 75:1037–1050.
68. Charo IF, Bekeart LS, Phillips DR. Platelet glycoprotein IIb-IIIa-like proteins mediate endothelial cell attachment to adhesive proteins and the extracellular matrix. J Biol Chem 1987; 262:9935–9938.
69. Kieffer N, Fitzgerald LA, Wolf D, Cheresh DA, Phillips DR. Adhesive properties of the β_3 integrins: comparison of GBIIb-IIIa and the vitronectin receptor individually expressed in human melanoma cells. J Cell Biol 1991; 113:451–461.
70. Sosnoski DM, Emanuel BS, Hawkins AL, et al. Chromosomal localization of the genes for the vitronectin and fibronectin receptors α subunits and for platelet glycoproteins IIb and IIIa. J Clin Invest 1988; 81:1993–1998.
71. Bray PF, Rosa J-P, Johnston GI, et al. Platelet glycoprotein IIb: chromosomal localization and tissue expression. J Clin Invest 1987; 80:1812–1817.

72. Bray PR, Barsh G, Rosa J-P, Luo XY, Magenis E, Shuman MA. Physical linkage of the genes for platelet membrane glycoprotein IIb and IIIa. Proc Natl Acad Sci USA 1988; 85: 8683–8687.
73. Zimrin AB, Gidwitz S, Lord S, et al. The genomic organization of platelet glycoprotein IIIa. J Biol Chem 1990; 265:8590–8595.
74. Burk CD, Newman PJ, Lyman S, Gill J, Coller BS, Poncz M. A deletion in the gene for glycoprotein IIb associated with Glanzmann's thrombasthenia. J Clin Invest 1991; 87:2706.
75. Nojima H, Sokabe H. Structure of a gene for rat calmodulin. J Mol Biol 1987; 193:439–445.
76. Prandini M-H, Denarier E, Frachet P, Uzan G, Marguerie G. Isolation of the human platelet glycoprotein IIb gene and characterization of the 5′ flanking region. Biochem Biophy Res Commun 1988; 156:595–601.
77. Luo X, Kim K-H. An enhancer element in the house-keeping promoter for acetyl-CoA carboxylase gene. Nucleic Acids Res 1990; 18:3249–3254.
78. Hickstein DD, Baker DM, Gollahon KA, Back AL. Identification of the promoter of the myelomonocytic leukocyte integrin CD11b. Proc Natl Acad Sci USA 1992; 89:2105–2109.
79. Corden J, Wasylyk B, Buchwalder A, Sassone-Corsi P, Kedinger C, Chambon P. Promoter sequences of eukaryotic protein-coding genes. Science 1980; 209:1405–1414.
80. Smale S, Baltimore D. The ''initiator'' as a transcription control element. Cell 1989; 57: 103–113.
81. Lanza F, Kieffer N, Phillips DR, Fitzgerald LA. Characterization of the human platelet glycoprotein IIIa gene. Comparison with the fibronectin receptor β-subunit gene. J Biol Chem 1990; 265:18098–18103.
82. Chen YQ, Gao X, Timar J, et al. Identification of the $\alpha_{II}\beta_3$ integrin in murine tumor cells. J Biol Chem 1992; 267:17314–17320.
83. Pytela R, Pierschbacher MD, Ruoslahti E. A 125/115-kDa cell surface receptor specific for vitronectin interacts with the arginine-glycine-aspartic acid adhesion sequence derived from fibronectin. Proc Natl Acad Sci USA 1985; 82:5766–5770.
84. Rodgers GP, Schechter AN, Noguchi CT, Klein HG, Nienhuis AW, Bonner RF. Microcirculatory adaptations in sickle cell anemia: reactive hyperemia response. Am J Physiol 1990; 258:H113–120.
85. Oyo R, Nakeff S, Huang SS, Ginsberg M, Deuel TF. New synthesis of platelet-specific protein: platelet factor 4 synthesis in megakaryocyte-enriched rabbit bone marrow system. J Cell Biol 1983; 96:515–520.
86. Debili N, Issaad C, Masse J-M, et al. Expression of CD34 and platelet glycoproteins during human megakaryocytic differentiation. Blood 1992; 80:3022–3035.
87. Orkin SH. GATA-binding transcription factors in hematopoietic cells. Blood 1992; 80:575–581.
88. Martin DIK, Zon LI, Mutter G, Orkin SH. Expression of an erythroid transcription factor in megakaryocytic and mast cell lineages. Nature 1990; 344:444–447.
89. Romeo P-H, Prandini M-H, Joulin V, et al. Megakaryocytic and erythrocytic lineages share specific transciption factors. Nature 1990; 344:447–449.
90. Zon LI, Mather C, Burgess S, Bolce ME, Harland RM, Orkin SH. Expression of GATA-binding proteins during embyonic development in *Xenopus laevis*. Proc Natl Acad Sci USA 1991; 88:10642–10646.
91. Pevny L, Simon CS, Robertson E, et al. Erythroid differentiation in chmaeric mice blocked by a targeted mutation in the gene for transcription factor GATA-1. Nature 1991; 349:257–260.
92. Uzan G, Prenant M, Prandini M-H, Martin F, Marguerie G. Tissue-specific expression of the platelet GPIIb gene. J Biol Chem 1991; 266:8932–8939.
93. Prandini M-H, Uzan G, Martin F, Thevenon D, Marguerie G. Characterization of a specific erythromegakaryocytic enhancer within the glycoprotein IIb promoter. J Biol Chem 1992; 267:10370–10374.

94. Lemarchandel V, Ghysdael J, Mignotte V, Rahuel V, Romeo P-H. GATA and Ets cis-acting sequences mediate megakayocyte-specific expression. Mol Cell Biol 1993; 13:668–676.
95. Bosselut R, Duvall JF, Gegonne A, et al. The product of the c-ets-1 proto-oncogene and the related Ets-2 protein act as transcriptional activators of the long terminal repeat of human T cell leukemia virus HTLV-1. EMBO J 1990; 9:3137–3144.
96. Ravid K, Beeler DL, Rabin MS, Ruley HE, Rosenberg RD. Selective targeting of gene products with the megakaryocyte platelet factor 4 promoter. Proc Natl Acad Sci USA 1991; 88:1521–1525.
97. Ravid K, Doi T, Beeler DL, Kuter DJ, Rosenberg RD. Transcriptional regulation of the rate platelet factor 4 gene: interaction between an enhancer/silencer domain and the GATA site. Mol Cell Biol 1991; 11:6116–6127.
98. Tapscott SJ, Weintraub H. MyoD and the regulation of myogenesis by helix-loop-helix proteins. J Clin Invest 1991; 87:1133–1338.
99. DeMarco L, Girolami A, Zimmerman TS, Ruggeri ZM. Von Willebrand factor interaction with the glycoprotein IIb/IIIa complex. J Clin Invest 1986; 77:1272–1277.
100. Savage B, Shattil SJ, Ruggeri ZM. Modulation of platelet function through adhesion receptors. J Biol Chem 1992; 267:11300–6.
101. Savage B, Ruggeri ZM. Selective recognition of adhesive sites in surface-bound fibrinogen by glycoprotein IIb-IIIa on nonactivated platelets. J Biol Chem 1991; 266:11227–33.
102. Weiss HJ, Turitto VT, Baumgartner HR. Platelet adhesion and thrombus formation on subendothelium in platelets deficient in glycoproteins IIb-IIIa, Ib, and storage granules. Blood 1986; 67:322–330.
103. Sims PJ, Ginsberg MH, Plow EF, Shattil SJ. Effect of platelet activation on the conformation of the plasma membrane glycoprotein IIb-IIIa complex. J Biol Chem 1991; 266:7345–7352.
104. O'Toole TE, Mandelman D, Forsyth J, Shattil SJ, Plow EF, Ginsberg MH. Modulation of the affinity of integrin αIIβb3 (GPIIb-IIIa) by the cytoplasmic domain of αIIb. Science 1991; 254:845–847.
105. Du X, Gu M, Weisel JW, et al. Long range propagation of conformational changes in integrin $\alpha_{IIb}\beta_3$. J Biol Chem 1993; 268:23087–23092.
106. Brass LF, Manning DR, Shattil SJ. GTP-binding proteins and platelet activation. Prog Hemost Thromb 1990; 10:127–174.
107. Golden A, Nemeth SP. Brugge JS. Blood platelets express high levels of the pp60c¯src-specific tyrosine kinase activity. Proc Natl Acad Sci USA 1986; 83:852–856.
108. Shattil SJ, Brugge JS. Protein tyrosine phosphorylation and the adhesive functions of platelets. Curr Opin Cell Biol 1991; 3:869–879.
109. Taniguchi T, Kitagawa H, Yasue S, et al. Protein-tyrosine kinase $p72^{svk}$ is activated by thrombin and is negatively regulated through Ca^{2+} mobilization in platelets. J Biol Chem 1993; 268:2277–2279.
110. Wilson DB, Neufeld EJ, Majerus PW. Phosphoinositide interconversion in thrombin-stimulated human platelets. J Biol Chem 1985; 260:1046–1051.
111. Brass LF, Joseph SK. A role for inositol triphosphate in intracellular Ca^{2+} mobilization and granule secretion in platelets. J Biol Chem 1985; 260:15172–15179.
112. Ferrell JE Jr, Martin GS. Platelet tyrosine-specific protein phosphorylation is regulated by thrombin. Mol Cell Biol 1988; 8:3603–3610.
113. Nakamura S, Yamamura H. Thrombin and collagen induce rapid phosphorylation of a common set of cellular proteins on tyrosine in human platelets. J Biol Chem 1989; 264:7089–7091.
114. Ferrell JE Jr, Martin GS. J Biol Chem 1989; 264:20723–20729.
115. Parise LV, Criss AB, Nannizzi L, Wardell MR. Glycoprotein IIIa is phosphorylated in intact human platelets. Blood 1990; 75:2363–2368.
116. Hillery CA, Smyth SS, Parise LV. Phosphorylation of human platelet glycoprotein IIIa (GPIIIa). J Biol Chem 1991; 266:14663–14669.

117. Hannun YA, Greenberg CS, Bell RM. Sphingosine inhibition of agonist-dependent secretion and activation of human platelets implies that protein kinase C is a necessary and common event of the signal transduction pathways. J Biol Chem 1987; 262:13620–13626.
118. Shattil SJ, Cunningham M, Wiedmer T, Zhao J, Sims PJ, Brass LF. Regulation of glycoprotein IIb–IIIa receptor function studied with platelets permeabilized by the pore-forming complement proteins C5b-9. J Biol Chem 1992; 267:18424–18431.
119. Lipfert L, Haimovich B, Schaller MD, Cobb BS, Parsons JT, Brugge JS. Integrin-dependent phosphorylation and activation of the protein tyrosine kinase $pp125^{FAK}$ in platelets. J Cell Biol 1992; 119:905–912.
120. Huang M-M, Lipfert L, Cunningham M, Brugge JS, Ginsberg MH, Shattil SJ. Adhesive ligand binding to integrin $\alpha_{IIb}\beta_3$ stimulates tyrosine phosphorylation of novel protein substrates before phosphorylation of $pp125^{FAK}$. J Cell Biol 1993; 122:473–483.
121. Burridge K, Turner CE, Romer LH. Tyrosine phosphorylation of paxillin and $pp125^{FAK}$ accompanies cell adhesion to extracellular matrix: a role in cytoskeletal assembly. J Cell Biol 1992; 119:893–903.
122. Akiyama SK, Yamada SS, Chen W-T, Yamada KM. Analysis of fibronectin receptor function with monoclonal antibodies; roles in cell adhesion, migration, matrix assembly, and cytoskeletal organization. J Cell Biol 1989; 109:863–875.
123. Chen W-T, Wang J, Hasegawa T, Yamada SS, Yamada KM. Regulation of fibronectin receptor distribution by transformation, exogenous fibronectin, and synthetic peptides. J Cell Biol 1986; 103:1649–1661.
124. Kornberg L, Earp HS, Parsons JT, Schaller M, Juliano RL. Cell adhesion or integrin clustering increases phosphorylation of a focal adhesion-associated tyrosine kinase. J Biol Chem 1992; 267:23439–23442.
125. Kornberg LJ, Earp HS, Turner CE, Prockop C, Juliano RL. Signal transduction by integrins: increased protein tyrosine phosphorylation caused by clustering of β_1 integrins. Proc Natl Acad Sci USA 1991; 88:8392–8396.
126. Phillips DR, Jennings LK, Edwards HH. Identification of membrane proteins mediating the interaction of human platelets. J Cell Biol 1980; 86:77–86.
127. Wheeler ME, Cox AC, Carrol RC. Retention of the glycoprotein IIb–IIIa complex in the isolated platelet cytoskeleton. Effects of separable assembly of platelet pseudopodal and contractile cytoskeletons. J Clin Invest 1984; 74:1080–1089.
128. Horwitz A, Duggan K, Buck C, Beckerle MC, Burridge K. Interaction of plasma membrane fibronectin receptor with talin-a transmembrane linkage. Nature 1986; 320:531–533.
129. O'Halloran T, Beckerle MC, Burridge K. Identification of talin as a major cytoplasmic protein implicated in platelet activation. Nature 1985; 317:449–451.
130. Gartner TK, Bennett JS. The tetrapeptide analogue of the cell attachment site of fibronectin inhibits platelet aggregation and fibrinogen binding to activated platelets. J Biol Chem 1985; 260:11891–11894.
131. Plow EF, Pierschbacher MD, Ruoslahti E, Marguerie G, Ginsberg MH. Arginyl-glycyl-aspartic acid sequences and fibrinogen binding to platelets. Blood 1987; 70:110–115.
132. Pytela R, Pierschbacher MD, Ginsberg MH, Plow EF, Ruoslahti E. Platelet membrane glycoprotein IIb/IIIa: member of a family of arg-gly-asp-specific adhesion receptors. Science 1986; 231:1559–1562.
133. Plow EF, Srouji AH, Meyer D, Marguerie G, Ginsberg MH. Evidence that three adhesive proteins interact with a common recognition site on activated platelets. J Biol Chem 1984; 259:5388–5391.
134. Farrell DH, Thiagarajan P, Chung DW, Davie EW. Role of fibrinogen α and γ chain sites in platelet aggregation. Proc Natl Acad Sci USA 1992; 89:10729–10732.
135. Bennett JS, Shattil SJ, Power JW, Gartner TK. Interaction of fibrinogen with its platelet receptor. Differential effects of α and γ chain fibrinogen peptides on the glycoprotein IIb–IIIa complex. J Biol Chem 1988; 263:12948–12953.

136. D'Souza SE, Ginsberg MH, Burke TA, Lam SC-T, Plow EF. Localization of an Arg-Gly-Asp recognition site within an integrin adhesion receptor. Science 1988; 242:91–93.
137. Gartner TK, Taylor DB. The amino acid sequence gly-ala-pro-leu appears to be a fibrinogen binding site in the platelet integrin, glycoprotein IIb. Thomb Res 1990; 60:291–309.
138. Charo IF, Nannizzi L, Phillips DR, Hsu MA, Scarborough RM. Inhibition of fibrinogen binding to GPIIb-IIIa by a GP IIIa peptide. J Biol Chem 1991; 266:1415–1421.
139. Du X, Plow EF, Freilinger AL III, O'Toole TE, Loftus JC, Ginsberg MH. Ligands "activate" integrin αIIbβ3 (platelet GPIIb-IIIa). Cell 1991; 65:409–416.
140. Parise LV, Helgerson SL, Steiner B, Nannizzi L, Phillips DR. Synthetic peptides derived from fibrinogen and fibronectin change the conformation of purified platelet glycoprotein IIb-IIIa. J Biol Chem 1987; 262:12597–12602.
141. Kouns WC, Hadvary P, Haering P, Steiner B. Conformational modulation of purified glycoprotein (GP) IIb-IIIa allows proteolytic generation of active fragments from either active or inactive GPIIb-IIIa. J Biol Chem 1992; 267:18844–18851.
142. Frelinger AL III, Lam SC-T, Plow EF, Smith MA, Loftus JC, Ginsberg MH. Occupancy of an adhesive glycoprotein receptor modulates expression of an antigenic site involved in cell adhesion. J Biol Chem 1988; 263:12397–12402.
143. Frelinger AL III, Cohen I, Plow EF, et al. Selective inhibition of integrin function by antibodies specific for ligand-occupied receptor conformers. J Biol Chem 1990; 265:6346–6352.
144. Brass LF. Ca^{2+} transport across the platelet plasma membrane. J Biol Chem 1985; 260: 2231–2236.
145. Rybak ME, Renzulli LA. Ligand inhibition of the platelet glycoprotein IIb-IIIa complex function as a calcium channel in liposomes. J Biol Chem 1989; 264:14617–14620.
146. Rybak ME, Renzulli LA, Bruns MJ, Cahaly DP. Platelet glycoproteins IIb and IIIa as a calcium channel in liposomes. Blood 1988; 72:714–720.
147. Powling MJ, Hardisty RM. Glycoprotein IIb-IIIa complex and Ca^{2+} influx into stimulated platelets. Blood 1985; 66:731–734.
148. Suldan Z, Brass LF. Role of the glycoprotein IIb-IIIa complex in plasma membrane Ca^{2+} transport: a comparison of results obtained with platelets and human erythroleukemia cells. Blood 1991; 78:2887–2893.
149. Fujimoto T, Fujimura K, Kuramoto A. Electrophysiological evidence that glycoprotein IIb-IIIa complex is involved in calcium channel activation on human platelet plasma membrane. J Biol Chem 1991; 266:16370–16375.
150. Handagama PJ, George JN, Shuman MA, et al. Incorporation of a circulating protein into megakaryocyte and platelet granules. Proc Natl Acad Sci USA 1987; 84:861–864.
151. Handagama P, Bainton DF, Jacques Y, Conn MT, Lazarus RA, Shuman MA. Kistrin, an integrin antogonist, blocks endocytosis of fibrinogen into guinea pig megakaryocyte and platelet α-granules. J Clin Invest 1993; 91:193–200.
152. George JN, Caen JP, Nurden AT. Glanzmann's thrombasthenia: the spectrum of clinical disease. Blood 1990; 75:1383–1395.
153. Gartner KT, Ogilvie ML. Peptides and monoclonal antibodies which bind to platelet glycoproteins IIb and/or IIIa inhibit clot retraction. Thromb Res 1988; 49:43–53.
154. Cohen I, Burk DL, White JG. The effect of peptides and monoclonal antibodies that bind to platelet glycoprotein IIb-IIIa complex on the development of clot tension. Blood 1989; 73:1880–1887.
155. Caen J. Glanzmann thrombasthenia. Clin Haematol 1972; 1:383–392.
156. Gu J-M, Xu W-F, Wang X-D, Wu Q-Y, Chi C-W, Ruan C-G. Identification of a nonsense mutation at amino acid 584-arginine of platelet glycoprotein IIb in pateints with type I Glanzmann thrombasthenia. Br J Haematol 1993; 83:442–449.
157. Kato A, Yamamoto K, Miyazaki S, Jung SM, Moroi M, Aoki N. Molecular basis for Glanzmann's thrombasthenia (GT) in a compound heterozygote with glycoprotein IIb gene: a

proposal for the classification of GT based on the biosynthetic pathway of glycoprotein IIb-IIIa complex. Blood 1992; 79:3212–3218.
158. Newman PJ, Seligsohn U, Lyman S, Coller BS. The molecular genetic basis of Glanzmann thrombasthenia in the Iraqi-Jewish and Arab populations in Israel. Proc Natl Acad Sci USA 1991; 88:3160–3164.
159. Wilcox DA, Wautier J-C, Pidard D, Newman PJ. A single amino acid substitution flanking the fourth calcium binding domain of α_{IIb} prevents maturation of the $\alpha_{IIb}\beta_{IIIa}$ integrin complex. J Biol Chem 1993; 269:4450–4457.
160. Bray PF, Shuman MA. Identification of an abnormal gene for the GPIIIa subunit of the platelet fibrinogen receptor resulting in Glanzmann's thrombasthenia. Blood 1990; 75:8818.
161. Simsek S, Heyboer H, Bruijne-Admiraael LG, Goldschmeding R, Cuijpers HTM, von dem Borne AEGK. Glanzmann's thrombasthenia caused by homozygosity for a splice defect that leads to deletion of the first coding exon of the glycoprotein IIIa mRNA. Blood 1993; 81: 2044–2049.
162. Chen Y, Djaffar I, Pidard D, et al. Ser-752-Pro mutation in the cytoplasmic domain of integrin β_3 subunit and defective activation of platelet integrin $\alpha_{IIb}\beta_3$ (glycoprotein IIb-IIIa) in a variant of Glanzmann thrombasthenia. Proc Natl Acad Sci USA 1992; 89:10169–10173.
163. Loftus JC, O'Toole TE, Plow EF, Glass A, Frelinger AL III, Ginsberg MH. A β3 integrin mutation abolishes ligand binding and alters divalent cation-dependent conformation. Science 1990; 249:915–918.
164. Bajt ML, Ginsberg MH, Frelinger AL III, Berndt MC, Loftus JC. A spontaneous mutation of integrin $\alpha_{IIb}\beta_3$ (platelet glycoprotein IIb-IIIa) helps define a ligand binding site. J Biol Chem 1992; 267:3789–3794.
165. Lanza F, Stierle A, Fournier D, et al. A new variant of Glanzmann's thrombasthenia (Strasbourg 1). J Clin Invest 1992; 89:1995–2004.
166. Bardwell JCA, Beckwith. The bonds that tie: catalyzed disulfide bond formation. Cell 1993; 74:769–771.

29

Platelet Integrins Other than Glycoprotein IIb-IIIa ($\alpha_{IIb}\beta_3$)

Samuel A. Santoro
Washington University School of Medicine, St. Louis, Missouri

I. INTRODUCTION

The integrins are a family of noncovalently associated heterodimeric adhesive receptors that play major roles in mediating the cell-substrate and cell-cell interactions of many different cell types, including platelets. General aspects of integrin structure are dealt with only briefly in this chapter. Several excellent general reviews are available (1–5). At least 14 distinct α subunits and 8 different β subunits have now been described. Ligand specificity appears to be determined in large part by the particular combination of α and β subunits, although other factors, such as alternative splicing and the degree of integrin activation, may contribute. As more members of the growing family of integrins have been characterized, it has become apparent that the integrin-mediated adhesive mechanisms exhibit a great deal of redundancy, even on a single cell type. Several integrins may all bind to a given ligand, that is, the $\alpha_1\beta_1$, $\alpha_2\beta_1$, $\alpha_3\beta_1$, $\alpha_6\beta_1$, and $\alpha_6\beta_4$ integrins may all bind laminin. On the other hand, multiple substrates may serve as ligands for a single integrin. For example, the $\alpha_V\beta_3$ integrin may bind vitronectin, fibrinogen, von Willebrand factor, thrombospondin, and osteopontin.

On platelets, integrins are important mediators of both the interactions that mediate platelet aggregation (platelet-platelet interactions) and the adhesion of platelets to components of the vascular subendothelium. Only a limited subset of the known integrins are expressed on the platelet surface. The most abundant of these, and the most extensively studied of the platelet integrins, is the $\alpha_{IIb}\beta_3$ integrin, which is also known as the platelet membrane glycoprotein (GP) IIb-IIIa complex (6,7). The $\alpha_{IIb}\beta_3$ integrin mediates the activation-dependent binding of fibrinogen to platelets that gives rise to platelet aggregation and is considered in detail elsewhere in this volume. The $\alpha_2\beta_1$ integrin, a collagen receptor on platelets; the $\alpha_5\beta_1$ integrin, a fibronectin receptor; the $\alpha_6\beta_1$ integrin, a laminin receptor; and the $\alpha_V\beta_3$ integrin, which, as noted, may bind several different adhesive proteins, are the other integrins expressed on platelets and are the subjects of this chapter (Table 1).

Table 1 Integrins Expressed on Platelets

Integrin	Platelet nomenclature	Ligands
$\alpha_{IIb}\beta_3$	IIb-IIIa	Fibrinogen, fibronectin, von Willebrand factor, vitronectin
$\alpha_v\beta_3$	Ic-IIIa	Vitronectin, thrombospondin, von Willebrand factor, fibrinogen, osteopontin, collagen?
$\alpha_2\beta_1$	Ia-IIa	Collagen[a]
$\alpha_5\beta_1$	Ic-IIa	Fibronectin
$\alpha_6\beta_1$	Ic-IIa	Laminin

[a]On platelets the $\alpha_2\beta_1$ integrin is a collagen-specific receptor. On some other cell types the $\alpha_2\beta_1$ integrin functions as both a collagen and a laminin receptor.

Many of these heterodimeric complexes have also been identified using a platelet nomenclature based upon the electrophoretic mobility of the constituent polypeptides (e.g., GPIIb-IIIa and GPIa-IIa). This nomenclature becomes problematical when applied to some platelet integrins because the α_5, α_6, and α_V subunits all comigrate at the position corresponding to platelet membrane glycoprotein Ic. The platelet nomenclature is also confusing when applied to these integrins, which are all expressed on other cell types in addition to platelets. For reference purposes, both the platelet nomenclature and the more general $\alpha\beta$ integrin subunit nomenclature are given in Table 1.

The chromosomal locations of each of the genes for the integrin subunits expressed on platelets have been determined. These are summarized in Table 2. With the exception of the α_{IIb} and β_3 integrin subunits (8,9), structures of the genes encoding the integrins expressed on platelets have not been determined.

II. INTEGRIN STRUCTURE

A. β Subunit Structure

The β_1 and β_3 integrin subunits are the only integrin β chains expressed on platelets. The β_3 subunit is paired with either the α_{IIb} or the α_V subunit, whereas the β_1 subunit is found in combination with the α_2, α_5, or α_6 chains. The amino acid sequences of both the β_1 and β_3 subunits have been deduced from the corresponding cDNA sequences (10–12). All integrin β chains share a number of common structural features, which are represented schematically in Figure 1. The β_1 subunit exhibits apparent molecular weights of 130,000 and 110,000 under reducing and nonreducing conditions, respectively. The β_3 subunit exhibits apparent molecular weights of 105,000 and 90,000 under reducing and nonreducing conditions. Integrin β subunits typically exhibit a short cytoplasmic domain composed of the most carboxyl-terminal 40 or so amino acid acid residues. The only notable exception to date is the β_4 subunit, which is not expressed on platelets and contains an exceptionally large cytoplasmic domain of greater than 1000 amino acids (13). Additional structural variability within the β_1 and β_3 cytoplasmic domains may arise from alternative splicing (see later).

Table 2 Chromosomal Localization of Genes for Integrin Subunits Expressed on Platelets

Integrin subunit	Chromosome	References
β_1	10	108, 109
β_3	17	11, 110
α_2	5	3
α_5	12	110
α_6	2	21
α_{IIb}	17	110, 111
α_v	2	110

All β subunits contain a single hydrophobic transmembrane domain of length sufficient to span the plasma membrane a single time. The large extracellular domains contain 56 conserved cysteine residues, 31 of which are clustered into four tandemly repeated segments, each of which is believed to contain a large number of internal disulfide bonds. The amino-terminal 45–50 kD also contains within it several other regions of highly conserved sequence. These regions in association with the α subunit may contribute to ligand binding activity. Cross-linking experiments with adhesive peptide ligands for the $\alpha_V\beta_3$ and $\alpha_{IIb}\beta_3$ integrins indicate that residues 100–200 are cross-linked to the adhesive peptides, thus implicating these regions of the β subunits in ligand binding activity (14,15).

Alternatively spliced forms of both the β_1 and β_3 subunits have been described. These variants differ in the structure of their cytoplasmic domains. The usual cytoplasmic domains of the β_1 and β_3 subunits are apparently encoded by two exons. Alternative splicing retains the intron located between these two exons (16,17). The presence of a stop codon near the 5′ end of the intron in both alternatively spliced mRNAs results in truncation of the β_1 and β_3 cytoplasmic domains in these variants. Languino and Ruoslahti (18) recently described a second alternatively spliced β_1 cytoplasmic domain. The variant, which they denote β_{1S}, is generated by the addition of a 116 base pair sequence between the two exons encoding the cytoplasmic domain of the β_1 subunit. Because the β_{1S} sequence is not present in the reported 5′ end sequence of the intervening intron, the sequence is presumably located farther 3′ in the intron. Although it represented only a minor fraction of total β_1 mRNA present, platelets were an exceptionally rich source of

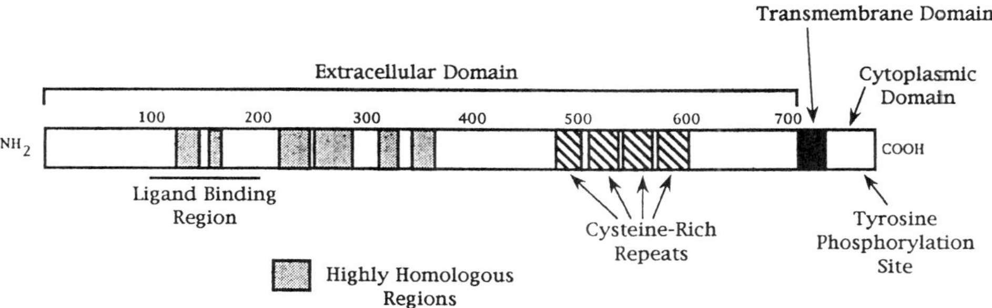

Figure 1 Structural features common to integrin β subunits.

β_{1S} mRNA. Immunoprecipitation experiments employing antiserum directed against the β_{1S} sequence indicated the presence of the subunit in human erythroleukemia cells. No data were provided regarding the expression of β_{1S} protein in platelets. The functional significance of the alternatively spliced β subunit cytoplasmic domains is at present unknown.

B. α Subunit Structure

The amino acid sequences of the integrin α_2, α_5, α_6, and α_V subunits have all been deduced from cDNA sequences (10,19–24). Because none of these cDNAs were derived from cells with megakaryocytic properties, the presence of structural variants of these subunits on platelets cannot be excluded. The biochemical and molecular properties of the α subunits of integrins expressed on platelets are summarized in Table 3. The α chains of these integrins, like all other integrin α chains, share common structural features that are represented schematically in Figure 2. The α subunits all have short cytoplasmic domains. Although the amino acid residues just inside the plasma membrane (GFFKR) are highly conserved among all integrin α chains, the remainder of the cytoplasmic domains differ considerably in sequence. Each of the α chains contains a single hydrophobic membrane spanning segment.

Some α subunits are cleaved posttranslationally to yield heavy- and light-chain components linked by a disulfide bond. The 25–30 kD light chain contains the membrane spanning segment, whereas the heavy chain is completely extracellular. The α_5, α_6, α_V, and α_{IIb} chains undergo such posttranslational cleavage; the α_2 subunit does not.

Integrin α subunits contain within the large extracellular domain a sevenfold repeat of an homologous segment. Depending upon the integrin α subunit, the last three or four such repeats contain the sequence Asp-*X*-ASP-*X*-Asp-Gly-*X*-*X*-Asp or a closely related sequence. Based upon the similarity of these sequences to those responsible for binding divalent cations in known metal binding proteins, it is thought that these sequences contribute to the divalent cation binding properties of integrins. Although the identity of

Table 3 Biochemical Properties of Integrin α Subunits Expressed on Platelets

Integrin subunit	Molecular weight[a] Reduced	Molecular weight[a] Non reduced	Posttrans-lational cleavage	I domain	Divalent cation binding sites
α_{IIb}	120,000 25,000	145,000	Yes	No	4
α_v	125,000 24,000	150,000	Yes	No	4
α_2	165,000	160,000	No	Yes	3
α_5	135,000 25,000	155,000	Yes	No	4
α_6	120,000 30,000[b]	155,000	Yes	No	4

[a]The molecular weights are those assigned by Hemler (3). Slightly different values have been reported by other investigators.

[b]A second α_6 light chain of 31,000 is believed to arise from alternative splicing and/or alternative posttranslational cleavage.

(a)

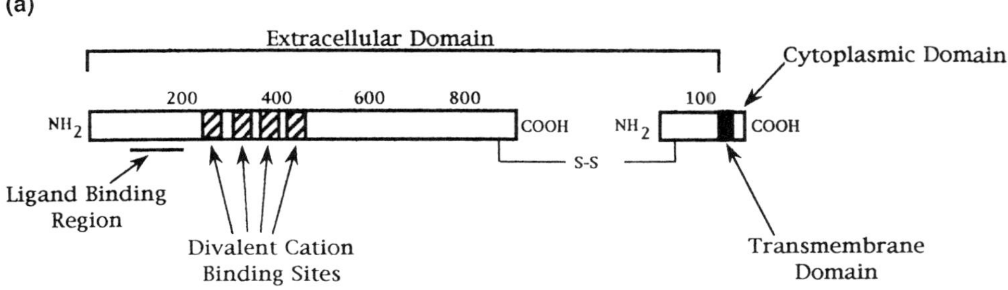

(b)

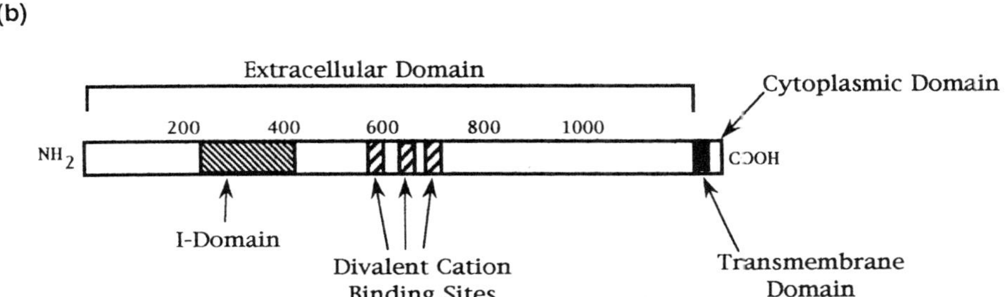

Figure 2 Structural features present within the integrin α subunits. (a) The presence of a post-translational cleavage site, the presence of four putative divalent cation binding sites, and the absence of an I domain are typical of the α_{IIb}, α_5, α_6, and α_V integrin subunits of platelets. (b) The absence of postranslational cleavage, the presence of three putative divalent cation binding sites, and the presence of an I domain are characteristic of the α_2 integrin subunit.

the divalent cation(s) varies from integrin to integrin, all known integrins require the presence of divalent cations for ligand binding activity. The α_5, α_6, α_V, and α_{IIb} subunits contain four putative divalent cation binding sites, whereas the α_2 integrin subunit contains only three.

The α_2 integrin subunit contains an additional segment of 191 amino acids that is inserted before the last five of the homologous repeats containing the divalent cation binding domains (19). This so-called I domain was originally observed in the α_M, α_L, and α_X integrin subunits, which pair with the β_2 subunit to form the leukocyte adhesion integrins (3). More recently, an I domain was shown to be present in the α_1 integrin subunit (25). The I domains are homologous to the collagen binding domains of von Willebrand factor and cartilage matrix protein and to domains within the complement factors B and C2. Because of the homology to other known collagen binding domains, it has been suggested that the I domain may contribute to the collagen binding activity of the $\alpha_2\beta_1$ and $\alpha_1\beta_1$ integrins (3,19). However, the β_2 integrins and the complement proteins, which also contain I domains, have not been shown to bind collagen. On the other hand, the I domain is inserted into a region of the integrin α chains thought to be involved in ligand binding. Cross-linking experiments using synthetic peptides containing recognition sequences for the $\alpha_{IIb}\beta_3$ and $\alpha_V\beta_3$ integrins suggest that regions of the α chain proximal to the divalent cation binding sites are in contact with the adhesive peptides (26,27).

Two of the integrin α subunits expressed on platelets, α_{IIb} and α_6, have been shown to exist in variant forms arising from alternative mRNA splicing (28–30). Alternative splicing gives rise to two forms of the α_6 subunit (α_{6A} and α_{6B}), which differ in the length and sequence of their cytoplasmic domains (29,30). The two variations of the α_6 cytoplasmic domain have in common only the conserved α subunit sequence CGFFKR proximal to the cell membrane. The 54 amino acid cytoplasmic domain of the α_{6B} subunit is 17 amino acid residues longer than that of the α_{6A} subunit and is considerably richer in charged amino acids. Although it seems likely that differences in α_6 integrin function are attributable to the alternative cytoplasmic domains, such differences in function have yet to be demonstrated. It is not known which forms of the α_6 subunit are present in platelets.

III. INTEGRINS AND PLATELET FUNCTION

A. The $\alpha_2\beta_1$ Integrin

Although several extracellular matrix molecules support platelet adhesion, the fibrillar collagens support not only adhesion but also subsequent platelet activation and aggregation. Baumgartner (31) concluded that fibrillar collagen was the most thrombogenic component of the vascular subendothelium. The nonfibrillar collagens, such as type IV collagen, a basement membrane collagen, effectively support adhesion but are less thrombogenic. The interactions of platelets with collagens are clearly complex, and the complexity of the problem has been reflected in the number of potential mediators of the platelet-collagen reaction proposed over the years. See Reference 32 for a review. Only over the last few years have sufficient data accumulated to provide compelling evidence for the function of any platelet surface component as a collagen receptor. It is now clear that the $\alpha_2\beta_1$ integrin functions as a collagen receptor not only on platelets but on many other cell types.

Recognition that platelet adhesion to collagen substrates was markedly enhanced in the presence of Mg^{2+} represented a key early step in elucidating the mechanism of platelet adhesion to collagen (33,34). Platelet adhesion to collagen types I–VIII under both static and flow conditions is supported by a Mg^{2+}-dependent mechanism, although the effectiveness of the different collagen types varies (34,35). The Mg^{2+}-dependent process supports only adhesion on monomeric collagen but leads to platelet activation on fibrillar substrates (34). In addition to Mg^{2+}, Mn^{2+}, Co^{2+}, Cu^{2+}, Fe^{2+}, and Zn^{2+} support platelet adhesion to collagen. Ca^{2+} not only failed to support adhesion but was found to inhibit adhesion supported by Mg^{2+}.

Several independent lines of investigation have established the role of the $\alpha_2\beta_1$ integrin as mediator of Mg^{2+}-dependent platelet adhesion to collagen. Santoro and colleagues (36,37) exploited the Mg^{2+} dependence of the adhesion process to purify by affinity chromatography on collagen-Sepharose from detergent-solubilized platelet membranes a heterodimeric complex composed of 160 and 130 kD polypeptides that bound to collagen in a Mg^{2+}-dependent manner. Electrophoretic analysis suggested that the components corresponded to platelet membrane glycoproteins Ia and IIa. Monoclonal antibody reactivity confirmed the identity of the 160 kD component as platelet membrane glycoprotein Ia (36).

Kunicki et al. (38) first suggested that platelet membrane glycoprotein Ia and IIa existed as a noncovalently associated heterodimeric complex within the membrane.

Pischel et al. (39) presented evidence that the platelet membrane glycoprotein Ia-IIa complex was physicochemically and immunochemically indistinguishable from the very late activation antigen 2 (VLA-2) expressed on T cells. Subsequent studies carried out in our laboratory (36,37) and independently by Kunicki et al. (40) and by Takada et al. (41) established the identity of the purified collagen binding complex not only with platelet membrane glycoprotein Ia-IIa complex with the VLA-2 complex but also with the ECMR II (extracellular matrix receptor II), which serves as a collagen receptor on fibroblasts (42). The complex is now most commonly referred to as the $\alpha_2\beta_1$ integrin, platelet glycoprotein Ia corresponding to the α_2 integrin subunit and glycoprotein IIa to the β_1 subunit.

When the purified $\alpha_2\beta_1$ integrin was incorporated into liposomes, the liposomes adhered to collagen substrates in a Mg^{2+}-dependent manner and exhibited the same collagen type specificity as intact platelets (37). The $\alpha_2\beta_1$-containing liposomes did not adhere to any of the several other extracellular matrix and adhesive proteins examined, indicating the specificity for collagen of platelet-derived $\alpha_2\beta_1$. Liposomes containing the purified $\alpha_2\beta_1$ integrin were also employed to explore the mechanism by which Ca^{2+} inhibited Mg^{2+}-dependent adhesion to collagen. The study revealed that the inhibition occurred via a simple, linear, noncompetitive mechanism, suggesting that separate classes of binding sites for Ca^{2+} and Mg^{2+} were present on the integrin (37). Peptide mapping experiments revealed that Ca^{2+} and Mg^{2+} stabilized distinct structures within the complex (43). Presumably when the Ca^{2+} binding sites are occupied, the receptor assumes a conformation that no longer supports collagen binding activity. The ligand binding activity of several other integrins, including the entire β_2 family and the $\alpha_V\beta_1$ integrin, is also inhibited by Ca^{2+}, raising intriguing possibilities about the ability of local concentrations of Ca^{2+} to regulate integrin function at, for example, sites of inflammation or bone remodeling (44,45).

A second important approach to establishing the role of the $\alpha_2\beta_1$ integrin as a platelet surface collagen receptor has been the use of specific inhibitory monoclonal antibodies directed against the complex. Several such antibodies that inhibit the Mg^{2+}-dependent adhesion of platelets to collagen or the ability of platelets to agglutinate collagen-coated beads have been described (37,40,46,47). Although the 6F1 antibody studied by Coller et al. (46) had only a modest effect on collagen-induced platelet aggregation in plasma, producing concentration-dependent prolongation of the lag time preceding the onset of aggregation, the 176D7 monoclonal antibody effectively inhibited collagen-induced platelet aggregation in plasma in addition to its ability to inhibit platelet adhesion to collagen (47). Recent studies with this antibody indicate that it is also an effective inhibitor of platelet deposition onto collagen substrates under conditions of flow using whole blood and a rectangular perfusion chamber (35,48). Quantitative binding studies using monoclonal antibodies directed against the $\alpha_2\beta_1$ integrin indicate that approximately 1000–2000 copies of the integrin are present on the platelet surface (39,46).

Parmentier et al. (50) reported a monoclonal antibody directed against the β_1 integrin subunit (glycoprotein IIa) that inhibited platelet adhesion to type III collagen, prolonged the lag time preceding collagen-induced platelet aggregation, and inhibited collagen-induced platelet aggregation and secretion.

The third avenue of investigation leading to the delineation of the collagen receptor function of the $\alpha_2\beta_1$ integrin has been the study of patients. Indeed, the initial suggestion that the $\alpha_2\beta_1$ integrin serves as a platelet surface collagen receptor came from the description by Nieuwenhuis et al. (51) of a patient with a mild bleeding disorder. The patient's platelets showed no response to collagen but exhibited normal responsiveness to other platelet agonists. Electrophoretic analysis of the platelets revealed a deficiency of glycoprotein Ia. Recently, a more detailed analysis of the adhesive properties of these platelets revealed marked diminution of Mg^{2+}-dependent platelet adhesion to collagen in both static and perfusion assays (48,49). Nieuwenhuis et al. (52) earlier described marked decrements in platelet adhesion to collagen substrates, as well as to substrates composed of arterial subendothelium under conditions of flow. In contrast to normal platelets, which spread upon the complex arterial subendothelium substrate, the few Ia-deficient platelets that adhered did not spread, suggesting that adhesion to collagen via other mechanisms and/or adhesion to subendothelial constituents other than collagen did not lead to platelet activation and spreading.

Kehrel et al. (53) described a patient with a long-standing bleeding disorder whose platelets were unresponsive to collagen. The platelets were found to be deficient in both glycoprotein Ia and thrombospondin. Surprisingly, at the onset of menopause, the bleeding disorder abated and the platelets were shown to express both thrombospondin and glycoprotein Ia and to respond normally to collagen. This interesting report suggests that an underlying derangement in hormonally regulated expression of glycoprotein Ia may have contributed to the disorder.

In studies reported to date only in abstract form, Handa et al. (54) and Nagai et al. (55) observed that patients with myeloproliferative disorders exhibiting prolonged bleeding times and impaired collagen-induced platelet aggregation were found on further analysis to show impaired platelet adhesion to collagen and partial or complete deficiency of glycoprotein Ia. Deckmyn et al. (56) described a patient with an acquired bleeding disorder and impaired collagen-induced platelet aggregation. An autoantibody reactive with glycoprotein Ia was apparently responsible for the bleeding disorder.

The many studies summarized here have provided compelling evidence for the function of the $\alpha_2\beta_1$ integrin as a collagen receptor on platelets. Although the data indicate that the $\alpha_2\beta_1$ integrin is an important mediator of early adhesive events along the pathway leading to collagen-induced platelet aggregation, studies carried out in several laboratories indicate that subsequent activation and/or aggregation requires interactions between the platelet surface and collagen in addition to those that give rise to adhesion (57–59). The identity of the platelet receptor(s) that mediate these later events remains to be determined.

Several integrins, such as $\alpha_5\beta_1$, $\alpha_V\beta_3$, and $\alpha_{IIb}\beta_3$, recognize the sequence RGD within their ligands, and the binding of these integrins to their ligands is inhibited by soluble RGD peptides (1–5). Platelet adhesion to collagen mediated by the $\alpha_2\beta_1$ integrin is not inhibitable by RGD-containing peptides (34). Studies with the $\alpha_2\beta_1$ integrin on platelets and other cells have revealed that the major $\alpha_2\beta_1$ recognition site on the α_1(I) collagen chain is located within the α_1(I)-CB3 fragment (60,61). The sequence DGEA appears to be a key determinant within the recognition site (62). Although the DGEA sequence is not present in all types of collagen recognized by the $\alpha_2\beta_1$ integrin, related sequences, such as DGES and DGET, which are active in experiments in vitro (Staatz and Santoro, unpublished), are present in homologous regions of other collagen chains. Vandenberg et al. (63) have presented evidence to support the presence of an $\alpha_2\beta_1$ integrin recognition site within the CB3 fragment of type IV collagen, a region that is not homologous to the CB3 fragment of type I collagen.

The $\alpha_2\beta_1$ integrin is widely distributed on various cell types (64), and studies from many laboratories have now documented its function as a collagen receptor on many diverse cell types, for the most part by use of inhibitory monoclonal antibodies and, to a lesser extent, affinity chromatography. Transfection of full-length α_2 integrin cDNA into cells that do not express the $\alpha_2\beta_1$ integrin but express other β_1 integrins results in expression of the $\alpha_2\beta_1$ integrin and acquisition of the ability to adhere to collagen substrates and to contract collagen gels (65,66). Interestingly, whereas on platelets and fibroblastic cells the $\alpha_2\beta_1$ integrin is a collagen receptor, on endothelial cells, melanoma cells, and many epithelial cell types it appears to function as both a collagen and a laminin receptor (67–69). A very recent report indicates that the differences in ligand specificity of the $\alpha_2\beta_1$ integrin on different cells are a consequence of cellular environment rather than any variation in primary sequence and that an activating β_1 antibody can convert the collagen-specific form of the integrin to the collagen/laminin binding form through conformational changes initiated at a binding site on the β_1 subunit that is not involved in ligand binding (70).

B. The $\alpha_5\beta_1$1 Integrin

Fibronectin within vascular subendothelium appears to play a significant role as a substrate for platelet adhesion following endothelial injury and exposure of the subendothelium to flowing blood (71,72). Preincubation of arterial subendothelium with an antiserum or $F(ab')_2$ fragments against fibronectin reduced platelet deposition onto segments of vascular subendothelium mounted in an annular perfusion chamber when the perfusions were performed at intermediate (800 s^{-1}) or high (1800 s^{-1}) shear rates (71). Similarly, antibodies directed against fibronectin also diminished platelet deposition onto the extracellular matrix produced by cultured cells and onto collagen (72). Early studies established the role of the $\alpha_{IIb}\beta_3$ integrin (GPIIb-IIIa) as an activation-dependent receptor on platelets for fibronectin. The binding of fluid phase, radiolabeled fibronectin to platelets is dependent upon platelet activation (73), does not occur with platelets from individuals with Glanzmann thrombasthenia deficient in the $\alpha_{IIb}\beta_3$ complex (74), and is inhibited by the monoclonal antibodies directed against the $\alpha_{IIb}\beta_3$ complex (75). The binding of the $\alpha_{IIb}\beta_3$ integrin to fibronectin has been demonstrated by affinity chromatography (76,77), by the adhesion to fibronectin of liposomes containing purified $\alpha_{IIb}\beta_3$ (77,78), and by the binding of radiolabeled fibronectin to liposomes containing the purified $\alpha_{IIb}\beta_3$ complex (79).

However, the early studies of Grinnell et al. (80) and Hynes et al. (81), subsequently confirmed by many investigators, demonstrated that unactivated platelets are also capable of adhering to fibronectin substrates. Platelets adherent to fibronectin undergo a profound morphological transition to a flattened and spread morphology on fibronectin substrates. Because platelet adhesion and spreading on fibronectin were observed even in the presence of inhibitors of platelet activation, such as prostaglandin E_1, it seemed unlikely that inadvertent activation and adhesion mediated by $\alpha_{IIb}\beta_3$ could account for the observed adhesion and spreading.

Giancotti et al. (82) observed that platelets expressed a heterodimeric receptor, distinct from the $\alpha_{IIb}\beta_3$ integrin, composed of 160 and 138 kD subunits, which was immunoprecipitated by an antiserum prepared against a 135 kD β subunit of the mouse fibronectin receptor ($\alpha_5\beta_1$ integrin). The data of Giancotti et al. (82) did not permit assignment of function to the complex. Piotrowicz et al. (83) employed an antiserum specific for band 3 (β_1 subunit) of the chicken embryo fibroblast receptor for fibronectin ($\alpha_5\beta_1$ integrin) to immunopurify a heterodimer composed of glycoprotein Ic and IIa from human platelets. The complex was found to be present not only on normal platelets but also on platelets from individuals with Glanzmann thrombasthenia. Piotrowicz et al. (83) went on to show that the adhesion to fibronectin of unactivated platelets (from either normal individuals or patients with Glanzmann thrombasthenia) was inhibited by soluble fibronectin, by antibodies against fibronectin, by peptides containing the RGD sequence, and by polyclonal antibodies specific for the β_1 integrin subunit. These observations indicated that the platelet membrane glycoprotein Ic-IIa complex functioned as an activation-independent receptor for fibronectin on platelets. Wayner et al. (84) subsequently established, with the use of monoclonal antibodies, that the platelet glycoprotein Ic-IIa complex described earlier was identical to the human fibroblast receptor for fibronectin isolated by Pytela et al. (85), to the VLA-5, and to the ECMR VI. This complex is now most commonly referred to as the $\alpha_5\beta_1$ integrin.

The accumulated data suggest that the complex serves as an activation-independent receptor for fibronectin on platelets and that the receptor mediates the spreading of platelets on fibronectin. The $\alpha_5\beta_1$ integrin recognizes the RGD sequence and adjacent flanking sequences within the cell binding domain of fibronectin (86,87). Recent studies indicate that an as yet undefined third receptor for fibronectin is required to account more fully for the observed adhesive interactions of platelets with fibronectin (88). A nonintegrin receptor appears to mediate divalent cation-independent adhesion to the gelatin binding domain of fibronectin, which, unlike adhesion mediated by the $\alpha_5\beta_1$ integrin, does not lead to platelet spreading (88).

C. The $\alpha_6\beta_1$ Integrin

Laminin may contribute to platelet deposition onto vascular subendothelium. Ill et al. (89) first described the adhesion of platelets onto surfaces coated with purified laminin and observed that adhesion to laminin differed significantly from adhesion to type I collagen or to fibronectin, in that adhesion to laminin resulted in neither platelet activation nor spreading. Subsequent platelet studies have contributed significantly to the delineation of the role of the $\alpha_6\beta_1$ integrin as a laminin receptor, not only on platelets, but also on many other cell types.

The GoH3 rat monoclonal antibody originally developed to study differentiation antigens on breast epithelial cells was shown to immunoprecipitate an heterodimeric complex from human platelets composed of polypeptides corresponding to platelet membrane glycoproteins Ic and IIa (90). Subsequent preclearing immunoprecipitation experiments, peptide mapping, and two-dimensional electrophoretic analysis distinguished the Ic-IIa complex immunoprecipitated by the GoH3 antibody from the Ic-IIa complex corresponding to the $\alpha_5\beta_1$ integrin (91). The new complex was designated VLA-6 and is now more commonly referred to as the $\alpha_6\beta_1$ integrin.

The role of the $\alpha_6\beta_1$ integrin as a laminin receptor was first established by the demonstration that the GoH3 antibody inhibited platelet adhesion to laminin substrates, but had no effect on the adhesion of platelets to collagen, fibronectin, or fibrinogen in static assays (92). Hindriks et al. (93) further established the role of the $\alpha_6\beta_1$ antibody as a laminin receptor on platelets by demonstrating that the GoH3 antibody also completely inhibited platelet adhesion to laminin under conditions of flow.

Further examination of platelet adhesion to laminin revealed that the process did not require platelet activation and that it was supported by Mg^{2+}, Mn^{2+}, and Co^{2+} but not by Ca^{2+}, Zn^{2+}, or Cu^{2+} (92). Adhesion to laminin was not inhibited by RGD-containing peptides or by peptides containing the YIGSR sequence, previously shown to inhibit the adhesion of some cells to laminin mediated by a 67 kD nonintegrin receptor (94). Recent studies indicate that the binding site for the $\alpha_6\beta_1$ integrin on laminin is located within the E8 fragment derived from the terminal half of the long arm of laminin (95–97).

Thus, it has become quite clear that the $\alpha_6\beta_1$ integrin is a major mediator of platelet interactions with laminin. A recent report indicates that the 67 kD nonintegrin laminin binding protein is also expressed on platelets and may contribute to platelet adhesive interactions with laminin (98).

D. The $\alpha_V\beta_3$ Integrin

Although relatively little information about the $\alpha_V\beta_3$ integrin has accrued from studies on platelets, most likely because of its very low level of expression (see later), the integrin has been extensively studied on other cell types. As noted earlier, the integrin binds many different ligands, including vitronectin, fibrinogen, von Willebrand factor, thrombospondin, osteopontin, and possibly collagen. The receptor appears to recognize RGD sequences within each of its ligands (99–102).

Binding studies with radiolabeled vitronectin indicated that approximately 5000 activation-dependent binding sites were present on platelets (103). The activation-dependent binding of vitronectin to platelets and the adhesion of activated platelets to vitronectin substrates were inhibited by RGD-containing peptides and by a monoclonal antibody directed against the $\alpha_{IIb}\beta_3$ integrin (103,104). These results suggested that the activation-dependent interactions of platelets with vitronectin were mediated by the $\alpha_{IIb}\beta_3$ integrin. Purified $\alpha_{IIb}\beta_3$ incorporated into liposomes has been shown to bind to vitronectin (77).

Lam et al. (105) provided the first direct evidence for the presence of the $\alpha_V\beta_3$ integrin on platelets. Lam et al. (105) detected the $\alpha_V\beta_3$ integrin following affinity chromatography of detergent extracts of ^{125}I surface-labeled platelets on columns containing immobilized KYGRGDS. The glycoprotein IIb-IIIa complex (the major species retained on the column) was eluted with soluble HHLGGAKQAGDV. A second heterodimeric complex was subsequently eluted from the column with soluble GRGDSP. The complex was composed of subunits with the electrophoretic mobilities of platelet glycoproteins Ic and IIIa. The heterodimer was not immunoprecipitated by a monoclonal antibody directed against the β_1 integrin subunit or by a monoclonal antibody directed against the α_6 integrin subunit. The purified complex was immunoprecipitated, however, by the LM609 monoclonal antibody directed against the $\alpha_V\beta_3$ integrin complex. The α chain of the platelet-derived material also reacted on western blots with an antiserum directed against the α_V integrin subunit. Lam et al. (105) furthermore determined that the amino-terminal amino acid sequence of the α chain from the platelet-derived material was identical to the amino-terminal sequences previously determined for α_V subunits of endothelial cell and fibroblast origin. Thus, the complex appears to be identical to the $\alpha_V\beta_3$ vitronectin receptor.

Lawler and Hynes (106) provided independent evidence for the presence of the $\alpha_V\beta_3$ integrin on platelets. These workers subjected detergent extracts of radiolabeled platelet membranes to affinity chromatography on columns of thrombospondin-Sepharose. Elution of the column with soluble GRGDSP yielded two distinct heterodimeric complexes. One was shown to be identical to the $\alpha_{IIb}\beta_3$ integrin, whereas the second contained the identical β_3 subunit and an α subunit that comigrated with endothelial cell-derived α_V.

The identity of the additional α subunit as α_V was ascertained by its monoclonal antibody reactivity.

The studies of Lam et al. (105) and Lawler and Hynes (106) suggested that the $\alpha_V\beta_3$ integrin is expressed in low abundance relative to other characterized platelet adhesive receptors. Coller et al. (107) determined the number of $\alpha_V\beta_3$ complexes on platelets using both α_V-specific and complex-specific monoclonal antibodies and concluded that normal platelets expressed between 50 and 100 $\alpha_V\beta_3$ receptors per platelet. Coller et al. (107) were also able to devise adhesion experiments that indicated that under at least some conditions the $\alpha_V\beta_3$ complexes could support the adhesion of unactivated platelets to vitronectin. It thus seems likely, but is yet unproven, that the low levels of the $\alpha_V\beta_3$ integrin on platelets contributes to platelet adhesion to any of the several matrix molecules that may serve as ligands for the complex.

REFERENCES

1. Hynes RO. Integrins: a family of cell surface receptors. Cell 1987; 48:549–554.
2. Albelda SM, Buck CA. Integrins and other cell adhesion molecules. FASEB J 1990; 4: 2868–2880.
3. Hemler ME. VLA proteins in the integrin family: structures, functions, and their role on leukocytes. Annu Rev Immunol 1990; 8:365–400.
4. Ruoslahti E. Integrins. J Clin Invest 1991; 87:1–5.
5. Hynes RO. Integrins: versatility, modulation, and signalling in cell adhesion. Cell 1992; 69: 11–25.
6. Bennett JS. The molecular biology of platelet membrane proteins. Semin Hematol 1990; 27:186–204.
7. Phillips DR, Charo IF, Scarborough RM. GPIIb-IIIa: the responsive integrin. Cell 1991; 65: 359–362.
8. Heidenreich R, Eisman R, Surrey S, et al. Organization of the gene for platelet glycoprotein IIb. Biochemistry 1990; 29:1232–1244.
9. Zimrin AB, Gidwitz S, Lord S, et al. The genomic organization of platelet glycoprotein IIIa. J Biol Chem 1990; 265:8590–8595.
10. Argraves WS, Suzuki S, Arai H, Thompson K, Pierschbacher MD, Ruoslahti E. Amino acid sequence of the human fibronectin receptor. J Cell Biol 1987; 105:1183–1190.
11. Rosa JP, Bray PF, Gayet O, et al. Cloning of glycoprotein IIIa cDNA from human erythroleukemia cells and localization of the gene to chromosome 17. Blood 1988; 72:593–600.
12. Fitzgerald LA, Steiner B, Rall SC, Lo SS, Phillips DR. Protein sequence of endothelial glycoprotein IIIa derived from a cDNA clone: identity with glycoprotein IIIa and similarity to integrin. J Biol Chem 1987; 262:3936–3939.
13. Suzuki S, Naitoh Y. Amino acid sequence of a novel integrin β_4 subunit and primary expression of the mRNA in epithelial cells. EMBO J 1990; 9:757–763.
14. Smith JW, Cheresh DA. The Arg-Gly-Asp binding domain of the vitronectin receptor. Photoaffinity cross-linking implicates amino acid residues 61–203 of the β subunit. J Biol Chem 1988; 263:18726–18731.
15. D'Souza SE, Ginsberg MH, Burke TA, Lam SC-T, Plow EF. Localization of an Arg-Gly-Asp recognition site within an integrin adhesion receptor. Science 1988; 242:91–93.
16. Altruda F, Cervella P, Tarone G, et al. A human β_1 integrin subunit with a unique cytoplasmic domain generated by alternative RNA processing. Gene 1990; 95:261–266.
17. Van Kuppevelt THMSM, Languino LR, Gailet JO, Suzuki S, Ruoslahti E. An alternative cytoplasmic domain of the integrin β_3 subunit. Proc Natl Acad Sci USA 1989; 86:5415–5418.

18. Languino LR, Ruoslahti E. An alternative form of the integrin β_1 subunit with a variant cytoplasmic domain. J Biol Chem 1992; 267:7116–7120.
19. Takada Y, Hemler ME. The primary structure of the VLA-2/collagen receptor α_2 subunit (platelet GPIa): Homology to other integrins and the presence of a possible collagen-binding domain. J Cell Biol 1989; 109:397–407.
20. Fitzgerald LA, Poncz M, Steiner B, Rall SC Jr, Bennett JS, Phillips DR. Comparison of cDNA-derived protein sequences of the human fibronectin and vitronectin receptor α-subunits and platelet glycoprotein IIb. Biochemistry 1987; 26:8158–8165.
21. Hogervorst F, Kuikman I, Van Kessel AG, Sonnenberg A. Molecular cloning of the human α_6 integrin subunit. Alternative splicing of the α_6 mRNA and chromosomal localization of the α_6 and β_4 genes. Eur J Biochem 1991; 199:425–433.
22. Tamura RN, Rozzo C, Starr L, Chambers J, Reichardt LF, Cooper HM, Quaranta V. Epithelial integrin $\alpha_6\beta_4$: complete primary structure of α_6 and variant forms of β_4. J Cell Biol 1990; 111:1593–1604.
23. Suzuki S, Argraves WS, Pytela R, et al. cDNA and amino acid sequences of the cell adhesion protein receptor recognizing vitronectin reveal a transmembrane domain and homologies with other adhesion protein receptors. Proc Natl Acad Sci USA 1986; 83:8614–8618.
24. Suzuki S, Argraves WS, Arai H, Languino LR, Pierschbacher M, Ruoslahti E. Amino acid sequence of the vitronectin receptor α subunit and comparative expression of adhesion receptor mRNAs. J Biol Chem 1987; 262:14080–14085.
25. Ignatius MJ, Large TH, Houde M, et al. Molecular cloning of the rat integrin α_1-subunit: a receptor for laminin and collagen. J Cell Biol 1990; 111:709–720.
26. D'Souza SE, Ginsberg MH, Burke TA, Plow EF. The ligand binding site of the platelet integrin receptor GPIIb-IIIa is proximal to the second calcium binding domain of its α subunit. J Biol Chem 1990; 265:3440–3446.
27. Smith JW, Cheresh DA. Integrin ($\alpha_V\beta_3$)-ligand interaction. J Biol Chem 1990; 265:2168–2172.
28. Bray PF, Leung CSI, Shuman MA. Human platelets and megakaryocytes contain alternatively spliced glycoprotein IIb mRNAs. J Biol Chem 1990; 265:9587–9590.
29. Tamura RN, Cooper HM, Collo G, Quaranta V. Cell type-specific integrin variants with alternative α chain cytoplasmic domains. Proc Natl Acad Sci USA 1991; 88:10183–10187.
30. Cooper HM, Tamura R, Quaranta V. The major laminin receptor of mouse embryonic stem cells is a novel isoform of the $\alpha_6\beta_1$ integrin. J Biol Chem 1991; 115:843–850.
31. Baumgartner HR. Platelet interaction with collagen fibrils in flowing blood. I. Reaction of human platelets with α-chymotrypsin digested subendothelium. Thromb Haemost 1977; 37: 1–16.
32. Santoro SA. Molecular basis of platelet adhesion to collagen. In: Jamieson GA, ed. Platelet Membrane Receptors: Molecular Biology, Immunology, Biochemistry, and Pathology. New York, Alan R. Liss, 1988:291–314.
33. Shadle PJ, Barondes SH. Adhesion of human platelets to immobilized trimeric collagen. J Cell Biol 1982; 95:361–365.
34. Santoro SA. Identification of a 160,000 dalton platelet membrane protein that mediates the initial divalent cation-dependent adhesion of platelets to collagen. Cell 1986; 46:913–920.
35. Saelman EUM, Nieuwenhuis HK, Hese KM, et al. Platelet adhesion to collagen types I–VIII under conditions of stasis and flow is mediated by GPIa/IIa ($\alpha_2\beta_1$-integrin). Blood 1994; 83:1244–1250.
36. Santoro SA, Rajpara SM, Staatz WD, Woods VL Jr. Isolation and characterization of a platelet surface collagen binding complex related to VLA-2. Biochem Biophys Res Commun 1988; 153:217–223.
37. Staatz WD, Rajpara SM, Wayner EA, Carter WG, Santoro SA. The membrane glycoprotein Ia-IIa (VLA-2) complex mediates the Mg^{++}-dependent adhesion of platelets to collagen. J Cell Biol 1989; 108:1917–1924.

38. Kunicki TJ, Nurden AT, Pidard D, Russell NR, Caen JP. Characterization of human platelet glycoprotein antigens giving rise to individual immunoprecipitates in crossed-immunoelectrophoresis. Blood 1981; 58:1190–1197.
39. Pischel KD, Bluestein HG, Wood VL Jr. Platelet glycoproteins Ia, Ic, are IIa and physicochemically indistinguishable from the very late activation antigens adhesion-related proteins of lymphocytes and other cell types. J Clin Invest 1988; 81:505–513.
40. Kunicki TJ, Nugent DJ, Staats SJ, Orchekowski RP, Wayner EA, Carter WG. The human fibroblast class II extracellular matrix receptor mediates platelet adhesion to collagen and is identical to the platelet glycoprotein Ia-IIa complex. J Biol Chem 1988; 263:4516–4519.
41. Takada Y, Wayner EA, Carter WG, Hemler ME. Extracellular matrix receptors, ECMRII and ECMRI, for collagen and fibronectin correspond to VLA-2 and VLA-3 in the VLA family of heterodimers. J Cell Biochem 1988; 37:385–393.
42. Wayner EA, Carter WG. Identification of multiple cell adhesion receptors for collagen and fibronectin in human fibrosarcoma cells possessing unique α and β subunits. J Cell Biol 1987; 105:1873–1884.
43. Staatz WD, Peters KJ, Santoro SA. Divalent cation-dependent structure in the platelet membrane glycoprotein Ia-IIa (VLA-2) complex. Biochem Biophys Res Commun 1990; 168:107–113.
44. Dransfield I, Cabañas C, Craig A, Hogg N. Divalent cation regulation of the function of the leukocyte integrin LFA-1. J Cell Biol 1992; 116:219–226.
45. Kirchhofer D, Grzesiak J, Pierschbacher MD. Calcium as a potential physiological regulator of integrin-mediated cell adhesion. J Biol Chem 1991; 266:4471–4477.
46. Coller BS, Beer JH, Scudder LE, Steinberg MH. Collagen-platelet interactions: evidence for a direct interaction of collagen with platelet GPIa/IIa and an indirect interaction with platelet GPIIb/IIIa mediated by adhesive proteins. Blood 1989; 74:182–192.
47. Gralnick HR, McKeown LP, Williams SS, et al. A murine monoclonal antibody that identifies a 157/130 kDa platelet collagen receptor. Circulation 1988; 78 (Suppl. II):308.
48. Saelman EUM, Hese KM, Nieuwenhuis HK, de Groot PG, Gralnick HR, Sixma JJ. Role of GPIa/GPIIa in platelet adhesion to collagen under static and flow conditions. Thromb Haemost 1991; 65:678.
49. de Groot PG, Agbanyo F, Beumer S, et al. Role of adhesive proteins and membrane glycoproteins in platelet adhesion. Thromb Haemost 1991; 65:744.
50. Parmentier S, Catimel B, McGregor L, Leung LLK, McGregor JL. Role of glycoprotein IIa (β_1 subunit of very late activation antigens) in platelet functions. Blood 1991; 78:2021–2026.
51. Nieuwenhuis HK, Akkerman JWN, Houdijk WPM, Sixma JJ. Human blood platelets showing no response to collagen fail to express surface glycoprotein Ia. Nature 1985; 318:470–472.
52. Nieuwenhuis HK, Sakariassen KS, Houdijk WPM, Nievelstein PFM, Sixma JJ. Deficiency of platelet membrane glycoprotein Ia associated with a decreased platelet adhesion to subendothelium: a defect in platelet spreading. Blood 1986; 68:692–695.
53. Kehrel B, Balleisen L, Kokott R, et al. Deficiency of intact thrombospondin and membrane glycoprotein Ia in platelets with defective collagen-induced aggregation and spontaneous loss of disorder. Blood 1988; 71:1074–1078.
54. Handa M, Nagai H, Ando Y, et al. Defect of GPIa-IIa heterodimer complex in a patient whose platelets showed no response to collagen. Blood 1988; 72:323a.
55. Nagai H, Handa H, Kamata T, et al. Defective platelet adhesion to collagen and lack of membrane glycoprotein Ia in patients with myeloproliferative disorders. Thromb Haemost 1989; 62:417.
56. Deckmyn H, Chew SL, Vermylen J. Lack of platelet response to collagen associated with an autoantibody against glycoprotein Ia: a novel cause of acquired qualitative platelet dysfunction. Thromb Haemost 1990; 64:74–79.

57. Karniguian A, Legrand YJ, Lefrancier P, Caen JP. Effect of collagen-derived octapeptide on different steps of the platelet/collagen interaction. Thromb Res 1983; 32:592–604.
58. Morton LF, Peachey AR, Barnes MJ. Platelet reactive sites in collagens type I and type III. Evidence for separate adhesion and aggregatory sites. Biochem J 1989; 258:157–163.
59. Santoro SA, Walsh JJ, Staatz WD, Baranski KJ. Distinct determinants on collagen support $\alpha_2\beta_1$ integrin-mediated platelet adhesion and platelet activation. Cell Regul 1991; 2:905–913.
60. Staatz WD, Walsh JJ, Pexton T, Santoro SA. The $\alpha_2\beta_1$ integrin cell surface collagen receptor binds to the α1(I)-CB3 peptide of collagen. J Biol Chem 1990; 265:4778–4781.
61. Gullberg D, Gehlsen KR, Turner DC, et al: Analysis of $\alpha_1\beta_1$, $\alpha_2\beta_1$, and $\alpha_3\beta_1$ integrins in cell-collagen interactions: identification of conformation dependent $\alpha_1\beta_1$ binding sites in collagen type I. EMBO J 1992; 11:3865–3873.
62. Staatz WD, Fok KF, Zutter MM, Adams SP, Rodriguez BA, Santoro SA. Identification of a tetrapeptide recognition sequence for the $\alpha_2\beta_1$ integrin in collagen. J Biol Chem 1991; 266:7363–7367.
63. Vandenberg P, Kern A, Ries A, Luckenbill-Edds L, Mann K, Kühn K. Characterization of a type IV collagen major cell binding site with affinity to the $\alpha_1\beta_1$ and the $\alpha_2\beta_2$ integrins. J Cell Biol 1991; 113:1475–1483.
64. Zutter MM, Santoro SA. Widespread histologic distribution of the $\alpha_2\beta_1$ integrin cell-surface collagen receptor. Am J Pathol 1990; 137:113–120.
65. Chan BMC, Matsuura N, Takada Y, Zetter BR, Hemler ME. In vitro and in vivo consequences of VLA-2 expression on rhabdomyosarcoma cells. Science 1991; 251:1600–1602.
66. Schiro JA, Chan BMC, Roswit WT, et al. Integrin $\alpha_2\beta_1$ (VLA-2) mediates reorganization and contraction of collagen matrices by human cells. Cell 1991; 67:403–410.
67. Elices MJ, Hemler ME. The human integrin VLA-2 is a collagen receptor on some cells and a collagen/laminin receptor on others. Proc Natl Acad Sci USA 1989; 86:9906–9910.
68. Languino LR, Gehlsen KR, Wayner E, Carter WG, Engvall E, Ruoslahti E. Endothelial cells use $\alpha_2\beta_1$ integrin as a laminin receptor. J Cell Biol 1989; 109 2455–2462.
69. Kirchhofer D, Languino LR, Ruoslahti E, Pierschbacher MD. $\alpha_2\beta_1$ Integrins from different cell types show different binding specificities. J Biol Chem 1990; 265:615–618.
70. Chan BMC, Hemler ME. Multiple functional forms of the integrin VLA-2 can be derived from a single α_2 cDNA clone: interconversion of forms induced by an anti-β_1 antibody. J Cell Biol 1993; 120:537–543.
71. Houdijk WPM, Sixma JJ. Fibronectin in artery subendothelium is important for platelet adhesion. Blood 1985; 65:598–604.
72. Bastida E, Escolar G, Ordinas A, Sixma JJ. Fibronectin is required for platelet adhesion and for thrombus formation on subendothelium and collagen surfaces. Blood 1987; 70:1437–1442.
73. Plow EF, Ginsberg MH. Specific and saturable binding of plasma fibronectin to thrombin-stimulated human platelets. J Biol Chem 1989; 256:9477–9482.
74. Ginsberg MH, Forsyth J, Lightsey A, Chediak J, Plow EF. Reduced surface expression and binding of fibronectin by thrombin-stimulated thrombasthenic platelets. J Clin Invest 1983; 71:619–624.
75. Plow EF, McEver RP, Coller BS, Woods VL Jr, Marguerie GA. Related binding mechanisms for fibrinogen, fibronectin, von Willebrand factor and thrombospondin on thrombin-stimulated human platelets. Blood 1985; 66:724–727.
76. Gardner JM, Hynes RO. Interaction of fibronectin with its receptor on platelets. Cell 1985; 42:439–448.
77. Pytela R, Pierschbacher MD, Ginsberg MH, Plow EF, Ruoslahti E. Platelet membrane glycoprotein IIb/IIIa: Member of a family of Arg-Gly-Asp-specific adhesion receptors. Science 1986; 231:1559–1562.

78. McGraw DJ, Santoro SA. Gangliosides enhance the adhesive activity of the platelet membrane glycoprotein IIb/IIIa complex. FASEB J 1990; 4:A1029.
79. Parise LV, Phillips DR. Fibronectin binding properties of the purified platelet glycoprotein IIb-IIIa complex. J Biol Chem 1986; 261:14011–14017.
80. Grinnell F, Feld M, Snell W. The influence of cold insoluble globulin on platelet morphological response to substrata. Cell Biol Int Rep 1979; 3:585–592.
81. Hynes RO, Ali IU, Destree AT, et al. A large glycoprotein lost from the surfaces of transformed cells. Ann NY Acad Sci 1978; 312:317–342.
82. Giancotti FG, Languino LR, Zanetti A, Peri G, Tarone G, Dejana E. Platelets express a membrane protein complex immunologically related to the fibroblast fibronectin receptor and distinct from GPIIb/IIIa. Blood 1987; 69:1535–1538.
83. Piotrowicz RS, Orchekowski RP, Nugent DJ, Yamada KM, Kunicki TJ. Glycoprotein Ic-IIa functions as an activation-independent fibronectin receptor on human platelets. J Cell Biol 1988; 106:1359–1364.
84. Wayner EA, Carter WG, Piotrowicz RS, Kunicki TJ. The function of multiple extracellular matrix receptors in mediating cell adhesion to extracellular matrix: preparation of monoclonal antibodies to the fibronectin receptor that specifically inhibit cell adhesion to fibronectin and react with platelet glycoproteins Ic-IIa. J Cell Biol 1988; 107:1881–1891.
85. Pytela R, Pierschbacher MD, Ruoslahti E. Identification and isolation of a 140 kD cell surface glycoprotein with properties expected of a fibronectin receptor. Cell 1985; 40:191–198.
86. Pierschbacher MD, Ruoslahti E. Cell attachment activity of fibronectin can be duplicated by proteolytic fragments and synthetic peptides which support fibroblast adhesion. Nature 1984; 309:30–33.
87. Obara M, Kang MS, Yamada KM. Site directed mutagenesis of the cell binding domain of human fibronectin: separable synergistic sites mediate adhesive function. Cell 1988; 53: 649–657.
88. Winters KJ, Walsh JJ, Rubin BG, Santoro SA. Platelet interactions with fibronectin: divalent cation-independent platelet adhesion to the gelatin-binding domain of fibronectin. Blood 1993; 81:1778–1786.
89. Ill CR, Envall E, Ruoslahti E. Adhesion of platelets to laminin in the absence of activation. J Cell Biol 1984; 99:2140–2145.
90. Sonnenberg A, Janssen H, Hogervorst F, Calafat J, Hilgers J. A complex of platelet glycoproteins Ic and IIa identified by a rat monoclonal antibody. J Biol Chem 1987; 262:10376–10383.
91. Hemler ME, Crouse C, Takada Y, Sonnenberg A. Multiple very late antigen (VLA) heterodimers on platelets. J Biol Chem 1988; 263:7660–7665.
92. Sonnenberg A, Modderman PW, Hogervorst F. Laminin receptor on platelets is the integrin VLA-6. Nature 1988; 336:487–489.
93. Hindriks G, Ijsseldijk MJW, Sonnenberg A, Sixma JJ, de Groot PG. Platelet adhesion to laminin: Role of Ca^{2+} and Mg^{2+} ions, shear rate and platelet membrane glycoproteins. Blood 1992; 79:928–935.
94. Graf J, Iwamoto Y, Sasaki M, et al. Identification of an amino acid sequence in laminin mediating cell attachment, chemotaxis, and receptor binding. Cell 1987; 48:989–996.
95. Aumailley M, Timpl R, Sonnenberg A. Antibody to integrin α_6 subunit specifically inhibits cell-binding to laminin fragment 8. Exp Cell Res 1990; 188:55–60.
96. Sonnenberg A, Linders CJT, Modderman PW, Damsky CH, Aumailly M, Timpl R. Integrin recognition of different cell binding fragments of laminin (P1, E3, E8) and evidence that $\alpha_6\beta_1$ but not $\alpha_6\beta_4$ functions as a major receptor for fragment E8. J Cell Biol 1990; 110: 2145–2155.
97. Hall DE, Reichardt LF, Crowley E, Holley B, Moezzi H, Sonnenberg A, Damsky CH. The α_1/β_1 and α_6/β_6 integrin heterodimers mediate cell attachment to distinct sites on laminin. J Cell Biol 1990; 110:2175–2184.

98. Tandon NN, Holand EA, Kralisz U, Kleinman HK, Robey FA, Jamieson GA. Interaction of human platelets with laminin and identification of the 67 kD laminin receptor on platelets. Biochem J 1991; 274:535–542.
99. Pytela R, Pierschbacher MD, Ruoslahti E. A 125/115-kDa cell surface receptor specific for vitronectin interacts with the arginine-glycine-aspartic acid adhesion sequence derived from fibronectin. Proc Natl Acad Sci USA 1985; 82:5766–5770.
100. Cheresh DA, Spiro RC. Biosynthetic and functional properties of an Arg-Gly-Asp-directed receptor involved in human melanoma cell attachment to vitronectin, fibrinogen, and von Willebrand factor. J Biol Chem 1987; 262:17703–17711.
101. Lawler J, Weinstein R, Hynes RO. Cell attachment to thrombospondin: the role of Arg-Gly-Asp, calcium, and integrin receptors. J Cell Biol 1988; 107:2351–2361.
102. Smith JW, Ruggeri ZM, Kunicki TJ, Cheresh DA. Interaction of integrins $\alpha_V\beta_3$ and glycoprotein IIb-IIIa with fibrinogen. Differential peptide recognition accounts for distinct binding sites. J Biol Chem 1990; 265:12267–12271.
103. Thiagarajan P, Kelly KL. Exposure of binding sites for vitronectin on platelets following stimulation. J Biol Chem 1988; 263:3035–3038.
104. Thiagarajan P, Kelly K. Interaction of thrombin-stimulated platelets with vitronectin (S-protein of complement) substrate: inhibition by a monoclonal antibody to glycoprotein IIb-IIIa complex. Thromb Haemost 1988; 60:514–517.
105. Lam SC-T, Plow EF, D'Souza SE, Cheresh DA, Frelinger AL, Ginsberg MH. Isolation and characterization of a platelet membrane protein related to the vitronectin receptor. J Biol Chem 1989; 264:3742–3749.
106. Lawler J, Hynes RO. An integrin receptor on normal and thrombasthenic platelets that binds thrombospondin. Blood 1989;74:2022–2027.
107. Coller BS, Cheresh DA, Asch E, Seligsohn U. Platelet vitronectin receptor expression differentiates Iraqi-Jewish from Arab patients with Glanzmann thrombasthenia in Israel. Blood 1991; 77:75–83.
108. Peters PM, Kamarck ME, Hemler ME, Strominger JL, Ruddle FH. Genetic and biochemical characterization of human lymphocyte cell surface antigens. The A-1A5 and A-3A4 determinants. J Exp Med 1984; 159:1441–1454.
109. Rettig WG, Dracopoli NC, Goetzger TA, et al. Somatic cell genetic analysis of human cell surface antigens: chromosomal assignments and regulation of expression in rodent-human hybrid cells. Proc Natl Acad Sci USA 1984; 81:6437–6441.
110. Sosnoski DM, Emanuel BS, Hawkins AL, et al. Chromosomal localization of the genes for the vitronectin and fibronectin receptors alpha subunits and for platelet glycoproteins IIb and IIIa. J Clin Invest 1989; 81:1993–1998.
111. Bray PF, Rosa J-F, Johnston GI, et al. Platelet glycoprotein IIb: chromosomal localization and tissue expression. J Clin Invest 1987; 80:1812–1817.

30

Thrombospondins

Jack Lawler
Brigham and Women's Hospital and Harvard University Medical School, Boston, Massachusetts

I. INTRODUCTION

The thrombospondins comprise a family of extracellular calcium binding proteins. Platelet thrombospondin is the first member of this family to be identified, and it is the most extensively characterized member because it is readily purified from the supernatant of thrombin-treated platelets. Amino acid sequence data indicate that platelet thrombospondin is a homotrimer of the thrombospondin-1 gene product. In the last 3 years, it has become apparent that there are at least four other genes that are evolutionarily related to thrombospondin-1. These have been designated thrombospondin-2, 3, and 4 and cartilage oligomeric matrix protein (COMP) (1–6). In addition, a superfamily of proteins that contain the thrombospondin type 1 repeats is rapidly growing.

Thrombospondin-1 has been shown to interact with proteoglycans, integrins, CD36, and other cell surface molecules. In some cases, multiple receptors act concurrently to bind thrombospondin-1. In addition, proteoglycans appear to be able to bind to multiple sites on the thrombospondin molecule. This complexity makes it difficult to correlate receptor occupancy with cellular phenotype. Thrombospondin-1 acts to modulate cell adhesion, migration, and proliferation. The molecule is expressed at the site of vascular, cutaneous, and muscle wounds. In addition, thrombospondin-1 is a major component of the proteins that are released from human blood platelets in response to thrombin. These data have led to the hypothesis that the thrombospondins are important in tissue genesis and remodeling and in the cell's response to injury.

A. Thrombospondin in Platelet Aggregation

Platelet thrombospondin is secreted from α granules in response to various stimuli. The secreted protein has been reported to become associated with the platelet surface in the presence of calcium (7). In the platelet, some of the thrombospondin is associated with the α granule membrane (8). These molecules may represent a significant portion of the protein that remains associated with the platelet after activation. Specific coclustering of thrombospondin and glycoprotein IIb-IIIa ($\alpha_{IIb}\beta_3$ integrin), as well as thrombospondin and fibrinogen, is observed on the surface of activated platelets (9). The evidence for a direct interaction of thrombospondin with $\alpha_{IIb}\beta_3$ is summarized later (Sect. V.C). Antithrombospondin antibodies have been reported to inhibit thrombin-, collagen- and A23187-induced platelet aggregation (see Ref. 10 for a review). These antibodies did not inhibit dense granule or α granule release as measured by the secretion of serotonin or β-thromboglobulin, respectively. Whereas the primary phase of ADP-induced aggregation is not inhibited, the secretion-dependent secondary phase is inhibited by antithrombospondin antibodies. These antibodies also caused a decrease in the affinity of fibrinogen for $\alpha_{IIb}\beta_3$ (11). Leung (11) proposed that thrombospondin functions in platelet aggregation by stabilizing the fibrinogen-$\alpha_{IIb}\beta_3$ complex. However, the expression of thrombospondin on the surface of activated platelets from thrombasthenic and afibrinogenemic patients is indistinguishable from that on activated normal control platelets (12–15). Platelet glycoprotein IV (CD36) has also been proposed to function as a receptor for thrombospondin (see Sect. V.B). Monoclonal antibodies to the NH_2-terminal or the COOH-terminal globular domains of thrombospondin inhibit platelet aggregation (Fig. 1) (15,16). Legrand et al. (15) observed that a monoclonal antibody to the heparin binding domain inhibits thrombin-induced aggregation and secretion of serotonin. This antibody also inhibited the binding of thrombospondin to fibrinogen and CD36 in a solid-phase binding assay. Dixit et al. (16) have shown that a monoclonal antibody directed against the COOH-terminal domain also inhibits platelet aggregation. These data indicate that multiple sites on the thrombospondin molecule are capable of interacting with multiple receptors of the platelet membrane.

B. Role of Thrombospondin in Clot Formation and Lysis

In addition to binding to the platelet surface, secreted thrombospondin also becomes associated with the fibrin clot (17). Thrombospondin is covalently bound to fibrin by

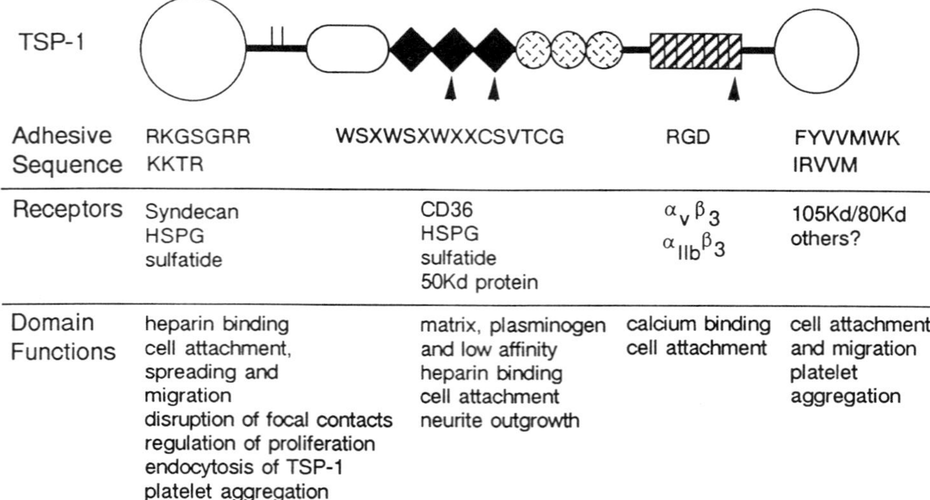

Figure 1 Structural and functional properties of thrombospondin-1.

factor XIIIa. Clots that form in the presence of thrombospondin are composed of more numerous and thinner fibers. In addition to affecting the structure of the clot, thrombospondin may function at the platelet surface to anchor the fibrin fibrils and may modulate fibrinolysis (see Ref. 18 for a review). Solid-phase binding assays have been used to demonstrate that thrombospondin binds to fibrinogen, plasminogen, histidine-rich glycoprotein, and both the single- and two-chain forms of urokinase. Thrombospondin can form trimolecular complexes with plasminogen and tissue plasminogen activator (19,20). The resulting complex is considerably more efficient in the generation of plasmin. The plasmin that remains bound to the thrombospondin is protected from inhibition by α_2-plasmin inhibitor. Similarly, when urokinase is bound to thrombospondin it is protected from inhibition by plasminogen activator inhibitor type 1. These data support the hypothesis that the thrombospondin that is incorporated into the fibrin clot acts to accelerate fibrinolysis. In striking opposition to this conclusion, Hogg and coworkers (23) reported that, in solution, thrombospondin is a potent inhibitor of plasmin. Thus, like much of the in vitro data that have been obtained on the function of thrombospondin, conflicting results have been obtained. Mice that have been genetically engineered to lack thrombospondin represent a critical model to test the in vivo relevance of much of the in vitro data.

II. THE THROMBOSPONDIN GENE FAMILY

The members of the thrombospondin gene family can be divided into two subgroups based on their protein structure (Fig. 2). Subgroup A includes thrombospondin-1 and 2. Thrombospondin-1 and 2 are very similar in overall organization (1–3,24). There are globular regions at the amino and carboxyl termini that are low in cysteine content and do not possess internal repeating structure or strong homology to other proteins in the database. In contrast, the center portions of the molecules are rich in cysteine and contain repeated sequence motifs that are found in other proteins. The type 1 repeats were first identified in thrombospondin-1 and have been designated thrombospondin repeats (TSR) by some investigators. The thrombospondin type 1 superfamily is an expanding group of heterogeneous proteins (Fig. 2). Multiple copies of the type 1 repeats are found in the complement factors, including C8, C9, and properdin (25,26). The properdin molecule is composed of six consecutive type 1 repeats, with short regions of distinct polypeptide at the amino and carboxyl termini. The TSR is also found in two proteins that occur in malaria-parasitized erythrocytes, the circumsporozoite protein and the thrombospondin-related anonymous protein (27). The most recent additions to this superfamily are two proteins that are involved in axon guidance, the unc-5 gene product of *C. elegans* and F spondin (the F indicates that the protein is found in the floor plate, and spondin indicates its relationship to thrombospondin) (28,29). The type 1 repeats appear to be involved in the interaction of thrombospondin with cell surfaces (see later). The type 1 repeats are found only in subgroup A of the thrombospondin genes.

Subgroup A contains three type 2 repeats, whereas subgroup B contains four type 2 repeats (16,24,30). These repeats are similar to epidermal growth factor (EGF) in that they contain six cysteines in a characteristic placement (31). Multiple copies of the EGF repeats are commonly found in cell adhesion molecules and adhesive glycoproteins. These repeats probably provide structurally stable spacers in regions in which the molecules are extended and accessible to the solvent.

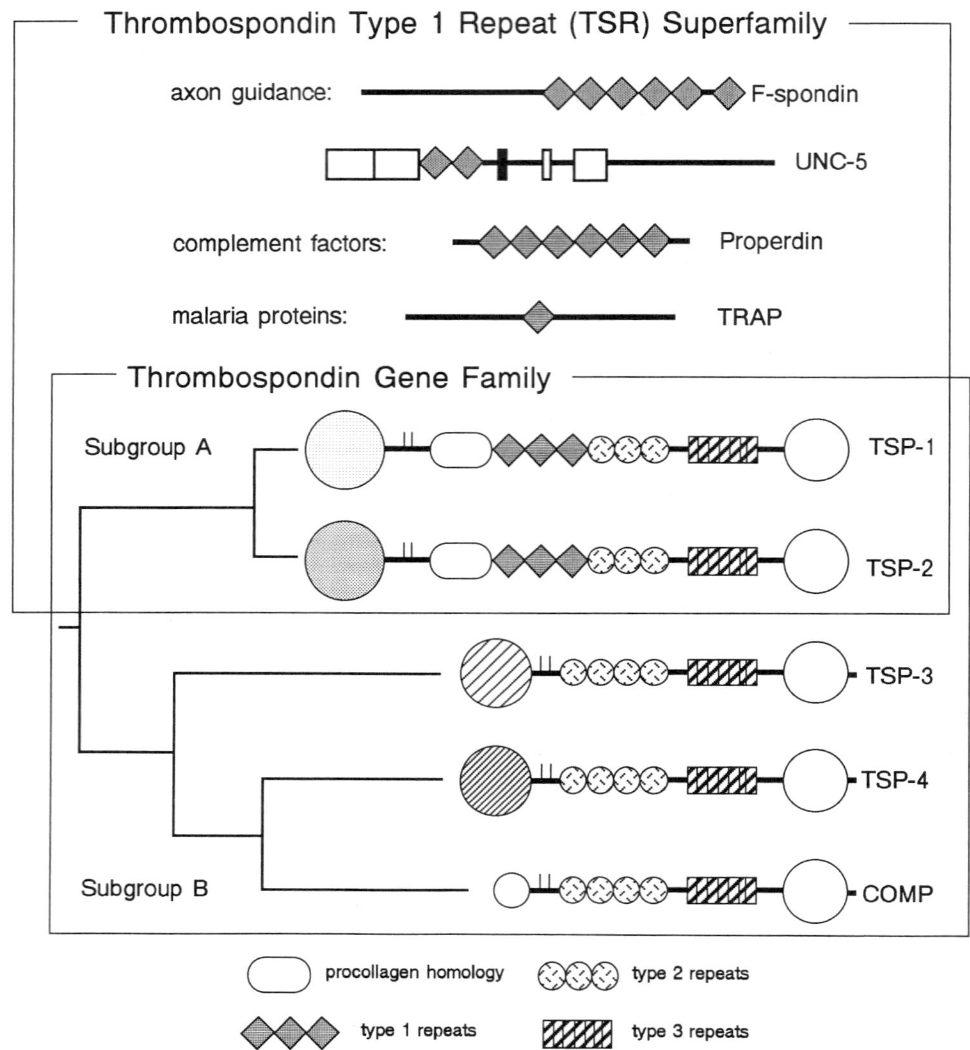

Figure 2 The thrombospondin gene family and the thrombospondin type 1 repeat (TSR) superfamily.

A gradient of amino acid sequence identity from the carboxyl terminal (highest) to the amino terminal (lowest) is a general feature of the members of the thrombospondin gene family. The carboxyl-terminal and the type 3 repeats represent the most conserved feature of all the thrombospondins, including COMP. The type 3 repeats represent a contiguous series of calcium binding sites (24). The calcium binding loops are rich in aspartic acid and are similar to the calcium binding pockets of calmodulin. Whereas the calcium binding sites of calmodulin are stabilized by adjacent α-helical polypeptide segments, the calcium binding sites of the thrombospondins are stabilized by disulfide bonds. Calcium induces a significant conformational change in thrombospondin-1 and has been shown to modulate the function of the molecule (see later) (32–37). If each of the type 3 repeats can bind calcium, then each subunit of thrombospondin could bind 12 calcium ions. This would correspond to 36 calcium ions for intact thrombospondin-1 and 2, which are trimers, and 60 calcium ions for COMP, which is a pentamer (38). The number of subunits in the intact thrombospondin-3 and 4 molecules has not been determined.

We used progressive sequence alignment algorithms to predict the evolution of the members of the thrombospondin gene family (Fig. 3) (39). The analysis indicates that subgroups A and B were produced by the duplication of a primordial thrombospondin gene more than 900 million years ago. Further duplications of these genes occurred around the time of the appearance of the first vertebrates to yield the current members of the thrombospondin gene family. Because the subgroup B members do not contain the region of homology with procollagen or the type 1 repeats, it has been proposed that these sequences were inserted by exon shuffling events that occurred after the first gene duplication, but before the gene duplication that produced thrombospondin-1 and 2. The evolutionary analysis indicates that the thrombospondin genes have existed during most of the evolution of the animal kingdom and probably serve a function that is common to most of its members.

III. GENE STRUCTURE AND CHROMOSOMAL LOCALIZATION

The structure of the mouse thrombospondin-1 gene and the known structure of the mouse thrombospondin-2 gene are shown schematically in Figure 4 (1,4,40,41). The thrombospondin-1 gene is approximately 18 kb in length and is composed of 22 exons (40,41). The human and mouse thrombospondin-1 genes have similar sequences in portions of the promoter, the coding sequence, and the 3′-untranslated regions (42,43). The thrombospondin-1 genes are similar to the housekeeping and growth control genes in that they contain SP1 and AP1 binding sequences and lack a CAAT box (44). Both the human and mouse thrombospondin-1 promoters contain a NF-κB binding sequence that has been reported to be involved in interleukin-1 (IL-1)-induced upregulation of transcription. The thrombospondin-1 gene has been reported to be induced by IL-1 (45). The human and mouse thrombospondin genes also contain regulatory elements for regulation by steroid hormones, cAMP, and serum- and platelet-derived growth factor (40–43). Thrombospondin-1 behaves as an immediate early response gene in that mRNA levels are rapidly induced in response to serum and superinduced in response to serum in the presence of cycloheximide (46). Two regions of the human promoter have been proposed to be involved in the response to serum (30). A serum-response element that has been characterized in the c-fos gene is found 1280 bp upstream of the start site and a NF-Y site that is 65 bp upstream of the start site. In addition, human and mouse thrombospondin-1 contain a sequence in the 3′-untranslated region that has been reported to be involved in the serum responsiveness of the JE gene (see later). This latter sequence has not been shown to be functional in the thrombospondin gene.

Whereas the overall organization of the thrombospondin-2 gene is similar to that of thrombospondin-1, the nucleotide sequence of the promoter is distinct (1). The mouse thrombospondin-2 promoter does not contain SP1 or AP1 sites within 1116 bp of the start site. Two sequences that are similar to CAAT box sequences are present upstream of a typical TATA box sequence. The thrombospondin-2 promoter does not contain a serum-response element in the 1116 bp that has been sequenced; however, such an ele-

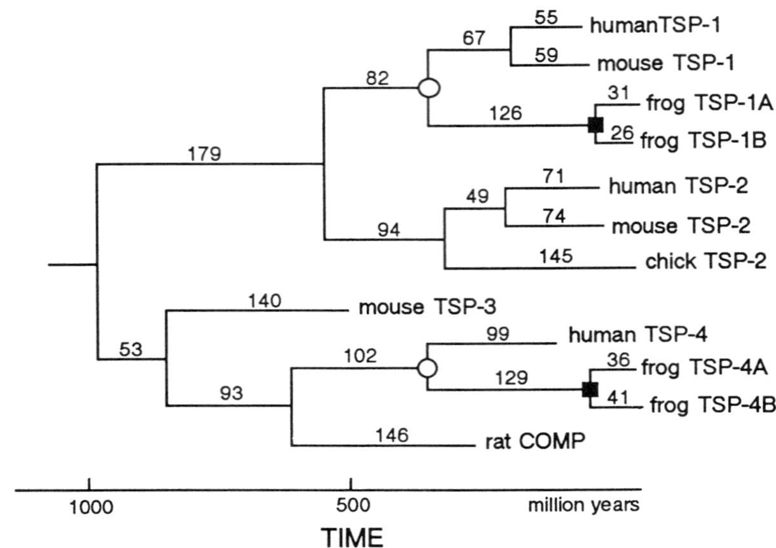

Figure 3 Phylogenetic tree of the thrombospondin gene family. The tree was constructed using progressive sequence alignment algorithms (39). The divergence of humans and frogs (open circles) is assumed to have occurred 350 million years ago, and the duplication of the *Xenopus laevis* genome (solid boxes) is assumed to have occurred 30 million years ago. The lengths of the mouse TSP-3 and rat COMP branches are artifactually short because only one sequence is available for these proteins (39).

ment may exist farther upstream. Conflicting results have been reported regarding the serum responsiveness of the thrombospondin-2. Laherty et al. (3) reported that the thrombospondin-2 mRNA increased rapidly in the first hour after serum stimulation and then decreased rapidly, returning to baseline levels in 4–8 h. These authors reported that thrombospondin-2 mRNA is superinduced when the cells are treated with serum in the presence of cycloheximide. By contrast, Bornstein et al. (1) reported that the thrombospondin-2 gene is not upregulated by serum.

The gene structure of thrombospondin-1 and 2 supports the hypothesis that the proteins are members of the thrombospondin type 1 repeat superfamily and the EGF repeat (thrombospondin type 2) superfamily. Each of the type 1 and type 2 repeats is encoded in a single exon. The phasing of these exons is such that most could be removed without disrupting the reading frame (47). When differences exist between thrombospondin-1 and 2, the unaligned DNA usually corresponds to an integer number of codons so that the phasing is preserved. By contrast, the exons that contain the type 3 repeats do not have symmetrical phasing and do not contain individual type 3 repeats (47). In addition, two of the larger exons (17 and 18) that contain some of the type 3 repeats of thrombospondin-1 are divided into two exons in thrombospondin-3 (4).

Thrombospondin-3 and 4 are distinct from thrombospondin-1 and 2 in that they have relatively short (150 bp) 3′-untranslated regions (4,6). The 3′-untranslated regions of thrombospondin-1 and 2 may serve regulatory functions. The approximately 2 kb of 3′-untranslated region have multiple copies of the ATTT and TATT sequences that have been reported to decrease message stability (48). In addition, the sequence TTTTGTA is

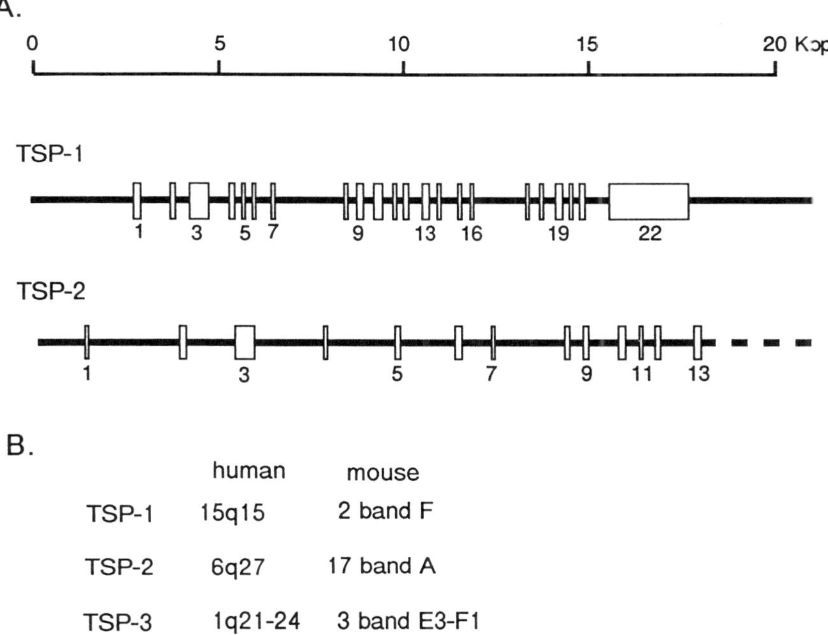

Figure 4 Structure and chromosomal location of the thrombospondin genes. (A) Genomic structures of murine thrombospondin-1 (TSP-1) and 2 (TSP-2) are shown schematically. The open boxes are exons. (B) Chromosomal location of the thrombospondin genes in the human and mouse genomes.

found in a region of 100 nucleotides near the polyadenylation site that is highly conserved in humans and mice. This sequence has been reported to be involved in the response of the JE gene to serum.

IV. TISSUE EXPRESSION OF THE THROMBOSPONDIN GENES

The thrombospondin genes are expressed throughout the tissues of the human and mouse (Table 1). Most tissues express more than one member of the gene family (1–4,6,49). Heart and skeletal muscle express thrombospondin-1, 2, 3, and 4, but not COMP (Fig. 5). Thrombospondin-4 and COMP show the highest level of lineage restriction in that they are found principally in muscle and cartilage, respectively. Each of the thrombospondins is expressed at high levels in adult tissues. Thrombospondin-3 and 4 are expressed at the highest levels in the adult lung and heart, respectively.

The expression of thrombospondin-1 and 2, has also been characterized in the developing mouse and in the chicken embryo (3,50–52). In the embryo, the level of expression varies within different tissues during development. In general, significant expression is observed at neurulation and during organogenesis. The brain and heart are consistently positive during organogenesis. By contrast, expression in the mouse liver goes up sharply at 15.5 days of development and then decreases rapidly (3).

From these data, several generalizations regarding the expression of the thrombospondin genes can be made. First, most vertebrate cell types synthesize and secrete at

Table 1 Tissue Expression of the Members of the Thrombospondin Gene Family[a]

Tissue	TSP-1	TSP-2	TSP-3	TSP-4	COMP
Liver	+/−	+/−	+/−	−	−
Lung	++	+/−	++	−	?
Spleen	+	−	−	?	?
Bone	++	+	+	?	−
Tail	++	++	+	?	?
Skin	+	+	+	?	?
Thymus	−	−	−	?	?
Muscle	+	+	+	+	−
Heart	+	+	+/−	++	−
Kidney	+	+/−	+	−	−
Brain	+	+	−	−	?
Placenta	+	+/−	+	−	?
Cartilage	?	?	?	?	++

[a]The +/− indicates conflicting results have been obtained. The data summarized here are from tissues of adult human or 4-week-old mouse, with the exception of the kidney, in which the positive signal for TSP-2 was observed in embryonic mouse tissue. The ++ indicates that these tissues displayed the highest level of expression.

least one member of the family. Of the organs that have been studied to date, the thymus is the only organ that appears to be negative for the known thrombospondins. Thus, the thrombospondins probably serve a common and universal function. Second, many cell types produce several thrombospondins. This raises the possibility that some thrombospondins are redundant in some tissues and that synthesis of the thrombospondins may be coordinately regulated. Thus, despite the broad distribution of some thrombospondins, deletion of a gene may not have a dramatic impact on the development or function of the adult tissue. For example, deletion of the thrombospondin-4 gene may not lead to defects in muscle tissue, because other thrombospondins are present. Third, the presence of multiple genes with variable expression has the potential for a great deal of biodiversity. O'Rourke et al. (53) showed that cells that synthesize thrombospondin-1 and 2 secrete homo and heterotrimers. Because thrombospondin-1 and 2 are trimers, eight different combinations can be produced. This may enable the cell to modulate subtly the function of the thrombospondin molecules produced by adjusting the subunit composition.

V. CELL SURFACE RECEPTORS FOR THROMBOSPONDIN-1

The cell binding activities of thrombospondin-1 have been extensively studied through the use of cell attachment assays. Thrombospondin-1 is bound to substrates, and cells are added in the presence or absence of inhibitors of known receptors for other adhesive proteins. The results of these studies have often appeared contradictory, different laboratories obtaining conflicting results. This seems to be partly because of the complexity of the interaction, in that multiple receptor systems are able to interact with various domains of the molecule (Fig. 1) (54–56). In addition, some receptors are able to bind to the molecule in more than one location. Various cell types express different sets of receptors, so that each has unique interactions with the molecule.

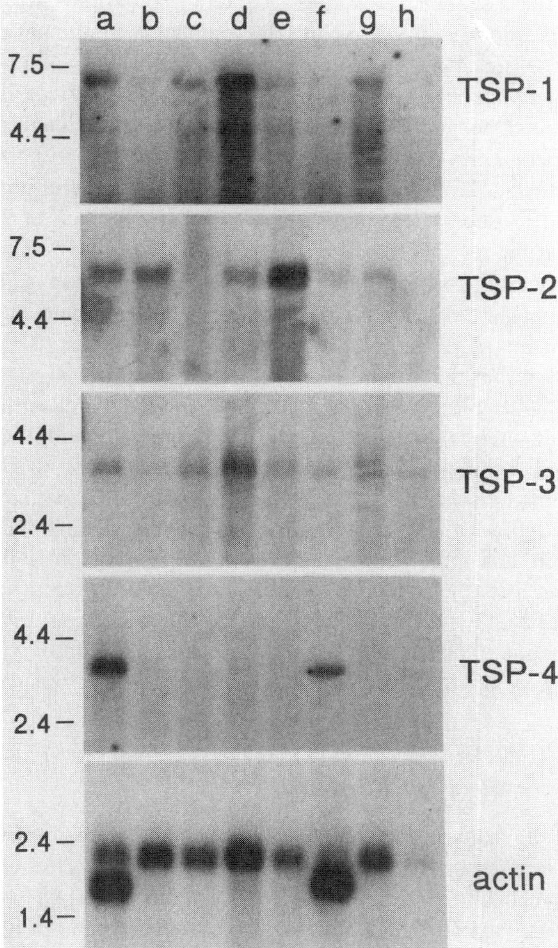

Figure 5 Expression of thrombospondin genes in adult human tissue. A northern blot of poly $(A)^+$ RNA from adult human heart (a), brain (b), placenta (c), lung (d), liver (e), skeletal muscle (f), kidney (g), and pancreas (h). The probe is indicated on the right, and the positions and sizes (kb) of the markers are indicated on the left.

A. Interaction of Thrombospondin with Proteoglycans

The NH_2-terminal domain of thrombospondin-1 contains a high-affinity binding site for heparin sulfate-containing proteoglycans (57). A specific interaction with syndecan has been demonstrated (58). This binding activity has been extensively characterized using soluble heparin and heparin-Sepharose. Quantitative studies indicate that thrombospondin-1 binds heparin with a dissociation constant of approximately 10 nM (59). Whereas heparin can be fractionated into low- and high-affinity pools on fibronectin or laminin, thrombospondin can bind all forms of heparin with high affinity (59). Site-directed mutagenesis has been used to demonstrate that two sequences that are rich in basic residues participate in the binding of thrombospondin-1 to heparin (57). Mutagenesis of either of these sequences resulted in a decreased affinity for heparin as measured by the salt concentration required to elute the mutated recombinant molecules from heparin-Sepharose. The presence of multiple sites on each subunit and the oligomeric nature of the molecule may account for the high-affinity binding to heparin.

Thrombospondin-1 has additional heparin binding sites of lower affinity in other portions of the molecule. Several groups have shown that the 140,000 dalton proteolytic fragment of thrombospondin that lacks the NH_2-terminal heparin binding domain retains the ability to bind to heparin-Sepharose (57,60,61). Haverstick et al. (60) reported that further cleavage at the COOH terminus decreased the heparin binding activity. Deletion of the type 1 or type 2 repeats by oligonucleotide-directed mutagenesis did not abolish the heparin binding activity of this fragment (57). These data indicate that a heparin binding site is present in the COOH-terminal globular domain. By contrast, Dardik and Lahav (61) reported that the 70,000 dalton proteolytic fragment that includes the type 1 and 2 repeats contains a heparin binding site. These data indicate that heparin binding sites are present at three distinct sites along the thrombospondin molecule.

Heparin binding sequences have been identified in the type 1 repeats (62–64). A detailed analysis of the sequence requirements has led to the definition of the minimum sequence that will bind to heparin and to the conclusion that flanking sequences are necessary to obtain maximum activity (63,64). The sequence WSPW is the shortest peptide that has heparin and sulfatide binding. The inclusion of upstream basic amino acids in the peptide KRFKQDGGSHWSPWSS resulted in an increase in the activity and made the peptide a selective inhibitor of heparin but not sulfatide binding (63,64). This peptide also supports the attachment of melanoma cells when it immobilized on plastic substrates.

B. Receptors for the Type 1 Repeats

A comparison of the sequences of the members of the TSR superfamily reveals that the sequence VTSG is conserved in many of its members (27). This sequence can promote adhesion of human hematopoietic cells (27). In addition, peptides that contain this sequence inhibit the attachment of various cell types to thrombospondin-coated substrates and inhibit platelet aggregation (27,62,65,66). Multiple copies of this sequence occur in thrombospondin-1 and 2, some species having a VTCG sequence in all three types 1 repeats (Fig. 1). Two cell surface receptors, CD36 and a 50,000 dalton polypeptide, have been reported to bind to thrombospondin through the VTCG peptide sequence (65,67).

CD36 is an 88,000 dalton glycoprotein expressed on many cell types, including platelets (glycoprotein IV), endothelial cells, monocytes, and many epithelial cells (for a review, see Ref. 68). Platelet CD36 functions as a receptor for collagen, and endothelial CD36 is involved in the sequestration of parasitized erythrocytes (69,70). The role of CD36 as a thrombospondin receptor is controversial because contradictory and seemingly unreconcilable results have been obtained. In support of an interaction of thrombospondin with CD36, it has been reported that (1) antibodies to CD36 inhibit the thrombospondin-dependent adhesion of monocytes to platelets and macrophages, (2) thrombospondin binds to CD36 in solid-phase binding assays, (3) liposomes that contain CD36 bind to thrombospondin, and (4) antithrombospondin antibodies immunoprecipitate a CD36-thrombospondin complex from activated platelets (71–77). Asch and coworkers (72) have provided data that CD36 is able to bind directly to the CSVTCG sequence. Leung et al. (78) proposed that the binding of thrombospondin to CD36 is a two-step process. The initial interaction results in a conformational change in thrombospondin that results in the exposure of a second high-affinity site on thrombospondin for CD36. Data suggesting that CD36 is not a receptor for thrombospondin include the observations that (1) COS cells that are transfected with CD36 do not bind thrombospondin, (2) anti-CD36 antibodies do not inhibit the adhesion of platelets to solid-phase thrombospondin, and (3) Nak^a-negative platelets that lack CD36 display normal thrombospondin binding (79–82). Greenwalt et al. (68) proposed that CD36 is not a receptor for thrombospondin, but it is in close proximity to a thrombospondin receptor. This proximity would result in coimmunoprecipitation of CD36 with the thrombospondin-receptor complex and steric hindrance of the thrombospondin-receptor interaction with anti-CD36 antibodies.

A second receptor for the CSVTCG sequence was recently identified by Tuszynski and coworkers (67). When detergent extracts of A549 human lung adenocarcinoma cells are applied to a column of CSVTCG-Sepharose, a 50,000 dalton protein is eluted. This protein produces bands of 50,000 and 60,000 daltons under reducing conditions. Like other receptors for thrombospondin, the purified protein binds to solid-phase thrombospondin in a calcium-dependent manner, and the binding is inhibited by heparin. Antibodies to this protein inhibit A549 cell attachment and spreading on thrombospondin-coated substrates.

C. Interaction of Thrombospondin with Integrins

Integrins are a family of cell surface glycoproteins that function as receptor for the adhesive glycoproteins (for a review see Ref. 83). Santoro (84) demonstrated that thrombin-treated platelets attach to thrombospondin-coated substrates. Relatively little attachment is observed with unactivated platelets. The attachment of activated platelets to thrombospondin-coated substrates is inhibited by the synthetic peptide GRGDSP. The control peptide (GRGESP) is not inhibitory. The inhibition of platelet adhesion by RGD-containing peptides indicates that an integrin is involved. The $\alpha_{IIb}\beta_3$ (platelet glycoprotein IIb/IIIa) integrin is expressed at very high levels on platelets. Several lines of evidence indicate that thrombospondin can interact with $\alpha_{IIb}\beta_3$. Antibodies to $\alpha_{IIb}\beta_3$ inhibit the binding of thrombospondin to activated platelets (85,86). The binding of thrombospondin to activated platelets is also inhibited by soluble fibrinogen (86). Neither fibrinogen nor anti-$\alpha_{IIb}\beta_3$ antibodies inhibited the binding of thrombospondin to resting platelets in the studies of Wolff et al. (86). These data are consistent with the observation that thrombospondin, fibrinogen, and $\alpha_{IIb}\beta_3$ colocalize on the surface of activated platelets (9). Karczewski et al. (87) used a solid-phase binding assay to study the interaction of thrombospondin with $\alpha_{IIb}\beta_3$. Specific binding is observed when either the thrombospondin or the $\alpha_{IIb}\beta_3$ is adsorbed by the substrate. As with intact platelets, the binding is calcium dependent and is inhibited by antibodies directed against $\alpha_{IIb}\beta_3$; however, this interaction is not significantly inhibited by the synthetic peptide GRGDSP (87). Affinity chromatography has also been used to characterize the interaction of thrombospondin with $\alpha_{IIb}\beta_3$. Platelet $\alpha_{IIb}\beta_3$ is retained on thrombospondin-Sepharose columns (87,88). When platelet membrane proteins are radiolabeled with ^{125}I-lactoperoxidase, solubilized in *n*-octyl glucoside, and applied to a column of thrombospondin-Sepharose, two integrins are bound to the column and specifically eluted with the peptide GRGDSP (88). Relatively small amounts of $\alpha_{IIb}\beta_3$ are eluted, suggesting that this interaction is weak. The second integrin shares the same β subunit but has a distinct α subunit, designated α_V. The $\alpha_V\beta_3$ integrin on thrombasthenic platelets has also been shown to be retained on thrombospondin-

Sepharose columns (88). In contrast to $\alpha_{IIb}\beta_3$, the platelets from thrombasthenic patients have normal or elevated levels of $\alpha_V\beta_3$ (89).

Tuszynski and Kowalska (81) reported that unactivated platelets are capable of attaching and spreading on thrombospondin-coated substrates. This interaction is also inhibited by EDTA and by GRGDSP. In addition, this adhesion is inhibited by antibodies directed against the $\alpha_2\beta_1$ integrin (platelet glycoprotein Ia-IIa complex) and the $\alpha_{IIb}\beta_3$ integrin. The amount of platelets from thrombasthenic patients that attach is approximately 40–50% of their normal counterparts. A similar amount of inhibition is observed when the normal platelets are treated with anti-$\alpha_{IIb}\beta_3$ antibodies. Antibodies to the $\alpha_2\beta_1$ integrin are more effective in blocking the adhesion of platelets from normal and thrombasthenic individuals to thrombospondin-coated substrates (81). The involvement of $\alpha_{IIb}\beta_3$ is consistent with the fact that the RGD-containing peptides inhibit attachment; however, $\alpha_2\beta_1$ has been reported to recognize a DGEA sequence in type 1 collagen (90). The DGEA sequence is not found in the thrombospondin sequence.

The role of integrins in the attachment of cultured cells has also been studied using peptides and antibodies that inhibit integrin function. In some studies, the RGDS peptide alone does not inhibited attachment of G361 melanoma cells; however, when RGDS is added with heparin, the attachment is inhibited (54). These results indicate that multiple receptor systems are functioning in parallel. Inhibition of attachment is observed only when all functional receptor systems are blocked, concurrently. Whereas other studies with G361 melanoma cells support the proposal that multiple receptors are functioning concurrently, in some cases an antagonist to a single receptor system is sufficient to block cell attachment (56).

Whereas Asch et al. (54) did not observe an effect of RGDS on C32 melanoma cells, Tuszynski et al. (91) observed a partial inhibition with GRGDS or GRGDSP. In the latter study, an antibody to the $\alpha_V\beta_3$ integrin was also found partially to inhibit C32 melanoma cell attachment to platelet thrombospondin-coated substrates. The synthetic peptide GRGDSP and a monoclonal antibody, designated LM609, that is directed against $\alpha_V\beta_3$ have also been reported to inhibit the attachment of human endothelial cells, smooth muscle cells, U937 cells, and normal rat kidney cells (92). Attachment is also inhibited by the peptide GRGDAC, the RGD-containing sequence that is found in thrombospondin (92). The attachment of all four cell types was dependent on the presence of divalent cations. When *n*-octyl glucoside extracts of ^{125}I-radiolabeled endothelial cells or smooth muscle cells are applied to a column of thrombospondin-Sepharose, the $\alpha_V\beta_3$ integrin is eluted with the GRGDS peptide (92).

Antibodies against $\alpha_V\beta_3$ and RGD-containing peptides also partially inhibit the attachment and spreading of human embryonic fibroblasts (55). In these studies, fucoidan was used to inhibit the heparin binding activity of thrombospondin. Fucoidan and antibodies to CD36 also blocked the attachment of embryonic fibroblasts to thrombospondin (55). Furthermore, an additive effect was observed when antibodies to $\alpha_V\beta_3$ were added along with either fucoidan or anti-CD36 antibodies. Thus, multiple receptors appear to be functioning on these cells concurrently.

D. Receptors for the COOH-Terminal Domain of Thrombospondin

The ability of G361 melanoma cells, human intestinal smooth muscle cells, epidermal keratinocytes, and MG-63 osteosarcoma cells to attach to thrombospondin in inhibited by antibodies (C6.7 and ESTs10) to the COOH-terminal domain of thrombospondin (Fig. 1) (56,93,94). In addition, monoclonal antibody C6.7 inhibits platelet aggregation (16). The epitope for monoclonal antibody C6.7 is contained in the amino acid sequence between residue 1030 and 1152, and the epitope for Ests10 is between residues 913 and 1152 (32,56). Kosfeld et al. (94) expressed amino acids 941–1152 as a bacterial fusion protein. This protein is able to support attachment of G361 melanoma cells.

Yabkowitz and Dixit (95) have shown that 11B squamous carcinoma cells bind to thrombospondin through receptors for the NH_2-terminal heparin binding domain and the COOH-terminal domain. When a detergent extract of the plasma membrane from 11B cells is applied to a column of thrombospondin-Sepharose, two peptides with molecular weights of 80,000 and 105,000 are eluted with buffer containing 5 mM EDTA and 5 mM EGTA. These polypeptides do not bind antibodies to β_1 or β_3 integrins or antibodies to CD36 (95). The peptide maps that are produced by V-8 digestion of the 80,000 and 105,000 dalton bands are distinct, indicating that these peptides are not products of the same gene.

VI. CONCLUSIONS

Since the discovery of thrombospondin in the α granules of human blood platelets, the understanding of scope and complexity of the functions of thrombospondin has continually increased. Approximately 10 years ago, it became apparent that thrombospondin has a wide tissue distribution. In the last 2 years, thrombospondin-1 has been recognized as a member of an emerging family of extracellular calcium binding proteins. The potential for mixing different thrombospondin subunits during the assembly of the intact molecule leads to the possibility of hundreds of distinct intact thrombospondin molecules. The existence of multiple cell surface receptors for the variety of assembled thrombospondin molecules leads to a great deal of complexity and the possibility of a wide range of phenotypic cellular responses. One way to reduce this complexity is to focus on the interaction of specific cell types with the specific thrombospondins that are produced. To accomplish this goal, several laboratories are studying the expression of the members of the thrombospondin gene family by in situ hybridization. These studies also require purified natural or recombinant proteins for each family member. A second useful approach to the function of specific members will be the production of thrombospondin-deficient mice by homologous recombination. The existence of multiple genes increases the chances that these mice will be viable. By producing various deficient lines, specific questions can be asked about (1) the correlation between expression and function, (2) the role of each thrombospondin in the function of various tissues, and (3) the possibility that the members of the thrombospondin gene family are coordinately regulated and can compensate for individual deficiencies. Thrombospondin-1-deficient mice will be a powerful tool for analyzing the role of thrombospondin-1 in (1) platelet aggregation, (2) thrombosis, (3) fibrinolysis, (4) wound healing, (5) development, (6) angiogenesis, and (7) metastasis.

ACKNOWLEDGMENTS

The manuscript was typed by Pamela Caffey and edited by Sami Lawler. The author is the recipient of National Institute of Health Grant HL 28749 and HL 42443 from the National Heart, Lung and Blood Institute.

REFERENCES

1. Bornstein P, Devarayalu S, Li P, Disteche CM, Framson P. A second thrombospondin gene in the mouse is similar in organization to thrombospondin-1 but does not respond to serum. Proc Natl Acad Sci USA 1991; 88:8636–8640.
2. Bornstein P, O'Rourke K, Wikstrom K, et al. A second, expressed thrombospondin gene (Thbs2) exists in the mouse genome. J Biol Chem 1991; 266:12821–12824.
3. Laherty CD, O'Rourke K, Wolf FW, Katz R, Seldin MF, Dixit VM. Characterization of mouse thrombospondin-2 sequences and expression during cell growth. J Biol Chem 1992; 267: 3274–3281.
4. Vos HL, Devaraylau S, de Vries Y, Bornstein P. Thrombospondin 3 (*Thbs3*), a new member of the thrombospondin gene family. J Biol Chem 1992; 267:12192–12196.
5. Oldberg A, Antonson P, Lindblom K, Heinegard D. COMP (cartilage oligomeric matrix protein) is structurally related to the thrombospondins. J Biol Chem 1992; 267:22346–22350.
6. Lawler J, Duquette M, Whittaker CA, Adams JC, McHenry K, DeSimone DW. Identification and characterization of thrombospondin-4, a new member of the thrombospondin gene family. J Cell Biol 1993; in press.
7. Phillips DR, Jennings LK, Prasanna HR. Ca^{2+}-mediated association of glycoprotein G (thrombin-sensitive protein, thrombospondin) with human platelet. J Biol Chem 1980; 255:11629–11632.
8. Gogstad GO, Hagen I, Korsmo R, Solum NO. Evidence for release of soluble, but not membrane-integrated proteins from human platelet α-granules. Biochim Biophys Acta 1982; 702:81–89.
9. Asch AS, Leung LLK, Polley MJ, Nachman RL. Platelet membrane topography: colocalization of thrombospondin and fibrinogen with the glycoprotein IIb-IIIa complex. Blood 1985; 66:926–934.
10. Lawler J. The structural and functional properties of thrombospondin. Blood 1986; 67:1197–1209.
11. Leung LLK. Role of thrombospondin in platelet aggregation. J Clin Invest 1984; 74:1764–1772.
12. Aiken ML, Ginsberg MH, Plow EF. Identification of a new class of inducible receptors on platelets: thrombospondin interacts with platelets via a GPIIb-IIIa-independent mechanism. J Clin Invest 1986; 78:1713–1716.
13. Boukerche H, McGregor JL. Characterization of an anti-thrombospondin monoclonal antibody (P8) that inhibits human blood platelet functions. Normal binding of P8 to thrombin-activated Glanzmann thrombasthenic platelets. Eur J Biochem 1988; 171:383–392.
14. Legrand C, Dubernard V, Kieffer N, Nurden AT. Use of a monoclonal antibody to measure the surface expression of thrombospondin following platelet activation. Eur J Biochem 1988; 171:393–400.
15. Legrand C, Thibert V, Dubernard V, Beqault B, Lawler J. Molecular requirements for the interaction of thrombospondin with thrombin-activated human platelets: modulation of platelet aggregation. Blood 1992; 79:1995–2003.
16. Dixit VM, Haverstick DM, O'Rourke KM, et al. A monoclonal antibody against human thrombospondin inhibits platelet aggregation. Proc Natl Acad Sci USA 1985; 82:3472–3476.
17. Bale MD, Westrick LG, Mosher DF. Incorporation of thrombospondin into fibrin clots. J Biol Chem 1985; 260:7502–7508.
18. Mosher DF, Misenheimer TM, Stenflo J, Hogg PJ. Modulation of fibrinolysis by thrombospondin. Ann NY Acad Sci 1992; 667:64–69.
19. Silverstein RL, Leung LLK, Harpel PC, Nachman RL. Complex formation of platelet thrombospondin with plasminogen: modulation of activation by tissue activator. J Clin Invest 1984; 74:1625–1633.
20. DePoli P, Bacon-Baguley T, Kendra-Franczak S, Cederholm MT, Walz DA. Thrombospondin interaction with plasminogen. Evidence for binding to a specific region of the kringle structure of plasminogen. Blood 1989; 73:976–982.

21. Silverstein RL, Harpel PC, Nachman RL. Tissue plasminogen activator and urokinase enhance the binding of plasminogen to thrombospondin. J Biol Chem 1986; 261:9959–9965.
22. Harpel PC, Silverstein RL, Pannell R, Gurewich V, Nachman RL. Thrombospondin forms complexes with single-chain and two-chain forms of urokinase. J Biol Chem 1990; 265: 11289–11294.
23. Hogg PJ, Senflo J, Mosher DF. Thrombospondin is a slow tight-binding inhibitor of plasmin. Biochemistry 1992; 31:265–269.
24. Lawler J, Hynes RO. The structure of human thrombospondin, an adhesive glycoprotein with multiple calcium-binding sites and homologies with several different proteins. J Cell Biol 1986; 103:1635–1648.
25. Robson KJH, Hall JRS, Jennings MW, et al. A highly conserved amino-acid sequence in thrombospondin, properdin and in proteins from sporozoites and blood stages of a human malaria parasite. Nature 1988; 335:79–82.
26. Goundis D, Reid KBM. Properdin, the terminal complement components, thrombospondin and the circumsporozoite protein of malaria parasites contain similar sequence motifs. Nature 1988; 335:82–85.
27. Rich KA, George FW IV, Law JL, Martin WJ. Cell adhesive motif in region II of malarial circumsporozoite protein. Science 1990; 249:1574–1577.
28. Klar A, Baldassare M, Jessell T. F-spondin: a gene expressed at high levels in the floor plate encodes a secreted protein that promotes neural cell adhesion and neurite extension. Cell 1992; 69:95–110.
29. Leung-Hagestijn C, Spence AM, Stern BD, et al. Unc-5, a transmembrane protein with immunoglobulin and thrombospondin type 1 domains, guides cell and pioneer axon migrations in *C. elegans*. Cell 1992; 71:289–299.
30. Bornstein P. Thrombospondins: structure and regulation of expression. FASEB J 1992; 6: 3290–3299.
31. Hanford PA, Mayhew M, Baron M, Winship PR, Campbell ID, Brownlee GG. Key residues involved in calcium-binding motifs and EGF-like domains. Nature 1991; 351:164–167.
32. Galvin NJ, Dixit VM, O'Rourke KM, Santoro SA, Grant GA, Frazier WA. Mapping of epitopes for monoclonal antibodies against human platelet thrombospondin with electron microscopy and high sensitivity amino acid sequence. J Cell Biol 1985; 101:1434–14441.
33. Lawler J, Simons E. Cooperative binding of calcium to thrombospondin: the effect of calcium on the circular dichroism and limited tryptic digestion of thrombospondin. J Biol Chem 1983; 258:12098–12101.
34. Lawler J, Derick LH, Connolly JE, Chen JH, Chao FC. The structure of human platelet thrombospondin. J Biol Chem 1985; 260:3762–3774.
35. Dixit VM, Galvin NJ, O'Rourke KM, Frazier WA. Monoclonal antibodies that recognize calcium-dependent structures in human thrombospondin. J Biol Chem 1986; 261:1962–1968.
36. Slane JMK, Mosher DF, Lai C-S. Conformational change in thrombospondin induced by removal of bound Ca^{++}: a spin label approach. FEBS Lett 1988; 229:363–366.
37. Vuillard L, Clezardin P, Miller A. Models of human platelet thrombospondin in solution: a dynamic light-scattering study. Biochem J 1991; 275:263–266.
38. Morgelin M, Heinegard D, Engel J, Paulsson M. Electron microscopy of native cartilage oligomeric matrix protein purified from the swarm rat chondrosarcoma reveals a five-armed structure. J Biol Chem 1992; 267:6137–6141.
39. Lawler J, Duquette M, Urry L, McHenry K, Smith TF. The evolution of the thrombospondin gene family. J Mol Evol 1993; in press.
40. Bornstein P, Alfi D, Devarayalu S, Framson P, Li P. Characterization of the mouse thrombospondin gene and evaluation of the role of the first intron in human gene expression. J Biol Chem 1990; 265:16691–16698.
41. Lawler J, Duquette M, Ferro P, Copeland NJ, Gilbert J, Jenkins NA. Characterization of the murine thrombospondin gene. Genomics 1991; 11:587–600.

42. Donoviel DB, Framson P, Eldridge CF, Cooke M, Kobayashi S, Bornstein P. Structural analysis and expression of the human thrombospondin gene promoter. J Biol Chem 1988; 263: 18590–18593.
43. Laherty CD, Gierman TM, Dixit VM. Characterization of the promoter region of the human thrombospondin gene. J Biol Chem 1989; 264:11222–11227.
44. Dynan WS. Promoters for house keeping genes. Trends Genet 1986; 2:196–197.
45. Donoviel DB, Bornstein P. The thrombospondin gene is inducible by basic fibroblast growth factor (bFGF) and interleukin-1 (IL-1). J Cell Biol 1988; 107:596a.
46. Majack RA, Cooke SC, Bornstein P. Control of smooth muscle cell growth by components of the extracellular matrix: autocrine role for thrombospondin. Proc Natl Acad Sci USA 1986; 83:9050–9054.
47. Wolf FW, Eddy RL, Shows TB, Dixit VM. Structure and chromosomal localization of the human thrombospondin gene. Genomics 1990; 6:685–691.
48. Hennessy SW, Frazier BA, Kim DD, et al. Complete thrombospondin mRNA sequence includes potential regulatory sites in the 3′ untranslated region. J Cell Biol 1989; 108:729–736.
49. Hedbom E, Antonsson P, Hjerpe A, et al. Cartilage matrix proteins. J Biol Chem 1992; 267: 6132–6136.
50. O'Shea KS, Dixit VM. Unique distribution of the extracellular matrix component thrombospondin in the developing mouse embryo. J Cell Biol 1988; 107:2737–2748.
51. Corless CL, Mendoza A, Collins T, Lawler J. Colocalization of thrombospondin and syndecan during murine development. Dev Dynamics 1992; 193:346–358.
52. Tucker RP. The in situ localization of tenascin splice variants and thrombospondin 2 mRNA in the avian embryo. Development 1993; 117: in press.
53. O'Rourke KM, Laherty CD, Dixit VM. Thrombospondin 1 and thrombospondin 2 are expressed as both homo- and heterotimers. J Biol Chem 1992; 267: in press.
54. Asch AS, Tepler J, Silbiger S, Nachman RL. Cellular attachment to thrombospondin: cooperative interactions between receptor systems. J Biol Chem 1991; 266:1740–1745.
55. Stomski FC, Gani JS, Bates RC, Burns GF. Adhesion to thrombospondin by human embryonic fibroblasts is mediated by multiple receptors and includes a role for glycoprotein 88 (CD36). Exp Cell Res 1992; 198:85–92.
56. Adams JC, Lawler J. Diverse mechanisms for cell attachment to platelet thrombospondin. J Cell Sci 1993; in press.
57. Lawler J, Ferro P, Duquette M. Expression and mutagenesis of thrombospondin. Biochemistry 1992; 31:1173–1180.
58. Sun X, Mosher DF, Rapraeger A. Heparin sulfate-mediated binding of epithelial cell surface proteoglycan to thrombospondin. J Biol Chem 1989; 264:2855–2889.
59. San Antonio JD, Slover J, Lawler J, Karnovsky MJ, Lander AD. Specificity in the interactions of extracellular matrix proteins with subpopulations of the glycosaminoglycan heparin. J Cell Biol 1991; 115:124a.
60. Haverstick DM, Dixit VM, Grant GA, Frazier WA, Santoro SA. Localization of the hemagglutinating activity of platelet thrombospondin to a 140,000-dalton thermolytic fragment. Biochemistry 1985; 23:5597–5603.
61. Dardik R, Lahav J. The structure of endothelial cell thrombospondin: characterization of the heparin-binding domains. Eur J Biochem 1987; 168:347–355.
62. Prater CA, Plotkin J, Jaye D, Frazier WA. The properdin-like type 1 repeats of human thrombospondin contain a cell attachment site. J Cell Biol 1991; 112:1031–1040.
63. Guo N-H, Krutzsch HC, Negre E, Zabrenetzky VS, Roberts DD. Heparin-binding peptides from the type 1 repeats of thrombospondin. J Biol Chem 1992; 267:19349–19355.
64. Guo N-H, Krutzsch HC, Negre E, Vogel T, Bleke DA, Roberts DD. Heparin- and sulfatide-binding peptides from the type 1 repeats of human thrombospondin promote melanoma cell adhesion. Proc Natl Acad Sci USA 1992; 89:3040–3044.

65. Asch AS, Silbiger S, Heimer E, Nachman RL. Thrombospondin sequence motif (CSVTCG) is responsible for CD36 binding. Biochem Biophys Res Commun 1992; 182:1208–1217.
66. Tuszynski GP, Rothman VL, Deutch AH, Hamilton BK, Eyal J. Biological activities of peptides and peptide analogues derived from common sequences present in thrombospondin, properdin and malarial proteins. J Cell Biol 1992; 116:209–217.
67. Tuszynski GP, Rothman VL, Papale M, Hamilton BK, Eyal J. Identification and characterization of a tumor cell receptor for CSVTCG, a thrombospondin adhesive domain. J Cell Biol 1993; 120:513–521.
68. Greenwalt DE, Lipsky RH, Ockenhouse CF, Ikeda H, Tandon NN, Jamieson GA. Membrane glycoprotein CD36: a review of its roles in adherence, signal transduction and transfusion medicine. Blood 1992; 80:1105–1115.
69. Tandon NN, Kralisz U, Jamieson GA. Identification of GPIV (CD36) as a primary receptor for platelet-collagen adhesion. J Biol Chem 1989; 264:7576–7583.
70. Howard RJ, Gilladoga AD. Molecular studies related to the pathogenesis of cerbral maloria. Blood 1989; 74:2603–2618.
71. Beiso P, Pidard D, Fournier D, Dubernard V, Legrand C. Studies on the interaction of platelet glycoprotein IIb-IIIa and glycoprotein IV with fibrinogen and thrombospondin: a new immunochemical approach. Biochim Biophys Acta 1990; 1033:7–12.
72. Asch AS, Barnwell J, Silverstein RL, Nachman RL. Isolation of the thrombospondin membrane receptor. J Clin Invest 1987; 79:1054–1061.
73. Silverstein RL, Asch AS, Nachman RL. Glycoprotein IV mediates thrombospondin-dependent platelet-monocyte and platelet-U937 cell adhesion. J Clin Invest 1989; 84:546–552.
74. McGregor JL, Catimel B, Parmentier S, Clezardin P, Dechavanne M, Leung LLK. Rapid purification and partial characterization of human platelet glycoprotein IIIb. J Biol Chem 1989; 264:501–506.
75. Kieffer N, Nurden AT, Hasitz M, Titeux M, Breton-Gorius J. Identification of platelet membrane thrombospondin binding molecules using an anti-thrombospondin antibody. Biochim Biophys Acta 1988; 967:408–415.
76. Legrand C, Pidard D, Beiso P, Tenza D, Edelman L. Interaction of a monoclonal antibody to glycoprotein IV (CD36) with human platelets and its effect on platelet function. Platelets 1991; 2:99–105.
77. Berk GL, Asch AS, Scollo AW, Silvestein RL, Nachman RL. Glycoprotein IIIb incorporated into liposomes binds thrombospondin. Blood 1989; 74:172a.
78. Leung LLK, Li W-X, McGregor JL, Albrecht G, Howard RJ. CD36 peptides enhance or inhibit CD36-thrombospondin binding. A two-step process of ligand-receptor interaction. J Biol Chem 1992; 267:18244–18250.
79. Oquendo P, Hundt E, Lawler J, Seed B. CD36 directly mediates cytoadherence of *Plasmodium falciparum* infected erythrocytes. Cell 1989; 58:95–101.
80. Tandon NN, Ockenhouse CF, Greco NJ, Jamieson GA. Adhesive functions to platelets lacking GPIV (CD36). Blood 1991; 78:2809–2813.
81. Tuszynski GP, Kowalska MA. Thrombospondin-induced adhesion of human platelets. J Clin Invest 1991; 87:1387–1394.
82. Kehrel B, Kronenberg A, Schwippert B, et al. Thrombospondin binds normally to glycoprotein IIIb-deficient platelets. Biochem Biophys Res Commun 1991; 179:985–991.
83. Hynes RO. Integrins: versatility, modulation and signalling in cell adhesion. Cell 1992; 69: 11–25.
84. Santoro SA. Thrombospondin and the adhesive behavior of platelets. Semin Thromb Hemost 1987; 13:290–297.
85. Plow EF, McEver RP, Coller BS, Woods VL Jr, Marquerie GA, Ginsberg MH. Related binding mechanisms for fibrinogen, fibronectin, von Willebrand factor, and thrombospondin on thrombin-stimulated human platelets. Blood 1985; 66:724–727.

86. Wolff R, Plow EF, Ginsberg MH. Interaction of thrombospondin with resting and stimulated human platelets. J Biol Chem 1986; 261:6840–6846.
87. Karczewski J, Knudsen KA, Smith L, Murphy A, Rothman VL, Tuszynski GP. The interaction of thrombospondin with platelet glycoprotein GPIIb-IIIa. J Biol Chem 1989; 264:21322–21326.
88. Lawler J, Hynes RO. An integrin receptor on normal and thrombasthenic platelets that binds thrombospondin. Blood 1989; 74:2022–2027.
89. Coller BS, Cheresh DA, Asch E, Seligsohn U. Platelet vitronectin receptor expression differentiates Iraqi-Jewish and Arab patients with Glanzmann's thrombasthenia in Israel. Blood 1991; 77:75–83.
90. Staatz WD, Fok KF, Zutter MM, Adams SP, Rodriquez BA, Santoro SA. Identification of a tetrapeptide recognition sequence for the $\alpha_1\beta_1$ integrin in collagen. J Biol Chem 1991; 266: 7363–7367.
91. Tuszynski GP, Karczewski J, Smith L, Murphy A, Rothman VL, Knudsen KA. The GPIIb-IIIa-like complex may function as a human melanoma cell adhesion receptor for thrombospondin. Exp Cell Res 1989; 182:473–481.
92. Lawler J, Weinstein R, Hynes RO. Cell attachment to thrombospondin: the role of Arg-Gly-Asp, calcium, and integrin receptors. J Cell Biol 1988; 107:2351–2361.
93. Roberts DD, Sherwood JA, Ginsburg V. Platelet thrombospondin mediates attachment and spreading of human melanoma cells. J Cell Biol 1987; 104:131–139.
94. Kosfeld MD, Pavlopoulos TV, Frazier WA. Cell attachment activity of the carboxyl-terminal domain of human thrombospondin expressed in *Escheriehia coli*. J Biol Chem 1991; 266: 24257–24259.
95. Yabkowitz R, Dixit VM. Human carcinoma cells express receptors for distinct domains of thrombospondin. Can Res 1991; 51:1645–1650.

31

Platelet Thrombin Receptor

Shaun R. Coughlin
Cardiovascular Research Institute, University of California at San Francisco, San Francisco, California

I. INTRODUCTION

Thrombin is a multifunctional protease generated at sites of vascular injury. Although thrombin's best studied actions have been those on soluble proteins (Chaps. 2, 15, and 20), it is also a powerful agonist for a variety of cellular responses (1). First and foremost, it is the most potent activator of platelets in vitro (2,3), and a number of pharmacological studies demonstrate that thrombin activity is critical for platelet-dependent arterial thrombosis in vivo (4–8). A host of other thrombin activities have been defined, mainly in vitro. Thrombin is chemotactic for monocytes (9) and is mitogenic for lymphocytes and mesenchymal cells, including vascular smooth muscle cells (10–12). Thrombin's actions upon the vascular endothelium include stimulating endothelial production of prostacyclin (13), platelet-activating factor (14), plasminogen activator inhibitor (15), and the potent smooth muscle cell mitogen platelet-derived growth factor (16). Thrombin also induces neutrophil adherence to the vessel wall by an endothelial cell-dependent mechanism (17), probably by causing surface expression of P-selectin on the endothelial surface (18). Teleologically, these disparate functions of thrombin may be unified by viewing thrombin as an orchestrator of the response to vascular injury or wounding, potentially mediating not only hemostatic but perhaps inflammatory and proliferative or reparative responses. However, although thrombin's critical role in hemostasis and thrombosis is well established, the in vivo importance of its proliferative and inflammatory actions remains to be defined.

This discussion begs an understanding of the mechanisms underlying thrombin's actions upon cells. How does thrombin, a protease rather than a classic ligand, activate platelets and other cells? The recent cloning of a platelet thrombin receptor has provided a framework for understanding how thrombin talks to cells (19) and, potentially, a new target for antithrombotic and other therapies (20).

II. GENETICS

The thrombin receptor was cloned only 2 years ago, and the genetics of this system is in its infancy. The structure of the thrombin receptor gene is under study and will no doubt be reported within the next year. The thrombin receptor cDNA predicts a 3.5 kb mRNA with 224 bases of 5′-untranslated sequence and nearly 2.0 kb of 3′-untranslated sequence (19). The latter contains several AU-rich sequences that are known to confer instability and regulability to other mRNAs (21).

No naturally occurring mutations associated with human disease have yet been described. Efforts to "knock out" the thrombin receptor gene in mice by homologous recombination are in progress in a number of laboratories.

III. STRUCTURE-FUNCTION RELATIONSHIPS

A. General Features of the Receptor Structure

The thrombin receptor deduced amino acid sequence revealed it to be a member of the seven transmembrane domain receptor family (19). Its primary sequence is most closely related to the receptors for neuropeptides, glycoprotein hormones, and proinflammatory mediators, such as C5a and interleukin-8 (Ref. 19 and data not shown). The thrombin receptor's predicted topology in the cell membrane is remarkable for a long extracellular amino-terminal extension that contains structures critical for receptor function (see Secs. IIIB and C).

Posttranslational modifications have not yet been rigorously studied. The receptor sequence reveals five consensus N-linked glycosylation sites in regions predicted to be extracellular (19), and in vitro translation studies suggest that the receptor is glycosylated (Gerszten and Coughlin, unpublished). A pair of adjacent cysteines potentially analogous to the palmitoylated cysteines of the β-adrenergic receptor and rhodopsin are also noted in the thrombin receptor's carboxyl tail (19,22,23).

B. Molecular Basis for Thrombin-Receptor Interaction

Within the receptor's amino-terminal exodomain is a putative thrombin cleavage site resembling the known thrombin cleavage site in protein C (Fig. 1). Carboxyl to this site is a sequence resembling the carboxyl tail of hirudin, a structure known to interact with thrombin's anion-binding exosite (24). These observation suggested that this region of the thrombin receptor serves as a thrombin substrate (19). Indeed, a variety of studies with mutant thrombin receptors and synthetic peptides have shown that cleavage of the receptor protein at this site is necessary and sufficient for receptor activation:

1. Mutation of the cleavage site such that it cannot be cleaved rendered the receptor unactivatable by thrombin (19).
2. Replacing the thrombin cleavage recognition sequence with that for enteropeptidase switched receptor specificity; cells (25) or oocytes (26) expressing this construct responded to enteropeptidase but not to thrombin.
3. Synthetic peptides mimicking the receptor's cleavage site were cleaved by thrombin (26), and uncleavable "mutant" peptides mimicking this region bound thrombin and inhibited its activity against synthetic substrates, fibrinogen, and its receptor (26–28).

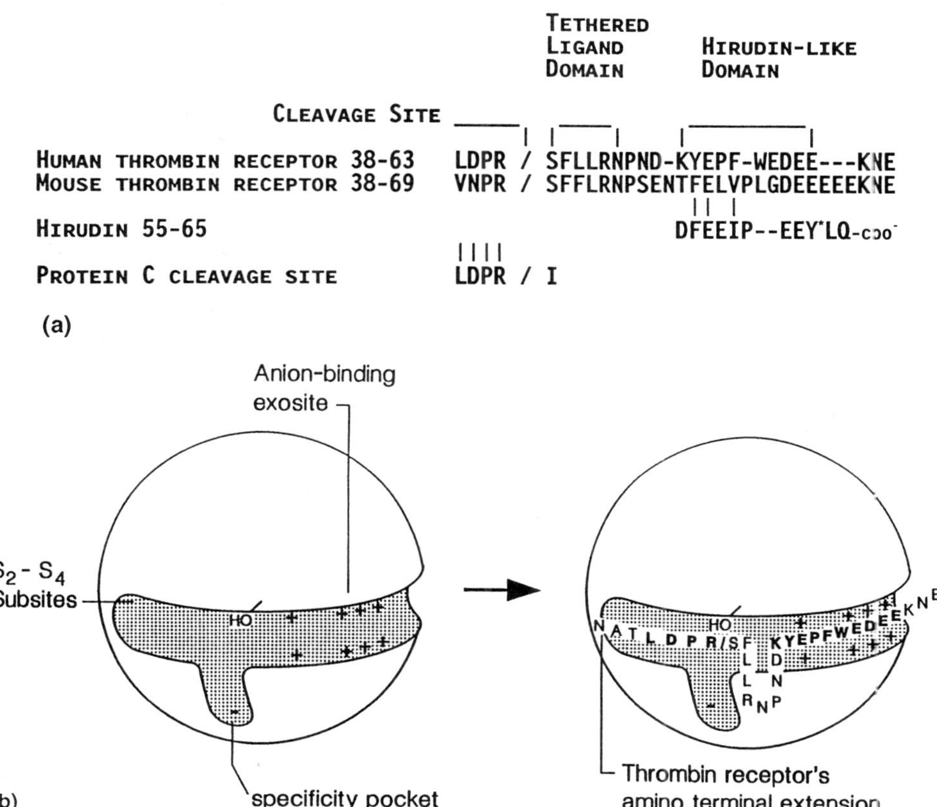

Figure 1 Thrombin-receptor interaction. (A) Functional domains within the thrombin receptor amino-terminal extension. The cleavage recognition sequence (LDPR), thrombin cleavage site, agonist peptide or tethered ligand domain, and anion-binding exosite binding domain (hirudin-like domain) as defined by structure-activity studies with the human receptor are shown. These are aligned with the murine thrombin receptor sequence and the anion-binding exosite binding sequence of the leech anticoagulant hirudin, as well as with the known thrombin cleavage site in protein C. (B) A model for interaction of these domains with thrombin. Thrombin has an extended substrate binding surface (represented by the canyon running laterally) that recognizes residues both amino and carboxyl to its substrate's cleavage site. Structure-function studies suggest that the receptor's hirudin-like domain (KYEPF) interacts with thrombin's anion-binding exosite, and its cleavage site (LDPR/S) interacts with thrombin's S_1–S_4 subsites. This model has important implications for the development of blocking antibodies and receptor peptide-based thrombin inhibitors. [Reprinted with permission from *Nature* (26).]

4. Recent studies utilized an epitope-tagged thrombin receptor to demonstrate that receptor cleavage on intact cells correlated with signaling (see later) (29).

In addition to the primary cleavage recognition sequence LDPR, the receptor's hirudin-like domain has been implicated in thrombin-receptor interaction. Specifically, the receptor's KYEPF sequence binds thrombin's anion-binding exosite in a manner grossly analogous to the DFEEI sequence in hirudin's carboxyl tail (Fig. 1) (26,27).

X-ray crystallographic studies of cocrystals of thrombin with receptor-based peptides have recently revealed the details of this interaction (58). Whether additional receptor domains participate in thrombin-receptor interaction remains an open question.

C. Proteolytic Unmasking of a Tethered Peptide Ligand: A Novel Mechanism of Receptor Activation

How might proteolysis within the thrombin receptor's amino-terminal extension activate the receptor? There are several precedents for a protease activating a target protein by unmasking a new amino terminus within that protein. In particular, proteolytic activation of the zymogen trypsinogen occurs when enteropeptidase cleaves it to unmask a new amino terminus that then binds intramolecularly to effect a conformational change and create an active trypsin molecule (30). A grossly analogous mechanism exists for the thrombin receptor (19). Synthetic peptides that mimick the new amino terminus created when thrombin cleaves its receptor are full agonists for receptor activation and bypass the requirement for receptor proteolysis (Figs. 1 and 2) (19). This observation suggests the model shown in Figure 2. Thrombin cleaves its receptor's amino-terminal extension to unmask a new amino terminus. This new amino terminus then functions as a tethered peptide ligand, binding to an as yet undefined site within the body of the receptor to effect receptor activation (19). As discussed later, synthetic peptides mimicking this "agonist peptide domain" (Fig. 1) provide a new tool for defining the role of the thrombin receptor in various cellular events.

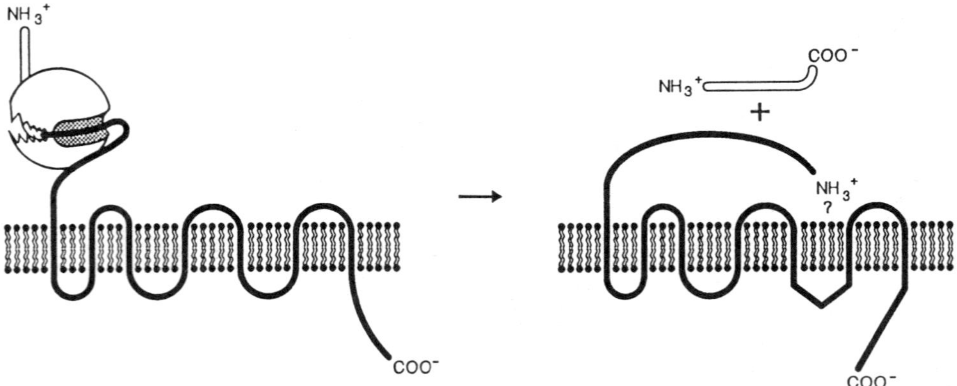

Figure 2 Model of thrombin receptor activation. Thrombin binds its receptor via the receptor's amino-terminal extension as in Figure 1; whether additional receptor domains participate in thrombin binding is unknown. After binding to the amino-terminal extension, thrombin cleaves the receptor at the LDPR/S cleavage site (junction between open and filled receptor segments), releasing an inactive fragment of the receptor's amino terminus (open fragment) and exposing a new amino terminus. This newly unmasked amino terminus then functions as a tethered peptide ligand, binding to an as yet undefined pocket to effect receptor activation. Synthetic peptides mimicking the tethered ligand function as full agonists for receptor activation. [Reprinted with permission from *Nature* (26).]

D. Mechanism of the "Proteolytic Switch": How Does the Tethered Ligand Remain Silent in the Uncleaved Receptor?

The model just described casts the thrombin receptor as a peptide receptor that contains its own ligand. This ligand remains cryptic until unmasked when thrombin cleaves the receptor. How might this "proteolytic switch" work? Recent structure-function studies of the receptor's agonist peptide domain suggest a possible answer. The first five amino acids of the receptor's agonist peptide domain (SFLLR in the single-letter code) are sufficient to specify agonist activity (31–33). The protonated amino group of Ser^1 and the Phe^2 side chain are vital for agonist function; Leu^4 and Arg^5 side chains play less important roles (32–34). The importance of the protonated amino group of Ser^1 is particularly appealing, because this group is created by receptor cleavage. This may explain in part how the agonist peptide domain's activity is masked when the receptor is in the uncleaved state (33). Steric and structural contributions to maintaining the agonist peptide silent in the uncleaved receptor remain to be defined.

E. Kinetics of Thrombin Receptor Cleavage and Relation to Signaling: How Does a Protease Elicit Concentration-Dependent Responses?

Like other important signaling molecules, thrombin effects concentration-dependent and graded responses in its target cells (3,35–39), a feature vital to normal homeostasis. Classic ligands effect concentration-dependent responses via graded receptor occupancy. How thrombin acting as a enzyme rather than a classic ligand might effect concentration-dependent responses has been a long-standing question (3,35–37). Specifically, one would predict that even low amounts of thrombin would eventually cleave and activate all cell surface receptors. How then can a concentration-dependent response be achieved? Recent studies utilized a thrombin receptor with an epitope-tagged activation peptide to distinguish immunologically the naive receptor from the cleaved and activated form (29). Examination of the kinetics of receptor cleavage on intact cells revealed that the rate of thrombin receptor cleavage was proportional to thrombin concentration over the physiological range, but low thrombin concentrations ultimately cleaved all cell surface thrombin receptors. Cumulative phosphoinositide hydrolysis in response to thrombin correlated precisely with absolute receptor cleavage, not with the integral of receptor cleavage as a function of time. These data strongly suggest that each activated thrombin receptor generates a quantum of second messenger, then "shuts off" (29). Thus, unlike the classic ligands, cells cannot utilize graded receptor occupancy to effect graded responses to thrombin. Rather, concentration-dependent graded responses to thrombin must be determined by the balance between rate of receptor activation and second-messenger clearance. This formulation suggests that thrombin receptor antagonists need only slow thrombin receptor activation enough that clearance of second messengers outstrips their generation to block signaling effectively (29). This notion may encourage attempts at antagonist development, which might otherwise be discouraged by the receptor's tethered liganding mechanism.

F. Transmembrane Signaling

Where the thrombin receptor's agonist peptide domain binds and how this binding event causes a transmembrane signal that allows the receptor to activate G proteins is unknown.

Ligand binding for seven transmembrane domain receptors is best studied for the β_2-adrenergic receptor (40). In this case, the catecholamine ligand interacts with residues predicted to reside within the transmembrane domains. The details of how this binding "switches on the receptor," and even whether peptide ligands bind in an analogous manner, is unknown.

G. Mechanisms of Receptor Shutoff: Did the Novel Activation Mechanism Beget a Novel Shutoff Mechanism?

The formulation of thrombin receptor signaling just outlined suggests that the thrombin receptor's shutoff mechanism is a critical determinant of the gain of the system and thereby of thrombin responsiveness (19,29). The mechanism of thrombin receptor shutoff has not been rigorously defined. By analogy with other seven transmembrane domain receptors (40), it is likely that receptor kinases play an important role in the immediate termination of thrombin receptor signaling. Indeed, recovery of receptor responsiveness to agonist peptide after desensitization was inhibited by phosphatase inhibitors (41), consistent with but not proving a role for phosphorylation in desensitization. Direct demonstration of receptor phosphorylation and mapping of phosphorylation sites remain to be accomplished.

Receptor internalization (sequestration) and degradation (downregulation) may also be involved in termination of thrombin receptor signaling. Recent work suggests that the thrombin receptor undergoes agonist-induced internalization (29); the relative contribution of internalization to termination of thrombin signaling is unknown.

A striking paradox has been noted in thrombin receptor signaling. Based on recent kinetic studies of signaling in receptor-transfected cell lines, it appears that the thrombin receptor stops signaling despite the continued presence of cleaved and "activated" receptors on the cell surface and at a time when cells are refractory to thrombin but sensitive to agonist peptide (29). These observations may be consistent with the earlier finding that responsiveness to agonist peptide recovers faster than responsiveness to thrombin in HEL cells (41). The finding of an agonist peptide-responsive/thrombin-unresponsive state, despite the continued presence of cleaved and activated thrombin receptor on the cell surface, is provocative. A trivial explanation of this observation is the existence of a receptor pool accessible to exogenous agonist peptide but not to thrombin. A more exciting possibility is that the receptor becomes modified such that the tethered agonist peptide domain cannot function but exogenous agonist peptide can. Thus an as yet uncharacterized and novel shutoff mechanism may have evolved to deal with the tethered ligand and with the obligate relationship of receptor activation to phosphoinositide hydrolysis.

IV. BIOLOGICAL ROLES OF THE CLONED THROMBIN RECEPTOR

Synthetic peptides mimicking the thrombin receptor's agonist peptide domain and the receptor cDNA itself have provided new tools for defining the role of the cloned receptor both in intracellular signaling events and in various thrombin-induced cellular functions. For the purpose of this chapter, discussion of the thrombin receptor's role in cellular events is largely confined to platelets.

A. Intracellular Signaling Events

1. Phosphoinositide Hydrolysis and Inhibition of Adenylyl Cyclase

Thrombin is known both to activate phosphoinositide hydrolysis and to inhibit adenylyl cyclase in platelets (42,43) and other responsive cells (39,44). In platelets, both these second-messenger events serve to promote aggregation (45). The thrombin receptor agonist peptide has been reported to elicit both these second-messenger events in fibroblasts (31,46), and transfection of the cloned thrombin receptor conferred both receptor-mediated phosphoinositide turnover and inhibition of adenylyl cyclase to Rat 1 cells (25), suggesting that the cloned receptor can mediate both signaling events. In the Rat 1 transfectants, receptor-mediated inhibition of cyclase was inhibited by pertussis toxin but phosphoinositide hydrolysis was not. These data suggest that the cloned receptor can interact with both pertussis-sensitive Gi-like G protein(s) to effect inhibition of cyclase and pertussis-insensitive Gq-like G protein(s) to mediate phosphoinositide hydrolysis (25). This scheme also accommodates the variable pertussis sensitivity of the phospholipase C-sensitive response across cell types (see later).

The relative importance of particular G proteins in thrombin signaling in different cell types may depend on their repertoire of both G proteins and effector molecules. For example, although inhibition of adenylyl cyclase is completely blocked by pertussis toxin in a variety of cell types, the reported sensitivity of thrombin-induced phosphoinositide hydrolysis has been variable. In CHO cells, thrombin-induced phosphoinositide hydrolysis is reported to be pertussis sensitive (47), in CCL-39 cells there appears to be both pertussis-sensitive and insensitive components (39,46), and in Rat 1 cells transfected with the cloned receptor, phosphoinositide hydrolysis was largely pertussis insensitive (25). Recent studies reveal that phosphoinositide hydrolysis may be mediated by distinct phospholipase Cs activated by $\beta\gamma$ subunits released when receptors activate G_i ($\alpha_i\beta\gamma$) (48,49) or by members of the recently described G_q α subunit pathway (50). Thus the pertussis-sensitive component of thrombin-induced phosphoinositide hydrolysis may be mediated via $\beta\gamma$ subunits released upon G_i activation, and the pertussis-insensitive component may be mediated by G_q. The relative importance of these pathways may depend on the abundance of specific G protein subunits and phospholipase subtypes expressed in a particular cell.

2. Other Thrombin-Induced Signaling Events

In addition to activation of phosphoinositide hydrolysis and inhibition of adenylyl cyclase, thrombin elicits a variety of other signaling events. At least some of these appear to be mediated by the cloned receptor. For example, the thrombin receptor agonist peptide has been reported to cause increases in cytoplasmic calcium in both receptor-transfected and naturally responsive cells (19,29,31,41,51) and to activate prostacyclin production in endothelial cells (51).

B. Cellular Events: Platelet Activation

The cloned thrombin receptor clearly plays an important role in mediating platelet activation by thrombin. Thrombin receptor agonist peptide causes platelet secretion and aggregation (19,32–34), and the potency of mutant agonist peptides for platelet activation parallels that for activation of the cloned receptor (33). The agonist peptide also causes platelet phosphoinositide hydrolysis (52). Moreover, peptide antibodies to the cloned receptor block platelet activation by thrombin (53,54). These data strongly suggest that the cloned receptor is both sufficient and necessary for platelet activation by thrombin. Inhibition of platelet function by receptor antisera was overcome by high concentrations of thrombin; thus available data do not rigorously exclude the existence of a second platelet thrombin receptor. It should be noted, however, that the inhibitory activity of the receptor antisera was also overcome by high thrombin concentrations even in a defined system in which the cloned receptor was expressed in *Xenopus* oocytes (53).

The relationship of the platelet thrombin receptor to platelet glycoprotein (GP) Ib, a distinct platelet surface thrombin binding site (Chap. 24), in thrombin-induced platelet activation remains to be defined. Clearly, thrombin signaling occurs in cells that do not express GPIb, and expression of the cloned receptor alone is sufficient to confer signaling to low concentrations of thrombin to cells as diverse as *Xenopus* oocytes and fibroblast-like cells (19,46). It has been proposed that thrombin binding to platelet GPIb serves to increase the local thrombin concentration, thereby promoting thrombin interaction with its receptor (55), and several forms of indirect evidence support such a role. However, direct demonstration of an adjunctive role for GPIb in thrombin receptor signaling by coexpression of these molecules in a null host has not yet been accomplished.

The in vivo role of platelet thrombin receptor activation in hemostasis and thrombosis remains to be defined. The receptor became accessible as a target for reagent and drug therapy only 2 years ago. One approach to defining the thrombin receptor's importance in signaling is the use of blocking antibodies (53,54), but antibodies sufficiently avid for in vivo use are not yet available. Similarly, modified analogs of the thrombin receptor's agonist peptide may function as receptor antagonists but are in their infancy.

C. Role in Disease

The role of thrombin receptor activation in human hemostasis and thrombosis remains to be rigorously demonstrated, but the receptor's clear role in mediating thrombin-induced platelet activation, the importance of thrombin in platelet-dependent models of arterial thrombosis (see earlier), and the efficacy of antithrombin therapy for unstable angina (56) all suggest that thrombin receptor activation will play an important role. Pharmaceuticals that block thrombin receptor function are needed. Meanwhile, several laboratories are screening platelets from patients with bleeding diatheses for selective refractoriness to thrombin or agonist peptide. To the author's knowledge, no candidate syndrome for a thrombin receptor genetic disease has been described.

Given thrombin's known actions on inflammatory and mesenchymal cells and its generation at sites of vascular injury, it is tempting to postulate a role for thrombin receptor activation in inflammatory and proliferative responses. Recently, robust thrombin receptor expression was noted in human atherosclerotic plaques, apparently by smooth muscle cells, mesenchymal-appearing cells of unknown origin, and macrophages (57). Little receptor was seen in the normal artery wall, except apparently, low-level endothelial expression (57). The finding of cells selectively expressing the thrombin receptor in atherosclerotic lesions compared with normal artery wall suggests a possible role for thrombin receptor activation in the genesis of such lesions. The role of thrombin itself in animal models of restenosis is currently being examined.

ACKNOWLEDGMENTS

This work was supported in part by Grants HL44907 and HL43322 from the National Institutes of Health and by the University of California's Tobacco Related Disease Research Program Grant 2RT19. Dr. Coughlin is an Established Investigator of the American Heart Association.

REFERENCES

1. Shuman MA. Thrombin-cellular interactions. In: Walz DA, Fenton JW II, and Shuman MA, eds. Bioregulatory functions of thrombin. Ann NY Acad Sci 1986; 485:228–239.
2. Davey MG, Luscher EF. Actions of thrombin and other coagulant and proteolytic enzymes on blood platelets. Nature 1967; 216:857–858.
3. Berndt MC, Phillips DR. Platelet membrane proteins: composition and receptor function. In: Gordon JL, ed. Platelets in Biology and Pathology. Amsterdam: Elsevier/North Holland Biomedical Press, 1981:43–74.
4. Eidt JF, Allison P, Nobel S, et al. Thrombin is an important mediator of platelet aggregation in stenosed canine coronary arteries with endothelial injury. J Clin Invest 1989; 84:18–27.
5. Hansen SR, Harker LA. Interruption of acute platelet-dependent thrombosis by the synthetic antithrombin PPACK. Proc Natl Acad Sci USA 1988; 85:3184–3188.
6. Fitzgerald DJ, Fitzgerald GA. Role of thrombin and thromboxane A_2 in reocclusion following coronary thrombolysis. Proc Natl Acad Sci USA 1989; 86:7585–7589.
7. Jang I-K, Gold HK, Ziskind AA, Leinbach RC, Fallon JT, Collen D. Prevention of platelet-rich arterial thrombosis by selective thrombin inhibition. Circulation 1989; 81:219–225.
8. Heras M, Chesebro JH, Penny WJ, Bailey KR, Badimon L, Fuster V. Effects of thrombin inhibition on the development of acute platelet-thrombus deposition during angioplasty in pigs. Circulation 1989; 79:657–665.
9. Bar-Shavit R, Kahn A, Wilner GD, Fenton JW II. Monocyte chemotaxis: stimulation by specific exosite region in thrombin. Science 1983; 220:728–731.
10. Chen LB, Teng NNH, Buchanan JM. Mitogenicity of thrombin and surface alterations on mouse splenocytes. Exp Cell Res 1976; 101:41–46.
11. Chen LB, Buchanan JM. Mitogenic activity of blood components. I. Thrombin and prothrombin. Proc Natl Acad Sci USA 1975; 72:131–135.
12. McNamara CA, Sarembok, IJ, Gimple LW, Fenton JW II, Coughlin SR, Owens GK. Thrombin stimulation of smooth muscle cell proliferation is mediated by a proteolytic, receptor-mediated mechanism. J Clin Invest 1992; 91:94–98.
13. Weksler BB, Ley CW, Jaffe EA. Stimulation of endothelial cell prostacyclin production by thrombin, trypsin, and the ionophore A23817. J Clin Invest 1978; 62:923–930.
14. Prescott SM, Zimmerman GA, McIntyre TM. Human endothelial cells in culture produce platelet-activating factor when stimulated by thrombin. Proc Natl Acad Sci USA 1984; 81: 3534–3538.
15. Gelehrter TD, Sznycer-Laszyk R. Thrombin induction of plasminogen activator-inhibitor in cultured human endothelial cells. J Clin Invest 1986; 77:165–169.
16. Daniel TO, Gibbs VC, Milfay DF, Garavoy MR, Williams LT. Thrombin stimulates c-*sis* gene expression in microvascular endothelial cells. J Biol Chem 1986; 261:9579–9582.
17. Zimmerman GA, McIntyre TM, Prescott SM. Thrombin stimulates neutrophil adherence by an endothelial cell-dependent mechanism. Ann NY Acad Sci 1986; 485:349–368.
18. Hattori R, Hamilton KK, Fugate RD, McEver RP, Sims PJ. Stimulated secretion of endothelial vWF is accompanied by rapid redistribution to the cell surface of the intracellular granule membrane protein GMP-140. J Biol Chem 1989; 264:7768–7771.
19. Vu T-KH, Hung DT, Wheaton VI, Coughlin SR. Molecular cloning of a functional thrombin receptor reveals a novel proteolytic mechanism of receptor activation. Cell 1991; 64:1057–1068.
20. Coughlin SR, Vu T-KH, Hung DT, Wheaton VI. Perspectives. Characterization of the cloned platelet thrombin receptor: issues and opportunities. J Clin Invest 1992; 89:351–355.

21. Wilson T, Treisman R. Removal of poly(A) and consequent degradation of c-fos mRNA facilitated by 3' AU-rich sequences. Nature 1988; 336(6197):393–399.
22. O'Dowd BF, Hantowich M, Caron MG, Lefkowitz RJ, Bouvier B. Palmitoylation of the human β_2-adrenergic receptor. J Biol Chem 1989; 264:7564–7569.
23. Karnik SS, Ridge KD, Bhattacharya S, Khorana HG. Palmitoylation of bovine opsin and its cysteine mutants in COS cells. Proc Natl Acad Sci USA 1993; 90:40–44.
24. Rydel TJ, Rabichandran KG, Tulinsky A, et al. The structure of a complex of recombinant hirudin and human alpha-thrombin. Science 1990; 249:277–280.
25. Hung DT, Wong YH, Vu T-KH, Coughlin SR. The cloned platelet thrombin receptor couples to at least two distinct effectors to stimulate both phosphoinositide hydrolysis and inhibit adenylyl cyclase. J Biol Chem 1992; 267:20831–20834.
26. Vu T-KH, Wheaton VI, Hung DT, Coughlin SR. Domains specifying thrombin-receptor interaction. Nature 1991; 353:674–677.
27. Liu L, Vu T-KH, Esmon CT, Coughlin SR. The region of the thrombin receptor resembling hirudin binds to thrombin and alters enzyme specificity. J Biol Chem 1991; 266:16977–16980.
28. Hung DT, Vu T-KH, Wheaton VI, et al. ''Mirror image'' antagonists of thrombin-induced platelet activation based on thrombin receptor structure. J Clin Invest 1992; 89:444–450.
29. Ishii K, Hein L, Kobilka B, Coughlin SR. Kinetics of thrombin receptor cleavage on intact cells: relation to signaling. J Biol Chem 1993; 268:9780–9786.
30. Bode W, Schwager P, Huber R. The transition of bovine trypsinogen to a trypsin-like state upon strong ligand binding. J Mol Biol 1978; 118:99–112.
31. Vouret-Craviari V, Van Obberghen-Schilling E, Rasmussen UB, Pavirani A, Lecocq J-P, Pouyssegur J. Synthetic α-thrombin receptor peptides activate G-protein coupled signalling pathways but are unable to induce mitogenesis. Mol Biol Cell 1992; 3:95–102.
32. Vassallo RR Jr, Kieber-Emmons T, Cichowski K, Brass LF. Structure-function relationships in the activation of platelet thrombin receptors by receptor-derived peptides. J Biol Chem 1992; 267:6081–6085.
33. Scarborough RM, Naughton M, Teng W, et al. Tethered ligand agonist peptides: structural requirements for thrombin receptor activation reveal mechanism of proteolytic unmasking of agonist function. J Biol Chem 1992; 267:13146–13149.
34. Coller BS, Ward P, Ceruso M, et al. Thrombin receptor activating peptides: importance of the N-terminal serine and its ionization state as judged by pH dependence, NMR spectroscopy, and cleavage by aminopeptidase M. Biochemistry 1992; 31:11713–11720.
35. Detwiler TC, Feinman RD. Kinetics of the thrombin-induced release of calcium (II) by platelets. Biochemistry 1973; 12:282–289 36. Martin BM, Feinman RD, Detwiler TC. Platelet stimulation by thrombin and other proteases. Biochemistry 1975; 14:1308–1314.
37. Martin BM, Wasiewski WW, Fenton JW II, Detwiler TC. Equilibrium binding of thrombin to platelets. Biochemistry 1976; 15:4886–4893.
38. Rittenhouse-Simmons S. Production of diglyceride from phosphatidylinositol in activated human platelets. J Clin Invest 1979; 63:580–587.
39. Paris S, Pouyssegur J. Pertussis toxin inhibits thrombin-induced activation of phosphoinositide hydrolysis and Na^+/H^+ exchange in hamster fibroblasts. EMBO J 1986; 5:55–60.
40. Dohlman HG, Thorner J, Caron MG, Lefkowitz RJ. Model systems for the study of seven transmembrane segment receptors. Annu Rev Biochem 1992; 60:653–688.
41. Brass LF. Homologous desensitization of HEL cell thrombin receptors. J Biol Chem 1992; 267:6044–6050.
42. Banga HS, Walker RK, Winberry LK, Rittenhouse SE. Platelet adenylate cyclase and phospholipase C are affected differentially by ADP ribosylation. Biochem J 1988; 252:297–300.
43. Brass LF, Lasposata M, Banga HS, Rittenhouse SE. Regulation of the phosphoinositide hydrolysis pathway in thrombin-stimulated platelets by a pertussis toxin-sensitive guanine nucleotide-binding protein. J Biol Chem 1986; 261:16838–16847.

44. Jones LG, McDonough PM, Brown JH. Thrombin and trypsin act at the same site to stimulate phosphoinositide hydrolysis and calcium mobilization. Mol Pharmacol 1989; 36:142–149.
45. Kroll MH, Schafer AI. Biochemical mechanisms of platelet activation. Blood 1989; 74:1181–1195.
46. Hung DT, Vu T-KH, Nelken NA, Coughlin SR. Thrombin-induced events in non-platelet cells are mediated by the novel proteolytic mechanism defined by the cloned platelet thrombin receptor. J Cell Biol 1992; 116:827–832.
47. Ashkenazi A, Peralta EG, Winslow JW, Ramachandran J, Capon DJ. Functionally distinct G-proteins selectively couple different receptors to PI hydrolysis in the same cell. Cell 1989; 56:487–493.
48. Camps M, Carozzi A, Schnabel P, Scheer A, Parker PJ, Gierschik P. Isozyme-selective stimulation of phospholipase C-β_2 by G protein $\beta\gamma$-subunits. Nature 1992; 360:684–686.
49. Katz A, Wu D, Simon MI. Subunits $\beta\gamma$ of heterotrimeric G protein activate β_2 isoform of phospholipase C. Nature 1993; 360:686–689.
50. Lee CH, Park D, Wu D, Rhee SG, Simon MI. Members of the Gq alpha subunit family gene family activate phospholipase C beta isozymes. J Biol Chem 1992; 267:16044–16047.
51. Ngaiza JR, Jaffe EA. A 14 amino acid peptide derived from the amino terminus of the cleaved thrombin receptor elevates intracellular calcium and stimulates prostacyclin production in human endothelial cells. Biochem Biophys Res Commun 1991; 179:1656–1661.
52. Huang R, Sorisky A, Church WR, Simons E, Rittenhouse SE. Thrombin receptor-directed ligand accounts for activation by thrombin of platelet phospholipase C and accumulation of 3-phosphorylated phosphoinositides. J Biol Chem 1991; 266:18435–18438.
53. Hung DT, Vu T-KH, Wheaton VI, Ishii K, Coughlin SR. The cloned platelet thrombin receptor is necessary for thrombin-induced platelet activation. Blocking antiserum to the thrombin receptor's hirudin-like domain. J Clin Invest 1992; 89:1350–1353.
54. Brass LF, Vassallo RR, Belmonte E, Ahuja M, Cichowski K, Hoxie JA. Structure and function of the human platelet thrombin receptor. Studies using monoclonal antibodies directed against a defined domain within the receptor N terminus. J Biol Chem 1992; 267:13795–13798.
55. Okamura T, Hasitz M, Jamieson GA. Platelet glycocalicin: interaction with thrombin and role as thrombin receptor on the platelet surface. J Biol Chem 1978; 253:3435–3443.
56. Theroux P, Ouimet H, McCums J, et al. Aspirin, heparin, or both to treat acute unstable angina. N Engl J Med 1988; 319:1105–1111.
57. Nelken NA, Soifer SJ, O'Keefe J, Vu T-KH, Charo IF, Coughlin SR. Thrombin receptor expression in normal and atherosclerotic human arteries. J Clin Invest 1992; 90:1614–1621.
58. Mathews II, Padmanabhan KP, Ganesh V, Tulinsky A., Ishii M, Chen J, Turck CW, Coughlin SR, Fenton JW II. Crystallographic structures of thrombin complexed with thrombin receptor peptides: existence of expected and novel binding modes. Biochemistry 1994; 33(11):3266–3279.

32

The Molecular and Cellular Biology of Thrombopoietin, the MPL Ligand

Kenneth Kaushansky
University of Washington, Seattle, Washington

Si Lok
ZymoGenetics, Inc., Seattle, Washington

I. INTRODUCTION

The process of platelet formation, or thrombopoiesis, is an enormous endeavor. Each day, the adult human produces approximately 1×10^{11} platelets, essential for the maintenance of normal hemostasis. Furthermore, this process can increase approximately 10-fold under conditions of increased demand. Hematopoiesis, the process of blood cell development, requires the interplay of stem and progenitor cells, marrow stromal elements, and polypeptide growth factors (1–5). Previous studies have suggested that a hierarchical system accounts for hematopoiesis in which the survival and proliferation of progenitor cells is supported by multiple *early*-acting growth factors [interleukin (IL)-3, c-*kit* ligand (KL), flt-3/flk-2 ligand (FL), granulocyte (G)-colony-stimulating factor (CSF) or IL-11] and their differentiation to mature blood cells supported by a single, lineage-specific *late*-acting cytokine [such as G-CSF for neutrophils, IL-5 for eosinophils, monocyte (M)-CSF for monocytes or erythropoietin (EPO) for red cells (6–10)]. Although many proteins have been shown to support the expansion of megakarocytic precursor cells [including IL-3 (11,12), GM-CSF (13,14), and KL (11,15,16)], identification of the late-acting, lineage-specific growth factor for platelet production, termed thrombopoietin, has remained elusive (reviewed in 17,18).

II. THROMBOPOIETIN, AN OPERATIONAL DEFINITION

By analogy to the other late-acting, lineage-specific hematopoietic growth factors, most investigators agree that the properties of a thrombopoietin would include lineage-specificity, direct action on maturing megakaryocytic precursors, induction of programs of terminal differentiation, such as the expression of platelet-specific surface membrane glycoproteins, and, most important, stimulation of platelet production in vivo and an inverse relationship with platelet levels. Pioneering studies of Levin, McDonald, Rosenberg, and others began to characterize substances from the plasma of thrombocytopenic animals that display several of these activities (17—22). However, progress was impeded by the very low levels of thrombopoietin in plasma, the complex nature of plasma or serum as a starting material for biochemical purification, and a very cumbersome in vivo assay of thrombopoietic activity. The failure to clone thrombopoietin as techniques in molecular biology and biochemistry advanced led some to doubt its existence as a distinct entity.

In recent years, three well-characterized cytokines have been proposed to function as thrombopoietins. IL-6, IL-11, and leukemia inhibitory factor (LIF) display a myriad of biological activities, including effects on hematopoiesis. For example, IL-6 acts synergistically with IL-3 and KL to augment the expansion of very primitive hematopoietic progenitor cells in blast-cell colony assays (2,23), and although not able to stimulate megakaryocyte progenitor cells alone, IL-6 acts together with IL-3 to promote CFU-Mk-derived colony formation (24–26). IL-11 can synergize with IL-3 to enhance the production of megakaryocytic colonies from marrow progenitors and shorten the G_0 period of very early hematopoietic progenitors (27,28). And like IL-6 and IL-11, LIF works with IL-3 to promote CFU-Mk expansion (29). Of note, all three of these cytokines are pleiotropic in effect, at odds with the commonly held assumption that thrombopoietin should be absolutely or, at the least, relatively tissue-specific. Despite concerns over their physiological relevance, the hematopoietic effects of these agents lead to preclinical and clinical trials for thrombocytopenia. Unfortunately, results have been disappointing. For example, in normal mice, even toxic doses of IL-6, IL-11, or LIF increase platelet counts by only 70% (29–35). Responses in humans have also yielded mixed results (36–38). Recent editorials argue persuasively that none of these agents are thrombopoietin (39–41).

III. MPL AND HEMATOPOIESIS

Although science usually proceeds logically (and somewhat predictably) from one conclusion to the next, occasionally a finding from one arena provides the critical insight into an apparently unrelated field. In 1986, Wendling and co-workers described a retroviral complex that induced a myeloproliferative disorder in recipient animals (42). The transducing virus was termed myeloproliferative leukemia virus (MPLV), and the transforming oncogene, v-*mpl* (43), and its cellular homolog, c-*mpl* (44), were identified and cloned. Sequence analysis of the two genes quickly revealed an important homology with a growing family of polypeptides, the hematopoietic cytokine receptors.

The hematopoietic cytokine receptor family is a rapidly expanding group of type I transmembrane proteins (at present, at least 20 distinct polypeptides have been identified), and is characterized by a 200-amino-acid extracellular motif consisting of four characteristically spaced cysteine residues, 14 beta strands, and a distinct juxtamembrane WSXWS sequence (45–47). The cytoplasmic domain of the polypeptides can be grouped according to their length and function, either long (180–400 residues), containing a ''Box 1/2'' motif (47), and capable of transmitting a proliferative signal, or short (40–100 residues) and requiring a second component for signal transduction. The c-Mpl receptor shares all of these structural features, but contains a cytoplasmic tail intermediate in length (122 residues). Thus, as its ligand was unknown, the c-*mpl* proto-oncogene appeared to encode an orphan cytokine receptor.

Shortly after its description, several lines of evidence were presented that strongly suggested the involvement of the c-Mpl receptor in megakaryocytic development. First, c-*mpl* expression was almost exclusively limited to normal cells that participate in the megakaryocytic lineage (e.g., $CD34^+$ cells, megakaryocytes, and platelets) or cell lines that display one or more features of megakaryocytic commitment (e.g., CMK, HEL, KU 812) (48). Second, when fused to the extracellular domain of the IL-4 receptor, the cytoplasmic domain of c-Mpl could transmit an IL-4-dependent proliferative signal to BaF3 cells, and IL-3-dependent cell line that can be supported by other cytokines only if first engineered to express the corresponding receptor (49). Similar results were reported for a GM-CSF/c-Mpl fusion receptor (50). And finally, down-modulation of c-*mpl* expression in $CD34^+$ human marrow cells led to the selective loss of megakaryocytic colony formation (48).

IV. CLONING OF THE MPL LIGAND (ML)

With the observations implicating that the ligand for c-Mpl is potentially an important regulator of megakaryocytopoiesis, and a likely candidate for either thrombopoietin or Meg-CSF, several groups raced to clone the receptor ligand. C-Mpl played an integral role in the cloning by effecting the purification of the ligand from biological fluids by affinity chromatography employing column immobilized receptor. In addition, the capacity of the receptor to transduce a proliferative signal when expressed in factor-dependent cell lines such as BaF3 (49,50) allowed the development of a sensitive bioassay to monitor the purification progress. This strategy was employed by scientists at Genentech (51) and Amgen (52) for the purification of c-Mpl ligand (ML) from aplastic plasma of irradiated pigs and dogs, respectively. From minute quantities of protein eluted from the receptor affinity column, both groups were able to obtain sufficient N-terminal amino acid sequence to enable the isolation of cDNAs encoding ML from a cDNA library.

The group from ZymoGenetics (Novo Nordisk A/S) adopted a very different cloning strategy (53). They extended the observation that transformation from factor-dependent to factor-independent growth in many hematopoietic cell lines is due to autocrine production of a growth factor to which the cells have a receptor (54,55). To exploit this strategy, a c-*mpl* cDNA was transfected into a murine IL-3-dependent pre-B-cell line to construct a cell line producing the receptor, thus rendering the cell line ML as well as IL-3 growth dependent. The cells were then treated with the mutagen EMS (55), causing single nucleotide base pair changes randomly throughout the genome, some of which activate expression of nearby genes. To select for a cell capable of autonomous growth due to the chance activation of the endogenous *ML* gene, the EMS-treated cells were grown in the absence of IL-3. Of the 18 autonomous clones obtained, one was found to produce and secrete ML. A cDNA expression library was prepared from the mutant-expressing cells. Library pools were transfected into baby hamster kidney cells and culture supernatant assayed for the capacity to support proliferation of the c-Mpl-bearing BaF3 cells (53). Two positive pools were identified and a cDNA encoding ML was isolated by conventional pool breakdown techniques.

Following the publication from ZymoGenetics (53), Genentech (51), and Amgen (52) on the successful cloning of ML encoding cDNA, Kirin Brewery Company (56) and a group led by Robert Rosenberg (57) announced that they also have isolated a ligand for c-Mpl. Unlike the other investigators, the Kirin and Rosenberg groups purified the ligand from aplastic plasma employing chromatographic steps without the use of c-Mpl. The

Kirin group was able to purify a sufficient quantity of protein from plasma of irradiated rats to obtain sequences to enable the isolation of a cDNA. Rosenberg's laboratory was able to show that partial sequences from their isolated protein were a match to the published sequence of c-Mpl ligand (51–53).

V. IN VITRO ACTIVITIES OF c-MPL LIGAND

As noted in the Introduction, thrombopoiesis is thought to be organized in a hierarchical manner, in which early-acting factors support the proliferation of megakaryocytic progenitors, and late-acting factors act to induce the full maturation of these cells to platelet-producing megakaryocytes. Thus, investigators have tested ML for activity in two types of assay. The first, designed to determine its capacity to stimulate progenitor cell proliferation, is termed a colony-forming assay. By immobilizing marrow-derived hematopoietic progenitor cells in semisolid medium, such assays determine the capacity of cytokine to induce clonal proliferation of individual progenitor cells, retrospectively scored by the presence of colonies of megakaryocytes. Such proteins are termed megakaryocyte colony-stimulating factors (Meg-CSFs). In addition, a number of experimental systems have been devised to assess megakaryocyte development. Most are initiated in suspension culture, using readouts of megakaryocyte size, polyploidy, or the expression of platelet-specific membrane glycoproteins.

Previous studies have shown that megakaryocyte colony formation is best supported by the plasma of aplastic animals. Using this observation to advantage, Wendling and colleagues have shown that irradiated pig serum could support the proliferation and differentiation of human megakaryocyte progenitors, and that this activity was eliminated by adsorbing the serum with a soluble c-Mpl receptor (58). These results suggest that the megakaryocyte colony-stimulating activity of serum might be identical with ML. Using the recombinant murine protein, we arrived at an identical conclusion (59). As it supports the formation of small megakaryocytic colonies in semisolid culture, like IL-3, GM-CSF, and KL, ML is a Meg-CSF (Fig. 1). However, at low concentrations of murine ML, the addition of IL-3 led to increased numbers of colonies and increased colony size (59). Furthermore, in cultures initiated with IL-3, delayed addition of ML was not detrimental to colony numbers. Taken together with the size of ML-induced megakaryocyte colonies, these data suggest that ML works relatively late in megakaryocytic progenitor cell development.

Many groups have now begun to determine the effects of recombinant ML on megakaryopoiesis in suspension culture. We have found that culture of murine marrow cells with ML increases the size of megakaryocytes 50% over that seen in the presence of IL-3, KL, IL-6, or IL-11, alone or in combination (59 and personal observations). Using megakaryocyte-specific monoclonal antibodies, fluorescence-activated cell sorting, and propidium iodide as a measure of cellular DNA content, two groups have now shown that ML shifts megakaryocyte ploidy to higher values (58,59). And using GP Ib or GP IIb/IIIa-specific monoclonal antibodies, several groups have shown that ML induces expression of these platelet-specific cell surface markers (51,52,58,59). However, at present, it is uncertain whether these effects are direct; it is possible that ML induces a general program of terminal differentiation and that stimulation of the promoters of GP Ib/V/IX and of GP IIb/IIIa is the province of other downstream factors. Direct studies of platelet-specific glycoprotein transcription are underway.

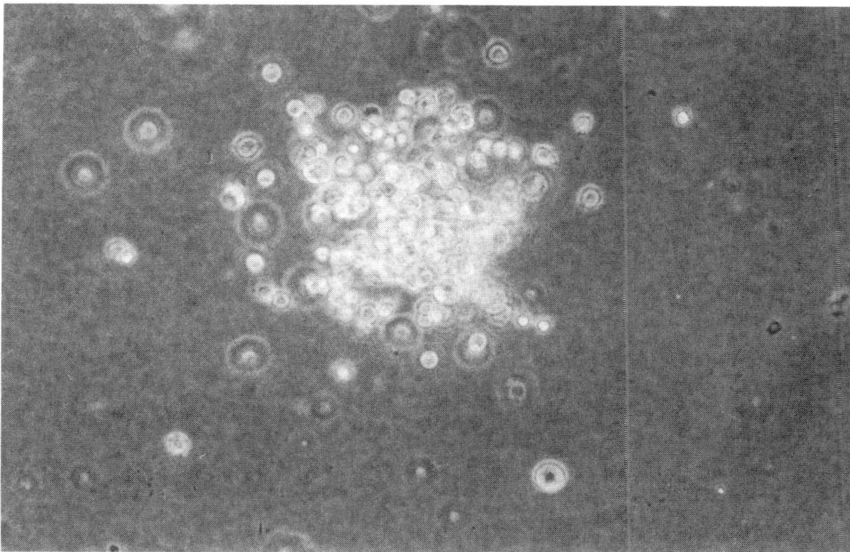

Figure 1 Large megakaryocyte colony, derived from normal murine marrow grown in the presence of recombinant murine Tpo. The colony contains approximately 200 megakaryocytes.

VI. IN VIVO ACTIVITIES OF ML

When administered to normal mice, reombinant ML induces a four- to fivefold increase in platelet level within 5 days (53). The level begins to climb on day 3, peaks at day 5, and remains elevated as long as daily administration continues. Platelet counts begin to fall 5–7 days following discontinuation of ML administration (personnel observations), in keeping with the 5-day life span of the murine platelet. This rise is due to a five- to 20-fold increase in marrow and splenic megakaryocytic progenitors, a 10-fold increase in marrow and splenic megakaryocytes, and a 70% increase in megakaryocyte size. Other groups have administered human ML to mice (51,56). Both have reported a modest (20–60%) increase in platelet formation. Although the human homolog has been shown to engage the murine c-Mpl receptor and to stimulate proliferation in various assays (51,52), the relative specific activities of human and murine ML are presently unknown. Moreover, Wendling and colleagues were unable to detect ML in human aplastic serum using a BaF3/murine c-Mpl proliferation assay (58). Therefore, it is likely that the discrepancy between our results and those of other workers is due to differences in the relative specific activity of the murine and human proteins on murine receptor–bearing cells. However, together with the in vitro results, the effects of ML on murine thrombopoiesis argue persuasively that ML is thrombopoietin (TPO), the critical regulator of platelet production.

VII. SITE OF EXPRESSION OF c-MPL LIGAND

Murine and human c-Mpl ligand are each encoded by a low-abundance, 1.8-kb mRNA species (51–53). The predominant organ where transcripts can be detected by Northern blot analysis is liver, with lesser amounts in the kidney. In mouse, detectable amounts of a 1.8-kb species can also be found in skeletal muscle, heart, brain, and spleen mRNAs, along with a 5-kb species that might correspond to an unprocessed form of the transcript (53). *EPO* mRNA synthesis is limited to specialized cells, the peritubular cells, of the kidney (60). In contrast, the apparent wide tissue distribution of murine c-*mpl* ligand mRNA may be a consequence of expression in a few specialized cell types present in the many tissues examined. Initial RT-PCR examinations of multiple cell lines and primary cells in culture support this support (personal observations).

VIII. HUMAN THROMBOPOIETIN GENE STRUCTURE AND CHROMOSOMAL LOCALIZATION

Southern blot analysis of genomic DNA with a *ML* cDNA probe produced a very simple hybridization pattern consistent with the presence of a single *TPO* locus in humans (61). The human *TPO* gene spans over 6 kb and has a structure similar to that of the *EPO* gene (61–65). All of the introns in the human and murine *TPO* and *EPO* genes occur in precisely the same relative position to the protein structure (Fig. 2). The similarity in exon structure argues that the two genes have evolved from a common ancestral sequence by gene duplication (61). The junction between the two TPO domains does not correspond to an intron position in the TPO gene. Interestingly, except for the first 130 nucleotides, the sequence of the TPO carboxy-terminal domain bears a weak similarity to the 3′ noncoding region of the *EPO* gene. This observation supports the theory that a portion of the TPO carboxy-terminal domain may have evolved from an ancestral sequence that may also have given rise to the present 3′ noncoding region of the *EPO* gene. During or following the gene duplication event, either a 130-bp sequence was inserted into the *TPO* locus creating the carboxy-terminal domain or a similar sequence may have been deleted in the *EPO* gene leaving 3′ noncoding sequences. Another feature of the Tpo locus is that it has a 5′ noncoding exon, which is not present in the *EPO* locus (61).

Fluorescence in situ hybridization mapped the *TPO* locus to chromosomal position 3q26 (61). By contrast, human *EPO* resides on chromosome 7 (66). Previously, a site encoding a regulator of megakaryopoiesis was proposed to reside on the long arm of chromosome 3. Chromosome 3q abnormalities have been reported in cases of abnormal thrombopoiesis associated with acute myeloid leukemia (AML) (67–69) and in blastic

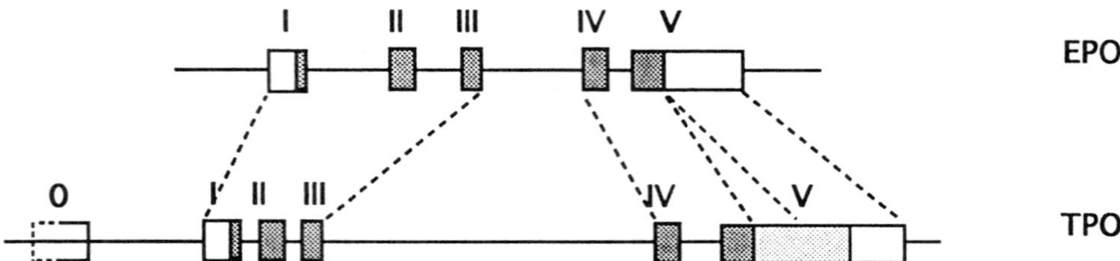

Figure 2 Human *EPO* and *TPO* gene structures. Exon sequences are indicated by boxed regions. The exon encoding the initiation codon is numbered exon I. Exon regions encoding EPO and the "EPO-like" domain of TPO are indicated with dark shading. Exon regions encoding the carboxyl-terminal glycosylation domain of TPO are indicated with light shading. Exon regions of noncoding sequences are unshaded. The proposed 5′ noncoding exon is indicated as a dashed box.

transformation of chronic myeloid leukemia (CML) (70,71). AML patients with the so-called 3q21q26 syndrome, the majority of whom have a chromosomal inversion of (3) (q21q26), have marked thrombocytosis. In one case examined, elevated serum TPO activity levels were detected in the patient's serum (67). With the characterization of the *TPO* locus, the molecular cytogenetics of these patients are open to investigation to determine whether disruption of the locus has a contributory effect to the disease.

Abnormalities at or near the 3q26 region have also been implicated in a number of other leukemias (72,73). Several potential oncogenes including *EVI 1* (74), a zinc finger protein whose perturbed expression may have contributed to oncogenesis, also maps to this region. Most recently, the role of the 3q26 region in AML and in CML blast crisis was further indicated by translocation of chromosome 21q22 sequences to the 3q26 region (75–77). The resulting translocation, t(3;21)(q26;q22), resulted in the production of fusion mRNAs (presumably also fusion proteins) that are initiated from the *AML1* gene located at 21q22 to several genetic loci in the 3q26 region. It should prove interesting to determine whether the *TPO* locus are also disrupted by similar translocation events.

IX. REGULATION OF THROMBOPOIETIN PRODUCTION

Initial studies suggested that the level of plasma thrombopoietin was inversely related to platelet demand (17,18). Using a proliferation assay based on BaF3 cells expressing the c-Mpl receptor, Wendling and colleagues have shown that the level of ML rises in animals made thrombocytopenic, either by irradiation or by use of an antiplatelet antiserum (58). Moreover, nearly all of the BaF3/mpl stimulatory activity could be removed by adsorbing either of the sera with a soluble c-Mpl/immunoglobulin fusion protein. Although immunologically based assays are not yet available, it appears likely that the level of ML will be inversely related to platelet demand. This information serves to reinforce the concept that ML is identical with TPO.

Despite clear evidence for an inverse relationship of thrombopoietin and platelet demand, the mechanisms underlying its regulation are uncertain. Specifically, it is unclear whether the rate of gene expression is altered in response to physiological stimuli or simply the protein level. Two theories have been proposed. One school suggests that the rate of *TPO* gene expression is constant; plasma levels of the protein are dependent entirely on the rate of receptor-mediated uptake and destruction by the end-product of its action, the platelets themselves (57). In this way, when platelet levels are high, TPO is rapidly destroyed and levels fall, thereby limiting megakaryocyte production. A similar mechanism has been invoked to explain the regulation of CSF-1 (M-CSF) by its end-product, the monocyte (78). In contrast, others believe that gene expression is a regulated event; feedback is provided by the platelet or the megakaryocyte and transcription is varied accordingly. The two theories are not mutually exclusive. This early conundrum in megakaryocyte regulation is likely to be solved shortly.

X. PROTEOLYTIC PROCESSING OF THROMBOPOIETIN

Several lines of evidence suggest that thrombopoietin may undergo proteolytic processing. Although the majority of the recombinant murine protein purified from serum free tissue culture medium runs with an apparent M_r of 70 kD, the material purified from irradiated rat, pig, sheep, or canine plasma was reported to be 18–32 kD (51,52,56,57). Of interest, immediately following the ''EPO-like'' domain of TPO, an Arg-Arg dipeptide is present, a sequence commonly found within the recognition sequence of a number of processing proteases. Moreover, recombinant forms of TPO, artificially truncated 20 residues downstream of this position (52), or following the first of the two arginine residues (51), are biologically active. Recombinant and natural EPO undergo proteolytic processing near their carboxyl termini (79). Although only a single residue is removed, this finding indicates that a processing signal might be shared between the two hormones. Therefore, it is possible that proteolytic processing plays an important role in the normal physiology of thrombopoietin.

XI. SUMMARY AND FUTURE DIRECTIONS

The regulation of platelet production is a complex process dependent on the interaction of hematopoietic progenitors, marrow stroma, and circulating growth factors. The critical regulator of platelet production, thrombopoietin, has been cloned, recombinant protein expressed, and many of its biological activities established. In addition to acting to induce megakaryocyte differentiation, thrombopoietin also supports the proliferation of megakaryocytic progenitor cells. In animals, the recombinant protein rapidly increases platelet production by affecting megakaryocyte number and size and by expanding the progenitor cell pool. It thus shows great promise to alleviate the thrombocytopenic complications of cytotoxic chemotherapy and in states of natural marrow failure.

Despite this progress, relatively little is now known of the molecular biology of thrombopoietin. Messenger RNA is expressed in the liver and kidney, and at lesser levels in the spleen, muscle, and marrow. The cellular origin of specific transcripts is unknown, and whether gene expression is regulated or is constitutive is also unclear. Thrombopoietin is organized into two distinct domains. Although it is clear that the erythropoietin-like amino terminal domain is responsible for the known biological activities of the protein, the function of the carbohydrate rich carboxyl-terminal region remains an enigma. Moreover, thrombopoietin purified from plasma is proteolytically processed; whether this is an artifact introduced during purification or is critical for biological activity remains an unsettled issue. Although having been defined some 36 years ago, the first steps, purification and cloning of thrombopoietin, have only now been accomplished; it is hoped that a more complete molecular characterization of its physiology will not require so long a period of time.

REFERENCES

1. Metcalf D. Hemopoietic regulators. TIBS 1992; 17:286–289.
2. Ogawa M. Differentiation and proliferation of hematopoietic stem cells. Blood 1993; 81: 2844–2853.
3. Debili N, Massé J-M, Katz A, Guichard J, Breton-Gorius J, Vainchenker W. Effects of the recombinant hematopoietic growth factors interleukin-3, interleukin-6, stem cell factor, and leukemia inhibitory factor on the megakaryocytic differentiation of CD34$^+$ cells. Blood 1993; 82:84–95.
4. Metcalf D. Lineage commitment of hemopoietic progenitor cells in developing blast cell colonies: influence of colony-stimulating factors. Proc Natl Acad Sci USA 1991; 88:11310–11314.

5. Dpsnhtufr HJ. Biological and clinical aspects of hematopoietic stem cells. Annu Rev Med 1994; 45:91–104.
6. Brandt JE, Bhalla K, Hoffman R. Effects of interleukin-3 and c-*kit* ligand on the survival of various classes of human hematopoietic progenitor cells. Blood 1994; 83:1507–1514.
7. Emerson SG, Thomas S, Ferrara JL, Greenstein JL. Developmental regulation of erythropoiesis by hematopoietic growth factors: analysis on populations of BFU-E from bone marrow, peripheral blood, and fetal liver. Blood 1989; 74:49–55.
8. Broxmeyer HE, Williams DE, Hangoc G, et al. Synergistic myelopoietic actions in vivo after administration to mice of combinations of purified natural murine colony-stimulating factor 1, recombinant murine interleukin 3, and recombinant murine granulocyte/macrophage colony-stimulating factor. Proc Natl Acad Sci USA 1987; 84:3871–3875.
9. Paul CC, Tolbert M, Mahrer S, Singh A, Grace MJ, Baumann MA. Cooperative effects of interleukin-3 (IL-3), IL-5, and granulocyte-macrophage colony-stimulating factor: a new myeloid cell line inducible to eosinophils. Blood 1993; 81:1193–1199.
10. McNiece I, Andrews R, Stewart M, Clark S, Boone T, Quesenberry P. Action of interleukin-3, G-CSF, and GM-CSF on highly enriched human hematopoietic progenitor cells: synergistic interaction of GM-CSF plus G-CSF. Blood 1989; 74:110–114.
11. Debili N, Massé J-M, Katz A, Guichard J, Breton-Gorius J, Vainchenker W. Effects of the recombinant hematopoietic growth factors interleukin-3, interleukin-6, stem cell factor, and leukemia inhibitory factor on the megakaryocytic differentiation of $CD34^+$ cells. Blood 1993; 82:84–95.
12. Segal GM, Stueve T, Adamson JW. Analysis of murine megakaryocyte colony size and ploidy: effects of interleukin-3. J Cell Physiol 1988; 137:537–544.
13. Kaushansky K, O'Hara RJ, Berkner K, Segal GM, Hagen FS, Adamson JW. Genomic cloning, characterization and multilineage growth-promoting activity of human granulocyte-macrophage colony-stimulating factor. Proc Natl Acad Sci USA 1986; 83:3101–3105.
14. Robinson BE, McGrath HE, Quesenberry PJ. Recombinant murine granulocyte macrophage colony stimulating factor has megakaryocyte colony stimulating activity and augments megakaryocyte colony stimulation by interleukin-3. J Clin Invest 1987; 79:1648–1652.
15. Avraham H, Vannier E, Cowley S, et al. Effects of the stem cell factor, c-*kit* ligand, on human megakaryocytic cells. Blood 1992; 79:365–371.
16. Briddell RA, Bruno E, Cooper RJ, Brandt JE, Hoffman R. Effect of c-*kit* ligand on in vitro human megakaryocytopoiesis. Blood 1991; 78:2854–2859.
17. McDonald TP. Thrombopoietin: its biology, purification, and characterization. Exp Hematol 1988; 16:201–205.
18. Hill RJ, Levin J. Regulators of thrombopoiesis: their biochemistry and physiology. Blood cells 1989; 15:141–166.
19. Evatt BL, Levin J. Measurement of thrombopoiesis in rabbits using ^{75}selenomethionine. J Clin Invest 1969; 48:1615–1626.
20. Shreiner DP, Levin J. Detection of thrombopoietic activity in plasma by stimulation of suppressed thrombopoiesis. J Clin Invest 1970; 49:1709–1713.
21. McDonald TP. Bioassay for thrombopoietin utilizing mice in rebound thrombocytosis. Proc Soc Exp Biol Med 1973; 144:1006–1012.
22. Tayrien G, Rosenberg RD. Purification and properties of a megakaryocyte stimulatory factor present both in the serum-free condition medium of human embryonic kidney cells and in thrombocytopenic plasma. J Biol Chem 1987; 262:3262–3268.
23. Ikebuchi K, Wong GG, Clark SC, Ihle JN, Hirai Y, Ogawa M. Interleukin-6 enhancement of interleukin 3-dependent proliferation of multipotential hemopoietic progenitors. Proc Natl Acad Sci USA 1987; 84:9035–9039.
24. Ishibashi T, Kimura H, Uchida T, et al. Human interleukin 6 is a direct promoter of maturation of megakaryocytes in vitro. Proc Natl Acad Sci USA 1989; 86:5953–5957.
25. Quesenberry PJ, McGrath HE, Williams ME, et al. Multifactor stimulation of megakaryocytopoiesis: effects of interleukin 6. Exp Hematol 1991; 19:35–41.

26. Kimura H, Ishibashi T, Uchida T, Maruyama Y, Friese P, Burstein SA. Interleukin 6 is a differentiation factor for human megakaryocytes in vitro. Eur J Immunol 1990; 20:1927–1931.
27. Paul SR, Bennett F, Calvetti JA, et al. Molecular cloning of a cDNA encoding interleukin 11, a stromal cell-derived lymphopoietic and hematopoietic cytokine. Proc Natl Acad Sci USA 1990; 87:7512–7516.
28. Musashi M, Yang Y-C, Paul SR, Clark SC, Sudo T, Ogawa M. Direct and synergistic effects of interleukin 11 on murine hemopoiesis in culture. Proc Natl Acad Sci USA 1991; 88:765–769.
29. Metcalf D, Waring P, Nicola NA. Actions of leukaemia inhibitory factor on megakaryocyte and platelet formation. Ciba Found Symp 1992; 167:174–187.
30. Ishibashi T, Shikama Y, Kimura H, et al. Thrombopoietic effects of interleukin-6 in long-term administration in mice. Exp Hematol 1993; 21:640–646.
31. Ishibashi T, Kimura H, Shikama Y, et al. Interleukin-6 is a potent thrombopoietic factir in vivo in mice. Blood 1989; 74:1241–1244.
32. Patchen ML, MacVittie TJ, Williams JL, Schwartz GN, Souze LM. Administration of interleukin-6 stimulates multilineage hematopoiesis and accelerates recovery from radiation-induced hematopoietic depression. Blood 1991; 77:472–480.
33. Neben TY, Loebelenz J, Hayes L, et al. Recombinant human interleukin-11 stimulates megakaryocytopoiesis and increases peripheral platelets in normal and splenectomized mice. Blood 1993; 81:901–908.
34. Yonemura Y, Kawakit M, Masuda T, Fujimoto K, Takatsuki K. Effect of recombinant human interleukin-11 on rat megakaryopoiesis and thrombopoiesis in vivo: comparative study with interleukin-6. Br J Haematol 1993; 84:16–23.
35. Metcalf D, Nicola NA, Gearing DP. Effedts of injected leukemia inhibitory factor on hematopoietic and other tissues in mice. Blood 1990; 76:50–56.
36. Schrezenmeier H, Späth-Schwalbe E, Dreschler S, et al. Phase I trials of interleukin-6 (IL-6) in patients with advanced renal cell carcinoma (RCC) and patients with aplastic anemia (AA): effects of long-term application and differences of hemopoietic response to IL-6 in patients with normal hemopoiesis versus patients with aplastic anemia. Blood 1993; 82(Suppl 1):368a.
37. Gordon MS, Battiato L, Hoffman R et al. Subcutaneously (SC) administered recombinant human interleukin-11 (Neumega™ rhl-11 growth factor; rhIL-11) prevents thrombocytopenia following chemotherapy (CT) with cyclophosphamide (C) and doxorubicin (A) in women with breast cancer (BC). Blood 1993; 82(Suppl 1):318a.
38. Orazi A, Cooper R, Tong J, et al. Recombinant human interleukin-11 (Neumega™ rhIL-11 growth factor; rhIL-11) has multiple profound effects on human hematopoiesis. Blood 1993; 82(Suppl 1):369a.
39. Williams N. Is thrombopoietin interleukin 6? Exp Hematol 1991; 19:714–718.
40. Hill RJ, Warren MK, Levin J, Gouldie J. Evidence that interleukin-6 does not play a role in the stimulation of platelet production after induction of acute thrombocytopenia. Blood 1992; 80:346–351.
41. Tsukada J, Misago M, Ogawa R, et al. Synergism between serum factor(s) and erythropoietin in inducing murine megakaryocyte colony formation: The synergistic factor in serum is distinct from interleukin-11 and stem cell factor (c-*kit* ligand). Blood 1993; 81:866–867.
42. Wendling F, Varlet P, Charon M, Tambourin P. A retrovirus complex inducing an acute myeloproliferative leukemia disorder in mice. Virology 1986; 149:242–246.
43. Souyri M, Vigon I, Penciolelli J-F, Heard J-M, Tambourin P, Wendling F. A putative truncated cytokine receptor gene transduced by the myeloproliferative leukemia virus immortalizes hematopoietic progenitors. Cell 1990; 63:1137–1147.
44. Vigon I, Mornon J-P, Cocault L, et al. Molecular cloning and characterization of *MPL*, the human homolog of the v-*mpl* oncogene: Identification of a member of the hematopoietic growth factor receptor superfamily. Proc Natl Acad Sci USA 1992; 89:5640–5644.

45. Cosman D. The hematopoietin receptor superfamily. Cytokine 1993; 5:95–106.
46. Miyajima A, Hara T, Kitamura T. Common subunits of cytokine receptors and the functional redundancy of cytokines. TIBS 1992; 17:378–382.
47. Kishimoto T, Taga T, Akira S. Cytokine signal transduction. Cell 1994; 76:253–262.
48. Methia N, Louache F, Vainchenker W, Wendling F. Oligodeoxynucleotides antisense to the proto-oncogene c-*mpl* specifically inhibit in vitro megakaryocytopoiesis. Blood 1993; 82:1395–1401.
49. Skoda RC, Seldin DC, Chiang M-K, Peichel CL, Vogt TF, Leder P. Murine c-*mpl*: a member of the hematopoietic growth factor receptor superfamily that transduces a proliferative signal. EMBO J 1993; 12:2645–2653.
50. Vigon I, Florindo C, Fichelson S, et al. Characterization of the murine *mpl* proto-oncogene, a member of the hematopoietic cytokine receptor family: molecular cloning, chromosomal location and evidence for a function in cell growth. Oncogene 1993; 8:2607–2615.
51. de Sauvage FJ, Hass PE, Spencer SD, et al. Stimulation of megakaryocytopoiesis and thrombopoiesis by the c-Mpl ligand. Nature 1994; 369:533–538.
52. Bartley TD, Bogenberger J, Hunt P, et al. Identification and cloning of a megakaryocyte growth and development factor that is a ligand for the cytokine receptor Mpl. Cell 1994; 77: 1117–1124.
53. Lok S, Kaushansky K, Holly RD, et al. Cloning and expression of murine thrombopoietin cDNA and stimulation of platelet production in vivo. Nature 1994; 369:565–568.
54. Stocking C. Bergholz U, Friel J, et al. Distinct classes of factor-independent mutant can be isolated after retroviral mutagenesis of a human myeloid stem cell line. Growth Factors 1993; 8:197–209.
55. Wilks AF, Kurban RR, Dunn AR. Direct demonstration of an autocrine mechanism in EMS-induced, tumorigenic mutants of growth factor-dependent hemopoietic cell line, FDC-P1. Growth factors 1989; 2:32–42.
56. Miyazaki H, Kato T, Ogami K, et al. Isolation and cloning of a novel human thrombopoietic factor. Exp Hematol 1994; 22:838.
57. Kuter DJ, Beeler DL, Rosenberg R. The purification of megapoietin: a physiological regulator of megakaryocyte growth and platelet production. Presented at the Workshop on Megakaryocytopoiesis and Platelet Production, National Institutes of Health Campus, Bethesda, MD, August 18–19, 1994.
58. Wendling F, Maraskovsky E, Debili N, et al. c-Mpl ligand is a humoral regulator of megakaryocytopoiesis. Nature 1994; 369:571–574.
59. Kaushansky K, Lok S, Holly RD, et al. Promotion of megakaryocyte progenitor expansion and differentiation by the c-Mpl ligand thrombopoietin. Nature 1994; 369:568–571.
60. Lacombe C, Da Silva J-L, Bruneval P, et al. Peritubular cells are the site of erythropoietin synthesis in the murine hypoxic kidney. J Clin Invest 1988; 81:620–623.
61. Foster DC, Sprecher CA, Grant FJ, et al. Human thrombopoietin: gene structure, cDNA sequence, expression and chromosomal localization. Proc Natl Acad Sci USA. In press.
62. McDonald JD, Link F-K, Goldwasser E. Cloning, sequencing, and evolutionary analysis of the mouse erythropoietin gene. Mol Cell Biol 1986:6:842–848.
63. Shoemaker CB, Mitsock LD. Murine erythropoietin gene: cloning, expression and human gene homology. Mol Cell Biol 1986; 6:849–858.
64. Jacobs K, Shoemaker CB, Rudersdorf R, et al. Isolation and characterization of genomic and cDNA clones of human erythropoietin. Nature 1985; 313:806–810.
65. Lin F-K, Suggs S, Lin C-H, et al. Cloning and expression of the human erythropoietin gene. Proc Natl Acad Sci USA 1985; 82:7580–7584.
66. Law ML, Cai G-Y, Lin F-K, et al. Chromosomal assignment of the human erythropoietin gene and its DNA polymorphism. Proc Natl Acad Sci USA 1986; 83:6920–6924.
67. Pinto MR, King MA, Goss GD, et al. Acute megakaryoblastic leukaemia with 3q inversion and elevated thrombopoietin (TSF): an autocrine role for TSF? Br J Haematol 1985; 61: 687–694.

68. Sweet DL, Golomb HM, Rowley JD, Vardiman JM. Acute myelogeneous leukemia and thrombocythemia associated with an abnormality of chromosome no. 3. Cancer Genet Cytogenet 1979; 1:33–37.
69. Bernstein R, Pinto MR, Mendelow B. Chromosome 3 abnormalities in acute nonlymphocytic leukemia (ANLL) with abnormal thrombopoiesis: report of three patients with a ''new'' inversion anomaly and a further case of homologous translocation. Blood 1982; 60:613–617.
70. Bernstein RA, Bagg A, Pinto M, Lewis D, Mendelow B. Chromosome 3q21 abnormalities associated with hyperactive thrombopoiesis in acute blastic transformation of chronic myeloid leukaemia. Blood 1986; 3:652–657.
71. Carbonell F, Hoelzer D, Thiel E, Bartl R. Ph^1-positive CML associated with megakaryocytic hyperplasia and thrombocythemia and an abnormality of chromosone no. 3. Cancer Genet Cytogenet 1982; 6:153–161.
72. Norrby A, Ridell B, Swolin B, Westin J. Rearrangement of chromosome no. 3 in a case of preleukemia with thrombocytosis. Cancer Genet Cytogenet 1982; 5:257–263.
73. Carroll AJ, Poon M-C, Robinson NC, Crist WM. Sideroblastic anemia associated with thrombocytosis and a chromosome 3 abnormality. Cancer Genet Cytogenet 1986; 22:183–187.
74. Morishita K, Parker DS, Mucenski ML, Jenkins NA, Copeland NG, Ihle JN. Retroviral activation of a novel gene encoding a zinc finger protein in IL-3-dependent myeloid leukemia cell lines. Cell 1988; 54:831–840.
75. Mitani K, Ogawa S, Tanaka T, et al. Generation of the AML1-EVI-1 fusion gene in the t(3; 21)(q26;q22) cause blastic crisis in chronic myelocytic leukemia. EMBO J 1994; 13:504–510.
76. Nucifora G, Begy CR, Erickson P, Drabkin HA, Rowley JD. The 3;21 translocation in myelodysplasia results in a fusion transcript between the AML1 gene and the gene for EAP, a highly conserved protein associated with the Epstein-Barr virus RNA EBER1. Proc Natl Acad Sci USA 1993. 90:7784–7788.
77. Nucifora G, Begy CR, Kobayashi H, et al. Consistent intergenic splicing and production of multiple transcripts between AML1 at 21q22 and unrelated genes at 3q26 in (3;21)(q26;q22) translocations. Proc Natl Acad Sci USA 1994; 91:4004–4008.
78. Bartocci A, Mastrogiannis DS, Migliorati G, Stockert RJ, Wolkoff AW, Stanley ER. Macrophages specifically regulate the concentration of their own growth factor in the circulation. Proc Natl Acad Sci USA 1987; 84:6179–6183.
79. Recny MA, Scoble HA, Kim Y. Structural characterization of natural human urinary and recombinant DNA-derived erythropoietin. J Biol Chem 1987; 262:17156–17163.

Index

About the Editors

KATHERINE A. HIGH is Director of Hematology Laboratories at the Children's Hospital of Philadelphia, Pennsylvania, and an Associate Professor of Pediatrics and Laboratory Medicine and a Member of the Institute for Human Gene Therapy at the University of Pennsylvania Medical Center, Philadelphia. She is a member of the International Society on Thrombosis and Hemostasis, the American Society of Hematology, and the American Society for Clinical Investigation, among other organizations. She is the author or coauthor of over 60 book chapters, professional papers, and abstracts, and serves on the editorial board of the *American Journal of Hematology*. Dr. High received the M.D. degree (1978) from the University of North Carolina at Chapel Hill School of Medicine.

HAROLD R. ROBERTS is Chief of the Division of Hematology and Sarah Graham Kenan Professor of Medicine and Pathology at the University of North Carolina at Chapel Hill School of Medicine. He is the Executive Director of the International Society on Thrombosis and Hemostasis, a Fellow of the American College of Physicians and the American Association for the Advancement of Science, and a member of the Society for Experimental Biology and Medicine, the American Society of Hematology, International Society of Hematology, the American Society for Clinical Investigation, and the Association of American Physicians, among other organizations. The editor or coeditor of nine books and the author or coauthor of over 350 professional papers and abstracts, he serves on the editorial boards of several journals. Dr. Roberts received the M.D. degree (1955) from the University of North Carolina at Chapel Hill School of Medicine.